# Physiology of the Eye

# STANDARD
# LOAN

UNLESS RECALLED BY ANOTHER READER
THIS ITEM MAY BE BORROWED FOR
## FOUR WEEKS

To renew, telephone:
01243 816089 (Bishop Otter)
01243 816099 (Bognor Regis)

1 8 MAY 2006

# Physiology of the Eye

**Hugh Davson** DSc (Lond)

Honorary Research Fellow, King's College, London and
Fellow, University College, London

FOURTH EDITION

CHURCHILL LIVINGSTONE
EDINBURGH LONDON AND NEW YORK 1980

CHURCHILL LIVINGSTONE
Medical Division of Longman Group Limited

First Edition 1949
Second Edition 1963
Third Edition 1972
Fourth Edition 1980

ISBN 0 443 01892 8

**British Library Cataloguing in Publication Data**
Davson, Hugh
   Physiology of the eye—4th ed.
   1. Eye
   I. Title
   612′.84     QP476     79-41016

Printed and bound in Great Britain by
William Clowes (Beccles) Limited, Beccles and London.

TO THE MEMORY OF
Dr Samuel L. Saltzman

# Preface

It was in 1948 that I called at the offices of Messrs Churchill with a brown paper parcel containing the manuscript of *Physiology of the Eye*, and offered it to them for publication, and the first edition appeared in 1949. A second edition appeared in 1963, and in the preface to this I wrote: 'The task (of revision) has now been completed and has resulted in about a 50 per cent increase in size, an increase that is the necessary consequence of the tremendous advances made in those aspects of the physiology of the eye that are dealt with in Sections I and II. Thus, it is true to say that in 1948 our knowledge of the vegetative physiology and biochemistry of the eye was rudimentary in the extreme. Now, thanks to the establishment of numerous ophthalmological research organizations—mainly in the U.S.A. —it is safe to say that our knowledge in this branch of ocular science is abreast of our knowledge in comparable fields of physiology and biochemistry. Although, in 1948, the field covered by the title *Mechanism of Vision* rested on a much firmer basis of accurate scientific investigation, the striking improvements in photochemical and electrophysiological techniques that have since taken place have permitted just as rapid progress in elucidating basic problems. Consequently, because I have tried to maintain the same sort of treatment as I had in mind when I wrote the first edition of this book— namely, to present a simple yet thorough exposition of the fundamental principles of ocular physiology—I have had to double the amount of space devoted to these first two sections. The muscular aspects covered by Section III have been expanded a little to take account of the important advances made possible by the application of modern methods to the analysis of the eye movements, pupillary function, and so on. Sections IV and V rest on the firm theoretical and practical foundation laid by the work of Helmholtz and other classical research workers in the realm of physiological optics, so that they have been tampered with least.'

In presenting this (fourth) edition, I feel that I must convey my apologies for a book that has grown well beyond the limits expected of a 'small textbook'. I must temper these apologies, however, with the mitigating plea that the advances in electrophysiology of the receptors, and of the central nervous pathways of vision and eye movement, demanded a great increase in space if a fair presentation was to be achieved. In the field of the vegetative physiology of the eye, the remarkable developments in separative techniques of the lens crystallins have likewise demanded an expansion of the treatment of the biochemical aspects of lens transparency. Mere expansion of the subject matter does not necessarily justify an expansion of treatment in a textbook; however, in writing these editions I have always kept in mind the desirability of showing how the facts of visual physiology were discovered, rather than epitomizing them in a succinct summary, so that new facts do, in fact, usually require new space for their description. By suppressing the detailed description of visual optics present in the first three editions, I have made room for more physiology and biochemistry, and thus rescued the book from an unconscionable increase in size.

London, 1980                                        Hugh Davson

# Acknowledgments

It is a pleasure to acknowledge, with grateful thanks, the assistance of the Leverhulme Trust Fund, which, by awarding me an Emeritus Fellowship, has contributed materially to defraying the costs of preparing this new edition. In addition I should like to thank my former colleague, Dr Malcolm Segal, for invaluable assistance in preparing the illustrations, and Mary Ann Fenstermacher for secretarial help. To the authors and publishers from whose works illustrations have been reproduced my sincere thanks for their permission to do this, and especially to H. Anderson, A. Bill, D. F. Cole, H. J. Hoenders, H. Inomata, T. Kuwabara, A. Palkama, M. Passatore, B. Philipson, Y. Pouliquen, A. Spector, R. C. Tripathi, and R. W. Young for generously providing originals of micrographs.

H.D.

# Contents

# The vegetative physiology and biochemistry of the eye

# 1 (a). Anatomical introduction
# (b). Aqueous humour and the intraocular pressure

THE THREE LAYERS

The globe of the eye consists essentially of three coats enclosing the transparent refractive media. The outermost, protective tunic is made up of the *sclera* and *cornea*—the latter transparent; the middle coat is mainly vascular, consisting of the *choroid, ciliary body*, and *iris*. The innermost layer is the *retina*, containing the essential nervous elements responsible for vision—the *rods* and *cones*; it is continued forward over the ciliary body as the *ciliary epithelium*.

DIOPTRIC APPARATUS

The dioptric apparatus (Fig. 1.1) is made up of the transparent structures—the cornea, occupying the anterior sixth of the surface of the globe, and the *lens*, supported by the *zonule* which is itself attached to the ciliary body. The spaces within the eye are filled by a clear fluid, the *aqueous humour*, and a jelly, the *vitreous body*. The aqueous humour is contained in the *anterior* and *posterior chambers*, and the vitreous body in the large space behind the lens and ciliary body. The posterior chamber is the name given to the small space between the lens and iris.

The iris behaves as a diaphragm, modifying the amount of light entering the eye, whilst the ciliary body contains muscle fibres which, on contraction, increase the refractive power of the lens (accommodation). An image of external objects is formed, by means of the dioptric apparatus, on the retina, the more highly specialized portion of which is called the *fovea*.

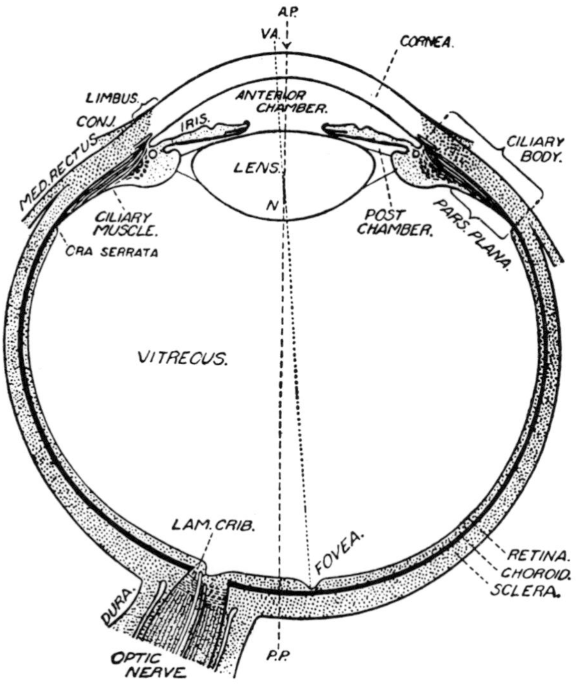

VISUAL PATHWAY

The retina is largely made up of nervous tissue—it is an outgrowth of the central nervous system—and fibres carrying the responses of visual stimuli lead away in the *optic nerve* through a canal in the bony orbit, the *optic foramen* (Fig. 1.2); the visual impulses are conveyed through the optic nerve and tract to the lateral geniculate body and thence to the cerebral cortex; on their way, the fibres carrying messages from the medial, or nasal, half of the retina cross over in the *optic chiasma*, so that the lateral geniculate body of the left side, for example, receives fibres from the temporal half of the left retina and the nasal half of the right. The nerve trunks, proximal to the chiasma, are called the *optic tracts*. This partial decussation may be regarded as a development associated with binocular vision; it will be noted that the responses to a stimulus from any one part of the visual field are carried in the same optic tract (Fig. 1.3), and the necessary motor response, whereby both eyes are directed to the same point in the field, is probably simplified by this arrangement. It will be noted also that a right-sided event, i.e. a visual stimulus arising from a point in the right half of the field, is associated with impulses passing to the left cerebral hemisphere, an arrangement common to all peripheral stimuli.

VASCULAR COAT

The nutrition of the eye is taken care of largely by the capillaries of the vascular coat; let us examine this structure more closely.

**Fig. 1.1** Horizontal section of the eye. P.P., posterior pole; A.P., anterior pole; V.A., visual axis. (Wolff, *Anatomy of the Eye and Orbit.*)

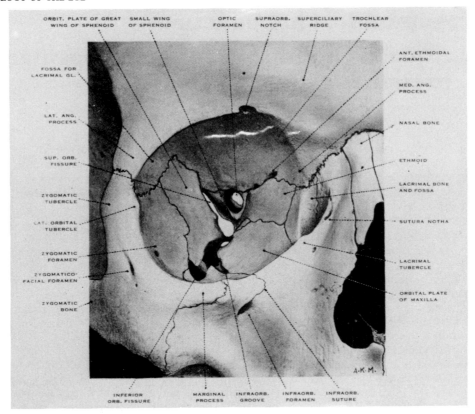

Fig. 1.2    The orbit from in front. (Wolff, *Anatomy of the Eye and Orbit.*)

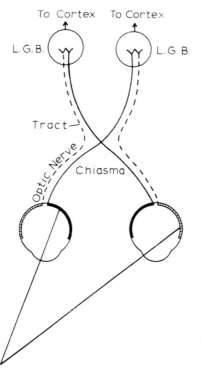

Fig. 1.3    The partial decussation of the optic pathway.

CHOROID

The choroid is essentially a layer of vascular tissue next to the retina; it is separated from this nervous tissue by two membranes—the structureless *membrane of Bruch* and the pigment epithelium. The retina comes to an end at the ciliary body, forming the *ora serrata* (Fig. 1.4), but the vascular coat continues into the ciliary body as one of its layers, the *vessel layer*, which is separated from the eye contents by membranes—the two layers of ciliary epithelium which, viewed embryologically, are the forward continuations of the retina and its pigment epithelium, and the *lamina vitrea*, which is the continuation forward of Bruch's membrane.

CILIARY BODY AND IRIS

The ciliary body in antero-posterior section is triangular in shape (Fig. 1.5) and has a number of processes (seventy) to which the zonule is attached; viewed from behind, these processes appear as radial ridges to which the name *corona ciliaris* has been given (Fig. 1.4). The relationship of the iris to the ciliary body is seen in Figure 1.5; the blood vessels supplying it belong to the same system as that supplying the ciliary body. Posteriorly, the stroma is separated from the aqueous humour by the *posterior epithelium*, a double, heavily

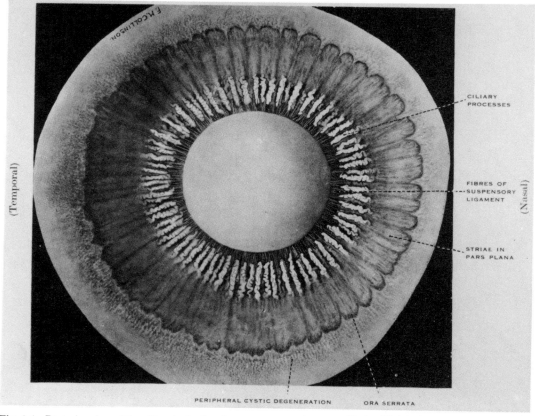

**Fig. 1.4**   Posterior view of the ciliary body showing ora serrata and corona ciliaris. (Wolff, *Anatomy of the Eye and Orbit.*)

pigmented layer which may be viewed as a prolongation of the ciliary epithelial layers.

## NUTRITION

The vessels of the vascular coat nourish the internal structures of the eye; so far as the lens is concerned, this process must take place by diffusion of dissolved material from the capillaries through the aqueous humour and vitreous body. The inner (nearest the vitreous) nervous elements of the retina, however, are provided for by a functionally separate vascular system derived from the *central artery of the retina*; this artery, a branch of the *ophthalmic*, enters the globe with the optic nerve and it is its ramifications, together with the *retinal veins*, that give the fundus of the eye its characteristic appearance. The choroid, ciliary body, and iris are supplied by a separate system of arteries, also derived from the ophthalmic—the *ciliary system of arteries*.

The anterior portion of the sclera is covered by a mucous membrane, the *conjunctiva*, which is continued forward on to the inner surfaces of the lids, thus creating the *conjunctival sac*. The remainder of the sclera is enveloped by *Tenon's capsule*.

## MUSCLE

Movements of the eye are executed by the contractions of the six extra-ocular muscles; the space between the globe and orbit being filled with the orbital fat, the movements of the eye are essentially rotations about a fixed point in space.

## NERVE SUPPLY

The essentials of the nerve supply to the eyeball are indicated in Figure 1.6; sensory impulses (excluding, of course, the visual ones) are conveyed through the *long* and *short ciliary nerves*. The long ciliary nerves are composed mainly of the axons of nerve cells in the *Gasserian ganglion*—the ganglion of the trigeminal (N. V); they convey impulses from the iris, ciliary body,

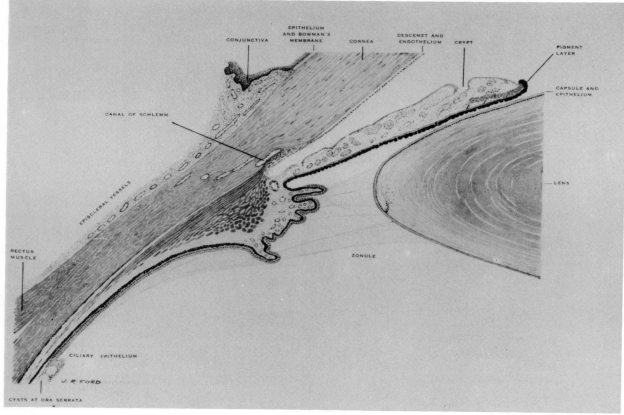

**Fig. 1.5**   Antero-posterior section through the anterior portion of the eye. (Wolff, *Anatomy of the Eye and Orbit.*)

and cornea. The short ciliary nerves also contain axons of the trigeminal; they pass through a ganglion in the orbit, the *ciliary ganglion*, into the *naso-ciliary nerve*; the fibres carry impulses from all parts of the eyeball, but chiefly from the cornea.

The voluntary motor nerve supply to the extraocular muscles is through the cranial nerves III (oculomotor), IV (trochlear) and VI (abducens). Parasympathetic motor fibres to the ciliary muscle and iris travel through the lower division of N. III as the *motor root* of the ciliary ganglion; post-ganglionic fibres to the muscles are contained in the short ciliary nerves.

Sympathetic fibres from the superior cervical ganglion enter the orbit as the sympathetic root of the ciliary ganglion and run in the short ciliary nerves to supply the vessels of the globe and the dilator fibres of the pupil. Other sympathetic fibres avoid the ciliary ganglion, passing through the Gasserian ganglion and

entering the globe in the long ciliary nerves, whilst still others enter the globe in the adventitia of the ciliary arteries.

ADDITIONAL NERVE SUPPLIES

Figure 1.6 may be said to represent the classical view of the ciliary ganglion with its three roots; however, the work of Ruskell (1970, 1974) has revealed the existence of ocular fibres from both the facial nerve (N. VII) and from the maxillary branch of the trigeminal, in addition to the nasociliary branch.

**Facial nerve**

This represents an ocular parasympathetic pathway derived from N. VII; the fibres are derived from the pterygopalatine ganglion, running as *rami orbitales* to the *retro-orbital plexus* (*top* of Fig. 1.7); the *rami oculares*

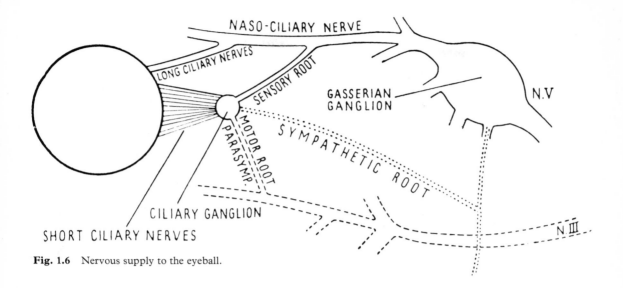

**Fig. 1.6** Nervous supply to the eyeball.

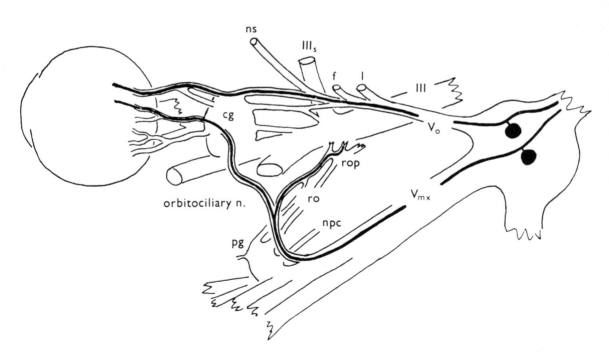

**Fig. 1.7** Distribution of trigeminal nerve fibres to the eyeball. cg, ciliary ganglion; f, frontal nerve; l, lacrimal nerve; npc, nerve of the pterygoid canal; ns, nasociliary nerve; pg, pterygopalatine ganglion; ro, rami orbitales; rop, retro-orbital plexus; $V_o$, ophthalmic nerve; $V_{mx}$, maxillary nerve; III, oculomotor nerve; $III_s$, superior branch of oculomotor nerve. (Ruskell, *J. Anat.*)

find their way into the globe mainly through the long ciliary nerves and along the ciliary arteries.

## Maxillary fibres
These are illustrated in Figure 1.7, some of them entering the retro-orbital plexus which had hitherto been assumed to be entirely autonomic, and the other division represents a maxillary root of the ciliary ganglion; degeneration experiments showed that the fibres passed into the globe through short ciliary nerves.

# 1 (b). Aqueous humour and the intraocular pressure

The aqueous humour is a transparent colourless fluid contained in the anterior and posterior chambers of the eye (Fig. 1.1, p. 3). By inserting a hypodermic needle through the cornea, the fluid may be more or less completely withdrawn, in which case it consists of the mixed fluids from the two chambers. As Seidel first showed, the posterior fluid may be withdrawn separately by inserting a needle through the sclera just behind the corneoscleral junction; because of the valve-like action of the iris, resting on the lens, the anterior fluid is restrained from passing backwards into the posterior chamber, so that subsequently another needle may be inserted into the anterior chamber and this anterior fluid may be withdrawn. In the cat Seidel found that the volume was some 14 per cent of the volume of the anterior fluid, whilst in the rabbit it is about 20 per cent, according to Copeland and Kinsey. The fluids from the two chambers are of very similar composition, so that we may consider the composition of the mixed fluid when discussing the relationships between aqueous humour and blood. Thanks largely to the studies of Kinsey, however, it is now known that there are characteristic differences between the two fluids, differences that reflect their histories, the posterior fluid having been formed first and therefore exposed to diffusional exchanges with the blood in the iris for a shorter time than the anterior fluid.

## ANATOMICAL RELATIONSHIPS

The nature of the aqueous humour is of fundamental interest for several reasons. There is reason to believe that it is formed continuously in the posterior chamber by the ciliary body, and that it passes through the pupil into the anterior chamber, whence it is drained away into the venous system in the angle of the anterior chamber by way of Schlemm's canal. As a circulating fluid, therefore, it must be a medium whereby the lens and cornea receive their nutrient materials. Furthermore, because it is being continuously formed, the *intraocular pressure* is largely determined by the rate of formation, the ease with which it is drained away, and by the forces behind the process of formation. To understand the nature of the fluid we must consider its chemical composition and variations of this under experimental conditions, whilst to appreciate the drainage process we must examine the detailed micro-anatomy of the structures that are concerned in this process. Since, moreover, the aqueous humour cannot be considered in isolation from the vitreous body, the posterior portion of the eyeball is also of some interest.

## THE VASCULAR COAT

The choroid, ciliary body and iris may be regarded as a vascular coat—the *uvea*—sandwiched between the protective outer coat—cornea and sclera—and the inner neuroepithelial coat—the retina and its continuation forwards as the ciliary epithelium (Fig. 1.8). This

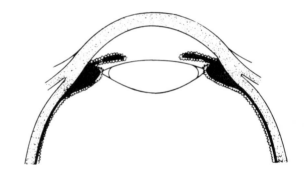

**Fig. 1.8** Illustrating the uvea, or vascular coat (black), sandwiched between the sclera on the outside and the retina, ciliary epithelium, etc., on the inside.

vascular coat is made up of the ramifications of the ciliary system of arteries that penetrate the globe independently of the optic nerve.

### CILIARY ARTERIES

The posterior ciliary arteries arise as trunks from the ophthalmic artery (Fig. 1.9), dividing into some ten to twenty branches which pierce the globe around the optic nerve. Two of them, the long posterior ciliary arteries, run forward through the choroid to the ciliary body where they anastomose with the anterior ciliary arteries to form what has been incorrectly called the *major circle of the iris*; incorrectly, because the circle is actually in the ciliary body. From this circle, arteries run forward to supply the ciliary processes and iris, and backwards—

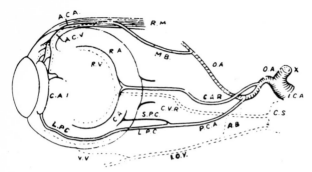

**Fig. 1.9**  Physiological plan of the circulation in man. OA, ophthalmic artery; MB, muscular branch; ACA, anterior ciliary artery; CAI, circulosus arteriosus iridis major; LPC, long posterior ciliary artery; SPC, short posterior ciliary artery; PCA, posterior ciliary arteries; CAR, central artery of the retina; RA, retinal artery; VV, vortex vein; IOV, inferior orbital vein; AB, anastomosing branch; CS, cavernous sinus. (Duke-Elder, *Brit. J. Ophthal.*)

as recurrent ciliary arteries—to contribute to the choroidal circulation.

By the *choroid* we mean the vascular layer between the retina and sclera; it is made up of the ramifications of the recurrent ciliary arteries, just mentioned, and of the short posterior ciliary arteries, which enter the globe around the posterior pole. The arteries break up to form a well-defined capillary layer—the *choriocapillaris* —next to the retina, but separated from this by a transparent glassy membrane—the *membrane of Bruch* or *lamina vitrea*. This capillary layer is certainly responsible for nutrition of the outer layers of the retina—pigment epithelium, rods and cones, and bipolar cells—and may also contribute to that of the innermost layer.

### VENOUS SYSTEM

The venous return is by way of two systems; the anterior ciliary veins accompany the anterior ciliary arteries, whilst the vortex veins run independently of

the arterial circulation; these last are four in number, and they drain all the blood from the choroid, the iris and the ciliary processes, whilst the anterior ciliary veins drain the blood from the ciliary muscle and the superficial plexuses, which we shall describe later. According to a study by Bill (1962) only 1 per cent of the venous return is by way of the anterior ciliary veins.

### RETINAL CIRCULATION

As indicated earlier, the innermost part of the retina is supplied directly, in man, by the ramifications of the central retinal artery, a branch of the ophthalmic that penetrates the meningeal sheaths of the optic nerve in its intraorbital course and breaks up into a series of branches on the inner layer of the retina, branches that give the fundus its characteristic appearance in the ophthalmoscope. As Figure 1.10 indicates, the central artery gives off branches that supply the optic nerve*; moreover, the posterior ciliary arteries may send off branches in the region of the optic disk, forming a circle—*circle of Zinn or Haller*—which may form anastomoses with the branches of the central artery of the retina. To the extent that these anastomoses occur, the uveal and retinal vascular systems are not completely independent. Venous return from the retinal circulation is by way of the central retinal vein which, for a part of its course, runs in the subarachnoid space of the optic nerve and is thereby subjected to the intracranial pressure. For this reason, an elevated intracranial pressure may be manifest as an engorgement of the retinal veins.

---

* Some authors, for example François and Neetens, speak of a *central artery of the optic nerve*, derived either as a branch of the central retinal artery or entering separately as a branch of the ophthalmic artery. Wolff describes them as *arteriae collaterales centralis retinae*. The literature relating to this and other points is well summarized in the careful studies of Singh and Dass (1960), whilst Hayreh (1969) has described the capillary supply to the optic disk; a large part of this—the prelaminar—is from cilio-retinal arteries.

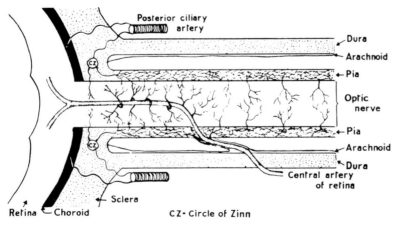

CZ - Circle of Zinn

**Fig. 1.10**  Schematic section of the optic nerve and eyeball showing course and branches of the central retinal artery. (Singh and Dass, *Brit. J. Ophthal.*)

So much for the general lay-out of the vascular circulation of the eye; as we have seen, the uveal circulation supplies the ciliary body and iris, the two vascular structures that come into close relationship with the aqueous humour. It is important now that we examine these structures in some detail.

## CILIARY BODY

In sagittal section this is a triangular body (Fig. 1.5, p. 6), whilst looked at from behind it appears as a ring, some 6 mm wide in man. Where it joins the retina—the *ora serrata*—it appears smooth to the naked eye, but farther forward the inner surface is ridged, owing to the presence of some seventy to eighty *ciliary processes*, and acquires the name of *corona ciliaris* (Fig. 1.4, p. 5). As the oblique section illustrated in Figure 1.11 shows, the ciliary processes are villus-like projections of the main body, jutting into the posterior chamber. According to Baurmann, the total area of these processes, in man, is some 6 cm² and we may look upon this expansion as subserving the secretory activity of this part of the ciliary body, since there is good reason to believe that it is here that the aqueous humour is formed. The vessel-layer of the ciliary body is similar to the choroid; it extends into each ciliary process, so that this is probably the most heavily vascularized part of the eye. Each process contains a mass of capillaries so arranged that each comes into close relationship, at some point in its course, with the surface of the epithelium. As stated above, blood from the ciliary processes is carried away by the vortex veins. In the rabbit, the ciliary processes are of two kinds—*posterior principal processes* projecting from the posterior region, and the *iridial processes*, located more anteriorly, and projecting towards the iris. These last have no zonular connections and may be those that are most closely concerned in production of the aqueous humour since they are the only processes to show significant morphological changes after paracentesis (p. 56), according to Kozart (1968).

### CILIARY EPITHELIUM

The innermost coat, insulating the ciliary body from the aqueous humour and vitreous body, is the *ciliary*

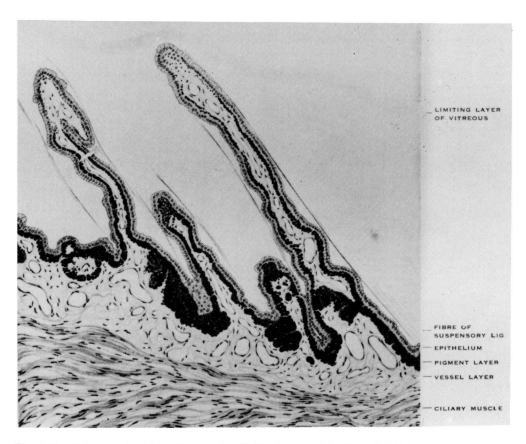

**Fig. 1.11**   Oblique section of the ciliary body. (Wolff, *Anatomy of the Eye and Orbit.*)

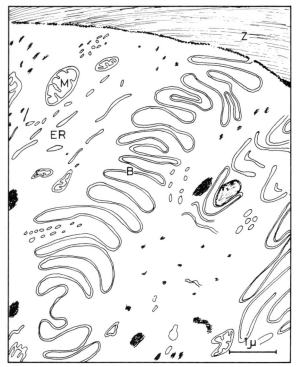

**Fig. 1.12**   A section of portions of two adjacent ciliary
epithelial cells of the rabbit showing complex interdigitations
(B) of their boundaries. Zonular fibres (Z) can be seen in close
approximation to the surface of these cells. ER, endoplasmic
reticulum; M, mitochondrion. (Drawing after an electron
micrograph by Pappas and Smelser, *Amer. J. Ophthal.*)

*epithelium*, a double layer of cells, the inner layer of
which, i.e. the layer next to the aqueous humour or
vitreous body, is non-pigmented, whilst the outer
layer is heavily pigmented, and represents the forward
continuation of the retinal pigment epithelium, whilst
the unpigmented layer is the forward continuation of
the neuroepithelium from which the retinal cells are
derived. The cells of this epithelium, which are
considered to be responsible for the secretion of the
aqueous humour, have been examined in the electron
microscope by several workers; a striking feature is the
interdigitation of the lateral surfaces of adjacent cells,
and the basal infoldings (Fig. 1.12), which are characteristic features of secretory epithelia concerned with
fluid transport. The relations of the two epithelial cell
layers are of importance, since the secreted aqueous
humour must be derived from the blood in the stroma
of the ciliary body, and thus the transport must occur
across both layers. The matter will be considered in
detail later in connexion with the blood-aqueous barrier,
and it is sufficient to indicate now that the cells of the
two layers face each other apex-to-apex, as a result of
the invagination of the neuroepithelial layer during
embryogenesis. Thus the *basal* aspects of the unpigmented epithelial cells face the aqueous humour, and
the *basal* aspects of the pigmented cells face the ciliary
stroma (Fig. 1.13).

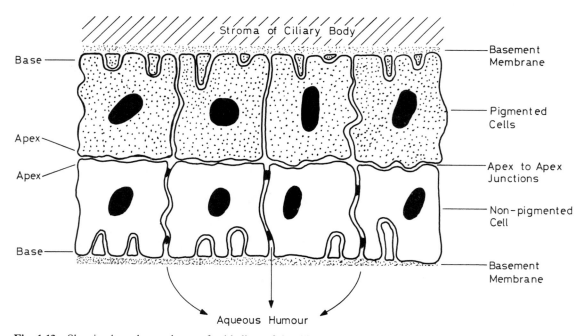

**Fig. 1.13**   Showing how the two layers of epithelium of the ciliary body face each other apex-to-apex.

## THE IRIS

### LAMINAE

This is the most anterior part of the vascular coat; in a meridional section it is seen to consist of two main layers or *laminae*, separated by a less dense zone, the *cleft of Fuchs*. The posterior lamina contains the muscles of the iris and is bounded by two layers of pigmented cells; the innermost, i.e. nearest the aqueous humour, represents the forward continuation of the unpigmented ciliary epithelial layer, and the outer is embryogenically equivalent to the pigmented epithelial layer; its cells, however, become differentiated into the muscle fibres of the dilator pupillae, so that the layer is often called *myoepithelium*. The blood vessels, derived from the major circle of the iris (p. 10), run in the loose connective tissue of the laminae. Both arteries and veins run radially; venous return is by way of the vortex veins.

### LININGS

On its anterior surface the iris is in contact with the aqueous humour, so that it is of some interest to determine exactly how the fluid is separated from the tissue. It is customary to speak of an *anterior endothelium*,

continuous with the corneal endothelium anteriorly and the posterior iris epithelium posteriorly, covering the anterior surface of the iris and thus insulating the aqueous humour from the underlying spongy tissue. The weight of recent evidence, however, bears against this so that it would seem that, in man at any rate, the

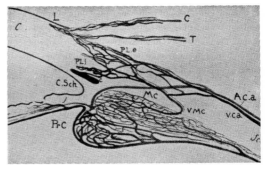

**Fig. 1.14** Illustrating the canal of Schlemm and its relation to the plexuses of the anteror segment of the eye. C, conjunctival plexus; T, Tenon's capsule plexus; PLe, episcleral plexus; PLi, intrascleral plexus; Mc, ciliary muscle; VMc, vein of ciliary muscle; Ac.a, anterior ciliary artery; v.ca, anterior ciliary vein; PrC, ciliary process; C, cornea; c, collector; L, limbus. (Maggiore, *Ann. Ottalm.*)

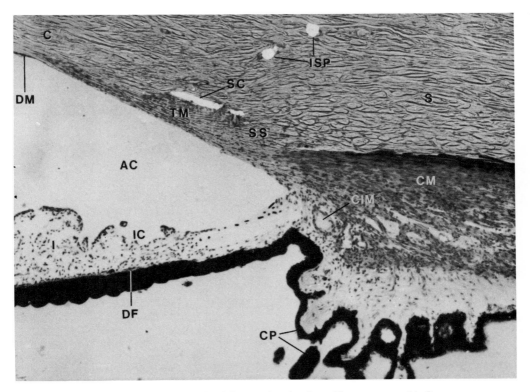

**Fig. 1.15** Meridional section of the angle of the anterior chamber of human eye. × 64. C, cornea; ISP, intrascleral plexus of veins; DM, Descemet's membrane; AC, anterior chamber; I, iris; IC, iris crypts; DF, dilator fibres of iris; CP, ciliary processes; CIM, circulus iridis major; CM, ciliary muscle; SS, scleral spur; TM, trabecular meshwork; SC, Schlemm's canal; S, sclera. (Courtesy R. C. Tripathi.)

aqueous humour has direct access to the stromal tissue. This is an important point when we come to consider the blood-aqueous barrier.

## THE DRAINAGE ROUTE

We have said that the aqueous humour is most probably secreted continuously by the cells of the ciliary epithelium; to make room for the new fluid there must obviously be a continuous escape, or drainage, and this occurs by way of the *canal of Schlemm*. This is a circular canal in the corneo-sclera of the limbus (Fig. 1.5, p. 6) which comes into relation with the aqueous humour on the one hand and the *intrascleral venous plexus* on the other, as illustrated by Figures 1.14 and 1.15.

### PLEXUSES OF THE ANTERIOR SEGMENT

The intrascleral plexus is largely made up of the ramifications of branches of the anterior ciliary veins, but, as Maggiore showed, there is also an arterial contribution, branches of the anterior ciliary arteries passing directly to the plexus and breaking up into fine vessels that come into direct continuity with the finer vessels of the plexus. It connects with the canal of Schlemm by some twenty fine vessels—*collectors*—which thus transport aqueous humour into the venous system. The blood from the plexus is drained away by large trunks into the *episcleral plexus*; from here large veins, also draining the more superficial plexuses—conjunctival and subconjunctival—carry the blood to the insertions of the rectus muscles in the sclera and accompany the muscular veins.

### CANAL OF SCHLEMM

The canal of Schlemm is usually represented as a ring lying in the corneo-sclera, which on cross-section (as in the meridional section of Fig. 1.15) appears as a circle or ellipse; its structure is, in fact, more complex, since it divides into several channels which, however, later reunite; hence the appearance of a cross-section varies with the position of the section. The channel, moreover, is partially interrupted by septa which probably serve a mechanical role in preventing obliteration of the canal when the pressure within the eye is very high.

The canal is essentially an endothelium-lined vessel, similar to a delicate vein; on the outside it rests on the scleral tissue whilst on the inside, nearest the aqueous humour, it is covered by a meshwork of endothelium-covered trabeculae—the *trabecular meshwork* or *pectinate ligament* (Fig. 1.16). Aqueous humour, to enter the canal, must percolate between the trabeculae and finally cross the endothelial wall

### TRABECULAR MESHWORK

On meridional section the meshwork appears as a series of meridionally orientated fibres (Fig. 1.15), but on tangential section it is seen to consist of a series of flat lamellae, piled one on top of the other, the fluid-filled spaces between these lamellae being connected by holes, as illustrated schematically in Figure 1.17. The meshwork has been divided into three parts with characteristically different ultrastructures; the innermost portion (1 of Fig. 1.18) is the *uveal meshwork*; the trabeculae making up the lamellae here are finer than those of the outer or *corneoscleral meshwork* (2) whilst the

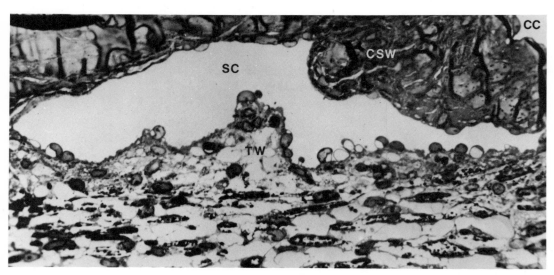

**Fig. 1.16**   Meridional section of Schlemm's canal and adjacent trabecular meshwork of rhesus monkey. Osmium fixed, toluidine blue stained. CC, collector channel; CSW, corneoscleral wall of Schlemm's canal; SC, Schlemm's canal; TW, trabecular wall of Schlemm's canal. Light micrograph × 750. (Courtesy, R. C. Tripathi.)

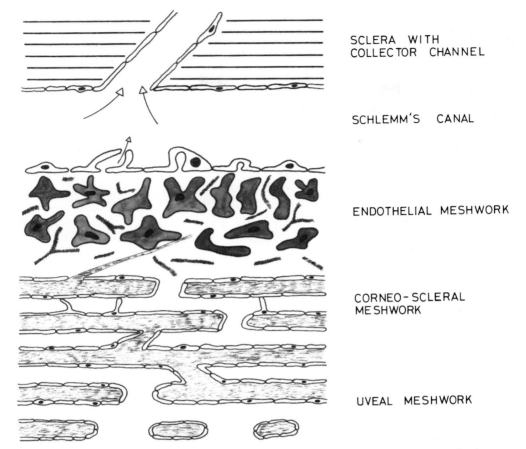

SCLERA WITH
COLLECTOR CHANNEL

SCHLEMM'S CANAL

ENDOTHELIAL MESHWORK

CORNEO-SCLERAL
MESHWORK

UVEAL MESHWORK

**Fig. 1.17**   Schematic representation of a section through chamber-angle tissue. (Bill, *Physiol. Rev.*)

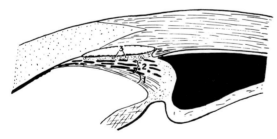

**Fig. 1.18**   Schematic drawing of the chamber-angle in primates, showing the three distinct parts of the trabecular meshwork. 1, uveal meshwork; 2, trabeculum corneosclerale; 3, inner wall, or pore area. Note trabeculum corneosclerale has been incorrectly shown as attaching to ciliary muscle. (Rohen, *Structure of the Eye.*)

meshes are larger. Each lamella of the uveal meshwork is attached anteriorly to the corneal tissue at the end of Descemet's membrane, constituting *Schwalbe's ring*,★ and posteriorly to the meridional fibres of the ciliary muscle. A pull of these fibres may well operate to open up the meshes of this trabecular structure, thereby

favouring passage of fluid. The third portion, lying immediately adjacent to the canal, was described by Rohen as the *pore area*, or *pore tissue*, but is now usually referred to as the *endothelial meshwork*.

**Ultrastructure**

As described by Tripathi (1977) the corneoscleral and outer uveal trabeculae are flattened perforated sheets orientated circumferentially parallel to the surface of the limbus. The inner 1 to 2 layers of uveal sheets are cord-like and orientated predominantly radially in a net-like fashion enclosing large spaces (Fig. 1.19). The beams constituting the meshwork are made up of collagen and of elastic tissue, covered by basement membrane material which, in turn, is completely or incompletely covered by a layer of flattened mesothelial cells. These cells exhibit both synthetic and phagocytic properties. As Figure 1.19 shows, the trabecular meshwork forms a labyrinth of intercommunicating intra-

★ *Iris processes* are made up of an occasional lamina of the uveal meshwork that crosses posteriorly to attach to the root of the iris.

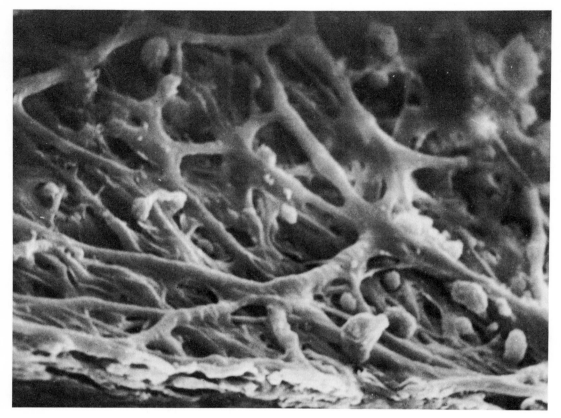

**Fig. 1.19** Scanning electron micrograph of the trabecular meshwork seen from the anterior chamber aspect. The rounded beams of the inner uveal meshwork, their endothelial covering and net-like arrangement and intercommunicating inter- and trans-trabecular spaces are seen in the foreground, whereas the deeper sheet-like trabeculae seen in the background are orientated circumferentially. (Approx. × 610) (Tripathi, *Exp. Eye Res.*)

and extra-trabecular spaces of variable size and shape, and may be likened to a coarse filter.

ENDOTHELIUM AND ADJACENT TISSUE

The endothelium of the canal consists in a single layer of cells whose lateral borders meet in either a narrow zone of apposition or in a more complex zone formed by cellular overlap or tongue-in-grove insertion (Grierson and Lee, 1975). Intercellular clefts exhibit junctional modifications which represent limited regions of apparent fusion between apposing plasma membranes, and are, in fact, gap-junctions in addition to adhering junctions.★ The cells rest on a basement membrane facing the *endothelial meshwork*, which may be considered to be the region lying between the basal aspect of the canal and the first layer of corneoscleral trabeculae; this is of variable thickness (8 to 16 μm).† The native cells constituting this region are usually orientated with their long axes parallel to the endothelium lining the canal, and form a network by process-contacts between neighbouring cells. In the outer corneoscleral meshwork, the adjacent trabeculae are separated by narrow intertrabecular spaces, and cellular contact across these spaces is achieved through cellular processes; once again, the regions of contact show gap-type and desmosome-type junctions. A characteristic feature of the endothelial cells lining the meshwork side of the canal is the prominent vacuolization of the cells; and its significance will be considered later in relation to the mechanism of drainage.

---

★According to Inomata *et al.* (1972), the junctional complexes include tight-junctions or *zonulae occludentes*.
†According to Inomata *et al.* (1972), the endothelial cells of the inner wall of Schlemm's canal, i.e. that facing the meshwork, have no basement membrane, the cells resting on very fine, unevenly distributed fibrillar material embedded in a homogeneous ground substance. This is in contrast to the cells of the outer layer, adjacent to the corneo-sclera, which rest on a well defined basement membrane. These cells lack vacuoles (p. 39).

## COMPARATIVE ASPECTS OF DRAINAGE

Earlier anatomists tended to distinguish between the primate angle of the anterior chamber, with its canal of Schlemm, and the drainage mechanism in lower mammals, e.g. Troncoso (1942). However, the physiological studies on drainage processes in non-primates,

this. It is generally located on the inner aspect of the anterior sclera and sclero-cornea, and is supported partly on the anterior chamber aspect or completely surrounded by the loose tissue of the angular meshwork. As in primates, the plexus forms a circumferential structure in lower mammals, birds and many reptiles, but is more localized in amphibians and fishes. Drainage

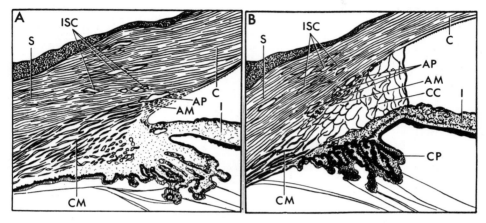

**Fig. 1.20** Diagrammatic representation of the angular region of a primate (A) and a lower placental (B) showing comparative morphological organizations. S, sclera; C, cornea; I, iris; AM, angular meshwork; AP, angular aqueous plexus; CM, ciliary muscle; CC, ciliary cleft; CP, ciliary processes; ISC, intrascleral collector channels. (Tripathi, *The Eye*.)

and the recent comparative anatomical studies of Tripathi (1974), have confirmed the essential uniformity in the mechanism and site of drainage, not only amongst mammals but also in lower vertebrates including fishes, although the morphological organization of the iridocorneal angle does show interesting variations.

ANGULAR AQUEOUS PLEXUS

Thus in primates, owing to the compact nature of the ciliary muscle, the angle is clearly defined (Fig. 1.20A), whereas in most amphibians, reptiles, birds and lower placentals it is ill defined owing to the wedge-shaped separation of the ciliary body into two leaves, one hugging the inner aspect of the sclera and the other being continuous with the fibrous base-plate of the ciliary body. The resulting *ciliary cleft* can be regarded as an extension of the anterior chamber (Fig. 1.20B); it is criss-crossed and bridged by fibro-cellular strands to form an *angular meshwork* or *pectinate ligament*★ (Fig. 1.21). The structure of this angular meshwork varies widely throughout the vertebrates. In all the species studied by Tripathi, an equivalent channel to Schlemm's canal could be discerned, and he suggested a generic term—*angular aqueous plexus*—to describe

of aqueous from this plexus takes place into intra- and epi-scleral plexuses of veins, and in some instances also by supraciliary and suprachoroidal plexuses of veins which connect directly to the angular plexus.

**Cellular lining**

As with the primate, the angular plexus is lined by mesothelium, the individual cells being linked by junctional complexes, whilst the large vacuoles described above for the primate eye are a universal feature, and which, on the basis of tracer studies, have been identified as representing the pathway into the channel. Thus, to quote Tripathi (1977) '. . . The normal presence of macrovacuolar structures in the mesothelial lining of the aqueous plexus in species as diverse as aquatic dogfish, arboreal pigeon and terrestrial man would suggest that the bulk flow of aqueous humour through the dynamic system of vacuolar transcellular pores is a fundamental biological mechanism of great antiquity.'

★The stout strands extending from the root of the iris to the inner aspect of the peripheral cornea (Fig. 1.21) have been likened to an inverted comb and hence the name pectinate ligament. These strands are most prominent in ungulates.

**Fig. 1.21**    The angular region of rabbit eye in meridional section. PL, pectinate ligament; CC, ciliary cleft, bridged by an accessory pectinate ligament (arrow); AM, angular meshwork; AP, angular aqueous plexus; C, peripheral cornea. Photomicrograph. $OsO_4$, 1 $\mu$m thick Araldite section. (Tripathi, *The Eye*.)

## THE AQUEOUS HUMOUR

The intraocular pressure depends primarily on the rate of secretion of aqueous humour, so that factors influencing this are of vital importance in understanding the physiology of the intraocular pressure. We may therefore pass to a consideration of the chemistry of the fluid and the dynamics of exchanges between it and the blood. These aspects, although apparently academic, are important for the full understanding of the experimental approaches to measuring the rate of secretion of the fluid and to controlling this.

## CHEMICAL COMPOSITION

The most obvious chemical difference between aqueous humour and blood plasma is to be found in the protein contents of the two fluids; in the plasma it is of the order of 6 to 7 g/100 ml, whereas in the aqueous humour it is only 5 to 15 mg/100 ml in man and some 50 mg/100 ml in the rabbit. By concentrating the aqueous humour

and then causing the proteins to move in an electrical field—*electrophoresis*—the individual proteins may be separated and identified. Such studies have shown that, although the absolute concentrations are so low, all the

**Table 1.1** Chemical composition of aqueous humour and blood plasma of the rabbit (after Davson, 1969)

|  | *Aqueous humour* | *Plasma* |
|---|---|---|
| Na | 143·5 | 151·5 |
| K | 5·25 | 5·5 |
| Ca | 1·7 | 2·6 |
| Mg | 0·78 | 1·0 |
| Cl | 109·5 | 108·0 |
| $HCO_3$ | 33·6 | 27·4 |
| Lactate | 7·4 | 4·3 |
| Pyruvate | 0·66 | 0·22 |
| Ascorbate | 0·96 | 0·02 |
| Urea | 7·0 | 9·1 |
| Reducing value (as glucose) | 6·9 | 8·3 |
| Amino acids | 0·17 | 0·12 |

Concentrations are expressed as millimoles per kilogramme of water.

plasma proteins are present in the aqueous humour, so that it is reasonable to assume that they are derived from the plasma rather than synthesized during the secretory process.

NON-COLLOIDAL CONSTITUENTS

As Table 1.1 shows, the crystalloidal composition of the aqueous humour is similar in its broad outlines to that of plasma, but there are some striking differences; thus the concentrations of ascorbate (vitamin C), pyruvate and lactate are much higher than in plasma, whilst those of urea and glucose are much less.

## Gibbs-Donnan equilibrium

The similarities between the two fluids led some earlier investigators to assume that the aqueous humour was derived from the blood plasma by a process of filtration from the capillaries of the ciliary body and iris; i.e. it was assumed that the aqueous humour was similar in origin and composition to the extracellular fluid of tissues such as skeletal muscle. If this were true, however, the ionic distributions would have to accord with the requirements of the Gibbs-Donnan equilibrium, according to which the concentrations of positive ions, e.g. $Na^+$ and $K^+$, would be greater in the plasma than in the aqueous humour, the ratio: $[Na]_{Aq}/[Na]_{Pl}$ being less than unity, whilst the reverse would hold with negative ions such as $Cl^-$ and $HCO_3^-$. Moreover, the actual values of the ratios: *Concn. in Aqueous Humour/Concn. in Plasma* would have to conform to those found experimentally when plasma was dialysed against its own ultrafiltrate. In fact, however, as Table 1.2 shows,

**Table 1.2** Values of the two distribution ratios; Concn. in Aqueous/Concn. in Plasma and Concn. in Dialysate/Concn. in Plasma. (Davson, Physiology of the Ocular and Cerebrospinal Fluids.)

|  | Concn. in Aqueous / Concn. in Plasma | Concn. in Dialysate / Concn. in Plasma |
|---|---|---|
| Na | 0·96 | 0·945 |
| K | 0·955 | 0·96 |
| Mg | 0·78 | 0·80 |
| Ca | 0·58 | 0·65 |
| Cl | 1·015 | 1·04 |
| HCO₃ | 1·26 | 1·04 |
| Urea | 0·87 | 1·00 |

the ratios show considerable deviations from what is demanded of a plasma filtrate, and the deviations are not attributable to experimental errors. Thus with a plasma filtrate the ratio: $[Na]_{Aq}/[Na]_{Pl}$ should be 0·945 whilst it is in fact 0·96, indicating that there is an excess of 1·5 per cent of sodium in the aqueous humour. More striking is the discrepancy of the bicarbonate ratio,

which in the rabbit is 1·26 compared with a value of 1·04 demanded by the Gibbs-Donnan equilibrium. It will be noted that the dialysis-ratio for calcium is 0·65; the theoretical ratio for a divalent cation is 0·92, and the discrepancy is due to the large amount of calcium held bound to plasma proteins in unionized form. With magnesium the discrepancy is not so large, due to a smaller degree of binding.

## Glucose

The concentration of glucose in the aqueous humour is less than in plasma, and the concentration in the vitreous body is even lower; these differences have been ascribed to consumption within the eye by the lens, cornea and retina; this utilization may be simply demonstrated by incubating the excised eye at 37°C; within two hours all the glucose has disappeared (Bito and Salvador, 1970). It is possible that a further factor is a low concentration in the primary secretion, since the concentration of glucose in the aqueous humour of cataractous eyes is not significantly lower than normal (Pohjola, 1966).

## Amino acids

The relative concentrations of amino acid in the plasma aqueous humour and vitreous body depend on the particular amino acid, so that the concentration in aqueous humour may be less than, equal to, or greater than that in plasma (Fig. 1.22); the concentration in the vitreous body is always very much less than in either of the other fluids (Reddy, Rosenberg and Kinsey, 1961; Bito et al., 1965). The low values in the vitreous body are not due to consumption within the eye since a non-metabolized acid, cycloleucine, shows a steady-state concentration in the vitreous body only 10 per cent of that in plasma after intravenous injection (Reddy and Kinsey, 1963), so that an active removal of amino acid by the retina has been postulated. When the dead eye is incubated, there is a large rise in concentration of amino acids in the vitreous body, due to escape from the lens, which normally accumulates these from the aqueous humour (Bito and Salvador, 1970).

## Ascorbic acid

The concentration of ascorbic acid in the rabbit's aqueous humour is some 18 times that in the plasma; the degree of accumulation varies with the species, being negligible in the rat and galago, but high in man, monkey, horse, guinea pig, sheep, and pig. As Kinsey (1947) showed, when the plasma level is raised, the concentration in the aqueous humour rises until a 'saturation level' is reached in the plasma, of about 3 mg/100 ml when the concentration in the aqueous humour reaches some 50 mg/100 ml. Raising the plasma level beyond the saturation level causes no further rise

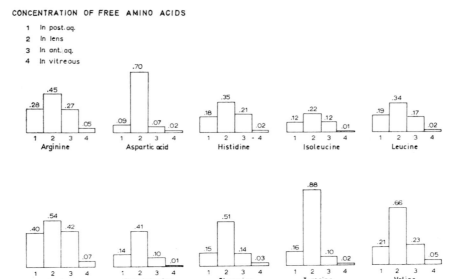

CONCENTRATION OF FREE AMINO ACIDS

1   In post. aq.
2   In lens
3   In ant. aq.
4   In vitreous

**Fig. 1.22**   Relative concentrations of free amino acids in lens, vitreous and aqueous humours of the rabbit eye. (After Reddy and Kinsey, *Invest. Ophthal.*)

in concentration in the aqueous humour. According to Linnér (1952) the ability of the eye to maintain this concentration of ascorbic acid depends critically on the blood-flow, so that unilateral carotid ligation causes a 17 per cent fall in concentration on the ligatured side. Hence, under non-saturation conditions, the concentration of ascorbic acid in the aqueous humour is some measure of the rate of blood-flow through the ciliary body. When the isolated ciliary body is placed in a medium containing ascorbic acid, the tissue accumulates this to concentrations considerably higher than in the medium; D-isoascorbate competed for accumulation but glucoascorbate, containing an additional CHOH-group, had no effect (Becker, 1967). The isolated retina also accumulates ascorbic acid (Heath and Fiddick, 1966); this is true of the rat, although in this species the ocular fluids contain a very low concentration.

AQUEOUS AS A SECRETION

Studies along these lines have left the modern investigator in no doubt that the fluid could not have been formed by such a simple process as filtration; in consequence the aqueous humour is called a *secretion*. What is implied by this is, essentially, that in order that it should be formed from the blood plasma, something more than a mechanical filtration process that 'skims' off the proteins is necessary; the cells of the ciliary epithelium must perform work—what is called *osmotic work*—in order to produce a fluid like the aqueous humour, which is not only different from plasma but also from a filtrate of plasma or dialysate. How this

happens is still a closed book; presumably the first step is the production of a plasma filtrate in the stroma of the ciliary processes; this is prevented from passing into the posterior chamber by the tightly packed cells of the ciliary epithelium. We may assume that, next, these cells abstract from this fluid the various ions and molecules that are ultimately secreted, and together with water they drive them into the posterior chamber. In making the selection, certain ions and molecules may be transferred across the cells to give higher concentrations in the secretion than in the original plasma filtrate, e.g. bicarbonate and ascorbate, whilst others are transferred at lower concentrations, e.g. calcium, urea, and so on.

ALTERATIONS OF THE PRIMARY SECRETION

Once formed, the composition of the fluid may well be modified by metabolic and other events taking place during its sojourn in the posterior and anterior chambers; thus the concentration of glucose could be low because the lens and cornea actually utilize it, whilst the relatively high concentrations of lactate and pyruvate could be the consequence of glycolytic activity by these structures. That this is probably true is shown by comparing the vitreous body with the aqueous humour; the vitreous body, being exposed to both lens and retina, might be expected to have an even lower concentration of glucose than that in the aqueous humour, and higher concentrations of lactate and pyruvate. In fact this is true. Moreover, if lactate and pyruvate are produced in this way, being stronger acids than carbonic acid, they will have decomposed some of the bicarbonate entering in the original secretion, so that the original concentration of bicarbonate must have been higher (and the chloride concentration lower) than that actually found, and recent studies of Kinsey and Reddy have shown that this is indeed true.

SPECIES VARIATIONS

The high concentration of bicarbonate in the aqueous humour shown in Table 1.1 is not a general characteristic common to all species; thus in horse, man, cat, rabbit and guinea pig the ratios: *Concn. in Aqueous Humour/Concn. in Plasma* are 0·82, 0·93, 1·27, 1·28 and 1·35 respectively. Hence in the horse and man the concentrations of bicarbonate are *less* than what would be expected of a filtrate of plasma. Study of the chloride composition shows an inverse relationship, as we might expect, the aqueous humours with high bicarbonates having low chlorides, and *vice versa*. It would seem that the secretion of a fluid with a high concentration of bicarbonate is concerned with the buffering needs of the eye, the bicarbonate being required to neutralize the large amounts of lactic and pyruvic acids formed by lens, retina and cornea. In animals like the rabbit and guinea pig the lens occupies a large proportion of the intraocular contents and so the buffering requirements are high; by contrast, in the horse and man the lens occupies a much smaller percentage of the contents (Davson and Luck, 1956). Wide species fluctuations in the concentration of ascorbate in the aqueous humour are also found; thus in the horse, man, rabbit, dog, cat and rat the concentrations are about 25, 20, 25, 5, 1 and 0 mg/100 ml respectively, although the concentrations in the plasma are all of the same order, namely about

1 mg/100 ml. Unfortunately the significance of ascorbate in the eye is not yet understood, so that the reason why the cat, dog and rat can manage with such low concentrations, whilst the rabbit and other species require high concentrations, is not evident. The suggestion that the lens actually synthesizes ascorbate and pours it into the aqueous humour has been disproved, at any rate as a significant source of the vitamin.

## THE BLOOD-AQUEOUS BARRIER

Our knowledge of the relationships between aqueous humour and plasma has been furthered by a study of the exchange of materials between the two fluids, studies of what has come to be known as the *blood-aqueous barrier*. If various substances are injected into the blood so as to establish and maintain a constant concentration in the plasma, then the rates at which the concentrations in the aqueous humour approach that of the plasma are usually very much less than occurs in such tissues of the body as skeletal muscle, liver, etc. It is because of this that we have come to speak of the 'blood-aqueous barrier', meaning that substances encounter difficulty in passing from the one fluid to the other.

### EXPERIMENTAL STUDIES

Figure 1.23 illustrates the penetration of some substances into the anterior chamber of the rabbit. In general, the substances fall into three categories. First there are the very large molecules like the plasma

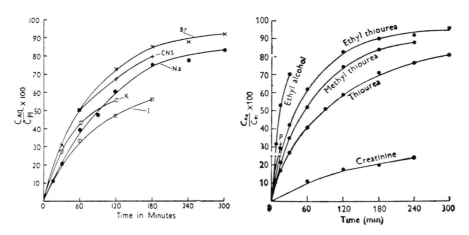

**Fig. 1.23** Penetration of some non-electrolytes (*left*) and ions (*right*) into the aqueous humour of the rabbit. The concentration of the given substance in the plasma ($C_{Pl}$) was maintained constant, and the concentration in the aqueous humour ($C_{Aq}$) was determined at different times. (Davson, *Physiology of the Ocular and Cerebrospinal Fluids*.)

proteins, dextrans and inulin which penetrate so slowly as to give the impression of an absolute barrier, although studies with isotopically labelled proteins leave us in no doubt that there is some, but very slow, penetration. Because the aqueous humour is constantly being drained away the concentration of these large molecules in the aqueous humour never rises very high and we may speak of a *steady-state* in which the ratio of concentrations is exceedingly low. This of course accounts for the low concentrations of plasma proteins in the aqueous humour. The second category of substance has a much smaller molecule, or may be an ion; the molecule is water-soluble, like urea, creatinine, *p*-aminohippurate, sucrose and so on. Penetration is considerably faster, usually, than that of the proteins. The third group are the lipid-soluble molecules which penetrate very rapidly; these include ethyl alcohol, various substituted thioureas, many sulphonamides. In general, the rate of penetration increases with increasing lipid-solubility, as measured by the oil-water partition coefficient. Water, although not strongly lipid-soluble, belongs to the category of rapidly penetrating substances. Again, the sugars such as glucose and galactose are insoluble in lipid solvents but they also penetrate the barrier rapidly.

## MECHANISM OF THE BARRIER

We may imagine that the cells of the ciliary epithelium continually secrete a fluid, which we may call the *primary aqueous humour*, that is emptied into the posterior chamber and carried through the pupil into the anterior chamber to be finally drained away out of the eye. If a foreign substance, or the isotope of a normally occurring constituent of the aqueous humour, e.g. $^{24}$Na, is injected into the blood it presumably equilibrates rapidly with the extracellular fluid in the ciliary processes, and thus gains ready access to the cells of the ciliary epithelium. If the substance penetrates these cells easily, then we may expect it to appear in the secretion produced by them relatively rapidly. The primary aqueous humour will therefore contain the substance in high concentration. This circumstance will favour penetration into the aqueous humour as a whole. On mixing with the fluid already present in the posterior chamber, the primary fluid will be diluted, so that it will take time for the aqueous humour as a whole to reach the concentration in the plasma, in the same way that if we were to add a coloured solution to a tank of water, removing fluid from the tank as fast as the coloured fluid was added, in time the concentration of dye in the tank would come up to that in the coloured solution, but the process would take time and be determined by the rate of flow.

### DIFFUSION FROM IRIS

While in the posterior and anterior chambers, however, the aqueous humour is subjected to diffusional exchanges with the blood in the iris, as illustrated in Figure 1.24, so that if these exchanges are significant

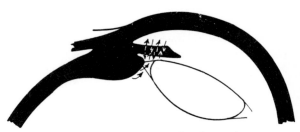

**Fig. 1.24**   Illustrating the diffusional exchanges between aqueous humour in posterior and anterior chambers on the one hand, and the blood plasma in the iris, on the other.

the rate at which the foreign substance comes into equilibrium with the aqueous humour is increased. There is no doubt that lipid-solubility is an important factor in determining the rapidity of these diffusional exchanges. The anterior surface of the iris is not covered by an intact epithelial or endothelial layer, so that restraint on diffusion from blood into the anterior chamber must depend on the capillaries of this tissue. That the iris offers considerable restraint to exchanges with the aqueous humour was demonstrated by Davson and Spaziani (1960), who infused artificial aqueous humour containing different solutes into the anterior chamber of the cat; under these conditions the pupil closed to a slit so that a large area of iris was exposed to the aqueous humour. If lipid-insoluble substances such as $^{24}$Na, para-aminohippurate or sucrose were injected, these escaped at a rate that could be completely accounted for by bulk flow through the canal of Schlemm; even lipid-soluble substances, such as propylthiourea, did not escape much faster than sucrose.

### Large molecules

On the basis of this scheme, we may interpret the blood-aqueous barrier phenomena qualitatively. Thus large lipid-insoluble molecules such as the plasma proteins will probably be held back partially by the capillaries in the ciliary processes; the permeability of ciliary body capillaries to proteins is unusually high, however, (Bill, 1968)★ so that the main restraint will be

---

★Thus Bill (1968, a, b), measured the penetration of labelled myoglobin, serum albumin and gammaglobulin from blood into the stroma of the ciliary body and choroid; the turn-over rates in the tissue were 25, 3·9 and 1·6 per cent per minute respectively, showing that the capillaries were exerting a sieve-like action. Bill estimated that the albumin concentration in the stroma was some 55 to 60 per cent of that in plasma, so that the oncotic pressure of the tissue is relatively high.

the clefts between the unpigmented ciliary epithelial cells (p. 27). Penetration into these cells and through the clefts will be minimal, and it may be that passage across this unpigmented layer is by a process of pinocytosis or engulfment of the proteins into small vacuoles, which subsequently empty themselves into the posterior chamber (Bill, 1975). During the fluid's sojourn in the posterior and anterior chambers, additional amounts may diffuse from the iris; once again, however, the vacuolar pathway seems likely, since the capillaries of the iris are probably completely impermeable to proteins (p. 29). Because of the sluggishness of all these processes, the concentration in the aqueous humour will never build up to that pertaining in the plasma; in fact, as indicated above, the concentration-ratio never rises above about 1/200 for the plasma proteins.★

### Ions and lipid-soluble molecules

With substances such as $^{24}$Na, the evidence suggests that the primary secretion contains this ion in the same concentration as in the plasma, whilst the contributions of diffusion from the iris are also significant. The overall rate of penetration into the aqueous humour is thus considerably greater than that of the proteins. With ethyl alcohol penetration is altogether more rapid because of the lipid-solubility of the molecules; this permits of very rapid exchanges across the iris.

*Facilitated transfer*

Sugars and amino acids are relatively large and very lipid-insoluble molecules so that their penetration into the aqueous humour, which occurs relatively rapidly (Davson and Duke-Elder, 1948; Harris and Gehrsitz, 1949) is doubtless due to facilitated transport, namely the development of 'carriers' or active membrane sites that specifically react with these solutes.

BLOOD-VITREOUS BARRIER

Early studies on the kinetics of the blood-aqueous barrier tended to ignore the influence of the vitreous body on the penetration of material into the aqueous humour. In general, the vitreous body comes into equilibrium with the blood much more slowly than the aqueous humour and we may speak of a *blood-vitreous barrier*. The vitreous body probably receives material from the choroidal and retinal circulations, but because, with many substances, equilibrium with plasma takes a long time to be achieved, it follows that the vitreous body must also receive material from the aqueous humour in the posterior chamber, the concentration tending to be higher in this fluid. The blood vitreous barrier is thus complex, and may be illustrated by Figure 1.25.

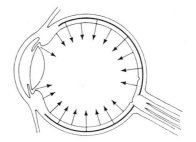

**Fig. 1.25**    The blood-vitreous barrier.

Since the vitreous body is exposed to diffusional exchanges with blood in the retinal circulation, as well as the choroidal, we must seek some reason for the slow rates of exchange of this body with blood. The pigment epithelium doubtless restricts passage from the choroid; as we shall see, the cells are sealed by tight junctions. Thus material escaping from the highly permeable choroidal capillaries is held up at the pigment epithelium. Treatment of rabbits with iodate, which has a specific effect in damaging the cells of the pigment epithelium (Noell, 1955; Grignolo *et al.*, 1966), causes a marked decrease in electrical resistance across the retina (Noell, 1963) and an increased permeability of the blood-vitreous barrier (Davson and Hollingsworth, 1972).

### Diffusion into vitreous body

A study of the penetration of, say, $^{24}$Na into the aqueous humour must take account of the circumstance that quite a considerable fraction of the $^{24}$Na passing from plasma to posterior chamber diffuses back into the vitreous body, instead of being carried into the anterior chamber. The extent to which this occurs depends largely on the relative rates at which the vitreous body and aqueous humour come into equilibrium with the plasma. With lipid-soluble substances like ethyl alcohol or ethyl thiourea the discrepancy is small, and the loss of material from the posterior chamber is not serious.† With $^{24}$Na and various other ions the losses are very significant.

MATHEMATICAL ANALYSIS

In consequence, the mathematical analysis of the process of equilibration becomes highly complex; moreover, the experimental study demands that we measure penetra-

---

★The concentrations of immunoglobulins in the tissues of rabbit and human eyes have been determined by Allansmith *et al.* (1971; 1973); these are remarkably high in the avascular cornea, and low in the retina and iris.
†With lipid-soluble substances, however, the penetration of material into the lens becomes a significant factor; in an animal like the rabbit, where the volume of the lens is about twice that of the aqueous humour, this can be a very serious factor.

tion into both anterior and posterior chambers separately, as well as into the vitreous body and sometimes the lens. It is not feasible to enter into the details of the mathematical analysis, which has been so elegantly carried out by Kinsey and Palm and Friedenwald and Becker. Figure 1.26 shows the experimental basis for one such analysis, namely the penetration of $^{24}$Na into

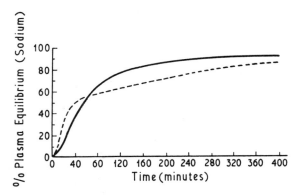

**Fig. 1.26**   The penetration of isotopically labelled sodium into the posterior (broken line) and anterior chambers (full line) of the rabbit. (After Kinsey and Palm, *Arch. Ophthal.*)

posterior and anterior chambers. As we should expect, the posterior chamber tends to come into equilibrium more rapidly than the anterior chamber; the interesting feature is the crossing over of the curves, so that at the later stages the anterior chamber is ahead; this is because the posterior chamber is losing material to the vitreous body. A similar situation is found with chloride, studied with the isotope $^{36}$Cl. The value of the mathematical analysis of the process of penetration is that it enables one to compute the probable character of the aqueous humour as it is primarily secreted (Kinsey and Reddy, 1959).

## POSTERIOR AND ANTERIOR CHAMBERS

In describing the chemistry of the aqueous humour we considered simply the fluid obtained by inserting a needle into the anterior chamber and withdrawing all the fluid; the aqueous humour so studied was a mixture

of posterior and anterior fluids. When the fluids are withdrawn separately and analysed, their comparative chemistry reveals, in some measure, their different histories. Thus the bicarbonate concentration in the posterior fluid is higher than that in the anterior fluid, in the rabbit, and this is because the freshly secreted fluid has a much higher concentration than in the plasma; while in the posterior chamber, some diffuses back to the vitreous body and some is decomposed by lactic and pyruvic acids; in the anterior chamber more is decomposed by the acids formed by lens and cornea, and some diffuses back into the blood in the iris. In a similar way we can account for the circumstance that the ascorbate concentration is higher in the posterior fluid; here, the diffusion from the anterior chamber into the blood in the iris is the determining factor. With chloride the concentration is less in the posterior fluid; this is because the freshly secreted fluid has a relatively low concentration of this ion; while in the anterior chamber, chloride diffuses from the blood into the anterior fluid. Some analyses, including the vitreous body, are shown in Table 1.3.

### POTASSIUM, MAGNESIUM AND CALCIUM

The relative concentrations of these ions in plasma and ocular fluids have not received a great deal of attention, due, presumably, to difficulties in chemical analysis. The matter is certainly of great interest since the concentrations of these ions are very important for the functioning of the central nervous system, and they are maintained in the cerebrospinal fluid and extra-cellular fluid of the central nervous tissue at levels that are different from those of a plasma filtrate. Since the retina is a part of the central nervous system, and since it is exposed to the vitreous body, the concentrations of these ions in the vitreous body are of great interest; since, moreover, the vitreous body is in free communication with the aqueous humour in the posterior chamber, the concentrations in the aqueous humour are also relevant.

### Potassium
The situation is complicated by species variability; for example, the concentration of $K^+$ in the anterior aqueous humour of the dog is considerably higher

**Table 1.3**   Relative compositions of plasma, aqueous humours and vitreous body in the rabbit. (Reddy and Kinsey, 1960)

|  | Na | Cl | Total CO$_2$ | Ascorbate | Lactate | Glucose |
|---|---|---|---|---|---|---|
| Plasma | 143·0 | 109·0 | 20·6 | 0·04 | 10·3 | 5·7 |
| Anterior aqueous | 138·0 | 101·0 | 30·2 | 1·11 | 9·3 | 5·4 |
| Posterior aqueous | 136·0 | 96·5 | 37·5 | 1·38 | 9·9 | 5·6 |
| Vitreous | 134·0 | 105·0 | 26·0 | 0·46 | 12·0 | 3·0 |

Concentrations are expressed in millimoles per kilogramme of water.

than that in a plasma dialysate, whereas in the cat and rabbit it is less. Some figures from a recent paper by Bito (1970) have been collected in Table 1.4. Perhaps the most striking feature is the existence of quite large differences in concentration between the anterior and posterior halves of the vitreous body; such differences indicate that the retina + choroid is acting as either a

gradients of concentration between posterior aqueous and anterior vitreous, and between anterior vitreous and posterior vitreous, indicate beyond doubt that the retina is actively removing amino acids from the vitreous body; and the process has been demonstrated experimentally by Reddy et al. (1977a,b). These authors have shown, for example, that treatment of the sheep with

**Table 1.4** Concentrations of certain ions (mequiv/kg $H_2O$) in ocular fluids and plasma dialysate. (After Bito, 1970)

| Species | Ion | Plasma dialysate | Aqueous Ant. | Post. | Vitreous Ant. | Post. |
|---------|-----|------------------|--------------|-------|---------------|-------|
| Dog | K | 4·06 | 4·74 | 5·32 | 5·24 | 5·11 |
| Cat | K | 4·53 | 4·15 | 4·70 | 5·15 | 4·99 |
| Rabbit | K | 4·37 | 4·26 | 4·69 | 4·69 | 4·62 |
| Dog | Mg | 1·18 | 1·07 | 1·06 | 1·28 | 1·38 |
| Cat | Mg | 1·26 | 0·99 | 0·89 | 1·01 | 1·08 |
| Rabbit | Mg | 1·49 | 1·41 | 1·50 | 1·99 | 2·16 |
| Dog | Ca | 3·05 | 2·96 | 2·92 | 3·34 | 3·56 |
| Cat | Ca | 3·02 | 2·70 | 2·78 | 3·04 | 3·22 |
| Rabbit | Ca | 4·12 | 3·48 | 3·48 | 3·90 | 3·94 |

source or sink for the ion. Thus, in the dog, the concentration of $K^+$ in the posterior half is less than in the anterior half, and this is less than in the posterior aqueous humour; these gradients could be due to the retina + choroid acting as a sink for $K^+$, secreted by the ciliary body at a higher concentration than in a plasma filtrate. A further complicating factor is the possibility that the lens may be actively taking up $K^+$ at its anterior surface and releasing it at its posterior surface (Bito and Davson, 1964) and this view is supported by Kinsey's studies on transport of this ion across the lens (pump-leak hypothesis, p. 129), and the observation of Bito et al. (1977) that, in the absence of the lens, the rabbit's vitreous has a concentration of $K^+$ considerably lower than in a dialysate of plasma.

## Magnesium

The gradients with $Mg^{2+}$, on the other hand, suggest that the retina plus choroid is acting as a source of this ion and this indicates that the actual concentration of $Mg^{2+}$ immediately adjacent to the retina is even higher than in the bulk of the posterior segment of vitreous. The concentration of $Mg^{2+}$ in the cerebrospinal fluid is likewise much higher than in a plasma dialysate (Davson, 1956) so that it seems that central nervous tissue requires a higher concentration of $Mg^{++}$ in its extracellular fluid than that provided by a simple filtrate of plasma.

### AMINO ACIDS

The low concentrations of amino acids in aqueous humour have been discussed earlier (p. 19); the

iodate, which attacks the pigment epithelium specifically, actually caused the steady-state concentration of cycloleucine in the vitreous body to increase, indicating that the iodate had inhibited the active transport of the amino acid out of the eye. These authors emphasized the species variation in the steady-state distribution of amino acids in the ocular fluids; thus in the sheep the concentrations in posterior and anterior aqueous humours were always above that in the plasma, indicating an active transport inwards across the ciliary epithelium, and possibly the iris. It is possible, moreover, that the low concentrations found in many species result from the vitreous 'sink', caused by active transport across the pigment epithelium.

### SUMMARY DIAGRAM

Figure 1.27 from Bito (1977) illustrates the gradients of concentration within the eye established by recent chemical studies.

### VITREOUS BODY AS EMERGENCY RESERVOIR

Bito (1977) has emphasized that, because of its relative bulk, the vitreous of, say, the rabbit, contains some 1·2 mg of glucose. In the enucleated eye maintained at body-temperature there is a rapid utilization of this (Davson and Duke-Elder, 1948), and this is due to utilization by both lens and retina; the retinal contribution to utilization was demonstrated by the gradients of concentration in the vitreous, and also by the continued utilization by the aphakic eye; in this case it amounted to some 10 μg/min (Bito et al., 1977). Thus

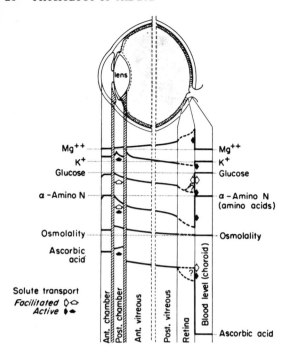

**Fig. 1.27**   Schematic representation of the gradients of some solute concentrations and total osmolality in the intraocular fluid system of a typical mammalian eye. The blood levels are indicated by the dotted line and by the value given for the choroid. There is active transport of magnesium into the eye across the blood–retinal barrier while potassium and most amino acids are transported outward. Potassium and ascorbic acid are actively transported into the posterior chamber, while amino acids and glucose transport can be either active or passive at this site. The facilitated transport of glucose and ascorbic acid across the blood–retinal barrier is indicated by a question mark since they have not been proven. The highest osmolality is found in the posterior chamber, and the fluids become more isotonic toward both the anterior chamber and the retina. (Bito, *Exp. Eye Res.*)

in acute metabolic stress the vitreous might act as a temporary supply, and as a 'sink' for accumulating lactate.

## WATER AND OSMOTIC EXCHANGES

The aqueous humour and blood plasma are approximately isosmolar, with the possibility that the aqueous humour is slightly hyperosmolar. Thus exchanges of water, apart from those concerned in the formation of the fluid, are normally negligible. When the osmolarity of the blood is altered experimentally, for example by intravenous injection of a hypertonic salt solution, then there is a rapid adjustment of the osmolarity of the aqueous humour, indicating that the barrier to water is relatively slight by comparison with the barrier to salts; and this is confirmed by direct studies with labelled

water. The loss of water resulting from the hypertonic injection brings about a fall in the intraocular pressure, and the effectiveness of various solutes in reducing the intraocular pressure is presumably related to the ease with which they may cross the blood-aqueous and blood-vitreous barriers. Thus, if the solute penetrates rapidly into the fluids a difference of osmotic pressure cannot be maintained for any length of time and so the loss of water will be small. For this reason sucrose or sorbitol is effective. The special effectiveness of urea, which may be used clinically to reduce intraocular pressure temporarily, is due to a relatively slow rate of penetration of the barriers, but also to the circumstance that it is not rapidly excreted by the kidneys, so that a high level may be maintained in the plasma for some time; finally, its molecular weight is low, so that from an osmotic point of view a 10 per cent solution of urea is as effective as a 60 per cent sucrose solution.

## THE MORPHOLOGY OF THE BARRIERS

The experimental studies outlined above have revealed a restraint on the passage of many solutes from blood into aqueous humour and vitreous body; and we must now enquire into the sites of the barriers that give rise to this restraint. Thus, as indicated earlier (p. 22), blood in the iris, ciliary body and choroid comes into close relations with the aqueous and vitreous humours. If relatively free and unselective exchanges are not to occur between blood and these fluids there are two possibilities; first, that a selectively permeable layer of epithelium be interposed between the stroma of the tissue and the fluids, or second, that the capillaries in the stroma be of a type similar to those found in the brain, exhibiting a high degree of selectivity and generally restraining the passage of most water-soluble compounds. In the eye, both mechanisms operate; thus the ciliary epithelium and retinal pigment epithelium separate aqueous humour and vitreous body from the stroma of the ciliary body and choroid respectively, whilst in the iris and retina the restraints to free exchanges with blood are imposed by the capillaries themselves.

## CILIARY EPITHELIUM

The blood capillaries of the ciliary body supplying the tissue immediately under the epithelium in the ciliary processes are large, like venules, and exhibit fenestrations that occupy the entire circumference of the endothelium. Capillaries with fenestrations are found to be highly permeable to all solutes, so that even plasma proteins escape into the stroma and tend to build up a high colloid osmotic pressure (Bill, 1975). Thus restraint to passage from the ciliary stroma must be determined by the ciliary epithelium and such a restraint may be visualized by injecting into the blood a tracer that escapes from the capillaries but yet is large enough to be visualized in the electron microscope. Such a tracer is horseradish peroxidase.* When Vegge (1971) injected

layers may fuse to give the tight junction, or *zonula occludens*, where the layers have been reduced to 5 (Fig. 1.28B); and it is found that electron-dense markers such as ferritin or horseradish peroxidase fail to pass through the intercellular cleft, which is therefore sealed. A membrane consisting of a layer of cells, each with its space occluded by a tight junction, would have a low

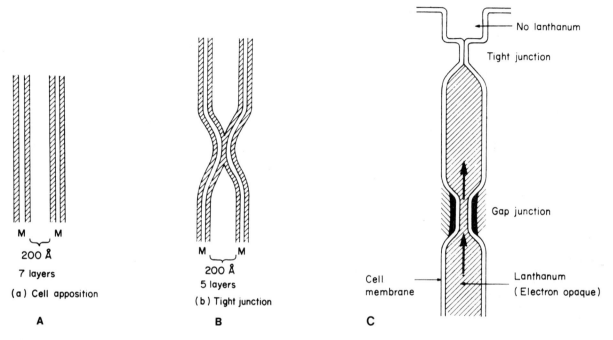

**Fig. 1.28**    Illustrating types of cell contact. (A) Simple apposition. (B) Tight junction where there has been fusion of adjacent laminae. (C) Tight junction plus gap-junction in series. This preparation has been treated with colloidal lanthanum from the blood side, as indicated by the cross-hatching. Lanthanum has passed through the gap-junction but has been restrained by the tight junction. (Davson and Segal, *Introduction to Physiology*.)

this into the monkey he found that it rapidly filled the stroma of the ciliary body and the intercellular spaces of the pigmented layer, but was held up at the non-pigmented layer passing along the intercellular clefts only to a restricted extent and was prevented from passing further by junctional complexes. A similar result was described by Shiose (1970) in the mouse and Uusitalo et al. (1973) in the rabbit.†

TIGHT JUNCTION

When examined in thin sections in the electron microscope, the individual cells of epithelial sheets are seen to be separated from each other by a space of about 20 nm; since each cell membrane appears, at high magnification, to consist of three layers, the region of cell apposition appears as a seven-layered structure (Fig. 1.28A). At certain regions these membranes may come into much closer apposition and their constituent

permeability to solutes, and this would be manifest as a high electrical resistance.

GAP-JUNCTION

Another type of junction, formerly confused with and included in the category of tight junction, is the *gap-junction*; here, the outer layers of the cell membranes come into very close juxtaposition, but when a small electron-dense marker, such as lanthanum hydroxide,

*The technique employing this as an electron-dense marker was developed by Karnovsky (1967); it has a molecular weight of 40 000 and a diameter of 50 to 60 A; by causing it to react with $H_2O_2$ and 3-3¹-diaminobenzidine, the reaction product reacts with osmium to produce a dense electron-staining.

† Uusitalo et al. (1973) point out that the ciliary epithelium on the ciliary processes and pars plana does allow some horse-radish peroxidase to pass between its non-pigmented cells, so that it is possible that the small amounts of protein entering the aqueous humour are carried through these regions.

is employed, it is seen to pass through a narrow inter-cellular space of about 4 nm diameter, the outer leaflets of the cell membranes not having fused as in the tight junction (Fig. 1.28C). Thus the gap-junction constitutes a pathway for ready passage of small solutes between cells, so that an epithelium whose cells were linked by this type of junction only would be very leaky and have a low electrical resistance. Another feature of the gap-junction is that it serves to connect the two adjacent cells, so that if a dye stuff is injected into one cell the adjacent cell soon becomes coloured, due to diffusion across the gap-junction; in this way the fibres of cardiac and smooth muscle are joined, so that the tissue behaves as a syncytium.

### DESMOSOME

A third main type of junction is the *desmosome*, or *macula adhaerens*; this serves only to hold the cells together mechanically.

### Junctional complex

In general these junctions do not occur singly but as a 'junctional complex' with the tight junction, or zonula occludens, closest to the apices of the cells.

### CILIARY JUNCTIONS

Figure 1.29 illustrates the types of junction in the ciliary epithelium; it is the apices of the unpigmented cells that contain the true zonulae occludentes that act as seals restricting diffusion between the cells, so that the whole layer behaves as a single sheet of membrane-covered cytoplasm rather than a discontinuous array of cell membranes and intercellular spaces (Raviola, 1971; 1977). The gap-junctions between adjacent pigmented cells, and between pigmented and unpigmented cells, allow the tissue to behave as a syncytium, in the sense that diffusion from one cell to another is relatively unrestricted. These gap-junctions have been demon-strated by Raviola (1971) and Smith and Rudt (1975) who showed that, whereas horseradish peroxidase (MW 40 000) was held up at the junctions, the smaller microperoxidase (MW 1900) did actually penetrate through the fine intercellular space at the gap-junctions between non-pigmented epithelial cells.

### THE IRIS BARRIERS

#### POSTERIOR EPITHELIUM

This layer seems to be functionally similar to its analogue in the ciliary body, namely the unpigmented layer, since its intercellular clefts are probably sealed by tight junctions; at any rate, according to Raviola (1977), horseradish peroxidase, which leaks out of the ciliary body capillaries into the stroma, diffuses into the root of the iris but is held up at the intercellular clefts of the posterior epithelium.

punctum
adhaerens

gap
junction

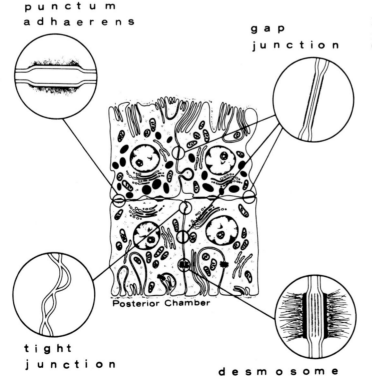

Posterior Chamber

tight
junction

desmosome

Fig. 1.29   Illustrating the intercellular junctions in the ciliary epithelium. (Raviola, *Exp. Eye Res.*)

## IRIS CAPILLARIES

The blood capillaries of the iris seem to be different from those of the ciliary processes, although they are derived from the same arterial circle; thus we have seen that the ciliary capillaries are fenestrated and allow the escape of large molecules such as horseradish peroxidase and plasma proteins. By contrast, horseradish peroxidase (Shiose, 1971; Vegge, 1971), microperoxidase (Smith and Rudt, 1975) and even fluorescein fail to escape from iris capillaries, and morphological studies have revealed tight junctions between the endothelial cells similar to those in brain capillaries (Vegge, 1972) mostly located close to the luminal end of the cleft. There seem to be some species variations, which have been summarized by Szalay et al. (1975); thus gap-junctions seem to be predominant in rat, cat and pig, whilst in mouse, monkey and man the junction is of the occluding type (Smith, 1971; Vegge and Ringvold, 1969). Thus it would appear that the maintenance of a barrier between the blood in the iris stroma and the aqueous humour in posterior and anterior chambers is achieved by restraint on passage out of the capillaries, just as in the brain; an additional barrier is provided by the posterior epithelium interposed between iris stroma and posterior chamber.

## IRIS ANTERIOR LINING

We have indicated that the consensus of opinion, based on light-microscopy, denied the existence of an epithelial or endothelial sheet, continuous with the corneal endothelium, and covering the anterior surface of the iris. Electron microscopy has, indeed, revealed a surface layer of cells, which, however, lack a basal lamina and are thus more similar to fibroblasts (Ringvold, 1975; Raviola, 1977). As seen in the scanning electron microscope, the cells are star-shaped, and their processes intertwine with those of neighbouring cells. Between the cell-bodies and processes large openings are present, traversed by small bundles of connective tissue fibres. The same openings are seen in freeze-fractured preparations so that they are unlikely to be due to shrinkage during fixation. These large holes between cells mean that there is no barrier to free diffusion between the stroma of the iris and the aqueous humour in the anterior chamber, so that the blood-aqueous barrier in this region is determined by the nature of the iris capillaries with their tight junctions.

# THE CHOROID AND PIGMENT EPITHELIUM

The choroidal capillaries are similar to those of the ciliary body, being fenestrated; and the intercellular contacts are peculiar and reminiscent of those in liver sinusoids. Studies on passage of macromolecules

from blood into the choroidal stroma indicate a high permeability (Bill, 1968), and horseradish peroxidase passes readily into these spaces (Shiose, 1970). Thus the choroid, like the ciliary body, contains vessels that are readily permeable to the solutes of plasma, so that the observed restraint on passage of such solutes from plasma into the vitreous body must be due to the inter-position of a selective membrane. At one time it was thought that the limiting membranes of the retina—inner and outer limiting membranes—constituted a barrier between vitreous and choroid.

## PIGMENT EPITHELIUM

However, when horseradish peroxidase was added to the vitreous body it was found to diffuse rapidly across these limiting membranes and to be brought to a halt at the tight junctions of the pigment epithelium (Peyman et al., 1971), whilst horseradish peroxidase, injected into the blood, fails to pass beyond the stroma of the choroid, being held up at tight junctions between the pigment epithelial cells (Peyman and Bok, 1972). The junctions between pigment epithelial cells have been described by Hudspeth and Yee (1973) in the eyes of a number of species, ranging from frog to cats and monkeys; they were similar in all species, so that the frog's epithelium may be treated as typical. The junctional complex occurred about half-way down the cell, rather than at the apex, and consisted of an apical gap-junction (nearest the vitreous) and a basal desmosome or macula adhaerens, and between and over-lapping these is the tight junction or zonula occludens. The presence of gap-junctions suggests electrical coupling between the cells, and this was proved by passing current through one cell and recording the electrical changes in cells at greater and greater distances away; the epithelium was clearly behaving as a syncytium.

# RETINAL CAPILLARIES

In those species that have a retinal circulation, in addition to the choroidal circulation, the maintenance of a blood-vitreous barrier must depend not only on a barrier between choroid and vitreous, exerted by the pigment epithelium, but also a barrier across the capillary endothelium of the retinal circulation. The retinal circulation is equivalent to the cerebral circulation, since the nervous layers of the retina are a part of the brain; in the cerebral circulation it has been demonstrated that the basis of the blood-brain barrier is the existence of tight junctions between the endothelial cells of the capillaries (Reese and Karnovsky, 1967). Experimental studies on the retinal circulation, using electron-dense markers, showed that horseradish peroxidase, for example, was prevented from leaving retinal capillaries by the tight junctions between endothelial cells (Shiose, 1970). Horseradish peroxidase has a molecular weight of 40 000 and a diameter of about 5 nm; the much smaller microperoxidase, with a molecular weight of only 1900 daltons, also fails to cross the retinal capillary endothelium (Smith and Rudt, 1975).

To conclude, then, the aqueous humour and vitreous body are insulated against free exchanges of solutes with blood by epithelial layers and, where these are lacking, by capillaries with an endothelium in which the intercellular clefts are closed by zonulae occludentes. A 'leak' in this system is probably provided by the anterior surface of the iris; solutes could diffuse out of the ciliary capillaries into the stroma of the ciliary body and from there they could diffuse to the iris root and thence into the anterior chamber. The same route in reverse is available for the uveo-scleral drainage process of Bill (p. 46).

## SECRETION OF AQUEOUS HUMOUR

### ACTIVE TRANSPORT

The continuous formation of a fluid that is not in thermodynamic equilibrium with the blood plasma requires the performance of work; such work is described as active transport, and the common basis for this, whether it be the production of saliva, the secretion of gastric juices, or the reabsorption of fluid from the kidney tubule, is accomplished by the transport of one or more ions from the blood side of the system to the secreted fluid side. The osmotic gradient caused by the transport of the ion or ions leads to flow of water, and, depending on the permeability characteristics of the limiting membranes across which the transport occurs, the fluid may ultimately be isotonic, hypertonic or hypotonic. Thus the process in the eye takes place across the ciliary epithelium; in the stroma of this tissue the blood is filtered by the capillaries to produce a filtrate comparable with tissue fluid in other parts of the body; because of the high permeability of the capillaries the filtrate is by no means free from protein. The unpigmented epithelium, however, acts as a diffusional barrier to this protein and to most of the other constituents of the filtrate, so that the transport is slowed and becomes selective.

### COLE'S HYPOTHESIS

The basic hypothesis, developed by Cole (1977), is that the unpigmented cells absorb selectively $Na^+$ from the stroma and transport it into the intercellular clefts which, as Figure 1.30 shows, are closed at the stromal side by tight junctions but are open at the aqueous humour side; the development of hyperosmolarity in the

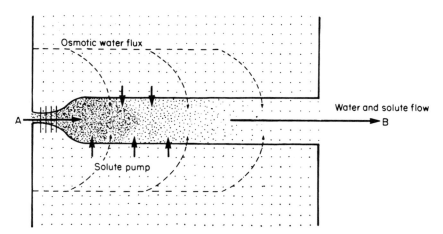

**Fig. 1.30**   Representation of standing gradient osmotic flow system. The system consists of a long, narrow channel, the entry to which is restricted at the left-hand end (A). The density of dots indicates the solute concentration. Solute is actively transported from the cells into the intercellular channel by a solute pump, thus making the channel-fluid hypertonic. As solute diffuses towards the open end of the channel at the right-hand side (B), water enters across the walls because of the osmotic differential. In the steady state a standing gradient will be maintained with the osmolarity decreasing from the restricted to the open end and with volume-flow directed towards the open end. In the limiting case, the emergent fluid approaches iso-osmolarity with the intracellular compartments. The arrow on the left-hand side (across the restriction) represents an element of hydraulic flow. (Cole, *Exp. Eye Res.* modified from House, *Water Transport in Cells and Tissues.*)

clefts leads to osmotic flow of water from the stroma and thus to a continuous flow of fluid along the clefts. The passage of other ions may also be governed by independent active processes, e.g. $Cl^-$, $HCO_3^-$, $K^+$, and so on, whilst others may diffuse passively down concentration gradients established by the primary process; the same may apply to sugars and amino acids.

## BASIC CONSIDERATIONS

### POTENTIAL

Active transport of an ion is often associated with the development of a potential; thus the frog-skin, which for simplicity we may represent as a single layer of cells as in Figure 1.31, actively transports NaCl from

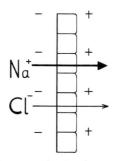

**Fig. 1.31** Active transport of $Na^+$ across a layer of epithelium can cause a potential due to acceleration of the positive ions with respect to the passively flowing negative ions.

outside to inside even though the concentrations on both sides are maintained the same, or, if the concentration is much lower outside when the animal is in fresh water. The transport of the $Na^+$-ion is equivalent to an increase in its mobility in relation to the negative $Cl^-$-ion, so that the side to which $Na^+$ is transported becomes positive; the difference of potential created by the separation of the ions accelerates the $Cl^-$-ions, which are said to follow passively in the wake of the actively transported $Na^+$-ions; thus in a given time, because of the establishment of this potential, equal numbers of ions pass from the bulk of the outside medium to the bulk of the inside medium, and the potential is confined to the membrane. Electrodes placed in the two media will therefore register the active transport potential, often described as a pump-potential since it results from the pumping action of metabolism on a given ion. Often the transport of $Na^+$ is accompanied by a transport of $K^+$ in the opposite direction, in which case the pump is called a $Na^+$-$K^+$-pump, the transport of $Na^+$ in one direction across the membrane being linked to the transport of $K^+$ in the reverse direction, not always in a one-for-one manner. Inhibition of the ion pump can be achieved through metabolic inhibitors, such as

dinitrophenol, or by substances that exert a more specific attack on the transport process, such as the cardiac glycosides. The effect will be to reduce, if not to inhibit entirely, the potential, but this will depend on the extent to which the measured potential is directly dependent on the ion-pump.

### SHORT-CIRCUIT CURRENT

Ussing and Zerahn (1951) drew attention to the importance of this current in the quantitative assessment of the active transport process. Figure 1.32 illustrates the

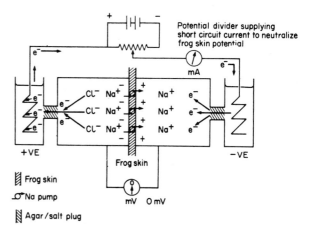

**Fig. 1.32** Apparatus suitable for measuring the short-circuit current across the frog skin. The active transport of $Na^+$ from left to right creates a potential across the membrane due to the slower movement of $Cl^-$-ions. If a potential is applied from a battery (top) opposing this potential, a current must flow in the external circuit such that for each ion of $Na^+$ transported one electron passes to neutralize it. Thus, electrons in the external circuit replace the $Cl^-$-ions that would have moved from left to right accompanying the actively transported $Na^+$ ions. If, under these short-circuit conditions, the only ion being transported is $Na^+$, the short-circuit current is a measure of the transport of $Na^+$. (Davson and Segal, *Introduction to Physiology.*)

procedure for measurement and the principle on which it is based. Ussing and Zerahn argued that, if a counter-potential was applied across the skin so as to make the potential zero, there would be no force driving $Cl^-$ from one side of the membrane to the other if there were equal concentration on either side in the first place. The counter-potential would presumably not affect the movement of $Na^+$ if it was being transported as an unionized complex, so that the neutralization of the positive charges brought on to the inside of the skin by the transported $Na^+$ would have to be done by electrons in the external circuit. Thus the current flowing in the external circuit—*the short-circuit current* —would be equivalent to the ions carried by the active transport mechanism, if this transport of $Na^+$ were the sole process generating the potential.

## TRANSPORT OF Na⁺

Cole blocked the escape route from the anterior chamber by inserting silicone oil into the angle (Fig. 1.33), and the posterior and anterior chambers were perfused with an isotonic mannitol solution; the $Na^+$, $K^+$ and $Ca^{2+}$ entering the fluid were estimated chemically. When a

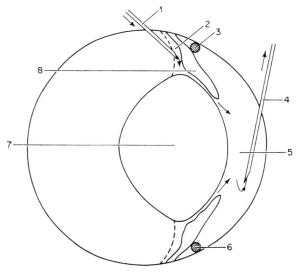

**Fig. 1.33** Diagrammatic transverse section of eye, showing positions of needles and oil during transfusion. 1, inflow needle (connected to motor-driven syringe; 2, posterior chamber; 3 and 6, oil or silicone fluid in drainage angle; 4, outflow needle to collecting system; 5, anterior chamber; 7, lens; 8, iris. (Cole, 1965.)

steady state was established, most of the ions entering the fluid were derived from the newly secreted aqueous humour, together with additional quantities derived from diffusion from the iris, cornea, etc. Treatment of the animal with DNP reduced the $Na^+$-influx by about a half, suggesting that about half the influx was due to the active secretory process. In general, Cole's studies suggested that a passive influx of $Na^+$, possibly by a process of filtration from the stroma across the tight junctions sealing the clefts, contributed perhaps a third of the total flow, the remainder being dependent on active transport inhibitable by DNP.

### POTENTIAL

Lehmann and Meesmann (1924) measured a potential of 6 to 15 mV between an electrode in blood and one in the aqueous humour, the aqueous humour being positive; subsequent work has confirmed this in the intact rabbit's eye and the isolated epithelium (Miller and Constant, 1960; Cole, 1961; Berggren, 1960); although there may be species differences respecting the orientation of the potential; thus Holland and Gipson

(1970) found values of 1·3 to 1·8 mV, the aqueous humour side negative.

### SHORT-CIRCUIT CURRENT

With an isolated iris-ciliary body preparation, separating two chambers of saline as in the Ussing-Zerahn set-up, Cole measured a p.d. of 5 mV and a short-circuit current of 46 $\mu A/cm^2$, using ox tissue, and 3·8 mV and 29·5 $\mu A\beta cm^2$ for rabbit tissue. In the cat isolated preparation, Holland and Gipson (1970) obtained a value equivalent to the transport of 1 $\mu E/cm^2/hr$.

### ATPase

The inhibition of active transport of $Na^+$ by the cardiac glycosides is related to the inhibition of an enzyme that catalyses the hydrolysis of ATP to ADP and phosphate; the enzyme is specifically activated by $Na^+$- and $K^+$-ions as well as $Mg^{2+}$, and its presence on cell membranes concerned with active transport has been visualized in the light- and electron-microscope. The localization of ATPase in the ciliary epithelium is illustrated by Figure 1.34 where it is seen to be associated with the

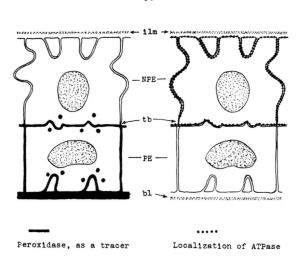

**Fig. 1.34** Localization of ATPase at cell borders of the ciliary epithelium. (Shiose, *Jap. J. Ophthal.*)

membranes of the unpigmented cells. Cole found that strophanthin G reduced influx of $Na^+$ into the anterior chamber by some 50 per cent. Figure 1.35 illustrates the results of a more recent study where blood-aqueous potential, aqueous humour formation, $Na^+$-influx and plasma sodium concentration are shown before, during, and after treatment with ouabain, another cardiac glycoside. When ouabain was injected into the vitreous humour, there was a fall in intraocular pressure, due presumably to lowered rate of secretion of aqueous

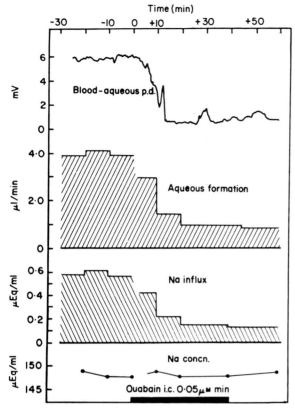

**Fig. 1.35**  Effects of close arterial injection of ouabain on several parameters of aqueous humour dynamics. (Cole, *Exp. Eye Res.*)

humour; and the time courses of the two processes ran parallel to each other (Bonting and Becker, 1964).

## FURTHER ASPECTS OF SECRETION

### ISOLATED CILIARY PROCESSES

Berggren (1964) examined the ciliary body *in vitro* when spread over a cylinder of perspex such that the processes projected outwards, with the stroma in contact with the cylinder. The whole was immersed in a bath and it was observed that, because of secretion, the ciliary epithelial cells shrank. Using this preparation, Berggren showed that secretion could be inhibited by removing $Na^+$ or $K^+$ from the medium, whilst typical inhibitors of active transport, such as ouabain, DNP, azide etc., were also effective in inhibiting shrinkage.

### ANION TRANSPORT

The effects of ouabain are strongly suggestive of the existence of a primary active transport of $Na^+$ across the ciliary processes; a probable linkage of this with an active transport of anions, such as chloride and bicarbonate, is suggested by the prominent action of

the carbonic anhydrase inhibitor, acetazolamide, or Diamox, on the rate of formation of aqueous humour, and also by the short-circuit studies of Holland and Gipson, which indicated that a part of the net flux of chloride could be accounted for by an active transport of this ion. Acetazolamide lowers intraocular pressure and reduces the rate of production of aqueous humour by some 40 to 50 per cent (Becker and Constant, 1955; Garg and Oppelt, 1970). The attack seems to be on an anion-secreting mechanism, and in view of acetazolamide's well known inhibition of carbonic anhydrase, it is reasonable to infer that the inhibition of synthesis of bicarbonate by the secreting cells represents the prime effect; as Maren (1977) has emphasized in relation to a similar inhibition of secretion of cerebrospinal fluid, the inhibition of formation of bicarbonate would lead to about a 50 per cent inhibition in the total rate of secretion of fluid. Thus the observed inhibition of sodium transport (Garg and Oppelt, 1970; Becker, 1959, Zimmerman *et al.*, 1976a,b), is consistent with inhibition of bicarbonate transport, a transport that is in some way linked to that of sodium (Maren, 1977).

### ULTRAFILTRATION

At one time it was argued that the sole process of formation of aqueous humour was by filtration from the blood vessels of the anterior uvea (Duke-Elder, 1932); later, the reverse extreme was adopted, namely that the process was governed entirely by secretion (e.g. Kinsey and Bárány, 1949). However, Davson (1953) showed that the penetration of two large water-soluble molecules, namely p-aminohippuric acid and raffinose, into the aqueous humour from the blood, took place at the same rate. Since these molecules would be unlikely to enter the primary secretion, it was argued that their penetration was through relatively large pores either in the iris or ciliary epithelium or both; if they entered by diffusion their rates would be proportional to the square roots of their molecular weights, i.e. the penetration of p-aminohippuric acid would be 1·6 times that of raffinose; if, on the other hand, the non-colloidal constituents of plasma were carried through by flow, i.e. by ultrafiltration, then, if the pores were large, there should be no difference in the rates, and this was found.

### Pressure-reduced flow

Further support for a contribution of passive filtration to flow has been provided by several studies, notably the demonstration by Bill and Bárány (1966) that a rise in pressure in the eye, artifically produced, caused a reduction in the computed rate of secretion of aqueous humour. This has been confirmed in the monkey by Bill (1968) who caused a hypertension by intracameral injection of erythrocytes; the suppression corresponded to 0·06 $\mu$l/min/mmHg rise in pressure; thus with a normal flow of 1·24 $\mu$l/min this means that a rise of some 20 mmHg would suppress aqueous humour formation altogether. This does not mean that secretion would be suppressed; far from it, since it is well established that secretory processes can proceed against very high pressures; Bill's observation simply indicates that, when pressure in the eye reaches a critical level, the influx by secretion is just balanced by an outflux through the filtration channels.*

## Site of filtration

The contribution of the iris to fluid flow into the aqueous humour seems small or negligible (Bill, 1974), so that the site of filtration may well be through the clefts in the ciliary epithelium; thus secretory epithelia have been categorized as 'tight'—e.g. that of frog skin—and 'leaky', such as that of gall bladder and choroid plexus. By a leaky epithelium is meant one that has a relatively high passive permeability to ions, and as a result, the establishment of large potentials during secretion is precluded. As we have seen, the potential across the ciliary body is small, about 5 mV; again, the electrical resistance is low (Cole, 1966) about 60 ohm/cm$^2$ compared with 2000 for frog skin. Thus the ciliary epithelium may be classed with the leaky epithelia, and a significant passive filtration across this is feasible, provided the pressure within the blood capillaries of the ciliary body is adequate. According to Cole (1977) the pressure is in the region of 25 to 33 mmHg; this would certainly permit filtration into the stroma, especially as the capillaries are highly permeable and allow escape of proteins so that the opposing colloid osmotic (oncotic) pressure is probably less than the theoretical oncotic pressure based on impermeability of the capillaries to proteins (Bill, 1968).

## ROLE OF THE PIGMENT EPITHELIUM

This layer constitutes the barrier between the blood, in the choroid, and the fluid immediately adjacent to and surrounding the cells of the retina, in particular the receptors with which it makes especially intimate relations (p. 200). Since there is free communication between the retinal extracellular spaces and the vitreous body, the pigment epithelium may be said to contribute to the blood-vitreous barrier, the retinal capillaries (e.g. in the holangiotic retina) representing the other locus. A number of experiments have indicated that the pigment epithelium, like the unpigmented ciliary epithelium, is more than a passive barrier to diffusion from the adjacent stroma, and contributes, by active transport processes, to control the ionic composition

of the medium bathing the retinal cells (Noell et al., 1965; Lasansky and De Fisch, 1966; Steinberg and Miller, 1973).

### IONIC MOVEMENTS

It is sufficient to describe, here, a recent study of Miller and Steinberg (1977) on the cold blooded frog's retina that permits the study of transport across a choroid-pigment epithelium preparation mounted in an Ussing-type chamber. The preparation shows a trans-epithelial potential of some 5 to 15 mV, with the apical (vitread) side positive, suggesting the active transport of a positive ion, such as $Na^+$; a short-circuit current of some 35 to 80 $\mu$A/cm$^2$/hr was measured, and this was equivalent to some 30 to 40 per cent of the measured net flux of $Na^+$, as measured with isotope exchanges across the preparation. Ouabain reduced the net sodium-flux by 50 to 60 per cent, and this was associated with an increase in the retina-choroid flux rather than a decrease in the choroid-retina flux. In addition to an active sodium transport, revealed by these measurements, there was a net flux of chloride, measured by exchanges of isotope, in the opposite direction to that of sodium, namely from retina to choroid; this varied with the season being highest in spring, when it amounted to 0·5 $\mu$Eq/cm$^2$/hr compared with some 0·23 $\mu$Eq for sodium. This chloride-flux was inhibited by acetazolamide, and by lowering the concentration of bicarbonate in the medium. Of considerable significance for the function of the photoreceptors was the demonstration of a net flux of $Ca^{++}$ into the retina of some 6 nM/cm$^2$/hr.

### ANALOGY WITH CHOROID PLEXUS

Miller and Steinberg emphasize the analogy between the choroid-pigment epithelium system of the eye and the choroid plexus of the brain, both of which actively transport ions, thereby exerting a control over the ionic environment of the neurones in their neighbourhood; to some extent this control may be extended to the composition of the vitreous body, but its large mass and close relations to the lens and ciliary body may obscure any effects of active processes in the pigment epithelium, which could well be confined to the fluid immediately adjacent to the receptors.

---

*That the pressure-suppressed inflow of aqueous humour is, indeed, formed by filtration is made very likely by Macri's (1967) finding that, when he raised the intraocular pressure by occluding anterior ciliary veins, there was no suppression; in this case the transmural pressure-difference across the blood capillaries was probably not decreased, whereas when the intraocular pressure was raised by increasing the pressure-head of saline during perfusion, there was a suppression of inflow of aqueous humour, as revealed by the phenomenon of pseudofacility.

# RATE OF FLOW OF AQUEOUS HUMOUR

The measurement of the rate of production (or drainage) of aqueous humour is of great interest for the study of the physiology and pathology of the eye, but unfortunately the measurement is not easy. Thus, simply placing a needle in the eye and collecting the fluid dropping out would cause a breakdown of the barrier, and the rate of production of the plasmoid aqueous humour would not be a measure of the normal rate of production. In consequence, essentially indirect methods, largely based on studies of the kinetics of penetration and escape of substances into and out of the eye, have been employed. Space will not allow of a description of the methods. In general, the principle on which most are based is the introduction of some material into the aqueous humour and to measure its rate of disappearance from the anterior chamber. The substance is chosen such that the rate of escape by simple diffusion into the blood stream, across the iris, is negligible by comparison with the escape in the bulk-flow of fluid through Schlemm's canal, for example, [131]I-labelled serum albumin, [14]C-labelled sucrose or inulin, and so on.

## INTRODUCTION OF MARKER

The main experimental difficulty is the introduction of the marker into the aqueous humour without disturbing the blood-aqueous barrier; and this is extremely difficult. Maurice applied a solution of fluorescein onto the cornea through which it diffused to establish a measurable concentration in the aqueous humour; the subsequent changes in concentration were measured with a slit-lamp fluorimeter and, with the aid of suitable equations, a flow-rate could be determined. The method is applicable to man. In animals Bill (1967), for example, has introduced needles into the anterior chamber and replaced the aqueous humour by an artificial one containing the labelled marker; by an ingenious arrangement of syringes this could be done without lowering the intraocular pressure. Oppelt (1967) has introduced cannulae into posterior and anterior chambers and perfused an artificial fluid through the system, measuring the dilution of the marker due to formation of new fluid; under these conditions, however, the eye was not normal since the perfusate contained a high concentration of protein.

## ESTIMATED VALUES

Estimates based on these principles agree on a value of some 1 to $1\frac{1}{2}$ per cent of the total volume per minute. Thus, in the cat, with a volume of about 1 ml, this would correspond to 10 to 15 $\mu$l/min, whilst in the rabbit, with a volume of about 0·35 ml, the absolute flow-rate would be less, namely 3·5 to 5·2 $\mu$l/min.

## PRESSURE-CUP TECHNIQUE

The outflow of aqueous humour may be blocked by applying a uniform positive pressure on the sclera at the limbus in such a way as to compress the drainage channels. The technique was developed by Ericson (1958) to permit an estimate of the rate of flow of aqueous humour, since, if rate of secretion is unaffected by this treatment, the rise in pressure, and calculated rise in intraocular volume, at the end of, say, 15 minutes' occlusion of the drainage channels should give the rate of formation. The blockage of outflow is achieved by placing a suction-cup on the globe (Fig. 1.36) fitting

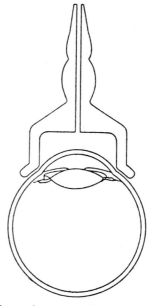

**Fig. 1.36** The suction-cup applied perilimbally with the contact surface blocking the anterior outflow passages of the aqueous humour (Ericson, *Acta Ophthal.*)

snugly round the limbus; a negative pressure in the cup leads to a positive pressure on the rim; thus a negative pressure of 50 mmHg caused a positive pressure of some 29 mmHg, which was adequate to block both the epi- and intra-scleral vessels. Langham (1963) obtained flow-rates between 0·8 and 1·9 $\mu$l/min in six human subjects, comparing with a mean of 0·85 $\mu$l/min found by Galin *et al.* (1961).

## TONOGRAPHY

In human subjects the established technique depends on measuring the resistance to flow in response to an applied pressure. If this is measured (p. 50) then the normal rate of flow under steady state conditions will be given by the equation:

$$F \times R = P_i - P_e \qquad (1)$$

where $F$ is rate of flow, $R$ is the resistance, $P_i$ is the pressure within the eye and $P_e$ is the pressure in the blood vessels into which the aqueous humour drains; this is taken as the pressure in the episcleral veins, which can be measured (Brubaker, 1967).

## ACTIVE TRANSPORT OUT OF THE EYE

It is common to think of the ciliary processes as organs for the production of secretion, in the sense of transporting material from blood to posterior chamber. It is therefore of considerable interest to discover that the ciliary processes are capable of actively removing certain substances from the aqueous humour into the blood. Thus Forbes and Becker showed that, if diodrast was injected into the vitreous body, it disappeared from this into the blood but did not appear in the aqueous humour, being apparently actively removed by the ciliary epithelium; studies on the isolated ciliary body confirmed that the epithelium was indeed able to accumulate diodrast and also iodide.

### ANALOGY WITH KIDNEY AND LIVER

Another substance that is actively carried out of the posterior chamber is p-aminohippurate (PAH), so that the ciliary body behaves similarly to the kidney in the sense that it actively eliminates this substance from the eye; a similar process was described in the cerebrospinal fluid system, in the sense that iodide, thiocyanate, p-aminohippurate are removed rapidly (Davson, 1955) and against concentration gradients (Pappenheimer et al., 1961; Pollay and Davson, 1963). In addition to this 'kidney'-type of active process, there is a 'liver'-type

that actively removes compounds like iodipamide, a substance typically used to measure the liver's capacity to excrete into the bile (Bárány, 1972). The significance of these active processes was obscure, since these compounds penetrate the eye very slowly from the blood and therefore are unlikely to present a problem to the eye in so far as there is little necessity to remove them actively.

### PROSTAGLANDIN TRANSPORT

However, Bito (1972) showed that the ciliary body is capable of accumulating prostaglandins (PG) when incubated in vitro in a manner comparable with the accumulation of p-aminohippurate. Again, when a mixture of labelled sucrose and prostaglandin was injected into the vitreous body of the rabbit, only the labelled sucrose appeared in the anterior aqueous humour (Fig. 1.37), indicating that the prostaglandin was removed directly into the blood rather than through the usual drainage channels. If the mixture was injected into the anterior chamber both substances disappeared at the same rate, showing that the iris is not responsible for active removal (Bito and Salvador, 1972). Bito suggested that the excretory mechanism, manifest in the active transport of p-aminohippurate, etc., from the eye was developed in association with a need to eliminate prostaglandins after their synthesis and liberation in the ocular tissues. Thus the eye is unlike lung in that it has a very low power of inactivation of prostaglandins (Fig. 1.38), so that any removal of liberated prostaglandin would have to be achieved by the ordinary bulk flow mechanism unless supplemented by this active secretory process. That the secretory process uses the same mechanism as that involved in elimination of

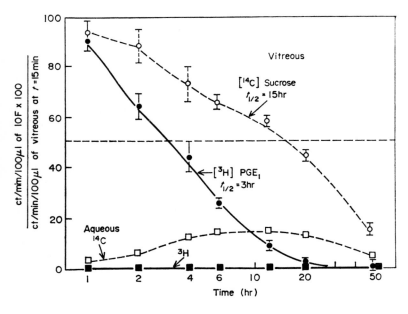

**Fig. 1.37**  A mixture of [3]H-labelled prostaglandin and [14]C-labelled sucrose was injected into the vitreous body. The curves show the changes in activity in aqueous humour (lower block) and vitreous humour (upper and lower blocks) as a function of time. The prostaglandin fails to appear in the aqueous humour, and disappears from the vitreous humour much more rapidly than does sucrose. (Bito and Salvador, *Exp. Eye Res.*)

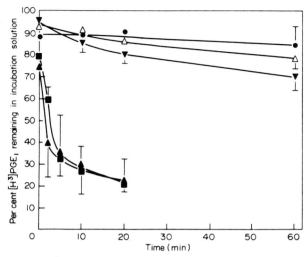

**Fig. 1.38** Metabolic degradation of prostaglandin $E_1$ in supernatant fractions of rabbit kidney, lungs and ocular tissues. Conjunctiva, △; lens, ●; iris/ciliary body, ▼; kidney, ▲; lung, ■. Each point is the mean of four experiments + s.e.m. (Eakins *et al., Exp. Eye Res.*)

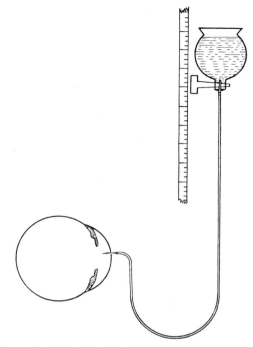

**Fig. 1.39** Illustrating Seidel's experiment.

*p*-aminohippurate is indicated by the fact that probenecid inhibits the process (Bito and Wallenstein, 1977).*

## THE DRAINAGE PROCESS

### EXPERIMENTAL STUDY

In 1921 Seidel established beyond reasonable doubt that the aqueous humour was continually formed and drained away. He inserted a needle, connected to a reservoir containing a solution of a dye, into the anterior chamber of a rabbit (Fig. 1.39). By lowering the reservoir below the intraocular pressure a little aqueous humour was drawn into the needle and then, when the reservoir was raised, this fluid, mixed with dye, was returned to the anterior chamber. Soon the dye mixed with the rest of the fluid. Seidel observed that the dye stained the blood vessels in the surface of the globe—episcleral vessels—indicating that aqueous humour was being carried into the venous system. This passage out would only occur, however, if the pressure in the eye, determined by the height of the reservoir, was at or above 15 mmHg; when the pressure was reduced below this it was presumed that drainage of aqueous humour ceased. Since the pressure within the eye is normally above 15 mmHg, we may conclude that fluid is being continuously lost to the blood; moreover, the loss presumably depends on the existence of a pressure-gradient between the aqueous humour and the venous system. Since the episcleral vessels are derived from the anterior ciliary system, we may conclude that the dye

was not being absorbed from the iris, the venous return from which is by way of the vortex veins which were uncoloured in these experiments. More recent studies, in which isotope-labelled material is introduced into the aqueous humour and the venous blood collected from both ciliary and vortex veins, have confirmed that at least 99 per cent of drainage is by the anterior route (Bill, 1962).

## AQUEOUS VEINS

Confirmation of Seidel's view was provided in a striking manner by Ascher in 1942; when examining the superficial vessels of the globe with the slit-lamp microscope, he observed what appeared to be empty veins but they turned out to be full of aqueous humour, and he called them *aqueous veins*. Usually one of these veins could be followed till it joined a blood vessel, in which event the contents of the two vessels, aqueous humour and blood, did not mix immediately but often ran in parallel streams forming a *laminated* aqueous vein, as illustrated schematically in Figure 1.40. If the blood-vein beyond the junction was compressed, one of two

---

*Ehinger (1973) incubated iris-ciliary body tissue in a medium containing $^3H$-$PGE_1$ and found an accumulation in the ciliary processes.

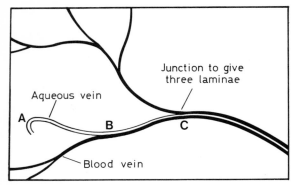

**Fig. 1.40** Aqueous veins. The aqueous vein emerges from the deeper tissue and is visible at A. At B it joins a blood-vein so that the latter has two strata. At C the stratified vessel joins another vein to give three laminae. (Ascher, *Amer. J. Ophthal.*)

things happened; either the blood drove the aqueous humour out of its channel or the aqueous humour drove the blood out.

### GLASS-ROD PHENOMENA

The latter situation was described as the *positive glass-rod phenomenon*, whilst the former, with the blood driving aqueous humour out, was called the *negative glass-rod phenomenon*. These superficial vessels, full of aqueous humour, are probably derived from the deeper vessels of the intrascleral plexus, the pressure-relationships in the collectors from Schlemm's canal being such as to favour displacement of blood throughout the course of the fluid to the surface of the globe as illustrated by Figure 1.41.

### FLUORESCEIN

Goldmann proved conclusively that these vessels contained aqueous humour by injecting fluorescein intravenously. This substance passes very slowly from blood into aqueous humour so that one would not expect

**Fig. 1.41** Ascher's view of the origin of an aqueous vein.

any vessel containing aqueous humour to fluoresce immediately after the injection, whilst vessels full of blood or plasma should do so; in fact only the blood vessels did fluoresce.

### CONNECTIONS

Later, Ashton identified an aqueous vein in a human eye that was about to be enucleated; a wire was tied round the vein and, after enucleation, the canal of Schlemm was injected with Neoprene. After digesting the tissue away a cast of the canal and its connecting vessels remained, and examination showed that the aqueous vein did, indeed, take origin from the canal of Schlemm. Subsequent studies of Jocson and Sears (1968), employing a silicone vulcanizing fluid, have shown that the connections of the canal of Schlemm with the vascular system are of two main kinds, large (aqueous veins) to the episcleral vessels and much finer connections to the intrascleral vessels. In addition, Rohen (1969) has described bridge-like channels running parallel with Schlemm's canal, which branch off and later rejoin it; these make connections with the intrascleral plexus.

## PHYSIOLOGY OF THE VASCULAR SYSTEM

The vascular circulation of the eye has attracted a great deal of interest because of its close relationship to the formation and drainage of aqueous humour. Space will not permit a detailed description of this aspect of ocular physiology, and we must be content with a few salient points.

### PULSES

With the ophthalmoscope an arterial pulse may be observed in the retinal vessels; if the intraocular pressure is increased by compressing the globe the pulse may be exaggerated, since the collapse of the artery during diastole becomes greater; if the pressure is raised sufficiently the pulse will cease, at which point the pressure in the eye is equal to the systolic pressure in the artery from which the retinal artery is derived, namely, the ophthalmic artery. The retinal veins also show a pulse; it is best seen where the large veins lie on the optic disk; it appears as a sudden emptying of this portion of the vein, progressing from the central end towards the periphery followed by a pronounced dilatation, beginning at the periphery and passing centrally. The pulse is not simply a reflexion of events taking place in the right atrium, and is closely related to ventricular systole. The rise in intraocular pressure associated with systole will tend to compress the veins; and a collapse of the latter will take place where the pressure is least, i.e. most centrally. The subsequent

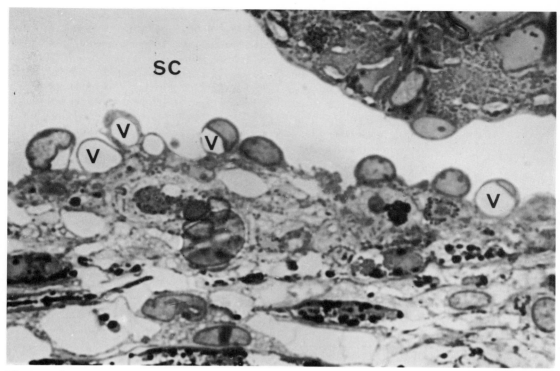

**Fig. 1.42**   Light photomicrograph of the trabecular wall of Schlemm's canal; the lining cells are characterized by prominent bulging nuclei and vacuolar structures (V). Note also the predominance of loosely organized cellular elements in the supporting tissue zone. SC, lumen of Schlemm's canal. Araldite section. (× 980) (Tripathi, *Exp. Eye Res.*)

refilling of the vein will occur in diastole, or late in systole, and will proceed in the reverse direction. The exact time-relationships of the pulse were established cinematographically by Serr.

PRESSURES

These were studied experimentally by Seidel and later by Duke-Elder; according to the latter author the following values (mmHg) are associated with an intra-ocular pressure of 20 mmHg:

| | | | |
|---|---|---|---|
| Ophthalmic artery | 99·5 | Retinal vein | 22·0 |
| Retinal arteries | 75·5 | Intrascleral vein | 21·5 |
| | | Episcleral vein | 13·0 |

Values in normal human subjects are included in Table 1.6 (p. 54).

## MORPHOLOGICAL BASIS OF DRAINAGE

GIANT VACUOLES

The connections between the canal of Schlemm and the vascular system have been well established since the work of Maggiore; the mechanism by which fluid passes from the trabecular meshwork into the canal is not so clear. McEwan (1958) showed on hydrodynamic principles that the flow could be accounted for on the basis of a series of pores in the inner wall of the canal, and recent electron-microscopical studies have lent support to this view. Holmberg (1959) made serial sections of the trabecular tissue immediately adjacent to the canal and found channels apparently ending in vacuoles enclosed within the endothelial cells of the canal, and he considered that these vacuoles opened into the canal. Subsequent work, notably that of Kayes (1967) Tripathi (1968–1977) and Bill and Svedbergh (1972) has confirmed the existence and emphasized the significance of these giant vacuoles which are formed by the cells of the trabecular wall and cells lining the septa, recognizable even at the light-microscopical level of magnification (Fig. 1.42), and there seems little doubt they are not artefacts of fixation.*

---

* Shabo *et al.* (1973) claimed that the vacuoles were, indeed, due to inadequate fixation, increasing with the delay between death and fixation; however Kayes (1975), while confirming the effects of pressure on vacuole formation (p. 43), were unable to find more than a few vacuoles in tissue maintained at zero pressure and left for as long as 30 minutes before fixation.

### TRANSCELLULAR CHANNELS

By studying serial sections in the electron microscope it could be shown that the majority of the vacuoles were, in fact, invaginations of the basal aspect of the endothelial cell and thus may be regarded as extensions of the extracellular space into the cell, with openings measuring up to 3·5 or 4 μm on the trabecular aspect.

At any given time a small proportion of these vacuoles show communications with the lumen of Schlemm's canal through smaller apical openings, in addition to the basal opening into the meshwork. Thus such a cell is, in effect, cloven into two by a channel communicating between aqueous humour in Schlemm's canal and aqueous humour in the endothelial mesh-

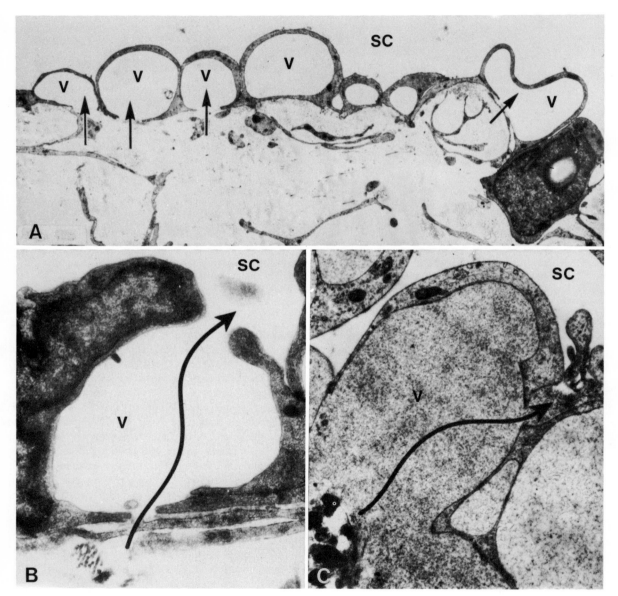

**Fig. 1.43** (A) A survey electron micrograph of the cellular lining of Schlemm's canal (SC) in section showing many giant vacuoles (V). In this plane of section, many vacuoles are seen to have communication (arrow) towards the spongy subendothelial tissue containing aqueous humour. (× 3810). (B) In a given plane of section, some giant vacuoles show openings on both sides of the cell thereby consituting a vacuolar transcellular channel (continuous arrow). N, cell nucleus; SC, Schlemm's canal. (× 20 000). (C) Such transcellular channels provide the pathway for the bulk outflow of aqueous humour as indicated by the passage of colloidal tracers (ferritin) shown in this Figure. The tissue was processed and electron micrograph taken following experimental introduction of the tracer into the anterior chamber of the eye. V, vacuolar transcellular channel filled with electron dense tracer material; SC, Schlemm's canal. (× 17 550). (Courtesy, R. C. Tripathi.)

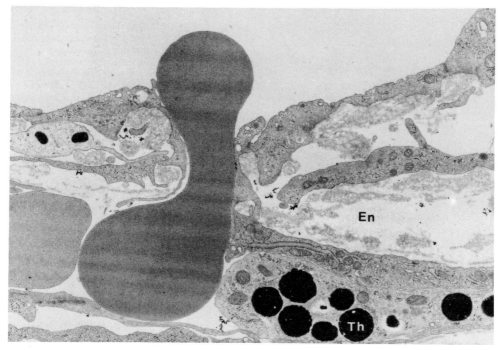

**Fig. 1.44**   An erythrocyte is seen passing through a pore in the flat portion of the endothelium (En) to the lumen of Schlemm's canal. The endothelium extends a funnel-like process toward the meshwork side. Th. thorotrast. ( × 2000). (Inomata *et al.*, *Amer. J. Ophthal.*)

work. An example of this communication is shown in Figure 1.43; and Figure 1.44 from Inomata *et al.* (1972) shows an erythrocyte actually on its way through a pore, the anterior chamber having been perfused with fluid containing these cells before fixation. Even more striking is the appearance of the inner (luminal) side the Schlemm's canal as seen in the scanning electron microscope; as Figure 1.45 shows, there are unmistakeable pores opening into the lumen.

VALVULAR ACTION

The relatively large basal openings compared with the smaller apical (luminal) openings, these last being located on the convex surface of the vacuole and usually not directly opposite the basal opening, could well provide the basis for a valvular function, flow being favoured by a pressure-gradient from aqueous humour towards the canal of Schlemm; reversal of this gradient might well cause closure of the apical openings and prevent reflux of aqueous humour into the anterior chamber. The basic concept of the structure of the angle is illustrated by Figure 1.46 from Inomata *et al.*, the endothelial lining of Schlemm's canal, as seen from the inside by scanning electron microscopy, exhibiting

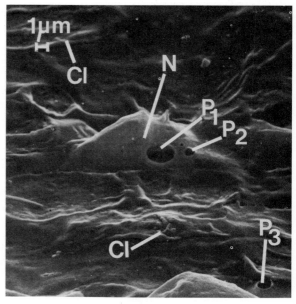

**Fig. 1.45**   Detail of the inner wall of Schlemm's canal. A partly collapsed bulging structure has two openings, P1 and P2. Part of the nucleus, N, can be seen through one opening. Structures most probably representing collapsed invaginations, CI, are seen at several places. P3 is also a pore. Freeze-dried preparation. (Bill and Svedbergh, *Acta Ophthal.*)

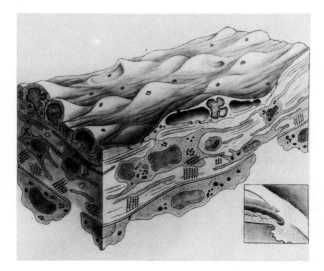

**Fig. 1.46**  Three-dimensional schematic drawing of endothelium lining the inner wall of Schlemm's canal and the endothelial meshwork. Endothelial cells of inner wall of Schlemm's canal are long and spindle-shaped and the nuclear portion, which often contains vacuoles, bulges into the lumen. Pores through endothelial cells are formed in flat portions of cells as well as in the wall of vacuoles opening both toward the meshwork and to the lumen of Schlemm's canal. Endothelial meshwork cells have long processes which contact those of adjacent cells forming a complicated network. These cells are often located close to the pores in the endothelial cell and are involved in 'sieving' of particles in the aqueous humour. A tiny rectangle in the Insert shows the part of the wall of Schlemm's canal that is illustrated. (Inomata *et al.*, *Amer. J. Ophthal.*)

depressions corresponding to the endothelial cell nuclei, and at the base of these depressions the openings of the pores, or giant vacuoles, are located.

VACUOLIZATION CYCLE

Tripathi suggested that the channel was just one stage in a cycle of vacuolization, as illustrated by Figure 1.47, in which event it is possible to envisage changes in the resistance to outflow in accordance with changes in the degree of vacuolization. Cole and Tripathi (1971) calculated that the normal facility of outflow could be accounted for if the life-span of a transcellular channel

occupied only one fiftieth of the total life-cycle of the vacuole as depicted in Figure 1.47.

When particular matter, such as Thorotrast, ferritin, etc. is introduced into the aqueous humour, it can be identified within the large vacuoles, and even intact blood cells can be seen in them; by contrast, no particulate matter is seen to leak through the intercellular clefts, a route for bulk outflow suggested by Shabo *et al.* (1973), and in fact Bill (1975) has calculated that these clefts would be unable to contribute as much as 1 per cent of the total fluid conductance across the endothelial lining of Schlemm's canal.

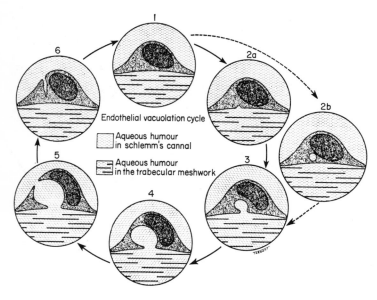

Endothelial vacuolation cycle
□ Aqueous humour in schlemm's cannal
▤ Aqueous humour in the trabecular meshwork

**Fig. 1.47**  Illustrating hypothetical cycle of vacuolization in an endothelial cell of Schlemm's canal resulting in creation of a temporary channel communicating between the canal and the trabecular meshwork. (Tripathi, *Exp. Eye Res.*)

PINOCYTOSIS

Finally, in this context, it can be argued that flow across the endothelium is achieved through pinocytosis by the endothelial cells, i.e. through engulfment (endocytosis) of aqueous humour in small pinocytotic vesicles and ejection of the contents at the opposite side of the cell (exocytosis). However, Tripathi's studies have shown that the pinocytotic vesicles within the endothelial cells do not accumulate particulate matter from the aqueous humour.

EFFECTS OF PRESSURE

The effects of pressure on the state of the drainage channels are of great interest; these have been studied in particular by Johnstone and Grant (1973), by Grierson and Lee (1974, 1975) and Kayes (1975) in primate eyes, using both scanning and transmission electron microscopy. To confine attention to the studies of Grierson and Lee on monkeys, they found that changes in intraocular pressure had negligible effects on the uveal meshwork and inner corneoscleral meshwork, so that the obvious effects were on the structure closer to Schlemm's canal. The effects may be summarized as a tendency for the meshwork to become compressed at low intraocular pressures, reducing the intertrabecular spaces, and for the meshwork to become progressively expanded as pressure was raised.

**Increasing pressure**

Thus at 8 mmHg the endothelial meshwork had become compact and was reduced to less than 10 $\mu$m in thickness, the native cells being arranged in two or three irregular layers; the narrow and tortuous extracellular spaces were filled with extracellular material as in normal tissue. As pressure was increased up to and past normal, the intertrabecular spaces in the outer corneoscleral meshwork became wider and the cells of the endothelial meshwork became separated, so that the thickness of this region became as much as 20 $\mu$m. Eventually as pressure rises further the endothelial lining of Schlemm's canal, adjacent to the endothelial meshwork, balloons into the canal to become apposed to the corneal layer, thus locally occluding the channel; this tendency is opposed by rigid structures such as the outer trabeculae which are strengthened by attachment to the scleral spur and cornea; this, together with the struts formed by the bridging septa within the canal, prevent complete closure even at pressures of 50 mmHg. At these high pressures the endothelial meshwork becomes still more distended to reach a thickness of 15 to 35 $\mu$m, so that cell-to-cell contacts were often broken although in other regions these were maintained through fine cytoplasmic processes connecting meshwork cells to each other and also connecting meshwork cells to endothelial cells of Schlemm's canal.

**Low pressure**

At zero pressure the endothelial cells and outer corneoscleral meshworks were very highly compressed. The absence of blood cells in the inner meshwork, although

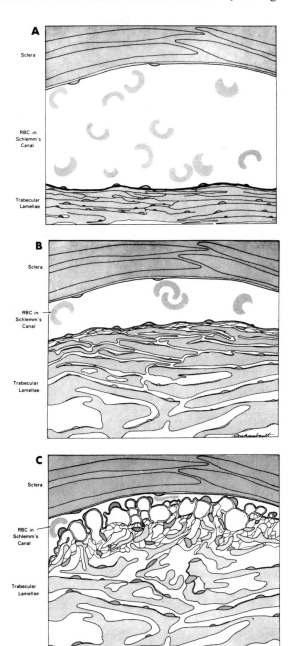

**Fig. 1.48**  Composite schematic drawing of the trabecular meshwork and Schlemm's canal of the human eye. (A) with intraocular pressure less than the episcleral venous pressure; (B) with equal intraocular and episcleral venous pressures; (C) with intraocular pressure 30 or 40 mmHg greater than episcleral venous pressure. (Johnstone and Grant, *Amer. J. Ophthal.*)

the canal of Schlemm was packed with them, suggests that the collapsed meshwork adjacent to the canal acts as a filter preventing reflux from the canal (Johnstone and Grant, 1973). The changes in appearance of the meshwork with pressure are illustrated schematically by Figure 1.48.

## Vacuoles

At the low intraocular pressure of zero there were no giant vacuoles (Johnstone and Grant, 1973), and as pressure was raised there was a progressive increase in the numbers of vacuolated cells, and an increase in size and length of the vacuoles within individual cells, so that it appears that vacuole formation does depend on a positive intraocular pressure; this, of course, does not prove that at high pressures there are more transcellular channels, since the increased vacuolation may not be accompanied by increased numbers of openings into the canal lumen. This is important since many studies have indicated a nearly linear relation between pressure and flow-rate, indicating a resistance to flow that was independent of pressure so that Poiseuille's Law would apply. If rising pressure were accompanied by lowered resistance due to increased numbers of channels, then flow would rise steeply with increasing pressure. At the highest pressure the number of vacuoles decreased, so that the increases in resistance to flow observed at high intraocular pressures by Ellingsen and Grant (1971) in experiments on enucleated human eyes, may have been due to the reduction in vacuolization as well as the herniation of the canal of Schlemm that also occurs.*

## Extracellular material

In the normal and low-pressure eyes the characteristic extracellular material covering the endothelial cells, and probably consisting of mucopolysaccharides, was distributed normally; at the raised pressures this tended to disappear; thus at 22 mmHg it became focally aggregated and usually in limited regions close to the endothelium lining the canal; at 30 mmHg the material was still further reduced and in some areas could not be detected.

### ESCAPE OF PARTICULATE MATTER

Particulate matter of diameter around 1·5 to 3·0 $\mu$m, when injected into the anterior chamber, escapes into the blood-stream; even erythrocytes; introduced into the anterior chamber with care not to alter the intraocular pressure, disappear and may be recovered in the blood when appropriately labelled; the rate of escape is not so rapid as that of serum albumin, showing that there is some obstruction to passage; moreover, if the cells are left in the anterior chamber and the eye is allowed to adopt its true pressure, this rises due to a partial obstruction of the outflow routes (Bill, 1968). The

passage of particulate matter has been visualized in the electron microscope, e.g. by Inomata *et al.* (1972) in the monkey. Thorotrast (10 nm diameter), 0·1 $\mu$m latex spheres and blood cells (about 6 to 7 $\mu$m) readily entered Schlemm's canal, but fewer 0·5 $\mu$m and 1·0 $\mu$m latex spheres were found, indicating some selection at these sizes. Thorotrast penetrated every space of the trabecular meshwork, except areas with very fine fibrils, and it entered each vacuole in the endothelium lining the canal; as Tripathi also found, Thorotrast particles did not enter micropinocytotic vesicles in the endothelial cells lining Schlemm's canal, nor did they pass between adjacent cells lining the canal. The large latex spheres of 0·5 to 1·0 $\mu$m diameter reached the endothelial meshwork adjacent to the canal but most were trapped by its long cytoplasmic processes, which thus acted as a sieve. The escape of erythrocytes is due to their great deformability, so that although 6 to 7 $\mu$m in diameter, they were able to squeeze through 2·5 to 3·5 $\mu$m diameter pores (Fig. 1.44, p. 41). The exit vacuole has a smaller diameter than the inlet (1·0 to 1·8 $\mu$m compared with 2·5 to 3·5 $\mu$m), so that erythrocytes that entered a pore tended to remain there.

## LOCUS OF OUTFLOW RESISTANCE

The actual locus of the main resistance to outflow from the anterior chamber has been implicitly accepted as the trabecular meshwork and endothelial lining of the canal, as opposed to the resistance to flow from the canal through the collector channels and vessels of the scleral plexuses. This viewpoint is supported by the experiments of Grant (1958; 1963) and Ellingsen and Grant (1971) on the trabeculotomized eye, the resistance in this eye being only some 25 per cent of normal. The change in resistance on trabeculotomy is strikingly illustrated by Figure 1.49; in the monkey some 83 to 97 per cent of the resistance was eliminated by trabeculotomy compared with some 75 per cent in human eyes (Ellingsen and Grant, 1971).

### COMPUTED RESISTANCE

Bill and Svedbergh (1972) examined the meshwork and Schlemm's canal of human eyes in the scanning electron microscope, in an attempt to estimate the

---

*These authors discuss in some detail the effects of pressure on resistance, as measured by cannulation of the eye and determining the relation between pressure and flow from a reservoir; an important pitfall is the variation in resistance with changing angle of the anterior chamber; thus if no deepening is allowed to take place, facility tends to decrease with increasing pressure, whereas if deepening occurs with raised pressure, then facility tends to remain nearly constant with increasing pressure.

probable number of open pores and their dimensions, and from this to compute a likely resistance to flow on hydrodynamic principles; they pointed out that, with conventional transmission electron-microscopy, the

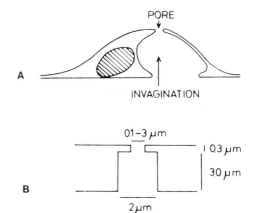

**Fig. 1.50** (A) Invagination and pore in an endothelial cell. (B) Simplified model used to calculate resistance to flow. (Bill and Svedbergh, *Acta. Ophthal.*)

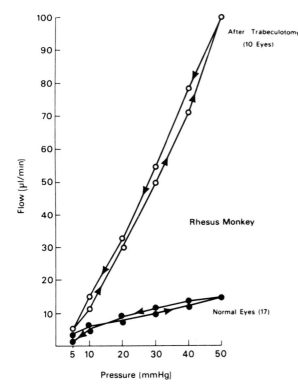

**Fig. 1.49** Steady-state intraocular pressures and corresponding aqueous perfusion flow rates in enucleated monkey eyes in which the anterior and posterior chambers were in free communication, showing the difference between normal control eyes and eyes that had internal trabeculotomy of one quadrant. The measurements were made at a series of steady states as the pressure was raised stepwise to 50 mmHg and again as pressure was reduced, in the sequence indicated by the arrowheads. (Ellingsen and Grant, *Invest. Ophthal.*)

chances of finding a pore-opening are relatively small compared with scanning electron microscopy with its smaller magnification and larger field. They measured an average of some 1840 pores per $mm^2$ of endothelial surface (comparing with $1200/mm^2$ in the monkey). Some of the pores are illustrated in Figure 1.45. Taking a value of 11 $mm^2$ for the total area of the wall (McEwan, 1958) and the dimensions illustrated schematically in Figure 1.50, they calculated the resistance per pore on the basis of a modified Poiseuille's relationship which took account of the shortness of the pore; and from the numbers of pores of given diameters

they computed the total resistance of the angle. This varied between 0·16 and 0·25 mmHg min $\mu l^{-1}$.

### ACTUAL RESISTANCE

Although the calculation is highly approximate it does indicate that the facility is more than adequate to account for flow; thus usual values for resistance in man are 3 to 5 mmHg min $\mu l^{-1}$ (Grant, 1958), i.e. some ten times that computed for flow through the pores. Hence the study of Bill and Svedbergh has shown that the pores actually observed could easily account for the flow under the existing difference of pressure between anterior chamber and the episcleral venous pressure, supposing that the pressure-drop occurred between the anterior chamber and canal of Schlemm.

### RESISTANCE IN DRAINAGE VESSELS

The study, of course, does not answer the question as to whether a large proportion of the resistance occurs along the collectors and vascular channels, as argued by Perkins (1955), who found the pressure in the canal of Schlemm equal to that in the anterior chamber. In view of the fact that the resistance can be strongly affected by drugs that cause a pull on the ciliary muscle, opening up the meshwork, it is difficult to believe that the meshwork, especially the corneo-scleral and endothelial regions, is not a major locus of resistance to flow.

### EFFECT OF HYALURONIDASE

Bárány and Scotchbrook (1954) showed that the facility of outflow through the perfused rabbit eye was increased by adding hyaluronidase to the medium, suggesting that a factor in resistance to flow was the mucopolysaccharide matrix to the trabecular meshwork, but it would seem from the study of Melton and De Ville (1960) that this does not happen in the cat and dog. Again, Grant (1963) found no effect on human eyes. In the monkey, Peterson and Jocson (1974) found decreases ranging from 24 to 65 per cent in resistance with a mean of 40 per cent; the remaining resistance could be decreased by a further 60 per cent by trabeculotomy, whilst hyaluronidase treatment after trabeculotomy only decreased resistance by some 25 to 30 per cent. Thus the authors considered that hyaluronidase might well be acting at some other locus than the meshwork.

## Glycosaminoglycans

However, there is no doubt that glycosaminoglycans are present in the meshwork; thus Grierson and Lee (1975b) and Armaly and Wang (1975) used colloidal thorium and colloidal iron as electron-dense stains that react with the glycosaminoglycan of the mucopolysaccharide complexes. These particles were found on the surface of the trabecular meshwork cells, on the surface of the endothelial cells of the canal, and in close association with extracellular connective-tissue elements. The giant vacuoles were lined by a continuous layer of polysaccharide, thus confirming their origin as invaginations of the cell membrane which carried with it its external coating of mucopolysaccharide; when an opening of a vacuole appeared in a section, the coating was continuous with that on the plasma membrane of the vacuolated endothelial cell. Thus the transmembrane channels through which aqueous humour presumably flows are apparently lined with mucopolysaccharide, and this may well influence fluid-flow.★

## UVEOSCLERAL DRAINAGE

Bill (1964) deduced from his studies on the passage of albumin from blood into the uveal tissue that there must be some route for drainage additional to that provided by Schlemm's canal. By perfusing the anterior chamber of the monkey's eye with a mock aqueous humour containing labelled albumin, he was able to divide the losses of protein into two components, the one appearing in the blood and doubtless having been drained through Schlemm's canal, and the other that remained in the uveal tissue, having presumably passed directly into the tissue of the anterior uvea. The average 'uveo-scleral drainage' was $0.44$ $\mu$l/min compared with a conventional drainage of $0.80$ $\mu$l/min; thus the rate of formation of aqueous humour, the sum of these, was $1.24$ $\mu$l/min In the cat, the rate of secretion was $14.4$ $\mu$l/min and

only 3 per cent of this passed by uveoscleral routes. In man, Bill (1971) demonstrated the uveoscleral flow by autoradiography, iodinated albumin passing through the anterior uvea and along the suprachoroidal space to reach, eventually, the posterior pole of the eye. In rabbits, Cole and Monro (1976) perfused eyes with fluorescein-labelled dextrans and were able to demonstrate that these high-molecular weight compounds found their way to suprachoroid, scleral limbus and iris, and, by quantitative analysis of the material in the dissected eye, it was shown that the amounts recovered were in this order.

### THE ACTUAL ROUTE

There is no complete endothelial layer lining the anterior surface of the ciliary body which faces the anterior chamber between the cornea and iris; there is also no delimitation of the spaces between the trabecular sheets and the spaces between the bundles of the ciliary muscle so that fluid can pass from the chamber-angle into the tissue spaces of the ciliary muscle, and these spaces in turn open into the suprachoroidal space from which fluid can pass across the sclera or along the perivascular spaces of the large vessels penetrating the sclera. Under normal conditions the pressure in the suprachoroid is less than that in the anterior chamber by a few mmHg (Bill, 1975), and thus the conditions for a flow along this 'unorthodox channel' are present.†

### Electron-microscopy

Inomata et al. (1972b) followed the passage of particulate matter (Thorotrast of 10 nm diameter and latex spheres of $0.1$, $0.5$ and $1.0$ $\mu$m diameter) from anterior chamber along the unconventional route. As others had found, the anterior surface of the iris allowed passage of all sizes of particles into the stroma; the pupillary zone appeared less permeable. All sizes of particles also passed into the intertrabecular spaces, penetrating the corneoscleral and uveoscleral meshwork and thence into the spaces between bundles of the ciliary muscle; the spaces between muscle cells allowed the smaller Thorotrast particles to pass, excluding the latex spheres. All sizes of particles appeared in the suprachoroid and even

---

★ Schachtschabel et al. (1977) have cultured explants of trabecular meshwork and shown that they synthesize glycosaminoglycans consisting of mainly hyaluronic acid and the sulphated glycans, chondroitin-4-sulphate and dermatan sulphate.

† The proportion of uveoscleral drainage seems to depend on the size of the ciliary muscle; thus in the cat it is considerable although less than in primates; in the rabbit with its very poor ciliary muscle it is negligible (Bill, 1966).

the largest particles passed along this as far as the macular region in three hours. The compact sclera, as one might expect, allowed only the smaller (Thorotrast) particles to penetrate, and these in relatively small amount. In a later study, the passage of particulate matter was followed through the sclera in the loose connective tissue surrounding the blood vessels piercing this coat (Inomata and Bill, 1977).

### EFFECTS OF CILIARY MUSCLE CONTRACTION

Bill and Wålinder (1966) found in monkeys that, although drugs such as pilocarpine, which cause contraction of the ciliary muscle, increase facility of outflow, they actually decrease uveoscleral flow, and this is clearly because of the reduction in the spaces between ciliary muscle bundles (Fig. 1.74, p. 71). A similar effect was found in human eyes (Bill, 1971).

## THE INTRAOCULAR PRESSURE AND FACILITY

On the basis of Seidel's studies, we may expect the flow of aqueous humour to be determined by the difference of pressure between the fluid within the eye— the *intraocular pressure*—and the blood within the episcleral venous system into which the fluid must ultimately flow to reach the surface of the globe. Thus, according to Seidel's formulation we should have:

$$\text{Flow} = (P_i - P_e)/R \qquad (2)$$

if the flow follows Poiseuille's Law, $R$ being a resistance term determined by the frictional resistance through the trabecular meshwork and along the various vessels through which the fluid flows. To understand the dynamics of flow of aqueous humour, then, we must consider in some detail the nature of the intraocular pressure, $P_i$, and its relationships with the other factors in the above equation, namely, the episcleral venous pressure, $P_e$, and the resistance term, $R$.

## MEASUREMENT OF THE INTRAOCULAR PRESSURE

On inserting a hypodermic needle into the anterior chamber the aqueous humour flows out because the pressure within is greater than atmospheric; we may define the intraocular pressure as that pressure required just to prevent the loss of fluid. Manometric methods have been developed that permit the measurement of this pressure with a minimal loss of fluid, this being

necessary since, as we shall see, loss of fluid *per se* may upset the normal physiology of the eye.

### MANOMETRY

The general principle of the manometric methods employed for measuring the intraocular pressure is illustrated by Figure 1.51. The fluid-filled chamber is connected to a reservoir, R, which is of variable height, and also to the hypodermic needle. The end of the

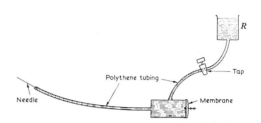

**Fig. 1.51** Illustrating the general principle on which manometric measurement of the intraocular pressure is based. The reservoir serves to fill the system with saline. When the needle is in the anterior chamber, and the tap is closed, changes in pressure cause movements of the membrane.

chamber is covered by a membrane, and it is essentially the movements of the membrane, caused by changes in pressure, that are recorded. If the membrane consists of a latex skin, it may be made to move a small mirror in contact with it, so that movements of the membrane can be magnified into movements of a spot of light reflected from the mirror. Alternatively, the membrane may be made of metal and act as the plate of a condenser, as in the Sanborn electromanometer, or it may be connected to a transducer valve, i.e. a valve whose anode may be moved from outside, so that a change in its position is converted into a change of voltage. The reservoir serves to fill the system with a saline solution and may be employed to calibrate the system; thus, with the needle stopped and the tap open, raising and lowering the reservoir will establish known pressures in the chamber.

### TONOMETRY

For studies on man, the introduction of a needle into the anterior chamber is rarely permissible, so that various *tonometers* have been developed permitting an indirect measure of the pressure within the eye. With the *impression type* of tonometer, such as the Schiøtz instrument, the depth to which a weighted plunger applied to the cornea sinks into the eye is measured, whilst with the *applanation tonometer* the area of flattening of the cornea, when a metal surface is applied with a controlled force, is measured. The two principles are illustrated schematically in Figure 1.52; the greater the intraocular pressure the smaller will be the depth of impression, or area of applanation.

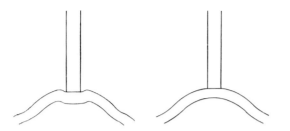

**Fig. 1.52** Illustrating the principles of impression tonometry (*left*) and applanation tonometry (*right*).

### Schiøtz tonometer

So widespread has been the use of this instrument, or its electronic adaptations, that it may profitably be described in some detail. An actual instrument is illustrated in Figure 1.53, whilst the mechanical features are shown schematically in Figure 1.54. The footplate, F, is curved to fit the average curvature of the human cornea. The weighted plunger, P, passes through the footplate; movements of this plunger operate on the hammer, H, which converts vertical movements into readings of the pointer on the scale. This scale reads from 0 to 20, the greater the indentation the lower the intraocular pressure and the *larger* the scale-reading.

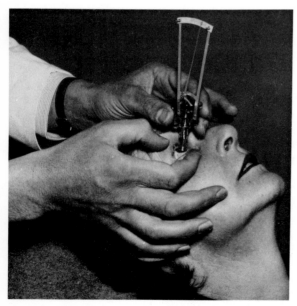

**Fig. 1.53** The Schiøtz tonometer applied to the eye (Department of Audiovisual Communication, Institute of Ophthalmology, London.)

The frame, Y, serves to hold the instrument upright on the eye, and because the footplate can move freely within the cylindrical part of the frame, the actual weight resting on the eye is that of the plunger, footplate, hammer and scale (17·5 g), the frame being held by the observer. With the smallest weight on the plunger, the total effective thrust on the eye amounts to 5·5 g, and the readings made with this are referred to as the '5·5 g readings'. By using heavier weights the thrust may be increased to 7·5 or 10 g, thereby allowing the instrument to give reasonable scale-readings at higher levels of intraocular pressure.

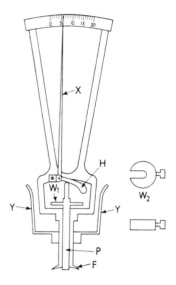

**Fig. 1.54** Diagrammatic illustration of the Schiøtz tonometer.

CALIBRATION

To calibrate the tonometer, the scale-reading, $R$, must be converted to the actual pressure within the eye before the tonometer was applied, $P_o$. The detailed procedure has turned out to be very complex and not completely satisfying;[*] it is impossible to enter into details of the procedures here, so we must content ourselves with the general principles. When the tonometer is on the eye, the pressure within it obviously rises because of the weight of the instrument; and the pressure within the eye with the tonometer on it is referred to as $P_t$. $P_t$ may be related to the scale-reading, $R$, by taking an excised eye, and inserting a needle connected to a reservoir and manometer into the anterior chamber. Different pressures may be established in the eye, and tonometer readings

---

[*]The literature on calibration scales and ocular rigidity is quite large and cannot be summarized here; some key references are Friedenwald, 1957; McBain, 1960; Drance, 1960; Perkins and Gloster, 1957; Hetland-Eriksen, 1966; Moses and Grodski, 1971.

corresponding to these may be made, Friedenwald found the relationship:

$$W/P_t = a + bR \qquad (3)$$

so that with $a$ and $b$ determined experimentally we have the relationship between $P_t$ and scale-reading, $R$. To convert $P_t$ to $P_o$, we require to know by how much the indentation, corresponding to the scale-reading, $R$, has raised the pressure; and this depends on the distensibility of the sclera, i.e. the extent to which a change in volume of the eye will raise the pressure. Thus, if we know the volume of displacement, $V_c$, associated with a given scale-reading, and if we can relate $V_c$ to a change in pressure of the eye, we can finally establish a relationship between $R$ and $P_o$. Friedenwald considered that there was a logarithmic relationship between pressure and volume of the globe, to give an equation:

$$\text{Log } P_2/P_1 = k(V_2 - V_1) \qquad (4)$$

whence he defined a *rigidity coefficient*, $k$, characteristic of a given eye. Experimentally he measured the volumes of displacement, $V_c$, for different scale-readings, and making use of an average rigidity coefficient he was able to construct a table relating scale-readings to values of $P_o$, the intraocular pressure corresponding to the scale-reading. As subsequent work has shown (Perkins and Gloster; Macri, Wanko and Grimes; McBain), the weakness in this calibration is the assumption that the rigidity coefficient is a constant characteristic of each eye, i.e. that it is independent of pressure; in fact values ranging from 0·003 to 0·036 could be obtained in a given eye according to the intraocular pressure. Moreover, there is no doubt that the distensibility of the globe varies from one individual to another, so that unless this can be allowed for, it is not very sound to use a single calibration scale.*

Because of these difficulties in accurate calibration, the Schiøtz instrument is slowly being superseded by an applanation type of manometer, since with this instrument the deformation of the eye during applanation is very small, with the result that calibration is independent of scleral extensibility (Goldmann, 1955; Armaly, 1960).

## Applanation tonometer

This measures, in effect, the force required to produce a fixed degree of applanation when a flat surface is pressed against the cornea. The earliest instrument (Maklow) consisted of a cylindrical piece of metal with a flat base weighing 5 to 15 g. A drop of a concentrated dye-solution was spread over the cornea, the instrument was allowed to rest a moment on the corneal surface and a print of the area of contact between cornea and metal was obtained by placing the foot of the instrument on a piece of paper. Goldmann's (1955) instrument is a refinement on this procedure, in which the degree of applanation is kept fixed and the instrument indicates the force required to produce this. There is no doubt that the calibration of this instrument involves far fewer uncertainties than does that of the Schiøtz indentation type. Other applanation instruments are the Mackay-Marg (1959) tonometer and the pneumatic applanation tonometer of Durham et al. (1965).

Because of the superiority of the Goldmann

applanation tonometer to the indentation type, it seems reasonable to calibrate the Schiøtz tonometer by measuring pressure on the same eye with both instruments successively, taking care that the head is in the same position for both measurements. The most recent calibration carried out in this way has shown that the pressure deduced from the Schiøtz measurement, using Friedenwald's calibration table, is lower than that measured by the Goldmann instrument (Anderson and Grant, 1970).

## THE NORMAL INTRAOCULAR PRESSURE

Clearly, the magnitude of the normal average intraocular pressure in man will depend on the validity of the calibration scale employed; early studies, using Schiøtz' original calibration, gave mean values in the neighbourhood of 20 mmHg, but later studies in which more modern calibration scales have been employed, e.g. that of Leydhecker et al., or in which a more accurate applanation technique was used (Goldmann), indicate a lower mean value in the range 15 to 16 mmHg. In general, Leydhecker et al.'s figures, based on 13 861 apparently healthy eyes, suggest that 95·5 per cent of all healthy eyes have a pressure lying between 10·5 and 20·5 mmHg. A pressure above 20·5 mmHg may thus be a sign of abnormality. Between the ages of 10 and 70 there is no significant change in mean intraocular pressure, and no difference between the sexes is found. In newborn infants likewise the pressure falls within the normal adult range, the mean value found by Hörven being 16·5 mmHg with limits of 10·9 and 24·2 mmHg. In experimental animals the intraocular pressure may well be of the same order; thus in cats Davson and Spaziani reported a mean value of 16·5 mmHg, whilst Langham found 20 mmHg in this animal and 20·5 mmHg in the rabbit.†

---

*If the pressure is measured on the same eye with different weights on the tonometer, the same result should be read off the calibration scales. If this did not happen it was attributed to the circumstance that the ocular rigidity was abnormal in the eye being studied and Friedenwald, and later Moses and Becker, actually devised a nomogram and tables from which the ocular rigidity could be estimated from paired readings on the same eye using different weights. It will be clear, however, that the scales must be rigidly consistent within themselves if deviations from them are to be interpreted as the correlates of the mechanical properties of the eye rather than as errors in the scales themselves. Subsequent studies summarized by McBain (1960) have shown that they were not sufficiently consistent to warrant their use in this way.

†Manometric methods usually demand the use of a general anaesthetic, so that the figures given may not be true of the normal unanaesthetized animal; according to Sears (1960) and Kupfer (1961), the normal unanaesthetized value for rabbits is 19 mmHg; with urethane it was 16·8 mmHg, with paraldehyde, 20·4 and aprobarbital, 17 mmHg.

### DIURNAL VARIATION

In man the intraocular pressure shows a slight but significant diurnal variation, being highest in the early morning and lowest at about midnight. The range of fluctuation is small, 2 to 5 mmHg. According to Ericson (1958) the cause is a diurnal variation in the rate of secretion of fluid, as measured by the suction-cup technique (p. 35).

## MECHANICAL FACTORS AFFECTING THE INTRAOCULAR PRESSURE

The simple relationship between flow-rate of aqueous humour, $F$, and the pressure-drop, mentioned above, may be rewritten:

$$P_i = F \times R \times t + P_e \qquad (5)$$

where $P_i$ is the intraocular pressure, $F$ is the volume of aqueous humour drained during time $t$, $R$ is the resistance and $P_e$ is the episcleral venous pressure. With the aid of this relationship some of the factors determining the intraocular pressure may be easily appreciated. Thus an increase in resistance to flow will increase the intraocular pressure, provided that the rate of formation of fluid does not decrease in proportion. The factors determining the rate of production of fluid are not completely understood. As we have seen, the fluid is formed by a process of secretion and, when comparable systems are studied, e.g. secretion of saliva, it is found that the process can operate against very high pressures, so that the secretion of aqueous humour may be assumed to occur at a constant rate independent of fairly wide fluctuations in the pressure. If a process of passive filtration from the vascular bed contributes to flow—and as we shall see this seems to be true of some species—then this contribution will be reduced at higher pressures (p. 52). We shall not go far wrong, however, if we conclude that intraocular pressure varies inversely with resistance to flow. Alterations in episcleral venous pressure, $P_e$, may likewise be expected to be reflected in predictable alterations in intraocular pressure.

### TONOGRAPHIC MEASUREMENT OF RESISTANCE

The essential principle on which this measurement is based is to place a weight on the cornea; this raises the intraocular pressure and causes an increased flow of aqueous humour out of the eye. Because of this increased flow, the intraocular pressure falls eventually back to its original value, and the rate of return enables the computation of the resistance to flow. Thus:

$$F_o = \frac{P_i - P_e}{R}(t_2 - t_1) \qquad (6)$$

$F_o$ being the normal volume flow, without the weight on the eye, during the period $t_2 - t_1$.

With the weight on the eye we have:

$$F_t = \frac{\frac{1}{2}(P_{t_2} - P_{t_1}) - P_e}{R}(t_2 - t_1) \qquad (7)$$

Where the average intraocular pressure during the period $t_2 - t_1$ with the weight on the eye is taken as: $\frac{1}{2}(P_{t_2} - P_{t_1})$.

The extra flow,

$$F_t - F_o = \left\{ \frac{\frac{1}{2}(P_{t_2} - P_{t_1}) - P_e}{R} - \frac{P_i - P_e}{R} \right\}(t_2 - t_1)$$

$$= \{\frac{1}{2}(P_{t_2} - P_{t_1}) - P_i\} \frac{t_2 - t_1}{R} \qquad (8)$$

$$\text{or } \Delta F = \frac{t_2 - t_1}{R} \cdot \Delta P \qquad (9)$$

$$\text{or } \Delta F = (t_2 - t_1) \cdot C \cdot \Delta P \qquad (10)$$

where $C$ is the reciprocal of $R$ and may be called the *facility of outflow*.

Experimentally the weight is applied to the eye in the form of an electronic tonometer so that it serves also to measure the intraocular pressure when appropriately calibrated. Since, moreover, the calibration of the tonometer had already given the change in volume of the eye associated with changes in $P_t$, $\Delta F$ was also known. Hence by measuring the change in $P_t$, $\Delta F$ was found and these two quantities could be inserted into the equation to give $C$, the facility of outflow. The average value found by Grant, who first described the technique, was $0.24$ mm$^3 \cdot$ min$^{-1} \cdot$ mmHg$^{-1}$, i.e. the flow when the difference of pressure was 1 mmHg was $0.24$ mm$^3$ per minute. Measurements on excised human eyes gave comparable results; in this case fluid was passed into the eye from a reservoir and the rate of movement measured at different pressures.

### EPISCLERAL VENOUS PRESSURE

The episcleral venous pressure, $P_e$, representing as it does the pressure of the blood into which the aqueous humour is draining, is an important experimental parameter. The measurement has been studied especially by Brubaker (1967), who has compared three different techniques, two based on the force required to make a vein collapse, and one involving direct cannulation of the vein and determining the pressure required to cause a change in direction of flow. The cannulation method is the most reliable, but for human subjects the technique employing a transparent pressure-chamber must be used, a membrane made of the toad pericardium being employed. With this method, an average value of $10.8 \pm 0.5$ mmHg was found for the rabbit. Values for man are included in Table 1.5 (p. 54).*

# FACTORS AFFECTING THE INTRAOCULAR PRESSURE

## VENOUS PRESSURE

In so far as the aqueous humour must drain into the episcleral venous system, either by way of the intrascleral plexus or more directly along the aqueous veins, the intraocular pressure will vary directly with the venous pressure in accordance with the simple equation given above. Certainly in man there is a good correlation between intraocular pressure and $P_e$, whilst in cats Macri showed the same thing when he varied the intraocular pressure experimentally. Acute changes of venous pressure will influence the intraocular pressure in a twofold manner. First, by affecting drainage as just indicated, and second, by a direct transmission of the changed venous pressure across the easily distensible walls of the intraocular veins. The expansion of the veins will be eventually compensated by a loss of aqueous humour from the eye, so that the rise due to this cause may be more or less completely compensated, but the effect on drainage should last as long as $P_e$ is elevated. Experimentally the effects of changed venous pressure may be demonstrated by administering amyl nitrite; the peripheral vasodilatation causes a general increase in venous pressure and leads to a considerable rise in intraocular pressure in spite of a lowered arterial pressure. Again, ligation of the vortex and anterior ciliary veins causes large rises in intraocular pressure, as we should expect. Finally, destruction of the aqueous veins, by restricting the outlets of aqueous humour, may cause quite considerable rises in intraocular pressure in rabbits, according to Huggert.

## ARTERIAL PRESSURE

The arterial pressure can influence the intraocular pressure through its tendency to expand the intraocular arteries; since these are not easily distensible, however, relatively large changes in arterial pressure will be required to produce measurable changes in intraocular pressure. That the arterial pressure does affect the intraocular pressure in this way is manifest in the *pulse* shown in records of the intraocular pressure; the amplitude of this pulse is small, about 1 mmHg, and coincides with cardiac systole. Thus a change of about 20 mmHg, which is the pulse-pressure in an intraocular artery according to Duke-Elder, causes a change of only 1 mmHg in the intraocular pressure.

In the rabbit, ligation of one common carotid artery lowers the intracranial arterial pressure on the side of the ligation, whilst the pressure on the other side rises above normal; hence we may study the effects of lowered arterial pressure on one eye, using the other as a control. Wessely in 1908 found a small lowering of the intraocular pressure on the ligated side and Bárány found that after 24 hours the pressure had returned to normal. The immediate effect of the ligation is undoubtedly the result of the fall in arterial pressure which leads to a reduced volume of blood in the eye. As Figure 1.55 shows, there is a tendency to compensate

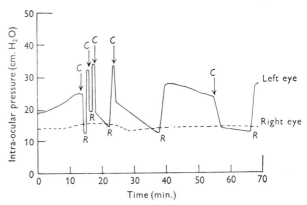

**Fig. 1.55**   Intraocular pressure of the rabbit. At points marked C the left common carotid artery was clamped, and at points marked R it was released. (Davson and Matchett, *J. Physiol.*)

for this since alternate clamping and unclamping of the artery leads to an overshoot on release of the clamp, suggesting that there has been some vasodilatation which would partly compensate for the reduced flow caused by the fall in pressure. The compensation is not complete, however, since Linnér found a 19 per cent decrease in the rate at which blood escaped from a cut vortex vein on the side of the ligature.

Bill (1970), working on the monkey, found at normal mean arterial pressure of 119 mmHg a rate of flow of aqueous humour of $2.64~\mu l/min$; this was unaffected by lowering the blood pressure, through haemorrhage, to 70 to 90 mmHg; however, when the pressure was reduced to 60 to 40 mmHg there was a sharp fall to as little as $0.5~\mu l/min$, presumably due to lowered blood-flow through the ciliary processes.

---

*The episcleral venous pressure will be a factor determining the intraocular pressure so that a relationship between these two variables may be expected. Weigelin and Lohlein (1952) gave the following: $P_e = 0.48 \times P_i + 3.1$, where $P_e$ is the episcleral venous pressure and $P_i$ is the intraocular pressure. In animals, too, Macri (1961) found a linear relationship between the two.

RATE OF FLOW OF AQUEOUS HUMOUR

Other things being equal, we may expect changes in rate of secretion of fluid to produce changes in intraocular pressure in the same sense, but the effects are not likely to be large. Thus, if the intraocular pressure, $P_i$ is 16 mmHg, and the episcleral venous pressure, $P_e$, is 12 mmHg, the pressure drop, $P_i - P_e$, is 4 mmHg. If the rate of flow is doubled the pressure-drop must be doubled too, in which case it becomes 8 mmHg, and if $P_e$ remains the same this means that the intraocular pressure, $P_i$ rises from 16 to 20 mmHg. Thus doubling the rate of secretion causes only a 25 per cent increase in intraocular pressure. In practice it is not easy to vary the rate of secretion; as we shall see, the carbonic anhydrase inhibitor, Diamox, reduces the rate and this is accompanied by a fall in intraocular pressure; the same is true of digitalis compounds, according to Simon, Bonting and Hawkins.

By injecting saline continuously into the anterior chamber we may experimentally increase the flow of fluid through the eye and test the applicability of the simple equation relating flow-rate to pressure-drop; Langham found that the equation did not hold, the rise in pressure for a given increase in flow being smaller than than predicted, and his results suggested some homeostatic mechanism that decreased the resistance to flow as the intraocular pressure rose. Section of the cervical sympathetic, or ligation of the common carotid, caused the relationship between flow and pressure to become linear, i.e. to follow the simple equation:

flow × resistance = pressure drop

and this suggested that the adaptation was largely vascular. The situation is apparently complicated, however, by the effects of pressure on the rate of production of aqueous humour. Thus Bill and Bárány (1966) found a linear relation between flow-rate and pressure, but they also showed that the rate of secretion diminished with higher pressures, so that the linear relation found between flow-rate out of the cannula and pressure means that the actual flow through Schlemm's canal was not linearly related to pressure, a rise in pressure producing too small a flow, and thus suggesting an increased resistance at the higher pressure.

## PSEUDOFACILITY

If raising the intraocular pressure reduces the production of fluid, this means that, when facility is determined by measuring the flow of fluid from a cannula in the anterior chamber under an applied pressure, a reduction in fluid production caused by the raised pressure will make room for some fluid from the reservoir and so give too high a measure of the extra flow through Schlemm's

canal; the extra fluid entering for a given rise in pressure can be expressed as a 'pseudofacility' (Bill and Bárány, 1966).

BÁRÁNY'S THEORETICAL TREATMENT

If a part of the production of fluid is pressure-dependent, then the relationships between pressure, resistance (or facility), flow-rate and vascular pressures become more complex. Bárány (1963) has developed an equation* to describe these relationships, defining the pseudofacility as the rate of decrease in flow, $dF/dP_i$ with increase in intraocular pressure, $\Delta P_i$. Thus $dF/dP_i = -A x_c$ where $A$ represents the hydraulic conductivity across the filtering membranes, e.g. the capillaries of the ciliary body, and $x_c$ represents the 'pressure index' corresponding to the filtering vessels and is related to the pressures at arterial and venous ends of the filtering systems, $P_a$ and $P_v$. An important relation emerges from this treatment, describing the change in intraocular pressure, $\Delta P_i$ when the episcleral venour pressure is changed by $\Delta P_e$:

$$\frac{\Delta P_i}{\Delta P_e} = \frac{C_{true}}{C_{true} + C_{pseudo}} \quad (11)$$

$C$'s being facilities, defined as before, $C_{gross}$ being that measured without regard to pseudofacility, and being equal to $C_{true} + C_{pseudo}$.

From this it follows that:

$$C_{pseudo} = C_{gross} \times \left(1 - \frac{\Delta P_i}{\Delta P_e}\right) \quad (12)$$

EXPERIMENTAL DETERMINATION

Brubaker and Kupfer (1966) determined $C_{gross}$ by perfusion of the monkey's eye at constant pressure. They then raised $P_e$ by inflating a cuff round the monkey's neck and determined $C_{gross}$ for the new steady state; $P_i$ and $P_e$ at the new pressures were also determined, to give $\Delta P_i/\Delta P_e$.

---

*Bárány's equation is:

$$P_i = \frac{S}{C + Ax_c} + P_e \frac{C}{C + Ax_c} + P_a \frac{Ax_c}{C + Ax_c} - P_{coll} \frac{A}{C + Ax_c}$$

$S$ is rate of secretion; $C$ is true facility, $P_i$ and $P_e$ are intraocular and episcleral venous pressures respectively, $P_{coll}$ is the colloid osmotic pressure of the plasma proteins, and $A$ represents the conductivity of the filtering part of the vascular tree, i.e. the flow produced by unit difference of pressure, and $x_c$, as indicated above, represents the pressure index, and is essentially a fraction of the difference between arterial and venous pressures within the eye. Thus $x_c \times (P_a - P_v)$ represents the available pressure favouring filtration and may be equated with the capillary pressure; as indicated, $Ax_c$ is the pseudofacility.

Mean values for the monkey were:

$P_i = 12.7$ mmHg

$P_e = 9.0$ mmHg

$C_{gross} = 0.63$ $\mu$l.min$^{-1}$ mmHg$^{-1}$

$C_{pseudo} = 0.19$ $\mu$l.min$^{-1}$ mmHg$^{-1}$

Thus in this species, pseudofacility was 30 per cent of the total. In human subjects Goldmann (1968) found a value of 20 per cent.

## EFFECTS OF BLOOD OSMOLALITY

Aqueous humour and blood plasma are approximately isosmolal, and this is to be expected in view of the rapid exchanges of water across the blood-ocular fluid barriers (Kinsey et al., 1942) compared with the much slower exchanges of solutes. Raising the osmolality of blood artificially causes a fall in intraocular pressure because of the osmotic outflux of water (Davson and Thomassen, 1950; Auricchio and Bárány, 1959); clinically an acute fall in intraocular pressure may be induced by oral urea, this being chosen because it is not rapidly excreted from the blood and its passage across the blood-ocular barriers is slow; mannitol is less effective presumably because its renal excretion is greater (Galin, Davidson and Pasmanik, 1963).

### REFLEXION COEFFICIENTS

As Pederson and Green (1973) have emphasized, the effectiveness of a hypertonic solution in reducing intraocular pressure depends on the reflexion coefficient of the solute, e.g. glycerol, employed to induce the hypertonicity. Thus the relatively small effects* obtained with many hypertonic solutions are probably due to the low reflection coefficient of the solute; for example, Pederson and Green calculate from their measurements that the reflection coefficient for salt, such as NaCl, is as low as 0.02, in which case the difference of osmotic pressure actually developed between blood and aqueous humour would be only one fiftieth of the theoretical pressure based on complete semipermeability of the lining membranes.

## RELATIVE CONTRIBUTIONS OF FILTRATION AND SECRETION

It has generally been assumed that the contribution of filtration to the net flow of fluid through the eye is small and perhaps negligible in some species (Bill, 1975). However, Weinbaum et al. (1972) and Pederson and Green (1973a,b) and Green and Pederson (1972) have developed Bárány's treatment of the fluid relations in the eye and illustrated how some of the required

parameters may be deduced from experimental studies. On this basis they conclude that the contribution of filtration is very much higher than that deduced, say, from the effect of inhibitors of sodium transport (Cole, 1966), in fact they conclude that only some 35 per cent of total formation is the result of active secretion, the remainder being due to filtration. This very high estimate of filtration is surprising and may well result from an erroneous estimate of the reflection coefficient for such solutes as Na$^+$ and sucrose which, surprisingly, Pederson and Green find the same, although the rate of penetration of sucrose into the aqueous humour from blood is very much less than that of $^{24}$Na (see, for example, Davson and Matchett, 1953).†

## HUMAN STUDIES

The study of intraocular pressure in man must be made with the tonometer; this permits both the measurement of pressure and facility of outflow. The measurement of episcleral venous pressure, $P_e$, in humans by Kupfer and his colleagues has permitted the assessment of the pseudofacility, an important parameter since, without a knowledge of the pressure-dependent flow into the eye—and out of it when the pressure is high—no accurate notion of the true facility is obtainable, and without this, of course, no true notion of rate of secretion is obtainable. Kupfer and Ross (1971) measured total facility, using a modified Grant equation that corrects for ocular rigidity, and pseudofacility was determined by measuring change in intraocular pressure, $\Delta P_i$, with change in episcleral venous pressure $\Delta P_e$, in accordance with Equation 11. The flow out of the eye, $F$, is given by the equation:

$$F = C_{true}(P_i - P_e) \qquad (13)$$

$C_{true}$ being equal to $C_{total} - C_{pseudo}$.

### THE NORMAL VALUES

Average normal values derived from ten determinations of right and left eyes of the same subject are given in Table 1.5, and average results on four separate human subjects are shown in Table 1.6. It emerged that the errors were not very high, so that it is practicable, using

---

* Care must be taken in interpreting the quantitative results of hypertonic solutions since the vitreous body represents a large unstirred mass of fluid and will only slowly be affected by the hypertonicity of the blood.

† Pederson and Green (1975) have developed their treatment further in relation to the effects of prostaglandin PGE$_2$ on aqueous flow; they make the important point that filtration and secretion might well cooperate in determining the efficacy of the flow process, the passive pressure-induced flow 'sweeping' the secreted fluid in the correct direction.

**Table 1.5** Parameters of the right and left eyes of a human subject. (Kupfer and Ross, 1971).

|  | $P_i$ | $P_e$ | $C_{total}$ | $C_{true}$ | $C_{pseudo}$ | $P_k$ | *Flow* |
|---|---|---|---|---|---|---|---|
| *Right eye* | 13·8 | 9·2 | 0·328 | 0·256 | 0·072 | 31·6 | 1·19 |
|  | ± | ± | ± | ± | ± | ± | ± |
| SEM | 0·36 | 0·14 | 0·014 | 0·010 | 0·006 | 2·5 | 0·13 |
| *Left eye* | 14·0 | 9·0 | 0·288 | 0·226 | 0·062 | 33·6 | 1·14 |
|  | ± | ± | ± | ± | ± | ± | ± |
| SEM | 0·33 | 0·12 | 0·016 | 0·012 | 0·006 | 2·3 | 0·11 |

Mean of ten determinations.
Pressure mmHg; $C$ in $\mu$l/min/mmHg. Flow in $\mu$l/min. (Kupfer and Ross, 1971).

**Table 1.6** Similar to Table 1.5, but results of four human subjects averaged. (Kupfer and Ross, 1971).

|  | $P_i$ | $P_e$ | $C_{total}$ | $C_{true}$ | $C_{pseudo}$ | $P_k$ | *Flow* |
|---|---|---|---|---|---|---|---|
|  | 13·3 | 8·4 | 0·285 | 0·224 | 0·061 | 31·0 | 1·09 |
|  | ± | ± | ± | ± | ± | ± | ± |
| SEM ± | 0·21 | 0·26 | 0·008 | 0·006 | 0·002 | 0·59 | 0·030 |

these techniques, to study the effects of drugs and to analyse them in terms of changed flow-rate and facilities.

THE CRITICAL INTRAOCULAR PRESSURE

Goldmann (1968) pointed out that if the intraocular pressure was caused to rise, e.g. by raised episcleral venous pressure, the pressure-dependent flow would decrease and finally reverse, and that there was a theoretical intraocular pressure, $P_k$, at which net flow through the eye would be zero, the secretory inflow being balanced by the pressure-dependent outflow. This becomes clear from Figure 1.56 where intraocular pressure, $P_i$, is abscissa and the ordinate is flow through the eye. Line II represents the constant (pressure-independent) secretion, $S$, and line I is the pressure-dependent flow, which is positive at lower values of $P_i$ but becomes negative at higher values. The relationship is probably linear and the slope measures the pseudofacility, $C_{pseudo} = \Delta F/\Delta P_i$. Line III is the sum of lines I and II and represents the net flow through the eye and becomes zero at $P_k$. It is clear that the slope of line III also represents the pseudofacility.

$P_k$ is given by:

$$P_k = \frac{P_{i_2}P_{e_1} - P_{e_2}P_{i_1}}{\Delta P_i - \Delta P_e} \quad (14)$$

where the numbered suffices indicate the pressures before and after a change in $P_e$.

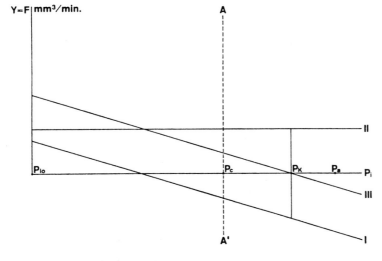

**Fig. 1.56** Illustrating significance of the theoretical pressure, $P_k$, such that flow through the eye is zero, the secretory inflow being balanced by the pressure-dependent outflow. $P_i$ is the abscissa and the rate of inflow is ordinate. Line II represents the pressure-independent influx (secretion), whilst Line I represents the pressure-dependent component, being positive (inflow) at lower values of intraocular pressure ($P_i$) and negative (outflow) at higher values. Line III is the algebraic sum of I and II. It is clear that total rate of flow is linearly dependent on intraocular pressure and becomes zero at $P_k$. (Goldmann, *Bibl. Ophthal.*)

## FLOW-CURVES

Experimentally, then, so-called flow-curves may be constructed based on tonographic measurements of flow-rate and the influence of altered $P_e$ on $P_i$. The effects of drugs etc. might then be revealed as an altered slope of the flow-curve, indicating a changed pseudofacility, whilst an altered flow-rate, with unchanged pseudofacility, would give a new line parallel to the first.

To check the linearity of flow versus $P_i$, Kupfer et al. (1971) raised $P_e$, the episcleral venous pressure, to successively higher values by inflating a cuff round the neck, measuring $P_e$ and $P_i$ at each steady state. Finally tonography was carried out at normal pressure to give total facility; from this the pseudofacility was computed, whence true facility was obtained:

$$C_{true} = C_{total} - C_{pseudo} \qquad (15)$$

Hence for each value of $P_i$ a computed flow rate given by:

$$F = (P_i - P_e) C_{true} \qquad (16)$$

was obtained. Values have been plotted in Figure 1.57 to give a good straight line.

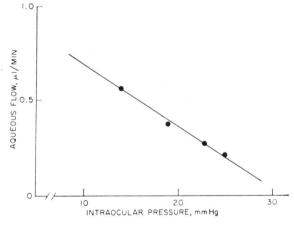

**Fig. 1.57** Relation of aqueous flow to steady-state intraocular pressure in one eye of a young normal volunteer. Sequential elevation of intraocular pressure obtained by sequentially greater inflation of a pressure cuff around the neck. Flow calculated using value for $C_{true}$ obtained at base-line pressure. The relation is very close to linear. (Kupfer et al., Invest. Ophthal.)

### Effects of Diamox

Figure 1.58 illustrates the effects of Diamox on the flow-curve; there has been a parallel shift of the line to the left indicating that, at any given value of intraocular pressure, the flow-rate is considerably smaller. This representation indicates how an underestimate in the

reduction in flow-rate can be made through ignoring pseudofacility. Thus the initial estimate of aqueous flow, $F_a = 1.24$ $\mu$l/min was obtained at the baseline intraocular pressure, $P_i = 13.1$ mmHg. After treatment, the pressure fell to $P_p = 9.6$ mmHg and total flow decreased to $F_p = 0.40$ $\mu$l/min, an observed difference of 0.84 $\mu$l/min. However, this observed difference does not take

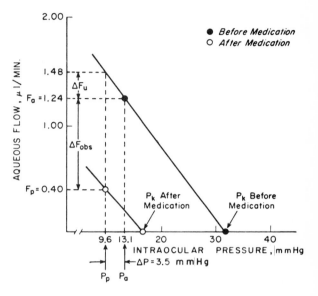

**Fig. 1.58** Flow curves for mean values of data obtained on four subjects before and after acetazolamide. Illustrates how experimentally observed drop of aqueous flow ($\Delta F_{obs}$) underestimates total effect of medication by magnitude of increase of ultrafiltration ($\Delta F_u$) when intraocular pressure becomes lower. Magnitude of total flow decrease due to acetazolamide would be $\Delta F_{obs} + \Delta F_u$. (Kupfer et al., Invest. Ophthal.)

into account the increased pressure-dependent flow due to the lowered intraocular pressure. This is represented as $\Delta F_u$ and its magnitude is given by the pseudofacility (0.07 $\mu$l min$^{-1}$ mmHg$^{-1}$) times the change of pressure, 3.5 mmHg = 0.24 $\mu$l/min). Thus the total change in flow due to the drug is $\Delta F_{obs} + \Delta F_u = 0.84 + 0.24 = 1.08$ $\mu$l/min.

### Effect of epinephrine

When epinephrine was applied to the eye there was a change in the slope of the flow-curve (Fig. 1.59). The intraocular pressure was decreased, and this was associated with a decrease in pseudofacility and a decrease of observed flow, which, however, was not significant. In these young normal subjects there was no significant increase in true facility, by contrast with glaucoma subjects where this occurs (see, for example, Weekers et al., 1954; Kronfeld, 1964; Krill et al., 1965).

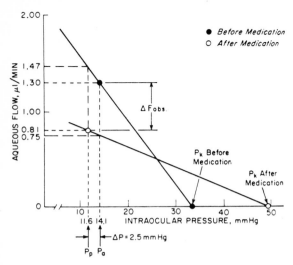

**Fig. 1.59** Flow curves for mean values of data obtained for the treated eye of four subjects, before and after topical L-epinephrine. Illustrates how the experimentally observed decrease of aqueous flow ($\Delta F_{obs}$) is related to change of slope (*pseudo-facility*) and change of intraocular pressure; it underestimates total effect of medication. The total effect of medication can be determined for any given value of intraocular pressure. (Kupfer *et al.*, *Invest. Ophthal.*)

## BREAKDOWN OF THE BLOOD-AQUEOUS BARRIER

Under certain pathological or experimental conditions, the blood-aqueous barrier breaks down, so that substances such as proteins, that are normally almost completely excluded from penetration, now appear in the aqueous humour in measurable amounts. Thus, on treatment of the eye with nitrogen mustard, there is a large increase in the protein content of the aqueous humour, and this is easily revealed in the slit-lamp microscope as a pronounced Tyndall beam (Fig. 1.60); associated with these two events there is a constriction of the pupil. The breakdown of the barrier not only allows proteins and other high molecular weight substances in the blood to appear in the aqueous humour in large amounts, but it is also associated with an increased permeability to smaller molecules, such as sucrose or *p*-aminohippurate.

### FLUORESCEIN TEST

The dye, fluorescein, normally penetrates the blood-aqueous barrier very slowly, but to a measurable extent because, in the intact eye, it may be detected by the fluorescence it causes in the slit-lamp beam. When the blood-aqueous barrier is broken down, the rate of appearance of fluorescein increases, and the change in rate, measured by some simple fluorimetric device such

as that devised by Amsler and Huber or Langham and Wybar, is used as a measure of the breakdown. Clinically this has been found useful in the diagnosis of uveitis, an inflammatory condition associated with breakdown of the barrier.★

## PARACENTESIS

When aqueous humour is withdrawn from the eye in appreciable quantities, the fluid is re-formed rapidly but it is now no longer normal, the concentration of proteins being raised. The fluid is described as *plasmoid*. The breakdown of the barrier following paracentesis was attributed by Wessely to the sudden fall in intraocular pressure, causing an engorgement of the blood vessels of the uvea that led to dilatation of the epithelial linings and escape of protein into the anterior or posterior chamber.

### ANTAGONISM OF BREAKDOWN

Wessely showed that the breakdown could be prevented by sympathetic stimulation or treatment of the eye with epinephrine, and more recently Cole (1961) showed that the anti-inflammatory agent, polyphoretin, which has a direct action on capillaries, also prevented breakdown. It must be emphasized, however, that the degree of breakdown after paracentesis varies with the species, being very pronounced in the rabbit but much less in primates.

### PHYSIOLOGICAL SIGNIFICANCE

According to Bito, the escape of proteins, including fibrinogen, into the anterior chamber serves to seal the

---

★A few minutes after an intravenous injection of fluorescein, a line of green fluorescence may be seen along the vertical meridian of the cornea. This is called the *Ehrlich line*, and it results from the thermal currents in the anterior chamber. The rate of appearance of the Ehrlich line was used as a rough measure of rate of flow of aqueous humour, on the assumption that the fluorescein passed first into the posterior chamber. More recently the 'fluorescein appearance time', determined with the slit-lamp microscope, has been recommended as a measure of rate of flow of aqueous humour, but the situation with regard to the mechanism of penetration of fluorescein into the aqueous humour is complex; thus the initial appearance seems definitely to be due to diffusion across the anterior surface of the iris (Slezak, 1969), a measurable concentration in the posterior chamber only building up later. There seems no doubt, moreover, from the work of Cunha-Vaz and Maurice (1967) that fluorescein is actively transported out of the eye, certainly by the retinal blood vessels and possibly, also, by the posterior epithelium of the iris. Thus appearance of fluorescein in the anterior chamber is determined primarily by diffusion from the surface of the iris rather than by the rate of secretion of aqueous humour.

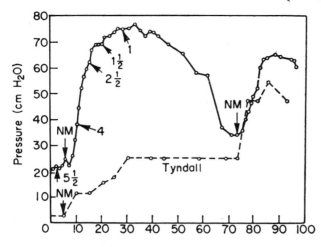

**Fig. 1.60** Rise in IOP and intensity of the Tyndall beam in the aqueous humour after subconjunctival injection of nitrogen mustard in the rabbit's eye. At points marked NM the injection was made. Numbers applied to the pressure curve indicate diameter of the pupil in mm. (Eakins, *Exp. Eye Res.* from Davson and Huber, *Ophthalmologica.*)

**Table 1.7** Experimental procedures producing a breakdown of the blood–aqueous barrier (Eakins, 1977)

| | |
|---|---|
| *Trauma*<br>   Mechanical injury to iris, lens<br>   Contusions<br>   Paracentesis | *Endogenous chemical mediators*<br>   Histamine<br>   Bradykinin<br>   Prostaglandins (and arachidonic acid)<br>   Serotonin<br>   Acetylcholine |
| *Chemical irritants*<br>   Nitrogen mustard<br>   Formaldehyde<br>   Acid burns<br>   Alkali burns | *Miscellaneous*<br>   Bacterial endotoxins<br>   X-ray irradiation<br>   Laser irradiation |
| *Nervous activity*<br>   Stimulation of trigeminal nerve | |
| *Immunogenic mechanisms*<br>   Bovine serum albumin | |

anterior chamber after puncture by clotting of the re-formed aqueous humour. The source of the newly formed plasmoid aqueous humour is presumably the ciliary body and iris;* certainly the rate of refilling of the eye is much greater than can be accounted for by a normal secretion, so that a large part is simply a plasma exudate; thus the concentration of chloride in the newly formed fluid approximates that in a plasma filtrate rather than in normal aqueous humour (Davson and Weld, 1941).

## PROSTAGLANDINS

Breakdown of the blood-aqueous barrier results from a number of treatments, which are summarized in Table 1.7, and it appears that a feature common to most of the insults is the liberation of prostaglandins into the eye. Ambache (1957) extracted a substance from the iris that

he called *irin*; this seemed to be responsible for the pupillary constriction that followed stroking the iris or mechanical stimulation of the trigeminal (Maurice, 1954); and in a later study Ambache, Kavanagh and Whiting (1965) showed that the effects were not due to cholinergic mechanisms, since they were not blocked by atropine, nor yet was histamine or 5–HT responsible since mepyramine and LSD did not block. By perfusing the anterior chamber they were able to obtain material—irin—that caused contraction of smooth muscle; increased amounts of the irin were obtained by collapse

---

* Scheie *et al.* (1943) found that the protein concentration in the fluid formed after paracentesis in the cat was reduced to about a half if the animal was completely iridectomized, and they attributed to the iris a major role in production of the newly formed fluid. However, as Unger *et al.* (1975) point out, iridectomy removes a major source of prostaglandins which, as we shall see, are mainly, if not entirely, responsible for the breakdown of the barrier.

of the anterior chamber, stroking the iris, or moving the lens. Chromatographic analysis of irin indicated that it consisted of one or more fatty acids of the prostanoic acid class called *prostaglandins* (Ambache *et al.*, 1966). The metabolic pathway of PGE$_2$ is illustrated in Figure 1.61. Waitzman and King (1967) injected prostaglandins E$_1$ and E$_2$ into the anterior chamber of the rabbit's eye and obtained a sustained rise in intraocular pressure accompanied by a contracted pupil; the effect on facility

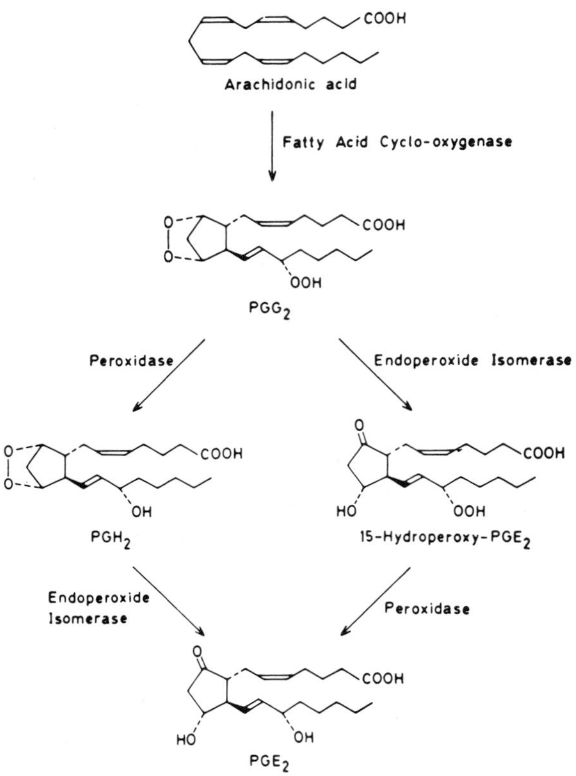

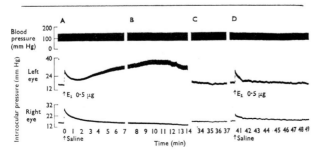

*Fig. 1.61* Pathways of biosynthesis of PGE$_2$ from arachidonic acid. (Eakins, *Exp. Eye Res.*)

of outflow was slight, so they concluded that the influence was on the production of fluid; these authors stated that the protein content of the aqueous humour was not raised significantly, but Beitch and Eakins (1969) found a considerable increase; Figure 1.62 shows

the course of the rise in intraocular pressure, the course being similar to that caused by nitrogen mustard; like Davson and Quilliam (1947), these authors found that application to one eye could have a contralateral effect on the other eye. As Cole had found with breakdown induced by nitrogen mustard, the effects of prostaglandin could be blocked by polyphloretin. Prostaglandin synthetic enzymes are present in the ocular tissues (Bhatterchee and Eakins, 1974) and the precursor of prostaglandins, arachidonic acid, is effective in causing breakdown of the barrier.

INHIBITORS OF PROSTAGLANDIN SYNTHESIS

The synthesis of prostaglandins is inhibited by aspirin and indomethacin (Ferreira *et al.*, 1971),\* and the breakdown of the barrier caused, for example, by paracentesis, may be largely inhibited by these agents (Unger *et al.*, 1975). An actual release of PGE$_2$-like activity into the rabbit's aqueous humour was described by Miller *et al.* (1973), and this, too, was prevented by treatment of the animal with aspirin. When different prostaglandins were compared for activity in raising the intraocular pressure, Beitch and Eakins (1969) found the order:

$$E > E_2 > F_{2\alpha} > F_{1\alpha}$$

SPECIES VARIATIONS

The cynomolgous monkey's eye is much less sensitive to PG's than the rabbit's, and in this species Kelly and Starr (1971) found that the changes in protein concentration in the aqueous humour were unrelated to the rise in intraocular pressure, suggesting that something more than a breakdown of the barrier was involved in the rise in pressure.

EXCEPTIONS

The rise in intraocular pressure and breakdown of the barrier caused by nitrogen mustard are not inhibited by pretreatment of the animal with aspirin, in contrast to

---

\* Most of the aspirin-like nonsteroidal anti-inflammatory drugs inhibit synthesis at an early stage and thus block the formation of the whole cascade of compounds formed from the essential fatty acids, such as arachidonic acid.

*Fig. 1.62* Effect of prostaglandin PGE$_1$ on intraocular pressure of rabbit. Right eye is the control. (A) Injection of 0·5 μg of PGE$_1$ into the anterior chamber of the left eye. (B) Shows peak response 9–12 min after the injection. (C) Shows return to normal after about 30 min and (D) shows failure of eye to respond to a second injection. (Beitch and Eakins, *Brit. J. Pharmacol.*)

the inhibition of the effects of paracentesis (Neufeld, Jampol and Sears, 1972). By contrast, the effects of nitrogen mustard could be partially inhibited by retrobulbar local anaesthetics, such as lidocaine, as indeed Davson and Huber (1950) had found, and the effects are completely blocked by retrobulbar alcohol, whereas these sensory blocks had no effect on the breakdown due to paracentesis (Jampol *et al.*, 1975). A similar discrepancy is found with the breakdown associated with stimulation of the trigeminel nerve; Perkins (1957) showed that stimulation of NV, especially if brought about mechanically, causes a marked rise in intraocular pressure accompanied by a dilatation of the intraocular blood vessels and a breakdown of the blood-aqueous barrier, and is presumably an antidromic response to sensory nerve stimulation causing a primary vasodilation in the anterior uvea. These effects are not blocked by aspirin (Cole and Unger, 1973); again, the breakdown due to formaldehyde is not inhibited by aspirin-like drugs (Cole and Unger, 1973).

## EFFECTS OF POLYPHLORETIN

We have seen that this anti-inflammatory agent blocks the effects of chemical irritants, such as nitrogen mustard and formaldehyde, an action originally attributed to its antiprostaglandin activity. However, with the demonstration that the breakdown of the blood-aqueous barrier by these drugs is not affected by inhibitors of prostaglandin synthesis, the mechanism of polyphloretin action is probably more direct. Eakins (1971) has shown that the low-molecular weight fraction of the mixture of polyphloretins employed in these studies has anti-prostaglandin action, whilst the high-molecular weight fraction has none; yet Cole (1974) has shown that it is this fraction that blocks the action of chemical irritants whilst the low molecular weight polymers failed to antagonize their action. Thus the polyphloretin action is apparently the result of a direct action on the blood vessels. The results of Cole's experiments are summarized in Figure 1.63, where it is seen that the mixed preparation and the high-molecular weight fraction reduce the effects on intraocular pressure, whilst the low-molecular eight fraction has little or no effect. None of the preparations affected the constriction of the pupil, indicating that this response to chemical trauma is likewise not induced by prostaglandins (Cole and Unger, 1973).

## LASER IRRADIATION

Laser irradiation of the eye is used clinically to cause lesions in the iris allowing communication between posterior and anterior chambers (laser iridotomy). Lesions obtained by directing the high energy of a ruby pulse laser on to the iris cause a transient breakdown of the blood-aqueous barrier with pupillary constriction (Unger *et al.*, 1974). The effects may be partly abolished by treatment with indomethacin, indicating the involvement of prostaglandins; the residual effect was not abolished by atropine and is thus analogous with the effects

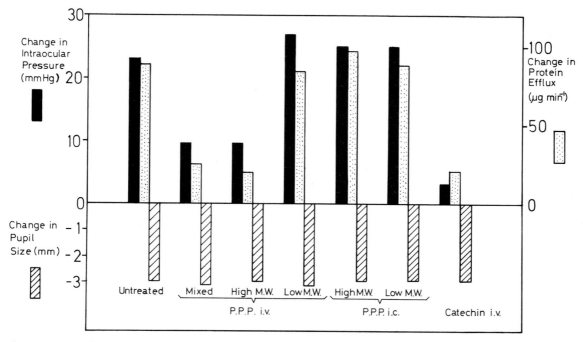

**Fig. 1.63** Effects of polyphloretin preparations on the change in intraocular pressure induced by formaldehyde. Solid columns represent mean changes in intraocular pressure in mmHg; shaded columns (above the line) represent mean changes of protein efflux in μg/min, and the shaded columns below the line represent mean changes in pupil diameter in mm. (Cole, *Exp. Eye Res.*)

of nitrogen mustard or formaldehyde. Since application of local anaesthetics abolished this residual effect, it would seem that there is some non-cholinergic, non-prostaglandin-mediated response to the laser irraciation; and this is presumably analogous with the response to stimulation of NV; the fact that tetrodotoxin, which blocks nervous conduction, also reduced the laser effect when given into the anterior chamber confirms the neurogenic character of part of the response to laser irradiation (Unger et al., 1977).

PARASYMPATHETIC BREAKDOWN

It is well established that cholinergic potentiators, such as DFP, cause a breakdown of the blood-aqueous barrier (v. Sallmann and Dillon, 1947) suggesting that the parasympathetic innervation of the eye might be involved. Stimulation of the intracranial portion of NIII caused an increase in protein concentration of the rabbit's aqueous humour (Stjernschantz, 1976), an effect that was not blocked with a curarine-type drug and so was not due to contraction of the extraocular muscles; it was also unaffected by indomethacin, thus ruling out involvement of prostaglandins, but it was prevented by treatment with biperiden, a drug that blocks muscarininc receptors.

## OSMOTIC AGENTS

Infusion of a hypertonic solution into the internal carotid and ophthalmic arteries causes a rapid and reversible breakdown of the blood-aqueous and blood-vitreal barriers (Latties and Rapoport, 1976), as manifest by penetration of fluorescein from the blood; an exception is the barrier in the iris. The effect is due to shrinkage of the cells constituting the barrier, e.g. endothelial cells of retinal capillaries or epithelial cells of ciliary epithelium, a shrinkage that would impose strains on the junctions tending to pull the cells apart. In the bilayered ciliary epithelium the hypertonic solutions produce permanent structural changes, by contrast with the pigment epithelium adjacent to the retina where increased fluorescein permeability is not accompanied by any obvious structural change. As illustrated by Figure 1.64, the pigmented layer of ciliary epithelium separates from the nonpigmented layer and eventually some 50 to 80 per cent of the cells in this layer are lost; the structure of the nonpigmented layer rapidly returns to normal after separating from the pigmented cells, being connected at their basal tight junctions. Figure 1.65 shows the changes in intraocular pressure; this is low immediately after the perfusion and ultimately returns to normal in 4 to 6 weeks. The fact that normal intraocular pressure returns although as much as 80 per cent of the pigmented layer of ciliary epithelium is lost, indicates that secretion of aqueous humour can proceed independently of the pigmented layer.

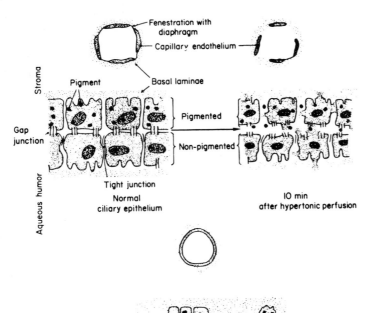

Fig. 1.64   Summary of structural changes in the ciliary epithelium as a consequence of an internal carotid perfusion of hypertonic solution for 20–30 seconds. (Rapoport, *Exp. Eye Res.*)

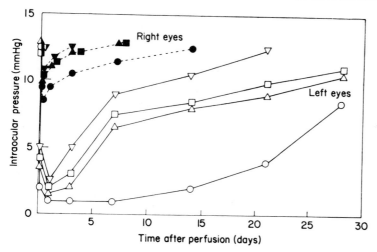

**Fig. 1.65**  Effects of intracarotid perfusion of hypertonic solution on intraocular pressure of monkeys. (Rapoport, *Exp. Eye Res.*)

## MORPHOLOGY OF THE BREAKDOWN

### CILIARY EPITHELIUM

Smelser and Pei (1965) and Bairati and Orzalesi (1966) examined the electron-microscopical appearance of the ciliary epithelium after paracentesis of the rabbit; the changes seemed to be confined to the pigmented layer, the intercellular spaces of which being dilated to form large vacuoles or cysts so that the pigmented cells were converted into slender cytoplasmic processes separated by large cavities filled with finely granular material; in spite of this, the cells retained their attachments to the basement membrane on the stroma side and to the non-pigmented cells on the other. In man, where the effects of paracentesis are not so large, the intercellular spaces showed only slight dilatations.

When particulate matter was in the blood before paracentesis, this apparently escaped mainly through the enlarged intercellular clefts; in particular Smelser and Pei (1965) noted that the interdigitations of the cells were reduced and the particulate matter became concentrated in the enlarged intercellular spaces. The basement membranes of epithelial and capillary endothelial cells offered no absolute restriction to movement of particles.

### IRIS

So far as the iris is concerned, it appears that there are species differences. Thus in the rat and cat the iris capillaries, normally impermeable to fluorescein, now become permeable to this, and also to much larger molecules, such as Thorotrast (10 nm diameter) and carbon particles (20 nm) (Shakib and Cunha-Vaz, 1966); by contrast, in the monkey Raviola (1974) found the iridial vessels impermeable to horseradish peroxidase (4 nm diameter) after paracentesis. The difference may be due to the fact that the endothelial cells of the monkey are sealed by tight junctions whereas those of the rat and cat have gap-junctions.

### RAT

The breakdown in the rat has been examined in some detail by Szalay *et al.* (1975); in normal young animals there was a little escape of fluorescein, and this was greater in older animals; in the old animals paracentesis was accompanied by increased escape from the iris capillaries. Carbon particles injected into the blood also escaped from the iris capillaries after paracentesis, and it seemed that the paracentesis had produced an inflammatory type of reaction in the iris capillaries leading to large gaps between endothelial cells through which the carbon particles escaped into the stroma. The escaped particles were rapidly phagocytosed by macrophages.

### RABBIT

In the rabbit, Cole (1974), employing vasodilator drugs to break the barrier down, concluded that escape of fluorescein occurred primarily into the posterior chamber, the fluorescein passing through the pupil (Fig. 1.66). When a suspension of carbon particles was given intravenously, the drugs caused 'carbon staining' of the blood vessels of the ciliary body but not of the iris; this carbon staining was due to escape of the carbon particles into the surrounding basement membranes of the capillaries and thence into the stroma.

### CILIARY PROCESSES AS MAIN SITE

Cole (1975) emphasized that this obvious increase in ciliary body capillary permeability is not the whole mechanism, since in the normal eye there is some escape of proteins through the ciliary capillaries (Bill, 1968), further progress being prevented by the ciliary

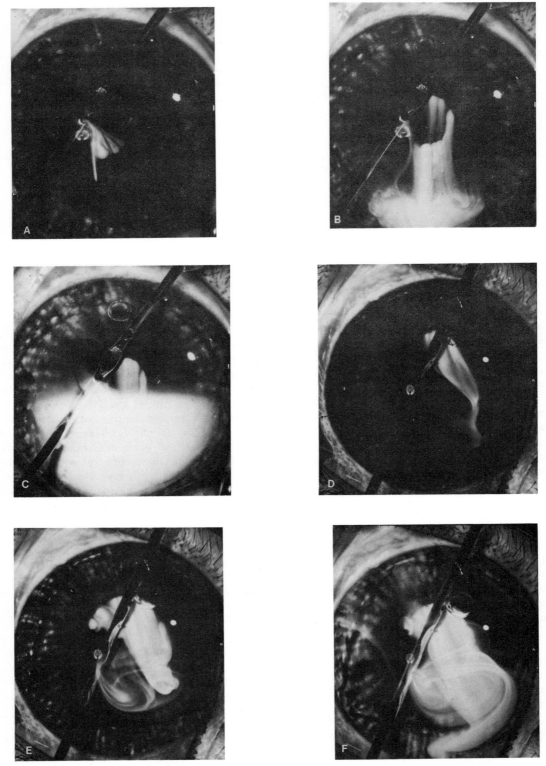

**Fig. 1.66** Effects of bradykinin on passage of fluorescein from blood into the anterior segment of the eye. (A) to (C), rabbit I, and (D) to (F) rabbit II. Nearly all the fluorescein enters the anterior chamber by transpupillary flow and only at a late stage (F) is there any indication of diffusion across the anterior surface of the iris. (Cole, *Exp. Eye Res.*)

epithelium. Thus a breakdown of the occluding junctions between ciliary epithelial cells is also necessary, and this may be the mechanical consequence of the oedematous condition of the ciliary processes, or it may be that the primary cause of the breakdown affects capillary endothelium and ciliary epithelium together. Extending this work to the situation of paracentesis, Unger *et al.* (1975), using the same techniques of fluorescein angiography and colloidal carbon labelling, showed that the breakdown was indeed primarily in the ciliary processes; these were highly congested, whilst carbon particles were deposited in the walls of their blood vessels, by contrast with those of the iris, which appeared normal.*

PROSTAGLANDINS AND SODIUM ARACHIDONATE

In strong contrast to the effects of paracentesis, these substances, applied topically to the eye, caused increased permeability of the iris vessels, so that fluorescein appeared to come primarily from the anterior surface of this tissue rather than from the posterior chamber. Thus the effects of paracentesis are probably twofold; first, a release of prostaglandins which attack mainly the iris vessels, and secondly, because of the mechanical strains due to the fall in intraocular pressure, the ciliary body vessels are affected (Bhattacherjee and Hammond, 1975).†

PARASYMPATHETIC STIMULATION

Uusitalo *et al.* (1974) studied the breakdown of the barrier, following N. III stimulation, in the electron microscope. Horseradish peroxidase injected into the blood escaped into the stroma of the ciliary body; it no longer remained accumulated here between the two epithelial layers but presumably, because of the electrical stimulation, was able to pass into the aqueous humour between the epithelial cells at points where the tight junctions had opened. There was no evidence of a vasodilatation in the ciliary body, in that there was no extravasation of red cells into the stroma, and the capillaries seemed to be intact; and this is consistent with the studies of Perkins (1957) and Bill (1962). The ciliary epithelium of the stimulated eyes showed striking changes; in the pigmented layer there were saccular dilatations filled with slightly electron-dense material; these dilatations appeared to be intracellular and thus were giant vacuoles which projected into the intercellular space between pigmented and non-pigmented cells, interrupting this. In the non-pigmented layer there were also dilatations at the lateral borders of the cells, dilatations that seemed to contain electron-dense material.

**Greeff vesicles**

In so far as it allowed the comparison of stimulated and unstimulated eyes in the same animal, this study, although not very instructive as to the nature of the dilatations, i.e. as to whether they were inter- or intracellular, does confirm the fact that breakdown of the barrier is, indeed, associated with the appearance of 'vesicles' probably the same as the Greeff vesicles described so long ago in the light-microscope. That nervous stimulation alone, apparently not producing vasodilatation, can produce these morphological changes is of great interest and could mean some nervous influence on the intactness of the blood-aqueous barrier.

## EFFECTS OF NERVOUS SYSTEM ON SECRETION AND PRESSURE

### INTRODUCTION

On theoretical grounds, nervous mechanisms may be expected to influence the steady-state intraocular pressure primarily by affecting either the resistance to outflow or the rate of secretion of aqueous humour. Temporary, non-steady-state, changes can be brought about by alterations in the volume of blood in the eye or mechanical compression brought about by the extraocular musculature. A direct influence of the nervous system on the secretory process is a possibility, whilst inhibition by drugs is well proven, as we shall see; alternatively, the secretory process may be limited by blood-flow in the ciliary body, in which event secretion could be controlled by an influence on the vasculature. The resistance of the angle of the anterior chamber could well be affected by the ciliary muscle, as originally suggested by Fortin, since the trabecular meshwork is the insertion of the meridional fibres; a contraction

---

*In the monkey, however, Okisaka (1976) found a characteristic engorgement of capillaries in the ciliary body following $PGE_1$-treatment; the channels between the two ciliary epithelial layers became enlarged and spaces between cells of both layers were expanded. The striking feature of the drainage angle was the large number of red blood cells in the canal of Schlemm, and the reduction in the number of giant vacuoles in the lining endothelial cells; no red blood cells found their way back into the meshwork.

† Raviola's (1974) study on the effects of paracentesis on the monkey's blood-aqueous barrier strikes a discordant note; she failed to observe any changes in the ciliary body, so that, as in the normal animal, horseradish peroxidase was blocked at the tight junctions of the non-pigmented epithelial layer, and the same was true of the iris, the tracer being blocked at the tight junctions of the capillaries. The large influx of peroxidase into the aqueous humour that she observed was apparently due to reflux from the canal of Schlemm, which entered between gaps in the endothelial cell lining. Whilst there is no reason to doubt this reflux from the canal of Schlemm and intrascleral blood vessels, the failure to observe changes in the ciliary body is surprising.

of these fibres should open up the meshwork and promote percolation of fluid through the holes in the lamellae. It seems unlikely that the nervous system can directly affect the meshwork, although the studies of the effects of ganglionectomy, to be described below, do suggest this; such an effect could be brought about by a shrinkage or swelling of the endothelial cells lining the trabeculae, and possibly through an alteration of the size of the vacuolar system in the inner wall of the canal of Schlemm. Finally, the pressure within the blood vessels into which the aqueous humour drains is under nervous control.

## SYMPATHETIC SYSTEM

Stimulation of the peripheral end of the cut cervical

immediate effect on the normal intraocular pressure (Greaves and Perkins; Langham and Taylor), although occasionally quite large rises, lasting for thirty minutes, can be observed (Davson and Matchett). Twenty-four hours after ganglionectomy the intraocular pressure is definitely lower than normal, and a study of the dynamics of flow of aqueous humour suggested that this was due to a diminished outflow resistance, since $P_e$ was said to be unaffected by the procedure. Tomar and Agarwal (1974) confirmed in rabbits the fall in pressure, reaching a maximum at about 24 hr; by 3 to 4 days, however, it had returned to normal. The fall in pressure was associated with an increased facility of outflow and the return to normal pressure was accompanied by a return to normal facility.

**Table 1.8** Blood-flow through eye of monkey on sympathectomized side, and effects of sympathetic stimulation (Alm, 1977).

| Tissue | Blood flow in whole tissue (mg/min) (sympathectomized side) | Blood flow (g/min/100 g tissue) (sympathectomized side) | Percent reduction of blood flow on stimulated side | Significance level |
|---|---|---|---|---|
| Retina | $33 \pm 2$ | | $8 \pm 4$ | |
| Iris | $4 \pm 1$ | | $52 \pm 7$ | $P < 0.001$ |
| Ciliary body | $79 \pm 4$ | | $22 \pm 3$ | $P < 0.001$ |
| Choroid | $558 \pm 59$ | | $30 \pm 6$ | $P < 0.005$ |
| Ciliary processes | | $120 \pm 18$ | $23 \pm 8$ | $P < 0.025$ |
| Ciliary muscle | | $176 \pm 10$ | $29 \pm 5$ | $P < 0.001$ |
| Anterior sclera | | $2 \pm 1$ | $52 \pm 21$ | $P < 0.05$ |

The weights for the ciliary processes and muscle are calculated from the dry weight on the assumption that dry weight/wet weight ratio is 0·20 (see Alm and Bill, 1972).

sympathetic trunk in the rabbit causes a constriction of the uveal vessels associated with a marked fall in the intraocular pressure. This is accompanied by a considerable fall in blood-flow through the uvea; thus Alm and Bill (1973), working on the cat, and Alm (1977) on the monkey, measured blood-flow with radioactive microspheres.* Results on the monkey are shown in Table 1.8, where a sympathectomized eye is compared with one where the cervical sympathetic on the non-sympathectomized side was stimulated for one minute. The effect on retinal blood-flow was small and probably insignificant, and this is consistent with a correspondingly small effect on cerebral blood-flow; presumably constriction of the retinal artery is compensated by the autoregulatory mechanism peculiar to the central nervous system.

## SECTION OF CERVICAL SYMPATHETIC

Section of the sympathetic, or extirpation of the superior cervical ganglion has, in general, remarkably little

## EFFECTS OF CATECHOLAMINES

### ALPHA- AND BETA-ACTION

The sympathetic activity in the body is mediated by norepinephrine which is liberated at the nerve terminals. The same catecholamine, as well as adrenaline (or epinephrine) is secreted by the adrenal gland. It is well established that there are at least two types of receptor for the catecholamines, in the sense that a given catecholamine may be preferentially bound at one site and exert a specific action whilst the same or another catecholamine may be bound to a different site and exert a different and usually opposite activity. Thus the neuropharmacologist speaks of $\alpha$- and $\beta$-receptors, and the normally occurring catecholamines, norepinephrine and epinephrine, are capable of exerting both $\alpha$- and $\beta$-action on a given effector, such as the pupil. Nor-

*Radioactive microspheres of about 15 $\mu$m diameter are injected into the left ventricle, and the animal is subsequently killed; the microspheres are unable to pass beyond the capillary circulation and their density in any tissue is a measure of the blood-flow at the time of injection.

epinephrine exerts a predominantly $\alpha$-action, whilst epinephrine exerts a mixed $\alpha$- and $\beta$-action. The $\beta$-action can be mimicked by isoproterenol (Fig. 1.67); usually the separate effects of, say, epinephrine can be deduced from the effects of specific inhibitors; thus phenoxybenzamine inhibits alpha-action, and when an animal is treated with this its pupillary response to

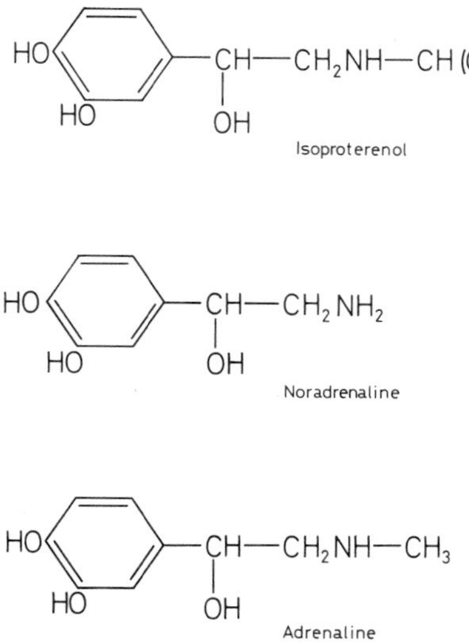

HO—⟨⟩—CH—CH$_2$NH—CH(CH$_3$)$_2$
HO        OH

Isoproterenol

HO—⟨⟩—CH—CH$_2$NH$_2$
HO        OH

Noradrenaline

HO—⟨⟩—CH—CH$_2$NH—CH$_3$
HO        OH

Adrenaline

**Fig. 1.67**  Structural formulae of some catecholamines.

epinephrine—midriasis—is abolished, indicating that the predominant effect of epinephrine on the iris is an $\alpha$-action. A specific inhibitor of $\beta$-activity is propanolol, and this barely affects the action of epinephrine on the pupil showing that there is little $\beta$-activity on the iris dilator muscle.

MYDRIASIS AND HYPOTENSION

The effects of catecholamines and their antagonists have been examined in detail by Langham, mainly on the rabbit; they are highly complex, indicating possible effects on resistance to flow and rate of secretion, effects that involve both $\alpha$- and $\beta$-actions. The most obvious effect or epinephrine, applied to the cornea, is a fall in intraocular pressure extending over about 24 hours; the same effect can be mediated by the almost pure $\alpha$-agonist norepinephrine and also by the $\beta$-agonist isoproterenol. A typical response is shown by Figure 1.68 which includes the changes in pupillary diameter taking place at the same time. The two effects— mydriasis and hypotension—do not follow the same time-course, suggesting that the effect on intraocular pressure is more complex, perhaps involving both $\beta$- and $\alpha$-activity. A $\beta$-blocker, propanolol, reduced the hypotensive effect by 17 per cent, indicating some $\beta$-action; a much larger reduction in the effect was produced by an $\alpha$-blocker, e.g. phenoxybenzamine, indicating the predominantly $\alpha$-action in the hypotensive response.

Salbutamol is a drug that has a purely $\beta$-action by contrast with the mixed effects of epinephrine; as Figure 1.69 shows, it had no effect on pupillary diameter, but it produced an initial rapid fall in intra-ocular pressure which was over in 5 hr, by contrast with the more prolonged action of epinephrine (Fig. 1.68). It would seem, then, that the powerful and prolonged effects of epinephrine are due to its combined $\alpha$- and $\beta$-actions working in the same direction. It is unusual for $\alpha$- and $\beta$-actions on an effector to work in the same direction, the one usually inhibiting the other; it would seem, then, that the catecholamines can act at two sites,

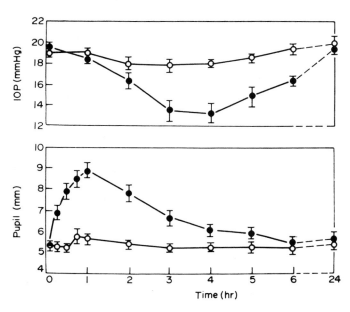

**Fig. 1.68**  The mean pupillary and intraocular pressure responses to a 2 per cent solution of epinephrine. 25 $\mu$l of solution was applied topically to 1 eye (●—●) of 8 conscious rabbits at T = 0, 2 and 4 min. The contralateral eyes (○—○) were untreated. The vertical bars represent $\pm$ standard error of the mean. (Langham *et al.*, *Exp. Eye Res.*)

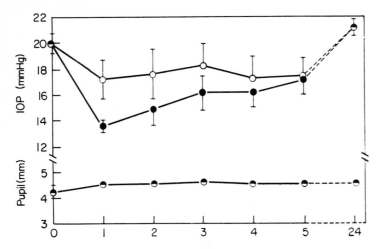

**Fig. 1.69**  The average time courses of the pupillary and intraocular pressure responses of conscious rabbits to salbutamol ascorbate applied topically. Twenty-five microlitres of a 2 per cent solution (pH 6·5) was applied to one eye (●—●) of individual rabbits at 0, 2 and 4 min. The contralateral eyes (○ - - ○) were not treated. (Langham and Diggs, *Exp. Eye Res.*)

presumably the rate of formation of fluid and the facility of outflow. Langham and Diggs (1974) concluded that the rapid $\beta$-action was, in effect, a reduction in secretion rate, whilst the more prolonged $\alpha$-action was due to an effect on facility. This could be achieved by reduction of blood-flow through the intrascleral plexus leaving more room for aqueous flow in the mixed channels.*

HYPERTENSIVE RESPONSE

The effects of epinephrine are even more complex than the above description suggests, since there is some evidence for a hypertensive response; this was observed by Kornfeld (1971), in his study of human subjects, as an initial response in some subjects. On average, however, the effect of single or repeated doses of epinephrine was a prolonged fall associated with an increase in facility from control values of 0·10 to 0·12 to 0·19. Again, Norton and Viernstein (1972), working on rabbits, found a transitory rise in intraocular pressure in rabbits if they used high concentrations (4 to 8 per cent) of epinephrine; this occurred in 75 per cent of animals, and the remainder showed only a continuous decline in pressure; the effect could be mimicked by the $\beta$-agonist, isoproterenol, and blocked by propanolol a $\beta$-blocker. These authors attributed the hypertensive effect to a decrease in facility.

BIPHASIC RESPONSE

Langham and Kriegelstein (1976) were able to bring this hypertensive action of epinephrine into prominence by studying the effects of repeated applications; Figure 1.70A shows the usual hypotensive response to a single application. B shows the response to a second application 24 hr later; there is now an initial rise in intraocular pressure, which gives way to a long-lasting hypotensive response so that after 24 hr the pressure is still below normal. Surprisingly, the rise in pressure could be

blocked by $\alpha$-blockers, indicating the participation of an $\alpha$-mechanism; and this was confirmed by showing that a similar effect of repeated application could be obtained with norepinephrine, which has a predominantly $\alpha$-action; it was blocked by phenoxybenzamine but not by propanolol.†

**Norepinephrine**

The same biphasic responses were obtained with noradrenaline by Langham and Palewicz (1977), and it was found, as with epinephrine, that both responses, namely the fall and rise in intraocular pressure, were accompanied by changes in outflow resistance, the rise in pressure being associated with an increased resistance and vice versa. This surprising opposite action of an $\alpha$-agonist on resistance was interpreted as being due to difference in sites of action on the intraocular vasculature, so that the effects being observed were the result of altered filling of the vascular networks with blood, leading to altered facility of flow of aqueous humour along channels shared with the blood.

PRIMATES

As indicated earlier, Kupfer *et al.* (1971) found, in normal human subjects, a decrease in intraocular

---

*A similar conclusion had been reached by Sears (1966) who ascribed the early fall in tension to reduced production; this overlapped with a later-occurring increase in facility.

†In general, the studies of Lamble (1977 and earlier) are in good agreement with those from Langham's laboratory; Lamble concludes that any effect on flow—measured quantitatively by clearance of inulin—are rapid and soon over, so that the main effect in inducing a lowered intraocular pressure is through increased facility, mediated by $\alpha$-receptors. The delay in fall in intraocular pressure observed by Lamble (2 hr) is attributed to an initial hypertensive response which temporarily negates the hypotensive effect of increased facility.

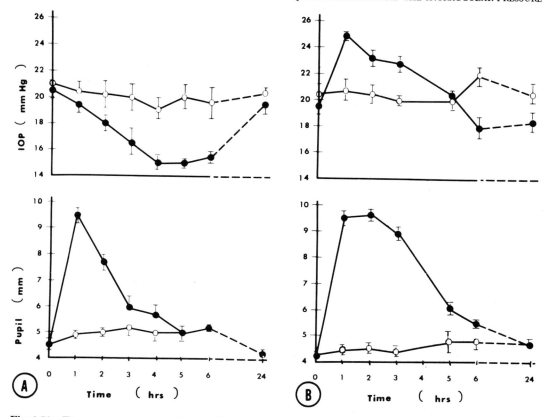

**Fig. 1.70**    The mean time courses of the pupillary and intraocular pressure responses of six conscious rabbits to single applications of 25 $\mu$l of 1 per cent epinephrine at T = 0 (*A*) and T = 24 hours (*B*). $\bigcirc$—$\bigcirc$ and $\bullet$—$\bullet$ represent the untreated and drug-treated eyes, respectively. (Langham and Krieglstein, *Invest. Ophthal.*)

pressure with epinephrine, and their treatment permitted an estimate of true and pseudofacility; they concluded that the increased facility was due entirely to pseudofacility, as might be expected of a vascular phenomenon.

With norepinephrine, the virtually pure $\alpha$-agonist, increased rate of flow being accompanied by a reduction in pseudofacility. Isoproterenol, the $\beta$-agonist, lowered intraocular pressure and decreased flow-rate without altering facility. The combined effects of norepinephrine and isoproterenol were equivalent to those of epinephrine, and they concluded that the acute effects of epinephrine were in fact due to the flow-reduction due to $\beta$-stimulation combined with the reduced pseudofacility due to alpha-action. (Gaasterland *et al.*, 1973).

**Monkey**

In the vervet monkey Bill (1970) measured both flow-rate and facility, separating flow into its two components, namely uveoscleral and conventional. Isoproterenol increased production of fluid and increased

gross facility and uveoscleral drainage; he estimated that flow had increased by 30 per cent and facility by 55 per cent. Norepinephrine had no significant effects on flow or facility. Neither drug had a significant effect on intraocular pressure. Thus in this species only the $\beta$-agonist seems to have a significant effect on aqueous humour dynamics, and the effects mainly cancel each other out.

**GANGLIONECTOMY**

The 'ganglionectomy-effect' has been the subject of a great deal of investigation largely in the laboratories of Langham and Bárány. The effect is definitely due to destruction of the postganglionic neurones, since preganglionic sympathotomy is without effect (Langham and Fraser, 1966). It is well known that denervation of a tissue causes hypersensitivity, and it was argued by Sears and Bárány that the resistance to flow was governed by an adrenergic mechanism which became hypersensitive after denervation, whilst the liberation of catecholamines from the degenerating terminals produced the adrenergic effect. The slow release of the catecholamines would account for the slow onset of the change, and the subsequent return of the intraocular pressure to normal would be accounted for by the eventual release of all stored catecholamines. The effects of reserpine and adrenergic blocking agents generally confirmed the hypothesis, although Langham and Rosenthal's observation that prolonged stimulation of the sympathetic had no effect on outflow resistance is in conflict with the theory.

## LANGHAM-HART MODEL

In general, then, the effects of the adrenergic system on intraocular dynamics are highly complex, and the reports are often contradictory; there are obvious species differences, whilst acute must be distinguished from chronic effects, and even the mode of administration of

the drug may be critical. If the effects are mediated entirely through the ocular vasculature—and this seems the most profitable working hypothesis—then the model built by Hart (1972), according to which facility of outflow is governed by the degree to which the intra- and episcleral blood vessels are perfused (Fig. 1.71), and the rate of secretion influenced by the blood supply to the ciliary processes, seems a reasonable basis for the interpretation of many of the effects. Thus according to the model of Figure 1.71, it is assumed that increases of intraocular pressure and outflow resistance are due to constriction of the aqueous and episcleral veins, and that the decreases in pressure are due to constriction of the blood supply to the intrascleral venous plexus. In addition, the changes in calibre of the episcleral veins induced by adrenergic drugs can alter outflow resistance; thus the blanching of these vessels observed after application of $\alpha$-agonists could be a cause of an increased resistance, taking place rapidly owing to the short distance required for the agent to travel.

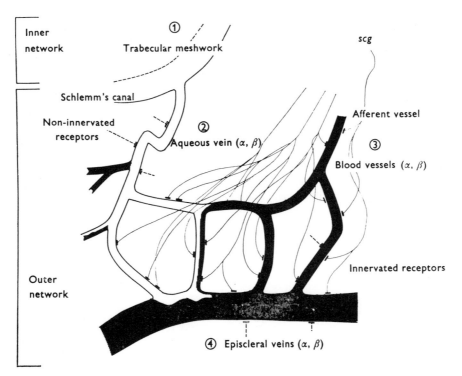

**Fig. 1.71**  The anatomical components of the inner and outer networks of the channels draining the aqueous humour from the eye. The inner network includes the trabecular meshwork and the layer of endothelial cells lining the inner wall of Schlemm's canal. It remains uncertain whether innervated receptors are present in the inner network (1). The outer network comprises the intrascleral vessels filled with aqueous humour (2), blood (3), or a mixture of blood and aqueous humour, and the recipient episcleral veins (4). Post-ganglionic adrenergic neurones from the superior cervical ganglia (*scg*) innervate the vessels of the outer network. In addition, vessels of the outer network show vasoconstrictor response to $\alpha$-adrenergic receptor agonists and vasodilatation to $\beta$-adrenergic agonists. (Langham and Palewicz, *J. Physiol.*)

## UVEOSCLERAL DRAINAGE

When uveoscleral drainage represents a significant proportion of the total, as in primates, then the state of the ciliary muscle becomes important (p. 71), whilst altered tendency to filtration in the anterior uvea, due to vascular changes, would also affect this drainage pathway and, of course, any flow of aqueous humour due to direct filtration from the vascular bed, a flow that reflects pseudofacility.

## CYCLIC AMP

It is generally agreed that $\beta$-adrenergic activity may be mediated through the second messenger, cyclic AMP, the neurotransmitter activating the enzyme adenyl cyclase which leads to the intracellular release of cAMP, which then initiates the cellular reaction, be this contraction of heart muscle, changed liver metabolism, and so on. The physiological response and the associated rise in cAMP concentration can usually be blocked by a $\beta$-adrenergic antagonist but they are little affected by $\alpha$-blockers. As Figure 1.72 shows, the effects of adrenergic agonists, both $\alpha$- and $\beta$- are accompanied by a rise in aqueous humour cAMP concentration; the falls in intraocular pressure run parallel with the rises in cAMP, suggesting a causal relationship; interestingly, the effects of the $\beta$-agonist, isoproterenol, were not blocked by propanolol; thus isoproterenol reduced intraocular pressure but the fall was not blocked by propanolol, whilst the rise in cAMP was also not blocked. These findings were essentially confirmed by Radius and Langham (1973) using norepinephrine, a predominantly $\alpha$-agonist, as the adrenergic agent; this caused a rise in cAMP in the aqueous humour associated with a fall in intraocular pressure; the $\alpha$-blocker

phenoxybenzamine itself actually increased the concentration of cAMP in the aqueous humour; a further increase was achieved by subsequent application of norepinephrine, whilst the pressure and pupillary responses to norepinephrine were blocked by phenoxybenzamine. Thus it would appear that both $\alpha$- and $\beta$-agonists are able to stimulate adenylcyclase activity in the eye and, through the raised cAMP, to exert a hypotensive effect, but of course the precise mechanism or mechanisms of the hypotensive effect remain to be discovered.*

## PARASYMPATHETIC

## EFFECTS OF STIMULATION

### INTRAOCULAR PRESSURE AND BLOOD-FLOW

The reports on the effects of stimulation of the parasympathetic supply to the eye on intraocular pressure are contradictory; and this may be largely due to the uncertainty as to the nerve fibres actually stimulated, e.g. when an electrode is placed on the ciliary ganglion (Armaly, 1959) or by stereotactic location intracranially (Stjernschantz, 1976). The various studies have been summarized by Stjernschantz and Bill (1976) who have described experiments on rabbits stimulated intracranially. The mean decrease of 1·8 mmHg in intraocular pressure was not statistically significant, and this was associated with a barely significant increase in

---

*Some of the characteristics of a particulate fraction with adenylcyclase activity obtained from rabbit ciliary process tissue have been described by Waitzman and Woods (1971); its activity could be stimulated by epinephrine, an effect that was blocked by propanolol, a $\beta$-blocker, but not by phenoxybenzamine. The prostaglandin $PGE_1$ was also a stimulator and added its effects with those of epinephrine.

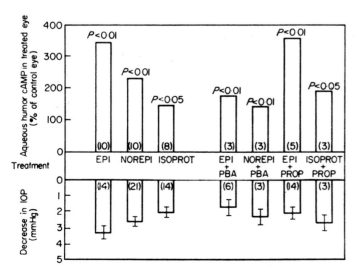

Fig. 1.72 The effects of adrenergic agonists and antagonists on intraocular pressure and cyclic-AMP in the aqueous humour. Agonists: EPI = 1 per cent l-epinephrine, NOREPI = 1 per cent l-norepinephrine, ISOP = 2 per cent l-isoproterenol, were administered topically to one eye, the control vehicle (1mM HCl) to the other eye. Antagonists: PBA = 30 mg/kg phenoxybenzamine, PROP = 5 mg/kg propanolol, were administered intravenously. The change in intraocular pressure is expressed as the mean ± S.E.M. of the difference between the two eyes. The cyclic-AMP content of the aqueous humour in the drug-treated eye is expressed as the mean per cent of the contralateral eye. The 'P' values were obtained from the mean ± S.E.M. of the cyclic-AMP levels of the drug-treated and control eyes. The numbers of observations are in parentheses. (Neufeld et al., Exp. Eye Res.)

outflow facility; the mean increase in flow-rate was not significant. Stimulation of NIII, as we have seen, causes a breakdown of the blood-aqueous barrier, which might be expected to raise intraocular pressure by increasing flow of abnormal aqueous humour from the ciliary body, and such an increased production of fluid has been described by Macri and Cevario (1973) in perfused enucleated cat's eyes. In the rabbit Stjernschantz and Bill (1976) found an actual decrease in blood-flow through the iris and ciliary body, as measured with radioactive microspheres; this was accompanied by an increased flow in the choroid. Thus the equivocal effects on intraocular pressure, as with catecholaminergic agents, are probably the results of opposing actions on the various parameters that go to make up the intraocular pressure.

### RESISTANCE TO OUTFLOW

Direct stimulation of the oculomotor nerve as it emerges from the brainstem in the monkey causes a fall in resistance to outflow, whilst cutting the nerve unilaterally causes the resistance to be higher on the same side (Tornqvist, 1970).

## PARASYMPATHOMIMETIC AGENTS

Drugs that mimic the effects of parasympathetic stimulation, e.g. pilocarpine and eserine, are used in the treatment of glaucoma, and their hypotensive effect has been regarded as largely exerted through the pull of the ciliary muscle on the trabecular meshwork. Atropine, the cholinergic blocking agent, may raise the pressure in glaucoma. In the normal human eye the effects of pilocarpine and atropine are usually negligible, but this does not mean that the drugs are without effects on all the parameters that go to determine the intraocular pressure.

### PILOCARPINE

Thus there seems little doubt that pilocarpine causes a decreased resistance to outflow, and Bárány (1967) has shown that the effect is so rapid that it occurs before sufficient of the drug has reached the aqueous humour to enable it to exert any direct effect on the meshwork; its action is therefore on the ciliary body and/or iris; and we may presume that it is exerted through the ciliary muscle. The effects obtained were just as large as

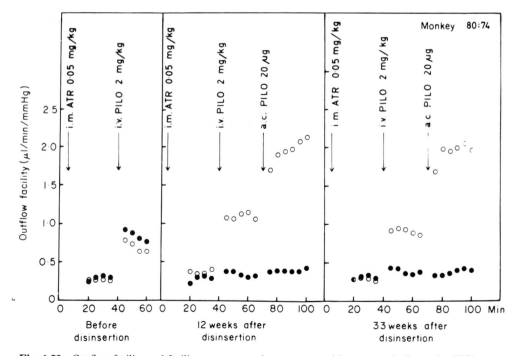

**Fig. 1.73** Outflow facility and facility responses to intravenous and intracameral pilocarpine-HCl (i.v. PILO; a.c. PILO) before and after unilateral ciliary muscle disinsertion in a typically bilaterally iridectomized monkey. Intramuscular atropsine sulphate (i.m. ATR) was given before each perfusion to minimize systemic effects of intravenous pilocarpine. Note absence of facility increase following intravenous and intracameral pilocarpine in the iridectomized + 'disinserted' eye (solid circles), as opposed to the large facility increases in the opposite iridectomized only eye (open circles). (Kaufman and Bárány, *Exp. Eye Res.*)

any obtained through intracameral injection, so it is unlikely that the drug influences the trabecular meshwork directly. Further proof that the increased facility is a mechanical response to contraction of the ciliary muscle is provided by Kaufman and Bárány (1976) who disinserted the monkey's ciliary muscle and showed that the acute effects of pilocarpine were abolished (Fig. 1.73).

## UVEOSCLERAL-SCHLEMM CANAL SWITCH

Some of the complexities in the study of the effects of drugs on intraocular pressure are well revealed by the studies of Bill and Wålinder (1966), who showed that, in macaque monkeys, pilocarpine increased the intraocular pressure; this was associated with a fall in rate of secretion of aqueous humour from a control value of 1·60 μl/min to 1·05 μl/min. The reason for the rise in pressure was an almost complete blockage of the uveoscleral drainage route, so that drainage was confined to the canal of Schlemm (Fig. 1.74).

### Blood-flow

We have seen that stimulation of NIII causes a decreased blood-flow through the anterior uvea (of rabbits); the effects of parasympathetic potentiators such as eserine and DFP are opposite, however, doubling the blood-flow through the anterior uvea

(Alm et al. 1973), whilst in humans Wilke (1974) found dilatation of superficial blood vessels associated with increased episcleral venous pressure.

### EXPERIMENTS ON PERFUSED CAT'S EYE

Macri and his colleagues, working on the isolated cat's eye, perfused through the ophthalmic artery, have studied the effects of various agents on the production of fluid. Under these conditions there is a production of fluid of some 8 μl/min, compared with about twice this in the normal animal. Macri and Cevario (1973) found a marked increase in rate of production of fluid if acetylcholine plus eserine were applied, an effect that was reduced by atropine. This increase could be abolished by reducing the arterial pressure (Macri et al., 1974) so that it is likely that the main phenomenon observed in the perfused cat's eye is the production of an intraocular fluid that is largely a filtrate from the plasma. Whether or not this production is a normal component of the cat's aqueous flow remains an open question; it is significant that the measured flow in the perfused eye is only half normal, so that it is possible that this pressure-sensitive fraction is, indeed, a normal component of the cat's secretion. This may well account for the small effects of Diamox (acetazolamide) on the cat's fluid formation (Macri and Brown, 1961) an effect that was attributed to vascular changes rather than inhibition of carbonic anhydrase; certainly Davson and Spaziani (1960), working on the intact cat, found no

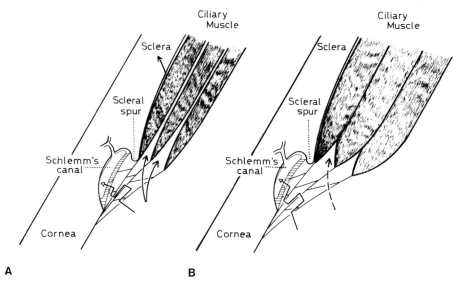

**Fig. 1.74**  (A) Ciliary muscle is relaxed, spaces between lamellae of uveal and corneoscleral meshwork are narrow, and interstitial spaces between muscle bundles of ciliary muscle are wide. Aqueous humour leaves the anterior chamber via Schlemm's canal and via interstitial spaces of ciliary muscle. (B) Ciliary muscle is contracted, spaces in different parts of meshwork are relatively wide, and interstitial spaces between muscle bundles have almost disappeared. Aqueous humour leaves the eye almost entirely via Schlemm's canal. (Bill, *Physiol. Rev.*)

evidence of a decrease in rate of secretion greater than about 20 per cent.

Macri (1971) and Macri and Cevario (1975, 1977) have brought forward a body of evidence suggesting that a part of the effects of acetazolamide and ouabain are due to their constrictive action on the ciliary body vasculature. Thus, working on the isolated perfused cat's eye, they demonstrated an indubitable vasoconstrictive action of these drugs; the effects could be blocked by hexamethonium, suggesting that acetazolamide and ouabain were exerting a pharmacological action on the blood vessels rather than specific inhibition of active transport. The action could be on a local ganglion in the uvea, sensitive to cholinergic drugs, since acetylcholine-eserine accelerated fluid production, and this would account for the action of the ganglion blocker, hexamethonium. The strongly hypotensive action of hexamethonium on the blood pressure precluded a direct study of its action on fluid production in the intact animal, but the ganglion-blocker phencyclidine, which does not have this hypotensive action, was tested, and it blocked the effects of acetazolamide on fluid production in the isolated eye preparation as well as the vasoconstrictive action. In intact monkeys phencyclidine alone had no effect on rate of production of aqueous humour, but it did reduce the inhibition caused by high doses of acetazolamide. It should be emphasized that Macri postulates a vasoconstrictive action that can accelerate fluid production—acetylcholine-eserine treatment—and a vasoconstrictive action that can decrease fluid production—acetazolamide, ouabain. He assumes that the situation in the ciliary processes is similar to that in a kidney glomerulus, so that constriction of an afferent blood vessel reduces fluid formation, and constriction of an efferent vessel increases it.★

# HYPOTHALAMIC CENTRE

## STEREOTACTIC STIMULATION

Attempts to locate a centre controlling the intraocular pressure have been made by v. Sallmann and Loewenstein and Gloster and Greaves, by inserting stereotactically controlled stimulating electrodes into different parts of the brain. The parts of the brain from which effects on intraocular pressure could be obtained were in the hypothalamus, but since this region contains many centres controlling the autonomic system generally, it is important to distinguish primary effects on the intraocular pressure from the secondary consequences of stimulating these autonomic centres. Thus many stimulated points gave rise to rises or falls in intraocular pressure that were obviously associated with corresponding rises or falls of arterial pressure. Nevertheless, other points could be found where a rise or fall of intraocular pressure, independent of general vascular changes, resulted from stimulation; the falls in intraocular pressure were probably mediated by the sympathetic, since they could be prevented by section of the sympathetic trunk. It is possible, then, that there is a co-ordinating centre in the hypothalamus capable of influencing the intraocular pressure, and subject to afferent impulses carried in the long ciliary nerves. The

evidence for this is not completely convincing, however, and we must bear in mind the possibility that there is no central control, local homeostatic mechanisms being adequate to keep the intraocular pressure within a narrow range.

If there is a 'centre' controlling the intraocular pressure, we may expect to find evidence of a sensory system capable of keeping the centre informed of the state of the intraocular pressure at any moment. Studies from the laboratories of Vrábec and v. Sallmann have shown that the trabecular meshwork is richly provided with fine nerve fibres.

# SENSORY INPUT

## TRIGEMINAL

Using electron microscopy and nerve degeneration, Ruskell (1976) showed that fibres from N. V terminate in the meshwork. (Sympathetic terminals were also identified by the presence of granular visicles in terminal varicosities, and these might well be concerned with some of the effects of adrenergic substances on intraocular dynamics.) Bergmanson has examined the innervation of the uvea in monkeys before and after lesions of the ophthalmic division of N. V. It appeared that the choroid was probably not innervated, the fibres running in the suprachoroid having terminals in the anterior uvea; similarly the origin of the ciliary muscle lacked sensory innervation, as also did the individual ciliary muscle fibres. The subepithelial stroma of the ciliary processes was well innervated and it seemed that the fibres terminated in the stroma as the adjacent tissue either had none or very few. It could be these terminals in the ciliary body stroma that gave rise to the action potentials described by Tower (1940) when pushing on the front surface of the lens in cats. According to Bergmanson's study, there are few or no sensory terminals in the monkey's iris; this is surprising in view of the painful nature of iridectomy without retrobulbar anaesthesia.

Thus the source of sensory input from the uvea is apparently largely the stroma of the ciliary processes and the iris root, and we may presume that if such afferent information is concerned in responses to raised or lowered intraocular pressure, then these fibres carry the information. The absence of sensory fibres in the ciliary muscle rules out the trigeminal nerve as a source of muscle proprioception.

---

★ In the monkey, Bill et al. (1976) have found evidence for a ganglionic type of intraocular vascular response to oculomotor nerve stimulation; these authors showed that the stimulation liberated both vasoconstrictor and vasodilator agents in the anterior uvea; the vasoconstrictor response was completely blocked by hexamethonium, and also, in the ciliary processes, by the α-adrenergic blocking agent, phentolamine.

AFFERENT DISCHARGES

Belmonte *et al.* (1971) recorded discharges in ciliary nerves of the cat; lowering the intraocular pressure by steps progressively reduced the discharge-frequency.

## EFFERENT IMPULSES

### FACIAL NERVE

Gloster (1961) has shown that stimulation of N. VII causes a rise in intraocular pressure; it is not merely due to a contraction of the extraocular muscles since gallamine did not abolish the effect. Cutting the nerve of the pterygoid canal causes a fall in intraocular pressure in the monkey (Ruskell, 1970b). This effect is consistent with Ruskell's (1970a) description of *rami orbitales* of the pterygopalatine ganglion containing parasympathetic fibres of N. VII; these supply the ciliary arteries, and if they are antagonistic to the sympathetic supply, the rise in intraocular pressure obtained by stimulation could be due to vascular dilatation. Cutting this nerve supply caused a prolonged fall in intraocular pressure (Ruskell, 1970b).

### CILIARY NERVES

Gallego and Belmonte (1974) have described discharges in the central end of the cut ciliary nerves, presumably carrying efferent messages to the internal structures of the eye; some increased their firing frequency when intraocular pressure was raised and others decreased their frequency. So far as motor terminals in the eye are concerned, there is no doubt from various studies (Nomura and Smelser, 1974; Ruskell, 1976) that sympathetic and probably parasympathetic fibres terminate in the angle, although the manner in which a nervous impulse could alter the meshwork directly, i.e. independently of action on the ciliary muscle, must remain obscure. They are located close to the insertion of the ciliary muscle in the scleral spur with no terminals in the more anterior parts of the meshwork, so it may well be that they are related more to the muscle than the meshwork as such (Nomura and Smelser, 1974). The autonomic innervation of the uveal vasculature is well established and physiologically this is revealed by the effects of nerve stimulation on blood-flow (p. 71).

## HORMONAL INFLUENCES ON INTRAOCULAR PRESSURE

### CORTICOSTEROIDS

François (1954) observed that the prolonged use of topical cortisone in the treatment of uveitis sometimes led to a rise in intraocular pressure; patients under-

going systemic corticosteroid therapy for rheumatic conditions also showed raised intraocular pressure. Bernstein and Schwartz (1962) showed that the cause was a decreased facility of outflow, which was reversible when the dose of corticosteroid was reduced. In human subjects, Anselmi, Bron and Maurice (1968) found no change in rate of secretion, although the intraocular pressure was high. In experimental animals, such as the rabbit, results have been contradictory (see, for example, Armaly, 1964, whose results were negative, and Jackson and Waitzman, 1965, with positive, but complex, results); the study of Oppelt, White and Halpert (1969), employing the posterior-anterior chamber perfusion technique in cats, indicated a dose-related fall in rate of secretion after intravenous hydrocortisone; associated with this, there was a 23 per cent decrease in facility of outflow, and presumably it would be this that caused the rise in intraocular pressure if results on cats may be transposed to man.

### TRABECULAR MESHWORK

Rohen *et al.* (1973) examined samples of trabecular meshwork removed from two human subjects who had developed glaucoma as a result of corticosteroid therapy;

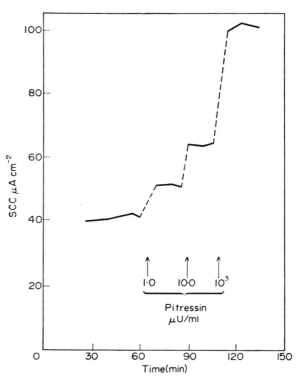

**Fig. 1.75** Effects of increasing doses of vasopressin (Pitressin) on short-circuit current across the isolated ciliary body. (Cole and Nagasubramanian, *Exp. Eye Res.*)

in the electron microscope the outer region of the mesh-work appeared swollen and very compact, and the inter-trabecular spaces were filled with amorphous and fibrous material embedded in a homogeneous substance. The endothelial lining of the canal of Schlemm, adjacent to the meshwork, seemed thin and was without vacuoles. Thus the evidence suggests that the resistance to flow through the meshwork and into the canal could have been seriously increased through the deposition of material.*

VASOPRESSIN

The antidiuretic hormone is an agent concerned in the regulation of blood osmolality, being secreted when this is reduced; a lowered blood osmolality probably increases aqueous humour production, in that a greater amount of water accompanies a given amount of active salt transport; thus if secretion of ADH increased salt transport, the decreased secretion that accompanies a lowered blood-osmolality would decrease salt transport into the eye and thus tend to compensate for the increased flow of aqueous humour caused by the lowered blood osmolality. To test this hypothesis, Cole and

Nagasubramanian (1973) measured the effects of ADH on short-circuit current in the isolated iris-ciliary body preparation; as Figure 1.75 shows, ADH increased this strikingly. To test the effects in the intact animal, the endogenous ADH in the rabbit was reduced by an infusion of ethyl alcohol; this was associated with a fall in intraocular pressure; when vasopressin was administered, in addition, however, the result was a rise in intraocular pressure. Against this view of the role of vasopressin, or ADH, we have the finding of Niederer et al. (1975) that this hormone actually decreased flow of aqueous humour, as measured by the rate of elimination of labelled protein introduced by injection into the anterior chamber; it is possible that breakdown of the blood-aqueous barrier occurred, however, and this would obscure the effects of hormones on the secretion of aqueous humour.

---

*Niederer et al. (1975) found that cortisol and prolactin, alone, had no significant effect on flow of aqueous humour in the rabbit; however, when combined, they increased production, as measured by the rate of elimination of labelled protein from the eye after injection into the anterior chamber.

## REFERENCES

Allansmith, M., Newman, L. & Whitney, C. (1971) The distribution of immunoglobulin in the rabbit eye. *Arch. Ophthal.* **86**, 60–64.
Allansmith, M. R., Whitney, C. R., McClellan, B. H. & Newman, L. P. (1973) Immunoglobulins in the human eye. *Arch. Ophthal.* **89**, 36–45.
Alm, A. (1977) The effect of sympathetic stimulation on blood flow through the uvea, retina and optic nerve in monkeys (*Macaca irus*). *Exp. Eye Res.* **25**, 19–24.
Alm, A. & Bill, A. (1972) The oxygen supply to the retina II. *Acta physiol. Scan.* **84**, 306–319.
Alm, A. & Bill, A. (1973) The effect of stimulation of the cervical sympathetic chain on retinal oxygen tension and on uveal, retinal and cerebral blood flow in cats. *Acta physiol. Scand.* **88**, 84–94.
Alm, A., Bill, A. & Young, F. A. (1973) The effects of pilocarpine and neostigmine on the blood flow through the anterior uvea in monkeys. A study with radioactively labelled microspheres. *Exp. Eye Res.* **15**, 31–36.
Ambache, N. (1957) Properties of irin, a physiological constituent of the rabbit's iris. *J. Physiol.* **135**, 114–132.
Ambache, N., Brummer, H. C., Rose, J. G. & Whiting, J. (1966) Thin-layer chromatography of spasmogenic unsaturated hydroxy-acids from various tissues. *J. Physiol.* **185**, 77–78.
Ambache, N., Kavanagh, L. & Whiting, J. (1965) Effect of mechanical stimulation on rabbits' eyes: release of active substance in anterior chamber perfusates. *J. Physiol.* **176**, 378–408.
Amsler, M. & Huber, A. (1946) Methodik und erste klinische Ergebnisse einer Funktionsprüfung der Blut-Kammerwasser-Schranke. *Ophthalmologica, Basel,* **111**, 155–176.
Anderson, D. R. & Grant, W. M. (1970) Re-evaluation of

the Schiøtz tonometer calibration. *Invest. Ophthal.* **9**, 430–446.
Anselmi, P., Bron, A. J. & Maurice, D. M. (1968) Action of drugs on the aqueous flow in man measured by fluorophotometry. *Exp. Eye Res.* **7**, 487–496.
Armaly, M. F. (1959) Studies on intraocular effect of orbital parasympathetic. *Arch. Ophthal., N.Y.* **62**, 117–124.
Armaly, M. F. (1960) Schiøtz tonometer calibration and applanation tonometry. *Arch. Ophthal., N.Y.* **64**, 426–432.
Armaly, M. F. (1964) Aqueous outflow facility in monkeys and the effect of topical corticoids. *Invest. Ophthal.* **3**, 534–538.
Armaly, M. F. & Wang, Y. (1975) Demonstration of acid mucopolysaccharides in the meshwork of the rhesus monkey. *Invest. Ophthal.* **14**, 507–516.
Ascher, K. W. (1942) Aqueous veins. Preliminary note. *Am. J. Ophthal.* **25**, 31–38.
Ascher, K. W. (1961) *The Aqueous Veins.* Springfield: Thomas.
Ashton, N. (1951) Anatomical study of Schlemm's canal and aqueous veins by means of Neoprene casts. *Br. J. Ophthal.* **35**, 291–303.
Auricchio, G. & Bárány, E. H. (1959) On the role of osmotic water transport in the secretion of the aqueous humour. *Acta physiol. Scand.* **45**, 190–210.
Bairati, A. & Orzalesi, N. (1966) The ultrastructure of the epithelium of the ciliary body. *Z. Zellforsch.* **69**, 635–658.
Bárány, E. H. (1947) The recovery of intraocular pressure, arterial blood pressure, and heat dissipation by the external ear after unilateral carotid ligation. *Acta Ophthal.* **25**, 81–94.
Bárány, E. H. (1963) A mathematical formulation of intraocular pressure as dependent on secretion, ultra-filtration, bulk outflow, and osmotic reabsorption of fluid. *Invest. Ophthal.* **2**, 584–590.

Bárány, E. H. (1967) The immediate effect on outflow resistance of intravenous pilocarpine in the vervet monkey, *Cercopithecus ethiops. Invest. Ophthal.* **6,** 373–380.

Bárány, E. H. (1972) Inhibition by hippurate and probenecid of *in vitro* uptake of iodipamide and o-iodohippurate. *Acta physiol. Scand.* **86,** 12–27.

Bárány, E. H. & Scotchbrook, S. (1954) Influence of testicular hyaluronidase on the resistance to flow through the angle of the anterior chamber. *Acta physiol. Scand.* **30,** 240–248.

Baurmann, M. (1930) Über die Ciliar-Fortsatz-Gefässystem. *Ber. dtsch. ophth. Ges.* **48,** 364–371.

Becker, B. (1959) Carbonic anhydrase and the formation of aqueous humour. *Amer. J. Ophthal.* **47,** Pt. 2, 342–361.

Becker, B. (1967) Ascorbate transport in guinea pig eye. *Invest. Ophthal.* **6,** 410–415.

Becker, B. & Constant, M. A. (1955) The effect of the carbonic anhydrase inhibitor acetazolamide on aqueous flow. *Arch. Ophthal. N.Y.* **54,** 321–329.

Beitch, B. R. & Eakins, K. E. (1969) The effects of prostaglandins on the intraocular pressure of the rabbit. *Brit. J. Pharmacol.* **37,** 158–167.

Belmonte, C., Simon, J. & Gallego, A. (1971) Effects of intraocular pressure changes on the afferent activity of ciliary nerves. *Exp. Eye Res.* **12,** 342–355.

Berggren, L. (1960) Intracellular potential measurements from the ciliary processes of the rabit eye *in vivo* and *in vitro. Acta physiol. scand.* **48,** 461–470.

Berggren, L. (1964) Direct observation of secretory pumping *in vitro* of the rabbit eye ciliary processes. *Invest. Ophthal.* **3,** 266–272.

Bergmanson, J. P. G. (1977) The ophthalmic innervation of the anterior uvea in monkeys. *Exp. Eye Res.* **24,** 225–240.

Bernstein, H. N. & Schwartz, B. (1962) Effects of long-term systemic steroids on ocular pressure and tonographic values. *Arch. Ophthal., N.Y.* **68,** 742–753.

Bhattacherjee, P. & Eakins, K. E. (1974) A comparison of the inhibitory activity of compounds on ocular prostaglandin biosynthesis. *Invest. Ophthal.* **13,** 967–972.

Bhattacherjee, P. & Hammond, B. R. (1975) Inhibition of increased permeability of the blood-aqueous barrier by non-steroidal anti-inflammatory compounds as demonstrated by fluorescein angiography. *Exp. Eye Res.* **21,** 499–505.

Bill, A. (1962) The drainage of blood from the uvea and the elimination of aqueous humour in rabbits. *Exp. Eye Res.* **1,** 200–205.

Bill, A. (1964) The albumin exchange in the rabbit eye. *Acta physiol. scand.* **60,** 18–29.

Bill, A. (1967) The effects of atropine and pilocarpine on aqueous humour dynamics in cynomolgus monkeys (*Macaca irus*). *Exp. Eye Res.* **6,** 120–125.

Bill, A. (1968a) The effect of ocular hypertension caused by red cells on the rate of formation of aqueous humour. *Invest. Ophthal.* **7,** 162–168.

Bill, A. (1968b) Capillary permeability to and extravascular dynamics of myoglobin, albumin and gammaglobulin in the uvea. *Acta physiol. Scand.* **73,** 204–219.

Bill, A. (1968c) A method to determine osmotically effective albumin and gammaglobulin concentrations in tissue fluids. *Acta physiol. Scand.* **73,** 511–522.

Bill, A. (1968d) The elimination of red cells from the anterior chamber in vervet monkeys (*Cercopithecus ethiops*). *Invest. Ophthal.* **7,** 156–161.

Bill, A. (1970a) Scanning electron microscopic studies of Schlemm's canal. *Exp. Eye Res.* **10,** 214–218.

Bill, A. (1970b) Effects of norepinephrine, isoproterenol and sympathetic stimulation on aqueous humour dynamics in vervet monkeys. *Exp. Eye Res.* **10,** 31–46.

Bill, A. (1970c) Scanning electron microscopy of the canal of Schlemm. *Exp. Eye Res.* **10,** 214–218.

Bill, A. (1970d) The effect of changes in arterial blood pressure on the rate of formation of aqueous humour in a primate (*Cercopithecus ethiops*). *Ophth. Res.* **1,** 193–200.

Bill, A. (1971) Uveoscleral drainage of aqueous humour in human eyes. *Exp. Eye Res.* **12,** 275–281.

Bill, A. (1974) The role of the iris vessels in aqueous humor dynamics. *Jap. J. Ophthal.* **18,** 30–36.

Bill, A. (1975) Blood circulation and fluid dynamics in the eye. *Physiol. Rev.* **55,** 383–417.

Bill, A. (1977) Basic physiology of the drainage of aqueous humour. *Exp. Eye Res.* Suppl. 291–304.

Bill, A. & Bárány, E. H. (1966) Gross facility, facility of conventional routes, and pseudofacility of aqueous humour outflow in the cynomolgus monkey. *Archs. Ophthal., N.Y.* **75,** 665–673.

Bill, A. & Svedbergh, B. (1972) Scanning electron microscopic studies of the trabecular meshwork and the canal of Schlemm—an attempt to locate the main resistance to outflow of aqueous humor in man. *Acta ophthal.* **50,** 295–320.

Bill, A., Stjernschantz, J. & Alm, A. (1976) Effects of hexamethonium, biperiden and phentolamine on the vasoconstrictive effects of oculomotor nerve stimulation in rabbits. *Exp. Eye Res.* **23,** 615–622.

Bill, A. & Wålinder, P. -E. (1966) The effects of pilocarpine on the dynamics of aqueous humour in a primate (*Macaca irus*). *Invest. Ophthal.* **5,** 170–175.

Bito, L. Z. (1970) Steady-state concentration gradients of magnesium, potassium and calcium in relation to the sites and mechanisms of ocular cation-transport processes. *Exp. Eye Res.* **10,** 102–116.

Bito, L. Z. (1972) Accumulation and apparent active transport of prostaglandins by some rabbit tissues *in vitro. J. Physiol.* **221,** 371–387.

Bito, L. Z. (1977) The physiology and pathophysiology of the intraocular fluids. *Exp. Eye Res.* **25,** Suppl. 273–289.

Bito, L. Z. & Davson, H. (1964) Steady-state concentrations of potassium in the ocular fluids. *Exp. Eye Res.* **3,** 283–297.

Bito, L. Z., Davson, H., Levin, E., Murray, M. & Snider, N. (1965) The relationship between the concentrations of amino acids in the ocular fluids and blood plasma of dogs. *Exp. Eye Res.* **4,** 374–380.

Bito, L. Z. & Salvador, E. V. (1970) Intraocular fluid dynamics. II. *Exp. Eye Res.* **10,** 273–287.

Bito, L. Z. & Salvador, E. V. (1972) Intraocular fluid dynamics. III. The site and mechanism of prostaglandin transfer across the blood intraocular fluid barriers. *Exp. Eye Res.* **14,** 233–241.

Bito, L. Z., Salvador, E. V. & Petrivonic, L. (1978) Intraocular fluid dynamics. IV. *Exp. Eye Res.* **26,** 47–53.

Bito, L. Z. & Wallenstein, M. C. (1977) Transport of prostaglandins across the blood-brain and blood-aqueous barriers and the physiological significance of these absorptive transport processes. *Invest. Ophthal.* Suppl. 229–243.

Bonting, S. L. & Becker, B. (1964) Inhibition of enzyme activity and aqueous humor flow in the rabbit eye after intravitreal injection of ouabain. *Invest. Ophthal.* **3,** 523–533.

Brubaker, R. F. (1967) Determination of episcleral venous pressure in the eye. *Arch. Ophthal.* **77,** 110–114.

Brubaker, R. F. (1970) The measurement of pseudofacility

and true facility by constant pressure perfusion in the normal rhesus monkey eye. *Invest. Ophthal.* **9**, 42–52.

Brubaker, R. F. & Kupfer, C. (1966) Determination of pseudofacility in the eye of the rhesus monkey. *Arch. Ophthal., N.Y.* **75**, 693–697.

Cole, D. F. (1960) Rate of entrance of sodium into the aqueous humour of the rabbit. *Br. J. Ophthal.* **44**, 225–245.

Cole, D. F. (1961) Electrochemical changes associated with the formation of the aqueous humour. *Br. J. Ophthal.* **45**, 202–217.

Cole, D. F. (1961) Prevention of experimental ocular hypertension with polyphloretin phosphate. *Br. J. Ophthal.* **45**, 482–489.

Cole, D. F. (1961) Electrical potential across the ciliary body observed *in vitro. Br. J. Ophthal.* **45**, 641–653.

Cole, D. F. (1965) Citrate cycle and active sodium transport in rabbit ciliary epithelium. *Exp. Eye Res.* **4**, 211–222.

Cole, D. F. (1966) Aqueous humour formation. *Doc. Ophthalmol.* **21**, 116–238.

Cole, D. F. (1974) The site of breakdown of the blood-aqueous barrier under the influence of vaso-dilator drugs. *Exp. Eye Res.* **19**, 591–607.

Cole, D. F. (1977) Secretion of the aqueous humour. *Exp. Eye Res.* Suppl., 161–176.

Cole, D. F. & Monro, P. A. G. (1976) The use of flurorescein-labelled dextrans in investigation of aqueous humour outflow in the rabbit. *Exp. Eye Res.* **23**, 571–585.

Cole, D. F. & Tripathi, R. C. (1971) Theoretical consideration on the mechanism of the aqueous outflow. *Exp. Eye Res.* **12**, 25–32.

Cole, D. F. & Unger, W. G. (1973) Prostaglandins as mediators for the responses of the eye to trauma. *Exp. Eye Res.* **17**, 357–368.

Copeland, R. L. & Kinsey, V. E. (1950) Determination of the volume of the posterior chamber of the rabbit. *Arch. Ophthal., N.Y.* **44**, 515–516.

Cunha-Vaz, J. G. & Maurice, D. M. (1967) The active transport of fluorescein by the retinal vessels and the retina. *J. Physiol.* **191**, 467–486.

Davson, H. (1953) The penetration of large water-soluble molecules into the aqueous humour. *J. Physiol.* **122**, 10P.

Davson, H. (1955) The rates of disappearance of substances injected into the subarachnoid space of rabbits. *J. Physiol.* **128**, 52–53P.

Davson, H. (1956) *Physiology of the Ocular and Cerebrospinal Fluids.* London: Churchill.

Davson, H. (1969) The intra-ocular fluids. The intra-ocular pressure. In *The Eye*, Ed. Davson, pp. 67–186, 187–272. London: Academic Press.

Davson, H. & Duke-Elder, W. S. (1948) The distribution of reducing substances between the intraocular fluids and blood plasma, and the kinetics of penetration of various sugars into these fluids. *J. Physiol.* **107**, 141–152.

Davson, H. & Hollingsworth, J. R. (1972) The effects of iodate on the blood-vitreous barrier. *Exp. Eye Res.* **14**, 21–28.

Davson, H. & Huber, A. (1950) Experimental hypertensive uveitis in the rabbit. *Ophthalmologica, Basel*, **120**, 118–124.

Davson, H. & Luck, C. P. (1956) A comparative study of the total carbon dioxide in the ocular fluids, cerebrospinal fluid, and plasma of some mammalian species. *J. Physiol.* **132**, 454–464.

Davson, H. & Matchett, P. A. (1951) The control of the intraocular pressure in the rabbit. *J. Physiol.* **113**, 387–397.

Davson, H. & Matchett, P. A. (1953) The kinetics of penetration of the blood-aqueous barrier. *J. Physiol.* **122**, 11–32.

Davson, H. & Quilliam, J. P. (1947) The effect of nitrogen mustard on the permeability of the blood-aqueous humour barrier to Evans blue. *Br. J. Ophthal.* **31**, 717–721.

Davson, H. & Spaziani, E. (1960) The fate of substances injected into the anterior chamber of the eye. *J. Physiol.* **151**, 202–215.

Davson, H. & Spaziani, E. (1961) The effect of hypothermia on intraocular dynamics. *Exp. Eye Res.* **1**, 182–192.

Davson, H. & Thomassen, T. L. (1950) The effect of intravenous infusion of hypertonic saline on the intraocular pressure. *Br. J. Ophthal.* **34**, 355–359.

Davson, H. & Weld, C. B. (1941) Studies on the aqueous humour. *Amer. J. Physiol.* **134**, 1–7.

Drance, S. M. (1960) The coefficient of scleral rigidity in normal and glaucomatous eyes. *Arch. Ophthal. N.Y.* **63**, 668–674.

Duke-Elder, W. S. (1927) The pressure equilibrium in the eye. *J. Physiol.* **64**, 78–86.

Duke-Elder, W. S. (1932) *Textbook of Ophthalmology.* Vol. 1. London: Kimpton.

Duke-Elder, S. & Wybar, K. C. (1961) *System of Ophthalmology.* Vol. II: *The Anatomy of the Visual System.* London: Kimpton.

Durham, D. G., Bigliano, R. P. & Masino, J. A. (1965) Pneumatic applanation tonometer. *Trans. Am. Acad. Ophthal. Oto-lar.* **69**, 1029–1047.

Eakins, K. E. (1971) Prostaglandin antagonism by polymeric phosphates of phloretin and related compounds. *Ann. N.Y. Acad. Sci.* **180**, 386.

Eakins, K. E. (1977) Prostaglandin and non-prostaglandin mediated breakdown of the blood-aqueous barrier. *Exp. Eye Res.* **25**, Suppl. 483–498.

Ehinger, B. (1973) Localization of the uptake of prostaglandin E in the eye. *Exp. Eye Res.* **17**, 43–47.

Ellingsen, B. A. & Grant, W. M. (1971) Influence of intraocular pressure and trabeculotomy on aqueous outflow in enucleated monkey eyes. *Invest. Ophthal.* **10**, 705–709.

Ericson, L. A. (1958) Twenty-four hourly variations of the aqueous flow. Examinations with perilimbal suction cup. *Acta ophthal.* Suppl. 50.

Ferreira, S. H., Moncada, S. & Vane, J. (1971) Indomethacin and aspirin abolish prostaglandin release from the spleen. *Nature, New Biol.* **231**, 237–239.

Forbes, M. & Becker, B. (1960) *In vivo* transport of iodopyracet (diodrast). *Am. J. Ophthal.* **50**, Pt. 2, 862–867.

Fortin, E. P. (1929) Action du muscle ciliare sur la circulation de l'oeil, etc. *C.r. Seanc. Soc. Biol.* **102**, 432–434.

François, J. (1954) Cortisone et tension oculaire. *Annls Oculist.* **187**, 805–816.

François, J. & Neetens, A. (1962) Comparative anatomy of the vascular supply of the eye in vertebrates. In *The Eye*, Ed. Davson, pp. 369–416. London: Academic Press.

Friedenwald, J. S. (1957) Tonometer calibration. An attempt to remove discrepancies found in the 1954 calibration scale for Schiøtz tonometers. *Trans. Am. Acad. Ophthal. Otolaryngol.* **61**, 108–123.

Friedenwald, J. S. & Becker, B. (1956) Aqueous humour dynamics. Theoretical considerations. *Am. J. Ophthal.* **41**, 383–398.

Gaasterland, D., Kupfer, C., Ross, K. & Gabelnick, H. L. (1973) Studies of aqueous humor dynamics in man. III. Measurements in young normal subjects using norepinephrine and isoproterenol. *Invest. Ophthal.* **12**, 267–279.

Galin, M. A., Baras, I. & Mandell, G. L. (1961) Measurements of aqueous flow utilizing the perilimbal suction cup. *Arch. Ophthal.* **66**, 65.

Galin, M. A., Davidson, R. & Pasmanik, S. (1963) An osmotic comparison of urea and mannitol. *Am. J. Ophthal.* **55**, 244–247.

Gallego, R. & Belmonte, C. (1974) Nervous efferent activity in the ciliary nerves related to intraocular pressure changes. *Exp. Eye Res.* **19**, 331–334.

Garg, L. C. & Oppelt, W. W. (1970) The effect of ouabain and acetazolamide on transport of sodium and chloride from plasma to aqueous humor. *J. Pharm.* **175**, 237–247.

Gloster, J. (1961) Influence of facial nerve on intraocular pressure. *Br. J. Ophthal.* **45**, 259–278.

Gloster, J. & Greaves, D. P. (1957) Effect of diencephalic stimulation upon intraocular pressure. *Br. J. Ophthal.* **41**, 513–532.

Goldmann, H. (1946) Weitere Mitteilung über den Abfluss des Kammerwassers beim Menschen. *Ophthalmologica, Basel,* **112**, 344–349.

Goldmann, H. (1955) Un nouveau tonomètre à aplanation. *Bull. Soc. franc. Ophtal.* **67**, 474.

Goldmann, H. (1968) On pseudofacility. *Bibl. Ophthal.* **76**, 1–14.

Grant, W. M. (1950) Tonographic method for measuring the facility and rate of aqueous outflow in human eyes. *Arch. Ophthal., N.Y.* **44**, 204–214.

Grant, W. M. (1958) Further studies on facility of flow through trabecular meshwork. *Arch. Ophthal., N.Y.* **60**, 523–533.

Grant, W. M. (1963) Experimental aqueous perfusion in enucleated human eyes. *Arch. Ophthal., N.Y.* **69**, 783–801.

Greaves, D. P. & Perkins, E. S. (1952) Influence of the sympathetic nervous system on the intraocular pressure and vascular circulation of the eye. *Br. J. Ophthal.* **36**, 258–264.

Green, K. & Pederson, E. (1972) Contribution of secretion and filtration to aqueous humor formation. *Amer. J. Physiol.* **222**, 1218–1226.

Grierson, I. & Lee, W. R. (1974) Changes in the monkey outflow apparatus at graded levels of intraocular pressure: a qualitative analysis by light microscopy and scanning electron microscopy. *Exp. Eye Res.* **19**, 21–33.

Grierson, I. & Lee, W. R. (1975a) The fine structure of the trabecular meshwork at graded levels of intraocular pressure. (1) Pressure effects within the near-physiological range (8–30 mm Hg) (2) Pressure outside the physiological range (0 and 50 mmHg). *Exp. Eye Res.* **20**, 523–530.

Grierson, I. & Lee, W. R. (1975b) Acid mucopolysaccharides in the outflow apparatus. *Exp. Eye Res.* **21**, 417–431.

Grignolo, A., Orzalesi, N. & Calabria, G. A. (1966) Studies on the fine structure and the rhodopsin cycle of the rabbit retina in experimental degeneration induced by sodium iodate. *Exp. Eye Res.* **5**, 86–97.

Harris, J. E. & Gehrsitz, L. B. (1949) The movement of monosaccharides into and out of the aqueous humour. *Amer. J. Ophthal.* **32**, Pt. 2, 167–176.

Hart, R. W. (1972) Theory of neural mediation of intraocular dynamics. *Bull. Math. Biophys.* **34**, 113–140.

Hayreh, S. S. (1969) Blood supply of the optic nerve head and its role in optic atrophy, glaucoma, and oedema of the optic disc. *Br. J. Ophthal.* **53**, 721–748.

Heath, H. & Fiddick, R. (1966) The active transport of ascorbic acid by the rat retina. *Exp. Eye Res.* **5**, 156–163.

Hetland-Eriksen, J. (1966) On Tonometry. I–IX. *Acta Ophthal.* **44**, 5–11, 12–19, 107–113, 114–120, 515–521, 522–538, 725–736, 881–892, 893–900.

Holland, M. G. & Gipson, C. C. (1970) Chloride ion transport in the isolated ciliary body. *Invest. Ophthal.* **9**, 20–29.

Holmberg, A. (1959) The fine structure of the inner wall of Schlemm's canal. *Archs. Ophthal. N.Y.* **62**, 956–958.

Hörven, I. (1961) Tonometry in newborn infants. *Acta ophthal. Kbh.* **39**, 911–918.

Hudspeth, A. J. & Yee, A. G. (1973) The intercellular junctional complexes of retinal pigment epithelia. *Invest. Ophthal* **12**, 354–365.

Huggert, A. (1958) Experimentally induced increase in intraocular pressure in the rabbit eye. *Acta ophthal. Kbh.* **36**, 750–760.

Inomata, H. & Bill, A. (1977) Exit sites of uveoscleral flow of aqueous humour in cynomolgus monkey eyes. *Exp. Eye Res.* **25**, 113–118.

Inomata, H., Bill, A. & Smelser, G. K. (1972a) Aqueous humour pathways through the trabecular meshwork and into Schlemm's canal in the cynomolgus monkey (*Macaca irus*). *Amer. J. Ophthal.* **73**, 760–789.

Inomata, H., Bill, A. & Smelser, G. K. (1972b) Unconventional routes of aqueous humor outflow in cynomolgus monkey (*Macaca irus*). *Amer. J. Ophthal.* **73**, 893–907.

Jackson, R. T. & Waitzman, M. B. (1965) Effect of some steroids on aqueous humor dynamics. *Exp. Eye Res.* **4**, 112–123.

Jampol, L. M., Neufeld, A. H. & Sears, M. L. (1975) Pathways for the response of the eye to injury. *Invest. Ophthal.* **14**, 184–189.

Jocson, V. L. & Sears, M. L. (1968) Channels of aqueous outflow and related blood vessels. I. *Macaca mulatta* (Rhesus). *Arch. Ophthal.* **80**, 104–114.

Johnstone, M. A. & Grant, W. M. (1973) Pressure-dependent changes in structure of the aqueous outflow system of human and monkey eyes. *Amer. J. Physiol.* **75**, 365–383.

Karnovsky, M. J. (1967) The ultrastructural basis of capillary permeability studied with peroxidase as a tracer. *J. Cell Biol.* **35**, 213–236.

Kaufman, P. L. & Bárány, E. H. (1977) Recent observations concerning the effects of cholinergic drugs on outflow facility in monkeys. *Exp. Eye Res.* (Suppl.) 415–418.

Kaye, G. I. & Pappas, G. D. (1965) Studies on the ciliary epithelium and zonule. III. The fine structure of the rabbit ciliary epithelium in relation to the localization of ATPase activity. *J. Microsopie,* **4**, 497–508.

Kayes, J. (1967) Pore structure of the inner wall of Schlemm's canal. *Invest. Ophthal.* **6**, 381–394.

Kayes, J. (1975) Pressure gradient changes on the trabecular meshwork of monkeys. *Amer. J. Ophthal.* **79**, 549–556.

Kelly, R. G. M. & Starr, M. S. (1971) Effects of prostaglandins and a prostaglandin antagonist on intraocular pressure and protein in the monkey eye. *Can. J. Ophthal.* **6**, 205–211.

Kinsey, V. E. (1947) Transfer of ascorbic acid and related compounds across the blood-aqueous battier. *Am. J. Ophthal.* **30**, 1262–1266.

Kinsey, V. E. & Bárány, E. (1949) The rate of flow of aqueous humour. II. Derivation of rate of flow and its physiological significance. *Amer. J. Ophthal.* **39** Pt. 2, 189–201.

Kinsey, V. E. & Palm, E. (1955) Posterior and anterior chamber aqueous humour formation. *Arch. Ophthal., N.Y.* **53**, 330–344.

Kinsey, V. E. & Reddy, D. V. N. (1959) An estimate of the ionic composition of the fluid secreted into the posterior chamber, inferred from a study of aqueous humour dynamics. *Doc. Ophthal.* **13**, 7–40.

Kinsey, V. E., Grant, M. & Cogan, D. G. (1942) Water movement and the eye. *Archs. Ophthal., N.Y.* **27**, 242–252.

Kozart, D. M. (1968) Light and electron microscopic study of regional morphologic differences in the processes of the ciliary body in the rabbit. *Invest. Ophthal.* **7**, 15–33.

Krill, A. E., Noell, F. W. & Novak, M. (1965) Early and long-term effects of levo-epinephrine on ocular tension and outflow. *Amer. J. Ophthal.* **59**, 833–839.

Kronfeld, P. C. (1964) Dose-effect relationships as an aid in the evaluation of ocular hypotensive drugs. *Invest. Ophthal.* **3**, 258–265.

Kupfer, C. (1961) Studies on intraocular pressure. I. *Arch. Ophthal., N.Y.*, **65**, 565–570.

Kupfer, C., Gaasterland, D. & Ross, K. (1971) II. Measurements in young normal subjects using acetazolamide and L-epinephrine. *Invest. Ophthal.* **10**, 523–533.

Kupfer, C. & Ross, K. (1971) Studies of aqueous humour dynamics in man. I. Measurements in young normal subjects. *Invest. Ophthal.* **10**, 518–522.

Lamble, J. W. (1977) Some effects of topically applied ( − )-adrenaline bitartrate on the rabbit eye. *Exp. Eye Res.* **24**, 129–143.

Langham, M. E. (1958) Aqueous humour and control of intra-ocular pressure. *Physiol Rev.* **38**, 215–242.

Langham, M. E. (1960) Steady-state pressure relationships in the living and dead cat. *Am. J. Ophthal.* **50**, Pt. 2, 950–958.

Langham, M. E. (1963) A new procedure for the analysis of intraocular dynamics in human subjects. *Exp. Eye Res.* **2**, 314–324.

Langham, M. E. (1977) The aqueous outflow system and its response to autonomic receptor agonists. *Exp. Eye Res.* **25**, Suppl. 311–322.

Langham, M. E. & Diggs, E. M. (1974) β-Adrenergic responses in the eyes of rabbits, primates and man. *Exp. Eye Res.* **19**, 281–295.

Langham, M. E. & Fraser, L. K. (1966) The absence of supersensitivity to adrenergic amines in the eye of the conscious rabbit following preganglionic cervical sympathotomy. *Life Sciences.* **5**, 1699–1705.

Langham, M. E. & Kriegelstein, G. K. (1976) The biphasic intraocular pressure response of rabbits to epinephrine. *Invest. Ophthal.* **15**, 119–127.

Langham, M. E. & Palewicz, K. (1977) The pupillary, the intraocular pressure, and the vasomotor responses to norepinephrine. *J. Physiol.* **267**, 339–355.

Langham, M. E. & Rosenthal, A. R. (1966) Role of cervical sympathetic nerve in regulating intraocular pressure and circulation. *Am. J. Physiol.* **210**, 786–794.

Langham, M. E. & Taylor, C. B. (1960) The influence of pre- and post-ganglionic section of the cervical sympathetic on the intraocular pressure of rabbits and cats. *J. Physiol.* **152**, 437–446.

Langham, M. E. & Taylor, C. B. (1960) The influence of superior cervical ganglionectomy on the intraocular pressure. *J. Physiol.* **152**, 447–458.

Langham, M. E. & Wybar, K. C. (1953) A fluorophotometer for the study of intraocular dynamics in the living animal. *J. Physiol.* **120**, 5P.

Lasansky, A. & De Fisch, F. W. (1966) Potential current and ionic fluxes across the isolated pigment epithelium and choroid. *J. gen. Physiol.* **49**, 913–924.

Laties, A. M. & Rapoport, S. (1976) The blood-ocular barriers under osmotic stress. *Arch. Ophthal.* **94**, 1086–1091.

Lehmann, G. & Meesmann, A. (1924) Über das Bestehen eines Donnangleichgewichtes zwischen Blut und Kammerwasser bzw. Liquor cerebrospinalis. *Pflüg. Arch.* **205**, 210–232.

Leydhecker, W., Akiyama, K. & Neumann, H. G. (1958) Der intraokulare Druck gesunder menschlicher Augen. *Klin. Mbl. Augenheilk.* **133**, 662–670.

Linnér, E. (1952) Effect of unilateral ligation of the common carotid artery on the blood flow through the uveal tract as measured directly in a vortex vein. *Acta physiol. Scand.* **26**, 70–78.

McBain, E. H. (1960) Tonometer calibration. III. *Arch. Ophthal., N.Y.* **63**, 936–942.

Macri, F. J. (1961) Interdependence of venous and eye pressure. *Arch. Ophthal., N.Y.* **65**, 442–449.

Macri, F. J. (1967) The pressure dependence of aqueous humor formation. *Arch. Ophthal., N.Y.* **78**, 629–633.

Macri, F. J. & Brown, J. G. (1961) The constrictive action of acetazolamide on the iris arteries of the cat. *Arch. Ophthal.* **66**, 570–577.

Macri, F. J. & Cevario, S. J. (1973) The induction of aqueous humor formation by the use of Ach-eserine. *Invest. Ophthal.* **12**, 910–916.

Macri, F. J. & Cevario, S. J. (1975) A possible vascular mechanism for the inhibition of aqueous humor formation by ouabain and acetazolamide. *Exp. Eye Res.* **20**, 563–569.

Macri, F. J. & Cevario, S. J. (1977) Blockade of the ocular effects of acetazolamide by phencyclidine. *Exp. Eye Res.* **24**, 121–127.

Macri, F. J., Cevario, S. J. & Ballintine, E. J. (1974) The arterial pressure dependency of the increased aqueous humor formation induced by Ach-eserine. *Invest. Ophthal.* **13**, 153–155.

Macri, F. J., Dixon, R. L. & Rall, D. P. (1965) Aqueous humour turnover rates in the cat. I. Effect of acetazolamide. *Invest. Ophthal.* **4**, 927–934.

Macri, F. J., Wanko, T. & Grimes, P. A. (1958) The elastic properties of the human eye. *Arch. Ophthal., N.Y.* **60**, 1021–1026.

Mackay, R. S. & Marg, E. (1959) Fast automatic, electronic tonometers based on an exact theory. *Acta Ophthal.* **37**, 495–507.

McEwen, W. K. (1958) Application of Poiseuille's law to aqueous outflow. *Arch. Ophthal., N.Y.* **60**, 290–294.

Maggiore, L. (1917) Struttura, comportamento e significato del canale di Schlemm nell'occhio umano, in condizioni normali e patologische. *Ann. Ottalm.* **40**, 317–462.

Maren, T. H. (1977) Ion secretion into the posterior aqueous humor of dogs and monkeys. *Exp. Eye Res.* **25**, (Suppl.) 245–247.

Maurice, D. M. (1954) Constriction of the pupil in the rabbit by antidromic stimulation of the trigeminal nerve. *J. Physiol.* **123**, 45–46P.

Melton, C. E. & De Ville, W. B. (1960) Perfusion studies on eyes of four species. *Am. J. Ophthal.* **50**, 302–308.

Miller, J. E. & Constant, M. A. (1960) The measurement of rabbit ciliary epithelial potentials *in vitro. Am. J. Ophthal.* **50**, Pt. 2, 855–862.

Miller, J. D., Eakins, K. E. & Atival, M. (1973) The release of $PGE_2$-like activity into aqueous humor after paracentesis and its prevention by aspirin. *Invest. Ophthal.* **12**, 939–942.

Miller, S. S. & Steinberg, R. H. (1977) Active transport of ions across frog retinal pigment epithelium. *Exp. Eye Res.* **25**, 235–248.

Moses, R. A. & Grodski, W. J. (1971) Theory and calibration of the Schiötz tonometer. I–V. *Invest. Ophthal.* **10**, 534–538, 539–543, 589–591, 592–600, 601–604.

Neufeld, A. H., Jampol, L. M. & Sears, M. L. (1972) Cyclic—AMP in the aqueous humor: the effects of adrenergic agents. *Exp. Eye Res.* **14**, 242–250.

Noell, W. K. (1955) Metabolic injuries of the visual cell. *Amer. J. Ophthal.* **40**, Pt. 2, 60–76.

Noell, W. K. (1963) Cellular physiology of the retina. *J. Opt. Soc. Am.* **53**, 36–48.

Noell, W. K., Crapper, D. R. & Paganelli, C. V. (1965) Transretinal currents and ion fluxes. In *Transcellular Membrane Potentials and Ionic Fluxes*. Ed. Snell, F. & Noell, W. K. New York: Gordon & Breach.

Nomura, T. & Smelser, G. K. (1974) The identification of adrenergic and cholinergic nerve endings in the trabecular meshwork. *Invest. Ophthal.* **13**, 525–532.

Okisaka, S. (1976) The effects of prostaglandin $E_1$ on the ciliary epithelium and the drainage angle of cynomolgus monkeys: a light and electronmicroscopic study. *Exp. Eye Res.* **22**, 141–154.

Oppelt, W. W. (1967) Measurement of aqueous humor formation rates by posterior-anterior chamber perfusion with inulin. *Invest. Ophthal.* **6**, 76–83.

Oppelt, W. W., White, E. D. & Halpert, E. S. (1969) The effect of corticosteroids on aqueous humor formation rate and outflow facility. *Invest. Ophthal.* **8**, 535–541.

Pappas, G. D. & Smelser, G. K. (1961) The fine structure of the ciliary epithelium in relation to aqueous humour secretion. In *The Structure of the Eye*, Ed. Smelser, pp. 453–467. New York: Academic Press.

Pappenheimer, J. R., Heisey, S. R. & Jordan, E. F. (1961) Active transport of Diodrast and phenolsulfonphthalein from cerebrospinal fluid to blood. *Amer. J. Physiol.* **200**, 1–10.

Pederson, J. E. & Green, K. (1973a) Aqueous humor dynamics: a mathematical approach to measurement of facility, pseudofacility, capillary pressure, active secretion and $x_c$. *Exp. Eye Res.* **15**, 265–276.

Pederson, J. E. & Green, K. (1973b) Aqueous humor dynamics: Experimental studies. *Exp. Eye Res.* **15**, 277–297.

Pederson, J. E. & Green, K. (1975) Solute permeability of the normal and prostaglandin-stimulated ciliary epithelium and the effect of ultrafiltration on active transport. *Exp. Eye Res.* **21**, 569–580.

Perkins, E. S. (1955) Pressure in the canal of Schlemm. *Br. J. Ophthal.* **39**, 215–219.

Perkins, E. S. (1957) Influence of the fifth cranial nerve on the intraocular pressure in the rabbit eye. *Br. J. Ophthal.* **41**, 257–300.

Perkins, E. S. & Gloster, J. (1957) Distensibility of the eye. *Br. J. Ophthal.* **41**, 93–102, 475–486.

Peterson, W. S. & Jocson, V. L. (1974) Hyaluronidase effects on aqueous outflow resistance. *Amer. J. Ophthal.* **77**, 573–577.

Peyman, G. A., Spitznas, M. & Straatsma, B. R. (1971) Peroxidase diffusion in the normal and photocoagulated retina. *Invest. Ophthal.* **10**, 181–189.

Peyman, G. A. & Bok, D. (1972) Peroxidase diffusion in the normal and laser-coagulated primate retina. *Invest. Ophthal.* **11**, 35–45.

Pohjola, S. (1966) The glucose content of the aqueous humour in man. *Acta Ophthal.* Suppl. 88.

Pollay, M. & Davson, H. (1963) The passage of certain substances out of the cerebrospinal fluid. *Brain* **86**, 137–150.

Radius, R. & Langham, M. E. (1973) Cyclic—AMP and the ocular responses to norepinephrine. *Exp. Eye Res.* **17**, 219–229.

Rapoport, S. I. (1977) Osmotic opening of blood brain and blood-ocular barriers. *Exp. Eye Res.* **25**, Suppl. 499–509.

Raviola, G. (1971) The fine structure of the ciliary zonule and ciliary epithelium. *Invest. Ophthal.* **10**, 851–869.

Raviola, G. (1974) Effects of paracentesis on the blood-aqueous barrier—an electron microscope study on *Macaca mulatta* using horseradish peroxidase as a tracer. *Invest. Ophthal.* **13**, 828–858.

Raviola, G. (1977) The structural basis of the blood-ocular barriers. *Exp. Eye Res.* **25**, Suppl. 27–63.

Reddy, D. V. N., Chakrapani, B. & Lim, C. P. (1977) Blood-vitreous barrier to amino acids. *Exp. Eye Res.* **25**, 543–554.

Reddy, D. V. N. & Kinsey, V. E. (1960) Composition of the vitreous humour in relation to that of the plasma and aqueous humour. *Arch. Ophthal., N.Y.* **63**, 715–720.

Reddy, D. V. N. & Kinsey, V. E. (1963) Transport of amino acids into intraocular fluids and lens in diabetic rabbits. *Invest. Ophthal.* **2**, 237–242.

Reddy, D. V. N., Rosenberg, C. & Kinsey, V. E. (1961) Steady-state distribution of free amino acids in the aqueous humors, vitreous body and plasma of the rabbit. *Exp. Eye Res.* **1**, 175–181.

Reddy, D. V. N., Thompson, M. R. & Chakrapani, B. (1977) Amino acid transport across the blood-aqueous barrier of mammalian species. *Exp. Eye Res.* **25**, 555–562.

Reese, T. S. & Karnovsky, M. J. (1967) Fine structural localization of a blood-brain barrier to exogenous peroxidase. *J. Cell Biol.* **34**, 207–217.

Riley, M. V. (1966) The tricarboxylic acid cycle and glycolysis in relation to ion transport by the ciliary body. *Biochem. J.* **98**, 898–902.

Ringvold, A. (1975) An electron microscopic study of the iris stroma in monkey and rabbit with particular reference to intercellular contacts and sympathetic innervation of anterior layer cells. *Exp. Eye Res.* **20**, 349–365.

Rodriguez-Peralta, L. (1962) Experiments on the site of the blood-ocular barrier. *Anat. Rec.* **142**, 273.

Rohen, J. (1961) Morphology and pathology of the trabecular meshwork. In *The Structure of the Eye*, Ed. Smelser, pp. 335–341. New York: Academic Press.

Rohen, J. W. (1965) *Die Struktur des Auges*. Stuttgart: Schattauer Verlag.

Rohen, J. W. (1969) New studies on the functional morphology of the trabecular meshwork and the outflow channels. *Trans. Ophthal. Soc., U.K.* **89**, 431–447.

Ruskell, G. L. (1970a) The orbital branches of the pterygopalatine ganglion and their relationship with internal carotid nerve branches in primates. *J. Anat.* **160**, 323–339.

Ruskell, G. L. (1970b) An ocular parasympathetic nerve pathway of facial nerve origin and its influence on intraocular pressure. *Exp. Eye Res.* **10**, 319–330.

Ruskell, G. L. (1974) Ocular fibres of the maxillary nerve in monkeys. *J. Anat.* **118**, 195–203.

Ruskell, G. L. (1976) The source of nerve fibres of the trabeculae and adjacent structures in monkey eyes. *Exp. Eye Res.* **23**, 449–459.

Sallmann, L. von & Dillon, B. (1947) The effect of di-isopropyl fluorophosphate on the capillaries of the anterior segment of the eye in rabbits. *Amer. J. Ophthal.* **30**, 1244–1262.

Sallmann, L. von, Fuortes, M. G. F., Macri, F. J. & Grimes, P. (1958) Study of afferent electric impulses induced by intraocular pressure changes. *Am. J. Ophthal.* **45**, Pt. 2, 211–220.

Sallmann, L. von & Loewenstein, O. (1955) Responses of

intraocular pressure, blood pressure and cutaneous vessels to electrical stimulation of the diencephalon. *Am. J. Ophthal.* **39**, Pt. 2, 11–29.

Schachtschabel, D. O., Bigalke, B. & Rohen, J. W. (1977) Production of glycosaminoglycans by cell cultures of the trabecular meshwork of the primate eye. *Exp. Eye Res.* **24**, 71–80.

Scheie, H. G., Moore, E. & Adler, F. H. (1943) Physiology of the aqueous in completely iridectomised eyes. *Arch. Ophthal., N.Y.* **30**, 70–74.

Sears, M. L. (1960) Outflow resistance of the rabbit eye: technique and effects of acetazolamide. *Arch. Ophthal., N.Y.* **64**, 823–838.

Sears, M. L. (1966) The mechanism of action of adrenergic drugs in glaucoma. *Invest. Ophthal.* **5**, 115–119.

Sears, M. L. & Bárány, E. H. (1960) Outflow resistance and adrenergic mechanisms. *Archs. Ophthal., N.Y.* **64**, 839–849.

Seidel, E. (1918) Experimentelle Untersuchungen über die Quelle und den Verlauf der intraokularen Saftströmung. *v. Graefes' Arch. Ophthal.* **95**, 1–72.

Seidel, E. (1921) Über den Abfluss des Kammerwassers aus der vorderen Augenkammer. *v. Graefes' Arch. Ophthal.* **104**, 357–402.

Serr, H. (1937) Zur Analyse der spontanen Pulserscheinungen in den Netzhautgefässen. *v. Graefes' Arch. Ophthal.* **137**, 487–505.

Shabo, A. L., Reese, T. S. & Gaasterland, D. (1973) Postmortem formation of giant endothelial vacuoles in Schlemm's canal of the monkey. *Amer. J. Ophthal.* **76**, 896–905.

Shakib, M. & Cunha-Vaz, J. G. (1966) Junctional complexes of the retinal vessels and their role in the permeability of the blood-retinal barrier. *Exp. Eye Res.* **5**, 229–234.

Shiose, Y. (1970) Electron microscopic studies on blood-retinal and blood-aqueous barriers. *Jap. J. Ophthal.* **14**, 73–87.

Shiose, Y. (1971) Morphological study on permeability of the blood-aqueous barrier. *Jap. J. Ophthal.* **15**, 17–26.

Simon, K. A., Bonting, S. L. & Hawkins, N. M. (1962) Studies on sodium-potassium-activated adenosine triphosphatase. II. Formation of aqueous humour. *Exp. Eye Res.* **1**, 253–261.

Singh, S. & Dass, R. (1960) The central artery of the retina. *Br. J. Ophthal.* **44**, 193–212, 280–299.

Slezak, H. (1969) Über Fluorescein in der Hinterkammer des menschlichen Auges. *v. Graefes' Arch. Ophthal.* **178**, 260–267.

Smelser, G. K. & Pei, Y. F. (1965) Cytological basis of protein leakage into the eye following paracentesis. *Investigative Ophthal.* **4**, 249–263.

Smith, R. S. (1971) Ultrastructural studies on the blood-aqueous barrier I. Transport of an electron-dense tracer in the iris and ciliary body of the mouse. *Amer. J. Ophthal.* **71**, 1066–1077.

Smith, R. S. & Rudt, L. A. (1975) Ocular vascular and epithelial barriers to microperoxidase. *Invest. Ophthal.* **14**, 556–560.

Speakman, J. S. (1960) Drainage channels in the trabecular wall of Schlemm's canal. *Br. J. Ophthal.* **44**, 513–523.

Steinberg, R. H. & Miller, S. (1973) Aspects of electrolyte transport in frog pigment epithelium. *Exp. Eye Res.* **16**, 365–372.

Stjernschantz, J. (1976a) Effect of parasympathetic stimulation on intraocular pressure, formation of the aqueous humour and outflow facility in rabbits. *Exp. Eye Res.* **22**, 639–645.

Stjernschantz, J. (1976b) Increase in aqueous humour protein concentration induced by oculomotor nerve stimulation in rabbits. *Exp. Eye Res.* **23**, 547–553.

Stjernschantz, J. & Bill, A. (1976) Effects of intracranial oculomotor nerve stimulation on ocular blood flow in rabbits: modification by indomethacin. *Exp. Eye Res.* **23**, 461–469.

Szalay, J., Nunziata, B. & Henkind, P. (1975) Permeability of iridial blood vessels. *Exp. Eye Res.* **21**, 531–543.

Tomar, V. P. S. & Agarval, B. L. (1974) Effect of ganglionectomy on intraocular pressure and outflow facility of aqueous humour. *Exp. Eye Res.* **19**, 403–408.

Theobald, G. D. (1960) The limbal area. *Am. J. Ophthal.* **50**, 543–557.

Törnqvist, G. (1970) Effect of oculomotor nerve stimulation on outflow facility and pupil diameter in a monkey (*Cercopithecus ethiops*). *Invest. Ophthal.* **9**, 220–225.

Tower, S. (1940) Units for sensory reception in the cornea; with notes on nerve impulses from sclera, iris and lens. *J. Neurophysiol.* **3**, 486–500.

Tripathi, R. C. (1968) Ultrastructure of Schlemm's canal in relation to aqueous outflow. *Exp. Eye Res.* **7**, 335–341.

Tripathi, R. C. (1969) Ultrastructure of the trabecular wall of Schlemm's canal. *Trans. Ophthal. Soc., U.K.* **89**, 449–465.

Tripathi, R. C. (1971) Mechanism of the aqueous outflow across the trabecular wall of Schlemm's canal. *Exp. Eye Res.* **11**, 111–116.

Tripathi, R. C. (1974) Comparative physiology and anatomy of the aqueous outflow pathway. In *The Eye*. Vol. 5, chap. 3 (Ed. Davson, H. & Graham, L. T.) New York: Academic Press.

Tripathi, R. C. (1977) The functional morphology of the outflow systems of ocular and cerebrospinal fluids. *Exp. Eye Res.* **25**, (Suppl.) 65–116.

Troncoso, M. U. (1942) The intrascleral vascular plexus and its relations to the aqueous outflow. *Amer. J. Ophthal.* **25**, 1153–1162.

Unger, W. G., Perkins, E. S. & Bass, M. S. (1974) The response of the rabbit eye to laser irradiation of the iris. *Exp. Eye Res.* **19**, 367–377.

Unger, W. G., Cole, D. F. & Hammond, B. (1975) Disruption of the blood-aqueous barrier following paracentesis in the rabbit. *Exp. Eye Res.* **20**, 255–270.

Unger, W. G., Cole, D. F. & Bass, M. S. (1977) Prostaglandin and neurogenically mediated ocular response to laser irradiation of the rabbit iris. *Exp. Eye Res.* **25**, 209–220.

Ussing, H. H. & Zerahn, K. (1951) Active transport of sodium as the source of electric current in the short-circuited isolated frog skin. *Acta physiol. Scand.* **23**, 110–127.

Uusitalo, R., Palkama, A. & Stjernschantz, J. (1973) An electron microscopical study of the blood-aqueous barrier in the ciliary body and iris of the rabbit. *Exp. Eye Res.* **17**, 49–63.

Uusitalo, R., Stjernschantz, J. & Palkama, A. (1974) Studies on para-sympathetic control of the blood-aqueous barrier in the rabbit. An electron microscopic study. *Exp. Eye Res.* **19**, 125–134.

Vegge, T. (1971) An epithelial blood-aqueous barrier to horseradish peroxidase in the ciliary processes of the vervet monkey (*Cercopithecus aethiops*). *Z. Zellforsch.* **114**, 309–320.

Vegge, T. (1972a) A study of the ultrastructure of the small iris vessels in the vervet monkey (*Cercopithecus aethiops*). *Z. Zellforsch.* **123**, 195–208.

Vegge, T. (1972b) A blood-aqueous barrier to small

molecules in the ciliary processes of the vervet monkey (*Cercopithecus aethiops*). *Z. Zellforsch.* **135**, 483–499.

Vegge, T. & Ringvold, A. (1969) Ultrastructure of the wall of the human iris vessels. *Z. Zellforsch.* **94**, 19–31.

Vrábec, F. (1954) L'innervation du système trabéculaire de l'angle irien. *Ophthalmologica, Basel.* **128**, 359–364.

Waitzman, M. B. & King, C. D. (1967) Prostaglandin influences on intraocular pressure and pupil size. *Am. J. Physiol.* **212**, 329–334.

Waitzman, M. B. & Woods, W. D. (1971) Some characteristics of an adenyl cyclase preparation from rabbit ciliary process tissue. *Exp. Eye Res.* **12**, 99–111.

Weekers, R., Prijot, E. & Gustin, J. (1954) Mesure de la résistance à l'écoulement de l'humeur aqueuse au moyen du tonomètre électronique. *Ophthalmologica,* **128**, 213–217.

Weigelin, E. & Löhlein, H. (1952) Blutdruckmessungen an den episkleral Gefässen des Auges bei kreislaufgesund Personen. *v. Graefes' Arch. Ophthal.* **153**, 202–213.

Weinbaum, S., Langham, M. E. & Goldgraben, J. R. (1972) The role of secretion and pressure-dependent flow in aqueous humor formation. *Exp. Eye Res.* **13**, 266–277.

Wessely, K. (1900) Über die Wirkung des Suprarenins auf das Auge. *Ber. deutsch. ophth. Ges.* pp. 69–83.

Wessely, K. (1908) Experimentelle Untersuchungen ü. d. Augendruck, sowie über qualitative und quantitative Beeinflussung des intraokularen Flussigkeitswechsels. *Arch. Augenheilk.* **60**, 1–48, 97–160.

Wilke, K. (1974) Early effects of epinephrine and pilocarpine on the intraocular pressure and the episcleral venous pressure in the normal human eye. *Acta ophthal.* **52**, 231–241.

Wolff, E. (1968) *Anatomy of the Eye and Orbit* (revised by R. J. Last). London: H. K. Lewis.

Zimmerman, T. J., Garg, L. C., Vogh, B. P. & Maren, T. H (1976a) The effect of acetazolamide on the movements of anions into the posterior chamber of the dog eye. *J. Pharmacol.* **196**, 510–516.

Zimmerman, T. J., Garg, L. C., Vogh, B. P. & Maren, T. H. (1976b) The effects of acetazolamide on the movement of sodium into the posterior chamber of the dog eye. *J. Pharmacol.* **199**, 510–517.

# 2. The vitreous body

This is a transparent jelly occupying, in man and most vertebrates, by far the largest part of the globe. As Figure 1.1, p. 3, shows, it is bounded by the retina, ciliary body and the posterior capsule of the lens. If the sclera is cut through round the equator, the posterior half with the attached retina comes away from the vitreous body, leaving this suspended from its firm attachments to the ciliary body and pars plana of the retina. To isolate it completely from these attachments careful cutting and scraping are necessary.

## CHEMISTRY

When discussing the dynamic relationships between aqueous humour and the vitreous body it was considered sufficient to treat the latter as an essentially aqueous medium; and this is justified by the observation that over 99 per cent of the material in this body is water and dissolved salts. The special features of the inorganic composition have been considered earlier, and here we are concerned with the composition in so far as it bears on the structure and stability of the vitreous body as a gel. If the vitreous body is held in a pair of forceps, or placed on a filter, it drips and after a time only a small skin of viscid material remains; this residue was called by Mörner the *residual protein*, and was considered to constitute the gel-forming material, in essentially the same way that fibrin constitutes the gel-forming material of a plasma clot. The liquid dripping away, or passing through the filter, contains the dissolved crystalloids together with a mucopolysaccharide, *hyaluronic acid*, that imparts to the solution a high viscosity by virtue of its high molecular weight and highly asymmetrical molecules. In addition there is a small percentage of soluble proteins, some of which are peculiar to the vitreous whilst others are identical with plasma proteins (Laurent, Laurent and Howe, 1962). In particular, the glycoproteins, containing sialic acid, are in much higher concentration than in plasma. Figures given for the colloidal composition of bovine vitreous body are: Vitrosin, 150 mg/litre, Hyaluronic acid, 400 mg/litre and Soluble proteins 460 mg/litre. Thus the colloidal material constitutes less than 0·1 per cent of the total body.

## RESIDUAL PROTEIN OR VITROSIN

Purified residual protein may be dissolved in alkali and subsequently precipitated by acid; when the precipitated material is examined in the electron microscope it appears fibrillar and very similar to collagen; this similarity is borne out by chemical analysis which reveals the characteristically high proportions of glycine and hydroxyproline; it differs from that in skin in having a high proportion of sugars, probably in the form of a polysaccharide firmly attached to the polypeptide skeleton (McEwan and Suran, 1960); in this respect it is similar to the collagens derived from Bowman's membrane and the lens capsule. The enzyme collagenase will liquefy the vitreous body.

## STRUCTURE

Tentatively we may suppose that the structure of the vitreous gel is maintained by a framework of branching vitrosin fibrils, in the meshes of which the watery solution of crystalloids, hyaluronic acid and soluble proteins is imprisoned. The concentration of residual protein is so small (0·01 per cent) that the fibrils must be very fine, and the meshwork relatively coarse, if such a small amount of material is to form a coherent jelly.

### COLLAGEN FIBRILS

This picture, which may be deduced on purely physico-chemical principles, is partly sustained by modern electron-microscopical studies, although there are still many points of uncertainty. The fibrillar basis is certainly well established by the examination of thin sections of frozen vitreous; this is illustrated by Figure 2.1, from a paper by Fine and Tousimis which confirms earlier studies, particularly those of Schwarz, in this respect. As predicted, the fibrils are very thin, being some 20 to 25 nm in diameter, and of indefinite length.*

---

* Schwarz (1961) gives an even smaller value, namely 4 to 10 nm with an average of 6·7 nm, whilst Grignolo (1954) described three types with diameters ranging from 15 to 80 nm. Olsen's (1965) filaments had a diameter of 12 to 13 nm and those of Smith and Serafini-Fracassini 11 nm on average, the finest being 8 nm. The important point is the fineness compared with those in the cornea.

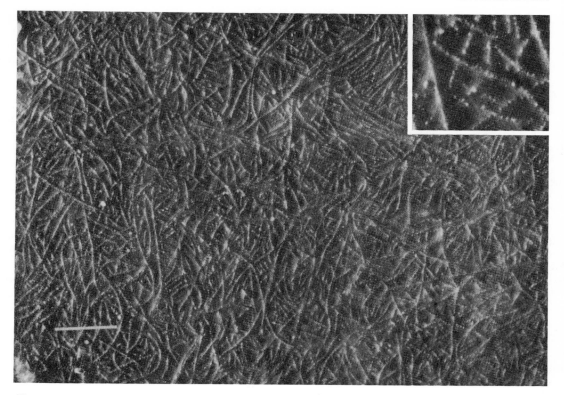

**Fig. 2.1**    Shadowcast section of posterior vitreous structure. The orientation of the filaments in the plane of the section suggests a tangential cut through a lamella. The filaments are uniform in width and show a distinct periodicity ( × 12 000). The enlarged inset ( × 38 000) demonstrates periodicity of filaments. (Fine and Tousimis, *Arch. Ophthal.*)

Olsen (1965) observed that some fibrils were made up of the parallel alignment of thin filaments, some 2 to 3 nm in diameter, and those could have represented chains of the fundamental unit of collagen, the tropocollagen molecule. The characteristic feature of collagen, prepared from most sources, is a cross-striation of period 64 nm; in general, the periodicity observed along the vitrosin fibrils is much smaller than this, namely 11 nm.

## PERIODICITY

The large period of 64 nm, usually observed in collagen, is due to the staggered side-by-side alignment of the individual tropocollagen molecules, so that at intervals certain polar groups come into register; these take up electron-stain and appear as dark bands on the large fibres made up of the aggregation of many of the fine fibrils. It is quite likely that the failure to observe the 64 nm banding is due to a less regular arrangement of filaments or, simply, that there are not sufficient in the very thin fibrils of vitrosin to render this periodicity visible; thus Smith and Serafini-Fracassini (1967) stated that, when these fibrils of diameter 14 nm were seen packed side-by-side, faint cross-bands of 64 nm periodicity were visible.

## HYALURONIC ACID

By treating vitreous preparations with hyaluronidase, which breaks down the large hyaluronic acid molecules into their constituent sugar molecules and uronic acid, the vitrosin fibrils become more distinct, suggesting that hyaluronic acid is normally in close association with the vitrosin; when Smith and Serafini-Fracassini stained the vitreous gel with lead, which precipitates the hyaluronic acid, they observed dark aggregates at regular intervals along the vitrosin fibres suggesting that, in the normal state, the hyaluronic acid was attached at regular points.

## LOCAL VARIATIONS IN STRUCTURE

The picture just presented, of a fibrous framework holding the fluid components of the vitreous in its meshes, would suggest that the vitreous body was essentially homogeneous throughout its bulk; a considerable amount of evidence indicates, however, the existence of differentiation to give microscopically visible structures; these are most probably super-imposed on the fundamental basis described above. To appreciate the significance of the differentiation we must review the embryonic history of the vitreous body.

## EMBRYONIC HISTORY

In the early stages, the optic cup is mainly occupied by the lens vesicle (p. 117); as the cup grows, the space formed is filled by a system of fibrillar material presumably secreted by the cells of the embryonic retina. Later, with the penetration of the hyaloid artery, more fibrillar material, apparently derived from the cells of the wall of the artery and other vessels, contributes to filling the space. The combined mass is known as *primary vitreous*. The *secondary vitreous* develops later and is associated with the increasing size of the vitreous cavity and the involution of the hyaloid system. The main hyaloid artery remains for some time but eventually disappears, and its place is left as a tube of primary vitreous surrounded by secondary vitreous; this tube is called *Cloquet's canal*, but is not a liquid-filled canal but simply a portion of differentiated gel. The *tertiary* vitreous is the name given to the grossly fibrillar material secreted by the inner walls of the optic cup in the region that will later become the ciliary body; this material eventually becomes the zonule of Zinn or suspensory ligament of the lens.

## PHASE-CONTRAST MICROSCOPY

When examined in the ordinary light-microscope the vitreous body is said to be optically empty; in the phase-contrast microscope, however, Bembridge, Crawford and Pirie described three characteristic differentiations; in the anterior portion definite fibres of relatively large diameter were found, whilst in the bulk of the body there were much finer, but still optically resolvable, fibres. Chemically the large fibres seemed to be different from the fine fibres; thus the enzyme collagenase left these large fibres intact whilst it dissolved the fine fibres; again, the enzyme trypsin or chymotrypsin dissolved the large fibres but not the others. Finally, on the surface, there appeared to be a hyaline membrane (Fig. 2.2). Since phase-contrast microscopy is carried out on perfectly normal fresh tissue it is difficult to rule out these structures as artefacts; nevertheless the electron-microscopical observations, indicating a fibril of the order of 10 to 20 nm by contrast with these optically visible fibres of the order of 1 to 2 $\mu$m, i.e. 1000 to 2000 nm, indicate that the fundamental basis of structure is probably the fine fibril, whilst the larger fibres must be regarded as essentially localized densifications of this primary structure. This is especially true of the zonular fibres proper; electron-microscopical observations of the vitreous in this region indicate large fibres made up of closely aggregated fine filaments, whilst the neighbouring vitreous shows only the fine filaments loosely scattered.

## LIMITING MEMBRANE

Again, in the electron-microscope there is little evidence for a structured limiting hyaline membrane belonging to the vitreous body; according to Tousimis and Fine the *internal limiting membrane* of the retina belongs to this tissue and cannot be regarded as a limiting membrane of the vitreous; when the vitreous is removed the membrane remains adherent to the cells of the retina

**Fig. 2.2** Hyaline membrane of vitreous. (Courtesey of Antoinette Pirie.)

and must be regarded as a basement membrane for these cells; as seen in the electron-microscope it is an acellular structure some 300 to 500 nm thick, and is essentially the basement membrane of the glial cells of the retina, being apparently continuous, embryologically, with the basement membrane of the ciliary epithelium, whose cell-bases face towards the vitreous body. It extends from the surface of the plasma membranes of the glial cells to the surface of the cortical layer of vitreous, which itself is covered by a delicate basal lamina made of the same basement-membrane material. The vitrosin fibrils run parallel with the surface, but make right-angle turns to connect with the basement membrane; the extent to which this happens varies with position, being greatest at the points of strong attachment, such as the ora serrata (Zimmerman and Straatsma, 1960).*

### CORTICAL ZONE

We must regard the hyaline layer, then, as a region in which the fibrils are more densely arranged, so as to give a sufficient difference of refractive index to create the appearance of a membrane in phase-contrast micro-scopy. This so-called *cortical zone* has been studied by Balazs who has emphasized the presence of numerous stellate cells buried in it; he has called these *hyalocytes* since he considers that they are responsible for synthesis of hyaluronic acid. Gärtner (1965) has made a careful study of the cells in the cortical layer in the rat embryo and young child; he characterizes the great majority of these cells as fibroblasts; in the region of the pars plana of the ciliary body they form a definite layer covering the epithelium; and the appearance of collagen-like fibrils in close association suggested that it was these cells that were responsible for secretion of vitrosin as well as hyaluronic acid.

### CHEMICAL DIFFERENCES

There is chemical evidence for some differentiation of the vitreous body; for example Bembridge *et al.* found that the proportions of vitrosin varied, being highest in the anterior portion where also the electron-microscopists had differentiated the highest density of fibrils. Again Balazs found a very high concentration of hyaluronic acid and vitrosin in the cortical layer, corresponding with its more compact character; finally the concentrations of acid glycoproteins and water-soluble proteins vary with the region analysed.†

### HYALURONIDASE

This enzyme, which breaks hyaluronic acid down to its constituent amino sugar and hexuronic acid molecules, occurs naturally, for example in tears, where it is called *lysozyme* (Sect. III). It was considered to be present in the intraocular tissues where it might exert an influence

on the fluidity of the vitreous body; recent studies have, however, ruled this out (Berman and Voaden, 1970).

### ORIGIN AND MAINTENANCE

It is customary to describe the vitreous as the product of secretion of ectodermal elements, namely the glial cells of the retina and the ciliary epithelium. The point has been discussed at some length by Gärtner (1965) in the light of his electron-microscopical studies; and he favours a mesenchymal origin for the primary vitreous and a connective tissue origin for the secondary vitreous. The tertiary vitreous, i.e. the zonule, he unequivocally attributes to fibroblasts, and his experimental evidence and arguments are convincing.

The eye reaches its full size within the first few years of life in man; after this there is no necessity for synthesis of new material for the vitreous body, unless there is an appreciable wear-and-tear. The problem as to whether vitreous may be replaced after loss is of considerable interest to the opthalmologist, but experimental and pathological studies provide little evidence in favour of this. It seems that the hyalocytes, separated from the vitreous by centrifugation, can incorporate radioactive precursors of hyaluronic acid, e.g. glucosamine, into high-molecular weight compounds, and it is interesting that acellular parts of the vitreous also have activity, suggesting the presence of extracellular enzymes. Studies on the intact animal, e.g. those of Österlin in the owl monkey, indicate that, whatever turnover there is in the hyaluronic acid moiety of the

---

*These regions of attachment have been examined with especial care by Hogan (1963); thus the vitreous fibrils attaching to the ciliary epithelium enter crypts and make contact with the basement membrane at the edges and depths of the crypts along with zonular fibres where they form a strong union. In eyes with vitreal detachment, the cortex of the vitreous separates from the basement membrane of the glial cells; the separation extends as far forward as the peripheral third of the retina and seems to be limited by the radial attachments of the vitreous fibrils to the retina in this area.

†During development there are marked changes in composition; for example the concentration of hyaluronic acid steadily increases. Where different species are compared some very large differences are found; for example the vitreous of the bush-baby (*Galago*) and of the owl monkey (*Aotus*) is liquid except for a thin 50 to 100 $\mu$ thick cortical layer; it is this relatively solid cortex, which contains all the collagen of the vitreous, that acts as a barrier to flow between the posterior chamber and the fluid vitreous. Thus on removing the aqueous humour by paracentesis, vitreous is not withdrawn; the fluid material contains most of the hyaluronic acid (Österlin and Balazs, 1968). By contrast the vitreous bodies of the frog and hen are very solid, and have very little hyaluronic acid and large amounts of collagen (Balazs *et al.*, 1959a,b).

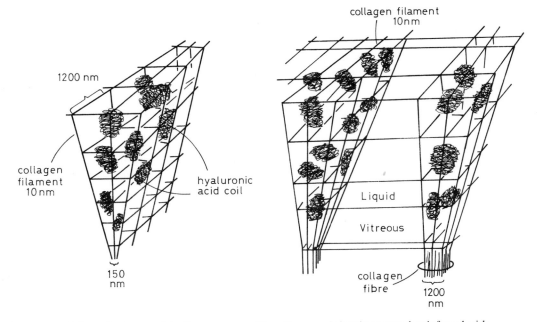

**Fig. 2.3** Schematic picture of the fine structure of the vitreous gel showing network reinforced with hyaluronic acid molecules. *Left :* Random distribution of the structural elements. *Right :* Formation of liquid pool and partial collapse of the network. (Balazs, in *Structure of the Eye.*)

vitreous, it is very slow (Österlin, 1968; Berman and Gombos, 1969).

## PROBABLE STRUCTURE

Modern studies on connective tissue indicate that both the collagen and mucopolysaccharide constituents contribute to the stability of the gel-structure that constitutes its basis. In the vitreous body the mucopolysaccharide is hyaluronic acid,* a polysaccharide constituted by alternating acetyl glycosamine and glucuronic acid molecules, as indicated on p. 95. The molecular weights of such polysaccharides may be very large, of the order of a million, and in the case of the vitreous it would seem that the material exists in the form of a variety of different sized molecules, i.e. different degrees of polymerization. The molecule probably exists as a randomly kinked coil entraining relatively large quantities of water in its framework so that in solution it occupies some 1000 times the volume it occupies in the dried state. The individual molecules probably entangle with each other to form a continuous three-dimensional network. In the vitreous body, then, such a network might act to stabilize the primary network formed by the vitrosin fibrils, as illustrated in Figure 2.3. As early postulated by Davson and Duke-Elder (1935), and later by Pirie, Schmidt and Waters (1948), the effects of environmental changes on the state of the vitreous body are largely exerted through the

hyaluronic acid moiety; and the same seems to be true of the cornea (p. 97 footnote).

### TRANSPORT THROUGH GEL

The gel structure formed in this way is doubtless adequate to prevent significant bulk flow of fluid through it, so that any posterior flow of aqueous humour, if it occurs, would take place meridionally (Fowlks, 1963) and not through 'Cloquet's canal'. Diffusion of molecules would, however, be quite unrestricted, so that from the point of view of penetration of material across the blood-vitreal barrier the vitreous may be treated as an unstirred liquid; this applies to the movement of such large molecules as albumin too (Maurice, 1959); it has been suggested that the more tightly packed cortical region offers some restraint to entry of colloidal material, e.g. from the retina, as indicated above, but the diffusion of protein within the much denser cornea casts some doubt on this.

### TRANSPARENCY

Finally, we may note that the presence of solid fibrils in the vitreous in no way impairs its transparency; this is presumably because the fibrils are thin and their

---

*An acidic glycosaminoglycan containing galactosamine and sulphate has been described in human vitreous by Breen *et al.* (1977); this appears to be covalently bound to the collagen.

concentration is low (p. 82). Pathological aggregates of colloidal material would cause opacities.

## THE ZONULE

### MACROSCOPIC APPEARANCE

The zonule, or suspensory ligament of the lens, may be regarded as an extreme instance of differentiation of the vitreous body, its fibrous parts being composed presumably of the same material as that making up the large fibres of the vitreous body, described by Bembridge *et al.*, since they are both dissolved by the enzyme trypsin; the difference between the two is essentially that the zonular fibres have found attachment to the lens capsule whilst the coarse fibres of the vitreous attach to the ciliary body. In the fresh specimen the zonule appears as a thick band-shaped structure attached internally to the equatorial region of the lens capsule; the anterior surface runs straight from the lens to the anterior region of the ciliary processes (Fig. 1.5, p. 6). The surfaces of the zonule show a marked striation running in a general sense from the lens to the ciliary body; these striations probably represent fibres of

highly condensed residual protein; between them is a perfectly homogeneous transparent gel which gives the surface a membrane-like appearance. The space between the anterior and posterior surfaces is crisscrossed by strands of fibrous material, the spaces so formed being once again filled by transparent homogeneous jelly.

### CHEMISTRY

As indicated above, the zonular fibres are attacked by trypsin but not by collagenase, and this indicates a difference in chemical composition, which has been confirmed by the chemical analyses of Buddecke and Wollensak (1966), who found negligible amounts of hydroxyproline and only a small percentage of glycine—amino acids that are present in characteristically high concentration in collagen. There was a remarkably high proportion of cysteine (cystine) in the zonular protein, and this may have pathological interest for the Marfan syndrome associated with homocystinuria, i.e. the incomplete formation or tearing of the zonular fibres leading to ectopia lentis. The extracted material contained some 5 per cent of carbohydrate which was not attacked by hyaluronidase, and it may be that the

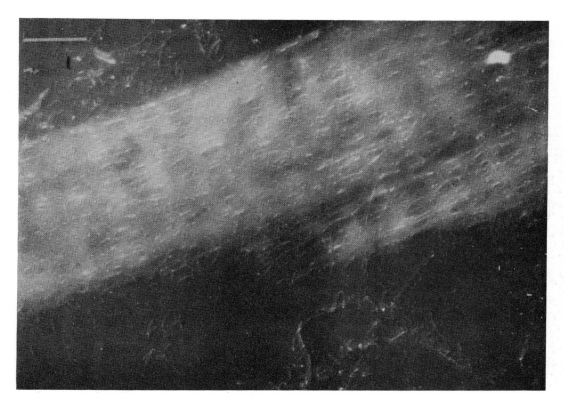

**Fig. 2.4**   Shadowcast section of adult human zonule and adjacent vitreous body framework. The widths of the filaments of both zonular and adjacent framework are similar; also the periodicities of the filaments (22 nm) ( × 13 000). (Fine and Tousimis, *Arch. Ophthal.*)

polysaccharide acts as a cover to the fibrils, protecting them against the attack of proteases in the anterior chamber. Thus in old people there is less of the polysaccharide, and it may be for this reason that their zonules are relatively easily attacked by trypsin, a procedure that is employed in cataract surgery.

ELECTRON-MICROSCOPY

In the electron microscope the zonular fibres appear as dense aggregations of the fine fibrils that constitute the framework of the surrounding vitreous (Fig. 2.4). Preparation of the ciliary epithelium made by Pappas and Smelser show that the fibres may be regarded as extensions of the basement membrane of the epithelial cells, which is itself fibrillar in nature. Thus the basement membrane invests the surface of the ciliary body with its processes very closely, so that the attachment of the zonular fibres to the ciliary body may be said to be similar to the 'attachment of a tight-fitting glove on one's hand', the mere close approximation of the two membranes—epithelial cell membrane and basement membrane—on an irregular contour, giving a very efficient attachment. No fibrils of the zonule entered the epithelial cells.

# REFERENCES

Balazs, E. A. (1961) Molecular morphology of the vitreous body. In *The Structure of the Eye*. Ed. Smelser, pp. 293–310. New York: Academic Press.

Balazs, E. A. (1968) Die Mikrostruktur und Chemie des Glaskörpers. *Ber. deutsch. ophth. Ges.* **68,** 536–572.

Balazs, E. A., Laurent, T. C. & Laurent, U. B. G. (1959a) Studies on the structure of the vitreous body. VI. *J. biol. Chem.* **234,** 422–430.

Balazs, E. A., Laurent, T. C., Laurent, U. B. G., De Roche, M. H. & Dunney, D. M. (1959b) Studies on the structure of the vitreous body. VIII. *Arch. Biochem. Biophys.* **81,** 464–479.

Bembridge, B. A., Crawford, C. N. C. & Pirie, A. (1952) Phase-contrast microscopy of the animal vitreous body. *Br. J. Ophthal.* **36,** 131–142.

Berman, E. R. & Gombos, G. M. (1969) Studies on the incorporation of U-$^{14}$C-glucose into vitreous polymers *in vitro* and *in vivo*. *Invest. Ophthal.* **8,** 521–534.

Berman, E. R. & Voaden, M. (1970) The vitreous body. In *Biochemistry of the Eye*. Ed. C. Graymore, pp. 373–471. London: Academic Press.

Breen, M., Bizzell, J. W. & Weinstein, H. G. (1977) A galactosamine containing proteoglycan in human vitreous. *Exp. Eye Res.* **24,** 409–412.

Buddecke, E. & Wollensak, J. (1966) Zur Biochemie der Zonulafaser des Rinderauges. *Z. Naturf.* **21b,** 337–341.

Davson, H. & Duke-Elder, W. S. (1935) Studies on the vitreous body. II. *Biochem. J.* **29,** 1121–1129.

Fine, B. S. & Tousimis, A. J. (1961) The structure of the vitreous body and the suspensory ligaments of the lens. *Arch. Ophthal., Chicago* **65,** 95–110.

Fowlks, W. L. (1963) Meridional flow from the corona ciliaris through the pararetinal zone of the rabbit vitreous. *Invest. Ophthal.* **2,** 63–71.

Gärtner, J. (1965) Elektronemikroskopische Untersuchungen über Glaskörperrindenzellen und Zonulafasern. *Z. Zellforsch.* **66,** 737–764.

Hogan, M. J. (1963) The vitreous, its structure, and relation to the ciliary body and retina. *Invest. Ophthal.* **2,** 418–445.

Laurent, U. B. G., Laurent, T. C. & Howe, A. F. (1962) Chromatography of soluble proteins from the bovine vitreous body on DEAE-cellulose. *Exp. Eye Res.* **1,** 276–285.

McEwen, W. K. & Suran, A. A. (1960) Further studies on vitreous residual protein. *Am. J. Ophthal.* **50,** 228–231.

Maurice, D. (1959) Protein dynamics in the eye studied with labelled proteins. *Am. J. Ophthal.* **47,** Pt. 2, 361–367.

Mörner, C. T. (1894) Untersuchungen der Proteinsubstanzen in den lichtbrechenden Medien des Auges. *Hoppe-Seyl. Z.* **18,** 61–106.

Olsen, B. R. (1965) Electron microscope studies on collagen. IV. *J. Ultrastr. Res.* **13,** 172–191.

Österlin, S. E. (1968) The synthesis of hyaluronic acid in vitreous. III. *Exp. Eye Res.* **7,** 524–533.

Österlin, S. E. & Balazs, E. A. (1968) Macromolecular composition and fine structure of the vitreous in the owl monkey. *Exp. Eye Res.* **7,** 534–545.

Pappas, G. D. & Smelser, G. K. (1958) Studies of the ciliary epithelium and the zonule. I. *Am. J. Ophthal.* **46,** Pt. 2, 299–318.

Pirie, A., Schmidt, G. & Waters, J. W. (1948) Ox vitreous humour. 1. The residual protein. *Br. J. Ophthal.* **32,** 321–339.

Schwarz, W. (1956) Elektronenmikroskopische Untersuchungen an Glaskörperschnitten. *Anat. Anz.* **102,** 434–442.

Smith, J. W. & Serafini-Fracassini, A. (1967) The relationship of hyaluronate and collagen in the bovine vitreous body. *J. Anat.* **101,** 99–112.

Zimmerman, L. E. & Straatsma, B. R. (1960) In *Importance of the Vitreous Body in Retina Surgery with Special Emphasis on Reoperations*. Ed. C. L. Schepens, pp. 15–28. St. Louis: Mosby.

# 3. The cornea

## STRUCTURE

The cornea is illustrated by the meridional section in Figure 3.1; the great bulk (up to 90 per cent of its thickness) is made up of the *stroma* which is bounded

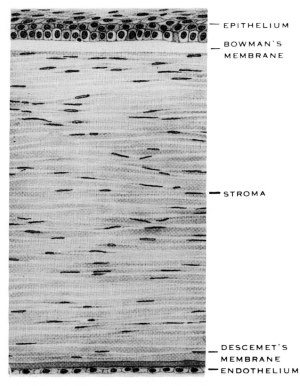

— EPITHELIUM

BOWMAN'S
MEMBRANE

— STROMA

DESCEMET'S
MEMBRANE
— ENDOTHELIUM

**Fig. 3.1** Meridional section through the human cornea. (Wolff, *Anatomy of the Eye and Orbit*.)

externally by Bowman's membrane and the epithelium, and internally by Descemet's membrane and the endothelium. The total thickness in man is just over 0·5 mm in the central region; towards the periphery it becomes some 50 per cent thicker.

## EPITHELIUM

This may be regarded as the forward continuation of the conjunctiva; it is a squamous epithelium consisting of some five or six layers of cells with a total thickness of 50 to 100 $\mu$m. The cells at the base are columnar, but as they are squeezed forward by new cells they become flatter so that three layers are recognized: *basal, wing-shaped*, and *squamous cells*. The more superficial cells do not show keratinization, as with skin, the process of shedding being accompanied by the breaking up of the cell into fragments (Teng). Replacement of cells occurs by mitotic division of the basal layer, the average life of a cell being some 4 to 8 days (according to Hanna and O'Brien). Junctional complexes between cells in the various layers seem to be mainly of the desmosome type, not involving fusion of adjacent membranes. Nevertheless, because the permeability of the corneal epithelium as a whole is low, we must postulate tight junctions or zonulae occludentes between the cells of at least one layer, possibly the basal layer; according to Pedler (1962) these cells show extensive lateral interdigitation of plasma membranes. The surface layer of cells may be examined by the scanning electron microscope; viewed in this way the anterior surfaces of this outermost layer are covered by villous projections (Blümcke and Morgenroth, 1967). Amongst the cells a special, possibly secretory, type has been described by Teng; it may be that it secretes the basement membrane upon which the epithelium rests; this latter is a layer some 10 to 30 nm thick resolved into an anterior apparently lipid layer and a reticular fibre meshwork in contact with Bowman's membrane.

### REGENERATION

When the epithelium is stripped off from the underlying Bowman's membrane there is a rapid regeneration, so that within twenty-four hours the surface may be completely covered by a new layer of regenerated epithelial cells derived from multiplication of the conjunctival cells at the periphery (Fig. 3.2). This single layer subsequently develops to give, within several weeks, a complete epithelium 4 to 6 cells thick. According to Thoft and Friend (1977), the epithelial cells making up this new epithelium retain the biochemical characteristics of conjunctival cells. Smaller lesions are repaired by sliding of the adjacent epithelial cells into the denuded region. Thus Kuwabara *et al.* (1976) made small cuts into the surface of the cornea and examined the behaviour of the adjacent epithelial cells over the subsequent several days. The cells begin to slide into the wound about one hour after the event, the main sliding cells being the wing cells, although the flat superficial (squamous) cells regain the cytoplasmic constituents of the younger cell and slide in a similar

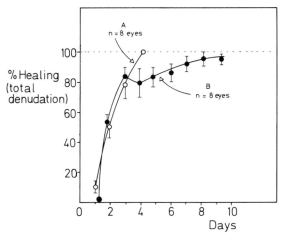

**Fig. 3.2**   Re-epithelialization of the completely denuded cornea either by knife blade (B) or by a rough stone (A). The response to denudation by knife was apparently biphasic. (Srinivasan *et al.*, *Exp. Eye Res.*)

fashion. The basal cell also participates, becoming flatter and sliding less. The tight interdigitations between cells are broken down but the cells are held together by desmosomes of extended cytoplasm and never become entirely free of attachments to underlying cells. The cells that have slid on to the surface of the stroma attach firmly to the adjacent cells, and some more distant cells slide over these recently attached cells. The cells on the surface of the denuded area begin to increase in height within 24 hours and acquire the appearance of basal cells. By the third day a basement membrane appears. Kuwabara *et al.* showed that the same sliding would occur in excised cornea *in vitro*.

**Epithelium vs. conjunctiva**

As to what proportion of corneal cells must remain after injury for the epithelium to be replaced by corneal cells rather than conjunctival ones, Maumenee and Scholz (1948) stated that as little as 5 to 10 per cent of residual uninjured epithelium was sufficient. It is not clear from the lesions of Pfister and Burstein (1976), which extended to a diameter of 12 mm, whether these were replaced by conjunctival, corneal, or both types of epithelium. When the new cells are of conjunctival origin, i.e. after complete denudation of Bowman's membrane, then they gradually acquire the morphology of epithelial cells although they may retain their con-

junctival biochemistry, e.g. relatively low concentration of glycogen, for more than six weeks (Thoft and Friend, 1977).

## STROMA AND BOWMAN'S MEMBRANE

### THE LAMELLAE AND FIBRILS

In meridional section the stroma appears as a set of lamellae, running parallel with the surface, and superimposed on each other like the leaves of a book; between the lamellae lie the *corneal corpuscles*, flattened in the plane of the lamellae; they are equivalent to fibrocytes of other connective tissue. The lamellae are made up of microscopically visible *fibres*, which run parallel to form sheets; in successive lamellae the fibres make a large angle with each other. In lower animals the lamellae extend over the whole surface of the cornea and the fibres in adjacent laminae are at right-angles to each other; in mammals and man the lamellae are said to be made up of bands some 90 to 260 $\mu$m wide; there is apparently no interweaving of these bands, however, so that essentially the structure consists of layers of fibres arranged parallel with themselves in any given lamella.

The lamellae in mln are some 1·5 to 2·5 $\mu$m thick, so that there are at least 200 lamellae in the human cornea. The fibrous basis of the lamellae is made up of collagen which constitutes some 80 per cent of the solid matter of the cornea; by submitting the cornea to ultrasonic vibrations in order to break down its structure into its elementary components, Schwarz (1953) observed typical collagen fibrils of 25 to 33 nm diameter, and having the characteristic 64 nm periodicity (p. 83); the fibrils isolated in this way were covered by a dense layer of cement substance which was presumably the mucopolysaccharide. The fibrils were remarkably homogeneous in regard to diameter, in marked contrast to those of the sclera where the diameter varied from 28 to 280 nm, and where the amount of ground substance was very much less. As we shall see, this uniformity in diameter is important for the transparency of the cornea by contrast with the opacity of the sclera. With the development of thin-sectioning techniques, the fibrillar structure of the individual laminae was firmly established, e.g. by Jakus (1961), the fibrils running parallel with each other in any lamina and at right-angles to those in adjacent laminae. This regularity is clear in the electron micrograph of Figure 3.3. The stromal cells in the human cornea appeared to lie within, rather than between, the laminae.

### COMPARISON WITH SCLERA

The basic difference in structure between cornea and sclera consists in the much larger and more variable diameters of the scleral collagen fibrils and their much

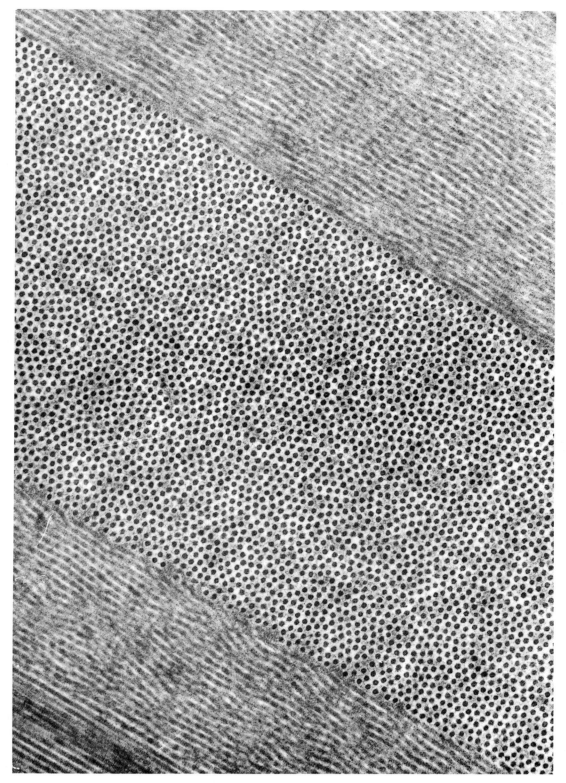

**Fig. 3.3**   Electron micrograph of normal human cornea. Note regularity of the fibres. (Giraud *et al.*, *Exp. Eye Res.*) (Courtesy of Pouliquen.)

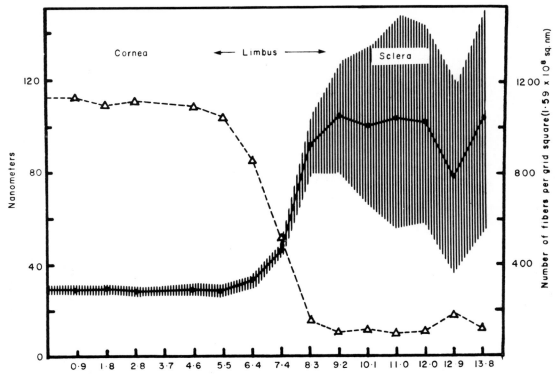

**Fig. 3.4** Comparison of the average number of collagen fibres per unit area ($\triangle$- - -$\triangle$) with the mean fibre diameter ($\bullet$—$\bullet$) $\pm 1$ s.d. (shaded area). Note steepness of change at limbus. (Borcherding *et al.*, *Exp. Eye Res.*)

less dense packing. These differences are strikingly shown by Figure 3.4, of interest is the transition zone, or limbus, where the changes occur very steeply. The regularity of fibre diameter in the cornea, and its gradual deterioration on passing through the limbus, are also manifest by the increasing size of the standard deviation shown by the vertical lines. The mean diameter in the central cornea found in this study (Boerscheding *et al.*, 1975) was 29·6 nm; in the limbal areas adjacent to the sclera the size was $105.4 \pm 36·5$ nm and in the sclera proper the mean was about the same.

BOWMAN'S MEMBRANE

Immediately next to the epithelium, the classical histologists described a layer with no resolvable structure, some 8 to 14 $\mu$m thick in man—*Bowman's membrane*★ (Fig. 3.1). In the electron-microscope it became evident that Bowman's membrane is only a less ordered region of the stroma; thus, according to Jakus, and Kayes and Holmberg, the stroma immediately beneath the basement membrane of the epithelium consists of collagen fibrils closely but randomly packed into a felt-like layer which is not sharply differentiated from the remainder of the stroma beneath it. It should

therefore be described as *Bowman's layer* rather than Bowman's membrane.

DESCEMET'S MEMBRANE

This is some 5 to 10 $\mu$m thick in man and is revealed as a characteristic differentiation of the stroma in the electron-microscope, although in man the organization is not so precise as in the ox cornea. The collagen is different, having an axial period of 117 nm in place of the usual banding at intervals of 64 nm, and the carbohydrate is also different but similar chemically to that in the lens capsule (Dohlman and Balazs). Descemet's membrane is secreted by the endothelium.

THE ENDOTHELIUM

This is a layer of flattened cells some 5 $\mu$m high and 20 $\mu$m wide. In the light-microscope the cells may be seen to form a regular mosaic (Fig. 3.5) with perfect contact between them.

★In many lower mammals, such as the rabbit and mouse, Bowman's membrane is absent, so that only the basement membrane, which is common to all species, separates the epithelium from the stroma proper.

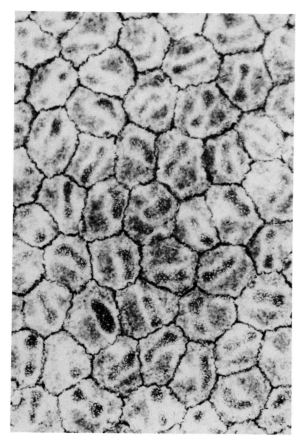

**Fig. 3.5** Aspect of normal rabbit cornea after silver impregnation. The cells form a regular mosaic, with perfect contact between them. (Hirsch *et al.*, *Exp. Eye Res.*) (Courtesy of Pouliquen.)

INTERCELLULAR JUNCTIONS

As we shall see, the endothelium acts as a permeability barrier capable of actively transporting ions, so that on physiological grounds we should expect the individual cells to be closely apposed, with their adjacent membranes forming junctional complexes comparable with those in the ciliary epithelium for example. When colloidal materials were introduced into the anterior chamber, such as $ThO_2$ or saccharated iron oxide, these failed to pass into the stroma, being blocked at the intercellular junctions, what little material appearing within the stroma being carried in by engulfment in vesicles (endocytosis) formed by the plasma membranes of the endothelial cells (Kaye and Pappas, 1962; Iwamoto and Smelser, 1965). The use of gross particles such as those of $ThO_2$ gives very little idea of the type of junction between adjacent endothelial cells, beyond indicating whether the spaces between the cells are large with no sealing off by fusion of membranes to form a zonula occludens (p. 27). Kaye *et al.* (1973) found

that horseradish peroxidase completely filled the clefts between endothelial cells and could be seen in the stroma after injection into the anterior chamber; this is a smaller molecule (40 000 mol. wt.) but a membrane permitting ready passage of this would be highly permeable unless the regions through which it could pass were very strictly limited.

**Gap-junctions**

Subsequent work, notably that of Leunberger (1973), of Kreutziger (1976) and Hirsch *et al.* (1977) has established that the main, or most significant, type of junction between adjacent endothelial cells is of the 'gap-type' representing a close apposition of adjacent membranes leaving a gap of about 30 Å or 3 nm. These junctions are found in both the apical and lateral regions of the cells.* The special feature of these junctions was that peroxidase, which would be unable to pass directly through the 3 nm-wide junction, does in fact infiltrate past it indicating that the junction does not form a complete belt round the cells, but is incomplete so that in limited regions the marker can pass through.

CELL DIVISION

Division of endothelial cells only occurs during growth and this is mitotic, in spite of claims to the contrary (v. Sallmann, Caravaggio and Grimes). During growth, multiplication of cells does not keep pace with the increase in area, so the cells become progressively flatter (Laing *et al.*, 1976).

REGENERATION

After damage to the endothelium the intact cells at the edge of the damaged region begin to migrate over those cells that have been destroyed, whilst an intense regeneration, or 'mitosis', area develops, providing new cells with which to cover the denuded Descemet's membrane. Experimentally a localized destruction of the endothelium may be achieved by application of a freezing probe to the cornea; this leads to complete destruction of the endothelium covering the frozen region.

**Morphological changes**

Hirsch *et al.* (1975) have described in detail the successive stages in regeneration of the rabbit's endothelium after contact for 10 sec with a probe of 1 cm diameter at the temperature of liquid nitrogen. They described during regeneration three

---

*Hirsch *et al.* (1977) have employed the freeze-fracture technique for elucidating the nature of the junctions between endothelial cells; in general, their pictures confirm that the junctions, both apical and lateral, are of the gap-type that permit free passage of solute and water between the cells and also provide some coupling between cells, so that they behave as a syncytium.

zones, namely the outermost *normal ring* that had not been frozen, a *zone of regeneration*, subdivided into a peripheral *transition zone* and a *front of intense regeneration* closest to the centre, and finally the *central zone* without covering. Twenty-four hours after the lesion (Day 1) these zones are easily distinguished; thus the transition zone is narrow but there are many cells showing complete regeneration, as indicated by the normal intercellular spaces. The regeneration front consisted of cells that had no established contact with each other, fusiform and stellate in shape with long protoplasmic protrusions and frequent mitosis. By Day 2 the transition zone had widened as well as the regeneration front, and by Day 4 Descemet's membrane was completely covered, with only the transition zone left in the central region. Finally by Days 5 to 6 the endothelium was apparently normal, with no mitoses and a hexagonal arrangement of the flat cells. However the number of cells per unit area was less than normal.*

## Relation to corneal thickness

Khodadoust and Green (1976) followed the regeneration of the endothelium, and related this to the deturgescence of the cornea (p. 98). Thus, after the damage, the cornea above the damaged endothelium swelled; although the damaged area was completely covered after 3 days, the corneal thickness was nearly twice normal, but as the number of cells in the denuded area increased over Days 3 to 8, oedema decreased, indicating that the large flattened cells that first replace the old endothelium are inadequate to function completely; only when the number has increased to permit a less flattened morphology is full physiological function possible. In a similar study, Hirsch *et al.* (1976) concentrated attention on the cellular contacts developing between the regenerating cells; after five days the junctions between adjacent cells are indistinguishable in the electron microscope from those in the normal cornea, i.e., they are of the gap-type and located both near the apex and laterally; application of horseradish peroxidase reveals the typical failure of the marker to fill the gap-junction, but regions of the intercellular cleft on each side of the junction are filled with the tracer. Stages in regeneration, as seen in the light-microscope, are shown in Figure 3.5 (p. 93).

## Primate cornea

It is generally recognized that the regenerative capacity of the rabbit's endothelium is greater than that of the primate, so that experimental studies on this class are preferable if results are to be extended to man. Van Horn and Hyndiuk (1975) made freezing lesions in monkey corneas, which led to denudation of Descemet's membrane over the frozen area; this was accompanied by rapid doubling of corneal thickness; as time progressed the cornea regained its normal thickness, indicating endothelial repair, but there was very little evidence of cell division so that the new endothelium was apparently created by migration and enlargement of the

uninjured more peripheral endothelial cells. The return to normal thickness was never complete, the value stabilizing at 0.6 to 0.7 mm compared with a normal thickness of 0.45 to 0.50 mm; thus the new endothelium was inadequate to support full activity of the fluid-pump. A similar failure is seen in aphakic bullous keratopathy when, due to trauma during cataract extraction, the endothelium is severely damaged (Dohlman and Hyndiuk, 1972).†

### CHANGES WITH AGE

Laing *et al.* (1976) employed the clinical specular microscope to photograph the endothelial cells in the living human eye; there was a pronounced increase in the area of individual cells with age, indicating a flattening and reduced total number.

## CHEMISTRY

The cornea contains some 22 per cent of solid matter, and is thus quite obviously built on a more rigid plan than the vitreous body; this solid matter is mainly collagen,‡ but in addition there are mucopolysaccharides, other proteins than collagen, and crystalloids of which salts make up the bulk.

Figures given by Maurice (1969) are as follows:

|  | per cent |
|---|---|
| Water | 78.0 |
| Collagen | 15.0 |
| Other proteins | 5.0 |
| Keratan sulphate | 0.7 |
| Chondroitin sulphate | 0.3 |
| Salts | 1.0 |

*Van Horn *et al.* (1977) have compared endothelial regeneration, after transcorneal freezing, in rabbit and cat; in the latter there was little regenerative activity so that the response was much more similar to that of primates than to that of the rabbit.
†Quoted by Van Horn and Hyndiuk (1975).
‡Collagen's molecule is built up of a triple helix with characteristic polypeptide chains making up the individual constituents of the helix; when different collagens are studied it is found that the chains are not always the same, and four distinct molecular species have been described. Thus type I collagen contains two identical polypeptide chains called α1 type I and a third called α2; this collagen is present in skin, bone and tendon. A second species contains three identical chains called α1 type II, and is present in a variety of cartilages; and so on. Trelstad and Kang (1974) have examined the collagens of lens, vitreous body, cornea and sclera from the chicken; sclera and cornea had the same collagen α1(I) α2 whilst vitreous collagen appeared to consist of entirely α1 chains, like that of cartilage.

# MUCOPOLYSACCHARIDES

This is the name commonly given to a group of polysaccharides containing an amino sugar, and widely distributed throughout the animal body where they may act as lubricants in the joints and saliva, as the anticoagulant called heparin, as blood-group substances on the surface of red cells, as ground substance of connective tissue, as cement between cells, and so on. Strictly speaking the term mucopolysaccharide should be applied, however, only to the combination between this type of polysaccharide and a protein, the reaction of the complex being predominantly polysaccharidic, whilst the term *muco-protein* is applied to a complex between protein and carbohydrate where the behaviour is predominantly protein in character.★

### HYALURANIC ACID

Perhaps the best known of the polysaccharides is *hyaluronic acid*, already described in relation to the vitreous body; it consists of large asymmetrical molecules (weight of the order of a million or more) made up by the polymerization of N-acetylglucosamine and glucuronic acid, linked together as follows:

Hyaluronic acid

Hyaluronic acid is specifically broken down by the enzyme, *hyaluronidase*, isolated from streptococci and pneumococci. Chondroitin sulphate was first found in the cornea but also occurs in cartilage; it is built up by the polymerization of N-acetylgalactosamine, glucuronic acid and sulphate, as below:

Chondroitin sulphate

### SULPHATED GLYCANS

Chondroitin is the desulphated chondroitin sulphate, whilst the latter has been found in three different forms, called chondroitin sulphate A, B and C, the differences being not so much due to differences in the sugar moieties as in the nature of the linkages.

### HISTOCHEMISTRY

Histochemically these polysaccharides are identified by their so-called *metachromatic staining* with toluidine blue, whereby the stain acquires a different colour than it had in free solution. The *periodic acid-Schiff test* (P.A.S.) is also characteristic for the chondroitins although there is some doubt as to whether hyaluronic acid reacts positively. The polysaccharides react easily with proteins to form *proteoglycans*, and it is probably in this form that they associate with the collagen fibres of connective tissue.

# EMBRYOLOGICAL ASPECTS

The manner in which the corneal collagen fibrils are laid down to give the regular spacing of about 600 Å (60 nm), with the orthogonal arrangement in the successive layers, is of some general interest, especially since it seems to be related to the type of glycosaminoglycan (GAG) present in the cornea; and it has been argued that, during the laying down of collagen, the arrangement is, in fact, governed by the nature of the GAG in association with the collagen, both being synthesized by the secreting cell and subsequently ejected into the extracellular space.

### DEVELOPMENT OF CHICK CORNEA

The process has been studied in detail by Coulombre and Hay and their collaborators in the chick. At 3 days of development, the newly formed lens of the developing chick induces the overlying epithelium to start producing the primary corneal stroma; and by four days (Stage 22) this cellular stroma consists of about 15 layers of collagen with the characteristic orthogonal arrangement of fibrils in successive layers; the fibrils are embedded in a glycosaminoglycan which has also been synthesized by the corneal epithelium. Thus the primary cornea, by contrast with the later-developed or secondary cornea, is a product of *epithelial*, or ectodermal, synthesis; a similar situation prevails in the vitreous body. Between 4·5 and 5 days the corneal endothelium moves into place and by 5·6 to 6 days (Stages 27 to 28) it forms a continuous layer separating the corneal stroma from the lens.

### MESENCHYMAL INVASION

The invasion of this stroma by mesenchymal tissue takes place during this period, and this is preceded by a swelling due to increased hydration of its hyaluronic acid matrix, secreted by the endothelium, a process that is presumably necessary to permit this invasion of mesenchyme. Subsequently thickness decreases as the cornea becomes transparent, and this is associated with a decrease in the hyaluronic acid, brought about by the transitory appearance of the enzyme hyaluronidase (Toole and Trelstad, 1971). Thus the function of hyaluronic acid seems to be that of favouring dispersal of collagen fibrils, and this is especially manifest in the vitreous body where the spacing is of microscopic dimensions.

**Collagen synthesis**

The secretion of the collagen by the epithelium and the nature of its GAG are of especial interest, since it seems that these primary layers act as the scaffolding, or template, on which successive layers are laid down by the invading mesenchymal tissue. Trelstad and Coulombre (1971) showed that the fully formed primary cornea is identical in pattern to the secondary cornea, and they suggested that the fusiform mesenchymal cells, synthesizing collagen, use the primary cornea as a scaffold, a process made easier by the primary collagen being organized in bundles, not sheets. Trelstad et al. (1974) showed that the outer primary epithelium produced sulphated gycosaminoglycans, chiefly chondroitin sulphate, which was distributed in the basement membrane of the basal epithelial cells in an ordered pattern. A second sulphated GAG was heparan sulphate (Meier and Hay, 1973).

---

★The term *glycosaminoglycan* is recommended for a polysaccharide containing a hexosamine; more specifically, *glucosaminoglycan* is a polysaccharide containing glucosamine; *galactosaminoglycan*, one containing galactosamine, and so on.

## Primary cornea as template

It would seem that the chondroitin sulphate appears at critical periods in morphogenesis, and, in the case of the cornea, when the primary cornea has to be laid down in a pattern that will act as a template for the later-formed material of mesenchymal origin. Thus the 'corneal epithelial cell, by controlling the stoichiometry of syntheses (of GAG and collagen), by the intracellular packaging and timing of the mode of excretion into the extracellular space would play a very direct role in extracellular matrix morphogenesis, and the self-assembly of the macromolecules in the extracellular space would occur, in part, because suitably interactive molecules were deposited at the correct morphogenetic site at the appropriate time' (Trelstad *et al.*, 1974).

### HUMAN CORNEA

In a study of the human fetus of 5 to 9 months gestation, Breen *et al.* (1972) compared the appearance of the different GAG'S in skin, sclera and cornea. In the human the differentiation of these tissues is virtually complete at 5 months, and after this the skin, for example, becomes more compact, and this is associated with a decrease in hyaluronic acid and an increase in sulphated GAG (dermatan sulphate). In the sclera there was very little change in the amounts of GAG's, the much tighter packing of collagen being associated with the presence of dermatan sulphate and some chondroitin sulphate. In the fetal cornea there was a striking change in the total GAG's by 33 per cent at 7 months, when it remained constant. In general, the main change from early fetal to newborn and adult was the increased proportion of keratan sulphate together with chondroitin.

### CORNEO-SCLERAL TRANSITION ZONE

The cornea and sclera differ markedly in structure and in glycosaminoglycan composition, as we have seen; it is therefore of interest to study the topographical variation in composition on moving from corneal centre to the limbus and sclera. Borscherding *et al.* (1975) analysed the acidic glycosaminoglycans in cornea and sclera at successive distances from the corneal centre, and found, as with corneal diameter that the major areas of transition lay between central and peripheral

cornea and between corneolimbus and sclerolimbus; in both cases the decrease in ordered arrangement and increase in size of collagen fibrils (Fig. 3.4, p. 92) coincided with decreases in the concentration of keratan sulphate and the appearance of more highly sulphated galactosaminoglycans (choindroitin sulphate and dermatan sulphate). Thus chondroitin is found primarily in the central cornea, and chondroitin sulphate in the periphery (Fig. 3.6), possibly because the central cornea is unable to derive sufficient sulphate to convert the precursor, chondroitin, to the sulphated form.

### THE GLYCOSAMINOGLYCAN COAT

It seems unlikely that the corneal collagen fibrils are held apart by covalently bound cross-linkages (Hodson, 1971) although ruthenium red-stainable filaments are seen between collagen fibrils of many other connective tissues, such as aorta (Myers *et al.*, 1973). As Borscherding *et al.* point out, however, the mere presence of a coating of a proteoglycan containing acidic glycosaminoglycans, especially the highly negatively charged keratan sulphate, might be sufficient. Their protein cores would orientate themselves perpendicular to the long axis of the collagen fibre, like bristles on a brush, so as to create a negative field around each fibre that would maintain the spatial separation; the greater negative charges of chondroitin sulphate in peripheral cornea would increase the interfibre distance. Sclera, in contrast to cornea, contains less proteoglycan of different type, molecular size and shape, and thus a different arrangement of fibres is achieved.

## HYDRATION AND TRANSPARENCY

### SWELLING

If an excised eye is kept at, say, 4°C in the refrigerator for 24 hours or more the cornea will be seen to have lost some of its transparency, becoming 'smoky' in appearance; associated with this optical change there is found an increase in thickness due to absorption of aqueous humour. Thus, with an ox eye, there is an increase of water-content from 77·3 to 82 per cent. Expressed in this way the increase in water-content is not impressive, but expressed as the amount of water per gramme of solid

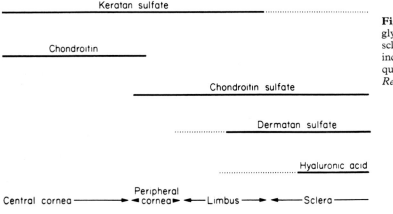

**Fig. 3.6** Distribution of acidic glycosaminoglycans from mid-cornea to sclera in the human eye. Dotted line indicates presence of detectable quantities. (Borcherding *et al.*, *Exp. Eye Res.*)

matter the increase is from 3·4 g $H_2O$/g solids to 4·55 g/g, i.e. an increase of 33 per cent. The increase in water is associated with an increase in the salt content and corresponds to an uptake of aqueous humour rather than the selective absorption of water (Davson, 1949). If an excised cornea is placed in saline the same increase in water-content, with absorption of salt, is observed; if the epithelium is removed the uptake is just the same, if somewhat more rapid, so that it is clear that we are dealing with an increased hydration of the stroma, which is behaving similarly to many of the inanimate gels studied by the colloid chemists; like these gels the cornea shows a swelling pressure, which may be measured by exerting a contrary pressure to prevent uptake of fluid; it amounts to 75 to 85 g/cm² in the normal cornea; as the cornea takes up water, the pressure it can exert decreases, whilst it increases as water is withdrawn (see, for example, Fatt and Hedbys, 1970).

EFFECT OF TEMPERATURE

This uptake of fluid by the dead cornea, and its associated loss of transparency, was considered by Cogan and Kinsey to result from the removal of the normal osmotic pressures of tears and aqueous humour that were supposed, in the living eye, to remove water continuously from the cornea as fast as it entered. In the first edition of this book the present author showed the theoretical inadequacy of this theory and subsequent experimental studies have confirmed this criticism. Thus, to quote a crucial experiment, on the basis of the Cogan-Kinsey theory the swelling of the dead cornea should be much more rapid and extensive if the eye were kept at body-temperature than in the refrigerator, yet if the two eyes of a rabbit are excised and the one is kept at 4°C and the other in a moist chamber at the normal temperature of the aqueous humour, namely 31°C, it is found that the eye kept in the warm is almost completely normal, with good transparency, and water-content only slightly different from the average normal figure. Clearly, by cooling the eye we are robbing the corneal tissue of its power of excluding fluid, and this is presumably due to reducing the metabolism of the eye. In other words, the cornea is probably using metabolic energy to resist the intrusion of fluid from the aqueous humour. To understand the factors concerned, let us turn now to the fluid and salt relationships between cornea, on the one hand, and aqueous humour and tears on the other.

## STROMA-AQUEOUS HUMOUR-TEARS

Analysis of the salt content of the cornea and aqueous humour suggests the simple Gibbs-Donnan type of equilibrium between the stroma and the outside fluids,

the stroma containing a high concentration of ionized colloid and being separated from the outside fluids, aqueous humour and tears, by membranes that allow movements of ions and water but not of the colloid. To quote Otori's figures for the rabbit, the concentrations of $Na^+$, $K^+$, and $Cl^-$ in meq/kg $H_2O$ are as follows:

|  | Aqueous humour | Stroma |
|---|---|---|
| $Na^+$ | 144·0 | 171 |
| $K^+$ | 4·9 | 22 |
| $Cl^-$ | 103·0 | 108 |

The high concentration of $K^+$ in the stroma is doubtless due to the presence of keratocytes, whilst the high concentration of $Na^+$ is presumably due to the presence of ionized collagen and mucopolysaccharide; on this basis, it is to be remarked, however, that the concentration of $Cl^-$ should be lower in the stroma than in the aqueous humour, and it may be that the bicarbonate ion is not at equilibrium (Hodson, 1971).

SWELLING TENDENCY

The general picture of the relationships between the corneal stroma and the aqueous humour and tears is illustrated schematically in Figure 3.7. The stroma,

Fig. 3.7 Illustrating the tendency for the stroma of the cornea to hydrate by virtue of the inward movement of water and salts from aqueous humour, tears and limbal capillaries.

being a mixture of collagen and its associated mucopolysaccharides, may be treated as a gel, and largely by virtue of the polysaccharide this gel has a tendency to take up water.* It is separated from the two fluids from which this water is obtainable by membranes, the epithelium and endothelium. These membranes are permeable to salts and water (Maurice, 1951) so that the

---

*It seems well proved that it is the mucopolysaccharide that is responsible for the hydration tendency; thus the fibrils of a swollen cornea are no larger than normal when examined in the electron-microscope (François, Rabaey and Vandermeersche, 1954); again, extraction of the mucopolysaccharides from the cornea reduces its tendency to take up water (Leyns, Heringa and Weidinger, 1940).

swelling tendency of the gel must be resisted by some active process on the part of either the endothelium or epithelium or both. Removal of the epithelium or endothelium in the intact eye causes a rapid uptake of fluid (Maurice and Giardini). As Figure 3.7 indicates, at the limbus the stroma may take up fluid and salts from the pericorneal capillary plexus, since there is no reason to believe that there would not be filtration from these capillaries as in other parts of the body.

### REVERSAL OF SWELLING

If the cornea of the excised eye is allowed to swell by keeping it for some time at 4°C, the process may be reversed by transferring the eye now to a moist chamber at 31°C (Davson, 1955) so that either the epithelium or endothelium has actively removed the salts and water that entered at the low temperature. Harris and Nordquist (1955) showed that this 'temperature-reversal effect' would still occur when the epithelium was removed, so that the 'pump' that removes fluid as it enters must be located in the endothelium. More recent studies have confirmed the importance of the endothelium; e.g. Green and Otori (1970) showed that the isolated cornea swelled up rapidly on removal of the endothelium, although when the endothelium was intact it could be maintained at normal thickness for hours provided glucose was in the incubation medium.

Again, if a layer of plastic is implanted intra-laminarly, it is the region *above* the implant that becomes oedematous whilst that below retains its normal thickness. Finally, Doane and Dohlman (1970) have actually maintained the cornea *in vivo* at normal thickness after total removal of the epithelium by sealing a plastic contact lens over the denuded surface. Maurice and Giardini had shown that removal of the epithelium caused a rapid increase in thickness of the cornea, and the fact that this does not occur when the surface is sealed means that an intact epithelium is important largely or only because it restrains the passive influx of salts and water from the tears and the limbal blood vessels, i.e. the influx becomes too great for the endothelium to deal with.

### ACTIVE PROCESS

As to the nature of the active process, it was early assumed that this could be achieved by the active transport of sodium from stroma to aqueous humour across the endothelial membrane, or from stroma to tears across the epithelium. The salt so transferred would carry with it its osmotic equivalent of water. Such an active transport, if it occurred, should be detectable by modern isotopic methods in which the fluxes of $^{24}$Na are measured together with the potential across the cornea.

### THE POTENTIAL

The potential between the tears and the aqueous humour sides of the cornea may be measured in the excised tissue using the cornea to separate two chambers of fluid containing electrodes, as in the Ussing technique (Fig. 1.32, p. 31), or it may be measured *in vivo* as the difference in potential between an electrode on the surface of the cornea and another dipping, say, in a pool of saline in the animal's ear, which makes effective electrical contact with the inside of the eye by way of the blood. In the isolated corneal system, Donn *et al.* (1959) found a potential of some 10 to 40 mV, the tears-side being negative to the aqueous humour-side, and similar potentials were found *in vivo* by Potts and Modrell (1957) and Maurice (1967). By splitting the cornea so that the potential across the epithelium alone could be measured, or by inserting a micro-electrode through the epithelium into the subjacent stroma, it was found that this transepithelial *potential* was the same, within experimental error, as the total transcorneal potential; thus the stroma appeared to be at the same potential as the aqueous humour (Fig. 3.17, p. 106).

### ACTIVE TRANSPORT

A potential of this polarity, with the stroma positive, would suggest an active transport of $Na^+$ from tears into the stroma, associated with a passive influx of negative ions; alternatively the potential could be the manifestation of an active transport of $Cl^-$ or $HCO_3^-$ out of the stroma into the tears, associated with passive flux of $Na^+$. In the latter event the active process might represent the salt-pump driving salt out of the stroma and as a consequence of the associated flow of water to maintain isotonicity, helping to maintain the cornea in its normal state of hydration. An active transport of $Na^+$ in the opposite direction would, on the other hand, contribute to corneal hydration and so be valueless as a fluid-pump. In fact, Donn *et al.* found that the measured fluxes of isotopic $Na^+$ indicated an active transport of this ion *into* the cornea, so that, in the rabbit at any rate, we must look to the endothelium for an active transport mechanism that would constitute the basis for the fluid-pump.

### THE ENDOTHELIAL PUMP

Maurice (1972) allowed a rabbit's cornea to swell by cooling for a definite period; after this, it was mounted in such a way that the aqueous humour-side could be perfused with an artificial aqueous humour* whilst the

---

*Dikstein and Maurice (1972), in their study of the most suitable medium for maintaining an excised cornea, found that glutathione considerably enhanced fluid transport; Anderson *et al.* (1974) have shown that the effective molecule is oxidized glutathione, and have discussed the effects in relation to the maintenance of adequate supplies of ATP in the endothelium.

epithelial side, after denuding it of epithelium, was covered with oil in spite of the absence of epithelium, the cornea lost fluid to the perfusing fluid and returned to its normal thickness (Fig. 3.8). An estimate of the rate of transport of fluid out of the swollen stroma was obtained by covering the outer—epithelium-denuded—surface of the cornea with a layer of artificial aqueous humour and placing above this a layer of silicone oil, as in Figure 3.9; the cornea, during its deturgescence, removes fluid from its stroma into the artificial aqueous humour and this is replaced by fluid from above, and the decrease in thickness of the layer of fluid can be measured with a microscope. A value of $5.0\ \mu l/hr/cm^2$ was found, and this occurred against a positive pressure in the artificial aqueous humour, and so must have resulted from an active transport of salt followed by water across the endothelium. The rate of transport was such that each endothelial cell was actually transporting its own volume every 5 minutes. When the cornea is treated with ouabain, which specifically inhibits active transport of $Na^+$, it is found to swell (Trenberth and Mishima, 1968) the rate of increase in thickness being some $45\ \mu m/hr$; the rate of fluid movement measured by Maurice corresponded to an ability to thin the cornea at a rate of $60\ \mu m/hr$, indicating that the active pump, before being inhibited, was capable of maintaining the cornea in its normal state, pumping fluid out as fast as it entered passively from the tears.

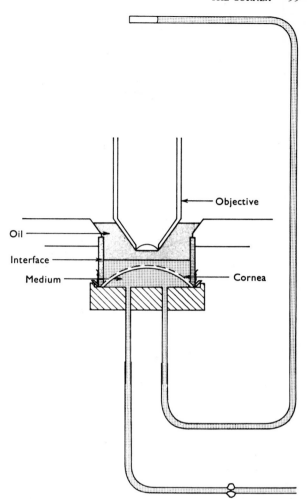

Fig. 3.9 Method of determining flux of fluid across corneal endothelium. The cornea is mounted after the removal of its epithelium and is just covered with the same medium that bathes its inner surface. A thick covering of silicone oil is then poured over the watery layer. Fluid movement is determined by focusing the microscope in turn on the oil–water interface and on the endothelial surface. (Maurice, *J. Physiol.*)

ENDOTHELIAL ACTIVE TRANSPORT

The transport of water in biological systems requires the transport of solute, usually a salt, as a primary step, the water following passively along osmotic gradients. Thus modern work has been concerned with the nature of the solute that must be actively transported out of the stroma, and the possibility that this transport gives rise to a difference of potential between stroma and aqueous humour. As we have seen, the transcorneal potential, i.e. that across tears to aqueous humour, is quantitatively equal to the transepithelial potential, i.e. that between tears and stroma, so that any difference of potential between stroma and aqueous humour, i.e. across the endothelium, must be smaller than the experimental

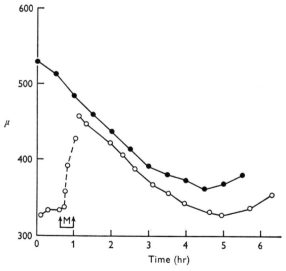

Fig. 3.8 Thinning of a cornea in the absence of epithelium. Open circles: cornea, without epithelium, swollen by application of perfusion fluid to bare surface of cornea at time M, first arrow. At second arrow perfusion fluid replaced by oil. Filled circles: temperature reversal of intact second eye of pair on next day. Vertical displacement of curves corresponds to thickness of scraped-off epithelial layer, $40\ \mu m$ (Maurice, *J. Physiol.*)

error in these studies. Fischbarg (1972) and a number of other workers have shown that the de-epthelialized rabbit cornea does, in fact, exhibit a difference of potential between stroma and aqueous humour, or other medium bathing the endothelium; the potential was small, of the order of 500 $\mu$V and was directed so that the stromal side was positive and the lens-side negative. Such a potential could be the expression of an ionic pump responsible for the transport of fluid out of the stroma, and in view of its polarity it might be expected to consist of an active transport of a negative ion, e.g. $Cl^-$ or $HCO_3^-$, from the stroma; an associated movement of $Na^+$ would tend to keep the pump 'neutral' but leaving a small excess of negativity on the lens-side of the endothelium (Fig. 3.10).*

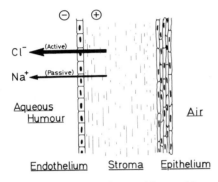

**Fig. 3.10** Showing polarity of potential caused by active transport of an anion from stroma into aqueous humour.

## Relation to potential

Hodson (1971) showed that the fluid-pump required, for its full activity in maintaining the cornea dehydrated, the presence of $Na^+$ and $HCO_3^-$, and subsequent studies, in which the fluid-transporting capacity of the endothelium was compared with the electrical or ionic-

pump characteristics, have shown a remarkably good correlation which definitely implicates the potential, or rather the ionic movements that create it, with the fluid-transport. For example Figures 3.11 and 3.12 compare the effects of reducing the concentration of $Na^+$ on the transendothelial potential and on the rate of fluid transport respectively. As a result of a thorough study of fluid transport and potential, Fischbarg and Lim (1974) constructed Table 3.1; it will be seen that the carbonic anhydrase inhibitors tended to block fluid-transport, just as with the aqueous humour, but they failed to influence the potential, suggesting that the electrogenic stage in the pumping process preceded another stage that could be inhibited by carbonic anhydrase and presumably involved hydration of $CO_2$. Fischbarg and Lim were reluctant to suggest a simple anion-pump, e.g. one based on active transport of bicarbonate linked to that of $Na^+$, since total removal of bicarbonate, although it abolished fluid-transport, left a residual potential; so they suggested a model requiring secretion of $H^+$-ions as well as active transport of $Na^+$ by an ouabain-sensitive mechanism.

### ION-EXCHANGE GEL

Hodson (1974) emphasized that the short-circuit current of his de-epithelialized corneas was equivalent to the transport of $0.27$ nEq/cm$^2$/sec, which is nearly identical with the measured transport of fluid multiplied by the isotonic concentration as found by Maurice (1972), namely $0.28$ nEq/cm$^2$/sec, but of course the transport of $Na^+$ would work in the wrong direction. He suggested, however, that the cornea would behave as an ion-exchange gel with fixed negative charges, and so the equilibria between it and the aqueous humour would be

*Fischbarg (1972) found an electrical resistance across the endothelium of 50 ohm cm$^2$; if the potential of $0.5$ mV operated across this resistance it would maintain a current equivalent to the carriage of $0.4$ $\mu$eq/hr/cm$^2$ of a univalent ion. The flow of solute due to the endothelial pump corresponds to $0.6$ $\mu$eq/hr/cm$^2$.

**Table 3.1** Relation between potential difference across endothelium and fluid transport. (Fischbarg and Lim, 1974)

| Factor | Effect on p.d. | Effect on fluid transport |
|---|---|---|
| [HCO$_3$] decrease | Reversibly decreased following behaviour similar to that of fluid pump | Decreases or arrests reversibly decreases |
| [Na$^+$] decrease | Reversibly decreases | Arrests |
| K$^+$ absence | Reversibly decreases | Reversibly arrests |
| [H$^+$] increase | Reversibly decreases | Arrests |
| Temperature decrease | Decreases | Arrests |
| Ouabain | Abolishes | Arrests |
| Carbonic anhydrase inhibitors | No effect | Reduces by half; reduces |
| Cyanide | Abolishes | Arrests |
| Iodoacetate | Abolishes | Arrests |
| Cytochalasin B | Reversibly decreases | Arrests? |

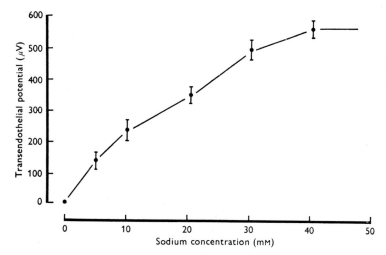

**Fig. 3.11** The effect of substituting choline for sodium in the bathing medium on the transendothelial potential (stromal side positive). The endothelial resistance remains nearly constant. (Hodson, *J. Physiol.*)

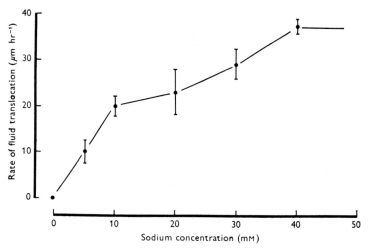

**Fig. 3.12** De-epithelialized corneas translocate fluid into their lens-side. The rate of fluid translocation is a function of the sodium concentration in the bathing medium. (Hodson, *J. Physiol.*)

modified in a predictable manner, so that even though $Na^+$ and $Cl^-$ or $HCO_3^-$ were being moved out of the stroma by an electrically neutral mechanism, a potential would develop due to passive leak back of salt. However, this plausible theory was put out of court by his more recent study (Hodson *et al.*, 1977) in which the potential was shown to be the same (400 to 700 $\mu V$) when the stroma had been stripped from the cornea leaving only the endothelium with its attached Descemet's membrane. As they emphasized, the problem remains as to what is the basis for the potential—if an anion-pump is active it cannot involve active transport of $Cl^-$, $SO_4^{2-}$ or $PO_4$ since removal of these does not affect the potential or fluid-pump; the sensitivity of the pump and potential to bicarbonate, however, strongly suggest that this ion is involved,[*] and a recent study of Riley (1977) lends further support. He demonstrated the presence of an ATPase in the rabbit's corneal endothelium that is dependent, for its activity, on the presence of anions in the medium, including bicarbonate; the enzyme was not inhibited by ouabain, but it was by thiocyanate and cyanate, both of which tended to reduce the ability of the endothelium to pump water out of the swollen cornea.

POTENTIAL AND WATER-FLUX

Barfort and Maurice (1974) measured transendothelial potential and water-flux in the excised cornea simultaneously and were thus able to correlate the two more accurately; as Figure 3.13 shows, there is a positive correlation between the magnitude of the potential and water-flux; and it is interesting that the potential measured by these authors was larger, from 0·6 to 1·7 mV with a median of 1·3 mV. As Fischbarg had found, the correlation extended to the effects of various inhibitors, including cooling, ouabain and antimycin, the blocking of fluid transport being accompanied by reduction of the potential to zero. These authors have

[*]The anions derived from glucose metabolism in the cornea are lactate and labellin, this latter having a molecular weight of 1500 daltons; Riley *et al.* (1977) have shown that the amounts of these formed are far too small to act as anions in the fluid transport system.

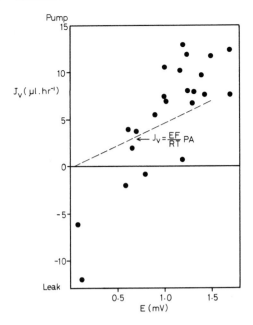

**Fig. 3.13** Illustrating correlation between magnitude of potential, E, across the corneal endothelium, and the magnitude of the water-flux. 'Pump' and 'Leak' represent water-flow against and with hydrostatic gradient. The line represents the flow when there is maximal coupling between ion pump and water. (Barfort and Maurice, *Exp. Eye Res.*)

discussed in some detail the way in which this potential could cause, or reflect, fluid transport. The obvious hypothesis, namely that active transport of an anion out of the stroma accompanied by a cation such as $Na^+$ would create the necessary osmotic flow, was tested by calculating the passive flux due to the potential from the relation:

$$J_{Na} = P_{Na}C_{Na}EF/RT$$

and the corresponding water flux, $J_v$, would be given by:

$$J_v = P_{Na}EF/RT$$

on the assumption that this occurred with maximum efficiency such that each mole of transported salt carried its osmotic equivalent of water. On this basis it appeared that the ionic flux would not be adequate to bring about the water-flow, so that additional, or alternative, hypotheses had to be considered.

### TEMPERATURE FLUCTUATIONS

The cornea can undergo much larger fluctuations in temperature than those in deeper structures of the eye. Hodson (1975) has discussed the effects of temperature on the corneal system; a lowering of temperature

decreases the efficiency of the endothelial pump; and this is seen by the slower rate of thinning of a previously swollen cornea when the temperature is reduced from 35°C to 25°C, but the same final condition of normal turgescence is reached. To compensate for the reduced pumping capacity we have the reduced tendency for the corneal stroma to swell, due to the lowered permeability of the endothelium to ions.

### HIBERNATION

An extreme example of the response to lowered corneal temperature is given by the hibernating animal. Bito *et al.* (1973) have drawn attention to the normal state of hydration of the hibernating wood-chuck in spite of its corneal temperature being maintained at 9·4°C for days on end; moreover, if the enucleated eyes are maintained at this temperature in a moist chamber, they increase their hydration in the same way that rabbit corneas do. Thus the swelling of the cornea that takes place in the cooled enucleated eye, or in the isolated cornea maintained in a Ussing-type chamber, apparently depends on something in addition to cooling, but what this extra factor is, it is difficult to say.

### MODEL OF ENDOTHELIUM

As indicated earlier, the morphological studies on the endothelium indicated that the clefts between endothelial cells were not completely sealed off, so that the endothelium would act as a rather leaky barrier between stroma and tears, exhibiting a rather low electrical resistance and relatively high permeability to ions compared with the epithelium, where the outer-most epithelial cells are sealed together by tight junctions. Fischbarg (1973) has measured the electrical resistance and capacity of the endothelium and, on the basis of these measurements, has constructed the picture illustrated by Figure 3.14, where the gap-junction★ constitutes a restraint on diffusion sufficient to account for the low resistance of the endothelium as a whole. Removal of $Ca^{2+}$ from the medium causes a large increase in permeability of the endothelium, and morphologically this is associated with abolition of the gap-junctions (Kaye *et al.*, 1973). Calculation of the rate of passive flow across the endothelium under the pressure-head that would occur *in vivo* in the absence of a pump, and employment of the dimensions of Figure 3.14, give a value of thickening of the cornea of 41 μm per hr, and this agrees well with values of 40 to 45

---

★ Kreutziger (1976) has emphasized the geometrical complexity of the cleft between corneal endothelial cells, since these interdigitate by means of cytoplasmic protrusions, whilst the clefts between these interdigitations have many gap junctions reducing their lumen to about 3 nm. Thus the pathway for a molecule, or ion, passing between cells is highly complex, and not so simple as envisaged by Figure 3.14.

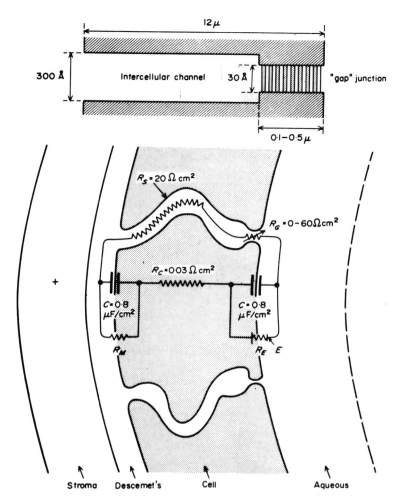

**Fig. 3.14** *Above :* Schematic diagram of the intercellular spaces and the terminal bars of 'gap junctions' in the endothelium. Dimensions obtained from electronmicrographs. *Below :* Electrical model for the corneal endothelium. $R_S$, resistance of the intercellular spaces; $R_G$, resistance of the gap junctions; $R_C$, resistance of the cell cytoplasm; $R_M$, resistance of the cell membrane; $R_E$, equivalent electrical resistance of the ionic pump; $\varepsilon$, equivalent electromotive force of the ionic pump; $C$, capacitance of the cell membrane. (Fischbarg, *Exp. Eye Res.*)

$\mu$m/hr found experimentally. In the absence of gap-junctions, as in the absence of $Ca^{++}$, the calculated flow would correspond to 800 $\mu$m/hr, and this once again agrees with the experimental finding.

## CORNEAL ATPASE

Active transport of $Na^+$ across cell membranes is usually inhibited by ouabain, which acts by inhibiting an enzyme that specifically hydrolyses ATP in the presence of appropriate proportions of $Na^+$ and $K^+$ and $Mg^{2+}$; it is generally referred to as a $Na^+$-$K^+$-activated ATPase. The enzyme's localization can be revealed by histochemical techniques, since fixation of the tissue does not inhibit the enzyme, but, as Tervo and Palkamo (1975) have emphasized in their study of the corneal enzyme, great care must be employed to avoid spurious results. According to these authors, who have summarized earlier studies, both corneal epithelium and endothelium of the rat contain the enzyme in association

with their cell membranes; in the epithelium the enzyme was located on the membranes, especially those of the basal cells in the region of their interdigitations, the surfaces facing the basement membrane being unstained. Similarly, the endothelial cell membranes were stained in the region of lateral cell apposition, the anterior and posterior faces being free from reaction-product (Fig. 3.15).

## COMPARATIVE ASPECTS

### ELASMOBRANCH

The elasmobranch cornea, for example that of the dogfish *Squalus acanthias*, does not swell even when both epithelium and endothelium are removed (Smelser, 1962); and this was attributed by Ranvier in 1881 to the presence of sutural fibres that ran at right-angles to the laminae, binding the basement membrane of the epithelium to Descemet's membrane. The electron

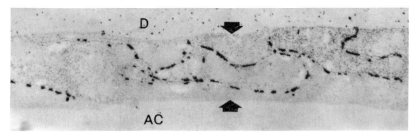

**Fig. 3.15**   NaK–ATPase reaction in the corneal endothelium. Note the enzyme activity located mainly in the adjoining cell membranes. On the other hand, no activity can be seen on those parts of the cell membrane which face either Descemet's membrane (D) or the anterior chamber (AC) (arrows) ( × 11 340). (Tervo and Palkama, *Exp. Eye Res.*)

microscopical study of Goldman and Benedek (1967) has confirmed this picture of antero-posteriorly directed fibres constituting a 'sutural complex', so that if these sutural fibres failed to stretch appreciably this would prevent swelling. Praus and Goldman (1970) have shown that the mucopolysaccharide of the shark cornea is more similar to that in cartilage than that found in the mammalian cornea, with very little keratan sulphate and a relatively high proportion of chondroitin sulphate.

**Resistance and potential**

Fischer and Zadunaisky (1977) found that the electrical resistance across the shark cornea was low, indicating high permeability of its epithelium; however, when corneal abrasions, normally suffered when the fish are in captivity, were prevented, high values of about 1000 ohm cm$^2$ were obtained. Only a very small transcorneal potential was observed, which was variable in polarity, so that, on average, the potential was close to zero; and this might well be consistent with the absence of a tendency for this cornea to swell under conditions, e.g. of cold, when that of the mammal and amphibian does.

TELEOSTS

Teleosts are more similar to the mammal insofar as, on cooling, the cornea swells; like the mammal there is a transcorneal potential with the outside negative, but this is small, ranging between 0·5 and 2·9 mV. It would seem, then, that the teleostean cornea requires a fluid-pump to maintain normal hydration and transparency, whereas that of the elasmobranch does not.

**THE TRANS-EPITHELIAL POTENTIAL**

The potential across the epithelium in the rabbit was found by Donn et al. to be accounted for by the active transport of Na$^+$, a test for this hypothesis being the measurement of the short-circuit current which, if the sole means of transport is through the active processes,

and if no other ion is being actively transported, may be equated with the transport of Na$^+$.

AMPHIBIAN CORNEA

In the frog the ionic transport mechanism seems to be different, and it may well be that an epithelial ionic pump, transporting net amounts of salt from stroma to tears, is the mechanism that maintains hydration constant. Zadunaisky (1966, 1969) measured a potential between tears and aqueous humour of 10 to 60 mV, the aqueous humour being positive; this was due to the active transport of Cl$^-$ from stroma to tears and could be abolished by removing this ion from the cornea. By splitting the cornea, the separate effects of epithelium and endothelium could be examined, and in this case it was the epithelium that reproduced the behaviour of the whole cornea. There was no evidence for active transport by the endothelium. When the transparency had been allowed to decrease by imbibition of fluid at a low temperature, the partial reversal of this by warming was accompanied by enhanced active transport of Cl$^-$; moreover the temperature-reversal effect was prevented by substitution of Cl$^-$ by SO$_4^{2-}$, and this also blocked the short-circuit current (Zadunaisky and Lande, 1971). Bromide would substitute for Cl$^-$ but not iodide.

Furosemide, a typical inhibitor of active chloride transport in the kidney, inhibited the frog's corneal chloride-pump when applied to the endothelial side (Candia, 1973). The existence of a Na$^+$-pump, transporting this ion from epithelium to stroma, as in the rabbit, was demonstrated by Candia and Askew (1968); its contribution to the total short-circuit current was only ten per cent.

OXYGEN-CONSUMPTION

Active transport involves the supply of metabolic energy; Reinach et al. (1977) have shown that inhibition of Cl$^-$-transport in the frog cornea, e.g. by removal of Cl$^-$, or treatment with furosemide, causes a reduction in $O_2$-consumption. A similar effect of inhibition of the $Na^+$-transport mechanism, by treatment with ouabain or removal of $Na^+$ from the medium, caused a decrease of 36 per cent. Since active transport of $Cl^-$ is dependent on the presence of $Na^+$, it is likely that the larger reduction of $O_2$-consumption by $Na^+$-blocking treatments is due to the associated blocking of $Cl^-$-transport.

### CYCLIC AMP

The chloride-pump of the amphibian cornea is activated by the cyclic AMP system, and this is revealed by the large effects of epinephrine, aminophylline, etc. on the short-circuit current (Chalfie *et al.*, 1972).

### CHLORIDE-PUMP IN RABBIT

Klyce *et al.* (1973) found that epinephrine and theophylline, placed on the lens-side of the isolated rabbit cornea, caused a rapid rise in short-circuit current, which suggested to them that a $Cl^-$-pump might be operating in addition to the $Na^+$-pump described by Donn *et al.*

In a chloride-free medium the responses to these agents were reduced, and Klyce *et al.* concluded that there was, indeed, a chloride-pump actively transporting this ion from stroma to tears and contributing to maintain normal hydration, i.e. it was operating in opposition to the $Na^+$-pump working in the opposite direction. That this anion-pump can contribute to corneal thinning was shown by Klyce (1975), using a cornea with endothelium removed, the endothelial side of the stroma being bathed with silicone oil; a swollen cornea treated in this manner showed some thinning, due to the intact epithelium, and this was increased by theophylline, the rate being predictable from the rate of salt transport. Klyce considered that under normal conditions the two epithelial pumps balanced each other; theophylline, by accelerating the $Cl^-$-pump and leaving the $Na^+$-pump unaffected, would cause net transport of salt and fluid out of the cornea and promote thinning.

### RELATIONS BETWEEN ANION AND CATION PUMPS

Fischer and Zadunaisky (1975) found that $Na^+$ was necessary for the epinephrine acceleration of the $Cl^-$-pump, suggesting a link between the two pumps; and this is supported by the observation that ouabain, a specific inhibitor of ATPase-mediated $Na^+$-$K^+$-linked active transport of $Na^+$, blocks a large part of the epinephrine stimulated $Cl^-$-transport. A tight coupling between the active transport of $Na^+$ and $Cl^-$ is unlikely, since the anion-pump is so dominant in the frog's cornea; moreover the stimulator of $Na^+$-pumps, amphotericin B, increased the short-circuit current due to $Na^+$ but had no effect on the $Cl^-$-transport (Candia *et al.*, 1974). Klyce and Wong (1977) suggest a model illustrated by Figure 3.16; although there are several layers of cells, making up the epithelium, we may treat these as a syncytium due to electrical coupling; and the whole structure may, to a first approximation, be treated as a giant cell with an inner—stroma-facing—and an outer—tears-facing—membrane. At the stromal side, $Cl^-$ is accumulated into the 'cell', a process linked to the extrusion of $Na^+$. The $Cl^-$ diffuses passively out of the cell into the tears down a gradient of electrochemical potential, since, although the intracellular con-

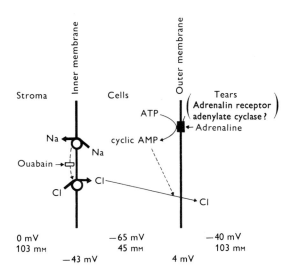

**Fig. 3.16** Model for adrenaline influence on corneal epithelial Cl transport. Potential and Cl concentrations [(Cl)] for the stroma, cells and tear solution are listed, as well as the Cl electrochemical potential gradients ($\Delta\tilde{\mu}_{Cl}$) across the inner and outer membranes. See text for additional details. (Klyce and Wong, *J. Physiol.*)

| | Stroma | Inner membrane | Cells | Outer membrane | Tears |
|---|---|---|---|---|---|
| Potential | 0 mV | | −65 mV | | −40 mV |
| [Cl] | 103 mM | | 45 mM | | 103 mM |
| $\Delta\tilde{\mu}_{Cl}$ | | −43 mV | | 4 mV | |

centration of $Cl^-$ is probably only 45 meq/litre and thus much less than that in tears, the negative potential between the inside of the 'cell' and the tears, of about 65 mV, is more than adequate to overcome this concentration difference, leaving an electrochemical potential favouring passage of $Cl^-$ to tears of about 4 mV.

## THE EPITHELIAL POTENTIAL PROFILE

By moving a micro-electrode from the tears into the epithelium by steps of a few microns at a time, the potential between the tears and successive depths of epithelium and stroma may be measured. Figure 3.17 shows the results of a study by Klyce (1972) where A

is a schematic outline of the structural make-up of the epithelium with the important transition points indicated, and B gives the potential, *referred to the stroma as zero*, at successive depths. C indicates the changes of resistance at critical sites. The total transcorneal potential is $V_c$ and here represents the potential between stroma and tears, i.e. before the micro-electrode has penetrated; in this example it is about 15 mV negative in respect to the stroma. On penetrating the squamous cell, $s$, there is a jump of negativity corresponding to the resting potential of the squamous cell, and this is indicated by $V_\alpha$, the membrane potential of the outer membrane of the squamous cell. On pushing the electrode further into the epithelium, no significant change occurs until the space $(wb)$, separating the inner wing cell $(w)$ and the basal cell $(b)$, has been crossed

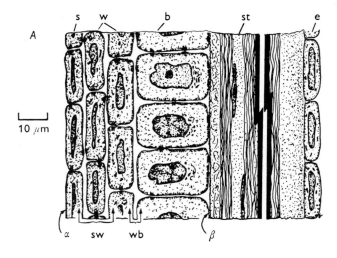

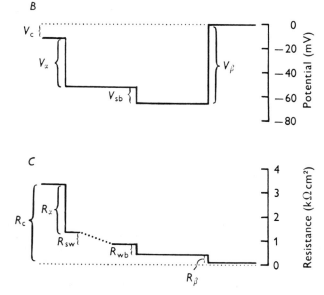

**Fig. 3.17**  (A) Schematic illustration of the cornea showing the epithelium, a portion of the stroma (st) and the endothelium (e). On the basis of the dye study, the following regions in the electrical profiles were identified: $\alpha$, the outer membrane of the squamous cell; $\beta$, the inner membrane of the basal (b) cell; sw, the region between squamous (s) and wing (w) cells; and wb, the transition region between wing and basal cells. (B) The average potential profile for the epithelium. The extracellular gradient in potential from the sub-epithelial stroma to the aqueous humour was less than 1 mV. (C) The average resistance profile for the epithelium. The stroma and endothelium contributed only a few per cent to total corneal resistance. (Klyce, *J. Physiol.*)

and the electrode has entered the basal cell. There is then a further increase in negativity, $V_{sb}$, and now the maximal negativity has been reached so that further movement through the basal cell produces no change. When the stroma-facing membrane of the basal cell has been penetrated, the potential swings to zero—the stromal potential—which is positive in relation to the tears. Approximately, then, the corneal potential can be described as that of a giant cell with tears-facing membrane, $\alpha$, and stroma-facing membrane, $\beta$, but the transition between the wing cell and basal cell shows that coupling between these cells is not perfect, by contrast with that between squamous and wing cells; and this is indicated by the resistance-profiles of Figure 3.17C.

## PERMEABILITY OF THE CORNEA

In a homogeneous medium a dissolved substance will pass from a region of high to one of low concentration by the process of diffusion. When the diffusion from one region to another is restricted by the presence of a membrane, the process is described as one of 'permeability', and the degree of restraint imposed on the migration of the solute molecules is measured inversely by the *permeability constant* of the membrane. In general, the ease with which a substance can penetrate cell membranes depends on its lipid-solubility; thus ethyl alcohol, which is highly lipid-soluble, penetrates cell membranes easily, whereas glycerol penetrates with difficulty. The membranes separating the stroma from its surrounding media are formed by layers of closely packed cells; lipid-soluble substances may be expected to pass easily into the cornea because they may pass easily into and out of the cells of these membranes; lipid-insoluble substances, including ions, may be expected to penetrate with difficulty; moreover, the endothelium, with its single layer of cells, might be expected to be the more permeable membrane.

### TIGHT JUNCTIONS

However, the essential factor in permeability of such cellular layers is the extent to which the intercellular clefts are closed by zonulae occludentes or tight-junctions. If these clefts, or the majority of them, are not closed by the fusion of adjacent membranes, we may expect a high degree of permeability extending to such large molecules as inulin and serum albumin, and this permeability will be relatively unselective because the large diameter of the intercellular clefts will not permit any selective sieve action on the various solutes commonly studied. In general, the qualitative studies of Swan and White, and Cogan and Hirsch, confirm these general principles. Where the substance is an organic base or acid, such as atropine or salicylic acid, the degree of dissociation is an important factor since the undissociated molecule penetrates more rapidly than the ion; hence the permeability of the cornea to these substances is influenced by pH.

### EPITHELIUM AND ENDOTHELIUM

More recent quantitative measurements, carried out mainly by Maurice, Mishima and Trenberth, and by Thoft and his collaborators, have allowed separate estimates of the permeabilities of the epithelium and endothelium. Thus *in vivo* Maurice (1951) showed that the endothelium was 100 times

more permeable to the $Na^+$-ion than the epithelium; this does not mean that the endothelium offers no restraint to the passage of this ion, however; it means, rather, that the permeability of the epithelium is very low, so that for many practical purposes the epithelium may be treated as a 'semi-permeable membrane', being permeable to water but effectively impermeable to the solutes of the tears. Maurice (1969) has calculated that, with $Na^+$, the endothelium offers some 1700 times the resistance to diffusion that would be offered by the same thickness of water; with sucrose it was 2000 times, with fluorescein, 4200, inulin 4400 and serum albumin greater than 100 000.

Thoft and Friend (1975) studied an amino acid analogue, $\alpha$-aminoisobutyric acid (AIB), a non-metabolizable sugar, 3-0-methylglucose, and mannitol. The calculated epithelial permeabilities were 0·00097, 0·0013 and 0·0015 cm/hr respectively; these values are equivalent to about $3.10^{-7}$ cm/sec, which are rather higher than the $1.10^{-7}$ cm/sec obtained by Maurice in his study. Freshly regenerated epithelium had rather higher permeabilities, 0·0044, 0·0021 and 0·0025 cm/hr respectively. As Maurice had found, endothelial permeability was some 100 times greater, values of 0·36, 0·17 and 0·08 cm/hr being obtained; the difference between mannitol, on the one hand, and the sugar and amino acid on the other, is consistent with the notion of facilitated transport for amino acids and sugars; such a carrier mechanism would not be available for mannitol. These authors stress the extreme inefficiency of any attempt to increase available sugar or amino acid to the cornea by application to the corneal surface.

### ENDOTHELIUM

Kim *et al.* (1971) and Mishima and Trenberth (1968) studied the permeability of the endothelium in isolation by removing the epithelium from the excised cornea; in effect, of course, they were measuring in addition the permeability of Descemet's membrane and the stroma, but these do not impose a significant barrier to diffusion, so that the process was governed by passage across the endothelium. The results of Kim *et al.*'s study are shown in Table 3.2. In general, permeability decreased as molecular weight or radius increased, in fact a linear relation between molecular radius and permeability was found, suggesting that the passage was through large water-filled pores, such as those constituted by the gap-junctions described earlier.

**Table 3.2** Permeability coefficients (cm/sec) of corneal endothelium. (Kim *et al.*, 1971)

| Solute | Permeability | Mol. wt |
|---|---|---|
| $Cl^-$ | $2 \cdot 2 \times 10^{-5}$ | 35·5 |
| Mannitol | $7 \cdot 0 \times 10^{-6}$ | 182 |
| Sucrose | $4 \cdot 5 \times 10^{-6}$ | 342 |
| Inulin | $1 \cdot 1 \times 10^{-6}$ | 5200 |
| Dextrin A | $5 \cdot 8 \times 10^{-7}$ | 15 000–17 000 |
| PVP | $2 \cdot 9 \times 10^{-7}$ | 40 000–50 000 |
| Dextrin B | $8 \cdot 1 \times 10^{-8}$ | 60 000–80 000 |

### Facilitated transport
Where metabolically important substances are concerned, however, permeability may be much greater than anticipated on the basis of molecular radius; thus glucose and amino acids are examples, the permeability of alpha-amino isobutryic acid being some $5.10^{-5}$ cm/sec, which is more than 112 times that of mannitol of comparable molecular radius, (Thoft and

Friend, 1972) and the same is true of glucose (Hale and Maurice, 1969). With these metabolically important substances the pathway is through the endothelial cells as well as through the gap-junctions, passage through these being significant because of the carrier-mediated facilitated transport across the cell membranes.*

## WATER

As with practically all cellular layers, the permeabilities of epithelium and endothelium to water are large; thus for epithelium Donn *et al.* (1963) found a value of $3.10^{-5}$ cm/sec, comparing with $1.10^{-7}$ cm/sec for $Na^+$; in the endothelium the value was $1·64.10^{-4}$ cm/sec comparing with $2·1.10^{-5}$ cm/sec for urea (Mishima and Trenberg, 1968). The high permeabilities are due to the relatively high permeability of cell membranes to water, and this permits transport over the whole surface of the cellular layer instead of confining it to the intercellular clefts.†

# TRANSPARENCY OF THE CORNEA

## SCATTERING OF LIGHT

By transparency we mean rather more than the ability of a material to transmit light, we are concerned with the manner in which it is transmitted; thus dark glasses may reduce the transmission to 10 per cent of the incident intensity yet the optical image remains perfect; on the other hand an opal screen may reduce the intensity much less, but because of the *scattering* of the light no image is possible. Thus diffraction, or scattering, of light is the important factor for transparency; and in this respect the cornea is very highly transparent, less than 1 per cent of the light being scattered. The question arises as to how this transparency is attained in an inhomogeneous medium consisting of fibres embedded in a matrix of different refractive index; and furthermore, why the transparency should be affected by increases in hydration of the cornea and changes in intraocular pressure, as when haloes are seen during an acute attack of glaucoma.

## LATTICE THEORY

In general, when light strikes a small discontinuity in the medium the latter acts as origin for new wavelets of light travelling in all directions; hence a large number of discontinuities in a transparent medium will give rise to wavelets of light passing backwards as well as forwards and thus making the medium opaque. A solution of very small particles, such as NaCl, remains transparent in spite of these discontinuities in the medium; this is because the discontinuities are small compared with half the wavelength of light. When the particles are of comparable size or greater, the diffraction becomes important, and the question arises as to why

the cornea, built up of fibrils with a diameter greater than the wavelength of visible light, should in fact be transparent. The problem has been examined in detail with considerable acumen by Maurice (1957) who has shown, from an analysis of the birefringence‡ of the cornea, that the size of the fibrils, and the difference of refractive index between them and the ground-substance in which they are embedded, are such as to lead one to expect a highly opaque structure, due to scattering of incident light. He has shown, nevertheless, that this scattering may be avoided if the fibres are of relatively uniform diameter and are arranged in parallel rows; in this event the diffracted rays passing forwards tend to cancel each other out by destructive interference, leaving the normal undiffracted rays unaffected. In this way the opacity of the sclera is also explained, since there is no doubt from Schwarz' study that the prime distinction between the two tissues is the irregularity in arrangement and diameter of the scleral fibres.

### Confirmation of theory

Although Maurice's explanation was questioned, e.g. by Smith, subsequent theoretical and experimental studies, notably those of Cox *et al.*, 1970; Benedik, 1971; Farrell *et al.*, 1973 and Twersky, 1976, have amply confirmed the contention that transparency depends on an ordered arrangement of the collagen fibrils; this ordering need not be so regular as that found in a crystal lattice, so that the short-range ordering actually found, in which regularity of spacing does not extend over many wavelengths in any given region, is adequate (Hart and Farrell, 1969).

### SWOLLEN CORNEA

An important tool in the analysis of the structure is the dependence of scattering on wavelength; thus the inverse third power dependence of scatter on wavelength found by these authors allowed them to conclude that

---

*According to Riley (1977) the transport of amino acids across the endothelium is not active, in the sense that fluxes are equal in both directions; the tendency for labelled amino acid to accumulate in the cornea is due to an active accumulation by the stroma cells.

†The permeability coefficient of the epithelium for $Na^+$ is about $1.10^{-7}$ cm/sec; the permeability of the capillary endothelium is about a thousand times bigger than this. The permeability of the epithelium to water is about $3.10^{-5}$ cm/sec comparable with that found in a variety of cells (Donn, Miller and Mallett, 1963).

‡The ordered arrangement of the collagen fibrils makes the velocity of light in one plane different from that in one at right-angles; this leads to double refraction or birefringence. From the magnitude and character of this birefringence important information regarding the orientation of the fibrils in the cornea may be obtained (Stanworth and Naylor, 1953; Maurice, 1957).

the ordering was short ranged. When the cornea was grossly swollen the relationship changed to a second power dependence indicating the presence of an inhomogeneous fibril distribution. Electron microscopy of the swollen corneas showing this relationship revealed lakes in which the fibrils had apparently disappeared. In analysing the basis for the loss of transparency, Farrell et al. (1973) considered the two main factors, namely the disruption in the ordering of the fibrils, and the decreased difference in refractive index between fibrils and surrounding medium, an effect that increases the 'scattering cross-section' of the fibril. They showed that the disorientation effect is by far the most significant.*

## METABOLISM

### PATHWAYS

The metabolic energy of the cornea is derived from the breakdown of glycogen and glucose; since the cornea utilizes oxygen, which it derives from the atmosphere, through the tears, and from the aqueous humour, at least some of the energy is derived from oxidative mechanisms; the high concentration of lactate in the cornea indicates that aerobic glycolysis is an important pathway for the breakdown of glucose. According to a recent study of Riley (1969) on the perfused cornea, 85 per cent of the glucose utilized is converted to lactate and only 15 per cent is oxidized. The lactate so formed is carried away into the aqueous humour and thus the cornea contributes to the high concentration of this anion in the fluid. A considerable fraction of the glucose apparently passes through the pentose phosphate shunt by which glucose-6-phosphate is converted to triose phosphate, a pentose phosphate being formed as an intermediate. The triose phosphate is metabolized to pyruvate, which can either be converted to lactate or enter the citric acid cycle to be oxidized by $O_2$. The pathways are indicated in Figure 3.18.

### OXYGEN TENSION

Langham (1952) used the concentration of lactate in the cornea as a measure of its 'oxygen-deficit', on the assumption that a greater availability of $O_2$ would be reflected in a lowered concentration of lactate in the cornea. In this way he was able to show that the corneal metabolism was affected by the tension of $O_2$ in the atmosphere, so that replacing air with $O_2$ decreased the lactate concentration in the cornea, whilst replacement with $N_2$ increased the concentration. When one eye was kept closed, the lactate concentration of this cornea increased, but not so much as in an eye exposed to $N_2$, so that it seems that the conjunctival blood vessels can provide some $O_2$.

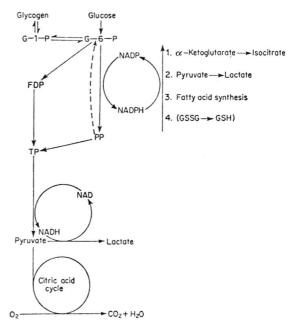

**Fig. 3.18** Pathways of glucose and hydrogen transfer in the corneal epithelium. PP, pentose phosphate; TP, triose phosphate; ——→ transaldolase and transketolase reactions. (Maurice and Riley, in *Biochemistry of the Eye*, New York: Academic Press.)

### Dependence on atmosphere

Because of this dependence on the atmosphere, under normal conditions, because of utilization, there is a gradient of oxygen-tension from tears, 155 mmHg, to the aqueous humour, 55 mmHg (Fatt and Bieber, 1968); thus the endothelium can apparently function normally at a relatively low oxygen tension. The question arises as to the extent to which the endothelial layer depends on the atmospheric supply, and the extent to which $O_2$ is derived from the aqueous humour. Barr and Roetman (1974) found a very steep gradient of $O_2$-tension sloping away from the endothelium when they measured the tension at different depths in the aqueous humour (Fig. 3.19), and this suggests that the cornea is, in fact, supplying the aqueous rather than vice versa. Barr et al. (1977), employing an ingenious experimental set-up in which the cornea was mounted on a layer of agar-gel, determined the tendency of $O_2$ to diffuse from aqueous

---

* It has been argued that most of the normal scattering in the cornea takes place at the epithelium and endothelium (Lindström et al., 1973); such a view might well vitiate the lattice theory of transparency, which implies that transparency and its loss depend on the structure of the stroma. McCally and Farrell (1976) have emphasized the instrumental and theoretical difficulties in assessing the variation of scattering with depth, and their own analysis confirms the notion that scattering depends on the stroma.

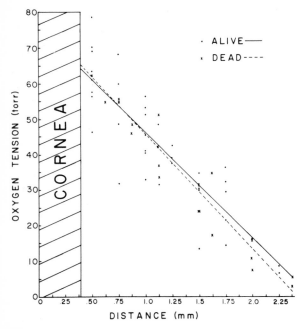

**Fig. 3.19** Oxygen tension in anterior chamber versus distance behind the corneal surface. Individual experimental values and least square analysis regression lines are shown for living (·—) and dead (x - - -) rabbits. (Barr and Roetman, *Invest. Ophthal.*)

humour to cornea as a function of gradient of $O_2$-tension in the agar, by a suitable mathematical derivation. They found that, when the cornea was exposed to air at an $O_2$-tension of 155 mm Hg, $O_2$ tended to diffuse out of the cornea when the $P_{O_2}$ in the aqueous humour was 95·9 mmHg and into the cornea when it was 107 mmHg, whilst there was zero net flux at an aqueous humour tension of 102·5 mmHg. Thus, under normal conditions, the tendency is for oxygen to diffuse out of the cornea so that all depths of the cornea receive their oxygen from the atmosphere.

**Contact lenses**

When access of oxygen to the outside of the cornea is prevented by a contact glass, or by keeping the eyes closed, the $O_2$-tension falls from 32 mmHg to 9 mmHg (Barr and Silver, 1973), and this is doubtless the cause for the corneal haziness that accompanies the wearing of contact lenses for some time (Smelser and Ozanics, 1953).

UTILIZATION BY COMPONENT LAYERS

Freeman (1972) developed a rapid polarographic technique for measuring $O_2$-consumption and applied it to the cornea, finding that the relative consumption-rates of endothelium, stroma, and epithelium were as 21:40:39. On the basis of volume of tissue, the epithelium

consumed $O_2$ ten times faster than stroma; interestingly, its rate was only a fifth of that of the endothelium, on this basis of tissue-volume.★

RESERVES

The excised cornea, or the cornea in the excised eye, can maintain normal metabolic function for many hours, and this is because of the reserves of glycogen in the epithelium as well as the lactate and glucose in the tissue (and aqueous humour of the intact eye).

SUPPLY OF MATERIALS

GLUCOSE

The supply to the cornea of materials required for metabolism, notably glucose, is theoretically possible by three routes; namely diffusion from the limbus, where there is a capillary network; diffusion from tears across the epithelium, and diffusion from aqueous humour. Maurice (1969) has examined the relative significance of these pathways and has shown that, because there is no significant flow of fluid through the stroma from limbus to central region, the transport by simple diffusion will be very inefficient. This is because, while the material diffuses slowly from limbus to central region, it will be exposed to concentration gradients from stroma to aqueous humour and tears (Fig. 3.20). Thus, with a small molecule or ion, if the concentration at the limbus is put equal to 100, the concentration at 6 mm from the limbus will be only 1 when a steady-state has been reached. On the other hand, where diffusion out into the aqueous humour and tears is restricted, as with plasma proteins, then a significant supply from the limbus is practicable, and would account for their presence in the cornea, as well as that of numerous enzymes. The supply of glucose, then, must be from aqueous humour and tears; however, the significance of the tears as a supply is probably negligible since, according to Giardini and Roberts, the concentration in this fluid is very small, whilst the permeability of the epithelium is low. A large number of experiments involving intralamellar implantation of plastic material in the cornea have shown that such a procedure leads to degeneration of the tissue above the layer of plastic, thereby emphasizing the importance of the diffusion of material from the aqueous humour; analysis of the epithelium showed that it had become deficient in glucose, glycogen and ATP (Turss, Friend and Dohl-

---

★ Herrmann and Hickman (1948) argued that the metabolisms of stroma and epithelium were interdependent, the epithelium requiring lactate from the stroma and the stroma requiring a phosphate-acceptor from the epithelium. More modern work has shown that all layers have the correct metabolic apparatus (Maurice and Riley, 1970).

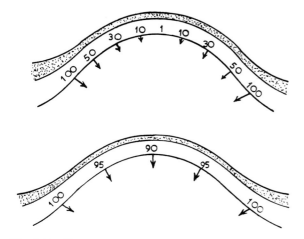

**Fig. 3.20** Illustrating the effects of diffusion from cornea into aqueous humour on the transport of material from limbus to corneal centre. *Above :* Diffusion into aqueous humour is rapid, so that at the corneal centre the effective concentration is low. *Below :* Diffusion out is slow, so that a relatively high concentration can be built up at the corneal centre.

man, 1970). Again, if the epithelium is removed we may place a cup on the stromal surface and measure diffusion of solutes from the aqueous humour into this fluid in the cup. Thoft *et al.* (1971b) measured a diffusion of glucose equal to 110 $\mu$g/hr/cm$^2$, which compares with a consumption of 38 to 83 $\mu$g/hr/cm$^2$. When the aqueous humour was replaced with silicone oil, the flux was decreased very considerably, indicating the importance of the aqueous humour as a source of glucose, in fact it was computed that the epithelium derived 80 per cent of its glucose from the aqueous humour and 20 per cent from the limbus (Thoft *et al.*, 1971b).

## CORNEAL VASCULARIZATION

Under conditions of metabolic stress, for example after injury, when the tissue may be invaded by leucocytes and fibrocytes, the nutritive supply and metabolic reserves may become inadequate, with the result that vessels sprout from the limbal plexus and grow into the stroma—*corneal vascularization.* It is likely that the effects of vitamin deficiency fall into the same category of metabolic stress; thus riboflavine deficiency is associated with corneal vascularization, whilst deficiencies of essential amino acids, such as tryptophan, lysine and methionine, will also produce the condition.

### STIMULUS TO VASCULARIZATION

The nature of the stimulus to ingrowth of vessels is a matter of some interest; it has been argued that the prime factor is the loosening of the tissue associated with the injury, i.e. the corneal oedema, which permits the vessels to grow into the tissue. It is certain that corneal vascularization is very frequently, if not invariably, associated with oedema, but the fact that, after tissue repair, the vessels may regress well before the subsidence of the oedema suggests that an additional factor is necessary (Langham, 1953). Campbell and Michaelson found that, if the injury to the cornea occurred too far from the periphery, i.e. at the pole of the cornea, vascularization failed to take place, and it was argued that a vascularization factor had to diffuse from the point of injury to the limbus to stimulate growth; if the distance from the point of liberation to the limbus was too great, sufficient would be lost to the aqueous humour to reduce its effectiveness.

### Tissue factor

Maurice, Zauberman and Michaelson (1966) concluded from their own experiments on rabbits that, whereas oedema was a necessary precondition for vascularization, an additional factor was required; and their observations were consistent with the notion of a factor, liberated by the cells of the tissue, that had to diffuse to the limbus to provoke ingrowth of the vessels. Thus they implanted fine tubes, running radially in the stroma, and found that an injury to the centre of the cornea would provoke the growth of vessels inside the tube provided this was open at both ends; if the end nearest the pole of the cornea had been occluded, there was no ingrowth. Again, vascularization was more certain to occur if the aqueous humour was replaced by silicone oil, and it could be argued that this was because loss of the factor had been reduced. Berggren and Lempberg (1973) considered that hyaline cartilage, which may be both vascular and avascular according to the age of the animal, might also contain a factor favouring or preventing vascularization; when they injected repeatedly extracts of calf cartilage into the central region of the rabbit's cornea, they obtained unequivocal vascularization; the layers of cartilage that proved most potent were not vascularized, which was the opposite to what they expected. However, if it is assumed that invading vessels inactivate the vascularizing factor, then the vascularized cartilage would have less factor to liberate, and be the less effective. Tissue injury causes the liberation of a number of pharmacologically active substances such as histamine, 5-hydroxytryptamine and bradykinin; and Zauberman *et al.* (1969) showed that the growth of vessels into tubes implanted in the cornea was more likely to occur if these tubes contained acetylcholine, histamine and 5-hydroxytryptamine, although bradykinin had no effect.

## NERVOUS SUPPLY

The innervation of the cornea is probably entirely sensory, mediated by the ophthalmic division of the trigeminal by way of the ciliary nerves. According to Zander and Weddell the fibres travel towards the limbus mainly in the perichoroidal space, and enter the substance of the cornea from three levels, scleral, episcleral and conjunctival. Most of the bundles entering the cornea are arranged radially and give rise, by subdivision, to a plexus in the stroma from which fibres supply the different regions. So great is the branching of any nerve fibre during its course in the cornea that a single axon may innervate an area equivalent to a quadrant. The endings are free, similar to those of pain fibres in other parts of the body; those reaching the epithelium are extremely fine, and send loops between the cells to within one cell-layer of the surface; similarly in the endothelium, the terminations are very fine.

### MODALITIES OF SENSATION

Since v. Frey's time it has been categorically stated that the cornea is sensitive only to pain, in the sense that all forms of stimulus in man arouse only this sensation. According to Lele and Weddell (1956), however, the sensation of touch, as distinct from that of pain, may be evoked by gentle application of a nylon thread; again, a copper cylinder, maintained at 1·5°C above or below the corneal temperature, evoked a sensation of warmth or cold. By recording from the ciliary nerves of the cat, Lele and Weddell (1959) found action potentials in the myelinated fibres in response to both thermal and mechanical stimuli, and it appeared that the same fibres responded to both types of stimulus. Thus, apparently in contradiction to Müller's Law of Specific Nerve Energies, the sensory fibres of the cornea were able to mediate both thermal and touch sense. According to Kenshalo, however, the sensations evoked by thermal stimulation of the human cornea are not those of warmth and cold but are of different types of 'irritation'; thus the nerve endings are *sensitive* to thermal change as well as mechanical, but the sensations evoked with the different stimuli are essentially of the same modality.

### REGENERATION OF FIBRES

When the cornea is cut through, as in making a section for cataract extraction, there is a loss of sensibility of the affected region, followed by a return associated with the growth of new fibres from the cut bundles proximal to the scar; there is no sprouting of adjacent uninjured neurones, however, since Rexed and Rexed repeatedly renewed a cut in the periphery and found no recovery of sensibility in the area supplied by the cut fibres.★

## RESPONSE OF THE CORNEA TO INJURY

### STROMAL REPAIR

The repair processes taking place in endothelium and epithelium have already been described; damage to the substantia propria is associated with the migration of new cells into the damaged area, either by chemotaxis from the limbal capillaries or, if the damage is severe, by the ingrowth of vessels into the stroma—the so-called corneal vascularization. The cells secrete new collagen fibres, but because the orientation of fibres is not so accurate as in normal tissue some degree of light-scattering remains in the scar. When the repair processes are over, the wandering cells that moved in settle down to become corneal corpuscles (Maumenee and Kornbluth).

### CORNEAL GRAFT

The cornea is peculiar in that it is possible to transplant a portion from one individual to another of the same species, i.e. to carry out a *homograft*. This may be the result of the low antigenic powers of the collagen of the cornea, or of the absence of a vascular supply through which the antigen-antibody response has to be mediated. An important element in the 'take' of the graft seems to be the sealing into place of the donor tissue by the migration of the host's corneal corpuscles to the cut edge of the trephine hole, and the extension of processes into the donor tissue (Hoffmann and Messier).

---

★A vegetative innervation of the cornea has been postulated and denied; the technique developed by Falck for identifying adrenergic fibres by virtue of their fluorescence has permitted the identification of this type of fibre in the anterior layers of the stroma of the embryonic cornea; in primates these fibres disappear after birth but in most other mammals they remain (Laties and Jacobowitz, 1964; Ehinger, 1966).

## REFERENCES

Andersen, E. I., Fischbarg, J. & Spector, A. (1974) Disulfide stimulation of fluid transport and effect on ATP level in rabbit corneal endothelium. *Exp. Eye Res.* **19**, 1–10.

Barfort, P. & Maurice, D. (1974) Electrical potential and fluid transport across the corneal endothelium. *Exp. Eye Res.* **19**, 11–19.

Barr, R. E., Hennessey, M. & Murphy, V. G. (1977) Diffusion of oxygen at the endothelial surface of the rabbit cornea. *J. Physiol.*, **270**, 1–8.

Barr, R. E. & Roetman, E. L. (1974) Oxygen gradients in the anterior chamber of anesthetized rabbits. *Invest. Ophthal.* **13**, 386–389.

Barr, R. E. & Silver, I. A. (1973) Effects of corneal environment on oxygen tension in the anterior chambers of rabbits. *Invest. Ophthal.* **12**, 140–144.

Benedek, G. B. (1971) Theory of transparency of the eye. *Appl. Optics* **10**, 459–473.

Berggren, L. & Lempberg, R. (1973) Neovascularization in the rabbit cornea after intracorneal injections of cartilage extracts. *Exp. Eye Res.* **17**, 261–273.

Bito, L. Z., Roberts, J. C. & Saraf, S. (1973) Maintenance of normal corneal thickness in the cold *in vivo* (hibernation) as opposed to *in vitro*. *J. Physiol.* **231**, 71–86.

Blümcke, S. & Morgenroth, K. (1967) The stereo ultrastructure of the external and internal surface of the cornea. *J. Ultrastr. Res.* **18**, 502–518.

Borscherding, M. S. *et al.* (1975) Proteoglycans and collagen fibre organization in human corneoscleral tissue. *Exp. Eye Res.* **21**, 59–70.

Breen, M., Johnson, R. L., Sittig, R. A., Weinstein, H. G. & Veis, A. (1972) The acidic glycosaminoglycans in human fetal development and adult cornea, sclera and skin. *Conn. Tissue Res.* **1**, 291–303.

Campbell, F. W. & Michaelson, I. C. (1949) Blood-vessel formation in the cornea. *Br. J. Ophthal.* **33**, 248–255.

Candia, O. A. (1973) Short-circuit current related to active transport of chloride in frog cornea: effects of furosemide and ethacrynic acid. *Biochim. biophys. Acta* **298**, 1011–1014.

Candia, O. A. & Askew, W. A. (1968) Active sodium transport in the isolated bullfrog cornea. *Biochim. biophys. Acta* **163**, 262–265.

Candia, O. A., Bentley, P. J. & Cook, P. I. (1974) Stimulation by amphotericin B of active Na transport across amphibian cornea. *Amer. J. Physiol.* **238**, 1438–1444.

Chalfie, M., Neufeld, A. H. & Zadunaisky, J. A. (1972) Action of epinephrine and other cyclic AMP-mediated agents on chloride transport of the frog cornea. *Invest. Ophthal.* **11**, 644–650.

Cogan, D. G. & Hirsch, E. D. (1944) The cornea. VII. Permeability to weak electrolytes. *Arch. Ophthal., N.Y.* **32**, 276–282.

Cogan, D. G. & Kinsey, V. E. (1942a) Transfer of water and sodium by osmosis and diffusion through the excised cornea. *Arch. Ophthal., N.Y.* **27**, 466–476.

Cogan, D. G. & Kinsey, V. E. (1942b) The cornea. V. Physiologic aspects. *Arch. Ophthal., N.Y.* **28**, 661–669.

Cox, J. L., Farrell, R. A., Hart, R. W. & Langham, M. E. (1970) *J. Physiol.* **210**, 601–616.

Davson, H. (1949) Some considerations on the salt content of fresh and old ox corneae. *Br. J. Ophthal.* **33**, 175–182.

Davson, H. (1955) The hydration of the cornea. *Biochem. J.* **59**, 24–28.

Dikstein, S. & Maurice, D. (1972) The metabolic basis to the fluid pump in the cornea. *J. Physiol.* **221**, 29–41.

Dohlman, C. H. & Balazs, E. A. (1955) Chemical studies on Descemet's membrane of the bovine cornea. *Arch. Biochem. Biophys.* **57**, 445–457.

Donn, A., Maurice, D. M. & Mills, N. L. (1959) The active transport of sodium across the epithelium. *Arch. Ophthal., N.Y.* **62**, 748–757.

Donn, A., Miller, S. & Mallett, N. (1963) Water permeability of the living cornea. *Arch. Ophthal., Chicago* **70**, 515–521.

Ehinger, B. (1966) Adrenergic nerves to the eye and to related structures in man and in the cynomolgus monkey (*Macaca irus*). *Invest Ophthal.* **5**, 42–52.

Farrell, R. A., McCally, R. L. & Tatham, P. E. R. (1973) Wave-length dependencies of light scattering in normal and cold swollen rabbit corneas and their structural implications. *J. Physiol.* **233**, 589–612.

Fatt, I. & Bieber, M. T. (1968) The steady-state distribution of oxygen and carbon dioxide in the *in vivo* cornea. I. *Exp. Eye Res.* **7**, 103–112.

Fatt, I. & Hedbys, B. O. (1970) Flow conductivity of human corneal stroma. *Exp. Eye Res.* **10**, 237–242.

Fischbarg, J. (1972) Potential difference and fluid transport across rabbit corneal endothelium. *Biochim. biophys. Acta* **288**, 362–366.

Fischbarg, J. (1973) Active and passive properties of the rabbit corneal endothelium. *Exp. Eye Res.* **15**, 615–638.

Fischbarg, J. & Lim, J. J. (1974) Role of cations, anions and carbonic anhydrase in fluid transport across rabbit corneal endothelium. *J. Physiol.* **241**, 647–675.

Fischer, F. H. & Zadunaisky, J. A. (1975) Sodium requirement for the chloride transport stimulation by epinephrine in the rabbit corneal epithelium. *Physiologist* **18**, 214.

Fischer, F. H. & Zadunaisky, J. A. (1977) Electrical and hydrophilic properties of fish cornea. *Exp. Eye Res.* **25**, 149–161.

François, J., Rabaey, M. & Vandermeersche, G. (1954) Etude de la cornée et de la sclérotique. *Ophthalmologica, Basel* **127**, 74–85.

Freeman, R. D. (1972) Oxygen consumption by the component layers of the cornea. *J. Physiol.* **225**, 15–32.

Giardini, A. & Roberts, J. R. E. (1950) Concentration of glucose and total chloride in tears. *Br. J. Ophthal.* **34**, 737–743.

Goldman, J. N. & Benedek, G. B. (1967) The relationship between morphology and transparency in the non-swelling corneal stroma of the shark. *Invest. Ophthal.* **6**, 574–600.

Green, K. & Otori, T. (1970) Studies on corneal physiology *in vitro. Exp. Eye Res.* **9**, 268–280.

Hale, P. N. & Maurice, D. M. (1969) Sugar transport across the corneal endothelium. *Exp. Eye Res.* **8**, 205–215.

Harris, J. E. & Nordquist, L. T. (1955) The hydration of the cornea. I. Transport of water from the cornea. *Am. J. Ophthal.* **40**, 100–110.

Hart, R. W. & Farrell, R. A. (1969) Light scattering in the cornea. *J. opt. Soc. Amer.* **59**, 766–774.

Herrmann, H. & Hickman, F. H. (1948) *Bull. Johns Hopkins Hosp.* **82**, 182, 225, 260.

Hirsch, M., Faure, J.-P., Marquet, O. & Payrau, P. (1975) Régénération de l'endothélium cornéen chez le lapin. *Arch. d'Ophthal.* **35**, 269–278.

Hirsch, M., Renard, G., Faure, J.-P. & Pouliquen, Y. (1976) Formation of intracellular spaces and junctions in regenerating rabbit corneal endothelium. *Exp. Eye Res.* **23**, 385–397.

Hirsch, M., Renard, G., Faure, J.-P., & Pouliquen, Y. (1977) Study of the ultrastructure of the rabbit corneal endothelium by the freeze-fracture technique: apical and lateral junctions. *Exp. Eye Res.* **25**, 277–288.

Hodson, S. (1971) Evidence for a bicarbonate-dependent sodium pump in corneal endothelium. *Exp. Eye Res.* **11**, 20–29.

Hodson, S. (1974) The regulation of corneal hydration by a salt pump requiring the presence of sodium and bicarbonate ions. *J. Physiol.* **236**, 271–302.

Hodson, S. (1975) The regulation of corneal hydration to maintain high transparency in fluctuating ambient temperatures. *Exp. Eye Res.* **20**, 375–381.

Hodson, S., Miller, F. & Riley, M. (1977) The electrogenic pump of rabbit corneal endothelium. *Exp. Eye Res.* **24**, 249–253.

Hoffmann, R. S. & Messier, P. E. (1949) *Arch. Ophthal., N.Y.* **42**, 140, 148.

Iwamoto, T. & Smelser, G. K. (1965) Electron microscopy of the human corneal endothelium with reference to transport mechanisms. *Invest. Ophthal.* **4**, 270–284.

Jakus, M. (1946) The fine structure of Descemet's membrane. *J. Biophys. biochem. Cytol.* **2**, Suppl., 243–252.

Jakus, M. (1961) The fine structure of the human cornea. In *Structure of the Eye.* Ed. Smelser, pp. 343–366. New York: Academic Press.

Kaye, G. I. & Pappas, G. D. (1962) Studies on the cornea. I. The fine structure of the rabbit cornea and the uptake and transport of colloidal particles by the cornea *in vivo. J. Cell Biol.* **12**, 457–479.

Kaye, G. I., Sibley, R. C. & Hoefle, F. B. (1973) Recent

studies on the nature of and function of the corneal endothelial barrier. *Exp. Eye Res.* **15**, 585–613.

Kayes, J. & Holmberg, A. (1960) The fine structure of Bowman's layer and the basement membrane of the corneal epithelium. *Am. J. Ophthal.* **50**, 1013–1021.

Kenshalo, D. R. (1960) Comparison of thermal sensitivity of the forehead, lip, conjunctiva and cornea. *J. app. Physiol.* **15**, 987–991.

Khodadoust, A. A. & Green, K. (1976) Physiological function of regenerating endothelium. *Invest. Ophthal.* **15**, 96–101.

Kim, J. H., Green, K., Martinez, M. & Paton, D. (1971) Solute permeability of the corneal endothelium and Descemet's membrane. *Exp. Eye Res.* **12**, 231–238.

Klyce, S. D. (1972) Electrical profiles in the corneal epithelium. *J. Physiol.* **226**, 407–429.

Klyce, S. D. (1975) Transport of Na, Cl, and water by the rabbit corneal epithelium at resting potential. *Amer. J. Physiol.* **228**, 1446–1452.

Klyce, S. D., Neufeld, A. H. & Zadunaisky, J. A. (1973) The activation of chloride transport by epinephrine and Db cyclic-AMP in the cornea of the rabbit. *Invest. Ophthal.* **12**, 127–139.

Klyce, S. D. & Wong, R. K. S. (1977) Site and mode of adrenaline action on chloride transport across the rabbit corneal epithelium. *J. Physiol.* **266**, 777–799.

Kreutziger, G. O. (1976) Lateral membrane morphology and gap junction structure in rabbit corneal endothelium. *Exp. Eye Res.* **23**, 285–293.

Kuwabara, T., Perkins, D. G. & Cogan, D. G. (1976) Sliding of the epithelium in experimental corneal wounds. *Invest. Ophthal.* **15**, 4–14.

Laing, R. A., Sandstrom, M. M., Berrospi, A. R. & Leibowitz, H. M. (1976) Changes in the corneal endothelium as a function of age. *Exp. Eye Res.* **22**, 587–594.

Langham, M. E. (1952) Utilization of oxygen by the component layers of the living cornea. *J. Physiol.* **117**, 461–470.

Langham, M. E. (1953) Observations on the growth of blood vessels into the cornea. *Br. J. Ophthal.* **37**, 210–222.

Langham, M. E., Ed. (1969) *The Cornea.* Baltimore: Johns Hopkins Press.

Laties, A. & Jacobowitz, D. (1964) A histochemical study of the adrenergic and cholinergic innervation of the anterior segment of the rabbit eye. *Invest. Ophthal.* **3**, 592–600.

Lele, P. P. & Weddell, G. (1956). The relationship between neurohistology and corneal sensitivity. *Brain* **79**, 119–154.

Lele, P. P. & Weddell, G. (1959) Sensory nerves of the cornea and cutaneous sensibility. *Exp. Neurol.* **1**, 334–359.

Leuenberger, P. M. (1973) Lanthanum hydroxide tracer studies on rat corneal endothelium. *Exp. Eye Res.* **15**, 85–91.

Leyns, W. F., Heringa, C. & Weidinger, A. (1940) Water binding capacity of the cornea. *Acta brev. Neerland. Physiol.* **10**, 25–26.

Lindström, J.-I., Feuk, T. & Tengroth, B. (1973) The distribution of light scattered from the rabbit's cornea. *Acta ophthal.* **51**, 656–669.

Maumenee, A. E. & Kornbluth, W. (1949) Regeneration of the corneal stromal cell. *Am. J. Ophthal.* **32**, 1051–1064.

Maumenee, A. E. & Scholz, R. O. (1948) The histopathology of the ocular lesions produced by the sulfur and nitrogen mustards. *Bull. J. Hopk. Hosp.* **82**, 121–147.

Maurice, D. M. (1951) The permeability to sodium ions of the living rabbit's cornea. *J. Physiol.* **122**, 367–391.

Maurice, D. M. (1957) The structure and transparency of the cornea. *J. Physiol.* **136**, 263–286.

Maurice, D. M. (1967) Epithelial potential of the cornea. *Exp. Eye Res.* **6**, 138–140.

Maurice, D. M. (1969) The cornea and sclera. In *The Eye.* Ed. Davson, Vol. I, pp. 489–600. London: Academic Press.

Maurice, D. M. (1972) The location of the fluid pump in the cornea. *J. Physiol.* **221**, 43–54.

Maurice, D. M. & Riley, M. V. (1970) The cornea. In *Biochemistry of the Eye.* Ed. Graymore, C., pp. 1–103. London: Academic Press.

Maurice, D. M., Zauberman, H. & Michaelson, I. C. (1966) The stimulus to neovascularization in the cornea. *Exp. Eye Res.* **5**, 168–184.

McCally, R. L. & Farrell, R. A. (1976) The depth dependence of light scattering from the normal rabbit cornea. *Exp. Eye Res.* **23**, 69–81.

Meier, S. & Hay, E. D. (1973) Synthesis of sulfated glycosaminoglycans by embryonic corneal epithelium. *Dev. Biol.* **35**, 318–331.

Mishima, S., Kaye, G. I., Takahashi, G. H., Kudo, T. & Trenberth, S. M. (1969) The function of the corneal endothelium in the regulation of corneal hydration. In *The Cornea.* Ed. Langham, M., pp. 207–235. Baltimore: Johns Hopkins Press.

Mishima, S. & Trenberth, S. M. (1968) Permeability of the corneal endothelium to nonelectrolytes. *Invest. Ophthal.* **7**, 34–43.

Myers, D. B., Highton, T. C. & Rayns, D. G. (1973) Ruthenium red-positive filaments interconnecting collagen fibrils. *J. Ultrastr. Res.* **42**, 87–92.

Otori, T. (1967) Electrolyte content of the rabbit corneal stroma. *Exp. Eye Res.* **6**, 356–367.

Pedler, C. (1962) The fine structure of the corneal epithelium. *Exp. Eye Res.* **1**, 286–289.

Pfister, B. R. & Burstein, N. (1976) The alkali burned cornea. I. Epithelial and stromal repair. *Exp. Eye Res.* **23**, 519–535.

Potts, A. M. & Modrell, R. W. (1957) The transcorneal potential. *Am. J. Ophthal.* **44**, 284–290.

Praus, R. & Goldman, J. N. (1970) Glycosaminoglycans in the nonswelling corneal stroma of dogfish shark. *Invest. Ophthal.* **9**, 131–136.

Reinach, P. S., Schoen, H. F. & Candia, O. A. (1977) Effects of inhibitors of Na and Cl transport on oxygen consumption in the bullfrog cornea. *Exp. Eye Res.* **24**, 493–500.

Rexed, B. & Rexed, V. (1951) Degeneration and regeneration of corneal nerves. *Br. J. Ophthal.* **35**, 38–49.

Riley, M. V. (1969) Glucose and oxygen utilization by the rabbit cornea. *Exp. Eye Res.* **8**, 193–200.

Riley, M. V. (1977) Anion sensitive ATPase in rabbit corneal endothelium and its relation to corneal hydration. *Exp. Eye Res.* **25**, 483–494.

Riley, M. V. (1977) A study of the transfer of amino acids across the endothelium of the rabbit cornea. *Exp. Eye Res.* **24**, 35–44.

Riley, M. V., Miller, F., Hodson, S. & Ling, D. (1977) Elimination of anions derived from glucose metabolism as substrates for the fluid pump of rabbit corneal endothelium. *Exp. Eye Res.* **24**, 255–261.

Schwarz, W. (1953) Elektronenmikroskopische Untersuchungen uber den Aufbau der Sklera und der Cornea des Menschen. *Z. Zellforsch.* **38**, 20–49.

Schwarz, W. & Keyserlingk, D. G. (1966) Uber die Feinstruktur der menschlichen Cornea, mit besonderer Berüchsichtigung des Problems der Transparenz. *Z. Zellforsch.* **73**, 540–548.

Smelser, G. K. (1962) Corneal hydration. Comparative physiology of fish and mammals. *Invest. Ophthal.* **1**, 11–21.

Smelser, G. K. & Ozanics, V. (1953) Structural changes in corneas of guinea pigs after wearing contact lenses. *Arch. Ophthal., N.Y.* **49**, 335–340.

Smith, J. W. (1969) The transparency of the corneal stroma. *Vision Res.* **9**, 393–396.

Stanworth, A. & Naylor, E. S. (1953) Polarized light studies of the cornea. *J. exp. Biol.* **30**, 164–169.

Swan, K. C. & White, N. G. (1942) Corneal permeability. *Am. J. Ophthal.* **25**, 1043–1058.

Teng, G. C. (1961) The fine structure of the corneal epithelium and basement membrane of the rabbit. *Am. J. Ophthal.* **51**, 278–297.

Tervo, T. & Palkama, A. (1975) Electron microscopic localization of adenosine triphosphatase (NaK-ATPase) activity in the rat cornea. *Exp. Eye Res.* **21**, 269–279.

Thoft, R. A. & Friend, J. (1972) Corneal amino acid supply and distribution. *Invest. Ophthal.* **11**, 723–727.

Thoft, R. A. & Friend, J. (1975) Permeability of regenerated corneal epithelium. *Exp. Eye Res.* **21**, 409–416.

Thoft, R. A. & Friend, J. (1977) Biochemical transformation of regenerating ocular surface epithelium. *Invest. Ophthal.* **16**, 14–20.

Thoft, R. A., Friend, J. & Dohlman, C. H. (1971a) Corneal glucose concentration. *Arch. Ophthal.* **85**, 467–472.

Thoft, R. A., Friend, J. & Dohlman, C. H. (1971b) Corneal glucose flux. II. *Arch. Ophthal.* **86**, 685–691.

Toole, B. P. & Trelstad, R. L. (1971) Hyaluronate production and removal during corneal development in the chick. *Dev. Biol.* **26**, 28–35.

Trelstad, R. L. & Coulombre, A. J. (1971) Morphogenesis of the collagenous stroma in the chick cornea. *J. Cell Biol.* **50**, 840–858.

Trelstad, R. L., Hayashi, K. & Toole, B. P. (1974) Epithelial collagens and glycosaminoglycans in the embryonic cornea. *J. Cell Biol.* **62**, 815–830.

Trelstad, R. L. & Kang, A. H. (1974) Collagen heterogeneity in the avian eye: lens, vitreous body, cornea and sclera. *Exp. Eye Res.* **18**, 395.

Trenberth, S. M. & Mishima, S. (1968) The effect of ouabain on the rabbit corneal endothelium. *Invest. Ophthal.* **7**, 44–52.

Turss, R., Friend, J. & Dohlman, C. H. (1970) Effect of a corneal fluid barrier on the nutrition of the epithelium. *Exp. Eye Res.* **9**, 254–259.

Twersky, V. (1976) Transparency of pair-correlated, random distributions of small scatterers, with application to the cornea. *J. opt. Soc. Amer.* **65**, 524–530.

Van Horn, D. L. & Hyndiuk, R. A. (1975) Endothelial wound repair in primate cornea. *Exp. Eye Res.* **21**, 113–124.

Van Horn, D. L., Sendele, D. D., Seideman, S. & Buco, P. J. (1977) Regenerative capacity of the corneal endothelium in rabbit and cat. *Invest. Ophthal.* **16**, 597–613.

v. Sallmann, L., Caravaggio, L. L. & Grimes, P. (1961) Studies on the corneal endothelium of the rabbit. *Am. J. Ophthal.* **51**, Pt. 2, 955–969.

Zadunaisky, J. A. (1966) Active transport of chloride in frog cornea. *Am. J. Physiol.* **211**, 506–512.

Zadunaisky, J. A. (1969) The active chloride transport of the frog cornea. In *The Cornea.* Ed. Langham, M. E., pp. 3–34. Baltimore: Johns Hopkins Press.

Zadunaisky, J. A., Lande, M. A. & Hafner, J. (1971) Further studies on chloride transport in the frog cornea. *Amer. J. Physiol.* **221**, 1832–1836.

Zander, E. & Weddell, G. (1951) Observations on the innervation of the cornea. *J. Anat. Lond.* **85**, 68–99.

Zauberman, H., Michaelson, I. C., Bergman, F. & Maurice, D. M. (1969) Stimulation of neovascularization of the cornea by biogenic amines. *Exp. Eye Res.* **8**, 77–83.

# 4. The lens

## DEVELOPMENT AND STRUCTURE

The lens is a transparent biconvex body (Fig. 4.1); the apices of its anterior and posterior surfaces are called the *anterior* and *posterior poles* respectively; the line passing through the lens and joining the poles is the *axis*, whilst lines on the surface passing from one pole to the other are called *meridians*. The *equator* is a circle on the surface at right-angles to the axis (Fig. 4.1). Structurally, the lens is divided into the *capsule*, an elastic envelope capable of moulding the lens substance during accommodative changes (Sect. III), and by which the attachment to the zonule is made; the *epithelium*, extending as a single layer of cells under the anterior portion of the capsule as far as the equator; and the *lens substance* made up of fibres and interstitial cement material. The significance of the intimate structure of the lens substance can only be appreciated with a knowledge of its embryonic and later development.

## EMBRYOLOGY

The lens vesicle, on separation from the surface ectoderm, is a sphere whose wall consists of a single layer of epithelium. On their outer aspect the cells of this epithelium secrete a structureless hyaline membrane, the capsule; the cells of the posterior wall and the equatorial region, moreover, undergo marked differentiation and from these the adult lens is formed. The cells of the posterior wall differentiate first, becoming primitive fibres by a process of elongation (Fig. 4.2):

their nuclei ultimately disappear and they remain as an optically clear *embryonic nucleus* in the centre of the mature lens. At the equator, meanwhile, epithelial cells likewise elongate to become fibres; the anterior end of each passes forward towards the anterior pole, insinuating itself under the epithelium, whilst the posterior end passes backwards towards the posterior pole. This process continues, and since each new fibre passes immediately under the epithelium it forces the earlier formed fibres deeper and deeper; successive layers thus encircle the central nucleus of primitive fibres.

## FIBRE FORMATION

The actual process whereby the developing lens fibres force themselves over less recently formed ones may now be described. The epithelial cells at the equator tend to become arranged in meridional rows (Fig. 4.3); they assume a pyramidal shape and finally their inner parts elongate and turn obliquely to run under the epithelium of the anterior capsule; the cell is now a young fibre and its inner end, which must now be referred to as its anterior end, grows forward whilst its outer (posterior) end is pushed backwards by a similarly developing cell immediately in front. As the cells grow into the lens substance, the nuclei gradually recede with them, becoming arranged in a wide bend which curves round anteriorly as the process of elongation proceeds; this configuration is termed the *nuclear* or *bow zone*.

At a certain stage the nucleus disappears—*denucleation*—so that the fibres in the less superficial layers are non-nucleated cells.

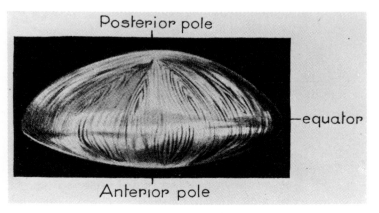

Posterior pole

equator

Anterior pole

Fig. 4.1   The human lens. (Duke-Elder, *Textbook of Ophthalmology*.)

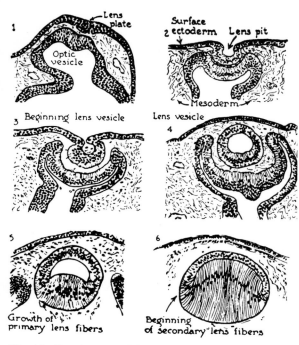

**Fig. 4.2**   Development of the human lens. (Bellows, *Cataract and Anomalies of the Lens.*)

pletely symmetrical, the meeting point would be the same for all cells, as illustrated schematically in Figure 4.4A. In fact, no such complete symmetry exists and the fibres meet each other in quite complicated figures which are called *sutures*; the simplest is probably the line-suture of the dogfish (Fig. 4.4B) and other lower vertebrates, whilst the human embryonic and adult nuclei have Y- and four-pointed stars respectively (Fig. 4.4C and D).

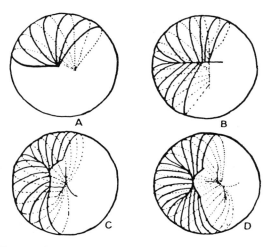

**Fig. 4.4**   Diagram showing the formation of lens sutures. A, simple arrangement with no sutures; B, linear sutures only; C, the Y-sutures of the human embryonic nucleus; D, four-pointed star of human adult nucleus. (Mann, *Development of the Human Eye.*)

## SUTURES

A recently formed lens fibre thus has its centre approximately at the equator, whilst the two limbs pass forward and backward towards the anterior and posterior poles respectively to meet the limbs of another cell developing from an 'antipodeal' position. If the system were com-

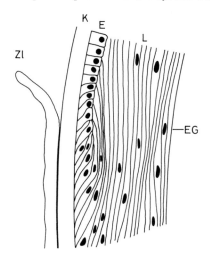

**Fig. 4.3**   Meridional section of adult human lens, illustrating the formation of meridional rows and the elongation of the epithelial cells. Zl, zonular lamella; K, lens capsule; E, lens epithelium; EG, level of epithelial border. (After Salzmann, *Anatomie u. Histologie d. menschlichen Augapfels.*)

## THE LENS FIBRES

The superficial lens fibres are long, prismatic, ribbon-like cells some 8 to 10 mm long, 8 to 12 $\mu$m broad, and 2 $\mu$m thick. An equatorial section of the lens cuts these fibres transversely, since they run in a generally antero-posterior direction; the appearance of such a section is shown in Figure 4.5, the orderly development of layer after layer of fibres from the cells of the meridional rows leading to a lamellar appearance, the so-called *radial lamellae*. A meridional section, on the other hand, cuts the fibres longitudinally, giving the so-called 'onion-scale' appearance (Fig. 4.3).

## CORTEX AND NUCLEUS

Since the oldest fibres become more sclerosed and less translucent than the more recently formed ones, the lens is not optically homogeneous; it may be broadly differentiated into a central core—the *nucleus*—and a softer *cortex*. With the slit-lamp the nucleus can be further differentiated into zones, which correspond with the periods at which their fibres were laid down, as in

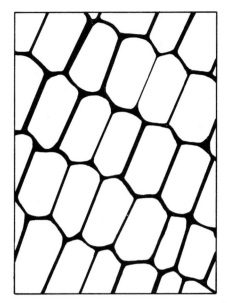

**Fig. 4.5** Drawing after phase-contrast micrograph of equatorial section of guinea-pig lens. Haematoxylin-eosin (× 3300). (Hueck and Kleifeld, *v. Graefe's Arch. Ophthal.*)

Figure 4.6. The lens grows throughout life, increasing from an average weight of 130 mg at 0 to 9 years to 255 mg at 80 to 89 years. The relative thickness of the cortex increases continuously until in old age it represents a third of the total. As a result of this growth, the curvature of the outer zone progressively decreases, i.e. the lens becomes flatter; the decrease in dioptric power that would otherwise ensue is prevented by the

simultaneous increase in the refractive index of the nucleus. When the increase in refractive index fails to keep pace with the flattening of the surface, the eye becomes hypermetropic; in early cataract the nucleus becomes excessively sclerosed so that the eye becomes myopic.

## DEVELOPMENT OF THE LENS FIBRES

Because of the continual growth throughout life, the lens cells differ in morphology, and the differences reveal the successive stages in the life-cycle of the lens cell from the relatively undifferentiated anterior lens cell—*epithelial cell*—through the young fibre immediately beneath the epithelium in the bow zone, to the mature cortical cells, whilst the innermost *nuclear* cells reveal the size and shape of the earliest formed embryonic and fetal fibres. In discussing the ultrastructure of the lens fibres, then, it is probably best to consider this in relation to the process of development.

### EPITHELIAL CELL

Kuwabara (1975) has described the changes in the epithelial cell that lead to its ultimate detachment from the basal lamina to become a lens fibre. In describing, first, the epithelium he points out that this layer is different from many other covering layers in that the apices of the cells face inwards, and that their main function is to produce lens fibres and synthesize the lens proteins, so that the rough endoplasmic reticulum

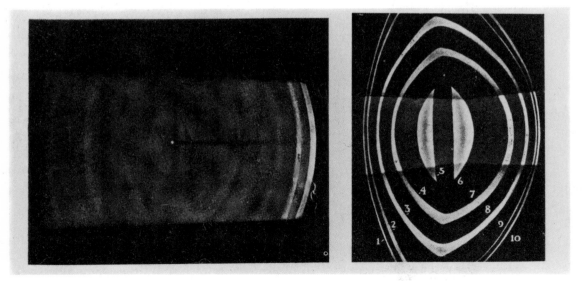

**Fig. 4.6** *Left :* Optical section of normal human crystalline lens as seen in the slit-lamp microscope. *Right :* Schematic representation of the bands of discontinuity of normal lens as seen in optical section. 1, anterior capsule; 2, anterior disjunctive band; 3, anterior band of adult nucleus; 4, anterior external band of foetal nucleus; 5, anterior internal band of foetal nucleus, and so on. (Grignolo, *Boll. d'Oculist.*)

is the only predominant organelle, whilst the profound lateral interdigitations may represent the anlage for the characteristic lateral junctions of the mature lens fibres, interdigitations that probably subserve a structural role. In its early developmental stage, the anterior cell begins to contain a fine granular substance, the lens proteins, and this feature is maintained throughout life.

### YOUNG CELLS IN BOW ZONE

The elongation of the anterior lens (epithelial) cells that have migrated to the equator leads to a bow configuration of the nuclei in nearby fibres, because the nuclei retain their central configuration whilst adjacent fibres show different degrees of elongation. The basal ends of these cells remain firmly attached to the basement membrane (capsule) and the apical ends extend towards the anterior of the lens. The cells in the innermost zone of the bow area appear to attach at the posterior pole with their anterior ends at the anterior pole.

### Sutures

Cells in the deeper cortex maintain apical junctions with cells extending from the other side, the junction line being the *anterior suture*. The basal ends of the cells detach from the basement membrane at the posterior pole when the cells expel their nuclei, and thus they are free to be pushed deeper into the lens. This area is seen as the Y-shaped *posterior suture*; the basal ends of the cells from opposite sides, forming the suture, seem to be only loosely joined, by comparison with the anterior apical junctions. In the monkey, the nucleated bow cells form a layer averaging 300 to 500 $\mu$m thick; the lateral interdigitations seen in the original epithelial cell become protrusions in the extended fibre, forming socket-and-ridge junctions.

### LOSS OF ORGANELLES

Maturation of the lens fibre is associated with loss of protein synthetic power, and this is revealed by a gradual loss of the typical organelles, such as ribosomes, mitochondria and rough-surfaced endoplasmic reticulum, which is manifest even in the cells of the meridional rows (Rafferty and Esson, 1974).

### DENUCLEATION

Kuwabara and Imaizumi (1974) followed the process of denucleation in the deeper zones of the bow; the loss of organelles goes hand in hand with the loss of chromatin from the nucleus, which gradually disappears with no signs of degeneration; the membrane finally breaks up into a set of vesicles and the outline becomes Feulgen-negative indicating the loss of all DNA. It would seem that the loss of organelles and nucleus is brought about by phagolysosomes (Gorthy *et al.*, 1971).

## MATURE LENS CELLS

These cells have lost their attachment to the basement membrane and become slightly flat hexagonal cylinders, or band-like fibres, with numerous knob-and-socket invaginations; the diameter of a knob is about 1 $\mu$m and the depth of a socket 1·5 $\mu$m, and gap-junctions are regularly present at the socket invaginations (Fig. 4.7). Cells in the deep cortex have curved band-like shapes; the thin edges are serrated with numerous socket junctions between adjacent fibres, whilst the flat sides are covered with densely distributed ridges (Fig. 4.8) which are probably engaged with those of adjacent cells. On cross-section it is seen that the cell membranes are finely reticulated, corresponding with the structure of the ridges. These mature cells have lost all their organelles so that ribosomes and microtubules, prominent in the cells of the superficial cortex, are absent, and we may presume that protein synthesis is complete.

### NUCLEAR CELLS

These, the oldest cells, are essentially similar in morphology to the cortical cells, but are smaller and more tightly packed; cell membranes can be seen although these appear poorly defined in the electron microscope. Figure 4.9 summarizes the changes in cell structure, as revealed by differences in location.

We may fill out this picture with a few more details derived from recent electron-microscopical studies.

### EPITHELIUM

This single layer of cells covers only the anterior surface of the lens and should be regarded more as a source of cells, produced by mitotic division near the equator, that produce the lens fibres, rather than as a covering epithelium devoted to selective transport. As seen in the electron microscope, the epithelial cells show marked lateral interdigitations (Fig. 4.10) on which characteristic epithelial junctional apparatuses are found; according to Cohen (1965) and Gorthy (1968) these include tight junctions or zonulae occludentes, that effectively limit diffusion along the intercellular clefts, although Kuwabara (1975) describes them as gap-junctions.

### Zones

According to Farnsworth *et al.* (1976), three zones may be distinguished exhibiting cells with different size and morphology, namely the equatorial, intermediate and central; thus the sizes varied from 8 × 9 $\mu$m in the equatorial region, to a more elongate form of 12 × 20 $\mu$m in the intermediate zone, to 20 × 45 to 30 × 75 $\mu$m in the intermediate-central zone. These differences are clearly related to the conversion from epithelial cell to lens fibre, mitotic activity being in the germinative equatorial intermediate zone, and inward migration in the intermediate-central zone.

### Relation to capsule

When viewed in the scanning electron microscope, the surfaces of the cells facing the basal lamina—capsule—show characteristic indentations, whilst the overlying basal lamina shows serpentine bodies on its surface which probably represent the

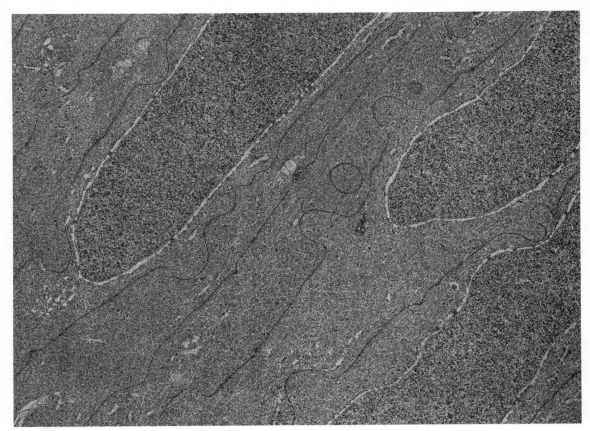

**Fig. 4.7** Cross-section of the deep bow zone illustrating the 'socket-and-junction' type of cell apposition. The cells are nucleated but the cytoplasm is devoid of micro-organelles except for a few microtubules ($\times$ 12 000). (Kuwabara, *Exp. Eye Res.*)

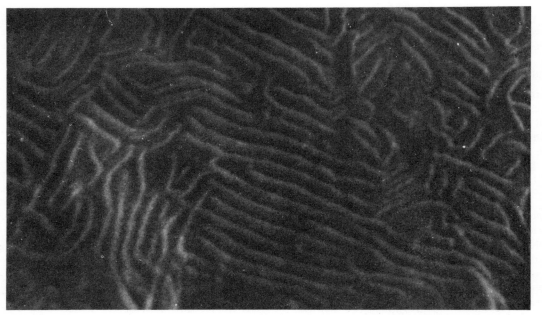

**Fig. 4.8** Illustrating the densely distributed ridges on the flat surface of the lens fibre ($\times$ 30 000). (Kuwabara, *Exp. Eye Res.*)

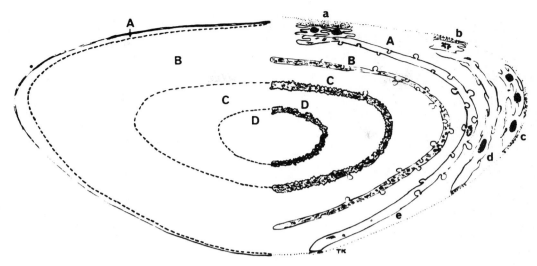

**Fig. 4.9**  Diagram to show distribution of the cells in various maturation stages and cytological appearances. Broken lines on the left indicate the areas proportionally. A, anterior cells and bow cells; B, cortex; C, deep cortex; D, lenticular nucleus. Sizes of cells shown on the right are not in proportion. Young lens cells, a–e, which occupy only a thin superficial layer, A, contain nuclei and attach to the basement membrane. Matured cells B and C occupy the majority of the lens. These cells have no basal attachment but they have unique apparatus in joining to each other by socket and fine ridge joints. Ridges become more numerous towards the centre. Older cells, D, form the lenticular nucleus. Transition between C and D is gradual. (Kuwabara, *Exp. Eye Res.*)

counterparts of these indentations, or craters, and doubtless these maintain adhesions between the cell and the basal lamina. When the basal lamina is peeled away and the cellular surface is compared with the lamina surface, filamentous processes can be seen to have been torn from the epithelial cells giving a pattern of the ridges of the cells indicating attachments at cell borders, perhaps similar to the inter-digitations between cells laterally. By contrast, the developing lens fibre, immediately beneath the epithelium, shows a

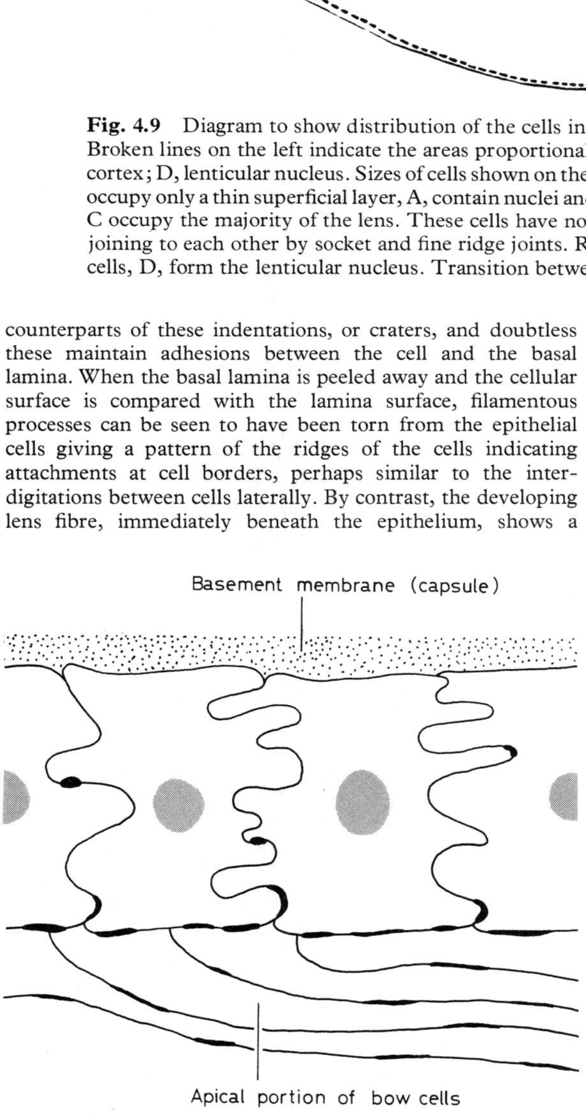

relatively smooth surface, so presumably during development the attachments to the basal lamina are broken. According to Kuwabara (1975) the epithelial cells are attached by numerous gap-junctions to superficial lens fibres.

INTERFIBRE CONNEXIONS

The connexions between fibres have been described by several workers using transmission and scanning electron microscopy. As indicated earlier, the characteristic feature is a ball-and-socket type; according to Dickson and Crock (1972) these occur all along, at or near, the six angular regions formed by adjoining flat surfaces of the fibres, and consist of short cytoplasmic stalks of 0·58 $\mu$m diameter expanding to form a 1·1 $\mu$m diameter dome that embeds in a complementary depression in the adjoining fibre. Also on the six flat surfaces there is a smaller type of tongue-and-groove connexion; the two are illustrated in Figure 4.11 on the basis of a scanning electron-microscopical view, whilst a cross-section of a ball-and-socket junction as seen by transmission electron-microscopy, is illustrated in Figure 4.7. These connexions presumably result in a very strong cohesion between lens fibres, tending to reduce the intercellular space, which may be as little as 8 nm (Philipson, 1969), a factor that may be important for transparency. At the suture zone, according to Wanko and Gavin (1961), there seem to be interdigitations between the tips of fibres from opposite sides (Fig. 4.12).

**Basement membrane (capsule)**

**Apical portion of bow cells**

**Fig. 4.10**  Diagrammatic illustrations of lateral interdigitations between lens epithelial cells, with junctional complexes. (Kuwabara, *Exp. Eye Res.*)

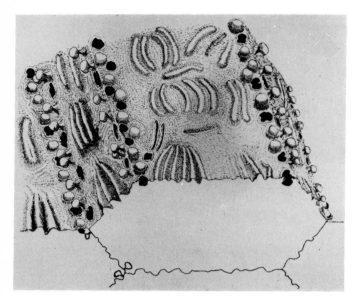

**Fig. 4.11** Diagram of primate lens fibres based on information derived from transmission and scanning electron micrographs. Rows of 'ball-and-socket' junctions are found at or near the six angular regions formed by adjoining flat surfaces. 'Tongue-and-groove' attachments cover the six flattened surfaces of the cells. In cross-section the former is composed of a narrow stalk with an expanded head region, embedded in an adjacent fibre. The latter is represented here in section by smaller finger-like projections of one fibre into its neighbour. (Dickson and Crock, *Invest. Ophthal.*)

### THE POSTERIOR SURFACE

Kuwabara has emphasized that the posterior surface of the lens is essentially similar to the anterior surface, being made up of epithelial cells, with the difference simply that the cells attached to the basement membrane at this side are elongated whilst their nuclei are situated only in the bow zone; they appear not to produce basement membrane to the same degree as the anterior (epithelial cells). The attachment of these posterior portions of the lens cells to the basal lamina (capsule) is rather loose, and their gap-junctions are sparse. By contrast, the anterior (epithelial) cells are attached to the underlying superficial cortical cells tightly by numerous gap-junctions.

### CELL CONTACTS

The width of the intercellular spaces between fibres of cortex and nucleus has been investigated with some care by Philipson *et al.* (1975); this seems to be characteristic of other tissues, giving a space in the region of 20 nm. At frequent points the space was reduced to form a junctional complex with a separation of membranes as small as 2 to 3 nm, and these junctions were described as gap-junctions by virtue of the penetration of lanthanum into the intercellular gaps. In the nucleus of the human lens, membranes were often hard to demonstrate, probably through difficulties in staining and fixation.

### CAPSULE

The light-microscopists described the capsule as a 'structureless membrane'; in the electron microscope, however, it appears as a regularly lamellated structure,

with a cement substance between the layers; and there is no doubt that the laminae arise from successive depositions of basement membrane by the underlying cells. This was demonstrated autoradiographically by Young and Ocumpaugh (1966) who administered $^3$H-

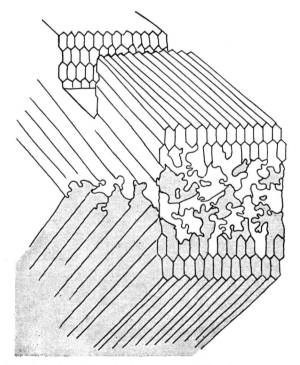

**Fig. 4.12** Lens fibres in the suture area. The diagram illustrates how the cortical fibres converge from opposite sectors at the lens suture, where they interlace and at the same time interdigitate with one another and also with fibres located in the same lens sector. (Wanko and Gavin, in *Structure of the Eye.*)

glycine and $^{35}$S-sulphate to young rats; labelling was at first exclusively in the epithelium and superficial fibres; an hour later there was some labelling of the capsule and after three to four hours practically all the label had been displaced to more anterior regions of the capsule, indicating that growth consisted of deposition from within. The capsule resembles Descemet's membrane and like this its basis is a collagen that shows characteristic differences from the 'ordinary collagen' of connective tissue both electron-microscopically and chemically. This difference in ultrastructure is probably related to the special need of this elastic envelope which it shares with Descemet's membrane; this need is presumably an extra resistance to internal slip of the fibres under stretch. Its strong positive periodic acid-Schiff reaction is due to the presence of a polysaccharide which is difficult to separate from the collagen (Pirie). The capsule is obviously divided into two layers, the zonular fibres being attached to the outer one, which is therefore called the *zonular lamella*. The features of this attachment, and the appearance of the zonular fibres, have been described by Farnsworth *et al.* (1976).

## TRANSPARENCY

The transparency of this densely packed mass of cells, containing very high protein concentrations that give the whole a high refractive index, probably results from a special character of the lens proteins—*crystallins*—which permits very high concentrations without separation of large aggregates that would cause scatter. In addition, the loss of nucleus and other intracellular organelles, and the dense packing of the fibres that reduces the intercellular spaces to very small values, are contributory factors. Thus the main discontinuities of refractive index are presumably the cell membranes separating the individual fibres; at any rate the X-ray diffraction pattern indicates a spacing of about 7 $\mu$m which might well correspond to the separation of membranes by the cytoplasm, and it is likely that the scattering of light by the lens, which makes it visible in the slit-lamp, is due to these discontinuities and could be responsible for the haloes around bright objects (Simpson, 1953).

### BIREFRINGENCE

The uniform distribution of protein in the lens fibre described by Philipson (1969; 1973) is consistent with an absence of optical anisotropy revealed by a negligible birefringence (Bettelheim, 1975); however, this author has suggested that the absence of birefringence is due to cancellation of a negative form-birefringence, governed by the orderly spacing of the lens membranes

at right-angles to the optic axis, and a positive intrinsic birefringence, due to an ordered arrangement of macromolecules in the cytoplasm; and it may be that the special feature of the lens crystallins that permits this orientation of aggregates is an adaptation that allows cancellation of the form birefringence to give increased transparency.

## THE LENS AS A FUNCTIONAL UNIT

### REGIONAL VARIATIONS

The essentials of the chemical composition of the lens are shown in Table 4.1. Because of its history, however, we may expect to find differences in composition according to (*a*) the total age of the lens and (*b*), in a given lens, according to the region selected. Thus the average water-content of the calf lens, containing a high percentage of young fibres, was found by Amoore, Bartley and van Heyningen to be 2·1 grammes of $H_2O$ per gramme of dry weight (g/g), comparing with 1·8 g/g

**Table 4.1** Chemical composition of the lens

|  | Young | Old |
|---|---|---|
| Water (per cent) | 69·0 | 64·0 |
| Protein (per cent) | 30·0 | 35·0 |
| Sodium (meq/kg $H_2O$) | 17·0 | 21·0 |
| Potassium | 120·0 | 121·0 |
| Chloride (meq/kg $H_2O$) | 27·0 | 30·0 |
| Ascorbic acid (mg/100 g) | 35·0 | 36·0 |
| Glutathione (mg/100 g) | 2·2 | 1·5 |

for ox lens. Again, by carefully peeling off concentric zones and analysing them, Amoore *et al.* found the following values from without inwards: 3·67, 2·86, 2·32, 1·79, 1·51 and 1·28 g/g respectively; thus the water-content of the innermost zone is less than half that of the outermost zone including the capsule. In the human, however, although there is a difference in water-contents between nucleus (63·4 ± 2·9 per cent) and cortex (68·0 ± 4·3 per cent) there seems to be no tendency for the water-content to decrease with age (Van Heyningen, 1972; Fisher and Pettet, 1973).

## SODIUM AND POTASSIUM

The lens is essentially a cellular tissue, so that we may expect a high concentration of potassium and a low concentration of sodium; this is generally true since the figures given by Amoore *et al.*, for example, indicate a value of approximately 20 m-molar for the concentration of sodium in the tissue-water and 120 m-molar for potassium. If this sodium is largely extracellular, whilst the potassium is largely intracellular, we may compute an extracellular space for the lens of about 13 per cent.

## EXTRACELLULAR SPACE

Direct measurements based on the distribution of large molecular weight extracellular markers, such as inulin, indicate a much smaller space, perhaps as low as 4 to 5 per cent for the mammalian lens (Thoft and Kinoshita, 1965; Paterson, 1970a), but the situation is complicated by the low diffusibility of inulin within the tissue compared with, say, mannitol, so that Yorio and Bentley (1976), for example, found values of 7 per cent and 12 per cent respectively using inulin and mannitol; and unless some mannitol enters the cells the value of 12 per cent should be the correct one. However, a great deal of this space seems to be provided by the capsule and perhaps a small pool of fluid immediately beneath, since, when Yorio and Bentley removed the capsule, the inulin-space fell to 1·3 per cent; a similar but smaller influence of removing the capsule was described by Paterson (1970a). A small value for the extracellular space is consistent with the histology of the tissue which shows the lens fibres to be tightly apposed to each other.

## Na+-K+-PUMP

In tissues such as nerve and muscle, where we also have a mass of cells with high internal potassium and high external sodium, these gradients of concentration are maintained by active processes, in the sense that a part of the metabolic energy of the cells is directed towards 'pumping out' any excess of sodium that penetrates into the cells; this Na+-pump is linked to transport of the K+-ion so that extrusion of Na+ from the cell is accompanied by penetration of K+, and a steady state is attained with the passive leaks of Na+ in and K+ out being compensated by active extrusion of Na+ and an accumulation of K+. This type of pump is specifically inhibited by cardiac glycosides, such as ouabain, which is also a specific inhibitor of an ATPase closely connected with the supply of metabolic energy for the pump.

## MEMBRANE POTENTIAL

The high internal concentration of K+ and low concentration of Cl− are associated with a resting, or membrane, potential across the cell membrane, as illustrated by Figure 4.13, the inside of the cell being negative, and under ideal conditions, being governed by the Nernst Equation:*

$$E = -\frac{RT}{zF}\ln\left(\frac{C_i}{C_o}\right)$$

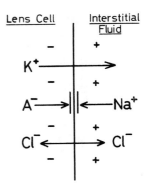

**Fig. 4.13**  Possible basis of potential across a lens fibre.

An important point to appreciate is that the K+ and Cl+ ions are, under these conditions, at equilibrium, although there are large concentration gradients; and this is because the potential operates in the opposite direction to the concentration gradient, so that the two ions are said to be at *equal electrochemical potential* across the membrane. By contrast, the Na+-ion is not in equilibrium, both the concentration gradient and the electrical potential favouring passage into the cell, and it is the pumping action that prevents the attainment of equilibrium, leaving the concentrations in their steady-state values.

## NON-ELECTROGENIC PUMP

An important point to appreciate is that the action of the Na+-pump does not give rise to the potential directly, but only secondarily by allowing the K+ to remain accumulated in the cell; thus, poisoning the pump will initially have no effect on the potential, and it will only fall in the course of time as the outside Na+ penetrates the cell and K+ leaks out in exchange, and the value of the Nernst-ratio becomes smaller. Thus the pump is said not to be *electrogenic*, by contrast with the corneal and frog-skin pumps where the active transport of Na+ causes a local accumulation of positive charge on one side of the skin, or cornea, that leads to a passive movement of an anion (Fig. 1.31, p. 31).

---

*The Nernst equation describes the potential for an ion at equilibrium; in most biological systems the given ions are rarely at equilibrium, and a more appropriate equation to describe the potential across a membrane is that of Goldman which takes into account the permeabilities of the membrane to other relevant ions, namely Na+ and Cl−, as well as K+:

$$E = \frac{RT}{F}\ln\frac{P_K[K]_o + P_{Na}[Na]_o + P_{Cl}[Cl]_i}{P_K[K]_i + P_{Na}[Na]_i + P_{Cl}[Cl]_o}$$

where the P's are relative permeability coefficients and the terms in square brackets are concentrations.

## LENS FIBRE

It may be that the situation in the lens is similar to that in muscle or nerve in that the tissue consists of a mass of fibrous cells each enclosed in a membrane capable of sustaining a resting potential associated with a high internal $K^+$-concentration and low $Na^+$-concentration; the situation would be represented schematically by Figure 4.14, where a single fibre represents the total; it is surrounded by extracellular fluid of about the same composition as the vitreous and aqueous humours, these latter fluids being presumably in some sort of dynamic equilibrium with the extracellular fluid since there is no doubt that the capsule is freely permeable to small crystalloids (Friedenwald).

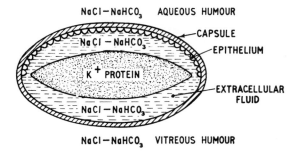

**Fig. 4.14** Illustrating the salt and water relationships of the lens fibres, extracellular space, aqueous humour and vitreous body. For simplicity the lens fibres are represented as a single one containing predominantly potassium and protein. The actual volume of the extracellular fluid may be very small indeed, since the fibres are very tightly packed.

## EPITHELIUM

A complicating feature is, of course, the layer of epithelial cells on the anterior surface, but it must be appreciated that any activities exerted by these in pumping ions into or out of the extracellular fluid of the lens will be largely vitiated by free exchanges across the posterior surface where there is no epithelial layer unless, of course, a layer of fibres on the posterior surface has become differentiated to permit it to function as an active membrane comparable with the epithelium.

## ACTIVE ION TRANSPORT

The importance of metabolism for the maintenance of the normal sodium, potassium and chloride relationships has been demonstrated by several studies; thus Langham and Davson (1949) found that within a few hours *post mortem* the chloride concentration in the lens began to rise, indicating, presumably, an absorption of aqueous humour; again, Harris and Gehrsitz (1951) found that cooling the lens *in vitro*, thereby inhibiting its metabolism, caused the potassium concentration to fall and that of sodium to rise, processes that could be reversed by re-warming.

An obvious interpretation of these findings is that the individual lens fibres maintain their high internal $K^+$-concentration and low $Na^+$ by virtue of active pumps; cooling deprives the pumps of metabolic energy and allows $K^+$ to escape, and $Na^+$ and $Cl^-$ to enter the fibres; the lost $K^+$ would diffuse from the extracellular space into the ambient medium.

## DUNCAN MODEL

Thus by treating the lens as a mass of cells insulated from each other by selectively permeable membranes, and surrounded by an extracellular fluid in free communication through the capsule with the aqueous and vitreous humours, we can account for the basic chemical composition of the lens and its response to environmental changes. However, a number of experimental studies have suggested that this picture is inadequate and that a better picture would be given by the system indicated schematically by Figure 4.15 in which the fibres are viewed as a syncytium, in the sense that there are gap-type junctions between adjacent fibres permitting free diffusion of ions from one cell to the next. On this basis, it has been argued that the outermost layers of lens fibres would constitute a cellular barrier, beneath the capsule, that would act as a single membrane sealing the contents of the lens fibres from the outside medium. This, which we may describe as the Duncan hypothesis, is a view that seemed to be demanded by the measured electrical potential between the inside of the lens and the adjacent vitreous humour, or aqueous humour, first described by Brindley.

## LENS POTENTIAL

### 'GIANT CELL'

Brindley (1956) impaled the amphibian lens with a microelectrode; as soon as the capsule was penetrated a potential, inside negative, of 63 to 68 mV, was registered; this was decreased if the external concentration of $K^+$ was raised. By fitting the Goldman equation to the potentials obtained when the external ion concentrations were varied, Brindley found that the potential could be accounted for by a system consisting of a 'giant cell' containing a high concentration of $K^+$ separated from aqueous or vitreous humour, containing low $K^+$ and high $Na^+$, by a membrane with relative permeabilities to $Na^+$ and $K^+$ in the ratio 0·03, comparable with that found in nerve. In the rabbit the potential was somewhat lower ($-66$ mV). Brindley found, moreover, that the potential across the anterior surface of the lens was the same as that measured across

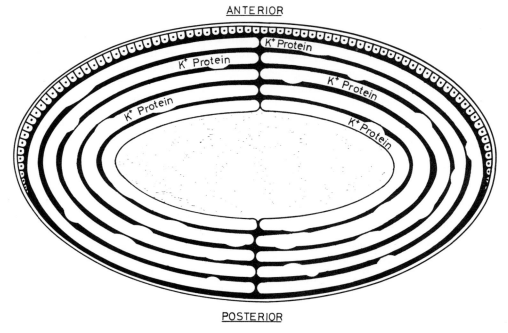

**Fig. 4.15**    Duncan's hypothesis to account for the lens potential based on a syncytial organization of the lens fibres so that the system acts as a giant cell with high internal concentrations of potassium and protein.

the posterior surface, so that it is unlikely that the potential was due to activity of the epithelial layer, i.e. it was unlikely to be due to an electrogenic pump comparable with that seen in the cornea.

### IMPORTANCE OF FIBRES

This point was subsequently stressed by Duncan (1969) who found that the potential of the toad lens, measured with a micro-electrode, remained the same however deep it penetrated. Furthermore, Andrée (1958) had found that the removal of the capsule and underlying epithelium only abolised the potential temporarily, so that after some hours it was re-established to reach a value of 68 mV compared with an initial one of 74 mV. Again, Duncan (1969) found that the changes in potential, measured between an electrode in the lens and one in the outside medium, caused by altering the concentration of $K^+$ in the outside medium, took place far too rapidly for the altered concentration of $K^+$ to have reached the internal electrode, so that the change of potential must have occurred at the surface of the lens and been transmitted electrically through the lens-substance to the electrode; if the electrode had been in a lens fibre insulated from adjacent fibres by a high-resistance membrane, then the effects of altered external concentration of $K^+$ would have had to wait on diffusion from the medium up to the fibre. If this fibre were connected to the surface through low-resistance path-

ways, as envisaged by the concept of a syncytial arrangement of Figure 4.15, then the effects would be transmitted rapidly, as indeed they are.

### IONIC CONCENTRATIONS AND PERMEABILITIES

In a more definitive study Delamere and Duncan (1977) have developed this concept of the lens potential further and applied it to both amphibian and mammalian lenses as well as to an invertebrate (*Sepiola*) lens. Table 4.2 shows the computed internal ion concentrations and also the Nernst potentials that would arise were the potential governed exclusively by the concentration difference of a given ion; thus if it were governed by $K^+$ alone it would be 90 mV in the frog compared with a measured value of 63 mV; the fact that the measured potential is less than the Nernst $K^+$-potential means that there is a significant permeability to $Na^+$ and $Cl^-$, and by measuring the effects of altered concentrations of all three ions on the potential and applying the Goldman or Kimizuka-Koketsu equations, estimates of the relative permeabilities of the limiting membrane responsible for the potential could be obtained. Since the electrical resistance of this membrane is a function of permeability to the ions and their concentrations, a function that is predictable on the basis of the above equations, estimates of the resistance or its reciprocal, conductance, were obtained and compared with those measured directly. As Table 4.3 shows, there is reason-

**Table 4.2** Concentrations of ions in lenses of different species (mmole/kg lens $H_2O$). (Delamere and Duncan, 1977.)

| Lens | Na | K | Cl |
|---|---|---|---|
| Frog | $14.7 \pm 1.1$ | $90.3 \pm 3.1$ | $17.7 \pm 2.0$ |
| Sepiola | $62.4 \pm 2.9$ | $124.9 \pm 2.8$ | $235.1 \pm 5.6$ |
| Bovine | $30.0 \pm 3.1$ | $142.0 \pm 3.4$ | $32.8 \pm 2.1$ |

The concentrations are given as the mean $\pm$ S.E. of thirty lenses in each case.

**Table 4.3** Computed internal ion concentrations, Nernst potentials and measured lens potentials. (Delamere and Duncan, 1977.)

| | $Na^+$ | $K^+$ | $Cl^-$ | $E_{Na}$ | $E_K$ | $E_{Cl}$ | $E_m$ (mV) |
|---|---|---|---|---|---|---|---|
| Frog | 9.4 | 94.9 | 12.8 | $+62$ | $-90$ | $-54$ | $-63.0 \pm 1.4$ |
| Sepiola | 39.9 | 131.0 | 219.3 | $+61$ | $-63$ | $-22$ | $-63.3 \pm 1.2$ |
| Bovine | 24.7 | 149.0 | 26.9 | $+43$ | $-84$ | $-44$ | $-23.1 \pm 0.9$ |

The ion concentrations (mmole/kg lens water) were calculated assuming a value of 5 per cent for the extracellular space.

able agreement between the two, suggesting that the basic concept of Duncan is correct.★

MEMBRANE RESISTANCE

A very important aspect of Duncan's treatment is the finding that, not only is the potential independent of position of the micro-electrode in the lens, but also that the resistance between an electrode inside the lens and one in the outside medium is also independent of position of the micro-electrode. Such an independence would be very unlikely were each individual fibre encased in a membrane of high electrical resistance, and this would be the case were the electrode within a lens fibre, or in the extracellular space, since the cells are so tightly packed together than the conducting pathway through the space would be tortuous and the resistance would be greater the greater the depth of the electrode in the tissue. Thus the independence of depth indicates an electrical coupling of the lens fibres such as that given by gap-junctions.

**Measurement**

In order to measure the resistance, Duncan placed a 'resistance electrode' within the lens and the potential between this and an external electrode was measured. A second 'current-passing' electrode was placed in the lens and the effect on the potential, due to a pulse of current through this current-passing electrode, was measured; and from this the resistance across the lens was computed. Later workers, e.g. Rae and Blankenship (1973) obtained conflicting results, but a thorough theoretical treatment of the problem of current-spread

within a cell from a micro-electrode, and a re-examination of the situation within the lens, has confirmed Duncan's conclusion (Eisenberg and Rae, 1976) to the effect that the resistance of the lens is, in effect, dominated by a single membrane at the surface; the resistance to flow of current within the tissue is rather high, namely 625 $\Omega$ cm compared with 130 $\Omega$ cm measured in cytoplasm, but the difference could be due to the resistance through the gap-junctions, i.e. it would depend on the effectiveness of the cell coupling.

CAPACITANCE

The estimate of membrane capacitance, $C_m$ was very high if the total area was equated to the area of the lens, so that it is likely that the area of membrane charged by the current pulses was much larger, and represented many individual fibre-membranes in the neighbourhood of the electrode.

POTENTIAL PROFILES

A basic question pertinent to the hypothesis of Duncan is 'Where is his micro-electrode located during measurements of potential and current flow?' If the individual fibres maintain high internal concentrations of $K^+$ they should individually have membrane potentials across them such that the inside of each fibre is negative to the extracellular space. If an electrode enters a given fibre the measured potential is, in effect, the resting potential of this fibre (Fig. 4.16), i.e. the potential between the inside and extracellular fluid which we may presume is in equilibrium with the aqueous humour and vitreous.

If the electrode within the fibre were pushed a little further so as just to emerge into the extracellular fluid, then a fall in potential should be recorded and, if there is free communication between the extracellular space and the surrounding medium, the potential should fall to zero. Rae (1974) argued that potential jumps should be measured on pushing a micro-electrode gently through the lens tissue, and furthermore that the frequency of the jumps should vary according to the angle at which the lens penetrated a given region; for example if it penetrated at right-angles to the long axis it would emerge from a fibre sooner than if it penetrated more obliquely. Rae's experimental study indicated that there were, indeed, successive changes in the frog's

★Paterson et al. (1974) have determined the activity of $K^+$ within the lens fibres by micro-electrode penetration, using a $K^+$-sensitive electrode; the activity was 83.6 mM which was converted to concentration by employing an activity coefficient of 0.77 to give 110 mM; this agreed with the values obtained by flame-photometry, assuming an extracellular space of 4.5 per cent, but it is higher than the value for frog shown in Table 4.3.

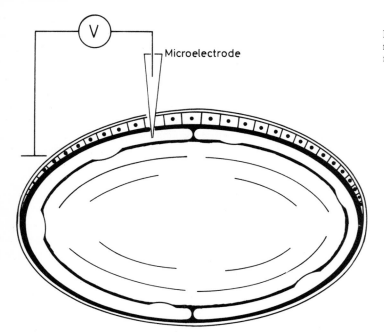

**Fig. 4.16** A micro-electrode is measuring the potential across the membrane of the fibre it is impaling.

potential as indicated by Figure 4.17 suggesting that each fibre does, indeed, maintain a potential difference between itself and the extracellular fluid. However, the potential does not fall to zero, indicating that the extracellular fluid is negative in respect to the bathing medium, but the general concept of a coupling between the fibres is sustained although not proved, and Rae emphasized that more direct evidence on this point would be valuable; thus he found (1974c) that the dye,

Procion yellow, injected into a single fibre, was not confined to this but passed from cell to cell, probably through gap-junctions similar to those that couple the photoreceptors (p. 259). Tentatively, then, we may assume that the potentials recorded by micro-electrodes normally register the potential within the fibre system; whether the difference of potential between the lens and its medium is entirely governed by the fibre-membrane potential of the surface-layers, as suggested by Duncan,

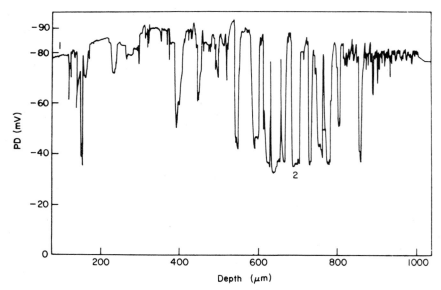

**Fig. 4.17** Indicating successive jumps in potential between a recording micro-electrode pushed into the lens substance, presumably resulting from penetration of successive fibres. (Rae, *Exp. Eye Res.*)

or whether there is an additional step leaving the extra-cellular fluid negative in respect to the outside medium, awaits further investigation.*

## ASYMMETRY POTENTIALS

When examined *in vivo*, or in the excised lens surrounded uniformly by a bathing medium, the lens shows the same difference of potential across both anterior and posterior faces; however Candia *et al.* (1970), by mounting a lens in an apparatus illustrated by Figure 4.18, in which the whole lens separates two compartments, found that there was a difference of potential between anterior and posterior faces of about 26 mV, the front face being positive with respect to the back face. Thus the separate trans-lens potentials

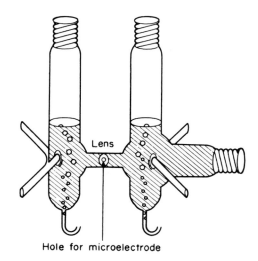

**Fig. 4.18** Measurement of asymmetry potential of lens. (Candia, *Exp. Eye Res.*)

were −57 mV for the anterior face and −31 mV for the posterior face, leading to this *asymmetry potential* of 26 mV. *In vivo* this difference of potential would be short-circuited by the suspensory ligament, and *in vitro*, when completely immersed, by the bathing solution. Thus the potential, as normally measured, would be governed by the separate potentials short-circuited, and from estimates of the appropriate electrical parameters Duncan computed that the toad potential would be −49 mV, considerably less than that actually found when totally immersed, namely 59 mV. More recently Duncan *et al.* (1977) have improved on the experimental study of the asymmetry potential, pointing out that the lens capsule itself represents a short-circuiting medium in the double-chamber system of Candia *et al.* This was avoided by immersing the lens in liquid paraffin and making separate electrical contacts with agar-bridges to obtain the trans-lens asymmetry potential,

while micro-electrodes inserted into the anterior and posterior faces measured the separate potentials. In the ox lens studied by them, the potentials were smaller, as indicated in Table 4.4, to give a trans-lens potential of only 6·3 mV. In general, each trans-membrane potential, i.e. that across the anterior and posterior faces respectively, responded to changed ion concentrations in the external medium, but not identically, so that the limiting membranes governing the separate potentials are not the same; and it is interesting that the posterior membrane, presumably made up of a layer of lens fibres, had the higher resistance although the epithelium on the anterior surface might have been expected to provide additional resistance.

**Table 4.4** Potential differences (mV) in bovine lens. (Duncan *et al.*, 1977.)

| Lens | I–A | I–P | A–P |
|------|-----|-----|-----|
| In situ | −30·0 ± 2·0 | −30·3 ± 2·0 | 0·3 ± 0·4 |
| Under oil | −28·7 ± 1·4 | −21·7 ± 1·4 | 6·3 ± 0·4 |

*I* refers to inside of lens; *A* = anterior medium; *P* = posterior medium. Note that the trans-lens potential difference *in situ* is not significantly different from zero, whereas the potential difference for the excised lens, measured under oil, is significantly different.

## THE PUMP-LEAK HYPOTHESIS

Kinsey and his colleagues have made a series of studies of transport of solutes from the external medium into the lens and interpreted them on the basis of active transport mechanisms, operating across the lens epithelium, associated with passive leak of the actively transported materials at the posterior surface where there is no epithelium. Thus the exchange of $K^+$ between medium and lens has been described by an active accumulation across the epithelium from the aqueous humour associated with a passive diffusion of the accumulated $K^+$ into the posterior chamber and vitreous body. The distributions of $K^+$ between these fluids are not necessarily inconsistent with this view (Bito and Davson, 1964), whilst a kinetic study of the uptake of $K^+$, $Rb^+$ and $Cs^+$ by Kinsey and McLean (1970) was also consistent with the pump-leak hypothesis, in the sense that the measured uptakes could be made to fit the solutions to simple equations based on a primary accumulation process at the epithelium, showing competitive inhibition between the three cations, and a passive diffusion process at the posterior capsule.†

---

* Rae (1974b) described two distinctly different compartments in the frog lens according as the potential measured in the one was high (− 70 mV) and the other low (− 30 mV). He has discussed in detail the possibility that the low potential represents that obtained when the electrode is in the extra-cellular compartment.

†The estimated transfer coefficients for the epithelium were 0·095, 0·056 and 0·024 hour⁻¹ for $K^+$, $Rb^+$ and $Cs^+$ respectively; this means that passage is not through a simple water-filled pore, which would be unable to exert this kind of selectivity.

## SODIUM

So far as the exchanges of $Na^+$ are concerned, Kinsey (1973) has applied the pump-leak hypothesis to experimental measurements of $Na^+$-fluxes on the basis of an assumed passive diffusion of $Na^+$ through the posterior surface of the lens and an active transport out through the anterior surface, mediated through the lens epithelium. As with the other ions studied, the curves of uptake and efflux of isotopic $Na^+$ could be made to fit theoretical ones by appropriate choice of parameters. However, as Duncan has emphasized, the difficulty with the pump-leak hypothesis is that it implies a low-resistance posterior membrane compared with that of the actively transporting anterior membrane (epithelium), but the separate study of resistances, and of variations of potential with ionic concentrations, all indicate a basic similarity between the two membranes, with the posterior membrane the more highly resistant. Nevertheless, the pump-leak hypothesis, by emphasizing the capabilities of the epithelium, has been of value, and it may well be that this active transport contributes to the postulated extracellular potential of Sperelakis and Potts (1959) and of Rae (1974b).

## EFFECTS OF OUABAIN

In this connexion, the effects of this inhibitor of $Na^+$-pumps on the lens potentials is of some interest. Kinsey and McGrady (1971) found that ouabain caused a relatively rapid fall in the anterior face potential, from $-40$ to $-20$ mV in the rabbit lens, this being followed by a much slower change towards zero. In the toad-lens, too, Bentley and Candia (1971) found a decrease in the short-circuit current, measured by short-circuiting the anterior and posterior faces, when ouabain was in contact with the anterior face, by contrast with a negligible effect on the posterior face. The effect was slow, as shown by Table 4.5. As we have indicated

**Table 4.5** Effects of ouabain on toad-lens short-circuit current (Bentley and Candia, 1971)

|  | Short-circuit current | | | | |
|  | 2 hr | 4 hr | 6 hr | 8 hr | 12 hr |
|---|---|---|---|---|---|
| Control | 93 | 81 | 73 | 68 | 70 |
| Ouabain ant. face | 70 | 45 | 36 | 29 | 9 |
| Ouabain post. face | 89 | 80 | 71 | 62 | 44 |

earlier, mere inhibition of a $Na^+$-pump in a tissue like muscle or nerve need not affect the potential immediately, since this depends primarily on the concentration differences of $K^+$. Inhibition of the $Na^+$-pump leads, after a time, to a leak of $K^+$ out of the cell and of $Na^+$ in, and thus the concentration difference of $K^+$ falls and hence the potential. Thus, if the lens potential were governed completely by the $K^+$ concentration gradient, ouabain should only have a slow effect as the concentration of $K^+$ fell.

## ELECTROGENIC PUMP

Hence the rapid decline described by Kinsey and McGrady, and later by Paterson et al. (1975), could be directly associated with inhibition of the $Na^+$-pump, so that the pump would be *electrogenic*, possibly accelerating $Na^+$ out of the lens across the epithelium, as suggested by Kinsey's pump-leak hypothesis; if this acceleration is associated with a slower passive diffusion of $Cl^-$ or $HCO_3^-$ we have the condition for an electrogenic pump comparable with the frog skin or corneal epithelium. Superimposed on this would be the activities of the lens membranes at anterior and posterior faces; these need not be electrogenic, the extrusion of $Na^+$ merely maintaining the high internal $K^+$-concentration which gives the potential. Duncan et al. (1977) measured the effects of ouabain on both faces of the lens; after 90 min the trans-lens asymmetry potential was abolished, due to the much slower fall in potential of the posterior surface than that of the anterior, and this would be consistent with the operation of an electrogenic pump by the epithelium, which would be rapidly inhibited; but Duncan et al. have suggested alternative explanations.*

## LOCALIZATION OF ATPase

Palva and Palkama (1974), using a specific test for localization of $Na^+$-$K^+$-activated ATPase found that the enzyme was confined to the epithelial cell membranes (Fig. 4.19) and no activity could be definitively associated with the fibres or capsule. The absence of activity on the fibre membranes was surprising, suggesting as it does that a $Na^+$-pump is not active in maintaining the intracellular concentrations of $Na^+$ at a low value, and it is interesting that in a later study (1976) cortical elongating fibres in the equatorial region *did* show marked activity.

## EFFECTS OF HYPEROSMOLALITY

Yorio and Bentley (1976) raised the osmolality of the bathing medium of the lens by 25 mM with NaCl or by 50 mM with a nonelectrolyte such as mannitol; there was a sharp fall in trans-lens short-circuit current and potential, the effect being confined to application to the anterior surface and presumably reflecting reduced active transport across the epithelium. When the potential across the anterior and posterior surfaces was measured separately, it was found that only the anterior potential was sensitive to altered osmolality of the medium. The basis for these effects is not easy to discover; the degree of hyperosmolality is not comparable with that used by Rapoport (p. 60) to break tight junctions, so that they are

---

*Hightower and Kinsey (1977) have again emphasized the electrogenic character of the $Na^+$-pump across the lens epithelium; thus they have shown a rapid fall in lens potential (10 to 15 min) from 65 to about 40 mV on cooling; this is more extensive than that due to ouabain. Since these treatments did not affect membrane permeability appreciably the fall in potential was undoubtedly due to inhibition of the $Na^+$-pump. We may note that Duncan (1970) found no immediate effect of ouabain on the active efflux of $Na^+$ from the amphibian lens; only after an incubation of 1·5 hr did this fall.

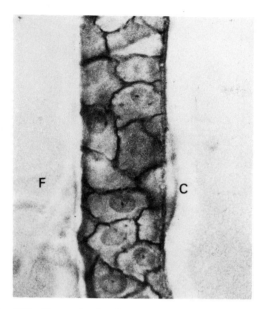

**Fig. 4.19**    Illustrating localization of ATPase activity on the cell membranes of the rat lens epithelium. A weak background staining of some nuclei is visible. The lens fibres (F) and capsule (C) are negative. ( × 440.) (Palva and Palkama, *Exp. Eye Res.*)

probably the result of a general osmotic efflux of water. Puzzling is the fact that the trans-membrane potential increased under these conditions.

## METABOLISM

### LENS CULTURE

The metabolism of the lens is most conveniently studied *in vitro*, and since Bakker in 1936 first maintained a lens in an artificial medium for several days with normal transparency, there have been frequent studies of the best medium for the purpose and the most suitable index of viability; outstanding in this respect has been the work of Kinsey and, more recently, Schwartz, and their collaborators. In general, the medium must contain a suitable ionic composition similar to that of the aqueous humour, whilst in order to allow for consumption of glucose and production of lactic and other acids, the medium must be changed regularly. As criterion of normal viability the best is probably the mitotic activity of the epithelium, the cells of which, it will be recalled, are continually dividing to produce new fibres (Kinsey *et al.*). Besides imitating the aqueous humour with regard to ionic make up, a variety of amino acids and vitamins, together with inositol, lactate and glutathione, were added. By using a constant flow of this solution, Schwartz kept a lens transparent for some 40 hours, but at the end of this time it was not completely normal,

having a higher lactic acid content, for example, than when it began.

## METABOLIC PATHWAYS

### GLYCOLYSIS

The lens utilizes glucose as its source of metabolic energy; since the $O_2$-tension of the aqueous humour is low, most of this metabolism is anaerobic, the $Q_{O_2}$ being only $0.1$; thus the main pathway for the utilization of glucose is by way of the Embden-Meyerhof or glycolytic reactions leading to pyruvic acid, which is largely converted to lactic acid; the latter is carried away in the aqueous humour. The enzyme lactic dedydrogenase is involved in this process, together with the coenzyme $NADH_2$:

$$Glucose + 2Pi + 2ADP \rightarrow 2 \ Lactate + 2ATP + 2H_2O$$

### PENTOSE PHOSPHATE SHUNT

As with the cornea, Kinoshita and Wachtl (1958) have shown that the pentose phosphate shunt contributes significantly to the oxidative breakdown pathway, in addition to the ordinary glycolytic pathway; this involves the coenzyme NADP.

### SORBITOL

Finally Van Heyningen (1959) has shown that the lens contains sorbitol and fructose, together with enzymes capable of converting glucose to fructose by way of sorbitol. Thus the metabolic events in the lens may be summarized by the scheme of Figure 4.20.

### ATP

Glycolysis, the pentose phosphate shunt and the Krebs cycle all lead to the generation of ATP which acts as substrate, or phosphate donor, in a variety of endergonic reactions, e.g. the synthesis of protein and fatty acids, the active transport of ions, and so on. The pentose phosphate shunt is concerned with the production of a variety of intermediates in synthesis, e.g. nucleic acids; it also leads to the production of $NADPH_2$, the coenzyme concerned with many syntheses, especially of fatty acids.

### SORBITOL PATHWAY

The significance of the sorbitol pathway, which converts glucose, via sorbitol, to fructose, is not clear; the enzymes concerned are aldol reductase, $NADPH_2$ and polyol reductase:

$$Glucose + NADPH_2 \rightarrow Sorbitol + NADP$$
Aldol reductase

$$Sorbitol + NAD \rightarrow Fructose + NADH_2$$
Polyol reductase

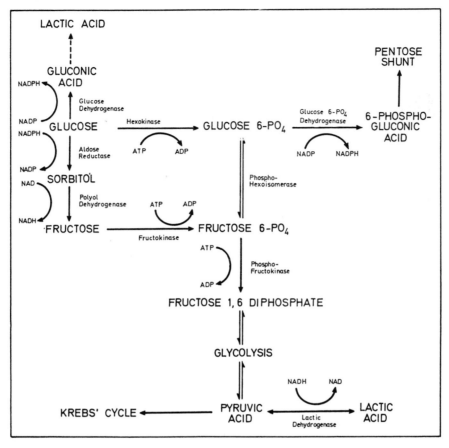

**Fig. 4.20**   Some of the possible metabolic routes of glucose in the lens. (After Spector, *Arch. Ophthal.*)

The same pathway allows xylose to be converted to xylitol, but dulcitol and arabitol, formed from galactose and arabinose by aldol reductase, are not metabolized further.

OXIDATIVE PHOSPHORYLATION

Molecular oxygen is used in biological reactions by way of the cytochrome system, which is contained in the mitochondria; this seems to be confined to the epithelium, according to Kinsey and Frohman, so that it may be that respiration, as contrasted with glycolysis, is confined to this cellular layer. Because of the relative insignificance of $O_2$ in metabolism,[*] keeping a lens anaerobically has no effect on transparency, nor on its power to retain its high $K^+$ and low $Na^+$; if glucose is absent from the medium, however, anaerobiosis causes a drastic deterioration in the state of the lens, presumably because, in the absence of glucose, $O_2$ is utilized to oxidize other substrates than glucose (Kinoshita, 1965).

CONTROL

Control over the rate at which a given metabolic process takes place is generally exerted through inhibition of one or more key enzymes; in the case of the lens these are hexokinase and phosphofructokinase, the former catalysing the conversion of glucose to glucose 6-phosphate and the latter the phosphorylation of fructose 6-phosphate to fructose 1,6-diphosphate:

Glucose + ATP → Glucose 6-phosphate + ADP
Fructose 6-phosphate + ATP
→ Fructose 1,6-diphosphate + ADP

Thus ATP inhibits phosphofructokinase activity whilst glucose-6-phosphate inhibits hexokinase activity; as a result, increasing the concentration of glucose in

[*] Because of the high efficiency with which ATP is produced through the oxidative phosphorylation mechanism, the amount produced by this pathway in the lens actually could be as high as 33 per cent; a definite Pasteur-effect can be demonstrated, moreover (Kern, 1962).

the lens does not increase lactate production or the level of glucose 6-phosphate. Increasing the level of glucose does, however, stimulate the sorbitol pathway, a finding of significance for the interpretation of diabetic cataract (p. 156).*

## PROTEIN SYNTHESIS

As indicated earlier, there is a small turnover of protein in the adult lens; in addition, new fibres are being laid down throughout life, requiring further synthesis. The total amount of protein synthesized per day is, nevertheless, small, and this, together with an extremely efficient amino acid accumulating mechanism that maintains the concentrations of amino acids in the lens far above those in the aqueous humour and blood plasma, is presumably why the lens remains normal and continues to grow when the level of amino acids in the blood is so low as to inhibit body-growth entirely.

### AMINO ACID ACCUMULATION

The active accumulation of amino acids by the lens was demonstrated by Kern (1962) and Reddy and Kinsey (1962), and is illustrated in Figure 1.22 (p. 20); the process seems similar to that described in other cells (Brassil and Kern, 1967) and ensures that protein synthesis will not be limited by availability of amino acids. We may note that the lens, like the brain, may synthesize certain amino acids from glucose, namely glutamic acid and alanine (see, e.g. Van Heyningen, 1965).

### NUCLEIC ACIDS

Protein synthesis occurs in association with nucleic acids by mechanisms that are now well understood and need not be repeated here. All the evidence indicates that synthesis in the lens is in every way similar; the loss of the nucleus from the lens fibre, like that of the reticulocyte, presumably demands that the template messenger RNA be longer-lived than that in nucleated cells; in this case protein synthesis will not be inhibited completely by actinomycin D as is, indeed, found (Zigman and Lerman, 1968).

## GLUTATHIONE

This tripeptide, glutamyl-cysteinyl-glycine, whatever its role, must be of importance for the lens since its concentration is some 1000 times that in the aqueous humour; studies on the incorporation of radioactive glycine into the lens show that glutathione is continually

being synthesized; in all forms of cataract the fall in concentration of glutathione is striking. Glutathione constitutes a redox system:

$$2\,GSH + \tfrac{1}{2}O_2 \rightleftharpoons GSSG + H_2O_2$$

In the lens there is an enzyme, *glutathione reductase*, catalysing the reduction of oxidized glutathione. Reoxidation can take place by a complex series of reactions with the intermediate formation of $H_2O_2$ and the intervention of the enzyme *glutathione peroxidase*. Glutathione can behave as coenzyme in the glyoxylase reaction. Its function in the lens is obscure: it has been argued that it prevents the oxidation of SH-groups in lens proteins. Of some significance is its very much greater concentration in the lens epithelial cell; when Giblin *et al.* (1976) oxidized the glutathione of the lens with tertiary butyl hydroperoxide they found reduced $Na^+$-$K^+$-activated ATPase in the lens, and this was accompanied by penetration of salt and water, suggesting a failure of the cation pumps. Another significant feature is the great decrease in GSH content of the lens in nearly all types of cataract.†

## ASCORBIC ACID

Ascorbic acid (vitamin C) is present in the aqueous humour in high concentration in many species; when this occurs it is also concentrated in the lens, but its significance is not clear since in scurvy the lens shows no defect; furthermore, the *in vitro* preparation seems quite stable without it.

## THE PROTEINS

The lens is highly transparent yet its cells contain a very high concentration of proteins, so high that unless these were specially adapted for their specific purpose, which is presumably to create a medium of high optical density and therefore high refractive index, they would impart to the lens a milkiness that would militate against the formation of distinct images. It is for this reason that the identification of the proteins and the determination of their structure have been investigated in great detail.

---

*The role of insulin in control of glucose utilization by the lens is a controversial topic; Farkas and Weese (1970) have shown that aqueous humour contains a stimulator to utilization which is enhanced by systemic administration of insulin in the presence of trace quantities of trivalent chromium.
†Waley (1969) has described a number of other S-containing compounds identified in the lens, e.g. *ophthalmic acid* or glutamyl-α-aminobutyryl-glycine, which inhibits the glyoxalase reaction; *norophthalmic acid*, γ-glutamylalanyl-glycine; *sulphaglutathione*, and so on.

ONTOGENY

As summarized by Clayton (1978), the function of the lens is to provide a transparent structure which focuses light on the retina and which will continue to do so accurately as the eye increases in size; this will require a fall in the refractive index of the lens from centre to periphery, which is associated with a parallel fall in protein concentration. The earliest formed, central fibres eventually have the highest protein concentration, and in order to sustain this and maintain transparency it appears that they have a different array of proteins from those in the outer fibres formed later. Certain of the crystallins are successively replaced during ontogeny, while other crystallins are synthesized throughout life, and it would seem that the high refractive index, and therefore high concentration of protein, of the nucleus are made possible by the preferential synthesis of proteins that are capable of entering into closer molecular packing than those in the more peripheral cells (Clayton, 1970; 1974).

Because of these special features of the lens proteins, or *crystallins* as they are called, their isolation and characterization have been studied in great detail, and any exhaustive review of these studies, which at any rate would be more appropriate to a textbook of the biochemistry of the eye, cannot be attempted here, so we must concentrate on those features that bear most directly on the changing composition with ontogeny and the relation to loss of transparency in cataract.

THE CRYSTALLIN FRACTIONS

The proteins were classically divided by Mörner into several fractions as shown in Table 4.6 where the

**Table 4.6** Protein fractions as described by Mörner

| Protein | per cent |
|---|---|
| Insoluble albuminoid | 12·5 |
| α-Crystallin | 31·7 |
| β-Crystallin | 53·4 |
| γ-Crystallin or albumin | 1·5 |
| Mucoprotein | 0·8 |
| Nucleoprotein | 0·07 |

percentages are those given by a more recent study by Krause, the 'insoluble albuminoid' being the residue after extraction of the crystallins etc. The modern classification tends to retain this terminology, the segregation of the crystallins into α-, β- and γ-fractions being brought about experimentally by such procedures as migration in an electric field—electrophoresis—sedimentation in the ultracentrifuge, gel-filtration, and so on. Table 4.7 enumerates the main features of these fractions as summarized in a recent review by Harding and Dilley (1976) on the basis of Waley's (1969) review. Figure 4.21 illustrates the separation of the soluble proteins on the basis of molecular weight by filtration through a column loaded with Sephadex. It will be seen that β-crystallin is resolved into two quite distinct components, $\beta_H$ and $\beta_L$. Within any class, modern refinements of technique have often permitted the isolation of sub-classes, whilst treatment with hydrogen-bond breaking agents, such as urea, has permitted the dissociation into small sub-units and their subsequent reassociation after removal of the dissociating agent.

LOW MOLECULAR WEIGHT PROTEINS

THE GAMMA-CRYSTALLINS

The γ-crystallin fraction is separated on the basis of its moving most slowly in electrophoresis and, more effectively, by gel-filtration on the basis of its low molecular weight, in the region of 20 000. Björk (1964a) has separated four distinct proteins from the γ-crystallin fraction; their molecular weights are very close, in the region of 19 000 to 20 000; three of the fractions were immunologically identical with each other whilst the fourth was only partially identical with these. Thus their differences reside in the amino acid composition, differences that presumably arose by mutations on specific parts of the gene during evolution, but since these mutations did not affect the transparency of the lens they were not eliminated by selection (Björk, 1964). The protein shares with the β-crystallin fraction the feature of a high cysteine content, which may be a significant feature in the genesis of a cataract if, as has

**Table 4.7** Characteristics of the main classes of soluble proteins from mammalian lens. (Harding and Dilley, 1976, after Waley, 1969.)

| Property | Class | | |
|---|---|---|---|
| | α | β | γ |
| Electrophoretic mobility (towards the anode, at pH 8–9) | High | Medium | Low |
| Range of isoelectric points shown by proteins | pH 4·8–5·0 | pH 5·7–7·0 | pH 7·1–8·1 |
| Molecular weight | Over $5 \times 10^5$ | $4$ to $20 \times 10^4$ | About $2 \times 10^4$ |
| Molecular form | Aggregates | Aggregates | Monomers |
| Thiol content | Low | High | High |
| N-terminal amino acid | Masked | Masked | Free; glycine or alanine |

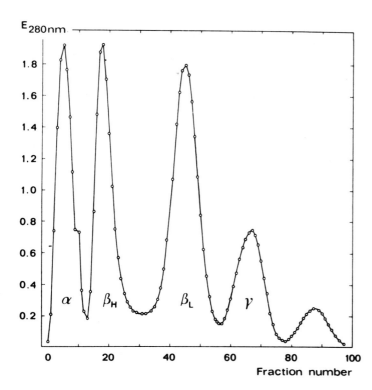

**Fig. 4.21**   Illustrating the different water-soluble fractions separated from an extract of calf lenses on a Sephadex column. (Bloemendal and Herbrink, *Ophthal. Res.*)

been suggested, this results from an oxidation of thiol-groups (Dische and Zil, 1951). Subsequent fractionation has shown that Björk's fractions III and IV could be separated into IIIa, IIIb and IVa, and IVb.

### Cortex and nucleus

Papaconstantinou (1965) found that the soluble proteins of the lens epithelium contained no $\gamma$-crystallin either in the adult or calf. In the fibres $\gamma$-crystallin was present in adult, calf, and embryonic lens, so that he concluded that it was the *initiation of differentiation* that was accompanied by synthesis of $\gamma$-crystallin. Thus the embryonic and adult nucleus both had $\gamma$-crystallin of similar electrophoretic properties; however, the cortical $\gamma$-crystallin of the adult lens was different, so that it seemed that the type of $\gamma$-crystallin that was synthesized during cell differentiation depended on the age of the animal.

### $\beta_s$-CRYSTALLIN

Thus Van Dam (1966) isolated from adult cow lenses, a protein with molecular weight 28 000 and a mobility between those of $\alpha$- and $\gamma$-crystallins, and it was named $\beta_s$-crystallin. Kabasawa *et al.* (1974) confirmed that the elution pattern of gel-filtration of calf and adult lens extracts differed strongly, the adult extract being more complex with the appearance of a peak corresponding to $\beta_s$-crystallin whilst the $\gamma$-fraction following this is different from the adult $\gamma$-fraction. In general, they

found a $\gamma$-crystallin of molecular weight 24 000 belonging to the adult lens cortex, and a $\gamma$-crystallin of molecular weight 20 000 common to the adult nucleus and both cortex and nucleus of the calf. Thus, according to this study, ageing was accompanied by the appearance of a new $\gamma$-crystallin of M.W. 24 000 as well as of $\beta_s$-crystallin. In a later study[*] these findings were confirmed and the molecular weight of $\beta_s$-crystallin of cattle cortex was given as 28 000, this being characterized by the absence of a terminal free amino acid; all three proteins were immunochemically distinct with different amino acid compositions. It thus appears that in the cattle lens cortex, $\gamma$-crystallin in the newer fibres is different immunochemically, in charge, and molecular weight from the $\gamma$-crystallins found in the young lens.

### $\beta$-CRYSTALLINS

This class constitutes a larger proportion of the total soluble lens proteins than either the $\alpha$- or $\gamma$-crystallins in the adult; in the embryonic and fetal lens, however, the $\alpha$-proteins are most prominent constituting 62 per cent of the total; this decreases sharply so that at 4 months prenatal it is only 35 per cent, the same as in the postnatal bovine lens; the $\beta$-crystallins represent

---

[*] Harding and Dilley (1976) had suggested that Kawabara *et al.* (1974) had confused $\beta_s$-crystallin with a $\gamma$-crystallin, but this later study fully confirms the earlier work.

only 17 per cent in very young embryonic lenses but increase to 40 per cent (Van Kamp *et al.*, 1974).

### HEAVY AND LIGHT FRACTIONS

When the proteins of the lens are separated by gel-filtration, four main fractions are obtained which have been named A, B, C and D; A and D are $\alpha$- and $\gamma$-crystallin respectively and B and C represent two classes of $\beta$-crystallins of greatly differing molecular weights, which are described as heavy—$\beta_H$—and light—$\beta_L$—respectively, with molecular weights of 210 000 and 52 000 (Zigler and Sidbury, 1973).

## Subunits

When treated with dissociating agents, such as urea, the molecular weights of both drop to the region of 25 000, and the component subunits may be separated by electrophoresis into a number of fractions, the most prominent being described as $\beta B_p$ being common to both heavy and light crystallins. The molecular weights are shown in Table 4.8. When the amino acid contents of the heavy and light fractions and of the subunit $\beta B_p$ were compared, together with $\beta_s$ mentioned above, strong similarities were obvious (Herbrink and Bloemendal, 1974; Van Dam, 1966) but the differences were sufficiently large to confirm the separate nature of the fractions.

**Table 4.8** Subunits derived from fractions B and C of the lens proteins. (Zigler and Sidbury, 1973.)

|  | Subunit molecular weight | % Composition |
|---|---|---|
| Fraction B | 35 000 | 8% |
|  | 31 000 | 9% |
|  | 27 500 | 35% |
|  | 24 000 | 48% |
| Fraction C | 27 500 | 60% |
|  | 24 000 | 40% |

Thus Herbrink *et al.* (1975) carried out a more elaborate fractionation of the heavy and light $\beta$-crystallins and found that, altogether, they shared, in addition to the $\beta B_p$, three other subunits or chains. The various subunits were indicated by the order in which they were eluted from DEAE, namely $\beta B_1$, $\beta B_2$, ..., the B referring to the basic character of the subunit; in addition, the acidic subunit, $\beta A$, was described. When amino acid composition (Table 4.9) and other properties of the fractions were compared, it was concluded that $\beta B_4$ having a molecular weight of 27 000 daltons, was identical with $\beta B_p$; and $\beta B_2$ and $\beta B_5$ were apparently identical, having a molecular weight of 30 000. $\beta A$ had a molecular weight of 25 000. When peptide maps were made, a small difference between $\beta B_2$ and $\beta B_5$ was found, which might have been due to a difference in

**Table 4.9** Amino acid analysis of the subunits of $\beta$-crystallins. (Herbrink *et al.*, 1975.)

| Amino acid | $\beta B_2$ $\beta B_5$ (mol %) | $\beta B_P$ $\beta B_4$ (mol %) | $\beta A$ (mol %) |
|---|---|---|---|
| Aspartic acid | 7·7 | 8·3 | 6·3 |
| Threonine | 2·0[a] | 3·4[a] | 3·7[a] |
| Serine | 7·1[a] | 8·4[a] | 8·4[a] |
| Glutamic acid | 13·6 | 15·8 | 15·4 |
| Proline | 5·0 | 6·8 | 3·4 |
| Glycine | 9·1 | 9·3 | 10·1 |
| Alanine | 6·0 | 4·2 | 5·8 |
| Cysteine | 1·0[b] | 1·0[b] | 1·9[b] |
| Valine | 6·2[c] | 7·5[c] | 6·8[c] |
| Methionine | 0·9 | 1·0 | 0·8 |
| Isoleucine | 2·8[c] | 3·1[c] | 2·3[c] |
| Leucine | 7·3 | 5·0 | 6·4 |
| Tyrosine | 3·8 | 4·4 | 3·6 |
| Phenylalanine | 4·2 | 4·0 | 6·6 |
| Tryptophan | 6·1[d] | 3·0[d] | 3·6[d] |
| Lysine | 4·2 | 6·0 | 3·2 |
| Histidine | 4·6 | 4·1 | 5·1 |
| Arginine | 8·5 | 5·0 | 7·0 |

[a] Extrapolated to zero time hydrolysis.
[b] Estimated as cysteic acid.
[c] Values for 72 hr hydrolysis.
[d] Determined according to Benzce–Schmid (1957).

two acidic groups. Thus, in all, there are at least 4 subunits and these are common to the heavy and light fractions, namely $\beta B_2$, $\beta B_4$, $\beta B_5$ and $\beta A$.

## $\alpha$-CRYSTALLIN

### HETEROGENEITY

This has the largest molecular weight of the crystallins, with values between 600 000 and 4 000 000; and this wide variation is due to a tendency for aggregation of subunits to greater or less extent. In the electron microscope this heterogeneity is especially manifest. By gel-filtration, or in the ultracentrifuge, the $\alpha$-crystallin may be separated into two main fractions, Fraction I called high-molecular weight $\alpha$-crystallin (HM$\alpha$), with weights varying between $0.9$ and $4.0 . 10^6$ daltons and Fraction II, or low-molecular weight (LM$\alpha$) $\alpha$-crystallin, with weights between $6 . 10^5$ and $9 . 10^5$ (Spector *et al.*, 1971) in the calf preparation.[*] A similar range of molecular weights was found by Horwitz (1976) in

[*] Actually the Peak I in gel-filtration containing high molecular weight material may be resolved into two populations; thus Spector *et al.* (1971) described, in all, three populations with molecular weight ranges $6 . 10^5$ to $9 . 10^5$, $0.9 . 10^6$ to $4 . 10^6$ and the third with molecular weights greater than $1 . 10^7$. In Hoenders' laboratory, similarly, three states of aggregation have been described, namely $\alpha_L$-crystallin with $s_{20,w}$ values ranging from 17 to 24 S $\alpha_H$-crystallin (30–55 S) and $\alpha_{VH}$-crystallin (100–190 S) (Van Kleef *et al.*, 1974a).

human lenses, in fact, when the low-molecular weight fractions were compared it was impossible to establish any difference with respect to Stokes radius, diffusion coefficient, and ultraviolet CD. In general, the low-molecular weight preparation, which constitutes some 55 per cent of the total α-crystallin, is similar in amino acid composition to the high molecular weight material, and it is customary in any study to treat the two as the same, with the exception of size, although, as we shall see, with the really large aggregates, β-crystallin seems also to be present.

TRANSFORMATION TO FIBRE

Alpha-crystallin is of special interest in the study of the physiology of the lens since it is the protein that seems to be specifically concerned with transformation of the epithelial cell into a lens fibre; this was made clear by Delcour and Papaconstaninou (1972), who calculated the rate of synthesis of the proteins of the cultured lens by measuring the amount of incorporation of labelled leucine in a given period. In the epithelial cell the rate of synthesis of α-crystallin was seven times higher than in a cortical fibre, indicating a large decrease in rate of synthesis on transformation to a fibre. However, the ratio for the non-α-crystallins was much higher, namely 41:1, so that although total protein synthesis falls, the proportional rate for α-crystallin increases, indicating the importance of α-crystallin for the fibre, so that fibrogenesis involves a quantitative biochemical specialization in addition to the qualitative specialization for initiation given by γ-crystallin synthesis.*

THE SUBUNITS

The preparation described by Bloemendal et al. (1962) had a molecular weight of some 810 000; on addition of hydrogen-bond breaking agents, such as 7 M urea, the molecular weight was reduced drastically to around 26 000, suggesting the breakdown into subunits. When the purified α-crystallin was examined in the electron microscope spherical molecules of about 150 Å diameter were observed, and these appeared to be made up of smaller subunits. Björk (1964b) examined the chemical character of the subunits and their re-aggregation brought about by removing the urea and increasing the ionic strength of the medium. The subunits were definitely not identical chemically; moreover, complete reaggregation to the original α-crystallin was never achieved, so that it is likely that there are several types of subunit arranged in a definite manner; breakdown of this structure and reaggregation would lead to the production of a variety of different combinations which would show only partial immunochemical identity with the native material.

## Acidic and basic

Subsequent studies have shown that the α-crystallin molecule is built up on a basis of four subunits, two acidic and two basic, and these have been designated (Waley, 1969) as $\alpha A_1$ and $\alpha A_2$ (the A referring to acidity), and $\alpha B_1$ and $\alpha B_2$ (the B referring to basicity). In bovine lens the subunits are found uniformly in the proportion of two A's to one B (Van Kamp et al., 1974).

## Amino acid sequences

The molecular weights of the subunits were similar at about 20 000, whilst the amino acid compositions of the acidic and basic pairs were apparently identical, their separation, being by virtue of electrophoretic mobility, indicating some small change. In fact it is now clear that probably $\alpha A_1$ is derived from $\alpha A_2$, so that there is no 'genetic' origin for $\alpha A_1$, there being no messenger RNA† for its synthesis; in fact, the embryonic lens does not contain $\alpha A_1$, which only appears later (Schoenmakers et al., 1969). The amino acid sequences of αA and αB subunits have been determined by Van der Ouderaa et al. (1973; 1974); the $\alpha A_2$ consists of a single chain of 173 residues giving a molecular weight of 19 832 daltons, and $\alpha B_2$ has 175 residues, giving a molecular weight of 20 070 daltons. As Figure 4.22 shows, there is a high degree of homology between the two, indicating an origin from a common ancestral α-crystallin gene.

## Evolutionary changes

Van der Ouderaa et al. (1974) computed, on the basis of the known rate of substitution of amino acids during evolution, which amounts to a substitution of 1 in 100 every 40 to 50 million years, that, with only 57 per cent homology, the production of the two α-subunits with separate genes would have occurred long before the appearance of the lowest vertebrates some 450 million years ago, so that we may expect to find invertebrate α-crystallins homologous with the vertebrate protein. So far as the conversion of $\alpha A_2$ to $\alpha A_1$ is concerned, the amino-acid sequencing revealed the conversion of one glutamine to glutamic acid, i.e. the change consisted of deamidation (Bloemendal et al., 1972).

---

*Approximately 60 per cent of the leucine incorporated into α-crystallin appeared in the $\alpha A_2$ fraction and 30 per cent in the $\alpha B_2$ fraction; since the relative proportions of these subunits in adult epithelial α-crystallin is as 60 to 30, this means that the newly formed α-crystallin has the same subunit pattern as that in the old.

†The messenger RNA for synthesis of α-crystallin has been isolated from the calf lens polyribosomes by Chen et al. (1974); analysis of the material synthesized in cell-free extracts under its influence indicated the presence of α-crystallin containing all four main subunits, so that the conversion of the $A_2$ and $B_2$ subunits to the $A_1$ and $B_1$ derivatives is rapid. In an earlier study Berns et al. (1973) had found only $B_2$ and $A_2$ chains.

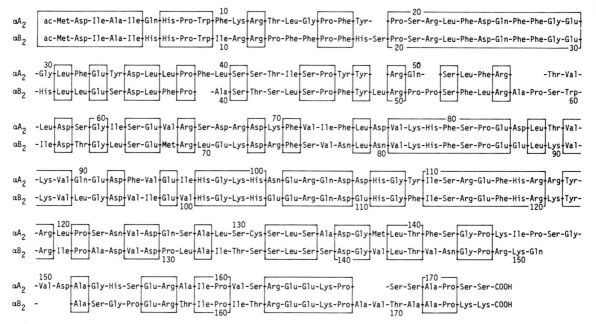

**Fig. 4.22** Illustrating the homology of the two major chains of α-crystallin. Identical residues are enclosed in rectangles. (Van der Ouderaa et al., *Eur. J. Biochem.*)

## Species variations

De Jong *et al.* (1975) have determined the amino-acid sequences of the A-chains of several mammalian species, comparing them with the previously determined sequence of bovine material. There were variations (Fig. 4.23), although the homologies were very strong, all substitutions being compatible with the change of a single base in the RNA-codon. Thus the leucine found in position 90 of the cat and dog preparations, in place of glutamine in all other species, suggests the common origin of these two carnivores. The rabbit, rat and monkey share four substitutions, as compared with other species, suggesting a common ancestry of the lagomorphs, rodents and primates.

### RE-ASSEMBLY OF α-CRYSTALLIN

As we have seen, removal of the dissociating agent permits the subunits to aggregate to form a higher molecular weight protein; in general, the work of Spector and his colleagues indicated that the re-association was never perfect, in the sense that the molecular weight of the aggregate was smaller than that of the 'native' α-crystallin from which the subunits had been derived; and Spector suspected that a low molecular weight material, formed during dissociation, was lost during the subsequent manipulations (Li and Spector, 1972). When Spector *et al.* (1971) allowed single subunits to re-aggregate, e.g. αB, proteins were formed, once again with a smaller molecular weight; and subse-

| Position | 3 | 4 | 13 | 61 | 90 | 91 | 127 | 133 | 146 | 147 | 148 | 150 | 153 | 155 | 162 | 168 | 172 |
|---|---|---|---|---|---|---|---|---|---|---|---|---|---|---|---|---|---|
| Cow | Ile | Ala | Thr | Ile | Gln | Glu | Ser | Leu | Ile | Pro | Ser | Val | Gly | Ser | Ser | Ser | Ser |
| Pig | Ile | Ala | Ala | Val | Gln | Glu | Ser | Leu | Val | Pro | Ser | Val | Gly | Ser | Ser | Ser | Thr |
| Horse | Ile | Ala | Ala | Ile | Gln | Glu | Thr | Val | Ile | Pro | Ser | Met | Gly | Ser | Ser | Gly | Ser |
| Dog | Ile | Ala | Ala | Ile | Leu | Glu | Ser | Leu | Val | Pro | Ser | Val | Gly | Ser | Ser | Ser | Ser |
| Cat | Ile | Ala | Ala | Ile | Leu | Glu | Ser | Leu | Val | Pro | Ser | Val | Gly | Ser | Ser | Ser | Ser |
| Rabbit | Val | Thr | Thr | Ile | Gln | Glu | Ser | Leu | Val | Gln | Ser | Leu | Gly | Ser | Ser | Ser | Ser |
| Rat | Val | Thr | Ala | Ile | Leu | Glu | Ser | Leu | Val | Gln | Ser | Leu | Gly | Ser | Ser | Ser | Ser |
| Rhesus monkey | Val | Thr | Thr | Ile | Gln | Asp | Ser | Leu | Ile | Gln | Thr | Leu | del | Thr | Ala | Ser | Ser |

**Fig. 4.23** Variable positions in the αA chains of eight mammalian species. In the monkey αA chain, a deletion occurs in position 153. (De Jong *et al.*, *Eur. J. Biochem.*)

quent studies, based on the re-aggregation of preparations of individual subunits, as well as of mixtures
of these, have all pointed to the possibility of producing
polymers of the type $(\beta A)_n$, $(\alpha B)_n$, or more complex
aggregates, although the sedimentation coefficient was
always less than that of the untreated $\alpha$-crystallin (Li
and Spector, 1973). When different mixtures of A and
B subunits were made, Van Kamp *et al.* (1974) found
that all the products, with differing proportions of A
and B in their make-up, had identical sedimentation
coefficients of 12 S, compared with 18 S for the starting
material: the re-aggregated products, however, did not
reflect the proportions of A and B present in the
mixture, and often pure B aggregates appeared out of
a mixture of A and B. Only when the ratio of A/B
was 4/6 or less were all B chains accepted completely;
as we have seen the ratio A/B in 'native' $\alpha$-crystallin
is 2.

### Re-assembly from dissociated mixtures of crystallins

When all the soluble proteins were dissociated together
and the resulting subunits were allowed to re-aggregate,
some hybridization between $\alpha$- and $\beta$-crystallin subunits

occurred, so that when the re-aggregated material was
submitted to a new gel-filtration most of the $\beta_H$ had
disappeared and the $\alpha$-peak had increased appropriately.
It seemed that $\gamma$-crystallin subunits did not participate
in this hybridization. In the electron microscope the
re-aggregated $\alpha$-fraction appeared as molecules of different size and shape. However, when the dissociated
protein preparation was strongly diluted, and association
allowed, the pattern obtained by gel-filtration was
virtually identical with that of the normal preparation
(Fig. 4.24). One difference was observed, however,
namely that the reconstituted $\alpha$-crystallin molecules
were smaller, as seen in the electron microscope, and
the difference was confirmed by ultracentrifugation
when it was found that, whereas the sedimentation
coefficients for the $\beta$- and $\gamma$-crystallins were unaffected
by dissociation and re-aggregation, that of $\alpha$-crystallin
had decreased from 18·2 to 12·3 Sverdberg units.

### The hybrids

Hybridization was recognized by reactions of the re-
aggregated material immunoelectrophoretically to both
$\alpha$- and $\beta$-antisera, and also by the large sedimentation
coefficients, ranging from 30 to 70 S; in the electron

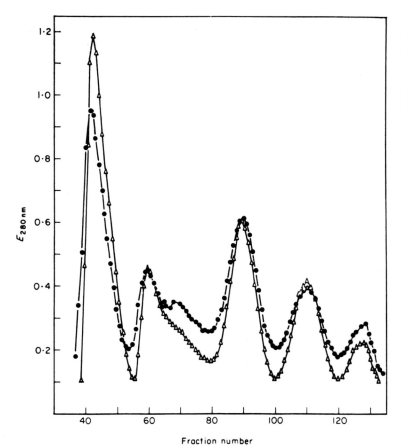

**Fig. 4.24**  Selective reassociation of s-
crystallins as revealed by gel filtration on
Sephadex G-200. Control, ●—●—●;
recombined crystallins, △—△. Fraction
1 is the first protein-containing fraction
behind the void volume. (Bloemendal *et
al., Exp. Eye Res.*)

Fraction number

microscope rod-like molecules made their appearance by contrast with the spherical shapes of $\alpha$- and $\beta$-crystallins; and it may be assumed that they represent aggregates of $\alpha$ and $\beta$ subunits.

## NEWLY SYNTHESIZED $\alpha$-CRYSTALLIN

By incubating a calf lens in a nutritional medium to which radioactively labelled leucine was added, Spector *et al.* were able to separate from the $\alpha$-crystallins a 'highly-labelled fraction' (NS) representing material that had been synthesized during the incubation. The molecular weight was remarkably homogeneous and corresponded with the minimum molecular weight of ordinary preparations; the amino acid composition was basically similar, with the exception of a greater percentage of free amino groups, suggesting that the terminal acetylation had not been completed. In a later study, Stauffer *et al.* (1974) found a molecular weight of $7{\cdot}4 . 10^5$ and only traces of $A_1$ and $B_1$ subunits; the actual relative amounts of the subunits are compared with those in LM$\alpha$-crystallin as follows:

| Subunit | NS$\alpha$ | LM$\alpha$ |
|---|---|---|
| $B_2$ | 32 | 24 |
| $B_1$ | 3 | 11 |
| $A_2$ | 56 | 40 |
| $A_1$ | 9 | 24 |

Very little radioactivity was associated with the $A_1$ and $B_1$ fractions, so that it is unlikely that any of these were synthesized directly. Thus it is reasonable to assume that LM$\alpha$ is directly derived from NS$\alpha$ through intracellular modification of the $A_2$ and $B_2$ chains.

### Human lens

When they extended this work to the human lens, Spector *et al.* (1976) found the isolated labelled $\alpha$-crystallin to contain 3 subunits, one equivalent to the bovine $B_2$ and two others slightly different from bovine A chains; the molecular weight was $4{\cdot}9 . 10^5$ compared with $7 . 10^5$ for the bovine material; the amino acid composition was similar. The proportion of A to B was 3:1 by comparison with 2:1 for bovine $\alpha$-crystallin.

## CAPSULAR PROTEIN

As indicated earlier, the structural protein of the capsule is collagen; it is hydrolysed by collagenase and has the characteristic proportion of hydroxproline residues. The high proportion of carbohydrate differentiates it from the collagen of skin, and the fluorescent antibody-staining experiments of Roberts (1957) show that the material is analogous with the basement membrane material characteristically seen surrounding capillaries;

embryologically it would seem to arise as a condensation of the basement membranes of lens cells (Cohen, 1965). The major sugar moiety is a sialohexosaminoglycan containing galactose, mannose, fucose and neuraminic acid; some sulphated esters are also present, an unusual feature for a connective-tissue glycan, possibly connected with the need for transparency of the capsule (Dische and Zelmenis, 1966).

## FM-CRYSTALLIN

This was called by Van Dam pre-$\alpha$-crystallin since it migrated in an electric field more rapidly than $\alpha$-crystallin; however, Van den Broek *et al.* (1973) have shown that it has no relation to $\alpha$-crystallin so far as amino acid content is concerned; its molecular weight is 14 500 daltons and it has been renamed 'fast-moving' or FM-crystallin.

## MEMBRANE PROTEIN

After extracting a lens homogenate with the usual saline solutions there is left a water-insoluble fraction or albuminoid (p. 134); most of this material can be solubilized with a hydrogen-bond dissociating agent such as 6 M urea, and the urea-insoluble residue represents mainly membrane material (Lasser and Balazs, 1972; Maisel, 1977). The protein associated with this material has been resolved by Bloemendal *et al.* (1977) into two components of molecular weight 26 000 and 34 000 designated as MP (membrane protein) 26 and MP 34 respectively.[*]

## CHICKEN LENS PROTEINS

### FISC OR $\delta$-CRYSTALLIN

Most work on lens proteins has been done on mammalian material; the study of the chicken is important experimentally since it is in this species that embryological studies are normally carried out. In general, the protein composition is very different, the only common protein being probably $\alpha$-crystallin; the major component is what Rabaey (1962) called FISC or *first important soluble crystallin* and is the avian counterpart

[*] The proteins associated with the cell membrane are probably incorporated into the basic lipid bilayer of the Danielli–Davson model; additional material may extend into the cytoplasm remaining attached to the membranes. The reader should read the paper of Lasser and Balazs (1972) for a combined morphological and biochemical study of the protein. Other studies are those of Alcala *et al.* (1975), Broekhuyse *et al.* (1976) on bovine material and Maisel *et al.* (1976) on chick lens material. Broekhuyse and Kuhlmann (1974) have presented a complete analysis of the lens plasma membrane previously treated with urea to remove cytoplasmic material.

of the mammalian γ-crystallin, predominating before birth and becoming concentrated in the nucleus in the adult. Thus it represents some 70 to 80 per cent of the lens proteins in the embryonic chick (Piatigorsky et al., 1972). Its molecular weight is, according to Piatigorsky et al. (1974), 200 000 daltons. As with the mammalian proteins FISC, or δ-crystallin as it has been called (Zwaan, 1966), is built up of subunits (Clayton, 1969) there being apparently four with a similar molecular weight of 45 000 to 50 000 daltons (Piatigorsky et al., 1974). The messenger RNA responsible for synthesis of δ-crystallin has been isolated by Zelenka and Piatigorsky (1976) and it would appear that the same RNA synthesizes the same subunit, so that differences found amongst the four are apparently due to post-translational changes.

### β-CRYSTALLINS

Chicken lens contains a group of crystallins that have been called γ-crystallins, because of their low molecular weight and the slow electrophoretic mobility of some of their members, or β-crystallins, because of their antigenic similarities to bovine β-crystallin. Truman and Clayton (1974) have summarized many of the basic facts regarding this group of chick proteins, which have molecular weights ranging from as low as 16 000 to as high as 50 000 (Truman et al., 1971). According to Clayton and Truman (1967), it is likely that the various proteins are heteropolymers, built up of a limited number of subunits, say 15 (Truman and Clayton, 1974), so that two given individual proteins, when dissociated, could produce the same subunit amongst their total (Truman et al., 1971). The presence of common subunits imparts an antigenic similarity amongst individual β-crystallins. However, of the total number of subunits, these were all immunologically distinct, no pair having exactly the same total number of antigenic determinants (Clayton and Truman, 1974).

### γ-CRYSTALLIN

Birds have a γ-crystallin with a low molecular weight that must represent a monomer (Rabaey et al., 1972), although the chick lens has a negligible amount.

### α-CRYSTALLIN

Bird lenses contain an α-crystallin, so that, in all, there are four crystallins in birds and reptiles (Clayton, 1974).*

## ORGAN SPECIFICITY

As long ago as 1903 Uhlenhuth showed that the immunological characteristics of the lens were unusual. Thus if we inject an extract of, say, ox muscle into a rabbit, the rabbit produces antibodies in its serum that will react with this extract if injected again, to produce the so-called immune reaction; if the reaction is carried out in vitro, there will be a precipitation of the proteins by the antiserum. This reaction is on quite a strictly species basis, so that the rabbit's serum will not react to sheep muscle for example. If an ox antiserum is prepared to ox lens, by injecting a lens extract into a rabbit, it is found that this antiserum reacts to lens extracts of all other mammalian species, as well as to those of fishes, amphibians and birds (but not to that of the octopus). The lens is said, therefore, to be organ—rather than species—specific and we may conclude that the lens proteins that are responsible for this antigenicity are all built on a very similar plan in the lenses of different species. It was thought for some time that it was only the α-crystallin that behaved as an antigen, but by the technique of immuno-electrophoresis, for example, it has been

shown that all three groups of crystallins are antigenic. Thus, when a rabbit antiserum was prepared to human lens, this serum was shown to react with some nine components of the lens proteins, two in the alpha-, four in the beta- and three in the gamma-crystallin groups. In the fish there were only five, and in the squid none. According to Manski, Auerbach and Halbert, the origin of this organ-specificity lies in the retention, by the lens, of its evolutionary history; thus the mammalian lens can produce anti-bird, anti-amphibian, and anti-fish antibodies because it contains proteins common to these classes; the fish lens, on the other hand, cannot produce anti-bird or anti-mammal antibodies. Clayton, Campbell and Truman (1968) have shown, along with earlier workers, that many of the antigens in lens are not organ specific in so far as they react with antibodies to other tissues; and they conclude that the real finding is that the concentrations of certain antigens are very much higher in the lens than in other tissues. This is understandable, since the lens contains many enzymes, such as lactic dehydrogenase, that are common to the whole body; thus organ specificity is a tendency rather than an absolute fact, a tendency promoted by the isolated position of the lens.†

### PHACOANAPHYLACTIC ENDOPHTHALMITIS

Early workers were unable to detect homologous lens reactions, i.e. they were unable to produce anti-lens antibodies in a rabbit by injecting an extract of rabbit lens, but with the aid of Freund's adjuvants, Halbert et al. were able to do this, so the possibility naturally arises that an animal may produce antibodies to its own lens and thus cause, perhaps, a localized precipitin reaction in its lens leading to cataract. In fact, however, the production of very high titres of anti-lens antibodies in the rabbit failed to cause cataract, nor yet could congenital cataract be induced in the progeny of females with high anti-lens titres during pregnancy. Furthermore, human cataract patients had no anti-lens antibodies in their serum. Nevertheless, it is possible to induce a localized anaphylactic response in rabbits by producing anti-lens antibodies in the normal way and then to traumatize the lens; the resulting inflammatory reaction in the eye has been called by Burky phacoanaphylactic endophthalmitis.‡

## PROTEASES

An excised lens, maintained under sterile conditions, eventually undergoes autolysis, i.e. non-bacterial decomposition. This is due to the activity of certain proteases that are normally suppressed. In general, because of the extremely slow turnover of the lens protein—not greater than 3 per cent per month according to Waley (1964)—proteolytic activity is not great; nevertheless the study of the lens proteinases, and the factors that normally hold them in check, is important from the point of view of the maintenance of the correct protein composition and the removal of precipitated material. Moreover it is likely that the subtle post-synthetic changes in the lens crystallins are brought about by proteases.

---

* The comparative aspects of the lens proteins have been described in masterly (mistressly?) fashion by Dr Ruth Clayton (1974).
† Mehta et al. (1964) found that it was the β- and γ-crystallins that showed species specificity, so that organ-specificity was peculiar to α-crystallin.
‡ Sandberg (1976) has found appreciable quantities of α-crystallin in human aqueous humour from cataractous eyes by radio-immuno-assay.

NEUTRAL AND AMINOPEPTIDASE

A neutral protease was prepared by Van Heyningen and Waley (1962) from bovine lens by precipitation at pH 5·0; the activity was associated with α-crystallin and the material was described as $α_1$-crystallin. In addition, a leucine aminopeptidase was obtained which could be removed by treatment with urea. It was suggested that the neutral protease broke the lens proteins, e.g. α-crystallin, down to polypeptides while the leucine aminopeptidase completed the hydrolysis to amino acids. Thus Blow *et al.* (1975) found that, when leucine aminopeptidase activity had been removed with urea, the product of autolysis of α-crystallin was peptides with an average number of 10 to 14 residues. However, Van Heyningen and Trayhurn (1976) found an average number of four residues when autolysis was allowed to proceed in the presence of both enzymes, and they suggest that some other enzyme than leucine aminopeptidase is required for final breakdown of the peptides. In this connexion we must note that there are very large species variations in the leucine aminopeptidase activity extractable from lens, so that Trayhurn and Van Heyningen (1976) found none in human and fish lenses, greatest in bovine lens and only a little in rabbit, rat, chicken and monkey.*

NORMAL INHIBITION

Under normal conditions proteolytic activity must be held in check, and this can be achieved by segregation of the enzymes in intracellular vesicles (lysosomes), by the presence of inhibitors, or by the requirement that the protein be submitted to a primary denaturation. That denaturation greatly accelerates the proteolysis of lens extracts is suggested by the finding that proteolysis is very much more rapid at 55°C than at 37°C, a difference that can be reduced by preliminary denaturation of the protein substrate (Blow, 1974). As to the presence of inhibitors, it has been found that autolysis of lens proteins occurs far more rapidly if the reaction proceeds in a dialysis sac that allows the products of digestion to escape into the outside medium (Blow, 1974; Van Heyningen and Trayhurn, 1976), so it has been suggested that these products are inhibitory. With regard to lysosomal segregation, there seems little doubt from the studies of Swanson (1966, 1977) and Roelfzema *et al.* (1974) that many of the typical enzymes required in tissue breakdown (acid phosphatase, cathepsin C, β-glucuronidase, etc.) are contained within cytoplasmic organelles within the lens epithelium and the lens fibres.

## CHANGES DURING DEVELOPMENT

EMBRYONIC AND FETAL LENSES

Some of the changes in relative amounts of the crystallins during development have been indicated earlier. Van Kamp *et al.* (1974) have examined the soluble proteins from bovine lenses from embryos and fetuses from the age of 6 weeks to 9 months; as Figure 4.25 shows, the proportion of α-crystallin is highest in the earliest stages (62 per cent) but as the proportion of β-crystallin increases, the proportion of α-crystallin falls to reach the postnatal value of 35 per cent at 4 months. It will be seen, also, that γ-crystallin occupies about the same percentage throughout prenatal life. So far as the subunit composition of the α-crystallin is

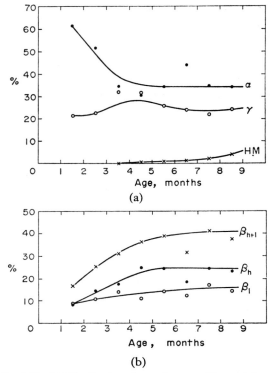

**Fig. 4.25**    (a) Distribution of α- and γ-crystallins and the high-molecular weight fraction (HM) in prenatal bovine lenses. (b) Distribution of β-crystallins in prenatal bovine lenses. (Van Kamp *et al., Comp. Biochem. Physiol.*)

concerned, remarkable changes take place; $αA_2$ and $αB_2$ were always present and presumably represent the 'essential' subunits; in the early embryonic lenses two specific subunits, not found in postnatal lenses to any extent, appear; these have been designated as $A_x$ and $B_x$; after 6 months $B_x$ has disappeared but $A_x$ is still present; in the 3 months post-natal lens only traces of $A_x$ are found. At $2\frac{1}{2}$ months $B_1$ appears and at $3\frac{1}{2}$ months the first trace of $A_1$ is seen.

## CHANGES IN ALPHA-CRYSTALLIN

We have seen that the α-crystallins are separated into a high- and low-molecular weight fraction, the dif-

*Swanson and Nichols (1971) reported the presence of a protease in human cataractous lens extracts with optimum activity at pH 5·2 activated by Co, Ni and Mn; Trayhurn and Van Heyningen were unable to find this, the only material in their extracts being maximally active in the neutral range pH 7 to 7·5 and activated by Mg; in the absence of Mg, which should enhance any acid activity, there was no pH dependence, and activity was uniformly low. The matter has been discussed further by Swanson and Beaty (1977) and Van Heyningen (1977); there is no doubt that there is in *bovine* lens, a proteolytic system active in the acid range (Hanson, 1962), but the existence in the human lens requires confirmation.

ference being solely one of aggregation of smaller units. When lenses of different ages were compared, Dilley and Harding (1975), working on human material, found that the percentage of the low-molecular weight fraction decreased from 66 per cent in the fetal lens to 16 per cent in the old lens. In the fetal lens there was no $\alpha B_1$ and only a trace of $\alpha A_1$, and in the neonatal lens there were very small amounts of these.

NUCLEUS VERSUS CORTEX

Since the nucleus is the older portion of any lens, the effects of age should be more obvious in this region than in the cortex. In the bovine lens Spector *et al.* (1971) found changes in proportions of the Peak I (high mol wt) to Peak II (low mol wt) in cortex and nucleus of calf and steer as follows:

|                 | Peak I | Peak II |
|-----------------|--------|---------|
| Calf periphery  | 30     | 70      |
| Calf nucleus    | 45     | 55      |
| Steer periphery | 43     | 57      |
| Steer nucleus   | 70     | 30      |

Thus in the adult animal the percentage of heavy material in the nucleus had become 70 compared with 45 in the calf nucleus and only 30 in the calf cortex.*

**Post-natal changes**

Figure 4.26 illustrates changes observed by Spector *et al.* (1974) in lenses taken from newborn to 80-year-old human subjects; the much more striking rise in the percentage of high molecular weight material in the nucleus compared with the cortex is especially clear.

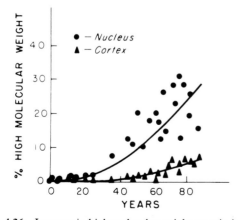

**Fig. 4.26** Increase in high molecular weight protein in soluble fraction of cortical and nuclear regions of human lenses as a function of ageing. (Spector *et. al., Invest. Ophthal.*)

These authors measured also the scattering of light by the lenses and found a good correlation between this and the percentage of high molecular weight material in the nucleus; in the cortex, however, scatter was much higher at all ages and there was only a small increase with age.

CHANGES IN THE SUBUNITS

Delcour and Papaconstantinou (1970) found mainly aggregates of $A_2$ and $B_2$ in the epithelium with only small amounts of $A_1$ and $B_1$ whereas in the fibres the $\alpha A_1$ and $\alpha B_1$ had become the major constituent of the $\alpha$-crystallin. The proportions are indicated below:

|            | $\dfrac{\alpha B_1}{(\alpha B_2 + \alpha B_1)} \times 100$ | $\dfrac{\alpha A_1}{(\alpha A_2 + \alpha A_1)} \times 100$ |
|------------|----------------------|----------------------|
| Epithelium | $8\cdot5 \pm 2$      | $10\cdot4 \pm 4$     |
| Fibres     | $29\cdot5 \pm 2$     | $42\cdot9 \pm 2\cdot6$ |

**Trends with depth**

By dividing the lens into nine layers from epithelium (I) to nucleus (IX) a trend towards increasing proportion of the $A_1$ and $B_1$ fractions became quite evident (Fig. 4.27A). In a similar way the sharp increase in molecular weight in the deepest layers was manifest (Fig. 4.27B). The very high molecular-weight material appearing in the more central regions resembled an aggregate composed of the polypeptide chains of both $\alpha$- and $\beta$-crystallin, and it seems highly likely that the insoluble albuminoid of the lens (p. 134) represents material resulting from progressive aggregation of the 'native' proteins of the lens involving not only $\alpha$-crystallin but probably all.

SHORTENED SUBUNITS

We have seen that the change from $\alpha A_2$ to $\alpha A_1$ results from removal of an amino group from glutamine. DeJong *et al.* (1974) found, during their determinations of the amino acid sequences of the subunits, that a small fraction lacked some amino acid residues at the C-terminal end, and they showed that this fraction was derived from the $\alpha A$ chain; it was called $\alpha A^{1-168}$ since it was missing the last 5 amino acids of the 173 residues in normal $\alpha A$. Both $\alpha A_2$ and $\alpha A_1$ exhibited the shortened version. The actual residues removed were: Ser-Ala-Pro-Ser. When layers of lens were analysed, the proportion of this degraded subunit increased progressively from epithelium to nucleus as we should expect of a process occurring after initial synthesis of the subunits; interestingly, though, the very high proportion that was found in early embryonic lenses

---

*These authors noted that a high glucose content was associated with the high molecular-weight fractions, and suggested that this was necessary for aggregation.

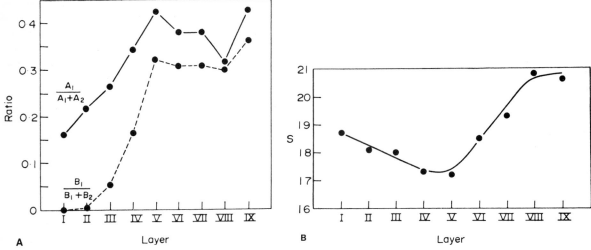

**Fig. 4.27**  (A) Ratios of polypeptide contents $A_1/A_1 + A_2$ and $B_1/B_1 + B_2$ in $\alpha$-crystallin from successive layers of lens from epithelium (I) to nucleus (IX). (B) Sedimentation coefficients of $\alpha$-crystallin isolated from successive layers of lens. (Hoenders *et al., Exp. Eye Res.*)

declined rapidly with increasing fetal and post-natal age; thus it seems that the $\beta A^{1-168}$ is further degraded during life, in fact Van Kleef has described a subunit missing the terminal 22 residues.

### VARIETY OF POLYPEPTIDE CHAINS

Using the technique of isoelectric focussing, which permits accurate resolution, Van Kleef *et al.* (1974) have identified nine chains which include $\alpha A_2^{1-151}$ and $\alpha A_1^{1-151}$ lacking two terminal amino acids (Van Kleef *et al.*, 1974a). They point out that, on the basis of subunit composition, there is as great a difference between $\alpha$-crystallin from calf cortex and cow cortex as there is between $\alpha$-crystallin from cortex and nucleus. When the subunit composition of the $\alpha LM$ and $\alpha HM$ fractions of any given extract were examined, e.g. cortex or nucleus, there was very little difference, indicating a direct aggregation of $\alpha LM$ to $\alpha HM$ in any given situation; by contrast there were large differences in subunit pattern when different extracts were compared, e.g. calf cortex, cow cortex, calf nucleus, cow nucleus. As these different extracts represent different ages, this means that subunit composition is very definitely a function of age of the synthesized protein.

## THE AGEING AND CATARACTOUS LENS

### AGGREGATION

Studies on the soluble extracts of lens proteins have indicated an increase, with age, of the proportion of

**Table 4.10** Percent high molecular weight proteins present in the soluble fraction of human normal lenses. (Jedziniak *et al.*, 1973.)

| Experiment number | Age | Concentration of heavy soluble protein / Concentration of total soluble protein × 100 |
|---|---|---|
| 1 | 96 | 10·98 |
| 2 | 79 | 11·36 |
| 3 | 76 | 9·30 |
| 4 | 76 | 4·33 |
| 5 | 75 | 5·18 |
| 6 | 71 | 6·08 |
| 7 | 68 | 7·69 |
| 8 | 65 | 5·75 |
| 9 | 64 | 6·07 |
| 10 | 64 | 1·89 |
| 11 | 55 | 4·97 |
| 12 | 46 | 6·46 |
| 13 | 46 | 2·01 |
| 14 | 42 | 5·35 |
| 15 | 28 | 3·12 |

high-molecular weight material, much of this representing aggregation of $\alpha$-crystallin polypeptide chains to larger and larger units, although the participation of other crystallins to give mixed aggregates also takes place. Studies on the human lens have exhibited the same phenomenon, namely the increase in the proportion of high-molecular weight material in the soluble fraction (Table 4.10) whilst amino acid analysis of this material suggests some difference between that of $\alpha$-crystallin and of the other pure crystallins, suggesting, in fact, an aggregation of subunits derived from more than one protein with perhaps some modification in the amino acid composition of the subunits (Jedziniak *et al.*, 1973).

## INSOLUBLE ALBUMINOID

When aggregation is carried beyond a certain point the product becomes insoluble in the usual salt solutions employed for extraction, and it could well be, as Spector has suggested, that the 'insoluble albuminoid' of the lens, i.e. the residue remaining after water extraction, represents an aggregation, during the course of the lens fibre's life, of proteins that were synthesized at a lower molecular weight.

## Urea-soluble material

The term 'insoluble albuminoid' was given by Mörner to the material that failed to dissolve in aqueous salt solutions; as Rao, Mehta and Cooper (1965) have shown, however, it may be solubilized by hydrogen-bond breaking agents, such as 7 M urea; and the resulting product shows marked similarities with α-crystallin as revealed immunoelectrophoretically. Certainly there is a very close reciprocal relationship between the insoluble albuminoid and α-crystallin content of lenses; thus in cataractous conditions the insoluble albuminoid fraction increases whilst the α-crystallin fraction decreases; similarly, in a given lens, the proportion of α-crystallin decreases on passing from cortex to nucleus whilst the proportion of insoluble albuminoid increases. We may conclude, then, that the molecules of α-crystallin aggregate together by the formation of hydrogen-bonds between chains; the bond-formation presumably reduces the number of hydrophilic groups in the surface and thus induces insolubility. Strong solutions of urea or guadinium chloride break hydrogen-bond linkages

and thus permit dissolution of the previously 'insoluble albuminoid'. Thus the urea-soluble (US) fraction of a lens constitutes this material, aggregated by hydrogen-bond linkages; any residue (urea-insoluble material) would be linked by stronger—covalent—bonds. Since the building up of large aggregates may well be related finally to a loss of transparency (Benedek, 1971),★ the investigation of the water-insoluble fractions, and the process of aggregation of the lens crystallins to larger units, has been carried out in considerable detail.

## SPECIES DIFFERENCES

Whereas the high molecular weight aggregates associated with the ageing lens appear to be, in the ox, largely derived from α-crystallin, this is by no means true of all species, so that in the rat and dogfish, for example, the main constituent is γ-crystallin; this is especially true of the urea-insoluble material, i.e. material that is covalently linked, mainly by disulphide bonds formed by cysteine residues (Zigman *et al.*, 1969).

## CHANGES IN INSOLUBLE PROTEIN WITH AGE

Figure 4.28A shows the change in the average percentage of the insoluble fraction of human lenses with age; it will be seen that this fraction is barely detectable in the first few years of life but increases to approximately 15 per cent of the total dry-weight of the lens by the age of 20; at the age of 60 there is a further sharp rise. As with the formation of high molecular weight soluble material, it was the nucleus that was responsible for most of the insoluble material. In Figure 4.28B

---

★ Although cataract is associated with increased formation of insoluble protein, it must be appreciated that the mere formation of albuminoid is not a cause of opacification since this material, although insoluble, probably exists, in the nuclear zone for example, as a transparent gel.

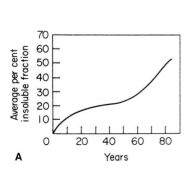

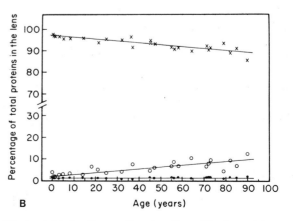

**A**    Years    **B**    Age (years)

**Fig. 4.28**   (A and B) Change in the average per cent of the insoluble fraction as a function of age. The dry weight contribution of the insoluble fraction to the dry weight of the total lens was determined on more than 40 lenses ranging in age from 3 months to 85 years. (Spector *et al.*, *Exp. Eye Res.*)

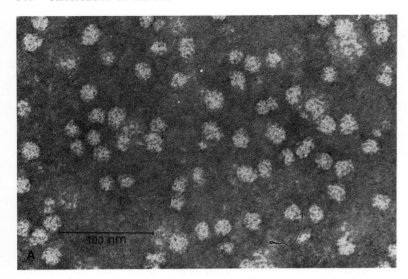

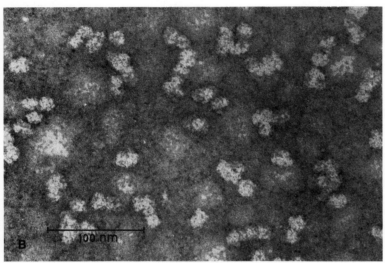

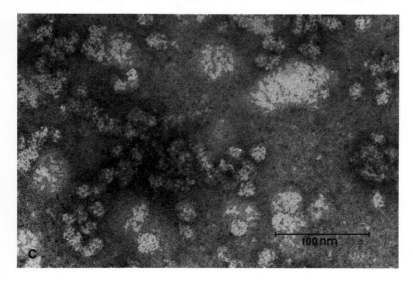

**Fig. 4.29** Electron micrographs of (A) α-crystallin showing spherical molecules with a diameter of approximately 17 nm, (B) cortical HM-crystallin showing mainly small aggregates, obviously consisting of several α-crystallin molecules linked together, (C) HM-crystallin showing irregularly shaped, giant aggregates with sizes reaching 500 nm. (Liem-The *et al.*, *Exp. Eye Res.* Courtesy Hoenders.)

there is shown a more elaborate analysis of the changes in normal human lenses with age; in this study (Coghlan and Augusteyn, 1977) the extracted proteins were categorized as soluble, urea-soluble; solubilized urea-insoluble, and finally urea-insoluble, this last being material that was not brought into solution with 8 M urea plus mercaptoethanol. The soluble protein fraction decreases by about 0·1 per cent per year and there is a corresponding increase in the urea-soluble fraction, so that by the age of 90 this represents 10 per cent. By contrast, the solubilized urea-soluble and the urea-insoluble proteins remain constant throughout life in the normal lens, but show a striking rise in senile nuclear cataracts, in fact the level could be used as a mode of classification of this condition.

THE TRANSITION TO HIGH MOLECULAR WEIGHT COMPOUNDS

Liem-The and Hoenders (1974a,b) have shown that the rabbit's lens proteins are essentially similar to those of the ox, although the $\alpha$-crystallins include a HM-fraction which appears to be an aggregate containing subunits of both $\alpha$- and $\beta$-crystallins, by contrast with the purely $\alpha$-crystallin make-up of bovine HM-fraction. When they compared the subunit composition of $\alpha$-crystallin, HM-fraction and the insoluble albuminoid, solubilized with urea (called urea-soluble material, US), they found sufficient similarities to justify the suggestion that HM is, indeed, a stage in aggregation of crystallins to albuminoid. Moreover, in the nucleus of the lens, the transition involved increasing numbers of $\beta$-crystallin chains, so that HM in the rabbit is not a simple aggregate of $\alpha$-crystallin. The urea-soluble material—albuminoid—revealed a complex pattern of $\alpha$- and $\beta$-crystallin subunits and three additional components

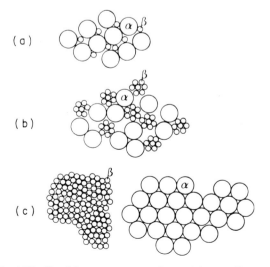

**Fig. 4.30** Possible arrangements of $\alpha$- and $\beta$-crystallins to form aggregates. (Liem-The *et al.*, *Exp. Eye Res.*)

with molecular weights of 52 000 to 63 000 daltons. In the electron microscope the HM extracted from the cortex consisted of relatively small aggregates of $\alpha$-crystallin units; thus Figure 4.29A shows a preparation of $\alpha$-crystallin consisting of spherical particles of diameter 17 nm; Figure 4.29B shows HM from the cortex consisting of relatively small aggregates obviously consisting of several units linked together. Figure 4.29C shows nuclear HM crystallin, containing irregularly shaped giant masses with sizes reaching up to 500 nm. The possible modes of aggregation have been discussed by Liem-The *et al.* (1975) as illustrated by Figure 4.30.

## Human lens

Similar studies on human lens proteins have been carried out by Spector and his colleagues. It seemed that, as with the rabbit, the insoluble material was simply a more highly aggregated form of HMW-protein, in fact the distinction between the HMW and insoluble albuminoid tends to vanish, the larger HMW material simply remaining in suspension when the still larger 'insoluble' material is removed from the preparation. Roy and Spector (1976) extracted the nuclei of cataractous human lenses and separated the material by centrifuging at different speeds; thus the material sedimenting with slow centrifugation (10 000 rev/min) was essentially insoluble albuminoid and had a molecular weight in the region of $50 . 10^6$. Fractions separated by higher centrifugal forces had lower molecular weights, but the amino acid composition of the three fractions was very similar, so that it appeared that the definition of HMW-crystallin was essentially operational, there being a continuous spectrum of material of larger and larger molecular weight but with little similarity to $\alpha$-crystallin. When the material was fractionated into subunits it became clear (Fig. 4.31) that the composition was complex; material of 45 000, 27 000, 22 000, 20 000 and 11 000 daltons was found in each preparation, indicating that the heavy material, whether isoluble albuminoid or HMW-crystallin, was a mixture of more than one crystallin; thus the 20 000 and 22 000 fractions are derived probably from $\alpha$-crystallin, but the 27 000 from $\beta$-crystallin.

ARTEFACTS

The urea-insoluble material, representing aggregates of crystallins linked by covalent bonds, seems to be partly artefactual, and results from oxidation of cysteine residues during extraction (Harding, 1969; 1972) so that in calf lens, for example, if steps are taken to prevent this oxidation only small amounts of urea-insoluble material are obtained and this is protein

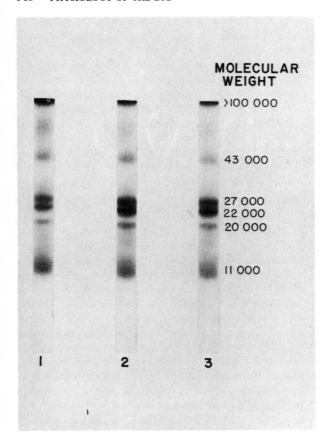

MOLECULAR
WEIGHT

>100 000

43 000

27 000
22 000
20 000

11 000

1          2          3

**Fig. 4.31**  Indicating heterogeneity of three fractions separated from extracts of cataractous human lens nuclei revealed as bands in polyacrylamide electrophoresis. Fractions 1, 2 and 3 were separated by 12, 20 and 30 thousand revolutions per minute respectively. (Roy and Spector, *Exp. Eye Res.*)

associated with the cell membranes of the lens. Nevertheless there is little doubt that the amount of the urea-insoluble material extractable from the lens increases with age and in the cataractous condition; and, according to Harding (1972) this represents, in effect, an increased susceptibility to oxidation of the native proteins. According to Harding (1972) this could have been due to a change in the conformation of the lens proteins whereby their thiol-groups became exposed and more susceptible to oxidation. Such an exposure might well account for a non-artefactual aggregation of protein molecules by oxidation within the lens fibres.

## CATARACT

By cataract we mean any opacification of the lens, which may be a highly localized 'spot' or a complete loss of transmission throughout the substance; it may occur as a senile change; as a result of trauma, which presumably injures the capsule and the underlying epithelium; as a result of metabolic or nutritional defects; or as a consequence of radiation damage.

## MORPHOLOGY

### NUCLEAR VS CORTICAL CATARACTS

The histology of the cataractous lens has been described in some detail by v. Sallmann (1951, 1957), and more recently by Cogan (1962). The *nuclear cataract* in which the nucleus of the lens becomes yellow or brown, is probably an exaggeration of the normally occurring senile changes. The cortical cataracts, on the other hand, represent opacities developing in the more recently laid-down lenticular fibres, and thus reflect the response to some abnormality in metabolism.

### CORTICAL CATARACTS

The gross morphology of these cortical cataracts consists in the appearance of aberrant forms of epithelial cells, which become large and round and are called 'balloon cells'; again, lens fibres may become fused into discrete masses, called morgagnian globules; if this process goes far enough the whole cortex becomes a milky liquid mass, and material may escape from the lens into the aqueous humour and vitreous body.

Of more interest, of course, are the early changes which, in the nature of things, must be examined in

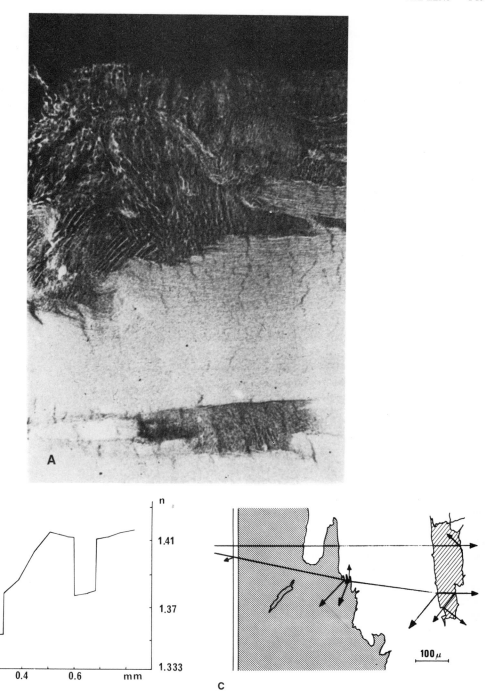

**Fig. 4.32** (A) Microradiogram from a human lens with cortical cataract, both subcapsular and supra-nuclear spokes. To the left is the anterior cortex with a reduced X-ray absorption. The capsule is torn off during the preparation procedure. About 0·6 mm from the periphery is another area with reduced X-ray absorption. This area has the same position as the spoke-like opacities, as revealed under slit lamp examination. Note the sharp and irregular interface between the regions with normal X-ray absorption and the region with reduced absorption. (×150); (B) The protein concentration and the refractive index determined along the marked line in the lens section shown in (A). Note the low concentration in the dark areas of the microradiogram and the sudden steep jumps in protein content and refractive index; (C) schematic drawing of the microradiogram in (A). Only the strong refractive interfaces are marked. Two rays of light are shown, and some of the main centres of light reflection or scattering are indicated. (Philipson, *Exp. Eye Res.*)

experimental forms of cataract in animals. In galactose cataract, Kuwabara, Kinoshita and Cogan (1969) have shown, in the electron microscope, that the first change is a vacuolization of the epithelial and superficial cortical lens cells; also the intercellular spaces between cortical fibres are expanded to form cysts. All these changes were reversible by returning the animals to a normal diet. After 7 days the lenses were opaque, and now the intercellular cysts had enlarged and become vacuoles; the membranes of individual lens fibres had broken and the cells had liquefied. After 10 to 15 days the lens was densely opaque, and the whole cortex had liquefied whilst the epithelium was irregular and showed regions of cell proliferation. Thus the early feature seems to be the appearance of vacuoles in the cells and enlargements of the extracellular space to form cysts; if this process is not held in check there is irreversible rupture of lens fibres.

### HUMAN CORTICAL CATARACT

In the human lens the cortical cataracts characteristically begin as localized points of opacity in surrounding transparent material in the subcapsular or supranuclear regions (Nordmann, 1962). In the electron microscope Philipson (1973) found essentially the same changes, consisting of an initial loss or damage to cell membranes leading to cellular dissolution with formation of large intercellular spaces or vacuoles. By microradiography, which permitted the estimate of local protein concentrations, the localized opacities were shown to be associated with a decrease in protein concentration (Fig. 4.32) accompanied by abrupt reduction in refractive index, and, as suggested by Figure 4.32, such localized refractive index transitions would act as scatter sources for light.

### NUCLEAR CATARACTS

Nuclear cataracts were completely different, the early stages being essentially a lenticular sclerosis accompanied by yellowing and a diffuse scattering of light with no large foci of scatter. In the advanced stages, usually associated with yellow-brown pigmentation, the electron-microscopical picture was normal so far as extracellular spaces were concerned, but in the dense whitish nuclear cataracts large aggregates of intracellular matrix of diameter 50 to 100 nm became obvious; such particles would be capable of scattering light and might account for the nuclear opacity (Benedek, 1971).

## SENILE CATARACT

### CLASSIFICATION

Pirie (1968) examined some 328 human cataractous lenses and classified them in accordance with their appearance as follows:

| Group I | Uniform pale yellow | 45% |
| II | Pale cortex with visible nucleus | 42% |
| III | Pale cortex with hazel brown nucleus | 11% |
| IV | Pale cortex with deep brown nucleus | 2 |

Since in many subsequent studies this classification has been adopted it is worth pointing out that Group I probably corresponds with Mach's Group 1 (*Cataracta nuclearis intumescens* et *cataracta corticale posteriore*), Group IV corresponds with Mach's Group 3 (*Cataracta brunescens*) and Groups II and III might be Mach's Group 2, showing both cortical and nuclear changes.

### PROTEIN CHANGES

We have seen that the characteristic feature of the ageing lens is the increase in percentage of 'insoluble albuminoid', which represents an aggregation of lower molecular weight species, with HMW proteins as an intermediate form on the way to complete water-insolubility. The increase is much more prominent in the nucleus so that there are appreciable amounts even in the young animal's nucleus.

In the cataractous lens the process of aggregation goes farther, so that the fraction of high-molecular weight material in the soluble fraction is higher than in normal lenses of the same age (Table 4.10). As Table 4.11 shows,

**Table 4.11** Effects of age on protein content of human lenses. (After Mach, 1963.)

| Age (years) | Weight (mg) | Protein content (mg/100 g lens) | | | %-Soluble |
|---|---|---|---|---|---|
| | | Total | Soluble | Insoluble | |
| 0–10 | 122 | 64·0 | 61·0 | 2·3 | 96·3 |
| 10–20 | 122 | 63·0 | 59·0 | 3·8 | 94·1 |
| 21–30 | 140 | 62·8 | 57·3 | 5·5 | 91·1 |
| 30–40 | 153 | 57·6 | 51·7 | 5·9 | 90·2 |
| 40–50 | 170 | 55·6 | 49·7 | 5·6 | 89·2 |
| 50–60 | 212 | 50·7 | 43·3 | 7·6 | 84·9 |
| 60–70 | 230 | 49·7 | 39·5 | 10·3 | 79·4 |
| 70–80 | 215 | 47·0 | 37·0 | 10·0 | 78·6 |
| >80 | 269 | 47·6 | 36·0 | 11·5 | 75·9 |
| *Cataract* | | | | | |
| Normal (40–86) | 297 | 50·2 | 41·4 | 9·1 | 81·7 |
| Cataract | 190 | 44·0 | 23·4 | 19·9 | 51·4 |

the ageing process is accompanied by a decrease in the absolute amount of protein in the lens from 64 to 47 mg/100 g; in cataract this process is enhanced so that in the age-group 40 to 86 the content is 50·2 mg/100 g compared with 44·0 mg/100 g in cataractous lenses, and the percentage of soluble proteins falls from 81 per cent to 51·4 per cent.

## γ-Crystallin

François, Rabaey and Stockmans (1965) have emphasized that the most significant feature of these changes
in composition is the decrease in the low-molecular
weight material, probably γ-crystallin, whilst the concentration of α-crystallin remains remarkably constant.

## Urea-insoluble material

We have seen that the studies from Spector's and
Bloemendal's laboratories indicate a progressive aggregation of crystallins, especially in the nucleus, so that
the soluble fraction contains materials of very high
molecular weight. There is no doubt, moreover, that
the urea-insoluble material increases at the expense of
urea-soluble material, i.e., that aggregates involving
covalent cross-linkages, and thus not disrupted in 8 M
urea, are formed in the cataractous lens. Thus, although
in many studies the appearance of urea-insoluble
material could have been due to artefactual formation
of S-S cross-linkages, the natural accumulation of high
molecular weight protein in some cataractous lenses,
formed by disulphide linkage and related to unfolding
of the lens proteins, has been established (Harding,
1973); these bonds are broken by mercaptoethanol so
that when a cataractous lens was treated with this
reagent* the gel-filtration profile showed a loss of the
high-molecular weight peak occurring in cataractous
lenses, but had no effect on a smaller peak in the normal
extract indicating that the normal high-molecular
weight fraction is not aggregated by S-S-linkages. Most
of this material was in the nucleus of these *nuclear
cataracts*, which were characterized as Group II (p. 150)
having a pale cortex and yellow nucleus.

AMINO ACIDS

Careful analysis of the amino acid composition of normal
and cataractous lens proteins, with a view to detecting
changes in critical groupings that might be concerned
in the formation of covalent cross-links, e.g. cysteine,
tryptophan and histidine, has revealed that the most
significant change is in the cysteine residues, so that,
according to Takemoto and Azari (1976), the free
cysteine decreased from 16 in normal to 1·9 residues/
1000 residues in advanced nuclear plus cortical cataract;
this was accompanied by a rise in the cystine residues
from 0·9 to 12·5/1000 indicating a process of auto-
oxidation of the SH-groups in crystallins associated with
cross-linkages (Table 4.12). There was no change in the
trytophan content.

## Non-disulphide-cross-linkages

Buckingham (1972) found, in advanced (Group III)
cataractous lenses, that 15 to 20 per cent of the protein
was in a form resistant to dissociation by hydrogen-bond
breaking agents and by reduction of the S-S-linkages
with dithiothreitol; and he concluded that this (nuclear)
material was linked by some other covalent link than
the disulphide bond. He observed that this covalently
linked material was associated with a strong absorption
of light at 310 nm giving it a yellow colour.

## YELLOW COLORATION

It is well established that ageing of the lens is accompanied by a yellow coloration, a coloration that is

---

*To prevent recombination of SH-groups the preparation is
carboxymethylated.

---

**Table 4.12** Cysteine (-SH) and half-cystine (-S-S-) content of proteins from normal and
cataractous human lenses. (Takemoto and Azari, 1976.)

| Type of lens | Age span (years) | Number analysed | Total cysteine[a] Mean ± s.e. | Free cysteine[b] Mean ± s.e. | Half-cystine[c] Mean ± s.e. |
|---|---|---|---|---|---|
| | | | Residue/1000 residues | | |
| Normal | 56–72 | 3 | 15·9 ± 0·07 | 16·0 ± 0·96 | 0·9 ± 0·43 |
| Posterior subcapsular cataract | 53–73 | 3 | 15·8 ± 0·90 | 13·0 ± 1·43 | 2·8 ± 1·03 |
| Nuclear cataract | 54–83 | 3 | 14·3 ± 0·43 | 5·9 ± 2·36 | 8·4 ± 2·33 |
| Nuclear plus cortical cataract | 53–80 | 3 | 14·4 ± 0·36 | 1·9 ± 2·53 | 12·5 ± 2·56 |

[a] The total cysteine content of lens is determined as CM cysteine after complete reduction of all
the disulphide bonds and alkylation of the original (free) and the newly liberated sulphydryl
groups.
[b] Determined as CM cysteine after carboxymethylation of lens proteins, prior to reduction of
cystine (-S-S-).
[c] The half-cystine values represent the cysteine residues which are involved in the formation of
disulphide bonds and are found indirectly from the difference between the total cysteine and the
free cysteine.

associated with the insoluble material of normal and cataractous lenses; this yellow material may well be related to the fluorescence associated with protein in cataractous lenses described by Dilley and Pirie (1974), a fluorescence that is not due to tryptophan. It is associated with the insoluble yellow protein of the nucleus of the cataractous lens although some is found associated with soluble protein of the high molecular weight fraction.

DISULPHIDE LINKAGES

Truscott and Augusteyn (1977) extracted cataractous lenses, typed in accordance with the coloration of their nuclei (Pirie, 1968), using hydrogen-bond and di-sulphide breaking agents; as Table 4.13 shows, the

N-terminal groups.† Yellowing of the lens, which is a normal age related feature of the human (Said and Weale, 1959), does not occur in the ox, and corresponding with this there is no non-tryptophan fluorescence in older ox lenses.

## MOLECULAR WEIGHTS OF THE AGGREGATES

So far as the molecular weights of the aggregates in cataractous lenses, before dissociation are concerned, these may be very high indeed; thus Takemoto and Azari (1977) found a range of 2 to $33 . 10^6$ daltons from advanced cataracts. As these authors point out, these

**Table 4.13** Distribution of protein fractions in the cortex and nucleus of cataractous lenses. (Truscott and Augusteyn, 1977.)

| Group | Dry weight (mg) Cortex | Nucleus | Soluble Cortex | Nucleus | Urea-soluble Cortex | Nucleus | Yellow Cortex | Nucleus | Insoluble Cortex | Nucleus |
|---|---|---|---|---|---|---|---|---|---|---|
| Type I | 36 | 31 | 36 | 33 | 10 | 18 | 1 | 1 | 2 | 0 |
| Type II | 34 | 33 | 37 | 26 | 11 | 19 | 1 | 4 | 2 | 0 |
| Type III | 42 | 30 | 31 | 13 | 17 | 12 | 2 | 22 | 3 | 1 |
| Type IV | 34 | 28 | 19 | 10 | 18 | 16 | 2 | 28 | 4 | 3 |

yellow protein increases markedly in amount in cataract, and is associated only with the nucleus.* The require-ment of dicaptoethanol suggests that disulphide linkages are primarily involved in the aggregation, and the fact that in brown cataracts of Classes III and IV the nucleus remains intact in solutions of urea and ioadacetamide, whereas Type I and II completely disintegrate, suggests that the cross-links in the more advanced nuclear cataracts are so extensive that they remain insoluble under these conditions. Molecular weight analysis showed that the de-aggregated protein consisted of particles of molecular weight varying from 130 000 to 20 000 daltons. The evidence indicated that the di-sulphide links were not formed during the extraction procedure; the very high percentage of this urea-insoluble material in the nucleus of the mature cataract (almost 60 per cent of the total) indicates that more than one crystallin is probably involved in the aggregation process.

FLUORESCENT POLYPEPTIDE

Spector et al. have isolated a polypeptide from old and cataractous human lens nuclei with which the fluor-escence is associated; its molecular weight was 43 000 daltons; it lacked N-terminal groups and it has been suggested that the polypeptide consists of two chains linked together by this fluorescent material through their

aggregates are not large enough to cause the scattering of light postulated by Benedek, which would require molecular weights of $50 . 10^6$ daltons. These authors emphasized that, although covalent aggregation is largely through S-S-links, a significant portion of the high molecular weight material in the cortex of advanced cataracts is linked in some other way.

## WATER AND ELECTROLYTE BALANCE

LEAKY MEMBRANES

The pathology of the cataractous lens suggests that the cell membranes have broken, and if this is true we may expect the normal electrolyte balance, with the high internal concentration of $K^+$ and low concentrations of $Na^+$ and $Cl^-$, to be upset as with any other tissue, so that initially $Na^+$ will leak into the cells and $K^+$ leak out; so long as the exchange is one-for-one there will be little change in internal osmotic pressure, but,

---

* Interestingly, the increase in urea-soluble material that occurs in cataractous lenses is confined to the cortex; thus the aggregates involving only hydrogen-bonding accumulate in the cortex by contrast with the urea-insoluble (covalently linked) material which is confined to the nucleus.
† The fluorescent substance seems to be a carboline (Dillon et al., 1975).

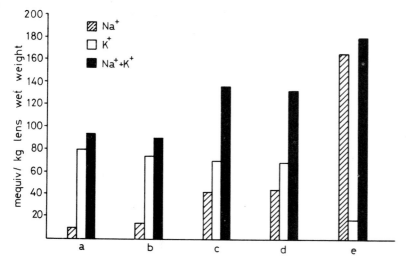

**Fig. 4.33**  Content of sodium and potassium ions (in mequiv/kg lens wet weight) of senile human normal and cataractous lenses: *a*, normal lenses; *b*, pure nuclear cataracts; *c*, nuclear cataracts with early cortical involvement; *d*, early cortical cataracts; *e*, complete cortical cataracts. (Maraini and Maraini, *Ciba Symp.*)

because of the Donnan-distribution imposed by a high internal concentration of negatively charged protein, ultimately the cells should swell and the final situation, if all the cells lost their membrane selectivity, would be a swollen lens with low internal $K^+$ and high internal concentrations of $Na^+$ and $Cl^-$ (Duncan and Croghan, 1969).

### CORTICAL VERSUS NUCLEAR CATARACTS

A number of early studies on cataractous lenses showed an increase in $Na^+$ and fall in $K^+$ with increase in percentage of water; however an important study of Maraini and Mangili (1973) emphasized the distinction between the cortical* and nuclear forms of cataract; it was only lenses with cortical opacities that showed an increased water-content, expressed as percentage of lens weight, the nuclear cataracts being normal in this respect (Table 4.14); as Figure 4.33 shows, the $K^+$-$Na^+$- and $Cl^-$-contents of purely nuclear cataracts are the same as in normal lenses, whereas the large rise in $Na^+$ and fall in $K^+$ are peculiar to the cortical cataract. In more elaborate studies, Duncan and Bushell (1975) found the same general picture, those cataractous lenses having

high $Na^+$ and low $K^+$, comparable with normal lenses, having mainly nuclear opacities; a smaller number of lenses had a moderately raised $Na^+$ and moderately lowered $K^+$, and another, large, group had very high $Na^+$ concentrations, considerably higher than that in the aqueous humour.

### THREE CATEGORIES

Thus three broad categories of cataract were established: A, B and C as in Table 4.15, Group A being purely nuclear, Group B a mixed type, and Group C containing mostly purely cortical cataracts. As Maraini and Mangili had found, the percentage of water was high in the cortical cataracts, which had also lower total solid matter per lens; thus the cortical cataract not only gains water but loses solid matter—protein.

---

*Dilley and Pirie (1974) have discussed the meaning to be attributed to the lens 'nucleus' in experimental studies; the term probably includes all the material formed before birth—embryonic and fetal—although biomicroscopically the infantile and adult regions are also distinguished (Fig. 4.6, p. 118). In practice the central approximately 100 mg of the mature human lens is employed for extraction of the proteins.

**Table 4.14** Wet weight, dry weight and water content of normal and senile cataractous human lenses. (Maraini and Mangili, 1973.)

|  | Normal lens | Posterior subcapsular cataract | Early cortical cataract | Intumescent cortical cataract | Complete cortical cataract | Nuclear cataract |
|---|---|---|---|---|---|---|
| Number of lenses | 7 | 6 | 15 | 7 | 10 | 9 |
| Wet weight (mg) | 236·5 ± 3·2 | 195·2 ± 11·5 | 207·5 ± 3·1 | 254·4 ± 14·5 | 196·7 ± 32·1 | 225·6 ± 20·6 |
| Dry weight (mg) | 72·9 ± 1·3 | 59·6 ± 2·2 | 68·6 ± 9·0 | 48·2 ± 2·7 | 45·2 ± 9·5 | 68·9 ± 8·5 |
| H₂O (mg) | 163·5 ± 9·7 | 135·6 ± 9·7 | 138·9 ± 24·1 | 206·1 ± 16·8 | 151·4 ± 28·7 | 156·6 ± 16·9 |
| H₂O (% lens wet weight) | 69·2 ± 0·7 | 70·7 ± 4·1 | 66·6 ± 2·7 | 81·0 ± 5·5 | 76·7 ± 4·4 | 69·4 ± 3·1 |

**Table 4.15** Ionic concentrations in lenses, given as mequiv/kg lens $H_2O$. The figures marked with an asterisk are taken from Maraini and Mangili. (Duncan and Bushell, 1975.)

| Lens type | Sodium | Potassium | Calcium | Magnesium | Chloride | % Water | Dry weight (mg) |
|---|---|---|---|---|---|---|---|
| Group A | 24·7 | 139·6 | 1·8 | 11·6 | 36·4 | 68·2 | 74·5 |
| Group B | 52·2 | 105·7 | 5·6 | 10·4 | 40·0 | 67·4 | 66·1 |
| Group C | 171·7 | 24·2 | 32·4 | 10·4 | 88·7 | 70·9 | 54·3 |
| Normal bovine | 25·0 | 120·0 | 0·6 | 10·4 | 35·0 | 67 | — |
| Normal human* | 14·5 | 113·5 | — | — | — | 70 | — |
| Nuclear* | 21·2 | 113·2 | — | — | — | 68 | — |
| Early cortical* | 67·7 | 106·1 | — | — | — | 67 | — |
| Complete cortical* | 210·6 | 21·4 | — | — | — | 79 | — |

HIGH $Na^+$-CONCENTRATION

The high concentration of $Na^+$ over that in the aqueous humour in cortical cataract may be accounted for by the high protein content of the tissue; a similar high $Na^+$-concentration occurs in cornea (Davson, 1949) and is due to a similar cause, but in the case of the lens this is normally prevented by the $Na^+$-pumps that exclude $Na^+$ and replace it with $K^+$; failure of the $Na^+$-pump allows an equilibrium to be established with a part of the $Na^+$-ions neutralized by protein anions. When the fibre membranes were damaged by freezing and thawing, a similar uptake of $Na^+$ and $Cl^-$ and loss of $K^+$ were observed (Duncan and Bushell, 1976), and the final levels reached were consistent with a Gibbs–Donnan equilibrium so far as these three ions were concerned.

## THE CALCIUM CONTENT

Great interest has attached to the calcium content of the lens since it is known that interference in the general calcium metabolism of the organism, such as parathyroidectomy, or in rickets, is associated with the development of cataract. Moreover, in aged, sclerotic or cataractous lenses the calcium content increases. It is unlikely, however, that the deposition of insoluble calcium salts is the primary change in any form of cataract; most probably the breakdown of the proteins and possibly of the lipids, as a result of metabolic disturbances, liberates sulphates, phosphates, and carbonates which are precipitated by calcium already present in the lens. Thus Bellows has shown that in galactose cataract the calcium deposition always follows the primary cataractous changes.

RELATION TO α-CRYSTALLIN AGGREGATION

Of great interest is the finding of Jedziniak et al. (1972) that the aggregation of bovine α-crystallin is favoured by high concentrations of calcium in the medium; this is illustrated by the curve of Figure 4.34 showing the

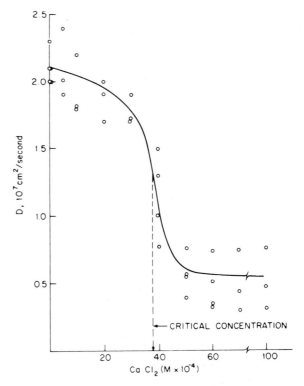

**Fig. 4.34** The effects of increasing the concentration of calcium ions on the diffusion coefficient of α-crystallin. (Jedziniak, *Invest. Ophthal.*)

diffusion coefficients of the aggregates against concentration of calcium in which the aggregation took place. At a critical concentration of $4 . 10^{-3}$ molar (4 mM) and above, aggregates as high as $300 . 10^6$ to greater than $2 . 10^9$ daltons molecular weight, computed from the diffusion coefficient, were obtained. The aggregation was not brought about by disulphide linkages, and belonged to α-crystallin and not to β or γ. Removal of the calcium with EDTA reversed the process, so that the turbid suspensions containing the highly aggregated material became transparent. In human lens material calcium also caused aggregation, but this was not reversible with EDTA or EGTA, indicating a much firmer binding (Jedziniak *et al.*, 1973).

## THE CALCIUM CONCENTRATIONS

As Duncan and Bushell have pointed out, the concentration of calcium within the lens water is very much less than that in the aqueous humour, so that the lens fibre shares with many other cells, such as the neurone, this feature of a low internal concentration of calcium, which must be maintained by an active pump since there is an electrical potential favouring influx of positive ions from aqueous humour to lens. Duncan and Bushell (1975) found that a high lens calcium was only associated with those cataractous lenses that had high $Na^+$ and low $K^+$, i.e. with the Group C lenses (Table 4.15). When Duncan and Bushell (1976) destroyed the fibre membranes by freezing and thawing the lens, the tissue now took up calcium, but the degree of accumulation was not comparable with that found in cataractous human lenses, and was prevented by treatment with EGTA; the rise in lens calcium under these conditions was accompanied by a loss of transparency, which was not due to the uptake of $Na^+$ or loss of $K^+$, since prevention of uptake of calcium by EGTA prevented the loss of transparency without affecting the changes in $Na^+$ and $K^+$ contents.

## ASSOCIATION WITH PROTEINS

The high internal concentration of calcium that builds up in the damaged lens is due, primarily, to failure of the outward pump mechanism, but on thermodynamic grounds such a failure could not lead to the very high concentrations actually found, especially in the cataractous lens, and a binding of the calcium in unionized form to the lens proteins must be invoked, similar to the binding to plasma proteins which, as we have seen, leads to a concentration about twice that found in a dialysate of plasma (p. 19). The extent of this binding, and its association with given protein fractions, have been studied in human cataractous lenses by Jedziniak *et al.* (1976) and Duncan and Van Heyningen (1977). Duncan and Van Heyningen used, as a measure of

binding, the rate at which calcium would diffuse out of a homogenate of lens retained in a dialysis-sac; and on this basis concluded that the insoluble fraction of the nucleus of the cataractous lens bound far more calcium per unit weight than the soluble fraction (Table 4.16).

Jedziniak *et al.* (1976) determined the calcium-contents of normal and cataractous lenses, relating the amount to the protein content; thus the average value for normal lenses was $0.14$ $\mu g/mg$ protein compared with $0.5$ $\mu g/mg$ protein, and they concluded that, on a statistical basis, the mean Ca concentration of the cataractous lens was some 2 to 13 times higher than that of the normal lens, but they emphasized that individual cataractous lenses could have a calcium-content within the normal range, or even less than this. Their results also indicated that, in the yellow type of cataract, the calcium associated with the HM-fraction of water-soluble protein was 2 to 3 times higher than that associated with the low-molecular weight protein, but their statistical analysis of results from other types of cataract emphasized the uncertainty with which associations of calcium with a protein fraction could be established.

## CALCIUM AND CATARACT

To conclude, then, it seems that failure of the membrane-pumps that occurs in mature cortical cataracts necessarily leads to uptake of calcium as well as $Na^+$ and $Cl^-$. The binding capacity of the lens proteins means that the concentration of calcium finally reached in the abnormal lens fibres is greater than would be expected of a passive distribution; however, the very high concentrations found in cataractous lenses do suggest that there has been an alteration in the binding capacity of the lens proteins in the cataractous condition. It is tempting to suggest, along with Spector, that the failure of the cataractous lens to exclude calcium leads to aggregation of lens proteins, an aggregation that leads to the much tighter binding of the calcium that is actually observed.

---

* Jedziniak *et al.* (1976) state that the concentration of calcium within the normal lens is higher than that required to cause aggregation, namely 4 mM; according to Duncan and Bushell (1975), however, the normal concentration is probably less than 1 meq/litre, i.e. less than 0.5 mM.

**Table 4.16** Calcium associated with the soluble and insoluble protein fractions of normal and cataractous human lenses. (Duncan and Van Heyningen, 1977.)

| Lens type | % Ca in soluble fraction | % Ca in insoluble fraction | Soluble/ insoluble ratio | Total Na (mM/kg water) | Total Ca (mM/kg water) |
|---|---|---|---|---|---|
| Normal | 20·2 | 13·8 | 1·3 | 54 | 1·8 |
| Group I | 10·0 | 13·5 | 0·83 | 68 | 4·2 |
| Group II | 8·2 | 30·1 | 0·39 | 105 | 9·8 |
| Group III | 6·1 | 35·3 | 0·20 | 184 | 12·5 |

## LENS POTENTIAL AND CATARACT

The asymmetry potential measured between the anterior and posterior faces of the lens disappears in cataractous lenses containing high $Na^+$ and low $K^+$; this is due to a fall in the trans-membrane potential at both faces to a value of $-15$ mV, the inside being negative with respect to the medium. Such a potential might be expected of a passive distribution of ions between the colloidal high-protein phase of the lens fibre and an outside medium free of colloid, i.e. it could result from the abundance of negatively charged sites on the fibre proteins that give rise to a Gibbs–Donnan distribution and appropriate potential difference. However, calculations suggest that this should only be of the order of 5 mV and comparable with that found in the freeze-dried lens.

## METABOLIC DEFECTS

The way in which a metabolic defect can lead to a loss of transparency is not clear, but an important factor must unquestionably be the amount and physical condition of the proteins in the lens fibres; the lens continually produces new proteins and the existing ones are in a continuous state of 'turnover', in the sense that they are broken down and resynthesized. These processes require energy which they derive from the metabolic breakdown of glucose, and it may be surmised that defects in the protein synthetic mechanism could cause cataract. Other factors may be an alteration in pH causing a precipitation of protein in the fibres; a failure of the ionic pumps that maintain the normal salt and water contents of the lens fibres; finally there is the death of the lens fibres, producing gaps in the ordered arrangement, or the partial death leading to vacuolization of the cells.

### TOXIC CATARACTS

A number of drugs will cause opacities in the lens; one that has been studied experimentally in some detail is naphthalene, which is converted in the body to, among other substances, naphthalene diol; according to Van Heyningen and Pirie (1967), this can be converted in the eye to 1-2-dihydroxy-naphthalene, which is probably responsible for the blue fluorescence in the eye of the naphthalene-fed rabbit. This is oxidized to $\beta$-naphthaquinone (Rees and Pirie, 1967) and is highly reactive; for example, it reacts with ascorbic acid to form $H_2O_2$ which is toxic. This reaction converts ascorbic acid into its oxidized product, dehydroascorbic acid; and it is interesting that the concentration of ascorbic acid in the lens actually increases in the naphthalene-fed rabbit; it has been suggested by Van Heyningen (1970) that the dehydroascorbic acid, which

readily penetrates the lens, is reduced there to ascorbic acid from which it cannot escape so easily, being less lipid-soluble than dehydroascorbic acid. The actual means whereby $\beta$-naphthaquinone causes cataract has not so far been clarified. In addition to naphthalene, many drugs, often used therapeutically at one time, such as dinitrophenol, myeleran, dimethyl sulphoxide, will cause cataracts but not necessarily in man in therapeutic doses; for example dimethyl sulphoxide has not reportedly caused eye damage in man although in experimental animals, such as the dog, doses of 5 g/kg will cause cataract, albeit of an unusual type (Rubin and Barnett, 1967). Such knowledge as we have of the mode of action of a variety of drugs has been summarized by Van Heyningen (1969), whilst Pirie (1968) has described the early pathology of naphthalene lesions in lens and retina.

### SUGAR CATARACTS

The essential factor in this form of cataract seems to be the high concentration of sugar in the aqueous humour, be it glucose, as in diabetes, or galactose and xylose in the experimental cataracts obtained by feeding these sugars to certain animals such as the rat and hamster. The only metabolic pathway common to all the cataractogenic sugars is the production of a polyhydric alcohol, e.g. sorbitol in the case of a high blood-glucose, dulcitol with high blood galactose, etc. Van Heyningen (1959) observed that feeding these sugars led to accumulation of the alcohols in the lens; and it was suggested by Kinoshita, Merola and Dikmak (1962) that the primary event was, indeed, the intracellular accumulation of the alcohol, which, not being able to escape from the lens fibres, caused an increased osmolality leading to osmotic swelling.

**Swelling of lens fibres**
Figure 4.35 shows changes in water- and dulcitol-content of rat lenses when the animals were maintained

**Fig. 4.35** The accumulation of dulcitol and water in lenses during development of galactose cataract in rats. (Kinoshita et al., *Exp. Eye Res.*)

on a galactose diet. *In vitro*, also, exposure of the lens to galactose caused corresponding increases in hydration and alcohol concentration. When an aldose reductase inhibitor was added to the galactose medium, the formation of dulcitol was prevented, and likewise the increase in hydration (Kinoshita *et al.*, 1968). As to the ultimate steps leading to the mature cataract, i.e. the opacity of the lens, the position is not clear; Patterson and Bunting (1966) consider that this does not result from the progressive swelling and disruption of the lens fibres, since this happens suddenly over a period of 24 hours, and up to this point the changes are reversible by removing the sugar from the diet; according to Kinoshita, the swelling induces secondary changes in the lens fibres that reduce the efficacy of their active transport mechanisms, e.g. the extrusion of $Na^+$, accumulation of $K^+$ and amino acids. Thus Kinoshita *et al.* (1969) were able to inhibit the loss of amino acids and myoinositol, found in galactose-exposed lenses, by preventing the swelling through addition of osmotic material to the medium. This last finding seems to be conclusive proof that the primary insult is the osmotic swelling; this presumably causes an increase in passive permeability to ions and metabolic substrates such as amino acids, thereby making the active transport mechanisms ineffective; the failure of the $Na^+$-pump would eventually lead to bursting of the cells, and failure to accumulate amino acids might reduce protein synthesis and turnover.*

### Genetic diabetes

Mice that are genetically diabetic, in contrast to other diabetic animals, do not develop cataracts; according to Varma and Kinoshita (1974) this is accounted for by a low concentration of the enzyme aldose reductase, responsible for formation of the polyalcohols. In a similar way it is difficult to cause accumulation of polyols in the mouse lens by feeding it galactose.

## RADIATION CATARACT

The lens is subject to damage by ionizing radiations because it is essentially only those cells that undergo division that are susceptible. If the radiation is confined to the central area of the lens, by screening the equatorial region, no cataract develops (Goldmann and Leichti, 1938). If only a sector of the lens is exposed, the damage in this region is much less severe than if the whole lens had been exposed, and it would seem that the undamaged region can exert a protective action on the remainder (Pirie and Flanders, 1957). As with other forms of radiation damage, injections of cysteine and certain other protective substances, e.g. cysteamine, reduce the severity of the lesion.

### BIOCHEMICAL CHANGES

The basic biochemical and morphological changes resulting from X-irradiation of the eye have been summarized by Liem-The *et al.* (1975), who have described their own studies on the rabbit. Earlier work had emphasized the reduction in reduced glutathione

---

* Lerman and Zigman (1967) found that the onset of galactose cataract was associated with a fall in the level of soluble $\gamma$-crystallin; they also found that $^3$H-sorbitol was bound to protein, in this condition, especially to $\gamma$-crystallin. Holt and Kinoshita (1968) have described alterations in the $\beta$- and $\gamma$-crystallin fractions in this form of cataract.

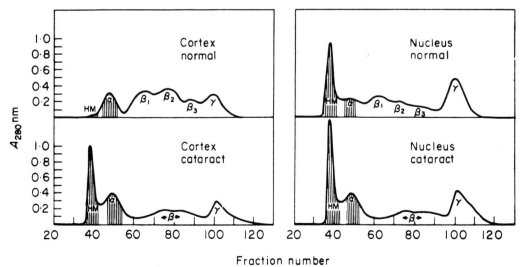

**Fig. 4.36** Column chromatography of the water-soluble fractions from normal and cataractous rabbit lens cortex and nucleus. Note that the cataractous lens cortex tends to approach the normal lens nucleus in its crystallin composition. (Liem-The *et al.*, *Exp. Eye Res.*)

(Pirie *et al*., 1953) and accumulation of insoluble albuminoid probably resulting from cross-linkage of crystallins through disulphide bonds (Dische, 1966). Liem-The *et al*. assumed that errors in metabolism might lead to synthesis of abnormal polypeptide chains causing aggregation to unusual high-molecular weight material. The content of the urea-soluble fraction increased some tenfold. In general, the cataractous lens cortex tended to approach the normal lens nucleus in its crystallin composition; this is shown by Figure 4.36 where a large peak of HM-crystallin appears in the cataractous cortex; a further feature is the decrease in $\beta$-globulin content of the cataractous lens.

The reduction in $\beta$-crystallin revealed by Figure 4.36 might well be accounted for by its incorporation into the HM-fraction.

### ANALOGY WITH SENILE LENS

The electron microscopical appearance indicated the appearance of large aggregates in cataractous cortex consisting presumably of $\alpha$-crystallin plus other crystallins since the subunit composition of the HM-material was not characteristic of a simple aggregation of $\alpha$-crystallin molecules. These findings confirm the general hypothesis that X-irradiation, by inhibiting production of new fibres, accelerates the normal ageing process, so that the cataractous lens has many similarities with the senile lens.

## GLASSBLOWER'S CATARACT

Exposure of the eyes to strong infra-red radiation causes cataract and this was shown by Goldmann in 1933 to be due to the high temperature to which the lens was raised by the absorption of heat; absorption of light by the ciliary body and iris, if sufficiently intense, will also raise the temperature of the lens and cause cataract (Langley, Mortimer and McCulloch).

## TRAUMATIC CATARACT

Trauma to the eye, or more specifically to the lens, can cause cataract, but the mechanism is obscure. Nelson and Rafferty (1976) have drawn attention to the rat's lens which seems especially immune to injury, so far as subsequent opacification is concerned (Rafferty, 1973). Using scanning electron-microscopy, they showed that the lens fibres adjacent to the injured region exhibited morphological changes (flattening) suggesting a specific reaction to injury, changes that were not permanent so that the fibre returned to its original shape, by contrast with the cataractogenic process in, say, the hereditary strain of mouse, which is permanent.

## HEREDITARY CATARACTS

Gorthy and Abdelbaki (1974) have described the morphological changes in lenses from rats of a strain bred from whole-body X-irradiated animals by Léonard and Maisin (1965). The primary defects originate in the cortex, and opacification is obviously related to formation of vacuoles. Hamai *et al*. (1974) have described the changes in a hereditary strain of cataracts in mice; an important feature was the defective denucleation, cells failing to show normal denucleation becoming hydropic; in fact Sakuragawa *et al*. (1975) have described a remarkable ballooning of the lens fibres in localized regions so that the volume of the swollen portion of a fibre may be 300 times greater than normal. The mechanism of these osmotically induced changes may well reside in a loss of $Na^+$-$K^+$-activated ATPase so that the $Na^+$-pump that maintains a normally low internal $Na^+$-concentration is out of action (Iwata and Kinoshita, 1971; Kinoshita, 1974).

### SULPHYDRYL GROUPS

Takemoto *et al*. (1975) found that the free SH-groups or cysteine residues in the lenses of rats with hereditary cataract decreased to zero when the cataract became mature; the changes are illustrated by Table 4.17.

**Table 4.17** Changes in SH- (cysteine) and SS- (cystine) groups in rats with hereditary cataract. (Takemoto *et al*., 1975.)

| Age (days) | | Total cysteine | Free cysteine | $\frac{1}{2}$ cystine |
|---|---|---|---|---|
| 29 | Normal | 26 | 26 | 0 |
| 30 | Cataractous (incipient) | 26 | 26 | 0 |
| 60 | Normal | 28 | 28 | 0 |
| | Cataractous (incipient) | 24 | 22 | 2 |
| 95 | Normal | 21 | 15 | 6 |
| | Cataractous (mature) | 22 | 0 | 22 |

The fall in free cysteine was associated with a rise in high molecular-weight protein, which could be disaggregated by SH-reagents. Interestingly, when the rat's lens was exposed *in vitro* to high pressures of pure oxygen, cataractous changes associated with loss of free SH-groups were observed.

## REVERSIBLE CATARACTS

### COLD CATARACT

Finally, we may mention two interesting forms of cataract that are reversible; when a young animal, such as a rat, is cooled, cataracts may appear, which disappear on rewarming, and they are due to a reversible precipitation of protein in the lens fibres. Zigman and Lerman (1965) isolated the precipitated material and showed that its sedimentation constant of 4 S was characteristic of a single low-molecular weight protein, although immunochemically it had characteristics of $\alpha$-, $\beta$-, and $\gamma$-crystallins. According to Balazs and Testa (1965) the material is $\gamma$-crystallin. Lerman and Zigman (1967) made the interesting observation that the $\alpha$- and $\beta$-crystallins, in high enough concentration, will inhibit cold-precipitation of $\gamma$-crystallin. Philipson found non-stained areas in electron-micrographs of calf nucleus made cataractous at less than $10°C$; their diameters ranged from 100 to 200 nm, and these regions might have represented aggregates of separated $\gamma$-crystallin.

## ALKALOIDS

The other form of opacity is hardly a true cataract; it follows the administration of morphia-type alkaloids to rodents and seems to be the consequence of a precipitation of the altered drug on the anterior surface of the lens; it develops in a few minutes and disappears within a few hours (Weinstock and Stewart).

## REFERENCES

Alcalá, J., Lieska, N. & Maisel, H. (1975) Protein composition of bovine lens cortical fibre cell membranes. *Exp. Eye Res.* 21, 581–595.

Amoore, J. E., Bartley, W. & van Heyningen, R. (1959) Distribution of sodium and potassium within cattle lens. *Biochem. J.* 72, 126–133.

Andrée, G. (1958) Uber die Natur des transkapsularen Potentials der Linse. *Pflüg. Arch. ges. Physiol.* 267, 109–116.

Bakker, A. (1936) Eine Methode, die Linsen erwachsener Kaninchen ausserhalb des Körpers am Leben zu erhalten. *v. Graefes' Arch. Ophthal.* 135, 581–592.

Balazs, E. A. & Testa, M. (1965) The cryoproteins of the lens. *Invest. Ophthal.* 4, 944.

Becker, B. & Cotlier, E. (1965) The efflux of [86]rubidium from the rabbit lens. *Invest. Ophthal.* 4, 117–121.

Bellows, J. G. (1944) *Cataract and Anomalies of the Lens.* London: Kimpton.

Benedek, G. B. (1971) Theory of transparency of the eye. *Appl. Optics.* 10, 459–473.

Bentley, P. J. & Candia, O. A. (1971) Effects of some metabolic inhibitors on the electrical potential difference and short-circuit current across the lens of the toad *Bufo marinus. Invest. Ophthal.* 10, 672–676.

Benzce, W. L. & Schmid, K. (1957) Determination of tyrosine and tryptophan in proteins. *Anal. Chem.* 29, 1193.

Berns, T. J. M., Schreurs, V. V. A. M., van Kraaikamp, M. W. G. & Bloemendal, H. (1973) Synthesis of lens proteins *in vitro. Eur. J. Biochem.* 33, 551–557.

Bettelheim, F. A. (1975) On the optical anisotropy of lens fiber cells. *Exp. Eye Res.* 21, 231–234.

Bito, L. & Davson, H. (1964) Steady-state concentrations of potassium in the ocular fluids. *Exp. Eye Res.* 3, 283–297.

Björk, I. (1964a) Studies on $\gamma$-crystallin from calf lens. II. *Exp. Eye Res.* 3, 254–261.

Björk, I. (1964b) Studies on the subunits of $\alpha$-crystallin and their recombination. *Exp. Eye Res.* 3, 239–247.

Björk, I. (1964c) Fractionation of $\beta$-crystallin from calf lens by gel filtration. *Exp. Eye Res.* 3, 248–253.

Bloemendal, H., Berns, T. J. M., Van der Ouderaa, F. & De Jong, W. W. W. (1972) Evidence for a 'non-genetic' origin of the A 1 chains of $\alpha$-crystallin. *Exp. Eye Res.* 14, 80–81.

Bloemendal, H., Berns, T., Zweers, A., Hoenders, H. & Benedetti, E. L. (1972) The state of aggregation of $\alpha$-crystallin detected after large-scale preparation by zonal centrifugation. *Eur. J. Biochem.* 24, 401–406.

Bloemendal, H. Bont, W. S. & Benedetti, E. L. (1965) On the sub-units of $\alpha$-crystallin. *Exp. Eye Res.* 4, 319–326.

Bloemendal, H., Bont, W. S., Jongkind, J. F. & Wisse, J. H. (1962) Splitting and recombination of $\alpha$-crystallin. *Exp. Eye Res.* 1, 300–305.

Bloemendal, H., Zweers, A., Benedetti, E. L. & Walters, H. (1975) Selective reassociation of the crystallins. *Exp. Eye Res.* 20, 463–478.

Bloemendal, H. *et al.* (1977) Nomenclature for the polypeptide chains of lens plasma membranes. *Exp. Eye Res.* 24, 413–415.

Blow, A. M. J. (1974) PhD. Thesis, Oxford.

Blow, A. M. J., Van Heyningen, R. & Barrett, A. J. (1975) Metal-dependent proteinase of the lens. *Biochem. J.* 145, 591–599.

Bonting, S. L., Caravaggio, L. L. & Hawkins, N. M. (1963) Studies on sodium-potassium-activated adenosinetriphosphate. VI. *Arch. Biochem. Biophys.* 101, 47–55.

Brassil, D. & Kern, H. L. (1967) Characterization of the transport of neutral amino acids by the calf lens. *Invest. Ophthal.* 7, 441–451.

Brindley, G. S. (1956) Resting potential of the lens. *Br. J. Ophthal.* 40, 385–391.

Broekhuyse, R. M. & Kuhlmann, E. D. (1974) Lens membranes. I. Composition of urea-treated plasma membranes from calf lens. *Exp. Eye Res.* 19, 297–302.

Broekhuyse, R. M., Kuhlmann, E. D. & Stols, A. L. H. (1976) Lens membranes. II. Isolation and characterization of the main intrinsic polypeptide (MIP) of bovine lens fiber membranes. *Exp. Eye Res.* 23, 365–371.

Buckingham, R. H. (1972) The behaviour of reduced proteins from normal and cataractous lenses in highly dissociating media: cross-linked protein in cataractous lenses. *Exp. Eye Res.* 14, 123–129.

Bullough, W. S. & Laurence, E. B. (1964) Mitotic control by internal secretion: the role of the chalone-adrenalin complex. *Exp. Cell. Res.* 33, 176–194.

Burky, E. L. (1934) Experimental endophthalmitis phacoanaphylactica. *Arch. Ophthal.* 12, 536–546.

Candia, O. A., Bentley, P. J., Mills, C. D. & Toyofuku, H. (1970) Asymmetrical distribution of the potential in the toad lens. *Nature* 227, 852–853.

Chen, J. H., Lavers, G. C. & Spector, A. (1974) Translation and reverse transcription of lens mRNA isolated by chromatography. *Exp. Eye Res.* 18, 189–199.

Clayton, R. M. (1969) Properties of the crystallins of the chick in terms of their subunit composition. *Exp. Eye Res.* 8, 326–339.

Clayton, R. M. (1970) Problems of differentiation in the vertebrate lens. *Curr. Top. Develop. Biol.* **5**, 115–180.

Clayton, R. M. (1974) Comparative aspects of lens proteins. In *The Eye* (Eds Davson, H. and Graham, L. T.). Vol. 5, pp. 399–494. Academic Press, New York.

Clayton, R. M. (1978) Divergence and convergence in lens cell differentiation: regulation of the formation and specific content of lens fibre cells. In *Stem Cells and Homeostasis*. Ed. Lord, Potten & Cole. pp. 115–138. C.U.P.

Clayton, R. M., Campbell, J. C. & Truman, D. E. S. (1968) A re-examination of the organ specificity of lens antigens. *Exp. Eye Res.* **7**, 11–29.

Clayton, R. M. & Truman, D. E. S. (1974) The antigenic structure of chick β-crystallin subunits. *Exp. Eye Res.* **18**, 495–506.

Cogan, D. G. (1962) Anatomy of lens and pathology of cataracts. *Exp. Eye Res.* **1**, 291–295.

Coghlan, S. D. & Augusteyn, R. C. (1977) Changes in the distribution of proteins in the aging human lens. *Exp. Eye Res.* **25**, 603–611.

Cohen, A. I. (1965) The electron microscopy of the normal human lens. *Invest. Ophthal.* **4**, 433–446.

Cremer-Bartels, G. (1962) A light-sensitive, fluorescent substance in bovine and rabbit lenses. *Exp. Eye Res.* **1**, 443–448.

Davson, H. (1949) Some considerations on the salt content of fresh and old ox corneae. *Br. J. Ophthal.* **33**, 175–182.

De Jong, W. W., Van Kleef, F. S. M. & Bloemendal, H. (1974) Intracellular carboxy-terminal degradation of the αA chain of α-crystallin. *Eur. J. Biochem.* **48**, 271–276.

De Jong, W. W., Van der Ouderaa, F. J. & Bloemendal, H. (1975) Primary structures of the α-crystallin A chains of seven mammalian species. *Eur. J. Biochem.* **53**, 237–242.

Delamere, N. A. & Duncan, G. (1977) A comparison of ion concentrations potentials and conductances of amphibian, bovine and cephalopod lenses. *J. Physiol.* **272**, 167–186.

Delcour, J. & Papaconstantinou, J. (1970) A change in α-crystallin subunit composition in relation to cellular differentiation in adult bovine lens. *Biochem. Biophys. Res. Comm.* **41**, 401–406.

Delcour, J. & Papaconstantinou, J. (1972) Synthesis and aggregation of α-crystallin subunits in differentiating lens cells. *J. biol. Chem.* **247**, 3289–3295.

Dickson, D. H. & Crock, G. W. (1972) Interlocking patterns on primate lens fibers. *Invest. Ophthal.* **11**, 809–815.

Dilley, K. J. & Harding, J. J. (1975) Changes in proteins of the human lens in development and aging. *Biochim. Biophys. Acta.* **386**, 391–408.

Dilley, K. J. & Pirie, A. (1974) Changes to the proteins of the human lens nucleus in cataract. *Exp. Eye Res.* **19**, 59–72.

Dische, Z. (1965) The glycoproteins and glycolipoproteins of the bovine lens and their relation to albuminoid. *Invest. Ophthal.* **4**, 759–778.

Dische, Z. (1966) Alterations of lens proteins as etiology in cataracts. In *Biochemistry of the Eye*. Symposium Tutzing Castle. p. 413. Karger, Basel/N.Y.

Dische, Z. & Zelmenis, G. (1966) The sialohexosaminoglycan of the bovine lens capsule. *Doc. Ophthal.* **20**, 54–72.

Dische, Z. & Zil, H. (1951) Studies on the oxidation of cysteine to cystine in lens proteins during cataract formation. *Amer. J. Ophthal.* **34**, 104–113.

Duncan, G. (1969a) Relative permeabilities of the lens membranes to sodium and potassium. *Exp. Eye Res.* **8**, 315–325.

Duncan, G. (1969b) The site of the ion restricting membranes in the toad lens. *Exp. Eye Res.* **8**, 406–412.

Duncan, G. (1969c) Kinetics of potassium movement across amphibian lens membranes. *Exp. Eye Res.* **8**, 413–420.

Duncan, G. (1970) Movement of sodium and chloride across amphibian lens membranes. *Exp. Eye Res.* **10**, 117–128.

Duncan, G. & Bushell, A. R. (1975) Ion analysis of human cataractous lenses. *Exp. Eye Res.* **20**, 223–230.

Duncan, G. & Bushell, A. R. (1976) The bovine lens as an ion-exchanger: a comparison with ion levels in human cataractous lenses. *Exp. Eye Res.* **23**, 341–353.

Duncan, G. & Croghan, P. C. (1969) Mechanisms for the regulation of cell volume with particular reference to the lens. *Exp. Eye Res.* **8**, 421–428.

Duncan, G., Juett, J. R. & Croghan, P. C. (1977) A simple chamber for measuring lens asymmetry potentials. *Exp. Eye Res.* **25**, 391–398.

Duncan, G. & Van Heyningen, R. (1977) Distribution of non-diffusible calcium and sodium in normal and cataractous human lenses. *Exp. Eye Res.* **25**, 183–193.

Farkas, T. G. & Weese, W. C. (1970) The role of aqueous humor, insulin and trivalent chromium in glucose utilization of rat lenses. *Exp. Eye Res.* **9**, 132–136.

Farnsworth, P. N., Mauriello, J. A., Burke-Gadomski, P., Kulyk, T. & Cinotti, A. A. (1976) Surface ultrastructure of the human lens capsule and zonular attachments. *Invest. Ophthal.* **15**, 36–40.

Fisher, R. F. & Pettet, B. E. (1973) Presbyopia and the water content of human crystalline lens. *J. Physiol.* **234**, 443–447.

François, J. & Rabaey, M. (1957) The protein composition of the human lens. *Am. J. Ophthal.* **44**, Pt. 2, 347–357.

François, J. & Rabaey, M. (1958) Permeability of the capsule for the lens proteins. *Acta ophthal., Kbh.* **36**, 837–844.

François, J., Rabaey, M. & Recoulès, N. (1961) A fluorescent substance of low molecular weight in the lens of primates. *Arch. Ophthal., N.Y.* **65**, 118–126.

François, J., Rabaey, M. & Stockmans, L. (1965) Gel filtration of the soluble proteins from normal and cataractous human lenses. *Exp. Eye Res.* **4**, 312–318.

Friedenwald, J. S. (1930) Permeability of the lens capsule. *Arch. Ophthal.* **3**, 182–193.

Fuchs, R. & Kleifeld, O. (1956) Uber das Verhalten des wasserlöslichen Linseneiweiss junger und alter Tiere bei papierelektrophoretische Untersuchung. *v. Graefes' Arch. Ophthal.* **158**, 29–33.

Fulhorst, H. W. & Young, R. W. (1966) Conversion of soluble lens protein to albuminoid. *Invest. Ophthal.* **5**, 298–303.

Giblin, F. J., Chakrapari, B. & Reddy, V. N. (1976) Glutathione and lens epithelial function. *Invest. Ophthal.* **15**, 381–393.

Goldmann, H. & Leichti, A. (1938) Experimentelle Untersuchungen über die Genese des Röntgenstars. *v. Graefes' Arch. Ophthal.* **138**, 722–736.

Gorthy, W. C. (1968) An electron microscope study of the rat lens epithelium. The central zone. *Exp. Eye Res.* **7**, 394–401.

Gorthy, W. C. & Abdelbaki, Y. Z. (1974) Morphology of a hereditary cataract in the rat. *Exp. Eye Res.* **19**, 147–156.

Gorthy, W. C., Snavely, M. R. & Berrong, N. D. (1971) Some aspects of transport and digestion in the lens of the normal young adult rat. *Exp. Eye Res.* **12**, 112–119.

Halbert, S. P., Locatcher-Khorazo, D., Swick, L., Witmer, R., Seegal, B. & Fitzgerald, P. (1957) Homologous immunological studies of ocular lens. I and II. *J. exp. Med.* **105**, 439–452, 453–462.

Hamai, Y., Fukui, H. N. & Kuwabara, T. (1974) Morphology of hereditary mouse cataract. *Exp. Eye Res.* **18**, 537–546.

Hanson, H. (1962) Proteolytic enzymes. *Exp. Eye Res.* **1**, 468–479.

Harding, J. J. (1969) The nature and origin of the urea-insoluble protein of human lens. *Exp. Eye Res.* **8**, 147–156.

Harding, J. J. (1972) Conformational changes in human lens proteins in cataract. *Biochem. J.* **129**, 97–100.

Harding, J. J. (1973) Disulphide cross-linked protein of high molecular weight in human cataractous lens. *Exp. Eye Res.* **17**, 377–383.

Harding, H. H. & Dilley, K. J. (1976) Structural proteins of the mammalian lens: A review with emphasis on changes in development, aging and cataract. *Exp. Eye Res.* **22**, 1–73.

Harris, J. E. & Becker, B. (1965) Cation transport of the lens. *Invest. Ophthal.* **4**, 709–722.

Harris, J. E. & Gehrsitz, L. B. (1951) Significance of changes in potassium and sodium content of the lens. *Am. J. Ophthal.* **34**, Pt. 2, 131–138.

Harris, J. E., Gehrsitz, L. B. & Nordquist, L. (1953) The *in vitro* reversal of the lenticular cation shift induced by cold or calcium deficiency. *Am. J. Ophthal.* **36**, 39–49.

Herbrink, P. & Bloemendal, H. (1974) Studies on β-crystallin. I. Isolation and partial characterization of the principal polypeptide chain. *Biochim. biophys. Acta.* **336**, 370–382.

Herbrink, P., Van Westreenen, H. & Bloemendal, H. (1975) Further studies on the polypeptide chains of β-crystallin. *Exp. Eye Res.* **20**, 541–548.

Hightower, K. R. & Kinsey, V. E. (1977) Studies on the crystalline lens. XXIII. Electrogenic potential and cation transport. *Exp. Eye Res.* **24**, 587–593.

Hoenders, H. J., Van Kamp, G. J., Liem-The, K. & Van Kleef, F. S. M. (1973) Heterogeneity, aging and polypeptide composition of α-crystallin from calf lens. *Exp. Eye Res.* **15**, 193–200.

Holt, W. S. & Kinoshita, J. H. (1968) Starch-gel electrophoresis of the soluble lens proteins from normal and galactosemic animals. *Invest. Ophthal.* **7**, 169–178.

Horwitz, J. (1976) Some properties of the low molecular weight α-crystallin from normal human lens: comparison with bovine lens. *Exp. Eye Res.* **23**, 471–481.

Iwata, S. & Kinoshita, J. H. (1971) Mechanism of development of hereditary cataract in mice. *Invest. Ophthal.* **10**, 504–512.

Jedziniak, J. A., Kinoshita, J. H., Yates, E. M. & Benedek, G. B. (1975) The concentration and localization of heavy molecular weight aggregates in aging normal and caractous human lenses. *Exp. Eye Res.* **20**, 367–369.

Jedziniak, J. A., Kinoshita, J. H., Yates, E. M., Hocker, L. O. & Benedek, G. B. (1972) Calcium-induced aggregation of bovine lens alpha crystallins. *Invest. Ophthal.* **11**, 905–915.

Jedziniak, J. A., Kinoshita, J., Yates, E. M., Hocker, L. O. & Benedek, G. B. (1973) On the presence and mechanism of formation of heavy molecular weight aggregates in human normal and cataractous lenses. *Exp. Eye Res.* **15**, 185–192.

Jedziniak, J. A., Nicoli, D. F., Yates, E. M. & Benedek, G. B. (1976) On the calcium concentration of cataractous and normal human lenses and protein fractions of cataractous lenses. *Exp. Eye Res.* **23**, 325–332.

Kabasawa, I., Kinoshita, J. H. & Barber, G. W. (1974) Aging effects on the bovine lens γ-crystallins. *Exp. Eye Res.* **18**, 457–466.

Kabasawa, I., Tsunematsu, Y., Barber, G. W. & Kinoshita, J. H. (1977) Low molecular weight proteins of bovine lenses. *Exp. Eye Res.* **24**, 437–448.

Kern, H. L. (1962) Accumulation of amino acids by calf lens. *Invest. Ophthal.* **1**, 368–376.

Kinoshita, J. H. (1965) Pathways of glucose metabolism in the lens. *Invest. Ophthal.* **4**, 619–628.

Kinoshita, J. H. (1974) Mechanisms initiating cataract formation. *Invest. Ophthal.* **13**, 713–724.

Kinoshita, J. H., Barber, G. W., Merola, L. O. & Tung, B. (1969) Changes in the level of free amino acids and myo-inositol in the galactose exposed lens. *Invest. Ophthal.* **8**, 625–632.

Kinoshita, J. H., Dvornik, D., Kraml, M. & Gabbay, K. H. (1968) The effect of an aldose reductase inhibitor on the galactose-exposed rabbit lens. *Biochim. biophys. Acta.* **158**, 472–475.

Kinoshita, J. H., Merola, L. O. & Dikmak, E. (1962) Osmotic changes in experimental galactose cataracts. *Exp. Eye Res.* **1**, 405–410.

Kinoshita, J. H. & Wachtl, C. (1958) A study of the $C^{14}$-glucose metabolism of the rabbit lens. *J. biol. Chem.* **233**, 5–7.

Kinsey, V. E. (1973) Studies on the crystalline lens. XIX. *Exp. Eye Res.* **15**, 699–710.

Kinsey, V. E. & Frohman, C. E. (1951) Distribution of cytochrome, total riboflavin, lactate and pyruvate and its metabolic significance. *Arch. Ophthal., N.Y.* **46**, 536–541.

Kinsey, V. E. & McGrady, A. V. (1971) Studies in the crystalline lens. XVII. *Invest. Ophthal.* **10**, 282–287.

Kinsey, V. E. & McLean, I. W. (1970) Characterization of active transport and diffusion of potassium, rubidium, and cesium. *Invest. Ophthal.* **9**, 769–784.

Kinsey, V. E. & Reddy, D. V. N. (1965) Studies on the crystalline lens. XI. *Invest. Ophthal.* **4**, 104–116.

Kinsey, V. E., Wachtl, C., Constant, M. A. & Camacho, E. (1955) Mitotic activity in the epithelia of lenses cultured in various media. *Am. J. Ophthal.* **40**, Pt. 2, 216–223.

Kuwabara, T. (1975) The maturation of the lens cell: a morphologic study. *Exp. Eye Res.* **20**, 427–443.

Kuwabara, T. & Imaizumi, M. (1974) Denucleation process of the lens. *Invest. Ophthal.* **13**, 973–981.

Kuwabara, T., Kinoshita, J. H. & Cogan, D. G. (1969) Electron microscopic study of galactose-induced cataract. *Invest. Ophthal.* **8**, 133–149.

Langham, M. & Davson, H. (1949) Studies on the lens. *Biochem. J.* **44**, 467–470.

Langley, R. K., Mortimer, C. B. & McCulloch, C. (1960) The experimental production of cataracts by exposure to heat and light. *Arch. Ophthal., N.Y.* **63**, 473–488.

Lasser, A. & Balazs, E. A. (1972) Biochemical and fine structure studies on the water-insoluble components of the calf lens. *Exp. Eye Res.* **13**, 292–308.

Leinfelder, P. J. & Schwartz, B. (1954) Respiration and glycolysis in lens transparency. *Acta XVII Conc. Ophth.* pp. 984–991.

Leon, A. E., de Groot, K. & Bloemendal, H. (1970) The molecular weight of the subunits of α-crystallin. *Exp. Eye Res.* **10**, 75–79.

Léonard, A. & Maisin, J. R. (1965) Hereditary cataract induced by X-irradiation of young rats. *Nature* **205**, 615–616.

Lerman, S. & Zigman, S. (1967) Metabolic studies on the cold precipitable protein of the lens. *Acta Ophthal.* **45**, 193–199.

Li, L.-K. (1974) Physical and chemical changes in alpha-crystallin during maturation of lens fibres. *Exp. Eye Res.* **18**, 383–393.

Li, L.-K. & Spector, A. (1972) Studies on the reaggregation of isolated subunits of calf lens α-crystallin. *Exp. Eye Res.* **13**, 110–119.

Li, L.-K. & Spector, A. (1973) The reaggregation of purified subunits of alpha-crystallins. *Exp. Eye Res.* **15**, 179–183.

Liem-The, K. N. & Hoenders, H. J. (1974a) Characterization of the soluble proteins from rabbit eye lens. *Exp. Eye Res.* **18**, 143–152.

Liem-The, K. N. & Hoenders, H. J. (1974b) HM-crystallin as an intermediate in the conversion of water-soluble into water-insoluble rabbit lens proteins. *Exp. Eye Res.* **19**, 549–557.

Liem-The, K. N., Stols, A. L. H. & Hoenders, H. J. (1975) Further characterization of HM-crystallin in rabbit lens. *Exp. Eye Res.* **20**, 307–316.

Liem-The, K. N., Stols, A. L. H., Jap, P. H. K. & Hoenders, H. J. (1975) X-ray induced cataract in rabbit lens. *Exp. Eye Res.* **20**, 317–328.

Mach, H. (1963) Untersuchungen von Linseneiweiss und Mikroelektrophorese von Wasserlöslichem Eiweiss im Altersstar. *Klin. Mbl. Augenheilk*, **143**, 689–710.

Maisel, H. (1977) The nature of the urea-insoluble material of the human lens. *Exp. Eye Res.* **24**, 417–419.

Maisel, H., Perry, M., Alcalá, J. & Waggoner, P. R. (1976) The structure of chick lens water-insoluble material. *Ophth. Res.* **8**, 55–63.

Mann, I. (1949) *The Development of the Human Eye.* London: B.M.A. (2nd ed.).

Manski, W., Auberach, T. P. & Halbert, S. P. (1960) The evolutionary significance of lens organ specificity. *Am. J. Ophthal.* **50**, Pt. 2, 985–991.

Maraini, G. & Mangili, R. (1973) Differences in proteins and in the water balance of the lens in nuclear and cortical types of senile cataract. In *The Human Lens in Relation to Cataract.* Eds. Elliott, K. M. & Fitzsimmons, D. W. Ciba Symp. No. 19. Elsevier Excerpta Medica: Amsterdam, London, N.Y. pp. 79–97.

Mehta, P. D., Cooper, S. N. & Rao, S. S. (1964) Identification of species-specific and organ-specific antigens in lens proteins. *Exp. Eye Res.* **3**, 192–197.

McEwan, W. K. (1959) The yellow pigment of human lenses. *Am. J. Ophthal.* **47**, Pt. 2, 144–146.

Mok, C.-C. & Waley, S. G. (1967) Structural studies on lens proteins. *Biochem. J.* **104**, 128–134.

Mörner, C. T. (1894) Untersuchung der Proteinsubstanzen in der lichtbrechenden Medien des Auges. III. *Z. physiol. Chem.* **18**, 233–256.

Nelson, K. J. & Rafferty, N. S. (1976) A scanning electron microscopic study of lens fibers in healing mouse lens. *Exp. Eye Res.* **22**, 335–346.

Nordmann, J. (1962) Acquisitions récentes dans le domaine de la biologie du crystallin. *Progr. Ophthal.* **12**, 1–264.

Palva, M. & Palkama, A. (1974) Histochemically demonstrable sodium-potassium activated adenosine triphosphatase (Na-K-ATPase) activity in the rat lens. *Exp. Eye Res.* **19**, 117–123.

Palva, M. & Palkama, A. (1976) Electron microscopical, histochemical and biochemical findings on the Na-K-ATPase activity in the epithelium of the rat lens. *Exp. Eye Res.* **22**, 229–236.

Papaconstantinou, J. (1965) The γ-crystallins of adult bovine, calf and embryonic lenses. *Biochim. biophys. Acta* **107**, 81–90.

Paterson, C. A. (1970a) Extracellular space of the crystalline lens. *Am. J. Physiol.* **218**, 797–802.

Paterson, C. A. (1970b) Sodium exchange in the crystalline lens. I. *Exp. Eye Res.* **10**, 151–155.

Paterson, C. A., Neville, M. C., Jenkins, R. M. & Cullen, J. P. (1975) An electrogenic component of the potential difference in the rabbit lens. *Biochim. biophys. Acta* **375**, 309–316.

Paterson, C. A., Neville, M. C., Jenkins, R. M. & Nordstrom, D. K. (1974) Intracellular potassium activity in frog lens determined using ion specific liquid ion-exchange filled microelectrodes. *Exp. Eye Res.* **19**, 43–48.

Patterson, J. W. & Bunting, K. W. (1966) Sugar cataracts, polyol levels and lens swelling. *Doc. Ophthal.* **20**, 64–72.

Philipson, B. (1969) Distribution of protein within lenses with X-ray cataract. *Invest. Ophthal.* **8**, 271–280.

Philipson, B. (1969) Galactose cataract: changes in protein distribution during development. *Invest. Ophthal.* **8**, 281–289.

Philipson, B. (1973) Changes in the lens related to the reduction in transparency. *Exp. Eye Res.* **16**, 29–39.

Philipson, B. T., Hanninen, L. & Balazs, E. A. (1975) Cell contacts in human and bovine lenses. *Exp. Eye Res.* **21**, 205–219.

Piatigorsky, J., Webster, H. De F. & Craig, S. P. (1972) Protein synthesis and ultrastructure during the formation of embryonic chick lens fibres *in vivo* and *in vitro.* *Dev. Biol.* **27**, 176–189.

Piatigorsky, J., Zelenka, P. & Simpson, R. T. (1974) Molecular weight and subunit structure of delta-crystallin from embryonic chick lens fibers. *Exp. Eye Res.* **18**, 435–446.

Pirie, A. (1951) Composition of the lens capsule. *Biochem. J.* **48**, 368–371.

Pirie, A. (1968) Color and solubility of the proteins of human cataracts. *Invest. Ophthal.* **7**, 634–650.

Pirie, A. (1968) Pathology of the eye of the naphthalene-fed rabbit. *Exp. Eye Res.* **7**, 354–357.

Pirie, A. & Flanders, H. P. (1957) Effect of X-rays on partially shielded lens of the rabbit. *Arch. Ophthal., N.Y.* **57**, 849–854.

Pirie, A., van Heyningen, R. & Boag, J. W. (1953) Changes in lens during the formation of X-ray cataract in rabbits. *Biochem. J.* **54**, 682–688.

Rabaey, M. (1962) Electrophoretic and immuno-electrophoretic studies on the soluble proteins in the developing lens of birds. *Exp. Eye Res.* **1**, 310–316.

Rabaey, M., Rikkers, I. & de Mets, M. (1972) Low molecular weight protein (γ-crystallin) in the lens of birds. *Exp. Eye Res.* **14**, 208–213.

Rae, J. D. (1974a) Potential profiles in the crystalline lens of the frog. *Exp. Eye Res.* **19**, 227–234.

Rae, J. L. (1974b) Voltage compartments in the lens. *Exp. Eye Res.* **19**, 235–242.

Rae, J. L. (1974c) The movement of procion dye in the crystalline lens. *Invest. Ophthal.* **13**, 147–150.

Rae, J. L., Hoffert, J. R. & Fromm, P. O. (1970) Studies on the normal lens potential of the rainbow trout (*Salmo gairdneri*). *Exp. Eye Res.* **10**, 93–101.

Rafferty, N. S. (1973) Experimental cataract and wound healing in mouse lens. *Invest. Ophthal.* **12**, 156–159.

Rafferty, N. S. & Esson, E. A. (1974) An electron-microscope study of adult mouse lens: some ultrastructural specializations. *J. Ultrastr. Res.* **46**, 239–253.

Rao, S. S., Mehta, P. D. & Cooper, S. N. (1965) Antigenic relationship between insoluble and soluble lens proteins. *Exp. Eye Res.* **4**, 36–41.

Reddy, D. V. N. & Kinsey, V. E. (1962) Studies on the crystalline lens. Quantitative analysis of free amino acids and related compounds. *Invest. Ophthal.* **1**, 635–641.

Rees, J. R. & Pirie, A. (1967) Possible reactions of 1,2-naphthaquinone in the eye. *Biochem. J.* **102**, 853–863.

Riley, M. V. (1970) Ion transport in damaged lenses and by isolated lens epithelium. *Exp. Eye Res.* **9**, 28–37.

Roberts, D. St. C. (1957) Studies on the antigenic structure of the eye using the fluorescent antibody technique. *Br. J. Ophthal.* **41**, 338–347.

Roelfzema, H., Broekhuyse, R. M. & Veerkamp, J. H. (1974) Subcellular distribution of sphingomyelinase and other acid hydrolases in different parts of the calf lens. *Exp. Eye Res.* **18**, 579–594.

Roy, D. & Spector, A. (1976) High molecular weight protein from human lenses. *Exp. Eye Res.* **22**, 273–279.

Roy, D. & Spector, A. (1976) Human α-crystallin: characterization of the protein isolated from the periphery of cataractous lenses. *Biochemistry*, **15**, 1180–1188.

Roy, D. & Spector, A. (1976) Human alpha-crystallin. III. Isolation and characterization of protein from normal infant lenses and old lens peripheries. *Invest. Ophthal.* **15**, 394–399.

Rubin, L. F. & Barnett, K. C. (1967) Ocular effects of oral and dermal application of dimethyl sulfoxide in animals. *Ann. N.Y. Acad. Sci.* **141**, 333–345.

Said, F. S. & Weale, R. A. (1959) The variation with age of the spectral transmissivity of the living crystalline lens. *Gerontologia*, **3**, 213–231.

Sakuragawa, M., Kuwabara, T., Kinoshita, J. H. & Fukui, H. N. (1975) Swelling of the lens fibers. *Exp. Eye Res.* **21**, 381–394.

Sandberg, H. O. (1976) The alpha-crystallin content of aqueous humour in cortical, nuclear and complicated cataracts. *Exp. Eye Res.* **22**, 75–84.

Schoenmakers, J. G. G. & Bloemendal, H. (1968) Subunits of alpha-crystallin from adult and embryonic cattle lens. *Nature* **220**, 790–791.

Schoenmakers, J. G. G., Gerding, J. J. T. & Bloemendal, H. (1969) The subunit structure of α-crystallin. *Eur. J. Biochem.* **11**, 472–481.

Schwartz, B. (1960) Initial studies of the use of an open system for the culture of the rabbit lens. *Arch. Ophthal., N.Y.* **63**, 643–659.

Simpson, G. C. (1953) Ocular haloes and coronas. *Brit. J. Ophthal.* **37**, 450–486.

Spector, A. (1969) Physiological chemistry of the eye. *Arch. Ophthal., N.Y.* **81**, 127–143.

Spector, A., Adams, D. & Krul, K. (1974) Calcium and high molecular weight protein aggregates in bovine and human lens. *Invest. Ophthal.* **13**, 982–990.

Spector, A., Freund, T., Li, L.-K. & Augusteyn, R. C. (1971) Age-dependent changes in the structure of alpha crystallin. *Invest. Ophthal.* **10**, 677–686.

Spector, A., Li, L.-K., Augusteyn, R. C., Schneider, A. & Freund, T. (1971) α-crystallin. The isolation and characterization of distinct macromolecular fractions. *Biochem. J.* **124**, 337–343.

Spector, A., Li, S. & Sigelman, J. (1974) Age-dependent changes in the molecular size of human lens proteins and their relationship to light scatter. *Invest. Ophthal.* **13**, 795–798.

Spector, A., Roy, D. & Stauffer, J. (1975) Isolation and characterization of an age-dependent polypeptide from human lens with non-tryptoplan fluorescence. *Exp. Eye Res.* **21**, 9–24.

Spector, A., Stauffer, J., Roy, D., Li, L.-K. & Adams, D. (1976) Human alpha-crystallin. I. The isolation and characterization of newly synthesized alpha-crystallin. *Invest. Ophthal.* **15**, 288–296.

Spector, A., Wandel, T. & Li, L.-K. (1968) The purification and characterization of the highly labeled protein fraction from calf lens. *Invest. Ophthal.* **7**, 179–190.

Sperelakis, N. & Potts, A. M. (1959) Additional observations on the bioelectric potentials of the lens. *Am. J. Ophthal.* **47**, 395–409.

Stauffer, J., Rothschild, C., Wandel, T. & Spector, A. (1974) Transformation of alpha-crystallin polypeptide chains with aging. *Invest. Ophthal.* **13**, 135–146.

Swanson, A. A. (1966) The identification of lysosomal enzymes in bovine lens epithelium. *Exp. Eye Res.* **5**, 145–149.

Swanson, A. A. (1967) Histochemical identification of lysosomal enzymes in lens epithelial cells. *Exp. Eye Res.* **6**, 351–355.

Swanson, A. A. & Beaty, H. B. (1977) Lens proteases. *Exp. Eye Res.* **24**, 89–91.

Swanson, A. A. & Nichols, J. T. (1971) Human senile cataractous lens protease. *Biochem. J.* **125**, 575–584.

Szent-Györgyi, A. (1955) Fluorescent globulin of the lens. *Biochim. biophys. Acta.* **16**, 167.

Tanaka, K. & Iino, A. (1967) Zur Frage der Verbindung der Linsenfasern im Rinderauge. *Z. Zellforsch.* **82**, 604–612.

Takemoto, L. J. & Azari, P. (1976) Amino acid composition of normal and cataractous human lens proteins. *Exp. Eye Res.* **23**, 1–7.

Takemoto, L. J., Azari, P. & Gorthy, W. C. (1975) Role of sulfhydryl groups in the formation of a hereditary cataract. *Exp. Eye Res.* **20**, 1–13.

Thoft, R. A. & Kinoshita, J. H. (1965) The effect of calcium on rat lens permeability. *Invest. Ophthal.* **4**, 122–128.

Trayhurn, P. & van Heyningen, R. (1976) Neutral proteinase activity in the human lens. *Exp. Eye Res.* **22**, 251–257.

Truman, D. E. S., Brown, A. G. & Rao, K. V. (1971) Estimates of the molecular weights of chick β- and δ-crystallins and their subunits by gel filtration. *Exp. Eye Res.* **12**, 304–310.

Truman, D. E. S. & Clayton, R. M. (1974) The subunit structure of chick β-crystallins. *Exp. Eye Res.* **18**, 485–494.

Truscott, R. J. W. & Augusteyn, R. C. (1977) Changes in human lens proteins during nuclear cataract formation. *Exp. Eye Res.* **24**, 159–170.

Uhlenhuth, P. T. (1903) Zur Lehre von der Unterscheidung verschiedener Eiweissarten mit Hilfe spezifischer Sera. In *Koch Festschrift*, p. 49. Jena: G. Fischer.

Van Dam, A. F. (1966) Purification and composition studies of $\beta_s$-crystallin. *Exp. Eye Res.* **5**, 255–266.

Van den Broek, W. G. M., Leget, J. N. & Bloemendal, H. (1973) FM-crystallin: a neglected component among lens proteins. *Biochim. biophys. Acta* **310**, 278–282.

Van der Ouderaa, F. J., De Jong, W. W. & Bloemendal, H. (1973) The amino-acid sequence of the αA₂ chain of bovine α-crystallin. *Eur. J. Biochem.* **39**, 207–222.

Van der Ouderaa, J., De Jong, W. W., Hilderink, A. & Bloemendal, H. (1974) The amino-acid sequence of the αB₂ chain of bovine α-crystallin. *Eur. J. Biochem.* **49**, 157–168.

Van Heyningen, R. (1959) Formation of polyols by the lens of the rat with 'sugar' cataract. *Nature, Lond.* **184**, 194–195.

Van Heyningen, R. (1965) The metabolism of glucose by the rabbit lens in the presence and absence of oxygen. *Biochem. J.* **96**, 419–431.

Van Heyningen, R. (1969) The Lens Metabolism and Cataract. In *The Eye*, Ed. Davson, H. Vol. I, pp. 381–488. London: Academic Press.

Van Heyningen, R. (1970) Ascorbic acid in the lens of the naphthalene-fed rabbit. *Exp. Eye Res.* **9**, 38–48.

Van Heyningen, R. (1972) The human lens. III. Some observations on the post-mortem lens. *Exp. Eye Res.* **13**, 155–160.

Van Heyningen, R. (1977) Lens proteases. *Exp. Eye Res.* **24**, 91.

Van Heyningen, R. & Pirie, A. (1967) The metabolism of naphthalene and its toxic effect on the eye. *Biochem. J.* **102**, 842–852.

Van Heyningen, R. & Trayhurn, P. (1976) Proteolysis of lens proteins (autolysis). *Exp. Eye Res.* **22**, 625–637.

Van Heyningen, R. & Waley, S. G. (1962) Search for a neutral proteinase in bovine lens. *Exp. Eye Res.* **1**, 336–342.

Van Kamp, G. J., Schats, L. M. M. & Hoenders, H. J. (1973) Characteristics of α-crystallin related to fiber cell development in calf eye lenses. *Biochim. biophys. Acta* **295**, 166–173.

Van Kamp, G. J., Struyker Boudier, H. A. J. & Hoenders, H. J. (1974) The soluble proteins of the prenatal bovine eye lens. *Comp. Biochem. Physiol.* **49B**, 445–456.

Van Kamp, G. J., Van Kleef, F. S. M. & Hoenders, H. J. (1974) Reaggregation studies on the polypeptide chains of calf lens α-crystallin. *Biochim. biophys. Acta* **342**, 89–96.

Van Kleef, F. S. M., Nijzink-Maas, M. J. C. M. & Hoenders, H. J. (1974a) Intracellular degradation of α-crystallin. *Eur. J. Biochem.* **48**, 563–570.

Van Kleef, F. S. M., Nijzink, M. J. C. M. & Hoenders, H. J. (1974b) Nine polypeptide chains in α-crystallin. *Exp. Eye Res.* **18**, 201–204.

Varma, S. D. & Kinoshita, J. H. (1974) The absence of cataracts in mice with congenital hyperglycaemia. *Exp. Eye Res.* **19**, 577–582.

v. Sallmann, L. (1951) Experimental studies on early lens changes after roentgen irradiation. *Arch. Ophthal., N.Y.* **45**, 149–164.

v. Sallmann, L. (1957) The lens epithelium in the genesis of cataract. *Am. J. Ophthal.* **44**, 159–170.

Voaden, M. J. (1968) A chalone in the rabbit lens? *Exp. Eye Res.* **7**, 326–331.

Voaden, M. J. & Froomberg, D. (1967) Topography of a mitotic inhibitor in the rabbit lens. *Exp. Eye Res.* **6**, 213–218.

Waley, S. G. (1964) Metabolism of amino acids in the lens. *Biochem. J.* **91**, 576–583.

Waley, S. G. (1965) The problem of albuminoid. *Exp. Eye Res.* **4**, 293–297.

Waley, S. G. (1969) Nomenclature of the polypeptide chains of α-crystallin. *Exp. Eye Res.* **8**, 477–478.

Waley, S. G. (1969) The lens: function and macromolecular composition. In *The Eye*. Ed. Davson, H. pp. 299–379. New York: Academic Press.

Wanko, T. & Gavin, M. A. (1961) Cell surfaces in the crystalline lens. In *The Structure of the Eye*, Ed. Smelser, pp. 221–233. New York: Academic Press.

Weinstock, M. & Stewart, H. C. (1961) Occurrence in rodents of reversible drug-induced opacities of the lens. *Br. J. Ophthal.* **45**, 408–414.

Yorio, T. & Bentley, P. J. (1976) Distribution of the extracellular space of the amphibian lens. *Exp. Eye Res.* **23**, 601–608.

Yorio, T. & Bentley, P. J. (1976) The effects of hyperosmotic agents on the electrical properties of the amphibian lens *in vitro. Exp. Eye Res.* **22**, 195–208.

Young, R. W. & Ocumpaugh, D. E. (1966) Autoradiographic studies on the growth and development of the lens capsule in the rat. *Invest. Ophthal.* **5**, 583–593.

Zelenka, P. & Piatigorsky, J. (1976) Molecular weight and sequence complexity of δ-crystallin mRNA. *Exp. Eye Res.* **22**, 115–124.

Zigler, J. S. & Sidbury, J. B. (1973) Structure of calf lens β-crystallins. *Exp. Eye Res.* **16**, 207–214.

Zigman, S. & Lerman, S. (1965) Properties of a cold-precipitable protein fraction in the lens. *Exp. Eye Res.* **4**, 24–30.

Zigman, S. & Lerman, S. (1968) Effect of actinomycin D on rat lens protein synthesis. *Exp. Eye Res.* **7**, 556–560.

Zigman, S., Schultz, J. & Yulo, T. (1969) Chemistry of lens nuclear sclerosis. *Biochem. Biophys. Res. Comm.* **35**, 931–938.

Zwaan, J. (1966) Sulfhydryl groups of the lens proteins of the chicken in embryonic and adult stages. *Exp. Eye Res.* **5**, 267–275.

# The mechanism of vision

# 5. Retinal structure and organization

In this section we shall be concerned with the peripheral mechanisms in the visual process. When light strikes the retina, physical and chemical changes are induced that lead eventually to a discharge of electrical impulses in the optic nerve fibres. The 'messages' in these nerve fibres are a record of the events taking place in the retina; because of the variety of sensations evoked by visual stimuli, these records must be of almost fantastic complexity when even a simple image is formed on the retina. The aim of investigators in this field of eye physiology is to disentangle the skein; their approaches are various but may be classed generally as psychophysical, chemical, and electrophysiological. Remarkable progress has been made, especially in recent years, but there is still a long way to go before anything approaching an exact picture of the retinal processes can be elaborated. The following description is designed more to show the modes of approach to the various problems rather than to present a theoretical picture of the 'mechanism of vision'.

## THE STRUCTURE OF THE RETINA

SENSORY PATHWAYS

The retina is a complex nervous structure made up of a number of layers (Fig. 5.1) whose significance is best understood by recalling that in other parts of the body, for example the skin, the nervous elements involved in the transmission of sensory impulses to the cerebral cortex consist of:

1. a specialized arrangement of cells called a receptor-organ, e.g. a Pacinian corpuscle sensitive to pressure
2. a neurone connecting the receptor to the spinal cord and medulla (neurone I) with its cell body in the posterior root ganglion
3. a connector neurone to the thalamus (neurone II) with its cell body in the nucleus cuneatus or gracilis
4. a neurone to transmit the impulses to the cortex (neurone III) with its cell body in the thalamus.

With the vertebrate eye, not only are the specialized receptor cells, the *rods* and *cones*, in the retina, but also the cell bodies and connecting fibres of neurones I and II so that the nerve fibres leading out of the eye, which

constitute the optic nerve, are the axons of neurone II; the majority of the fibres run to the lateral geniculate body where the impulses, initiated in the rods and cones, are relayed to the occipital lobe of the cortex. The analogy between the two systems is demonstrated in Figure 5.2.

THE RETINAL LAYERS

We may thus expect to differentiate in the retina a layer of receptors (*layer of rods and cones*), a layer of cell bodies of neurone I, the *bipolar cells* (*inner nuclear layer*) and a layer of cell bodies of neurone II, the *ganglion cells* (*ganglion layer*) (Fig. 5.1). The rods and cones are highly differentiated cells, and their orderly arrangement gives rise to a *bacillary layer*, consisting of their outer segments; an *outer nuclear layer* containing their cell bodies and nuclei; and an *outer plexiform layer* made up of their fibres and synapses with the bipolar cells. The region of synapse between bipolar and ganglion cells is the *inner plexiform layer*. Two other types of nerve cell are present in the retina, namely the *horizontal* and *amacrine* cells, with their bodies in the inner nuclear layer; the ramifications of their dendritic and axonal processes contribute to the outer and inner plexiform layers respectively; their largely horizontal organizations permits them to mediate connections between receptors, bipolars and ganglion cells. Besides the nerve cells there are numerous *neuroglial cells*, e.g. those giving rise to the radial fibres of Müller, which act as supporting and insulating structures.

THE RODS AND CONES

The photosensitive cells are, in the primate and most vertebrate retinae, of two kinds, called *rods* and *cones*, the rods being usually much thinner than the cones but both being built on the same general plan as illustrated in Figure 5.3. The light-sensitive pigment is contained in the *outer segment*, o, which rests on the pigment epithelium; the other end is called the *synaptic body*, s, and it is through this that the effects of light on the receptor are transmitted to the bipolar or horizontal cell. In the rods this is called the *rod spherule*, and in the cone the *cone pedicle*. The human rod is some 2 $\mu$ thick and 60 $\mu$ long; the cones vary greatly in size and

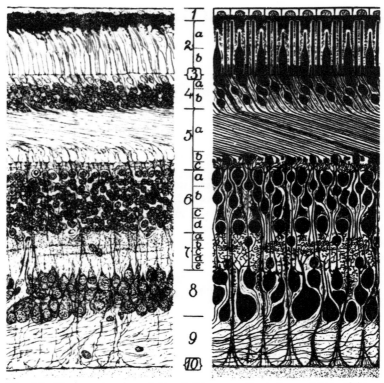

**Fig. 5.1**    Cross-section through the adult human retina in the periphery of the central area, showing details of stratification at a moderate magnification. The left-hand figure shows the structures as they appear when stained with a non-selective method (e.g. haematoxylin and eosin). The right-hand figure reproduces schematically the same view as it would appear from a study with an analytical method such as Golgi's or Ehrlich's, using many sections to get all details. (1) Pigment epithelium; (2) bacillary layer; (2-*a*) outer segments and (2-*b*) inner segments of the thinner rods and the thicker cones; (3) outer limiting membrane; (4) outer nuclear layer; (4-*a*) outer zone of the outer nuclear layer chiefly composed of the cone nuclei; (4-*b*) inner zone of the outer nuclear layer composed exclusively of the rod nuclei; (5) outer plexiform layer; (5-*a*) outer zone of the outer plexiform layer composed of the inner rod and cone fibres and of the corresponding portions of Müller's 'radial fibres' enveloping the first (here only one shown, beginning with the outer limiting membrane encasing the body and the fibre of a cone and continuing through ventral layers down to the inner limiting membrane); (5-*b*) middle zone of the outer plexiform layer composed of the rod spherules and cone pedicles; (5-*c*) inner zone of the outer plexiform layer made up of the expansions of the nerve cells and of Müller's supporting 'radial fibres' whose bodies and nuclei reside in the sixth layer; (6) inner nuclear layer with its four zones: 6-*a*, 6-*b*, 6-*c* and 6-*d*; (7) inner plexiform layer with its five zones: 7-*a*, 7-*b*, 7-*c*, 7-*d*, and 7-*e*; (8) layer of the ganglion cells; (9) layer of the optic nerve fibres; (10) inner limiting membrane, next to the vitreous. (Polyak, *The Retina.*)

shape with their position in the retina, becoming long and thin and not easily distinguished from rods in the most central portion (Fig. 5.4).

## SYNAPTIC RELATIONSHIPS

If it is the rod or cone that absorbs the light-energy of the stimulus, it is the bipolar cell that must transmit any effect of this absorption to higher regions of the central nervous system; consequently the relationship

between the two types of cell is of great interest. They have been examined in the electron microscope in some detail.* In general, the relation between receptor and

---

* Since Golgi's and Polyak's classical studies, developments on the primate retina have been due largely to Missotten (1965), Sjöstrand (1969), Dowling and Boycott (1966, 1969), Kolb (1970); of special interest is the study of Dowling and Werblin (1969) on the retina of the mudpuppy *Necturus*, since this has been accompanied by a parallel study of the electro-physiology of its connections (p. 251).

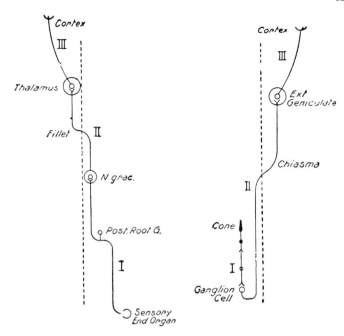

**Fig. 5.2** Illustrating the analogy between a typical cutaneous afferent pathway and the visual pathway. Note the cone has not been treated as a neurone.

neurone is of the synaptic type, in the sense that the dendrite of the bipolar cell, for example, comes into very close contact with the rod spherule or cone pedicle, and fusion of membranes does not occur, a definite synaptic cleft being maintained. The contacts are of two main types, namely when the dendritic process penetrates deep into an invagination of the pedicle or spherule, and when the process makes a simple contact with the surface.

THE TRIAD

Figure 5.5 illustrates a characteristic form of the invaginating synapse called by Missotten (1962) the *triad*; the figure shows a portion of the cone pedicle, CP, containing many vesicles reminiscent of synaptic vesicles as seen elsewhere in the central nervous system. The triad is formed by the penetration into the invagination of a central dendritic process from a bipolar cell, flanked by two processes from horizontal cells. In the same figure, a simpler *contact synapse* can be seen formed by a dendritic process from another bipolar cell on the cone pedicle. A *dyad* form of synapse is found in the inner plexiform layer (Fig. 5.6); here a bipolar axon and an amacrine process make adjacent synaptic contacts with a ganglion cell dendrite, and it seems that the amacrine cell is being both postsynaptic and presynaptic in its relations with the bipolar cell. Thus the bipolar cell involved in the dyad, like the cone pedicle involved in the triad, contains a linear portion of heavily staining material surrounded by vesicles, the so-called *ribbon*

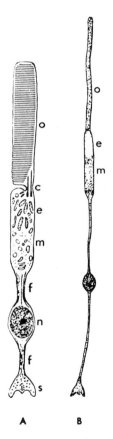

**A      B**

**Fig. 5.3** Diagrammatic representation of retinal photo-receptor cells. A is a schematic drawing of a vertebrate photoreceptor, showing the compartmentation of organelles as revealed by the electron microscope. The following subdivisions of the cell may be distinguished: outer segment (*o*); connecting structure (*c*); ellipsoid (*e*) and myoid (*m*), which comprise the inner segment; fibre (*f*); nucleus (*n*); and synaptic body (*s*). The scleral (apical) end of the cell is at the top in this diagram. B depicts a photoreceptor cell (rod) from the rat. (Young, *J. Cell Biol.*)

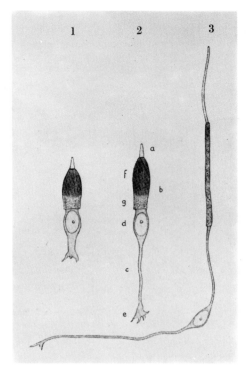

**Fig. 5.4**  Human cones. 1, Near ora serrata; 2, at equator; 3, at macula. *a*, outer segment; *b*, inner segment; *c*, cone fibre; *d*, cell body and nucleus; *e*, cone foot; *f*, ellipsoid; *g*, myoid. (Duke-Elder, *Textbook of Ophthalmology*.)

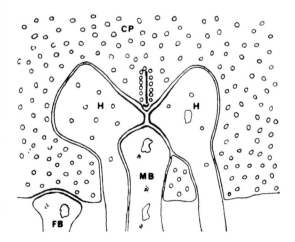

**Fig. 5.5**  Synaptic junction of the cone pedicle, CP, called a triad. Two processes of horizontal cells, H, and a midget bipolar dendrite invaginate into the cytoplasm of the cone pedicle. (Dowling and Boycott, *Proc. Roy. Soc.*)

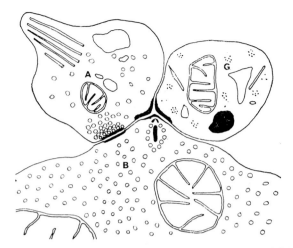

**Fig. 5.6**  Illustrating the dyad type of synaptic contact of the bipolar terminal, *B*, with ganglion cell dendrite, *G*, and amacrine process, *A*. Frequently the amacrine cell process makes a synaptic contact backwards on to the bipolar terminal, forming a reciprocal contact between bipolar terminal and amacrine process. (Dowling and Boycott, *Proc. Roy. Soc.*)

## CHARACTERIZATION OF THE SYNAPSE

One of the most important and difficult points to establish is the synaptic relations between the processes of any two cells of the retina, i.e. to determine whether, at a given synapse, a horizontal cell, for example, is transmitting information to a receptor or bipolar cell (whether it is acting *pre*synaptically) or whether it is receiving information (or acting *post*synaptically).

### SYNAPTIC VESICLES

As Sjöstrand (1976) has emphasized, the main criterion is still the accumulation of vesicles in the cytoplasm of a process at the region of junction, the process with the accumulation presumably liberating transmitter at the junction and behaving presynaptically. Sjöstrand has made the interesting observation, in the rabbit's outer plexiform layer, of processes of horizontal cells within a synaptic complex, which, as illustrated by Figure 5.7, show accumulations of vesicles within their cytoplasm in one region whilst the cytoplasm of the rod spherule in which they are embedded contains many vesicles that would permit it to act presynaptically on the horizontal cell processes, except in the region of the synaptic ribbon complex. Thus the rod could transmit information to the horizontal process at one point and receive information at another. Such an arrangement would provide a mechanism for lateral inhibition, light falling on the rod would excite it and the effect would be carried vertically along the bipolar-ganglion cell pathway; excitation of the horizontal cell could cause a lateral action on neighbouring receptors, inhibiting

*synapse*; the conventional type of synapse is illustrated by the relation of the amacrine cell to the bipolar cell in Figure 5.6; here there are dense lines on the apposing membranes, and opposite them in the presynaptic cell—amacrine—is a collection of synaptic vesicles.

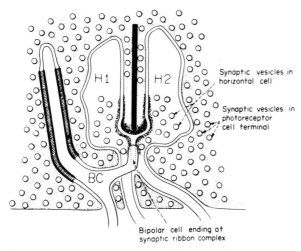

**Fig. 5.7** Illustrating the synaptic ribbon complex. (Sjostrand, *Vision Res.*)

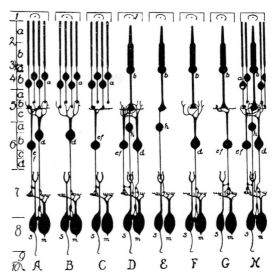

**Fig. 5.8** Some of the types of receptor-bipolar synapses in the retina. *a*, rod; *b*, cone; *d*, *e*, *f*, bipolars that make several synaptic relationships with receptors; *h*, midget bipolar; *s*, midget ganglion cell; *m*, parasol ganglion cell (used here to typify the type that makes several synaptic relationships with bipolars). (Polyak, *The Retina*.)

these. It will be seen that the bipolar cell also sends a process to the complex, and if this is, too, postsynaptic, the horizontal cell could inhibit this, exerting an effective *negative feedback* on the rod that is receiving the light.

## ORGANIZATION OF THE RETINA

### LATERAL SPREAD

The main organization of the retina is on a vertical basis, in the sense that activity in the rods and cones is transmitted to the bipolar cells and thence to the ganglion cells and out of the eye into higher neural centres—the lateral geniculate body, and so on. However, a given receptor may well activate many bipolar cells, and a single bipolar cell may activate many ganglion cells, so that the effect of a light-stimulus may spread horizontally as it moves vertically. The connections brought about by the horizontal and amacrine cells further increase the possibilities of horizontal spread and interaction, so that the problem of the analysis of the pathways of activity in the retina is highly complex, and only a beginning has so far been made in its solution.

### VERTICAL ORGANIZATION

The basic picture of a vertical retinal organization, as conceived by Polyak, is illustrated by Figure 5.8; according to this, the bipolar cells are of two main classes, namely *midget bipolars*, *h*, which connect only to a cone, and the diffuse type, which makes synaptic relations with several receptors, *d*, *e* and *f*. Similarly, the ganglion cells fall into two classes, the *midget ganglion cell*, connecting only with a single midget bipolar cell, and the diffuse type connecting with groups

of bipolar cells. On this basis, then, convergence of receptors on ganglion cells would be largely achieved by the diffuse types of bipolar and ganglion cells, whilst a virtual one-to-one relationship would exist between those cones that made a connection with a midget bipolar cell. As the figure indicates, however, this relationship is not strictly one-for-one, since a cone, connected to a midget bipolar, also connects with a diffuse type of bipolar, and so its effects may spread laterally. The more recent studies of Boycott and Dowling (1969) on the primate retina have largely confirmed Polyak's view but have corrected some errors; their general picture is indicated in Figure 5.9.

### BIPOLAR CELLS

According to this study, the bipolar cells related to cones are of two types; the *midget*, making a unique type of synapse with the cone, and the *flat* cone bipolar, making connections with some seven cones; the synapses are characteristically different, the dendritic process from the midget bipolar cell penetrating into an invagination of the cone pedicle, whereas the process from the flat cone bipolar makes a more simple type of contact (Fig. 5.9, *top left inset*). The third type of bipolar is also diffuse, and serves to collect messages from rods, being connected to as many as fifty of these; the synaptic arrangement is one of invagination, as seen in Figure 5.9, *top right inset*. Bipolars that synapse with both rods and cones, as indicated in Polyak's figure, almost certainly do not occur.

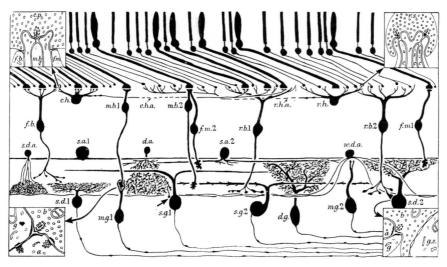

**Fig. 5.9**  The main kinds of nerve cell identified in the primate retina. *Inset, top left* shows how the cones (c.p.) contact their nerve cells. The *inset top right* shows how nerve cell processes invaginate into a rod spherule (r.s.) The *inset, lower left* represents a dyad in which a bipolar cell *b* is presynaptic to a ganglion cell dendrite *g* and an amacrine cell process *a*. The amacrine cell process has a reciprocal contact back on to the bipolar terminal and an amacrine to ganglion cell dendrite synapse. The inset *lower right* shows that the rod bipolar cell axosomatic contacts may consist of 'close contacts' on to the ganglion cell perikaryon (g.s.) (though which types of ganglion cell is unknown). This inset also shows that these same bipolar terminals have dyad contacts. *c.h.* and *r.g.* are horizontal cells considered to connect cones and rods to each other respectively; more recent work shows that horizontal cells only connect cones to rods, however; *f.b.*, flat bipolar making contact with several cones; *m.b.*1 and *m.b.*2 are midget bipolars; *f.m.*1 and *f.m.*2 are flat midget bipolars, making contact with only the base of the cone pedicle by contrast with the *m.b.*'s which invaginate into the triad. *r.b.*1 and *r.b.*2 are rod bipolars whose dendrites make the central elements of the rod spherule. (*Upper right inset*). *s.a.*1 and *s.a.*2 are unistratified amacrine cells whose processes ramify immediately under the inner nuclear layer, *d.a.* and *s.d.a.* are amacrine cells with branches extending much deeper; *d.a.* may be seen making an axo-somatic contact with the ganglion cell *s.g.*1 *w.d.a.* indicates a wide-field diffuse amacrine cell, with little or no ramification of the main branches as they run to the ganglion cell perikarya. Midget ganglion cells, *m.g.*1 and *m.g.*2 have a single dendrite branching at its tip to embrace the terminal of a midget bipolar cell. *s.g.*1 and *s.g.*2 are unistratified ganglion cells, the dendrites branching in a single plane, while the diffuse ganglion cell, *d.g.* has dendrites branching in planes throughout the inner plexiform layer. *s.d.*1 and *s.d.*2 are stratified diffuse ganglion cells corresponding to Polyak's 'parasol ganglion cell'. (Boycott and Dowling, *Phil. Trans.*)

## Midget bipolar

The midget bipolar has been examined in serial sections by Kolb (1970), and she has shown that a single bipolar may make some twenty-five connections with a single cone pedicle, providing single central dendritic processes for the twenty-five invaginations. Thus, since these twenty-five invaginations constituted the total for the cone pedicle, this proved that the bipolar cell made synapses with only a single cone. In the peripheral retina there were some branched midgets, probably making contact with two cones; whilst in some cases it seemed likely that the pedicle accommodated processes from two midget bipolars. In addition to this connection with a midget, each cone also connected with one or more flat bipolars (Boycott and Dowling, 1969). Kolb

also showed that the midget bipolar occurred in two types, the *invaginating midget* as described above, and also a *flat midget*, which at first was mistaken for the invaginating type; in fact, however, its dendritic 'tree-top' consists in a large number of very fine processes that do not enter the invaginations of the pedicle but only make a superficial contact with the base. Thus a cone pedicle contained twenty-five invaginating contacts and some forty-eight superficial contacts. Like the invaginating midget, the flat midget makes contact with only a single cone pedicle.

### HORIZONTAL CELLS

The dendritic processes of these cells make synaptic contacts with the cone pedicles as lateral processes of the triads (Fig. 5.5); they were considered by Boycott and Dowling (1969) to be of two types, Type A making dendritic connection with cones and Type B with rods (Fig. 5.9); however, Kolb (1970) has shown that both types actually make dendritic connection only with cones; and the only differentiation consists in the numbers of cones with which a single horizontal cell makes contact, the 'small-field' contacting some seven cones and the 'large-field' some twelve.

## Cone pedicles

A more systematic study (Boycott and Kolb, 1973), in which the synaptic terminals of cone pedicles on the dendrites of horizontal cells were examined, confirmed the unitary character of the primate horizontal cell, but there was a systematic increase in the number of terminal groups on an individual horizontal cell from about 10 near the foveal pit to as many as 40 in the periphery. Figure 5.10 illustrates a reconstruction of a horizontal cell with its associated cone pedicles, the black spots on the outlines of the pedicles indicating the number of stained lateral elements of the triads actually observed. In the region of retina from which this cell was taken, each pedicle had some 25 triads or 50 lateral elements

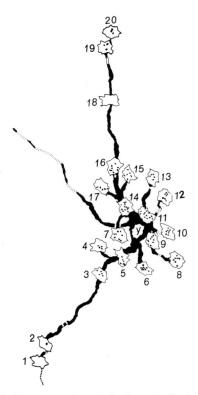

**Fig. 5.10**  Reconstruction of a horizontal cell with its associated cone pedicles. The cell is reconstructed from serial sections cut horizontally. The view is of sectioned cone pedicles with the perikaryon and the dendrites of the horizontal cell below. (Boycott and Kolb, *J. comp. Neurol.*)

from horizontal cells; thus only a portion of the 50 lateral elements in any cone triad are from this particular horizontal cell.

## Rod and cone mixing

Of greatest interest, of course, is the manner in which the axons of the horizontal cells terminate; thus their arrangement with the cones is post-synaptic, i.e. they receive messages from these; according to Kolb the axon processes terminate exclusively in rod spherules, constituting lateral elements in triads; each rod spherule thereby receives input from two horizontal cells. Thus the horizontal cells in the primate retina apparently mix the rod and cone messages.

### OUTER PLEXIFORM LAYER

In the light of this work, the outer plexiform layer may be illustrated by Figure 5.11, whilst the triad structures of rod and cone are shown in Figure 5.12; it will be noted that the structure of the cone triad is much more complex than at first thought, containing, in addition to the invaginating midget bipolar process (IMB) and the lateral horizontal processes (HC 1 and HC 2), more superficially placed processes from flat midget bipolars (FMB) and from flat cone bipolars FB. The possibilities for spread of information are thus quite large.

### INNER PLEXIFORM LAYER

Essentially this process of vertical and horizontal connections is repeated in the inner plexiform layer, which consists of axons and terminals of bipolar cells, processes of amacrine cells, and dendrites of ganglion cells. The ganglion cells fall into two main types, namely the midget ganglion cell which connects with a midget bipolar cell and thus has a very close relationship with a single cone, providing the latter with an exclusive pathway as far as the inner plexiform layer; the other type is diffuse, being connected with many bipolar cells; this type falls into several categories according to the manner in which their dendrites branch. The extent to which a ganglion cell may relay only from rod bipolars or cone bipolars is not clear from the anatomical evidence; physiological evidence indicates that a ganglion cell may respond to both rod and cone stimulation, whilst another may respond exclusively to cone stimulation; the latter is probably of the midget type.

### AMACRINE CELLS

The amacrine cells are of several types with respect to the nature and extent of their ramifications; they may, apparently, receive influences from bipolar cells and from other amacrines, whilst they may transmit influences to other bipolar terminals and to ganglion

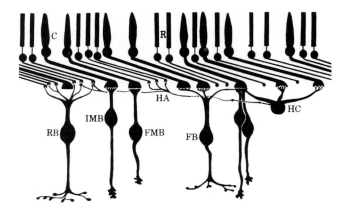

**Fig. 5.11** Illustrating organization in the outer plexiform layer of the primate retina. The cells are drawn diagrammatically and do not, except for the midget bipolar cells, show the full extent of their dendritic spread. In general the drawing does not show the overlap of cell types onto their respective receptors, the exception being the third cone from the right, where all the cell types connecting with a single cone pedicle are shown. The rod bipolar (RB) makes contact with rod spherules only. Two or three rod bipolars contact an individual rod, but a single rod bipolar only makes one dendritic contact per rod. Horizontal cell axon terminals (HA) end in red spherules only. Probably only two axons contribute a terminal to an individual rod. The axon is presented as a dotted line because the direct evidence for axon terminals being connected to the horizontal cell is not yet available.

The invaginating midget bipolar (IMB) is connected to a single cone. The flat midget bipolar (FMB) is also exclusive to a single cone pedicle. The flat bipolar (FB) is a diffuse cone bipolar and contacts six to seven cones. Probably six or seven flat bipolars overlap onto a single cone pedicle. The horizontal cell (HC) has dendrites ending in cone pedicles only. (Kolb, *Phil. Trans.*)

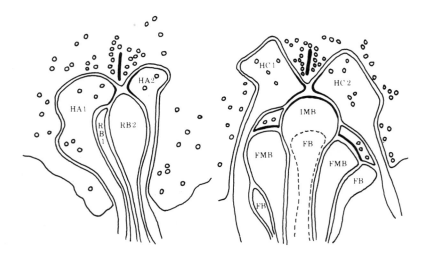

**Fig. 5.12** Illustrating synaptic arrangements in rod spherule (left) and cone pedicle (right) as deduced from electron-microscopy.

*Left :* The lobed lateral elements are horizontal cell axon terminals, each from a different horizontal cell (HA1 and HA2). The central elements are rod bipolar dendrites from different rod bipolars (RB1 and RB2).

*Right :* The horizontal cell dendrites, usually from different horizontal cells (HC1 and HC2) form the lateral elements of the invagination. The invaginating midget bipolar (IMB) pushes up into the invagination to lie between the two lateral elements. The synaptic ribbon and synaptic vesicles point into the junction of the three processes. The flat midget bipolar (FMB) dendrites lie alongside the invaginating midget bipolar dendrite and push part of the way up into the triad, but are in contact with the cone pedicle base. The flat bipolar (FB) terminals are also clustered around the invaginating bipolar dendrite and make superficial contact on the cone pedicle base. (Kolb, *Phil. Trans.*)

cells. Certain ganglion cells have extremely wide receptive fields (p. 293) in the sense that they may respond to light falling on retinal points that are far remote from each other; this means that the ganglion cell is collecting information from a wide area of rods and cones, and the manner in which their effects are funnelled into the ganglion cell is probably through the amacrine cells. This *convergence* of the retinal message is also assisted by the diffuse type of bipolar and ganglion cells, and it is not surprising that the total number of optic nerve fibres emerging from the human retina, namely about 1 million, is far less than the total number of receptors (7 million cones and 70 to 150 million rods). It is interesting, nevertheless, that in the central retina, there are three times as many bipolar cells as cones.

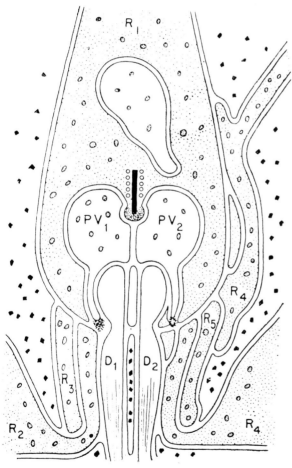

**Fig. 5.13** Schematic drawing showing the relationship between the ends of the lateral processes from $\beta$-type receptor cells ($R_2$, $R_3$, $R_4$, $R_5$) and the synaptic body of one $\alpha$-type cell ($R_1$), as well as between these processes and the end branches of bipolar cells ($D_1$, $D_2$) in the guinea pig retina. (Sjöstrand, *The Retina*. University of California Press.)

INTER-RECEPTOR CONTACTS

In the rabbit and guinea pig Sjöstrand described two types of rod, differentiated by the nature of their synaptic bodies; in the $\alpha$-type cells this is ovoid whilst in the $\beta$-type it is conical with the base of the cone facing the neuropil of the outer plexiform layer; the $\beta$-type rods make lateral contacts with the $\alpha$-type cells on the lateral aspects of their triad synaptic bodies as illustrated by Figure 5.13; in addition, the processes from these $\beta$-type cells make synaptic contacts with the vitread pole of several surrounding $\alpha$-type cells. Sometimes the processes are long so that contacts are made with cells some distance away. In addition, the $\beta$-type cells are interconnected with each other. The lateral processes of the $\beta$-type cells always contain vesicles. Sjöstrand (1969) has suggested that these contacts mediate some form of lateral inhibition analogous with that described by Hartline (p. 245).

CENTRIFUGAL PATHWAYS

Polyak described centrifugal bipolar cells that transmitted messages backwards to the receptors, but the more recent work of Dowling and Boycott on the primate retina, and of Brindley and Hamasaki (1966) on the cat retina, has failed to confirm their existence in these higher vertebrates; the same may well apply for centrifugal pathways from the higher centres of the brain back to the retina, at any rate so far as primates are concerned. In the pigeon, Dowling and Cowan (1966) confirmed Cajal's description of centrifugal fibres terminating in the outer edge of the inner plexiform layer; destruction of the isthmo-optic nucleus caused characteristic terminations on amacrine cells to lose their synaptic vesicles and finally to disappear, so that it would seem that the avian brain exerts control over transmission in the retina through effects on amacrine cells, which seem to be presynaptic to the terminals of bipolar cells.* In the amphibian there are efferent fibres to the retina, e.g. in the frog (Branston and Fleming, 1968) and in the phylogenetically intermediate group, the turtles, Marchiafava (1976) has demonstrated, from intracellular recordings, that stimulation of the optic nerve causes synaptic potentials in the amacrine cells (EPSP's) by contrast with the predominantly antidromic spikes recorded from ganglion cells.†

**Reflex arc**

It would seem that the optic nerve fibres acting centrifugally on retinal cells represent the efferent limb

* Spinelli and Weingarten (1966) observed single units in the optic nerve of the cat that were selectively activated by non-visual, e.g. auditory, stimuli, and it was assumed that these were efferent fibres.
† Ganglion cells show, in addition to the antidromic spike, a later synaptic potential which must be due to the amacrine activation occurring in response to activation of the centrifugal fibres; thus ganglion cells do not send collaterals to other ganglion cells so that a synaptic antidromic invasion is not possible, as, for example, that following antidromic excitation of a motor neurone.

of a reflex pathway by which control is exerted on the same restricted retinal areas from which the afferent impulses originate, carried by optic nerve fibres *via* the optic tectum to the isthmo-optic nucleus. So far as the pigeon is concerned, it appears that the centrifugal fibres originating in the isthmo-optic nucleus terminate solely on amacrine cells (Dowling and Cowan, 1966); from here the centrifugal influence can be transmitted to both bipolar and ganglion cells to which amacrine cells are extensively connected.★

## THE RETINAL ZONES

The retina is divided morphologically into two main zones, the *central area* (containing Regions I–III) and the *extra-areal periphery* (containing Regions IV–VII).

### FOVEA

Region I, the *central fovea* (Fig. 5.14), is a small depression, shaped like a shallow bowl with a concave floor, on the vitreal face of the retina, caused by the bending away of layers 5 to 9 and a thickening of the bacillary and outer nuclear layers. From edge to edge it is some 1500 $\mu$ across in man and thus subtends about 5° at the nodal point of the eye; the floor of the fovea is about 400 $\mu$ across. The fovea, described in this way, is generally referred to as the *inner fovea*, the word 'inner' being used to denote the aspect from which the depression is regarded; the *outer fovea*, some 400 $\mu$ across, represents a

depression in the opposite sense, i.e. a cupping of the outer limiting membrane from outwards inwards (Fig. 5.14), and corresponds roughly in extent to the floor of the inner fovea; it is caused by the lengthening of the cones in the bacillary layer, the cone proper increasing in length more than the remaining portion. The outer fovea is thus the central portion of the inner fovea; here the receptor elements consist entirely of cones which, as we have seen, are longer and thinner than elsewhere in the retina, being some 70 $\mu$ long and 1·5 $\mu$ thick at the base and 1 $\mu$ thick at the tip. The '*rod-free territory*' is actually rather greater in extent than the outer fovea, being some 500 to 600 $\mu$ across and containing some 34 000 cones. Beyond this limit rods appear and their proportion to cones increases progressively. The total number of cones in the territory of the inner fovea amounts to 115 000. As we shall see, the most central, rod-free, portion of the retina is concerned with the highest degree of visual acuity since it is here that the obstruction to light caused by the nerve fibres and other layers is reduced to a minimum and the density of receptors is highest; vision mediated by this portion of the retina is thus spoken of as 'foveal vision'; it must be emphasized, however, that the fovea proper embraces both this zone and the walls of the inner fovea where obstructions to the passage of light are not removed, where the cone density is considerably less, and where rods are present, so that it is inaccurate to speak of foveal vision when what is meant

★ Cowan and Powell (1963) have shown that the avian optic nerve fibres relay in the optic tectum; they run in the tractus isthmo-tectalis to the isthmo-optic nucleus; here they relay and centrifugal fibres run through the lateral geniculate nucleus to the bipolar cells of the retina.

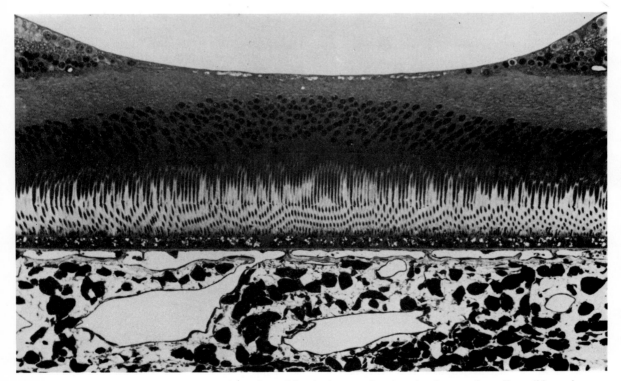

**Fig. 5.14** Transverse section through the primate fovea. Note in the central region, the absence of ganglion cell layer, inner nuclear layer and retinal capillaries, the thickening of the outer nuclear layer and the long slender cones (G.A./OsO$_4$ fixed, toluidine blue stained ×285; courtesy of R.C. Tripathi).

is vision mediated by the rod-free area or perhaps even a more restricted central zone (the 'central bouquet of cones', for example, some 50 to 75 $\mu$ across).

## MACULA

Mention may be made here of the *yellow spot* or *macula lutea*; this is an ill-defined area of the central retina characterized by the presence of a yellow pigment in the nervous layers; in actuality the region extends over the whole of the central area (Regions I–III) but the intensely pigmented part is limited to the fovea (Region I); the central portion of the fovea (outer fovea) is only slightly pigmented in man, since the pigment is associated with layers beyond number 4 which, as we have seen, are very much reduced or absent here.

## PARAFOVEA AND PERIFOVEA

Regions II and III are the *parafoveal* and *perifoveal* regions respectively, the outer edge of the former being 1250 $\mu$ from the centre of the fovea and that of the latter 2750 $\mu$ from the centre, the diameter of the entire central area being thus about 5000 to 6000 $\mu$. The cone density is considerably diminished in these regions whilst that of the rods is increased, so that in Region III there are only some twelve cones per 100 $\mu$ (compared with some fifty per 100 $\mu$ in the outer fovea) with two rods between each pair of cones. The *extra-areal periphery* (Regions IV to VII) is associated with a further decrease in the density of cones and an increase in the proportion of rods. In the whole human retina there are said to be about 7 million cones and 75 to 150 million rods.

## REFERENCES

Boycott, B. B. & Dowling, J. E. (1969) Organization of the primate retina: light microscopy. *Phil. Trans.* **225,** 109–184.
Branston, N. M. & Fleming, D. G. (1968) Efferent fibers in the frog optic nerve. *Exp. Neurol.* **20,** 611–623.
Brindley, G. S. & Hamasaki, D. I. (1966) Histological evidence against the view that the cat's optic nerve contains centrifugal fibres. *J. Physiol.* **184,** 444–449.
Cowan, W. M. & Powell, T. P. S. (1963) Centrifugal fibres in the avian visual system. *Proc. Roy. Soc. B.,* **158,** 232–252.
Dowling, J. E. & Boycott, B. B. (1966) Organization of the primate retina: electron microscopy. *Proc. Roy. Soc. B,* **166,** 80–111.
Dowling, J. E. & Cowan, W. M. (1966) An electron microscope study of normal and degenerating centrifugal fiber terminals in the pigeon retina. *Z. Zellforsch.* **71,** 14–28.
Dowling, J. E. & Werblin, F. S. (1969) Organization of retina of the mud-puppy *Necturus maculosus*. I. Synaptic structure. *J. Neurophysiol.* **32,** 315–335.

Kolb, H. (1970) Organization of the outer plexiform layer of the primate retina: electron microscopy of Golgi-impregnated cells. *Phil. Trans.* **258,** 261–283.
Marchiafava, P. L. (1976) Centrifugal actions on amacrine and ganglion cells in the retina of the turtle. *J. Physiol.* **255,** 137–155.
Missotten, L. (1965) *The Ultrastructure of the Retina.* Brussels: Arscia. Uitgaven N.V.
Polyak, S. L. (1941) *The Retina.* Chicago: University Press.
Polyak, S. L. (1957) *The Vertebrate Visual System.* Chicago: University Press.
Sjöstrand, F. S. (1969) The outer plexiform layer and the neural organization of the retina. In *The Retina,* Ed. Straatsma, B. R. *et al.,* pp. 63–100. Los Angeles: University of California Press.
Sjöstrand, F. S. (1976) The outer plexiform layer of the rabbit retina, an important data processing center. *Vision Res.* **16,** 1–14.
Spinelli, D. N. & Weingarten, M. (1966) Afferent and efferent activity in single units of the cat's optic nerve. *Exp. Neurol.* **15,** 347–362.

# 6. Measurement of the stimulus and dioptrics of the human eye

## PHOTOMETRY

### VISIBLE LIGHT

Visible light consists of electromagnetic vibrations limited to a certain band of wavelengths—the *visible spectrum*. The limits are given classically as 400 nm and 760 nm, wavelengths shorter than 400 nm being described as ultraviolet, and longer than 760 nm as infrared.* As we shall see, however, the retina is sensitive to wavelengths shorter and longer than these limits. Light is a form of energy and, as such, is amenable to quantitative measurement in absolute units, namely ergs. For physiological purposes, however, that subjective quality of light we call 'brightness' or 'luminosity' is the one that is of greatest interest, and since this is not uniquely determined by the energy content of the light, but varies widely with the wavelength (p. 327), it has been considered desirable to base measurements of light on a somewhat arbitrary 'luminosity basis.' Thus, on a luminosity basis, two sources of light are said to be of equal intensity when they produce equal sensations of brightness when viewed by the same observer under identical conditions.

### LUMINOUS INTENSITY

The luminous intensity ($I$) of a source of light is measured in terms of an arbitrarily chosen standard source of light-energy. Classically, this was a sperm-wax candle made to burn at a certain rate, and it gave rise to the term *candle-power*, a source of light appearing $x$ times as bright as the standard candle having $x$ candlepower. The modern *candela* (abbreviated cd) is fixed in terms of a standard filament lamp. It is such that one square centimetre of the surface of a full radiator at the temperature of solidification of platinum has a luminous intensity of 60 cd in the direction normal to the surface. In effect it has about the same value as the classical candle.

If two sources of light illuminate a screen under identical conditions, to produce identical degrees of illumination they are said to be of equal intensity. An observer may easily determine whether two identical screens, side by side, are equally illuminated; if they are not equally illuminated (i.e. if one appears brighter than the other) he is unable to state exactly how much more intensely illuminated one is than the other, so that in comparing intensities of illumination recourse is had to the photometer, one type of which utilizes as its basis the law that the intensity of illumination of a screen, placed normal to the rays from a point source of light, varies inversely as the square of its distance from the source. Thus, if we have two sources of light, of intensities $I_1$ and $I_2$ and at distances $r_1$ and $r_2$ from a screen such that they illuminate it equally, we can say:

$$I_1/I_2 = r_2^2/r_1^2.$$

Hence, if one of the light sources ($I_2$) is the standard candle, the intensity of the other is given by:

$$I_1 = r_2^2/r_1^2 \text{ candelas.}$$

### INTENSITY OF ILLUMINATION

In introducing the measurement of the intensity of a light source, we have mentioned the *degree of illumination*; this is, of course, determined entirely by the amount of light falling on a surface and hence by the intensity of the light source, the distance of this source from the object, and the angle the surface makes with the direction of the incident light. Thus, if we have a screen perpendicular to the direction of the light incident from a point source of intensity $I$, we may define the *intensity of illumination*, $E$, by the equation:

$E = I/r^2$, where $r$ is the distance from the source.

When $I$ equals unity, i.e. one candela, and $r$ equals 1 foot, $E$ is unity and is called 1 *foot-candela*. If $r$ is expressed in metres, the unit of illumination is the *metre-candela* or *lux*; expressed in words, it is the illumination of a surface, normal to the incident rays from a source of 1 candela one metre away. The *centimetre-candela* is called a *phot*. If a surface, normal to the incident rays from a source, has an illumination $E$, its illumination when the surface is inclined at an angle $\theta$ to the rays is given by $E \cos \theta$.

### LUMINANCE

The illumination of the surface does not necessarily define its intensity as a stimulus, since the latter depends on the light entering the eye and so is determined by the light reflected from the surface and the direction of view. If, however, we imagine a white surface that reflects all the light it receives and is, moreover, a 'perfect diffuser' in the sense that it appears equally bright from all directions, then the *luminance* of a surface may be defined in terms of its illumination from a standard source. Thus, on the so-called *Lambert system*, a perfectly diffusing surface, normal to the incident light from a source of 1 candela 1 centimetre away, has a luminance of 1 *lambert*. This unit is too large for most purposes and the *millilambert* (mL) represents a thousandth of a lambert; for studies of light thresholds (p. 185) the log micro-micro-lambert scale is often used. A luminance of 3 log micro-microlamberts means a luminance of $10^3 \times 10^{-12} = 10^{-9}$ lamberts or a millionth of a millilambert; 2 log micro-micro-

---

* The units employed in the earlier literature to define wavelengths were the Ångstrom Unit (Å), equal to $10^{-8}$ cm or $10^{-10}$ m, and the 'millimu' (m$\mu$) equal to $10^{-7}$ cm or $10^{-9}$ m. The modern unit is the *nanometre*, equal to $10^{-9}$ m. Thus 7600 Å = 760 m$\mu$ = 760 nm.

lamberts are equivalent to $10^2 \times 10^{-12} = 10^{-10}$ lamberts, and so on. The *equivalent foot-candela*, or *foot-lambert*, is the luminance of a similar screen distant 1 foot from the source.

## SURFACE INTENSITY

This mode of defining the luminance of a surface in terms of the light incident on it is limited by the ideal nature of a perfectly diffusing surface; a definition that does not involve this would be superior from many points of view. Such a definition is given by describing the luminance in terms of the light emitted by the surface, i.e. we can treat the surface as a luminous source, whether it is self-luminous as in the case of a hot body, or whether its luminosity depends on reflected light. Thus if 1 square foot of a surface emits the same amount of light in a given direction as a source of 10 candelas, for instance, we may define the *luminous intensity of the surface* as 10 candelas; the subjective brightness of the surface and hence its intensity as a stimulus will clearly depend on the area from which this light emanates; if it all comes from a small area the brightness will be greater than if it came from a large area. The luminance is thus defined in terms of the candle-power of the surface per unit area, e.g. candelas per square foot or candelas per square centimetre. We may then calculate the luminance, in candelas per square foot, of a perfectly diffusing surface illuminated by a source of 1 candela at a distance of 1 foot; this turns out to be $1/\pi$ candelas per square foot, i.e. $1/\pi$ *candelas per square foot = 1 equivalent foot-candela*. Similarly $1/\pi$ candelas per square centimetre = 1 lambert.

## EFFECTIVELY POINT SOURCE

The definition of the luminance of a surface in terms of its candlepower *per unit area* has an obvious physiological basis. By the luminance of a surface we wish to express in a quantitative manner the intensity of the stimulus when its image falls on the retina. An illuminated surface of given area forms an image of a definite area on the retina and the sensation of brightness, or rather the stimulus, depends on the amount of light falling on this area of the retina. If the surface is moved closer to the eye, the amount of light entering the eye is increased in accordance with the inverse square law, but the surface does not appear brighter because the size of the image has increased, also in accordance with an inverse square law, and consequently the actual *illumination* of the retina has not altered. Thus the subjective brightness of an extended source is independent of its distance from the observer. It may be argued that the brightness of a lamp decreases as it is moved away from the eye and eventually becomes so faint that it disappears if it is moved far enough away. This, however, is a consequence of the fact that at a certain distance from the eye the lamp becomes an 'effectively point source', i.e. its image on the retina is of the same order of size as that of a receptor element; consequently, when it is moved further away still, although its image on the retina becomes smaller, the amount of light per receptor element now begins to fall in accordance with the inverse square law. Thus the definition of luminance in terms of candles per unit area of necessity implies that the surface has a finite area; when this surface becomes effectively a point from the aspect of the size of its image on the retina, its power of stimulating the retina will be determined only by its total luminous intensity, i.e. its candlepower, and its distance from the observer. The limiting angle, subtended by an object at the eye, below which it behaves as an effectively point source depends on the particular visual effect being studied and the conditions of observation; for the study of absolute thresholds in the dark it is as big as 30 to 60 minutes of arc.

## RETINAL ILLUMINATION

We have indicated that the characterization of the luminance of an extended surface by its candlepower per unit area is a sound one physiologically in so far as it indicates the illumination of the retina; the retinal illumination is not, however, completely defined by the candlepower per unit area of the surface being viewed, since the amount of light falling on unit area of the retina depends also on the pupil size; thus from Figure 6.1 it is

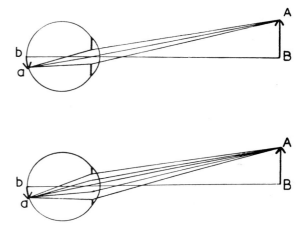

**Fig. 6.1** Illustrating the importance of the size of the pupil in defining the stimulus. *Above*: The pupil is small and only a thin pencil of light from A enters the eye. *Below*: The pupil is large, permitting a large pencil to enter the eye. AB = object. ab = image.

clear that the pencils of light that form the image of AB on the retina are larger in the case of the larger pupil. Hence the stimulus required to produce a definite sensation in the eye must be defined in terms of both the luminance, in candles per unit area, and pupillary area. The unit of retinal illumination suggested by Troland* and now given his name is defined by the statement: A troland is that intensity of stimulation that accompanies the use of a pupillary area of one square

*This unit of retinal illumination was originally called the *photon*, a term that had already been applied to the quantum of light energy. We may note that, in order to convert the luminance of an extended surface, expressed in millilamberts, to trolands of 'retinal illumination', it must be multiplied by $5d^2/2$, where $d$ is the pupillary diameter in millimetres. Because of the different sensitivities of the eye to the different wavelengths, according as it is light- or dark-adapted (p. 327), the equivalent stimuli for white light are not the same under photopic and scotopic conditions, so that we must speak of a *photopic* and *scotopic* troland; the theory behind the estimation of the conversion factor (1 scotopic lumen = 1·4 photopic lumens at a colour-temperature of 2750°C) has been discussed by Schneider and Baumgardt (1966). In terms of energy flux, a scotopic troland is equivalent to 4·46 quanta at 507 nm per square degree of visual field per second, incident on the eye.

millimetre and an external stimulus surface luminance of one candela per square metre. It is to be noted that although the troland is reffered to as a unit of 'retinal illumination,' it is not truly so, but is rather an indication of this quantity that is independent of the size of the observer's pupil. To define a stimulus in terms of true retinal illumination we must know the magnification and transmission of the optical media of the eye. On the basis of a value of 16·7 mm for the distance of the nodal point from the retina, the true retinal illumination is given by:

$$D = 0.36 \, \tau_\lambda \, s \, B$$

where $B$ is the luminance of the extended source of light, $s$ is the area of the pupil in $cm^2$ and $\tau_\lambda$ is the *transmission factor*, being the ratio of the intensities of light falling on the retina over the light falling on the cornea. This depends on wavelength, varying from 0·1 in the extreme violet where absorption by the yellow pigment of the lens is important, to 0·7 in the deep red; on average with white light it is about 0·5 indicating 50 per cent loss of light (Ludvigh and McCarthy).

When the stimulus in question is due to an effectively point source the pupillary area is of course important, and the stimulus is defined by the candlepower of the source, its distance from the eye and the pupillary area, i.e. by the total light flux entering the eye, which is given by the illumination of the pupil multiplied by its area.★

In view of the importance of the adequate definition of the stimuli used in experimental studies of vision, it is worth recapitulating the substance of the last few paragraphs. The luminance of an extended source is defined as its candlepower per unit area; its value as a stimulus is independent of its area and of its distance from the observer. For a given pupillary area it is an indication of the retinal illumination; the troland is a measure of retinal illumination which is independent of the observer's pupillary area; the true retinal illumination may be derived from the troland if the magnification and transmission characteristics of the eye are known. If the stimulus is due to an effectively point source, it is sufficient to characterize it by the total light flux entering the eye.

## LIGHT FLUX

### LUMEN

For the sake of simplicity, we have so far made our definitions without introducing the concept of *light flux*; this is defined as the amount of visible light energy passing any plane in unit time; its relationship to the luminous intensity of a point source will be clear from the following.

Suppose we have a point source of light of $I$ candlepower; let us imagine the point to be the centre of a sphere of radius $r$ (Fig. 6.2). The light energy from a point source travels in

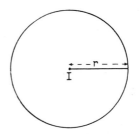

**Fig. 6.2**   A source of $I$ candlepower at the centre of a sphere of radius $r$. $4\pi I$ lumens are emitted per sec. and spread over a surface of $4\pi r^2$; the flux density is $I/r^2$.

all directions uniformly through space, and we may think of the sphere being occupied by energy passing through it at a definite rate. The flux, considered thus, is defined in relation to the unit of luminous intensity by saying that $4\pi$ times the candlepower is the flux emitted by the source, expressed in *lumens*:

$$4\pi \times I \text{ (candelas)} = \text{flux (lumens)}$$

$E$, the intensity of illumination of the inner surface of the sphere, as defined before, is:

$$E = I/r^2$$
$$= 4\pi I/4\pi r^2.$$

Now $4\pi I$ represents the number of lumens of flux falling on the surface of the sphere, and $4\pi r^2$ is the area of its surface. Hence $E$ may be expressed as the ratio of the flux emitted by the source of light over the area on which it falls: flux per unit area, or *flux density*. The units of illumination may therefore be, besides the foot- or metre-candela, a lumen per square foot or a lumen per square metre. Thus:

$$1 \text{ foot-candela} = 1 \text{ lumen/sq ft}$$
$$1 \text{ metre-candela} = 1 \text{ lumen/sq m}$$

**Table 6.1**   Some recommended illuminations

|  | foot-candela |
| --- | --- |
| General stores | 2–4 |
| Classroom | 5–10 |
| Billiard table | 20 |
| Type-setting | 35–50 |
| Operating table | 100 |

In Table 6.1 the illuminations recommended for different tasks are shown; in Table 6.2 some luminance levels commonly encountered are presented, and in Tables 6.1 and 6.2 some conversion factors for different units of illumination and luminance.

★Stiles and Crawford (p. 360) have shown that pencils of light entering the eye obliquely are less effective as stimuli than those entering the pupil centrally; this effect is not due to aberrations in the optical system but is most likely related to the orientation of the receptors in the retina. A correct definition of the stimulus must therefore take into account the directions of the pencils of light entering the eye.

**Table 6.2** Some commonly encountered luminances

|  | *candela/sq ft* |
|---|---|
| Starlit sky | 0·00005 |
| Moonlit sky | 0·002 |
| Full moonlight on snow | 0·007 |
| Ploughed land at noon in average daylight | 130 |
| Clear sky at noon | 1 000 |
| Sun's disk | 150 000 000 |

## PHOTOPIC AND SCOTOPIC LUMENS

The lumen is an arbitrary unit of power, based on the candela; it may be converted into absolute units on the basis of the light-energy emitted by the standard source of light; if this is the standard black body of absolute temperature T = 2042°K, then the conversion factor is 679·6 lumens per watt. It must be appreciated that the lumen is a unit in whose measurement the human eye is involved, so that the spectral sensitivity curve of the human eye must be taken into account, the luminance of the standard source being governed by the subjective luminances of the gamut of wavelengths of which the source is made up.

This is because the concept of the lumen as a unit of stimulus has been extended to cover any source of light; for example we may have a lumen of violet light in a narrow wavelength band close to 420 nm and a lumen of green light in a narrow wavelength band centred on 530 nm. If the two lumens are to be equal from the point of view of their subjective effects it is not correct for their flux-energies, in absolute units, to be equal; if they were, they would not match in the human eye, the green lumen being considerably brighter than the violet lumen; to be the same, we must know the relative sensies of the eye for the different wavelengths, i.e. we must have the *photopic luminosity curve* (Fig. 6.3) for a 'standard observer' and adjust the energy equivalents of the lumen in the different bands of the spectrum. Thus the conversion factor for the lumen, based on the standard white source, is 679·6 lumens per watt of energy-flux, and is the sum of the contributions of all the wavelengths contributing to stimulate the 'standard human eye', weighted according to their contributions to subjective brightness. This is called a *photopic lumen* because the relative brightness scale is based on a high level of illumination. A *scotopic lumen* would be different because the relative sensies of the eye to different wavelengths are different when viewing very weak lights (p. 327), and the *scotopic luminosity curve* (Fig. 6.3) must be used for assessing the contributions to the standard source of white light. The conversion factor for the *scotopic lumen* is 1745 lumen per watt. In effect this means that the standard source would appear equally bright to two eyes, the one with relative sensitivities to the different wave-

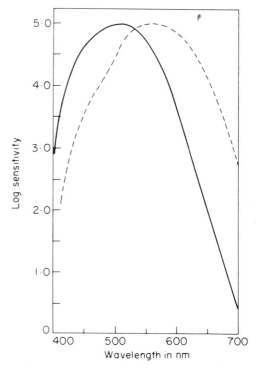

**Fig. 6.3** Internationally standardized spectral sensitivity curves for the dark (————) and light (— — — —) adapted eye respectively. The curves were recognized as standard in 1951 and 1931 respectively. (Ripps and Weale, *The Eye.*)

lengths governed by the photopic luminosity curve, and the other by the scotopic luminosity curve, provided the energy fluxes from the light source were in the proportion of 1745:679·6. For convenience the two luminosity curves, whose significance will be discussed later (p. 327), are presented here. Tables 6.3 and 6.4 provide useful conversion factors for units of illumination and luminance.

## DIOPTRICS OF THE HUMAN EYE

### REFRACTING SURFACES

The basic dioptrics of the eye have often been likened to those of a simple convex lens, and this is true to the extent that the formation of an image of external objects can be approximately described by drawing rays from a point on the object through a 'centre' and another parallel to the optic axis which passes through a 'focus'.

**Table 6.3** Conversion factors. Illumination. (Value in unit in left-hand column times conversion factor = value in unit at top of column; e.g. 3 foot-candle = 3 × 1·08 × 10 = 32·4 metre-candle.)

|  | *Foot-candle* | *Centimetre-candle* | *Metre-candle* |
|---|---|---|---|
| Foot-candle | 1 | 1·08 × 10$^{-3}$ | 1·08 × 10 |
| Cm-candle (Phot) | 9·29 × 10$^2$ | 1 | 1 × 10$^4$ |
| Metre-candle (Lux) | 9·29 × 10$^{-2}$ | 10$^{-4}$ | 1 |

**Table 6.4** Conversion factors. Luminance. (Value in unit in left-hand column times conversion factor = value in unit at top of column.)

|  | Candela/ft² | Equiv. fc. | Candela/cm² | Millilambert |
|---|---|---|---|---|
| Candela/ft² | 1 | 3·14 | $1·08 \times 10^{-3}$ | 3·38 |
| Equiv. fc. | $3·18 \times 10^{-1}$ | 1 | $3·43 \times 10^{-4}$ | 1·076 |
| Candela/cm² | $9·3 \times 10^{2}$ | $2·92 \times 10^{3}$ | 1 | $3·14 \times 10^{3}$ |
| Millilambert | $2·96 \times 10^{-1}$ | $9·29 \times 10^{-1}$ | $3·18 \times 10^{-4}$ | 1 |

In fact, of course, refraction by the eye is the result of bending light at several surfaces, in particular as it passes from air to the cornea, from aqueous humour to lens, and from lens to vitreous body, since it is at these surfaces that significant changes of refractive index take place.

SCHEMATIC EYE

A system of several refracting surfaces, sharing a common axis, can be analysed by the Gaussian treatment, which gives two *principal planes*, two *nodal points*, and two *principal foci*; and the result is described as the *schematic eye* based on measurements of the curvatures of the surfaces (cornea, lens) refractive indices and positions of the surfaces (Fig. 6.4).

REDUCED EYE

Since the principal planes and nodal points of the human eye are quite close to each other, a mean position for each is taken, to give a single principal plane 1·5 mm behind the anterior surface of the cornea, and a single nodal point, 7·2 mm behind the anterior surface of the cornea. This gives the 'reduced eye' of Figure 6.5, with an anterior focal length of 17·2 mm, measured from the single principal plane, and a posterior focal length of 22·9 mm, measured from the same plane. In the

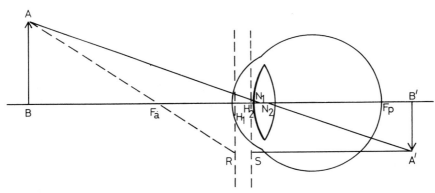

**Fig. 6.4**  The schematic eye. $F_a$ and $F_p$ are the principal foci; $N_1$ and $N_2$ the nodal points, and $H_1$ and $H_2$ are the principal planes. A ray from A strikes the first principal plane at R and leaves the second principal plane at S.

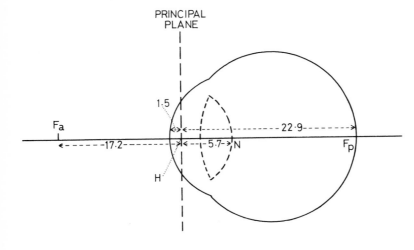

**Fig. 6.5**  The reduced eye. $F_a$ and $F_p$ = anterior and posterior principal foci respectively. N = nodal point. The optic axis intersects the principal plane at H.

'emmetropic' eye the pole of the retina coincides with the second principal focus, $F_p$.

## Image construction

On the basis of this reduced eye, image-formation may be carried out by the usual constructions, the undeviated ray of a pencil passing undeviated through the single nodal point, and parallel rays converging to the second focal plane, or retina (Fig. 6.6).

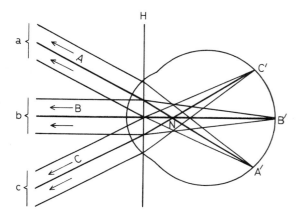

**Fig. 6.6** The image of a distant object, ABC, forms an image on the retina, A'B'C' of the emmetropic eye. H = principal plane. N = nodal point. a, b and c = pencils of light from A, B and C respectively.

## Accommodation

When the object is closer than infinity, the usual image-construction leads to the formation of the image behind the retina (Fig. 6.7), and this means that the reduced eye of Figure 6.5 is based on the dioptric characteristics of an eye 'focussed for infinity', or 'unaccommodated'. Focussing the image of distant objects on the retina clearly requires a shortening of the focal length, so that the posterior focus moves to $F_p'$, i.e. it requires an increase in dioptric power of the system. The mechanism of this accommodation will be described in Section III.

## HYPERMETROPIA AND MYOPIA

Failure of the second or posterior focus to coincide with the pole of the retina leads to a blurring of the image of distant points; in myopia (Fig. 6.8) the focal length is too short and in hypermetropia it is too long. The myopic eye is said to be too powerful for its axial length, and this may be due to a more-than-average curvature of the refracting surfaces, or to a longer-than-average axis; and the opposite conditions describe the hypermetropic eye.

### FAR POINT

The myopic eye can focus images of near objects distinctly on the retina, and a *far point* or *Punctum Remotum* is defined as that position on the optic axis that allows the formation of a distinct image on the retina of the unaccommodated eye (Fig. 6.9a). The myopia can be corrected by reducing the vergence of rays reaching the cornea by interposing a diverging lens, as in Figure 6.9b; now rays from infinity are made to diverge before reaching the cornea, and if they

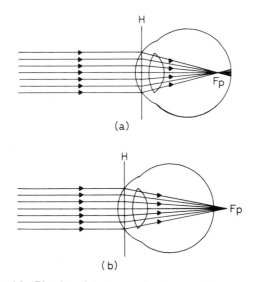

(a)

(b)

**Fig. 6.8** Blurring of the image of a distant point due to myopia (a) and hypermetropia (b). H = principal plane. $F_p$ = posterior principal focus.

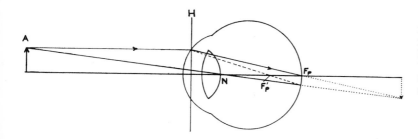

**Fig. 6.7** With an object not 'infinitely far away' the image would fall in a plane behind the retina. H = principal plane. N = nodal point. $F_p$ = posterior principal focus in relaxed eye; $F_p'$ is that for the accommodated eye.

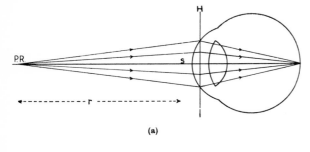

(a)

diverge from the Far Point they will come to a focus on the retina of the relaxed, unaccommodated eye.

## HYPERMETROPIA

The hypermetropic eye requires an increase in power, which is provided by a converging lens (Fig. 6.10). The hypermetropic subject can, of course, increase the dioptric power of the eye by accommodation, so that it is usually only in middle age, when accommodative power has been reduced or lost entirely, that correction for distant vision becomes necessary.

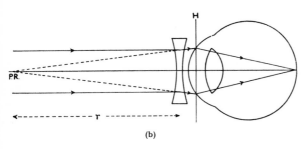

(b)

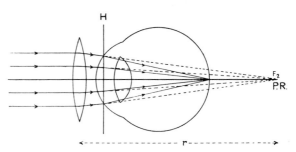

**Fig. 6.9** (a) The Far Point in myopia is the point in front of the eye at which an object must be placed for its image to fall on the retina. (b) Correction of myopia by placing a concave lens in front of the eye such that rays from a distant object will diverge from the Far Point. r = the focal length of the lens placed in front of the eye to achieve the necessary correction. PR = Punctum remotum or Far Point. H = principal plane. S = spectacle point.

**Fig. 6.10** The Far Point (P.R.) in hypermetropia is a point behind the retina to which rays from a distant object must converge in order that the image may fall on the retina. H = principal plane. $F_2$ = second principal focus of the convex lens which coincides with the Far Point. r = focal length of the convex lens.

## REFERENCES

Ludvigh, E. & McCarthy, E. F. (1938) Absorption of visible light by the refractive media of the human eye. *Arch. Ophthal.*, *Chicago* **20**, 37–51.

Schneider, H. & Baumgardt, E. (1966) Sur l'emploi en optique physiologique des grandeurs scotopiques. *Vision Res.* **7**, 59–63.
Walsh, J. W. T. (1953) *Photometry*. London: Constable.

# 7. Some general aspects of vision

## PHOTOCHEMICAL REACTION

The mechanism by which the energy contained in visible light is transmitted to the nervous tissue of the retina in such a way as to produce that change in its structure which we call a stimulus is a fascinating problem and one that lends itself to physiochemical study. There is a group of chemical reactions that derive a part, or all, of their energy from absorbed light—the *photochemical reactions*—a biological instance of this type of reaction is the synthesis of carbohydrate by the green leaves of plants, the pigment chlorophyll absorbing light and transmitting the energy to the chemical reactants. It is thus reasonable to inquire whether there is an analogous chemical reaction at the basis of the response of the retina to light, i.e. whether or not there are substances in the retina that can absorb visible light and, as a consequence of this absorption of energy, can modify the fundamental structure of the rod or cone to initiate a propagated impulse. One such substance, visual purple, was discovered in the retina as long ago as 1851 by H. Muller, but we owe the early knowledge of its function as an absorber of light to the researches of Kühne, described in his classical monograph of 1878. He was the first to obtain this pigment in solution by extracting the retina and he called it 'Sehpurpur', which has been translated as *visual purple*. As we shall see, thanks to the recent studies of Dartnall and Wald, the term visual purple must now be used to describe a class of visual pigments rather than a single entity.

## CATEGORIES OF VISION

Before the detailed study of the photochemical and electrical changes in the retina associated with the visual process is entered into, it would be profitable to discuss some of the more elementary phenomena of vision itself and to indicate the experimental methods available for its study. The visual perception of an object is a complicated physiological process, and attempts have been made to resolve it into simpler categories. Thus, it is common to speak of the *light sense*, defined as the appreciation or awareness of light, as such, and of modifications in its intensity. Secondly, the *colour sense*, which enables us to distinguish between the qualities of two or more lights in terms of their wavelengths, and finally the *form sense* that permits the discrimination of the different parts of a visual image. For the

present we shall preserve this simple classification for its convenience, but modern psychophysicists would hesitate to use the term 'sense' in relation to the discrimination of form.

## THE LIGHT SENSE

### THE ABSOLUTE THRESHOLD

The simplest way to study the light sense is to measure the *threshold* to white light, i.e. the minimum light-stimulus necessary to evoke the sensation of light. There are many ways of measuring this, and the general technique will be discussed later. In general, the subject is placed in a completely dark room and the illumination of a test-patch is increased until he is aware of its presence. The patch may be illuminated all the time and its luminance steadily increased, or it may be presented in flashes of varying luminance.

### DARK-ADAPTATION

Early attempts to measure the threshold soon showed that, in order to obtain other than very irregular results, the observer must be allowed to remain in the dark for at least half an hour (i.e. he must be *dark-adapted*) before any attempt is made to obtain a definite value. If this is not done, it is found that the threshold luminance of the patch is very high if the test is made immediately after the subject has entered the dark room, and that the threshold falls progressively until it reaches a fairly constant value which may be as small as one ten-thousandth of its initial figure. This effect is demonstrated by Figure 7.1, in which the illumination of the test patch (in microlux) required to produce the sensation of light in a subject, who had gazed at a bright light for some minutes before entering the dark room, is plotted against time in the dark; the curve is referred to as a *dark-adaptation curve*.

This extraordinary increase in sensitivity of a sense organ on removing it from its normal background of stimuli has not been met in the study of the other senses, and an inference to be drawn from the dark-adaptation curve is that a second mechanism of vision is being brought into play requiring half an hour, or more, to be established at its full level of efficiency. This inference represents the basis of the *Duplicity*

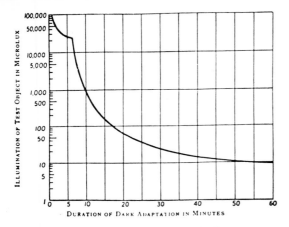

**Fig. 7.1** Typical dark-adaptation curve. Ordinate: illumination of the test-patch in microlux. Abscissa: duration of dark-adaptation in minutes. (Kohlrausch, in *Handbuch der Normalen und Pathologischen Physiologie*.)

*Theory of Vision,* a theory which assumes that at luminance levels below a certain value (about 0·01 mL) vision is mediated primarily by one mechanism—*scotopic vision*, involving the rods—and above this value by another mechanism—*photopic vision*, involving the cones; the transition is not abrupt so that there is a range in which the two mechanisms operate together, the *mesopic range*.

## DUPLICITY THEORY

### ROD VISION

The duplex nature of vision is readily appreciable under conditions of very low illumination. Thus, whereas in daylight, or photopic, vision the subject sees best by direct fixation, so that the image of the object or luminous point he wishes to see falls on the fovea, under night, or scotopic, conditions his best vision is *peripheral*; he must look away from the object he is trying to distinguish so that its image falls on the peripheral retina. Thus, in looking at a dim star one must, as the astromer Arago said, not look at it; if one fixates it, it disappears. This gives the clue to the anatomical basis of the duplicity theory; the fovea, in its central region specially, contains only cones. Since the fovea is apparently blind at low intensities of illumination, we may conclude that it is the rods that are functional under these conditions. That the rods are indeed responsible for night vision is now well recognized; it was Schultze who put forward this hypothesis on the basis of his comparative studies of animal retinae; thus he found that the nocturnal owl had a preponderance of rods whilst a diurnal bird, such as the falcon, had a preponderance of cones in its retina.

### Achromaticity

Another feature of vision at low luminance is its *achromaticity*, i.e. its failure to distinguish colours. Suppose, for example, that the threshold for vision is measured using different wavelengths of light; it is found that these thresholds are different, in the sense that more yellow energy, for instance, is required to give the sensation of light than blue energy, but the 'yellow' and 'blue' lights are both colourless. Only when the intensities are raised well above threshold do colours appear, i.e. when the threshold for stimulation of the *cones* is reached. Thus, on a moonless night colours are not distinguished, whereas in the *mesopic range*, corresponding to around full moonlight, colours may be appreciated but not very distinctly.

### Low visual acuity

A very obvious difference between scotopic and photopic vision is the low visual acuity at low luminances; in other words, in spite of the high sensitivity of the eye to light under scotopic conditions, the ability to discriminate detail (e.g. to read print) is very low.

Other features of the duplicity theory will be considered later; for the moment we may say that it is now a well established *principle* rather than a theory. In essence it states that at high levels of luminance the relatively insensitive cones are able to operate, whilst the highly sensitive rods are out of action (perhaps through overstimulation). Vision mediated by the cones is coloured and exhibits high powers of discrimination of detail. When the light available is restricted, the insensitive cones no longer operate whilst the rods, after a period of dark-adaptation as shown by Figure 7.1, become operative, mediating an achromatic, peripheral, low-acuity type of vision. Over an intermediate range, which may extend up to 1000 times the threshold for cone vision, both rods and cones are operative—the *mesopic range*.

## THE DIFFERENTIAL THRESHOLD OR L.B.I.

Instead of using the minimum luminance of a test patch as a means of studying the light sense, we may measure the *differential threshold*—the amount an existing stimulus, which itself produces a sensation of light, must be increased or decreased to produce a change in the sensation.

### MEASUREMENT

Thus the subject may be presented with a luminous field divided into two, as in Figure 7.2a, the luminances

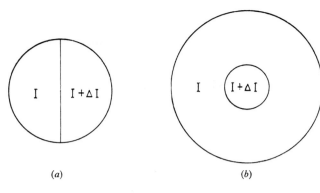

(a)                    (b)

of either being varied independently. Beginning with them both equal at I, the right-hand one may be increased to value $I + \Delta I$, such that the observer can just detect a difference of subjective brightness. The increment, $\Delta I$, is called the *liminal brightness increment* (l.b.i.) or *differential threshold*. Alternatively, as in Figure 7.2b, the subject may look at a large circular field of luminance $I$, and a smaller concentric field may be thrown on to this with a variable luminance greater than $I$, With this experimental technique, the measurements need not be confined to those intensities of illumination in which only scotopic vision is operative, but rather the whole range of intensities may be used.

WEBER-FECHNER LAW

The differential threshold, or l.b.i., varies with the initial value of the prevailing luminance, $I$; for example if $I$ is 100 mL then $\Delta I$ may be in the region of 2 mL; if $I$ is 1000 mL then $\Delta I$ is about 20 mL. Over a certain range, however, it may be expressed as a constant fraction of the prevailing luminance, so that the ratio: $\Delta I/I$ is constant. This is the basis of *Weber's Law of Sensation* which applies approximately to all sensory modalities, namely that the difference-limen is proportional to the prevailing stimulus. For example, in the case of the sense of touch or pressure, Weber showed that when two weights resting on the skin differed by about 1 in 30 they could be discriminated as different weights. Thus if an individual could discriminate between a 30 and a 31 gm weight he could also discriminate between a 3 and 3·1 gm weight; in other words, the *Weber fraction*, $\Delta S/S$, is constant, where $S$ is the prevailing stimulus and $\Delta S$ its liminal difference for perception.

FECHNER FRACTION

Fechner extended Weber's work to the study of visual discrimination, and the fraction $\Delta I/I$ is frequently called the Fechner-fraction. Figure 7.3 represents a more modern study, where the Fechner-fraction, $\Delta I/I$, has been plotted against the intensity level, $I$. At high intensities, between about 0·1 and 1000 mL, this fraction

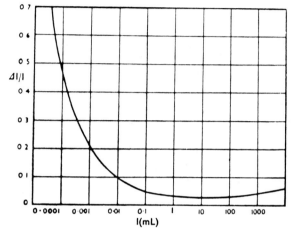

**Fig. 7.3** Intensity discrimination curve. The ratio $\Delta I/I$ is the liminal intensity ($\Delta I$) increment divided by the intensity ($I$). If the Weber-Fechner Law held, this ratio would be independent of the value of $I$. The curve shows that, over a certain range (0·1 to 1000 mL), this is approximately true. In the scotopic region, however, the ratio increases greatly, i.e. the power to discriminate changes in luminance falls off at low levels.

is fairly constant at about 0·02 to 0·03, indicating that a change of two to three per cent in luminance can be detected; at low luminances the power to discriminate falls off sharply so that at 0·0001 mL two patches must differ by about 50 per cent in luminance if they are to be discriminated.

## THE COLOUR SENSE

By means of the colour sense we become aware of those qualitative characteristics of light that depend on its spectral composition. On passing white light through a prism it is resolved spatially into its component wavelengths, and the result is the well-known spectrum; this resolution shows that the sensation of white is evoked by the summated effects of stimuli which, when operating alone, cause entirely different sensations, namely, those of red, orange, yellow, green, etc.

## HUE

The spectrum shows a number of characteristic regions of colour: red, orange, yellow, green, blue, indigo and violet; these regions represent large numbers of individual wavelengths; thus the red extends roughly from 750 to 650 nm; the yellow from 630 to 560 nm; green from 540 to 500 nm; blue from 500 to 420 nm, and violet from 420 to 400 nm. Within these bands of colour there are thus a variety of subtle variations, and a spectral colour or *hue*, has been defined somewhat idealistically as the sensation corresponding to a single wavelength. This is too rigid a definition, both because of the limitations to discrimination of the human eye, and the experimental difficulty in obtaining light confined to a single wavelength.

Thus coloured lights may be obtained by the use of

## HUE-DISCRIMINATION

The physiological limit to the differentiation of hues is given by the *hue-discrimination curve*, shown in Figure 7.4. The subject is presented with a split field, the two halves of which may be illuminated independently by lights of equal subjective intensity but differing wavelength. Starting with, say, a red of 700 nm in the left-hand field the wavelength of the light of the right-hand field is shortened till the subject can just discriminate a change in hue, say at 680 nm; the hue-limen is thus 20 nm. With both fields at 680 nm, the righthand field wavelength is again shortened and a new hue-limen obtained. Figure 7.4 shows the results for a normal human observer; discrimination is best at about 500 nm in the blue-green and at 600 in the yellow, but even here the smallest detectable change in wavelength is about 1 nm.*

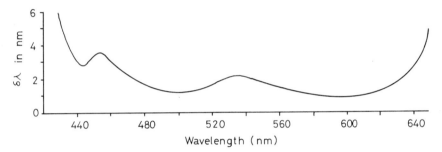

**Fig. 7.4**   Hue-discrimination curve. At any given wavelength, indicated on the abscissa, there is a certain change in wavelength, $\delta\lambda$, that must be made before the change in hue can be discriminated. (Wright, *Researches on Normal and Defective Colour Vision.*)

filters, which cut out many wavelengths and selectively transmit others; these filters are never sufficiently selective to confine the transmitted light to a single wavelength and at best they transmit a *narrow band* about a mean position. Hence by a green filter of 530 nm is meant a filter that transmits mainly green light with a peak of transmission at 530 nm; transmission at nearby wavelengths is also considerable, however, whilst even at remote wavelengths there may be a little transmission.

### MONOCHROMATOR

A more precise method is by the use of the monochromator, the essential principle of this technique being to select, from a spectrum, just that narrow band of wavelengths that one requires. In practice the band is selected by a movable slit that passes over the spectrum; and its width, in terms of the range of wavelengths, can be reduced by making the slit narrower and narrower. Unfortunately, this decreases the amount of light-energy transmitted, so that once again the experimenter must be content with a band of wavelengths rather than a single one.

### SATURATION

In practice, then, coloured lights are not monochromatic, being generally a mixture of different wavelengths; the hue is then said to be *impure*, and in the case when the impurity can be accounted for by an admixture of white light, it is said more specifically to be *unsaturated*, the actual colour being called a *tint* of the hue that has been thus diluted. The percentage saturation of a hue is the percentage contributed by the monochromatic light, estimated on a luminosity basis. By mixing white light with a spectral hue, the latter actually changes; thus red becomes pink, orange becomes yellow, yellow becomes green, green becomes yellow, and violet becomes salmon-pink; the yellow-green does, in effect, remain constant and the blue changes very little.

### LIMITS OF THE VISUAL SPECTRUM

The limits are usually given as 400 nm and 760 nm, wavelengths shorter than 400 nm and longer than

*Cornu and Harlay (1969) have examined the effects of luminance on hue discrimination, and have summarized the recent literature on this point.

760 nm being characterized as *ultraviolet* and *infrared* respectively. In fact, however, the retina is sensitive to wavelengths well beyond these limits; for example, Wald showed that the insensitivity of the eye to ultra-violet up to 350 nm is entirely due to absorption by the lens; aphakic subjects are able to see easily with these wavelengths, calling the sensation blue or violet. Again, the failure of early workers to evoke sensation with wavelengths greater than 760 nm was due entirely to their inability to obtain intense enough sources, the retina being so insensitive to red light. Griffin, Hubbard and Wald were able to evoke sensation with wavelengths as long as 1000 to 1050 nm, and Brindley has found that, when compared on an equi-energy basis (p. 327), the far reds are much more orange than the near reds; for example, a red of 887 nm could be matched by an orange of 640 nm.

Just as with white light, we may refer to the luminosity of a given hue and measure it in the same units, and although it may seem somewhat unreal to speak of comparative luminosities of widely different hues, by making use of a suitable colour photometer it is possible to use a scale of luminosity which applies to all the spectral colours in such a way that if all the luminosities of the hues of a spectrum are determined separately, the luminosity of the white light obtained by re-combining the spectrum is equal to the sum of these individual spectral luminosities.

### THE FLICKER PHOTOMETER

If it is desired to measure the relative luminosities of two differently coloured sources, the most satisfactory instrument to use is the *flicker photometer*. This is represented diagrammatically in Figure 7.5. It consists of a white screen, A, and a Maltese Cross, C, which is placed at such an angle that light from one of the sources ($S_2$) is reflected down the viewing tube, T. The screen, A, is likewise placed so that light from the other source ($S_1$) is reflected down the tube. When the cross is rotated, the eye receives light alternately from $S_2$ and $S_1$ by reflection from C and A respectively. If the luminances of the

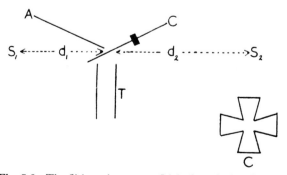

**Fig. 7.5** The flicker-photometer. Light from $S_1$ is reflected from A down the viewing tube, T. Light from $S_2$ is reflected from the rotating cross, C. When A and C are equally illuminated flicker ceases. (After Barton.)

two surfaces are different, a sensation of flicker is experienced which can be made to disappear by adjusting the distances of the two sources from the surfaces which they illuminate. When flicker disappears, the luminances of the two surfaces are equal. If A and C make the same angle with the direction of the incident light, and if their reflection factors are equal, the luminous intensities of $S_1$ and $S_2$ will be in the ratio of $d_2^2 : d_1^2$.

### VARIATION OF HUE WITH INTENSITY

We have seen that hue and saturation are not absolutely independent, since altering the saturation of a yellow, for example, can actually make it into a green. In a similar way, hue is not independent of luminosity. Thus, as the intensity of a coloured light is increased its subjective hue makes a characteristic shift, and ultimately all wave-lengths, if made intense enough, appear the same, evoking a yellowish-white sensation. These shifts are described as the *Bezold-Brücke phenomenon*, after their discoverers. On lowering the intensity there are also characteristic changes which will be discussed later in connection with the Purkinje phenomenon, and ultimately the whole spectrum becomes achromatic, in the sense that at sufficiently low intensity levels all wavelengths appear white and indistinguishable from each other qualitatively. It is therefore important to define the level of luminance at which studies of colour vision are made; the value generally chosen is of the order of 2 to 3 mL.

### COLOUR MIXTURES

If two colours, not too far apart in the spectrum—say a red and a yellow—are mixed, the resulting sensation is that produced by a single spectral hue of intermediate wavelength. The actual position of the resultant hue in the spectrum is determined by the relative luminosities of the two hues that are mixed; if they are equal, the resultant lies midway between; if the red predominates, the resultant lies towards the red end. As the distance apart of the two hues is increased, the resultant hue is found to become pro-gressively unsaturated, i.e. it can only be matched with an intermediate spectral hue provided that white light is added to the latter. Eventually a point is reached when the mixture results in white; any two hues that, when combined, give white light are called *complementary colours*. As a result of the classical researches of Young and Helmholtz and more recently those of Wright, the laws of colour mixture have been put on a quantitative basis. In general, it may be stated that any colour sensation, within an average range of intensities, can be produced by a mixture of not more than three *primary spectral wavelengths*, with the reservations that (1) in many cases white light must be added to the comparison colour to obtain a match in saturation as well as in hue. Thus, if it is required to match a given standard colour, say a bluish green, then by varying the intensities of the three primaries it will usually be found that it is impossible to make a perfect match with this standard, the mixture appearing unsaturated by comparison. We must therefore add white to the standard until the two match or, as it is often put, we must subtract white from the mixture to obtain a perfect match in hue and saturation. The three primaries are chosen so that any one cannot be produced by mixing the other two; thus red, green and blue are primaries.*

A given hue may thus be expressed in terms of three coefficients, namely:

$$\text{Colour} = \alpha R + \beta G + \gamma B$$

where the latter determine the relative quantities of these arbitarily chosen primaries. Light is measured on a luminosity basis in most work, so that it would be natural to measure the contributions of the primaries to any given hue in terms of luminosity units, remembering that the luminosity of the hue will be equal to the sum of the separate luminosities of the primaries required to make a match in hue and luminosity. Since the resultant mixture of the primaries will not necessarily correspond in saturation with the given hue to be matched, it is necessary to subtract a certain quantity of white—that is, certain quantities of the three primaries—and this may lead to negative coefficients. There is no particular theoretical objection to negative coefficients, but they can be avoided by choosing certain hypothetical primaries which do not correspond to any spectral hues. Thus we have seen that a mixture of three spectral primaries gives an unsaturated spectral hue; if the amounts of the primaries which must be subtracted from this unsaturated spectral hue, in order to saturate it, are subtracted from the primaries themselves, we obtain a set of hypothetical primaries which give positive coefficients. Thus Wright used hypothetical primaries, R', G', B', defined by the relationship:

$$R' = 1.015\,R - 0.015\,G$$
$$G' = -0.90\,R + 2.00\,G - 0.10\,B$$
$$B' = -0.05\,G + 1.05\,B$$

where R, G, and B were spectral hues of 650, 530, and 460 nm respectively.

## Trichromatic units

For ease in manipulation, the quantities of the primaries are often expressed in terms of their contributions to colour quality rather than to luminosity; thus the coefficients are defined by the statement that equal quantities of the primaries make white:

White = 0.333 R + 0.333 G + 0.333 B, the units defined in this way being referred to as *trichromatic*

*units*, and since the sum of the coefficients is equal to unity, the equation defining a colour is called a *unit colour equation*. It is a relatively simple matter to transpose from one set of coefficients to another.

### THE CHROMATICITY CHART

The unit colour equation is useful because it enables us to express graphically the composition of any hue in terms of the primaries; thus if we construct retangular co-ordinates so that the ordinates represent green coefficients and the abscissae red coefficients, then any hue will be given by a point representing its red and green coefficients; since $\alpha + \beta + \gamma = 1$, the blue coefficient will be given by subtracting the red and green coefficients from unity. The locus of spectral hues will be a line (Fig. 7.6) and white, if defined by the equation above, will be represented by a point whose ordinate and abscissa are 0.333.

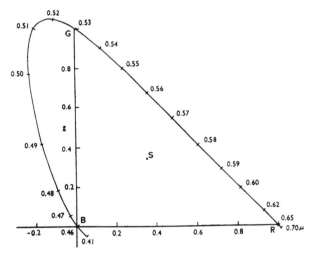

**Fig. 7.6**  Chromaticity chart. Units of green are indicated on the vertical line; units of red on the horizontal line. The amount of blue in a given mixture is given by the unit colour equation, R + G + B = 1. The line joining R to G and continuing as a curved loop to B is called the *spectral locus* and shows how the individual spectral hues can be matched by the three primaries. S is standard white. (Wright, *Researches on Normal and Defective Colour Vision*.)

* There are several ways of expressing the 'Laws of Colour Mixture'. A common one is as follows: given any four lights, it is possible to place three of them in one half of a photometric field and one in the other half, or else two in one half and two in the other, and by adjusting the intensities of three of the four lights to make the two halves of the field indistinguishable to the eye. This appears different from the statement in the text, but is actually the same in so far as it is another statement of the *trivariance* of colour matching, the three variables being the intensities of the three colours that are varied, the fourth remaining constant. The laws governing colour-matching are called *Grassmann's Laws*.

## FORM

The form or structure of the visual field is determined by the power to discriminate between two or more separate stimuli in regard to intensity and spatial extension; we perceive the shape of an object by virtue of our ability to integrate our responses to a large number of separate stimuli of varying intensity arising from different positions on its surface.

## RESOLVING POWER AND VISUAL ACUITY

The power of the eye to discriminate on a spatial basis is studied experimentally by measurements of the *resolving power*, the smallest angle subtended at the nodal point of the eye by two points or lines such that they are appreciated as separate, i.e. the so-called *minimum separable*. The visual acuity is defined as the reciprocal of this angle when it is measured in minutes of arc. Thus, under good experimental conditions, resolving powers of the order of 30 seconds of arc, or half a minute, may be measured, corresponding to a visual acuity of $1/0.5 = 2$.

### MEASUREMENT

The most commonly used test-objects for measuring resolving power are the Landolt-C and the grating (Fig. 7.7). The Landolt-C can be rotated into eight

**Fig. 7.7**  Snellen letter, grating and Landolt-C. In each case the angle subtended at the eye by the distance, *s*, apparently determines whether the gap will be recognized.

separate positions which the subject is asked to identify; C's of diminishing size are presented and the errors made in indicating the positions are recorded. Experimentally it is found, as with all psychophysical studies of thresholds, that there is no sharp division between the size of C that can be discriminated and that which cannot; near the 'end-point' we will have a range of sizes where the average number of errors in any set of trials increases with decreasing size of C until eventually the answers are purely random. We must therefore somewhat arbitrarily choose an end-point; Lythgoe chose that size of C with which 4.5 out of eight answers were correct, and the actual end-point was computed from the errors obtained with several sizes of C by the application of statistical theory. So far as

the resolving power of the eye is concerned, it would appear that the ability to detect the position of the break in the C, which distinguishes it from a complete circle, depends on the size of the break, *s*; in fact, the angle subtended at the eye by this distance, *s*, at the end-point, is indeed the measure of the resolving power of the eye. Thus at a distance of 6 metres, under given conditions, the size of the break, *s*, was 3.6 mm, at the end-point; this distance subtended 2 minutes at the eye and so the visual acuity was 0.5.

### Grating acuity
If the grating of Figure 7.7, consisting of equally spaced black and white lines, is removed sufficiently far from the observer, it appears uniformly grey; at a certain point the lines can be just discriminated and the angle subtended at the eye by the distance *s* once again is said to indicate the resolving power of the eye. According to Shlaer (1937) the limiting size for resolution subtends some 28 seconds of arc at the eye under optimal conditions, corresponding to an acuity of 2.12.

### Snellen test
The clinical *Snellen test* is essentially a resolution test; it is assumed that the average individual can resolve the details of a letter if they subtend one minute of arc at the eye, and the test letters are constructed so that their details (e.g. the gap in a C) subtend this angle, whilst the whole letter subtends five minutes of arc (Fig. 7.7), when they are at a definite distance from the eye. The chart has several rows of these letters, the sizes in the different rows being graded so that their details subtend one minute at different distances, e.g. the smallest letters subtend this angle at 4 metres whilst the largest do so at 60 metres. The subject is placed 6 metres from the chart and the smallest row of letters that can be read correctly is determined. The visual acuity is expressed as the ratio of the distance of the individual from the chart to the distance at which this row of minimum legible letters should be read, assuming a minimum visual angle of one minute. Thus at 6 metres if the individual can read down to the letters whose details subtend one minute when viewed from a distance of 18 metres his visual acuity is 6/18; if he can read letters so small that they subtend the same angle at only 5 metres his visual acuity is 6/5, and so on.

### EXPERIMENTAL RESULTS

Factors influencing visual acuity will be dealt with in detail later; for the present we need only note that it is markedly influenced by the illumination of the test object, the state of adaptation of the eye, and the pupil size. Under optimal conditions it has a value in the region between 2 and 3 (i.e. a minimum separable of about 30 to 20 seconds of arc) when a grating or Landolt-C is used as test-object.

### Visual acuity in animals
Visual acuity can be measured in animals by training them to choose between alternate pathways marked with

a grating or Landolt-C of variable visual angle; selection of the required pathway is rewarded with food. When the choice becomes random the animal is unable to resolve the grating or Landolt-C. Farrer and Graham (1967) found a limiting resolution of 0·5–1·0 minutes of arc in a monkey, indicating a visual acuity as good as that of man; in the rabbit, Van Hof (1967) found values between 10 and 20 minutes, and this corresponds with the absence of a fovea in this animal.

In the cat, Jacobson *et al.* (1976), employing an avoidance technique, found resolving powers as good as 3·5 minutes of arc, comparing with about 5 minutes in Smith's (1936) study. It is possible that the eagle's acuity is, indeed, better than man's (Schlaer, 1972).

## INDUCTION

The effect of light falling on a given portion of the retina is generally not confined to the retinal elements stimulated; moreover, the characteristics of those elements that have been stimulated are altered for an appreciable time, so that their response to a second stimulus is modified. The influence of one part of the retina on another is described as *spatial induction*; by *temporal induction* is meant the effect of a primary stimulus on the response of the retina to a succeeding stimulus.

### SPATIAL INDUCTION

Spatial induction, or *simultaneous contrast*, can be demonstrated by comparing the appearance of a grey square, when seen against a black background, with its appearance when seen against a white background; against the black background the grey appears almost white, whereas it appears quite dark when seen against the white background. The white surround apparently depresses the sensitivity of the whole retina, thereby making the grey appear darker; the net effect of this depression in sensitivity is to accentuate the contrast between the grey and the white surfaces. Where coloured lights are concerned the same principle of a reduction in sensitivity of the retina applies, but now this reduction is confined to certain wavebands, and the subjective result is to enhance the appearance of the complementary colour. Thus, if a white and blue light patch are placed side-by-side, the white patch appears yellow; apparently the blue light falling on one half of the retina has depressed the sensitivity of the whole retina to blue light; the white light falling on the retina is now losing some of its 'blueness' because of this loss of blue-sensitivity; removing blue from white leaves a yellow colour, since blue and yellow are complementaries.

## TEMPORAL INDUCTION

In a similar way, if the whole retina is stimulated with white light the sensitivity to a given stimulus following this is reduced, so that the second stimulus appears less bright than it would have done had there been no 'conditioning stimulus', as this preliminary, adapting, stimulus is called. This form of induction, which has been called *light-adaptation*, has been studied in detail by Wright in England, and Schouten in Holland.

### BINOCULAR MATCHING

The technique employed was that of binocular matching, based on the principle that the states of adaptation of the two eyes are independent. Thus the observer looks through a binocular eye-piece, each eye seeing the half of a circular patch of light, and these halves have to be matched for equal subjective brightness. If, say, the right eye is exposed to a bright light, we may determine the changes in sensitivity of this eye during its exposure to the bright light by matching its half-field with the left half-field, as seen by the dark-adapted eye. This means varying the intensity of the left half-field until the match is made. It is found that the intensity of the left half-field has to be continuously reduced during exposure of the right eye, until it levels off at a constant value which may be about one-sixteenth of the value at first. This means that the right eye

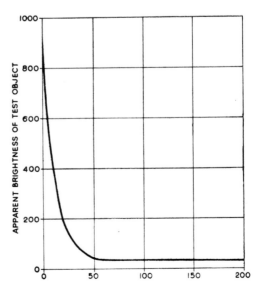

**Fig. 7.8**   Curve showing the fall in sensitivity of the fovea during exposure to a bright light. The right eye views the bright light whilst the dark-adapted left eye views a patch that must be matched in brightness with the sensation experienced through the right eye. The matching intensity decreases, indicating that the apparent brightness of the light, as seen through the right eye, becomes less and less. (Wright, *Trans. Illum. Eng. Soc.*)

becomes rapidly less sensitive to light from the moment it is exposed to it; this is the phenomenon of light-adaptation and is illustrated by Figure 7.8. The phenomenon is, of course reversible, so that if the two eyes are now allowed to match fields of moderate luminance, the left field will, at first, have to be made much weaker than the right because the left eye has remained dark-adapted and has remained the more sensitive; very rapidly, however, this matching field will be made stronger until they finally become equal when both eyes are completely dark-adapted.*

PRACTICAL DEMONSTRATION

This adaptation is very easy to demonstrate qualitatively; thus if an observer is instructed to stare at a brilliantly illuminated screen for a few minutes with one eye closed, and then this is replaced quickly by one of moderate intensity, he finds, on comparing the subjective brightness of the screen as seen through either eye separately, that the eye that had been kept closed sees the screen as very much brighter. We may note here that the phenomenon of *dark-adaptation*, as revealed by the progressive decrease in the absolute threshold, is a manifestation of temporal induction, in so far as the phenomenon reveals a change in the retina following the cessation of a stimulus, namely of the original bright light to which the observer was exposed before being placed in the dark.

ADAPTATION TO COLOURED LIGHTS

The same process of light-adaptation, or successive contrast, is seen with respect to colours; thus with the same binocular matching technique, except that the two fields must be matched for colour rather than subjective brightness, the effects of adapting to a given wavelength of light can be found. The same principle emerges, namely that adaptation of the one eye to, say, green, reduces the sensitivity of this eye to this wavelength, and in order that the other eye may make a match it has to employ much less green than before. Thus two half-fields may be made to match, the right eye viewing a yellow and the left a mixture of red and green, their intensities being adjusted to give the match with the right eye. The right eye is now exposed to an adapting green light for a few minutes and it is found

that, on viewing the two fields, the match no longer holds, the right field appearing reddish by comparison with the yellow of the left field. The left field must now be made less green to make the match, and the extent to which the green must be reduced in intensity in this field is a measure of the reduction of sensitivity of the right eye to green.

AFTER-IMAGES

When the eye is exposed to a brief flash of light the subjective sensations last for much longer than the stimulus, and are described as a succession of *after-images*. The phenomena may be classed as inductive, in so far as they reveal changes in the retina or its higher pathways following a primary stimulus; although they have attracted a lot of interest, little of fundamental value has emerged from their study and it would be a pity to spend much space on them. As classically described, the responses to a single flash are illustrated schematically in Figure 7.9 where upward movements indicate sensations of brightness (positive after-images) and downward movements sensations of blackness (negative after-images). If a coloured primary stimulus is used, some of the after-images are coloured, often with the complementary. The events depicted by Figure 7.9 occur rapidly and are exceedingly difficult to resolve, except by such devices as Bidwell's disc, illustrated by Figure 7.10. Here a disc, with a narrow slit in it, is rotating in front of a bright opal screen, and the after-images are projected on to the black disc to give the appearance of *Bidwell's ghost*. A more slowly developing and longer-lasting after-image is that observed after staring for a few minutes at a bright source and then transferring the gaze to a dimly lit screen. In a little while a 'negative' image of the bright source appears; if the source was coloured this has the complementary colour, e.g. magenta after looking at green; yellow after looking at blue, and so on.

---

*These studies of light-adaptation and their reversal are made with foveal vision and are essentially studies of cone vision, i.e. although comparisons are made with a dark-adapted eye, the intensities of the half-fields to be compared are always maintained well above cone-thresholds.

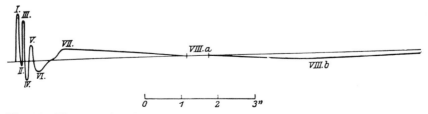

**Fig. 7.9**  The successive after-images said to be experienced after viewing a bright flash. Time-scale in seconds. (Tschermak, in *Handbuch der Normalen und Pathologischen Physiologie.*)

**Fig. 7.10** Bidwell's ghost. A black disc with a sector cut out rotates in front of a brightly illuminated opal screen. The after-images of the sector are projected to the left to give the appearance shown. (McDougall, *Brit. J. Psychol.*)

## INTERMEDIATE EXPOSURES

The after-effects of exposures intermediate in duration between the brief flash and the lengthy gaze have been studied under well controlled conditions by Trezona. Thus, with an exposure of the order of 8 seconds, the after-effects may be classified as follows:

1. *Persistence of vision*, this being the apparent continuation of the stimulus after it has been cut off, and lasting about 1 second at rapidly diminishing subjective brightness.
2. *After-colours*; these may last for seconds or minutes, changing colour frequently. The threshold for these after-colours was very high, of the order of 100 000 trolands of retinal illumination.
3. *After-blueness*; this was a characteristically different effect from the after-colours mentioned above; it lasts for some seconds and remains the same colour. The threshold was much lower, of the order of 100 to 1000 trolands of retinal illumination, and the effect could only be obtained with peripheral vision, suggesting mediation by the rods.
4. *After-blackness*. This was favoured by long exposures, and consisted of several appearances of the black image of the source.

### LIMINAL BRIGHTNESS INCREMENT

If a subject was exposed to pairs of bright flashes in the two halves of a test-field, the after-images could be distinguished if the intensities of the flashes differed, the Fechner-fraction, $\Delta I/I$, comparing with that found for direct vision; this was in spite of the fact that with direct vision the two halves of the field could not be distinguished because of the extreme brightness and shortness of the flashes. When the intensities of the flashes became too high, however, there was a fall in discrimination, i.e. a rise in the Fechner-fraction, and this was probably due to excessive bleaching of the cones (Brindley, 1959).

### PERIPHERAL ORIGIN

This finding, together with the study of the applicability of Bloch's Law (p. 241), strongly suggests that after-images have a peripheral origin, i.e. are determined by photochemical events in the receptors. Support for this is given by Craik's experiment, in which the eye was temporarily blinded by exerting pressure on the globe. On exposure of the eye to a bright light nothing was seen, but on restoring the circulation an after-image appeared. The possibility that some of the phases have a central origin must not be ruled out, however; thus Brindley was unable to find any regularity in behaviour of his after-images unless he ignored those appearing within the first 15 seconds; after this time they conformed to rules predictable on photochemical grounds (p. 241).

## VALUE OF CONTRAST

Simultaneous and successive contrast are essentially manifestations of an inductive process that tends to reduce the sensitivity of the retina to an existing stimulus and to enhance its sensitivity to a stimulus of a different kind, e.g. the sensitivity to the complementary of the existing, coloured, stimulus. As a result of simultaneous contrast the contours of an image become better defined and, as a result of successive contrast, the confusion that would result from the mixing of successively presented images is reduced.

## REFERENCES

Brindley, G. S. (1955) The colour of light of very long wavelength. *J. Physiol.* **130**, 35–44.

Brindley, G. S. (1959) The discrimination of after-images. *J. Physiol.* **147**, 194–203.

Cornu, L. & Harlay, F. (1969) Modifications de la discrimination chromatique en fonction de l'éclairement. *Vision Res.* **9**, 1273–1287.

Farrer, D. N. & Graham, E. S. (1967) Visual acuity in monkeys. *Vision Res.* **7**, 743–747.

Griffin, D. R., Hubbard, R. & Wald, G. (1947) The sensitivity of the human eye to infra-red radiation. *J. opt. Soc. Amer.* **37**, 546–554.

Jacobson, S. G., Franklin, K. B. J. & McDonald, W. I. (1976) Visual acuity of the cat. *Vision Res.* **16**, 1141–1143.

Kohlrausch, A. (1931) Tagessehen, Dämmersehen, Adaptation. *Handbuch der normalen und pathologischen Physiologie.* Bd. XII/2, pp. 1499–1594. Berlin: Springer.

Lythgoe, R. J. (1932) The measurement of visual acuity. *M.R.C. Sp. Rep. Ser.*, No. 173.

McDougall, W. (1904) The sensations excited by a single momentary stimulation of the eye. *Brit. J. Psychol.* **1**, 78–113.

Schouten, J. F. & Ornstein, L. S. (1939) Measurement of direct and indirect adaptation by means of a binocular method. *J. opt. Soc. Amer.* **29**, 168–182.

Shlaer, S. (1937) The relation between visual acuity and illumination. *J. gen. Physiol.*, **21**, 165–188.

Trezona, P. (1960) The after-effects of a white light stimulus. *J. Physiol.* **150**, 67–78.

Tschermak, A. (1929) Licht- und Farbensinn. *Handbuch der normalen und pathologischen Physiologie*. Bd. XII/1, pp. 295–499. Berlin: Springer.

Van Hof, M. W. (1967) Visual acuity in the rabbit. *Vision Res.* **7**, 749–751.

Wald, G. (1945) Human vision and the spectrum. *Science* **101**, 653–658.

Wright, W. D. (1939) The response of the eye to light in relation to measurement of subjective brightness and contrast. *Trans. Illum. Eng. Soc.* **4**, 1–8.

# 8. Photochemical aspects of vision

The receptors that mediate vision at night are the rods which, as mentioned earlier, contain a pigment—*visual purple* or *rhodopsin*—responsible for the absorption of light in the primary photochemical event leading to sensation. According to the photochemist, then, the visual process consists of the absorption of light by a specialized molecule, visual purple or rhodopsin. The absorption of light provides the rhodopsin molecule with a supply of extra energy and it is said, in this state, to be 'activated'. In this activated state it is highly unstable and so it will change to a new form, i.e. the molecule will undergo some kind of chemical change by virtue of this absorption of energy. The effects of this change will be to cause an 'excited' condition of the rod as a whole and it will be this excited condition that will ultimately lead to the sensation of light.

## PHOTOCHEMICAL AND THERMAL CHANGES

As we shall see, the effects of light-absorption on the rhodopsin molecule are complex because the new molecule formed is not stable but immediately undergoes a succession of changes, which are called *thermal reactions* to distinguish them from the primary *photochemical* change. From an economic point of view we may expect the photochemical and other changes taking place in the retina to be at least partly reversible, otherwise enormous quantities of rhodopsin would have to be synthesized continuously to keep pace with the material destroyed by light.

### REGENERATION

In fact, the early investigators who observed the bleaching of visual purple in the retina also observed that it would regenerate if the intact retina was allowed to remain in the dark. Subsequent studies on solutions of rhodopsin have indeed shown that, provided the necessary enzymes are present, the photochemical and subsequent chemical changes are largely reversible. Tentatively, then, the visual process may be represented schematically as follows:

Rhodopsin ⇌ Activated State ⇌ Bleached Products

As we shall see, the 'bleached products' are the results of a highly complex series of changes, so that the rhodopsin molecule alters its 'colour', i.e. its light-absorbing properties, from the original magenta through orange to yellow and ultimately to white, when it has become vitamin A and a protein, *opsin*. In general, when the retina has been kept away from light for some time— i.e. when it has been completely dark-adapted—we may expect to find only rhodopsin in the rods, whilst when the retina is being exposed to light we may expect to find a mixture of rhodopsin with many of its 'bleached products', the relative amounts of each product being determined by the intensity of light and the duration of exposure.

## ULTRASTRUCTURE OF THE RODS

### OUTER SEGMENT

Let us consider the structure of the rod in more detail, as revealed by the electron-microscope. As Figure 5.3, p. 169, showed, the rod may be divided into outer and inner segments; since rhodopsin has been shown to be confined to the *outer segment*, the structure of this region is of special interest. Figure 8.1 shows that this part consists essentially of a pile of flattened sacs contained within the limiting plasma membrane, as illustrated schematically by Figure 8.2. These sacs are formed by deep invaginations of the plasma membrane of the outer segment, the invaginations being ultimately nipped off so that the membranous sac is free in the cytoplasm of the outer segment. As indicated in Figure 8.3, the invaginations at the base of the outer segment remain continuous with the plasma membrane. When outer segments are treated with trypsin the outer plasma membrane of the segment is digested, liberating the intact discs (Trayhurn and Habgood, 1975).

### INNER SEGMENT

The structure of the inner segment is quite different and is comparable with that of many cells, having a nucleus, Golgi apparatus, mitochondria and ribosomes. This part of the rod is thus concerned with the energy-requiring metabolic events, whilst the outer segment is concerned with the absorption of light, since it may be shown, by

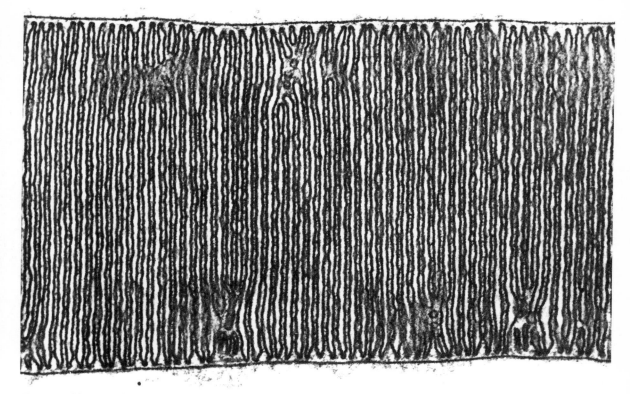

**Fig. 8.1** Electron micrograph of a section of rod outer segment of human retina showing the parallel arrangement of disc lamellae. Osmium fixed. ( × 64 000) (Courtesy, R. C. Tripathi.)

separating these from the body of the rod, that they contain the light-absorbing pigment, rhodopsin. The outer segment is connected to the inner segment by a cilium (Fig. 8.2B) which suggests that the rod has developed from a primitive ciliated photosensitive cell.

### CILIUM

The cilium has been examined in some detail by Richardson (1969) who has shown that, in the mammalian rod, it is not the sole connecting link between inner and outer segments; in addition there is a bridge of cytoplasm, and it is through this narrow neck that the products of synthesis within the inner segment may be conveyed to the outer segment.

## Protein synthesis

Thus studies involving radioautography carried out mainly in Young's laboratory (e.g., Young, 1976) have shown that, when tritiated amino acid is incorporated into the rod, it enters into protein synthesis within the myoid region and, within a short time, the radioactive protein may be seen migrating through the connecting cilium and ultimately appearing in the outer segment.

### CHEMICAL COMPOSITION

The outer segments of rods may be separated easily from the parent rods, and suspensions of these may be purified and analysed. Approximately 60 per cent of the separated membrane material is protein and 40 per cent is lipid, mainly phospholipid. Of the protein about 80 per cent is the visual pigment rhodopsin which, as we shall see, is a chromolipoprotein with retinal linked to it as a chromophore group.

### MEMBRANE STRUCTURE

The basic structure of the plasma membrane is essentially that postulated by Danielli and Davson (1935), and illustrated schematically in Figure 8.4, the bilayer of lipid molecules being orientated with their polar groups facing the watery phase and their hydrophobic tails orientated away from the watery phase and attracted to each other. Attached to each outer layer there is a layer of protein molecules.

### RHODOPSIN

In the case of the discs, the protein attached to the membrane is rhodopsin and X-ray studies of Blasie and Worthington (1969) have indicated that these rhodopsin molecules are arranged in a liquid-like array in the

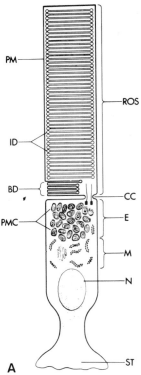

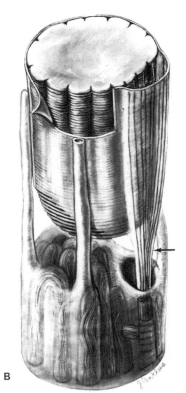

**Fig. 8.2** (A) Diagram of frog red rod photoreceptor. ROS, rod outer segment; PM, outer segment plasma membrane; ID, isolated outer segment discs; BD, basal outer segment discs which are continuous with the plasma membrane; CC, connecting cilium; E, mitochondria-rich ellipsoid; PMC, perimitochondrial cytoplasm; M, myoid region containing endoplasmic reticulum and Golgi apparatus; N, nucleus; ST, synaptic terminal (Basinger *et al.*, *J. Cell Biol.*) (B) Three-dimensional illustration of junction between inner and outer segments of rod. Note scalloped appearance of rod discs. (Young, *Invest. Ophthal.*)

plane of the disc membrane; the X-ray diffraction data indicate an asymmetry, suggesting that the rhodopsin is located only on the inward facing side of the lipid bilayer (Worthington, 1973) (Fig. 8.5). There is reason to believe that the rod plasma membrane, and hence that of the discs, is unusually fluid so that the rhodopsin molecules are not anchored firmly in the lipid bilayer but diffuse readily. This was shown very neatly by Poo and Cone (1973) who measured the change in absorbance of light by single rods when bleached from one side only. At first there was no change in absorbance of the side not exposed to light, but after a time this decreased, indicating diffusion of bleached rhodopsin from the exposed to the non-exposed side of the rod. By contrast, there was no diffusion in a longitudinal direction along the outer segment, and this is understandable since it is well established from Young's work that a rhodopsin molecule, once incorporated in a sac, never leaves this until the sac is finally discarded at the tip of the segment. Since rhodopsin may diffuse within the membrane-material, we may expect it to appear in the outer limiting plasma membrane of the outer segment, diffusing from the innermost set of discs which,

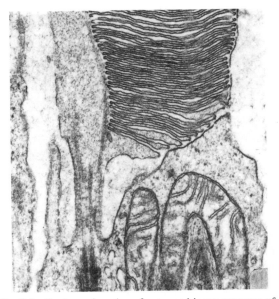

**Fig. 8.3** Region at junction of outer and inner segments of rod. Note that the invaginations at the base of the outer segment remain continuous with the plasma membrane. (Young, *Invest. Ophthal.*)

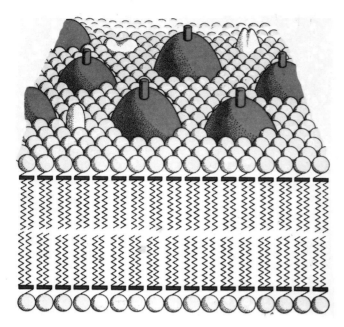

**Fig. 8.4**  Simplified model of rod outer segment membrane. The basic Danielli-Davson bilayer with its layers of protein on each side constitutes the essential structural element. A perspective view of large rhodopsin molecules, partly immersed in the bilayer, is also shown. The small cylindrical stem attached to each opsin molecule represents its carbohydrate chain. (Young, *Invest. Ophthal.*)

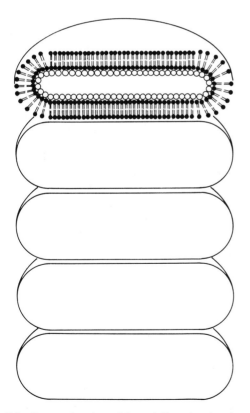

**Fig. 8.5**  Perspective view of the rod discs showing that the rhodopsin molecules are attached to the layer facing the inside of the hollow disc.

as we have seen, are continuous with the outer membrane (Fig. 8.2).

## Leucine labelling

When Basinger *et al.* (1976) labelled rods with tritiated leucine, they found heavy activity in the plasma membrane as well as the basal discs—this could very clearly be seen when, during fixation, the membrane became separated widely from the discs. When the radioactive material was separated and analysed it was over 90 per cent rhodopsin.* We shall see that the ultimate effect of light-absorption on the receptor is a change in electrical potential along its length; and it may be that the rhodopsin in the outer membrane, or in the most basal discs with which it is continuous, is important for some phase of the electrical change (p. 255).

### BIREFRINGENCE

The arrangement of the discs in the rod imparts a form-birefringence to the outer segment, revealed as a difference in refractive index according as the direction of the electric vector of the plane-polarized light is parallel with, or at right-angles to, the long axis of the rod (Fig. 8.6). This *form-birefringence* depends on the difference of refractive index of the disc and

---

*Papermaster *et al.* (1975) have followed the synthesis of rhodopsin and its migration from inner segment to outer segment; at no time in its life was the rhodopsin unattached to membrane, so that the transport could have occurred by movement through the plasma membrane of the rod, or by movement through the cytoplasm in the form of membrane-bound photopigment.

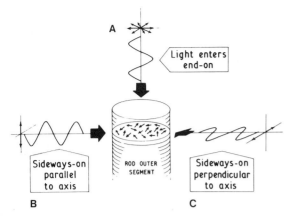

**Fig. 8.6** Diagram showing the different ways in which polarized light can pass through a receptor outer segment, and its reaction with the absorption vectors of the pigment molecules (arrows) lying in a particular membrane. (A) Light passing down the receptor is absorbed equally whatever the orientation of its plane of polarization. (B) Light entering the side of the receptor with its plane of polarization *parallel* to the long axis of the receptor is perpendicular to the plane containing the absorption vectors, and is not absorbed. (C) If the plane of polarization is *perpendicular* to the axis, it will also lie in the plane containing the absorption vectors and is strongly absorbed. (Knowles and Dartnall, *The Eye*.)

the cytoplasm in which it is suspended, and may be abolished by soaking the rod in a solution of refractive index equal to that of the discs. Because the molecules within individual discs are arranged in an ordered fashion, the long lipid molecules being orientated at right-angles to the plane of the disc, i.e. along the long axis of the rod, the outer segment is said to have, in addition, an *intrinsic birefringence*, and this will remain after abolishing the form-birefringence as described above. The measured birefringence will be the algebraic sum of the negative form- and the positive intrinsic birefringences. Liebman *et al.* (1974) found that, for frog outer segments, the refractive index for light incident side-on, with electric vector parallel to the cell axis, $n_{\parallel}$, exceeded that for light polarized perpendicular to this axis, $n_{\perp}$. Thus for wavelengths greater than 650 nm, the outer segments immersed in saline had a net positive birefringence, $\Delta n = n_{\parallel} - n_{\perp}$ of $+0.0010$. The intrinsic component was $+0.0050$ and the form $-0.0040$. The measurement becomes of interest when attempting to ascertain the structural consequences of the absorption of light (p. 212) by the rod.

### DICHROISM

Because the light-absorbing pigment molecules are orientated with their light-absorbing regions predominantly in the plane of the sacs, the absorption of plane-polarized light will be greater when its electric vector is perpendicular to the axis of the rod than when parallel with this, giving rise to what is called *dichroism*. This dichroism was demonstrated by Denton (1959), and when the magnitude of the dichroism, as measured by the optical density difference for light parallel to, and at right angles to, the planes of the discs, was plotted against the wavelength of the light, the curve corresponded with the total retinal optical density as a function of wavelength, indicating that most of the light-absorbing material was, indeed,

responsible for the dichroism. When bleaching had been maintained for some time, to give retinol as the final product, the sense of the dichroism reversed, so that it would seem that the retinol (vitamin A) molecules, liberated from rhodopsin, remained in an orientated position but with their axes sticking out of the plane of the disc. A similar dichroism has been measured in single goldfish cones by microspectrophotometry (p. 341), the density for side-illumination of the rods being 2 to 3 times greater when the electric vector of the light was in the plane of the discs than when it was at right-angles (Harosi and MacNichol, 1974).

### TURNOVER OF THE DISCS

In embryonic development, the formation of discs proceeds from the base, so that new-formed discs are added successively from below, displacing the first-formed discs towards the apex of the segment (Nilsson, 1964). Since the earliest-formed discs are smaller than those formed later, this presumably accounts for the conical shape of many outer segments—e.g. in the cones of most species—but the rod-like shape of the rod outer segments, in which the discs are of equal size, suggests a subsequent modification. In fact Young's studies on the incorporation of radioactive amino acids and other compounds involved in synthesis of the discs have indicated that the structure of the rod is not permanent, new discs being formed continuously throughout life at the base of the outer segment, whilst the oldest are shed at the apex by a process of nipping off, the 'autophaged'

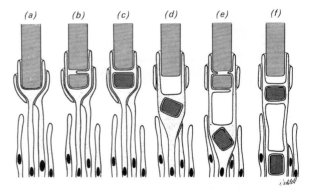

**Fig. 8.7** Phagocytosis. Long villus-like pigment epithelial apical processes reach toward the cone from the pigment epithelium. The processes that reach the cone (two are arbitrarily shown) expand to ensheath a portion of the outer segment (*a*). The sequence (*a*)–(*f*) shows the successive phagocytosis of two packets of discs by one apical process. It is assumed that separation of the packet and resealing of outer-segment plasma membrane occurred *before* phagocytosis, (*b*) and (*e*). The forms assumed by apical process pseudopodia —(*b*) and (*e*)—were arbitrarily selected to illustrate several possibilities. The discs in the packets are shown becoming more osmiophilic than those in the outer segment, after they have been phagocytosed. Notice that as the first phagosome ascends in the process, (*d*)→(*f*), the sheath-like region of the process lengthens. Phagocytosis of the second packet of discs might also occur by a different process which, for some reason, displaces the first process. (Steinberg *et al.*, *Phil. Trans.* Drawn by J. Weddell.)

discs being subsequently phagocytosed by the pigment epithelium as illustrated in Figure 8.7. This renewal of discs was recognized by the migration of radioactive material along the length of the outer segment as a dense line (Fig. 8.8) representing those discs that had

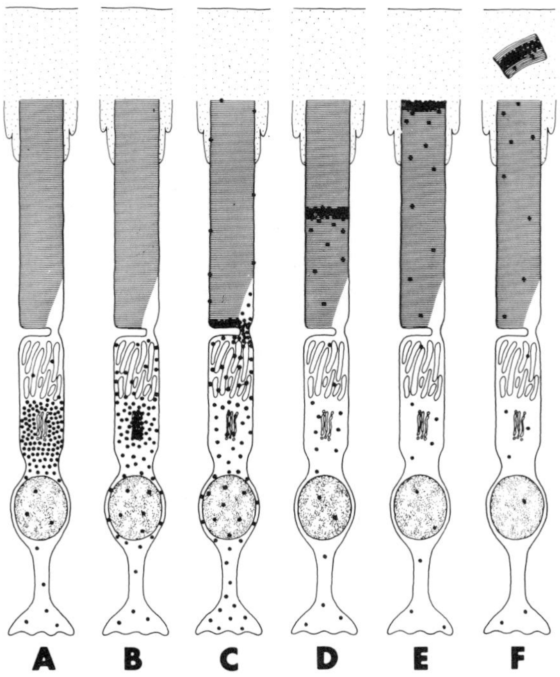

**Fig. 8.8** Diagram illustrating the renewal of protein in rod visual cells as it is revealed in autoradiograms after administration of radioactive amino acids. New protein is at first concentrated at its major site of synthesis, the myoid zone of the inner segment (A). The protein molecules then scatter throughout the cell, many of them migrating by way of the Golgi complex, where they are modified by the addition of carbohydrate (B). Much of the new protein traverses the connecting cilium, and is incorporated into growing membranes at the base of the outer segment. Some of the protein diffuses from the new membranes into the outer cell membrane with which they are continuous (C). Detachment of the disc-shaped double membranes from the outer membrane traps labelled protein within the discs. Repeated formation of new membranes displaces the labelled discs along the outer segment (D). Eventually, they reach the end of the cell (E) from which they are shed in small packets. These are phagocytosed by the pigment epithelium (F). (Young, *Invest. Ophthal.*)

incorporated radioactive amino acid during the period of exposure to this (Young and Droz, 1968). A similar migration of radioactive material was observed when radioactive glycerol was incorporated; this molecule constitutes the skeleton of the phospholipids that go to make up the plasma membrane and discs; and, as with protein, the radioactive glycerol became concentrated

first in the myoid. Young and Droz calculated that the discs were replaced at a rate of some 25 to 36 per day according to the species.

**Turnover of retinal**
When H-labelled retinal is injected into dark-adapted rats and the animals are maintained in the dark, thereby preventing any photic separation of retinal from its opsin linkage, it is found that the retinal that is subsequently extracted from the bleached retina becomes radioactive, increasing in amount for as long as 14 days. This 'turnover', manifest as an incorporation of exogenous retinal into the rhodopsin, is probably a reflexion of the renewal of rod discs, which requires, of course, resynthesis of rhodopsin (Bridges and Yoshikami, 1969).

## PIGMENT EPITHELIUM-RECEPTOR RELATIONS

The pigment epithelium is morphologically very intimately related to the photoreceptors. Thus, as described by Young (1971) in the monkey, the surface of the epithelium, facing the outer segments, presents a fringe of delicate cytoplasmic processes, and the rod outer segment is inserted deeply into this meshwork. At or near their origin from the cell body the pigment epithelial processes form an irregularly fitting cap of flimsy cytoplasm presenting a sort of socket into which the end of the outer segment is loosely fitted. The two cells—receptor and epithelial—are closely apposed but not apparently sealed by any junctional complexes. In the human retina the epithelial processes are of two types, thin slender and broad rampart-like villous extensions encompassing the tips of the outer segments to a depth of about 4 μm (Spitznas and Hogan, 1970).

### PHAGOCYTOSIS

The process of nipping off and subsequent phagocytosis of the rod discs has been followed in the electron microscope; a group of discs at the tip curls or rolls up at the edges, arching in such a way that the convexity faces the remaining rods; the rod cell membrane folds into the zone of separation so formed and finally dissects, or nips off, the packet of discs which remains in the extracellular space wrapped in the outer-segment membrane. Later the packet is seen within the cytoplasm of a pigment epithelial cell as a phagosome. Each pigment epithelial cell engulfs some 2000 to 4000 discs daily (Young, 1971). The process is illustrated schematically in Figure 8.7 taken from a later study by Steinberg *et al.* (1977) on human cones.

**Diurnal burst of disc shedding**
La Vail (1976) noted the paucity of phagosomes seen in pigment epithelial cells when regard is had to the rapid rate of disc shedding described by Young, and he considered that this might be due to a diurnal variation in rate of shedding. In fact he observed a burst of

shedding, with consequent large numbers of phagosomes, some half hour after the onset of light; within a
few hours the rate of shedding and the number of
phagosomes observed had fallen to a low value.

## ULTRASTRUCTURE OF THE CONES

These are built on the same plan, being made up of
an outer and inner segment (Fig. 8.9) but in the cone
the sacs are not separated from the plasma membrane,
as they are in rods, but remain continuous with this
and are thus analogous with the most basal sacs in the
rod, remaining, in essence, deep invaginations of the
plasma membrane. Development of the outer segment
follows the same pattern as that described for the rod,
but each invaginated sac moves outwards by a shift of
the point of invagination rather than by a movement
of the isolated sac (Nilsson, 1964).

### DISC RENEWAL

When Young applied his autoradiographic technique to
the study of turnover of the cone sacs he found a very

**Fig. 8.9**  Electron micrographs of outer segments of rhesus monkey cone (*left*) and
rod (*right*). ( × 40 000; Courtesy of Young.)

different situation; the labelled amino acid never became concentrated in a band that migrated along the length of the cone outer segment; instead, the radioactivity became diffusely distributed throughout the outer segment. It was surmised, therefore, that the structure of the cone outer segment was more permanent, the incorporation of radioactivity into it representing a renewal of membrane material, including photopigment, in a more piecemeal fashion, old molecules being cast off and replaced with new throughout the whole structure, a process that Young called *molecular replacement* by contrast with *membrane replacement* in the rods.

## Phagocytosis

Nevertheless, a number of studies have demonstrated phagocytosis of complete discs derived from cones; thus Hogan and Wood (1974) showed that human cones obviously shed bundles of discs that were enveloped in the plasma membrane of the cone outer segment and, outside this, the membrane of phagosome. Again Anderson and Fisher (1976) have shown that, in several species of squirrel retina, a portion of the outer segment contains, beyond the midpoint, discs that are free-floating in the cytoplasm, as with rods, and in the electron microscope there was clear evidence of a nipping off of bundles of discs by processes from pigment epithelial cells (Fig. 8.10); thus, since there is only diffuse labelling of the outer segment, the absence of a sharp band of labelling is no evidence for the absence of continual replacement of discs.

Finally, Steinberg *et al.* (1977), working on the human retina, observed phagocytosis of cone discs; the process began, as with rods, in a curling of the discs; the packet could be seen still enclosed in plasma membrane of the outer segment, but this membrane pinched at one side. The apical pigment epithelial processes extended along both sides of the outer segment but gave no indication that they might soon phagocytose the terminal packet. In other preparations, however, it appeared as though the long apical processes were sending in pseudopods that separated a bundle of terminal discs from the rest, but this process might have been preceded by a pinching off by the outer segment's plasma membrane. Since these processes occurred in eyes spanning a wide age it is unlikely that the phagocytosis was a pathological phenomenon.

Thus diffuse labelling of the discs seems not to be an absolute criterion for determining whether there is casting off and replacement; in this connexion we must note that, in the rod, there is some diffuse labelling similar to, but by no means as intense as, that in the cones (Bok and Young, 1972) indicating, presumably, exchange of newly synthesized protein with that already present in the discs or inter-disc space of the outer segment. Again Ditto (1975) saw no discrete bands of labelling in the developing salamander's outer segments yet there must have been a progressive formation of new discs from the base that would be expected to give discrete bands of labelling.

## EXTRACTION AND CHARACTERIZATION OF PHOTOPIGMENT

### SOLUBILIZATION

As mentioned earlier, Kühne observed that visual purple in the intact retina when first yellow and then white when exposed to light; on allowing the eye to remain in the dark the changes were reversed. The problem of the photochemist is to extract from the retina the visual purple, and to elucidate its structural chemistry and the changes it undergoes during exposure to light. Extraction of the pigment is achieved by the use of a 'solubilizing agent' such as digitonin or bile salts; these form complexes with the pigment that permit it to form a homogeneous solution. The amount of pigments obtainable from a single retina is so small that special methods for characterizing them, and the changes they undergo, must be employed; of these the study of absorption spectra has been easily the most profitable.

### DENSITY SPECTRA

A solution of a pigment is coloured because the pigment molecules absorb wavelengths more strongly than others, so that white light, after transmission through the solution, has had some of its wavelengths selectively removed. Thus chlorophyll solutions are green because its molecules selectively remove red and blue wavelengths, and hence allow the green and yellow parts of the spectrum to pass through. Visual purple is magenta coloured because its molecules selectively absorb the middle region of the spectrum—blues and greens— allowing a mixture of red and violet lights to pass through with relatively little absorption. The selective absorption of visual purple solutions can be measured by passing a beam of light, whose wavelength can be varied throughout the spectrum at will, through the solution and measuring, with a suitable photocell, the amount of light-energy transmitted. This may be compared with that transmitted through water alone, and the difference gives the light-energy absorbed. The results may be plotted as a density spectrum, as in Figure 8.11, where the optical density of the solution is plotted against wavelength. It will be seen that the spectrum has a maximum at about 5000 Å in the blue-green. It is this *wavelength of maximal absorption*, indicated by $\lambda_{max}$, that we may use to characterize a given pigment extracted from the retina.

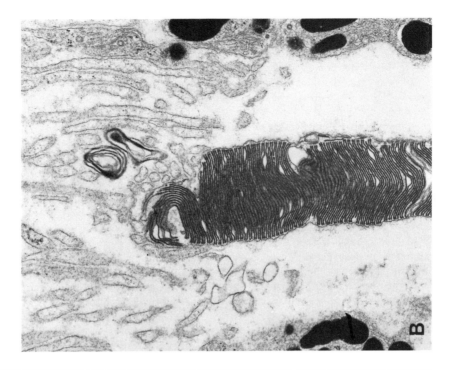

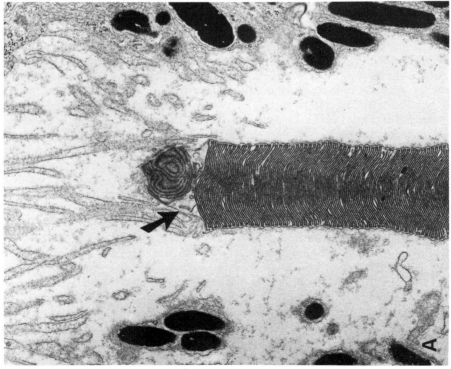

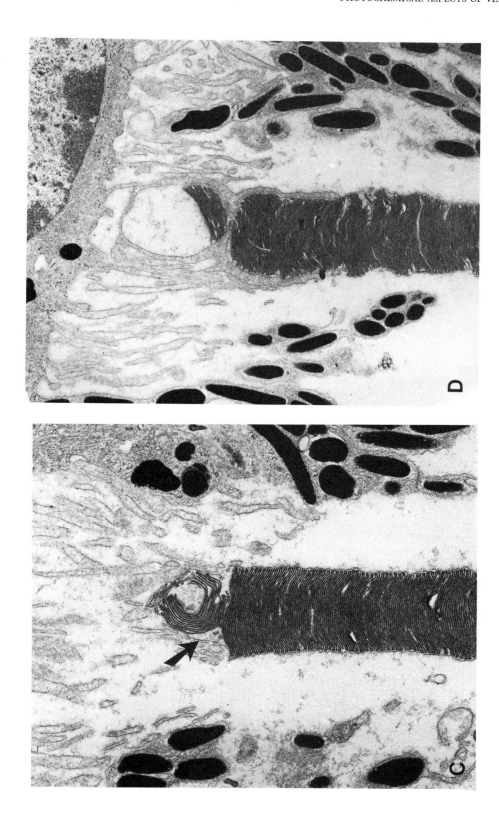

**Fig. 8.10** Phagocytosis of cone outer segments of 13-lined ground squirrel. The curling of the packets of discs takes place before any significant deformations of the outer cell membrane are obvious (arrows, A and C). In D pigment epithelial processes have segregated a small disc packet from the rest of the outer segment. (A–C, × 15 000; D, × 18 000.) (Anderson and Fisher, *J. Ultrastr. Res.*)

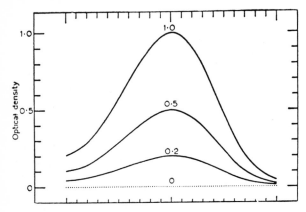

**Fig. 8.11** Density spectra of different concentrations of visual purple. Note that as the concentration is reduced the curve becomes flatter although the wavelength of maximum absorption remains the same. (Dartnall, *The Visual Pigments*.)

OPTICAL DENSITY

If the absorption of light obeys Beer's Law, it is determined by the length of the path through which the light travels, and the concentration of pigment, in accordance with the equation:

$$\log_e I_{inc}/I_{trans} = \alpha_\lambda \times c \times l$$

where $I_{inc}$ and $I_{trans}$ are the intensities of the incident and transmitted beams respectively, $c$ is the concentration, $l$ the path-length, and $\alpha_\lambda$ is the *extinction coefficient* for this particular wavelength. This extinction coefficient is the measure of the tendency for the molecules to absorb light of a given wavelength, so that by plotting $\alpha_\lambda$ against wavelength the *extinction spectrum* so obtained is a measure of the absorption of the different wavelengths. If logarithms to base 10 are used in the above equation, and the concentration is expressed in moles per litre, the *optical density* or 'absorbance', $A_\lambda$, of a solution is defined as:

$$A_\lambda = \log_{10} I_{inc}/I_{trans} = \varepsilon_\lambda \times c \times l$$

where $\varepsilon_\lambda$ is the *molar extinction coefficient*. For a given solution the optical density may be used as the measure of absorption of a given wavelength, and the curve obtained by plotting optical density against wavelength is called a *density spectrum** (Fig. 8.11). It will be clear from the above equations that the extinction coefficients will be independent of concentration of pigments so that extinction spectra will also be independent of this variable; the optical density, on the other hand, increases with increasing concentration, and the shape of the density spectrum will, indeed, vary according to the concentration of the pigment in the solution being studied. Where the concentration of absorbing material is not directly determined, as in studies on the intact retina, then the shape of the density spectrum gives a clue to the actual concentration present. We may note, finally, that if the density for any given wavelength is expressed as a percentage of the maximum density (in the case of rhodopsin 500 nm) then the density spectrum does become independent of concentration, and it may be used to characterize the pigment.

DIFFERENCE SPECTRA

It usually happens that visual pigment solutions are contaminated by light-absorbing impurities; if these are unaffected by light, in the sense that they retain their light-absorbing qualities unchanged after exposure, then the true absorption spectrum of the visual pigment may usually be deduced from the *difference spectrum*, given by measuring the absorption spectrum of the solution before and after bleaching with white light; thus, before the bleaching, the absorption will be due to visual pigment plus impurities; after bleaching it will be due to impurities alone if the visual pigment has become colourless. In many cases, of course, the visual pigment does not become colourless, but allowance can usually be made for this residual absorption.

PURITY

Retinal extracts may contain mixtures of visual pigments, in which case it becomes important to decide when any extract is homogeneous or not. Dartnall has developed a technique of partial bleaching with selectively chosen wavelengths that permits of an unequivocal demonstration whether or not the preparation contains more than one visual pigment.

PHOTOSENSITIVITY

The rate at which a simple photochemical reaction proceeds is governed primarily by the rate of absorption of quanta of light and the efficiency with which an absorbed quantum reacts. The rate of absorption will be given by the product of $I$, the light intensity expressed in quanta per second per unit area, $\alpha$, the extinction coefficient, and $\gamma$, the quantum efficiency of the reaction.

$$\text{Rate of reaction} = \alpha . \gamma . I$$

The product $\alpha . \gamma$ is called the *photosensitivity*; it has units of $cm^2$ and may be regarded as the effective cross-sectional area of the molecule when bombarded by photons. The *molar photosensitivity*, based on the molar extinction coefficient, $\varepsilon_\lambda$, can be calculated by multiplying by Avogrado's No. $6.023 . 10^{23}$, by $0.4342$ (normal to decadic logs) and by $10^{-3}$ ($cm^3$ to litre). $\varepsilon . \gamma = 2.615 . 10^{20} \alpha\gamma \, M^{-1} \, cm^{-1}$.

# DENSITY SPECTRUM AND SCOTOPIC SENSITIVITY

HUMAN SCOTOPIC SENSITIVITY

The density spectrum of the pigment extracted from the rods of the dark-adapted retina may be used to determine whether this pigment is, indeed, responsible for night vision, since the efficiency of any wavelength in evoking a sensation of light must be closely related to the efficiency with which it is absorbed by the rods. Clearly, if only one per cent of orange light falling on the retina were absorbed, whilst 90 per cent of the blue light were absorbed, we should have to use ninety times as much orange light-energy to evoke the sensation of light. In Figure 8.12 two things have been plotted; the smooth curve shows the absorption spectrum of visual purple extracted from a dark-adapted human eye, whilst

---

* The *absorption spectrum* is given by plotting the percentage of light absorbed against wavelength. Terminology is loose in this respect, however, so that absorption spectrum and density spectrum are terms used interchangeably.

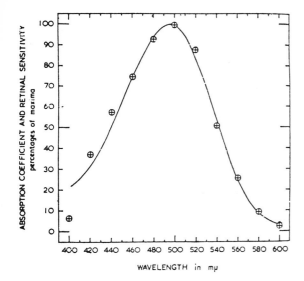

**Fig. 8.12** Comparison of human retinal scotopic sensitivity at different wavelengths, as determined by Crawford, with the absorption spectrum of human visual purple. The curve represents the absorption spectrum whilst the plotted points represent human scotopic sensitivity. (Crescitelli and Dartnall, *Nature*.)

the plotted points are the sensitivities of the dark-adapted human eye to the different wavelengths, the so-called action spectrum of the human eye. As indicated earlier, the dark-adapted eye is achromatic at light intensities in the region of the threshold; the different wavelengths do not evoke the sensation of colour but simply of light, and the experiment consisted of measuring the thresholds for the different wavelengths; the reciprocals of these, measured in quanta, represent the visual efficiencies of the wavelengths. The agreement between the two measurements is striking.

## SENSITIVITY IN ANIMALS

In animals, also, the scotopic sensitivity of the eye can be measured; for example, by determining the intensities of different wavelengths required to evoke a *b*-wave of fixed magnitude in the electroretinogram, or to evoke a discharge of fixed frequency in a single optic nerve fibre. The reciprocals of these values give the sensitivities of the retina at the different wavelengths and, when plotted against wavelength, give, usually, the characteristic 'rhodopsin-type' of curve with a maximum in the region of 500 nm. If the light-adapted retina is employed, on the other hand, there is a characteristic shift in the maximum towards, the longer wavelengths to give a maximum in the yellow-green at 560 nm. This is the *Purkinje shift* and represents the change in spectral sensitivity of the eye due to passing from rod to cone vision. Under these conditions, of course, the eye appreciates the colours of the stimuli.

# THE NATURE OF RHODOPSIN

## LIPID COMPONENT

The rhodopsin extracted from the retina is a chromoprotein; as characteristically prepared in a digitonin extract, a considerable proportion of the material is phospholipid, but the extent to which this is a genuine part of the rhodopsin molecule is difficult to state since, as we have seen, the molecule is embedded in the membranes of the outer segment, the material of the membrane being largely phospholipid; solubilization by digitonin brings most of the membrane lipid into the micelles. A carbohydrate fraction has also been identified (Heller, 1968). When preparations of rhodopsin are made with a view to characterization of its molecular weight, the lipid fraction is separated, for example by use of a detergent such as Triton X and subsequent removal of the detergent and lipid with toluene (Zorn and Futterman, 1971); in this way the amount of lipid associated with the extracted material is reduced to about 5 per cent of its original amount.

## Effects of lipid removal

By employing enzymatic treatments, the amount of lipid can be reduced still more, but when this is reduced to the point that the amounts of phosphatidyl ethanolamine and phosphatidyl serine are reduced to below 0·1 per molecule of rhodopsin there is a loss of absorbance of about 36 per cent at the wavelength of maximum absorbance, namely 500 nm (Borggreven *et al.*, 1972). In fact, considerably smaller losses of lipid are sufficient to affect the absorbing power of the rhodopsin so, although the absorption is due to the chromophore-protein complex, the shape of the complex is presumably stabilized by its association with lipid, as, of course, is the situation in the rod outer segment. The protein would be said to be partially denatured by removal from its normal lipid environment; this denaturation, or loss of shape, presumably affects the chemical interaction between the chromophore and the protein, an interaction that is responsible for its ability to absorb visible light.

## MOLECULAR WEIGHTS

The purified preparations of Heller and of Shichi *et al.* gave molecular weights of 27 000 to 30 000 for bovine rhodopsin, and these are consistent with the amino acid compositions. Amino acid analysis of frog and squid rhodopsin proteins gave larger values—40 000 and 48 500 respectively (Robinson *et al.*, 1972; Hagins, 1973).

## CARBOHYDRATE

The carbohydrate attached to the rhodopsin molecule contains three N-acetyl glucosamine as well as three mannose residues; thus it is a glucosaminoglycan and attaches to an aspartic acid residue. This behaves as an end-group of a protein amino-acid chain, and is accessible to concanavalin A, which reacts with this form of end-group; thus it apparently projects from the surface of the molecule. The evidence indicates that it is some distance from the point at which the chromophore group attaches (Renthal *et al.*, 1973) and when it is oxidized by periodate the 500 nm absorption band, and its regeneration after bleaching, are not materially impaired, indicating that the carbohydrate does not partake in the absorption of light-energy.

## Mannose incorporation

When isolated rod outer segments are incubated with radioactive mannose the radioactivity appears in the rhodopsin molecules after a delay which presumably represents the time required for incorporation into rhodopsin and its migration from the inner to the outer segment (O'Brien, 1977).

THE CHROMOPHORE AND VISUAL CYCLE

Each molecule has one light-absorbing chromophore group on it. The nature of this chromophore group, responsible for the colour of the molecule, and thus of its characteristic absorption, and the nature of its linkage to the protein moiety, are the keys to the photochemistry of vision.

When a retina is exposed to light the magenta colour of the visual pigment, rhodopsin, is lost—the retina is said to be bleached, and it was shown by Wald that,

during this bleaching process, the chromophore could be split off, and it was identified as *retinene*, which is vitamin A aldehyde (Fig. 8.13) or, as it has been subsequently called, *retinal*. The remaining colourless protein was called *opsin*, and it was the combination between the two molecules that transformed the yellow retinal into the magenta or purple rhodopsin. Within the retina, retinal could be, and is, indeed, reduced to vitamin A by an alcohol dehydrogenase—retinal reductase. Under appropriate conditions the retinal and opsin could be made to recombine to *regenerate* rhodopsin.

Thus the basic visual cycle defined by Wald was:

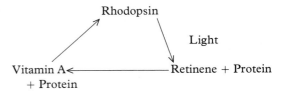

THE SHAPE OF THE MOLECULE

Wu and Stryer (1972) found that they were able to attach certain fluorescent markers to the rhodopsin molecule in what seemed to be specific sites, which they called A, B and C, and which were all different from the site of attachment of the retinal. These fluorescent markers were able to transfer their absorbed light-energy to retinal and cause it to emit this, since the $\lambda_{max}$ of rhodopsin was fairly close to the emission bands of the fluorescent markers. From the efficiency of energy-transfers they computed that the sites were 7·5,

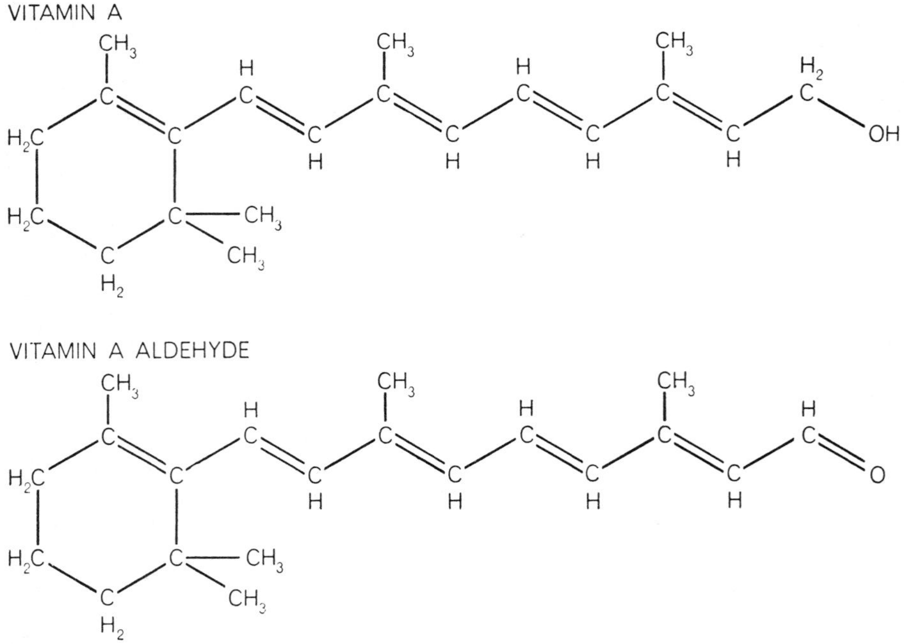

**Fig. 8.13**   Structural formulae of vitamin A or retinol, and its aldehyde, retinal (retinene).

5.5 and 4·8 nm distant from retinal, an arrangement that would correspond to an elongated shape corresponding to Figure 8.14. Thus the rhosopsin molecule might well occupy most of the thickness of a disc membrane.

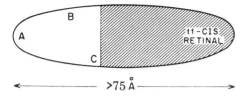

**Fig. 8.14** A model of the rhodopsin molecule based on observed proximity relationships. Measurements of energy transfer between label molecules bound at sites A, B and C and the chromophoric group allow calculation of distances from A, B and C to the chromophoric group of 7·5, 5·5 and 4·8 nm respectively. This leads to the conclusion that the molecule is cigar-shaped and at least 7·5 nm long. (Wu and Stryer, *Proc. Nat. Acad. Sci.*)

## THE RETINAL-OPSIN LINK

According to the studies of Collins and Morton, retinal forms a Schiff-base with amines; thus methylamine reacts with retinal to give retinal methylimine:

$$C_{19}H_{27}.CHO + CH_3.NH_2$$

$$\rightarrow C_{19}H_{27}C = N.CH_3 + H_2O$$
$$\hspace{1.4cm}|$$
$$\hspace{1.4cm}H$$

or more generally:

$$C_{19}H_{27}.C = O + H-N-R$$
$$\hspace{1.6cm}| \hspace{1.6cm}|$$
$$\hspace{1.6cm}H \hspace{1.6cm}H$$
$$\hspace{1cm}\text{Retinal} \hspace{2cm}\text{Amine}$$

$$\rightleftharpoons C_{19}H_{27} - C = NR + H_2O$$
$$\hspace{3cm}|$$
$$\hspace{3cm}H$$
$$\hspace{1cm}\text{Aldimine} \hspace{2cm}\text{Water}$$

The reactions are reversible so that the breakdown of the aldimine to retinal and amine may be described as hydrolysis. The aldimines in neutral solution are pale yellow but in acid solution, when they take on the protonated form, they give a red solution:

$$C_{19}H_{27}C = NCH_3 + H^+ \rightleftharpoons C_{19}H_2.C \overset{+}{=} N-CH_3$$
$$\hspace{1.2cm}| \hspace{5cm}| \hspace{0.5cm}|$$
$$\hspace{1.2cm}H \hspace{5cm}H \hspace{0.4cm}H$$

The protonated form is more stable than the neutral form in the sense that it is not hydrolysed in acid reaction.

### N-RETINYLIDENE OPSIN

By comparing the characteristics of these artificial aldimines with those of partially bleached rhodopsin, where the chromophore group was still attached to the opsin, Collins and Morton concluded that rhodopsin could be described as N-retinylidene opsin (NRO) with the retinal linked to an amino-group of the opsin molecule. Although this conclusion was based on studies of a rhodopsin that had been changed chemically through absorption of light, subsequent studies have confirmed that in its native unbleached state rhodopsin contains retinal linked in the above manner. For example Mathies *et al.* (1976) analysed the Raman spectrum and deduced from the line at 1660 cm$^{-1}$ the presence of a protonated Schiff base.

### ISOMERS OF RETINAL

Early studies on regeneration of rhodopsin from mixtures of the opsin and retinal formed by bleaching, indicated a far less efficient and complete regeneration than that obtained in the intact retina, and the main cause for this inefficiency turned out to be the necessity for a special isomer of retinene for the combination with opsin (Hubbard and Wald, 1952).

Retinal and vitamin A contain five double-bonds joining C-atoms, and the presence of these gives the possibility of numerous *cis-trans* isomers, due to the fact that rotation about a double bond is not possible, so that a molecule illustrated by Figure 8.15a is different from that illustrated by Figure 8.15b. Of the various

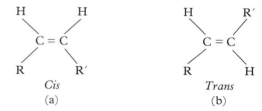

**Fig. 8.15** *Cis-trans* isomerism.

possible isomers obtained by rotating the molecule about different double bonds Wald found two that were capable of condensing with opsin to give photosensitive pigments. One, neo-b retine or 11-*cis* retinene (Fig. 8.16), when incubated with opsin gave rhodopsin, whilst the other, isoretinal-a or 9-*cis* retinal (Fig. 8.16) gave a photosensitive pigment, *isorhodopsin*, with $\lambda_{max}$ of 487 nm.

## Isomerization

The reason why reversal of the changes undergone by bleaching is not always possible is because the retinal that splits off from rhodopsin has the *all-trans* form, which is not active for resynthesis; thus retine in the rhodopsin molecule is in the 11-*cis* of neo-b form,

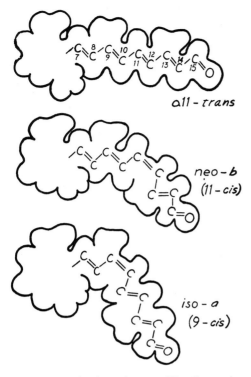

**Fig. 8.16** Shapes of retinene isomers. The all-*trans* isomer is straight; a *cis*-linkage introduces a bend into the molecule. In addition, the 11-*cis* linkage encounters steric hindrance and therefore twists the molecule out of the plane in a manner not shown in the diagram. (Hubbard and Kropf, *Ann. N.Y. Acad. Sci.*)

but when split off is in the all-*trans* form. In order to be of further use the all-*trans* must be isomerized, a process that can take place in the retina in the presence of an enzyme, *retinal isomerase*, although mere exposure to light also causes isomerization—*photo-isomerization*.

## INTERMEDIATE STAGES IN BREAKDOWN TO RETINAL AND OPSIN

It was originally thought that the primary action of light was to split retinal from opsin, but subsequent studies have emphasized that the photochemical reaction, induced by the absorption of a quantum of light, is only the beginning of a cascade of reactions that ultimately lead to this splitting off of the chromophore, reactions that are no longer dependent on the absorption of light and so are described as thermal. Many of these reactions involve very small changes in internal energy of the various intermediates and may occur so rapidly at body-temperature* that their products would escape detection even when employing the changes of absorption spectrum as a means of identifying the various stages.

### LOW TEMPERATURE

However, by carrying out the initial photochemical reaction at a very low temperature it is possible to arrest the subsequent thermal changes, and the history of the analysis of these thermal changes has depended on the degree of cooling that was employed. Thus Lythgoe and Quilliam were able to identify an intermediate compound, which they called *transient orange*, because they cooled the rhodopsin to 0°C during irradiation. When the solution was subsequently warmed, a yellow product, which they called *indicator yellow* and is now known as *N-retinylidine opsin*, was produced; subsequent work showed that a further thermal reaction resulted in the splitting of the Schiff base that bound retinal to opsin in N-retinylidine opsin to give retinal and opsin as separate molecules.

### BATHORHODOPSIN

When liquid nitrogen (temperature $-195°C$) was employed to cool the solution of rhodopsin before and during exposure to light, and the mixture was subsequently warmed in stages, the series of products indicated by Figure 8.17A was obtained (Abrahamson and Ostroy, 1967); thus the only reaction requiring light-energy was the conversion of rhodopsin to pre-lumirhodopsin (subsequently called bathorhodopsin), recognized by the change in $\lambda_{max}$ from 498 nm to 543 nm.

### LUMIRHODOPSIN AND METARHODOPSINS

By warming to $-140°C$ bathorhodopsin is converted to lumirhodopsin, which remains stable at temperatures up to $-40°C$, above which it is converted to a series of *metarhodopsins*, differing in their $\lambda_{max}$, then to N-retinylidene opsin (NRO) previously called *indicator yellow* because of its change of colour with pH, and finally NRO is hydrolysed to retinal and opsin. In all but the last of these successive changes, the rhodopsin molecule has retained its attached chromophore group, retinal, so that the alterations have been essentially changes in the configuration of the whole molecule. The final splitting of the chromophore group from the protein moiety represents the breaking of the Schiff-base link.

### HYPSORHODOPSIN

When liquid helium was employed (temperature $-270°C$) Yoshizawa identified an earlier protoproduct, which has been called hypsorhodopsin with a $\lambda_{max}$ at 430 nm; as indicated by the scheme of Figure 8.17B, this can be converted by a thermal reaction to bathorhodopsin by warming to $-250°C$.

---

* According to Oseroff and Callender (1974) the formation of bathorhodopsin takes place in less than 6 psec.

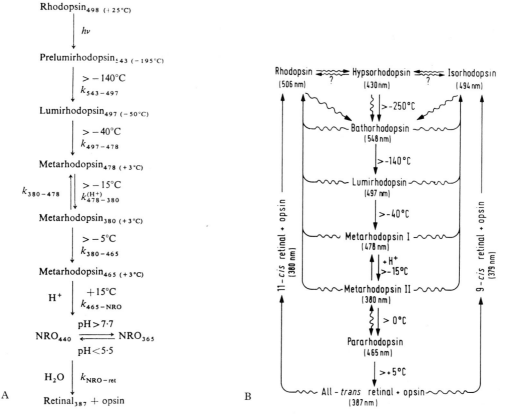

**Fig. 8.17** (A) Resections in the thermolysis of rhodopsin. (Abrahamson and Ostroy, *Progr. Biophys.*) (B) Stages in the bleaching of rhodopsin. Photoreactions are symbolized by wavy lines, thermal reactions by straight ones. $\lambda_{max}$ for rhodopsin, hypsorhodopsin, isorhodopsin and bathorhodopsin measured at $-268°C$ and for metarhodopsin I at $-65°C$; other pigments near $0°C$. (Yoshizawa, *Handbook of Sensory Physiology.*)

### PHOTO-ISOMERIZATION

The situation becomes much more complex, however, when the effects of light on the reaction mixture at different stages are taken into account, since light can itself bring about many of these conversions. For example, light converts hypsorhodopsin to bathorhodopsin at the temperature of liquid helium; moreover, light can cause regeneration of both rhodopsin and isorhodopsin so that at this low temperature there will be a mixture of four compounds, rhodopsin, isorhodopsin, hypso- and batho-rhodopsin in relative amounts depending on the wavelength of the irradiating light; thus a wavelength of 436 nm favours conversion to bathorhodopsin, whereas at 600 nm there is almost complete reversal to rhodopsin.

### ISORHODOPSIN

As we have indicated above, there are two isomers of retinal, namely 11-*cis* and 9-*cis*, capable of combining with opsin to produce photosensitive pigments, called rhodopsin and isorhodopsin respectively. As Figure

8.17B shows, photo-isomerization of the products of photolysis of rhodopsin can lead to production of isorhodopsin ($\lambda_{max}$ 496) in the retina. The requirement for the initial photo-isomerization is that the chromophore be attached to the opsin moiety. Examination of the products of exposing the eye to very bright and very short flashes of light—*flash photolysis*—has shown that isorhodopsin may, indeed, be formed under these conditions (Bridges, 1961; Frank 1969; Ripps and Weale, 1969). We may note also that this tendency for photoreversal of the light-induced changes leads to a considerable reduction in the efficiency of bleaching by short flashes of light; thus Hagins (1957) found that brief flashes of light, however intense, were incapable of bleaching all the rhodopsin in the retina, and more recently Pugh (1975), using a xenon flash, found that no more than 60 per cent of the rhodopsin in the human eye could be bleached.

### METARHODOPSINS

Metarhodopsin I is stable up to about $-15°C$, above which it comes into tautomeric equilibrium with metarhodopsin II, a change that involves a large change in $\lambda_{max}$ from 478 to 380 nm, and involves the uptake of an $H^+$-ion. On warming to a higher temperature, MR II is transformed into *N-retinylidene opsin* (NRO or *indicator yellow*) either directly or through Meta III, which has been called *pararhodopsin*, and is probably equivalent to Lythgoe's and Quilliam's transient orange. N-retinylidene opsin (NRO) has indicator qualities and may be hydrolysed to release the chromophore from opsin.

## THE NATURE OF THE CHANGES

Absorption of a quantum of light brings the light-sensitive molecule into an 'excited state' of high potential energy; in this state strains may be introduced in the molecule leading to, perhaps, the rearrangement of chemical bonds or the rupturing of these. These effects are accompanied by changes in the absorbance of the molecule, revealed as a shift in $\lambda_{max}$, and also in the shape, of the absorption spectrum. The most important finding of Wald was that the retinal released from rhodopsin had changed its stereo-isomeric configuration from the 11-*cis* to the all-*trans* form; this was certainly true of the released retinal, but the evidence indicated that the very first change following absorption of light consisted in this photo-isomerization of the retinal chromophore *in situ*. Thus, the first effect of the absorption of energy might be regarded as the removing of a linch-pin that held the retinal molecule in the 11-*cis* configuration, the linch-pin consisting of suitable secondary chemical interactions between the retinal and opsin molecules that held the retinal in this particular configuration (Dartnall, 1957).

### HYPSORHODOPSIN AND BATHORHODOPSIN

The conversion of rhodopsin to hypsorhodopsin, or bathorhodopsin, producing considerable shifts in $\lambda_{max}$, must be interpreted in terms of a primary conversion of the retinal group to the all-*trans* form, leading to different side-chain reactions between the prosthetic group and opsin; thus they occur at so low a temperature that it is considered unlikely that any change in the opsin molecule would take place.

### Strains in the chromophore

According to Abrahamson and Wiesenfeld (1972), the position of the $\lambda_{max}$ of hypsorhodopsin at the short wavelength of 430 nm indicates the absence of significant secondary interaction between the retinal molecule, in its all-*trans* state, and the opsin molecule,

since it is these secondary interactions that cause strains in the retinal molecule that enable it to absorb light of longer wavelengths. The position of the $\lambda_{max}$ of bathorhodopsin, at the longer wavelength of 548 nm, indicates, on the other hand, that new strains have been placed on the retinal molecule owing to secondary interaction with the opsin molecule. Thus, if hypsorhodopsin is an earlier intermediate than bathorhodopsin, the process of photo-isomerization may be accounted for as follows: In the excited state, the chromophore, because of the breakdown of secondary linkages with the opsin residue, is potentially able to undergo rotations about both single and double bonds, rotations that would be prevented sterically by its secondary interactions. For example, the rotation about the 11–12 double bond necessary for conversion of the 11-*cis* to the all-*trans* structure, might be preceded by rotation about a single bond (from S-trans to S-cis or *vice versa*) in order to avoid interference with the opsin structure. Such an isomerization could result in very little interaction with opsin, and this may be the basis for hypsorhodopsin. On absorption of a further photon at the liquid helium temperature, or by raising the temperature to 23°K (—250°C) in the dark, the chromophore might again undergo a rotation about the same, or another, single bond with the consequent establishment of a new side-chain that interacted strongly with opsin to give a highly strained and distorted molecule—bathorhodopsin—absorbing at long wavelengths. The changes through lumirhodopsin to metarhodopsin I, occurring at higher temperature, are thought to involve changes in interaction between the prosthetic group and opsin, this time owing to changes in conformation of the opsin molecule, changes that have become possible through the primary change in the shape of the prosthetic group from the 11-*cis* to the all-*trans* orientation.

### HUBBARD-KROPF MODEL

In Figure 8.18 is illustrated the hypothesis of Hubbard and Kropf that accounts, in a general way, for the successive changes; thus retinal is combined stably with rhodopsin because its 11-*cis* shape enables it to fit into a suitable pocket on the surface of opsin; absorption of light provides the prosthetic group with sufficient energy to break loose from this pocket and thereby assume the all-*trans* configuration, giving rise to hypso and bathorhodopsins. Subsequent changes are due to alterations in the shape of the opsin pocket that, among other things, reveal chemical groupings that were formerly shielded from the solution, e.g. SH-groups and an $H^+$-binding group, processes that are manifest as changes in the reactivity of the rhodopsin to SH-reagents and changes in pH of the medium.

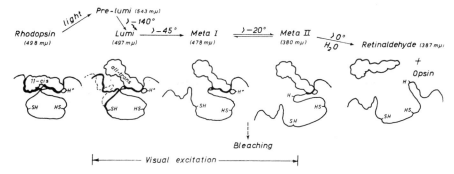

**Fig. 8.18** Hubbard-Kropf model of stages in the bleaching of rhodopsin. Rhodopsin has, as a chromophore, 11-*cis* retinal, which fits closely a section of the opsin structure. The only action of light is to isomerize retinal to the all-*trans* configuration (prelumirhodopsin). Then the structure of opsin opens in stages (lumirhodopsin, metarhodopsin I and II), until finally the retinaldehyde is hydrolysed away from opsin. Bleaching occurs in the transition from metarhodopsin I and II; and by this stage visual excitation must also have occurred. The opening of the opsin structure exposes new groups, including two —SH groups of one $H^+$-binding group. The absorption maxima shown are for prelumirhodopsin at $-190°C$, lumirhodopsin at $-65°$, and other pigments at room temperature. (Wald, *Proc. Int. Congr. Physiol.*)

### Changes in pH

Thus Wong and Ostroy (1973) measured the changes in light absorbance of metarhodopsin at 380 nm during its conversion to metarhodopsin II at $-15°C$, and correlated these with the change in pH of the medium as revealed by the change in colour of bromcresol green measured by a change in absorbance at the long wavelength 612·5 nm. As Figure 8.19 shows, there is a good correlation between the absorbances due to metarhodopsin and change in uptake of a proton ($H^+$).

### Borohydride reaction

Another example of changed reactivity is given by treating rhodopsin or metarhodopsin II with borohydride;

this substance is a reducing agent that adds hydrogen to the Schiff-base of N-retinylidene opsin:

$$C_{19}H_{27}CH{=}N.Opsin + NaBH_4 \rightarrow$$
$$C_{19}H_{27}.CH_2.NH.Opsin$$

It was found that this reduction did not occur with rhodopsin but only when the metarhodopsin II state had been reached, presumably because the $CH{=}N$ group had become exposed at this stage, being buried in the opsin molecule up to this point.

### SH-groups

Again, opsin will react with *p*-chloromercuribenzoate (PCMB) through its SH-groups; Wald and Brown

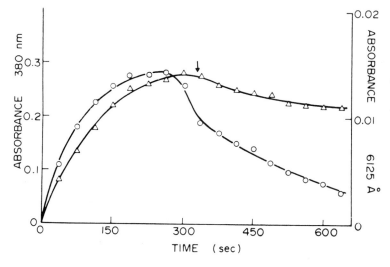

**Fig. 8.19** Hydrogen-ion changes, revealed by changes in absorbance of bromcresol green, during irradiation. Circles indicate absorbance of bromcresol green at 612·5 nm. Arrow indicates cessation of radiation. (Wong and Ostroy, *Arch. Biochem. Biophys.*)

(1952) found that rhodopsin did not react with PCMB, although when rhodopsin is bleached in the presence of the reagent an irreversible change takes place that now prevents the regeneration with 11-*cis* retinal.*

### THE METARHODOPSINS

The transitions from lumirhodopsin to NRO are thought to be accompanied by changes in the opsin molecule revealed by changes in the degree of chemical interaction of the chromophore with opsin, and thus changes in the absorption spectrum, and also by changes in the reactivity of the molecule according as reactive groups are exposed. The similarity in absorption spectra between Meta I and rhodopsin indicates a strong degree of interaction between opsin and chromophore in Meta I but some loosening of the opsin molecule is indicated by its susceptibility to attack by hydroxylamine. The change from Meta I to Meta II is considered to be the key reaction that involves sufficient change in entropy to permit it to alter the rod structure sufficiently to induce chemical changes. The basic reaction is only the addition of an $H^+$-ion but the entropy change is far greater than would be expected of this. One indication of chemical change is the finding that Meta II can be reduced by sodium borohydride whilst Meta I cannot, presumably because the aldimine bond has now become accessible by unfolding of protein.

### Metarhodopsin III

This product of bleaching, with a $\lambda_{max}$ of about 470 to 475 nm, has been observed in retinas of frog, skate and several mammals; it has also been called *pararhodopsin* and may be the same as the transient orange of Lythgoe and Quilliam (1938). As indicated in the scheme of Figure 8.17B, it is derived from Metarhodopsin II by thermal conversion, but the process may be apparently bypassed so that Meta II can be converted directly to retinal (Fig. 8.20). Exposure of

Meta III to light causes a photoreversal to Meta II, and this effect of light led to the suggestion that Meta III contained a different retinal isomer from the all-*trans*, which was photoisomerized to all-*trans* retinal in Meta II. It was suggested that the isomer was 13-*cis*. However, it is clear from the work of Reuter (1976) that Meta III is an all-*trans* intermediate, like Meta II, and that the effect of light on Meta III is to convert its retinal to the 9-*cis* and 11-*cis* configurations, and thus to regenerate isorhodopsin and rhodopsin, which are then converted to Meta II by absorption of a further photon. It is questionable whether this photo-isomerization is significant normally, since Meta II absorbs very little of the light reaching the retina, and of the remaining photoproducts only Meta III is long-lived enough for any photoconversion to be important, but the amount that accumulates when only a little rhodopsin is bleached is small because there is apparently a bypassing of the route through Meta III (Donner and Hemila, 1975) under these conditions. Thus, as Donner and Hemila point out, the strong bleaching that is employed experimentally to study the production and decay of Meta II is not likely to occur under natural conditions. Their own study of the kinetics suggested that the scheme of decay proposed by Baumann, and illustrated in Figure 8.20a, should be modified as in Figure 8.20b. Certainly there is evidence that retinal and retinol are orientated within the membrane of the rod discs.

### LATEST STAGES

The decay of Meta II to Meta III, and NRO, and finally the splitting of retinal from opsin are complex

---

*DeGrip *et al.* (1973) have re-examined the exposure of SH-groups during bleaching; when this is done with intact rod outer segment membranes, there is no increase in reactivity with DTNB or N-ethylmaleimide; only if the rhodopsin is solubilized with detergent is the increase, described by Wald and Brown, observable. We may note that the inhibition of regeneration of rhodopsin, from bleached rhodopsin and added 11-*cis* retinal, consequent on treatment with PCMB, described by Wald and Brown, is apparently not due to blockage of SH-groups since Zorn (1974) blocked these with N-ethylmaleimide, instead of PCMB, and found a high degree of regenerability. Thus PCNB has a specific effect—independent of its SH-blocking activity.

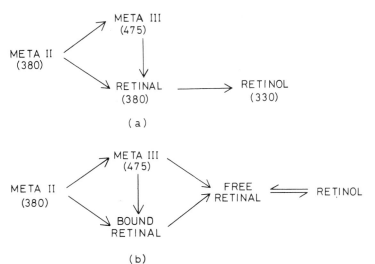

**Fig. 8.20** Alternate pathways for conversion of Metarhodopsin II to retinol.

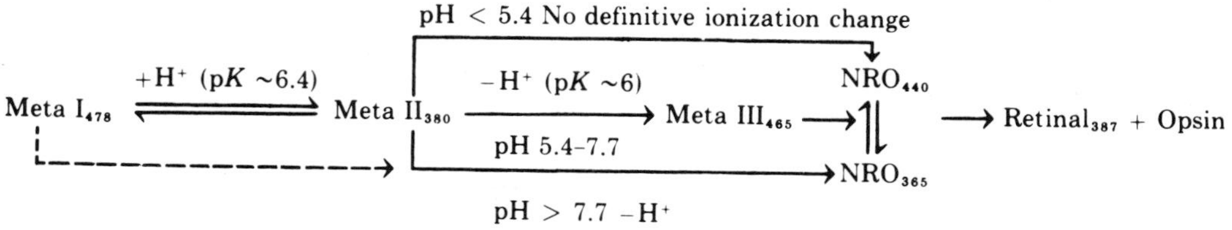

$$\text{pH} < 5.4 \text{ No definitive ionization change}$$

$$\text{Meta I}_{478} \underset{}{\overset{+\text{H}^+ \ (\text{p}K \sim 6.4)}{\rightleftharpoons}} \text{Meta II}_{380} \xrightarrow[\text{pH } 5.4-7.7]{-\text{H}^+ \ (\text{p}K \sim 6)} \text{Meta III}_{465} \rightarrow \begin{array}{c} \text{NRO}_{440} \\ \updownarrow \\ \text{NRO}_{365} \end{array} \rightarrow \text{Retinal}_{387} + \text{Opsin}$$

$$\text{pH} > 7.7 \ -\text{H}^+$$

**Fig. 8.21** Alternate pathways for decay of Metarhodopsin II. (Ostroy, *Arch. Biochem. Biophys.*)

processes by no means clearly understood. As indicated in Figure 8.21, from a study of Ostroy (1974), there are alternate pathways of decay, the one involving Meta III or pararhodopsin and a more direct one to NRO. According to Ostroy's study, the route taken is strongly pH-dependent, so that at pH 7·7 the major process consists in a release of a $\text{H}^+$-ion to give the non-protonated form of NRO, a form that is readily hydrolysed to give retinal and opsin. It would seem from the study of Kimble and Ostroy (1973) that the decay of Meta II is coincident in time with the appearance of two reactive sulphydryl groups.

## N-Retinylidene opsin (NRO)

This is the 'indicator yellow' of Lythgoe, changing from an orange ($\lambda_{max}$ 440 nm) in acid solution to a yellow ($\lambda_{max}$ 363 nm) in alkaline solution. At physiological pH the yellow form will predominate and, since this is rapidly hydrolysed, the presence of indicator yellow under normal conditions is difficult to establish, the final product being retinal plus opsin.

An interesting point emphasized by Knowles and Dartnall (1977) is that the absorption spectra of the NRO's from a variety of rhodopsins are essentially similar although their opsins are different; hence in NRO there is no secondary interaction between opsin and chromophore in this molecule.

## LINKAGE OF RETINAL

### LYSINE

The treatment of bleached rhodopsin with borohydride to give the reduced compound:

$$C_{19}H_{27}CH{=}N.\text{Opsin} \xrightarrow{\text{NaBH}_4} C_{19}H_{27}CH_2NH.\text{Opsin}$$

serves to 'freeze' the chromophore on the opsin moiety, and in this state it is feasible to establish the position of the link. Bownds (1967) showed that this link was through the ε-amino group of a lysine reside of opsin, so that unless retinal migrates during the few msec required for conversion of rhodopsin to Meta II, this is the mode of linkage in the unbleached state (Rotmans *et al.*, 1974).

### BINDING TO PHOSPHOLIPID

However Poincelot *et al.* (1969) concluded, from their studies on extracting rhodopsin with the lipid solvent methanol, that the binding site in rhodopsin is on the phospholipid moiety as N-retinylidene-phosphatidyl-ethanolamine, and that during bleaching of meta-rhodopsin I to II the chromophore was transferred from the lipid to the amino-acid skeleton of opsin.

### CONFIRMATION OF LYSINE-LINKAGE

There is little doubt from the work of Fager *et al.* (1972) and DeGrip *et al.* (1973) that the link is to a lysine residue. Fager *et al.* treated rhodopsin with cyano-borohydride, a more specific compound for reacting with protonated Schiff bases and giving a more stable product at acid pH, and found only a lysine-N-retinylidene complex during subsequent chromato-graphy, there being no evidence for the presence of a phosphatido-ethanolamine compound. Again DeGrip *et al.* (1973) amidinated 50 out of the total of 52 primary amino groups (including that attached to retinal) with methylacetimidate:

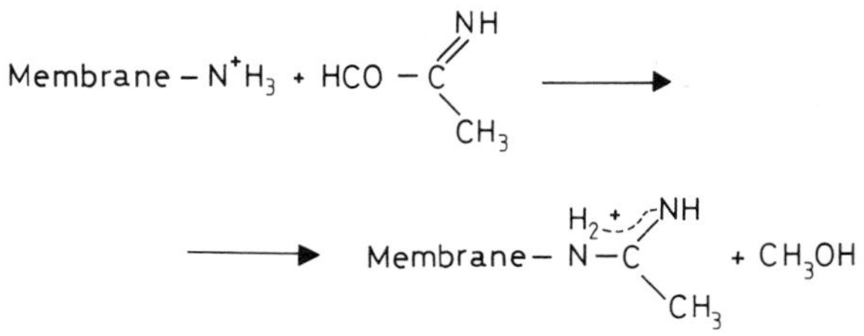

$$\text{Membrane} - \text{N}^+\text{H}_3 + \text{HCO} - \overset{\displaystyle \overset{\text{NH}}{\|}}{\underset{\displaystyle \text{CH}_3}{\text{C}}} \longrightarrow$$

$$\longrightarrow \text{Membrane} - \overset{\displaystyle \overset{\text{H}_2{}^+\,\nearrow\text{NH}}{}}{\underset{\displaystyle \text{CH}_3}{\text{N}-\text{C}}} + \text{CH}_3\text{OH}$$

In this condition the rhodopsin was normal to the extent that some 70 per cent could be bleached and regenerated. On subsequent denaturing of this modified rhodopsin, leading to release of retinal, one further ε-lysyl-group of opsin became free; if the rhodopsin was exposed to light before amidination, then one more amido group was taken up and now denaturation did not cause the appearance of one more amino group. DeGrip *et al.* conclude from this study that retinal is unequivocally bound to opsin through a lysyl residue in the unbleached state, and that it is held in this position at least during the important changes of structure—notably those leading to Meta II—during which the visual event presumably takes place.

## PROTEIN-LIPID INTERCHANGE

It would seem that a very facile transfer of the retinal from protein to lipid, or in the reverse direction, can take place, and it is possible that some such retinal-lipid linkage is important in the transfer of liberated retinal to its isomerizing or oxidizing site.

## IN VIVO CHANGES

The various compounds formed as intermediates in the bleaching of rhodopsin have been identified by optical means in solutions of the pigment or in isolated outer segments, and they have only been identified by the expedient of cooling and thus reducing, or inhibiting, the thermal reactions that occur very rapidly at body-temperature. In view of the extreme lability of the intermediates identified in this way it seems unlikely that significant concentrations of many of the inter-mediates will build up in the normal retina when exposed to light. There have been a variety of studies on the intact retina, and it would seem that it is meta-rhodopsin II and pararhodopsin (meta III) that are the main long-lived photoproducts (see, for example, Frank, 1969; Ripps and Weale, 1969; Knowles and Dartnall, 1977. Ch. 9).*

A situation approximating some way towards the normal is given by the study of the changes taking place in suspensions of rod outer segments (ROS) where the rhodopsin and its decomposition products remain within the discs and attached, presumably, to the membranes. In general, suspensions of rod outer segments reveal the same sequence of changes, although the stability of the intermediates, e.g. batho- and lumirhodopsins, is less in the receptor than in the digitonin micelles.

## CHANGES IN BIREFRINGENCE

Analysis of the changes in birefringence that take place after a flash of light indicates an initial small rapid decrease in the intrinsic component, symptomatic of a loss of crystallinity, i.e. a tendency for the molecules to lose their regular arrange-ment in the disc membranes (Liebman et al., 1974). The change corresponds in time to the formation of Meta II, and is complete within a few msec. Following this rapid change there is a slower loss in intrinsic birefringence followed by recovery and an overshoot, so that the intrinsic birefringence is greater than before bleaching. The recovery seemed to be associated with the disappearance of Meta II, whilst the final overshoot is assumed to be due to the appearance of free retinol which would orientate itself in the lipid membranes of the discs with its long axis parallel with the long axes of the phospholipid molecules of the membrane, i.e. at right-angles to the disc surface (Fig. 8.6, p. 200). A similar conclusion respecting a change in crystallinity was reached by Tokunaga et al. (1976), who compared the changes in absorbance in intact outer segments with those taking place in digitonin suspensions. Thus the orientation of the absorbing pigment molecules in the plane of the discs produces a characteristic change in the absorption spectrum, and an analysis of this difference, when

the systems were exposed to light at low temperatures so as to produce bathorhodopsin, lumirhodopsin and meta-rhodopsin, indicated that the chromophore group in lumi-rhodopsin tended to project out of the plane of the disc at an angle of 33° compared with 18° for rhodopsin; the shift to Meta I slightly reduced the angle to 27°. The authors suggest that this shift might be sufficient to alter the physical characteristics of the disc membrane, to act as trigger for the visual process.

# REGENERATION OF RHODOPSIN

In the retina the products of bleaching—retinal and opsin—are converted to rhodopsin in the dark—*regeneration*—and this is the basis for dark-adaptation, the recovery of sensitivity to light during the dark (p. 185). This regeneration process must be distinguished from the renewal process that occurs independently of exposure to light or dark, during which packets of rod discs are absorbed by the pigment epithelium whilst new discs, demanding the synthesis of opsin and retinal, are manufactured in the outer segments. Here we need only consider the former, regeneration process, although the continued engulfment of intact rhodopsin molecules by the pigment epithelium will clearly influence the amount of 11-*cis* retinal within its cells at any moment. A basic scheme for the regeneration process, which includes all possible reactions, is illustrated in Figure 8.22. Thus, as we have seen, retinal, when liberated, is converted to vitamin A or retinol and the relative amounts of retinal and retinol in the retina are governed by the proportions of NADP and NADPH, the co-enzyme that cooperates with retinol dehydrogenase in the reversible conversion of retinal to retinol; in the retina the conditions favour production of retinol, so that regeneration from retinol demands a preliminary oxidation of retinol to retinal.

## REGENERATION FROM ADDED 11-CIS RETINAL

### BREAKAGE OF THE RETINAL OPSIN LINK

When 11-*cis* retinal is added to a suspension of bleached rod outer segments, rhodopsin is rapidly regenerated (Amer and Akhtar, 1972); a similar regeneration was obtained with bleached isolated retinae (Amer and Akhtar, 1973), An important aspect of this regenera-tion in the dark is the apparent necessity for the retinal to be completely split off the opsin molecule, so that if there is a delay in this final step, e.g. if, after illumination,

---

*The lifetime of metarhodopsin II at room-temperature is of the order of milliseconds; it has been identified in the intact eye by Hagins (1956) using reflexion-densitometry (p. 231). Pugh (1975), in his study of flash photolysis in the human eye, calculated a half-life of 1·4 msec.

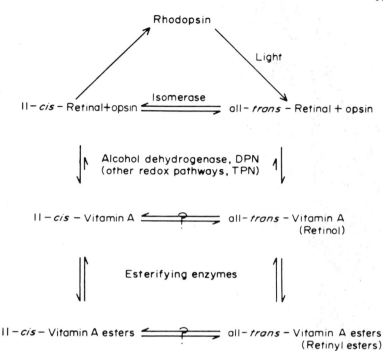

**Fig. 8.22**   The classic rhodopsin cycle. (Daemen *et al.*, *Exp. Eye Res.*)

some of the bleached rhodopsin is in the form of meta-rhodopsins, then regeneration through added 11-*cis* retinal will be delayed until the metarhodopsins have finally been converted to all-*trans* retinal plus opsin (Rotmans *et al.*, 1974; Paulsen *et al.* 1975). Thus Rotmans *et al.* (1974) measured the time-course of release of retinal from the bleached rhodopsin, which reached a plateau level in some 30 to 60 minutes. They also measured the amounts of rhodopsin that could be regenerated by adding pure 11-*cis* retinal to the bleached preparation at different times after the bleach, and they found that the per cent recombination increased with time after the bleach, following the course of the release of the chromophore manifest as loss of metarhodopsin II. This dependence on removal of chromophore doubtless explains the variety of kinetics of regeneration of rhodopsin observed in the intact eye.

## ISOMERIZATION

In the isolated retina, detached from its pigment epithelium, or in suspensions of rod outer segments, or in solutions of bleached rhodopsin, regeneration has been found either not to occur at all or very inefficiently compared with the intact eye; and the cause of this inefficiency is undoubtedly related to the separation of the receptors from the pigment epithelium. The require-ment for regeneration is the availability of a supply of 11-*cis* retinal to replace the all-*trans* retinal split off

from the rhodopsin molecule during bleaching. If the all-*trans* retinal is to be re-utilized it must be isomerized to 11-*cis* retinal; moreover, since conversion of retinal to retinol is rapid and may even occur before the final splitting off of the chromophore, then retinol must first be oxidized to retinal by retinol dehydrogenase; finally retinol is esterified so that, as indicated in Figure 8.22, hydrolysis back to retinol may be necessary. If one or more of these events requires the cooperation of the pigment epithelium, regeneration in the isolated retina will be impaired and rely on local stores of intermediate.

### PHOTO-ISOMERIZATION

If the chromophore is attached to the opsin molecule, the action of light can cause a photo-isomerization, and thence reversal of the light- and thermally induced changes, to regenerate rhodopsin. During continued light-exposure there is evidence of some regeneration due to this photo-isomerization. The process can lead to the formation of isorhodopsin, due to the conversion of the all-*trans* retinal to the 9-*cis* form; and under experimental conditions isorhodopsin can be demon-strated in the retina. Thus, using very brief flashes—flash-photolysis—permits some of the unstable products of the flash to absorb light before they break down and thus to isomerize. For example, Dowling and Hubbard (1963) found a mixture of 5:1 rhodopsin and iso-rhodopsin in the flash-irradiated rat retina. Under normal conditions, however, photo-isomerization is a relatively small factor, although it may account for regeneration observed in the frog's isolated retina free from pigment epithelium (Baumann, 1970).

## RETINAL ISOMERASE

By far the most important factor for regeneration, however, is the presence of an enzyme, *retinal isomerase*; its presence in the pigment epithelium of the frog, for example, probably accounts for the great fall in power of regeneration observed when the retina was detached from the pigment epithelium, so that it is generally believed that isomerization of the chromophoric group takes place outside the photoreceptor in the pigment epithelial cell, whence it is transferred back to the receptor, either in the form of 11-*cis* retinol or of 11-*cis* retinal.*

## RETINOL

### MIGRATION TO PIGMENT EPITHELIUM

An important factor is the rapid conversion of retinal to vitamin A, or retinol, in the retina; this must be converted to retinal before it can react with opsin, and this is carried out by an alcohol dehydrogenase in the visual cells. Retinol formed in the receptor does not remain there but migrates to the pigment epithelium. Thus Jancsó and Jancsó profited by the fluorescence of vitamin A in ultra-violet light to demonstrate that after maximal light-adaptation there is a high concentration of the vitamin in the pigment epithelium; during full dark-adaptation none could be demonstrated. Again, the chemical studies of Hubbard and Colman and of Dowling have confirmed this migration of vitamin A during light- and dark-adaptation, and have shown that the total amount of vitamin A in the eye remains constant, if we include under this title the retinene bound to rhodopsin.

### RETINAL–RETINOL SYSTEM

The changes in retinal and retinol in retina and pigment epithelium in the rat's eye are illustrated in Figure 8.23

from a study by Dowling (1960); in the rat, retinol is not stored in the pigment epithelium, by contrast with the frog; thus changes in retinal and retinol content are easy to demonstrate. The Figure shows that in the dark-adapted eye there is very little retinol in either pigment epithelium or retina, whilst retinal (as rhodopsin) is in high concentration. With light-adaptation the retinal falls and the retinol contents of the retina and pigment epithelium rise; soon the retinol in the retina falls, but that in the pigment epithelium continues to rise. During dark-adaptation the fall in pigment epithelial retinol parallels the rise in retinal in the retina (i.e. rhodopsin content). A similar cycle has been described more recently by Zimmerman (1974).

## Oxidation of retinol

The reduction of retinal to retinol and the reverse oxidation of retinol to retinal are catalysed by a retinene reductase-alcohol dehydrogenase enzyme, requiring NADP as co-enzyme. Daemen *et al.* (1974) and Zimmerman *et al.* (1975) have shown that bovine retinol dehydrogenase only acts on all-*trans* retinol, and if this is true the retinol transported from the pigment epithelium must arrive in the all-*trans* form, be oxidized in the outer segment and then isomerized to 11-*cis* retinal. However, Bridges (1976) was able to show that frog outer segments were capable of regenerating rhodopsin from added 11-*cis* retinol and were thus capable of oxidizing the retinol; this author points to the difficulty in explaining the utility of a small pool of 11-*cis* retinol in the dark-adapted rod outer segments described by

---

*Hubbard's (1956) preparation of retinene isomerase has been called into question by Plante and Rabinovitch (1972) who could find no evidence for an enzyme-controlled regeneration; Amer and Akhtar (1972) have described a preparation from rod outer segments that, when added to bleached rod outer segments, allows regeneration of rhodopsin from all-*trans* retinal; activity was lost by heating at 100°C. The complexities of dark-regeneration of rhodopsin have been discussed by Knowles and Dartnall (1977).

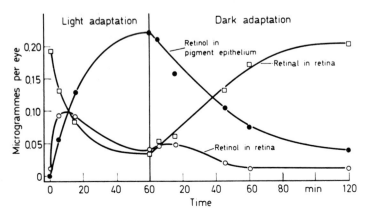

Fig. 8.23   Topographic distribution of retinal and retinol in the eyes of albino rats during light- and dark-adaptation. (Baumann, *Handbook of Sensory Physiology*. From Dowling.)

him, or yet the decrease in 11-*cis* retinol in the outer segments that takes place during dark-adaptation, unless the outer segments can utilize 11-*cis* retinol.

## ESTERIFICATION OF RETINOL

Retinol is stored in cells in oil droplets in the form of retinyl esters; in the frog only a few per cent of the total vitamin A is in the form of retinol (Bridges 1976), and over half of this is in the outer segments, which contain only 0.4 per cent of the total retinyl ester. In dark-adapted frogs between 35 to 47 per cent of the retinyl ester is in the *cis*-form, the remainder being all-*trans*. During light-adaptation the retinol formed passes into the pigment epithelium and is found in the oil-droplets as retinyl ester; thus some 90 per cent of that in the pigment epithelium is esterified and all that in the oil droplets; the esterification apparently takes place in the pigment epithelium so that it is retinol that leaves the outer segments. In fact the outer segment is the predominant site of free retinol; when contact with the pigment epithelium is lost, retinol accumulates in the outer segments which thus lack the ability to esterify retinol.

## Site of isomerization

During dark-adaptation a great deal of the rhodopsin regenerated comes from all-*trans* retinyl ester, and the site of the isomerization has been taken to be the pigment epithelium on the grounds that regeneration of rhodopsin is severely restricted in its absence. Bridges (1976) has discussed this assumption in some detail and raised, also, the question as to the form in which the isomerized material is transferred, i.e. is the returned material 11-*cis* retinal, 11-*cis* retinol, or 11-*cis* retinyl ester. In this last event, the outer segment would have to hydrolyse the ester, and subsequently oxidize it; this latter is certainly possible, as Bridges has shown, whilst Bibb and Young (1974) have observed a transfer of labelled fatty acids from pigment epithelium to outer segments, suggesting a hydrolysis of retinyl esters when they reach the rod outer segment. On the basis of this reasoning, Bridges has suggested a visual cycle illustrated by Figure 8.24. During exposure to light, all-*trans* retinol flows to the pigment epithelium where it is esterified and mixes with pre-existing stores of all-*trans* and 11-*cis* ester. Both isomers then return, as *esters*, to the outer segment where the all-*trans* ester is isomerized to 11-*cis* and is then hydrolysed to 11-*cis* retinol which is then oxidized to 11-*cis* retinal. Thus the loss of the pigment epithelium is manifest as a loss of esterification of the liberated retinol.*

---

*According to Zimmerman (1974), the vitamin A compounds in the pigment epithelium after complete dark-adaptation represent phagocytosed rod discs. Opsin also is present in corresponding amount (Sichi, 1973).

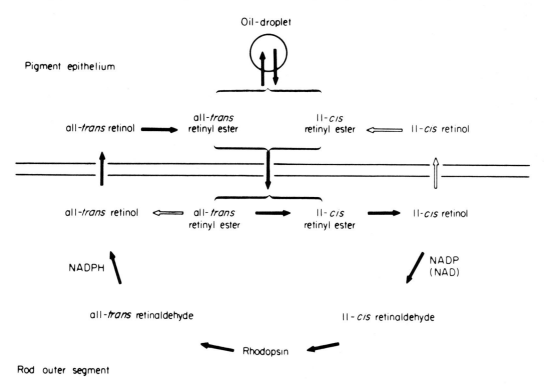

**Fig. 8.24** Suggested visual cycle in the frog eye. (Bridges, *Exp. Eye Res.*)

## SIGNIFICANCE OF THE SHUTTLE

As Azuma *et al.* (1977) point out, the outer segment-pigment epithelium 'shuttle' makes sense in that detached prosthetic groups are remotely stored for later use, being taken out of the danger zone where they might otherwise combine with opsin at 'wrong' sites. So far as the rate of transport is concerned, if we accept Azuma's figure of a regeneration capability of 12 per cent per minute, it is easy to show from Bridges' (1976) measurements of the traffic between pigment epithelium and rod outer segments, namely 130 molecules per sec per $\mu m^2$ of surface, that this would be adequate to maintain regeneration at its maximum rate.

## REGENERATION IN ISOLATED RETINA

Cramer and Sickel (1975) studied regeneration in an isolated perfused retina of the frog and were able to extract regenerated 11-*cis* rhodopsin, regenerated both thermally and by photo-isomerization. Using the same preparation but optical methods for identifying the regeneration process, Azuma *et al.* (1977) demonstrated a rapid and efficient regeneration provided that only small amounts of rhodopsin were bleached at a time; with this restriction, repeated bleaches followed by regeneration could be carried out. Figure 8.25 shows the time-course of regeneration, from which a rate-constant of $0.12 \ min^{-1}$ can be computed. Under these conditions, since any retinal would be rapidly lost, we must assume that the all-*trans* retinal is still attached to the opsin at one of the metarhodopsin stages; their spectral analysis indicated that it was metarhodopsin $II_{380}$ from which the regeneration took place. The spectral analysis indicated that, under these conditions

metarhodopsin $III_{465}$ did not appear in significant amounts. We may presume, then, that the regeneration observed in the isolated perfused retina is due to the retention of the retinal on the opsin molecule; its final hydrolysis would result in its rapid escape and require an intact pigment epithelium for its return to the outer segments.

## UPPER LIMIT TO BLEACHING

Hagins (1956) observed that, when using very short and very bright flashes of light—flash photolysis—an upper limit of 50 per cent bleaching was obtained, however intense the light. Williams (1974) has discussed the phenomenon, which he attributes to the photo-isomerization of the early products of bleaching, leading to re-formation of rhodopsin and of isorhodopsin; the essential feature is the time required by the metarhodopsins, produced by the bleaching light, to be broken down to compounds incapable of photo-isomerization, so that at a low temperature photo-isomerization is favoured and the bleaching is reduced, and the higher the temperature the greater the bleaching by the flash. Williams found a maximal bleaching of 88 per cent. *In vivo*, Pugh (1975) demonstrated that a 600 $\mu$sec xenon flash delivering as many as 15 rod-equivalent quanta per rhodopsin molecule, obtained a maximum of 40 to 50 per cent bleaching, as determined by reflexion-densitometry.

### TRANSPORT OF RETINOL

The visual cycle is not completely closed so that some of the liberated retinol is lost to the general circulation; moreover new supplies are required for replacement of that lost by phagocytosis of rod discs by the pigment epithelium. Retinol is transported in the blood stream attached to a retinol-binding protein (Kanai *et al.*, 1968; Muto and Goodman, 1972), and,

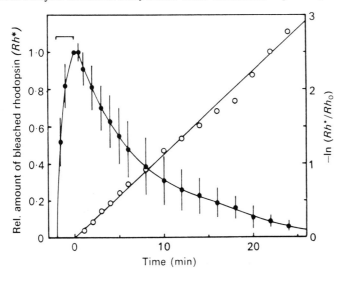

**Fig. 8.25** Time-course of regeneration of rhodopsin in the isolated perfused retina. Filled circles represent relative amount of bleaching (left-hand ordinate) and the open circles represent the function $-\ln (Rh^\star)/(Rh_o)$ (right-hand ordinates). (Azuma *et al.*, *J. Physiol.*)

in order that the retina may take up this blood-bound retinol, the tissue itself must contain retinol binding sites of sufficient affinity for retinol to permit a transfer from the plasma-binding protein to the retinol binding site.

### Pigment epithelial cell receptors

Bok and Heller (1976) injected labelled retinol-binding protein (RBP) into young rats and identified the labelled material attached to the pigment epithelial cells; the points of attachment presumably represented sites at which retinol would be first trapped from the circulating blood, and they were restricted to the basal and lateral parts of the plasma membrane. Once attached, the labelled retinol-binding protein remains where it is, so that addition of unlabelled material displaces it from its position. Thus this autoradiographic study suggests that the retinol in the blood is first trapped by attachment of the carrier protein, so that subsequent translocation to the interior of the cell requires transfer to a transport-protein belonging to the pigment epithelial cell.

### Tissue retinol-binding proteins

Homogenization of the retina should liberate material exhibiting this binding capacity. In fact retinol-binding proteins, different from plasma RBP have been isolated; thus Wiggert and Chader (1975) separated from homogenized pigment epithelium or retina a protein of about 19 000 daltons molecular weight; this appeared in the embryonic retina before development of the receptors and may well play a role during assembly of the outer segment membranes. This protein might serve not only to store retinol in the pigment epithelium but also in the translocation process from epithelium to receptor. Heller and Bok (1976) isolated soluble intracellular lipoproteins,* capable of binding retinol, from living pigment epithelial cells and receptor outer segments; these were similar to plasma lipoproteins, but differed from these in several respects; moreover the two lipoproteins from the outer segments and pigment epithelium also differed, suggesting different roles.

They suggest that these cytoplasmic carrier proteins are responsible for transport of the blood-borne retinol, trapped at the surface of the pigment epithelial cell, within the cell and thence to the receptor outer segment in accordance with the scheme of Figure 8.26; according to this, the cytosol-binding protein would be secreted into the interstitial space by the pigment epithelial cell whence it would pass to the receptor, be recognized there, and form a complex on its surface, a binding that would facilitate transfer of the retinol into the cytoplasm of the pigment epithelial cell, where it would be immediately taken up by another carrier protein and carried to the cell interior. A reverse process, whereby retinol is transported from pigment epithelial cell to receptor, is also envisaged. If the binding proteins of pigment epithelial cell and receptor are different, independent control over transport is easily envisaged.

## COMPARATIVE PHYSIOLOGY OF THE PIGMENTS

### RHODOPSIN AND PORPHYROPSIN

Köttgen and Abelsdorff (1896) observed that the 'visual purples' extracted from mammals, birds and amphibians were different from those extracted from various fresh-water fishes; the former had the usual maximal

---

* Saari et al. (1977) have emphasized that the cells of pigment epithelial preparations, obtained by brushing the eye-cup, are not intact, so that loss of low molecular weight proteins may occur. Their own studies indicated the presence of a binding protein of molecular weight 17 000 and emphasized the very high concentrations in the pigment epithelium vis à vis the retina. So far as this tissue was concerned, about 10 per cent of its binding protein was in the rod outer segments. Retinoic acid-binding protein is largely confined to the retina, probably in the inner segments.

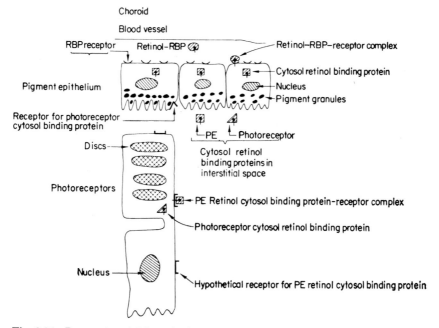

Choroid
Blood vessel
RBP receptor — Retinol-RBP — Retinol–RBP–receptor complex
Cytosol retinol binding protein
Pigment epithelium — Nucleus
Pigment granules
Receptor for photoreceptor cytosol binding protein
PE — Photoreceptor
Discs — Cytosol retinol binding proteins in interstitial space
Photoreceptors
PE Retinol cytosol binding protein-receptor complex
Photoreceptor cytosol retinol binding protein
Nucleus — Hypothetical receptor for PE retinol cytosol binding protein

**Fig. 8.26** Proposed model for retinol transport. (Heller and Bok, *Exp. Eye Res.*)

absorption near 500 nm whilst the fish pigments had a maximal absorption near 540 nm, and thus appeared more violet. Wald showed that the retinae of freshwater fishes gave retinene (retinal) and vitamin A (retinol) on bleaching, but it became evident that these were chemically different, having a double bond in the beta ionone ring:

Retinal

3 - dehydroretinal

The pigments were called porphyropsins, and the chromophore and its vitamin were called $retinene_2$ and vitamin $A_2$ to distinguish them from the $retinene_1$ and vitamin $A_1$ of the rhodopsins. With the use of the terms retinal and retinol in the place of $retinene_1$ and vitamin $A_1$, the corresponding compounds for the porphyropsins are called dehydroretinal and dehydro-retinol.

### MARINE AND FRESHWATER FISH

In general, the retinae of marine fish contain rhodopsins based on retinal, and we may view the development of porphyropsins, based on dehydroretinal, as an adaptation to permit the fish to make use of the longer wavelengths of light found in freshwater by comparison with sea-water. Euryhaline fish, spending part of their time in freshwater and part in sea-water, contain mixtures of rhodopsin and porphyropsin, the proportion varying with the environment.

## MULTIPLICITY OF PIGMENTS

Wald's statement that there were only two scotopic types of pigment, rhodopsin and porphyropsin, has been shown, mainly by Dartnall's work, to be incorrect. It is true that all the known pigments are based on only two retinals, but presumably because of the widely differing opsins, a very wide range of pigments with characteristic $\lambda_{max}$ ranging from 563 nm in the yellow to 430 nm in the violet has now been demonstrated; in general those with $\lambda_{max}$ in the long wavelength region— so-called porphyropsins—are based on $retinene_2$, but there is considerable overlap, so that the carp pigment based on $retinene_2$ has a $\lambda_{max}$ of 523 nm, whilst the gecko pigment, based on $retinene_1$ has a $\lambda_{max}$ of 524 nm. Furthermore, some

retinae gave extracts containing at least two photosensitive pigments of the scotopic type. Even the typical rhodopsin, moreover, differs significantly according to the species of retina from which it is extracted; thus cattle rhodopsin has a $\lambda_{max}$ of 499 nm, human 497 nm, squirrel and frog 502 nm.

### ADAPTATION TO ENVIRONMENT

The significance of the wide variations in absorption characteristics of the scotopic visual pigments, found in fish especially, is probably to be sought in the requirements of the animal's habitat. Thus Denton & Warren and Munz independently showed that deep-sea fishes, such as the conger eel, had golden coloured pigments that were apparently adapted for the reception of the predominantly blue light of their habitat, and subsequent study has indeed shown that the $\lambda_{max}$ of bathypelagic fishes range between 478 and 490 nm, and are based on $retinene_1$. As Munz has suggested, since some of these deep-sea fish have light-emitting organs, it is probable that their pigments are adapted to correspond with the $\lambda_{max}$ for *emission* of these organs.*

## VARIATIONS IN PIGMENT CONTENT

There is no doubt that with many species the pigment composition of a retina is not an invariable characteristic; this is revealed in the study of diadromous fishes, in the metamorphosis of larvae of some amphibians, and in the variation with the time of the year.

### DIADROMOUS FISH

Typical of these are the eels, which are spawned in sea-water, migrate to Europe where they are transformed to elvers or glass eels and begin the ascent of rivers, during which time they become known as yellow eels. Here they undergo a second metamorphosis preparatory to their 3000 to 4000 mile journey back to their

---

*Munz has established some very clear correlations between the spectral quality of the light available in different habitats and the $\lambda_{max}$ of the predominant visual pigment. Thus in clear oceanic waters the $\lambda_{max}$ of transmitted light at 200 metres is 475 nm and this corresponds to that of the absorption maxima of bathypelagic fishes; closer to the surface the $\lambda_{max}$ for transmission is about 485 to 495 nm corresponding to the $\lambda_{max}$ of the visual pigments of surface pelagic fishes. With rocky coasts the transmitted light is different from that in the turbid waters near sandy beaches, and there is a corresponding difference in the visual pigments (Munz, 1958). J. N. Lythgoe (1968) has pointed out, however, that for detection of a target, sensitivity to light is not the entire story since it is rather the contrast that is the determining factor, and better contrast may, in some circumstances, be obtained with a wavelength of light that is not the same as the wavelength of maximal absorption of the visual pigment. This is because the spectral composition of the light reflected from an object is different from that reaching the eye from the background.

spawning grounds, and they become silver eels. Both the American and European species of *Anguilla* have rhodopsin-porphyropsin pigment systems, with $\lambda_{max}$ at 501 and 523 nm respectively, sharing the same opsin. The immature yellow eels have mixtures of the two pigments in different proportions, but before the migration from the rivers to the sea, all the porphyropsin has been lost, but interestingly, some of the rhodopsin has shifted its $\lambda_{max}$ towards the short wavelengths to become a 'chrysopsin' of the deep-sea fish. Thus, as Beatty (1975) has emphasized in his study of an American species, *Anguilla rostrata*, the eels are peculiar in making use not only of the rhodopsin-porphyropsin paired system, based on the same opsin and different chromophores, but also, in its transition to sea-water it has utilized the variant employed by marine fishes in their adaptations to deep and shallow water, namely the modification of rhodopsin by a change in its opsin.

## Salmonids

The salmonids undergo the opposite change in salinity for breeding purposes. Ocean-caught salmon have predominantly rhodopsin in their retinas, but as Figure 8.27 shows, there is a progressive increase in the proportion of porphyropsin during the pre-spawn return to freshwater. It would seem that, with both eels and salmon, the changeover of pigment takes place before the actual migration to the new environment and is presumably initiated by some cyclical hormonal action.

### FROGS

Wald (1946) showed that in the frog, *Rana catesbiana*, the photopigment was rhodopsin* whereas in the tadpole it was porphyropsin, and Wilt (1959), studying the same species, showed that in the tadpole there was a mixture of rhodopsin and porphyropsin based on retinal and dehydroretinal, with a predominance of

porphyropsin which is absent in adult frogs. Figure 8.28 illustrates the results of a study on *Rana temporaria* by Muntz and Reuter (1966). When individual rods were examined by microspectrophotometry, the interesting observation was made that individual receptors from tadpoles contained mixtures of the pigments.

### SYNTHESIS OF VITAMIN A

It seems generally established that the vitamin A employed by the eye in synthesizing its pigment is determined by synthetic mechanisms in the eye rather than the liver, so that the vitamin A of the liver of freshwater fish is retinol although that of the porphyropsin system is dehydroretinol; the critical enzyme that catalyses the formation of dehydroretinol from retinol is retinol dehydrogenase, the transformation occurring in the isolated eye *in vitro* (Ohtsu *et al.* 1964), a transformation that is inhibited by thyroxine; thus the hormone that promotes metamorphosis acts directly on the synthetic mechanisms of the eye so far as retinol is concerned. The possible routes of interconversion of porpyropsin and rhodopsin are indicated in Figure 8.29, the fundamental reaction being the dehydrogenation of retinaldehyde to dehydroretinaldehyde, rather than the dehydrogenation of retinol.

### DEHYDROGENATION OF RETINOL

Owing to the rapid conversion of retinal to retinol, any dehydrogenation of retinol would have to occur in the outer segments; the migration of retinol to the pigment epithelium would require the dehydrogenation of this

---

*In the frog *Rana pipiens*, Liebman and Entine (1968) found either a rhodopsin ($\lambda_{max}$ 502) or a porphyropsin ($\lambda_{max}$ 527) in the frog or tadpole stages respectively, but there was never any evidence of a mixture, before the final stages of metamorphosis.

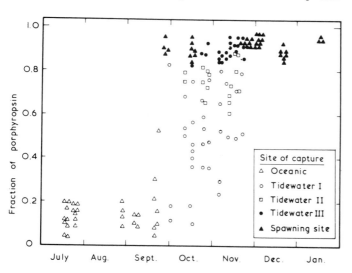

**Fig. 8.27** The proportion of porphyropsin in the retinas of individual adult coho salmon, *Oncorhynchus kisutch*, during their return to freshwater spawning grounds. The appearance of the Tidewater Group I fish is identical to ocean-caught fish; Group III fish are nearly identical with those found at the spawning grounds, while Group II are intermediate in appearance. (Knowles and Dartnall, *The Eye*, from Beatty.)

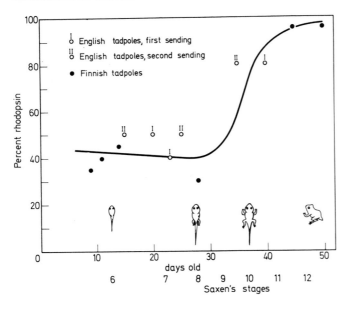

**Fig. 8.28** Variations in the proportion of rhodopsin in retinal extracts of *Rana temporaria* during ontogenesis. (Bridges, *Handbook of Sensory Physiology*, from Muntz and Reuter.)

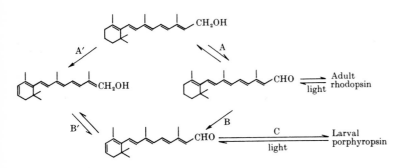

**Fig. 8.29** Alternative steps in the conversion of retinol into the 3-dihydroretinaldehyde prosthetic group of porphyropsin. (Bridges, *Handbook of Sensory Physiology*.)

to take place there. We may assume that thyroxine inhibits the dehydrogenase at steps A′ or B; this would limit the supply of dehydroretinaldehyde leading to the predominance of rhodopsin.

TEMPORAL VARIATIONS

Dartnall *et al* (1961) found that, in a given population of rudd, the proportions of the two retinal pigments based on retinal and dehydro-retinal varied with time of year, and this was due to the variations in the length of the days; in winter the dominant pigment was that with $\lambda_{max}$ 543 nm based on dehydroretinal$_2$, whilst in summer it was that with $\lambda_{max}$ 510 based on retinaldehyde. By artificially varying the environmental illumination in an aquarium, the proportions of the pigments could be varied, and it became clear that it was the total duration of exposure to light that determined the magnitude of the shift. Figure 8.30 illustrates the seasonal variations in rhodopsin and porphyropsin contents of the freshwater cyprinid *Notemigonus* studied by Bridges (1965), and the important point emerges

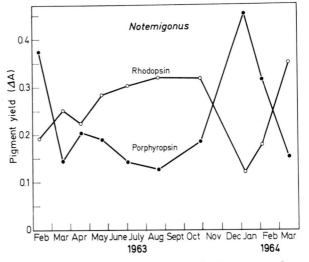

**Fig. 8.30** Seasonal reciprocal variation in the amounts of rhodopsin and porphyropsin in *Notemigonus* retinas. ○, rhodopsin; ●, porphyropsin. (Bridges, *Handbook of Sensory Physiology*.)

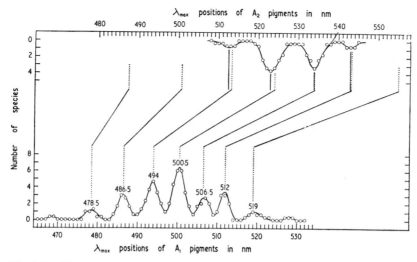

**Fig. 8.31** The relationship between the spectral distributions of $A_1$ and $A_2$ pigments in teleost fishes. The data for the $A_1$ pigments are derived from those fishes that possess *only* $A_1$ pigments; the data for the $A_2$ pigments are not restricted: *i.e.* they include all fishes that possess $A_2$ pigments, some of which have $A_1$ pigments as well. (Dartnall and Lythgoe, *Vision Res.*)

that there is a reciprocal relation between the contents, suggesting that the one is made at the expense of the other.

PHYLOGENETIC ASPECTS

Wald considered that the dehydroretinal pigments, his porphyropsins, were exclusively found in fresh-water fishes; and this was the basis for some interesting phylogenetic generalizations. However, there is no doubt from Schwanzara's (1967) study that many freshwater fishes have a retinal pigment, in fact 35 per cent of those examined had this, unmixed with a dehydroretinal pigment; moreover, mixtures commonly were found in fishes that were not euryhaline. Again, marine fishes did not necessarily have exclusively retinal pigments. Schwanzara found a dominance of dehydro-retinal pigments in freshwater fishes of temperate waters and of retinal pigments in tropical freshwater fishes; and he attributed this to the different spectral qualities of the light in tropical and temperate fresh-water.

RETINAL AND OPSIN VARIATIONS

In general, we can expect two ways of varying the character of visual pigment; first with a given retinene we may vary the opsin part of the molecule; with retinal this gives quite a wide range of possibilities permitting pigments that enable the honey-bee, for example, to make full use of ultra-violet light ($\lambda_{max}$ 440 to 450 nm); the deep-sea fish to have golden retinae with $\lambda_{max}$ in the region of 480 nm; mammals to have pigments with $\lambda_{max}$ around 500 and so on up to the chicken's pigment, iodopsin, with $\lambda_{max}$ on the long-wavelength side in the region of 560 nm. Another series of pigments may be obtained by varying the opsins attached to dehydroretinal Figure 8.31 shows the results of studies by Dartnall and Lythgoe (1965) and of Bridges (1964) on the $\lambda_{max}$ of pigments extracted from a variety of fishes based on dehydroretinal (above) and retinal (below). The figure illustrates, first, that there is considerable overlap between the two types, and it also shows that there is a tendency for the $\lambda_{max}$ to cluster

around certain wavelengths. Dartnall and Lythgoe established an empirical relationship between the $\lambda_{max}$ of pigments with the same opsin, and retinal or dehydroretinal attached; with the aid of this, they predicted the shifts in $\lambda_{max}$ of pigments containing retinal that would be obtained if the same opsin were attached to dehydroretinal. It will be seen that the predictions, indicated by the positions of the upper vertical dotted lines, agree with actual pigments very well.

OPSINS

With retinas containing mixtures of porphyropsins and rhodopsins it might be assumed that the same opsin would serve for both pigments although this would be difficult to prove; Figure 8.32 shows the relationship between the $\lambda_{max}$ of

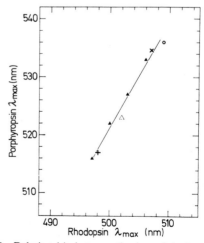

**Fig. 8.32** Relationship between the $\lambda_{max}$ of rhodopsins and their corresponding porphyropsins. Different symbols represent pairs from different species. + represents a synthetic pair based on cattle opsin. (Bridges, *Handbook of Sensory Physiology.*)

corresponding pairs from various species, and it is remarkable that when the $\lambda_{max}$ of one of a pair is plotted against the $\lambda_{max}$ of the other, a straight line is obtained which includes a synthetic pair namely a rhodopsin 497 and a porphyropsin 517 derived from cattle opsin.

## PHOTOPIC PIGMENTS

By means of reflexion densitometry Rushton and Weale have shown that both foveal and peripheral cones undergo a bleaching process on exposure to light; and the action spectrum for this bleaching corresponds with the photopic luminosity curve, with a maximum at longer wavelengths than for scotopic vision (*cf.* Ripps and Weale, 1970). The cones serve not only to respond to light at high intensities but also to discriminate its spectral quality; they are the receptors for colour vision. We shall see that this discrimination in man and many other species requires that the cones be of at least three kinds, containing different pigments; hence we need not expect to extract a single pigment with the absorption spectrum corresponding with the photopic spectral sensitivity curve. In fact, as we shall see, there is strong spectrophotometric evidence for the existence of three separate pigments in the cones of fishes and primates.

### IODOPSIN

When the relatively pure cone retina of the chicken was extracted Wald (1937) did, indeed, obtain a pigment with $\lambda_{max}$ 562 nm which he called *iodopsin*. It is based on retinal$_1$ and undergoes the same cycle of changes undergone by rhodopsin and thus differs from the latter in the opsin part of the molecule (Yoshizawa and Wald, 1967); by condensing dehydroretinal$_2$ with the chicken opsin, Wald *et al.* (1955) obtained a new pigment which they called *cyanopsin* with $\lambda_{max}$ of 620 nm; this has so far not been extracted from any retina.

### Artefact?

If it is suggested that iodopsin is the photopigment for the red-absorbing cones of the retina responsible for the red-contribution to chromatic sensation (p. 341), then it should only represent the pigment for a portion of the total cones, but it seems unlikely that the high proportion of iodopsin found in extracts of chicken retina could have been derived from only a fraction of the total population (Knowles, 1976); moreover, by microspectrophotometry it can be shown that the specific red-absorbing pigment of chicken retina has a $\lambda_{max}$ of 569 nm (Bowmaker and Knowles, 1977), and thus longer than that of iodopsin. According to Knowles (1976), extracts made in the presence of chloride ions did not contain a 560 nm pigment; when chloride was added to an extract, absorbance was lost around 500 nm and gained at 560 nm, suggesting that the 560 nm material was derived from a rhodopsin with $\lambda_{max}$ of about 500 nm. Thus it could be that iodopsin is an artefact of preparation.

## REFERENCES

Abrahamson, E. W. & Ostroy, S. E. (1967) The photochemical and macromolecular aspects of vision. *Progr. Biophys.* **17**, 181–215.

Abrahamson, E. W. & Wiesenfeld, J. R. (1972) The structure, spectra, and reactivity of visual pigments. *Hdb. Sensory Physiol.*, 7/1, 69–121.

Akhtar, M. & Hirtenstein, M. D. (1969) Chemistry of the active site of rhodopsin. *Biochem. J.* **115**, 607–608.

Amer, S. & Akhtar, M. (1972) The regeneration of rhodopsin from all-trans retinal: Solubilization of enzyme system involved in the completion of the visual cycle. *Biochem. J.* **128**, 987–989.

Amer, S. & Akhtar, M. (1973) Studies on the regeneration of rhodopsin from all-*trans* retinal in isolated rat retinae. *Nature* **245**, 221–223.

Anderson, D. H. & Fisher, S. K. (1976) The photoreceptors of diurnal squirrels; outer segment structure, disc shedding and protein renewal. *J. Ultrastr. Res.* **55**, 119–141.

Azuma, K., Azuma, M. & Sickel, W. (1977) Regeneration of rhodopsin in frog rod outer segments. *J. Physiol.* **271**, 747–759.

Basinger, S., Bok, D. & Hall, M. (1976) Rhodopsin in the rod outer segment plasma membrane. *J. Cell Biol.* **69**, 29–42.

Baumann, C. (1970) Regeneration of rhodopsin in the isolated retina of the frog (*Rana esculenta*). *Vision Res.* **10**, 627–637.

Beatty, D. D. (1975) Visual pigments of the American eel *Anguilla rostrata*. *Vision Res.* **15**, 771–776.

Bibb, C. & Young, R. W. (1974) Renewal of glycerol in the visual cells and pigment epithelium of the frog retina. *J. Cell Biol.* **62**, 378–389.

Bittar, E. E. (1966) Effect of inhibitors and uncouplers on the Na pump of the *Maia* muscle fibre. *J. Physiol.*, **187**, 81–103.

Blasie, J. K. & Worthington, C. R. (1969) Planar liquid-like arrangement of photopigment molecule in frog retinal receptor disk membranes. *J. mol. Biol.* **39**, 417–439.

Bok, D. & Heller, J. (1976) Transport of retinol from the blood to the retina: an autoradiographic study of the pigment epithelial cell surface receptor for plasma retinol-binding protein. *Exp. Eye Res.* **22**, 395–402.

Bok, D. & Young, R. W. (1972) The renewal of diffusely distributed protein in the outer segments of rods and cones. *Vision Res.* **12**, 161–168.

Borggreven, J. M. P. M., Rotmans, J. P., Bonting, S. L. & Daemen, F. J. M. (1971) The role of phospholipids in cattle rhodopsin studied with phospholipase C. *Arch. Biochem. Biophys.* **145**, 290–299.

Bowmaker, J. K. & Knowles, A. (1977) The visual pigments and oil droplets of the chicken retina. *Vision Res.* **17**, 755–764.

Bownds, D. (1967) Site of attachment of retinal in rhodopsin. *Nature* 216, 1178–1181.

Bridges, C. D. B. (1961) Studies on the flash photolysis of visual pigments. *Biochem. J.* 79, 128–134.

Bridges, C. D. B. (1964) The distribution of visual pigments in freshwater fishes. *Abstr. Fourth Internat. Congr. Photobiol.*, Oxford, p. 53. Bucks: Beacon Press.

Bridges, C. D. B. (1976) Vitamin A and the role of the pigment epithelium during bleaching and regeneration of rhodopsin in the frog eye. *Exp. Eye Res.* 22, 435–455.

Bridges, C. D. B. & Yoshikami, S. (1969) Uptake of tritiated retinaldehyde by the visual pigment of dark-adapted rats. *Nature* 221, 275–276.

Cohen, A. I (1968) New evidence supporting the linkage to extracellular space of outer segment saccules of frog cones but not rods. *J. Cell Biol.* 37, 424–444.

Cohen, A. I. (1970) Further studies on the question of the patency of saccules in outer segments of vertebrate photoreceptors. *Vision Res.* 10, 445–453.

Collins, F. D. & Morton, R. A. (1950) Studies on rhodopsin. I–III. *Biochem. J.* 47, 3–9, 10–17, 18–24.

Cramer, C. & Sickel, E. (1968) Measuring regeneration of rhodopsin by an extraction technique in perfused vertebrate retina. *Pflug. Arch. ges. Physiol.* 355, R 110.

Crescitelli, F. & Dartnall, H. J. A. (1953) Human visual purple. *Nature* 172, 195–196.

Daemen, F. J. M., Rotmans, J. P. & Bonting, S. L. (1974) On the rhodopsin cycle. *Exp. Eye Res.* 18, 97–103.

Danielli, J. F. & Davson, H. (1935) A contribution to the theory of permeability of thin films. *J. cell. comp. Physiol.* 5, 495.

Dartnall, H. J. A. (1957) *The Visual Pigments.* London: Methuen.

Dartnall, H. J. A., Lander, M. R. & Munz, F. W. (1961) Periodic changes in the visual pigment of a fish. In *Progress in Photobiology*, pp. 203–213. Amsterdam, Elsevier.

Dartnall, H. J. A. & Lythgoe, J. N. (1965) The spectral clustering of visual pigments. *Vision Res.* 5, 81–100.

DeGrip, W. J., Bonting, S. L. & Daeman, F. J. M. (1973) The binding site of retinaldehyde in cattle rhodopsin. *Biochim. biophys. Acta* 303, 189–193.

DeGrip, W. J., Van de Laar, G. L. M., Daemen, F. J. M. & Bonting, S. L. (1973) Biochemical aspects of the visual process. XXIII. *Biochim. biophys. Acta* 325, 315–322.

Denton, E. J. (1954) On the orientation of molecules in the visual rods of *Salamandra maculosa*. *J. Physiol.* 124, 17 P.

Denton, E. J. (1959) The contribution of the photosensitive and other molecules to the absorption of whole retina. *Proc. Roy. Soc., B.* 150, 78–94.

Denton, E. J. & Warren, F. J. (1956) Visual pigments of deep sea fish. *Nature* 178, 1059.

Ditto, M. (1975) A difference between developing rods and cones in the formation of the outer segment membranes. *Vision Res.* 15, 535–536.

Donner, K. O. & Hemila, S. (1975) Kinetics of long-lived rhodopsin photoproducts in the frog retina as a function of the amount bleached. *Vision Res.* 15, 985–995.

Dowling, J. E. (1960) Chemistry of visual adaptation in the rat. *Nature* 188, 114–118.

Dowling, J. E. (1965) Foveal receptors of the monkey retina: fine structure. *Science* 147, 57–59.

Dowling, J. E. & Hubbard, R. (1963) Effect of brilliant flashes on light and dark adaptation. *Nature* 199, 972–975.

Fager, R. S., Sejnowski, P. & Abrahamson, E. W. (1972) Aqueous cyanohydridoborate reduction of the rhodopsin chromophore. *Biochem. Biophys. Res. Comm.* 47, 1244–1247.

Frank, R. N. (1969) Photoproducts of rhodopsin bleaching in the isolated, perfused frog retina. *Vision Res.* 9, 1415–1433.

Frank, R. N. & Dowling, J. E. (1968) Rhodopsin photoproducts: effects on electroretinogram sensitivity in isolated perfused rat retina. *Science* 161, 487–489.

Hagins, F. M. (1973) Purification and partial characterization of the protein component of squid rhodopsin. *J. biol. Chem.* 248, 3298–3304.

Hagins, W. A. (1956) Flash photolysis of rhodopsin in the retina. *Nature* 177, 989–990.

Hagins, W. A. (1957) Rhodopsin in a mammalian retina. Thesis, University of Cambridge.

Harosi, F. I. & MacNichol, E. F. (1974) Visual pigments of goldfish cones. Spectral properties and dichroism. *J. gen. Physiol.* 63, 279–304.

Heller, J. (1968) Purification, molecular weight, and composition of bovine visual pigment. *Biochem.* 7, 2906–2913.

Heller, J. & Bok, D. (1976) Transport of retinol from the blood to the retina: involvement of high molecular weight lipoproteins as intracellular carriers. *Exp. Eye Res.* 22, 403–410.

Hogan, M. J. & Wood, I. (1974) Phagocytosis by pigment epithelium of human retinal cones. *Nature* 252, 305–307.

Hubbard, R. (1954) The molecular weight of rhodopsin and the nature of the rhodopsin–digitonin complex. *J. gen. Physiol.* 37, 373–379.

Hubbard, R. (1956) Retinene isomerase. *J. gen. Physiol.* 39, 935–962.

Hubbard, R. & Colman, A. D. (1959) Vitamin A content of the frog eye during light and dark adaptation. *Science* 130, 977–978.

Hubbard, R. & Kropf, A. (1959) Molecular aspects of visual excitation. *Ann. N.Y. Acad. Sci.* 81, 388–398.

Hubbard, R. & Wald, G. (1952) Cis-trans isomers of vitamin A and retinene in the rhodopsin system. *J. gen. Physiol.* 36, 269–315.

Jancsó, N. v. & Jancsó, H. v. (1936) Fluoreszenmikroskopische Beobachtung der reversiblen Vitamin-A Bildung in der Netzhaut wahrend des Sehaktes. *Biochem. Z.* 287, 289–290.

Kanai, M., Razh, A. & Goodman, De W. S. (1968) Retinol-binding protein: the transport protein for vitamin A in human plasma. *J. clin. Invest.* 47, 2025–2044.

Kimble, E. A. & Ostroy, S. E. (1974) Kinetics of the reaction of the sulfhydryl groups of rhodopsin. *Biochim. biophys. Acta* 325, 323–331.

Knowles, A. (1976) The effect of chloride upon chicken visual pigments. *Biochem. Biophys. Res. Comm.* 73, 56.

Knowles, A. & Dartnall, H. J. A. (1977) *The Photobiology of Vision.* Vol. 2B, *The Eye* (Ed. Davson, H.). Academic Press, New York and London.

Köttgen, E. & Abelsdorff, G. (1896) Absorption und Zersetzung des Sehpurpurs bei den Wirbeltieren. *Z. Psychol. Physiol. Sinnesorg.* 12, 161–184.

La Vail, M. M. (1976) Rod outer segment disc shedding in relation to cyclic lighting. *Exp. Eye Res.* 23, 277–280.

Liebman, P. A. & Entine, G. (1968) Visual pigments of frog and tadpole (*Rana pipiens*). *Vision Res.* 8, 761–775.

Liebman, P. A., Jagger, W. S., Kaplan, M. W. & Bargoot, F. G. (1974) Membrane structure changes in rod outer segments associated with rhodopsin bleaching. *Nature* 251, 31–36.

Lythgoe, J. N. (1968) Visual pigments and visual range. *Vision Res.* 8, 997–1011.

Lythgoe, R. J. (1937) Absorption spectra of visual purple and visual yellow. *J. Physiol.* 89, 331–358.

Lythgoe, R. J. & Quilliam, J. P. (1938) The relation of transient orange to visual purple and indicator yellow. *J. Physiol.* **94**, 399–410.

Mathies, R., Oseroff, A. R. & Stryer, L. (1976) Rapid flow resonance Raman spectroscopy of photolabile molecules: rhodopsin and isorhodopsin. *Proc. Nat. Acad. Sci. Wash.* **73**, 1–5.

Muntz, W. R. A. & Reuter, T. (1966) Visual pigments and spectral sensitivity in *Rana temporaria* and other European tadpoles. *Vision Res.* **6**, 601–618.

Munz, F. W. (1958) Photosensitive pigments from the retinae of certain deep-sea fishes. *J. Physiol.* **140**, 220–235.

Munz, F. W. (1958) The photosensitive retinal pigments of fishes from relatively turbid coastal waters. *J. gen. Physiol.* **42**, 445–459.

Muto, Y. & Goodman, De W. S. (1972) Vitamin A transport in rat plasma. *J. biol. Chem.* **247**, 2533–2541.

Nilsson, S. E. G. (1964) Receptor cell outer segment development and ultra-structure of the disk membranes in the retina of the tadpole (*Rana pipiens*). *J. Ultrastr. Res.* **11**, 581–620.

O'Brien, P. J. (1977) Incorporation of mannose into rhodopsin in isolated bovine retina. *Exp. Eye Res.* **24**, 449–458.

Ohtsu, K., Naito, K. & Wilt, F. H. (1966) Metabolic basis of visual pigment conversion in metamorphosing *Rana catesbiana. Dev. Biol.* **10**, 216–232.

Oseroff, A. R. & Callender, R. H. (1974) Resonance Raman spectroscopy of rhodopsin in retinal disk membranes. *Biochemistry* **13**, 4243–4248.

Ostroy, S. E. (1974) Hydrogen ion changes of rhodopsin. *Arch. Biochem. Biophys.* **164**, 275–284.

Papermaster, D. S., Converse, C. A. & Siu, J. (1975) Membrane biosynthesis in the frog retina: opsin transport in the photoreceptor cell. *Biochemistry* **14**, 1343–1352.

Paulsen, R., Miller, J. A., Brodie, A. E. & Bownds, M. D. (1975) The decay of long-lived photoproducts in the isolated bullfrog rod outer segment: relationship to other dark reactions. *Vision Res.* **15**, 1325–1332.

Plante, E. O. & Rabinovitch, B. (1972) Enzymes in the regeneration of rhodopsin. *Biochem. Biophys. Res. Comm.* **46**, 725–730.

Poincelot, R. P., Millar, P. G., Kimbel, R. L. & Abrahamson, E. W. (1969) Lipid to protein chromophore transfer in the photolysis of visual pigments. *Nature* **221**, 256–257.

Poo, M. M. & Cone, R. A. (1973) Lateral diffusion of rhodopsin in *Necturus* rods. *Exp. Eye Res.* **17**, 503–510.

Pugh, E. N. (1975) Rhodopsin flash photolysis in man. *J. Physiol.* **248**, 393–412.

Renthal, R., Steinemann, A. & Stryer, L. (1973) The carbohydrate moiety of rhodopsin: lectin binding, chemical modification and fluorescence studies. *Exp. Eye Res.* **17**, 511–515.

Reuter, T. (1976) Photoregeneration of rhodopsin and isorhodopsin from metarhodopsin III in the frog retina. *Vision Res.* **16**, 909–917.

Ripps, H. & Weale, R. A. (1969) Flash bleaching of rhodopsin in the human retina. *J. Physiol.* **200**, 151–159.

Ripps, H. & Weale, R. A. (1970) The photophysiology of vertebrate colour vision. *Photophysiol.* **5**, 127–168.

Robinson, W. E., Gordon-Walker, A. & Bownds, D. (1972) Molecular weight of frog rhodopsin. *Nature New Biol.* **235**, 112–114.

Rotmans, J. P., Daemen, F. J. M. & Bonting, S. L. (1974) Biochemical aspects of the visual process. XXVI. Binding site and migration of retinaldehyde during rhodopsin photolysis. *Biochim. biophys. Acta* **357**, 151–158.

Saari, J. C., Bunt, A. H., Futterman, S. & Berman, E. R. (1977) Localization of cellular retinol-binding protein in bovine retina and retinal pigment epithelium. . . . *Invest. Ophthal.* **16**, 797–806.

Schichi, H., Lewis, M. S., Irreviere, F. & Stone, A. L. (1969) Purification and properties of bovine rhodopsin. *J. biol. Chem.* **244**, 529–536.

Schwanzara, S. A. (1967) The visual pigments of freshwater fishes. *Vision Res.* **7**, 121–148.

Sjöstrand, F. S. (1953) The ultrastructure of the outer segments of rods and cones of the eye as revealed by electron microscopy. *J. cell. comp. Physiol.* **42**, 45–70.

Spitznas, M. & Hogan, M. J. (1970) Outer segments of photoreceptors and the retinal pigment epithelium. *Arch. Ophthal.* **84**, 810–819.

Steinberg, R. H., Wood, I. & Hogan, M. J. (1977) Pigment epithelial ensheathment and phagocytosis of extrafoveal cones in human retina. *Phil. Trans.* **277**, 459–476.

Tokunaga, D., Kawamura, S. & Yoshizawa, T. (1976) Analysis by spectral difference of the orientational change of the rhodopsin chromophore during bleaching. *Vision Res.* **16**, 633–641.

Trayhurn, P. & Habgood, J. O. (1975) The effect of trypsin on the retinal rod outer segments: trypsin digestion as a means of isolating viable discs. *Exp. Eye Res.* **20**, 479–487.

Wald, G. (1935) Carotenoids and the visual cycle. *J. gen. Physiol.* **19**, 351–371.

Wald, G. (1937) Photo-labile pigments of the chicken retina. *Nature* **140**, 545–546.

Wald, G. (1939) The porphyropsin visual system. *J. gen. Physiol.* **22**, 775–794.

Wald, G. (1946) The chemical evolution of vision. *Harvey Lectures* **41**, 148–152.

Wald, G. (1960) The distribution and evolution of visual systems. In *Comparative Biochemistry*, vol. I. New York: Academic Press.

Wald, G. & Brown, P. K. (1952) The role of sulphydryl groups in the bleaching and synthesis of rhodopsin. *J. gen. Physiol.* **35**, 797–821.

Wald, G., Brown, P. K. & Smith, P. H. (1955) Iodopsin. *J. gen. Physiol.* **38**, 623–681.

Wiggert, B. O. & Chader, G. J. (1975) A receptor for retinol in the developing retina and pigment epithelium. *Exp. Eye Res.* **21**, 143–151.

Williams, T. P. (1974) Upper limits to the bleaching of rhodopsin by high light intensities. *Vision Res.* **14**, 603–607.

Wilt, F. H. (1959) The differentiation of visual pigments in metamorphosing larvae of *Rana catesbiana. Dev. Biol.* **1**, 199–233.

Wong, J. K. & Ostroy, S. E. (1973) Hydrogen ion changes of rhodopsin. I. Proton uptake during the metarhodopsin $I_{478}$ metarhodopsin $II_{308}$ reactions. *Arch. Biochem. Biophys.* **154**, 1–7.

Worthington, C. R. (1973) X-ray analysis of retinal photoreceptor structure. *Exp. Eye Res.* **17**, 487–501.

Wu, C.-W. & Stryer, L. (1972) Proximity relationships in rhodopsin. *Proc. Nat. Acad. Sci. Wash.* **69**, 1104–1108.

Yoshizawa, T. (1972) The behaviour of visual pigments at low temperatures. *Hdb. Sensory Physiol.* 7/1, 146–179.

Yoshizawa, T. & Wald, G. (1967) Photochemistry of iodopsin. *Nature* **214**, 566–571.

Young, R. W. (1967) The renewal of receptor cell outer segments. *J. Cell Biol.* **33**, 61–72.

Young, R. W. (1971) Shedding of discs from rod outer

segments in the rhesus monkey. *J. Ultrastr. Res.* **34,** 190–203.

Young, R. W. (1976) Visual cells and the concept of renewal. *Invest. Ophthal.* **15,** 700–725.

Young, R. W. & Droz, B. (1968) The renewal of protein in retinal rods and cones. *J. Cell Biol.* **39,** 169–184.

Zimmerman, W. F. (1974) The distributions and proportions of vitamin A compounds during the visual cycle in the rat. *Vision Res.* **14,** 795–802.

Zimmerman, W. F., Lion, R., Daemen, F. J. M. & Bonting,

S. L. (1975) Distribution of specific retinol dehydrogenase activities in sub-cellular fractions of bovine retina and pigment epithelium. *Exp. Eye Res.* **21,** 325–332.

Zorn, M. (1974) The effect of blocked sulfhydryl groups on the regenerability of bleached rhodopsin. *Exp. Eye Res.* **19,** 215–221.

Zorn, M. & Futterman, S. (1971) Properties of rhodopsin dependent upon associated phospholipid. *J. biol. Chem.* **246,** 881–886.

# 9. Dark-adaptation and the minimum stimulus for vision

## THE DARK-ADAPTATION CURVE

We have seen that, in the dark, the sensitivity of the human eye to light increases until a maximum is reached after about 30 minutes; this gives the dark-adaptation curve showing the progressive fall in threshold. We have already concluded that the curve represents the course of recovery of function of the rods. This curve, however, is not smooth, but shows a 'kink' so that adaptation consists in an initial rapid phase of recovery of sensitivity, which is finished in about 5 to 10 minutes, and a slower phase.

### CONE AND ROD PORTIONS

The first rapid phase has been shown, by many experiments, to represent the increase in sensitivity of the cones in the dark, so that the threshold during the first few minutes in the dark is determined by the cones, the thresholds of the rods still being higher than those of the cones. Numerous lines of evidence support this conclusion. For example, if the area of the light-stimulus used for measuring the threshold is small, and is concentrated on the fovea, the curve of dark-adaptation consists only of the initial part. As the size of the stimulus is increased, so that overlap on to the parafovea, containing rods, occurs, the dark-adaptation curve shows the typical kink with cone and rod portions (Fig. 9.1). Again, if coloured stimuli are used to test the threshold, during the early portion of the curve the subject appreciates the colour of the threshold stimulus, whilst at the later stages the light is said to be white. Yet again, by varying the state of light-adaptation of the subject before he is placed in the dark, the type of curve may be changed. After a period of weak light-adaptation the cones are nearly at their maximum sensitivity, so that the threshold on going into the dark is very soon determined by the rods; hence the kink comes early or not at all (Fig. 9.2). If the wavelength of the adapting light is varied, the relative extents to which the thresholds of the rods and cones are raised will be different, so that on placing the subject in the dark after adapting him to red light, for instance, his rods will be relatively active—because red light barely influences these—whilst the cones will have a high threshold; the curve for dark-adaptation will therefore have no kink, the rods determining the threshold from the beginning.

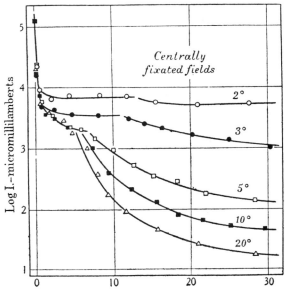

**Fig. 9.1** Dark-adaptation curves with centrally fixated fields subtending larger and larger angles at the eye. With the larger fields peripheral vision comes into play giving the characteristic break in the curve. (Hecht, Haig and Wald, *J. gen. Physiol.*)

### ROD MONOCHROMAT

Finally, a human subject defective in cone vision, the so-called *rod monochromat*, has a dark-adaptation curve with no kink (Fig. 9.3). This type of subject is useful since he permits the study of the rod thresholds without interference from the cones; thus, immediately on placing the normal light-adapted subject in the dark his threshold is determined by the cones; in other words the rods are less sensitive than the cones and so we cannot tell what their threshold is. In the rod-monochromat, however, we can measure the rod threshold as soon as he is placed in the dark and this is found to be very high; nevertheless the rods are not incapable of being stimulated, so that in ordinary daylight, although their thresholds are higher than those of the cones, they may be responding to light. However, there seems no doubt from Aguilar and Stiles' study, that the rods become 'saturated' at high levels of illumination (2000 to 2500 trolands or 100 to 300 cd/m²) in the sense that they become incapable of responding to changes of illumination (the l.b.i. is

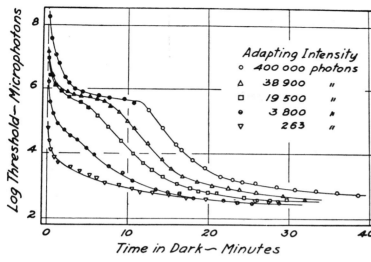

**Fig. 9.2**  The course of dark-adaptation as measured with violet light following different degrees of light-adaptation. The filled-in symbols indicate that a violet colour was apparent at the threshold, while the empty symbols indicate that the threshold was colourless. (Hecht, Haig and Chase, *J. gen. Physiol.*)

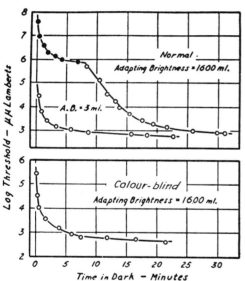

**Fig. 9.3**  Dark-adaptation curves for a normal subject and a rod monochromat. After adaptation to 1600 mL, the normal subject shows a rapid cone adaptation (solid circles) followed by a slow rod adaptation (open circles). The colour-blind shows only a rod curve of the rapid type. (Hecht, Shlaer, Smith, Haig and Peskin, *J. gen. Physiol.*)

infinitely high); the same was shown by Fuortes, Gunkel and Rushton on their rod-monochromat who was virtually blind at high levels of illumination; the rods were responding to light, but not to changes, so that there was no awareness of contrast.

## REGENERATION OF RHODOPSIN

Light-adaptation is accompanied by the bleaching of rhodospin as shown by experiments involving extraction

of pigment from eyes before and after exposure to light. It seems reasonable to suppose, therefore, that the process of dark-adaptation has as its basis the regeneration of rhodopsin in the rods; and the early studies of Tansley and Peskin certainly showed that the time-course of regeneration, as measured by extraction of frog's retinae after different periods in the dark, corresponded with the time-course of dark-adaptation. By the technique of *reflexion densitometry* the changes in concentration of pigment in the retina have been followed in the intact eye in man and lower animals.*

### REFLEXION DENSITOMETRY

The technique is based on the principle of measuring the light reflected from the fundus of the eye; if the retina contains rhodopsin it will absorb blue-green light of 500 nm in preference to red and violet light, hence the light reflected from the fundus will contain less light of 500 nm and more of these other wavelengths. The loss of intensity of light, moreover, will give a direct measure of the degree of absorption, and thus of the optical density of the rhodopsin in the retina. Careful analysis of the changes in amount of light after reflexion showed that they were indeed due, in the dark-adapted eye, to the presence of rhodopsin (Campbell and Rushton), the degree of absorption of the different wavelengths corresponding approximately with the extinction spectrum of rhodopsin. As Figure 9.4 shows, exposure to lights of increasing intensity caused decreases in the amount of rhodopsin.

In a more recent study Alpern and Pugh (1974) have compared the action-spectrum for bleaching some

---

*Brindley and Willmer (1952) iniated this ingenious technique, and it has been applied by both Rushton (Rushton and Cohen, 1954; Campbell and Rushton, 1955) and Weale (1953, 1962) to human and animal eyes respectively.

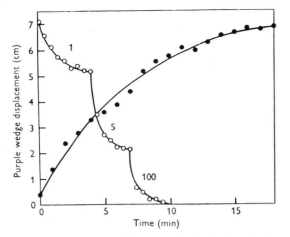

**Fig. 9.4**  ○—○, effect of bleaching the dark-adapted retina with lights of intensities 1, 5 and 100 units (1 unit = 20,000 trolands).
●—●, course of regeneration of rhodopsin. The ordinate indicates the degree of absorption of light in terms of displacement of an optical wedge. (Campbell and Rushton, *J. Physiol.*)

10 per cent of the rhodopsin in a human observer's retina with the scotopic sensitivity of the same observer. Figure 9.5 shows the comparison where the triangles represent the action spectrum for the bleach and the brackets determine the geometric mean of fifteen determinations of the absolute threshold at the corresponding wavelengths.

### PHOTOSENSITIVITY

An important characteristic of the retinal photopigment *in situ* must be its photosensitivity which defines, in effect, the rate at which bleaching will take place for a given intensity of retinal illumination (p. 206). The

determination *in situ*, as well as that of the optical density of the pigment, given by $Log_{10} I_i/I_t$, can be carried out by the technique of reflexion-densitometry. The optical density was, on average in human subjects, 0·3, indicating that one out of two quanta incident on the retina was absorbed. The photosensitivity was of the order of 100 000 l.(cm.mole)$^{+1}$, which was two to three times higher than values obtained *in vitro* or in the isolated animal eye, and Alpern and Pugh (1974) suggest that light is being funnelled into the outer segments of rods by the inner segments (Stiles-Crawford effect), although such a funnelling is predominantly a cone phenomenon. We may note that the optical density of rhodopsin in human rods is comparable with that of cone photopigments in the human retina deduced likewise on the basis of reflexion-densitometry.*

### RATES OF REGENERATION

After complete bleaching the curve of regeneration is slow, the total regeneration requiring 30 minutes, and thus being equivalent to the time for complete dark-adaptation in man. In the cat, where dark-adaptation is considerably slower, the regeneration, measured in a similar way, takes much longer (Weale). The half-times (min) for regeneration, determined by reflexion densitometry, in different species are as follows (Bonds and McLeod, 1974):

| Man | 4·75 | Rat | 40 |
|---|---|---|---|
| Cat | 11 | Guinea pig | 60 |
| Rabbit | 24 | Skate | 60 |
| Frog | 25–30 | | |

*Bonds and MacLeod (1974) found an *in vivo* photosensitivity of about 50 000 l.(cm.mole)$^{-1}$ in the cat.

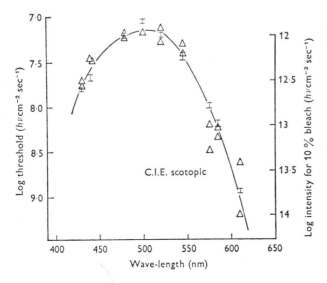

**Fig. 9.5**  Comparison of the action spectrum for bleaching rhodopsin (triangles and right-hand ordinate) with scotopic sensitivity of the same subject (brackets and left-hand ordinate). The brackets define the geometrical mean ±1 standard error of fifteen determinations of the absolute threshold in a single run. (Alpern and Pugh, *J. Physiol.*)

## RHODOPSIN CONCENTRATION AND RETINAL SENSITIVITY

Although there is a correlation between pigment regeneration and increase in scotopic sensitivity, the quantitative relationship between the two is not the simple linear one that might be predicted on the basis of photochemistry. Thus, exposure of the dark-adapted eye to a flash that only bleached 1 per cent of its rhodopsin caused an increase of tenfold in its threshold. Clearly, absorption of light by rhodopsin has reduced its sensitivity out of all proportion to the amount of bleaching; again, the final 1 per cent of regeneration during the last fifteen minutes of dark-adaptation brings about a disproportionate increase in sensitivity of the eye.

Nevertheless, of course, there is a relationship between the amount of rhodopsin in the retina and visual threshold; thus Rushton (1961) varied the amount of bleaching by different exposures of humans to light and found that the threshold for rod vision always occurred when 92 per cent of the rhodopsin had regenerated; before this had happened the measured threshold in the dark was always the cone threshold; the remainder of the dark-adaptation curve, i.e. beyond the 'kink' (Fig. 7.1, p. 186) corresponded to the regeneration of the remaining 8 per cent. Rushton was led to the belief that the relationship between threshold and rhodopsin concentration would be logarithmic, so that on plotting the log threshold against rhodopsin concentration a straight line should be obtained. Since, with a normal human subject in the dark, the threshold measured is that of the cones during the initial portion of dark-adaptation, it is not possible to find out what the condition of the rods is in the early stages, i.e. we cannot measure their threshold by simply exposing the eye to light, because it is higher than that of the cones. Thus we cannot pursue the relationship between rod threshold and rhodopsin concentration below concentrations corresponding to less than 92 per cent of the value in the completely dark-adapted eye.

### LOGARITHMIC RELATIONSHIP

In a study on a rod-monochromat, a subject with defective cones so that the threshold was determined at all states of adaptation apparently by the rods, Rushton showed that the logarithmic relationship applied over a large range of rhodopsin concentration. Independently Dowling, using the time required for the b-wave of the ERG to appear after bleaching the rhodopsin in the rat's eye, showed that the logarithm of the threshold was indeed linearly related to the rhodopsin concentration measured chemically by extraction; this is shown in Figure 9.6, where the amount of rhodopsin in the retina was varied, either by light-adaptation or by vitamin-A deficiency.*

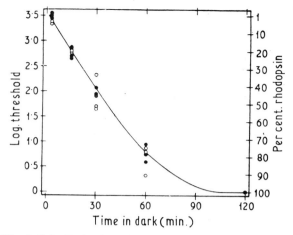

**Fig. 9.6(a)**   Dark-adaptation in the rat. As ordinates are plotted the threshold for producing an ERG (left) and the amount of rhodopsin extractable from the retina as a percentage of the dark-adapted value (right). Following light-adaptation the log threshold falls whilst the percentage of rhodopsin rises. ● Log threshold; ○ rhodopsin.

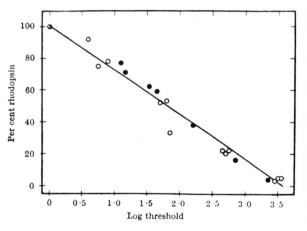

**Fig. 9.6(b)**   The relation between rhodopsin content of the retina and the visual threshold in animals dark-adapted after exposure to bright light, and in animals night-blind due to vitamin A deficiency. In both cases the log threshold varies linearly with the per cent of rhodopsin in the retina. ●, night-blindness; ○, dark-adaptation. (Dowling, *Nature*.)

---

*Granit, Munsterhjelm and Zewi (1939) were unable to establish any relationship between height of b-wave and rhodopsin concentration, but Dowling considers that this is because the *height* of the wave is not the parameter that should be measured. Wald (1954) has suggested an ingenious explanation for the experimental fact, namely that the absorption of a little light has an inordinate effect on threshold in the completely dark-adapted eye. According to this the rhodopsin is compartmented in such a way that the absorption of a quantum of light by any molecule in a compartment prevents all the other molecules from absorbing.

Finally, Weinstein, Hobson and Dowling (1967) measured the changes in concentration of rhodopsin in the isolated retina and correlated these with visual sensitivity, as measured by the *b*-wave of the electro-retinogram. The isolated retina remains functional in an artificial medium, but it does not regenerate rhodopsin after bleaching since it has lost its attachment to the pigment epithelium; thus the preparation is ideal for making controlled changes in the amount of unbleached rhodopsin. A perfect straight-line relationship between percentage of rhodopsin in the retina and the logarithm of the sensitivity was obtained.

NERVOUS EVENTS

The fact that the sensitivity of the eye to light is not related in a simple linear way to the amount of visual pigment in its receptors suggests that the sensitivity is determined by other factors than this; and a great deal of evidence supports this. Thus Arden and Weale found that the rate of adaptation depended on the size of the test-field, the larger the field the greater the rate, and they concluded that this was because, during dark-adaptation, the size of the receptive field was actually increasing, i.e. the power of the retina to summate was improving. The importance of the neural aspect of adaptation was emphasized by Pirenne's and Rushton's experiments, showing that it was apparently un-necessary to expose a given portion of the retina to light in order to reduce this portion's sensitivity to light. Rushton exposed the eye to an adapting light made up of a black and white grating of uniform stripes subtending 0.25° at the eye; the test flash, used to measure the threshold, was another identical grating, and the threshold was measured first when the two gratings were in phase, i.e. when white fell on white, and then when they were out of phase, white falling on black. It was found that there was no difference between the thresholds, the striped background raising the threshold to the same extent in both cases.*

## 'Dark light'

Thus we may speak of an effect of 'dark light', the sensitivity of part of the retina having been reduced in exactly the same way as if light had actually fallen on it. This concept, originally developed by Crawford in 1937, showed that the threshold for vision in the dark at any period of dark-adaptation after an initial bleach of the retina could be expressed in terms of the background illumination required to give the same threshold. Thus, if, after say 5 minutes in the dark, the threshold had fallen to a luminance of $I$ cd/m$^2$, then a background of $I - \Delta I$ could be found that would give an increment threshold, $\Delta I$, equal to the measured threshold. In this way the eye, at any given state of

adaptation, could be considered to be looking at its test-spot through a 'veiling luminance' or through dark light.†

## 'Retinal noise'

Barlow (1964) has developed this view still further and has suggested that the dark light is essentially an expression of 'retinal noise', i.e. of unnecessary retinal signals on to which a stimulus-light has to be imposed in order to be appreciated. Thus, when a subject has been exposed to light and his threshold is being determined in the dark, we may ask whether the high threshold we observe can be considered to be the high value he would have were he looking through a dark filter, or alternatively were he looking through dark light. The situations are different intrinsically; in the former case the signals from the stimulated rods are attenuated, whilst in the latter they are enhanced. Barlow provided convincing arguments against the filter hypothesis; he asked, however, why, if the rods in this light-adapted state are highly active, the subject is not aware of light when placed in the dark? The situation should be of someone looking at the test spot through an after-image of the adapting light. However, after-images fade rapidly, and this is presumably because they are, in effect, stabilized on the retina; thus if we wish to prove that the after-image of the adapting light is really reducing the sensitivity of the retina to a test spot, we must stabilize the test-spot on the retina and compare its subjective brightness with that of the after-image. Barlow and Sparrock (1964) carried out this experiment; the eye was exposed to a bright annulus which caused bleaching of the retina; then, in the dark, a stabilized image of a spot, seen in the centre of the annulus, was compared with the after-image, and its luminance adjusted until a match was obtained. As time progressed this fell, and the typical dark-adaptation curve was obtained (Fig. 9.7, solid circles). In separate experiments the increment-threshold relationship for the test lights used was determined, against a background provided by real light, and the equivalent background for each of the matching spots was obtained and plotted against time on the same curve (Fig. 9.7, open circles). The coincidence between the two demonstrates that when one measures the threshold in a partially adapted eye,

---

*Barlow and Andrews (1967) have been unable to repeat this finding exactly; they found the threshold on those parts of the retina exposed to the dark bars of the grating lower than on those exposed to the bright bars, but higher than if there had been no illumination at all.
†The validity of this concept of equivalent background has been confirmed and extended by Blakemore and Rushton (1965) in a rod monochromat whose scotopic visual function could be examined over a wide range of luminance without contamination by cones.

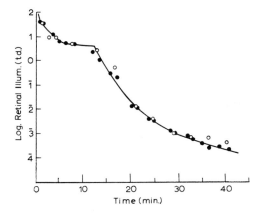

**Fig. 9.7**  Dark-adaptation curves. Solid circles are the equivalent luminance of a positive after-image obtained from the stabilized image that matched it. Open circles represent the equivalent background luminance that would be required to give the actual threshold, measured, now, as a liminal brightness increment, as postulated by Crawford. (Barlow and Sparrock, *Science*.)

one is measuring, in effect, the incremental threshold when the test light is seen against a background corresponding to a certain after-image luminance.

ADAPTATION POOL

Rushton (1965b) has developed the concept of an 'adaptation pool' which, at any given state of adaptation, measures the extent to which spatial summation of rod responses is possible; during complete dark-adaptation this is maximal and after complete rod bleaching it is minimal; in some way, then, the presence of bleached rhodopsin in the rod causes it to signal to the pool and control the degree of summation and hence the threshold.

Thus according to Rushton's view, the bleaching signals due to the presence of unregenerated rhodopsin in the rods are of a different kind from the signals indicating the action of light; for Rushton, signals generated by light enter an automatic gain-control device through the input, become attenuated by the gain-control, emerge at the output (after attenuation) to signal the light and are fed back into the gain-control to regulate sensitivity. On the other hand bleaching signals enter the gain-control, not at the input but through the feedback. They regulate visual sensitivity in this way by attenuating the gain but are not themselves regulated by it.

ESSENTIAL NIGHT-BLINDNESS

By contrast with the night-blindness associated with vitamin A deficiency, there is no deficiency in the amount of rhodopsin in the retina of the subject with this congenital visual defect, nor yet are the kinetics of bleaching or regeneration defective (Carr *et al.* 1966). The 'cone-type' of dark-adaptation curve shown in Figure 9.8 is expressive of normal cone function but failure of the retina to signal the bleaching of rhodopsin. The site of failure is thought to be in the retina since abnormalities in the ERG have been described in this condition; for example Carr *et al.* (1966) observed a depression in the *b*-wave and in one subject the ERG failed to appear at levels of luminance that gave well marked records in the normal. Carr *et al.* thought that the condition might represent a failure of spatial summation but this was not true, and their studies indicated that there was some defect in transmission of signals from both rods and cones, the impairment of rod-function being the more severe; thus one subject did reveal a scotopic sensitivity curve with maximum at 500 nm; however a shoulder on the curve at 580 nm indicated the intervention of cone function.

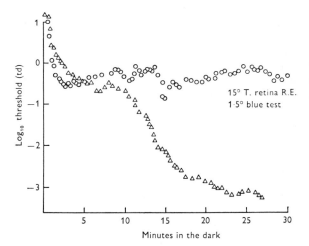

**Fig. 9.8**  Dark-adaptation curves of night-blind subject (open circles) and a normal subject (triangles) tested under identical conditions for the 15° peripheral retina with a 1·5° test target. Even after 30 min in the dark, the night-blind subject saw the blue colour of the test at threshold. (Alpern *et al.*, *J. Physiol.*)

PUPIL

It is interesting that the pupils of the essentially night-blind do respond to light under scotopic conditions, so that after bleaching the retina and allowing the subject to remain in the dark the recovery of pupil-size follows the course of regeneration of rhodopsin in accordance with the simple equation:

$$D = 4·5\,(1 - p)$$

where $p$ is the fraction of rhodopsin unregenerated, a relation found earlier for normal subjects by Alpern and Ohba (1972). Alpern *et al.* (1972) have concluded however that the pupil responses in the dark are not completely normal, and they cite an experiment that suggests that the signals providing the 'equivalent background of the bleach', and which are set up by the amount of rhodopsin ungenerated in the rods, are active, behaving as a background illumination and causing the pupils to constrict, whereas the rod light signals, governed by the bleaching of rhodopsin, are inactive. If their deductions are correct—and the argument is rather tenuous—it follows that Rushton's concept of two types of signal, the one indicating true light and another indicating 'background of the bleach', are different, by contrast with Barlow's view of the two signals as being equivalent.

## THE ABSOLUTE THRESHOLD

### FREQUENCY-OF-SEEING

In the fully dark-adapted state, the minimum stimulus necessary to evoke the sensation of light is called the absolute threshold. It must be appreciated, however, that this threshold is not a fixed quantity but changes from moment to moment, so that if a subject is presented with a luminous screen in the dark, we may find a range of luminance over which there will be some probability, but not certainty, of his seeing it. Experimentally, therefore, the screen must be presented frequently at different luminances and the experimenter must plot the 'frequency-of-seeing' against the luminance, to give the typical 'frequency-of-seeing' curve shown in Figure 9.9. From this curve it is a simple matter to compute a threshold on the basis of an arbitrarily chosen frequency, e.g. 55 per cent.

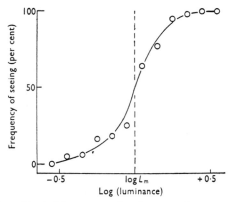

**Fig. 9.9** Typical frequency-of-seeing curve for dark-adapted human observer. (Pirenne, Marriott and O'Doherty, *M.R.C. Sp. Rep.*)

### MINIMUM RETINAL ILLUMINATION

The threshold *stimulus* may be defined in several ways according to the experimental method of measurement.

Thus we may express it as the luminance of a large test-screen, exposed for a relatively long time, e.g. one second. Since the size of the pupil is important, this must be measured, and the threshold expressed in trolands; furthermore the spectral composition of the light must be defined so that, if necessary, the number of quanta falling on unit area of the retina in unit time may be derived. This definition of the threshold stimulus is thus a statement of the *retinal illumination*. Experimentally this type of threshold has been found to be of the order of $0.75 \cdot 10^{-6}$ cd/m$^2$ varying between $0.4$ and $2.0 \cdot 10^{-6}$ (Pirenne, Marriott and O'Doherty). This is equivalent to a retinal illumination of $4.4 \cdot 10^{-5}$ scotopic trolands, and corresponds to the absorption of some 2300 quanta per second by some 13.4 million rods over which the large test-patch fell. Thus, in each second of time, only one rod in 5800 absorbs a single quantum when the eye is being stimulated at threshold intensity.

### MINIMUM FLUX OF ENERGY

If an effectively point source of light is used, the image is concentrated on to a point on the retina so that the concept of retinal illumination loses its application; instead, then, we define the threshold as the *minimum flux* of light-energy necessary for vision, i.e. the number of lumens per second or, better, the number of quanta per second entering the eye. Experimentally Marriott, Morris and Pirenne obtained a value of 90 to 144 quanta/second.

### MINIMUM AMOUNT OF ENERGY

Finally, if a very brief flash of light is used as the stimulus (less than 0.1 second) we may express the threshold as simply the total number of quanta that must enter the eye to produce a sensation. Thus there are three thresholds, the *minimum retinal illumination*, the *minimum light flux* and the *minimum amount of energy*. Of these, it is the last that interests us most since from this we are able to compute just how many quanta of light a single receptor must absorb to be excited.

## MINIMUM STIMULUS

This was determined in 1942 by Hecht, Shlaer and Pirenne by the method of presenting short flashes and determining the frequency-of-seeing curves. With a 60 per cent frequency as the end-point for threshold, the amount of energy striking the cornea was 54 to 148 quanta, depending on the observer; after allowing for scattering and absorption by the ocular media the range of useful light-energy was some 5 to 14 quanta. The image of the flash fell on about 500 rods, so that it was unlikely that one rod received more than a single quantum. This might suggest, therefore, that the minimum stimulus consisted of the simultaneous stimulation of some 11 rods, each absorbing a single quantum, their effects being transmitted to bipolar and ganglion cells by convergence. On this basis, then, the excitation of a single rod at threshold levels is not sufficient for vision, the excitatory effect being insufficient to permit conduction through the various synapses on the visual pathway. Only if several rods are excited will their effects be sufficient to break down these synaptic barriers and permit the passage of the excitation along the visual pathway to higher centres in the brain.

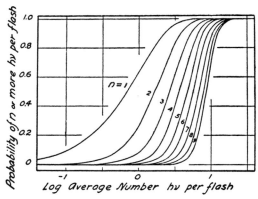

Fig. 9.10    These are probability curves based on quantum theory, showing the chance that a given flash, with average energy shown on the abscissa, will have $n$ quanta in it or more. For example, if we employ curve belonging to $n = 2$, the chance of having 2 or more quanta in a flash is about 0·2 (i.e. 1 in 5) if the average energy of the flashes is 1 quantum (log 1 = 0). (Hecht, Shlaer and Pirenne, *J. gen. Physiol.*)

## QUANTUM FLUCTUATIONS

The threshold varies from moment to moment and, according to Hecht, Shlaer and Pirenne, this must be due partly, if not wholly, to fluctuations in the number of quanta that a given flash contains when it is repeated. Thus, when we are dealing with very small amounts of light-energy, the actual number of quanta in any given flash will vary. Hence, according to the *uncertainty principle*, when we speak of a flash having an energy, $a$, we mean that this is the *average energy* the flash would have if it were repeated a large number of times, and the total number of quanta actually measured were divided by the number of flashes. For any given flash, with a mean energy-content of $a$ quanta, we can only speak of the *probability* that it will contain a given number $n$, or more, quanta; and this probability may be simply calculated. On raising the stimulus-strength, measured by $a$, over a range of, say, tenfold, the probability that a flash will have a given number of quanta will increase, and finally become unity. The curve obtained by plotting this probability against the stimulus-strength has a characteristic shape, the steepness depending on the value of $n$ as Figure 9.10 shows.

### FREQUENCY-OF-SEEING

Now, if the seeing of a flash depends only on the flash's containing the right number of quanta, say $n$, clearly the frequency-of-seeing will increase as the strength of the stimulus, measured by $a$, increases, because the probability that there will be $n$ quanta in a flash increases with increasing value of $a$. Moreover, the shape of the frequency-of-seeing curve should be similar to the shape of the probability curve. As Figure 9.11 shows, there is remarkably good agreement between the two, the range of stimulus-strength over which a flash is never seen and always seen being about a factor of tenfold, corresponding with the range over which the probability that a flash has $n$ quanta in it varies from 0 to unity. We may note that the slope of the probability curve depends on the number of quanta, $n$, required, becoming steeper the greater the number, so that in order that our frequency-of-seeing curve shall match the probability curve we must choose a suitable value of $n$, the number of quanta required for vision. Hecht *et al.* found that, with different observers, $n$ varied between 5 and 7, a fact suggesting that the absorption of these numbers of quanta corresponded with the stimulus for threshold vision. Direct measurements of the mean amount of quanta in a threshold stimulus gave values ranging from 6 to 17 quanta actually absorbed; since there is reason to believe that only one in two of absorbed quanta are effective—the *quantum efficiency* of the process is said to be 0·5—this would correspond then to 3 to 8 effective quanta.

### LATER DEVELOPMENTS

Since the publication of this study of Hecht *et al.* there have been a number of discussions on the actual minimum number of quanta required to excite vision. There is no doubt that the absorption of one quantum *can* excite a rod, but the

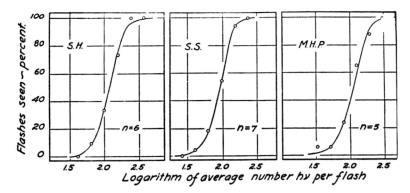

Fig. 9.11    Frequency-of-seeing curves for three human subjects. The number of times the flash was seen is plotted against the logarithm of the average energy per flash. The curves are theoretical ones, based on the theory of quantum fluctuations in the stimulus, and the fact that the experimental points fall on the curves suggests that the failure to see a given flash was largely due to its not having the required number of quanta in it. (Hecht, Shlaer and Pirenne, *J. gen. Physiol.*)

phenomena of summation, both temporal and spatial, prove that the excitation of more than one rod is necessary to evoke *sensation*. Thus, it has been found that, within certain limits of area, a given number of quanta in a flash are equally effective if they are concentrated on a small area of retina or spread over a larger one. If only one absorbed quantum was necessary to act as a stimulus, spreading these quanta over a large area should reduce the chance of any one rod absorbing this quantum, so that the threshold should rise. The fact that, within an area subtending some 10 minutes of arc, the stimulus is equally effective, however spread out, means that at threshold more than one rod must be stimulated, their combined effects, transmitted to a bipolar cell, constituting the minimum stimulus for vision. Pirenne inclines to the belief that $n$ is either equal to or greater than 4 although Bouman considers that his own and van der Velden's observations are better fitted by a value of 2.

## SIGNAL THEORY

Barlow and Levick (1969) have examined the responses of single ganglion cells of the cat to incremental light-stimuli, with a view to defining a threshold in terms of alteration in the background spike-discharge, the idea being to obtain, from the recorded spike activity before, during, and after a flash, a measurable quantity that would enable the cat to distinguish between a change in retinal illumination and a casual fluctuation in the background spike activity. The prime factor was undoubtedly the quantum/spike ratio, i.e. the extra quanta absorbed required to produce an extra spike in the sampling time, $\tau$. Additional factors are the time-course of the response, so that alterations in this may affect the threshold, as in Bloch's Law; and finally the statistical distribution of the impulses that occur in the time-interval $\tau$ in the absence of a stimulus.

## COUNTING EVERY QUANTUM

Barlow (1950) discussed the criterion employed for decision as to whether threshold stimulation had been achieved, as against a random variation in 'noise', and emphasized that if the criterion were lowered, i.e. if the observer were to say 'yes' at a trial when he was not absolutely certain of having seen a flash, the threshold would, of course, be lower when the responses of many tests were analysed statistically, but of course the rate of 'false positives' would be increased and the observer would be dismissed as 'untrained' or 'unreliable'. He concluded that at least 2 and probably 10 to 20 excited rods were needed to give a sensation of light and that noise in the optic pathway—e.g. thermal decompositions of rhodopsin molecules—provided the limit to sensitivity. Sakitt (1972) has developed this theme further and emphasized that the choice between 'seen' and 'not seen' is one of several possible choices. Thus, in order to be certain of seeing a flash, an observer chooses a larger signal (or absorbed number of quanta) than if he were allowed to give a less unequivocal statement; his false statements may be more frequent, but statistically he will still be giving meaningful answers and they will be more frequently correct than if answering at random. Thus in 'frequency-of-seeing' tests, subjects are chosen who, in fact, are trained to give no false responses and thus are being trained to use a large signal (large number of absorbed quanta), so that when they deny seeing a flash they may well be ignoring signals from the flash because they are not sufficiently strong—but they would be signals. Sakitt presented observers with three types of stimulus namely a blank, a flash corresponding to 66 photons at the cornea, or one of 55 photons, and the observer was allowed to make 6 types of rating from 0, meaning that he did not see anything, up to 6, indicating a very bright light. When ratings were plotted against number of quanta as in Figure 9.12 a linear relation was found, suggesting that the ratings might actually represent numbers of quanta effectively absorbed to give rod-signals greater than the background noise. On this basis he deduced a quantum efficiency of seeing of 0·0274, this being the fraction of quanta falling on the cornea that are actually effective after allowing for absorption in the eye media and the probability of about a half that an absorbed quantum will actually isomerize a molecule of rhodopsin (Hagins, 1955). If the notion that ratings indicated quantum events was correct, the variance in a given rating should follow a Poisson distribution and the result should be a set of 'frequency of seeing curves' as in Figure 9.13, each being made on the basis of a single, double, triple etc. rod signal. As we should expect, the curve becomes steeper and like the more classical frequency-of-seeing curve when the rating is high; thus if the observer used a criterion of 3 the absolute threshold (50 per cent probability) would correspond to a rod signal of 2·7 and, from the linear relation between this and quanta reaching the

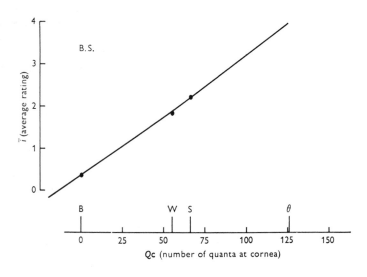

**Fig. 9.12** The average rating, $i$, plotted against $Q_c$, the average number of quanta at the cornea for one subject. The straight line through the points follows the equation: $i = 0\cdot0274\,Q_c + 0\cdot36$. (Sakitt, *J. Physiol.*)

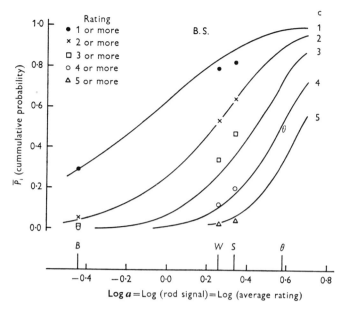

**Fig. 9.13** The experimental points are the cumulative probabilities $\bar{P}_1$ for a subject giving a rating $i$ or more. The abscissa is the log of the rod signal which is the same as the log of the average rating. The points labelled $B$, $W$ and $S$ are at the values of $a$ for the blank, weak and strong stimuli respectively. The symbol $\theta$ refers to the absolute threshold. The smooth curves are theoretical cumulative Poisson probabilities $\bar{P}(c, a)$ that $c$ or more rod signals occur when $a$ is the average number occurring. (Sakitt, *J. Physiol.*)

cornea,* this would correspond to a value of 87 quanta and only one false positive would be made in a hundred trials. If a weaker criterion 2 had been chosen, $a$, the average number of rod signals, would be 1·7 giving 50 quanta as the threshold stimulus, but 5 per cent of the answers would be false positives. A criterion of 1 would mean a response to as few as 12·6 quanta and would lead to a false-positive rate of 29 per cent, which would be experimentally unacceptable, but the observer would not be answering at complete random. Sakitt concluded, therefore, that his observers were, indeed, 'counting every quantum'.

## CAT'S SENSITIVITY

Bonds and MacLeod (1974) have studied the *in vivo* bleaching and regeneration of rhodopsin in the cat and have commented on the 'quantum catching' in this animal compared with that in man. They computed that a cat's retina absorbed some 24 per cent of the quanta incident on the cornea compared with only 9 per cent in man, so that the cat's quantum catch is 2·7 times man's on this account; its fully dark-adapted pupil has about 3·2 times the area of that in man, so that man must catch nine times as many quanta as the cat to perceive light. In fact, Gunter's behavioural study indicated that the cat's absolute threshold was some sixth to a seventh of man's.

## SPATIAL AND TEMPORAL SUMMATION

The minimum stimulus has been defined in three ways, as we have seen, but the significance of the differences in definition may not be clear unless the importance of spatial and temporal summation is appreciated. It is because of spatial summation that the size of the

test-stimulus must usually be stated, and because of temporal summation that the duration of the stimulus acquires significance.

## RECEPTIVE FIELDS

Studies on single optic nerve fibres (p. 248) have shown that they have quite extensive 'receptive fields' in the sense that a small stimulus applied over a large number of receptor cells can evoke a response in this single fibre. The anatomical basis for this is the convergence of receptors on bipolar cells and the bipolar cells on ganglion cells, so that a single ganglion cell may have synaptic relationships ultimately with many thousands of rods and cones. Because of this convergence, two spots of light falling within the receptive field may be expected to be more effective than a single one of the same light-energy, so that if we are measuring thresholds, the luminance of a single test-patch must be greater than that of two test-patches of the same area falling within the receptive field of a single ganglion cell. Our threshold luminance, expressed in terms of retinal illumination, will be smaller, then, the larger the area. If summation is perfect, moreover, we may expect the following relationship to hold:

$$Intensity \times Area = Constant$$

the greater the area of the stimulus the smaller the luminous intensity required to excite.

---

*Thus the relation between $a$, the average number of rod signals, and $Q_c$, the average number of quanta falling on the cornea during a flash was:

$$a = 0.0274\, Q_c + 0.36$$

0·36 is equivalent to the average number of noise events during the flash (Barlow, 1956).

## RICCO'S LAW

This is *Ricco's Law*. Expressed in terms of the actual number of quanta of energy reaching the retina this means that the threshold is independent of the area of the stimulus; thus if we have two stimuli forming images of 1 mm$^2$ and 2 mm$^2$ on the retina, according to Ricco's Law the luminous intensity of the smaller image will be four times that of the larger one at threshold. The *amounts* of energy falling on the two areas of retina in unit time will be equal, however, since the greater area of the one stimulus exactly compensates for its smaller retinal illumination. The area over which Ricco's Law holds varies with position on the retina, being about 30 minutes of arc in the parafoveal region with eccentricity of 4 to 7°, and increasing to as much as 2° at an eccentricity of 35° (Hallett, Marriott and Rodger, 1962). Over these areas, then, there is complete summation, in the sense that there is no loss of efficiency on spreading the available light-energy over a certain area instead of concentrating it on a much smaller one. When we are determining the minimum amount of energy necessary to excite the sensation of light, then, we must clearly ensure that the area of stimulus is within the limit over which Ricco's Law applies, since beyond this area the summation is only *partial*, in the sense that the larger stimulus will require more quanta than the smaller one, and will therefore have to have a higher luminous intensity. The existence of this partial summation will be indicated by the fact that the *luminous intensity* of the larger stimulus will be less than that of the smaller one, whilst its partial nature will be indicated by the fact that the *number of quanta* required to excite vision will be greater the greater the area over which they are spread.

As we shall see, there are some 'anomalies' in areal summation within the Ricco area that indicate that total summation of quanta falling on a given area is more the exception than the rule; for example, Sakitt (1971) found that presentation of a two-spot stimulus within the Ricco-area defined for a single-spot stimulus did not give total summation, the amount depending on the spatial separation of the spots.

## PIPER'S LAW

By Piper's Law is meant the statement that over a certain range of areas, where summation is partial, the relationship:

$$\sqrt{Area} \times Intensity = Constant$$

holds.* In practice partial summation extends over large areas, up to about 24°. Beyond this area there is no further summation, so that increasing the area of the stimulus requires no diminution in its intensity for threshold excitation; beyond this area then, the *luminous intensity* of the test-patch is independent of the area. The *amount of energy* entering the eye, of course, increases as the square of the size of the retinal image so that the threshold, expressed as the number of quanta per second required to excite, will increase as the area of test-patch increases.†

## PROBABILITY SUMMATION

Since partial summation extends over large areas of retina, it seems unlikely that convergence can be the sole basis for it, and much more likely that it results from an increased probability that the stimulus will have the necessary number of quanta to stimulate the retina. Thus, we may measure the threshold first with a single patch of light and next with two patches such that their images fall widely apart. If there is any summation at all, the two patches will have a lower luminance than the single one. If we suppose that the two patches are too far apart to permit any convergence of receptors on the same optic nerve fibres, we may nevertheless explain the greater effectiveness of the two patches on a probability basis. Thus, with a single patch the probability of seeing it has a certain value, $p$, so that the chance that it will not be seen is $(1 - p)$. The chance of *not seeing* two patches is $(1 - p)(1 - p)$ which is smaller, so that a lower luminance is necessary for these two patches to have a given 'frequency-of-seeing' than for the single one whose probability of not seeing is only $(1 - p)$. We may thus speak of *physiological summation*, when this is due to convergence of retinal elements on optic nerve fibres, and *probability summation*, when the parts of the retina on which the summation occurs may be treated as independent of each other.‡

---

*According to Baumgardt (1959) this relationship follows from probability theory, but there seems little doubt from Barlow's (1958) studies that no simple Law will cover all the phenomena; thus the degree of spatial summation depends on the duration of the stimulus whilst the degree of temporal summation depends on the size of the stimulus. Some of the limitations of studies on area are considered by Hallett *et al.* (1962).

†At the fovea threshold measurements are those for stimulation of the cones and, as we have seen, in the completely dark-adapted condition the thresholds are much higher. This may well be due, not so much to a lower intrinsic sensitivity of the cone as compared with the rod, but to the much smaller degree of convergence exhibited by cones, so that the area over which complete summation is possible is very small. Hence, as ordinarily measured, the threshold will be in the region of partial or no summation. By reducing the size of the test-stimulus to a subtense of 2·7 minutes of arc Arden and Weale (1954) found the threshold intensities to be the same for fovea and parafovea, but their contention that the sensitivities of foveal cones and peripheral rods are equal is probably unsound (Pirenne, 1962, p. 106).

‡An extreme example of probability summation is given by measuring the frequency of seeing a flash, when two observers are employed, with the frequency of seeing when only a single observer is used; from these we may compute thresholds in the usual way and it is found that the threshold for the 'double observer' is less than that for the single.

TEMPORAL SUMMATION. BLOCH'S LAW

By temporal summation we mean that a stimulus lasting a short time must be stronger in luminance than a stimulus lasting a long time; thus the stimulus with the longer duration can be considered to be made up of a lot of successively repeated stimuli, and it is these successive stimuli that apparently add their effects. It was found by Bloch in 1885 that, over a period of roughly 100 msec the relationship:

$$I \times t = Constant$$

holds, where $I$ is the threshold luminance in the dark-adapted eye, and $t$ is the duration of presentation. This is the *Bunsen-Roscoe Law* of photochemical reactions and indicates that up to this period of about 100 msec the effective stimulus is only the number of quanta entering the eye, and we may say that the summation is *complete*. On a photochemical basis this complete summation is what one might expect, since the limiting factor in the threshold is presumably the absorption of a definite number of quanta, and the chances that this will happen will increase proportionately with the number of quanta entering the eye.

**Partial summation**

Beyond a certain time, which has been called the *action time*, the absorption of a second quantum of light will be of no use to help the effects of the absorption of the first, because these effects, e.g. production of an electrical change, will subside. We can therefore understand the limited period over which the Bunsen-Roscoe Law applies. Beyond this critical period during which summation is complete there is a period of partial summation during which two successive stimuli will add their effects, but not completely, so that some other relationship between $I$ and $t$ holds. On theoretical grounds this may be expected, since the longer the stimulus the greater the probability it will contain a certain minimal number of quanta, and according to Baumgardt (1959) there should be a square-root relationship between threshold energy and duration of the stimulus. Over a certain range, Baumgardt found the experimental results agreed with this square-root relationship.

DETECTABILITY OF SIGNAL

On the basis of quantum fluctuation theory, a signal of incremental flux, $E$, above a background flux, $B$, due to a distant uniform emitting source subtending area, $A$ and lasting time, $T$, is given by:

$$d' = \frac{EAT}{(BAT)^{1/2}}$$

or if only a fraction, $F$, of the available photons are used in the task:

$$d' = \frac{FEAT}{(FBAT)^{1/2}}$$

Implicit in the equation are the deVries-Rose Law ($E \propto B^{1/2}$), Piper's Law ($E \propto A^{1/2}$), and Pieron's Law ($E \propto T^{1/2}$). However, Weber's Law, Ricco's Law and Bloch's Law all indicate the limited applicability of quantum fluctuation theory, although the successful application to the absolute threshold and other measurements indicates that these fluctuations constitute a factor that must be taken into account. Recently Cohn (1976) has applied the theory to detection of foveal luminance changes and shown that results accord with prediction, from which he deduced a quantum efficiency of only $2 . 10^{-4}$, i.e., only one in 10 000 of the quanta falling on the cornea are effective in foveal vision. However, when the size of the stimulus was decreased, efficiency increased by a factor of ·10, and thus only 3 to 10 times less than that reported for scotopic vision.

REFERENCES

Aguilar, M. & Stiles, W. S. (1954) Saturation of the rod mechanism in the retina at high levels of illumination. *Optica Acta* **1,** 59–65.

Alpern, M., Holland, M. G. & Ohba, N. (1972) Rhodopsin bleaching signals in essential night blindness. *J. Physiol.* **225,** 457–476.

Alpern, M. & Ohba, N. (1972) The effect of bleaching and background on pupil size. *Vision Res.* **12,** 943–951.

Alpern, M. & Pugh, E. N. (1974) The density and photosensitivity of human rhodopsin in the living retina. *J. Physiol.* **237,** 341–370.

Arden, G. B. & Weale, R. A. (1954) Nervous mechanisms and dark-adaptation. *J. Physiol.* **125,** 417–426.

Barlow, H. B. (1956) Retinal noise and absolute threshold. *J. opt. Soc. Amer.* **46,** 634–639.

Barlow, H. B. (1958) Temporal and spatial summation in human vision at different background intensities. *J. Physiol.* **141,** 337–350.

Barlow, H. B. (1964) Dark-adaptation: a new hypothesis. *Vision Res.* **4,** 47–58.

Barlow, H. B. & Andrews, D. P. (1967) Sensitivity of receptors and receptor 'pools'. *J. opt. Soc. Am.* **57,** 837–838.

Barlow, H. B. & Levick, W. R. (1969) Three factors limiting the reliable detection of light by retinal ganglion cells of the cat. *J. Physiol.* **200,** 1–24.

Barlow, H. B. & Sparrock, J. M. B. (1964) The role of afterimages in dark adaptation. *Science* **144,** 1309–1314.

Baumgardt, E. (1959) Visual spatial and temporal summation. *Nature* **184,** 1951–1952.

Blakemore, C. B. & Rushton, W. A. H. (1965) Dark adaptation and increment threshold in a rod monochromat. *J. Physiol.* **181,** 612–628.

Bonds, A. B. & MacLeod, D. I. A. (1974) The bleaching and regeneration of rhodopsin in the cat. *J. Physiol.* **242,** 237–253.

Bouman, M. A. (1955) Absolute threshold conditions for visual perception. *J. opt. Soc. Am.* **45,** 36–43.

Brindley, G. S. & Willmer, E. N. (1952) The reflexion of light from the macular and peripheral fundus oculi in man. *J. Physiol.* **116**, 350–356.

Campbell, F. W. & Rushton, W. A. H. (1955) Measurement of the scotopic pigment in the living human eye. *J. Physiol.* **130**, 131–147.

Carr, R. E., Ripps, H. & Siegel, I. M. (1966) Rhodopsin and visual thresholds in congenital night blindness. *J. Physiol.* **186**, 103–104P.

Crawford, B. H. (1937) The change of visual sensitivity with time. *Proc. Roy. Soc. B* **123**, 69–89.

Dowling, J. E. (1960) Night blindness, dark adaptation and the electroretinogram. *Am. J. Ophthal.* **50**, 875–889.

Dowling, J. E. (1960) Chemistry of visual adaptation in the rat. *Nature* **188**, 114–118.

Fuortes, M. G. F., Gunkel, R. D. & Rushton, W. A. H. (1961) Increment thresholds in a subject deficient in cone vision. *J. Physiol.* **156**, 179–192.

Granit, R., Munsterhjelm, A. & Zewi, M. (1939) The relation between concentration of visual purple and retinal sensitivity to light during dark adaptation. *J. Physiol.* **96**, 31–44.

Hagins, W. A. (1955) The quantum efficiency of bleaching rhodopsin *in situ*. *J. Physiol.* **129**, 22–23P.

Hallett, P. E., Marriott, F. H. C. & Rodger, F. C. (1962) The relationship of visual threshold to retinal position and area. *J. Physiol.* **160**, 364–373.

Hecht, S., Shlaer, S. & Pirenne, M. H. (1942) Energy, quanta and vision. *J. gen. Physiol.* **25**, 819–840.

Marriott, F. H. C., Morris, V. B. & Pirenne, M. H. (1959) The minimum flux of energy detectable by the human eye. *J. Physiol.* **145**, 369–373.

Pirenne, M. H. (1943) Binocular and uniocular thresholds of vision. *Nature* **152**, 698–699.

Pirenne, M. H. (1962) Visual function in man. In *The Eye*, vol. II, chapters 1–11. Ed. Davson. London: Academic Press.

Pirenne, M. H., Marriott, F. H. C. & O'Doherty, E. F. (1957) Individual differences in night-vision efficiency. *Med. Res. Council Sp. Rep. Ser.*, No. 294.

Rushton, W. A. H. (1961a) Dark-adaptation and the regeneration of rhodopsin. *J. Physiol.* **156**, 166–178.

Rushton, W. A. H. (1961b) Rhodopsin measurement and dark-adaptation in a subject deficient in cone vision. *J. Physiol.* **156**, 193–205.

Rushton, W. A. H. (1965a) Bleached rhodopsin and visual adaptation. *J. Physiol.* **181**, 645–655.

Rushton, W. A. H. (1965b) Visual adaptation. *Proc. Roy. Soc. B* **162**, 20–46.

Rushton, W. A. H. & Cohen, R. D. (1954) Visual purple level and the course of dark adaptation. *Nature* **173**, 301–304.

Rushton, W. A. H., Fulton, A. B. & Baker, H. D. (1969) Dark adaptation and the rate of pigment regeneration. *Vision Res.* **9**, 1473–1479.

Rushton, W. A. H. & Henry, G. H. (1968) Bleaching and regeneration of cone pigments in man. *Vision Res.* **8**, 617–631.

Rushton, W. A. H. & Westheimer, G. (1962) The effect upon the rod threshold of bleaching neighbouring rods. *J. Physiol.* **164**, 318–329.

Sakitt, B. (1971) Configuration dependence of scotopic spatial summation. *J. Physiol.* **216**, 513–529.

Sakitt, B. (1972) Counting every quantum. *J. Physiol.* **223**, 131–150.

Wald, G. (1954) On the mechanism of the visual threshold and visual adaptation. *Science* **119**, 887–892.

Weale, R. A. (1953) Photochemical reactions in the living cat's retina. *J. Physiol.* **122**, 322–331.

Weale, R. A. (1962) Further studies of photo-chemical reactions in living human eyes. *Vision Res.* **1**, 354–378.

Weinstein, G. W., Hobson, R. W. & Dowling, J. E. (1967) Light and dark adaptation in the isolated rat retina. *Nature* **215**, 134–138.

# 10. Electrophysiology of the retina

## THE SENSORY MESSAGE

When a receptor is stimulated, a succession of electrical changes takes place in its conducting nerve fibre. These changes are measured by placing electrodes on the nerve fibre and connecting them to a device for recording rapidly occurring potential differences, an *oscillograph*. A series of electrical variations, called *action potentials* or *'spikes'*, is obtained when the receptor is stimulated. The duration of an individual spike is very short (it is measured in milliseconds) and, as the strength of the stimulus is increased, the *frequency* of the discharge (the number of spikes per second) increases, but the size of the individual spikes remains unaltered (*'all-or-none' effect*).

### SPIKE POTENTIALS

Thus the message sent to the central nervous system is in a *code* consisting of a series of spike potentials apparently all with the same magnitude, the frequency being the variable that indicates how strongly the receptor has been stimulated. When the nerve fibres from different receptors are compared, e.g. touch, temperature, vision, hearing, the action-potentials are very similar in magnitude and time-course; and it has not been possible to correlate any given type of action potential with the type of sensation that it is mediating. In other words, it appears that the elements of the message, like the dots and dashes of the Morse code, are the same for all, whether they indicate a change in temperature or the flash of a light, and we must conclude that it is the places where the messages are carried in the brain—the *central terminations*—that determine the character of the sensation. The intensity of the sensation, as indicated above, is determined by the frequency of the spikes, i.e. the number that pass under our recording electrode in unit time.

### RECORDING THE MESSAGE

The accurate study of the events taking place when a receptor is stimulated demands that the recording electrode should be on or within the receptor or, if this cannot be achieved, on the sensory nerve that makes direct connection with this receptor. Moreover, if a recording from a nerve is to be made, the leads should be taken from a single fibre, otherwise the record of events taking place in the many thousands of fibres of a whole nerve would be a very confused and attenuated picture of the events taking place in the single fibres. With the vertebrate eye, these requirements are very difficult to meet, because the receptors—rods and cones—and the first sensory neurones, the bipolar cells, are buried in the depth of the retina; hence if we wish to record from a reasonably accessible nerve, namely the optic nerve, we will be recording from a second-order neurone, namely the ganglion cell. In its passage through the ganglion cell, the original message from the rod or cone and the bipolar cell has been submitted to a process of what the neurophysiologist calls *re-coding*; unnecessary parts of the message being dropped and other parts being accentuated at their expense. Thus, to understand what happens in the receptor cell, it would be a help if we could place our electrode on or within a rod or cone, or failing that, on a bipolar cell, which makes such intimate relationships with the rods and cones. For this reason the invertebrate eye, built on a simpler plan neurologically, was studied by Hartline; and it is to his pioneering studies that we owe so much of our knowledge of sensory mechanisms. Hartline studied the compound eye of the horse-shoe crab *Limulus*, the fibres of its optic nerve being, apparently, processes leading away from the retinula cells on which the light impinges. Thus the study of the electrical changes in these fibres brings us close to the electrical changes occurring directly in response to the impinging of light on a light-sensitive cell.

## THE INVERTEBRATE EYE

## OMMATIDIUM

Essentially the compound eye of insects and crustaceans is an aggregation of many unit-eyes or *ommatidia*, each of these units having its own dioptric apparatus that concentrates light on to its individual light-sensitive surface, made up of specialized *retinula cells*. The facetted appearance of the compound eye is thus due to the regular arrangement of the transparent surfaces of these unit-eyes (Fig. 10.1*a*). The *ommatidium* is illustrated schematically in Figure 10.1*b*; light is focused by the *crystalline cone* on to the highly refractile *rhabdom*, which constitutes an aggregation of specialized regions of the *retinula cells*. These retinula cells are wedge-shaped and are symmetrically arranged around the axial canal

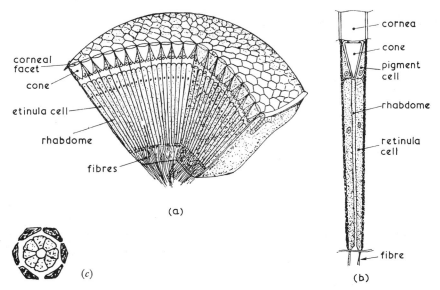

Fig. 10.1   The facetted eye. (a) Cut-away view showing the packing of the individual ommatidia; (b) longitudinal section of an ommatidium; (c) cross-section of ommatidium.

like the segments of an orange (Fig. 10.1c); the rhabdom is made up of the apical portions of these wedge-shaped cells, the separate portions being called *rhabdomeres*. As illustrated in Figure 10.2, which shows a cross-section of the *Limulus* (horse-shoe crab) ommatidium, the regions of junction of the retinula cells are characterized by dense staining, so that a cross-section appears like the spokes of a wheel, the axle being the central canal. This canal contains the process of the *eccentric cell*; this is apparently a neurone, with its cell body squeezed between the retinula cells at the base of the ommatidium; its long spear-like process fills the central canal and may be regarded as a dendrite; the central end of the cell consists of a long process that passes out of the ommatidium in the optic nerve.

Fig. 10.2   Cross-section of ommatidium of *Limulus*, showing individual retinula cells with the rhabdomeres represented in white. (After Miller, *Ann. N.Y. Acad. Sci.*)

ECCENTRIC CELL

The eccentric cell thus constitutes the nerve fibre of the ommatidium, whilst we may regard the retinula cells, with their specialized rhabdomeres, as the cells responsible for absorbing the incident light. Provisionally, then, we may say that the light-stimulus falls on the rhabdom; as a result of chemical changes taking place in this region, electrical changes are induced in the eccentric cell that culminate in the passage of impulses along the axon-like process of this cell.

RETINULA PROCESSES

Actually each retinula cell also has a long proximal process leading out of the ommatidium in the optic nerve; consequently the optic nerve is made up of some 10 to 20 retinula processes to every one eccentric cell process. The significance of these retinula processes is not clear; they do not, apparently, carry action potentials, so that it would appear that the great majority of the 'optic nerve fibres' of the compound eye are not nerve fibres at all.

The electron-microscope has contributed a lot to the understanding of the structure of the rhabdomere; the spokes illustrated in Figure 10.2 are apparently made up by innumerable interdigitating villi from the surfaces of contiguous retinula cells; they thus constitute a vast increase in the surface area of the retinula cells and it is here that the photosensitive pigment, responsible for absorbing the energy of the light-stimulus, is apparently concentrated.

SINGLE FIBRE RESPONSES

In Figure 10.3 are shown Hartline's oscillograms of the amplified potential changes in a single optic nerve fibre of *Limulus* when the eye is exposed to a steady illumination of two separate intensities; that for the upper record is ten times greater than that for the lower.

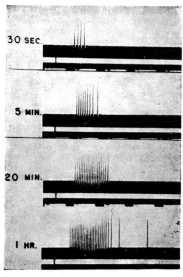

Fig. 10.3  Oscillogram of the amplified potential changes in a single *Limulus* optic nerve fibre due to steady illumination of the eye. Intensity of illumination with upper record 10 times greater than for lower record. Full length of each record corresponds to 1 second. (Hartline, *J. Opt. Soc. Amer.*)

The figure shows that, as with other receptors, the response to a continuous stimulus is intermittent, and that the effect of increasing the intensity of the stimulus is to increase the frequency of the action potentials, but not to modify the magnitude of the individual spikes.

DARK-ADAPTATION

The phenomenon of dark-adaptation has already been described; in Figure 10.4 are shown the electrical discharges of a *Limulus* single fibre in response to light stimuli of the same duration and intensity at intervals of 30 seconds, 5, 20 and 60 minutes after a light-adapting exposure of the eye to a bright light. It is quite clear the frequency of the response increases with the period of dark-adaptation. This suggests that the increase in sensitivity that occurs during a stay in the dark is a feature of the receptors, i.e. the rods and cones of the vertebrate eye, rather than of events in the central nervous system. As we shall see, however, there are many features of adaptation that indicate the intervention of higher nervous activity.

Fig. 10.4  The effect of dark-adaptation on the response of the eye of *Limulus* to a flash of constant intensity. As the eye dark-adapts, the frequency and duration of the response increase. (Hartline, *J. Opt. Soc. Amer.*)

LATENCY

When the stimulus is a brief flash of light, as in Figure 10.5, the discharge occurs entirely after the stimulus is over. The weaker the flash the longer this latency, as Figure 10.5 shows; the latency represents the time elapsing between the absorption of light by the rhabdom and the induction of an electrical change in the nearby eccentric cell, i.e. it is the time required for some photochemical change to take place in the rhabdom. The stronger the flash of light the more rapid will be this photochemical reaction and thus the shorter the latency.

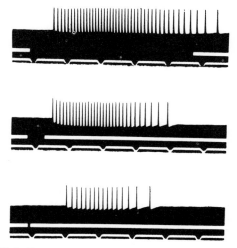

Fig. 10.5  Responses to flashes of light of decreasing duration as indicated by the size of the break in the upper white line. Note that in the lower records the response occurs entirely after the stimulus; also the long latency in the lowest record. (Hartline, Wagner and MacNichol, *Cold Spr. Harb. Symp. quant. Biol.*)

INHIBITION

It was considered in the early work on ommatidia that they were entirely independent of each other so far as the discharges in the optic nerve fibres were concerned; recently, however, Hartline, Wagner and Ratliff have shown that adjacent ommatidia may inhibit each other's response. An example is shown by Figure 10.6. This inhibition must have resulted from an interaction between the responses in the nerve fibres from individual ommatidia, taking place presumably in the plexus of fibres just outside the ommatidia.

This plexus, or *neuropil*, has been examined in the electron microscope by Whitehead and Purple (1970); regions of close contact between processes were common, and these were associated with the presence of electron-dense material and vesicles in the cytoplasm suggestive of a synaptic arrangement. In the more complex eyes of vertebrates we may expect inhibitory effects to be more marked because of the greater

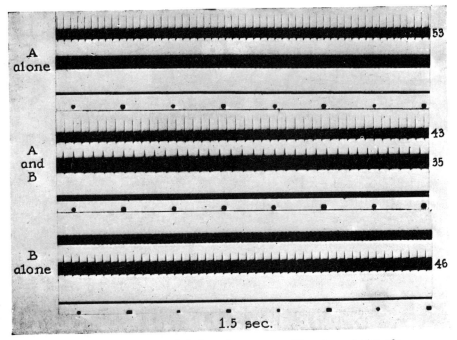

**Fig. 10.6**    Illustrating inhibition of discharge in one ommatidium by excitation of a neighbouring ommatidium. Ommatidium A, when illuminated alone, gives a frequency of 53 spikes/sec; ommatidium B gives a frequency of 46/sec; when A and B are illuminated together, A has a frequency of 43, and B one of 35/sec, i.e. they have mutually inhibited each other. (Hartline, Wagner and Ratliff, *J. Gen. Physiol.*)

possibilities of interaction between individual pathways resulting from the well defined system of interneuronal connections at the level of the bipolar and ganglion cells.

### Significance of inhibition

The importance of inhibition in the sensory pathway may not be immediately evident but will become so when we discuss such phenomena as flicker and visual acuity; for the moment we need only note that it is just as important to prevent a discharge—inhibit—as to initiate one—excite. Thus there is some evidence that a single cone can be excited separately by a spot of light, although we know on optical principles that the image of this spot must cover several cones. In fact several cones are stimulated, in the sense that light falls on them, but only one may actually be excited because the activities of adjacent ones are inhibited by the one that receives the most light, i.e. the cone in the centre of the image of the spot of light where the image is most intense. It is through inhibition then, that unnecessary responses fail to reach the higher levels of the central nervous system. The example shown in Figure 10.6 reveals inhibition at the lowest level; in the vertebrate retina it may occur at the level of the bipolar cells and ganglion cells and higher still in the geniculate body and cortex.

## MECHANISM OF ACTIVATION OF THE SENSORY NEURONES

### GENERATOR POTENTIAL

The electrical events following stimulation have been described as a series of spikes, brief changes in potential taking place across the nerve fibre's membrane. Those concerned with the way in which light, or some other stimulus, can provoke this electrical change have studied the events leading up to this spike discharge; and in general it appears that the common feature is the establishment in a sensory cell of what is called a *generator potential*; this is a more slowly developing and longer-lasting electrical negativity of the cell than the brief spike, and is essentially a 'preliminary change' that, if carried far enough, sets off the discharge. The development of this generator potential in the *Limulus* eye was followed by Fuortes, who placed an electrode inside the eccentric cell of an ommatidium; when light fell on this ommatidium the energy was presumably taken up by the rhabdom which, as we have seen (p. 244) surrounds the process of the eccentric cell. As a result of the chemical changes taking place in the rhabdom, the eccentric cell developed a generator potential, becoming negative internally; superimposed on this negativity were the much more rapid and abrupt changes that we call the spikes. This is illustrated

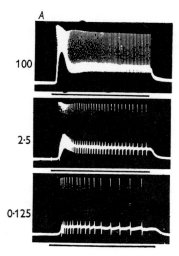

**Fig. 10.7** Generator and spike potentials from impaled eccentric cell of *Limulus* ommatidium. Duration of illumination indicated by black lines below records. Intensity of illumination decreases from above downwards. Note that the frequency of the spikes increases with the size of the generator potential. (Fuortes, *J. Physiol.*)

by Figure 10.7 where the ommatidium has been stimulated with three intensities (100, 2·5 and 0·125 units); the generator potential is indicated by the rise in the baseline, and the spikes by the strokes taking off from this; with increasing intensity the generator potential increases, and the spike frequency too.

### Retinula and eccentric cells

Behrens and Wulff (1965) were able to insert microelectrodes into two cells of the same ommatidium and, after recording their responses to illumination, they injected dye so as to be able to identify the types of cell from which records were made. In general, as others had found before, the responses fell into two classes: (*a*) a slow wave of depolarization with superimposed small spikes, and these could be identified with retinula cells; (*b*) a slow wave with large spikes identified with the eccentric cell; they showed that the retinula responses always preceded, by some 7 to 30 msec, the eccentric cell response; furthermore, the retinula responses, including the small spikes, were always synchronous with each other when two retinula cells were recorded from. Thus the retinula cell's slow potential wave seems to be the initial reaction to light; since the small spikes in this cell are not propagated along its nerve fibre, we may look on these as transmitted effects from the large spikes developed in the eccentric cell. It is possible, then, that the effect of light on the retinula cells is, after some photochemical change, to develop a slow depolarization which acts as a generator potential for the development of a depolarization of the

eccentric cell, which finally leads to spike activity. The fact that the electrical changes recorded from the retinula cells of the same ommatidium were always synchronous, indicates an electrical coupling between them, as suggested by Tomita, Kukuchi and Tanaka (1960) and proved by Smith, Baumann and Fuortes (1965).

### Summation

With a subthreshold stimulus there is only a small generator potential, too small to initiate spikes. However, this subthreshold generator potential will add with a second potential, so that one factor in the summation of subthreshold stimuli, observed in psychophysical studies, must be the ability of the generator potentials to add, provided, of course, that the interval between stimuli is not too long.

CHEMICAL TRANSMITTER

According to Fuortes and Hodgkin the electrical events can be best described on the assumption that the retinula cells (rhabdom) liberate a substance (transmitter) after absorbing light, and it is this chemical transmitter that activates the sensory neurone, in this case the eccentric cell.★

## VERTEBRATE RESPONSES

It is now time to turn our attention to the messages developed in the vertebrate retina, a system that poses more difficulties to the experimenter because the easily accessible fibres are those from the ganglion cells, whose responses are only distantly related to the initial effects of light on the receptors. It will be recalled that the receptors synapse in the outer plexiform layer with the bipolar and horizontal cells; the bipolar cells, in their turn, make complex synaptic relations with ganglion cells and amacrine cells in the inner plexiform layer. It is the axons of the ganglion cells, emerging from the eye, that constitute the optic nerve. Logically it would be reasonable to begin this account with the responses in the receptors, but since it is only very recently that such responses have been unequivocally identified, the more historical approach is preferable.

---

★In the thoroughly dark-adapted eye of *Limulus*, Werblin (1968) has examined discrete potentials in the retinula cells that occur in response to a weak light stimulus; these are of two types, namely small discrete 'quantum bumps', perhaps the result of absorption of a quantum of light, and larger potentials that have many of the features of regenerative potentials described in nerve, although they are not abolished by tetrodotoxin. Werblin suggests that this regenerative type has its origin in the retinula cell and represents an amplification of the 'quantal bump' permitting the eventual activation of the eccentric cell.

## FROG OPTIC NERVE

Single fibres of the frog's optic nerve were first studied by Hartline. In general, three types of fibre were found, differing in their responses to a flash of light. In the first place there is a group of fibres (representing on the average some 20 per cent of the total) which behave in essentially the same way as *Limulus* fibres, showing a response which starts almost immediately after the light has been switched on and ceasing when the light has been switched off (Fig. 10.8A); these are called

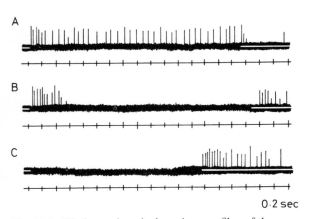

**Fig. 10.8** Discharges in a single optic nerve fibre of the vertebrate (frog) eye. Type A: Pure ON-element responding at ON with a discharge maintained during illumination. Type B: ON-OFF-element responding with discharges at ON and OFF but silent during the main period of illumination. Type C: Pure OFF-element responding only at OFF. Signal of the exposure to light blackens out white line above time-marker. Time marked in 0·2 second. (Hartline, *J. Opt. Soc. Amer.*)

ON-fibres; secondly, there is a group (some 50 per cent of the total) which respond with an initial outburst of impulses when the light is switched on and again when the light is switched off (Fig. 10.8B); these are called ON-OFF-fibres; during the intermediate period no impulses are recorded. Finally, there is a group of fibres (some 30 per cent of the total) which respond only when the light is switched off (Fig. 10.8C); these are called OFF-fibres. The group B, responding to the switching on and off of the light, shows a remarkable sensitivity to changes in the intensity of the stimulus and to the slightest movement of the image on the retina, the greater the change in intensity and the more rapid and extensive the movement of the image, the greater the frequency of the resultant discharge. These OFF-discharges probably represent a release from *inhibition*, the ganglion cells giving rise to them presumably being subjected to an inhibitory discharge by bipolar and other retinal neurones (horizontal and amacrine cells?) during the period of exposure to light; when the light

is switched off this inhibition ceases and the ganglion cells discharge, the phenomenon being called more generally 'inhibitory rebound'.

## MAMMALIAN EYE

Hartline's work was extended by Granit, and more recently by Kuffler and Barlow, to the mammalian eye where the same pattern of ON-OFF-, and ON-OFF-responses was revealed, although the phenomena were even more complex. The mammalian retina shows a background of activity in the dark, so that ON- and OFF-effects were manifest as accentuations or diminutions of this normal discharge. In general, ON-elements showed an increased discharge when the light was switched on, and an inhibition of the background discharge when the light was switched off. Similarly, an OFF-element showed an inhibited discharge when the light was switched on but gave a powerful discharge when the light was switched off. The OFF-effect is thus quite definitely a *release from inhibition*.★

## RECEPTIVE FIELDS

The number of fibres in an optic nerve is of the order of one million, whilst the number of receptors is of the order of 150 million. This means, in neurophysiological terms, that there is a high degree of *convergence* of the receptors on the ganglion cells, so that a single optic nerve fibre must have connections with many bipolar cells which themselves have connections with many rods and cones. In consequence, the area of retina over which a light-stimulus may evoke a response in a single optic nerve fibre may be quite large; the *receptive field* of a single fibre may be plotted by isolating a single fibre and, with a small spot of light, exploring the retina for those parts that will cause a discharge in this fibre. As Figure 10.9 shows, the field increases with the strength of stimulus, so that in order that a light-stimulus falling at some distance from the centre of the field may affect this particular fibre it must be much more intense than one falling on the centre of the field. This shows that some synaptic pathways are more favoured than others.

### SUMMATION

As we shall see, this convergence is the anatomical basis for the phenomena of summation: thus a single spot of

---

★In psychophysical terms the beginning of the ON-discharge may well correspond to the primary sensation of light when the eye is exposed to a flash; the so-called Hering after-image, that follows rapidly on the primary sensation, would then correspond to the activation of the ON-OFF-elements, whilst the dark-interval following this would represent activation of the OFF-elements. Finally, the appearance of the Purkinje after-image, rather later, could be due to a secondary activation of the ON-elements described by Grüsser and Rabelo (1958).

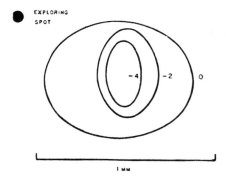

**Fig. 10.9**   Showing the retinal region with which a single optic nerve fibre makes connections. Each curve encloses the retinal region within which an exploring spot (relative size shown at upper left) produced responses in the fibre at an intensity of illumination whose logarithm is given in the respective curve. Note that the greater the stimulus the greater is the receptive field of the fibre. (Hartline, *J. Opt. Soc. Amer.*)

light falling on the periphery of the receptive field may be too weak to provoke a discharge in the ganglion cell; several spots each too weak to excite will, if presented together in different regions of the field, provoke a discharge. In neurophysiological terms we may say that the ganglion cell requires a certain bombardment by discharges from the bipolar cells, with which it makes synaptic contacts, before it discharges; this bombardment can be provided by a few bipolar cells discharging strongly or by many bipolar cells discharging more weakly. The electrical basis for this summation of effects is essentially the same as that for the summation in the receptors; the ganglion cells develop generator potentials which are this time called *synaptic potentials*; they are not all-or-none, and so can add their effects and ultimately build up to a sufficient height to initiate a spike discharge.

### CENTRE AND PERIPHERY

A careful study of the receptive fields in the cat's retina by Kuffler (1953) showed that they were not so simple as had been thought, the more peripheral part of the field giving the opposite type of response to that given by the centre. Thus, if at the centre of a receptive field the response was at ON, the response farther away, in the same fibre, was at OFF, and in an intermediate zone it was often mixed to give an ON-OFF-element as illustrated in Figure 10.10. In order to characterize an element, therefore, it was necessary to describe it as an *ON-centre* or *OFF-centre* element, meaning thereby that at the centre of its receptive field its response was at ON or OFF respectively. By studying the effects of small spot-stimuli on centre and periphery separately and together, Kuffler demonstrated a mutual inhibition between the two.

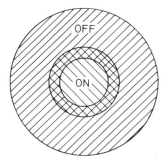

**Fig. 10.10**   Centre-surround organization of receptive field for a ganglion cell; the centre is excited at ON and the surround is excited at OFF. An intermediate zone may show ON-OFF characteristics.

### ANATOMICAL BASIS OF CENTRE-SURROUND

The anatomical basis for the centre-surround organization of the receptive field may well be that the centre corresponds with the dendritic field of the ganglion cell whilst the surround represents the influence of horizontally conducting elements like the horizontal cells. Brown and Major (1966) found that the dendritic fields of cat ganglion cells fell within two groups with diameters of 70 to 200 $\mu$m and 400 to 700 $\mu$m, and suggested that these corresponded with the centres of the receptive fields. Leicester and Stone (1967) found that the commonest centre-surround fields had centres with diameters of 110 to 880 $\mu$m, whilst dendritic fields of the deep multidentritic type of ganglion cell were in the range 70 to 710 $\mu$m, sufficiently close to make the suggestion of Brown and Major reasonable. However, Schwartz (1973) impaled individual ganglion cells of the turtle that gave ON-OFF responses to spots of light in the centre of their receptive fields; the periphery had an inhibitory action on the centre's response, and this peripheral inhibition could be achieved without any change in the membrane potential of horizontal cells. Schwartz concluded that peripheral inhibition of these ON-OFF elements, some of which may have been amacrine cells, was due to bipolar rather than horizontal cells. This was not to deny that horizontal cells influenced ganglion cell activity, since they certainly do so, but their large receptive fields apparently preclude them from exerting the quite localized inhibitory activity described by Schwartz.

### NON-RESPONSE TO UNIFORMITY

This opponence between centre and surround of the receptive field, although not universal in the retinae that have been examined exhaustively, is by far the predominant feature. It is continued to further stages in the visual pathway; thus the geniculate neurones exhibit the opponence to the point that centre and surround usually balance each other's effects exactly, so that uniform illumination of the retina usually fails to excite geniculate cells. The physiological significance of this opponence becomes clear if we appreciate that under ordinary conditions of illumination all receptors might be expected to be responding to light, and if these responses were all transmitted to the central nervous system along the million odd optic nerve fibres, this would mean a tremendous bombardment of the higher centres with information; the centre-surround organization means that for the great majority of ganglion cells uniform illumination of their receptive field will produce no,. or only a small, effect, the influence of light on the receptors that lie in the centre of the field being balanced by the opposite influence on the receptors that lie in the surround. On the other hand, a spatial discontinuity in the field, given by a bar on a white field, for instance, would activate centre and surround to different extents and so give rise to a message that would enable the perception of form. Since there is every reason to believe that the responses of the receptors to light are the same, the opposite responses of a ganglion cell to light falling on receptors in the two zones is the consequence of the differing synaptic relations of the receptors with the intermediate neurones. Thus, as Figure 10.11 shows, for a hypothetical ON-centre-OFF-surround ganglion cell, the

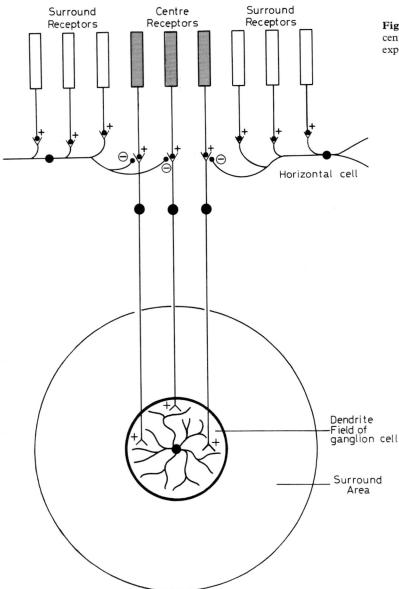

Fig. 10.11  Anatomical basis of the centre-surround organization. For explanation see text.

excitatory response of the receptors, whatever it may be, is transmitted to bipolar cells and the ganglion cell, whereas the excitatory response of the surround receptors causes the horizontal cells to liberate inhibitory transmitters on the bipolar cells, suppressing their normal state of excitation, which is reflected in an inhibitory influence on the ganglion cell. Alternatively, the amacrine cell might convert excitatory activity in bipolar cells to inhibitory action on the ganglion cell.

## MORE PERIPHERAL RESPONSES

The responses recorded from ganglion cells or optic nerve fibres represent activity in third-order cells; as such they provide us with an accurate picture of the messages sent to the brain, but they tell us little of the mode of transduction of the photochemical change in the receptors into an electrical event, the transmission of this change to bipolar and horizontal cells, and finally the integration of the messages from these second-order cells by the ganglion cells, aided by the amacrine cells. For a long time attempts at intracellular recording of the electrical changes in vertebrate receptors were unsuccessful and the experimenter's only clues to events at this level were provided by analysis of the electro-retinogram (p. 273), but even when this was recorded with microelectrodes placed at different depths in the retina the interpretation was by no means unequivocal. The first records purporting to represent intracellular records of responses to receptors were the S-potentials of Svaetichin, taken from the fish retina on account of its large cones. These will be considered later, but they turned out to be recordings from horizontal cells. With the refinement of micro-electrode techniques and especially with the development of the Procion yellow method for intracellular staining of the cell from which records were taken, enormous strides have been made in the understanding of the primary event in the receptors, especially in those of the fish, turtles and amphibians. The pioneering, and now classical, study is that of Werblin and Dowling (1969) on the retina of the mud-puppy, *Necturus*, and we may profitably begin with a brief description of this.

## INTRACELLULAR RECORDS FROM NECTURUS RETINA

### 'WIRING DIAGRAM'
The synaptic structures and relations between retinal cells in *Necturus* are similar to those of the primate retina, but the relations between horizontal cells and receptors and bipolar cells would appear to be different. The 'wiring diagram' of the *Necturus* retina is illustrated schematically in Figure 10.12, and it will be seen that

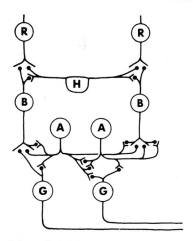

**Fig. 10.12**  Schematic 'wiring diagram' of the mud-puppy retina. R, receptor; H, horizontal cell; B, bipolar cell; A, amacrine cell; G, ganglion cell. Note that in this retina the horizontal cell influences both receptor and bipolar cell, and the amacrine cell influences both bipolar cell and ganglion cell. (Dowling and Werblin, *J. Neurophysiol.*)

the horizontal cell is not a mere mixer of rods and cones, since it forms presynaptic relations with bipolar cells, i.e. it transmits from receptor to bipolar cell and thus contributes to vertical transmission. Probably more important, however, is the feedback that allows activity in one receptor to influence activity in its neighbours, thus providing an anatomical basis for surround-on-centre antagonism. In a similar way, the amacrine cell not only links bipolars horizontally but also transmits from bipolars to ganglion cells. Thus there are two pathways from rod to bipolar, and two from bipolar to ganglion, either direct or through a horizontal or amacrine cell.

### RECORDING
Microelectrodes were inserted into different regions of the retina and intracellular recordings of the responses to light stimuli were made; the stimuli consisted of a small spot or of an annulus, either 250 $\mu$m or 500 $\mu$m wide, either singly or in combination with the spot. In this way the existence, or otherwise, of opponent responses typical of those described for ganglion cells, could be ascertained. After the recordings had been made, the position of the electrode was identified by ejecting Niagara blue from the electrode and subsequent histological examination. The results are summarized in Figure 10.13.

### RECEPTOR RESPONSE
The striking feature of the receptor response is that it consists of a hyperpolarization of the cell, rather than a depolarization as found in the invertebrate receptor.

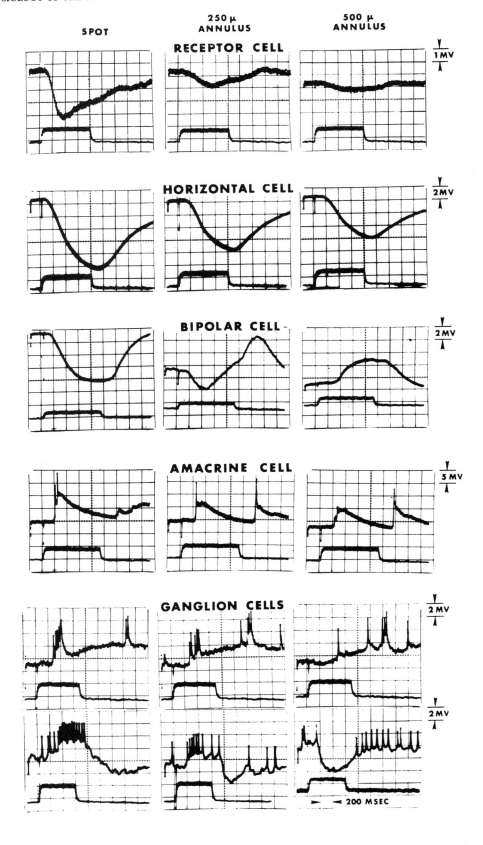

SPOT    250 μ ANNULUS    500 μ ANNULUS

RECEPTOR CELL    1 MV

HORIZONTAL CELL    2 MV

BIPOLAR CELL    2 MV

AMACRINE CELL    5 MV

GANGLION CELLS    2 MV

2 MV

◄ 200 MSEC

The response is graded, like the invertebrate generator potential, but there is no all-or-none spike activity. Annular stimulation did not affect the element, nor did it affect the response to a spot stimulus; thus there was no opponent organization of centre and surround, as found in the ganglion cell.

### HORIZONTAL CELL

This cell's response, too, is hyperpolarizing, and comes on with a longer latency than that of the receptor; it is graded and does not lead to spikes, and the effects of light are summated over quite a wide field since a 500 $\mu$m annulus of light gave the same response as a 250 $\mu$m annulus containing the same number of quanta of energy (Fig. 10.13).

### BIPOLAR CELL

Once again, this responds with a slow potential change, graded according to the strength of the stimulus and giving no spikes.* The receptive field, however, is concentrically organized in an opponent fashion so that if the centre hyperpolarizes, the periphery causes depolarization, and *vice versa*. In the middle column of Figure 10.13 the small annulus stimulated both centre and periphery of the receptive field so that both hyper- and de-polarizing responses were observed; at the right the large annulus was combined with the central spot; now the central spot has been suppressed and the response is purely depolarizing. The central response always preceded the antagonizing surround response; hence when both centre and surround are illuminated, and then the intensity is changed, there is always a transient central response even if the contrast between centre and surround is unchanged.

### AMACRINE CELL

Responses in these are recorded from the inner plexiform layer; they consist typically of one or two spikes superimposed on a transient depolarizing potential (Fig. 10.13). The transient responses occurred at both ON and OFF. In these cells the effect of variation of intensity of the stimulus was essentially a change in the latency, the number of spikes being independent of the intensity. In some units the receptive field was organized in an opponent fashion, but in others the responses were uniform over the whole area.

### GANGLION CELLS

These responded with a spike discharge taking off from an initial depolarization; the frequency of discharge was proportional to the degree of membrane depolarization. ON-, OFF- and ON-OFF-responses were obtained in different cells. Centre-surround antagonism was also a feature of some cells and was similar to that of the bipolar cells; moreover, the extents of the fields were very similar.

### RETINAL ORGANIZATION

The central part of the receptive field of the bipolar cell is about 100 $\mu$m wide and is surrounded by an annulus of about 250 $\mu$m. The dendritic field of a bipolar cell in *Necturus* retina is about 100 $\mu$m in diameter, so that the large annular area, stimulation of which gives responses in a single bipolar cell, must

---

*Kaneko and Hashimoto (1969) recorded spike activity from cells in the inner nuclear layer of the carp's retina, which they attributed to bipolar cells; Murakami and Shigematsu (1970) have also observed these, in the frog retina. Tetrodotoxin, which abolishes spike activity but not slow potential changes, allowed some form of transmission to ganglion cells in response to light, so that it was argued that the bipolar cells could transmit their slow potential changes to the ganglion cell as well as their spike activity. It may be, however, that the spike activity occurring normally was, in fact, recorded from amacrine cells, as indicated by Werblin and Dowling. Certainly Kaneko (1970) found no spike activity in bipolar cells of the goldfish retina.

---

◀ **Fig. 10.13**   Responses of different retinal cells to a spot-stimulus (extreme left) and to annular stimuli of 250 $\mu$ and 500 $\mu$ radius. Receptors have relatively narrow receptive fields, so that annular stimulation evokes very little response. The horizontal cell responds over a broader region of the retina, so that annular illumination with the same total energy as the spot (left column) does not reduce the response significantly (right columns). The bipolar cell responds by hyperpolarization when the centre of its receptive field is illuminated (left column). With central illumination maintained (right trace; note lowered base line of the recording and the elevated base line of the stimulus trace in the records) annular illumination antagonizes the sustained polarization elicited by central illumination, and a response of opposite polarity is observed. In the middle column the annulus was so small that it stimulated the centre and periphery of the field simultaneously. The amacrine cell was stimulated under the same conditions as the bipolar cell, and gave transient response at both the onset and cessation of illumination. Its receptive field was somewhat concentrically organized, giving a larger ON-response to spot illumination, and a larger OFF-response to annular illumination of 500 $\mu$ radius. With an annulus of 250 $\mu$ radius, the cell responded with large responses at both ON and OFF. The ganglion cell shown in the upper row was of the transient type and gave bursts of impulses at both ON and OFF. Its receptive-field organization was similar to the amacrine cell illustrated above. The ganglion cell shown in the lower row was of the sustained type. It gave a maintained discharge of impulses with spot illumination. With central illumination maintained, large annular illumination (right column) inhibited impulse firing for the duration of the stimulus. The smaller annulus (middle column) elicited a brief depolarization and discharge of impulses at ON, and a brief hyperpolarization and inhibition of impulses at OFF. (Werblin and Dowling, *J. Neurophysiol.*)

transmit its effects through another type of cell that synapses with the bipolar. This other type is presumably the horizontal cell; this receives influences from the receptors in the surrounding annulus and transmits them to bipolar cells connected with the central

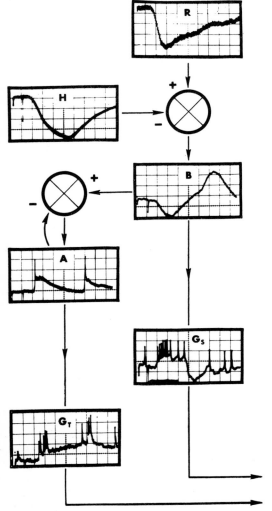

**Fig. 10.14** Summary diagram of synaptic organization of the retina. Transformations taking place in each plexiform layer are represented here by summing junctions. At the outer plexiform layer the direct input to the bipolar cell from the receptor is modified by input from the horizontal cells. At the inner plexiform layer the bipolar cell drives some ganglion cells directly. These ganglion cells generate a sustained response to central illumination, which is inhibited by additional annular illumination: and thus these cells follow the slow, sustained changes in the bipolar cell's response. Bipolar cells also drive amacrine cells, and the diagram suggests that it is the amacrine-to-bipolar feedback synapse that converts the sustained bipolar response to a transient polarization in the amacrine cell. These amacrine cells then drive ganglion cells which, following the amacrine cell input, respond transiently. (Werblin and Dowling, *J. Neurophysiol.*)

receptors, exerting an opposite type of action; thus, if the central receptors hyperpolarize the bipolar, the horizontal cell depolarizes it. Hence it is the horizontal cells that organize the contrast-detection, as exhibited in the centre-surround antagonism. The further transformation of the message, which gives it its transient character, is probably done through the amacrine cells; these respond transiently, by contrast with the more sustained responses of bipolar cells, and they thus impose a transient type of response on the ganglion cells; the ganglion cells, driven in this way, presumably are able to carry information about *changes* in the visual field, whilst other ganglion cells, driven by bipolar cells directly, would be better able to carry information about the static character, i.e. contrast. These principles are illustrated by Figure 10.14.

## THE HYPERPOLARIZING RESPONSE

The most important finding is clearly the hyperpolarization of the receptor by light, a hyperpolarization that is transmitted to the two types of cell that make synaptic relations with the receptor, namely the bipolar and horizontal cell. If, as is the general rule, depolarization of a nerve cell causes release of a transmitter, we may postulate that the receptor in the dark is continually releasing a transmitter on to the synaptic surfaces of bipolar and horizontal cells, thereby maintaining these in a depolarized state. Light-induced hyperpolarization of the receptor would decrease, or inhibit entirely, the release of transmitter, thereby allowing the bipolar and horizontal cells to become less depolarized, i.e. to *hyperpolarize*. We have therefore to enquire into the mechanism of the hyperpolarizing response in the receptor. The depolarizing responses seen typically in the generator potentials of invertebrate visual cells, or in mechanoreceptors such as the Pacinian corpuscle, are attributable to an increase in permeability of the excitable membrane to certain ions, which is reflected in a decrease in membrane resistance.

### NERVE RESTING POTENTIAL

Thus, in nerve, the resting membrane potential depends on a high internal concentration of $K^+$ and low concentration of $Na^+$, the situation being maintained by a low permeability of the membrane to $Na^+$. Under ideal conditions the potential is given by the Nernst equation:

$$E = \frac{RT}{nF} \ln \frac{[K]_i}{[K]_o}$$

and is described as a $K^+$-potential and we may represent the membrane by a simple electrical circuit in which the potential is considered to be due to the action of a battery—the $K^+$-potential—operating across a capacity and a resistance, or its reciprocal a conductance.

## SODIUM BATTERY

If the ideal situation does not prevail, and the membrane has a significant permeability to $Na^+$, the membrane potential is lower because the difference in concentrations of $Na^+$ is capable of acting like a battery working in the reverse direction, so that we must represent the situation as being rather more complex, with two batteries working against each other giving a smaller potential—the membrane is said to be depolarized by the $Na^+$-battery (Fig. 10.15). If, now, the effect of light is to reduce the permeability to $Na^+$ by 'blocking $Na^+$-channels', to use the terminology of Baylor *et al.* (1974), the cell will hyperpolarize.

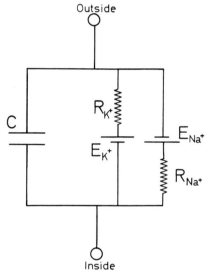

**Fig. 10.15** Electrical representation of an excitable cell with batteries due to potassium and sodium differences of concentration working in opposition.

## GRADIENT OF POTENTIAL

The observation that the retina exhibited in the dark a gradient of potential such that the tips of the receptors were negative in relation to the synaptic region, and that with light this potential tended to be reversed, suggested to Bortoff and Norton (1967) that the dark-depolarization of the receptors might be restricted to the outer segments, so that, as indicated by Figure 10.16, there would be a gradient of potential along the receptor, the tip being less positive (or more negative) than the inner segment, and there would be a flow of positive current through the inter-receptor space, from inner to outer segment, taking place in the dark. Light, by increasing the resistance of the outer segment, would reduce this current and the outer segment would hyperpolarize.

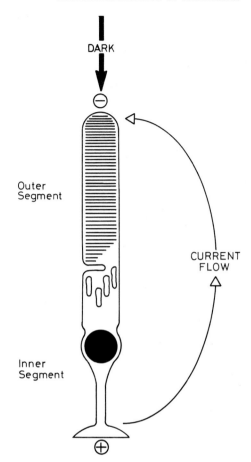

**Fig. 10.16** Dark current due to difference of potential between outer and inner segments of a rod.

### LIGHT-INDUCED CHANGES

Penn and Hagins (1969) and Hagins *et al.* (1970) placed electrodes at different depths in the retina alongside the receptors and showed that the voltage did, in fact, vary with depth, the electrode becoming progressively more positive as it moved from the receptor tip to the rod synapse, whilst the light-induced voltage increased in parallel. They showed, moreover, when the rod was partially illuminated, that the photocurrent was generated in the zone of the rod outer segment that was illuminated, so that the unilluminated region acted as a sink of current, thus confirming that the effect of light was to reduce a standing dark-current inward at the outer segment and outward along the remainder of each receptor cell. Figure 10.17 shows the membrane current produced by two flashes of light of different intensities at different depths in the retina; the flow of current induced by light is outwards at the outer segments, and this corresponds with a reduction in the dark-current, which occurs in the opposite direction (Fig. 10.16). At the junction of outer and inner segments

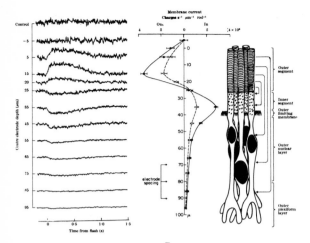

**Fig. 10.17**   Spatial distribution $\overline{\overline{i}}_m$ of membrane photocurrent density along rods. Left panel: photocurrent transients against time at electrode depths shown on scale, corrected for radial variations in retinal resistance as described. Each tracing is the average of sixteen responses. Stimuli: $1 \times 10^{11}$ hV cm$^{-2}$ at 560 nm. Centre panel: peak photocurrent against radial depth of centre electrode of recording triplet. Results from two slices stimulated with $1 \times 10^{11}$ hV cm$^{-2}$ ($\bigcirc$) and $4 \times 10^{11}$ hV cm$^{-2}$ ($\bullet$) are shown. Error bars indicate rms noise in current tracings. Right panel: retinal rods of the rat drawn to scale. Pattern of flow of photocurrent indicated by arrows. Outer and inner segments, outer limiting membrane, and outer nuclear and plexiform layers are labelled. (Penn and Hagins, *Nature*.)

the current reverses in sign, as we should expect if the flow of current were confined to the receptor, outward flow at one point being accompanied by inward flow at another. The studies of Hagins were on the rods of the rat; when the larger receptors of *Necturus* or the gekko were examined by Toyoda *et al.* (1969), the postulated increase in membrane resistance could be measured with intracellular electrodes.

**Ionic currents**
In terms of ionic movements, the flow of current would represent a net penetration of Na$^+$ into the outer

segments and a flow of K$^+$ out of the inner segments; if this persisted indefinitely, of course, the concentration gradients would disappear and the batteries would run down. We must postulate, as with other neurones, a Na$^+$—K$^+$ linked pump that, at the expense of metabolic energy, removes Na$^+$ and accumulates K$^+$; such a pump would be inhibited by oubain, and it is interesting that Frank and Goldsmith (1969) found that this ATPase inhibitor rapidly blocked the ERG. Further proof relating to the postulated mechanism of the photocurrents was given by Korenbrot and Cone (1972), who showed that the permeability to Na$^+$ of isolated rod outer segments was actually decreased by light. That it was the outer (plasma) membrane of the outer segment, rather than the disc membrane, that was affected was shown by Korenbrot *et al.* (1973) and they computed a permeability constant for Na$^+$ of about $2 \cdot 8 \cdot 10^{-6}$ cm/sec. The dependence of the hyperpolarizing response on external Na$^+$ was demonstrated by Brown and Pinto (1974) who substituted this ion by choline; as Figure 10.18 shows, shortly after the replacement the membrane hyperpolarized and the responses to light-stimuli were abolished; after restoration of Na$^+$ to the medium, the membrane potential returned to its depolarized value, and responses to light returned.

CALCIUM
It has been suggested that the effect of light on the receptor membrane is mediated through release of some substance within the discs, and Ca$^{2+}$ is a candidate; Brown and Pinto lowered the external Ca$^{2+}$ of the toad's retina, and this depolarized the membrane in the dark and increased the maximal response evoked by light, the magnitude of the increase being equal to the magnitude of increased depolarization. When the outside concentration of Ca$^{2+}$ was raised, there was a steady hyperpolarization and decreased light-response. Thus, if the lowered outside concentration causes depletion of Ca$^{2+}$ in the receptor, and if Ca$^{2+}$ tends to decrease the permeability to Na$^+$, then the loss of internal Ca$^{2+}$ would be expected to depolarize the receptor, but of

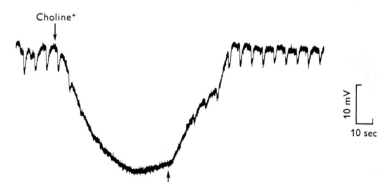

**Fig. 10.18**   Dependence of the hyperpolarizing response to light on external Na$^+$. At the first arrow the Na$^+$ in the medium was replaced by choline; shortly after this substitution the membrane hyperpolarized, and the receptor potential was abolished. These effects were reversed by restoration of Na$^+$ to the medium (second arrow). (Brown and Pinto, *J. Physiol.*)

course it is by no means certain that changes in external $Ca^{2+}$ are reflected in corresponding changes* in internal $Ca^{2+}$.

### TURTLE CONES

Baylor and Fuortes (1970) impaled the large cones of the turtle, *Pseudemys elegans*, and obtained hyperpolarizing responses that increased in size to a maximum of about 12 mV bringing the resting potential from about 25 mV (inside negative) to about 37 mV; the peak potential change at any intensity of illumination, $I$, was given by the Michaelis-type equation:

$$\frac{V_i}{V_{max}} = \frac{I}{I + \sigma}$$

$V_{max}$ being the potential change recorded at very bright lights and $\sigma$ the light-intensity when $V_i = \frac{1}{2}V_{max}$. Typical records are shown in Figure 10.19. The potentials were very similar to the S-potentials derived from horizontal cells in the fish, and they followed a similar equation (Naka and Rushton, 1966). When the size of the spot illumination was increased, there was an apparent summation which the authors considered

could be accounted for by scatter from the larger area, but as we shall see, summative interaction between cones does, in fact, occur.

### Changed membrane resistance

If a current is passed through a cone, the change in its potential is governed, other things being equal, by the resistance across its membrane; Baylor and Fuortes observed an increase in this resistance when the light was switched on, and this would correspond to a diminished permeability to $Na^+$-ions at the outer segment.

---

*The role of $Ca^{2+}$ in the response of invertebrate receptors to light is better established experimentally, especially in those of *Limulus*. Most recently Brown *et al.* (1977) have employed a very sensitive indicator of ionized $Ca^{2+}$ to measure the light-induced changes in the internal concentration. The estimated internal concentration in the light-adapted eye was $2 - 6 . 10^{-4}$ M; a flash of light produced a transient increase in internal concentration, as Brown and Blinks (1974) had found with a less quantitative method; with a prolonged light-stimulus, the internal concentration rose to a peak and fell to a lower level during the maintained stimulus; with onset of darkness the concentration fell still farther.

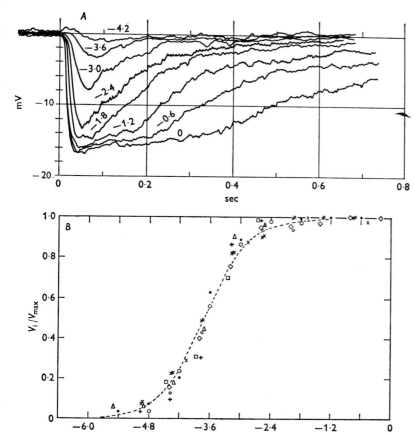

Fig. 10.19  Cone response to flashes of different intensities. *A*. Superimposed tracings of responses to 10 msec flashes of increasing intensity, as indicated (log units), applied at time zero. Potential drop on penetration of cell was 22 mV. Downward deflexion represents hyperpolarization.

*B*. Peak height of response plotted as a function of light intensity. The symbols plot normalized voltages obtained from ten different cells. As different cells had different sensitivities to light, some shift along the abscissa was necessary to superimpose all points. The largest shift was about 0·6 log unit. The interrupted line follows the Michaelis-type equation. (Baylor and Fuortes, *J. Physiol.*)

## Equivalent circuit

Baylor and Fuortes emphasized that the response to light could be represented by a similar circuit to that proposed to describe the end-plate potential, as in Figure 10.20, where the membrane is represented by a

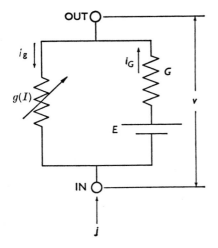

**Fig. 10.20** Equivalent circuit of cone membrane. The membrane is represented by a battery having a fixed e.m.f., $E$, and an internal conductance $G$. In parallel with this limb is a conductance $g$ whose value is reduced by light. The voltage across the membrane is denoted by $v$. $j$ denotes a constant applied current, and $i_G$ and $i_g$ denote currents through conductances $G$ and $g$ respectively. (Baylor and Fuortes, *J. Physiol.*)

fixed battery, $E$, a fixed conductance $G$, and another conductance, $g$, which varies with light. In the dark, $g$ shunts the battery and causes the voltage, $v$, across the membrane to be less than $E$. Light reduces $g$ and allows $v$ to approach $E$. On this basis, it is easy to demonstrate that the light-response will increase with hyperpolarizing, and decrease with depolarizing currents, passed across the receptor membrane. It could be deduced that the response would be zero when the depolarization was 29 mV, i.e. when the cone had been depolarized to give about zero resting potential.

### COMPUTED CONE RESPONSE

Baylor and Hodgkin and their colleagues have made measurements of the hyperpolarizing responses of turtle cones to flashes of light, and have sought to explain the curve, showing the development of the response with time, in terms of a model. The basic feature of this model is that light liberates a substance, possibly calcium ions, after a certain delay, which blocks sodium channels in the outer segment. Experimentally it was found that the response is shortened if the stimulus is increased in intensity or superimposed on a light background, so that an additional postulate, that the blocking molecules were removed by an autocatalytic process, was made. The model envisaged the formation of blocking particles through some six steps, whilst the breakdown occurs through four.

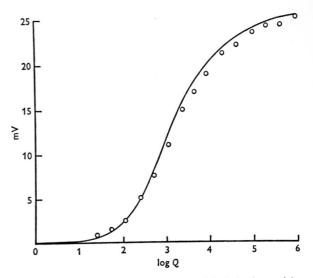

**Fig. 10.21** Relation between intensity of flash (on log scale) and peak amplitude of response for real and model cones. The circles are for the real cone and the continuous curve is for the model. The abscissa is the logarithm of the number of photoisomerizations per cone calculated on the basis that the flash sensitivity is 25 $\mu$V per photoisomerization. (Baylor *et al.*, *J. Physiol.*)

The model allows the prediction of a number of features of the responses by choice of appropriate parameters; thus Figure 10.21 shows the predicted relation between peak amplitude of the response, in mV, and the intensity of the flash computed in terms of number of isomerizations per cone; the circles are experimental points. As the authors emphasize, it is unlikely that the same equations would describe the behaviour of the rods by simply slowing down the cone model by a factor. The equivalent electrical circuit envisaged is shown in Figure 10.22; here we have the battery $E$ responsible for maintaining the resting potential and $g$ is the fixed conductance; $g_i$ is the variable conductance of the outer segment which is blocked, channel by channel, by a product of the absorbed light, and $g_f$ is a voltage-dependent

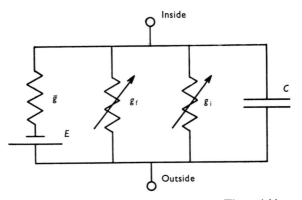

**Fig. 10.22** Equivalent circuit of model cone. The variable conductance $g_f$ increases reversibly and with a delay when the cell is hyperpolarized; $g_i$ is the light-sensitive conductance. (Baylor *et al.*, *J. Physiol.*)

conductance presumed to belong to the inner segment, and is introduced to account for the sagging of the hyperpolarization from its peak to a plateau. Essentially, then, the model must describe the change in $g_i$ in terms of production and decay of blocking particles.*

### REGENERATIVE RESPONSES

The spread of electrical change from the rod outer segment to the inner segment and thence to the synaptic body involves propagation over quite a large distance, and one is tempted to look for some form of regenerative action that assists this propagation, comparable with the regenerative depolarization observed in conducting cells such as nerve and muscle fibres. Werblin (1975) has applied voltage-clamps to *Necturus* rods and shown that the membrane resistance was indeed voltage-dependent, and when the membrane at the cell body was hyperpolarized and clamped at this hyperpolarized state a transient outward current was measured associated with a decrease in membrane resistance with a similar time-course. This outward current can be described as regenerative since it is initiated by hyperpolarization and its effect is to hyperpolarize the membrane still further.

## COUPLING OF RECEPTORS

Stimulation of turtle cones adjacent to that from which records are taken produces responses in this (unstimulated) cone; this spread of effect has a double cause, namely a direct coupling between adjacent cones, through gap-type junctions that permit an electrical change induced in one to be transmitted rapidly to another, and secondly to the cone's synaptic relations with horizontal cells.

The direct cone-cone interaction is demonstrated by passing current into one cone and recording a corresponding, if attenuated, change in nearby cones (Baylor *et al.*, 1971); this interaction extends to a radius of about 50 $\mu$m and is probably mediated through lateral processes that extend from one cone pedicle to another, since these have a maximum length of this order (Lasansky, 1971). If a small-radius spot is moved

from a central position over an impaled cone, the intensity of the response falls off to each side defining a receptive field which itself defines the extent of the cone-cone interaction (Fig. 10.23), since a small spot-stimulus is unable to exert an appreciable effect on the horizontal cells.

### MORPHOLOGICAL BASIS OF COUPLING

The physiological studies on inter-receptor coupling suggest that it is electrical through cell-to-cell junctions that permit passage of current from one to another; morphologically these junctions have been characterized—e.g. in epithelia—as gap-junctions and, as such, they have a characteristic appearance in freeze-fractured specimens. Fain (1975) demonstrated a very powerful interaction between rods of the toad such that a rod could be excited by light that failed to fall on it. The points of contact responsible for this coupling are probably located on the 'fins' or radial processes of the rods, which come into close relations with fins of other receptors. In the electron-microscope rod-rod junctions at these fins are large to give a total area of as much as 1 $\mu m^2$ and by freeze-fracture they were shown to have the appearance of characteristic gap-junctions (Fain *et al.*, 1975). These large areas of contact were peculiar to those between red-rods; much smaller points of contact were seen between adjacent rods and cones, and between red and green rods; these apparently do not contribute significantly to mixing of signals from different types of receptor since the spectral sensitivity of red rods corresponded remarkably closely with that of the toad's rhodopsin, whereas if there had been intermixture with green rods or cones there would have been significant deviations.

In the turtle retina, the cone pedicles have basal processes that may extend laterally as far as 40 nm; these processes end in dyad-structures of other cone pedicles and may well constitute the basis for electrotonic coupling (Lasanksy, 1971).

### SIGNIFICANCE OF COUPLING

Coupling of receptors, so that they have lost their independence as single units, militates against spatial discrimination (p. 315) although it increases sensitivity to large fields by permitting summation. According to

---

*Cervetto *et al.* (1977) have analysed their results on toad rods on the basis of a similar model.

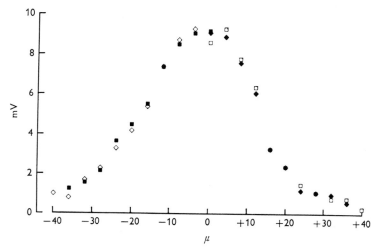

**Fig. 10.23** The receptive field of a cone. The peak amplitude of the response is plotted against distance from axis of stimulation light. A flash covering a circle of 4 $\mu$ radius was centred on the receptor. The intensity of the flash was adjusted to give a response of about half maximum amplitude. The spot was then moved away from the centre ($\square$ $\diamond$) and then back toward the centre ($\blacklozenge$ $\blacksquare$) in 4 $\mu$ steps. In this experiment, the retina was quite transparent (decreasing scatter) and at 40 $\mu$ from the centre the responses were quite small. (Baylor *et al.*, *J. Physiol.*)

Lamb and Simon (1976), the 'noise' of a single cone, i.e. the spontaneous changes in potential taking place without measurable change in the stimulus conditions, is decreased by exposing the retina to light (Fig. 10.24), a surprising result since one might expect random fluctuations in the number of quanta falling on the retina to introduce more rather than less noise. Lamb and Simon showed that there was a good correlation between the noise, or intrinsic voltage variance, and the tightness of coupling between cones, measured by the so-called space-constant of a given cone. Thus when a slit of light is moved laterally from its initial position covering the cone there is a falling off in the response, as indicated in Figure 10.25. If the cone were not coupled to adjacent cones there would be a very steep falling off as the slit passed off the surface of the cone, a falling off governed entirely by the spread of scattered light to the cone. In fact, the falling off may be quite gentle, due to the coupling between cones so that an electrical change induced in cones, lateral to that from which records are taken, is propagated to this by electrotonic spread of current.

## Space-constant

As with spread along an axon or muscle fibre, this spread can be indicated by a space-constant, $\lambda$, and in the case of coupled cells the magnitude of $\lambda$ is a measure of the tightness of coupling. As Figure 10.26 shows, there is a strong correlation between noise and coupling, the smooth curve being a theoretical one based on a model in which the cones are spaced on a square grid with individual uncoupled cones having a voltage variance of $0.4$ mV$^2$ and a cone-separation of 15 $\mu$m; this computed separation corresponds fairly closely to the anatomical value of 17.4 $\mu$m (Baylor and Fettiplace, 1975).

## Signal-to-noise ratio

Lamb and Simon considered that an important function of coupling was the improved 'signal-to-noise' ratio;

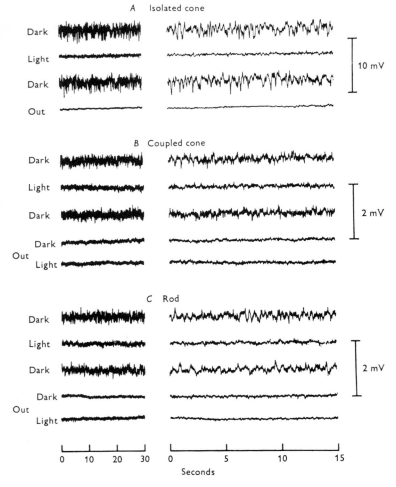

Fig. 10.24  Reduction of photoreceptor noise by bright illumination. Voltages were recorded both intracellularly and after withdrawing the electrode from the cell ('out'). Note the different ordinate scales. A, isolated red-sensitive cone (cell 13). Stimulus was a 105 $\mu$m diameter circle with photon flux of $8.39 \times 10^5$ photons $\mu$m$^{-2}$ sec$^{-1}$ at 639 nm and evoked a response of 11.5 mV at peak and about 6 mV in the steady state. B, coupled red-sensitive cone (cell 20); 105 $\mu$m spot; $8.39 \times 10^5$ photons $\mu$m$^{-2}$ sec$^{-1}$ at 639 nm; peak response 17.5 mV, steady response 11 mV. Stimulus intensity outside the cell was $8.5 \times 10^7$ photons $\mu$m$^{-2}$ at 639 nm. C, rod (cell 31); 210 $\mu$m spot; $9.31 \times 10^4$ photons $\mu$m$^{-2}$ sec$^{-1}$ at 558 nm; peak response 14 mV, steady response 4.5 mV. Outside the cell the stimulus was $5.47 \times 10^6$ photons $\mu$m$^{-2}$ sec$^{-1}$ at 559 nm, 1150 $\mu$m spot. (Lamb and Simon, *J. Physiol.*)

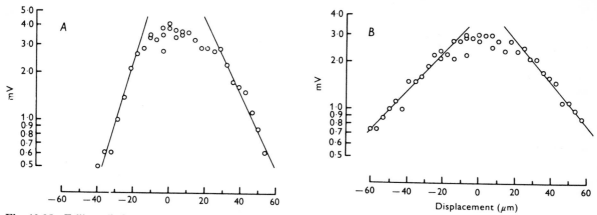

**Fig. 10.25**  Falling off of response of a cone as a slit of light is moved laterally from a position immediately over it. *A* is a red-sensitive cone and *B* a green-sensitive cone stimulated by 643 nm and 559 nm lights respectively. (Lamb and Simon, *H Physiol.*)

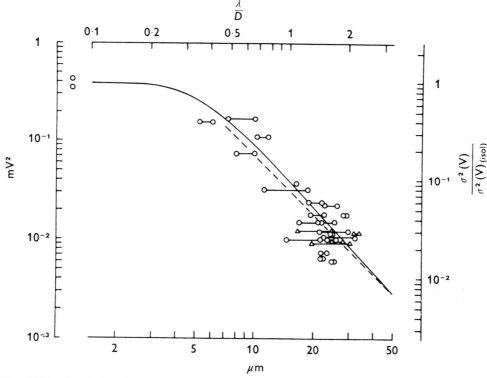

**Fig. 10.26**  Correlation of noise (intrinsic voltage variance) with length constants representing falling off of response with distance from cone-centre. Values for red-sensitive ($\bigcirc$) and green-sensitive ($\triangle$) cones are plotted on double logarithmic coordinates. Horizontal lines connect $\lambda_+$ and $\lambda_-$ for each cell. Continuous line is predicted from the square-grid model, and interrupted curve from distributed model. (Lamb and Simon, *J. Physiol.*)

thus for diffuse illumination the signal is independent of $\lambda$ but the root mean square noise is inversely proportional to $\lambda$, implying that the signal-to-noise ratio increases directly as $\lambda$. For a very small spot activating only one cone this is not true since the signal now decreases more rapidly with increasing $\lambda$ than does the square root of the variance.

**BIPOLAR CELLS**

It must be emphasized that this averaging, which reduces noise, could be performed, and is, by the bipolar cell which receives, in the peripheral retina where these measurements have been made, input from many cones. It has been well established that the dark noise in the bipolar cell is inhibited by light, but in this case its mechanism is the hyperpolarization of the receptor which reduces the release of transmitter. Thus in the dark the depolarized receptor is continually releasing transmitter, and spontaneous variations in this release would be the cause of the noise comparable with the miniature end-plate potential produced by a motor nerve terminal; hyperpolarization, be it of the motor nerve terminal or of the receptor terminal, would decrease the amplitude and frequency of the miniature potentials.

PHYSICAL BASE OF RECEPTOR NOISE

Schwartz (1977) and Lamb and Simon (1977) have investigated the features of noise in turtle rods and cones respectively. Schwartz described a voltage-dependent component of noise as that which was reduced by hyperpolarizing the receptor, and a light-sensitive noise, being the additional reduction in noise when the hyperpolarized cell was stimulated by light. The light-sensitive noise was concluded to be the result of random bleaching of molecules of photopigment, whilst the voltage-sensitive noise was due to random alterations in the permeability characteristics of the receptor membrane. So far as the magnitude of the noise is concerned, this, represented as a standard deviation of the membrane potential, was 271 $\mu$V comparing with 30 $\mu$V for the response to absorption of a single photon of light; thus the noise in the receptor is some nine times the response to a photon, so that we may expect the synapse at the bipolar cell to serve as a filter, removing noise and allowing the signal to pass (Baylor and Fettiplace, 1975). Schwartz estimated that the most advantageous transmission would be if the rod contacted a very limited number of rods through its dendritic field, since the larger the number of rods included the greater will be the noise transmitted to the bipolar cell, and he notes that the dendritic spread of bipolars receiving from rods is, in fact, smaller than those receiving from cones (Scholes, 1975). Lamb and Simon (1977) calculate the 'dark light' that would be equivalent to the observed noise in turtle cones; it amounts to some 2000 isomerizations per sec per cone and is such as to reduce the cone's sensitivity by a half. According to their model, photo-

isomerization of a pigment molecule leads to an active state of the receptor membrane allowing $Ca^{2+}$ to enter the cone and block $Na^+$-channels leading to hyperpolarization. This state lasts 40 msec. Removal of $Ca^{++}$ occurs with a time-constant considerably shorter than 40 msec. The mean time in the active state is reduced by hyperpolarization or by an increase in intracellular $Ca^{++}$, and in these ways the noise may be reduced.

THE RODS AS A SYNCYTIUM

Schwartz (1973) impaled single rods of the turtle, *Chelydra serpentina*; Figure 10.27 shows the hyperpolarizing responses to increasing intensities of a spot-stimulus of 100 $\mu$m diameter. The gain* was some four times that of cones under comparable conditions; this seems a small difference when the much greater sensitivity of the dark-adapted eye is considered, and suggests that in some way the rods can interact with each other, either directly, or through horizontal cells. This interaction was revealed by the greater response obtained by a large 1000 $\mu$m field than that of a 25 $\mu$m spot, in spite of the fact that both stimuli were adjusted to ensure that the same number of quanta fell on the retina. Fields between 200 $\mu$m and 800 $\mu$m produced no increase with area so that the summation area in this case is about 200 $\mu$m. Unlike the situation with turtle cones, where lateral interaction is probably mediated by horizontal cells, which are certainly activated, the rod interaction was probably mediated by direct rod-rod connections; and a more elaborate study (Schwartz, 1974; 1976) confirmed this, the treatment

*The gain is a measure of the response in terms of voltage and the duration, that permits a comparison between rods and cones when the time-course of the potential change is accurately determined (Schwartz, 1976). Another measure is the *scaled amplitude*, equivalent to the *sensitivity* of Fuortes and Hodgkin, 1964).

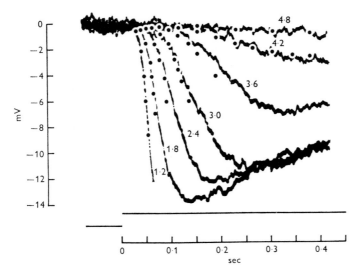

Fig. 10.27   Responses of turtle rods to steps of light of increasing intensity as indicated on the records. The filled circles represent theoretical values calculated on the basis of a model. (Schwartz, *J. Physiol.*)

of the retina with cobalt, which abolishes synaptic activity, being without effect on the response to a large spot of light.

## Enhancement

It was observed that the effect of a large spot of light could never match that of a small spot, however the relative intensities of the spots were varied; this indicated that, besides a summation, described as *enhancement*, there was some other effect of the more distant rods, and it was described as *disenhancement*, once again not due to synaptic action through horizontal cells but to some alteration in the coupling between the rods as a result of the light-stimulus. The enhancement, namely the increased effectiveness of a light-stimulus applied to one rod by light applied to neighbouring rods, is easily explained on the basis of a network of rods coupled electrically. Thus the rod has a low membrane conductance and the current-flow, after a light-stimulus, from the outer segment to the inner segment, which constitutes the primary electrical change, tends to be dissipated by flow into adjacent rods through the electrical coupling. When a large spot of light is employed, all the adjacent rods will develop the same voltage so that no flow of current between rods will occur, and thus the response of any individual rod to the light-flux will be greater.

## Disenhancement

Disenhancement is manifest as a distortion of the response of a rod by illumination of adjacent rods, and is considered to be due to a change in membrane conductance in addition to the light-sensitive change that is the cause of the hyperpolarization. The two phenomena of enhancement and disenhancement are illustrated in the records of Figure 10.28 which shows

the responses of a single rod when stimulated by 25 $\mu$m and 500 $\mu$m diameter flashes, each delivering $10^3$ photons/$\mu$m$^2$ at the centre of their images. It will be seen that the larger spot causes a greater voltage initially but that this falls sufficiently rapidly that, after 0·3 sec, the hyperpolarization is less than that produced by the smaller spot.

## Resolution

The coupling between receptors obviously militates against spatial resolution, since the resolution depends on the ability of the receptor to transmit its message independently of responses in its neighbours (p. 312). Thus this coupling of the rods into a syncytium, extending perhaps over the whole retina, might be expected to destroy resolution; however, Schwartz (1976) calculates that, because of the decay of the potential transmitted laterally to adjacent rods, the effects of a light-stimulus on one rod will be reduced to about 10 per cent at a distance of 20 $\mu$m; moreover, the scattering of light within the retina causes considerable spread of the stimulus to adjacent receptors, so that even if the receptors were not electrically coupled they would be affected by a light-stimulus designed to stimulate a single rod; as Figure 10.29 shows, the spread of the light-stimulus through scatter is comparable in magnitude to the spread of the electrical response of a single receptor through the syncytium.

### ROD-CONE INTERACTION

The studies on the rods described above were made using a dim green light, calculated to activate rods with a porphyropsin photopigment maximally; when different wavelengths of light were compared, the typical action-spectrum for porphyropsin was obtained pro-

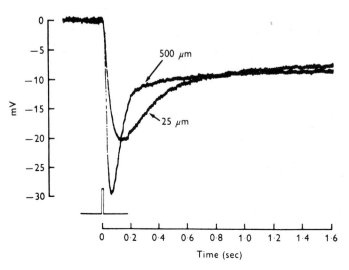

**Fig. 10.28** Illustrating enhancement and disenhancement of rod responses. Superimposed responses to bright lights covering either a small or large area. The larger spot produced a larger hyperpolarization, but this falls so rapidly that after 0·3 sec the hyperpolarization is less than that produced by the smaller spot. (Schwartz, *J. Physiol.*)

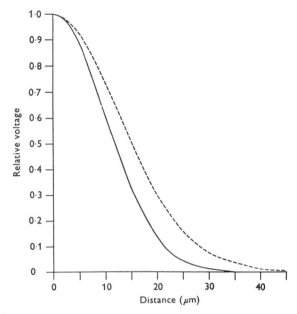

**Fig. 10.29** Showing that the scattering of light from a spot source would produce a falling off in response with distance comparable with that predicted with no light-scatter but electrotonic spread through a system of coupled receptors. Ordinate: potential as a fraction of the potential at the site of illumination; abscissa, distance in μm. The continuous line is the result predicted if the rods were not coupled; the dashed line is the result predicted if the rods are coupled. (Schwartz, *J. Physiol.*)

vided small spots of light were employed; with large spots there was an increased sensitivity to long wavelengths suggesting that cones were producing an effect (Schwartz, 1975). This effect could be mediated by horizontal cells but a direct examination of responses in these made this unlikely, and it was suggested that the rod-cone interaction might be direct.

### CONE-CONE INTERACTION

This is in marked contrast to the cone-cone interaction, which takes two forms, as illustrated by Figure 10.30;

within a radius of 50 μm they enhance their responses but they also receive a recurrent inhibition from horizontal cells which are electrically coupled so that their receptive fields are very large. The fact that it is the red-sensitive cones that are linked to the rods means that the sensitivity of the retina to red light at dim illuminations is enhanced.

### INTERCONNECTION OF RODS

Copenhagen and Owen (1976), also working on the snapping turtle, have examined the summative effects between adjacent rods in considerable detail. They too have ruled out a feedback from horizontal cells. As Figure 10.31 shows, the amplitude of the response to a flash increased with the size up to a maximum of some 300 μm diameter, and they computed that, when the whole retina was illuminated, the rod's response was only 4 per cent due to absorption of light by itself, the remainder being the result of summative interaction with a total of some 200 rods. In consequence, a rod should, and does, hyperpolarize in response to light absorbed by neighbouring rods even though it fails to absorb light itself. Copenhagen and Owen have suggested that the interaction is brought about by synaptic connexions through long basal processes extending from the synaptic terminals in the rod spherule, contacting other rods through swellings that might well contain synaptic vesicles, or else the contact points could be gap-junctions mediating an electrical coupling.

## THE HORIZONTAL CELL

### FEEDBACK ON CONES

Before discussing the responses of bipolar cells it would be profitable to review some of the properties of horizontal cells. The normal response of a horizontal cell to light falling on the receptors is one of hyperpolarization, and this hyperpolarization *feeds back* on the receptors. Thus when Baylor *et al.* (1971) hyper-

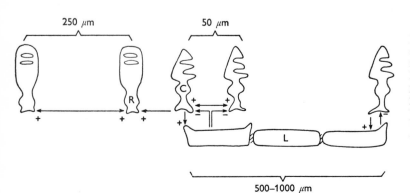

**Fig. 10.30** Diagram of the synaptic interactions between rods, R; red-sensitive cones, C; and luminosity horizontal cells, L. + indicates that the pathway preserves the polarity of the presynaptic voltage change; − indicates that it is reversed. The interaction between cones and rods may not be direct but may involve an interneurone.

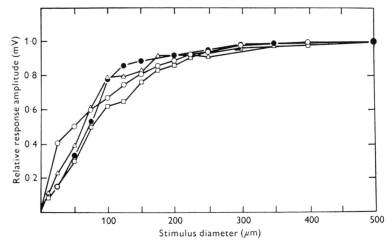

**Fig. 10.31** The relative variation in peak amplitude of the rod response as a function of stimulus diameter. Stimuli were flashes of white light, 200 msec in duration, at an intensity approximately 3 log units below saturation. For each of four rods, results are plotted as fractions of the amplitude of the response elicited by the stimulus 500 $\mu$m in diameter. (Copenhagen and Owen, *J. Physiol.*)

polarized a horizontal cell by the injection of current, they found that an impaled cone would exhibit depolarization, indicating an inhibitory action, or negative feedback. Similarly, if the response of a cone to a flash of light was measured, this was depressed by hyperpolarization of the horizontal cell. This horizontal cell-cone interaction extends over wide areas of the retina owing to the junctions between horizontal cells and is of opposite sign to that of direct cone-cone interaction. Thus, if a small spot of light is flashed on

to the retina, there is the hyperpolarizing response of the cone, whilst an impaled horizontal cell produces little or no response (Fig. 10.32); as the size of the spot increases the hyperpolarizing response in the cone increases—enhancement—up to about a radius of 70 $\mu$m; when the radius of the spot is increased further, there is an increased response in the horizontal cell, whilst the hyperpolarizing response in the cone shows a depolarizing deflexion indicating a depolarizing feedback on the cone.

### RESPONSE OF CONE

Thus the response to a large circle of light is complex, and can be described as the '*direct response*' of the cone, an *enhancement* due to coupling of cones with each other,[*] and a *negative feedback* due to a depolarizing action of the horizontal cells, this last effect extending over a large radius owing to the electrical coupling of the horizontal cells (Fuortes *et al.*, 1973).

### BIPOLAR SURROUND

We have seen (p. 253) that the bipolar cell of the mudpuppy exhibits a centre-surround antagonism, like the ganglion cell; and it is likely that the horizontal cell is involved in the activity of the surround, exerting an opposite effect on a group of receptors to that of the light-stimulus falling on these. If it is appreciated that the dendritic field of a bipolar cell rarely exceeds 100 $\mu$m, and that bipolar responses can be affected by light falling on the retina at far greater distances, it is clear that we must invoke the horizontal cell to explain these centre-surround interactions.

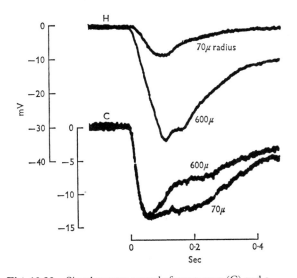

**Fig. 10.32** Simultaneous records from a cone (C) and a horizontal cell (H). Superimposed records of responses to stimuli covering different areas. A flash (delivered at a time 0) covering a circle of 70 $\mu$ radius produced a small response of the horizontal cell and a large, smooth response of the cone. A second flash of the same intensity but covering a circle of 600 $\mu$ radius was then given. The response of the horizontal cell was increased. The response of the cone was the same in its initial phase and peak. Later, however, it developed a depolarizing inflexion. (Baylor *et al.*, *J. Physiol.*)

[*]The relations between rods and horizontal cells of the toad *Bufo marinus* have been emphasized by Normann and Pochobradsky (1976), who observed strong oscillations in membrane potential in both types of retinal cell occurring spontaneously and suggesting some feed-back from horizontal cell to rod.

## S-POTENTIALS

Historically the horizontal cell is of interest since it was the first cell to be punctured in intracellular recording from the retina (Svaetichin, 1956), and at the time the hyperpolarizing and depolarizing potentials recorded from them were considered to be the responses of receptors. They have been called S-potentials and their main interest at the time was their colour-coding in the sense that some, the C- or chromatic cells, responded with depolarization with one wavelength of light and hyperpolarization with another—opponent—wavelength. Other cells, the L- or luminosity, cells, responded in the same way to all wavelengths.

## LOCATION

The horizontal cells of the fish and turtle retina occupy a very prominent part of the inner nuclear layer and may constitute as much as two-thirds of the whole; they are organized in layers probably in accordance with function; thus in the goldfish the external layer makes exclusive connexions with cones and the intermediate layer with rod spherules, whilst the internal layer apparently has no distally directed dendrites. In some higher vertebrates horizontal cells receive input from both rods and cones.

## SYNCYTIAL BEHAVIOUR

The striking feature of the horizontal cells of the fish, demonstrated by Naka and Rushton, was their behaviour as a syncytium, they behaved in effect electrically as if they were equivalent to a laminar conducting medium limited by a pair of high-resistance membranes and the whole was thus described as the S-space, extending over the entire retina. As with the syncytium of receptors, the potential at any distance, $x$, from an applied potential, $V_0$, was given by:

$$V = V_0 e^{-x/\lambda}$$

where $\lambda$ is the space-constant, being some 0·25 mm in the tench and 0·5 to 1·0 mm in the catfish.

## ELECTRICAL STIMULATION OF HORIZONTAL CELL

It is reasonable, then, to look for the interpretation of the centre-surround organization of the receptive field of the bipolar cell in the behaviour of the horizontal cells. Thus Naka and Witkowsky (1972) showed that, when they penetrated a dogfish horizontal cell with an electrode and applied an extrinsic current through it, either de- or hyper-polarizing it, a slow potential developed in a bipolar cell penetrated at the same time, and this could be associated with a discharge in a ganglion cell. The response-pattern depended on the type of polarizing current passed through the horizontal cell, so that in effect ON- or OFF- responses in bipolar and ganglion cells could be obtained.

## SYNAPTIC TRANSMISSION FROM RECEPTORS

We may assume that, in the dark, the depolarized receptor is releasing an excitatory or depolarizing transmitter at the receptor-horizontal cell synapse (Trifonov, 1968); light, by hyperpolarizing the receptor reduces the transmitter released and thus hyperpolarizes the horizontal cell, a change associated with an increase in resistance indicating, as with the receptor, a decreased permeability to ions such as $Na^+$ (Toyoda et al. 1969). The responses of a carp horizontal cell to flashes of light superimposed on different background-intensities are shown in Figure 10.33. Large hyperpolarizations are induced, bringing the membrane potential to as high as 80 mV; it will be seen that the steady background caused a sustained hyperpolarization, and imposing on this a flash caused a further hyperpolarization; with a background of maximal intensity, the flash produced only a negligible response presumably because, now, the release of transmitter was completely blocked so that further light could have no effect.

## Effects of cobalt

Synaptic transmission is blocked by $Co^{2+}$, and Figure 10.34 shows the effect of this ion on the responses of a horizontal cell to flashes of light before, and during,

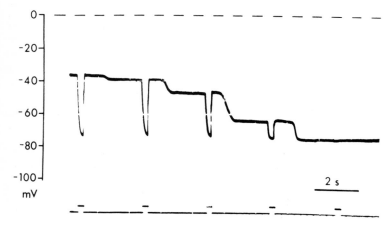

**Fig. 10.33** Intracellular records from a horizontal cell. In complete darkness the membrane potential was −35 mV and a flash gave a hyperpolarizing response to reach −75 mV; as the background was illuminated at successively higher values, the steady potential increased (hyperpolarization) and light-flashes superimposed on these backgrounds smaller hyperpolarizing responses. (Kaneko and Shimazaki, Cold Spr. Harb. Symp. Quant. Biol.)

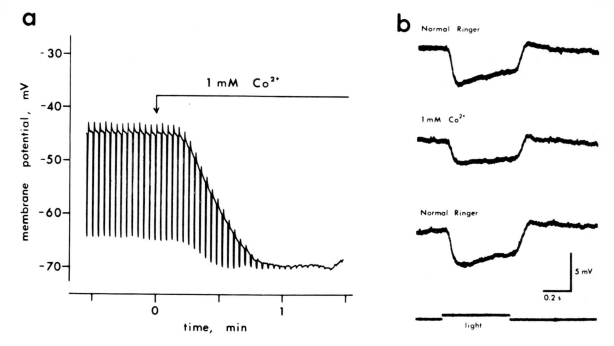

**Fig. 10.34** Effect of $Co^{++}$ on the membrane potential of a cone and a horizontal cell. (*a*) Intracellular recording from a horizontal cell. The perfusing medium was switched, at 0 min, from the standard Ringer to a test solution containing 1 mM $Co^{++}$. (*b*) Responses of a red-sensitive cone in normal Ringer (*top*), 2 min after application of $Co^{++}$ Ringer (*middle*), and 2 min after washing with standard Ringer (*bottom*). Note that horizontal cell responses have already been strongly suppressed 1 min after $Co^{++}$ application. Red light (620 nm, $3 \times 10^{12}$ photon/mm$^2 \cdot$s) illuminated the retina diffusely. (Kaneko and Shimazaki, *Cold Spr. Harb. Symp. Quant. Biol.*)

application to the retina. In records *b* the responses of a *cone* to light are shown; these are not blocked, confirming the synaptic effect of the cobalt. $La^{3+}$ facilitates transmitter release, and this caused a strong depolarization of horizontal cells.

TYPES OF L-CELL

Horizontal cells, producing only luminosity responses, apparently fall into two classes in lower vertebrate retinas. Thus Lasansky, on morphological grounds, separates one with an obvious soma and long thin dendritic branches, and the other with broad trunk-like main branches with no definite soma. So far as function goes, Simon (1973) described a Type I identified as that with the trunk-like axons, and giving a hyperpolarizing response increasing with the area of the stimulus up to 833 $\mu$m radius, and a Type II, with a smaller receptive field of 300 $\mu$m and corresponding to the cell with soma and thin dendrites. Type I cells were usually coupled, but there was no coupling between Type I and II; he was never able to penetrate two Type II cells and so could not determine whether coupling occurred here. Simon concluded that the morphology of the cell was not the important factor so far as the size of receptive field was concerned, this being governed by the degree of coupling, the Type II cells being presumably poorly coupled and thus not able to operate over large distances. Interestingly, both types had the same action-spectrum with maximal response at about 600 nm, indicating a predominant input from 'red-cones'.

ROD L-CELL

In the carp, Kaneko and Yamada differentiated L-cells that were obviously related synaptically with rods, their responses appearing in the dark-adapted eye and showing spectral sensitivity corresponding with absorption by the rod pigment, porphyropsin; adaptation with coloured lights failed to influence the shape of the sensitivity curve, indicating the absence of cone input. The penetrated cells were located in the intermediate layer of horizontal cells, and this conforms with the anatomical study of Stell (1967) on the goldfish retina, who found that, whereas the external layer made synaptic contact with cones, the intermediate layer made contacts with rods only. Thus, at the horizontal cell level in these fish, the input from rods is kept separate from that from cones.

CAT RETINA

An interesting feature of some horizontal cells of the cat is the apparent separation of function of two parts of the same cell; thus the cell body and dendrites receive signals predominantly from cones, whilst the terminal arborizations (Fig. 10.35) receive predominantly from rods; the long fine process connecting these two parts of the cell neither generates impulses nor allows significant electrotonic conduction between them. Both parts respond to light of all wavelengths by hyperpolarization, but when responses to different wavelength stimuli, matched so as to bleach equal amounts of rhodopsin, were measured, it was found that the responses could be matched at the terminal arborization but not at the

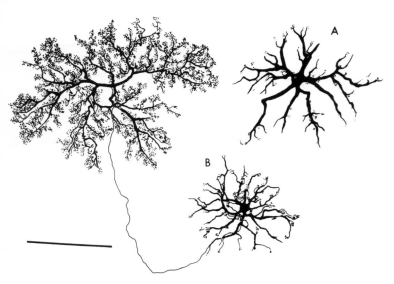

**Fig. 10.35** Horizontal cells of the cat retina drawn from Golgi-stained preparations. *A*, A-type or axonless horizontal cell. *B*, B-type or axon-bearing horizontal cell. Dendrites emanating from the cell bodies of both A- and B-type horizontal cells contact only cones, whereas the extensive axon terminal (upper left) of the B-type cell contacts only rods. Calibration, 100 μm. (Nelson *et al.*, *Invest. Ophthal.*)

cell body, and it was concluded that some 80 per cent of the peak response at the terminal arborization was due to rods and 20 per cent to cones, whereas at the cell body the figures were 42 per cent and 58 per cent. Nelson *et al.* emphasize that, in teleosts entirely separate horizontal cells receive inputs from rods or cones whereas in mammals one type receives input from both, but in the case described, because of the absence of significant interaction between the two parts, the inputs could be kept separate. The intermixture of rod and cone inputs to these separate parts is apparently due to junctions between rods and cones since the terminal arborization is known only to contact rods and the cell body and dendritic tree only contact cones. Another horizontal cell, called A-type (Nelson *et al.*, 1976) is apparently axonless, and its dendrites only contact cones.

## THE BIPOLAR CELL

### CENTRE-SURROUND ANTAGONISM

The classical study of Werblin and Dowling (1969) showed that the bipolar cells exhibited a centre-surround antagonism, so that a given bipolar cell might respond to a central spot with depolarization and to an annulus with a hyperpolarization—it would be called an ON-centre cell, whilst another bipolar cell might behave in the opposite fashion and be described as an OFF-centre cell, hyperpolarizing in response to a spot and depolarizing in response to an annulus. Figure 10.36

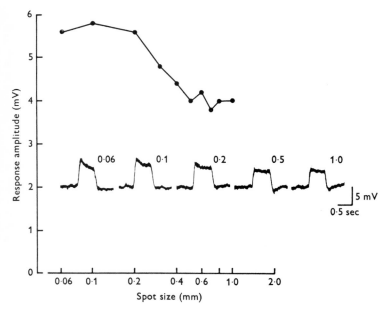

**Fig. 10.36** Spatial summation curve of an ON-centre bipolar cell. Response amplitude was measured from the resting level to the peak. Inset: response traces to light spots, sizes of which are indicated in mm on the right upper corner of each trace. D.c. recording, positivity upward. (Kaneko, *J. Physiol.*)

shows the variation in response-amplitude of an impaled goldfish ON-centre cell as a function of spot-size; as the spot-diameter was increased beyond 100 μm there was little change up to 200 μm but beyond this the response declined, indicating the effect of the antagonistic surround, the outer margin of which was of the order of 1 mm. Histological evidence indicated that the size of the centre of the receptive field corresponded with the spread of the bipolar's dendritic tree, whereas the size of the surround obviously exceeds that of any dendritic tree and Kaneko (1973) concluded that horizontal cells were responsible.

In later studies on turtle bipolars by Schwartz (1974) and Richter and Simon (1975), additional features of the bipolar cells' responses have been established. Thus Schwartz showed that the depressant effect of a surround annulus on the response of the centre was associated with a hyperpolarizing response in horizontal cells; moreover he, and Richter and Simon, demonstrated the far greater responses of the bipolar cell to a light stimulus than that in the associated receptors; this is illustrated by Figure 10.37, which shows the effects of three different sized stimulating spots on a bipolar cell (open symbols) and of a small 100 μm spot on cones (filled circles). Thus at an irradiance of about 54 quanta $\mu m^{-2}$ $sec^{-1}$ the cones produced a response of 1 mV and the bipolar a response greater than 10 mV.

## Two types of response

The bipolar cell quite clearly responds in one of two different manners to a central spot of light, namely by depolarization or hyperpolarization, whilst the surrounding annulus has an opposing influence, reducing the central depolarization or hyperpolarization. We have argued that receptors liberate a depolarizing transmitter in the dark when they are themselves depolarized, so that the onset of light reduces the liberation of transmitter from the receptor and thus causes a hyperpolarizing response in the bipolar cell. How, then, are we to explain the opposite, depolarizing, (or ON-centre) response in the bipolar cell? Either the receptors giving rise to this response liberate a hyperpolarizing transmitter, or all receptors liberate the same transmitter but the post-synaptic membrane of the bipolar cells is different, so that some bipolars are hyperpolarized by the same transmitter that depolarizes others (Fig. 10.38 from Tomita, 1976). Certainly there is evidence that ON- and OFF-centre bipolar cells respond differently to a given treatment; thus aspartate, which apparently mimics transmission in

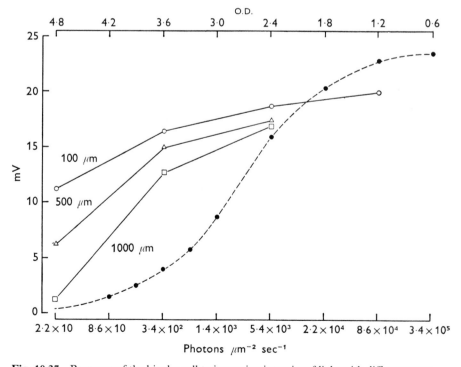

**Fig. 10.37**  Response of the bipolar cell to increasing intensity of light with different areas of test-spot. Note that for a given intensity of stimulus the larger spot produced a smaller response. The filled circles represent the average peak responses for a 100 μm stimulus for ten consecutively penetrated red-sensitive cones with maximum responses greater than 20 mV. (Schwartz, *J. Physiol.*)

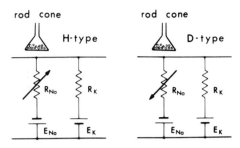

**Fig. 10.38**  Electrical models of two types of subsynaptic membrane in the ON-centre bipolar cell (left, H-type) and OFF-centre bipolar cell (right, D-type). (Tomita, *Vision Res.*)

the retina, depolarizes OFF-centre bipolar cells and hyperpolarizes ON-centre cells (Murakami *et al.*, 1975). Aspartate would, of course, be affecting the receptors responsible for the surround as well as the centre, so Murakami *et al.* assume that the effect on the receptors giving rise to the centre response is dominant, and this seems to be generally true of the light-stimulus, a diffuse illumination, exciting both centre and surround, usually giving rise to a centre-effect. Kaneko and Shimazaki (1975) have also observed differences in behaviour of impaled ON- and OFF-centre bipolar cells, suggesting that they are differently affected by the same treatment; thus a trans-retinal current that caused a hyperpolarization of an ON-centre cell caused a depolarization of an OFF-centre cell.

HORIZONTAL CELLS AND SURROUND

The influence of the surround stimulus on a bipolar cell is probably mediated through horizontal cells; this could be through a direct synaptic input from horizontal cell to the bipolar cell, or it could be mediated by feedback from horizontal cell to receptors, a view favoured by Schwartz. Whatever the mechanism, we may explain the OFF-centre responses to a surround-stimulus as a negative feedback so that the bipolar cell, instead of being hyperpolarized by light, is depolarized, i.e. is behaving as if it were in the dark.

## THE AMACRINE CELL

GOLDFISH RETINA

Goldfish amacrine cells were classified as transient and sustained types; the transient type always responded with depolarization at both ON and OFF, whereas the sustained type gave a steady hyperpolarization to white light. Only the sustained type showed colour-coding, in the sense that a green stimulus would depolarize the cell and a red stimulus would hyperpolarize (p. 344). When different coloured stimuli were used for the

transient type, the responses were qualitatively the same, but of course different wavelengths produced different magnitudes of response in accordance with the efficacy of absorption; the wavelength of maximum sensitivity was 620 nm, suggesting that these amacrine cells received input primarily, if not exclusively, from 'red cones'. In contrast to bipolar cells, there was no centre-surround antagonism. The very large receptive fields encountered, namely 2·5 mm, suggests that there are interconnexions between amacrine cells of similar types (Witkowsky and Dowling, 1969).

### Centre-surround of ganglion cells

Both bipolar and amacrine cells feed into ganglion cells, and the similarity in the centres of the receptive fields of ganglion cells and bipolar cells, both with respect to size and chromatic behaviour, leads Kaneko to assume that a bipolar cell feeds directly to the ganglion cell giving it its centre properties, its size being governed by the size of the 'bipolar' cell's dendritic tree. The very large surround of the ganglion cell—up to 5 mm according to Daw (1968)—and the large receptive fields of the amacrine cells, suggest that they are responsible for the surround input—chromatically they behave somewhat similarly.*

## THE GANGLION CELL

EXCITATION THROUGH CONES

Baylor and Fettiplace (1975) excited turtle cones electrically and observed the responses in ganglion cells. In this species (*Pseudemys scripta*) the ganglion cells show no spontaneous discharge, and most respond with either a transient discharge at ON or at OFF; the transitoriness of the response is due to the fact that the excitatory synaptic potentials evoked are not maintained during the stimulus. As Figure 10.39 shows, injection of hyperpolarizing current into a cone could be made to mimic the effects of light; moreover, the response of a

---

*About half of the amacrine cells of the carp exhibit centre-surround antagonism, being either OFF-centre or ON-centre (Toyada *et al.*, 1973). Injection of current into horizontal cells produced responses in about half the amacrine cells studied, the responses being similar to the surround responses to light. These surround effects must be mediated by bipolars, since horizontal cells make no direct connexions with amacrine cells. The changes in resistance suggested that the only input to the amacrine cells was excitatory, no evidence for an inhibitory postsynaptic potential being found. Thus the opposite responses of depolarization and hyperpolarization are brought about by onset and cessation of depolarization. ON-OFF amacrine cells were assumed to have an input from ON- and OFF-bipolar cells, whose effects would tend to cancel out with steady illumination, and thus would only signal at ON and OFF.

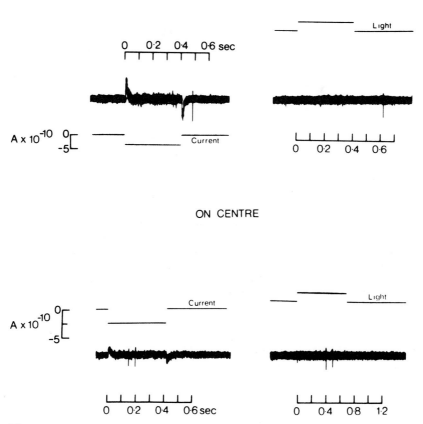

**Fig. 10.39** Duplication of the responses of two ganglion cells to light by injection of hyperpolarizing current into red-sensitive cones. The dim, 640 nm, stimulating spot was 60 μm in diameter above and 120 μm in diameter below. (Baylor and Fettiplace, *Cold Spr. Harb. Symp. Quant. Biol.*)

ganglion cell to light could be abolished by simultaneous injection of a depolarizing current into a cone, or its effect could be potentiated by injecting a hyperpolarizing current. Of most interest were the time-relations between application of the electrical stimulus to the receptor and the discharge in the ganglion cell; There was a delay of some 75 to 125 msec between switching the current on and the peak probability of obtaining a discharge; the probabilities then fell off rapidly, indicating the absence of any sustained discharge.

**Integration and differentiation**
Baylor and Fettiplace thus define two processes, an *integration* process, during which the effects of the stimulus are allowed to build up, and a *differentiation* process that puts a stop to transmission. The existence of these two processes is reminiscent of the build-up of

the excitation of nerve or muscle to give a strength-duration curve, and Figure 10.40 shows the effects of reducing the stimulating current on the time required for a response to be evoked for rods and cones. Analysis of these curves indicated that the time-constants for rods and cones differed by a factor of about ten, so far as the process of differentiation was concerned.

**Minimum stimulus**
The minimum threshold current in a red-sensitive cone, required to evoke a response in a ganglion cell, was $1.5.10^{-11}$ amp; this compares with Baylor and Hodgkin's estimate of the current generated by one photo-isomerization as $1.5.10^{-13}$ amp, so that the minimum cone stimulus would be equivalent to 100 photo-isomerizations. For rods the value might be 10 photo-isomerizations.

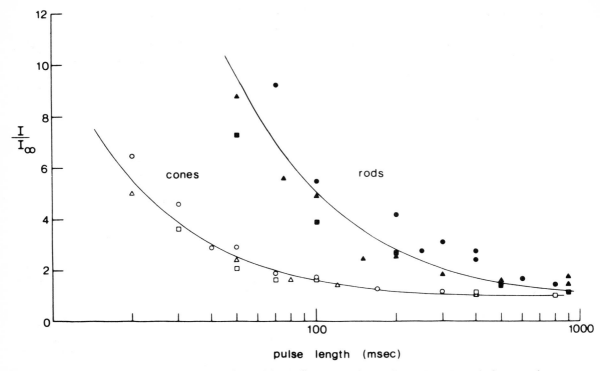

**Fig. 10.40**   Collected strength-duration curves for eliciting 'off' responses in ganglion cells at the end of rectangular hyperpolarizing currents injected into cones (○, □, △) or rods (●, ■, ▲). Threshold current divided by rheobase current plotted against pulse duration. Temperature, 19·5–21·2°C. The smooth curves were calculated with a differentiating time constant of 100 ms for the cones and 450 ms for the rods. (▲) From a snapping turtle preparation. (Baylor and Fettiplace, *Cold Spr. Harb. Symp. Quant. Biol.*)

## Relation to light responses

In continuing this study Baylor and Fettiplace (1977) showed that the type of ganglion cell response to electrical stimulation of a cone was related to its response to light: thus an ON-sensitive ganglion cell responded to negative-going electrical current pulses, so that it responded at the make of a hyperpolarizing current or the break of a depolarizing current; OFF-sensitive ganglion cells responded to positive going currents, and

ON-OFF cells to both negative- and positive-going currents. The responses to hyperpolarizing currents are indicated in Table 10.1. In general, then, the effects of a hyperpolarization of the receptor can be equated with those of light, and of depolarization as antagonistic to the effects of light. When the effects of a background light on electrical sensitivity were determined, in rods this was equivalent to a simple rise in threshold without change in the membrane resistance, as measured

**Table 10.1** Patterns of ganglion cell response to light and to hyperpolarizing current in a receptor (Baylor and Fettiplace, 1977).

| | Ganglion cell response | | |
|---|---|---|---|
| *Receptor* | *To light* | *To hyperpolarizing current* | *n* |
| Red-sensitive cone | On | Make | 7 |
| | Off | Break | 4 |
| | On and off | Make and break | 9 |
| | On and off | Break | 7 |
| Rod | Off | Break | 7 |

*n* is the number of pairs tested. In the experiments on cones the stimulating spots were at 640 nm and were 46 to 140 $\mu$m in diameter; the currents were 0·4 sec long. In the rod experiments the light was at 520 nm with spots 320 to 690 $\mu$m in diameter; the currents were 0·5 or 0·7 sec long.

by the current-voltage relationship. In cones, the effect depended on the ON- or OFF-character; thus the threshold of receptors utilizing the ON-pathway was raised by the background light, whereas that of receptors utilizing the OFF-pathway was lowered.

## Circuits

In general, the experimental results could be best explained on the basis of the circuits illustrated by Figure 10.41; according to these, an ON-responsive

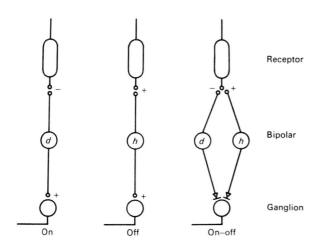

**Fig. 10.41** Schematic diagram of proposed connexions between red-sensitive cones and ON, OFF, and ON-OFF ganglion cells. Bipolars depolarized by central illumination denoted by *d*, hyperpolarizing bipolars by *h*. The ± indicates a synapse with sign reversal, + a synapse with sign preservation, — ▷| a rectifying synapse which transmits depolarizations but not hyperpolarizations to the post-synaptic cell. (Baylor and Fettiplace, *J. Physiol.*)

ganglion cell would be connected to a bipolar cell that depolarized in response to the receptor's hyperpolarizing response (synaptic reversal); this depolarization would trigger the ganglion cell through a synaptic depolarization. Thus light at ON would hyperpolarize the receptor, and this would depolarize the bipolar cell giving an ON-response. Bipolar cells responding by hyperpolarization to light (synaptic sign preservation)

would inhibit the ganglion cell during ON and excite it at OFF. The ON–OFF ganglion cell would be related to both types of bipolar cell, but an additional factor would have to be introduced if the opposing effects of the two bipolar cells were not to cancel; this additional effect would be a 'rectifying synapse', which transmitted depolarizations but not hyperpolarizations to the post-synaptic cell.★

## THE ELECTRORETINOGRAM

The electroretinogram† (ERG) is a record of comparatively slow changes of potential that take place across the retina in a radial direction when light falls on it. As classically studied in the intact eye, one electrode is placed on the cornea whilst the other, indifferent, electrode may be placed in the mouth or, in experimental animals, on the optic nerve. A better record may be obtained in experimental animals by placing the active electrode in the vitreous body. It is essentially because the retina is built up of large numbers of extremely well orientated units, arranged in layers, that the successive states of depolarization or hyperpolarization in these units lead to definite changes of polarity across the whole structure.

### TYPICAL RECORD

In Figure 10.42 is shown a typical example; the first sign of activity is shown as an *a*-wave corresponding to a decrease in the normal positivity of the corneal

★ In their most sensitive pairs of receptor and ganglion cell, the current required to excite was about $2.10^{-11}$ amp, corresponding to some 130 isomerizations. For rods, the minimum number of isomerizations detectable by a ganglion cell would be about 50, according to this study. Thus detection of a single quantum event is out of the question and the ganglion cell must summate the effects of light falling on the many cones constituting the centre of its receptive field.
† The 'retinal currents' now called the ERG were first described by Holmgren in 1865 but were independently discovered by Dewar and McKendrick. Kühne and Steiner (1880, 1881) showed that the same electrical changes could be obtained from the retina removed from the eye, whilst the remaining sclera, choroid and pigment epithelium were unresponsive.

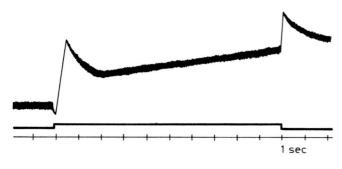

**Fig. 10.42** Typical electroretinogram (ERG). Lowest record, time scale; middle record, stimulus; top record, galvanometer deflexion. (Drawn from Hartline, *Amer. J. Physiol.*)

1 sec

electrode; this is conventionally recorded as a downstroke on the record, and is called a *negative* wave; it lasts less than 0·1 second. The *a*-wave is followed by a reversal in polarity so that the cornea now becomes strongly positive, and this gives the sharp upward *b*-wave conventionally described as a *positive* wave; the *c*-wave is a much slower rise in corneal positivity and is terminated, when the light is switched off, by the *d*-wave, or OFF-effect.

## ADAPTATION

According to the type of retina, and the state of adaptation of the eye, the records are qualitatively different; and this is because of the predominance of cones or rods in producing the net record. Thus, when the cones dominate the record, e.g. in the light-adapted mixed retina of the frog or the pure cone retina of the squirrel, the slow rise of the *c*-wave disappears, and the *a*-wave and OFF-effect become prominent. In the dark-adapted state, or in the virtually pure rod retinae of many nocturnal animals, the *a*-wave is not so pronounced whilst the *c*-wave is prominent.

## ORIGIN OF THE ERG

It is well established that the ERG is the record of changes taking place in the retina, and we may ask what is the cause of these fluctuations in potential. Essentially they are changes in the potential difference across the retina since they may be best recorded by a pair of electrodes one on the inner face and the other on the scleral face of the retina. We may recall that a neurone may develop a rapid spike which is reflected in a very brief phase of negativity of an electrode placed exactly at the site of this activity. This change is so rapid, however, that it is unlikely that spike activity on the part of the retinal neurones will have much to do with the ERG.

### SLOW POTENTIALS

Neurones will also, however, develop a much more slowly occurring change of potential; an example of this is the negativity which has been called the generator potential. In other cases we may record an increase in the normal membrane potential—a hyperpolarization—so that the neurone is said to become *positive*; this change is associated with inhibition of activity. We may assume, therefore, that it is this type of change, rather than spike activity, that is at the basis of the ERG and the questions that present themselves are (*a*) What are the actual changes taking place in the retina? and (*b*) Where are they occurring, i.e. in what layer or layers? Partial or complete answers to these questions have been provided by the studies of Tomita and Brown and

their colleagues, based largely on the insertion of microelectrodes to different depths of the retina and noting the changes in the records when the retina was exposed to light. Before describing these, we may consider the classical analysis of Granit.

### GRANIT'S ANALYSIS

Granit considered that the ERG was the record of several events taking place very nearly simultaneously, so that the measured potential changes were the algebraic sum, at any instant, of these. On the basis of experiments designed to block some processes, and thus allowing others to appear in a less obscure form, Granit deduced the presence of three 'components', namely PI, PII and PIII. PI is a slowly developing positive component; with strong stimuli, or after light-adaptation, it tends to disappear; it accounts for the secondary rise in positivity, the *c*-wave. PII is likewise positive but is much more rapid in development and is chiefly responsible for the *b*-wave. PIII is the negative component; it comes on rapidly and is responsible for the initial *a*-wave. The *d*-wave, or 'off-effect', is considered to represent an interference phenomenon resulting from the rapid swing of the negative PIII component back to the baseline; if this occurs more rapidly than the fall of the PI and PII components, it will clearly be reflected in a wave of positivity, the *d*-wave. Figure 10.43 illustrates the analysis for the dark-

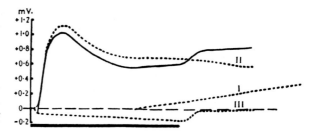

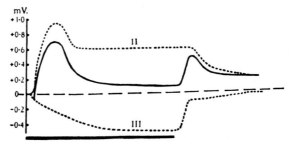

**Fig. 10.43**   Analysis of the ERG. Above: Dark-adapted frog's eye; PIII component and OFF-effect small. Below: Light-adapted frog's eye; PIII component and OFF-effect large; PI is missing so that the *c*-wave is absent. The black line indicates the duration of the stimulus (0·2 second). (Granit and Riddell, *J. Physiol.*)

adapted (above) and light-adapted eye of the mixed frog retina. The dotted lines indicate the probable courses of the pure components acting in isolation, and the full lines the algebraic sums.

### E- AND I-RETINAE

The results of numerous experiments by different workers have shown that there is a strong correlation between the magnitude of the negative *a*-wave and the magnitude of the 'off-effect'; thus in the light-adapted frog's eye the *a*-wave and 'off-effect' are both marked (Fig. 10.43); in the dark-adapted frog's eye, or the cat's eye, the *a*-wave and 'off-effect' are small, in fact the latter is just shown as a slight check in the downward trend of positivity. Granit has divided the retinae studied into two groups, *E-* and *I-retinae*, according principally to the predominance of the negative PIII component; thus the retinae of the cat, rat, and guinea pig are all E-retinae with small PIII components in their electroretinograms; the pigeon and light-adapted frog retinae belong to the I-type, with pronounced PIII characteristics. Since those retinae that give the E-type of response contain predominantly rods, whilst the I-type of response is shown either by retinae with predominantly cones, as with the pigeon or tortoise, or by retinae with both types of receptor in which rod function is suppressed (light-adapted frog), it is reasonable to assume that the negative PIII component is mainly associated with cone activity. The available evidence suggests that it is connected with *inhibition* of impulses, as opposed to their initiation.

### EXCITATION

The PII component, on the other hand, is closely associated with the passage of impulses along the optic nerve, i.e. with excitation; thus when the PII component is blocked by ether anaesthesia, simultaneous records of the electrical changes in the optic nerve show that the action potentials here are likewise blocked. So strong, indeed, is the correlation that the height of the *b*-wave has been used as an index to the intensity of the optic nerve response. Thus in Figure 10.44 we have the ERG analogue of a dark-adaptation curve, obtained by plotting the intensity of the light stimulus necessary to evoke a *b*-wave of fixed height against the time of dark-adaptation of a frog's retina. Dark-adaptation causes the sensitivity of the retina to increase; this is reflected in a decrease in the stimulus necessary to evoke the *b*-wave.

### PRE-EXCITATORY INHIBITION

Granit refers to the PIII component as a process of '*pre-excitatory inhibition*'; the correlations between the ERG and the sensory phenomena of vision indicate that the first effect of stimulation of the light-adapted retina is one of inhibition; it is as though the slate were being

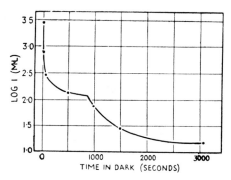

**Fig. 10.44** The ERG analogue of the dark-adaptation curve. The logarithm of the stimulus necessary to evoke a *b*-wave of fixed magnitude is plotted against time in the dark. As the eye adapts, the stimulus necessary becomes smaller. (Re-drawn from Riggs.)

wiped clean in preparation for a new picture. The component is probably associated with those single fibres of the optic nerve that show a discharge at 'Off' (Fig. 10.8C). The comparative rarity of this type of fibre in the mammalian retina is probably due to the fact that mainly E-type retinae have been studied (cat, rat, guinea pig).

## FURTHER DEVELOPMENTS

In general, Granit's analysis remains essentially valid, in the sense that the record is one of independently generated potential changes, although modifications have been introduced in the light of microelectrode studies.

### RECORDS FROM A DISTANCE

Thus the conventional ERG electrode on the cornea is recording events at a considerable distance from where they take place, and the actual record must be considerably attenuated so that it is, at first thought, surprising that it is not completely extinguished. It is only because the cells of the retina are all so well orientated that a record is possible. Thus, we may suppose the situation to be that illustrated by Figure 10.45, where the rectangles indicate radially orientated retinal cells; they have become active in the sense that a part of them has become negative in relation to other parts and the rest of the retina. This part would be described as a *sink of current* into which positive current would flow, and it may be shown that a distant electrode, e.g. one on the cornea, or closer still, in the vitreous, would record a phase of *positivity*. If the position were reversed, i.e. if later this cell, or another, became positive, we should expect to record negativity. If different cells showed opposite effects simultaneously

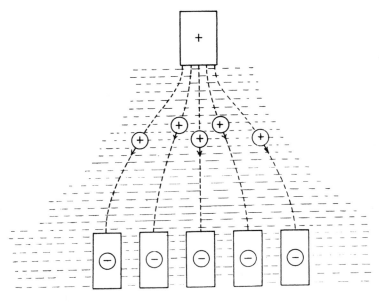

**Fig. 10.45** The orientated cells, indicated by the rectangles, have become negative in relation to their surroundings and thus become what is known as a *sink* for positive current; the electrode at the top of the figure, at some distance from the sink, thus becomes positive, i.e. a *source* of positive current which flows into the sink.

we should expect cancellation of effects, so that the magnitude of the changes recorded at a distance might be much smaller than if the electrode were close to the origins of activity. Thus, on causing our electrode to penetrate into the retina we may expect to find larger potential changes; furthermore, when the electrode comes very close to the cells the record changes polarity; the electrode becomes negative when the cell does, and positive when it becomes positive. At the site of electrical activity, then, we may expect to find a *reversal of the ERG*, or perhaps, if the events take place in different layers, reversal of some components only.

THE PIII COMPONENT

We may begin with the PIII component, which is the first to manifest itself in the record as a negative downstroke. This was earlier shown to be localized in the region of the outer segments of the receptors (Brown and Wiesel, 1961); a microelectrode was inserted to successive depths, which were later identified; the *a*-wave reached a maximum when the active electrode was some 120 $\mu$ from the scleral surface of the retina and this corresponds to the layer of receptors. Again Murakami and Kaneko (1966), using a rather more elaborate recording procedure, identified the origin of the *a*-wave at some 120 $\mu$ from the surface, but they noted that at this level there was a considerable *a*-wave, recorded from the receptor layer and the inner surface of the retina, i.e. they identified what they called a proximal component due to activity in the outer plexiform layer. Sillman, Ito and Tomita (1969) showed that this component, as well as the *b*-wave, was blocked

by aspartate, so that, in the presence of this inhibitor, they were able to record what was probably the pure receptor potential, i.e. the potential change taking place in the outer segments of the receptors. This is illustrated by Figure 10.46, where the top record shows the normal record, inverted because the retina had been inverted. The lower records show the *a*-wave in isolation after the *b*-wave had been blocked by two concentrations of aspartate.

**Isolated a-wave**

Brown and Watanabe (1962) were able to isolate the *a*-wave by making the retina anoxic by compressing the retinal vessels; in this case activity in all cells proximal to the receptors was abolished and thus the *b*-wave was suppressed; in this way they were able to compare the *a*-wave derived from cones—over the fovea—with that from both rods and cones—periphery—and they observed that the pure cone response from the fovea came on and off rapidly, whilst the mixed response consisted of two steps, indicating the presence of a much slower rod component.

**Intracellular records**

The unequivocal localization of electrical activity is given by insertion of the recording electrode into the cells responsible for this. We have seen that the response of the receptor of the mud-puppy, *Necturus*, is a wave of hyperpolarization; a similar type of response was found by Tomita (1965) in the large cones of the carp, and by Baylor and Fuortes (1970) in the turtle. Such a change in potential is consistent with the polarity of the *a*-wave, which indicates that the outermost layer of

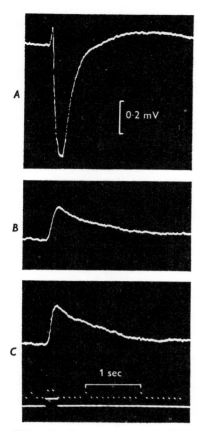

adjacent neurones when these are excited, so that it has been postulated that the responses of the distal retinal neurones to light induce depolarizing responses in the glial cells. Thus Granit's finding that the *b*-wave is associated with nervous activity is consistent with this hypothesis.

PI COMPONENT

This is defined as that responsible for the *c*-wave, and is prominent in the dark-adapted state; microelectrode studies have shown that this, as well as the *a*-wave, is generated close to the scleral surface of the retina; Noell (1954) showed that iodate selectively reduced the *c*-wave and destroyed the pigment epithelium, and he concluded that the *c*-wave was generated by these cells. Brown and Wiesel (1961) concluded from their intracellular recording that the *c*-wave was, indeed, derived from pigment epithelium cells. Brindley (1960) pointed out that the pigment of these cells is melanin so that we might expect the action spectrum for the *c*-wave to be that of the absorption spectrum of this pigment, whereas it is, in fact, similar to that of rhodopsin (Granit and Munsterhjelm, 1937).

**Bleaching signal**

Since it depends on the intactness of the relation between pigment epithelium and the receptor, the electrical change denoted as the *c*-wave must be the reflexion, in the pigment epithelial cell, of electrical changes induced in the rod by bleaching of pigment. Recently Lurie (1976) has developed this viewpoint in relation to the frog's ERG, where the *c*-wave is more simple. According to Lurie, the electrical event with which the *c*-wave is related is a change in the standing potential across the retina; thus Figure 10.47a shows the response to a flash of light; the *c*-effect is a wave rising after the *b*-wave and slowly declining; in *b* the light has been maintained for several minutes and the *c*-effect is now manifest as a slow rise continuing from the *b*-wave until a plateau is reached. Thus in steady illumination the *c*-effect is now a changed *standing potential* rather than a *c*-wave, and when the light is switched off the standing potential declines slowly. The decline in this steady potential relates quite closely to the regeneration of rhodopsin. Since the receptors themselves do not show any change in membrane potential with dark-adaptation, it is this pigment epithelial change that acts, in some way, as a modifying signal that provides the message—if such there be—signalling the degree of bleaching of visual pigment in the retina (Rushton, 1965).

D.C. COMPONENT

Brown and Wiesel attribute the OFF-effect, as seen in the cat's ERG, essentially to the cessation of what they

**Fig. 10.46** Effect of sodium aspartate on the ERG, recorded from the isolated, inverted frog retina. A, control, before treatment with aspartate; B, after washing the retina with 10 mM aspartate-Ringer; and C, with 110 mM aspartate-Ringer. (Sillman, Ito and Tomita, *Vision Res.*)

the retina becomes positive in relation to the indifferent electrode outside the eye. The transmission of such an effect to the bipolar cell might well inhibit spontaneous activity so that the first response to light could well be what Granit had described as *pre-excitatory inhibition*.

PII COMPONENT

The study of Brown and Wiesel (1961) located the maximum amplitude of the *b*-wave at a more proximal site, about half-way through the retina near the inner nuclear layer; by reducing the strength of the stimulus, the interfering effects of the *a*- and *c*-waves were removed. Since it could be abolished by local anaesthetics and anoxia, it was thought to represent synchronized neural activity. Rather surprisingly Miller and Dowling (1970) have shown that the time-course of the depolarizing responses of Muller glial cells to light is similar to that of the *b*-wave; glial cells, in general, do not generate action potentials but they do become depolarized as a result of the escape of $K^+$ from

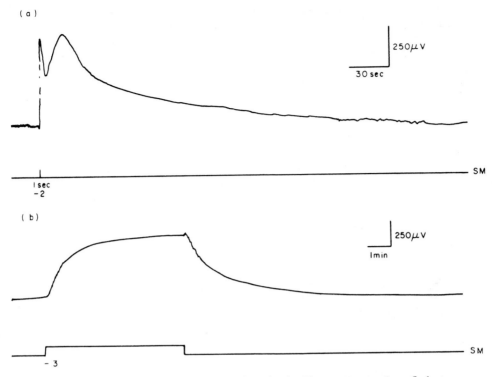

**Fig. 10.47** (*a*) The d.c.-ERG evoked in a dark-adapted animal in response to a 1-sec flash at an intensity 6 log units above that necessary to evoke a just-perceptible b-wave. Note the extremely long time course of the c-wave. (*b*) The d.c.-ERG of a dark-adapted eye in response to a step of illumination. (Lurie, *Exp. Eye Res.*)

called the d.c. component, a phase of negativity that comes on immediately with the light-stimulus and remains at a constant level during this, to fall abruptly when the light is switched off. Thus the d.c. component is a part of what Granit included in his PII, which gives rise not only to the b-wave but the sustained negativity during maintained stimulation. Steinberg (1969) has recently analysed both b-wave and d.c. component, and his results tend to confirm the distinction between the two.

## EARLY AND LATE RECEPTOR POTENTIALS

Brown and Murakami (1964), when making their ERG records with a microelectrode in the retina, found that the a-wave was preceded by a diphasic potential change with no measurable latency; this was obtained with the tip of the electrode in the region of the junction of the outer and inner segments of the receptors. We may therefore speak of an *early* and *late* receptor potential (ERP and RP), the latter being the PIII component of the electroretinogram, and the former being over before this is normally recorded.

EARLY RECEPTOR POTENTIAL

A comparison of the effectiveness of different wavelengths of light showed that, in the virtually all-rod retina of the albino rat, the pigment responsible was rhodopsin (Cone, 1964); however, the early receptor potential is only obtained with intense flashes, the relative amount of energy for late and early potentials being in the ratio of one to a million. The potential is diphasic, beginning with a cornea-positive change ($R_1$) followed by a cornea-negative change ($R_2$). $R_1$ occurs at sub-zero temperatures which abolish $R_2$. All the evidence indicates that both phases have nothing in common with bioelectric potentials as usually recorded; thus they can be recorded from fixed tissue, and, as indicated, at sub-zero temperatures. Some organization of retinal structure is necessary, however, since the potential vanishes if the eye is heated above 58°C, and this is accompanied by a measurable disorientation of the rhodopsin (Cone and Brown, 1967).

### Dipole change

The currently accepted hypothesis is that the potential arises from a change in the orientation, or magnitude, of the electric dipole of the visual pigment molecule dependent on ionized groups. For this change to appear between electrodes in the inside and outside of the retina, it is essential for it to take place within the thickness of the high-resistance membrane separating the receptor interior from the extracellular space, the result being a transient flow of current between the outer and inner segments in one direction and in the opposite direction within the interior of the receptor. The difference in potential measured in the extracellular fluid is the ERP. The first effect of light is to isomerize the 11-*cis* retinal on the rhodopsin molecule (p. 210); in order that rotation of a double-bond may occur, this must be broken, and this will result in the transfer of electrons, and if the transfer is such as to result in a transmembrane movement of current this would account for the difference of potential occurring between outer and inner segments.

## Intracellular records

Murakami and Pak (1970) compared intra- with extracellular recording of the ERP in the gekko, axolotl and mud-puppy retinae; the amplitude of the intracellular record was 1·7 mV on average, compared with only 0·09 mV extracellularly. The time-course of the intracellular events was slower. These findings are consistent with an initial transfer of electronic charge across the outer segment membrane followed by the spread of potential towards the inner segment, in the same way that an applied potential spreads along the surface of an excitable tissue such as a nerve—the spread is governed by the cable-properties of the receptor membrane— in fact computations of membrane capacitance and resistance, derived from these measurements of receptor potential, are consistent with those in other cells (ca 1 k$\Omega$m$^{-2}$ and 1 Fm$^{-2}$).

## Rods and cones

In the mixed retina, it is the cones that determine the early receptor potential in spite of the predominance of the rod pigment; thus the action-spectrum for the monkey retina has a maximum between 535–570 nm (Carr and Siegel, 1970), whilst the recovery of the potential after a flash is at a rate commensurate with cone pigment regeneration rather than that of rhodopsin (Goldstein and Berson, 1969).

It appears from the work of Zanen (1973) and Zanen and Debecker (1975) that the magnitudes of $R_1$ and $R_2$ are strongly dependent on the wavelength of light, suggesting that the relative contributions of rods and cones can vary; thus $R_1$ may well be due to cones entirely, since it is not obtained by violet flashes, whilst $R_2$ may represent a mixture of rod and cone responses, although the differing sensitivities of the components to temperature would suggest a more fundamental difference in origin.

## Polarized light

That rods and cones make separate contributions is also suggested by the effects of light polarized in different planes. Thus, in the cones, the membrane developing the potential is presumably the entire plasma membrane together with that covering the discs which is continuous with it, so that light with its electric vector at right-angles to the discs, i.e., parallel to the outer segments, should be more effective than light with its electric vector at right-angles to this direction. Using a photopic cone-stimulus, the maximal ERP was, indeed, found with the electric vector at right-angles to the discs, giving a dichroic ratio of 1:2; when the retina was bleached with deep red light, the ERP disappeared, indicating a cone phenomenon. When the cone effect was removed, then an opposite dichroic ratio was obtained, and this was presumably by activation of rods, since the active membrane in this case cannot be that in the discs, which are separate, and is probably the plasma membrane, which runs parallel with the outer segments. In the guinea pig, which behaves as a pure rod retina, only one ERP-generating mechanism was obtained and in this the maximal response was to light with its electric vector parallel with the outer segments. If this viewpoint is correct, the predominance of the cone contribution to the mixed retina's ERP is due to its special anatomical feature of continuity of plasma membrane with its infolded discs (Govardovskii, 1975).

## Dependence on retinal pigment

If the ERP depended on an electron charge transfer due to absorption of light, it might depend on the state of the visual pigment at the time of exposure to the intense flash, and Cone and Cobbs (1969) showed that the ERP changed its form at successive intervals following a bleaching exposure, and these changes were correlated with the appearance of Metarhodopsins I, II and pararhodopsin (p. 212). Subsequent work of Zanen (1973) along the same lines certainly indicates that the potentials developed by flashes following an intense bleaching exposure are dependent on the nature of the predominant intermediary of bleaching.*

TURTLE'S ERP

Hodgkin and O'Bryan (1977) have extended their studies of intracellular recording of single cone responses to the ERP, which, like those described by Murakami and Pak, consisted of a depolarizing and hyperpolarizing phase. They were able to separate the

*As we might expect of a process dependent on the absorption of light by photopigment, the appearance of the ERP in embryonic development coincides with the appearance of rhodopsin in the developing chick retina, namely on the 13th day (Hanawa *et al.*, 1976).

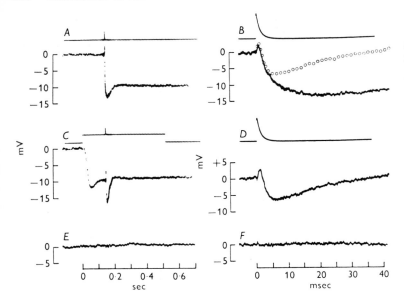

**Fig. 10.48** Use of a conditioning step of light to separate ERP from LRP, recorded from a red-sensitive cone with an internal electrode. *A*, effect of unattenuated xenon flash without conditioning light (slow time base). *B*, same but on faster time base. The response is a mixture of ERP (shown by circles obtained from *D*) and LRP. *C*, effect of applying same xenon flash on conditioning step of light which saturates the LRP mechanism. The response to the xenon flash in this record and in *D* is considered to be ERP only. *D*, same as *C* but showing effect of xenon flash on faster time base-expansion of 0·13–0·18 sec of record *C*; the zero on the voltage scale corresponds to − 10 mV in *C*. *E*, *F*, same stimulus as in *C* and *D* but recorded with electrode outside cone. (Hodgkin and O'Bryan, *J. Physiol.*)

ERP from the late receptor potential, i.e. from the hyperpolarizing response to light, by imposing the very bright flashes, required to evoke the ERP, on a light of sufficient intensity to saturate the late response. This is possible without serious bleaching of the cone pigment, so that the ERP appears as potential changes imposed on the plateau-response to the conditioning light (Fig. 10.48). A striking difference between the early and late responses was the finding that the shape of the early response was independent of the intensity of the flash, so that, by scaling, responses to weak and strong flashes could be superimposed; this is in contrast with the later response.

**Separate current-generating mechanism**

All the evidence pointed to the existence of two entirely independent current-generating mechanisms in the early and late receptor potentials; according to theory, the early current is generated by each photo-isomerization of cone pigment whereas the later is an indirect result of the liberation of a blocking molecule which closes ionic channels and hyperpolarizes the cell to a saturating value. So far as the two components of the ERP are concerned, it may be as Cone and Cobbs suggested, that the initial depolarizing phase corresponds to the photochemical events leading to the production of metarhodopsin I (or its equivalent cone pigment) and the later hyperpolarization to the conversion of metarhodopsin I to metarhodopsin II.

In contrasting the two receptor potentials, early and late, Hodgkin and O'Bryan emphasize the basic simplicity of the ERP, which increases linearly with the number of isomerizations brought about by the flash and may be due to a simple molecular rearrangement in

the receptor membrane; by contrast, the late potential changes in shape and size as the intensity of the flash is increased in a complex way. Again, there is an enormous difference in the number of photo-isomerizations required to saturate the responses; for the ERP it is half-maximal at $7.10^7$ photo-isomerizations per cone, and for the late receptor potential it is half-maximal at only 1000 photo-isomerizations per cone.

**Transmitter**

Finally, it seems likely that a transmitter action is unnecessary with the ERP, whereas the fact that absorption of a photon by any one of about $10^8$ molecules of photopigment can lead to the blockage of about one thousandth of the ionic channels in the outer segment requires some internal transmitter.

**Role of pigment epithelium**

Illumination of the eye-cup with the retina removed from its subjacent pigment epithelium gives rise to changes in potential analogous in their time-course to those taking place in the intact eye; Brown and Crawford (1967) showed that these were generated by the pigment epithelium cells with action spectra corresponding to the absorption of light by melanin. The potentials produced in this way will necessarily contribute to the potentials recorded in the intact eye; their contribution has been described non-committally as the *fast photochemical voltage (FPV)* and may be assessed by employing sufficiently intense light that effectively bleaches all the visual pigment; the 'photostable residue' of activity is due to the pigment epithelium. Zanen and Debecker (1971), working on human cadavers, showed that the ERP and the photostable fraction occurred in different layers of the retina whilst in the living human eye Debecker and Zanen (1975) found the ERP proportional to flash-intensity up to 130 to 150 μV; they never obtained light-saturation due to the photostable contribution, but they concluded that up to 100 μV the ERP is purely due to visual pigment.

# REFERENCES

Baylor, D. A. & Fettiplace, R. (1975a) Light path and photon capture in turtle photoreceptors. *J. Physiol.* **248**, 434–464.

Baylor, D. A. & Fettiplace, R. (1975b) Transmission of signals from photoreceptors to ganglion cells in the eye of the turtle. *Cold Spr. Harb. Symp. quant. Biol.* **40**, 528–536.

Baylor, D. A. & Fettiplace, R. (1977) Transmission from photoreceptors to ganglion cells in turtle retina. *J. Physiol.* **271**, 391–424.

Baylor, D. A. & Fuortes, M. G. F. (1970) Electrical responses of single cones in the retina of the turtle. *J. Physiol.* **207**, 77–92.

Baylor, D. A., Fuortes, M. G. F. & O'Bryan, P. M. (1971) Receptive fields of cones in the retina of the turtle. *J. Physiol.* **214**, 265–294.

Baylor, D. A. & Hodgkin, A. L. (1973) Detection and resolution of visual stimuli by turtle photoreceptors. *J. Physiol.* **234**, 163–198.

Baylor, D. A., Hodgkin, A. L. & Lamb, T. D. (1974) The electrical response of turtle cones to flashes and steps of light. *J. Physiol.* **242**, 685–727.

Baylor, D. A., Hodgkin, A. L. & Lamb, T. D. (1974) Reconstruction of the electrical responses of turtle cones to flashes and steps of light. *J. Physiol.* **242**, 759–791.

Behrens, M. E. & Wulff, V. J. (1965) Light-initiated responses of retinula and eccentric cells in the *Limulus* lateral eye. *J. gen. Physiol.* **48**, 1081–1093.

Bortoff, A. & Norton, A. L. (1967) An electrical model of the vertebrate photoreceptor cell. *Vision Res.* **7**, 253–263.

Brindley, G. S. (1960) *Physiology of the Retina and Visual Pathway* (1st ed.). London: Arnold.

Brown, J. E. & Blinks, J. R. (1974) Changes in intracellular free calcium during illumination of invertebrate photoreceptors detected with aequorin. *J. gen. Physiol.* **64**, 643–665.

Brown, J. E., Brown, P. K. & Pinto, L. H. (1977) Detection of light-induced changes of intracellular ionized calcium concentration in *Limulus* ventral photoreceptors using arsenazo III. *J. Physiol.* **267**, 299–320.

Brown, J. E. & Major, D. (1966) Cat retinal ganglion cell dendritic fields. *Exp. Neurol.* **15**, 70–78.

Brown, J. E. & Pinto, L. H. (1974) Tonic mechanism of the photoreceptor potential of the retina of *Bufo marinus*. *J. Physiol.* **236**, 575–591.

Brown, K. T. & Crawford, J. M. (1967) Intracellular recording of rapid light-evoked responses from pigment epithelium cells of the frog eye. *Vision Res.* **7**, 149–163.

Brown, K. T. & Crawford, J. M. (1967) Melanin and the rapid light-evoked responses from pigment epithelium cells of the frog eye. *Vision Res.* **7**, 165–178.

Brown, K. T. & Murakami, M. (1964) A new receptor potential of the monkey retina with no detectable latency. *Nature* **201**, 626–628.

Brown, K. T. & Watanabe, K. (1962) Isolation and identification of a receptor potential from the pure cone fovea of the monkey retina. *Nature* **193**, 958–960.

Brown, K. T., Watanabe, K. & Murakami, M. (1965) The early and late receptor potentials of monkey cones and rods. *Cold Spr. Harb. Symp. quant. Biol.* **30**, 457–482.

Brown, K. T. & Wiesel, T. N. (1958) Intraretinal recording in the unopened cat eye. *Am. J. Ophthal.* **46**, Pt. 2, 91–98.

Brown, K. T. & Wiesel, T. N. (1959) Intraretinal recording with micropipette electrodes in the intact cat eye. *J. Physiol.* **149**, 537–562.

Brown, K. T. & Wiesel, T. N. (1961) Analysis of the intraretinal electroretinogram in the intact cat eye. *J. Physiol.* **158**, 229–256.

Brown, K. T. & Wiesel, T. N. (1961) Localization of origins of electroretinogram components by intraretinal recording in the intact cat eye. *J. Physiol.* **158**, 257–280.

Carr, R. E. & Siegel, I. M. (1970) Action spectrum of the human early receptor potential. *Nature* **225**, 89–90.

Cervetto, L., Pasino, E. & Torre, V. (1977) Electrical responses of rods in the retina of *Bufo marinus*. *J. Physiol.* **267**, 17–51.

Cone, R. A. (1964) Early receptor potential of the vertebrate retina. *Nature* **204**, 736–739.

Cone, R. A. & Brown, P. K. (1967) Dependence of the early receptor potential on the orientation of rhodopsin. *Science* **156**, 536.

Cone, R. A. & Cobbs, W. H. (1969) Rhodopsin cycle in the living eye of the rat. *Nature*, **221**, 820–822.

Copenhagen, D. R. & Owen, W. G. (1976) Functional characteristics of lateral interactions between rods in the retina of the snapping turtle. *J. Physiol.* **259**, 251–282.

Daw, N. W. (1968) Colour-coded ganglion cells in the goldfish retina: extension of their receptive fields by means of new stimuli. *J. Physiol.* **197**, 567–592.

Debecker, J. & Zanen, A. (1975) Intensity functions of the early receptor potential and of the melanin fast photovoltage in the human eye. *Vision Res.* **15**, 101–106.

Dowling, J. E. & Werblin, F. S. (1969) Organization of retina of the mudpuppy *Necturus maculosus*. I. Synaptic structure. *J. Neurophysiol.* **32**, 315–335.

Fain, G. L. (1975a) Interactions of rod and cone signals in the mudpuppy retina. *J. Physiol.* **252**, 735–769.

Fain, G. L. (1975b) Quantum sensitivity of rods in the toad retina. *Science*, **187**, 838–841.

Fain, G. L., Gold, G. H. & Dowling, J. E. (1975) Receptor coupling in the toad retina. *Cold Spr. Harb. Symp. quant. Biol.* **40**, 547–651.

Frank, R. N. & Goldsmith, T. H. (1967) Effects of cardiac glycosides on electrical activity in the isolated retina of the frog. *J. gen. Physiol.* **50**, 1585–1606.

Fuortes, M. G. F. (1959) Initiation of impulses in visual cells of *Limulus*. *J. Physiol.* **148**, 14–28.

Fuortes, M. G. F. & Hodgkin, A. L. (1964) Changes in time scale and sensitivity in the ommatidium of *Limulus*. *J. Physiol.* **172**, 239–263.

Fuortes, M. G. F., Schwartz, E. A. & Simon, E. J. (1973) Colour-dependence of cone responses in the turtle retina. *J. Physiol.* **234**, 199–216.

Goldstein, E. B. & Berson, E. L. (1969) Cone dominance of the human early receptor potential. *Nature*, **222**, 1272–1273.

Govardovskii, V. I. (1975) The sites of generation of early and late receptor potentials in rods. *Vision Res.* **15**, 973–980.

Granit, R. (1947) *Sensory Mechanisms of the Retina*. London: Oxford University Press.

Granit, R. (1955) *Receptors and Sensory Perception*. Newhaven: Yale University Press.

Granit, R. & Munsterhjelm, A. (1937) The electrical responses of dark-adapted frogs' eyes to monochromatic stimuli. *J. Physiol.* **88**, 436–458.

Grüsser, O. -J. & Rabelo, C. (1958) Reaktioner retinaler Neurone nach Lichtblitzen. *Pflüg. Atch.* **265**, 501–525.

Hagins, W. A., Penn, R. D. & Yoshikami, S. (1970) Dark current and photocurrent in retinal rods. *Biophys. J.* **10**, 380–412.

Hanawa, I., Takahashi, K. & Kawamoto, N. (1971) A correlation of embryogenesis of visual cells and early receptor potential in the developing retina. *Exp. Eye Res.* **23**, 587–594.

Hartline, H. K. (1941–2) The neural mechanisms of vision. Harvey Lectures, Series 37.

Hartline, H. K. (1940) The nerve messages in the fibres of the visual pathway. *J. opt. Soc. Amer.* **30**, 239–247.

Hartline, H. K., Wagner, H. G. & MacNichol, E. F. (1952) The peripheral origin of nervous activity in the visual system. *Cold. Spr. Harb. Symp. quant. Biol.* **17**, 125–141.

Hartline, H. K., Wagner, H. G. & Ratliff, F. (1956) Inhibition in the eye of *Limulus. J. gen. Physiol.* **39**, 651–673.

Hodgkin, A. L. & O'Bryan, P. M. (1977) Internal recording of the early receptor potential in turtle cones. *J. Physiol.* **267**, 737–766.

Kaneko, A. (1970) Physiological and morphological identification of horizontal bipolar and amacrine cells in goldfish retina. *J. Physiol.* **207**, 623–633.

Kaneko, A. (1973) Receptive field organization of bipolar and amacrine cells in the goldfish retina. *J. Physiol.* **235**, 133–153.

Kaneko, A. & Hashimoto, H. (1969) Electrophysiological study of single neurons in the inner nuclear layer of the carp retina. *Vision Res.* **9**, 37–55.

Kaneko, A. & Shimazaki, H. (1975) Synaptic transmission from photoreceptors to bipolar and horizontal cells in the carp retina. *Cold Spr. Harb. Symp. quant. Biol.* **40**, 537–546.

Kaneko, A. & Yamada, M. (1972) S-potentials in the dark-adapted retina of the carp. *J. Physiol.* **227**, 261–273.

Korenbrot, J. I., Brown, D. T. & Cone, R. A. (1973) Membrane characteristics and osmotic behaviour of isolated rod outer segments. *J. Cell Biol.* **56**, 389–398.

Korenbrot, J. I. & Cone, R. A. (1972) Dark ionic flux and the effects of light in isolated rod outer segments. *J. gen. Physiol.* **60**, 20–45.

Kuffler, S. W. (1953) Discharge patterns and functional organization of mammalian retina. *J. Neurophysiol.* **16**, 37–68.

Kühne, W. & Steiner, J. (1880) Uber das electromotorische Verhalten der Netzhaut. *Unt. physiol. Inst. Univ., Heidelberg,* **3**, 327–377.

Kühne, W. & Steiner, J. (1881) Uber electrische Vorgänge im Sehorgane, *loc. cit.* **4**, 64–168.

Lamb, T. D. & Simon, E. J. (1977) Analysis of electrical noise in turtle cones. *J. Physiol.* **272**, 435–468.

Lasansky, A. (1971) Synaptic organization of cone cells in the turtle retina. *Phil. Trans.* **262**, 365–381.

Leicester, J. & Stone, J. (1967) Ganglion, amacrine and horizontal cells of the cat's retina. *Vision Res.* **7**, 695–705.

Lurie, M. (1976) Some observations on the C-wave of the electroretinogram in the intact frog eye. *Exp. Eye Res.* **23**, 197–207.

Miller, R. F. & Dowling, J. E. (1970) Intracellular responses of the Müller (glial) cells of mudpuppy retina: their relation to b-wave of the electroretinogram. *J. Neurophysiol.* **33**, 323–341.

Murakami, M. & Kaneko, A. (1966) Differentiation of PIII subcomponents in cold blooded vertebrate retinas. *Vision Res.* **6**, 627–636.

Murakami, M., Ohtsuka, T. & Shimazaki, H. (1975) Effects of aspartate and glutamate on the bipolar cells in the carp retina. *Vision Res.* **15**, 456–458.

Murakami, M. & Pak, W. L. (1970) Intracellularly recorded early receptor potential of the vertebrate photoreceptors. *Vision Res.* **10**, 965–975.

Murakami, M. & Shigematsu, Y. (1970) Duality of

conduction mechanism in bipolar cells of the frog retina. *Vision Res.* **10**, 1–10.

Naka, K. I. & Rushton, W. A. H. (1966) S-potentials from colour units in the retina of fish (*Cypridinae*). *J. Physiol.* **185**, 536–555.

Naka, K. I. & Witkovsky, P. (1972) Dogfish ganglion cell discharge resulting from extrinsic polarization of the horizontal cells. *J. Physiol.* **223**, 449–460.

Nelson, R., Kolb, H., Famiglietti, E. V. & Gouras, P. (1976) Neural responses in the rod and cone systems of the cat retina: intracellular records and Procion stains. *Invest. Ophthal.* **15**, 946–953.

Nelson, R., Lutzow, A. V., Kolb, H. & Gouras, P. (1975) Horizontal cells in cat retina with independent dendritic systems. *Science* **189**, 137–139.

Noell, W. K. (1954) The origin of the electroretinogram. *Am. J. Ophthal.* **38**, 78–90.

Normann, R. A. & Pochobradsky, J. (1976) Oscillations in rod and horizontal cell membrane potential: evidence for feed-back to rods in the vertebrate retina. *J. Physiol.* **261**, 15–29.

Penn, R. D. & Hagins, W. A. (1969) Signal transmission along retinal rods and the origin of the electroretinographic a-wave. *Nature* **233**, 201–205.

Richter, A. & Simon, E. J. (1975) Properties of centre-hyperpolarizing, red-sensitive bipolar cells in the turtle retina. *J. Physiol.* **248**, 317–334.

Rushton, W. A. H. (1959) A theoretical treatment of Fuortes's observations upon eccentric cell activity in *Limulus. J. Physiol.* **148**, 29–38.

Rushton, W. A. H. (1965) Visual adaptation. *Proc. Roy. Soc. B* **162**, 20–46.

Scholes, J. H. (1975) Colour receptors, and their synaptic connexions, in the retina of a cyprinid fish. *Phil. Trans. B*, **270**, 61–118.

Schwartz, E. A. (1973) Organization of ON-OFF cells in the retina of the turtle. *J. Physiol.* **230**, 1–14.

Schwartz, E. A. (1974) Responses of bipolar cells in the retina of the turtle. *J. Physiol.* **236**, 211–224.

Schwartz, E. A. (1975a) Responses of single rods in the turtle. *J. Physiol.* **232**, 503–514.

Schwartz, E. A. (1975b) Rod-rod interaction in the retina of the turtle. *J. Physiol.* **246**, 617–638.

Schwartz, E. A. (1975c) Cones excite rods in the retina of the turtle. *J. Physiol.* **246**, 639–651.

Schwartz, E. A. (1976) Electrical properties of the rod syncytium in the retina of the turtle. *J. Physiol.* **257**, 379–406.

Schwartz, E. A. (1977) Voltage noise observed in rods of the turtle retina. *J. Physiol.* **272**, 217–246.

Sillman, A. J., Ito, H. & Tomita, T. (1969) Studies on the mass receptor potential of the isolated frog retina. I. *Vision Res.* **9**, 1435–1442.

Simon, E. J. (1973) Two types of luminosity horizontal cells in the retina of the turtle. *J. Physiol.* **230**, 199–211.

Smith, T. G., Baumann, F. & Fuortes, M. G. F. (1965) Electrical connexions between visual cells in the ommatidium of *Limulus. Science* **147**, 1446–1447.

Steinberg, R. H. (1969) Comparison of the intraretinal b-wave and d.c. component in the area centralis of cat retina. *Vision Res.* **9**, 317–331.

Stell, W. K. (1967) The structure and relationships of horizontal cells and photoreceptor-bipolar synaptic complexes in goldfish retina. *Amer. J. Anat.* **121**, 401–424.

Svaetichin, G. (1956) Spectral response curves from single cones. *Acta physiol. Scand.* **39**, Suppl. 134, 17–46.

Tomita, T. (1970) Electrical activity of vertebrate photoreceptors. *Quart. Rev. Biophys.* **3,** 179–222.

Tomita, T. (1976) Electrophysiological studies of retinal cell function. *Invest. Ophthal.* **15,** 171–187.

Tomita, T., Kikuchi, R. & Tanaka, I. (1960) Excitation and inhibition in lateral eye of horseshoe crab. In *Electrical Activity of Single Cells*, pp. 11–23. Ed. Katsuki, Y. Tokyo: I. Shoin.

Toyoda, J., Nosaki, H. & Tomita, T. (1969) Light-induced resistance changes in single photoreceptors of *Necturus* and *Gekko. Vision Res.* **9,** 453–463.

Toyoda, J. I., Hashimoto, H. & Ohtsu, K. (1973) Bipolar amacrine transmission in the carp retina. *Vision Res.* **13,** 295–307.

Trifonov, Y. A. (1968) Study of synaptic transmission between photoreceptors and horizontal cells by means of electric stimulation of the retina. *Biophysika, Moscow* **13,** NS (In Russian).

Werblin, F. S. (1975) Regenerative hyperpolarization in rods. *J. Physiol.* **244,** 53–81.

Werblin, F. S. & Dowling, J. E. (1969) Organization of the retina of the mudpuppy, *Necturus maculosus.* II. Intracellular recording. *J. Neurophysiol.* **32,** 339–354.

Whitehead, R. & Purple, R. L. (1970) Synaptic organization in the neuropile of the lateral eye of *Limulus. Vision Res.* **10,** 129–133.

Zanen, A. (1973) Contribution à l'étude électrophysiologique des mécanismes protorecepteurs de l'oeil normale. Thesis, Univ. Brussels.

Zanen, A. & Debecker, J. (1971) Visual pigments and melanin contributions to the fast photovoltage of the human eye. *Vision Res.* **11,** 169–172.

Zanen, A. & Debecker, J. (1975) Wavelength sensitivity of the two components of the early receptor potential (ERP) of the human eye. *Vision Res.* **15,** 107–112.

# 11. Electrophysiology of the retina— further aspects

## EFFECT OF ADAPTATION ON THE CENTRE-SURROUND ORGANIZATION

A striking feature of the ganglion cell's receptive field is that it is altered in the dark-adapted state; as illustrated in Figure 11.1 from Barlow *et al.* (1957), the surrounding area of opposite activity becomes ineffective in the dark-adapted state. In this sense, therefore, the receptive field shrinks during dark-adaptation, but as it is a reduction in inhibitory effect between centre and periphery it means, in fact, that the effective field actually increases, i.e. the regions over which summation can occur. As Hammond has emphasized, the failure of the surround mechanism after dark-adaptation is largely due to the failure of the cones in the surround of the ganglion cell to respond; this does not mean, however, that rods are not active in the cat's surround, but merely that their presence can only be demonstrated by suprathreshold stimuli (Andrews and Hammond, 1970).

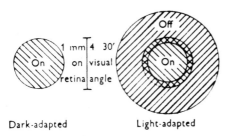

**Fig. 11.1**  Effect of dark-adaptation on receptive field. (Barlow *et al.*, *J. Physiol.*)

## DETERMINATION OF FIELD SIZES

Because of the effects of dark-adaptation, the sizes of centre and surround should be carried out using supra-threshold conditions; by increasing the size of the spot, the response of an ON-centre unit will increase due to spatial summation, and the diameter of the spot giving optimal discharge is the diameter of the field-centre; as the size of the spot increases further, the inhibitory action of the surround becomes manifest. The size of the field-surround is best determined by projecting annuli of increasing diameter on to the receptive field; the effects of the annulus on the discharge caused by a

small spot projected on to the centre is a measure of the effect of the annulus; if it increases the discharge, the annulus is within the field-centre, if it decreases it, it is within the surround. Figure 11.2 shows a typical example where a discharge of 100 represents that due to the central spot alone; the centre of the field is clearly some 2 degrees in diameter and the surround extends beyond 9 degrees, showing a maximal inhibitory action at about 3·5 degrees.★

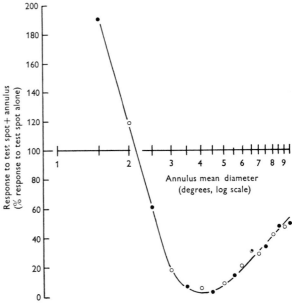

**Fig. 11.2**  Determination of the size of the receptive field. An annular stimulus is projected on to the retina and the discharge and its effects on the discharge caused by a small centred spot are recorded. A discharge of 100 represents that due to the central spot alone; when combined with annuli of increasing diameter the discharge decreases; annuli of smaller diameter than that of the test spot cause an increased discharge. The diameter of the centre-excitatory field is clearly about 2°, whilst the inhibitory surround extends to beyond 9°. (Hammond, *J. Physiol.*)

---

★ The actual unit shown here is one from the lateral geniculate body the neurones of which have a well defined centre-surround organization (p. 562). Experimentally, annuli of internal and external radii differing by 1 degree were used, and the abscissa is the mean of these two radii.

# QUANTITATIVE ASPECTS OF CENTRE-SURROUND INTERACTION

## EFFECTS OF FIELD SIZE

Rodieck and Stone (1965) measured the responses of centre and periphery of cat ganglion cell receptive fields to small spots of light, using, as a measure of the response, the summated spikes during the interval of summation; the summation was carried out over small intervals of time and the result presented as a response histogram (Fig. 11.3). Special

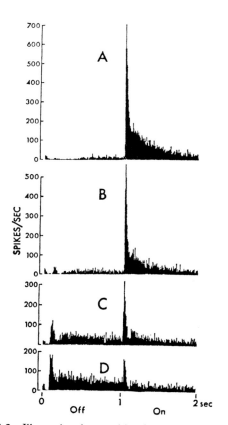

**Fig. 11.3** Illustrating the transition from a pure ON-response in A to an ON-OFF response in D. A small spot of light was flashed for 1 sec off and 1 sec on. A shows the prominent ON-response of the centre with suppression of background during the OFF-phase. At B to D the flashes were on more and more peripheral parts of the receptive field. (Rodieck and Stone, *J. Neurophysiol.*)

attention was paid to the sizes of receptive fields, summation within the field, and interaction between centre and surround. In general, the fields were radially symmetrical, and the maximal response of the central region was usually at the geometrical centre, but not always. The sizes of the central areas varied from 0·5 to 4 degrees. Beyond the centre, the strength of the surround-response increased quickly to a maximum and then gradually decreased until no response could be obtained, the limiting size of the peripheral field being determined by the intensity of the spot stimulus. When the retina was stimulated by a large uniform field, the response

of a given unit was that of its centre, i.e. surround and centre did not neutralize each other completely; as Barlow *et al.* had found, dark-adaptation abolished the surround effect. Usually a region in the receptive field could be found that gave an ON-OFF response and this was confined to a narrow ring; spot of light flashed 1 second off and 1 second on; it is an ON-centre field and the spot is projected first on the centre (A) and then more and more peripherally, and the transition from a pure ON-response to the ON-OFF response in D is clearly seen. A moving spot was a most effective stimulus, and the response to a small movement varied with the initial position in the receptive field and the direction of the movement; for an ON-centre unit, centripetal movement of the spot anywhere within the border of maximum OFF-surround caused excitation and outside it, inhibition. The converse was true for OFF-centre fields, Rodieck and Stone, 1965a, pp. 162–163.

### Calculated responses

Rodieck and Stone, on the basis of studies of this sort, computed the effects of stimulating a receptive field by a moving pattern; to do this it was necessary to make assumptions as to the way in which the different regions of a given field interacted, i.e. the manner in which similar portions summated and opposite portions subtracted. They accepted the suggestion of Wagner, MacNichol and Wohlbarsht (1963) that the surround field, did, in fact, extend into the centre, so that the response to a spot on the centre was really the algebraic sum of centre and 'surround' responses; the situation is illustrated in Figure 11.4, the curves indicating the relative effectiveness of stimuli falling on different portions of the centre and surround. The centre mechanism has a steep outline whilst that of the surround is altogether broader; the diagrams below indicate the hypothetical results of interaction when stimuli fall in more and more peripheral parts of the field, leading to typical centre-type response, no response, ON-OFF response, and surround-type response. The response of a unit to the movement of a contoured body passing across it will be given by the responses to movement of its leading and trailing edges, and Rodieck (1965) was able to predict these, on the basis of his simple theoretical treatment. Thus Figure 11.5 illustrates the calculated responses to moving bars of increasing width across the receptive field for two types of unit, those giving a centre-activated (CA) and centre-suppressed (CS) response; these agree remarkably well with the experimental responses (Rodieck and Stone, 1965a).*

# DIRECTIONAL SENSITIVITY

As we shall see, neurones within the cerebral cortex were early shown to be sensitive to the direction in which a moving spot passed over the retina; that a single neurone may respond so selectively to the retinal stimulus requires a considerable degree of integration and selection during the transmission of information; thus it has to ignore messages coming from receptors

---

*Varela and Maturana (1970) have shown that the latency of the centre is always less than that of its surround; moreover, it is possible to abolish the effect of stimulating one region by a stimulus applied to the other if the timing is chosen correctly.

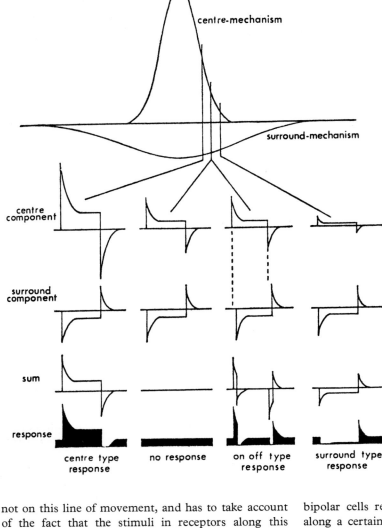

centre-mechanism

surround-mechanism

centre
component

surround
component

sum

response

centre type
response

no response

on off type
response

surround type
response

**Fig. 11.4** At the top the theoretical responses of the centre and surround mechanisms are depicted, in this case an ON-centre unit with positive response in the centre and a smaller negative response over the whole field. Near the centre of the field, as the left column of hypothetical responses shows, the centre mechanism predominates; a little farther out, centre and surround cancel completely; still farther out the response becomes ON-OFF, and in the extreme periphery of the field the response is that of the surround only. (Rodieck and Stone, *J. Neurophysiol.*)

not on this line of movement, and has to take account of the fact that the stimuli in receptors along this preferred direction are sequential, and so on. Barlow, Hill and Levick (1964) showed that, already at the ganglion cell level in the rabbit, this discriminative power is present; they observed many cells with the typical centre-surround arrangement with either ON- or OFF-responses in the centre and OFF- or ON-responses in the surrounds. Others, however, resembled the ON-OFF units of the frog over the whole field, and these* showed directional sensitivity in the sense that their maximal response occurred when a moving spot traversed a certain 'preferred' direction, whilst there was no response at all to movement in the opposite, 'null', direction. That inhibition was concerned was suggested by the observation that movement in the null direction would inhibit spontaneous activity; again, very slow motion in the null direction would evoke a response. Thus we may imagine that the

bipolar cells related to a series of receptors, located along a certain direction in the retina, all feed their responses into a ganglion cell; these responses are excitatory and inhibitory, the inhibitory effect being transmitted, after a small latency, to the ganglion cell.

INHIBITION

The picture that fitted best with the experimental facts is illustrated by Figure 11.6. Activity roused at either ON or OFF in the receptors C, B, etc., is passed laterally to inhibit the bipolar cell B. ~C', A . ~B' and, so on. This inhibition prevents the excitatory effect on B and A from getting through when the motion of the spot is in the null direction, but arrives too late if motion is in the preferred direction. If the movement is very

* In the visual streak area of the retina, characterized by densely packed ganglion cells, Levick (1967) found that the ON-OFF-units were not all direction-sensitive; many were local-edge detectors (p. 291).

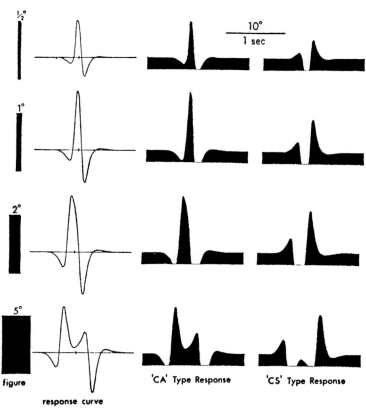

**Fig. 11.5** Calculated response curves for moving bars of increasing width across the receptive field of a cat's ganglion cell. (Rodieck, *Vision Res.*)

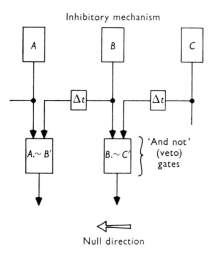

**Fig. 11.6** Hypothetical mechanism for direction-sensitive ganglion cell. The receptor elements A, B, C have a direct excitatory influence on bipolar cells and an inhibitory action on adjacent bipolar cells in the null direction, the inhibitory action occurring after a latency of $\Delta t$. These are said to act as 'And not' veto gates. A spot of light passing in the null direction, i.e. from right to left, would tend to inhibit its influence on successive receptors. (Barlow and Levick, *J. Physiol.*)

slow we can imagine easily that the inhibition may not last sufficiently long to inhibit the responses of adjacent elements. On this basis, two stationary stimuli, falling on regions of the retina separated along the preferred direction in sequence, will give a large response when the sequence corresponds with motion along the preferred direction and a smaller response for the null direction. In fact this was found, provided the separation was small; in the null direction sequence, the second stimulus could give no discharge at all (see also, Michael, 1968). Barlow and Levick (1965) postulated that the inhibitory action was mediated by the horizontal cells, and a study of Werblin (1970) on the responses of the different retinal cells—receptors, bipolars, etc.— to moving spots of light are consistent with this inhibitory role of the horizontal cells.

### THE PREFERRED DIRECTIONS

Oyster and Barlow (1967) determined the preferred directions in many units of the rabbit retina and found that they fell into four groups, and these corresponded accurately with the directions of movement of the fixation axis when the four rectus muscles were allowed to act separately; these were not purely horizontal and vertical, but inclined to these by a few degrees. Thus if the retina is to act as a servo-system correcting movements of the eyes, the direction-sensitive units fulfil an obvious role here.*

## CLASSIFICATION OF GANGLION CELL TYPES

Since Kuffler's demonstration of the centre-surround organization of the ganglion cell's receptive field, a variety of subdivisions of these ganglion cells have been made on the basis of special features of their responses in addition to the main classification of ON-Centre—OFF-surround and OFF-Centre—ON-surround. Additionally, other classes of cells have been described lacking the centre-surround antagonism, such as direction-sensitive units, responding to a spot of light only when it was moved across the retina in a specific direction, and so on.

### X- AND Y-CELLS

The first main subdivision was that of Enroth-Cugell and Robson (1966) on the basis of linear versus non-linear spatial summation over the receptive field, those showing linear summation being described as X-cells and those showing non-linear summation as Y-cells (p. 292).

### SUSTAINED AND TRANSIENT

Later, Cleland *et al.* (1971, 1973)† classified the cells in accordance with the duration of their responses, *sustained* and *transient*, and these were probably equivalent to X- and Y-cells respectively of Enroth-Cugell and Robson.

Figure 11.7 shows the difference in response to a spot of light centred on the receptive fields of a sustained and transient cell respectively. The responses to gratings, moved steadily across the receptive field, are correspondingly different; thus there were bursts of response as each bar moved across the receptive field; as the spatial frequency of the grating, was increased, i.e. the fineness of the pattern (p. 317), the difference

between the two became manifest; the sustained unit continued to modulate its discharge about a mean level until the lines became too fine to allow discrimination, whereas the transient unit's modulations soon ceased, giving an unmodulated increase in discharge while movement continued, and returning to the unstimulated level when movement ceased. With the finest patterns there was only a brief transient discharge when the grid was jerked into motion. Clearly the sustained unit is specifically suited to respond to the spatial pattern of the stimulus, whereas the transient unit seems to represent the initial stage in development of a specific sensitivity to motion, its main responsiveness being to the passage of an object suddenly crossing or moving within the receptive field.

### BRISK AND SLUGGISH

Finally Cleland and Levick (1974) introduced a further dichotomy of *brisk* and *sluggish*. Thus, within the Centre-Surround units there are four subclasses, as indicated in Table 11.1, so that, in all, we can characterize eight classes of ganglion cell in accordance with the type of receptive field, the briskness or sluggishness of the response and the transience or sustained nature of the response. A battery of tests were devised to permit the characterization of a given unit; thus the response to standing contrast was measured by

---

*These direction-sensitive ganglion cells are presumably analogous in some respects to Grüsser *et al.*'s (1967) Class 2 movement-sensitive units in the frog's retina; a number of quantitative aspects of their responses to moving stimuli, e.g. area of the stimulus, angular velocity, and so on, have been examined, and a model, based on convergence of excitatory and inhibitory influences from bipolar cells on to the ganglion cell, has been constructed to account for the experimental findings (Grüsser, Finkelstein and Grüsser-Cornehls, 1968). An interesting attempt to relate the electrophysiological findings to the frog's fly-catching activities is contained in the paper by Butenandt and Grüsser (1968).
†Fukada and Saito (1971) distinguished Type I, responding transiently, from Type II, responding for some time to a spot-stimulus; Type I had the larger conduction velocity and receptive field. There were characteristic differences in their responses to flickering stimuli.

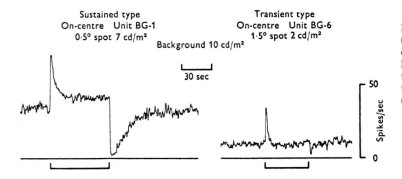

Sustained type
On-centre   Unit BG-1
0·5° spot 7 cd/m²

Transient type
On-centre   Unit BG-6
1·5° spot 2 cd/m²

Background 10 cd/m²

30 sec

50

Spikes/sec

0

**Fig. 11.7**   Sustained (left) and transient (right) responses to a spot of light. In each case a centred test spot, adding a luminance of about 10 × threshold, was turned on for the duration indicated by the thick bar beneath each record. (Cleland *et al.*, *J. Physiol.*)

**Table 11.1**  Receptive-field types and latencies of 960 cat retinal ganglion cells. Latencies are the ranges of antidromic conduction times from optic tract stimulus site (Cleland and Levick, 1974)

|  |  |  | Latency (msec) |
|---|---|---|---|
| *Concentrically organized* | 887 | (92%) |  |
| A. Brisk | 774 | (80%) |  |
|    1. Transient | 243 | (25%) | 1·0–2·4 |
|       (i) On-centre | 115 |  |  |
|       (ii) Off-centre | 128 |  |  |
|    2. Sustained | 531 | (55%) | 2·5–5·9 |
|       (i) On-centre | 271 |  |  |
|       (ii) Off-centre | 260 |  |  |
| B. Sluggish | 113 | (12%) |  |
|    1. Sustained[a] | 44 |  | 4·6–24·0 |
|       (i) On-centre | 22 |  |  |
|       (ii) Off-centre | 22 |  |  |
|    2. Transient[a] | 27 |  | 6·1–18·7 |
|       (i) On-centre | 13 |  |  |
|       (ii) Off-centre | 14 |  |  |
|  |  |  | Latency (msec) |
| *Not concentrically organized* | 73 | (8%) |  |
| 1. Local edge detector | 45 | (5%) | 6·6–15·9 |
| 2. Direction selective | 11 | (1%) | 6·1–12·4 |
| 3. Colour coded | 6 | (<1%) | 3·8–14·2 |
| 4. Uniformity detector | 5 | (<1%) | 8·7–13·9 |
| 5. Edge inhibitory off-centre | 3 | (<1%) | 3·9–6·6 |
| 6. Unclassified[b] | 3 | (<1%) |  |

[a] In a further forty-two there were insufficient observations to diagnose sustained or transient (twenty on-centre, twenty-two off-centre).
[b] Insufficient observations to reach a conclusion.

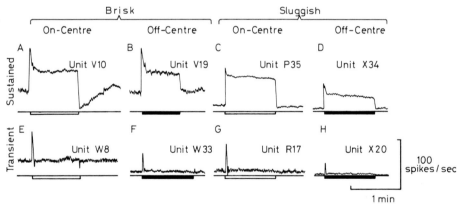

**Fig. 11.8**  Responses to standing contrast for all eight types of concentrically organized ganglion cells. A projected spot of light was turned on (A, E, G) or a contrasting disk was unmasked at the centre of the receptive field for the duration shown by the signal bar below each base line (□ brightening, ■ darkening); time calibration at lower right. The jagged line traces the mean discharge rate (calibration at lower right). The base line of each record corresponds to zero spikes/sec.
A, 0·5° spot, 10 cd/m²; B, 0·5° black disk; C, 0·8° white disk; D, 1·2° black disk; E, 1·0° spot, 6 cd/m²; F, 0·5° black disk; G, 0·5° spot, 20 cd/m²; H, 0·8° black disk. (Cleland and Levick, *J. Physiol.*)

projecting a spot of light for ON-centre cells, or unmasking a black disc for OFF-centre cells centred on the receptive field for 20 sec or longer and recording the mean impulse rate. An example is shown in Figure 11.8 for all eight types of concentrically organized ganglion cells. The response of the sustained brisk cell rose from a moderate level of maintained discharge sharply to a peak, declined a little and was maintained at a high plateau. When the spot was turned off, the discharge ceased but returned slowly to the original maintained discharge. The sluggish sustained ON-centre unit had a lower maintained discharge, which was very regular; when the light was turned on, the discharge accelerated and, with a very small overshoot, reached a plateau which only declined slightly. A characteristic feature of the sluggish maintained ON-centre cell is that the strongest responses were obtained with 1 to 2° spots, whereas the corresponding brisk cells required diameters of 0·3 to 0·7°. This standing contrast-test failed to bring out differences between sluggish and brisk cells of the transient class, so that the use of other tests was necessary. For example, the speed of motion of a spot across the retina was a useful stimulus-parameter, sluggish cells being primarily sensitive to slow movements (about 3°/sec) whilst brisk cells always responded well to rapid movements (*ca* 100°/sec or greater).

This sensitivity to motion is also manifest in the 'periphery effect' of McIlwain (1964); transient units are weakly excited by stimuli outside the conventionally defined receptive field; the response to the continual movement of a large pattern is an augmentation of the maintained discharge with no relation to the actual movements.

CONDUCTION VELOCITIES

There is little doubt that the transient units have the higher conduction velocities, so that Cleland *et al.* (1971), for example, found a conduction time of 2 to 3 msec to the lateral geniculat body compared with 3·4 to 6 msec for sustained cells.★

RECEPTIVE FIELDS

Both centre and surround areas of sustained units are smaller than those of transient units.

RELATION TO GANGLION CELL MORPHOLOGY

Boycott and Wässle (1974) characterized the cat's retinal ganglion cells morphologically on the basis of perikaryon size and mode of branching of the dendritic tree; three classes were derived on this basis, namely α, β and γ; it was important to appreciate that there were progressive changes mainly in size as one passed from the central area to more peripheral regions, so that comparison of types must be made at a given retinal point, in fact the variation in, say, the dendritic field size with retinal eccentricity for a given type of cell can give an important clue to its electrophysiological behaviour. Thus Figure 11.9A shows how the variation in centre-receptive field size of brisk sustained cells correlates with the size of dendritic fields of β cells, leading Cleland and Levick to equate the two. In B there is a corresponding plot for brisk transient cells and α-cells, and in C for sluggish and γ cells. On the basis of this comparison, the

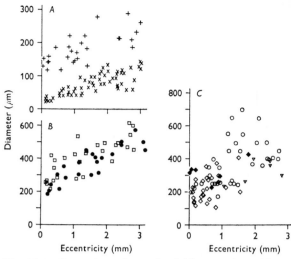

**Fig. 11.9** Comparison of receptive field centre sizes with dendritic fields as functions of eccentricity from the *area centralis*. A, Brisk-sustained centres ( + ) and dendritic fields of β cells ( × ). B, brisk-transient (□) and α cells (●). C, sluggish (◇, ◆) and γ cells (○▽); ◆ signifies members of the subclass: on-centre sustained; ▽ signifies members of the δ subclass. (Cleland and Levick, *J. Physiol.*)

**Table 11.2** Receptive fields related to morphology of retinal ganglion cells. (Levick, 1975)

| Receptive field | Morphology |
|---|---|
| 1. Concentric-surround | |
| (a) Brisk | |
| (i) Transient | α |
| (ii) Sustained | β |
| (b) Sluggish | |
| (i) Transient | |
| (ii) Sustained | |
| 2. Non-concentric | γ(δ) |
| (a) Local edge-detector | |
| (b) Direction-Selective | |
| (c) Colour-Coded | |
| (d) Uniformity-Detector | |
| (e) Edge-Inhibiting OFF-Centre | |

correlations indicated in Table 11.2 have been made, sluggish concentric cells corresponding apparently with the γ-cells.†

UNITS WITH NON-CONCENTRIC SURROUNDS

Cleland and Levick found some 73 units out of 960 with receptive fields departing from the common centre-

★The fast conducting optic tract fibres send branches to the superior colliculi and may constitute the afferent basis for high-speed direction-sensitivity of many collicular neurones (McIlwain and Buser, 1968).
†Boycott and Wässle described a subclass of γ-cells, the δ cells; their perikaryal diameters were larger, and their dendritic morphology was rather similar to that of smaller α-cells.

surround pattern, and these have been classified as in Table 11.1, and their features may be briefly described.

### Local edge detectors

These have also been called 'excited by contrast' units (Stone and Hoffmann, 1972) and are distinguished from concentric field units by their responses to centripetal and centrifugal movement of small contrast targets. Figure 11.10 illustrates some responses to a white spot; it is moved first into the centre of the field, producing a burst of spikes; on removing it, in the same direction, a second burst occurs, a form of behaviour different from that of an ON-centre unit. A black disc had the same effects. Large spots were much less effective, indicating that an important feature of the stimulus was the establishment of a contrast on the receptive field; when slowly moving targets were used, responses occurred as the leading or trailing edges passed across the receptive field. The surround-field produced no discharge when an annulus of light was presented, but it apparently exerted a strong inhibitory action on the centre, and this meant that discharge could be evoked by moving edges in the centre provided that edges were not moved at the same time in the surround—hence the terms 'local edge detector'.

### Direction-sensitive units

These are much less prominent and selective as to direction than the rabbit units described by Barlow *et al.* (1964) and Barlow and Levick (1965), and will not be discussed further (see p. 565).

### Colour-coded units

The responses of units in the visual pathway to different wavelengths of light will be discussed more fully in the section on colour vision. The basis of colour-coding

is that the character of the response should vary according to the wavelength of light, a character that does not depend on mere intensity of stimulation. In the cat, Cleland and Levick found very few with this property; they had in common an ON-response to blue light in the centre of their receptive fields and an OFF-response to green and red light. By varying the wavelength of the stimulating light, a 'cross-over point' could be found at which the excitatory (ON) effect of short wavelength light could be balanced by the inhibitory (OFF) effect of longer wavelength light; this occurred at about 509 nm (Fig. 11.11).

### Uniformity detectors

These cells had a brisk maintained discharge which was reduced or abolished by all forms of visual excitation, whether white spots on a dark background or black spots on a light background; moving a grating across the receptive field caused an immediate cessation of discharge, which began again soon after the movement stopped; by continuously moving the grating, the discharge could be suppressed indefinitely. Thus the message conveyed to the brain by discharge in these units is simply: 'No recent temporal or spatial discontinuities over the region covered by the receptive field' and it probably corresponds to the 'suppressed-by-contrast' unit of Rodieck (1967) but is probably different from similarly entitled units described by Stone and Fukuda (1974).

### Edge-inhibitory OFF-centre

This is similar to the 'suppressed by contrast' units of Stone and Fukuda (1974), since a grating suppressed discharge not only when moving but also when maintained stationary. Analysis of its receptive field with small spot-stimuli showed that it differed from the

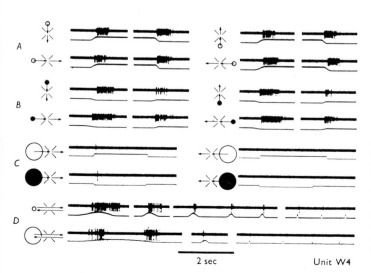

**Fig. 11.10** Local edge detector: responses to moving targets. A, 1° white disk moved into centre of receptive field (alignment marks on diagrams to the left of records), stopped, then moved out in the same direction; all four directions (horizontal and vertical) tested. In each record, upper trace is spike record (positivity down) of the cell; lower trace is the output of a photomultiplier receiving light via an optical system from a region 2° in diameter centred on the receptive field (increasing light caused upward deflexion). B, same for 1° black disk. C, 4° white and black disks used. D, 1° white disk (upper row) moved completely across the receptive field at increasing speeds (range 2·3°/sec at left, 75°/sec at right); 4° white disk (lower row) moved similarly (2·2°/sec at left, 430°/sec at right). (Cleland and Levick, *J. Physiol.*)

2 sec          Unit W4

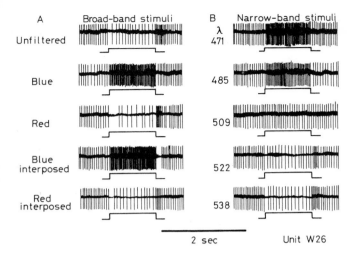

A   Broad-band stimuli     B   Narrow-band stimuli

Unfiltered

Blue

Red

Blue
interposed

Red
interposed

λ
471

485

509

522

538

2 sec          Unit W26

**Fig. 11.11** Responses of a colour-coded cell. For all records the stimulus was a large patch of light projected on to a grey background of luminance 4 cd/m². A, broad-band stimuli. The top record shows the response to flashing unfiltered white light. In the next two records a blue and a red filter were present in the beam during the flash. In the last two records the shutter was held open constantly and the blue and red filters flipped into the beam briefly. B, spectral cross-over point. Narrow-band interference-filters (peak wave-length shown at left of each record) were located in the beam during the flash to determine the filter for which the sustained excitatory and inhibitory actions were approximately balanced ( ∼ 509 nm). (Cleland and Levick, *J. Physiol.*)

classical OFF-centre concentric unit, in fact its field could be considered as being made up of three concentric processes, a central patch which, stimulated in any fashion, caused decrease or abolition of the maintained discharge; a surrounding zone of OFF-responsiveness, and an outermost annulus of ON-responsiveness.

### MORPHOLOGICAL CORRELATION

As suggested in Table 11.2, the sluggish and non-concentric units have morphological counterparts in the γ-cells of Boycott and Wässle.★

### INFORMATION PROCESSING

The conclusion we may draw from these studies of receptive field types is that the information received by the receptors as patterns of light and dark, with or without movement, has been considerably processed within the retina so that a given neurone's discharge

provides the higher centres of the brain with a significantly coded message.

### ANATOMICAL BASIS FOR ON- AND OFF-CENTRE UNITS

Famiglietti and Kolb (1976) have shown that ganglion cells may be differentiated on the basis of the layer in the inner nuclear layer in which their dendrites ramify, as illustrated by Figure 11.12. In layer *a* are to be found the endings of flat cone bipolars and in layer *b* the endings of the invaginated

---

★ In relating their classification to earlier ones, Cleland and Levick assumed that the brisk-transient and brisk-sustained cells are Y- and X-cells respectively on the basis of Enroth-Cugell and Robson's classification. The class of W-cells, described by Stone and Hoffmann (1972) and Stone and Fukuda (1974), represents some 40 per cent of the units encountered by them and obviously overlaps many of the categories described by Cleland and Levick; thus 60 per cent of W-cells had concentric fields, some were directionally sensitive, and others suppressed by contrast. Cleland and Levick conclude that W-cells are those that cannot be classed as X- or Y-cells.

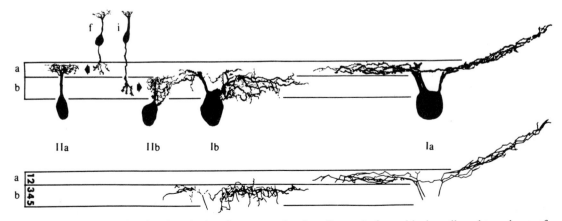

**Fig. 11.12** Camera lucida drawings (top) and computer drawings (bottom) of cone bipolar cells and two classes of ganglion cells. Top: Sublamina *a* contains the terminals of flat cone bipolar cells, *f*, and the dendrites of Type Ia and IIa ganglion cells. Sublamina *b* contains terminals of invaginating cone bipolar cells, *i*, and dendrites of Type Ib and Type IIb ganglion cells. For significance see text. (Famiglietti and Kolb, *Science.*)

cone bipolars; ganglion cells of Type I (Boycott and Wässle's α) can therefore be subdivided into Ia and Ib according to the layer in which their dendrites ramify; similarly Type II (β) are IIa and IIb. The overlap of dendrites and bipolar terminals in the respective layers is very striking, so that IIa ganglion cells receive major input from flat cone bipolars and IIb from invaginating cone bipolars. From various correlations between the populations of retinal regions and the type of ganglion cell response, the authors concluded that ganglion cells branching in layer *a* will be OFF-centre and in *b* OFF-centre. Since flat cone bipolars receive an excitatory synapse from cones their response will be hyperpolarizing and lead to OFF-excitation of ganglion cells with which they synapse. Invaginating cone bipolars apparently do not receive excitatory synapses from cones, these synapses appearing to be of the depolarizing type (Raviola and Gilula, 1975), so that they might well mediate an ON-centre response.

## CENTRE AND SURROUND MECHANISMS

It will be recalled (p. 286) that Rodieck and Stone were able to analyse the responses of a ganglion cell to stimuli impinging on different parts of its receptive field on the basis of overlapping excitatory and inhibitory regions; Hammond (1975) has extended this sort of analysis to sustained and transient cells, basing his study largely on the selective adaptation of the centre by virtue of its greater sensitivity. Hammond concluded that with both types of cell the centre and surround mechanisms were co-extensive but that the surround mechanisms differed in shape, as illustrated by Figure 11.13a, there being a

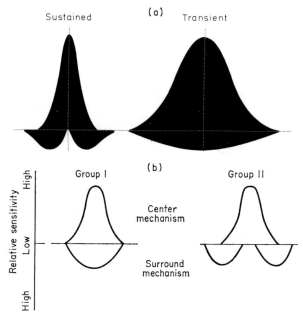

**Fig. 11.13** (a)    Schematic illustration of the difference in receptive fields of sustained and transient ganglion cells. In both types of cell the centre and surround mechanisms are co-extensive, but they differ in shape. (Hammond, *Exp. Brain Res.*) (b) An essentially similar description of the receptive fields of Group I and Group II cells. (Hickey *et al.*, *Vision Res.*)

region in the centre of the sustained cell's receptive field where only the centre mechanism was active. A rather similar conclusion had been reached by Hickey *et al.* (1973) as illustrated by Figure 11.13b.

## 'PERIPHERY' OR SHIFT EFFECT

### THE DISINHIBITORY SURROUND

Ikeda and Wright (1972a) have shown that there is a region beyond the classical inhibitory surround of an ON-centre ganglion cell which they describe as disinhibitory. Figure 11.14 illustrates its demonstration in a transient-type unit. A spot of light is flashed on the centre of the ON-centre receptive field to produce a maximal response; as the spot is made to fall on more peripheral parts of the field, the response decreases and at about 2 to 2·5° an OFF-response appears, although the ON-response is larger; on moving to more eccentric positions, the OFF-response disappears' and the ON-response increases, and it is this range, extending from about 3 to 5°, that is described as the disinhibitory zone. If the spike-count for stimulus-ON is plotted against eccentricity, we get the curve of Figure 11.15, the hump between about 2 and 4° representing the disinhibitory surround. Although harder to demonstrate, a similar surround was found in sustained cells, but because of the much smaller receptive fields of these cells, the degree of eccentricity was much less—namely about 0·6°.

### Centre-surround interaction

According to Ikeda and Wright the disinhibitory surround of the transient cell may be regarded as part of a zone, concentric with the field-centre, in which interaction between periphery and centre takes place, a zone in which a spot of light will increase the sensitivity of the central spot. Since the transient cell analyses the dynamic aspects of the visual stimulus, this disinhibition might be valuable in signalling movement of the visual stimulus over the retina permitting the fixation-reflex response. With the sustained cells, whose function is concerned with analysis of the spatial characteristics of the visual field, a zone of peripheral disinhibition may contribute to spatial frequency-tuning, but the exact manner in which this can be brought about is by no means clear.

### PERIPHERY EFFECT

This extension of the surround beyond its classical borders is probably merely one example of the more general 'periphery effect' first described by McIlwain (1964, 1966) who stimulated the centre of a receptive field with weak flashes and moved a black disc in the periphery 90° away from the centre of the field; in

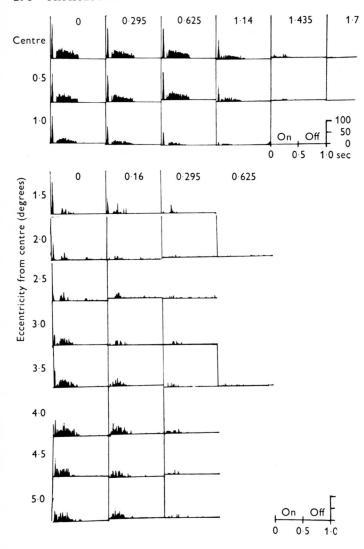

**Fig. 11.14**   The disinhibitory surround. Post-stimulus histograms of response of an ON-centre transient ganglion cell to a spot of light falling on the centre of the receptive field or at different degrees of eccentricity, indicated by the numbers at the left-hand side of the columns of histograms. The duration of ON and OFF is indicated at the right of the figure. The columns headed 0, 0·295, etc. represent different intensities of spot-stimulus, the numbers indicating the densities of the neutral filters employed. Note that an OFF-response occurs at an eccentricity of 2–2·5°; at eccentricities greater than this the OFF-response is suppressed (disinhibition). (Ikeda and Wright, *J. Physiol.*)

this case there was an increase in firing rate. The essential feature of the peripheral stimulus was not the change in retinal illumination but the edge of the contrast, and by using a large peripheral grating as in Figure 11.16, which was moved just the width of a bar of the grating, very powerful discharges were obtained from a ganglion cell whose field had not been exposed to the grating. When the grating was shifted the ON-centre unit was illuminated at its centre with its surround in darkness and there was a large increase in discharge. If the centre was in darkness and the surround illuminated there was a suppression of discharge (Fischer and Kruger, 1974). With OFF-centre cells the reverse occurred.

**Horizontal and amacrine cells**
The large extension of the receptive field of ganglion

cells (and geniculate cells, McIlwain, 1964) beyond their classical limits clearly indicates the influence of horizontal and amacrine cells which, by synaptic or electrotonic connexions, can transmit horizontally over large distances; and it is interesting that Ikeda and Wright (1972) pointed out that the periphery effects described by McIlwain and Fischer and Kruger were only seen in ganglion cells whose receptive fields did show the continuous expansion with increasing spot size.*

---

* Ikeda and Wright (1972b) distinguished two aspects of the periphery effect; the *modulated effect* is essentially that described by McIlwain, and is best brought about by moving a grating in the periphery; each shift of the grating produces a modulation in discharge synchronous with the shift. The *unmodulated effect* reflects a more static change in excitability extending over large distances.

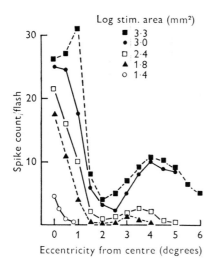

**Fig. 11.15** The results shown in Fig. 11.14 have been plotted to show the variation in spike-response to a spot of light with eccentricity from the centre. (Ikeda and Wright, *J. Physiol.*)

## PERCEPTION OF EYE MOVEMENT

Ikeda and Wright considered that the periphery or 'shift' effect was purely a function of transient Y-cells, but Barlow *et al.* (1977) could demonstrate it in sustained X-cells (Fig. 11.16), and they have suggested that, because it depends quantitatively on the size of retinal field stimulated, it is concerned with signalling a movement of the eye since under these conditions the image of the whole field moves suddenly. Thus, when the eye moves, there is a transient discharge from all ganglion cells that depends on the change in illumination and not on its value at the new position, so that with each movement the pattern of impulse-frequency depends on both the luminous patterns existing in the present position as well as in the new one. Perceptually such a confusing situation is not manifest, and this could be due to the shift-effect which could act to suppress responses during movement thus 'wiping the slate clean' for the next image. During a saccade there is undoubtedly a rise in threshold, and it could be

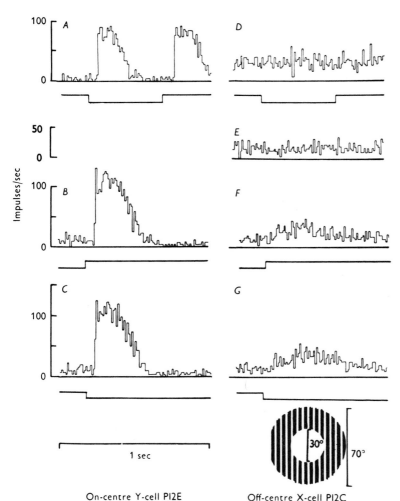

**Fig. 11.16** Illustrating the 'periphery effect'. Responses of a typical transient Y-cell (A–C) and a sustained X-cell (D–G) to a shift of the grating shown at the bottom right. The receptive fields were centred in the central 30° disc where the view of the grating was occluded. The grating was moved just the width of a bar, so that the stimulus is essentially a movement of the striped pattern in the far periphery of the receptive field. Note that the sustained X-cell shows some response although much smaller than that of the transient Y-cell. (Barlow *et al.*, *J. Physiol.*)

that the movement of the retinal image is the effective stimulus as MacKay's work suggested (p. 509).

## ROD AND CONE INTERACTION

Psychophysical studies suggest a great deal of independence of rod and cone activities when thresholds are measured; when the sensation of brightness is studied, however, the Purkinje phenomenon (p. 327) indicates that the rods may modify colour sensation, and this is presumably through ganglion cells that make neural connexions with both rods and cones, possibly through the horizontal cells. Many studies of ganglion cells in lower mammals have indicated a convergence of the two types of receptor; in the monkey, too, Gouras and Link (1966) demonstrated this, and the interaction of rod and cone influences at this level; thus the shorter latency of the cone response permits its differentiation from that of the rod, and by using red flashes to excite cones, and violet flashes to excite rods, intervals between the two may be found such that the rod response is occluded by the cones or *vice versa*. In general, with both threshold and suprathreshold stimuli, rods and cones do not contribute simultaneously to a ganglion cell's response; because the cone system is so much faster it manages to control ganglion cell function whenever stimuli are sufficient to excite it.

## REFERENCES

Andrews, D. P. & Hammond, P. (1970) Mesopic increment threshold spectral sensitivity of single optic tract fibres in the cat: cone-rod interaction. *J. Physiol.* 209, 65–81.

Barlow, H. B., Derrington, A. M., Harris, L. R. & Lennie, P. (1977) The effects of remote retinal stimulation on the responses of cat retinal ganglion cells. *J. Physiol.* 269, 177–194.

Barlow, H. B., Fitzhugh, R. & Kuffler, S. W. (1957) Change of organization of the receptive fields of the cat's retina during dark adaptation. *J. Physiol.* 137, 338–354.

Barlow, H. B., Hill, R. M. & Levick, W. R. (1964) Retinal ganglion cells responding selectively to direction and speed of image motion in the rabbit. *J. Physiol.* 173, 377–407.

Barlow, H. B. & Levick, W. R. (1965) The mechanism of directionally sensitive units in rabbit's retina. *J. Physiol.* 178, 477–504.

Boycott, B. B. & Wässle, H. (1974) The morphological types of ganglion cells of the domestic cat's retina. *J. Physiol.* 240, 397–419.

Butenandt, E. & Grusser, O.-J. (1968) The effect of stimulus area on the response of movement detecting neurons in the frog's retina. *Pflug. Arch. ges. Physiol.* 298, 283–293.

Cleland, B. G., Dubin, M. W. & Levick, W. R. (1971) Sustained and transient neurones in the cat's retina and lateral geniculate nucleus. *J. Physiol.* 217, 473–496.

Cleland, B. G. & Levick, W. R. (1974) Brisk and sluggish concentrically organized ganglion cells in the cat's retina. *J. Physiol.* 240, 421–456.

Cleland, B. G. & Levick, W. R. (1974) Properties of rarely encountered types of ganglion cells in the cat's retina and an overall classification. *J. Physiol.* 240, 457–492.

Cleland, B. G., Levick, W. R. & Sanderson, K. J. (1973). Properties of sustained and transient ganglion cells in the cat retina. *J. Physiol.* 228, 649–680.

Enroth-Cugell, C. & Robson, J. G. (1966). The contrast sensitivity of retinal ganglion cells of the cat. *J. Physiol.* 187, 517–552.

Famiglietti, E. V. & Kolb, H. (1976) Structural basis for ON- and OFF-center responses in retinal ganglion cells. *Science* 194, 193–195.

Fischer, B. & Krüger, J. The shift-effect in the cat's lateral geniculate neurons. *Exp. Brain Res.* 21, 225–227.

Fukada, Y. & Saito, H.-A. (1971) The relationship between response characteristics to flicker stimulation and receptive field organization in the cat's optic nerve fibres. *Vision Res.* 11, 227–240.

Gouras, P. & Link, K. (1966) Rod and cone interaction in dark-adapted monkey ganglion cells. *J. Physiol.* 184, 499–510.

Grüsser, O.-J. et al. (1967) A quantitative analysis of movement detecting neurons in the frog's retina. *Pflug. Arch. ges. Physiol.* 293, 100–106.

Grüsser, O.-J., Finkelstein, D. & Grüsser-Cornehls, U. (1968) The effect of stimulus-velocity on the response of movement sensitive neurons of the frog retina. *Pflug. Arch. ges. Physiol.* 300, 49–66.

Hammond, P. (1972) Chromatic sensitivity and spatial organization of LGN neurone receptive fields in cat: cone-rod interaction. *J. Physiol.* 225, 391–413.

Hammond, P. (1975) Receptive field mechanisms of sustained and transient retinal ganglion cells in the cat. *Exp. Brain Res.* 23, 113–128.

Hickey, T. L., Winters, R. W. & Pollack, J. G. (1973) Centre-surround interactions in two types of on-centre retinal ganglion cells in the cat. *Vision Res.* 13, 1511–1526.

Ikeda, H. & Wright, M. J. (1972a) The outer disinhibitory surround of the retinal ganglion cell receptive field. *J. Physiol.* 226, 511–544.

Ikeda, H. & Wright, M. J. (1972b) Functional organization of the periphery effect in retinal ganglion cells. *Vision Res.* 12, 1857–1879.

Levick, W. R. (1967). Receptive fields and trigger features of ganglion cells in the visual streak of the rabbit's retina. *J. Physiol.* 188, 285–307.

Levick, W. R. (1975) Form and function of cat retinal ganglion cells. *Nature* 254, 659–662.

McIlwain, J. T. (1964) Receptive fields of optic tract axons and lateral geniculate cells: peripheral extent and barbiturate sensitivity. *J. Neurophysiol.* 27, 1154–1173.

McIlwain, J. T. (1966) Some evidence concerning the physiological basis of the periphery effect in the cat's retina. *Exp. Brain Res.* 1, 265–271.

Michael, C. R. (1968) Receptive fields of single optic nerve fibres in a mammal with an all-cone retina; I. Contrast-sensitive units. II. Directionally sensitive units. *J. Neurophysiol.* 31, 249–256; 257–267.

Oyster, C. W. & Barlow, H. B. (1967) Direction-selective units in rabbit retina: distribution of preferred directions. *Science* 155, 841–842.

Raviola, E. & Gilula, N. B. (1975) Intramembrane organization of specialized contacts in the outer plexiform layer of the retina. *J. Cell Biol.* **65,** 192–222.

Rodieck, R. W. (1965). Quantitative analysis of cat retinal ganglion response to visual stimuli. *Vision Res.* **5,** 583–601.

Rodieck, R. W. (1967) Receptive fields in the cat retina: a new type. *Science* **157,** 90–92.

Rodieck, R. W. & Stone, J. (1965a) Response of cat retinal ganglion cells to moving visual patterns. *J. Neurophysiol.* **28,** 819–832.

Rodieck, R. W. & Stone, J. (1965b) Analysis of receptive fields of cat retinal ganglion cells. *J. Neurophysiol.* **28,** 833–849.

Stone, J. & Fukuda, Y. (1974) Properties of cat retinal ganglion cells: a comparison of W-cells with X- and Y-cells. *J. Neurophysiol.* **37,** 722–748.

Stone, J. & Hoffmann, K.-P. (1972) Very slow conducting ganglion cells in the cat's retina: a major new functional type? *Brain Res.* **43,** 610–616.

Varela, F. G. & Maturana, H. R. (1970) Time courses of excitation and inhibition in retinal ganglion cells. *Exp. Neurol.* **26,** 53–59.

Wagner, H. G., MacNichol, E. F. & Wohlbarsht, M. L. (1963) Functional basis for 'on'-centre and 'off'-centre receptive fields in the retina. *J. opt. Soc. Amer.* **53,** 66–70.

Werblin, F. S. (1970) Response of retinal ganglion cells to moving spots: intracellular recording in *Necturus maculosus. J. Neurophysiol.* **33,** 342–350.

# 12. Flicker

## CRITICAL FUSION FREQUENCY

The sensation of 'flicker' is evoked when intermittent light stimuli are presented to the eye; as the frequency of presentation is increased a point is reached—the *critical fusion frequency*—at which the flicker sensation disappears to be replaced by the sensation of continuous stimulation. The study of flicker has turned out to be a valuable method of approach to the fundamental problems of visual phenomena—it is amenable to fairly accurate measurement and it represents a perceptual process intermediate in complexity between intensity discrimination and form perception.

### THE TALBOT-PLATEAU LAW

This generalization states that, when the critical fusion frequency has been reached, the intensity of the resultant sensation, i.e. the brightness of the intermittently illuminated, but non-flickering, patch, is the mean of the brightness during a cycle. Thus if the subject's view of an illuminated patch, of luminance 5 mL, is interrupted by rotating an opaque disc in front of it from which a sector has been removed, the resultant sensation will be equal to that given by a continuous stimulus of luminance equal to 5 × Area of Sector/Total Area of Disc. This law finds a useful application in many experimental studies in which it is desired to cut down the illumination by accurately determined amounts; by varying the size of the sector in the rotating disc, different average intensities of illumination per cycle may be achieved and, so long as the critical fusion frequency is exceeded, these will appear as different steady luminances.

## EFFECT OF LUMINANCE ON FUSION FREQUENCY

In general, the greater the luminous intensity of the flickering light, the higher must be its frequency to attain fusion. This is illustrated by Figure 12.1 where the critical fusion frequency has been plotted against the logarithm of the retinal illumination in trolands. With foveal observation, the relationship is linear over a wide range, between 0·5 and 10 000 trolands, and this is the basis of the so-called *Ferry-Porter Law* which states that

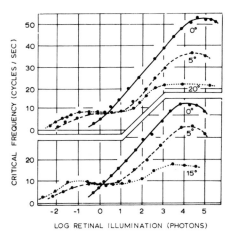

**Fig. 12.1** Dependence of critical fusion frequency on the intensity of the stimulus. With foveal fixation (0°), the Ferry-Porter Law holds over a wide range. With eccentric fixation, the curves show discontinuities indicating the activity of the rods. (Hecht, Shlaer and Smith, *Cold Spr. Harb. Symp. Quant. Biol.*)

the critical fusion frequency is proportional to the logarithm of the luminance of the flickering patch. At very high luminances the fusion frequency passes through a maximum in the region of 50 to 60 cycles/sec. At very low luminances, in the scotopic range, the fusion frequency is remarkably low, of the order of 5/sec, so that under these conditions the temporal resolution of the fovea is extremely small, a separation of 200 msec between successive stimuli not being discriminated.

### PERIPHERAL VISION

With peripheral vision, where use of both rods and cones is possible, there is an obvious discontinuity in the graph, indicating the limited application of the Ferry-Porter Law; by plotting the logarithm of the fusion frequency against the logarithm of the luminance, as in Figure 12.2, the break in continuity becomes more obvious and may be seen to occur in the region of a retinal illumination of 10 trolands (log 10 = unity), and corresponds to the transition from predominantly cone to predominantly rod vision. In Figure 12.2 the different curves correspond to the use

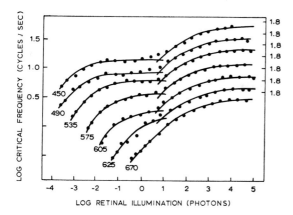

**Fig. 12.2** The log of the fusion frequency has been plotted against the log of the stimulus intensity. With red light, no break occurs in the curve. The ordinates to the left apply to the topmost curve; the others have been moved down in steps of 0·2 log units, and their exact positions are indicated to the right. (Hecht, Shlaer and Smith, *Cold Spr. Harb. Symp. Quant. Biol.*)

of different wavelengths of light, and it is seen that no break occurs with a red light of 670 nm, as one would expect since this wavelength tends to stimulate only cones.

### ROD VISION

The low fusion frequencies obtainable with rod vision might suggest that these receptors would not give as high fusion frequencies as the cones at high luminances, if they could be studied in isolation. This view of the rods as less discriminating from the point of view of temporal resolution has been amply confirmed by many experiments. Thus Hecht *et al.* (1948) studied a rod monochromat, a subject with apparently non-functional cones. The curve obtained by plotting fusion frequency against log luminance did, indeed, show a break, suggesting the operation of two mechanisms, but it was argued that these corresponded to two types of rod, since the break occurred with red light;* the highest fusion frequency measured was only 21/sec so that the rods do, apparently, have less powers of mediating temporal resolution than the cones.

### SIGNIFICANCE OF FERRY-PORTER LAW

The experimental finding—that to produce fusion the frequency of stimulation must be increased when the intensity of the light stimulus is increased—is, at first thought, surprising. The fact that fusion is possible is clearly dependent on the existence of an 'after-effect' resulting from a single stimulus, so that a succeeding stimulus can fall on the retina while it is still respond-

ing positively to the first; on purely photochemical grounds one might expect the after-effect to last longer the more powerful the primary stimulus, since a greater quantity of light-sensitive substance would be decomposed; if this were the case, clearly the minimum time elapsing between this stimulus and the next, such that the latter arrives while the eye is still responding strongly to the first stimulus, will be greater with the more powerful stimulus and therefore the critical fusion frequency should be lowered. The response of the eye to illumination, however, is far too complex to be described simply in terms of a primary photochemical reaction—the latter is the first stage only in a process that is elaborated in the nervous structures of the retina and higher visual centres. The study of the ERG has thrown valuable light on the mechanism of flicker and has emphasized the importance of inhibition, besides excitation, in determining this form of resolution.

## THE ERG AND THE INTERPRETATION OF FLICKER

The ERG in response to a single flash of light is a characteristic sequence of potential changes considerably outlasting the duration of the flash; a second flash, falling on the eye during these changes, starts off a new series of events, but the extent to which it will modify the existing state of the retinal potential will depend on how soon it follows; and we may envisage a condition, with rapidly repeated stimuli, such that the potential never falls to its baseline and, moreover, if the frequency is high enough, such that the record appears unbroken and smooth. From the point of view of the ERG we may call this the critical fusion frequency but this need not mean, at any rate without further proof, that the subjective sensation is likewise smooth. Figure 12.3 from Granit and Riddell illustrates the phenomenon of fusion in the ERG of the light-adapted frog; as the frequency of the intermittent light-stimuli increases, the record becomes smoother, whilst the amplitude of the potential change decreases.

### IMPORTANCE OF INHIBITION

In the dark-adapted frog's eye, or the rod-dominated retinae of the cat and guinea pig, the fusion frequency was low, and below fusion successive stimuli seemed merely to result in successive *b*-waves imposed on the declining limbs of the previous *b*-wave. In the light-adapted frog eye, or the *I*-type of retina (pigeon, squirrel, etc.), a much higher fusion frequency was

---

*The features of vision by the rod monochromat are discussed further on page 335.

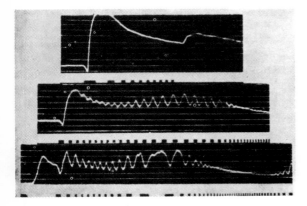

**Fig. 12.3** Flicker as shown by the frog's ERG. The light stimuli are shown at the bottom of each record as white rectangles. Top: Dark-adapted eye with no appreciable flicker. Middle and bottom: Light-adapted eye showing flicker with fusion when the stimuli are sufficiently frequent. (Granit and Riddell, *J. Physiol.*)

obtained, and an analysis of the ERG at frequencies below fusion indicated that the fluctuations of potential with successive stimuli were now characteristically different, the results of successive stimuli being to impose a negative *a*-wave on the preceding *b*-wave, i.e. the succession of events was *a-b-a-b-a* as opposed to *b-b-b-b*. If the *a*-wave really is a measure of pre-excitatory inhibition, then the high frequency of flicker is possible with the cone-dominated *I*-retinae because, before the excitatory effects of each flash, there is a preliminary inhibition of the excited state remaining from the previous flash. Since the size of the *a*-wave, i.e. of the inhibitory P III component, increases with increasing stimulus-strength, we have a reasonable explanation of the effects of luminance on fusion frequency.*

ROD-TYPE ERG

Furthermore, the great difference between rod and cone vision might be explained on the basis that the rods are less susceptible to inhibition than the cones. Thus in the rod-dominated retina of the cat at ordinary stimulus intensities the flicker is determined, according to Granit, by a different process; here the negative, inhibitory, component is relatively unimportant, and the sensation of flicker is dependent on the primary excitation (the *b*-wave in the ERG) dying down sufficiently to allow the next stimulus to evoke a new *b*-wave which will appear as a notch on the descending limb. In the absence of a pronounced P III component, this will necessitate a considerable pause between individual light-stimuli if they are to be appreciated as distinct, i.e. the fusion frequency will be low.†

SUBJECTIVE AND ERG-FLICKER

Subsequent studies of the flicker-ERG have tended to confirm Granit and Riddell's view, but we must remember that the explanation begs the question as to the justification of considering the *a*-wave as peculiarly inhibitory. Moreover, we cannot be certain that unevenness on the record necessarily means a flicker sensation, nor can we be certain that a smooth record corresponds to subjective fusion. This is well exemplified by the study of the human ERG when the subjective sensation can be compared with the electrical record. Subjectively, as we have seen, we may obtain fusion frequencies of the order of 60/sec or more by suitably increasing the luminance of the flickering patch; in the early studies of the human ERG, however, the fusion frequency rose to a maximum at about 25/sec, further increases in luminance causing no increase in fusion frequency. The human ERG, under photopic conditions, was apparently similar to that given by a rod-dominated retina, so that a smooth record did not correlate with subjective sensation. This could be because the ERG is, in general, the pooled response of all the receptors in the retina, and since rods are in such a huge majority in the human retina the ERG-response to flickering stimuli should be a rod-response. The *subjective sensation*, however, might be determined by the cones, in spite of their numerical inferiority, presumably because, under photopic conditions, their pathways to higher centres are more favoured, or because of their greater cortical representation.

CONE ACTIVITY

If this were true it ought to be possible, by adequate amplification of the ERG, and by increasing the intensity of the light stimuli, to bring out the cone activity in the ERG when the rod activity was suppressed. How can we suppress the rod activity, however? The flicker-ERG gives us just this opportunity since, as we have seen, the ERG record becomes smooth at a luminance of about 10 lux with a fusion frequency of about 25/sec. Thus the rod-response can no longer follow the stimuli and any irregularity in the record obtained by increasing the luminance must now be due to cone activity. Dodt

---

*The falling off in fusion frequency with very high luminances seems to be due to the intensification of the *a*-wave under these conditions, the flicker-ERG consisting solely of *a*-waves (Heck, 1957).

†We may note that in the typical rod-ERG the response to the first flash of a series of intermittent flashes is always very much larger than the succeeding responses; this is apparently not due to the light-adaptation that takes place to some extent during repeated exposure of the retina to flash stimuli; the relative amplitudes of first and successive responses depend critically on the duration of the dark phase between the two, as though the retina had to recover from an inhibitory condition imposed by the first flash (Arden, Granit and Ponte, 1960).

in 1951 was able to demonstrate just this; he increased the luminance of the flashes to far greater levels than earlier workers had used, and by employing an opal contact lens that enabled large areas of retina to be stimulated he found that, beyond 10 lux, flicker again appeared on the record, and the curve of fusion frequency against luminance corresponded with the subjective sensation. Fusion frequencies as high as 70/ sec were obtained as Figure 12.4 from Dodt and Wadensten shows. Even in the cat, moreover, Dodt and Walther were able to demonstrate a cone-type of ERG at sufficiently high luminance.

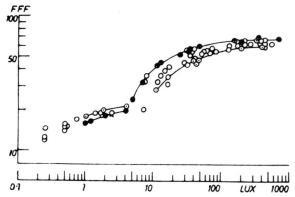

**Fig. 12.4** The critical fusion frequency (FFF) as a function of stimulus-intensity as indicated by the human ERG. Ordinate: flashes/second; Abscissa: luminance in lux. Note logarithmic scales. (Dodt and Wadensten, *Acta ophthal. Kbh.*)

THE CONE-ERG

The capacity of the rods to have their contribution to the ERG suppressed by use of an intermittent stimulus has permitted the study of the contribution of cones in some detail, and from thence it has been a simple step to the analysis of cone function in the intact retina. For example, by studying the effectiveness of different wavelengths in producing a cone-ERG under intermittent stimulation, Dodt and Walther found a typical photopic sensitivity curve with maximum in the yellow-green at 550 nm compared with the scotopic maximum of about 500 nm determined by measuring sensitivity of the retina in the dark-adapted state. This is the characteristic Purkinje shift (p. 327). Again, the characteristics of a single cycle of potential change, i.e. a single ERG, can also be studied by choosing frequencies that are well below fusion of the cone-ERG but of course above those for rod fusion; in general the *a*-wave is well defined and may be resolved into *two*, whilst the *b*-wave has *three*, components, suggesting the activities of different types of cone (Heck). In the completely colour-blind (rod-mono-chromat) there was no cone-ERG at all.

## GANGLION CELL DISCHARGES

We may ask what takes place in the optic nerve during flicker and fusion; The point was examined in detail by Enroth.* It will be recalled that single cells, or optic nerve fibres, respond to a flash of light in different manners and are described as Pure ON-, Pure OFF- and ON-OFF-elements according as they give a burst of spikes at ON, OFF or both at ON and OFF. In the cat the ON-OFF ganglion cells are in the majority so that these were studied most, but the ON-response of such a cell was essentially the same as a pure ON-response from the point of view of flicker, so that the same general principles of behaviour were valid; the same applied to the OFF-responses.

FUSION

To consider the pure ON-element, the first few flashes usually caused a simple spike discharge with no silent period between; this corresponds with the non-flickering period of the ERG. Soon each flash caused a burst of spikes in step with the stimulus (Fig. 12.5); as the frequency of flashing increased, the bursts became shorter but remained in step. At a certain frequency of flashing fusion occurs, in the sense that now the spike discharges, if they occur at all, bear no relationship with the stimuli. The unit shown in Figure 12.5 gave a dis-

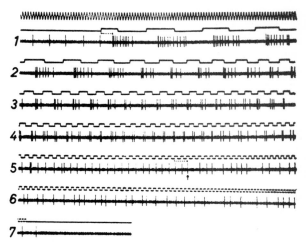

**Fig. 12.5** Response of ON-element to repetitive flash stimuli leading to fusion, in the sense that when the flash-frequency reaches a certain value, marked by the arrow, the response is a discharge that bears no relationship to the frequency of flash-stimulation. Before this phase is reached the spike bursts are synchronous with the light flashes, beginning after a latency indicated by the dotted lines on the records marked 1 and 5. Upward movement of signal indicates onset of flash. (Enroth, *Acta physiol. Scand.*)

*The main results were described by Enroth in her thesis (1952); they have been extended by Dodt and Enroth (1954) and Grüsser and Rabelo (1958).

charge corresponding to what would be seen with steady illumination, but others (usually OFF- and ON-OFF-elements) remained completely silent during fusion. Decreasing the flash-rate brought about an immediate flicker in the record. With ON-OFF elements, at low flickering rates the ON- and OFF-discharges behaved independently, each 'flickering' with the ON- and OFF-phases of the light cycle. Usually fusion of the one response occurred before that of the other, so that if, for example, the ON-element fused first, the record was the same as that for a flickering OFF-element.

### FERRY-PORTER LAW

The remarkable feature emerging from this study was the relationship between the fusion frequency for a given ganglion cell and the frequency of its spike discharge in response to a single flash; if the spike discharge had a high frequency the fusion frequency was high, and *vice versa*. Since, usually, spike discharge frequency increases with increasing intensity of light, we have an electrophysiological basis for the Ferry-Porter Law relating fusion frequency with light intensity.

### PRE-EXCITATORY INHIBITION

The mechanism of fusion is doubtless related to the pre-excitatory inhibition of the succeeding stimulus on the excitatory effect of the preceding one, and we may assume that the actual period between flashes that permits fusion is determined by the latency of this pre-excitatory inhibition and the latency of the excitatory response. Thus, to consider an OFF-element; at the end of the first flash it discharges, after a latency of a few msec. The second flash will inhibit this discharge provided the latency of its inhibitory effect is not too long, and as Figure 12.6 shows, it will be when the dark

period is equal to the difference between the two latencies. These latencies are variable, so that simple relationships between them and flicker-rate are not to be expected, although in general there is, indeed, a strong correlation between the length of the dark period and the difference of latencies (Grüsser and Rabelo).

### POST-EXCITATORY INHIBITION

With ON-elements, the significant factor is probably the post-excitatory inhibition that brings to an end the discharge in response to the onset of the flash. This inhibition is seen when the element is stimulated with a single flash of light; the response is a *primary activation* lasting about 20 to 70 msec and consisting of a burst of spikes; this is followed by a *discharge pause* of 80 to 250 msec and then a *secondary activation*. The discharge pause represents post-excitatory inhibition, and fusion will presumably occur when the flashes are so timed that the primary activation of the one flash is due to fall in the discharge pause of the preceding one. According to Grüsser and Rabelo, the latency of onset of the primary activation is the determining factor.

### SPIKE FREQUENCY ANALYSIS

By using modern methods of analysis of the responses in single ganglion cells of the cat, Ogawa, Bishop and Levick (1966) have made a more precise analysis of critical fusion frequency and the factors determining it. Figure 12.7 shows

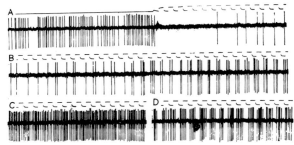

**Fig. 12.7** Development of steady-state discharge pattern by a dark-adapted OFF-centre unit in response to an 8 cycles/sec flashing spot. Upper trace: phases of flash-cycle; downward, light off. A: immediately before and after the onset of the flashing light; B: continued from A; C: 25 sec after B; D: 2·5 min after C. (Ogawa, Bishop and Levick, *J. Neurophysiol.*)

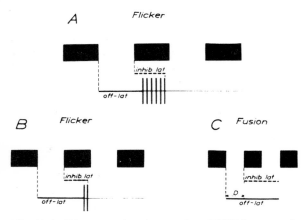

**Fig. 12.6** Diagrammatic representation of OFF-latency and pre-excitatory inhibition latency during flicker of an OFF-element. Flashes are indicated by black rectangles. A, the dark interval is too long for the inhibitory effect of the light flash to affect the OFF-response; B, the interval is shorter, but the OFF-response has not been completely inhibited; C, there is fusion because the dark period is equal to the difference between the OFF-latency and the inhibitory latency. (Enroth, *Acta physiol. Scand.*)

the response of a single unit to a flashing spot of light at 8 cycles/sec, i.e. well below fusion frequency; after an initial inhibition of the spontaneous discharge, because the unit examined is an OFF-centre type, spikes occur in response to the flashes, but in the early stages, e.g. in C, there is a virtually continuous discharge with little relation to the flash-rate; after a time, D, the discharges become grouped in accordance with the flash-rate. Subjectively this doubtless corresponds with the observation of Granit and Hammond (1931) that when a flickering light is first presented it may appear continuous, the sensation of flicker developing later. Ogawa *et al.* used as a criterion of fusion the point when the record, presented as an 'interval histogram', became identical with that of the spontaneous discharge during constant illumination. They found that the Ferry-Porter Law applied, the typical rod-cone break occurring at luminance levels, some 100-fold those at which it occurs in man. The fusion frequencies at high luminance were high, of the order of 70 cycles/sec which is much higher than that obtained in man, but may well correspond with that in the intact cat, since Kappauf (1937) obtained frequencies as high as these in some cats using a behavioural method of study. When the steady discharges during fusion were measured at different intensities of illumination, these showed no obvious relation although the Talbot-Plateau Law would seem to demand this; however, the optic nerve message indicating luminance need not necessarily be the frequency of discharge in a given ganglion cell, in fact in view of the ON-OFF characteristics of so many, this is unlikely; it may be that it is the small movements of the eyes, provoking ON-OFF discharges, that are important, and these were absent in the experimental preparation.

## THE VISUALLY EVOKED RESPONSE

When the sensation of fusion is attained, the organism is unable to utilize, or else it ignores, the information reaching it at the receptor level; an obvious reason for this could be the failure of the receptors, or elements in the conducting pathway, to follow the repetitive stimuli with discrete responses, so that it is of some interest to find where the failure to transmit information occurs. The work with the ERG and ganglion cells suggests that fusion occurs at the retinal level, but a number of studies in which the cortical responses to visual stimuli have been compared with subjective sensation in man, or behavioural responses in animals, indicate that the sensation of fusion may intervene long before the cortical response becomes flat.[*] Thus a repetitive response in the retina is transmitted to the cortex, and may be recorded from its surface as a series of potential changes, of the same frequency as the flash-rate, but this does not necessarily evoke a continuous sensation. The cortical response is referred to as the *visually evoked response (VER)*, and will be described in more detail in Section IV (p. 601). In the rabbit, Schneider (1968) recorded the integrated VER at different frequencies; he found that frequencies that gave fusion, so far as the animal's behaviour was concerned, still gave VER's in phase with the stimuli; only when the flash-rate was some 20 cycles/sec greater than the behavioural critical fusion frequency did the record become smooth.

The amplitude of the VER diminished as the flash-rate increased, and it is interesting that behavioural fusion always occurred when the VER had attenuated to 10 to 20 per cent of its maximal value. The attenuation of the VER is due to failure of a proportion of the cortical cells to follow, so that their responses become equivalent to those induced by steady illumination; and examination of individual cortical cells shows that these vary greatly in their power to follow intermittent stimuli (Ogawa *et al.*, 1966). Thus it seems to require a definite number of cortical cells to respond in phase with the stimuli, for flicker to be perceived, or more correctly, a definite amount of cortical activity.

## EFFECT OF DARK-ADAPTATION

On the basis of Granit's analysis we may expect the subjective phenomenon of flicker to show marked variations according as the human eye is light- or dark-adapted, i.e. according as the cone or rod mechanism predominates. This is, indeed, the case but the effects are complex for reasons that will become evident.

Let us suppose that a subject is placed in a box and that he looks at a flickering test-patch through a small window; the walls of the box may be illuminated to any desired intensity and thus the subject's state of light- or dark-adaptation may be controlled accurately. The flickering patch may likewise be given any desired luminance, so that we may make two kinds of measurement of the effects of dark-adaptation, i.e. we can make the walls of the box darker and darker and keep the luminance of the test-patch at a level such that rods only will be stimulated; alternatively we may make the walls of the box darker and darker as before, thereby increasing the dark-adaptation, but we can keep the luminance of the flickering test-patch at a level above the threshold for cone vision. It is very important to keep in mind this distinction in experimental approach as the effects of dark-adaptation are greatly different in the two cases. A third method, extensively adopted in the earlier work on this theme, consists in placing the subject in a dark room and varying the luminance of the flickering test-patch from the very highest levels to the lowest, threshold ones; in this case, however, it is very difficult to control the adaptation of the eyes—at the low levels of illumination of the test-patch the eyes will be more or less completely dark-adapted, but at the high levels they will be only partially so, and to different degrees, owing to the effects of the test-patch.

### ROD VISION

If the subject is placed in the box and is allowed to view a flickering patch of luminance below the cone threshold, then the effects of dark-adaptation are precisely those we should expect on the basis of the Ferry-Porter Law; as the dark-adaptation proceeds, the test-patch appears brighter and brighter and consequently we may expect the fusion frequency to increase with dark-adaptation—this is the case as Figure 12.8 (lower curve) shows; the general level of fusion frequency is low (3 to 12 cycles/sec).

---

[*] Walker *et al.* (1943) compared the responses to repetitive photic stimuli in the optic nerve, lateral geniculate body and cortex of the monkey; the optic nerve and lateral geniculate body could be 'driven' up to rates of 62 and 59 cycles/sec respectively, whereas the cortex gave fusion at 34 cycles/sec. The critical fusion frequency for the monkey is probably about 35 cycles/sec, so that here fusion is determined by the cortex rather than the retina. Brindley (1962) has shown that, when the eye is intermittently stimulated both electrically and photically, it is easy to obtain beats by adjusting the phase-relations appropriately; these beats consist of a repetitive emphasis of some feature of the visual field. The fact that beats could be obtained when the photic stimulation rate (111 to 125 cycles/sec) was far above the critical fusion frequency (76 to 87 cycles/sec) indicated that the photoreceptors were responding repetitively at these high frequencies.

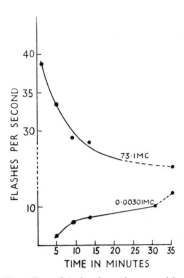

Fig. 12.8  The effect of dark-adaptation on critical fusion frequency. When the patch is of photopic intensity, dark-adaptation decreases c.f.f. When the patch is of scotopic intensity, dark-adaptation increases c.f.f. (10° eccentric fixation). (After Lythgoe and Tansley, *M.R.C. Sp. Rep.*)

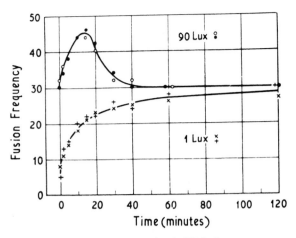

Fig. 12.9  Fusion frequency of cat's ERG. Lower curve shows that with a weak stimulus (1 lux) the fusion frequency increases with dark-adaptation to reach a maximum, characteristic of the rods. Upper curve shows that, with a moderately high intensity of light (90 lux), dark-adaptation causes an increase of fusion frequency, due to cone activity, but after a time the fusion frequency falls to that characteristic of rod vision, presumably because the rods are inhibiting the cones. (Dodt and Heck, *Pflüg. Arch.*)

### PHOTOPIC VISION

If, on the other hand, the flickering patch has a high luminance, so that both rods and cones are stimulated, the effect of dark-adaptation (decreasing the illumination of the walls of the box) is in the opposite sense, the critical fusion frequency falling from a value of nearly 40 cycles/sec in the light-adapted eye to about 25 cycles/sec in the completely dark-adapted condition (Fig. 12.8, upper curve). The curves shown in Figure 12.8 were obtained with 10° eccentric view, i.e. both rods and cones were being stimulated by the test patch; if central vision alone is used, the effects of dark-adaptation are said by Granit to be very small, hence the peripheral cones are apparently behaving differently from foveal cones. The peripheral cones differ in their organization from the foveal ones, in the sense that many share the same bipolar cell with rods, and it is just possible that these effects of dark-adaptation are due to the increased activity of rods that takes place.

### ROD-CONE INTERACTION

In other words, the rods may well inhibit the cones. A particularly good example of this rod-cone interaction was described by Dodt and Heck in the cat, using the ERG as the criterion of flicker and fusion. Figure 12.9 shows the effects of dark-adaptation on fusion frequency; with a scotopic stimulus, below the cone threshold (1 lux), the fusion frequency increases to a limiting rate of about 30/sec, the highest value for rod vision in the cat. This is the effect we have already seen

in man. When a moderately high stimulus-intensity is employed (90 lux), the fusion frequency corresponds at first to that of the rods; during dark-adaptation, however, the cones apparently take over, since the fusion frequency rises above the rod limit. After 15 minutes of dark-adaptation, nevertheless, the fusion frequency falls and eventually becomes equal to that obtained with pure rod stimuli; presumably the rods have taken over again and suppressed cone activity.

### ROD-CONE CANCELLATION

MacLeod (1972) has shown that rod and cone signals can interact during the appreciation of flicker in such a way as to cancel the flickering sensation evoked by the one type of receptor. He projected a flickering 3° patch of light on to the parafovea where both rods and cones could be excited; the patch could be illuminated by blue-green light, exciting mainly rods, and yellow light exciting rods and cones. At an alternating-frequency of 7·5 Hz flicker was visible at scotopic levels of intensity and again at photopic levels, but not at the mesopic level when both rods and cones were being excited; the effect was less pronounced at 6·5 Hz and absent at 4·5 Hz. It was postulated that the rod and cone responses were reaching the cortical integrating centre 180° out of phase so that there was no interval of non-stimulation reaching the centre, the cessation of the cone response coinciding with the onset of the rod response. To test this he projected green and deep red light to the test patch; at an alternating frequency of 7 Hz and

100 per cent contrast-modulation, flicker could be appreciated if either coloured light was projected separately, but not if both were projected together; if now a phase-drift of 180° was introduced, so that the maximum red luminance coincided with the minimum green luminance, flicker was obtained.

## SIZE OF PATCH AND THE LIGHT-DARK RATIO

### GRANIT-HARPER LAW

In general, if the size of the flickering patch is increased, its luminance being held constant, the fusion frequency increases; the effects can be large, so that a retinal area of 10 mm$^2$ may have a fusion frequency of say 60 cycles/sec whilst an area of 0·001 mm$^2$ of the same luminance may have just about half this. According to the so-called Granit-Harper Law, the fusion frequency is proportional to the logarithm of area. Varying the stimulus area, however, causes alterations not only in the total number of retinal elements stimulated, but in the relative extents to which peripheral and foveal elements enter into the response, and it would only be accidental if any simple relationship were found to hold over the whole retina. As Landis states: 'Area as a determinant of fusion frequency is a complicated affair. Not only is the size of the retinal area a determinant, but its position on the retina, its shape, whether it is discrete or composed of several parts, all enter into the areal effect.'

## SINUSOIDAL WAVE-FORM

### SQUARE-WAVE STIMULUS

These classical studies on flicker have been carried out using what is called a square-wave flickering stimulus, the observed patch being illuminated suddenly and as abruptly quenched, usually by rotating a sectored 'chopper-disc' in front of a constant source. The stimulus is illustrated schematically (Fig. 12.10) and the time-average of the luminance is indicated by the dotted line. This time-average will clearly be governed, not only by the intensity of light falling on the patch—the traditionally employed luminance when discussing

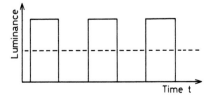

**Fig. 12.10** Periodic rectangular waveform used in traditional flicker experiments. The dotted line represents the time-average luminance of the stimulus. (Kelly, *Hdb. Sensory Physiol.*)

the effect of luminance on flicker as in the Ferry-Porter Law—but by the light-dark ratio or more correctly the pulse-to-cycle fraction,* or PCF. Thus variation in the PCF will clearly influence the adaptive condition of the eye, and a variety of studies have indicated that the effects of variation of the PCF on c.f.f. are highly complex and apparently unpredictable.

### SINUSOIDAL STIMULUS

A more promising approach is that of presenting a sinusoidal wave-form whose amplitude varies as well as the mean intensity of the light during a cycle in a manner comparable with that employed in visual acuity studies (p. 316) where a contrast-threshold was determined for a given spatial-frequency of the stimulus. It is impossible to generate a simple sine-wave flicker of the form sin $\omega t$, as this would involve negative values and radiant power is a non-negative quantity, but a closely related wave-form:

$$f(t) = L(1 + m \, . \, \text{Sin} \, \omega t)$$

can be generated and is illustrated by Figure 12.11

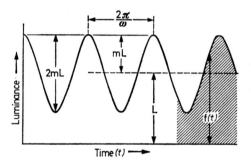

**Fig. 12.11** More modern wave-form of stimulus. This is the sum of a sinusoidal component, of amplitude $mL$, and a constant component, $L$. The time-average luminance, $L$, is often called the 'adaptation level', and the dimensionless ratio, $m$, is called the modulation. (Kelly, *Hdb. Sensory Physiol.*)

where a steady luminance, $L$, is presented, and this is modulated in a sine-wave fashion, with amplitude $m$ and frequency $\omega$. The time-average of the luminance is called the *adaptation level* and is the luminance parameter that is of much more interest than, say, the maximum luminance. The dimensionless $m$ is called the 'modulation'.

### C.F.F. VERSUS MODULATION

Experimentally, then, we may fix on a mean luminance or adaptation level and measure the critical fusion

---

*Gibbins and Howarth (1961) point out that the pulse-to-cycle or light-time fraction (PCF or LTF) is sometimes misleadingly called the light/dark ratio.

frequency (c.f.f.) for various modulations, the greater the modulation the greater the frequency of the wave-form necessary for fusion. Figure 12.12 shows typical values for three different adaptation levels; they show a peak sensitivity to modulation at 5 to 10 Hz, in the sense that, when the luminance is oscillating at these

frequencies, the degree of modulation of light required for the oscillation to be detected is minimal. At lower adaptation levels the peak disappears and the high-frequency cut-off shifts towards lower frequencies.

FIRST FOURIER COMPONENT

De Lange varied the wave-form of the stimulus, as indicated in the Figure, but because the Fourier equivalent wave-forms were, in fact, identical, there was no difference in modulation-threshold, the essential features being the first-component modulation and average luminance, i.e. the fundamental was the same, and it is this that determines threshold.

PULSE TO CYCLE FRACTION

Because it is the first Fourier component that governs threshold and critical fusion frequency, it is possible to predict the responses to various wave-forms, and these predictions have a bearing on the question of the pulse-to-cycle fraction (PCF). Thus the wave-forms in Figure 12.13 are, according to Kelly (1964), all equivalent because the product $m_1 L$ is constant*. Thus if we pass diagonally from left to right we have three wave-forms with very different PCF's, wave-forms that have been used in many of the classical studies designed to study the effect of PCF. On the basis of this analysis of equivalent wave-forms and estimates of average luminance, Kelly was able to predict the effects of light-dark ratio or PCF on critical flicker frequency.

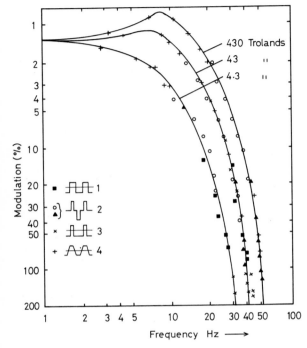

**Fig. 12.12** Modulation-threshold curves using the stimulus wave-form illustrated by Fig. 12.11. As the frequency of the wave-form increases the modulation necessary just to perceive flicker is greater. The curves show a peak of sensitivity between 5 and 10 Herz with a gradual decrease of sensitivity at lower frequencies. (Kelly, *Hdb. Sensory Physiol.*)

*More correctly, the stimuli in the horizontal rows are equivalent since the values of $m_1$ and $L$, the first Fourier component, are equal; at high frequencies, the product $m_1 L$, the absolute modulation measured in trolands, becomes the dominant factor and this is equal for all the wave-forms shown in Figure 12.13.

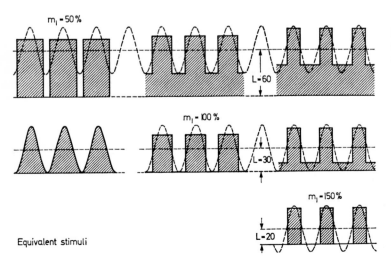

Equivalent stimuli

**Fig. 12.13** All these wave-forms are equivalent because the product $m_1 L$ is constant. (Kelly, *Hdb. Sensory Physiol.*)

Thus Figure 12.14 shows his predicted effects of PCF on c.f.f. on the basis of analysis of sine-wave flicker results; the experimental results of Bartley et al. (1961) are shown as vertical lines indicating the scatter. Thus the PCF effects are apparently completely predictable from the wave-form and the sine-wave flicker thresholds.

### FERRY-PORTER LAW

So far as the Ferry-Porter Law is concerned, this turns out to be somewhat of an accident; thus we may plot log

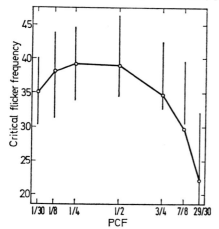

**Fig. 12.14** Predicted effects of varying the pulse-to-cycle fraction (PCF) on the critical fusion frequency. The points are calculated whilst the vertical lines represent the total spread of Bartley et al.'s (1961) PCF results for five observers. (Kelly, *Hdb. Sensory Physiol.*)

c.f.f. against the time-averaged luminance, $L$, using $m$, the modulation, as a parameter, and obtain a set of curves; a straight-line relation occurs over a range of about 2 log units of $L$ when $m$ is high. Thus with high contrast modulation, typical of the black-white alteration in the classical experiments, a logarithmic relation between luminance and c.f.f. is predictable.

### STEADY AND FLUCTUATING COMPONENTS

Gibbins and Howarth (1961) emphasized that there were two components in a flickering stimulus, namely a steady component equivalent to the mean intensity of light in a cycle, called $I_M$, and a fluctuating component, $I_F$. $I_M$ is the background adaptation and has the effect of masking $I_F$ and thus reducing the fusion-frequency when $I_F$ is held constant. $I_F$ is the effective stimulus, and for constant $I_M$, increasing $I_F$ increases the critical fusion frequency. If the mean luminance, $I_M$, were constant, the fusion frequencies at complementary pairs of pulse-cycle fraction would be equal, e.g. for P.C.F.'s of 0·1 and 0·9, 0·3 and 0·7, etc. By adding background light to keep $I_M$ constant, with different P.C.F.'s, this condition can be established and, as Figure 12.15 shows, this simple complementarity is then manifest, with maximum fusion frequency at a fraction of 0·5. The Figure shows that even in the uncompensated condition, complementarity can very nearly occur, depending on the light intensity and background conditions; with a very bright light on zero background, however, the relation becomes highly asymmetrical unless the compensating background is projected. Gibbins and Howarth discuss the Ferry-Porter Law and show that the limited region of light-intensities over which it applies is consistent with their approach and is due to the dominant influence of $I_F$ by comparison with $I_M$ over the region.

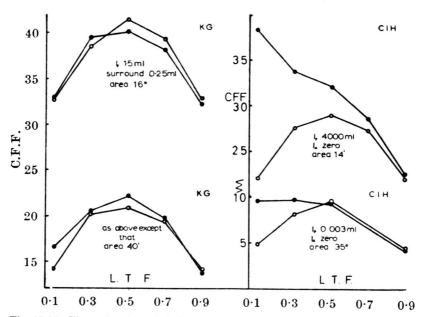

**Fig. 12.15** Illustrating the complementarity in critical fusion frequencies at certain values of the Light/Total Duration fraction (LTF), e.g. 0·3 and 0·7; 0·1 and 0·9. (Gibbins and Howarth, *Nature*.)

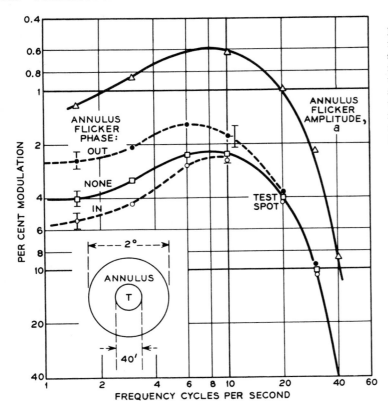

**Fig. 12.16** Effects of a subliminal flickering annulus on the modulation-sensitivity of a central spot as illustrated in the lower part of the Figure. The uppermost curve indicates flicker amplitude used in the annulus. When the two modulations are in phase (curve marked IN) there is a clear inhibition of sensitivity to flicker at the lower frequencies. (Levinson, *Doc. Ophthal.*)

SURROUND INHIBITION

Levinson (1964) measured the effect of a subliminal flickering annulus on the modulation-sensitivity of a flickering central spot, varying the phase-relations between the two flickers. The amplitude of the annular modulation was kept below threshold, and it was found that the modulation threshold of the test-spot varied with the relative phase of the annulus in such a way as to indicate inhibition at lower frequencies (Fig. 12.16) when the two were in phase. The fact that the inter-action only occurs at low frequencies indicates that the channels for spatial interaction integrate more slowly than the channels for transmission of the flicker signal to the brain.

## SUBJECTIVE ASPECTS

### BRÜCKE-EFFECT

v. Brücke in 1864 observed that the subjective bright-ness of a flickering patch of light could be higher than that experienced when the patch was presented continu-ously; the best effect was obtained with a flicker-rate of some 10 cycles/sec. It will be recalled that, according to the Talbot-Plateau Law, the sensation at fusion is the mean of the sensation during the light and dark intervals,

hence the subjective brightness of the flickering light is very considerably greater than what would be experienced at the critical fusion frequency, since with a light-dark ratio of 0·5 this would be one half that experienced with steady presentation of the bright patch. According to Bartley the optimum condition for eliciting this Brücke-effect is when the bright phase lasts one third the time of the dark phase. Under these conditions, then, the after-effect of the first stimulus accentuates the effect of the second, a finding that probably explains the observation of Strughold that visual acuity may be improved by use of a flickering light of some 5 to 6 cycles/sec. It must be appreciated that when a very short set of flashes is presented, at a rate high enough to give fusion, the Talbot-Plateau Law need not apply, in fact there is reason to believe that it will not, and Nelson and Bartley (1964) have shown that, when the duration of the presentation of the set of flashes gets very short, the sensation of brightness increases inversely with the duration, and may go above that for steady illumination. This seems to be an expression of the Broca-Sulzer effect, i.e. the initial rapid increase in visual sensation when a light is turned on, followed by a decline corresponding with the α-adaptation of Schouten (p. 363). As with the Broca-Sulzer effect, the 'Brücke-Bartley effect' is dependent on the wavelength of the light; moreover there is a very definite shift in subjective hue and of saturation (see, for example, Van der Horst and Muis, 1969).

SUBJECTIVE FLICKER-RATE

Another interesting subjective observation, recorded by Bartley, is that the actual flicker sensation, at the instant when flicker is about to disappear, does not correspond at all with the objective flash-rate. Thus at low levels of luminance it is possible to obtain fusion at 4 flashes/sec, nevertheless the flicker sensation at a flash-frequency just below this is that corresponding with a flashing light in the region of 10 to 20/sec. Conversely, at a high luminance, such that fusion is attained with 40 flashes/ sec, the flicker *sensation*, as it gives place to fusion, has about the same subjective frequency as that observed with an objective rate of 4 flashes/sec, i.e. in the region of 10 to 20/sec. Thus the final flickering sensation, obtained just before flicker disappears, seems to be independent of the actual flicker-rate.

## ELECTRICAL EXCITATION OF THE EYE

If a current is passed through an electrode on the surface of the eye and another, indifferent electrode, a sensation of light is evoked at make and break—the *electrical phosphene*—and Brindley (1955) has shown that this is most probably due to excitation of radially arranged retinal elements, either receptors or bipolar cells but almost certainly not of optic nerve fibres. When the current is alternated at, say, 50 cycles/sec, a flickering sensation is evoked. Wolff *et al.* (1968) have described in some detail the patterned sensations produced by much higher frequencies; these consisted of the appearance of blue patches of light with yellow lines radiating from them; as the frequency of alternation increased beyond 190 cycles/sec the blue disappeared and grey lines appeared looking like smoke-rings. At greater than 210 cycles/sec all patterns disappeared. When a green and red light were presented alternately at around the flicker fusion frequency, the predominance of either component in the fused sensation could be altered by electrical stimulation at the same frequency, the actual effect being determined by the phase-relations of the electrical and photic stimuli. This was presumably due to either a weakening or strengthening of the visual response by the electric currents and, depending on the phase relations, this could preferentially affect the response to red or green (Brindley, 1964).

## REFERENCES

Arden, G., Granit, R. & Ponte, F. (1960) Phase of suppression following each retinal b-wave in flicker. *J. Neurophysiol.* 23, 305–314.
Bartley, S. H. (1941) *Vision: A Study of its Basis.* New York: Nostrand.
Bartley, S. H., Nelson, T. M. & Ranney, J. E. (1961). The sensory parallel of the reorganization period in the cortical response in intermittent retinal stimulation. *J. Psychol.* 52, 137–147.
Brindley, G. S. (1955) The site of electrical excitation of the human eye. *J. Physiol.* 127, 189–200.
Brindley, G. S. (1962) Beats produced by simultaneous stimulation of the human eye with intermittent light and intermittent or alternating electric current. *J. Physiol.* 164, 157–167.
Brindley, G. S. (1964) A new interaction of light and electricity in stimulating the human retina. *J. Physiol.* 171, 514–520.
Broca, D. & Sulzer, A. (1903) Sensation lumineuse en fonction du temps pour les lumières colorées. *C.r. hebd. Séanc. Acad. Sci. Paris* 137, 944–946, 977–979, 1046–1049.
DeLange, H. (1954) Relationship between critical flicker frequency and a set of low-frequency characteristics of the eye. *J. opt. Soc. Amer.* 44, 380–389.
Dodt, E. (1951) Cone electroretinography by flicker. *Nature* 168, 738.
Dodt, E. & Enroth, C. (1954) Retinal flicker response in cat. *Acta physiol. Scand.* 30, 375–390.

Dodt, E. & Heck, J. (1954) Einfüsse des Adaptationszustandes auf die Rezeption intermittierender Lichtreize. *Pflüg. Arch.* 259, 212–225.
Dodt, E. & Wadensten, L. (1954) The use of flicker electroretinography in the human eye. *Acta ophthal. Kbh.* 32, 165–180.
Dodt, E. & Walther, J. B. (1958) Der photopischer Dominator im Flimmer-ERG der Katze. *Pflüg. Arch.* 266, 175–186.
Enroth, C. (1952) The mechanism of flicker and fusion studied on single retinal elements in the dark-adapted cat. *Acta physiol. Scand.* 27, Suppl. 100, p. 67.
Gibbins, K. & Howarth, C. I. (1961) Prediction of the effect of light-time fraction on the critical fusion frequency: an insight from Fourier analysis. *Nature* 190, 330–331.
Granit, R. & Hammond, E. L. (1931) The sensation time curve and the time-course of the fusion frequency of intermittent stimulation. *Am. J. Physiol.* 98, 654–663.
Granit, R. & Riddell, H. A. (1934) The electrical responses of the light- and dark-adapted frog's eyes to rhythmic and continuous stimuli. *J. Physiol.* 81, 1–28.
Grüsser, O. J. & Rabelo, C. (1958) Reaktionen retinaler Neurone nach Lichtblitzen. *Pflüg. Arch.* 265, 501–525.
Hecht, S., Shlaer, S. & Smith, E. L. (1935). Intermittent light stimuli and the duplicity theory of vision. *Cold Spr. Harb. Symp. quant. Biol.* 3, 237–244.
Hecht, S., Shlaer, S., Smith, E. L., Haig, C. & Peskin, J. C. (1948) The visual functions of the complete colorblind. *J. gen. Physiol.* 31, 459–472.
Heck, J. (1957) The flicker electroretinogram of the human eye. *Acta physiol. Scand.* 39, 158–166.

Kappauf, W. E. (1937). The relation between brightness and critical frequency for flicker discrimination in the cat. Ph.D. Thesis, Rochester. N.Y. University of Rochester. (Quoted by Ogawa *et al.*, 1966.)

Kelly, D. H. (1972) Flicker. *Hdb. Sensory Physiol.* VII/4. 273–302.

Landis, C. (1954) Determinants of the critical flicker-fusion threshold. *Physiol. Rev.* **34**, 259–286.

Levinson, J. (1964) Nonlinear and spatial effects in the perception of flicker. *Docum. Ophthal.* **18**, 36–55.

Lythgoe, R. J. & Tansley, K. (1929) The adaptation of the eye: its relation to the critical frequency of flicker. *Med. Res. Council Sp. Rep. Ser.* No. 134.

MacLeod, D. I. A. (1972) Rods cancel cones in flicker. *Nature* **235**, 173–174.

Nelson, T. M. & Bartley, S. H. (1964) The Talbot-Plateau law and the brightness of restricted numbers of photic repetitions at CFF. *Vision Res.* **4**, 403–411.

Ogawa, T., Bishop, P. O. & Levick, W. R. (1966) Temporal characteristics of responses to photic stimulation by single ganglion cells in the unopened eye of the cat. *J. Neurophysiol.* **29**, 1–30.

Schneider, C. W. (1968) Electrophysiological analysis of the mechanisms underlying critical flicker frequency. *Vision Res.* **8**, 1233–1244.

Van der Horst, G. J. C. & Muis, W. (1969) Hue shift and brightness enhancement of flickering light. *Vision Res.* **9**, 953–963.

Walker, A. E., Woolf, J. I., Halstead, W. C. & Case, T. J. (1943) Mechanism of temporal fusion effect of photic stimulation on electrical activity of visual structures. *J. Neurophysiol.* **6**, 213–219.

Wolff, J. G., Delacour, J., Carpenter, R. H. S. & Brindley, G. S. (1968) The patterns seen when alternating electric current is passed through the eye. *Quart. J. exp. Psychol.* **20**, 1–10.

# 13. Visual acuity

The perception of flicker demands the resolution of two stimuli separated in time; by visual acuity is generally meant the power of the eye to resolve two stimuli separated in space, i.e. it is fundamentally related to the spatial relationships between receptor elements as opposed to the temporal characteristics of the response of a single element.

## CENTRAL AND PERIPHERAL VIEWING

As with the other aspects of vision that we have so far considered in detail, the resolving power of the human eye differs markedly according as the conditions of view-

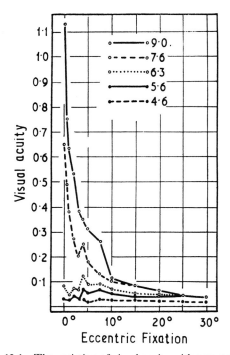

**Fig. 13.1** The variation of visual acuity with eccentricity of vision as measured along the horizontal meridian of the temporal retina. Five intensity levels, varying from 9·0 to 4·6 log micromicrolamberts were employed (9·0 log micromicrolamberts = $10^9$ micromicrolamberts = 1 mL). At 25° and 30° all the curves, except for 4·6 log micromicrolamberts, practically run together and are designated by single points. (Mandelbaum and Sloan, *Am. J. Ophthal.*)

ing favour rod or cone vision. Thus, visual acuity in the dark-adapted eye with parafoveal vision and scotopic luminous intensities is only a small fraction of that obtained with foveal vision and high luminous intensities. Nevertheless it would be wrong to consider this as a difference in rod and cone functions *per se*, since peripheral vision under photopic conditions gives a low order of visual acuity. This is illustrated by Figure 13.1, which shows visual acuity as a function of angle of view, foveal acuity being put equal to unity. Thus at 5° the acuity is only about a quarter of that found with foveal vision, and at 15° only about one-seventh.

### CONE DENSITY

The fall in visual acuity correlates fairly well with the fall in density of distribution of cones, which become more and more rare as we progress to the periphery of the retina. Density of receptors *per se*, however, is also not the determining factor since the rod density is very high throughout the retina yet visual acuity under scotopic conditions is always less than that under photopic conditions as Figure 13.1 shows except, perhaps, in the extreme periphery. As we shall see, it is the synaptic organization of the receptors that is the important factor.

## EFFECTS OF LUMINANCE OF TEST-OBJECT AND ADAPTATION

If the luminance of the test-object is such as to stimulate the cone mechanism, Lythgoe has shown that the curve showing the variation of visual acuity with the luminance of the test-object varies according to the experimental conditions; if the subject is placed in a dark box and is allowed to see the test-object through a small window, he can be considered to be in a state of dark-adaptation even though the luminance of the test-object is high enough to stimulate cone vision. In this case the acuity increases up to a maximum at approximately 10 e.f.c. and then begins to fall (Fig. 13.2, A); if the walls of the box are given a luminance of 0·011 e.f.c., i.e. if the subject is only partly dark-adapted, the acuity continues to increase to a maximum at approximately 50 e.f.c. but then falls off (Fig. 13.2, B). If the luminance of the box is continuously adjusted so that it is equal to that of the

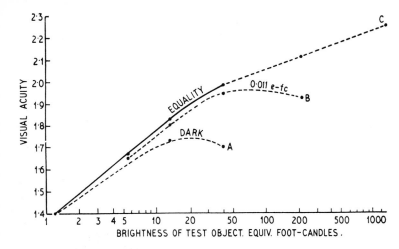

**Fig. 13.2** Illustrating the effects of the luminance of the surround (i.e. state of adaptation of the eye) on visual acuity. A, black surround: visual acuity increases with increasing luminance of test-object only up to a certain point. B, surround luminance of 0·11 e.f.c. C, surround luminance varied continuously so as to be equal to that of test-object. (Lythgoe, *Med. Res. Council Sp. Rep.* By permission of the Controller of H.M. Stationery Office.)

test-object, it is found that visual acuity increases progressively with the luminance of the test-object (Curve C). If the luminance of the box is made greater than that of the test-object, generally the acuity is lower than when the luminances are equal. Thus with visual acuity we are confronted again with some form of retinal interaction in so far as the acuity of the central retina is influenced by the general illumination of the eye, i.e. by light-adaptation.

ADAPTATION

We have seen that dark-adaptation increases visual acuity under scotopic conditions; this effect is doubtless due to the increased sensitivity of the rods; the curves in Figure 13.2 indicate that light-adaptation, under photopic conditions, increases visual acuity, and consequently dark-adaptation decreases it; since dark-adaptation increases the sensitivity of the cones (although not to the same degree as that of the rods) it is clear that the effect of dark-adaptation on photopic visual acuity is not closely connected with changes in sensitivity of the cones *per se.*

# NORMAL VISUAL ACUITY

We have so far considered variations in the visual acuity, but have said little about its absolute magnitude. It is customary to think of the visual acuity being determined by the 'grain' of the retinal mosaic of cones—thus the Snellen letters are constructed on the assumption that the average person can resolve points separated by 1 minute of arc, and this limit has been considered to be roughly determined by the diameter of a single cone. Thus if we consider two white points on a black field, if these are so close together that their images fall on a pair of adjacent cones it is clear that they would not be appreciated as separate since a single, larger, white point

would produce the same effect; only if one unstimulated cone is between the two images should the points appear as discrete. If we take Polyak's figure of 1·5 μm for the diameter of the base of a foveal cone, this would permit a resolving power of approximately 20 seconds of arc; the best figure for visual acuity indicates a resolving power of this order. When, however, visual acuity is measured by the power to detect a break in a contour, e.g. with a vernier scale adjustment, or to detect the presence of a single line on a uniform background, a far greater resolving power is found (of the order of 4 seconds of arc for the contour break and 0·5 second for the line), and numerous explanation have been put forward to explain the apparent anomaly.

THE 'ONE-TO-ONE' RELATIONSHIP

If the individual retinal cones are to be considered as the units on the basis of which resolving power is to be determined, as opposed to groups of cones, there must be a so-called 'one-to-one' relationship between cones and optic nerve fibres; in other words, two adjacent cones must be able to convey their impulses along independent paths; if two cones, close together, were connected to the same bipolar cell, the impulses relayed from this cell would be characteristic of neither cone separately, so that it would be difficult to conceive of a 'local sign' being attached to either cone, and consequently resolving power could not be so fine as to permit the differentiation of stimuli separated by one unstimulated cone.

## Midget cells
The organization of the retina has been discussed earlier, and we have seen that the midget bipolar, whose significance was first emphasized by Polyak (1941), makes a special type of contact with the cone pedicle; furthermore, a given midget bipolar only receives messages from one cone. The same exclusive relation-

ship is found with the midget ganglion cells, so that in this sense a given cone may have an exclusive pathway to the lateral geniculate body, a pathway not shared by another cone. Strictly speaking, however, there is no absolute 'one-to-one' relationship between cones, bipolars, and ganglion cells owing to the fact that cones, even in the central area, make connections with diffuse bipolars besides with a midget; similarly, the same midget bipolar that makes a 'private' synapse with a midget ganglion also makes contact with the diffuse varieties of ganglion cells. The 'one-to-one' relationship thus exists only by virtue of a peculiar synaptical arrangement between a cone and its related midget bipolar and ganglion cells; although a share of the electrical change excited in a cone is directed into collateral channels (diffuse bipolars, horizontal and amacrine cells) the main or specific cone influence seems to remain restricted to this special channel.

## Spatial summation

Thus spatial summation of stimuli, such as is observed with rod vision, could only be mediated in the central (cone) area by means of the subsidiary synaptic connections and it is therefore not surprising that very little evidence of this type of summation by the cones has been obtained. Lythgoe suggested that the increased visual acuity obtained by light-adaptation (see, e.g. Fig. 13.2) is due to an inhibition of these subsidiary nervous paths so that the cones, during light-adaptation, tend to react more and more as single units.

### DIFFRACTION AND CHROMATIC ABERRATION

Having established the existence of a 'one-to-one' relationship between cone and optic nerve fibre for the central region of the retina, albeit restricted in scope, we can postulate a certain local sign attached to any foveal cone, so that we have a basis for the differentiation by the higher centres between the stimulation of two adjacent cones and two separated by one unstimulated cone. In the absence of diffraction and aberration effects, then, a figure of 20 seconds of arc could be taken as the anatomical limit of resolution of two points by the most central portion of the fovea. Diffraction and chromatic aberration, however, must certainly be taken into consideration; it is all very well to assume that the image of a grating or Landolt-C will be a clear-cut replica of the original, but let us suppose that the white break in the Landolt-C at the limit of resolution produces an image two, three, or four times its 'geometrical size'; clearly it will stimulate two, three, or four cones so that it will be impossible to localize the break exactly; moreover, the black line of the Landolt-C has the same thickness as its break; if the white points on the inside and outside of the black contour all produce images extending over one or

more cones there will be no unstimulated cone on the retina and, on the simple view so far taken, a Landolt-C whose black line subtends only 20 seconds of arc should be invisible.

### DIFFRACTION RINGS

As a result of diffraction, a point of light produces an image of finite size on the retina; the image consists of a central bright spot surrounded by successive dark and bright rings; with a pupillary diameter of 3 mm and with light of wavelength 555 nm, the central spot will have a diameter of rather more than $3.8\ \mu$m, subtending some 47 seconds of arc at the nodal point of the eye; that is, there should be a definite overlapping of cone stimulation even when the stimulus arises from a theoretical point source. In general, the light falling in the bright rings concentric with the central bright disc only amounts to 16 per cent of the whole, so that the much greater overlapping due to these rings can probably be ignored.

## Chromatic aberration

The effects of chromatic aberration are even more serious; the effect is to make the image of a point source into a large disc covering some four cones.

## Retinal image

As a result of these combined effects the image of a grating, for example, will not consist of a series of bright and black patches on the retina; long before the resolution limit of such a grating has been reached, the image will be such that *no cone will be unstimulated*; this is shown by Figure 13.3 from Hartridge where the

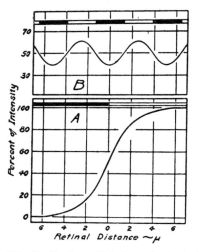

**Fig. 13.3**   The distribution of light intensity in the image of an edge (A) and a grating (B) on the retina as calculated by Hartridge. (Shlaer, *J. Gen. Physiol.*)

relative luminances of the different parts of the image of a grating on the retina are plotted; in the middle of the geometrical shadow of the black lines, the intensity is some 40 per cent of what it would be if there were no black lines at all; beneath the image of the white bar the intensity is only 60 per cent. In the same figure the distribution of light in the region of the geometrical edge of an extended object is shown; it is seen that there is no clear-cut shadow, light falling on the retina several cones' width within the geometrical shadow.

CRITERION FOR RESOLUTION

Clearly the classical conception of the basis of visual acuity—which demands that two resolvable points should be such that their images have an unstimulated cone between them—must be abandoned, since it fails to fit the observed facts of visual acuity. If we adopt Hartridge's criterion of the limit of resolution of two points, namely that their images must be sufficiently far apart so that there is a cone between that is *less stimulated than the rest*, we are not only provided with a way round our difficulty, but we can also explain the extremely high order of visual acuity exhibited in the detection of the break in a contour. Thus although a cone, falling under the geometrical image of a black bar, is stimulated, it receives less light than adjoining cones falling under the geometrical images of the white bars, and there is thus a peripheral basis for the resolution of points closer together than might otherwise have been expected.

**Induction effects**

According to Hartridge's analysis, the limit to resolution is determined almost exactly by the 'grain' of the retinal mosaic of cones, i.e. by the diameter of a central retinal cone. It may be mentioned here that local inductive effects, whereby the sensitivity of cones adjacent to the most strongly stimulated one is depressed (pp. 192, 373), assist in the resolution process. i.e. the inductive effects *amplify* the differences in response of adjoining cones on which the image of a grating falls.

DETECTION OF BREAKS IN CONTOURS

The high degree of resolution exhibited in the recognition of breaks in contours, where the limit is of the order of 4 seconds of arc—'vernier acuity'—appears, at first sight, to present greater difficulties. On the classical view, the limit to detection of a break in a contour would be determined by the break of such a size that the image of the line above the break fell on a row of cones whilst the image of the line below the break fell on another row (Fig. 13.4); the break in the

**Fig. 13.4** The break in the contour is recognized, on the classical view, by the fact that the lower cones in the row are stimulated whilst the upper ones are not.

contour would be recognized by the existence of some stimulated and some unstimulated cones in a row. Such a view demands, however, a cone size much too small for the observed resolution.

**Aberrations**

When diffraction and chromatic aberration are taken into account, the image of the contour extends to a considerable distance beyond the geometrical shadow, but the *break* in the contour is accurately reproduced, even though at a considerable distance (Fig. 13.5). Let

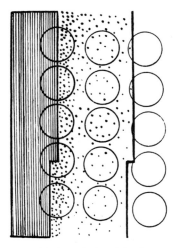

**Fig. 13.5** Illustrating how the break in the contour is reproduced accurately in the diffraction shadow, which extends to some distance to the right of the geometrical shadow. Above the break, the shadow is slightly deeper than below, hence the cones above are slightly less strongly stimulated than those below.

us imagine that the edge of the shadow falls along a row of cones (Fig. 13.5); at the break in the contour the cones below the break receive less illumination than those above it, but it is impossible, of course, that any number of cones in a row could be unstimulated whilst others were stimulated. The basis for the recognition of the break in the contour must be that in the region of the break some cones will be slightly less strongly stimulated than others. Hartridge has calculated that the variation of luminance in the shadow is sufficiently great to permit a significant difference in stimulus-intensity for the cones in the region of the break, even when the latter only subtends an angle of a few seconds at the eye; i.e., by moving along the retina a distance corresponding to 4 seconds of arc, a sufficient change in illumination is obtained to have an effect on the response of the cones. On this view, the edge of the shadow does not have to fall on any line of cones; the shadow varies continuously in luminance from left to right; at any point above the break, the luminance will be always rather less than at any point in the same vertical line below the break. So long as that difference in luminance is adequate to cause a differential stimulation of cones at the two points, we have the basis for the recognition of the break in the contour. Diffraction and aberration, by causing a progressive decrease in luminance of the image of the line to left and right, thus provide the basis for discrimination which would be impossible with a geometrical image.

## Alignment of dots

Vernier acuity is most commonly described in terms of separation of lines, but there seems little doubt that the same high degree of resolution is obtained when the subject is asked to determine the alignment of two dots; thus Westheimer and McKee (1977) obtained a resolution of 2 to 5 sec of arc under these conditions. Hence a process of averaging the local signs of receptors along a line is unnecessary, the local signs of single receptors being adequate for the discrimination.

## Moving target

A further interesting point brought out by Westheimer and McKee is that the same degree of 'hyperacuity' remains if the target moves over the visual field at a rate of 4°/sec. Thus it would seem that over a certain area of the retina the signals indicating spatial separation can be integrated over a time-period. The authors (1977a) show, moreover, that the phenomenon cannot be explained on a simple basis of sweeping the pattern over a simple 'template' that matches the stimulus offset. Thus the basic feature is the slight difference in illumination of adjacent receptors as the image falls on the retina; if the stimulus-pattern is moving, then different pairs of receptors will continue the signal until the integrated signals are ultimately interpreted.

## RESOLUTION OF A SINGLE LINE

Hecht and Mintz have shown that a line may be detected against a uniformly bright background when it subtends only 0·5 seconds of arc at the eye; the image of such a line on the retina consists of nothing more than a slight variation in luminance to left and right of the geometrical image; this is so slight that even along the line corresponding to the geometrical shadow the luminance is only 1 per cent less than that in regions unaffected by the line; nevertheless this difference of luminance is considered to be adequate to cause a differential cone stimulation that could be interpreted as a localized change in luminance, i.e. the increment corresponds to the value of $\Delta I/I$ found for the prevailing level of luminance. On this view, it would appear that the limiting factor in the detection of a line is the liminal brightness increment.

## EFFECT OF LUMINANCE OF TEST-OBJECT

We have seen that visual acuity increases progressively with illumination of the target; the explanation for this is probably not simple, since this improvement is only achieved up to a point unless the surround illumination is increased too (Fig. 13.2, p. 312). This influence of the surround illumination suggests that neuronal interaction in the retina is also affected by luminance. Hecht considered that increasing the luminance brought into operation more and more cones, and so the effective mosaic became finer and finer and thereby permitted better and better resolution. In the simple form in which it was presented the theory was inadequate, but as modified recently by Pirenne and his colleagues, it may well account for many of the phenomena.

### MULTIPLE UNIT HYPOTHESIS

Their point of view is best illustrated by a diagram (Fig. 13.6). We have seen that the requirement for

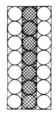 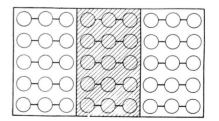

**Fig. 13.6**  Illustrating the concept of retinal units in defining resolving power. *Left:* The cones behave in a perfect one-to-one manner, so that the effective retinal mosaic for discriminating a grating is constituted by rows of single cones. *Right:* Because of convergence in the visual pathway, three cones behave as units, hence the grating at limiting resolution must be three times as wide.

maximal visual acuity is the effective one-to-one relationship between cones and optic nerve fibres. Any deviations from this relationship must reduce resolving power. Thus, as Figure 13.6 shows, a perfect three-to-one relationship would give a visual acuity of one third the maximum, since the effective size of the receptor is three times as large. The visual acuity, other things being equal, must therefore depend on the size of the effective retinal units than can be brought into operation, and Pirenne, Marriott and O'Doherty have suggested that it is mainly because increasing the luminance of the test-object brings into operation smaller units, i.e. smaller groups of receptors converging on a single optic nerve fibre, that the basic relationship between visual acuity and luminance pertains.

## Limiting light-flux

Thus the large units are able to summate light-stimuli falling over large areas and therefore their threshold will be lower than that of the smaller units in which summation will be more limited. On this basis we might expect that the flux of light-energy reaching the eye from the white break in the Landolt-C, corresponding to maximal visual acuity, would be the same for the various luminances. Thus under scotopic conditions we might expect it to be equal to the minimum flux necessary for

vision, of the order of 150 quanta/sec, and according to Pirenne *et al.* this is roughly true. Similarly, with photopic vision, the minimum flux to excite the cones is of the order of $2 \cdot 5 \cdot 10^{-12}$ lumens/sec, and if the fluxes emitted from the breaks of the Landolt-C's are calculated for different luminances these correspond approximately up to visual acuities of about 2. Beyond this, however, the amount of light entering the eye from the break is well above the threshold flux. As Pirenne has indeed pointed out, it is highly unlikely that this can be a complete explanation since threshold measurements are made under vastly different conditions from measurements of visual acuity, where large parts of the retina are stimulated by the surroundings of the Landolt-C.

### CONTRAST

In considering the luminance of the test object there is an unfortunate tendency to ignore the real factor, namely the contrast between the black details on the white background, or *vice versa*; there is a tendency to treat the black as absolutely black, and not reflecting light from its surface, thereby implying that increasing the luminance will increase the contrast in a parallel fashion. This is not necessarily true, so that it is more logical to express the lighting conditions of the target in

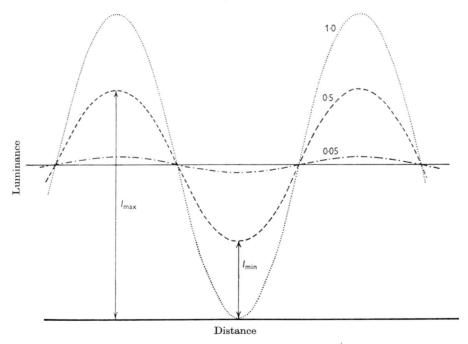

**Fig. 13.7** Illustrating the contrast-ratio for a grating generated by a sinusoidal variation in luminance on a cathode-ray screen. Contrast is defined as $(I_{max} - I_{min})/(I_{max} + I_{min})$. Three ratios of $1 \cdot 0$, $0 \cdot 5$ and $0 \cdot 05$ are shown. Note that the mean luminance level remains constant for different contrast-ratios. Spatial frequency is defined as the reciprocal of the angular distance between successive maxima in the sinusoidal distribution. (Campbell and Green, *J. Physiol.*)

terms of its *contrast*. A useful experimental method of establishing gratings with accurately defined contrast is to generate the grating on a cathode-ray oscilloscope screen by means of an appropriate oscillator; in the set-up described by Campbell and Green (1965) the intensity of emitted light from the screen varied in a sinusoidal manner with distance, as illustrated in Figure 13.7; the contrast is given by: $(I_{max} - I_{min})/(I_{max} + I_{min})$, and three different contrast-ratios are illustrated for the same grating-frequency. The advantage of this system, besides that of accurately defining the important experimental parameter, namely contrast, is that the average luminance of the screen during a cycle remains the same independently of altered contrast-ratios. We may define the *grating-*, or *spatial, frequency* as the angle subtended at the eye by the distance between two peaks of intensity; when the subject is just able to discriminate the grating, this grating-frequency, in cycles per degree, gives a measure of the resolving power, as usually defined, and its reciprocal is the visual acuity.★

## Contrast-sensitivity

The subject is presented with a given grating-frequency, and the contrast below which resolution is impossible indicates the threshold; the reciprocal of this contrast is called the *contrast-sensitivity*. The curve through the open circles in Figure 13.8 shows the relation between contrast-sensitivity and grating-frequency; at low grating-frequencies, i.e. with wide gratings, the contrast-sensitivity is high, i.e. small degrees of contrast are necessary.

---

★The resolving power of the eye is measured by the angle subtended at the eye by the minimum separable, i.e. by the line of a grating; the grating-frequency of the system illustrated by Figure 13.7 is given in cycles per degree subtended at the eye, the angular measure of the cycle being that between peaks of intensity, i.e. twice the angular width of the lines of the grating. Thus with a grating or spatial frequency of 60 cycles/degree, the distance between peaks is one minute of arc, and the grating-width is 30 seconds. Thus a resolving power of 60 cycles/degree is equivalent to one of 30 seconds of arc, or a visual acuity of two.

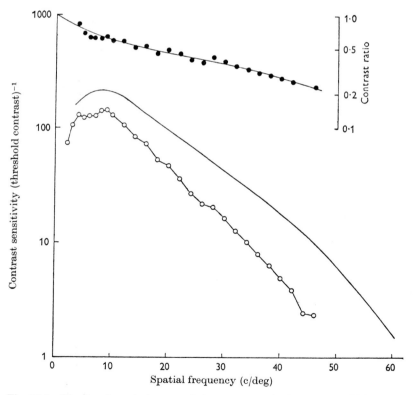

**Fig. 13.8**  The line through the open circles represents the 'contrast-sensitivity function' of a human observer. As ordinates are plotted the reciprocals of the contrast required to make gratings, of frequencies indicated as the abscissae, just resolvable. The continuous curve is a corresponding function for gratings that have been formed by interference on the retina and thus have suffered no optical defects. The filled circles indicate the ratios of the contrast-sensitivities for the two modes of presentation of the gratings. (Campbell and Green, *J. Physiol.*)

## Optical defects in image

Campbell and Green have applied this technique to assess the extent to which optical defects in the image contribute to a loss of visual acuity; to do this it was necessary to project on to the retina a grating of known contrast on the retina rather than of known contrast on the screen, since the effect of diffraction and aberrations is, by spreading light over the geometrical image of a black line, to reduce contrast. This was achieved by a technique developed by LeGrand in 1937; monochromatic light from two coherent sources was projected into the eye in such a way as to form interference fringes with calculable contrast-ratios, and the width and contrast of these fringes were varied until the subject failed to resolve them. The results have been drawn in Figure 13.8 as the continuous line, whilst the ratios of the contrast-sensitivities are plotted above in solid circles. Essentially these contrast-ratios are a measure of how the optics of the system reduce visual acuity.

### EFFECT OF ORIENTATION

The resolution of gratings is better if the lines are arranged vertically or horizontally (Taylor, 1963); Campbell, Kulikowski and Levinson (1966) have shown that the effects are similar with high and low spatial-frequency gratings, when they are assessed in terms of contrast-sensitivity (p. 322), so that it is unlikely that optical effects are the cause; and this was confirmed by finding similar orientational effects when the optical system of the eye was by-passed through the formation of interference fringes on the retina.

## Vernier acuity

This form of 'hyperacuity' may also be studied by the contrast-sensitivity technique as adapted by Tyler (1973); thus the typical vernier alignment task consists in a measure of the accuracy with which two line-segments can be aligned, and a periodical version of this employs as a stimulus a wavy line of different spatial frequencies as in Figure 13.9, and the requirement is that the subject decide whether the line is a straight line or whether it is wavy, and the subject is required to increase the amplitude of the waves until this discrimination can be made. Thus 'line modulation sensitivity' is determined at different spatial frequencies to give characteristic curves, as in Figure 13.10, the ordinate depicting the threshold amplitude of line modulation; as the scale is marked with displacement amplitude units increasing downwards, sensitivity increases upwards. Peak sensitivity was reached at a spatial frequency of between 2 and 6 c/degree and varied between 6 and 9 seconds of arc. As with the grating-acuity, there was a strong orientational bias, so

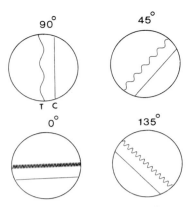

**Fig. 13.9** Periodic vernier stimuli, T, are shown with four different spatial frequencies but the same peak-to-peak amplitude of the sinusoid. Clearly, the task of deciding whether the line stimulus is wavy becomes more difficult as spatial frequency is increased. The subject varies the amplitude of the sinusoid waves until discrimination of waviness is just achieved. The Figure shows four orientations of the stimulus; a straight line C is situated about 2° to one side of the vernier stimulus acting as a comparison stimulus. (Tyler and Mitchell, *Vision Res.*)

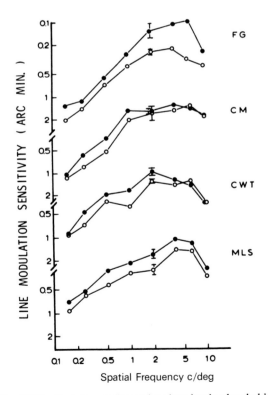

**Fig. 13.10** Periodic vernier acuity given by the threshold peak-to-peak amplitude of the sinusoid as a function of the spatial frequency of line modulation. Vertical and horizontal stimuli, filled circles; oblique stimuli, open circles. (Tyler and Mitchell, *Vision Res.*)

that sensitivity was reduced by about 30 per cent when oblique patterns were viewed (Tyler and Mitchell, 1977).

## ADAPTATIONAL EFFECTS

Gilinsky (1968) observed that the grating-acuity is lowered by adaptation to an identical pattern and bar-width but of higher mean luminance, an effect that was decreased if the orientation of the two was altered in relation to each other. Blakemore and Campbell (1969) showed that the contrast-threshold for a grating increased by 0·5 logunits after viewing for one minute a high-contrast grating; within a further minute the threshold returned to normal; the greater the contrast of the adapting grating, the greater the rise in threshold. When the effect of a given adapting grating on the contrast-sensitivity function (Fig. 13.8, p. 317) was studied, it was found that the adaptive effect was limited to a certain range of spatial frequencies, as illustrated in Figure 13.11 where the effect of a grating of 7·1 cycles/degree is seen to depress contrast sensitivity over only a part of the range of sensitivity, so that when the relative elevation of the threshold is plotted against spatial frequency, a narrow curve, peaking at about 7 cycles/degree, is obtained suggesting the preferential depression of some spatial frequency-sensitive channels. When different adapting grating-frequencies were used, different peaks were obtained, except at low spatial frequencies, i.e. wide gratings, when there was only a depression of sensitivity without any change of the peak frequency; thus at 2·5, 1·8, 1·5 and 1·3 cycles/degree the peaks all occurred at 3 cycles/degree. Again, Campbell and Kulikowski (1966) showed that simultaneous viewing of one grating reduced the sensitivity to another, the effect depending on the angles made by the two with each other, falling off to nearly zero over a range of 15 to 20 degrees. An interesting observation was that, not only does adaptation alter the contrast-sensitivity, but it also changes the

perceived spatial frequency, so that, after adaptation to a given grating, another grating, of about the same spatial frequency appears different, one with bars narrower than those of the adaptation grating appearing even narrower and one with bars wider appearing even wider. Such an effect is predictable if the cortical neurones showing the adaptation effect actually encode the information regarding spatial frequency of the stimulus (Blakemore, Nachmias and Sutton, 1970).

In general, then, it appears that the human visual system is organized in such a way as to give a number of channels tuned to a peak spatial-frequency, and presumably they would operate as size-detectors.[*]

## SIMPLE AND COMPLEX GRATINGS

Quick and Reichert (1975) presented subjects with simple gratings, i.e. of a single spatial frequency, and complex ones made up of two spatial frequencies; they estimated the contrast-sensitivity for these when presented alternately. A relative sensitivity of unity indicated that the complex grating was no more visible that the simple, and one of 2 that this was twice as visible

---

[*] Studies were usually carried out with gratings designed so that the luminous intensity along the grating varied in a sinusoidal fashion; when a square-wave form was employed, Campbell and Robson (1968) showed that the eye behaved as though it responded to the first higher harmonic, as well as to the fundamental spatial frequency; this is the third, with three times the spatial frequency of the fundamental and one third of the amplitude; it is interesting that Blakemore and Campbell (1969) found that the third harmonic of a square wave did, in fact, produce a substantial elevation of the threshold for a sine-wave grating, in addition to that produced by the fundamental, when the eye was adapted to this square-wave grating.

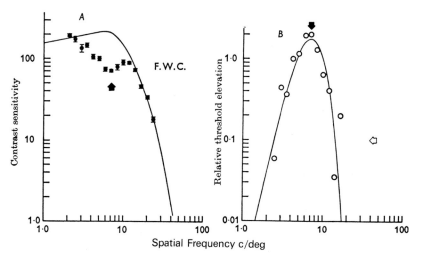

**Fig. 13.11**  Effect of adaptation to a given grating-frequency on the contrast-sensitivity function. A. The curve shows the contrast-sensitivity function for the observer F.W.C. The plotted points show the effects of adapting to a grating-frequency of 7·1 cycles/degree. B. The relative elevation of the threshold has been plotted against grating-frequency, giving a peak at about 7 cycles/degree. (Blakemore and Campbell, *J. Physiol.*)

(contrast-threshold $= \frac{1}{2}$). By plotting the relative sensitivity against the difference in spatial frequency, $\Delta f$, the curve of Figure 13.12 was obtained, showing that with a value of $\Delta f$ greater than about 2 cycles/degree there was little difference in visibility, suggesting that the detection of spatial frequency is quite sharply tuned. They concluded that a fixed area of retina, subtending about 1° of arc, is used for detection, the size being independent of the frequency of the channel. As they point out, this is at variance with results on

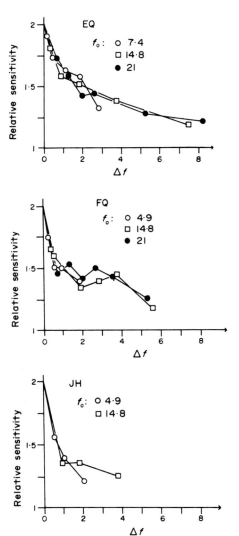

**Fig. 13.12** Contrast sensitivity for two-component gratings relative to contrast sensitivity for a one-component grating. A relative sensitivity of unity indicates that the two components together were no more visible than either presented alone. A relative sensitivity of 2 indicates that the two-component grating is twice as visible as either presented alone. $f$ is the difference in spatial frequency between the two components; when this is zero, relative sensitivity is 2 because both components are identical. (Quick and Reichert, *Vision Res.*)

adaptation (Blakemore and Campbell, 1969) suggesting that the adaptation properties of the channels are very different from the properties of the channels at near threshold without adaptation.

## IMPORTANCE OF TARGET-WIDTH

Below a certain value, the number of cycles of a grating presented for discrimination is important in defining the stimulus; for example, McCann *et al.* (1974) found that with low-spatial frequency gratings of 0·1 to 3 cycles/degree, the visibility depended only on the number of cycles presented and not on the spatial frequency. Hoekstra *et al.* (1974) found that the 'critical number' of cycles necessary for optimal resolution depended on the mean luminance, changing from 2 towards 5 when luminance was increased from 2 to 25 cd/m$^2$ and at 600 cd/m$^2$ it reached 10. The matter is of some significance since it has been argued that the fall in contrast-sensitivity at very low spatial-frequencies (Figure 13.8, p. 317) is an artefact due to the failure to present sufficient width of target and thus the critical number of bars for discrimination (Savoy and McCann, 1975). However, as Kelly (1975) has pointed out, Enroth-Cugell obtained the same effect on ganglion cells with targets covering the whole of the receptive field, and the existence of a minimum number of cycles of a truncated target that it must possess for it to act effectively as an infinitely wide target can be calculated from diffraction theory with no reference to the particular imaging system, provided only that there is a considerable attenuation of the contrast in the image at the frequency in question. He quotes the study of Barakat and Lerman (1967) which showed that a 3-bar test-object was a poor approximation to an infinite target whereas a 7-bar target is an excellent approximation if the central region of the target is used, simply because on optical grounds there are a few bars at the centre where the local contrast: $(B_{max} - B_{min})(B_{max} + B_{min})$ is essentially the same as if the number were infinite.[*]

## SCOTOPIC VISUAL ACUITY

The essential features of scotopic visual acuity have already been described, and are illustrated by Figure 13.1 (p. 311); the generally low order is presumably due to the synaptic organization of the rods such that the functional units are large. A study of the completely colour-blind subject with defective cone vision confirms this limitation of the rods; however great the luminance

---

[*]Van den Brink and Bilsen (1975) have expressed their disagreement with the argument.

employed the visual acuity could not be improved beyond 0·17, i.e. a value only about one-tenth that found for cone vision in the normal eye. Under mesopic conditions, and with peripheral vision, both rods and cones are active, yet Mandelbaum and Sloan found that usually the rods dominated the picture, presumably because they were operating far above their thresholds by comparison with the cones.

## PUPIL-SIZE, WAVELENGTH AND EYE MOVEMENTS

### THE EFFECT OF PUPIL-SIZE

Decreasing the size of the pupil will decrease the effects of aberrations on the optical image, and thus should improve visual acuity; on the other hand, when the pupil becomes sufficiently small, diffraction effects become important. Finally, the smaller the pupil the smaller the amount of light that may enter the eye; under scotopic conditions, at any rate, this must be an important factor. Experimentally it is found that with pupil-sizes below 3 mm the visual acuity is independent of pupil-size because the improvement in the retinal image due to reducing aberrations is compensated by the deleterious effects of diffraction. Increasing the pupil-size beyond 3 mm decreases visual acuity at high intensities of illumination, presumably because of the aberrations in the optical image. If a human subject is placed in a room that is darkened steadily the size of the pupil naturally increases, and it was shown by Campbell and Gregory that the size attained was, in fact, optimal for visual acuity at this particular luminance. Thus, by using artificial pupils they determined the optimum sizes for given levels of luminance, and they compared these with the actual values that the pupils took up under the same luminances. The results are shown in Figure 13.13 where the plotted points indicate the optimal

pupil size for different luminance levels, whilst the broken lines indicate the natural pupil-size under the same conditions.

### Diffraction

The spread of light from a point source, due to diffraction, is given by $\sin \theta = 1·22 \, \lambda a$, $a$ being the diameter of the entrance-pupil, $\lambda$ the wavelength of light and $\theta$ the angle subtended at the nodal point of the eye by the distance from the centre of the diffraction pattern to the first zero, i.e. the edge of Airy's disc. Westheimer (1964) has discussed the effects of diffraction on the theoretical visual acuity to be expected with varying pupil diameters; his theoretical values are given in Table 13.1, and they represent the limits of a grating that would provide a contrast-ratio between the images of the black and white lines greater than zero. This is obviously a theoretical upper limit, since the ability of the retina to detect this difference must be considered, whilst aberrations in the optical system have been ignored.

**Table 13.1** Theoretical Maximal Visual Acuity and Pupil Diameter (Westheimer, 1964)

| Entrance-pupil diameter (mm) | Visual acuity |
|---|---|
| 0·5 | 0·5 |
| 1·0 | 1·0 |
| 1·5 | 1·5 |
| 2·0 | 2·0 |
| 2·5 | 2·5 |
| 3·0 | 3·4 |

### INFLUENCE OF WAVELENGTH

In general, the use of different coloured lights has little effect on visual acuity provided their luminance be adequately maintained; the use of monochromatic, as

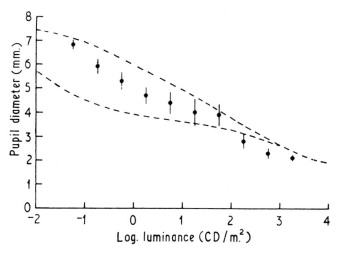

**Fig. 13.13** The broken lines indicate the natural pupil-size corresponding with the luminances shown on the abscissa; the points indicate the optimum pupil-sizes appropriate to these luminance levels. (Campbell and Gregory, *Nature*.)

opposed to white, light should abolish chromatic aberration so the fact that monochromatic light is not preferable indicates that physiological mechanisms, leading to suppression of the coloured fringes, come into play; in fact Hartridge found that artificially doubling the chromatic aberration of the human eye failed to influence visual acuity. If has been argued that, because colour vision is mediated by at least three types of cone, the use of monochromatic light would impair visual acuity since, with green light, say, only the green-sensitive cones would be operative and so the retinal mosaic would be much grosser, the intervening red-sensitive and blue-sensitive cones being functionless in this respect. Whilst under certain critical conditions it is possible to isolate one cone-mechanism from the rest, e.g. a blue-sensitive one, and when this is done to find that the visual acuity is indeed reduced, nevertheless under the ordinary conditions of measuring visual acuity it seems very likely that all types of cone would be operating, and so effects of wavelength would be negligibly small.

## Vernier acuity

Foley-Fisher (1968) has examined the effects of varying the colour of the illumination on vernier acuity and compared his results with earlier work. The acuity of different subjects varied between 4 and 10 seconds of arc, and the subjects fell into two groups, the one with a peak performance in the red and the other in the yellow-green. The best performance was never obtained with white light.

### THE EFFECT OF EYE MOVEMENTS

The question as to whether involuntary movements of the eyes play a role in determining visual acuity has been agitated repeatedly. As we shall see, the eyes are never absolutely still, so that the contours of the retinal image are repeatedly falling on new sets of receptors; and we may expect the OFF-effects to be just as important as the ON-effects, if not more so, in the interpretation of the retinal image by the brain. Troxler noted that when fixation was maintained by an effort, peripheral images tended to disappear, to reappear immediately on moving the eyes; this is the *Troxler phenomenon*, and suggests that, for peripheral vision at any rate, the eyes must move if the perception of contours is to be maintained.

## Stabilized retinal image

With foveal fixation it requires a special optical device to demonstrate the same effect; Ditchburn and Ginsborg and Riggs *et al.*, working independently, showed that with a 'stabilized retinal image', such that however the eyes moved the image of a fixated object remained on the same part of the retina, the image would disappear within a few seconds and, if it was of a fine line, would fail to reappear. The actual involuntary movements made by the eye are small, so that it is presumably the renewal of the stimuli at the contour that determines the retention of the visual percept.[*]

### SIGNIFICANCE OF EYE MOVEMENTS

Experimental studies on visual acuity, however, indicate that these movements have little influence on visual acuity. Mandelbaum and Sloan, in the study illustrated by Figure 13.1, used flash-exposures of their Landolt-C's lasting 0·2 second, so that it is very unlikely that eye movements could have contributed appreciably, yet their recorded visual acuities are not unusually low. Furthermore, if visual acuity were better with long exposures than with short this could be attributed rather to the extra time given the subject to make several tries at resolution rather than to the eye movements; these extra tries would improve resolution on a purely statistical basis. Direct experiments designed to test this point have been carried out by Riggs *et al.* (1953) and Keesey (1960), and both sets of experiments have shown that the involuntary movements are unimportant. Thus Keesey measured visual acuity when the targets were exposed from 0·02 to 1·28 seconds; in one series of experiments the target was viewed normally and in the other as stabilized images; as Figure 13.14 shows, the curves relating visual acuity to exposure-time are identical.

## Contrast-sensitivity

We have seen that when visual acuity is measured by the contrast-sensitivity function, there is an optimum spatial frequency at which contrast-sensitivity is maximal, i.e. the threshold contrast is lowest; as spatial frequency is increased, i.e. as the gratings become finer, the contrast-threshold rises, but also as the spatial frequency becomes *smaller*, and below about 10 cycles per degree, contrast-threshold rises. Tulunay-Keesey and Jones (1976) suggested that at these low spatial frequencies movements of the eyes might be important.

---

[*] The stabilized retinal image has acquired a large literature; the *relative visibility factor* is defined as $t/T \times 100$ per cent, where $t$ is the time during which the image is visible and $T$ is the total viewing time. If the fading of the image is due to failure to stimulate ON-OFF elements, then a flickering image should increase the visibility factor (Yarbus, 1960) or moving the target in a controlled manner (Ditchburn, Fender and Mayne, 1959). West (1968) has studied the effects of target-size, and of position on the retina, on the optimal flicker frequency necessary to give maximal relative visibility factor. Sparrock (1969) has shown that the logarithmic relation between increment threshold and background (Weber's Law) is unaffected by stabilization, so that although the subjective sensation of brightness is reduced the power to discriminate changes has not been affected.

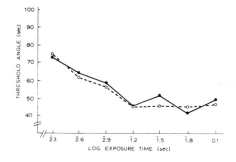

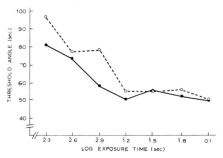

**Fig. 13.14** Showing unimportance of involuntary eye movements for visual acuity. Threshold visual angle has been plotted as a function of viewing time, the longer the time the lower the angle subtended by the grating elements at the eye. Dashed lines refer to normal viewing, the full lines to viewing with a stabilized retinal image. Curves for two subjects are shown. (Keesey, *J. Opt. Soc. Am.*)

Figure 13.15 shows the contrast-sensitivity functions of an observer for different exposure-times; as with the more classical study of acuity, reducing exposure-time from 1000 msec to 6 msec caused a rise in threshold, or decrease in sensitivity, but the graphs bring out a significant change in the shape of the sensitivity func-

tion, short exposure-times tending to flatten the curves at low spatial frequencies. Thus the sensitivity to coarse gratings tends to remain the same. Once again a stabilized image is without significant influence on contrast-sensitivity.

## BLOCH'S LAW

The variation of contrast-sensitivity with exposure-time, such that increasing the exposure increases the probability of detection of a grating, can be viewed as a temporal summation of stimuli comparable with the temporal summation for threshold luminance leading to the relation: $I \times t$ = constant, or Bloch's Law. When log-threshold contrast was plotted against exposure duration as in Figure 13.16, a critical duration of about 80 to 100 msec was revealed, below which the log-threshold is linearly related to log-duration; a slope of unity for this relationship would indicate perfect temporal summation, and the lines have been drawn with unit slope passing through the time of 70 msec, which we may take as the critical duration. At longer durations there is partial summation and after 1000 msec there is no improvement in sensitivity. The actual slopes to give best fit varied from 0·82 to 0·94 according to the spatial-frequency being studied, and the average for all was 0·88 indicating less than perfect temporal summation. Image stabilization was not completely without effect on contrast-sensitivity; with coarse gratings there was a decrease in sensitivity whereas with gratings of spatial frequencies between 2 and 7·5 c/deg sensitivity increased.*

---

*When the subjects were allowed to view the stabilized image indefinitely, contrast-sensitivity was uniformly decreased.

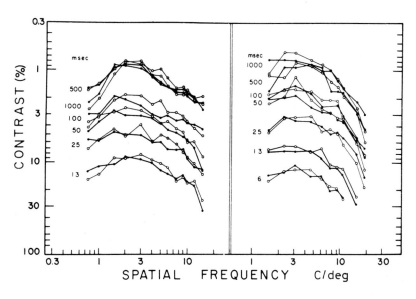

**Fig. 13.15** Effects of duration of stimulus on contrast-sensitivity functions for two observers (left and right panels). The closed symbols indicate the unstabilized condition and the open circles the stabilized condition. (Keesey and Jones, *Vision Res.*)

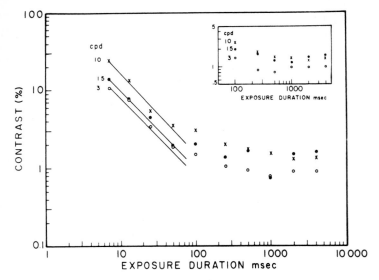

**Fig. 13.16** Threshold contrast for recognition of gratings of three spatial frequencies, 1·5, 3 and 10 c/degree. Lines of slope −1 indicating application of Bloch's Law, are drawn through 75 msec. The inset shows the results of an auxiliary experiment in which thresholds were obtained for 1·5, 3 and 10 c/degree and exposures ranging from 100 to 4000 msec. (Keesey and Jones, *Vision Res.*)

In general, however, the differences were small indicating that movements of the retinal image are of little value in detecting pattern, although the relief of local adaptation of the retina, preventing disappearance of the image, serves an obvious function.

### DEPENDENCE ON LEVEL OF LIGHT-ADAPTATION

In general dark-adaptation increases the critical duration time, indicating a longer time over which temporal summation can take place; according to the study of Brown and Black (1976), the effect is not due to a transition from cone to rod vision since a similar effect is observed when the luminance and spatial frequency of the target are such as to require cone vision.*

## Off-effects

According to Pirenne (1958) visual acuity under scotopic conditions is, indeed, improved if the subject is allowed to make quite large movements of the eyes, so that it can be argued that it is the OFF-effects, resulting from these movements, that are significant, rather than the continuous ON-effects from the white break in the C. This view is illustrated by Figure 13.17 where the Landolt-C is large enough for resolution at the given level of illumination. The C has been moved and the receptors that have been covered by the movement have indicated this by their OFF-discharges. At the break, however, there has been no OFF-effect, and it is the absence of OFF-effect in the region of the break that determines the perception of its position. If the break were too small, or the illumination too low, then the C would behave as an O, in the sense that the receptors

**Fig. 13.17** Illustrating the effects of an eye movement on the retina on which is projected the image of a Landolt-C. The initial position is in black; cones with crosses on them experience an OFF-effect whilst the remainder remain unaffected; it may be that it is the absence of OFF-effect at the break that signals the presence of the break.

under the break would not be stimulated, and movement of the C would cause OFF-effects to take place over the whole contour. For scotopic vision the movements would have to be large and greater than the involuntary movements whose maximum extent is only of the order

---

*The Bloch Law applies to the sensation of light and this will be governed by the amount of light-energy entering the eye until further increments fail to summate their effects with the initial input; to compare a discriminatory task with this simple appreciation of brightness is probably misleading; admittedly, the extra light-energy entering the eye with a longer exposure of a grating will be of value, but it is best to view the effect of duration as largely one of increasing the chance of making successive estimates of the visual problem. Baron and Westheimer (1973) showed that, when various gratings were viewed under equal-energy input situations, there was still an improvement of resolution with increased time of exposure.

of 10 minutes of arc; with photopic vision, on the supposition that OFF-effects were also of importance, the involuntary movements might well be adequate.★

## BINOCULAR VISUAL ACUITY

If the two eyes had indentical visual acuities we might, at first thought, expect to find no difference in visual acuity according as monocular or binocular vision were employed. Measurements of acuity are carried out, however, at the very limit of the subject's resolving power, the observer investigating how frequently a correct answer is made in the discrimination of breaks in successively presented Landolt-C's; there is thus no definite size of Landolt-C that can be invariably resolved whilst a slightly smaller one invariably cannot—the investigator chooses a certain 'end-point', e.g. five correct answers out of eight, so that he really estimates a probability that the subject will correctly resolve a Landolt-C of a given size. Bárány (1946) has shown that merely on a probability basis the apparent visual acuity should be increased in binocular vision, and that

the experimentally determined increase conforms to the predictions derived from probability theory. Campbell and Green (1965) have considered the matter from the point of view of a genuine summation of the responses from the two eyes, these being fed into a central integrating circuit. If these outputs are 'noisy', the standard error of the sum of $n$ independent measurements of a random noisy process decreases as $\sqrt{n}$; an observer using two eyes would make $\sqrt{2n}$ measurements, so the ratio of visual acuities in binocular and monocular vision should be $1/\sqrt{2}$; they did, indeed, find experimentally that the ratio was 1·414.

★When the test-chart is caused to move across the field of vision the visual acuity falls, and it is customary to speak of the *dynamic* visual acuity under these conditions, as opposed to the normally measured *static* acuity. With a movement of 40 degrees/sec, the acuity is about half the static value, and at 80 degrees/sec it is about one third (Methling, 1970). At these angular velocities the pursuit mechanism (p. 400) is insufficient to maintain a steady picture on the retina, and it is presumably the blurring of the retinal image that reduces acuity.

## REFERENCES

Bárány, E. H. (1946) A theory of visual acuity and an analysis of the variability of visual acuity. *Acta ophthal. Kbh.* **24**, 63–92.

Baron, W. S. & Westheimer, G. (1973) Visual acuity as a function of exposure duration. *J. Opt. Soc. Amer.* **63**, 212–219.

Blakemore, C. & Campbell, F. W. (1969) On the existence of neurones in the human visual system selectively sensitive to the orientation and size of retinal images. *J. Physiol.* **203**, 237–260.

Blakemore, C., Nachmias, J. & Sutton, P. (1970) The perceived spatial frequency shift—evidence for frequency-selective neurones in the human brain. *J. Physiol.* **210**, 727–750.

Brown, J. L. & Black, J. E. (1976) Critical duration for resolution of acuity targets. *Vision Res.* **16**, 309–315.

Campbell, F. W. & Green, D. G. (1965a) Optical and retinal factors affecting visual resolution. *J. Physiol.* **181**, 576–593.

Campbell, F. W. & Green, D. G. (1965b) Monocular versus binocular visual acuity. *Nature* **208**, 191–192.

Campbell, F. W. & Gregory, A. H. (1960) Effect of size of pupil on visual acuity. *Nature* **187**, 1121–1123.

Campbell, F. W. & Kulikowski, J. J. (1966) Orientational selectivity of the human visual system. *J. Physiol.* **187**, 437–445.

Campbell, F. W., Kulikowski, J. J. & Levinson, J. (1966) The effect of orientation on the visual resolution of gratings. *J. Physiol.* **187**, 427–436.

Campbell, F. W. & Robson, J. G. (1968) Application of Fourier analysis to the visibility of gratings. *J. Physiol.* **197**, 551–566.

Ditchburn, R. W., Fender, D. H. & Mayne, S. (1959) Vision with controlled movements of the retinal image. *J. Physiol.* **145**, 98–107.

Ditchburn, R. W. & Ginsborg, B. L. (1952) Vision with a stabilised retinal image. *Nature* **170**, 36–37.

Enroth-Cugell, C. & Robson, J. G. (1966) The contrast sensitivity of retinal ganglion cells of the cat. *J. Physiol.* **187**, 517–552.

Foley-Fisher, J. A. (1968) Measurements of vernier acuity in white and coloured light. *Vision Res.* **8**, 1055–1065.

Gilinsky, A. S. (1968) Orientation-specific effects of patterns of adapting light on visual acuity. *J. opt. Soc. Am.* **58**, 13–18.

Hartridge, H. (1947) The visual perception of fine detail. *Phil. Trans.* **232**, 519–671.

Hecht, S. (1937) Rods, cones and the chemical basis of vision. *Physiol. Rev.* **17**, 239–290.

Hecht, S. & Mintz, E. U. (1939) The visibility of single lines at various illuminations and the retinal basis of visual resolution. *J. gen. Physiol.* **22**, 593–612.

Hoekstra, J. H., van der Goot, J., van den Brink, G. & Bilsen, F. A. (1974) The influence of the number of cycles upon the visual contrast threshold for spatial sine wave patterns. *Vision Res.* **14**, 365–368.

Keesey, U. T. (1960) Effects of involuntary movements on visual acuity. *J. opt. Soc. Am.* **50**, 769–774.

Kelly, D. H. (1975) How many bars make a grating. *Vision Res.* **15**, 625–626.

Lythgoe, R. J. (1932) The measurement of visual acuity. *Med. Res. Council Sp. Rep. Ser.* No. 173.

Mandelbaum, J. & Sloan, L. L. (1947) Peripheral visual acuity. *Am. J. Ophthal.* **30**, 581–588.

McCann, J. J., Savoy, R. L., Hall, J. A. and Scarpetti, J. J. (1974) Visibility of continuous luminance gradients. *Vision Res.* **14**, 917–927.

Methling, D. (1970) Sehschärfe bei Augenfolgebewegungen in Abhängigkeit von der Gesichtsfeldleuchtdicke. *Vision Res.* **10**, 535–541.

Pirenne, M. H. (1957) "Off" mechanisms and human visual acuity. *J. Physiol.* **137**, 48–49 P.

Pirenne, M. H. (1958) Some aspects of the sensitivity of the eye. *Ann. N.Y. Acad. Sci.* **74**, 377–384.

Pirenne, M. H. (1962) Visual acuity. In *The Eye*, vol. II, Chapter 9. Ed. Davson. London: Academic Press.

Pirenne, M. H., Marriott, F. H. C. & O'Doherty, E. F. (1957) Individual differences in night-vision efficiency. *Med. Res. Council Sp. Rep. Ser.* No. 294.

Polyak, S. L. (1941) *The Retina.* Chicago: Chicago University Press.

Polyak, S. L. (1957) *The Vertebrate Visual System*. Chicago: Chicago University Press.

Quick, R. F. & Reichert, T. A. (1975) Spatial-frequency selectivity in contrast detection. *Vision Res.* **15,** 637–643.

Riggs, L. A. (1965) Visual acuity. In *Vision and Visual Perception*. Ed. Graham, C. H., pp. 321–349. New York: Wiley.

Riggs, L. A., Ratliff, F., Cornsweet, J. C. & Cornsweet, E. F. (1953) The disappearance of steadily fixated visual test objects. *J. opt. Soc. Am.* **43,** 495–501.

Savoy, R. L. & McCann, J. J. (1975) Visibility of low spatial-frequency line-wave targets; dependence on number of cycles. *J. opt. Soc. Am.* **65,** 343–350.

Shlaer, S. (1937) The relation between visual acuity and illumination. *J. gen. Physiol.* **21,** 165–188.

Sparrock, J. M. B. (1969) Stabilized images: increment thresholds and subjective brightness. *J. opt. Soc. Am.* **59,** 872–874.

Taylor, M. M. (1963) Visual discrimination and orientation. *J. opt. Soc. Am.* **53,** 763–765.

Tulunay-Keesey, U. & Jones, R. W. (1976) The effect of micromovements of the eye and exposure duration on contrast sensitivity. *Vision Res.* **16,** 481–488.

Tyler, C. W. (1973) Periodic vernier acuity. *J. Physiol.* **228,** 637–647.

Tyler, C. W. & Mitchell, D. E. (1977) Orientation differences for perception of sinusoidal line stimuli. *Vision Res.* **17,** 83–88.

Van den Brink, G. & Bilsen, F. A. (1975) The number of bars that makes a grating for the visual system: a reply to Dr. Kelly. *Vision Res.* **15,** 627–628.

West, D. C. (1968) Flicker and the stabilized retinal image. *Vision Res.* **8,** 719–745.

Westheimer, G. (1964) Pupil size and visual resolution. *Vision Res.* **4,** 39–45.

Westheimer, G. & McKee, S. P. (1977a) Integration regions for visual hyperacuity. *Vision Res.* **17,** 89–93.

Westheimer, G. & McKee, S. P. (1977b) Spatial configurations for visual hyperacuity. *Vision Res.* **17,** 941–947.

Yarbus, A. L. (1960) Perception of images of variable brightness fixed with respect to the retina of the eye. *Biophysics* **5,** 183–187.

# 14. Wavelength discrimination and the theory of colour vision

The problem of colour vision has attracted the interest and excited the speculation not only of physiologists but of physicists, anatomists, philosophers, and poets with the inevitable result that the subject has abounded with theories; in the present treatment we shall confine ourselves to the main experimental findings and show how they are related to the Young-Helmholtz trichromatic theory.

## THE PURKINJE PHENOMENON

We may introduce this account by studying the alterations in visual characteristics that take place on passing from scotopic to photopic vision.

## SCOTOPIC AND PHOTOPIC SPECTRAL SENSITIVITY CURVES

### SCOTOPIC SENSITIVITY CURVE

As we have seen, the scotopic sensitivity curve is obtained by measuring the absolute threshold of the dark-adapted eye for different wavelengths, and plotting the reciprocals of these thresholds, which measure sensitivity, against the wavelength. In a sense, then, this curve represents the relative luminosities of the different wavelengths so that, if we had a spectrum containing equal energies in all the wavelengths, the blue-green at about 500 nm would appear brightest, and the red and violet darkest. The spectral sensitivity curve is thus sometimes described as the *luminosity curve for an equal-energy spectrum*.

### PHOTOPIC SENSITIVITY CURVE

The photopic sensitivity curve is the measure of the relative effectiveness of the different wavelengths in exciting the light sense at levels well above threshold; at these levels the lights excite the sensation of colour as well and, since this is mediated by the cones, it would not be surprising if the curve of spectral sensitivities really indicated the relative sensitivities of the cones to different wavelengths. Thus we may take two patches, A and B, of light of different wavelengths and adjust their intensities until they appear equally bright by the method of flicker-photometry (p. 189). We may then compare B with a new patch of light, C, of another

wavelength, and so on through the spectrum. The reciprocals of these amounts of energies will be measures of the sensitivity of the eye to the different wavelengths and they give the curve of Figure 14.1. The maximum occurs at 555 nm in the yellow-green instead of at 500 nm in the blue-green for scotopic vision. On comparing the two sensitivity curves, as in Figure 14.2, it will be seen that there are characteristic changes in sensitivity of the eye on passing from scotopic to photopic vision, which are given the general name of *Purkinje shift*.

### SHIFT IN SENSITIVITY

Purkinje in 1823 observed that the relative luminosities of hues changed as the luminance was reduced (the *Purkinje shift*), and the nature of the change can be predicted from the curves of Figure 14.2. Thus let us

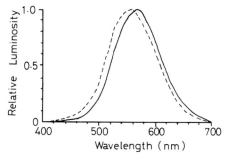

**Fig. 14.1**  Photopic luminosity curves. The curve in full line is that shown by Jainski whilst that in broken lines is the C.I.E. curve. (Le Grand, *Optique Physiologique*.)

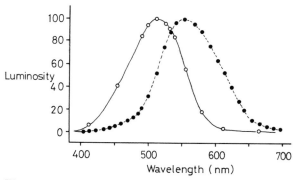

**Fig. 14.2**  The scotopic luminosity curve (open circles) compared with the photopic curve (full circles). (After Hecht.)

consider an orange of 650 nm and a yellow-green of 550 nm; under photopic conditions their luminosities are roughly in the ratio of 13:100, the yellow-green being some 8 times brighter than the orange; on referring to the scotopic curve it is seen that the same colours have luminosities in the ratio: 2:56, i.e. the yellow-green is now some 28 times brighter than the orange. In general, the short wavelengths will become brighter in comparison with the long wavelengths as the luminance level is reduced. It should be noted that the Purkinje phenomenon becomes apparent at levels where colour is still perceived, in the mesopic range where both rods and cones are operative; the shift in the apparent luminances of coloured lights is thus due to the superimposition of a rod effect—stimulating only the light sense—on the cone effect. The photopic and scotopic sensitivity curves are thus curves for cones and rods acting essentially alone; at intermediate luminances, where both are operative, the sensitivity curve lies between these limits.

### RED LIGHT AND DARK-ADAPTATION

We may note from the luminosity curves that red light is much less luminous to the dark-adapted than to the light-adapted eye in comparison with other wavelengths; in fact it is now generally considered that the rods are barely sensitive to the extreme red end of the spectrum, so that the dark-adapted eye may be exposed to fairly high luminance levels of deep red light—bright enough to permit reading with ease and hence to allow the photopic mechanism to function—without serious loss of dark-adaptation. This principle was applied extensively during the war; chartrooms were illuminated with red light and night-fighter pilots wore special red dark-adaptation goggles which permitted them to read in a well-illuminated room and yet maintain a considerable degree of dark-adaptation.

### IODOPSIN

Photochemical studies have demonstrated the remarkable correspondence between the absorption spectrum of rhodopsin and the scotopic sensitivity curve, from which it has been concluded that scotopic vision is mediated by this single pigment, rhodopsin. May we conclude that the photopic sensitivity curve corresponds to the presence of a cone-pigment with a maximum in the yellow-green at about 550 nm? This question is not easy to answer because cone vision is obviously more complex than rod vision, changes in wavelength being accompanied by changes in qualitative sensation that we call colour; and, as we shall see, wavelength discrimination requires the presence of several pigments, contained in separate types of receptor. The chicken retina contains practically only cones and Wald was able to extract from this a pigment, which he called *iodopsin*, with an absorption spectrum not greatly

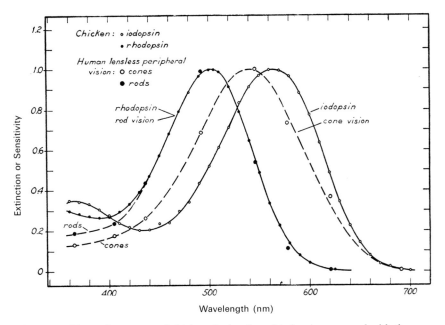

**Fig. 14.3** Absorption spectra of chicken rhodopsin and iodopsin compared with the spectral sensitivity of human rod and cone vision. The scotopic (rod) sensitivity agrees with the absorption spectrum of rhodopsin over most of its course. The photopic (cone) sensitivity curve does not coincide exactly with the absorption spectrum of iodopsin, being displaced some 20 nm toward the blue from this absorption spectrum. (Wald, Brown and Smith, *J. gen. Physiol.*)

different from the photopic sensitivity curve. This is illustrated by Figure 14.3, where both scotopic and photopic sensitivity curves have been drawn, whilst the plotted points represent absorption spectra of rhodopsin and iodopsin respectively. The agreement between rhodopsin absorption and scotopic sensitivity is good, as we have seen before; the maximum absorption of iodopsin, at 562 nm, is some 20 nm closer to the red end of the spectrum than the photopic sensitivity maximum measured, in this case, on a subject with a lens-less eye to reduce effects of absorption by the ocular media.

## ELECTROPHYSIOLOGICAL ASPECTS

### FROG ERG

We may expect the Purkinje effect to be manifest in all mixed retinae, whilst in pure rod (guinea pig) or pure cone (squirrel) retinae it should be absent. The electro-retinogram has been a useful tool in the study of this phenomenon; thus we may use the reciprocals of the relative amounts of light-energy to evoke a b-wave of fixed magnitude as the index to sensitivity of the retina to different wavelengths; if the light- and dark-adapted eyes are compared in the frog, for example, there is the same characteristic shift in sensitivity to longer wavelengths on passing from dark- to light-adaptation (Fig. 14.4).

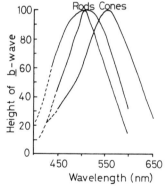

**Fig. 14.4** The Purkinje phenomenon in the frog's retina. The curves are luminosity curves and it is seen that the wavelength of maximum luminosity changes from 560 nm to 500 nm on passing from photopic to scotopic conditions. Double contour of rod graph indicates limits of variability. (Granit and Wrede, *J. Physiol.*)

### FLICKER ERG

Even in the cat, whose retina is very much dominated by rods, a shift in spectral sensitivity can be obtained by employing the flicker method of analysis of the ERG. It will be recalled that the rods are unable to follow a light-stimulus flickering at more than about 25 cycles/ sec. With a very strong light, flickering at, say, 50 cycles/ sec the responses on the record are thus cone responses,

and we may measure the effectiveness of the different wavelengths in producing an electrical effect of fixed magnitude. Under these conditions the maximum sensitivity shifts from 500 nm to 550 nm (Dodt and Walther). In the pure cone eye of the squirrel there is no Purkinje shift so that only high critical fusion frequencies of flicker are found and the spectral sensitivity shows a maximum in the yellow-green that is barely affected by dark-adaptation (Arden and Tansley).

### SINGLE FIBRES

Studies with single fibres of the optic nerve, or of large ganglion cells in the retina, show essentially the same sort of Purkinje shift but, as we shall see, this change belongs to a certain type of element which Granit has called the *dominator*.

## PHOTOCHROMATIC INTERVAL

The threshold stimulus for the different wavelengths gives us the scotopic sensitivity curve which, as we have seen, refers only to the light sense since no colour sensation is evoked (except possibly in the deep red). If the threshold stimulus for a given wavelength is increased, at a certain value the sensation of colour is reached (the *specific threshold*). The interval in luminance between the absolute light threshold and the specific threshold is called the *photochromatic interval*. In Figure 14.5 some results obtained by Dagher, Cruz

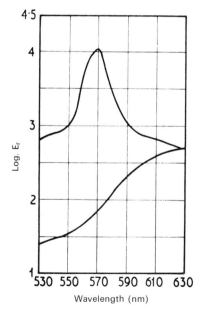

**Fig. 14.5** Illustrating the photochromatic interval. The lower curve indicates the thresholds for the sensation of light (absolute thresholds) whilst the upper curve indicates the thresholds for the sensation of colour. Note that the thresholds coincide in the orange-red. (After Dagher *et al.*, *Visual Problems of Colour.*)

and Plaza on a human subject are shown for the interval between 530 and 630 nm, i.e. over the green-orange-red part of the spectrum. The lower curve represents the absolute thresholds whilst the upper curve represents the specific thresholds. It will be seen that the interval becomes negligibly small at the red end of the spectrum, which means that the threshold for red light corresponds to the threshold for cone vision. At the fovea, too, the photochromatic interval is very small or nonexistent.

## CONE THRESHOLDS

### DARK-ADAPTATION CURVES

As stated above, if the photopic sensitivity curve represents the sensitivities of the cones, then if we can determine the thresholds for cone vision at different wavelengths we may obtain a spectral sensitivity curve which should be similar to that obtained by simple matching techniques. The problem is, of course, to be sure of stimulating only cones when measuring a threshold; in the completely dark-adapted eye this is impossible, except at the extreme red end of the spectrum, because the rods are the more sensitive. Wald made use of the fact that immediately after light-adaptation the first portion of the dark-adaptation curve (Fig. 7.1, p. 186) represents cone sensitivity i.e., under these conditions we may actually measure cone thresholds. By using different wavelengths for the light-stimulus he was able to plot a cone sensitivity curve which had a maximum at 555 nm and was similar in essentials to the C.I.E. curve of Figure 14.1.

### FOVEAL THRESHOLDS

Again, if vision is confined to the fovea by making use of flash stimuli of small extent, the thresholds should be determined by the cones. Under these conditions the subject usually states that he perceived the colour of the stimulus. By comparing results obtained in this way with those obtained by the dark-adaptation technique, we may compare the sensitivities of foveal and peripheral cones. As Figure 14.6 shows, the curves are very similar, the foveal cones being less sensitive, except in the red part of the spectrum where they are about the same.

### INCREMENT THRESHOLDS

Finally, we may note that measurement of the liminal brightness increment is a measure of the sensitivity of the retina under particular conditions of exposure to light; and according as to whether the rods or cones are the more sensitive under the particular conditions, the one or other type of receptor will determine this increment, or differential threshold as we have called it. By choosing suitable conditions of light-adaptation it is possible to ensure that the cones will be the more sensitive (e.g. by adapting with blue light to which the rods are so much more sensitive than the cones) and therefore it is possible to measure their thresholds with different wavelengths of light and obtain spectral sensitivity curves. This technique was developed by Stiles and Crawford and has been exploited widely in the measurement of spectral sensitivity curves. As we shall see, the measurements may be so refined that we may determine not only the thresholds of the cones in general, but thresholds of different types of cone (p. 334).

## WAVELENGTH DISCRIMINATION

As emphasized earlier, the difference between the scotopic and photopic sensitivity curves is not just the

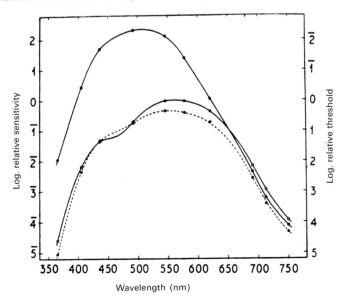

**Fig. 14.6** Sensitivity curves for foveal and peripheral cones compared with that for rods. Top curve: rods. Bottom curves: cones. The broken line refers to peripheral cones. (After Wald, *Science*.)

shift in maximal sensitivity to a different wavelength described as the Purkinje effect; there is the difference that vision becomes chromatic under photopic conditions, and the subject is able to state not only that light is present but that it belongs to a certain part of the spectrum. Before studying the mechanism of this wavelength discrimination, we may ask why it is that the retina is achromatic under scotopic conditions. The answer is that under scotopic conditions only one type of receptor is operating, the rods, and that these rods are physiologically identical in that they all have the pigment rhodopsin as their photopigment. Such a functionally homogeneous retina cannot act as wavelength discriminator.

### HOMOGENEOUS RETINA

To make this clear we may consider the responses in the optic nerve fibres of *Limulus* which, we may recall, are derived directly from the photoreceptors. A study of their responses to different wavelengths gives a typical sensitivity curve, with maximal sensitivity at 520 nm corresponding to the wavelength of maximum absorption of its photopigment. If we were to plot the intensity of response in a single fibre against wavelength we would obtain a curve roughly like Figure 14.7. In

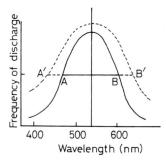

**Fig. 14.7**  Theoretical wavelength-response curve for a single receptor of *Limulus* retina. The maximum response will occur at 520 nm, the absorption maximum of the retinal pigment. With wavelengths of light corresponding to the points A and B, the responses will be identical, so that no discrimination between these two wavelengths is possible.

a sense, then, this receptor, and thus the eye, is showing some wavelength discrimination, in that it is responding with different frequencies of spike discharge to different wavelengths. A little consideration will show, however, that this receptor could not discriminate between changes of wavelength and changes of intensity; and even if we could keep intensity constant its discrimination is limited. Thus, at the points A and B on the curve, the responses are the same but the wavelengths are widely different; by drawing other lines parallel to the abscissa, cutting the curve in two points, many other pairs of spectral hues could be obtained producing

identical responses. Moreover, this curve was obtained by keeping the energy content of the stimulus the same; new curves would be obtained by varying the energy content, as indicated by dotted lines, so that in fact with a single wavelength we would be able to mimic the responses of all other wavelengths of the spectrum simply by varying the intensity of this single wavelength. Hence a retina containing a single type of receptor is unable to discriminate changes in both wavelength and intensity.

### MULTIPLE RECEPTOR HYPOTHESIS

The position becomes different if we have several types of receptor, with different photosensitive pigments; these would have different response curves and it is not difficult to show that the effects of varying wavelength on the pattern of discharges in the several types might be unique and not matchable by a change in intensity. Thus, for example, with six receptors a green light of wavelength 530 nm might evoke spike responses in these having say, 15, 34, 65, 20, 4, 2 spikes/sec. A yellow light of 600 nm might give a different set of responses, say 35, 65, 45, 20, 15, 7 spikes/sec; and it would be these *patterns* of discharges that would be characteristic of the respective wavelengths for this particular energy level of the stimuli. Increasing the energy would increase the frequencies of discharge, but it could well be that they would be able to maintain a unique pattern that could not be mimicked by other lights however much we varied the intensity. Clearly, the more types of receptor we have, the more certain we will be of having a unique type of response for a given combination of wavelength and intensity. As we shall see, Nature has economized in this respect, since it appears that only three types of receptor suffice for human wavelength discrimination.

## THE YOUNG-HELMHOLTZ THEORY

### TRIVARIANCE OF COLOUR VISION

We have already seen that colour-matching experiments have led to the generalization that we may match any colour by a mixture of three primary spectral wavelengths provided we admit that the match is not perfect so far as saturation is concerned. Thus if we wish to match a given colour *X*, we may place it in one half of a photometric matching field and in the other half we may project a mixture of red, green and blue lights; by varying the relative intensities of these we may match our colour from the point of view of chromaticity, but the mixture that we have made will be unsaturated, so we must add some white to *X* to make the match perfect in respect to both chromaticity and saturation. Thus our mixing equation would take the form:

$$\alpha R + \beta G + \gamma B = X + x\ White$$

or, $\qquad \alpha R + \beta G + \gamma B - x\ White = X$

But, since white may be represented by a mixture of red, green and blue, say: $\alpha'R + \beta'G + \gamma'B$, we may subtract these from the left-hand side of the equation, and we are left with an equation that defines our colour, $X$, in terms of only three variables:

$$(\alpha - \alpha')R + (\beta - \beta')G + (\gamma - \gamma')B = X$$

It is in this sense, then, that we are able to say that colour vision is a *trivariant phenomenon*; we can, in essence, describe any colour-stimulus in terms of three variables; normal human colour vision is thus *trichromatic*.

### SPECTRAL MIXTURE CURVES

The trichromaticity of colour vision tells us that *any* colour can be matched in terms of three primary stimuli. If we confine ourselves to matching the spectral colours we may represent this process by the chromaticity chart of Figure 7.6, p. 190; alternatively we may plot the individual contributions of our red, green and blue primaries to the different spectral colours as in Figure 14.8. As abscissae we have the wavelengths

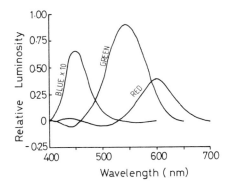

**Fig. 14.8** Spectral mixtures curve indicating the relative amounts of the three primaries that are required for matching any spectral wavelength. For example, with a blue of 500 nm the relative contributions are about 0·33 Green, 0·1 ( ÷10) Blue, and 0·05 Red. (Wright, *Researches on Normal and Defective Colour Vision.*)

of the spectral hues we wish to match, and as ordinates are plotted the amounts of each individual primary required for each match. These *spectral mixture curves* show us at a glance how much of a given primary is required to match a spectral hue; thus a yellow of 600 nm requires some 0·25 units of green and some 0·4 units of red with no blue. A blue-green of 500 nm requires about 0·4 units of green, 0·01 unit of blue* and some 'negative red'.

## THREE-RECEPTOR HYPOTHESIS

These curves lead naturally to the hypothesis that any colour sensation may be evoked by stimulating three fundamental receptors to varying extents. Thus we see that a yellow sensation may be created either by passing yellow light of, say, wavelength 600 nm into the eye, or we may create exactly the same sensation by passing into the eye a mixture of quite different lights, lights that by themselves would have excited the sensations of green and red respectively. Thus the colour mixing phenomena are essentially phenomena of confusion or failure to distinguish; the light of wavelength 600 nm is physically different from the lights of 530 nm and 650 nm that are used for the match, but the eye cannot distinguish between the light of 600 nm and the mixture. The phenomena illustrate too the economy of Nature; if an eye is to discriminate single wavelengths from mixtures then a large number of receptors must be necessary, but it is very questionable whether this discriminatory power would have much biological survival value, and instead the trivariance of colour sensation suggests that Nature 'makes do' with a very limited number of receptors.

## THE FUNDAMENTAL MECHANISMS

The three-receptor hypothesis is the essential basis of the Young-Helmholtz trichromatic theory of colour vision; in essence it has come to mean today that the retina contains three types of cone that respond differently to the same wavelengths; thus a 'blue' cone would respond most prominently to the short end of the spectrum, and only a little to lights of the middle region and perhaps not at all to very long wavelengths; a 'green cone' would respond preferentially to the middle regions, and so on. Each mechanism, or cone-type, may therefore be regarded as having its own spectral sensitivity curve, and we may look on the spectral mixture curves of Figure 14.8 as being related to the spectral sensitivity curves of the separate mechanisms. Thus, to refer to the figure, we might say that the yellow light of wavelength 600 nm had excited the green mechanism by 0·25 units and the red mechanism by 0·4 units and had failed to excite the blue mechanism at all. Thus the match occurred because the yellow light did exactly what the green and red lights would have done if they had fallen simultaneously on the retina. On this basis, then, we could plot three sensitivity curves for the three mechanisms, using as ordinates the frequency of discharge, say, in the nerve fibres leading from these fundamental mechanisms as in Figure 14.9.

---

* The contribution of blue to any mixture is very small, in terms of luminosity, so that the values have been multiplied by ten in the Figure.

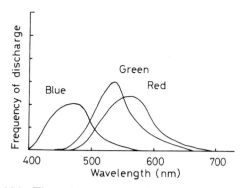

**Fig. 14.9** Theoretical wavelength-response curves for three hypothetical receptors maximally sensitive to red, green and blue respectively.

### SPECTRAL SENSITIVITIES

The next problem, of course, is to decide on the true spectral characteristics of these mechanisms; the colour-mixing phenomena tell us we can use any three primaries, so we can have an infinite number of sets of spectral mixture curves. Which, if any of these, is the correct set? Various methods were used to derive the true fundamental response curves, as they were called.

## Adaptation

For example, Wright used as his criterion the effects of adaptation. He has shown that if the eye is adapted to white light the sensitivity to all wavelengths is depressed in proportion; however, if the adapting light is coloured, there is a certain preferential depression of sensitivity to this colour, so that a subject who has been adapted to red, for instance, requires more red to make a match with white than he did before the adaptation. If the adapting light is yellow, both the red and green sensations are depressed, and so on. Thus if there are, indeed, three fundamental sensations, it should be possible to adapt the eye to at least one or two hues such that sensitivity to these alone is depressed, since these alone should be stimulated. Wright found that a blue of 460 nm and a red of 760 nm behaved in this way, and he suggested that these were two of the primary sensations; violet he concluded was definitely not a fundamental sensation since it was considerably modified by adaptation of the eye to red light. The wavelength corresponding to the third sensation, which from its position cannot be stimulated alone, was deduced to have the value of 530 nm, i.e. in the green. On the basis of this reasoning Wright drew a set of spectral sensitivity curves.

## Colour-blind

Other workers studied the so-called colour-blind, and on the assumption that their colour defect was the consequence of one of the mechanisms' being absent

(p. 335), they were able to deduce, from the colour-blind person's matching phenomena, the characteristics of the spectral sensitivity curve of the missing mechanism, and thence that of the remainder.

## TWO-COLOUR THRESHOLD TECHNIQUE

### LIMINAL BRIGHTNESS INCREMENT

Stiles developed an extremely ingenious technique for the identification of separate cone mechanisms in the human eye. The method is based on the measurement of increment thresholds, or liminal brightness increments (l.b.i.). It will be recalled that the l.b.i. for white light is measured by the extra luminance of a test patch, $\Delta I$, necessary for it to be distinguished against its white background of luminance $I$. By successively increasing the background luminance, $I$, we get a series of steps, $\Delta I$, and by plotting log $\Delta I$ against log $I$ we find that the graph has a characteristic shape. Thus, over the range of scotopic vision it may be represented by Figure 14.10; at very low luminance, the increment threshold, $\Delta I$, is approximately constant and equal to the absolute threshold; at a certain point the graph becomes a straight line of slope unity,[*] and finally there is a point where the rod mechanism is said to be saturated.

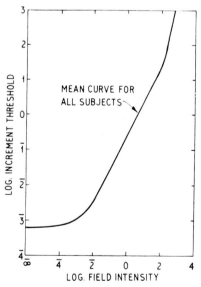

**Fig. 14.10** Showing the variation of the liminal brightness increment (l.b.i. or increment threshold) with field intensity. For details see text. (After Aguilar and Stiles, *Optica Acta.*)

---

[*] It is over this range that the Fechner fraction, $\Delta I/I$, is constant; so that we have the relationship:

$$\log \Delta I = \log I + Constant$$

so that with a logarithmic plot we should have a straight line with slope equal to unity.

ROD AND CONE MECHANISMS

This type of graph is only obtained, however, if a single mechanism is operating over the whole range of luminance, i.e. if the rods only are active; and the type of curve shown in Figure 14.10 could only be obtained by using special techniques to keep the cones from coming into play. In general, then, when the luminance comes up to a certain point the cones become the more sensitive and they now determine the l.b.i., and the graph of log $\Delta I$ against $I$ shows a characteristic break, or kink, indicating the taking over of the new 'mechanism', in this case the taking over of the threshold by cones in place of rods. This is illustrated by Figure 14.11. In this way we have identified, then, the

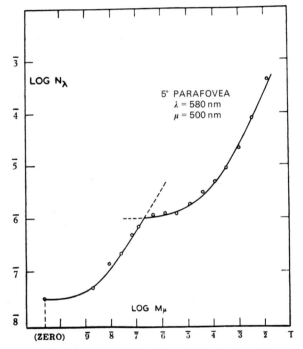

**Fig. 14.11**  The curve obtained by plotting the logarithm of the liminal brightness increment ($N\lambda$) against logarithm of the background intensity ($M\mu$). In this case the background was blue-green and the central patch a yellow of 580 nm. The left part of the curve represents increment thresholds for rod vision; at the break and beyond, the increment thresholds are determined by the cones. $N\lambda$ and $M\mu$ are expressed as erg sec$^{-1}$ (degree of arc)$^{-2}$. (Stiles, *Proc. Nat. Acad. Sci. Wash.*)

operation of a new mechanism. By varying the wavelength of the background, and of the test-stimulus, we may create conditions that favour one or the other mechanism and so study it over a large range of luminance; e.g. by using a red background and measuring the liminal brightness increment with a blue test-stimulus we may suppress cone activity and study rod

activity. In this way, then, we may not only identify the presence of rod and cone mechanisms but we may also measure their spectral sensitivities, using the l.b.i. as a measure of threshold for any given wavelength, care being taken that the background and test-stimulus are such that only one of the two mechanisms is operative.

CONE MECHANISMS

The same technique can be applied to identify different cone mechanisms, and Stiles found, indeed, that under appropriate conditions of background and test-stimulus the presence of three mechanisms could be established with maximum sensitivities in the red, green and blue. Subsequent studies indicated that the blue mechanism is complex, and can be described as being made up of three, $\pi_1$, $\pi_2$, and $\pi_3$, whilst the green and red mechanisms are called $\pi_4$ and $\pi_5$ respectively.*

### Composite nature of the mechanisms

Since the trivariance of colour vision is so well established, and since, as we shall see, direct experiments on retinal cones indicate unequivocally the presence of only three cone types, the Stiles mechanisms cannot be equated with the fundamental receptors, but rather with a feature of the integrating system. Pugh has examined the blue-sensitive mechanisms in detail and has shown that, over the short wavelength range of its action-spectrum, it behaves as one would expect of a single blue-sensitive type of cone; over longer wavelengths Grassman's law of additivity fails, indicating the mixing of activity of another type of cone with its region of maximal sensitivity at longer wavelengths. Pugh suggests that the composite $\pi_1$ mechanism is generated in the shoft wavelength-sensitive (blue) cones, but its signal must pass through two gain-stages, the gain in the first stage being controlled by the blue-cones alone and the gain in the second stage being controlled by a signal originating in the middle (green) or long wavelength (red) cones, or both.

## COLOUR-DEFECTS

Since the study of the colour-defective human subject (usually described somewhat inaccurately as the colour-blind) has thrown some light on the more general aspects

---

* Stiles' increment-threshold technique is one of several that have been employed to isolate the individual mechanisms; Brindley's studies on adaptation to very strong monochromatic stimuli are described later (p. 377); the blue mechanism may be identified by virtue of its much lower critical fusion frequency for flicker (Brindley, Du Croz and Rushton, 1966), and the greater area over which Ricco's Law applies, (12–18′ compared with 4′ for red and green), suggesting that the blue-sensitive cones are connected to a more diffuse type of bipolar (Brindley, 1954).

of colour discrimination, it is right that we devote some space to the main experimental findings at this juncture. In general, a colour-defective subject is one whose powers of discrimination between lights of different wavelengths are more limited than normal; the extent of this defect may range from complete failure of wavelength discrimination—*monochromatism*—through *dichromatism*, where the discrimination is so limited that we must postulate the absence of one of the three fundamental mechanisms, to the *colour-anomalous*, where the defect is a difference from average behaviour, in the sense that the protanomalous uses more red than the average in matching a yellow with a mixture of red and green.

## ROD MONOCHROMATISM

The more frequently occurring type of completely colour-blind is one whose cones are apparently pathologically defective; the subject is therefore called a *rod monochromat*, and we have already seen that his visual functions are largely what one might expect on the basis of the duplicity theory; thus he shows no Purkinje phenomenon, having a spectral sensitivity curve corresponding only to the scotopic type. Such monochromats occur with a frequency of about one in thirty thousand.

### VISUAL CHARACTERISTICS

There is by no means complete agreement as to the essential feature, if there is one, of the so-called rod monochromat; being more common than the cone monochromat, the rod monochromat is called 'typical', but various studies indicate that all the features of this condition cannot be forced into a single type. Walls and Heath have emphasized the complexities in some of the visual characteristics of the 'typical' achromat; thus we have already seen that the plot of critical fusion frequency against luminance shows a break, as in normals, suggesting the operation of two types of receptor, and sometimes the dark-adaptation curve also shows a 'kink'. These findings led v. Koenig and later Hecht to postulate the presence of 'photopic rods' in the retina of the monochromat. In general, we may regard 'typical' achromatopsia as a retinal dystrophy with the following signs, not all of which need be present in any subject: (a) achromatic vision; (b) low visual acuity, of the order of 0·1 to 0·3, better at moderate luminance than at high; (c) no Purkinje shift; (d) avoidance of very bright lights; (e) nystagmus; (f) some signs of macular dystrophy. The post-mortem study of Larsen showed that cones were indeed present in the retina, with most of them apparently normal.*

### THREE TYPES

Walls and Heath suggested that some cases could be accounted for on the assumption of the absence of functional cones together with the presence of rods in the fovea, whilst others could be described as containing, in addition to rods, the 'blue-type' cones postulated by trichromatic theory, the 'red' and 'green' cones being non-functional or absent. More recently Blackwell and Blackwell have reported on a study of nine subjects whose psychophysical findings allowed them to be placed in three categories: (a) typical rod monochromats with all visual functions determined by rods; (b) a group with a scotopic luminosity curve very similar to normal except in the blue where sensitivity was low, possibly due to the presence of the yellow xanthophyll macular pigment (they were using central fixation); this suggested that these achromats possessed rods in the fovea. At photopic luminances the sensitivity curve shifted towards the short-wave end of the spectrum, to give a maximum at about 440 nm, suggesting the coming into play of 'blue cones'. The third group were intermediate between (a) and (b).

## CONE MONOCHROMATISM

The cone, or atypical, monochromat is extremely rare, occurring with a frequency of only one in a hundred million; in this condition visual acuity is perfectly normal, but there is no discrimination of coloured lights if their luminosities are made equal. It might be thought that the retina, in this condition, contained only one type of cone, but if this were so we should expect the photopic spectral sensitivity curve to be abnormal. Thus, under photopic conditions the sensitivity of the eye to the different wavelengths must be the average sensitivity of the three or more cone mechanisms in the retina; removal of two of the mechanisms would leave the retina with only one type, and now the spectral sensitivity curve should coincide with the sensitivity curve of a single mechanism. In fact, Weale found the photopic sensitivity curve of the cone monochromat to lie within the normal range. When Weale (1959) applied his foveal bleaching techniques to a cone monochromat (p. 338), he found that the difference-spectrum for light reflected from the fundus indicated the presence of at least two cone mechanisms, and Gibson (1962), using Stiles' incremental threshold technique, found that three mechanisms were operative. Finally Fincham (1953) showed that a cone monochromat was able to make use of coloured aberration fringes in the accommodation reflex (p. 485), so that all the evidence indicates that the defect is not in foveal pigments but in the central integrating mechanism.

## DICHROMATISM

### COLOUR-MATCHING

The dichromat is one who can match all colours with suitable mixtures of only two, instead of three, primaries. The colour mixing equation will therefore be of the form:

$$C = \alpha A + \beta B$$

---

* Glickstein and Heath (1975) found no cones in the fovea of their monochromat with very few in the periphery.

where $A$ and $B$ are the primaries; instead of a pair of rectangular co-ordinates being required to represent their colour mixing data (p. 190), the latter may be exhibited in the form of a straight line where any point on this line corresponds to a definite mixture of the two primaries. It will be clear, without any mathematical analysis, that the range of colours that a dichromat can appreciate as distinct will be very restricted compared with the range of the normal individual, since it is obviously possible to obtain more variations by mixing three things than two.

### MISSING MECHANISMS

According to the classical interpretation, the *protanope* is missing the fundamental red sensation—he is red-blind; the *deuteranope* lacks the green sensation and the *tritanope* the blue. If this viewpoint is correct, we may anticipate the appearance of the spectrum to the protanope, for example, from a consideration of the curves in Figure 14.8, and imagining that the curve corresponding to the red sensation has been removed. In the region between 760 and about 650 nm no colour sensation should be experienced, i.e. the protanope's spectrum should be shortened; between 650 and about 500 nm the sensation should be that of green only (although it may be called yellow); with the introduction of the blue sensation at about 500 nm, changes in hue should be appreciated and colour discrimination should be approximately normal.

### HUE-DISCRIMINATION CURVES

Thus the curve, obtained by plotting the difference in wavelength necessary for a dichromat to appreciate a change in hue, as ordinates, against the wavelength, as abscissae (the hue discrimination curve), should be much simpler than that of a trichromat; in Figure 14.12 are shown typical curves for the protanope and deuteranope (that of the normal trichromat is shown in Figure 7.4, p. 188); it is seen that the maximum discrimination occurs in the region 480 to 500 nm where the blue mechanism is operative. Discrimination falls away rapidly on both sides of this point as we should expect.

### NEUTRAL POINT

The dichromat can match all hues (spectral and non-spectral and including white) by appropriate mixtures of two primaries; since white can be formed by a certain mixture of these primaries it follows that there will be a certain point in the spectrum which will appear to the dichromat as white—this point is called the *neutral point* and occurs at 495 nm with the protanope and at 500 nm with the deuteranope.

### ISO-COLOUR CHART

We have seen that for the normal trichromat the locus of spectral colours is given by a curved line in the colour chart, where each point represents by its co-ordinates the coefficients in the unit colour equation; by joining the two ends of this curve, to make a closed figure, we get the locus of certain

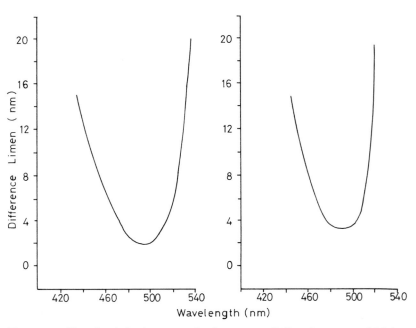

**Fig. 14.12** Hue-discrimination curves for deuteranope (left) and protanope (right). The smallest change in wavelength, necessary to produce a change in the sensation of hue, has been plotted against the wavelength. Greatest sensitivity is in the blue-green region (Wright and Pitt.)

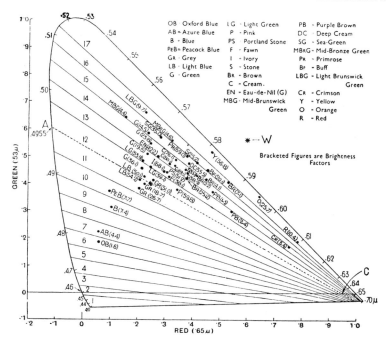

**Fig. 14.13** Iso-colour chart for protanope. (Pitt, *Med. Res. Council Sp. Rep.* By permission of the Controller of H.M. Stationery Office.)

non-spectral hues obtained by mixing blue and red, the purples (Fig. 7.6, p. 190). All points within this closed figure represent non-spectral hues appreciated by the trichromat. The locus of spectral colours for the dichromat is a straight line only, and since this cannot enclose space, all colours appreciated by the dichromat are spectral hues. This means that for all those numerous hues lying within the normal trichromatic locus, which are appreciated by the trichromat as different, there will correspond only a few spectral colours in the dichromat's colour range; in other words, to the dichromat, for any given spectral hue there will be a large number of non-spectral hues that are indistinguishable from it.

A determination of the colour-mixing data for dichromats permits the construction of iso-colour charts; in Figure 14.13 that for a protanope is given. The normal trichromatic spectral locus is drawn in the usual manner, and trichromatic white is indicated by a star. The neutral point of the protanope, i.e. the point on his spectral locus where the mixture of his two primaries gives white, lies at 495 nm; and an iso-colour line can be drawn through this point A, and the star representing trichromatic white, since the latter is appreciated as white by the protanope too. The line cuts the red axis at C, which corresponds to a very reddish purple. All colours represented by this line, ranging from blue to reddish purple, appear to the protanope as identical and white. From the data on the spectral hue discrimination of the protanope (Fig. 14.12), it is possible to mark off points on the spectral locus corresponding to just discriminable differences in hue, and the directions of the iso-colour lines passing through these points may be calculated from colour-mixing data (it will be remembered that between 530 and 760 nm all spectral colours appear the same to the dichromat). The iso-colour lines are drawn in the figure, and they divide the trichromatic colour area into a series of zones; all trichromatic hues represented by points in any one zone are indistinguishable by the dichromat, provided the luminosity is kept constant. For the protanope there are seventeen zones, whilst the deuteranope has some twenty-seven. The iso-colour charts permit one to

determine whether any pair of hues can be discriminated by a dichromat, and provide a rational interpretation of the numerous and hitherto uncorrelated facts of dichromatic matching and errors in discrimination.

### LUMINOSITY CURVES

If dichromatism does, indeed, represent the loss of one of the cone mechanisms we may expect the photopic luminosity curve to be different from normal in dichromatism. The regions of maximum luminosity are, in fact, shifted; that for the protanope lying at about 540 nm and for the deuteranope at about 570 nm. Furthermore we should expect the loss of the red mechanism to cause a shortening of the visible spectrum, and in the protanope this is indeed found, the reds at the end of the spectrum appearing grey. Again the loss of one of the 'colour mechanisms' presumably corresponds to the dysfunction on one group of cones; in this case we may expect a loss of absolute luminosity, i.e. the thresholds for vision in the dichromat should be generally higher than for the trichromat. With the protanope this is very well established in the red end of the spectrum and, according to Hecht, both protanopes and deuteranopes have a general loss of sensitivity over most of the spectrum, the average rise in threshold of the protanope being 49 per cent and that of the deuteranope 39 per cent. Subsequent studies on a uniocular deuteranope, i.e. on a subject with one eye normal and the other deuteranopic, showed that this difference in threshold was quite definite (Graham and Hsia, 1958).

### DEUTERANOPE

Pitt's analysis of deuteranopia suggested, however, that this could not be simply accounted for by the loss of the green mechanism; rather the deuteranope behaved as though he had, in place of the red and green mechanisms, a single one corresponding to a weighted mean of the normal red and green mechanisms; the deuteranope, on this basis, is indeed a dichromat, in the sense that his colour-mixing data can be expressed as a function of only two variables, but he behaves as though his red and green mechanisms are in some way linked so that they cannot act independently. More recent work, however, has shown that the deuteranope does, indeed, lack the green mechanism (p. 339).

### TRITANOPE

The tritanope is rare, constituting, according to Wright (1952), only one in 13 to 65 000 of the population by comparison with one in a hundred males for protanopia and deuteranopia. The tritanope seems to lack the blue mechanism, so that his discrimination is good in the red-green part of the spectrum but bad in the blue-green region; the neutral point is in the yellow at 575 nm. By employing Stiles' increment-threshold technique, Cole and Watkins (1967) showed that the tritanope lacked all three of the 'blue mechanisms', namely $\pi_1$, $\pi_2$ and $\pi_3$.

## ANOMALOUS TRICHROMATS

Here the subject requires three primaries to match all colours, but uses different proportions from those of the average; frequently this deviation from the average or 'normal' is associated with defective discrimination, but not always, according to Pickford. The abnormalities may be roughly classified as *protans*, *deuterans* and *tritans*; thus a protan is a trichromat who requires more red to match a yellow than does a normal trichromat. The anomalous trichromat differs from the dichromat in that he will not accept matches that the normal makes; thus the protanope matches a blue-green with white, but he will also accept as white a mixture of this blue-green with red. The protan will match a spectral yellow with an orange, i.e. he requires more red for his match than normal, but he will not accept the normal's red-green mixture as a yellow. Anomalous trichromasy is common, amounting to nearly 6 per cent of the male population.

## CONE PHOTOPIGMENTS

We have seen that, by the method of reflexion-densitometry, Campbell and Rushton and Weale were able to demonstrate the presence of rhodopsin in the human and animal eye, and to measure the changes that occurred during exposure to light. In the fovea, too, these authors have demonstrated the presence of two pigments absorbing preferentially in the red (*erythrolabe*) and green (*chlorolabe*) parts of the spectrum. The existence of a cone pigment was first established by exposing the eye to a strong orange light, which might be expected to bleach this pigment in preference to rhodopsin; by subsequently measuring the absorption of yellow light after reflexion from the fovea, a curve for regeneration of cone pigment was obtained. This regeneration was much faster than that of rhodopsin, being complete in 5 to 6 minutes.

### ERYTHROLABE AND CHLOROLABE

A study of the effects of bleaching with different wavelengths indicated the presence of two pigments, and these were isolated by studying a protanope who, according to classical theory, is missing the red mechanism and thus might be expected to have one fewer pigment. The curve of Figure 14.14 shows the

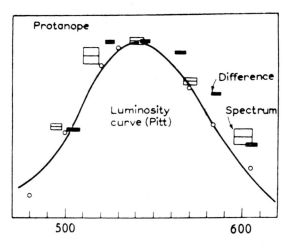

**Fig. 14.14**  The black and white rectangles represent the difference spectrum of the foveal pigments in the protanope, and presumably therefore correspond to the absorption spectrum of the green-sensitive pigment, chlorolabe. The curve represents the luminosity curve of the protanope as determined by Pitt, whilst the circles represent the actual luminosity curve of one of the subjects studied by reflexion densitometry. (Rushton, *Progress in Biophysics.*)

absorption spectrum of what is apparently a single photopigment in the protanope's fovea, and corresponds remarkably well with the photopic spectral sensitivity curve of the protanope, as determined by Pitt.

The peak absorption of this green-sensitive pigment was near 540 nm, and was called *chlorolabe*. In effect the retina of the protanope, under the conditions

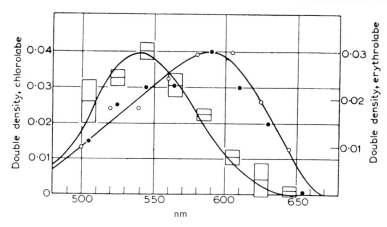

**Fig. 14.15**  Difference spectra of two pigments in the normal fovea. The left curve represents chlorolabe and the other erythrolabe. (Rushton, *Progress in Biophysics.*)

studied, was behaving as though only one pigment were present, and it was found that lights of different wavelengths, matched by the protanope as equally bright by means of flicker photometry, were, indeed, equally effective in bleaching the foveal photosensitive pigment. The failure to detect blue-sensitive pigment was presumably due to the relatively small contribution the blue mechanism makes to luminosity.

The deuteranope, like the protanope, fails to discriminate hues in the red-green part of the spectrum, and his colour-mixing data suggest that he lacks a green mechanism, so that, like the protanope, he has only a single mechanism, or pigment, operating in the red-green part of the spectrum. Rushton (1965) showed that, over this range, only a single pigment absorbed light significantly, and its absorption maximum corresponded reasonably well with the maximum of the deuteranope's luminosity curve. Rushton showed that the effects of bleaching the retina with different coloured lights were not consistent with the presence of a mixture of red- and green-absorbing pigments, as had been at first thought to be the case, so that we can conclude that the deuteranope is red-green blind because he lacks a green-absorbing pigment, *chlorolabe*, and not because, in some way, his red and green cones have become 'mixed'. In the normal eye we may expect to demonstrate the presence of both erythrolabe and chlorolabe; and this was in fact done by Baker and Rushton (Fig. 14.15) with the use of selective bleaching techniques.

### REDUCTION HYPOTHESIS

This so-called 'reduction hypothesis' to account for dichromacy has been generally accepted, although an alternative mechanism of fusion of the red- and green-sensitive mechanisms, either neurally or by their physical presence in the same cone, has been maintained, e.g. by Hurvich (1972). For example, certain deuteranopes have a normal luminosity curve which

might be interpreted on this fusion hypothesis. However, Alpern *et al.* (1968) showed that, when a normal luminosity curve was encountered in a deuteranope this was due to a deficiency in macular pigment.

### CORRECT DIAGNOSIS

Alpern[*] has re-investigated this matter, emphasizing that the important point to prove by retinal densitometry is that there is only one pigment active in the medium and long-wavelength regions of the spectrum, compared with two in the normal trichromat. An important point raised by him is the correct diagnosis of dichromacy; the Nagel anomaloscope, by which a subject matches a yellow with different proportions of red and green light, is highly sensitive in discovering dichromacy and anomalous trichromacy but, by itself, is insufficient, and two additional tests, including determining the neutral point, are recommended.

### VARIABILITY IN RED-GREEN SENSITIVITY

A further point, revealed by the experiments (Alpern and Wake, 1977) is the variability amongst individual protanopes and deuteranopes with respect to the ratio of spectral sensitivities to red and green, as determined on the anomaloscope. Examination of the absorption spectra of the protanope's erythrolabe and the deuteranope's chlorolabe shows that the relative sensitivity to green and red will be very different, the ratio $V_{645}/V_{535}$ being greater for the erythrolabe-containing eye (deuteranope) than for the chlorolabe-containing eye, (protanope). The fact that this ratio varies amongst the dichromats, as indicated in Figure 14–16, suggests that the absorption-spectrum of the pigment can vary with individuals, unless, of course, the variation is due to such causes as the presence of

[*]Alpern and Wake (1977), Alpern and Pugh (1977), and Alpern and Moeller (1977).

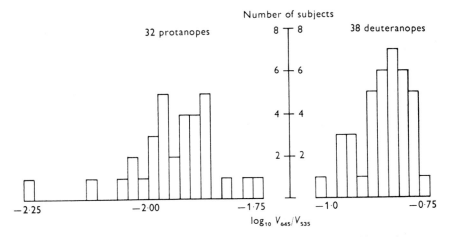

**Fig. 14.16** Histogram of the logarithm of the ratio of the spectral sensitivities at the anomaloscope primaries (645 nm and 535 nm) of seventy red-green dichromats. To condense the entire distribution to a single Figure, the hiatus in the abscissa between the two clusters has been drawn to a scale reduced by 0·22 that of the rest of the Figure. (Alpern and Wake, *J. Physiol.*)

colour-filtering pigments in the eye, or to the fact that in the dichromat there are, in fact, varying proportions of the two pigments in the individual cones.*

DIFFERENCE-SPECTRA

In Figure 14.17 are shown densitometrically determined difference-spectra indicating the effects of bleaching with red light (circles), with white light (triangles), and by bleaching first with red and then applying a further bleach with green. The deuteranopes are revealed as showing no difference-spectrum after bleaching with red followed by green, indicating that only one pigment is sensitive in the region of the spectrum studied; after a red bleach, sufficient of the pigment remained to give a transmission curve from which the difference-spectrum was computed (circles) showing a maximum at about 565 to 605 nm, the absorption-maximum of the postulated erythrolabe or red-sensitive pigment in the deuteranope's retina. With the normal trichromat, of course, with its two pigments absorbing over the medium and long-wavelength bands of the spectrum, bleaching with red and subsequently with green will leave a difference spectrum corresponding to the difference in absorbing qualities of the red- and green-sensitive pigments.

VARIABILITY IN ABSORPTION SPECTRUM

This study thus confirms the absence of chlorolabe in the deuteranope. The variability in the dichromat's colour mixing, and other parameters correlated with this, was investigated in some detail and it was suggested that it could well be due to a variability in the absorption spectrum of one of the visual pigments, analogous with the variability observed with rhodopsin

from different frogs (Bowmaker *et al.*, 1975). Such a variability would, of course, be expected in the absorption spectra of the cone pigments in the normal trichromat and would account for variability in colour mixing amongst individuals.

ANOMALOUS TRICHROMACY

The generally accepted view of this condition is the presence of one 'abnormal' pigment together with two normal ones; e.g. an 'abnormal' chlorolabe in the deuteranomalous trichromat, the 'deuteranolabe' of Piantanida and Sperling (1973). However, Alpern's study on the colour-matching and hue-discrimination of deuteranomalous subjects indicated that a deuteranomalous subject may have, in both green-sensitive and red-sensitive cones, the same pigment—erythrolabe —but differing in its extinction spectrum. In other words, instead of regarding the visual pigments as invariant molecules with fixed extinction spectra, Alpern suggests the existence of a 'cluster' of erythrolabes, a cluster of chlorolabes and a cluster of cyanolabes. Thus, on this basis, deuteranomaly and deuteranopia would represent the same defect but the deuteranomalous subject would retain a greater sensitivity to the middle wavelengths by virtue of cones containing erythrolabe with $\lambda_{max}$ closer to the average of the chlorolabe $\lambda_{max}$.

---

* Individual variations in colour-mixing function, whether encountered in normal trichromats or in the colour-defective, can often be explained on the basis of variations in transmissivity of the ocular media, etc. (see, for example, Pokorny *et al.*, 1973). The elaborate studies of Alpern, however, have ruled out such an explanation for the variations in computed action-spectra, and therefore of extinction spectra, observed in his colour-defective subjects.

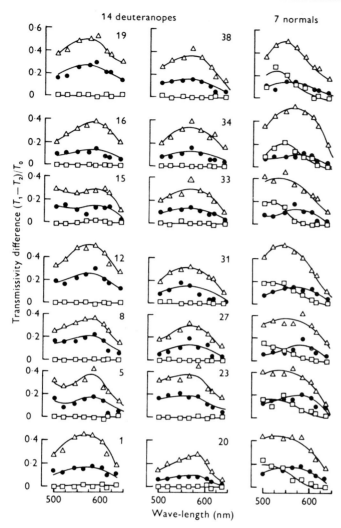

**Fig. 14.17** Difference spectra resulting from bleaching with red light (circles), white light (triangles) and with first red and then green (squares). The number on each deuteranope's graph indicates his position in the distribution of Figure 14.16. (Alpern and Wake, *J. Physiol.*)

Thus it is argued that red-green dichromats are merely deutans or protans in whom the long wavelength ('Red') cones have the same pigment as the medium wavelength ('Green') cones, and anomalous trichromats are subjects in whom the spectra of these pigments (always from the same cluster) differ, if ever so slightly.

## ABSORPTION SPECTRA OF CONES

A more precise characterization of the types of pigment in cones is given by measuring the absorption spectra of individual cones exposed to a minute pencil of light on a microscope slide. The pioneering study was made by Hanoaka and Fujimoto in 1957, using the large fish cones, and it was shown that different cones had different absorption spectra, with maxima in the regions of 500, 535, 575 and 620 nm. More recent studies of

Marks (1965), also on fish cones, gave maxima at 455, 530 and 625 nm, i.e. in the blue, green and red parts of the spectrum. Subsequent refinements of technique have permitted the study of the smaller human and monkey cones; Figure 14.18 shows the presence of three types of cone with $\lambda_{max}$ at 450, 525 and 555 nm (Brown and Wald, 1964); essentially similar results were found by Marks, Dobelle and MacNichol (1964), their maxima occurring at 445, 535 and 570 nm. These studies thus confirm in a striking manner the predictions of the trichromatic theory derived from colour mixing.

## EXTRACTED CONE PIGMENTS

Iodopsin with a $\lambda_{max}$ of 576 nm has generally been accepted to represent a cone pigment, presumably corresponding to the 'red receptor' of birds, since

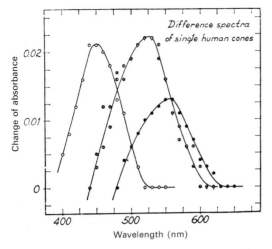

**Fig. 14.18** Difference spectra of the visual pigments in single cones of the human parafovea. In each case the absorption spectrum was recorded in the dark, then again after bleaching with a flash of yellow light. They involve one blue-sensitive cone, two green-sensitive cones, and one red-sensitive cone. (Brown and Wald, *Science*.)

**Table 14.1** Proportions of the three photopigments extracted from the retina of *Coryphaena* in light- and dark-adapted retina.

|  | Light-adapted |  | Dark-adapted |  |
|---|---|---|---|---|
| $\lambda_{max}$ | Per cent | | $\lambda_{max}$ | Per cent |
| 521·1 | 54 | | 521·1 | 19·8 |
| 498·9 | 31·6 | | 498·9 | 73·4 |
| 469·2 | 14·4 | | 469·2 | 68·0 |

## EVOLUTIONARY SIGNIFICANCE

The development of colour vision has survival value in so far as it permits the resolution of contrasts that are not apparent to the monochromat animal. To the predator fish there are two main visual problems according as it hunts from above its prey or not. If viewing its prey from above it is seen in contrast with the down-welling light reaching the depth at which it is hunting; if the absorption of its visual pigment is matched to this surrounding light the contrast provided by its prey is at a maximum and in this way we can explain the shift of $\lambda_{max}$ to shorter wavelengths in fishes in an environment in which the light is bluish (p. 222). If viewing the target horizontally the target may be brighter or darker than the surroundings, according as the surface is reflecting or non-reflecting, so that a visual pigment with $\lambda_{max}$ offset from that of the surrounding light will give better contrast for a reflecting surface, whilst a pigment matching the light is better for the non-reflecting prey. The development of a red-sensitive visual pigment may thus be regarded as an adaptation that increases the contrast of a prey that is reflecting the blue light of the underwater environment. These points are illustrated by Figure 14.19, which shows the contrasts between prey and environment under different conditions and as manifest to eyes with blue-sensitive or red-sensitive visual pigments.

As McFarland and Munz have pointed out, the development of several photopic visual pigments in the tropical fishes not only represents an evolutionary step towards increasing the perception of contrast but it represents the evolutionary process leading to colour-vision, as such.

Liebman (1972) found with his microspectrophotometric measurements red-sensitive cones exhibiting $\lambda_{max}$ of 570 nm.

### FISH PIGMENTS

Munz and McFarland (1975) have made a systematic study of extracted pigments from fish retinae in the hope of identifying photopigments responsible for their wavelength-discrimination. They studied more than a hundred marine diurnal tropical fish, extracting photo-sensitive pigment from light-adapted eyes and thus presumably free from the scotopic visual pigment. In all but two of the species studied, the mixture was a binary one, containing a red-sensitive ($\lambda_{max}$ 518 to 541 nm) and a less red-sensitive pigment ($\lambda_{max}$ 500 nm) differing in $\lambda_{max}$ by some 20 to 40 nm, the 'less-red' pigment probably being derived from rods. In two species the mixture was ternary containing, in addition, a more blue-sensitive pigment ($\lambda_{max}$ 480 to 500 nm) which was presumably a second cone-pigment. By cutting the retinae, containing these ternary mixtures, into quadrants, containing different proportions of rods to cones, it was established that the cone-poor regions had more of the 'middle pigment' i.e. that with $\lambda_{max}$ of about 500 nm and between that of the red-sensitive and the blue-sensitive pigments. If the middle pigment is, indeed, a rod pigment, its proportion should be much higher in the dark-adapted retina, and this was indeed the case, as Table 14.1 shows:

## ELECTROPHYSIOLOGY OF WAVELENGTH DISCRIMINATION

It is not, of course, sufficient to demonstrate the presence of 'mechanisms' by either subjective or by photochemical methods; the physiologist will only be satisfied when he can tap the messages whereby these

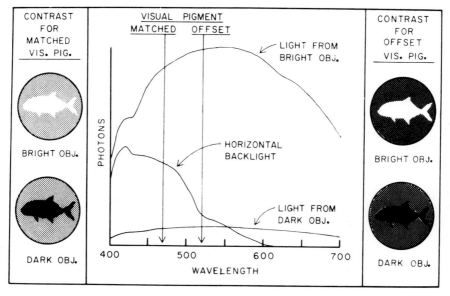

**Fig. 14.19** Comparisons of the relative photocontrast of 'bright' (reflective) and 'dark' (nonreflective) objects viewed along the horizontal line of sight at 1 m beneath the surface by fishes with visual pigments matched to ($P_{469}$), or offset from ($P_{521}$), the backlight. Both bright and dark targets are assumed to be grey. Light reflected into the horizontal field includes direct reflections of sunlight. Close to the bright objects, reflected light will be 50 to 100 times more intense than the radiance of the horizontal background (scaling not indicated on the ordinate). The insets portray the relative photocontrasts that would be produced by each visual pigment. The matched pigment enhances the contrast of a dark object and the offset pigment results in contrast enhancement of a bright target. (McFarland and Munz, *Vision Res.*)

mechanisms transmit their separate information. Thus, if there are red-, green- and blue-sensitive cones, these must respond in specific fashions to light of any given wavelength falling on the retina, and somehow these responses must be transmitted to bipolar cells and ganglion cells and pass along the optic nerves to the lateral geniculate bodies. The messages will, of course, be altered, but it is essential to appreciate that the neurones through which they are carried will be unable to *add* anything; the wavelength discrimination must have been carried out at the moment of reception of the light-stimulus, and subsequent alterations in the messages can only be described as a *re-coding* of the information, with perhaps the scrapping of some details and alteration of emphasis of the different parts. Thus, to consider the 'blue mechanisms', there might be three types of 'blue cone', but if their separate messages were all pooled in a single type of bipolar cell, the extra information would be lost. The fact that these mechanisms are shown to exist by psychophysical methods, however, proves that their special messages have indeed survived the re-coding processes so that it is difficult to escape the conclusion that there are several types of cone that are specially sensitive to blue light.

## INTRACELLULAR RECORDS FROM SINGLE UNITS

### S-POTENTIALS

In discussing the electrophysiology of the retina earlier, it was indicated that several workers had claimed to have recorded from individual cones by inserting a micro-electrode into the retina; some of these showed some interesting chromatic responses, namely the so-called *S-potentials* of Svaetichin in the fish retina*; subsequent work has shown that these are derived from horizontal cells, i.e. one step later than the receptors themselves. In general these responses to light were not spike discharges but graded potential responses consisting of either an increase of the normal resting potential, a *hyperpolarization*, or a decrease, *depolarization*. The responses were graded in the sense that the

---

*The fish retina has been studied with micro-electrodes because the cones are very large so that intracellular recording becomes practicable; moreover, the colour vision of fishes has been studied very extensively from a behavioural point of view, and there is no doubt that their colour discrimination is good and similar to that of man (see Svaetichin and MacNichol, 1958, for references).

change of potential increased with increasing intensity, whereas a spike potential is all-or-none in character.

## L- AND C-TYPES

According to the depth of the electrode in the retina the responses were of two main types; an L-type which responded in the same way with all wavelengths and therefore gave a 'luminosity message' (Fig. 14.20) and a

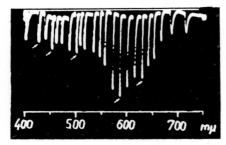

**Fig. 14.20** L-type of response of unit in fish retina; the polarity of the electrical change does not alter with varying wavelength of stimulus. (Svaetichin and MacNichol, *Ann. N.Y. Acad. Sci.*)

C-type which gave opposite electrical effects according as the retina was stimulated with red and green or blue and yellow lights (Fig. 14.21a, b). The messages carried by this second type thus indicated the chromatic nature of the stimulus, and the units were called *chromatic*, or C-units, being subdivided into *R-G* (red-green) and *Y-B* (yellow-blue) categories. Thus the C-potentials exhibit an 'opponent colour' basis so that a given element may be hyperpolarized by one band of wavelengths and depolarized by another band; if hyperpolarization indicates inhibition, and depolarization activation, then these opposite responses may indeed be described as opponent. An interesting characteristic of the S-responses was the very large receptive fields extending across the whole retina; in these fields summation was complete so that Ricco's Law applied perfectly (Norton *et al.*, 1968).

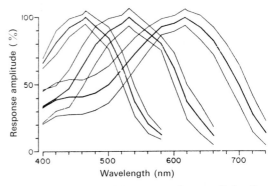

**Fig. 14.22** Responses of three types of cone to light of different wavelengths indicating maximum sensitivity in the blue, green and red parts of the spectrum. Thin lines indicate standard deviation curves. (Tomita, Kaneka, Murakami and Pautler, *Vision Res.*)

## CONE RESPONSES

### CARP RETINA

Of more interest are the responses of individual cones; and these have been described by Tomita *et al.* (1967) in the carp retina. The response to light was always a hyperpolarization, and when units were examined with light of different wavelengths, regions of maximum sensitivity were obtained in three broad regions in the blue, green and red. Figure 14.22 illustrates the averaged results; the peaks occurred in the following regions:

| Blue | Green | Red | |
|------|-------|-----|---|
| $462 \pm 15$ | $529 \pm 14$ | $611 \pm 23$ | Carp retina |
| $455 \pm 15$ | $530 \pm 5$ | $625 \pm 5$ | Goldfish |

these are compared with the absorption maxima of goldfish cones described by Marks (p. 341). The authors point out that their results provide no evidence for cones with maximal responses in the region of 680 nm, as deduced by Naka and Rushton from their analysis of S-potentials.

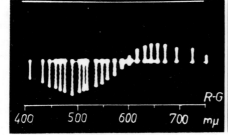
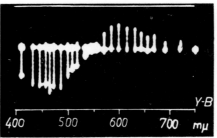

**Fig. 14.21** (*a*) C-type of unit, of the red-green variety, giving responses of opposite polarity according as the wavelength of the stimulus lies in the red or green regions of the spectrum. (*b*) C-type, of yellow-blue variety. (Svaetichin and MacNichol, *Ann. N.Y. Acad. Sci.*)

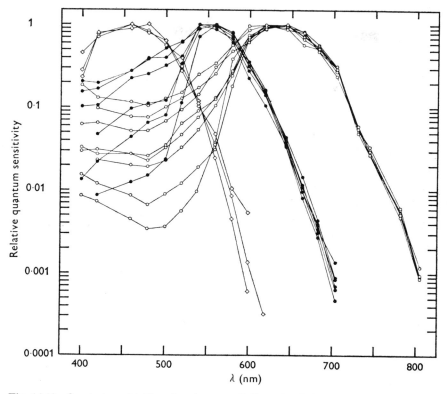

**Fig. 14.23**   Spectral sensitivities of turtle cones. Ordinate is relative quantum sensitivity; abscissa is wavelength. Collected results from seventeen cells. (Baylor and Hodgkin, *J. Physiol.*)

## TURTLE RETINA

Baylor and Hodgkin (1973) found three distinct classes of cones on the basis of their spectral sensitivities with maximal sensitivity at 630 nm (red-sensitive or red cones), 550 nm (green) and 460 nm (blue) respectively (Fig. 14.23). The maximum sensitivity of the green cones, namely 550 nm, is different from that of the rods (520 nm), although both receptors may well have the same porphyropsin photopigment based on dehydro-retinal with $\lambda_{max}$ of 518 nm (Liebman and Granda, 1971).

### Oil droplets

The shift of $\lambda_{max}$ to the red end of the spectrum in the cone is probably due to the orange oil-droplets in the green cones. Again, variations in wavelength-sensitivity amongst groups of red- and green-sensitive cones could well be due to the extent to which filtering by oil-droplets takes place.

### Univariance

When small spots of light were used as stimuli, the responses in a given cone were said to be *univariant*, meaning that the response at any wavelength and intensity could be matched by any other wavelength provided the intensity was adjusted appropriately, or more generally, the response at $I'$ and $\lambda_1$ could be matched by light of $kI'$ and $\lambda_2$ where $k$ is the same for all values of $I'$. With large spots of light, deviations were observed indicating some coupling of cones.

### COUPLING OF CONES

We have seen that cones are electrically coupled so that the stimulation of one may produce a similar response in cones within a radius of up to about 40 $\mu$m; this coupling is manifest in several ways, and Baylor and Hodgkin used the variation in response of a cone to changes in the position of the stimulating light as the criterion. If the cone acts in complete isolation the falling off in response as the light-spot is moved to one or other side of the central position should be symmetrical, whereas coupling results in bumps on the curves; thus Figure 14.24A shows the sensitivity-profile of a green-sensitive cone, and B that of a red-sensitive cone, and the bump in the profile at $+25$ $\mu$m in the one and at $-25$ $\mu$m in the other is evidence for

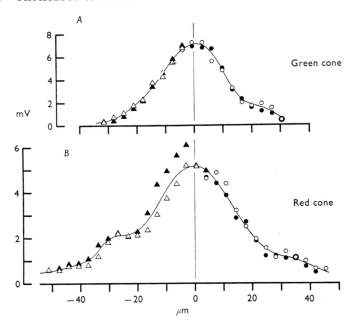

**Fio. 14.24** Coupling between cones. A is the profile of spatial sensitivity of a green cone being the response amplitude to a 7 µm spot of light as the spot is moved away from the central position. Light of 559 nm. There is a suggestion of a bump at +25 µm. B is profile for red-sensitive cone. There is a bump in the profile at −25 µm. White light. (Baylor and Hodgkin, *J. Physiol.*)

coupling with other cones. When the spectral characteristics of the responses were examined, it was concluded that coupling of cones must have been amongst cones of the same spectral sensitivity, i.e. green cones with green, red with red, and so on; this is reasonable since it would be pointless to develop specifically sensitive receptors and mix their responses at the very beginning. These studies involved generally small spots of light so that cone interaction was essentially due to electrical coupling; if spots greater than about 200 µm are employed, feedback from horizontal cells provides a complication as Baylor *et al.* (1971) showed (Fig. 10.32, p. 265).

FEEDBACK FROM HORIZONTAL CELLS

The feedback from horizontal cells tends to act in opposition to the direct action, so that the effect is not immediately evident until the directly effected hyperpolarization has decayed sufficiently for the depolarization to show (Fig. 10.32, p. 265). However, the direct effect can be eliminated by adapting the retina to a coloured light.

For example, red light is absorbed poorly by green-sensitive cones but is very effective on L-cells (p. 344), so that red light, applied over a large area, would activate red-sensitive cones directly but have little direct action on green-sensitive cones. This is illustrated by Figure 14.25 which shows the responses of a red cone, an L-cell, and a green cone to large fields of red (above) and green (below) light; the purely depolarizing action of the red light on the green cone is striking. Another mode of demonstrating feedback is to employ an annular

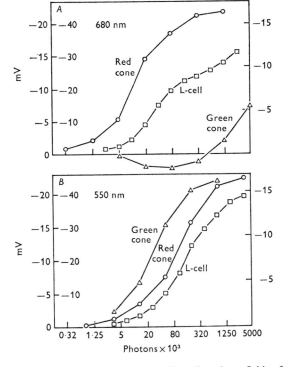

**Fig. 14.25** Responses of cones and L-cells to large fields of 1250 µm radius. The L-cell generated large responses which grew differently with stimulus strength for the two colours. The green cone produced depolarizing responses (points below the abscissa in *A*) for red flashes of moderate intensities but hyperpolarizing responses for brighter red flashes. The responses of the red cone to either colour and of the green cone to the green light were slightly larger than with the small field. (Fuortes *et al.*, *J. Physiol.*)

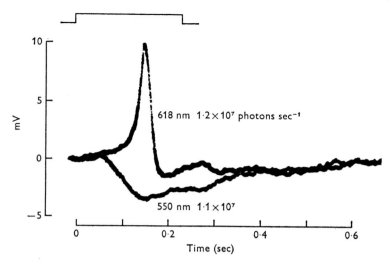

**Fig. 14.26** Responses of a green cone to annuli. Steps (monitored at top) of red or green light were applied to an annulus of 1250–500 μm radii. The green cone at the centre developed a large depolarizing transient for the red stimulus and was hyperpolarized when the light was green. Photo flux over 50 μm² is indicated near each record. (Fuortes et al., J. Physiol.)

stimulus; if this consists of an annulus of red light surrounding a green-sensitive cone, then if scatter of red light in the eye is small, the cone will be activated by the L-cells and exhibit depolarization. With green light in the annulus, on the other hand, probably because of scatter, the green-sensitive cone exhibits a hyperpolarizing response which is in marked contrast to the sharp depolarization caused by the red annulus (Fig. 14.26).

### DOUBLE CONES

In some vertebrates there are, in addition to the three cone types—red- green- and blue-sensitive—and rods, a fifth type, the double cone. This consists of two tightly joined elements; the 'chief' cone of the turtle, for example, contains an unknown photopigment and an oil droplet in its inner segment and the 'accessory' cone contains pigment $518_2$, i.e. the pigment found in the green-sensitive turtle cone (Liebman, 1972). Electrophysiological studies have shown that some retinal cones share electrical properties with both red and green sensitive cones, and since earlier work had shown that direct coupling of cones with each other was only between cones of similar type, it seemed that the receptors responsible for these mixed effects were the double cones. Richter and Simon (1974) made intracellular records from turtle cones, subsequently identifying them as double cones by injection of Procion yellow, which was found in both accessory and principal members. By contrast with single cones, the principle of univariance was not observed indicating the pooling of responses by receptors with different photopigments. The effects of variation of the wavelength of the stimulating light indicated green- and red-sensitive units, and the relative sensitivity to blue light was consistent with the absence of an oil-drop in the

accessory cone; the result was a very broad band of spectral sensitivity with peaks in the red and green regions (Fig. 14.27); by adapting with red light, the red peak could be suppressed, and with green light the green peak. The basis for the coupling of the cones could not be established, so that it might be through a chemical synapse or through a gap-junction.

### OIL-DROPLETS

In some species, of which the diurnal birds may be chosen as examples, the retinal cones contain, in their inner segments, coloured oil-droplets in such a position that they act as light-filters and therefore modify the spectral characteristics of the light reaching the outer segments. In the chicken there are at least three types, red, orange-yellow and yellow-green, when light-microscopy is employed, but Bowmaker and Knowles (1977), employing microspectrometric measurements, identified in all some six types of oil-droplets.

### Spectral characteristics

Figure 14.28 illustrates the spectral characteristics of five of these droplets, and it is evident that, with the exception of the clear droplet, they show a sharp increase in absorbance at specific regions, in fact they are behaving as cutoff filters for shorter wavelengths of light, Type A having a $\lambda_{T-50}$ (the wavelength at which 50 per cent transmission occurs) at about 454 nm, type C at about 520 nm and the red at about 585 nm; type $B_1$ has a shoulder and a $\lambda_{T-50}$ of 497 nm. The absorption spectrum of a given cone will depend on the visual pigment it contains together with the cone oil-droplet; according to Bowmaker and Knowles (1977) the chicken cones contain two photopigments with $\lambda_{max}$ at 569 and 497 nm respectively, so that the absorbance of a given cone may be predicted from the photopigment and oil-

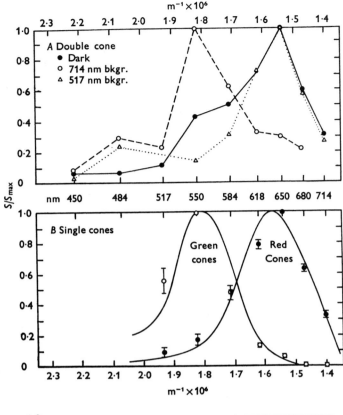

Fig. 14.27 Spectral sensitivities of single and double cones. A. Double cone. The responses to a flash applied from darkness (filled circles), over a 714 nm background (open circles) and over a 517 nm background (triangles). Note peak sensitivity in green and red elicited in this way. B. Spectral sensitivities of green-sensitive and red-sensitive single cones. (Richter and Simon, *J. Physiol.*)

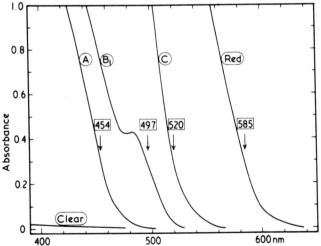

Fig. 14.28 Spectral absorbance of five types of oil-droplet from chicken cones. 'Clear', A, C and Red droplets occur in single cones; $B_1$ occurs in the chief member of the double cone. (Bowmaker and Knowles, *Vision Res.*)

droplet it contains. It must be appreciated, moreover, that many cones are double, namely 45 per cent, and these contain the $B_1$-droplet in the chief member and $B_2$ in the accessory member; the combined response will presumably be governed by the different absorption characteristics of the cone oil-droplets.

## Photopigment and droplet

There was some specificity regarding the association of

oil-droplet with photopigment; thus $P_{497}$ contained only Type C droplets, whilst $P_{569}$ cones were associated with red, type A, clear, or type B droplets. On no occasion was either the $P_{569}$ photopigment found associated with type C droplet or the $P_{497}$ photopigment with any other oil-droplet than the Type C. The computed effects of the oil-droplets on the absorptances of photopigment-oil droplet combinations are shown in Figure 14.29; thus the curve marked *Red* is the

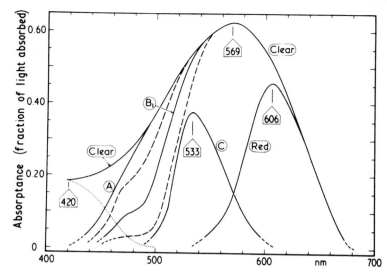

**Fig. 14.29** Computed absorptance for the photopigment-oil droplet combinations in the five recognized cone-types. Clear: absorptance of a cone having the clear oil droplet with $P_{569}$ in the outer segment; A: Type A droplet plus $P_{569}$; B: Type $B_1$ droplet plus $P_{569}$ (the dashed lines indicate the effect of the range of $B_1$ droplets found); C: Type C plus $P_{497}$; Red: Red droplet plus $P_{569}$. The dotted line indicates the possible response, maximal at 420 nm, derived by interaction between the 'clear' and 'Type A' cones. (Bowmaker and Knowles, *Vision Res.*)

absorptance due to a red oil-droplet plus $P_{569}$ giving a $\lambda_{max}$ at 606 nm, i.e. a shift in sensitivity towards the red. Curve C with $\lambda_{max}$ at 535 nm is derived from photopigment $P_{497}$ and Type C oil-droplet, once again showing a shift towards the red. A blue-sensitive photopigment has not been identified in the chicken retina, so that blue-sensitivity may depend on interaction of clear oil-droplet cone with Type A cones, a $\lambda_{max}$ of 420 nm can be predicted if both contain $P_{569}$.

## Relation to colour vision

This study of Bowmaker and Knowles on the chicken and another[*] on the pigeon (Bowmaker, 1977) prove conclusively that colour vision in the bird is not mediated by a single cone photopigment with variable droplet contents of individual cones; thus the oil-droplets *assist* colour discrimination (Muntz, 1972) to the extent that they sharpen the absorptance spectra of the cones, but they are unable to substitute for photopigments.

## OPPONENT COLOUR RESPONSES IN TURTLE HORIZONTAL CELLS

We have seen that turtle cones show an interaction that depends on feed-back from horizontal cells, this feed-back taking the form of a depolarization i.e. an electrical change that opposes the primary hyperpolarization in response to the action of light on the cone. As with the fish retina, horizontal cells could be classed as L-cells, showing little variation in response with wavelength of stimulating light, and C-cells which could be classified as R/G (65 per cent) and G/B (35 per cent).

L-CELL RESPONSES

As Figure 14.30 shows, the response is a smooth hyperpolarizing one, and when the total number of quanta were adjusted to be the same, the responses varied little with the wavelengths of stimulating light. The responses were not strictly invariant with respect to wavelength, and in some the variation was quite prominent, this was especially true of bright flashes. When the peak response was plotted against wavelength the sensitivity curve was similar to that of red-sensitive cones indicating that the L-cells were pre-dominantly connected with these. By exposing a cone to a background of light that it absorbs strongly its response becomes smaller, faster and often diphasic and its contribution to exciting the L-cell becomes correspondingly smaller. Using this means of sup-pressing the action of given cones, it was shown that the main input is indeed from red cones but some inter-mixture of possibly the double cones consisting of two tightly associated cones each containing one pigment-type (Liebman, 1972).

C-CELL RESPONSES

With regard to the C-cells, green or blue flashes applied to R/G cells gave smooth hyperpolarizing responses,

[*]In the pigeon Bowmaker (1977) identified by microspectro-metry three cone photopigments, $P_{567}$, $P_{514}$ and $P_{461}$, in addition to the scotopic rod pigment, $P_{503}$. By combining the absorbances of the photopigments and oil-droplets peculiar to given cone-types, he predicted response-curves to different wavelengths that coincided quite accurately with the modulator curves of Donner (1953), using Granit's single optic nerve fibre technique. Govardovskii and Zueva (1977), employing a technique of selective bleaching and its effect on the early receptor potential, have provided some evidence for four cone pigments in chicken and pigeon retina, these had $\lambda_{max}$ at 413, 467, 507 and 562 nm, but no allowance seems to have been made for droplet absorption.

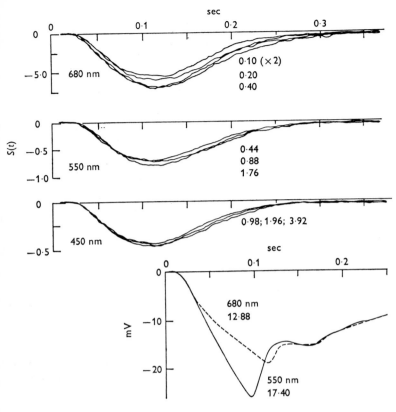

**Fig. 14.30** Responses of an L-cell to monochromatic flashes of three different intensities and three different wavelengths. The response is a smooth hyperpolarization and when stimuli are adjusted for equal numbers of quanta there is little variation with wavelength of the stimulating light. Inset: large responses of an L-cell to brighter flashes are clearly different for stimuli of 680 nm (dashes) and 550 nm (continuous). (Fuortes and Simon, *J. Physiol.*)

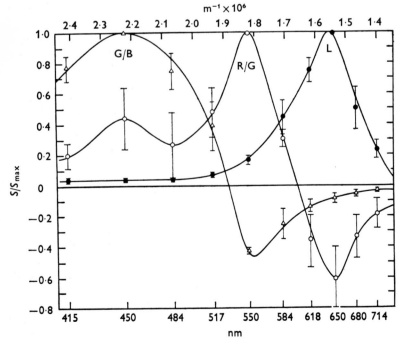

**Fig. 14.31** Spectral sensitivity curves of horizontal cells. Hyperpolarizing responses are plotted up and depolarizing responses are down. Results on 23 L-cells, sixteen R/G C-cells and four G/B C-cells. (Fuortes and Simon, *J. Physiol.*)

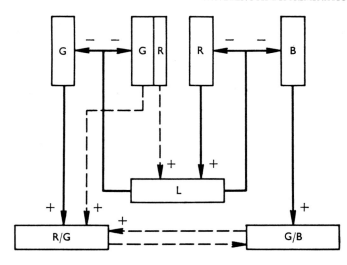

**Fig. 14.32** Diagram of the relations between cones and horizontal cells. The basic connexions responsible for the main properties of the cells are indicated by solid lines; dashed lines represent additional modifying interactions. The symbols + and − denote transmission without or with inversion of polarity. G, R and B are green, red and blue cones respectively; horizontal cells are identified by the usual notation; and the two adjoining R and G receptors represent the double cones. Additional outputs of C-cells could go to receptors, L-cells or bipolar cells. (Fuortes and Simon, *J. Physiol.*)

whereas red flashes gave depolarizing responses; in G/B cells, blue flashes elicited the hyperpolarizing waves and green flashes evoked depolarizing waves. The spectral sensitivities of the three main types of horizontal cell are indicated in Figure 14.31.

MODEL CIRCUIT

In general, it would appear that the chromatic responses of all types could be expressed in terms of a model circuit illustrated by Figure 14.32, where it is seen that each type of cone is connected with a corresponding type of horizontal cell and is responsible for its hyperpolarizing response, with sensitivity peaks around 640, 550 and 450 nm, as in red, green and blue cones respectively. Because of the recurrent loops from L-cells back to receptors, the responses of the horizontal cells will change. As an example, red light will hyperpolarize red cones producing negligible direct responses in blue and green cones; however, the excitation of L-cells causes a depolarizing impingement back on the red, green and blue cones; this will reduce the hyperpolarization of the red cones without changing the polarity, but the effects on green and blue cones will be one of pure depolarization, and this will be reflected in depolarization of the R/G and G/B cells. Green light is able to evoke hyperpolarization of red cones, L-cells and green cones but not of blue cones, hence the blue cones will be depolarized by green light as well as by red and thus also the G/B cells. Finally blue light hyperpolarizes all cones and consequently all horizontal cell types. In the scheme of Figure 14.32, an interaction between the two C-cells has been postulated to account for some of the discrepancies from theory that appear without this; thus the depolarizing action of the R/G cell would help to account for the observed depolarizing

action of green light in the G/B cell. This scheme, of course, is just a beginning and gives no clue as to the inputs into bipolar cells.

HORIZONTAL CELL ORGANIZATION IN GOLDFISH

S-Potentials in cyprinidae, such as the carp, have been classified as monophasic or L-type, hyperpolarizing in response to all wavelengths; biphasic or $C_1$ which hyperpolarize to light of shorter wavelengths and depolarize to longer wavelengths, and $C_2$ or triphasic which hyperpolarize to light of both longer and shorter wavelengths but depolarize to light of intermediate wavelengths. Stell and Lightfoot (1975) have examined the cone-horizontal cells connexions histologically and related these to the types of cones in the goldfish retina. As Figure 14.33 shows, these consist of double cones, comprising a long and short, two kinds of single long cones, one short and one miniature; as indicated, these have specific colour sensitivities, and it will be their connexions with horizontal cells that determine the S-potentials. As with other teleostean retinae, several types of horizontal cell could be distinguished on the basis of dendritic spread, density of contacts and so on, but in the goldfish a definite grouping of types in laminae was not observed. The types were designated as $H_1$, making contact with red + green + blue cones, $H_2$ with green + blue, and $H_3$ with blue. In order to account for the chromatic properties of the horizontal cells, synaptic connections between the horizontal cells had to be postulated as in Figure 14.34.

OPTIC NERVE FIBRE RESPONSE

DOMINATORS AND MODULATORS

Granit was the first to study the spectral sensitivity of optic nerve fibres. With the dark-adapted frog eye, and scotopic stimuli, the responses were the same whatever unit was picked up, the maximal response corresponding to the wavelength of maximal absorption of rhodopsin, about 500 nm. In the light-adapted frog eye the wavelength-response curves varied with the unit picked up; they fell into two categories described as

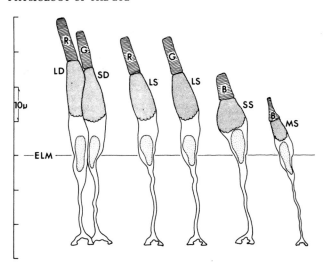

Fig. 14.33 Scale drawings of cones from central region of goldfish retina. The double cone comprises two members, long (LD) and short (SD). The single cones comprise probably two structurally similar long varieties (LS), one short (SS) and one miniature (MS) variety. The probable spectral regions of maximal absorption are indicated in the outer segments. (Stell and Lightfoot, *J. comp. Neurol.*)

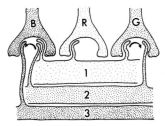

Fig. 14.34 Diagram summarizing contacts observed between cones believed to be red-, green- and blue-sensitive (R, G, B) and cone horizontal cells (1, 2, 3) in goldfish retina. Arrows indicate direction of information transfer which may account in part for known spectral response properties of horizontal cells. (Stell and Lightfoot, *J. comp. Neurol.*)

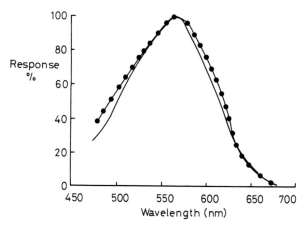

Fig. 14.35 Wavelength-response curves in single fibres of the frog (continuous line) and snake (interrupted by dots) optic nerve. The greatest response has been put equal to 100. The dominator. (Granit, *Nature.*)

*dominators* and *modulators*. The dominators were the most common and gave broad sensitivity curves with a maximum at the wavelength for maximal photopic sensitivity as found by the ERG, namely at about 555 nm (Fig. 14.35). The modulators had relatively narrow spectral sensitivity curves confined to one portion of the spectrum, so that it was possible to speak of a red-sensitive modulator whose peak of maximal sensitivity occurred at, say, 600 nm; a green-sensitive modulator with a peak at 530 nm, and so on. Figure 14.36 illustrates the modulators isolated by this electrophysiological method from various eyes.

THE OPTIC NERVE MESSAGE

Thus, so far as the *message* carried by the ganglion cells to the lateral geniculate bodies is concerned, we may say that it is organized apparently on a dual basis; the dominators indicate, by their frequency of discharge, the intensity of the prevailing light; under scotopic conditions their sensitivities correspond to the absorption of the scotopic pigment; rhodopsin in the case of the frog and mammals, and porphyropsin in the case of many fresh-water fishes. In the light-adapted mixed retina the dominator has the sensitivity corresponding to the photopic sensitivity curve. The second half of the message conveys information about the spectral composition; thus the red modulators would respond preferentially to red light falling on the retina, whilst the green and blue modulators would be much less affected, hence the message corresponding to the illumination of the retina with red light at photopic intensity might be a small discharge in the dominator, high discharge in the red modulator and small or negligible discharges in the green and blue modulators.

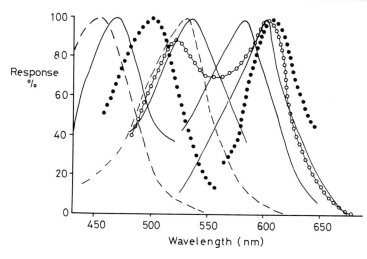

**Fig. 14.36** Wavelength-response curves in single fibres of the optic nerve of the rat (dots), guinea pig (broken lines), frog (lines in full) and snake (line interrupted by circles). The modulators. (Granit, *Nature.*)

## OPPONENT-TYPE RESPONSES

We have seen that the receptive fields of optic nerve fibres or of ganglion cells are not simple, but usually organized on a concentric basis, and subsequent work on wavelength-sensitivity of ganglion cells has shown that this concentric organization extends to colour sensitivity of the two components of the receptive field, so that the fields are organized on an opponent basis. Some examples of the types of receptive field are illustrated schematically in Figure 14.37; thus Type C

is relatively simple with no concentric organization, but the type of response, whether ON or OFF, depends on the wavelength of stimulating light. This was called a Type-P by Daw (1968), representing about 5 per cent of the goldfish ganglion cells; the majority, designated as Type-O, gave opponent responses between centre and surround. Where the centre-surround opponence is between complementary colours, the effects of uniform illumination of the field with white light is to cancel out any response, and this may be a means of reducing the

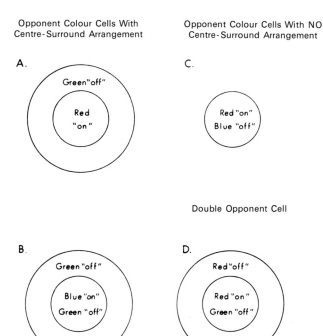

**Fig. 14.37** Examples of receptive fields of some colour-coded cells. (A) Centre-surround arrangement with spatial separation of the two colour components. (B) Centre-surround arrangement with overlap of the two colour components. (C) No centre-surround arrangement; complete overlap of the two colour components. (D) Double opponent cell; colour opponency in both centre and surround of the receptive field. (Daw, *Physiol. Rev.*)

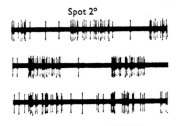
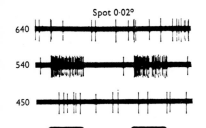

**Fig. 14.38** Responses from a colour-opponent ganglion cell with centre-surround organization to spots of monochromatic lights in the presence of an adapting white background. Bars indicate duration of light-stimulus. (Monasterio and Gouras, *J. Physiol.*)

total number of messages sent to the brain when the retina is exposed to uniform white light.

### MONKEY GANGLION CELLS

Hubel and Wiesel (1960), in their study of the receptive fields of monkey ganglion cells, found a small percentage with opponent colour responses, but Gouras (1968) found that the proportion was much higher and consistent with the high proportion of geniculate cells with a similar receptive field organization (p. 562). Gouras (1968, 1969) found two distinct types of ON-centre ganglion cells. Type I received signals from two cone mechanisms, Red and Green, as determined by the effects of selective adaptation to coloured lights. The responses were phasic, and both cone mechanisms excited the centre and surround. They could not be described as colour-coded, since they responded similarly to all wavelengths.

### Colour-coded cells

The other type was colour-coded, a single cone mechanism Red, Green or Blue exciting the centre with inhibition from the periphery through another cone-mechanism. These units were of the sustained type and were most common near the fovea and so may represent the midget ganglion cells of Polyak; their conduction-velocities were 1·8 m/s compared with 3·8 m/s for the phasic cells, suggesting that colour information was carried by small ganglion cells. In a more elaborate study De Monasterio and Gouras (1975) defined three general classes of monkey ganglion cells, with sub-divisions of these as shown in Table 14.2. The largest class were those with concentrically organized fields and colour opponence, in the sense that the centre responded specifically to one band of the spectrum and the surround to another, e.g. G + /R −, where the symbols indicate that the centre gives an excitatory ON-response to green and the surround an inhibitory OFF-response to red. An example of the responses of this type of cell is illustrated in Figure 14.38. By chromatic adaptation of centre and surround it was possible to determine the probable cone-inputs, and usually the centre had a single type whilst the surround could have a double type. The responses were of the sustained type, and in some cells the concentric arrangement was lacking.

### Broad-band cells

A second main type had concentric fields but no colour-opponency; they were described as *broad-band*, because their action-spectra were similar to the photopic luminosity function with maximum at about 560 nm (i.e. they could be described as 'dominators'). The action-spectra for centre and surround were identical, indicating the absence of colour opponence, whilst the effect of adaptation to different wavelengths (Fig. 14.39)

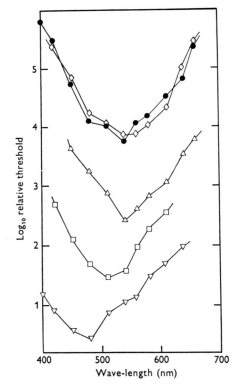

**Fig. 14.39** Broad-band cells. The action-spectrum is similar to the photopic luminosity function with maximum at about 560 nm. There is no colour-opponency so that action-spectra for a 0·08° spot (◇) and an annulus of 2° inner, 4° outer diameter (●) on a white background are superimposed. Adaptations with blue (△), red (□) and yellow (▽) backgrounds indicate the operation of at least two cone-mechanisms. (De Monasterio and Gouras, *J. Physiol.*)

indicated that at least two cone mechanisms were operative. The responses were of the transient type. A sub-class of these had a broad-band action-spectrum in its centre but the surround was more specifically activated, e.g. by the red plus blue cone mechanisms.*

SIGNIFICANCE OF SPATIAL OPPONENCE

The spatial opponent organization of colour-coded cells may well have little significance for conveying chromatic information to the brain and may simply represent Nature's method of transmitting both contour and colour information; thus the concentric organization of the receptive field of retinal and geniculate neurones, with its further development in simple cortical cells (p. 566), must be regarded as a development permitting the accurate assessment of moving or stationary contours; however, the same cones that are conveying this feature of the environment must be employed in conveying the colour as well; if all types of cone were to converge on single ganglion cells, the centre would

lose its chromatic selectivity so that it is understandable that these centres will have preferably a single type of cone input; the input to the surrounds, presumably operating through horizontally organized retinal cells, is usually chromatically opponent so that the response to any spot of light covering the whole receptive field— and this is the usual situation—will be governed by the opposing influences of centre and surround. In the special situation when the receptive field is trichromatic-

---

* In general, this study has revealed a larger number of types of ganglion cells in the monkey retina than had been suspected; the authors remark that it is generally considered that the ganglion cells of cat and monkey are simple by comparison with the large variety seen in frogs, pigeons, ground squirrels and rabbits; however, the greater diversity revealed here and in the cat (e.g. Cleland and Levick, 1974) suggests that differences amongst species are more quantitative than qualitative, so that evolution in this respect may work by allowing one type to multiply while others—perhaps representing the ancestral roots of the species—are either neglected or suppressed.

**Table 14.2** Retinal distribution of 460 units from monkey retina. B, G, R stand for cone types containing blue-, green- and red-sensitive pigments. $G+/R-$ means centre excitatory to Green, surround inhibitory to Red, and so on. Percentage refers to total sample. (De Monasterio and Gouras, 1975).

| Class | Type | (0–0.5°) | (0.5–2°) | (2–10°) | (10–40°) |
|---|---|---|---|---|---|
| 1. Colour-opponent concentric, (61%) | G+/R− | 22 | 33 | 35 | 5 |
| | G+/M− | 1 | 1 | 2 | — |
| | G−/R+ | 5 | 10 | 15 | — |
| | G−/M+ | — | — | 3 | — |
| | R+/G− | 7 | 24 | 45 | — |
| | R+/C− | — | 1 | 2 | — |
| | R−/G+ | 4 | 13 | 10 | 1 |
| | R−/C+ | — | — | 2 | — |
| | Y+/B− | 2 | 4 | — | — |
| | Y−/B+ | 4 | 4 | 2 | — |
| | B+/Y− | 2 | 5 | 11 | 1 |
| | B−/Y+ | — | 2 | 1 | 1 |
| 2. Colour-opponent, non-concentric (2%) | B+ Y− | 2 | 2 | 1 | — |
| | B− Y+ | — | 1 | — | — |
| | R+ G− | 2 | 1 | — | — |
| 3. Broad-band non-opponent (24%) | on/off | 2 | 4 | 36 | 27 |
| | off/on | 2 | 5 | 26 | 8 |
| 4. Broad-band, colour-opponent (4%) | Y+/R− | 1 | 6 | 4 | — |
| | Y+/G− | — | 3 | 1 | — |
| | W+/M− | — | 1 | 1 | — |
| | Y−/R+ | 1 | 1 | — | — |
| | Y−/G+ | — | 1 | — | — |
| 5. Non-concentric, phasic (6%) | on | 1 | 8 | 1 | — |
| | off | — | 2 | 1 | — |
| | on-off | 2 | 5 | 7 | — |
| 6. Non-concentric, motion-sensitive (3%) | ? | 2 | 3 | 5 | 4 |

ally organized, such that centre and surround are sensitive to complementary colours, e.g. Green-Magenta, Blue-Yellow, Red-Cyan, the ganglion cell may well be irresponsive to white light except when this falls on only a part of the receptive field, as with a contour. Some cortical and geniculate units certainly exhibit trichromatic opponency, and in the retina De Monasterio et al. (1975) identified GM, RC and BY cells, indicating that trichromaticity at the geniculate and cortical levels is not necessarily due to convergence of dichromatic or monochromatic inputs from the lower level.

## CONCEALED OPPONENCE

De Monasterio et al. (1975b) have emphasized that many units in the retina, and probably in geniculate and cortex, that are apparently non-opponent, do in effect have some concealed chromatic opponence, which may be revealed by appropriate chromatic adaptation, and fails to be normally evident by the extreme weakness of the surround response on a neutral background. Thus, on a neutral background, these cells behave as luminosity units without capacity to transmit chromatic information.

## RESPONSES IN THE LATERAL GENICULATE BODY

The optic nerve fibres relay in an ordered fashion to the cells of the lateral geniculate body (p. 542); De Valois was the first to record from these third-order neurones, showing that the messages sent to the cortex in the monkey, an animal with colour-vision comparable with that of man, were colour-coded. As with the ganglion cells and optic nerve fibres, there were pure ON-elements that responded by increasing frequency of discharge on illuminating the retina; the responses were sensitive to changes of wavelength so that their wave-length-response curves seemed to be typical modulators; these modulator type elements were especially numerous in the region to which the foveal ganglion cells projected.

### ON- OR OFF-ELEMENTS

A second type of element, the most common, was the ON or OFF, responding with increased discharge at ON or else responding with inhibition during ON and a discharge at OFF. As to whether the cell behaved as an ON- or OFF-element depended on the wavelength of the stimulating light. Thus cells could be found to respond at ON with red stimulation and at OFF with green stimulation, so that if the response, measured by the number of spikes, is plotted against wavelength for equal energy stimuli, we get the two curves of Figure 14.40. From this we see that with a green of 530 nm there is an inhibition of the existing spontaneous discharge followed by a large discharge at OFF; with a red of 640 nm there is a discharge at ON whilst at OFF there is some inhibition of spontaneous activity. With other elements the opposing wavelength bands were in the blue and yellow. It is clear that units of this type should be hardly affected by white light since this may be regarded as the combination of two of the complementary opponent stimuli, red-green, blue-yellow, etc. In fact, up to very high intensities, white light produced no response in these elements.

### PURE OFF-ELEMENTS

Finally pure OFF-elements were of the dominator type with maxima at 510 and 550 nm, presumably corresponding to scotopic and photopic pigments. At any rate Jacobs (1964), studying the squirrel monkey in which behavioural experiments had given the scotopic and photopic sensitivity curves, found reasonable concordance with the corresponding wavelength-sensitivity curves of dominator type units of the lateral geniculate body.

### SPATIAL OPPONENCE

De Valois used large light-stimuli and so he examined the overall effects of stimulating all the receptors in the receptive field of a given geniculate cell. Wiesel and Hubel (1966) showed that, like the ganglion cell units, the geniculate units were also spatially organized, having opposing centre and periphery effects with white light; when spectral sensitivity was studied a number of types were identified. The Type I unit gave an ON-response at its centre with a limited range of wavelengths whilst the surround gave an OFF-response to the complementary band, e.g. the commonest was a Red ON-centre, Green OFF-surround. These corresponded with De Valois'

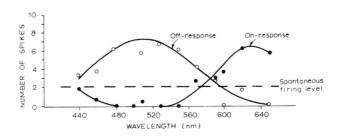

Fig. 14.40   The responses of a Red-ON, Green-OFF element to different wavelengths of light, the energy-contents of the stimuli being the same. (De Valois, *J. gen. Physiol.*)

ON-OFF elements since he illuminated the whole receptive field. At a certain wavelength, the *neutral point*, the opposing effects of centre and surround just balanced; thus at 540 nm a large spot of light had no effect. There were relatively few Blue ON-centre-Green-OFF-surround, and no Blue-Red units. With this Type-I, then, it appears as though the geniculate cell is connected to two types of cone, one of which inhibits and the other excites, or *vice versa*. When the intensity of the light-stimulus was reduced, a Purkinje-shift in sensitivity could be obtained with some Type I units, indicating connections with rods as well as cones.

The Type II units showed no centre-surround organization and consisted of a receptive field of some $\frac{1}{2}$ degree; the responses to different wavelength-bands were opposite in character, e.g. ON to blue and OFF to green, with a neutral point of 500 nm, ON to red and OFF to green with a neutral point of 600 nm, and so on. This type obviously receives input from two populations of cones, one excitatory and the other inhibitory, and the two types are distributed uniformly throughout the $\frac{1}{2}$ degree field. No Purkinje-effect was observed with these.

Type III units revealed no opponent colour effects; they had the centre-surround organization, the periphery usually suppressing the centre, but the responses were similar at all wavelengths. In these the representation of cone types was the same for inhibitory and excitatory areas of the field.

All these types were found in the dorsal layers of the lateral geniculate body; in the ventral layers an additional type was found, Type IV; they were organized on a centre-surround basis and were characterized by the very effective inhibition exerted by the surround.

**Luminosity units**
Like the ganglion cells with wide spectral sensitivity curves, similar to the luminosity curves of the intact animal, Wiesel and Hubel's Type III units may be classed as luminosity units. However, a much more thorough examination of the responses to different wavelengths under different degrees of chromatic adaptation would be necessary to decide whether there is no opponence in the responses. This point was emphasized by Padmos and Norren (1975) who showed, by selective adaptation, that many geniculate units that would have been classed as non-opponent did, in fact, have some concealed opponence with antagonistic green-red cone interactions, as revealed by the altered responses when adapted to different wavelength bands. This opponence might have been suspected from the deviation of the spectral sensitivity curve from the mean CIE curve; however, when many such units are averaged the average sensitivity is that of the CIE observer so that, as Padmos and Norren have argued, it could be that luminosity is not signalled by specific ganglion or geniculate cells, but rather through averaging this type of unit with antagonistic green and red inputs. In a similar way many cells that would have been characterized as having dichromatic input, e.g. Green-ON-Red-OFF by their response patterns on a neutral background could be shown to have a trichromatic input, in conflict with De Valois' generalization that all monkey geniculate cells had only dichromatic input.

## CORTICAL RESPONSES

The receptive fields of cerebral cortical neurones are much more complex than those of ganglion and geniculate neurones, so that it is advisable to postpone a description of the chromatic features of cortical neurone discharge until Section IV (p. 588).

## REFERENCES

Aguilar, M. & Stiles, W. S. (1954) Saturation of the rod mechanism at high levels of illumination. *Optica Acta* **1**, 59–65.

Alpern, M. & Moeller, J. (1977) The red and green cone visual pigments of deuteranomalous trichromacy. *J. Physiol.* **266**, 647–675.

Alpern, M. & Pugh, E. N. (1977) Variation in the action spectrum of enythrolabe among deuteranopes. *J. Physiol.* **266**, 613–646.

Alpern, M. & Wake, T. (1977) Cone vision in human deutan colour vision defects. *J. Physiol.* **266**, 595–612.

Arden, G. B. & Tansley, K. (1955) The spectral sensitivity of the pure cone retina of the squirrel (*Sciurus carolinensis leucotis*). *J. Physiol.* **127**, 592–602.

Baker, H. D. & Rushton, W. A. H. (1965) The red-sensitive pigment in normal cones. *J. Physiol.* **176**, 56–72.

Baylor, D. A., Fuortes, M. G. & O'Bryan, P. M. (1971) Receptive fields of cones in the retina of the turtle. *J. Physiol.* **214**, 265–294.

Baylor, D. A. & Hodgkin, A. L. (1973) Detection and resolution of visual stimuli by turtle photoreceptors. *J. Physiol.* **234**, 163–198.

Blackwell, H. R. & Blackwell, O. M. (1961) Rod and cone mechanisms in typical and atypical congenital achromatopsia. *Vision Res.* **1**, 62–107.

Bowmaker, J. K. (1977) The visual pigments, oil droplets and spectral sensitivity of the pigeon. *Vision Res.* **17**, 1129–1138.

Bowmaker, J. K. & Knowles, A. (1977) The visual pigments and oil droplets of the chicken retina. *Vision Res.* **17**, 755–764.

Bowmaker, J. K., Loew, E. R. & Liebman, P. A. (1975) Variation in the $\lambda_{max}$ of rhodopsin from individual frogs. *Vision Res.* **15**, 997–1003.

Brindley, G. S. (1954) The summation areas of human colour-receptive mechanisms at increment threshold. *J. Physiol.* **124**, 400–408.

Brindley, G. S., Du Croz, J. J. & Rushton, W. A. H. (1966) The flicker fusion frequency of the blue-sensitive mechanism of colour vision. *J. Physiol.* **183**, 497–500.

Brown, P. K. & Wald, G. (1964) Visual pigments in single rods and cones of the human retina. *Science* **144**, 45–52.

Cleland, B. G. & Levick, W. R. (1974) Properties of rarely encountered types of ganglion cells in the cat's retina and an overall classification. *J. Physiol.* **240**, 457–492.

Cole, B. L. & Watkins, R. D. (1967) Increment thresholds in tritanopia. *Vision Res.* **7**, 939–947.

Dagher, M., Cruz, A. & Plaza, L. (1958) Colour thresholds with monochromatic stimuli in the spectral region 530–630 $\mu$. In Visual Problems of Colour. *N.P.L. Symp.* No. 8, pp. 389–398.

Daw, N. W. (1968) Colour-coded ganglion cells in the

goldfish retina: extension of their receptive fields by means of new stimuli. *J. Physiol.* **197**, 567–592.

Daw, N. W. (1973) Neurophysiology of color vision. *Physiol. Rev.* **53**, 571–611.

Daw, N. W. & Perlman, A. L. (1970) Cat colour vision: evidence for more than one cone process. *J. Physiol.* **211**, 125–137.

De Valois, R. L. (1960) Colour vision mechanisms in the monkey. *J. gen. Physiol.*, **43**, Suppl. 115–128.

De Valois, R. L., Abramov, I. & Mead, W. R. (1967) Single cell analysis of wavelength discrimination at the lateral geniculate nucleus in the macaque. *J. Neurophysiol.* **30**, 415–433.

Dodt, E. & Walther, J. B. (1958) Der photopischer Dominator im Flimmer-ERG der Katze. *Pflüg. Arch.* **266**, 175–186.

Donner, K. O. (1953) The spectral sensitivity of the pigeon's retinal elements. *J. Physiol.* **122**, 524–537.

Fincham, E. F. (1953) Defects of the colour-sense mechanism as indicated by the accommodation reflex. *J. Physiol.* **121**, 570–580.

Fuortes, M. G. F., Schwartz, E. A. & Simon, E. J. (1973) Colour-dependence of cone responses in turtle retina. *J. Physiol.* **234**, 199–216.

Fuortes, M. G. F. & Simon, E. J. (1974) Interactions leading to horizontal cell responses in the turtle retina. *J. Physiol.* **240**, 177–198.

Gibson, I. M. (1962) Visual mechanisms in a cone-monochromat. *J. Physiol.* **161**, 10–11 P.

Glickstein, M. & Heath, G. G. (1975) Receptors in the monochromat eye. *Vision Res.* **15**, 633–636.

Gouras, P. (1968) Identification of cone mechanisms in monkey ganglion cells. *J. Physiol.* **199**, 533–547.

Gouras, P. (1969) Antidromic responses of orthodomically identified ganglion cells in monkey retina. *J. Physiol.* **204**, 407–419.

Govardovskii, V. I. & Zueva, L. V. (1977) Visual pigments of chicken and pigeon. *Vision Res.* **17**, 537–542.

Graham, C. H. & Hsia, Y. (1958) Colour defect and colour theory. *Science* **127**, 675–682.

Granit, R. (1943) A physiological theory of colour perception. *Nature, Lond.* **151**, 11–14.

Granit, R. & Wrede, C. M. (1937) The electrical responses of light-adapted frog's eyes to monochromatic stimuli. *J. Physiol.* **89**, 239–256.

Hanoaka, T. & Fujimoto, K. (1957) Absorption spectrum of a single cone in carp retina. *Jap. J. Physiol.* **7**, 276–285.

Hecht, S. (1947) Colourblind vision. I. Luminosity losses in the spectrum from dichromats. *J. gen. Physiol.* **31**, 141–152.

Hubel, D. H. & Wiesel, T. N. (1960) Receptive fields of optic nerve fibres in the spider monkey. *J. Physiol.* **154**, 572–580.

Hurvich, L. M. (1972) Color vision deficiencies. In *Hdb. Sensory Physiol.* VII/4, 607. Springer, Berlin.

Jacobs, G. H. (1964) Single cells in squirrel monkey lateral geniculate nucleus with broad spectral sensitivity. *Vision Res.* **4**, 221–232.

Larsen, H. (1921) Demonstration mikroskopischer Präparate von einem monochromatischen Auge. *Klin. Mbl. Augenheilk.* **67**, 301–302.

Le Grand, Y. (1948) *Optique Physiologique.* Vol. 2. Paris: Editions de la 'Revue D'Optique'.

Liebman, P. A. (1972) Microspectrophotometry of photoreceptors. In *Hdb. Sensory Physiol.* VII/1, 482–528. Berlin, Springer.

Liebman, P. A. & Granda, A. M. (1971)

Microspectrophotometric measurements of visual pigments in two species of turtle (*Pseudemys scripta* and *Chelona mydas*). *Vision Res.* **11**, 105–114.

MacNichol, E. J. & Svaetichin, G. (1958) Electric responses from the isolated retinas of fishes. *Am J. Ophthal.* Pt. 2 **46**, 26–46.

Marks, W. B. (1965) Visual pigments of single goldfish cones. *J. Physiol.* **178**, 14–32.

Marks, W. B., Dobelle, W. H. & MacNichol, E. F. (1964) Visual pigments of single primate cones. *Science* **143**, 1181–1183.

McFarland, W. N. & Munz, F. W. (1975) The evolution of photopic visual pigments in fishes. *Vision Res.* **15**, 1071–1080.

Michael, C. R. (1968) Receptive fields of single optic nerve fibers in a mammal with an all-cone retina. III. Opponent color units. *J. Neurophysiol.* **31**, 268–282.

Muntz, W. R. A. (1972) Inert absorbing and reflecting pigments. *Hdb. Sensory Physiol.* VII/1, 529–565.

Munz, F. W. & McFarland, W. N. (1975) Presumptive cone pigments extracted from tropical marine fish. *Vision Res.* **15**, 1045–1062.

Naka, K. I. & Rushton, W. A. H. (1966a) S-potentials from colour units in the retina of fish (*Cypridinae*) *J. Physiol.* **185**, 536–555.

Naka, K. I. & Rushton, W. A. H. (1966b) An attempt to analyse colour reception by electrophysiology. *J. Physiol.* **185**, 556–586.

Naka, K. I. & Rushton, W. A. H. (1966c) S-potentials from luminosity units in the retina of fish (*Cypridinae*). *J. Physiol.* **185**, 587–599.

Norton, A. L., Spekreijse, H., Wohlbarsht, M. L. & Wagner, H. G. (1968) Receptive field organization of the S-potential. *Science* **160**, 1021–1022.

Padmos, P. & Norren, D. V. (1975) Increment spectral sensitivity and colour discrimination in the primate studied by means of graded potentials from the striate cortex. *Vision Res.* **15**, 1103–1113.

Piantanida, T. P. & Sperling, H. G. (1973) Isolation of a third chromatic mechanism in the deuteranomalous observer. *Vision Res.* **13**, 2049–2058.

Pickford, R. W. (1951) *Individual Differences in Colour Vision.* Routledge and Kegan Paul, London.

Pitt, F. H. G. (1935) Characteristics of dichromatic vision. *Med. Res. Council Sp. Rep. Ser.*, No. 200.

Pokorny, J., Smith, V. C. & Katz, I. (1973) Derivation of the photopigment absorption spectra in anomalous trichromats. *J. opt. Soc. Am.* **63**, 232–237.

Pugh, E. N. (1976) The nature of the $\pi_1$ colour mechanism of W.S. Stiles. *J. Physiol.* **257**, 713–747.

Richter, A. & Simon, E. J. (1974) Electrical responses of double cones in the turtle retina. *J. Physiol.* **242**, 673–683.

Rushton, W. A. H. (1959) Visual pigments in man and animals and their relation to seeing. *Progr. Biophys.* **9**, 239–283.

Rushton, W. A. H. (1963) A cone pigment in the protanope. *J. Physiol.* **168**, 345–359.

Rushton, W. A. H. (1965) A foveal pigment in the deuteranope. *J. Physiol.* **176**, 24–37.

Stell, W. K. & Lightfoot, D. O. (1975) Color-specific interconnections of cones and horizontal cells in the retina of the goldfish. *J. comp. Neurol.* **159**, 473–502.

Stiles, W. S. (1959) Colour vision: the approach through increment-threshold sensitivity. *Proc. Natl. Acad. Sci. Wash.* **45**, 100–114.

Svaetichin, G. (1956) Spectral response curves from single cones. *Acta physiol. Scand.* **39**, Suppl. 134, 17–46.

Svaetichin, G. & MacNichol, E. F. (1958) Retinal mechanisms for chromatic and achromatic vision. *Ann. N.Y. Acad. Sci.* **74,** 385–404.

Tomita, T., Kaneka, A., Murakami, M. & Pautler, E. L. (1967) Spectral response curves of single cones in the carp. *Vision Res.* **7,** 519–531.

Wald, G. (1945) Human vision and the spectrum. *Science* **101,** 653–658.

Wald, G., Brown, P. K. & Smith, P. H. (1955) Iodopsin. *J. gen. Physiol.* **38,** 623–681.

Walls, G. L. & Heath, G. G. (1954) Typical total colour blindness reinterpreted. *Acta ophthal. Kbh.* **32,** 253–297.

Weale, R. A. (1953) Cone monochromatism. *J. Physiol.* **121,** 548–569.

Weale, R. A. (1959) Photosensitive reactions in fovea of normal and cone-monochromatic observers. *Optica Acta,* **6,** 158–174.

Wiesel, T. N. & Hubel, D. H. (1966) Spatial and chromatic interactions in the lateral geniculate body of the rhesus monkey. *J. Neurophysiol.* **24,** 1115–1156.

Wright, W. D. (1946) *Researches on Normal and Defective Colour Vision.* London: Kimpton.

Wright, W. D. (1952) The characteristics of tritanopia. *J. opt. Soc. Am.* **42,** 509–521.

# 15. Stiles-Crawford effect; adaptation; photopic sensitivity curves

We may conclude this account of modern research devoted to the elucidation of the mechanism of vision with a few special aspects of phenomena that have already been discussed in a more elementary fashion.

## DIRECTIONAL SENSITIVITY OF THE RETINA

### PUPILLARY DIAMETER

Stiles and Crawford decided to develop an indirect method of measuring pupillary diameter. They argued that the larger the pupil the greater the amount of light that would enter the eye and hence the greater the apparent brightness of a luminous field; the amount of light would increase in proportion to the area of the pupil and this in proportion to the square of its diameter. They found that the calculated diameter never increased beyond 5·5 mm although the maximum diameter actually measured was about 8 mm. This discrepancy was shown to be due to the circumstance that, as the pupil became larger, the retina received relatively more rays of light that struck it obliquely; if these oblique rays were less efficient than those entering through the centre of the pupil, then the discrepancy could be explained. The effects are large, so that if the subjective brightness of a pencil entering at the very periphery of the pupil is compared with one entering centrally, it may be only 15 per cent of this.

### CONE PHENOMENON

It was shown that the phenomenon was peculiar to the cones, both foveal and peripheral, and this has provided a valuable method for determining whether rods or cones are concerned in a given visual event. Thus Donner and Rushton, studying the responses of a single ganglion cell in the frog, showed that when it behaved as a scotopic dominator its responses were independent of the direction of incidence of the pencil of light; when it behaved as a photopic dominator, after light-adaptation, it showed directional sensitivity; moreover, in the mesopic range when a hump appears on the sensitivity curve in the blue, this 'blue hump' was shown to be insensitive to direction, and thus to be due to activity of rods, probably of a special type called 'green rods' (Denton and Wyllie).

### VARIATION WITH WAVELENGTH

The magnitude of the directional effect varies with wavelength (Stiles, 1937); moreover, if monochromatic light is used there is a change in its subjective hue as the angle of incidence of the rays varies; this is called the Stiles-Crawford effect of the second kind. Brindley showed that colour-matches could be upset by changing from direct to oblique incidence, so that it would seem that the forms of the fundamental response curves of the cone mechanisms responsible for wavelength discrimination change with angle of incidence of the radiation. Since the shape of an absorption spectrum depends on the density of the pigment in solution (p. 206), and since the shape of the sensitivity curve presumably depends on that of the absorption spectrum of the cone-pigment, we can conclude that the effective optical density of pigment in the cone varies with direction of incidence.

## WAVE-GUIDE THEORY

The most plausible explanation is that put forward by Wright and Nelson (1936) to the effect that light entering directly along the axis of the cone is more likely to be retained by a series of total internal reflexions than light entering obliquely.

The dense packing of the transverse discs makes the refractive index of the inside of the receptor segment greater than that of the outside medium, and thus promotes total internal reflexion. In addition, the stout outer segments act as light-collectors, funnelling the light into the thinner inner segments (Winston and Enoch, 1971). Such an arrangement would ensure that the pigment was maximally exposed to the incident quanta of light whilst providing the optical isolation of the receptors necessary for resolution (Kirschfeld and Snyder, 1976).

### EXPERIMENTS ON TURTLE CONE

The theoretical electromagnetic treatment of Snyder and Pask (1973) has confirmed the possibility, whilst recent electrophysiological studies have provided quantitative information respecting its wavelength-dependence in the turtle cone. This cone is interesting because of the presence of oil-droplets in its inner segment, and Baylor and Hodgkin (1973) noticed, in their study of hyperpolarizing responses in individual cones, that there were very wide variations in maximal sensitivity amongst any group, suggesting that the cones were being illuminated at different angles; moreover, there was a positive correlation between the absolute sensitivity of a cone and the reduction in sensitivity when the wavelength was changed from $\lambda_{max}$ towards the short wavelength. This positive correlation could be explained on the basis of Figure 15.1; at the left, light

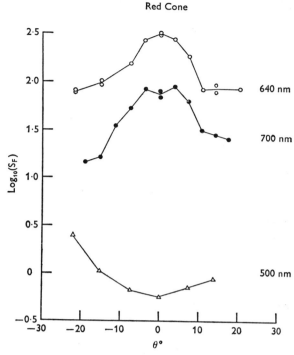

Fig. 15.2 Profiles of angular sensitivity of a red-sensitive turtle cone. The ordinate is the logarithm of the flash-sensitivity to light of the wavelengths indicated on the graphs. Note that with blue light oblique stimuli are more effective (sensitivity greater), and this is because the light avoids the yellow oil-droplets. (Baylor and Fettiplace, J. Physiol.)

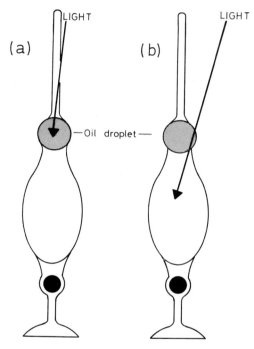

**Fig. 15.1** Showing how variation in angle of incidence of light on turtle cone can influence spectral response.

enters axially so that it is efficiently absorbed and the cone has a high sensitivity; however, the light passes through the oil-droplet, which absorbs short wavelength light and so there will be a large falling off in sensitivity with short wavelengths. By contrast, at the right the light falls obliquely, less is absorbed by the photosensitive pigment and less passes through the oil-droplet, so that the cut-off at short wavelengths is not so strong. Thus sensitivity and short wavelength cut-off run parallel, as found. In general, many cones showed sharp profiles of sensitivity versus angle of incidence of stimulating light, but this depended very strongly on the wavelength of the stimulating light, as we should expect. Thus with a red-sensitive cone (Fig. 15.2) orange and red lights gave sharp profiles but blue-

green light not only showed a low sensitivity but an inverted profile; light entering obliquely, and tending thereby to avoid the blue-absorbing (yellow) droplet, being more efficient.

**Orientation of receptors**

Histological study of the orientation of the receptors revealed that, although the receptors in a given region were aligned parallel with each other, the orientation of the inner segments varied, depending on retinal location, so that at the pole they were at right-angles to the retinal surface and thus pointing into the pupil whereas at the periphery the deviation from perpendicularity could be as great as 60°.

**Collection efficiency**

By measuring the amount of light passing through the cones experimentally, a collection efficiency of 0·55 ± 0·05 was obtained where the efficiency $E$ is given by

$$E = \frac{A_1 I_1}{A_1 I_1 + A_2 I_2}$$

$A_1$ being the cross-sectional area of the spot falling on the base of the outer segment with mean relative intensity $I_1$, and $A_2$ is the area of the corresponding inner segment minus $A_1$, and $I_2$ is the mean relative intensity in the dark region round the bright spot.

Figure 15.3 shows the directional sensitivity of red-sensitive cones compared with a theoretical curve based on a model cone computed by Winston and Enoch (1971).*

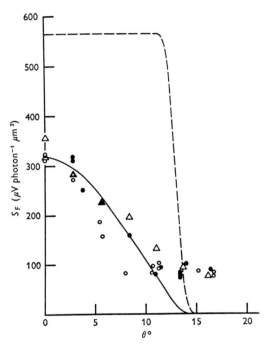

**Fig. 15.3** Directional sensitivity of a red-sensitive cone compared with a theoretical curve derived by Winston and Enoch (1971) for a cone with inner and outer segments with diameters in the ratio of 4·5. The dashed line shows the performance of an ideal collector with total internal reflexion and 100 per cent efficiency of collection of light not exceeding $\theta_{max}$ of 13°. (Baylor and Fettiplace, *J. Physiol.*)

ROD VISION

Psychophysical studies on man have usually failed to reveal a significant Stiles-Crawford effect for rod vision (see, for example, Flamant and Stiles, 1948) although optical studies on the retinae of several species have indicated wave-guide properties in the rods which, in the goldfish, may be as effective as those of the cones (Tobey *et al.*, 1975). As Van Loo and Enoch (1975) state, the limited angle of acceptance of the rod as a wave-guide, its cylindrical shape, orientated photopigment, its separation from neighbouring cells by interdigitating fibrils from the pigment epithelium, and the higher index of refraction of the cells relative to the

surrounding interstitial matrix, indicate that directionality should be present. These authors have re-examined the matter in three human subjects and, as Figure 15.4 shows, there is directionality, although small compared with that for cone vision.†

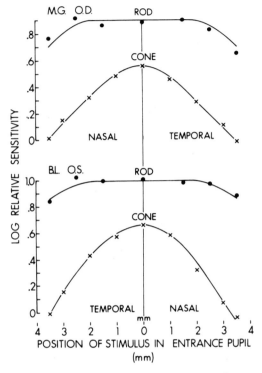

**Fig. 15.4** Rod and cone Stiles-Crawford curves for two human observers. (Van Loo and Enoch, *Vision Res.*)

---

*In addition to its capacity as a wave-guide the retinal cone may be described as an 'ideal light collector' by virtue of the ellipsoid, whose structure is such as to favour the funnelling of light from outer to inner segment. An ideal light collector has a completely reflecting inner surface; hence, because the cone relies on total internal reflexion, the collection is reduced on this account (Winston and Enoch, 1971). Tobey and Enoch (1973) passed light backwards through rods and showed that it emerged in a tight cone, so that, on the principle of reversibility of rays, there is a corresponding limitation to the acceptance of light.
† Brindley (1967) examined the 'phosphene' that results from exerting a small deforming force on the sclera; when the retina is illuminated under this condition, the phosphene appears as a dark patch and is presumably due to defective light-catching by the deformed part of the retina. Brindley argued that if this explanation were correct, the effect should be much stronger in cone-vision where the cones funnel light much more effectively than rods. In fact, this was found, the deformation-phosphene matching a paper with 68 per cent reflectance in rod vision but one of only 16 to 28 per cent in cone vision, depending on the size of the pupil.

# ADAPTATION AND INTERACTION OF LIGHT STIMULI

## ROD ADAPTATION

The main phenomena of adaptation have been already described throughout this Section, and here we shall discuss only some special aspects. By dark-adaptation we mean the recovery of sensitivity of the retina following exposure to light; so far as rod vision in man is concerned, there is little doubt that this process is largely determined by the regeneration of rhodopsin, although the relationship between retinal sensitivity and concentration of rhodopsin in the retina is not so simple as had been at first thought (p. 234), so that the concept of a neural brightness control mechanism has been introduced, a mechanism that behaves like 'dark light' and is apparently governed by signals generated, in the first place, by the amount of unregenerated photo-pigment in the receptors.

## FOVEAL ADAPTATION

With human foveal, i.e. cone, vision, Wright has established the existence of an essentially similar process that takes place far more rapidly however, presumably because cone pigments regenerate more rapidly (p. 338). The reverse process, which we may call foveal light-adaptation, constitutes a decrease in sensitivity of the retina during continued exposure to light; it was studied by Wright with his binocular matching technique, the change in sensitivity of one eye, during exposure to light, being measured by comparing the subjective sensations in the two eyes. The curve of decrease in sensitivity falls over a period of some 50 seconds to reach a plateau with a sensitivity about one sixteenth of the original value (Fig. 7.8, p. 192).

### ALPHA-ADAPTATION

A much more rapid adaptive effect was described by Schouten and Ornstein, who measured the effect of a small glare-source of light, stimulating a peripheral part of the retina, on the brightness-sensation experienced by the fovea. Thus the two eyes would view independently the two halves of an illuminated field, and these would be adjusted in luminosity to give a match. The peripheral glare-source, entering one eye only, would now be switched on, and immediately the match would be disturbed, the half of the field seen by the eye with the glare-source appearing much darker. By raising its luminance till a match was obtained the decrease in sensitivity could be determined. This *α-adaptation*, as Schouten and Ornstein called it, is complete within a fraction of a second and was thought to differ from the $\beta$-adaptation of Wright in that its effect extended to the whole retina, even though only a small part was stimulated. Provided exposure to the glare-source was short, recovery of sensitivity was rapid. Thus long before Wright, in his binocular matching studies, had made his first measurement of retinal sensitivity, this had presumably decreased by a factor of about five because of this rapid $\alpha$-adaptive process.

**Veiling luminance**
Subsequent work, notably that of Fry and Alpern has shown, however, that the phenomenon is not one of retinal interaction but is due to the scattered light from the glare-source. The effect of this scattered light is to behave as a 'veiling luminance', decreasing the sensitivity of the whole retina. We may regard the phenomenon as one of very rapid adaptation, but not, as originally thought, of inhibition of the whole retina by a stimulus falling on only a part. The rapid change in sensitivity of the retina to light is a very real phenomenon and spatial interaction, although not as marked as originally considered by Schouten and Ornstein, is likewise of considerable importance, and we shall now consider some aspects of temporal and spatial interaction.

## ROD-CONE INTERACTION

Where the effects of adapting lights on sensitivity of a given type of receptor are concerned, it is well established that the rod and cone systems of the human eye behave independently, in the sense that an adapting background affecting rods does not influence cone sensitivity, and vice versa. However, recent work summarized by Latch and Lennie (1977) has described some exceptions to this rule, which have been tentatively explained on the basis that independence is confined to retinal events, whereas adaptive effects manifest in higher stages of the visual pathway are not independent.

### IMPORTANCE OF FIELD SIZE

Latch and Lennie examined the situation when the threshold for rod vision in the peripheral retina is raised by a steady background acting through the cones, and showed that the important feature of the stimulus-conditions was the size of the stimulus or the adapting light; thus a large adapting field affected threshold for rods to the extent that the adapting field affected the rods, whereas with a small adapting field, threshold was elevated more by a long-wavelength background than a short-wavelength background although their intensities had been matched for equal effects on rods. Thus the long-wavelength background, stimulating cones, as well as rods, was the more effective background. In a reverse way, a small background-adapting light, affecting rods, could affect also the threshold for a stimulus seen through the cone mechanism. Latch and Lennie suggested that rod-cone interaction took place at a level in the visual pathway where image-size was discriminated, the phenomena described by them being comparable with those in which prolonged exposure to targets of a restricted size leads to reduced sensitivity to targets in this range of sizes but not to targets of different size.

## TEMPORAL ASPECTS

### BLOCH'S LAW

The limited application of Bloch's Law has already been described; to recapitulate, over a certain action-time two successive stimuli summate completely, so that the product: $I \times t = Constant$ applies. At longer intervals, summation is only partial and at still longer intervals may be characterized as *probability summation*, the two flashes being separated temporally to the extent that they may be considered as independent events, so that there is no question of summation of quanta.

#### CRITICAL DURATION

It becomes of interest, therefore, to ascertain with some precision the critical duration over which Bloch's Law of complete summation holds, and the effects of such stimulus variables as area, eccentricity and wavelength. So far as the minimum period over which the Law applies, Brindley (1952) found that it held down to flash-durations of extreme brevity, at least as short as $4 \cdot 1 . 10^{-7}$ sec when the retinal illumination was $3 . 10^8$ trolands. In the same study Brindley found that the greatest duration was of the order of 30 msec. The subject has been investigated recently by Saunders (1975) who varied size and eccentricity as well as duration of the flashes required to excite at threshold. Figure 15.5 shows the relative sensitivity of the dark-adapted eye to flashes of increasing duration, where relative sensitivity is the reciprocal of the threshold intensity. When plotted on a log-log basis in this way, the points fall on two straight lines; the initial line represents the application of Bloch's Law and the change to a new relation occurs at the arrow, namely at 80 msec.

### Size and eccentricity

The two sets of results apply to different sizes of stimulus and the break occurs at the same point indicating no effect of size; similarly, varying the eccentricity of the stimulus was without effect.

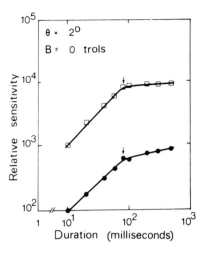

**Fig. 15.5** The relative sensitivity of the dark-adapted eye (B = 0 trols) to achromatic stimuli of increasing duration. (●) 3·2 minutes of arc stimulus; (□) 32 minutes of arc. Arrows indicate critical duration below which Bloch's Law applies. (Saunders, *Vision Res.*)

### Background luminance

When the flash was viewed against a bright background, so that the luminance-increment was being measured, rather than absolute threshold, the critical duration decreased on increasing the stimulus-area, as Barlow (1958) had found. The reason why at high background luminance the critical duration is decreased for large areas of stimulus, i.e. the effective temporal summation is reduced, is probably related to the effects of background on *spatial* summation. Thus the area over which there is complete spatial summation (Ricco's Law) becomes smaller at high background luminance (Barlow, 1958; Glezer, 1965), so that in order to achieve maximal temporal summation the stimulus-area must be smaller than Ricco's summation area when increment threshold measurements are made, by contrast with the situation at absolute threshold when summation-area becomes unimportant. When the Stiles colour mechanisms were studied, using increment thresholds against backgrounds designed to isolate a given mechanism (p. 333), a critical duration of 80 msec was obtained similar to that for achromatic stimuli at threshold, but as we might expect, this was only true of small areas of stimulus as with achromatic increment-thresholds.

### SUCCESSIVE CONTRAST

It is thought that the phenomena of successive contrast, or temporal induction, are associated with the slower $\beta$-adaptation described by Wright; here we are dealing with the effects of one stimulus on a succeeding one, both of them falling on the same part of the retina; the phenomena are essentially of an adaptive nature in the sense that the first stimulus depresses the sensitivity of the retina for a succeeding, like stimulus.

## META-CONTRAST

By *meta-contrast* is meant the inductive effect of a primary light stimulus on the sensitivity of the eye to a previously presented light stimulus on an adjoining part of the retina; it is a combination of temporal and spatial induction, or successive and simultaneous contrast. It is often called the 'flash after-effect' and has been studied by Baumgardt and Ségal (1946) and Alpern (1953). Baumgardt and Ségal illuminated the two halves of a circular patch consecutively for a brief duration. If the left half only, for example, is illuminated for 10 msec, it produces a definite sensation of brightness. If, now, both are illuminated for the same period, but the right half some 20 to 50 msec later, the left half of the field appears much darker than before, and near the centre may be completely extinguished. The left field has thus been inhibited by the succeeding, nearby, stimulus.

### PARA-CONTRAST

The right field, moreover, appears darker than when exposed alone—it has been inhibited by the first stimulus (*para-contrast*). Baumgardt and Ségal have shown that, by varying the interval between the presentation of the two halves of the illuminated field, a variety of intermediate degrees of meta-contrast may be achieved; moreover, these may be reproduced as after-images, but in the latter process the time relationships are altered so that if, for example, the left half of the field was dark and the right half bright in the primary sensation, the reverse relationship could be found in the after-image.

### BINOCULAR MATCHING

Alpern (1953) applied the binocular matching technique, whereby the left eye viewed a standard patch and the right eye a test-patch whose luminance could be adjusted by the observer to match the brightness of the left-eye patch. Besides the test-patch, an additional 'contrast-inducing' patch was presented to the right eye at varying intervals before or after the presentation of the test-patch. Figure 15.6 shows the change in apparent brightness of the test-patch as a function of the exposure-asynchrony, a value of zero representing simultaneous presentation. The ordinate is the luminance of the test-patch required for it to appear as bright as the fixed comparison standard seen by the other eye, so that increases represent diminished

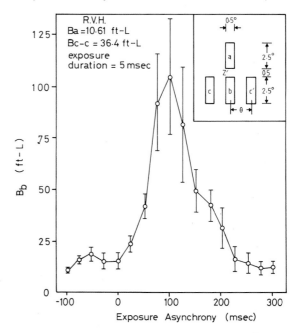

**Fig. 15.6** Showing change in apparent brightness of a test-patch as a function of the exposure asynchrony of two flashes presented in succession. Ordinate is luminance of the test-patch required to give a brightness-match with the comparison standard when the two contrast-inducing patches were flashed at various times before and after the exposure of the test- and comparison-flashes. Abscissa is time-interval between the onset of the test and comparison flashes and that of the contrast-inducing patches. *Inset*: Stimulus pattern. *a* is comparison standard seen with the left eye; *b* is the test patch seen with the right eye; *c-c'* are the two contrast-inducing patches also seen by the right eye. The fixation point, Z', is seen singly and binocularly. (Alpern, *J. opt. Soc. Amer.*)

sensitivity of the right eye. It will be seen that there is a little paracontrast with negative exposure asynchronies (the contrast-inducing patch appearing before the test-patch), and that the maximal meta-contrast occurs at an interval of about 50 msec. The meta-contrast effect operated over only a small area, so that the best effect was achieved if the edges of test and contrast-inducing fields coincided; no effect could be found in the fovea. Alpern's experiments suggested that the paracontrast effect was due to the light scattered within the eye by the preceding contrast-inducing flash.

## Cone-rod interaction

In a later study (1965) Alpern considered the possibility that the effect was due to cone-rod interaction, the test patch exciting rods and the contrast-inducing patch exciting cones and, because of the shorter latency of the cone response, the two stimuli being effectively simultaneous. He concluded, however, that meta-contrast was manifest with both rod and cone vision, but that there was no cone-rod interaction, a rod-stimulus being incapable of affecting the cone threshold, and *vice versa*.

## π-mechanisms

Alpern and Rushton (1965) showed, moreover, that when purely cone excitation was involved in both test- and contrast-inducing stimuli, the cone-cone interaction was highly specific, so that a given Stiles π-mechanism (p. 333) did not affect a test-flash exciting a different π-mechanism.

This degree of specificity is somewhat surprising since we are measuring an adaptive decrease in retinal sensitivity that presumably results from activation of a gain-control mechanism. Foster (1976) has re-examined the possibility of rod-cone interaction by using a cone-stimulating red test-flash with a rod-stimulating con-trast-inducing surround of green; as Figure 15.7 shows, there is a very definite elevation of threshold for contrast-inducing stimuli that both precede (para-contrast) and follow (metacontrast) the test stimulus.★

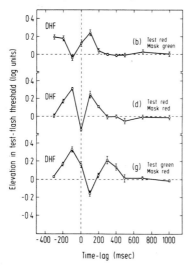

**Fig. 15.7** Elevation of threshold for a test-flash by masking stimuli preceding or following the test-flash. (Foster, *Vision Res.*)

## BROCA-SULZER EFFECT

Broca and Sulzer (1902) found that a brief light flash may appear brighter than a steadily maintained light of equivalent luminance.

### GROWTH OF SENSATION

Thus Figure 15.8 is a curve of brightness-sensation against time during which a bright light is regarded by

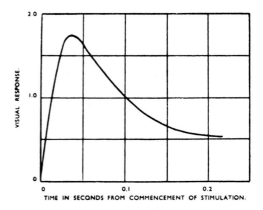

**Fig. 15.8** Actual sensation curve during the first 200 msec of stimulation by a bright light. (Wright, from Schouten.)

an observer, and was redrawn from the classical study of Broca and Sulzer. In effect, the ordinate is the luminance of a long-sustained flash of light necessary to match a brief flash of increasing duration. A number of studies have confirmed and extended this work, e.g. Katz (1964). If we accept, then, that the sensation of brightness builds up rapidly and passes through a maximum, it will be clear from Figure 15.8 that, if the light had been switched off after 50 msec the sensation, just before switching off, would have been far greater than if it had been maintained for 200 msec or more.

If, then, we expose one eye to flashes of increasing duration, and we assess the subjective brightness of each flash by matching it with a variable luminance presented to the other eye—the binocular matching technique—the first effect of increased duration will be the temporal summation described by Bloch's Law (p. 364); beyond a critical duration, sensation will not increase appreciably with further duration of the stimulus, and in fact the Broca-Sulzer effect becomes manifest as an actual *decrease* in brightness-sensation until, finally, extending the duration has no effect. This gives the impression that short flashes located in the transition zone between complete temporal summation and duration-independence are brighter than longer flashes.

### DEPENDENCE ON INTENSITY

The location in time of the Broca-Sulzer effect depends on the flash-intensity, so that Aiba and Stevens (1964), for example, found the maximum effect in the dark-adapted eye decreased from 300 msec to less than 100 msec as the flash-intensity increased.

---

★Bowen *et al.* (1977) showed that metacontrast masking failed if, instead of the test- and masking flashes being luminance-transients, i.e. increase or decrease in luminance referred to a steady background, they were colour-transients, i.e. hue-substitutions.

## ROD VISION

By reducing sensitivity of the cones with a red back-ground, the responses of rods could be examined in effective isolation; under these conditions a Broca-Sulzer effect could be demonstrated (White *et al.* 1976), the maximum effect occurring around 200 to 400 msec. As with photopic vision, the Broca-Sulzer effect depended on the temporal pattern of the flashes so that if both began together there was no effect, but there was one if they finished together or were presented successively.

## FINITE RESPONSE-TIME

The growth of visual sensation revealed by Figure 15.8 indicates in effect the finite response-time of the receptors and other elements in the visual pathway; operating against the growth is the development of adaptation or inhibition; the actual range over which the Broca-Sulzer effect will be manifest will be a function of the response-time, and this may be why the rod effect can be seen maximally later than the cone effect.

## BACKWARD MASKING

It has been known for a long time that a test flash can appear less bright, i.e. be masked, when a more intense flash follows it by a certain period, i.e. there is a retro-active brightness depression. Crawford (1947) examined the phenomenon in some detail; he found that the threshold for a 10 msec 0·5° test-flash was considerably increased if it was followed up by a 524 msec brighter flash subtending 12°, the test flash being within the larger conditioning or masking flash. The period within which this retroactive effect could be elicited was about 100 msec.

## DELAY-PERIOD

Subsequent studies (reviewed by Raab, 1963) have extended the work and the general principle has emerged that there is a delay-period between the two stimuli below which there will be a masking effect, and that this delay-period is governed by the difference in luminance between the two flashes, the dimmer the test-flash in relation to the bright (succeeding) flash, the longer the delay-period.

## LATENCIES OF FLASHES

If the bright flash is exerting its retroactive effect by an adaptive process—reducing the sensitivity of the retina to the test-flash—then we must assume that its effects reach some brightness-integrating centre before the effects of the weaker test-flash have finished, and the masking effect will be governed by the latencies of the two flashes; thus a weak flash has a longer latency than a brighter one, and this accounts for the fact that the period during which masking can occur is determined by the luminance-difference of test- and bright-flashes. It is interesting that the cortical evoked potential of the masked test-flash is inhibited so that it is, in effect, replaced by the subsequent conditioning flash's response (Donchin *et al.*, 1963).

## ROD-CONE INTERACTIONS

The Purkinje shift in the scotopic sensitivity curve at mesopic ranges of luminance is an example of rod-cone interaction whereby the relative brightnesses of different wavelengths are influenced by the brightness sensation evoked by rod vision. Frumkes *et al.* (1973) have examined the possible interactions when flash-stimuli are succeeded at different intervals, one flash being such as to stimulate rods only (420 nm) and the other to stimulate cones (680 nm). If, as it seems, the depression of brightness-sensation due to a conditioning flash

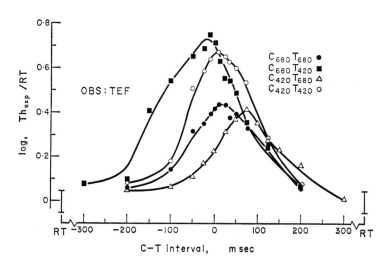

**Fig. 15.9** Test threshold as a function of the interval between the test stimulus and the conditioning stimulus (C-T interval). A negative interval indicates that the test precedes the conditioning flash onset. The test and conditioning stimuli are designed to stimulate rod and cones separately. (Frumkes *et al.*, *Vision Res.*)

depends on the difference in latencies for the test and conditioning flashes, then since rods have longer latencies than cones, the maximal effect might be obtained if the rod test-flash preceded the cone flash. As in earlier studies, the threshold began to rise when the test-flash preceded the conditioning flash by about 100 msec, and the effect reached a maximum at an interval of + 10 msec and then it returned to control values. As Figure 15.9 shows, the effect is manifest when conditioning and test-flashes stimulate different mechanisms. The authors considered that it was unlikely that the interaction occurred at the level of the horizontal cells, since primate horizontal cells receive input from cones but deliver their output to rods, and not *vice versa*, whereas the interaction described here is reciprocal.

## INCREMENTS AND DECREMENTS OF LUMINANCE

### INTERVAL VARIATION

Ikeda presented the observer with increments or decrements of luminance against a photopic background luminance, and the intervals between successive presentations were varied. Thus Ikeda was examining temporal summation. The two stimuli could be increments (+, +) or both decrements (−, −) or they could be an increment followed by a decrement (+, −) or *vice versa* (−, +). In the + + case there was complete temporal summation up to about 15 msec and this decreased, i.e. the threshold compared with that for simultaneous presentation increased up to an interval of 53 msec when summation was that expected of probability summation of two independent events. During this period there was evidence for some mutual inhibition of the two stimuli, so that threshold was higher than for a single presentation. As the luminance was decreased, the inhibitory phase was shifted towards longer times and with zero background no inhibition was observed. With the +, − type of stimulation, complete summation was revealed as an absence of sensation of light, and once again the period for complete summation was limited; it emerged that cancellation was good in the time-region where summation of like-stimuli (+, +; −, −) was large; cancellation was also large in the time-region where +, + stimuli tended to inhibit each other. Ikeda concluded that the response to a flash was diphasic, and it was the mutual interaction of these out-of-phase diphasic responses that accounted for the various manifestations. Ikeda's technique was employed by Rashbass (1970); Figure 15.10 shows the decrease in threshold as the duration of a single increment or decrement of luminance was increased; the line has a slope of unity indicating complete summation,

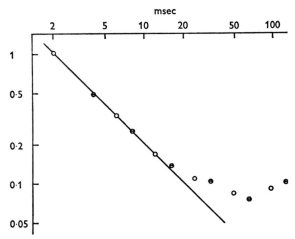

**Fig. 15.10** The threshold intensity of rectangular changes of luminance as a function of their duration. Plotted on log-log coordinates, the unit of intensity is the threshold of a 2 msec positive-going flash. The threshold intensity of a 2 msec flash is about 0·18 times the base-line intensity. Circles are positive-going incremental flashes; circles with bars across are negative-going decremental flashes. The line sloping down at 45°, drawn through the 2 msec flash threshold, represents the condition of total summation. (Rashbass, *J. Physiol.*)

falling off after about 20 msec. At 64 msec, threshold passes through a minimum suggesting some inhibition. In this experiment the change in luminance was brought on abruptly and maintained, so the condition is different from Ikeda's who presented the stimuli as separate flashes; when the same technique was employed Ikeda's result was obtained (Fig. 15.11), indicating failure of complete summation at a much earlier interval, about 10 msec, so once again we see that a patterning of the stimulus—this time a temporal discontinuity—tends to raise threshold.

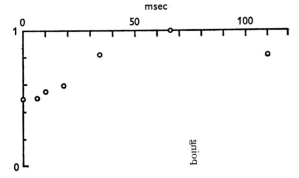

**Fig. 15.11** The threshold intensity of two equal 2 msec flashes, as a function of the interval between them. Plotted on linear co-ordinates, the unit of intensity is the threshold of a single 2 msec flash. Thresholds are calculated as the intensity for seeing with a probability of 0·5. Observations on positive-going and negative-going pairs of flashes are grouped together as there is no significant difference between them. (Rashbass, *J. Physiol.*)

## BANDPASS FILTER

Rashbass considered that the visual system contained a bandpass filter that accepted changes in the stimulus occurring at a certain range of frequencies but tended to attenuate those at high frequencies (short intervals between stimuli) and, to a lesser extent, at longer frequencies; on this basis he derived a model that described the phenomena reasonably well.

## SUBJECTIVE ESTIMATES

An interesting point is the subjective unawareness of the direction of luminance-change, so that the subject could not state that a brief decrement in luminance was a decrease or increase in subjective brightness; the stimulus was recognized as a change without sign. A change lasting 2 msec needs to be four times more intense to obtain any idea that it represents an increase or decrease from the baseline luminance, and as the duration increases the required change becomes smaller.

## BRIGHTNESS ENHANCEMENT

Donchin and Lindsley (1965) have described a subjective enhancement of brightness by a flash succeeding the test-flash by longer intervals, i.e. after the depressing effect has subsided. The brightness of the test-flash increased with delays between 100 to 150 msec and, as the delay was shortened, the increase became bigger until the masking effect began; the intervals at which enhancement occurred ranged from 150 to 20 msec depending on the luminance of the test-flash, the lower this the longer the period of enhancement. A satisfactory explanation of this enhancement is not so easy to find; the subject sees the test-flash as a contour different from the succeeding brighter flash and recognizes the period of dark between the two; as Donchin and Lindsley point out, two recognitions are made, namely the contour of the test-flash and its apparent brightness, and what is happening is that the brightness of the later flash is being attributed partly, or wholly, to the contour of the test-flash; thus an asynchrony in the brightness- and contour-messages could well result in the phenomenon.

## SPATIAL EFFECTS

## SIMULTANEOUS CONTRAST

At the time when the effects of the glare-source described by Schouten and Ornstein were interpreted as being due to an adaptive effect of one part of the retina on the rest, it was natural to assume that this α-adaptation lay at the basis of simultaneous contrast,

whereby the apparent brightness of a grey patch is diminished by surrounding it with a white background. However, with the demonstration that the glare-source effect was due to scattered light within the eye, this interpretation becomes suspect. Nevertheless, the fact of simultaneous contrast remains, in the sense that a patch of light appears much darker if surrounded by a white background than if surrounded by a black one; so it may well be that the stimulation of the surrounding retina does, indeed, reduce the sensitivity of the central retina to light.

## STEADY-STATE COMPARISONS

The results of an experiment by Heinemann (1955) are illustrated in Figure 15.12, where the inset shows the experimental arrangement. The test-field was a spot, T, surrounded

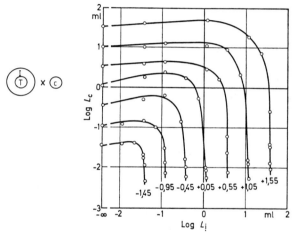

**Fig. 15.12** Effects of a surround-field on subjective brightness of a central spot. The experimental arrangement is shown on the left, the central test-spot, T, being matched with a comparison spot, C. The surround luminance is varied. The parameters on the curves are the luminances of the test fields in Log mL. (Heinemann, *Hdb. Sensory Physiol.*)

by an 'inducing field', I, and the subjective brightness of the test-field was estimated by adjusting its luminance, $L_t$, to equal that of the comparison spot, $C(L_c)$ when the luminance of I, $L_i$, was varied. As the luminance of the inducing field, $L_i$, is increased, the curves at first rise slightly, indicating that the brightness of the test-field increases—i.e. there is *enhancement*. For values of $L_i$ approximately half that of the test-field, the brightness of the test-field is depressed—the classical simultaneous contrast. Near the point where $L_i = L_t$, small increases in $L_i$ make necessary very large decreases in matching luminance, $L_c$, and when $L_i$ exceeds $L_t$ by only a few hundredths of a log-unit, the brightness of the test-field is so much reduced that a match becomes impossible. It must be emphasized that a depression in test-field brightness can also be induced by an inducing field of smaller luminance than that of the test.

**Scattered light**
It would seem, however, from the study of Fry and Alpern (1953) that a great deal of the effects of the inducing stimulus

are attributable to the scattered light producing a veiling glare that depresses sensitivity; thus the effect depends strongly on the separation of test and inducing fields, which would be expected of light-scatter. Heinemann (1972) concludes that physiologically transmitted effects at the fovea would not be observed at separations greater than 1·5 to 2°.

### INHIBITORY INTERACTION

The studies of Westheimer (1965, 1967) on the effects of surround-brightness on the threshold have demonstrated an inhibitory interaction. Westheimer caused a subject to view a test spot through an illuminated annulus and to adjust the luminance of the spot so that it could be discriminated from its surround; i.e. the value of $\Delta I$, or the *incremental threshold*, was obtained. When the size of this surround was increased, the threshold first increased, but beyond a certain point, namely 0·75 degrees, in scotopic vision it decreased by as much as one log unit up to a diameter of 2·5 degrees. The initial threshold-raising effect with increasing area is simply an expression of Ricco's Law; the annular surround is increasing its subjective brightness by spatial summation; at the point where the threshold falls a new factor is becoming evident, and this Westheimer attributes to an inhibitory action of the peripheral part of the annular surround on the more central part, so that the subjective sensation from the annulus is reduced and thus requires a smaller value of luminance of the central test spot for a match. In the fovea, too, there is a similar effect of increasing surround-area, the limit beyond which Ricco's Law fails is smaller, namely 5 minutes of arc.*

### SPATIAL SUMMATION

#### RICCO'S LAW

The existence of an area on the retina—the Ricco area—over which there is complete summation of quanta, has been discussed earlier; the phenomenon may be looked at as a special case of retinal interaction.

#### RICCO AREA AND STIMULUS PATTERN

Hallett (1963) and Sakitt (1971) have summarized the literature relating to the magnitude of Ricco's area; and the important fact emerges that this is strongly dependent on the pattern of the stimulus. Thus Bouman (1950, 1953) found complete summation for spots of light up to a diameter of 30 minutes of arc, whereas for line-shaped targets the area extended to 32 minutes. Again, when two separate spots are employed, the area in which they must fall for complete summation is much smaller than the Ricco area defined for a single spot. Sakitt measured the threshold as a function of area of stimulus, using a circular spot of increasing diameter; Figure 15.13 shows the linear relation between log threshold and log area; the slope of the line is − 1 for targets less than ½ degree in diameter, indicating complete summation up to a diameter of 30 minutes, the Ricco area for this subject.

When the thresholds for two small squares, separated by a variable distance, were measured it was found that summation was much more limited. Thus the logarithm

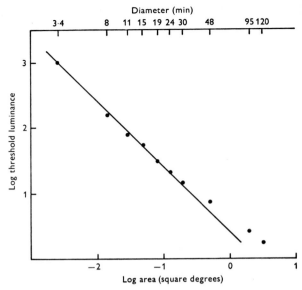

**Fig. 15.13** Illustrating spatial summation. The slope of the line has a value of − 1 indicating complete summation up to an area of about 30 min of arc. (Sakitt, 1972).

of the ratio of the two thresholds has been plotted against separation in minutes of arc (Fig. 15.14), for complete summation the ratio of thresholds would be 2 and the logarithm 0·3, and this only applies to the first point on the curve at a separation of centres that was so small that the two squares touched to form a continuous rectangle, whereas the same subject showed complete summation for circular targets up to 30 minutes of arc.

### Probability summation

If two spots excite independent neural detectors, the probability that the two spots together do not produce a sensation $(1 - f_2)$ is equal to the probability that the upper spot does not produce a sensation $(1 - f_1)$ times the probability that the lower spot does not produce a sensation, $(1 - f_1)$, i.e.

$$1 - f_2 = (1 - f_1)^2$$

where $f_1$ and $f_2$ are the frequencies of seeing the one spot and two spot targets respectively. As Table 15.1

---

* Sakmann, Creutzfeld and Schleich (1969) have shown that the area for summation of surround adaptive effects is not the same as that for excitation of a test spot in the cat's retina, in fact the receptive field of a ganglion cell for adaptive effects shows no centre-surround antagonism. Easter (1968) has measured adaptive effects of light stimuli on goldfish retinal ganglion cells, using cone-stimuli. The receptive field for the adaptive raising of threshold was certainly much larger than that for excitation of the ganglion cell, so that we may speak of an 'adaptation pool's field'; this may well be mediated by horizontal and amacrine cells.

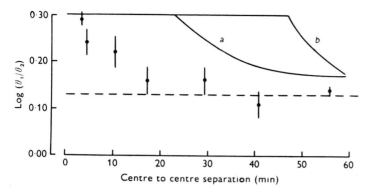

**Fig. 15.14** Illustrating reduction in summation when the test-stimuli are separated spatially. As ordinate the log of the ratio of absolute threshold luminance for one 3·4′ square, $\theta_1$, to the absolute threshold luminance of two such squares, $\theta_2$, has been plotted as a function of centre-to-centre separation of the squares. The continuous curves are predicted employing weighting functions. Summation is approximately complete at very small separations, the ratio being 2 and its log 0·3. (Sakitt, *J. Physiol.*)

**Table 15.1** Frequencies of seeing one 3·4′ square ($f_1$) or two such squares ($f_2$) separated by 20 or 23′ for three subjects. Test duration was 16 msec and the filament image was 1 mm at the pupil. Each entry is based on 120 trials and $s_1$ and $s_2$ are the standard errors of the means of $f_1$ and $f_2$ (Sakitt, 1971)

| Subject | Separation | $f_1$ | $f_2$ | $s_1$ | $s_2$ | $(1 - f_1)^2$ | $(1 - f_2)$ |
|---------|-----------|-------|-------|-------|-------|---------------|-------------|
| L.F. | 23′ | 0·48 | 0·68 | 0·05 | 0·04 | 0·27 | 0·32 |
| B.S. | 23′ | 0·18 | 0·28 | 0·04 | 0·04 | 0·67 | 0·72 |
| L.F. | 23′ | 0·46 | 0·68 | 0·05 | 0·04 | 0·29 | 0·32 |
| C.L. | 23′ | 0·41 | 0·68 | 0·05 | 0·05 | 0·35 | 0·32 |
| C.L. | 20′ | 0·39 | 0·64 | 0·05 | 0·05 | 0·37 | 0·36 |
| C.L. | 20′ | 0·47 | 0·73 | 0·05 | 0·04 | 0·28 | 0·27 |
| C.L. | 20′ | 0·33 | 0·61 | 0·05 | 0·05 | 0·44 | 0·39 |

shows, when the targets were separated by 20 or 23 minutes of arc this relationship between the frequencies of seeing holds, indicating probability summation; the dashed line in Figure 15.14 represents the log of the thresholds for one and two targets when the threshold for two targets is governed by probability summation, and it will be seen that neural, as opposed to probability, summation ceases at 20 to 30 minutes separation. By increasing the number of spots, more complex patterns of stimulus could be presented and threshold-ratios analysed; and Sakitt concluded that the spatial summation at absolute threshold was configuration-dependent but not uniquely related to a single variable, such as distance of separation. The important point to appreciate is that the introduction of a pattern into the stimulus tends to increase its absolute threshold.

### INCREMENTS AND DECREMENTS

This point was further emphasized by Cohn and Lasley (1975) who presented a subject with two spots of light that at arbitrarily chosen separations (Fig. 15.15). The luminance of pairs of spots could both be increased ( + + ), both decreased ( − − ) or there could be a paired

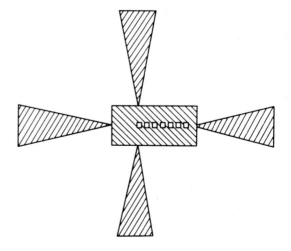

**Fig. 15.15** Illustrating the experimental stimulus-conditions. A row of seven light-emitting diodes is continuously lit at about 30 foot-lamberts. The left-most diode (*first spot*) is the fixation point and black arrows point to it to aid fixation. The left-most (first) spot and one of the other six (*second spot*) are the sites of any one of eight possible luminance modulations. (Cohn and Lasley, *Vision Res.*)

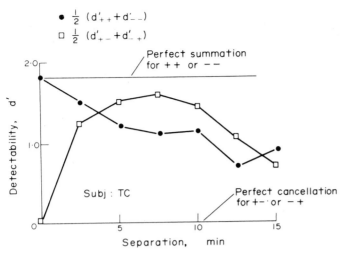

Fig. 15.16 Results of experiments using the experimental set-up of Fig. 15.15. Detectability of double increments or decrements (+ + and − −) and paired increment-decrement (+ − and − +) in luminance of spots one and two are plotted as a function of the separation of the spots. (Cohn and Lasley, *Vision Res.*)

increment-decrement (− +; + −). Perfect energy summation when an increment is presented at two separate spots would lead to the predication that detectability would be independent of the separation, and would be the same as if the spots were super-imposed; similarly for a pair of decrements. Secondly, if an increment occurred at one spot and a decrement at another, there should be cancellation and there should be no detectability regardless of separation. In fact there was no evidence for perfect summation and the results for double increments (+ +, − −) are shown as filled circles in Figure 15.16, and for paired increment-decrement (+ −, − +) as open squares, detectability being plotted as a function of separation. From these results a *summation index* $\psi$ was computed whose values would vary from − 1·0 if the two events to be detected were independent of each other (probability summation) to 0 if the detectability were governed by complete summation of the two stimuli. As Figure 15.17 shows, there was never complete summation, and this decreased up to a separation of about 4 minutes of arc, when

probability summation was reached; at larger separations the index had a larger negative value than − 1, indicating mutual inhibition between the two stimuli.

## Summing and differencing mechanisms

Cohn and Lasley interpreted their results on the plausible basis of a summing mechanism that averaged photon flux over the retinal area studied, and a differencing mechanism that detected local contrast. With superimposed stimuli, the differencing mechanism would not be active and summation would be perfect; with increasing separation the differencing mechanism would come into operation. Both mechanisms would be influenced by separation, and at large separations both might become independent (a) because, at the fovea especially, summation of photons is limited, and (b) because spatial difference-detection is governed by the receptive field of a ganglion cell. Thus, at large separations, the two component stimuli are signalled independently and $\psi$ is unity; at smaller separations

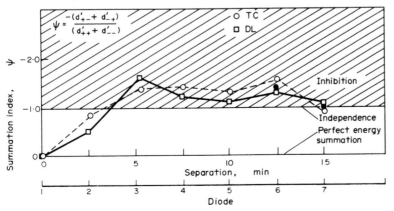

Fig. 15.17 Summation index versus separation plotted for two subjects. Ordinate: Summation index, $\psi$, computed from $-(d'_{+ -} + d'_{- +})/(d'_{+ +} + d'_{- -})$. Negative values plotted upwards. Abscissa: Separation of the two spots in minutes of arc. Shaded area indicates inhibition, $\psi < − 1$. d's are detectabilities. (Cohn and Lasley, *Vision Res.*)

there is interaction and it is because of the differencing channel that perfect summation is prevented. In this way we can account for the effects of pattern, for example Kristofferson and Blackwell found lower thresholds for non-circular targets.★

MACH BANDS

A manifestation of simultaneous contrast or spatial inductive effects is given by the so-called *Mach bands*. If a picture, such as Figure 15.18, is rotated above the

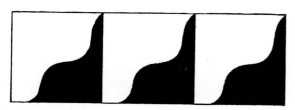

**Fig. 15.18** Illustrating the formation of Mach bands. If this figure is rotated on a drum, at a rate above the critical fusion frequency, the subject sees a grey surface with alternate light and dark stripes, although there are no actual sharp transitions of luminance on the figure. (Graham, *Vision and Visual Perception*. John Wiley Inc.)

critical fusion frequency in front of the eye, the subject sees a grey surface whose objective luminance varies in accordance with the Talbot-Plateau Law. However, although Figure 15.18 exhibits no sharp transitions of luminance, the subject sees alternate light and dark stripes on the height axis, adjacent stripes bordering each other at fairly clear lines. The light stripes correspond to the parts of the light-distribution curve that are concave to the height-axis (Fig. 15.19) and the

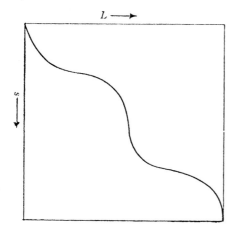

*L* ⟶

**Fig. 15.19** Illustrating the variation of luminance of the picture presented to the eye by Fig. 15.18, when rotated so as to cause fusion of successive images. The luminance is determined, under these conditions, by the Talbot-Plateau Law. (Graham, *Vision and Visual Perception*. John Wiley Inc.)

dark to those that are convex. We may presume that in the regions of fairly rapid change of luminosity, inhibition and facilitation occur, leading to the sharpening of the transition from one level of subjective sensation of greyness to another.

## RETINAL ELECTROPHYSIOLOGY AND ADAPTATION

The basic concepts of adaptation have been derived from psychophysical experiments and it becomes of interest to determine to what extent the characteristics of the ganglion cells of the retina determine adaptation phenomena, i.e. the extent to which the retina has processed the spatial and temporal features of the stimulus. Thus the centre-surround antagonism observed with many ganglion cells reveals the possibility of spatial interaction whereby light falling on a part of the retina may inhibit activity in another, but of course this interaction is on a relatively small spatial scale and is unlikely to explain the phenomena described, for example, by Westheimer.

## FIELD ADAPTATION AND ITS MECHANISM

Rushton described the change in sensitivity of the eye following exposures to light that had only a small bleaching action on the retina as field adaptation; it represents a diminished sensitivity of the visual mechanism to light, or to speak in engineering terms, there has been an automatic gain-control.

I/Q RATIO

Enroth-Cugell and Shapley (1973) used as a measure of gain, or rather its reciprocal, the ratio: $I/Q$, the impulse/quantum ratio, i.e. the number of impulses generated over the background discharge divided by the number of quanta of light effectively absorbed, a ratio which Barlow and Levick (1969) considered to be the most important parameter in determining a response. Figure 15.20 shows a typical increment-sensitivity curve, indicating the extra retinal flux (left-hand ordinates) required to evoke a criterion response of 30 imp/sec at increasing levels of background retinal luminance. The right-hand ordinates give the computed $I/Q$ ratios, which fall with increasing adaptive illumination of the retina. At the lowest levels the curve is flat, i.e. the $I/Q$ ratio is constant and is equivalent to the condition of dark-adaptation. At a particular background level, the $I/Q$ ratio begins to decrease sharply,

★Quoted by Cohn and Lasley (1975).

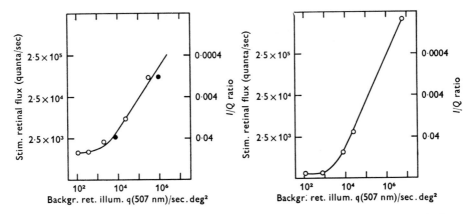

**Fig. 15.20**   Increment-sensitivity curves of retinal ganglion cells indicating the extra retinal flux (L.H. ordinate) required to evoke a criterion response of 30 imp/sec at increasing levels of background luminance. Right-hand ordinate gives calculated $I/Q$ ratios as defined in text. Open circles are for blue-green stimuli on a red background; filled circles for white stimuli on white background. (Enroth-Cugell and Shapley, *J. Physiol.*)

and the slope has a value of about $-0.9$ for the log-log plot, indicating that the sensitivity was proportional to the 0.9 power of the background, a relation close to the value expected for strict application of the Weber Law (p. 187). The level of retinal illumination at which the adaptive effect occurs is low, representing absorption of only 0.01 quanta per rod per sec; such absorption could have little effect in saturating the rod from a photochemical point of view, and in fact psychophysical studies of Aguilar and Stiles (p. 230) on man, and equivalent studies on the cat, showed that rod-saturation began only when retinal illumination was of the order of 7500 quanta rod$^{-1}$ sec$^{-1}$. Thus this field adaptation is certainly a neural rather than a photochemical phenomenon.

HORIZONTAL CELL FEEDBACK

Enroth-Cugell and Shapley presented a model of the adaptive process based on inhibitory feedback to the receptors through horizontal cells,* the hyperpolarizing potential of this cell acting as a control on transmission from receptors to bipolars, as illustrated schematically in Figure 15.21.

IMPORTANCE OF DENDRITIC FIELD

An important conclusion reached by these authors, was that the sensitivity, or reciprocal of $I/Q$, for a particular ganglion cell was governed by the amount of light-flux effectively gathered by it through its dendritic field; this is illustrated by Figure 15.22 which shows the relation between $I/Q$ and background retinal illumination for two ganglion cells with very different central fields, fields that are almost certainly governed by the size of the dendritic spread of the cell. It will be seen that, for the large central area, the transition point occurs at a

lower background illumination; and for any given illumination the $I/Q$ ratio is much larger, indicating lower sensitivity.

ADAPTATION POOL

Thus the 'adaptation pool' of the cat retina seems to be the receptors that transmit to the ganglion cell, whereas those that are responsible for the surround effects, presumably mediated through amacrine cells, do not contribute to this pool.

If the state of adaptation depends on the size of the receptive field of the ganglion cell, this means that, under many conditions, some cells will be more adapted than others; in general, cells with small fields will be the more sensitive (less adapted) and thus more suited to the detection of small targets.

CHARACTERISTICS OF THE SURROUND

The surround generally antagonizes the central spot-stimulus when the illumination of the retina is diffuse, so that the response to uniform illumination of the retina is small, and such as it is, is due to the different latencies of the centre and surround effects. Enroth-Cugell and Lennie (1975) emphasized the difficulties in determining the true effects of the surround because of the scatter of light from it which may be greater in effect than that on the surround, so that employment

---

*Enroth-Cugell and Shapley (1973) discuss some apparent objections to this thesis, but the weight of evidence seems to favour their conclusion that adaptation represents a feedback on the receptors through horizontal cells; thus adaptive inputs from many rods are pooled within horizontal cells which feedback on the rods to set the adaptation level.

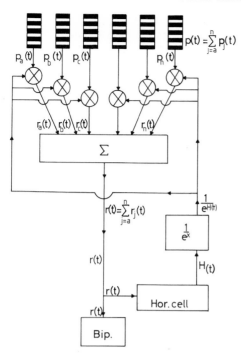

**Fig. 15.21**    Sketch of the proposed nonlinear feedback model.

$$p_a(t),\quad p_b(t),\quad \ldots,\quad p_n(t)$$

are the photocurrents in each of those individual rods which constitute the first stage of the central response mechanism of the ganglion cell. The amount of receptor transmission to bipolars and horizontals is signified by $r_a(t),\ldots,r_n(t)$. $H(t)$ is the horizontal cell potential due to its input

$$r(t) = \sum_{j=a}^{m} r_j(t).$$

An exponential function of the horizontal cell potential

$$\exp \{H(t)/H_{trig}\}$$

is what controls the sensitivity of the receptor transmission, $r(t)$, i.e.

$$r(t) = p(t) \times 1/\exp \{H(t)\}.$$

(Enroth-Cugell and Shapley, *J. Physiol.*)

of a surround in isolation, i.e. an annulus, to measure its effect on the centre gives results of very doubtful significance. Thus it is necessary to examine the effects of the surround on the centre when the latter is being excited by light, so that the effects of additional scattered light are negligible. Enroth-Cugell and Lennie steadily increased the size of a spot-stimulus until an optimum discharge was obtained; thus, as the size increased, because of summation, the threshold fell until a point was reached when the summational effect of added area was just cancelled by an inhibitory effect from the surround. Having obtained this optimum-spot, they measured the effects of diffuse illumination on the

response, but of course the diffuse illumination fell on the centre as well as the surround, so that allowance for this had to be made. The authors found that surround and centre responded in an additive fashion so that the contribution of the surround could be estimated by comparing the ganglion cell's response to diffuse illumination of its receptive field with that to an equiluminous spot which optimally stimulated the centre. When examined in this way, the surround's behaviour with regard to adaptation, etc., was often very similar to that of the centre, although working in opposite directions, with the result that centre and surround tended to balance each other over a wide range of luminances.

DARK ADAPTATION

However, it will be recalled that Barlow had found that centre-surround antagonism tended to disappear in dark-adaptation, and this is due to different stimulus-response functions of centre and surround at very low luminances, as revealed in Figure 15.23. Thus in the dark-adapted eye the surround does, indeed, contribute much less than the centre, but the difference is quantitative, not absolute. What is important is that in dark-adaptation the surround is unable to increase the centre discharge of an OFF-centre unit rather than to decrease the discharge of an ON-centre unit at ON.

BASIS OF THE OFF-DISCHARGE

According to Enroth-Cugell and Lennie (1975) the OFF-discharge is largely the result of delay in establishment of appropriate gain-control in centre and periphery; thus, under steady conditions, centre and surround are fairly closely matched and diffuse illumination produces no significant effect on the maintained discharge. On shifting to a new level, the more sluggish surround fails to adjust its gain in time with the centre, and it is the delay in adjustment that leads to OFF-responses. The same mechanism leads to 'overshoots', in the sense that an ON-centre unit at ON with diffuse illumination gives an initial sharp increase in discharge which declines over some hundreds of milliseconds.

CENTRE-SURROUND OVERLAP

According to the treatment of Rodieck and Stone (p. 285) the surround overlaps the centre extensively or completely, so that the rigid separation of the two, employed by Enroth-Cugell, is clearly not absolutely correct; as she points out, however, at any rate where X-cells are concerned, her test of studying the changes in spot-size until qualitatively different responses were obtained meant that, when this qualitative change occurred, she was observing an effect of surround, whereas up to this point the response was entirely due

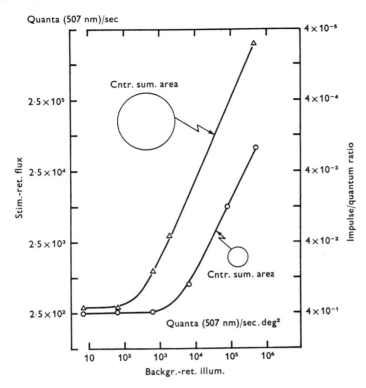

**Fig. 15.22** Increment-sensitivity curves for two ganglion cells with different centre-sizes for their receptive fields, as indicated by their summation areas. For the cell with the large central area the transition point occurs at a lower background illumination and for any given illumination the $I/Q$ ratio is much larger, indicating lower sensitivity. (Enroth-Cugell and Shapley, *J. Physiol.*)

to, or predominated by, the centre. With Y-cells the situation was more complex, the effects of their surrounds being more efficiently masked by those of their centres.

EFFECTS OF STEADY ILLUMINATION OF SURROUND

Westheimer's studies indicated that illumination of one part of the retina depressed the sensitivity of another; the question arises as to whether steady illumination of the surround affects the sensitivity of the centre of a concentric retinal ganglion cell. Sakmann *et al.* (1969) found that, when a receptive field was obtained on the basis of surround antagonism, there was no adaptive centre-surround antagonism, the effects of light on the centre being as effective as on the surround for adaptive suppression of response. This work was confirmed and extended by Enroth-Cugell *et al.* (1975) who measured the sensitivity of the centre of the receptive field to a small spot of flashing light when the field was exposed to a background spot of steadily increasing size from that of the small flashing spot until it extended well beyond the whole centre-plus-surround. Sensitivity decreased (the $I/Q$ ratio fell) by more than a factor of ten as the area of the background spot increased up to the value of the central summation area, $A_t$, beyond which there was no change. The fall in sensitivity up to $A_t$ was due to light-adaptation which summates within $A_t$, as shown earlier; the absence of further change indicates the absence of any effect, whether of

inhibition or facilitation, of the surround on the sensitivity of the centre.

**Sensitivity**

It must be appreciated that the surround was influencing the ganglion cell, in fact it decreased the maintained discharge; the point is that it failed to alter the *sensitivity* of the centre, as indicated by the number of quanta required to increase the discharge over the steady level by a fixed number of impulses ($I/Q$). The failure to influence sensitivity indicates that surround signals will either add to, or subtract from, those evoked by the central stimulus. It should be noted that when the surround stimulus was strong, an enhancement of apparent sensitivity was obtained, but this was spurious and due to the reduction of the inhibitory effect of the surround by the adapting surround light; the same effect can be produced by depressing the sensitivity of the centre by an adapting spot.

## ADAPTATION AND THE COLOUR SENSE

So far we have considered the effects of adaptation on the light and form senses; we know from the phenomena of spatial and temporal induction that the colour sense can be appreciably affected by adaptation, but it is only comparatively recently that serious quantitative measurements of the specific effects of any given wavelength have been made.

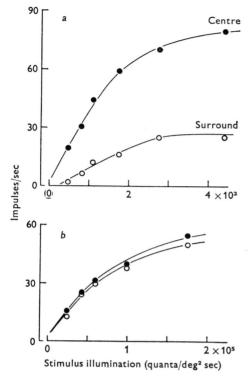

**Fig. 15.23** Stimulus-response functions for centre and surround of a ganglion cell under conditions of dark-adaptation (top) and light-adaptation (bottom). Note that under scotopic conditions the response functions of centre and surround are very different. (Enroth-Cugell and Lennie, *J. Physiol.*)

COLOUR MATCHES

Wright has applied his binocular matching technique to the problem. For example, the right eye was made to view a yellow patch; the left eye was presented with a patch which could be illuminated by the subject with three primary colours, red, green, and blue, and their proportions and intensities could be varied so that a match, both in hue and intensity, could be made with the yellow viewed by the right eye. The right eye was then exposed to an adapting light for three minutes and both eyes once again viewed their respective test and comparison patches. As a result of the adaptation it was found that the previous match no longer held in respect to either brightness or hue; and the change was measured by adjusting the intensities of the primaries as quickly as possible until a correct match was made. By repeating the matches every thirty seconds the course of recovery from adaptation could be plotted as three separate curves corresponding to the recoveries of the red, green, and blue sensations. In general, it was found that adaptation to any given wavelength had a general and specific effect; there was a general reduction in

sensitivity of the eye to all wavelengths, together with a specific reduction in sensitivity to the adapting wavelength. Consequently, a match that holds for one state of adaptation of the eye does not necessarily hold good for another. Thus with a yellow test-colour, a match was made with equal amounts of red and green before adaptation of the right eye to a white light; after adaptation, a match was made with red and green in the proportions of 38:60, i.e. far more green was required by the dark-adapted left eye to match the sensation experienced by the light-adapted right eye, a fact which indicates that an adapting white light depresses the sensitivity to red more than to green.

COLOUR-MIXING EQUATIONS

This observation does not, however, indicate that the colour-mixing equations, applicable for one level of luminance, are inapplicable for another; in fact, there is some reason to expect that, in the same eye, matches will hold independently of the condition of this eye, at any rate over a fairly wide range of light-adaptation. Thus if we take a mixture of red and green in one half of a field and match it with a mixture of yellow and white in the other, then the fact that there is a match means that the retina is responding in essentially the same way in the two halves of the field; any change in the retina, caused by adaptation to any wavelength, will influence the receptors receiving both halves of the field in the same way, so that although the subjective sensation will be altered the two halves may be expected to appear the same, and the match will be said to hold in spite of the adaptation of the retina. Thus, presentation of the two halves of the field to the two eyes separately, as in the binocular matching technique, would indicate a failure of the match, but that is because we are comparing two eyes in different states of adaptation. In practice, colour-matches do, indeed, hold under different conditions of adaptation.

ARTIFICIAL COLOUR-DEFECTIVENESS

Nevertheless, if very high adapting luminances are employed, Wright showed that uniocular colour matches may be upset, and the phenomenon was investigated in some detail by Brindley. On the basis of the trichromatic theory, the maintenance of colour matches in spite of adaptation implies that the forms of the three spectral response curves remain unaltered, since it is essentially because of their characteristics that the laws of colour matching are what they are. Thus, to consider the Red and Green response curves and their contribution to Yellow, the matching of Red + Green to equal yellow is represented by Figure 15.24; here the yellow stimulus matches the mixed Red and Green stimuli because it stimulates the Red by 3 units and the Green by 2 units. If we now adapt to red light so that the

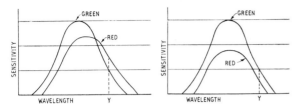

**Fig. 15.24** Showing that, if the forms of the fundamental response curves are altered by an adapting light, colour matching will be upset. *Left*: the yellow stimulus, Y, produces responses in the green and red mechanisms in the ratio of 2:3. *Right*: the same yellow stimulus produces responses in the ratio 1:2 because selective adaptation has altered the shape of the red mechanism's response curve.

shape of the Red curve is altered, then now our yellow radiation is stimulating the Red and Green mechanisms in different proportions and the match may no longer hold. By using very extreme conditions of adaptation such as might be expected to bleach the photopigment specifically from one type of cone completely, we might expect to achieve a state of artificial dichromatism, or, if we put two types of receptor out of action, one of monochromatism. In fact, all these disturbances of colour-matching may be produced (Brindley, 1953). For example, by adapting to a very bright violet light, an artificial tritanopia may be produced so that all spectral wavelengths may be matched by suitable mixtures of only two primaries, a red and a blue. With a very bright red or blue-green light this artificial tritanopia may be converted into an artificial·monochromatism, in which large parts of the spectrum can be matched with a single wavelength; thus a 'red monochromatism' was obtained by strong violet and blue-green adaptation, presumably because only the red mechanism was operative. Under these conditions all spectral colours from 500 nm to 700 nm could be matched by suitably varying the intensity of a yellow stimulus of 578 nm. The sensation was that of a very unsaturated red or pink. After adaptation with violet and red, a 'green mono-chromatism' was obtained, so that all wavelengths between 480 nm and 620 to 630 nm could be matched with a yellow, both fields appearing this time as an unsaturated blue-green.* Finally, a 'violet mono-chromatism' was obtained by adaptation to a bright yellow light; over the range of 400 to 500 nm all wavelengths were matched with violet of 447 nm, both fields appearing as a very saturated violet. Presumably, then, Brindley was putting the fundamental mechanisms out of action selectively, but of course the remaining mechanisms could hardly be called normal since they would have been subjected to quite considerable bleaching, so that the density of pigment in them would be less than that pertaining with normal trichromatic vision.

## BEZOLD-BRÜCKE PHENOMENON

When the intensity of a monochromatic light is increased considerably, there is a change of its subjective hue so that it can be matched with a different monochromatic light of lower intensity; this is the Bezold-Brücke phenomenon. The shifts are illustrated in Figure 15.25,

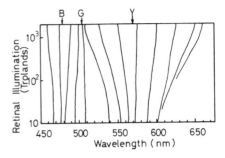

**Fig. 15.25** Wavelengths having the same hue at different levels of retinal illumination. For example, an orange of about 600 nm shifts towards the red when the retinal illuminance is increased from 10 to 1000 trolands. The arrows indicate the invariant hues. (Walraven, *J. opt. Soc. Amer.*)

where the lines join wavelengths of the same apparent hue when viewed at ten, and over a thousand, trolands; the arrows indicate the invariant hues, namely a blue of 476 nm, a blue-green of 508 nm and a yellow of 570 nm; thus oranges and reds at high intensities are equivalent to a yellow at moderate intensity, and so on. Peirce in 1877 suggested that these shifts could be explained on the basis of the Young-Helmholtz theory, with the assumption that the three sensitivity curves altered relatively to each other when the intensity-level of stimulation increased. Thus, if an orange at moderate intensity stimulated the red mechanism very strongly and the green mechanism only weakly, then increasing the intensity of the orange might be expected to increase the green's response more strongly than that of the red, and thus to shift the subjective sensation towards the green end of the spectrum, i.e. to make an orange approach a yellow. On this basis, the invarient hues would correspond to the crossing-over points of the three sensitivity curves. This is true of the invariant yellow at 507 nm if we accept Pitt's sensitivity curves, but as Walraven has shown, further assumptions are necessary to account for the other two invariant colours. As presented, the theory of Peirce, and its modification by Walraven, are obviously too naive, but the general

* In a paper published in the *Philosophical Transactions* in 1898, Burch described these 'red' and 'green' mono-chromatisms after adapting to strong violet + green or violet + red lights. It should be noted that these states last only for a short time, of the order of 10 seconds.

principle that the relative shapes of the response curves of the cone mechanisms will alter with intensity is in accordance with what would be expected on electro-physiological grounds.

## PHOTOPIC SENSITIVITY CURVES

The curves shown in Figure 14.1, p. 327, give the relative sensitivities of the light-adapted eye to different wavelengths, i.e. under conditions where the cones apparently determined the sensation. As we have seen, this sensation is mediated by three types of cone, so that one might expect to find discontinuities in the curve in regions where one type of cone tends to 'take over' from another. In fact, as we have seen, Stiles was able to create conditions where the 'taking over' by the separate mechanisms could be demonstrated with some precision, but this required conditions that ensured that he was measuring the threshold for the new mechanism, i.e. by light-adaptation with carefully chosen wave-lengths, the mechanisms with which he was not concerned could be depressed in sensitivity sufficiently to allow the single mechanism to manifest itself. Under the ordinary conditions of measurement of photopic sensitivity, where essentially what we are doing is to match different wavelengths for brightness sensation, thereby finding the relative energies required for equal sensation, we may expect all mechanisms to be operative to some extent, and thus to obtain a smooth curve of spectral sensitivity.

### 'HUMPS' ON THE CURVES

By reducing the size of the test-patch, or by using peripheral parts of the retina, however, some evidence for discontinuities, in the form of 'humps' on the sensitivity curves, can be obtained; at these humps the retina is presumably specially sensitive to a selected region of wavelengths. For example, Thomson found evidence for extra sensitivity in three regions of the spectrum, one at 460 nm, another at 520 nm and a third between 580 and 610 nm. Again, by studying the periphery, Weale found two humps indicating extra sensitivity in the blue at 460 nm and in the green at 540 nm. Using a rather different technique. Auerbach and Wald identified what they called a violet receptor with maximum sensitivity at 435 nm.

### SQUIRREL

Humps on the sensitivity curve are not peculiar to man; thus Figure 15.26 shows the mean results for a number of different species of squirrel, animals that are interesting because their retina is a pure cone one. There are two obvious peaks at 535 and 490 nm respectively, and

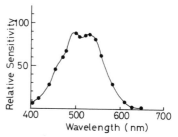

**Fig. 15.26** Mean spectral sensitivity curve for several species of squirrel, determined electroretinographically. Ordinate: percentage sensitivity. Abscissa: wavelength in nm of stimulating light. Intensity was adjusted so that at each wavelength the same number of quanta/sec fell on to the retina. (Tansley, Copenhaver and Gunkel, *Vision Res.*)

here it is interesting that neither of these coincides with the absorption peaks for the pigments that can be extracted from the squirrel retina, which is actually a rhodopsin with $\lambda_{max}$ of 502 nm (Dartnall). By selectively adapting the retina with blue or green light the two humps could be selectively depressed, and this suggests that there are different pigments mediating blue and green sensitivity.

## COLOUR MATCHING

Not only may the sensitivity curves be 'abnormal' under certain conditions but also the phenomena of colour-matching.

### FOVEAL TRITANOPIA

König, in 1894, discovered that the central fovea is tritanopic, blues and greens, and whites and yellows being confused if the matching fields were kept very small and central fixation was accurately maintained. Later Willmer rediscovered the phenomenon and suggested that the blue receptor—apparently lacking in the central fovea—was the rod, which behaved as a colour receptor under photopic conditions. However, this tritanopia is probably due, as Hartridge showed, to the use of very small fields *per se*, since a similar dichromacy could be observed in the periphery, provided the field was made small enough. This loss of blue sensitivity may be due to a relative rarity of the 'blue' cones, so that unless the field is large enough an insufficient number are stimulated to permit their effects to be manifested.

### FOVEA AND PERIPHERY

Frequently, however, an increased sensitivity to blue may be demonstrated under suitable conditions; in general, according to Weale, the visual functions of the different regions of the retina may be summarized as follows:

| Retinal area | Extent from centre nasally | Type of vision |
|---|---|---|
| Foveal centre | 50' | Tritanopia |
| Fovea | 2°30' | Trichromatic |
| Parafovea | 4°10' | Trichromatic, with enhanced blue sensitivity |
| Perifovea | 9°10' | Anomalous trichromacy with enhanced blue sensitivity |
| Inner periphery | 14°10' to | |
| Intermediate periphery | 24°10' | |
| External periphery | 77°35' | Dichromacy with enhanced blue sensitivity |

Thus, in the periphery, the sensitivity to blue seems to be *enhanced*.

## MESOPIC SENSITIVITY CURVES

The Purkinje shift has been described as a change in the luminosity curve, or spectral sensitivity function, as one passes from scotopic to photopic levels of luminance; at the intermediate ranges the sensitivity function becomes intermediate between that for rods and that for cones, and it is as though both were contributing to the sense of luminosity at the different wavelengths. When the quantitative aspects of these changes are investigated, however, it may be shown that a given mesopic luminosity curve cannot be constructed by linear addition of the pure scotopic and photopic curves, so that we must envisage interaction between rods and cones at some level in the visual pathway (Hough, 1968). It is interesting that when one studies the successive changes in the luminosity curve as the luminance is increased, it is as though Stiles' blue cone mechanism were coming into play with greater and greater effect, rather than all three cone-types. This was strikingly confirmed by Hough and Ruddock (1969), who found that in the tritanope, whose blue cone mechanism is defective, the Purkinje shift is very defective, the luminosity curve at a mesopic level of 10 photopic trolands being identical with the scotopic luminosity curve, whereas in a normal trichromat the

maximum sensitivity had changed from the scotopic 505 nm to some 540 nm. In view of the mixing of rod and cone responses that may occur as a result of the connections of horizontal cells (p. 173), it may well be that the mesopic luminosity curve reveals this interaction.[*]

### PLATEAU-PERIOD

During the plateau-period of the dark-adaptation curve of Figure 7.1, p. 186 the threshold remains constant although the rods are adapting but have not yet come to threshold. During this period, then, we may expect visual function to be governed purely by cones in peripheral vision, so that the mesopic range comes on later than this when cones may be activated by intensities that are sufficient to excite rods, i.e. beyond the break in the curve. Stabell and Stabell (1975, 1976) have shown that the colour-matching functions, and the luminosities of photopic stimuli measured during the period immediately following light-adaptation, remain invariant during the plateau period; thus only when the rod-threshold is exceeded is there rod-cone interaction.

---

[*]Dodt (1967) has described a Purkinje shift in the pure rod eye of the bush-baby, *Galago crassicaudatus*, when the visual response was measured by the height of the *b*-wave of the ERG.

## REFERENCES

Aiba, T. S. & Stevens, S. S. (1964) Relation of brightness to duration and luminance under light- and dark-adaptation. *Vision Res.* **4**, 391–401.

Alpern, M. (1953) Metacontrast. *J. Opt. Soc. Amer.* **43**, 648–657.

Alpern, M. (1965) Rod-cone independence in the after-flash effect. *J. Physiol.* **176**, 462–472.

Alpern, M. & Rushton, W. A. H. (1965) The specificity of the cone interaction in the after-flash effect. *J. Physiol.* **176**, 473–482.

Auerbach, E. & Wald, G. (1954) Identification of a violet receptor in human colour vision. *Science*, **120**, 401–405.

Barlow, H. B. (1958) Temporal and spatial summation in human vision at different background intensities. *J. Physiol.* **141**, 337–350.

Barlow, H. B. & Levick, W. R. (1969) Three factors limiting the reliable detection of light by ganglion cells of the cat. *J. Physiol.* **200**, 1–24.

Baumgardt, E. & Ségal, J. (1946) La localisation du mécanisme inhibiteur dans la perception visuelle. *C. R. Soc. Biol., Paris*, **140**, 431–432.

Baylor, D. A. & Fettiplace, R. (1975) Light path and photon

capture in turtle photoreceptors. *J. Physiol.* 248, 433–464.

Baylor, D. A. & Fuortes, M. G. F. (1970) Electric responses of single cones in the retina of the turtle. *J. Physiol.* 207, 77–92.

Baylor, D. A. & Hodgkin, A. L. (1973) Detection and resolution of visual stimuli by turtle photoreceptors. *J. Physiol.* 234, 163–198.

Bowen, R. W., Pokorny, J. & Cacciato, D. (1977) Metacontrast masking depends on luminance transients. *Vision Res.* 17, 971–975.

Brindley, G. S. (1952) The Bunsen-Roscoe law for the human eye at very short durations. *J. Physiol.* 118, 135–139.

Brindley, G. S. (1953) The effects on colour vision of adaptation to very bright lights. *J. Physiol.* 122, 332–350.

Brindley, G. S. (1967) The deformation phosphene and the funnelling of light into rods and cones. *J. Physiol.* 188, 24–25P.

Broca, A. & Sulzer, D. (1902) La sensation lumineuse en fonction du temps. *J. Physiol. Path. gén.* 4, 632–640.

Cohn, T. E. & Lasley, D. J. (1975) Spatial summation of foveal increments and decrements. *Vision Res.* 15, 389–399.

Crawford, B. H. (1947) Visual adaptation in relation to brief conditioning stimuli. *Proc. Roy. Soc., B.* 134, 283–302.

Denton, E. J. & Wyllie, J. H. (1955) Study of the photosensitive pigments in the pink and green rods of the frog. *J. Physiol.* 127, 81–89.

Dodt, E. (1967) Purkinje-shift in the rod eye of the bush-baby, *Galago crassicaudatus*. *Vision Res.* 7, 509–517.

Donchin, E. & Lindsley, D. B. (1965) Retroactive brightness enhancement with brief paired flashes of light. *Vision Res.* 5, 59–70.

Donchin, E., Wicke, J. D. & Lindsley, D. B. (1963) Cortical evoked potentials and perception of paired flashes. *Science* 141, 1285–1286.

Donner, K. O. & Rushton, W. A. H. (1959) Rod-cone interaction in the frog's retina analysed by the Stiles-Crawford effect and by dark-adaptation. *J. Physiol.* 149, 303–317.

Easter, S. S. (1968) Adaptation in the goldfish retina. *J. Physiol.* 195, 273–281.

Enoch, J. M. (1960) Waveguide modes: are they present, and what is their possible role in the visual system. *J. opt. Soc. Am.* 50, 1025–1026.

Enoch, J. M. (1961) Visualization of wave-guide modes in retinal receptors. *Amer. J. Ophthal.*, Pt. 2 51, 1107–1118.

Enroth-Cugell, C. & Lennie, P. (1975) The control of retinal ganglion cell discharge by receptive field surrounds. *J. Physiol.* 247, 551–578.

Enroth-Cugell, C., Lennie, P. & Shapley, R. M. (1975) Surround contribution to light adaptation in cat retinal ganglion cells. *J. Physiol.* 247, 579–588.

Enroth-Cugell, C. & Shapley, R. M. (1973) Adaptation and dynamics of cat retinal ganglion cells. *J. Physiol.* 233, 271–309.

Enroth-Cugell, C. & Shapley, R. M. (1973) Flux, not retinal illumination, is what cat retinal ganglion cells really care about. *J. Physiol.* 233, 311–326.

Flamant, F. & Stiles, W. S. (1948) The directional and spectral sensitivities of the retinal rods to adapting fields of different wavelengths. *J. Physiol.* 107, 187–202.

Foster, D. H. (1976) Rod-cone interaction in the after-flash effect. *Vision Res.* 16, 393–396.

Frumkes, T. E. *et al.* (1973) Rod-cone interaction in human scotopic vision. I. Temporal analysis. *Vision Res.* 13, 1269–1282.

Fry, G. A. & Alpern, M. (1953) The effect of a peripheral glare source upon the apparent brightness of an object. *J. opt. Soc. Am.* 43, 189–195.

Glezer, V. D. (1965) The receptive fields of the retina. *Vision Res.* 5, 497–525.

Hagins, W. A. (1955) The quantum efficiency of bleaching of rhodopsin *in situ*. *J. Physiol.* 129, 22–23P.

Hallett, P. E. (1963) Spatial summation. *Vision Res.* 3, 9–24.

Heinemann, E. G. (1972) Simultaneous brightness induction. *Hdb. Sensory Physiol.* VII/4, 146–169.

Hough, E. A. (1968) The spectral sensitivity functions for parafoveal vision. *Vision Res.* 8, 1423–1430.

Hough, E. A. & Ruddock, K. H. (1969) The parafoveal visual response of a tritanope and an interpretation of the $V_\lambda$ sensitivity functions of mesopic vision. *Vision Res.* 9, 935–946.

Ikeda, M. (1965) Temporal summation of positive and negative flashes in the visual system. *J. opt. Soc. Amer.* 55, 1527–1534.

Katz, M. S. (1964) Brief flash brightness. *Vision Res.* 4, 361–373.

Kirschfeld, K. & Snyder, A. W. (1976). Measurement of a photoreceptor's characteristic waveguide parameter. *Vision Res.* 16, 775–778.

Latch, M. & Lennie, P. (1977) Rod-cone interaction in light adaptation. *J. Physiol.* 269, 517–534.

Raab, D. H. (1963) Backward masking. *Psychol. Bull.* 60, 118–129.

Rashbass, C. (1970) The visibility of transient changes in luminance. *J. Physiol.* 210, 165–186.

Sakitt, B. (1971) Configuration dependence of scotopic spatial summation. *J. Physiol.* 216, 513–529.

Sakitt, B. (1972) Counting every quantum. *J. Physiol.* 223, 131–150.

Sakmann, B., Creutzfeldt, O. & Schleich, H. (1969) An experimental comparison between the ganglion cell receptive field and the receptive field of the adaptation pool in the cat retina. *Pflüg. Arch. ges. Physiol.* 307, 133–137.

Saunders, R. McD. (1975) The critical duration of temporal summation in the human central fovea. *Vision Res.* 15, 699–703.

Schouten, J. F. & Ornstein, L. S. (1939) Measurements on direct and indirect adaptation by means of a binocular method. *J. opt. Soc. Am.* 29, 168–182.

Snyder, A. W. & Pask, C. (1973) The Stiles-Crawford effect—explanation and consequences. *Vision Res.* 13, 1115–1137.

Stabell, U. & Stabell, B. (1975) The effect of rod activity on colour matching functions. *Vision Res.* 15, 1119–1123.

Stabell, U. & Stabell, B. (1976) Absence of rod activity from peripheral vision. *Vision Res.* 16, 1433–1437.

Stiles, W. S. (1937) The luminous efficiency of monochromatic rays entering the eye pupil at different points and a new colour effect. *Proc. Roy. Soc. B* 123, 90–118.

Stiles, W. S. & Crawford, B. H. (1933) The luminous efficiency of rays entering the eye pupil at different points. *Proc. Roy. Soc. B* 112, 428–450.

Thomson, L. C. (1951) The spectral sensitivity of the central fovea. *J. Physiol.* 112, 114–132.

Tobey, F. L. & Enoch, J. M. (1973) Directionality and waveguide properties of optically isolated rat rods. *Invest. Ophthal.* 12, 873–880.

Tobey, F. L., Enoch, J. M. & Scandrett, J. H. (1975) Experimentally determined optical properties of goldfish cones and rods. *Invest. Ophthal.* 14, 7–23.

Van Loo, J. A. & Enoch, J. M. (1975) The scotopic Stiles-Crawford effect. *Vision Res.* 15, 1005–1009.

Walraven, P. L. (1961) On the Bezold-Brücke phenomenon. *J. opt. Soc. Am.* **51**, 1113–1116.

Weale, R. A. (1953) Spectral sensitivity and wavelength discrimination of the peripheral retina. *J. Physiol.* **119**, 170–190.

Westheimer, G. (1965) Spatial interaction in the human retina during scotopic vision. *J. Physiol.* **181**, 881–894.

Westheimer, G (1967) Spatial interaction in human cone vision. *J. Physiol.* **190**, 139–154.

White, T. W., Collins, S. B. & Rinalducci, E. J. (1976) The Broca-Sulzer effect under scotopic viewing conditions. *Vision Res.* **16**, 1439–1443.

Willmer, E. N. (1946) *Retinal Structure and Colour Vision.* Cambridge: Cambridge University Press.

Winston, R. & Enoch, J. M. (1971) Retinal cone receptor as an ideal light collector. *J. Opt. Soc. Amer.* **61**, 1120–1121.

Wright, W. D. (1939) The response of the eye to light in relation to the measurement of subjective brightness and contrast. *Trans. Illum. Eng. Soc.* **4**, 1–8.

Wright, W. D. (1946) *Researches on Normal and Defective Colour Vision.* London: Kimpton.

Wright, W. D. & Nelson, J. H. (1936) The relation between the apparent intensity of a beam of light and the angle at which the beam strikes the retina. *Proc. Phys. Soc.* **48**, 401–405.

# The muscular mechanisms

# 16. The extraocular muscles and their actions

Apart from its importance in the treatment of squint, an adequate knowledge of the movements of the eyes, and their neuromuscular control, is of considerable value in understanding the physiological mechanisms concerned with the maintenance of posture; moreover, it would seem that our perceptions of objects, and particularly of their spatial relationships, are determined in part by the laws governing the movements of the eyes.

## THE MUSCLES

### THE RECTI

The muscles concerned with the eye movements are six in number: the *medial* and *lateral recti*, which determine sideways movements; the *superior* and *inferior recti* causing primarily upward and downward movements, and the *superior* and *inferior obliques*, which, when acting alone, primarily cause the eye to twist around its antero-posterior axis.

The insertions of the rectus muscles are indicated in Figure 16.1, and since the muscles arise from a common origin at the apex of the orbit, their main actions of elevation, depression, adduction (i.e. pulling towards the nose) and abduction, are not difficult to appreciate.

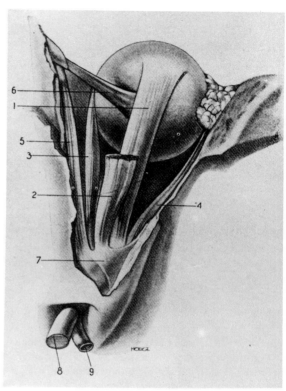

**Fig. 16.2** Extrinsic muscles of the eye from above. 1, superior rectus; 2, levator palpebrae superioris; 3, medial rectus; 4, lateral rectus; 5, superior oblique; 6, reflected tendon of the superior oblique; 7, annulus of Zinn; 8, optic nerve; 9, ophthalmic artery. (Duke-Elder, *Textbook of Ophthalmology*.)

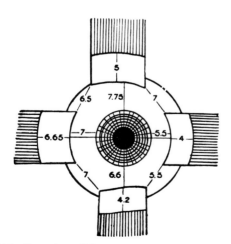

**Fig. 16.1** Insertions of the rectus muscles. The medial rectus is to the right of the diagram. (Duke-Elder, *Textbook of Ophthalmology*.)

### THE OBLIQUES

The superior oblique (Fig. 16.2) likewise arises from the apex of the orbit but after running above the medial rectus almost to the orbital margin, it becomes tendinous as it passes through a cartilaginous pulley, the trochlea, and turning sharply backward, downwards, and laterally over the eye it is inserted into the postero-lateral aspect of the sclera. Since the muscle pulls from above on an attachment behind the equator of the eye, it acts as a depressor.

The inferior oblique (Fig. 16.3) arises from the anterior part of the orbit, in the antero-medial corner

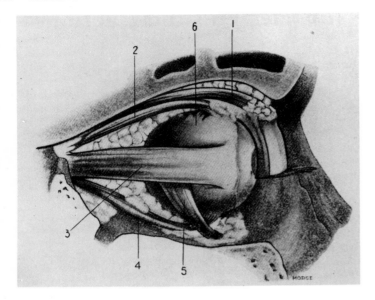

**Fig. 16.3** Extrinsic muscles of the eye. Lateral view. 1, levator palpebrae superioris; 2, superior rectus; 3, lateral rectus; 4, inferior rectus; 5, inferior oblique; 6, superior oblique. (Duke-Elder, *Textbook of Ophthalmology*.)

of its floor. It passes laterally and backwards beneath the inferior rectus and is inserted into the postero-lateral aspect of the globe. Pulling from below on an attachment behind the equator the inferior oblique thus acts as an elevator.

## DEFINING THE MOVEMENTS

### AXES OF ROTATION

In describing the actions of the muscles, we are concerned with changes in the direction in which the eye is 'pointing' i.e. the orientation of the *fixation axis* (defined as a line joining the point of fixation with the centre of rotation of the eye; Fig. 16.4). As a *primary position* or starting point, we may consider that position in which the eye is looking straight ahead when the head is erect. If the fixation axis swings horizontally, the eye is said to *adduct* or *abduct*, according as it swings

towards or away from the nose; if the fixation axis swings vertically, the eye is *elevated* or *depressed*. After any purely horizontal or vertical movement of the fixation axis from the primary position, the eye is said to have reached a *secondary position*. If the eye executes a movement which causes both a horizontal and vertical displacement of the fixation axis (e.g. if the eye looks up and to the right) it is said to adopt a *tertiary position*.

Experiment has shown that the movements of the eye in general can be approximately described as rotations about a fixed point,\* the *centre of rotation*, situated some 13·4 mm behind the anterior surface of the cornea. To define a movement it is, of course, not sufficient to indicate the centre of rotation, but an *axis*, passing through the centre, about which the eye turns, must be given; thus adduction or abduction is achieved by rotation about a vertical axis, the *z-axis* (Fig. 16.5); elevation or depression by rotation about a transverse horizontal axis, the *x-axis*. The fixation axis can be made to point in any direction by rotating the eye first about one axis and then about the other—alternatively the necessary position of the eye can be attained by a single rotation about an axis intermediate between the *x*- and *z*-axes. It is worth noting that when the eye is in the primary position the *x*- and *z*-axes are in a vertical frontal plane—called *Listing's plane* (Fig. 16.5)—and that *any* direction of the fixation axis can be obtained by rotating the eye about an axis lying in this plane.

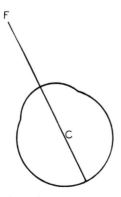

**Fig. 16.4** The fixation axis.

\* This is, indeed, an approximation; it would seem from the most recent work on this subject that the 'centre of rotation' varies within a range of about 2 mm according to the nature of the eye movement; in Van der Hoeve's view, this variability permits a more accurate adjustment of the eyes than would be possible with a fixed centre of rotation.

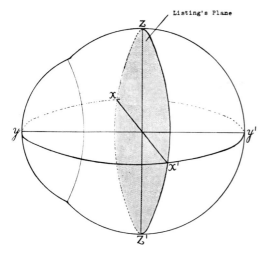

**Fig. 16.5** The primary axes of rotation and Listing's Plane. *xx'*, transverse horizontal, giving elevation and depression; *yy'*, antero-posterior, giving rolling or torsion; *zz'*, vertical, giving adduction or abduction.

## TORSION OR ROLLING

The physiologist is concerned not only with the direction of the fixation axis—which tells him where the eye is pointing—but also with the orientation of a certain fixed plane in the eye—the *retinal horizon* (Fig. 16.6).

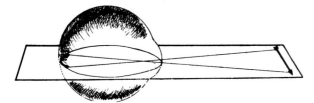

**Fig. 16.6** The retinal horizon. The image of the arrow, fixated in the primary position of the eye, falls on the horizontal meridian.

Imagine the eye in the primary position looking straight ahead; a horizontal plane through the pole of the cornea is the retinal horizon and the image of any point in space lying on this plane falls on a horizontal line on the retina—the *horizontal meridian*. The retinal horizon, being fixed in relation to the eye, moves with it. Now we shall see that the appreciation of direction—of a line for example—depends in large measure on the directions the images of external objects make on the retina, and if the orientation of the retina is upset, estimates of direction can become erroneous. Thus a horizontal line is appreciated as such because its image falls on, or is parallel to, the horizontal meridian; if the retina is twisted round, the horizontal line appears to be inclined unless the twist is compensated psychologically (p. 512).

We have thus to take into account the influence of movements of the eye on the orientation of the retina, or what amounts to the same thing, the orientation of the retinal horizon. So long as this plane keeps a definite orientation in space the images of horizontal lines in a frontal plane fall on the horizontal meridian or are parallel to it; the problem is how to define a position of the retinal horizon which we may consider normal—i.e. where there is no 'twist' of the retina; for the primary position of the eye this is easy, the retinal horizon must be horizontal and, if the eye rotates about its antero-posterior axis (the *y*-axis) from this position, the retinal horizon becomes inclined to the horizontal and the eye is said to have *rolled* or to have attained an *angle of rolling* or *torsion*. For other positions of the eye we cannot use the horizontal as our reference plane, e.g. if the eye is directed vertically upwards the retinal horizon is not twisted round but it is not horizontal. The normal position of the retinal horizon can be defined in relation to a fixed plane in space—the *median plane*, a vertical plane through the nose (Fig. 16.7); so long as

**Fig. 16.7** The median plane.

the retinal horizon remains perpendicular to this, its position is normal and there is no torsion. Thus in Figure 16.8 the eye is represented as looking up and to the right; the retinal horizon is inclined upwards, but it remains at right-angles to the median plane; there is no torsion and a horizontal line on the wall is parallel with the retinal horizon. In Figure 16.9 the retinal horizon has been twisted, i.e. there has been torsion; it is now no longer at right-angles to the median plane and a

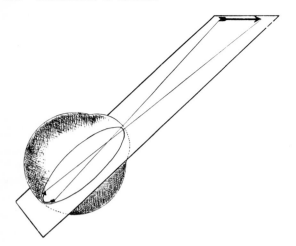

**Fig. 16.8** The eye is inclined up and to the right. No rolling has occurred, and the arrow produces an image coinciding with the horizontal meridian.

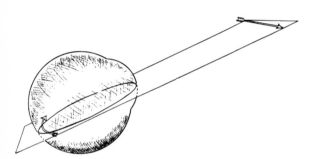

**Fig. 16.9** The eye is inclined up and to the right. Torsion has occurred, so that the horizontal arrow produces an image inclined across the horizontal meridian.

horizontal line on the wall is not parallel with the retinal horizon; its image on the retina is therefore not parallel with the horizontal meridian.*

### ROTATIONS ABOUT INTERMEDIATE AXES

Rolling is brought about by rotation of the eye around its antero-posterior axis; it is not difficult to see that a combined elevation and rolling can be achieved by rotating the eye first about the $x$-axis and then about the $y$-axis, or alternatively by rotating about a horizontal axis intermediate between the two. A motion consisting of all three types, e.g. elevation, adduction, and rolling, can be obtained by choosing an axis with components along each of the three primary $x$-, $y$-, and $z$-axes.

## ACTIONS OF THE INDIVIDUAL MUSCLES

We are now in a position to analyse the actions of the individual muscles.

### THE MEDIAL AND LATERAL RECTI

In Figure 16.10 is shown a horizontal meridional section through the eye; the medial and lateral recti are indicated by the heavy lines; they run over the surface of the globe but the pull takes place where the muscle is tangential to this surface, at T, the *tangential point*, and the direction of pull is long the *tangential line of force*, TT'. The *muscle plane* is defined as the plane containing the tangential line of force, TT', and the centre of rotation C, i.e. it is the plane of the paper in Figure 16.10. If C, the centre of rotation, is to remain

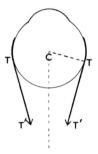

**Fig. 16.10** The actions of the medial and lateral recti. T, tangential point; TT', tangential line of force; C, centre of rotation. The muscle plane is the plane of the paper.

fixed in space, a pull along TT' must result in a rotation about an axis through C projecting vertically through the paper; in general we may state that *any muscle will rotate the eye about an axis perpendicular to the muscle plane concerned*. The muscle planes of the medial and lateral recti may be taken as horizontal and they thus rotate the eye about a vertical axis through C which has already been defined as the $z$-axis. Such a rotation leads, as we have seen, to a simple movement of adduction or abduction.

### THE SUPERIOR AND INFERIOR RECTI

The muscle planes of the superior and inferior recti are vertical and approximately identical; however they do not pass through the antero-posterior axis of the eye but are inclined at about 23° to it. The axis of rotation for these muscles is perpendicular to this plane and therefore lies horizontally but, as a result of this inclination, it does not correspond with the transverse horizontal $x$-axis but is inclined 23° away from it. The position is indicated in Figure 16.11, which shows a horizontal section through a right eye. The vertical muscle plane intersects this section in the line MM' inclined at an angle of 23° to the antero-posterior axis, $yy'$. The axis of rotation is the line AA' inclined at an angle of 23° to the transverse horizontal axis, $xx'$.

---

*The definition of the primary position can now be modified to take into account the orientation of the retina; it is the position when the eye is looking straight ahead with head erect and the retinal horizon horizontal.

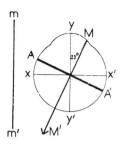

**Fig. 16.11** Actions of the superior and inferior recti (right eye). The muscle plane intersects the horizontal section of the eye in MM′. AA′ is the axis of rotation mm′ is the intersection of the median plane with the plane of the paper.

Rotation of the eye about $xx'$ causes pure elevation or depression; rotation about $yy'$ causes pure rolling; hence rotation about the intermediate axis AA′ causes a movement involving both these effects; thus the superior rectus causes a rotation such that M rises out of the paper and its action is therefore *elevation and inward rolling*, the movement of the retinal horizon being clockwise; the inferior rectus causes *depression and outward* rolling.

## Adduction

Besides causing elevation (or depression) and rolling, the superior and inferior recti act as adductors, causing the fixation axis to move towards the median plane. This is not immediately evident since it is clear that there can be no rotation about the vertical $z$-axis, the axis of rotation AA′ being perpendicular to it; this third action results from referring the orientation of a fixed direction in the eye—the fixation axis—to a plane outside the eye—e.g. the median plane through the head.

Thus in Figure 16.11 let us imagine that the superior rectus acts; $yy'$ corresponds to the fixation axis, which in the primary position remains parallel to the median plane, $mm'$; when the eye is elevated, $y$ moves out of the plane of the paper and moves through a circle which is clearly not parallel to $mm'$; hence $y$ becomes closer to $mm'$ whilst $y'$ moves farther away, and this can only result in the line $yy'$, the fixation axis, being directed towards the median plane, $mm'$, i.e. the superior rectus causes adduction. The inferior rectus causes $y$ to move downwards, but it still moves towards $mm'$ whilst $y'$ moves away; hence both superior and inferior recti cause adduction.

The difficulty arises from the physiological necessity of referring a direction in the eye, the fixation axis for example, to a plane outside it, the median plane; if the axis of rotation is parallel with, or perpendicular to, this plane of reference the inclination of this direction in the eye is unchanged during rotation; thus rotation about $xx'$ gives pure elevation because the axis of rotation is at right angles to the reference plane, $mm'$; similarly rotation about $yy'$ causes pure rolling because $yy'$ is

parallel to $mm'$; AA′, the axis of rotation for the superior and inferior recti, is neither parallel with, nor perpendicular to $mm'$, hence the inclination of the fixation axis to $mm'$ must alter during rotation, i.e. adduction must occur.

## Model experiment

The problem becomes very much clearer with a model; a tennis ball may be used; a point is marked to represent the pole of the cornea and a knitting needle is inserted through the centre to represent any desired axis of rotation. If the needle is passed vertically through the ball, rotation causes adduction or abduction; a horizontal needle through the equator permits elevation and depression; a horizontal one through the pole of the cornea causes pure rolling. A horizontal needle passed in a direction intermediate between these two positions, which gives us the axis of rotation for the superior and inferior rectus muscles, causes both rolling and elevation or depression, and it will be seen also that the direction in which the pole of the cornea points, referred to a vertical plane parallel to the original direction of the fixation axis, changes during the rotation. It becomes quite clear that this change of direction is not brought about by rotation about a vertical axis since the knitting needle is held horizontally all the time.

### THE SUPERIOR AND INFERIOR OBLIQUES

The muscle planes of the superior and inferior obliques may be considered to be vertical, as with the recti, but to run at an angle of about 50° to the antero-posterior axis, and on the opposite side, as in Figure 16.12. The axis of rotation, AA′, is likewise inclined at

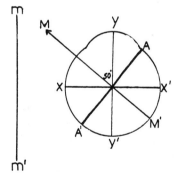

**Fig. 16.12** Actions of the obliques. (Right eye.) AA′ is the axis of rotation.

50° to the transverse horizontal axis. Contraction of the inferior oblique causes rotation about AA′ such that $y$ moves up out of the paper; the eye is thus elevated and rolled outwards; moreover, since the axis of rotation is inclined towards the median plane, the direction of the fixation axis, $yy'$, does not remain constant in respect to this median plane; as $y$ comes up out of the paper it clearly moves away from $mm'$ and thus the inferior oblique causes *abduction*; the superior oblique causes depression, inward rolling and abduction.

### SUMMARY

The individual actions of the muscles, thus deduced, may be summarized as follows:

| Muscle | Main action | Secondary action |
|--------|-------------|------------------|
| Lateral rectus | Abduction | None |
| Medial rectus | Adduction | None |
| Superior rectus | Elevation | Inward rolling and adduction |
| Inferior rectus | Depression | Outward rolling and adduction |
| Inferior oblique | Outward rolling | Elevation and abduction |
| Superior oblique | Inward rolling | Depression and abduction |

## COMBINED ACTIONS OF MUSCLES

As we shall see, a given movement involves activity in, or inhibition of, several if not all of the six extraocular muscles. It is therefore important to consider their combined actions. Furthermore, we have so far only discussed the direction of pull of a muscle from the primary position of the eye; since this must change frequently as the eye moves out of the primary position, the theoretical axis of rotation must also alter.

### EFFECT OF CHANGED POSITION OF EYE

The simplest example of the effects of moving out of the primary position on the action of a muscle is given by Figure 16.13 which illustrates the effects of abduction of the right eye on the direction of pull of the superior rectus. Obviously the rolling and adduction will decrease as the antero-posterior axis of the eye approaches the plane of the muscle, whilst the elevating action will increase. An exactly reverse relationship must exist with the obliques. The variations in the degree of elevation or depression exerted by the superior and inferior rectus, according as the eye is caused to adduct or abduct before the muscle acts, are shown in Figure 16.14 after Boeder. Here we imagine that we are looking into a sphere concentric with the left eye, and the lines are those that would be traced out by the fixation axis when the superior or inferior rectus acted alone, and caused a rotation of 50° about the axis of rotation. It is seen that, as the initial position becomes more and more one of adduction, the elevation caused by a given degree of rotation about the axis of rotation becomes smaller and smaller, whilst it increases with abduction.*

### DIAGNOSTIC DIRECTIONS OF GAZE

These essentially theoretical considerations on the change in the direction of pull with adduction and abduction, according to which the superior and inferior recti become more efficient in abduction whilst the inferior and superior obliques become more efficient in adduction, have led to the clinical concept of the *diagnostic directions of gaze*. In most movements of elevation and depression the obliques and the recti cooperate; thus a pathologically reduced ability to elevate could be due to paresis of either the superior rectus or the inferior oblique. If, however, the defect

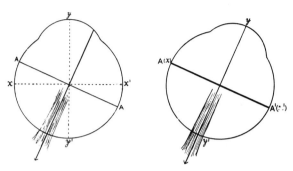

**Fig. 16.13** Illustrating the effect of abduction on the action of the superior rectus. Abduction causes the *x*-axis of the eye to approach the axis of rotation, AA′, so that rolling and adduction are decreased.

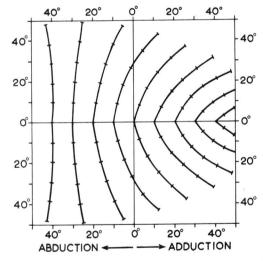

**Fig. 16.14** Trace of line of fixation on a sphere concentric with the eye when the superior and inferior recti act individually from positions of adduction or abduction. If we imagine we are viewing inside the sphere, the diagram corresponds to a left eye. Each complete curve represents rotation of 50° about the axis of rotation. Small cross-lines indicate cyclotorsion with respect to the objective vertical at every 10° of rotation. (After Boeder, *Amer. J. Ophthal.*)

*This assumes of course that the insertion of the muscle remains the same during abduction; in fact the effective insertion of an extraocular muscle varies very little because it is wide with its tendinous fibres fanning out; hence the axis of rotation tends to remain in the same *position in space* and therefore different in respect to the antero-posterior axis of the eye.

is accentuated when the eye is caused to abduct to the extent that the elevating action of the inferior oblique is very small, then the superior rectus is probably the defective muscle. As Boeder has shown, however, because the axes of rotation of the eyes, due to single muscle-pairs, vary so considerably as elevation or depression takes place, this simple view may well be illusory, so that in fact the superior and inferior recti probably always contribute more to elevation and depression than their synergistic obliques. According to Boeder's calculation, their proportionate contractions remain always in the ratio of about three to two. This is borne out by the observations of Breinin (1957) and of Tamler, Jampolsky and Marg (1959) that when the eye is looking upwards in strong adduction and then moves horizontally to abduction, e.g. if the left eye starts by looking up and to the right and then looks up and to the left, the superior rectus actually shows increased activity in the electromyogram although, on the classical view, the reverse would be expected since the eye is passing into a position where the action of the superior rectus becomes more and more efficient and therefore, presumably, requires less powerful contraction of its fibres.

## TORSION AND LISTING'S LAW

We have indicated (p. 387) that a rolling movement of the eye—e.g. by rotation about its antero-posterior $y$-axis, would upset the individual's estimate of direction in space unless psychologically compensated. It has been seen that the associated actions of the superior oblique with the inferior rectus tend to cancel out any rolling motion, and in actual fact a study of the movements of the eyes shows that they are executed in such a way as, in general, if not to exclude rotations about the antero-posterior axis, to reduce them to a minimum. This is the basis of *Listing's Law* which will be enunciated more exactly later; for the moment we need only note that since rotations about this third, antero-posterior, axis are apparently excluded, there are virtually only two degrees of freedom in the motions of the eyes, i.e. the motions can be resolved into rotations about the horizontal $x$- and vertical $y$-axes. Can we conclude, therefore, that there is no torsion during all movements of the eyes, so that the image of a horizontal line in the frontal plane always falls on the horizontal meridian or parallel with it? We have seen, in the analysis of the actions of the superior and inferior recti, that the latter adduct the eye even though they rotate it about an axis perpendicular to the vertical, i.e. in spite of the fact that there is no rotation about the vertical $z$-axis, and it can be shown that rolling or torsion will occur in spite of the exclusion of rotations about the antero-posterior axis.

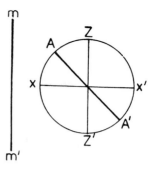

**Fig. 16.15**  Illustrating torsion. A vertical frontal section through the right eye. Rotation about AA' brings $z$ nearer to $mm'$ and takes $z'$ farther away.

### TORSION IN THE TERTIARY POSITION

In Figure 16.15 a vertical section through the equator of the eye in the primary position is shown, so that the plane of the paper corresponds to Listing's plane; $xx'$ is the transverse horizontal axis, rotation of the eye about which causes pure elevation or depression; $zz'$ is the vertical axis, rotation about which causes adduction or abduction. $xx'$ coincides in direction with the horizontal meridian and may be considered as the intersection of the retinal horizon with the plane of the paper; it is at right-angles to the median plane, $mm'$, coming out of the plane of the paper; in this, the primary position, there is no torsion and, so long as the retinal horizon, passing through $xx'$, remains at right-angles to this plane, torsion will be absent.

Rotation about $xx'$ obviously does not alter its inclination, and therefore that of the retinal horizon, to $mm'$, so that with pure elevation from the primary position there is no torsion; similarly, rotation about $zz'$, causing adduction or abduction, does not change the inclination of the retinal horizon to the plane $mm'$.

Rotation about the intermediate axis, AA', lying in the plane of the paper (Listing's plane), causes a combined movement of adduction and depression or abduction and elevation; during this rotation the point $x'$, for example, moves in a circle which is obviously not parallel with the plane $mm'$ nor at right angles to it, consequently the retinal horizon passing through $xx'$ becomes inclined to the plane $mm'$ and the eye experiences torsion in spite of the fact that the axis of rotation, AA', is in Listing's plane, at right-angles to the antero-posterior axis. If we imagine that we are looking into the eye from behind, a rotation that brings $x'$ up out of the paper corresponds to looking up and to the right; in this event the inclination of the retinal horizon towards $mm'$ follows an anti-clockwise direction and if the eye is a right eye it may be said to have undergone inward rolling. (It is perhaps easier to visualize the events by considering a median plane through $zz'$, corresponding to the vertical meridian of the retina; on rotation about AA′ z clearly moves closer to *mm′*.)

## Torsion and rolling

Thus rotation of the eye about an axis intermediate between the horizontal and vertical, whereby the eye passes from the primary position to a tertiary one, results in an angle of torsion; it is customary, but confusing, to draw a distinction between the two general ways of producing this deviation of the retinal horizon from perpendicularity to the median plane; if this is caused by rotation about the antero-posterior axis, or an axis with a component along this direction, it is said to be due to *rolling*; if it is a consequence of rotation about an axis in Listing's plane, it is called *torsion* or *false torsion*. No one has suggested a similar differentiation between adduction or abduction caused by rotation about a vertical axis and that caused by rotation about an axis perpendicular to this (p. 389), and the differentiation between the two processes is entirely unnecessary; the physiologist is chiefly concerned with the position of the retinal horizon in respect to the median plane, and the degree to which it deviates from perpendicularity may be called the angle of torsion or angle of rolling indifferently. Rather than increase the already existing confusion, however, it is probably advisable to retain the distinction; thus, with Quereau (1955) we may define rolling as a rotation about an antero-posterior axis and torsion (or false torsion as it is often called) as the consequence of oblique rotations about an axis in Listing's Plane.

So long as the head is vertical and no torsion has occurred, the retinal horizon will intersect a vertical frontal plane, e.g. the wall of a room, in a horizontal straight line (Fig. 16.8); if the eye fixates any horizontal line in this plane, therefore, its image will fall on the horizontal meridian of the retina so long as no torsion has been experienced during the movement of fixation.

### DONDERS' LAW

The position of the horizontal meridian of the eye in relation to the median plane, i.e. the degree of torsion present, can be ascertained by after-image experiments. If the eye, in the primary position, is directed at a horizontal straight line, the image of the latter falls on the horizontal meridian; if the stimulus is strong, an after-image may be evoked and the sensation is projected outwards, so that wherever the gaze is directed the line is perceived in the position it must have if its image is to fall on the horizontal meridian. If this projected after-image coincides with a true horizontal line, it can be said that the truly horizontal line produces an image on the horizontal meridian, i.e. that no torsion is present. If, on the other hand, torsion has occurred, the projection of the horizontal meridian appears inclined to a truly horizontal line, and the angle between the after-image and the horizontal gives the angle of torsion.

With the eye in the primary position, a black horizontal line is fixated for a time and the eye is then moved over a screen with horizontal lines on it. If the eye moves vertically upwards or downwards, or horizontally to the left or right, the after-image is found to coincide with the lines on the screen, indicating that there is no torsion. If the eye is moved to a tertiary position, the after-image no longer coincides with the horizontal lines; e.g. if the eye looks up and to the right, or down to the left, the after-image is inclined over to the left, as we should expect if rotation took place about an intermediate axis in Listing's plane. The actual directions, in which the projected after-image lies for different movements are indicated in Figure 16.16. It is found that no matter how the eye is moved to any position—whether in several steps or as a result of a single rotation—the torsion is the same; this is the basis of *Donders' Law* which states: '*When the position of the line of fixation is given with respect to the head, the angle of torsion will have a perfectly definite value for that particular adjustment; which is independent not only of the volition of the observer but of the way in which the line of fixation arrived in the position in question.*'

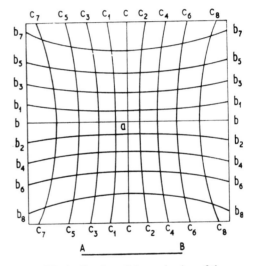

**Fig. 16.16** The inclination of the projection of the after-image of a horizontal line, fixated in the primary position of the eye, on moving the latter into various other positions. *a* is the original fixation point and AB represents the relative distance of the eye from this point. (Helmholtz, *Physiological Optics*.)

### PROJECTION OF VERTICAL LINES

With vertical lines, the position is more complicated; a vertical after-image is found to be inclined as though the eye had twisted in the opposite sense, i.e. on looking up and to the right the vertical line after-image is tilted over to the right instead of to the left; since the eye cannot exhibit torsion in two ways at once, and since the twist of the horizontal line agrees with theory, this apparent torsion in the opposite sense must be due to some other cause. Suppose now the eye is moved from the primary position upwards and to the right, and the torsion is corrected by an appropriate rolling movement.

The after-image of a horizontal line, fixated in the primary position, now corresponds in direction with the truly horizontal line through the new point of fixation; the vertical meridian of the retina is at right angles to the retinal horizon so that we might expect that its projection outwards would be at right angles to a horizontal line on the wall, coinciding with a vertical line. This, however, is not true; the retinal horizon is inclined upwards and to the right, and a line perpendicular to it—which represents the projection of the vertical meridian in the absence of torsion—is inclined over to the right when projected on the screen (Fig. 16.17).

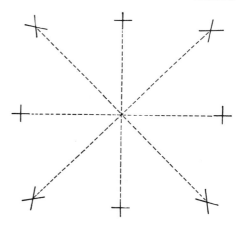

**Fig. 16.18** Approximate appearances of the projections of the after-image of a cross, fixated in the primary position of the eye, for various directions of the fixation axis.

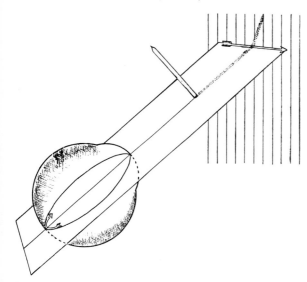

**Fig. 16.17** Illustrating the projection of the vertical meridian of the retina. The eye is looking up and to the right, and it is supposed that torsion has been corrected. The image of the pencil thus falls on the vertical meridian of the eye; its projection against the frontal plane is nevertheless inclined over to the right.

We are now in a position to explain the opposite appearances of horizontal and vertical after-images; due to torsion the horizontal and vertical meridians are tilted over to the left in the case examined. Owing to the characteristics of the projection of the vertical meridian on a vertical frontal plane in space, the after-image of a vertical line, fixated in the primary position, appears tilted over to the right, and more so than the tilt to the left due to torsion.

As a result of these combined effects, therefore, the after-image of a cross, fixated in the primary position, will have the appearances shown in Figure 16.18 when the eyes are moved into various tertiary positions.

LISTING'S LAW

It is clear that if, during a rotation of the eyes from the primary position, there is to be no component of rolling about the antero-posterior axis, then the axis of rotation

must be perpendicular to this axis, i.e. it must lie in Listing's plane (Fig. 16.5). According to Listing this is, indeed, true so that movements of the eye are confined to two, instead of three, degrees of freedom. Given this limitation—which is the basis of Listing's Law—the experimental finding of Donders that the torsion of the eye is characteristic for its position however this was achieved, follows at once, since the torsion due to rotation about an axis in Listing's plane is predictable on geometrical grounds. More recent experimental studies on the subject have shown that Listing's Law is only an approximation; thus, when the eye is made to look up and outwards, Quereau found that some rotation about an antero-posterior axis occurred so that the resulting torsion was not that predictable on geometrical grounds on the assumption of a rotation about an axis in Listing's plane.★

**BINOCULAR MOVEMENTS**

HERING'S LAW

The movements of the two eyes are so co-ordinated as to permit them to operate as a single entity; this unity in behaviour is the basis for *Hering's Law of the Ocular Movements* which states: '*The movements of the two eyes are equal and symmetrical.*'

A movement is said to be *conjugate* when the fixation axes are parallel and the movements of the separate eyes are equal in all respects (Fig. 16.19a). When the fixation

★This account of torsion is a slightly abbreviated version of that in the first edition of this book; in writing it I relied almost exclusively on Helmholtz's account. Since then several writers have contributed papers designed to clear up the confusion that enters once one departs from Helmholtz (Quereau 1954, 1955; Marquez, 1949; Moses, 1950). Since I find it very difficult to believe that Helmholtz was wrong in his geometry I have not been tempted to alter my first version.

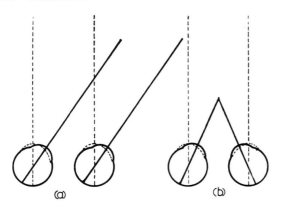

**Fig. 16.19**  Conjugate (*a*) and disjunctive (*b*) movements of the eyes.

axes are not parallel, as in convergence or divergence, the movement is said to be *disjunctive* (Fig. 16.19*b*); in this case the movements of the two eyes are symmetrical, and, as we shall see, they may well be equal in extent even though it can appear that one eye does not move at all (e.g. when converging on a point directly in front of one eye).

## THE POSITION OF REST

We have so far defined a primary position of the eyes as a reference point from which movements may be described; when the eyes are free from any stimulus determining their mutual orientation (accommodation, fusion of images, etc.) they adopt the so called *fusion-free position* or *physiological position of rest* and it generally happens that in this position the fixation axes diverge slightly (physiological heterophoria) and are thus not in their primary positions; hence, even for a distant point, the act of fixation requires a tonic action of the extraocular muscles whereby the fixation axes are maintained parallel. As we shall see, this 'tonic' action is the result of the operation of the fixation reflex that ensures that both eyes will fixate a given point accurately and thereby prevent the diplopia that would occur were this accurate binocular fixation not maintained (p. 520). To determine this fusion-free position various optical devices are employed to remove the tendency to binocular fixation; thus if a red spot on a green background is viewed by one eye through a red glass and by the other through a green one, the two eyes see quite different pictures and there is no tendency to fuse these, so that the fixation reflex is not operative and the eyes adopt their 'natural positions'. The muscle balance of the eyes would be said to be perfect if the two eyes maintained accurate alignment with the same point; if they tended to converge they would be said to be *esophoric*; if they diverged *exophoric*, and so on.

# CONJUGATE MOVEMENTS

## CONJUGATE INNERVATION

In conjugate binocular movements both eyes must move to the same extent if diplopia is not to result from any shift of the gaze, and it is therefore reasonable to suppose that there is some sort of conjugate innervation of the principal actors in any binocular movement, in the sense that the passage of impulses to one is accompanied by the facilitated passage of impulses to the other. Broadly speaking we may expect that for a conjugate deviation to the right the left medial rectus will be associated with the right lateral rectus; for a movement up and to the right, the right superior rectus (the elevator with the stronger action in abduction) will be associated with the left inferior oblique (the elevator with the stronger action in adduction). The muscles of the two eyes may thus be classified in pairs, or 'associates', such that in any general type of movement a pair may be designated as the chief actors. The clinician makes use of this classification in order to assess a muscle deficiency; the 'diagnostic directions of gaze', classified below, are chosen so that any defect in binocular fixation may be related to the muscle associates chiefly acting:

| Diagnostic direction | Chief actors | |
|---|---|---|
| | R.E. | L.E. |
| Right | L. rectus | M. rectus |
| Left | M. rectus | L. rectus |
| Up and right | S. rectus | I. oblique |
| Up and left | I. oblique | S. rectus |
| Down and right | I. rectus | S. oblique |
| Down and left | S. oblique | I. rectus |

That this linkage is real is proved by the phenomenon of the secondary deviation of a squinting eye. For example, if the left lateral rectus is partially paralysed, when the eyes look left the eyes squint because the left eye fails to make the necessary movement (Fig. 16.20*b*). If, however, a special effort is made by the left eye to fixate the point, then the right eye, by overaction of its medial rectus, squints, giving what is called a *secondary deviation* (Fig. 16.20*c*).

## LIMITS OF MOVEMENTS

For practical purposes, the limits of binocular conjugate movements are given by the *field of binocular single vision*, i.e. one determines the position, for any direction of the gaze, at which one eye can no longer keep pace with the other, as evidenced by an insuperable diplopia. Duane's measurements have shown that this field extends to at least 40° in any given direction and usually up to 50° or more, in fact he states that most individuals retain binocular single vision up to the very limits of the excursions of the eyes, so that it may happen that the binocular field of single vision is larger than either monocular field of fixation taken separately. Thus it appears that, no matter what the maximum excursion of each eye separately is, the excursion of both together is, in normal cases, such that one eye keeps pace with the other and diplopia is avoided even in extreme deviations of the eyes.

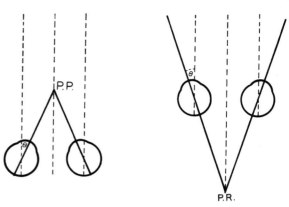

**Fig. 16.21**   The near point (P.P.) and far point (P.R.) of convergence.

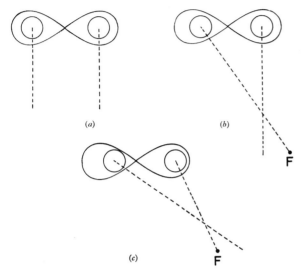

**Fig. 16.20**   Illustrating primary (*b*) and secondary deviations due to paresis of left lateral rectus.

## DISJUNCTIVE MOVEMENTS

### CONVERGING POWER

When a near object is fixated, the eyes converge as a result of the combined contractions of the medial recti. There is, of course, a limit to which the fixation axes can be deviated inwards so that if, for example, a pencil in the median line is moved closer and closer to the eyes, a point is reached where the two eyes can no longer fixate the pencil and it appears double. This point, the nearest consistent with single vision, is called the *near point of convergence* (Fig. 16.21); in 80 per cent of normal individuals it lies between 5 and 10 cm in front of the eyes and is not greatly affected by age as with the near point of accommodation. The *far point of convergence* is given by the point of intersection of the fixation axes in the physiological position of rest; since the eyes in this condition are usually divergent, this point is generally behind the eyes (Fig. 16.21), its distance being given a negative value.

### Metre-angle

By the *converging power* of the eyes, for any given point of fixation in the median line, is meant the angle through

which *either* fixation axis moves from the primary position when the convergence takes place (Fig. 16.21); this angle can, of course, be represented in degrees but Nagel introduced the *metre-angle* as a convenient unit. The metre-angle is the angle through which each eye has rotated from the primary position in order to fixate an object 1 metre away in the median line. If the eyes converge on an object 2 metres away, clearly the angle is approximately halved, and the convergence of each eye is said to be 0·5 metre-angle; if the object is 10 cm away the angle is approximately ten times as large and the convergence is 10 metre-angles. In general, the converging power of each eye is given by the reciprocal of the distance of the fixation point in metres. For an average interpupillary distance of 6 cm the metre-angle corresponds to the convergence angle of 1·7°. It will be noted that the accommodative power in dioptres for any point is given by the same reciprocal (p. 478), so that in some respects this is a convenient notation; nevertheless, because the actual angle of convergence depends on the interpupillary distance, this equality can be misleading. Thus for an emmetrope converging on a point $\frac{1}{3}$ metre away the convergence and accommodation are respectively 3 metre-angles and 3 dioptres. With interpupillary distances of 5, 6 and 7 cm the actual convergences are, however, 15, 18 and 21 prism-dioptres respectively. This is an important point when considering accommodative convergence (Breinin, 1957).

### Amplitude of convergence

The *amplitude of convergence* is given by the difference in the converging powers of the eyes for the near and far points; hence with a near point of + 8 cm and a far point of − 1 metre, the amplitude is:

$$1/0\cdot08 - 1/\text{-}1 = 12\cdot5 + 1 = 13\cdot5 \text{ metre-angles.}$$

A prism alters the apparent direction in which an object is seen; if it is base-out, the object appears to be deviated inwards and the eye adducts in order to fixate it (a base-out prism is called an *adducting prism*). Theoretically, then, the near point of convergence may be determined by placing pairs of base-out prisms in front of the eyes until a distant object can no longer be seen single (Fig. 16.22). The strength of the prisms required to achieve this could be converted into metre-angles of convergence by the formula:

*Strength of prism in dioptres = No. of metre-angles × ½ interpupillary distance in cm.*

Thus with a near point of 12·5 m.a. and an interpupillary distance of 6 cm the prism strength would be given by:

Prism strength = 12·5 × 3 = 37·5Δ (or 18·75°d).*

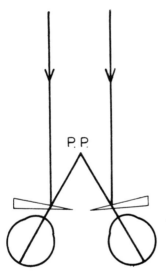

**Fig. 16.22** The use of adducting prisms to cause maximum convergence. (P.P., near point of convergence.)

In practice, however, this would be an unreliable measure of convergence-power since the ability to converge in response to prisms varies from moment to moment, and with practice.

**Fusion supplement**

It is stated that the desire for fusion permits a stronger convergence than that obtained without this stimulus; if an object is fixated at the near point of convergence and one eye is then covered up, the other deviates a little; the extra degree of convergence obtained when the stimulus for fusion is present is called the *fusion supplement*.

'SEE-SAW' MOVEMENTS

Disjunctive vertical movements—'see-saw' movements —in which one eye moves up and the other down,

are not normally produced voluntarily; in the interests of single vision, however, the eyes are able, on the average, to overcome the effects of a 2Δ prism base-up or base-down in front of one eye, i.e. they can exhibit a 'supervergence' of 1°.

CONVERGENCE AS A FUNCTION OF THE TWO EYES

It must be appreciated that disjunctive movements are a function of both eyes so that if a distant object in the median line is fixated and the strongest tolerated adducting prism is put in front of one eye, on placing a prism, however weak, in front of the other, diplopia results, i.e. no further convergence is possible. The near point of convergence is thus measured by half the strength of the strongest prism tolerated by one eye (when the other has none) or the strength of equal prisms tolerated by both eyes. The explanation for this, according to Hering, was that both eyes shared in the convergence although the direction of one fixation axis might remain finally unchanged. Thus in Figure 16.23 the fixation axes of the left and right eyes were initially parallel; a prism has been placed in front of the left eye so that it adducts maximally towards P. The right eye, during this process, was said not to remain fixed and the motions of the two eyes were analysed into an initial convergence to a point P′, such that the convergence of each eye is half the angle necessary for the left fixation axis to move through to fixate P, followed by a conjugate deviation to the right through the same angle; this leaves the right fixation axis pointing in its original direction, whilst the left is directed at P.

In a similar way the asymmetrical convergence to a point P, on the fixation axis of one eye when it is looking straight ahead (Fig. 16.24), was said to be achieved by an initial convergence to a point P′ in the median line, followed by a conjugate deviation. Since convergence is equally shared by the two eyes, we can obtain the convergence of *each* eye for any position of the fixation point by halving the sum of the angles moved through by the two fixation axes from their primary positions.

**Experimental evidence**

The asymmetrical convergence illustrated by Figure 16.24 need not take place in the gross steps illustrated, and the process may consist in a series of convergence and conjugate lateral movements that may escape observation in the absence of an accurate means of measuring either the movements of the eyes or activity in their muscles. Alpern and Wolter, employing the electro-

*The terminology of convergence is so varied that it is also necessary to be able to convert a convergence distance into an angle of convergence; the formula supplied by Duane is:

$$\text{Convergence angle} = \frac{\text{I.P.D. (mm)} \times 25}{\text{Convergence distance}} + 1.5°.$$

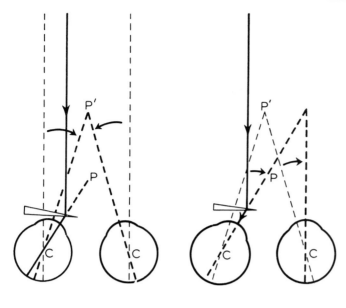

**Fig. 16.23**  The fixation axes were originally parallel. A strong adducting prism has been placed in front of the left eye so that it must be directed to P if it is to maintain fixation of the distant point. Both eyes converge to P′ and then make a conjugate deviation to the right. As a result, the left eye is directed to P, whilst the right eye re-assumes its original direction of fixation.

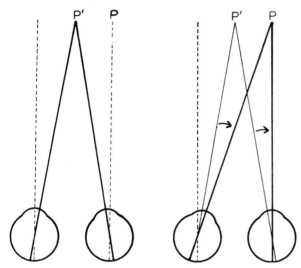

**Fig. 16.24**  Illustrating asymmetrical convergence to point P. The two eyes converge on the point P′, and then make a conjugate movement to the right. The direction of the right fixation axis is thus finally unchanged.

view exactly, the asymmetrical convergence consisting of a rapid 'saccade', i.e. a conjugate version, followed by convergence.*

## ROLLING

When the eyes are in the primary position they cannot be rolled voluntarily; however, the stimulus for fusion can cause such a movement. It is possible, with the aid of 90° prisms placed in front of one eye, to cause the image of an object on the retina to rotate about the fixation axis, e.g. the image of a cross may be tilted over. The two eyes are thus presented with different images that can only be fused if the eyes are rolled in opposite directions; the stimulus for fusion causes such a rolling, which may amount to as much as 10° to 14°, divided between the two eyes, but is generally much smaller.†  Convergence is said to be associated with a certain degree of outward rolling of the two eyes, so that their vertical meridians make an angle with each other which, in extreme instances, may amount to as much as 25°. This rolling must produce a rotation of the two retinal images in respect to each other, so that points in space

oculogram (EOG, p. 398) to record the eye movements, found that they were, indeed, compound, but apparently the conjugate movement preceded the convergence, i.e. the eyes tended to look to the right of P and subsequently to converge on it. As Alpern and Wolter emphasized, the speed of the conjugate type of movement, namely about 400°/sec, is so much greater than that of convergence (about 25°/sec) that the former might be missed. By measuring the eye movements by photographing the light reflected from the cornea, Westheimer and Mitchell were able to confirm Hering's

---

*As summarized by Zuber (1971) the evidence on this point is conflicting, and a great deal depends on the mode of presentation of the asymmetrical convergence stimulus, especially on whether or not the fusional mechanism is activated.
†According to Kertesz and Jones (1970), however, when cyclovergence is measured objectively under these conditions, it is negligible so that, in fact, fusion of the rotated images has taken place without rolling of the eyes, i.e. disparate images are fused.

formerly seen single might now be seen double (p. 520); on converging on a line of print, for example, diplopia should result from this asymmetrical torsion; in practice this does not occur so that it would seem, as Harms has pointed out, that when there is a stimulus for fusion, rolling does not take place during convergence; only when the object fixated has adequate symmetry about the fixation axis, so that its retinal image is not seriously modified by rotation about this axis, does rolling occur. For example, if an individual fixated a circle binocularly, rolling of one or both eyes would make no difference to the two retinal images; with a cross, on the other hand, the images would tilt in opposite directions and the individual would have to fuse dissimilar images; the rolling that occurs during convergence is therefore probably largely countered by a corrective fusion movement. As we shall see, the eyes execute a conjugate rolling movement when the head is tilted towards the shoulder; this is a reflex response to the altered pull of gravity on the head, and to the stretch of neck muscles (p. 437).

## ANALYSIS OF THE MOVEMENTS

It was Dodge in 1903 who made the first serious quantitative attempt to analyse the movements of the eyes; subsequent work has depended for its fruitfulness on improvements in the technique for recording eye movements. Probably the most satisfactory method consists of reflecting light from the cornea, or a contact glass fitting snugly on the eye, and focusing it on to a moving photographic film. By suitable calibration the movements of the reflected light may be converted into angular movements of the eye quite accurately.

### ELECTRO-OCULOGRAM
When an electrode is placed on the cornea and another on the posterior pole of the eye or elsewhere on the head, a 'standing potential' is recorded of about 1 mV, the cornea being positive. Illumination of the eye gives alterations of this standing potential which we call the electroretinogram (ERG); thus the ERG is really super-imposed on the 'standing potential'. Because of this potential the eye behaves like a dipole orientated along its antero-posterior axis, and if electrodes are placed at right-angles to this axis, e.g. in the two canthi, then movements of the eye will cause changes of potential between the electrodes because the dipole's orientation with respect to these electrodes changes. This is the basis of *electro-oculography* (EOG), an indirect method of recording eye movements, which can be employed even with the eyes closed. Methods specially adapted to record the small involuntary movements that take place

during maintained fixation are those of Rashbass (1960) and Ginsborg and Maurice (1959).

In experimental animals, such as the monkey, Fuchs and Robinson (1966) have implanted a fine wire coil under the conjunctiva; this was joined to a connector fixed to the skull. The animal's head was exposed to two alternating magnetic fields in spatial and phase quadrature, and signals could be made to indicate horizontal and vertical eye motions.

## TYPES OF MOVEMENT

The conjugate movements of the eyes may be classified as *saccades* and *smooth following*, or *tracking*, movements. Saccade was the name originally given by Dodge to the rapid movements between fixation-pauses that occur in reading, but it is now applied more generally to the conjugate shifts of gaze from one fixation point to another. The very small saccades, occurring during steady fixation, as described later, are now called *micro-saccades* or *flicks*. The movements of the eyes during convergence take place so much more slowly than the conjugate movements that a separate effector system has been postulated (Alpern and Wolter, 1956).

## THE SACCADE

The mechanics of the saccade were first seriously investigated by Westheimer (1954), the subject being told to fixate a light whose position could be suddenly changed. He found a reaction-time of some 120 to 180 msec, after which both eyes simultaneously moved; the maximum velocity attained varied with the magnitude of the saccade, being 500°/sec for a 30° movement and 300°/sec for 10°.

### PRE-EMPHASIS
Subsequent studies of Robinson (1964), in which he measured both the eye movements and the tensions developed in the extraocular muscles, showed that the eye was impulsively driven in a saccade by a brief burst of force much greater than necessary to maintain the eye in its final position; it is essentially this 'pre-emphasis', or excess force, that allows the saccade to be executed so rapidly, permitting a 10° saccade to occur within 45 msec after its beginning.

### DAMPING
The deceleration of the eye that brings it to rest at the end of the saccade is not achieved by any checking action of the antagonistic muscle, but relies entirely on the viscous-elastic damping of the muscles and orbital

connective tissue, a finding that has been amply confirmed by EMG and single unit studies of the eye motor
nuclei (p. 414). It is this orbital stiffness, of course, that
requires that the muscles contract when the eyes deviate
from their primary positions, the greater the angular
deviation the greater the force of contraction.★

### ACCURACY

In general, the saccadic movement matches that of the
target to about 0·2° (Rashbass, 1961) but sometimes it
does not, in which case it is followed, after a new
reaction-time, by a second saccade; in some subjects
there may be three or four saccades, each separated by
one reaction-time, before the final fixation is reached.
Rashbass also observed that there was a threshold
target displacement, so that a movement of 0·25 to 0·5°
is usually not accompanied by a response.

### CONTROL-MECHANISM

Young and Stark (1963) showed that the control
mechanism during the saccadic type of target-following
could be described as a 'sampled data system', in the
sense that the brain made discontinuous samples of the
position of the eyes in relation to the target, i.e. of the
'error', and corrected this. Proof of such discontinuity
was provided by Westheimer who showed that, if the
target was caused to step aside and return to its original
position within 100 msec, the eye nevertheless made a
saccade 200 msec later.†

## Large saccades

According to Becker and Fuchs (1969) a saccade of over
15° occurs in two pre-packaged stages; the first saccade
brings the eyes short by about 10 per cent of the target,
and after a reduced latency (130 instead of 230 msec)
a second saccade brings the eyes on target.

## Error-signal

It seems unlikely, however, that the secondary saccade,
subserving the obvious function of correcting the first,
should be executed in a pre-programmed fashion but
rather in the light of visual information that provides
the error-signal. Prablanc and Jeannerod (1975) showed
that the secondary corrective saccade failed if the visual
input was no longer present, i.e. if the target was
extinguished before the end of the primary saccade.
Interestingly, they showed that the error for the primary
saccade increased as the duration of presentation of the
target decreased, so that it was only 1° for a 200 msec
presentation and was 3° for a 20 msec presentation. Thus
time is required to weigh up the situation and
programme the saccade with accuracy.

## Latency

So far as the latency of the second corrective saccade was
concerned, Prablanc and Jeannerod found that it
depended on conditions; when they presented the signal
twice at the same place, e.g. two 20 msec pulses
separated by greater than 50 msec, then after the first
saccade there was a small corrective saccade with a
latency of 176 msec, which is less than that for the
primary saccade of some 217 msec. An important
condition for this reduced latency is the error in the
primary saccade; if this is greater than 4°, then the
latency is not reduced, presumably because a completely
new decision has to be made; with a smaller error it
appears that the execution of the first saccade in some
way speeds up the computation of the motor discharge
necessary for the secondary saccade.

## Visual input

Hallett and Lightstone (1976) studied the responses of
subjects to presentation of a target during the execution
of a saccade, to determine the extent to which information derived from the eyes could be utilized during the
eye movement. A target was presented, inducing a
saccade, $S_0$, and the eye-movement triggered a new
target movement before $S_0$ was completed; after the
target had made its new movement it was extinguished
so that the new saccade, $S_1$, required to fixate it, was
made while it was invisible. It was found that the
saccade, $S_0$, occurred with its normal latency, usually
undershooting its target, but it was never followed by
the expected corrective saccade, showing that this
requires visual input; after a further more or less normal
latent period, the usual response was a primary saccade,
$S_1$, towards the target in its second position; even
though it was not lit throughout the latent period, the
timing and accuracy were more or less normal. Thus
the cancellation of the corrective saccade and the
occurrence of $S_1$ indicate the effectiveness of visual
input during the saccade.

---

★The mechanical studies of Childress and Jones (1967), in
which they caused adduction by a mechanical pull on the eye
and observed the responses to quick releases, have generally
confirmed Robinson's assessment of the eye in its socket as a
high-inertia heavily damped system. Fuchs (1967) has studied
the eye movements of the monkey, which are qualitatively
similar in so far as the differentiation between saccades and
pursuit movements is concerned; quantitatively the monkey
differs in being able to follow more rapidly moving targets,
and in executing the saccades more rapidly.
†Fuchs (1967) has discussed the length of the sampling time;
he suggests that the sample may be taken over half a reaction
time, a position-correction decided on the basis of the information collected, and the correction effected without change half
a reaction time later. Such a system would be described as
*semi-discrete*, to distinguish it from smooth pursuit behaviour.

## SMOOTH PURSUIT OR TRACKING MOVEMENTS

Under these experimental conditions the fixated spot of light is caused to move at a constant velocity in a given direction; if this movement does not exceed a certain speed, namely about 30 to 40°/sec, the eyes move with the same speed, so that the image remains effectively stationary on the retina. As we should expect, the smooth pursuit movement is usually, itself, compound; thus, because of the reaction-time of some 150 msec before the smooth movement begins, and because it takes at least 400 msec for the eye to reach the target speed, the pursuit movement, by itself, would inevitably leave the point of fixation of the eye trailing behind the target unless a saccade intervened; and this in fact happens. Figure 16.25 from Robinson (1965) shows the

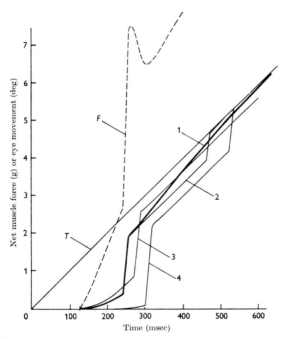

**Fig. 16.25** A composite graph to show the various types of eye-movement responses to a 10°/sec ramp target motion, T. Response 1 is the mean of fourteen responses of the most common type, and F is the net isometric muscle tension in the horizontal recti associated with it. Responses 2, 3 and 4 illustrate variations in overshoot and temporal spacing between smooth and saccadic components. (Robinson, *J. Physiol.*)

four main types of following movement, that marked 1 being the most common; here, after a delay of 125 msec, the eye accelerates at a mean value of 60°/sec² to reach a velocity of 6·1°/sec and a displacement of 0·38° in 112 msec before the interruption of a saccade; this

has an amplitude of 1·24°, so that an error of 0·7° remains. The eye leaves the saccade at a smooth velocity of 12·2°/sec, i.e. it has overshot the target velocity; it maintains this excessive velocity until the error is nearly zero, when the velocity slackens off to become equal to that of the target. Figure 16.25 also illustrates the development of tension in the horizontal recti associated with the movement; as with the saccade,. there is an initial 'pre-emphasis', this time on the rate of change of tension.

CONTROL

Young and Stark considered that smooth pursuit movements were also controlled by a sampled data system, the data utilized being the positions of the retinal image only. The studies of Rashbass and of Robinson, however, indicate that the control is continuous, presumably of the feed-back type, the eye responding to disparity, *per se*, and to the rate of change of position on the retina, i.e. there is response to both position and velocity, the latter being the dominant influence. Thus Rashbass' experiments have provided a number of interesting examples of the independence of the systems of control of saccades and pursuit movements. For example, if a target is given an initial rapid displacement of, say, 3° to the left, followed immediately by movement at uniform velocity to the right (step-ramp stimulus), a saccade is made to the left, after which the pursuit movement to the right occurs. Hence the saccade, because of the reaction-time, begins during the smooth movement of the target in the opposite direction; in this case, then, the position stimulus has taken precedence over the velocity stimulus. If the initial step-displacement is reduced to 1°, then the saccade fails, and the eye starts moving, after a longer-than-normal latency of 150 msec, in the same direction as the target. In this case the eye begins a smooth pursuit movement in response to target velocity even though the movement is away from the target position at this time. Thus the pursuit movement is primarily responsive to target velocity rather than position.

Young and Stark assumed that a retinal image-error exceeding 0·2° was adequate stimulus for a saccade, but the step-ramp experiment of Rashbass shows the inadequacy of this picture; similarly in curve 1 of Figure 16.25, after the initial saccade, an error of 0·7° remained but no saccade was elicited, the velocity overshoot being adequate for ultimate correction. Thus, the decision as to whether a saccade is to be made, as well as the determination of its magnitude, are functions of both retinal image-error and its rate of change, and it may be that the ganglion cells described by Barlow, Hill and Levick (p. 285), sensitive to direction of image-motion, are involved here.

## DISJUNCTIVE MOVEMENTS

### CONVERGENCE

The disjunctive movement of convergence is altogether slower than the conjugate movement, although the reaction-time of about 160 msec is of the same order. The movement takes place with a constant velocity, which is a function of the magnitude of the stimulus;* this velocity falls off asymptotically, with a total movement-time of about 800 msec. Even with a sudden change of vergence of the target, it is possible for control to be continuous rather than of the sampled data type; and according to Rashbass and Westheimer (1961a) this is true, so that, when momentary changes in the vergence stimulus are introduced during the execution of a movement, these are responded to after a reaction-time; thus disjunctive eye movements can be modified during their progress, and information concerning disparity can be assimilated during the reaction-time as well as during movement.

### VERSION PLUS VERGENCE

When the following of a target involves simultaneous conjugate and disjunctive movements, then, along whatever path the target moves, the movement of the eyes is resolved into two components corresponding to mean target position and target vergence, respectively, and appropriate responses to these components are made by two independent systems of control (Rashbass and Westheimer, 1961b).

## MOVEMENTS DURING STEADY FIXATION

It has been known for a long time that the eyes do not maintain a steady fixation of a point for any length of time; instead, fixation is interrupted by rapid jerks at irregular intervals. The actual movements that take place have been analysed by several workers using modern methods of recording, usually employing the reflexion of light from the cornea.†

## THREE TYPES

They have been resolved into three types, as follows:

1. Irregular movements of high frequency (30 to 70/sec) and small excursion, namely some 20 sec of arc; these are described as *tremor*.
2. Flicks or microsaccades of several minutes of arc, occurring at irregular intervals of the order of 1 sec.
3. Between flicks there are slow irregular drifts extending up to 6 minutes of arc. Figure 16.26 illustrates the region in which the fixation point would be 68 per cent of the time during a 30 seconds

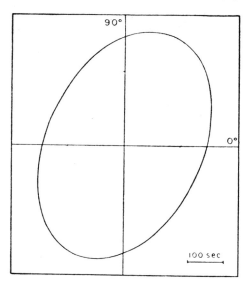

**Fig. 16.26** Indicating probable range of movements of the eye over a period of steady fixation. The solid line bounds the region within which a point retinal image will be found projected in space 68 per cent of the time during 30-sec fixation intervals. (After Nachmias, *J. Opt. Soc. Amer.*)

fixation period, and it will be seen that the range of excursion is not large, certainly not large enough to bring the image of the fixated point off the fovea.

### CORRECTIVE SACCADE

Most workers are agreed that the flick is a corrective movement that brings the image back when, as a result of a drift or previous saccade, it is in danger of moving off the central fovea. This is deduced from statistical studies that show that the chance of a flick occurring in a given direction correlates well with the degree of deviation from central fixation.‡

---

*Crone (1969) has emphasized the kinetic aspects of the retinal disparity that provokes a convergent fusional movement; the rate at which a fusional movement is brought about is, indeed, a linear function of the magnitude of the disparity.

†Of the contributors to this subject mention may be made of Riggs and his colleagues (Ratliff and Riggs, 1950; Cornsweet, 1956; Riggs and Niehl (1960); Krauskopf, Cornsweet and Riggs, 1960); Barlow (1952); Ditchburn and Ginsborg (1953); Drischel and Lange (1956); Nachmias (1959).

‡Cunitz and Steinman (1969) have pointed to some remarkable similarities between the flicks that occur during steady fixation and the larger saccades occurring during reading. Thus the intersaccadic intervals are remarkably similar; furthermore, during the reading pauses, the number of flicks is very small compared with a similar period of steady fixation, and they have argued that the flick, like the reading saccade, is a scanning mechanism. They point out that when a subject is asked to 'hold' rather than to fixate a target, flicks are very rare (Steinman *et al.*, 1967). Zuber, Stark and Cook (1965) have shown that, when maximum velocity is plotted against amplitude, the values for flicks fall on the same graph as those for the larger saccades in voluntary fixation movements.

## BINOCULAR FIXATION

When the involuntary movements in the two eyes are studied together during binocular fixation, it is found that the slow drifts take place independently in the two eyes, leading to small degrees of fixation-disparity; i.e. if the eyes are looking at a distant point the fixation axes will tend to converge or diverge at random. On the other hand, the rapid saccades take place simultaneously and in the same direction, and corrections for convergence-errors result from the circumstance that the saccades in the two eyes are not always equal in extent. According to Krauskopf, Cornsweet and Riggs, the eyes apparently behave as independent detectors of fixation error, and in correcting their own errors they cause conjugate movements of the other eye that may not necessarily be in the right direction for correction of this eye's error. Since the chance of initiating a saccade by either eye is proportional to the magnitude of its fixation error, we may expect the eye with the larger error to initiate the corrective saccade; if the movement of the other eye in this saccade is a little smaller than that of the initiator, then on a statistical basis we may expect a reasonably accurate binocular fixation. Thus it appears that it is not the error in convergence that is detected, and corrected for, but simply the error of fixation by either eye acting as an independent unit from the viewpoint of detection.

A more exhaustive study by St.-Cyr and Fender (1969) of both monocular and binocular fixation, in which the corrective effects of flicks and drifts were examined statistically suggests, however, that retinal disparity, as such, can act as a stimulus for corrective movements; furthermore, the drifts seem to be the movements responsible for these vergence corrections. Another point brought out by this work is the large individual variations in the character of the fixation movements, and the dependence of their direction on the type of target fixated; thus a vertical line tended to cause the flicks to take an upward direction.

## STEADY FIXATION IN THE CAT AND RABBIT

### THE CAT

Winterson and Robinson (1975) found that cats, with their heads fixed, when allowed to fixate objects of their own choosing, maintained a steady fixation by a slow control mechanism exclusively, in the sense that eye-drift, which occurred, was not corrected by micro-saccades, as frequently seen in man, but by equal and opposite slow drifts, which ensured that the mean eye position did not change. Thus, when looking about, they

make saccades, but these are maintained by the above slow control process. The stability of fixation was remarkably good, in fact as good as that of a human when told to fixate a given point. The stability relied on visual input since it was lost in the dark. Its basis is apparently the optokinetic reflex (p. 429), manifest experimentally in the slow movement of optokinetic nystagmus. Thus the drifts, which occur spontaneously, are being corrected by slow optokinetically induced compensatory movements. The latency is about 150 msec, and since a feedback system tends to oscillate with a frequency given by the reciprocal of twice the delay, the prominent frequency component of the drifts of 3/sec is intelligible on the basis of visual correction. Winterson and Robinson emphasize that this slow control is not peculiar to lower animals, in fact, when fixation patterns of man, rabbit and monkey were studied, a similar slow control was observed, the slow drifts being corrected by opposite slow drifts. So far as man is concerned, this does not mean that the microsaccades described above are not corrective, but that when an effort is made to prevent the microsaccade a slow drift-type of correction can replace this.

### RABBIT

The remarkable immobility of the rabbit's eye, when its head is fixed, is only apparent, and in fact it is the consequence of a stabilized fixation process, operating through visual feedback since, as soon as the rabbit is placed in the dark, large drifts take place. Collewijn and Van der Mark (1972) experimentally separated visual input from oculomotor response by the expedient of clamping one eye (the seeing eye) and covering the other (moving) eye. Under these conditions the seeing eye was providing no feedback because it was unable to drift, so that the covered, moving, eye exhibited the instability seen when both eyes were in the dark. By providing the seeing eye with suitable information regarding the moving eye, it was possible to enable the rabbit to stabilize its moving eye. This was achieved by using a servo-driven striped *optokinetic* drum whose rotation was steered by the position-signal of the moving eye; thus any movement of the eye caused a movement of the striped drum which induced an optokinetic correction in the moving, unseeing, eye. By making the drum rotate in the opposite direction to that which permitted stability, there was a positive feedback leading to very much enhanced instability, the drift being enhanced and interspersed with fast movements so that the movements assumed the character of a nystagmus. It is interesting that when single step-movements of the drum were made, as opposed to continuous rotations, there was no response of the eyes, so that the servo-system was velocity-sensitive rather than position-sensitive.

# REFERENCES

Alpern, M. & Wolter, J. R. (1956) The relation of horizontal saccadic and vergence movements. *Arch. Ophthal.*, **56**, 685–690.

Barlow, H. B. (1952) Eye movements during fixation. *J. Physiol.*, **116**, 290–306.

Becker, W. & Fuchs, A. F. (1969) Further properties of the human saccadic system: eye movements and correction saccades with and without visual fixation points. *Vision Res.*, **9**, 1247–1258.

Boeder, P. (1961) The co-operation of extraocular muscles. *Am. J. Ophthal.*, **51**, 469–481.

Breinin, G. M. (1957) Electromyographic evidence for ocular proprioception in man. *Arch. Ophthal.*, **57**, 176–180.

Breinin, G. M. (1957) Quantitation of extraocular muscle innervation. *Arch. Ophthal.*, **57**, 644–650.

Breinin, G. M. (1957) The nature of vergence revealed by electromyography. *Arch. Ophthal.*, **58**, 623–631.

Childress, D. S. & Jones, R. W. (1967) Mechanics of horizontal movement of the human eye. *J. Physiol*, **188**, 273–284.

Cornsweet, T. N. (1956) Determination of the stimuli for involuntary drifts and saccadic eye movements. *J. Opt. Soc. Am.*, **46**, 987–993.

Crone, R. A. (1969) The kinetic and static function of binocular disparity. *Invest. Ophthal.*, **8**, 557–560.

Cunitz, R. J. & Steinman, R. M. (1969) Comparison of saccadic eye movements during fixation and reading. *Vision Res.*, **9**, 683–693.

Ditchburn, R. W. & Ginsborg, B. L. (1953) Involuntary eye movements during fixation. *J. Physiol.*, **119**, 1–17.

Drischel, H. & Lange, C. (1956) Uber unwillkürliche Augapfelbewegungen bei einäugigen Fixieren. *Pflug. Arch.*, **262**, 307–333.

Fuchs, A. F. (1967) Saccadic and smooth pursuit eye movements in the monkey. *J. Physiol.*, **191**, 609–631.

Fuchs, A. F. & Robinson, D. A. (1966) A method for measuring horizontal and vertical eye movement chronically in the monkey. *J. appl. Physiol.*, **21**, 1068–1070.

Ginsborg, B. L. & Maurice, D. M. (1959) Involuntary movements of the eye during fixation and blinking. *Br. J. Ophthal.*, **43**, 435–437.

Hallett, P. E. & Lightstone, A. D. (1976) Saccadic eye movements towards stimuli triggered by prior saccades. *Vision Res.*, **16**, 99–106.

Hallett, P. E. & Lightstone, A. D. (1976) Saccadic eye movements to flashed targets. *Vision Res.*, **16**, 107–114.

Helmholtz, H. von (1925) *Physiological Optics*, vol. III, ed. Southall. Optical Soc. Am.

Kertesz, A. E. & Jones, R. W. (1970) Human cyclofusional response. *Vision Res.*, **10**, 891–896.

Krauskopf, J., Cornsweet, T. N. & Riggs, L. A. (1960) Analysis of eye movements during monocular and binocular fixation. *J. opt. Soc. Am.*, **50**, 572–578.

Marquez, M. (1949) Supposed torsion of the eye around the visual axis in oblique directions of gaze. *Arch. Ophthal.*, **41**, 704–717.

Moses, R. A. (1950) Torsion of the eye on oblique gaze. *Arch. Ophthal.*, **44**, 136–139.

Nachmias, J. (1959) Two-dimensional motion of the retinal image during monocular fixation. *J. opt. Soc. Am.*, **49**, 901–908.

Prablanc, C. & Jeannerod, M. (1975) Corrective saccades: dependence on retinal reafferent signals. *Vision Res.*, **15**, 465–469.

Quereau, J. V. D. (1954) Some aspects of torsion. *Arch. Ophthal.*, **51**, 783–788.

Quereau, J. V. D. (1955) Rolling of the eye around its visual axis during normal ocular movements. *Arch. Ophthal.*, **53**, 807–810.

Rashbass, C. (1960) New method for recording eye movements. *J. opt. Soc. Am.*, **50**, 642–644.

Rashbass, C. (1961) The relationship between saccadic and smooth tracking eye movements. *J. Physiol.*, **159**, 326–338.

Rashbass, C. & Westheimer, G. (1961*a*) Disjunctive eye movements. *J. Physiol.*, **159**, 339–360.

Rashbass, C. & Westheimer, G. (1961*b*) Independence of conjugate and disjunctive eye movements. *J. Physiol.*, **159**, 361–364.

Ratliff, F. & Riggs, L. A. (1950) Involuntary motions of the eye during monocular fixation. *J. exp. Psychol.*, **40**, 687–701.

Riggs, L. A. & Niehl, E. W. (1960) Eye movements recorded during convergence and divergence. *J. opt. Soc. Am.*, **50**, 913–920.

Robinson, D. A. (1964) The mechanics of human saccadic eye movement. *J. Physiol.*, **174**, 245–264.

Robinson, D. A. (1965) The mechanics of human smooth pursuit eye movement. *J. Physiol.*, **180**, 569–591.

Robinson, D. A. (1966) The mechanics of human vergence movements. *J. Pediat. Ophthal.*, **3**, 31–37.

St. Cyr, G. T. & Fender, D. H. (1969) The interplay of drifts and flicks in binocular fixation. *Vision Res.*, **9**, 245–265.

Steinman, R. M., Cunitz, R. J., Timberlake, G. T. & Herman, M. (1967) Voluntary control of microsaccades during maintained monocular fixation. *Science*, **155**, 1577–1579.

Tamler, E., Jampolsky, A. & Marg, E. (1958) An electromyographic study of asymmetric convergence. *Am. J. Ophthal.*, **46**, pt. 2, 174–182.

Tamler, E., Jampolsky, A. & Marg, E. (1959) Electromyographic study of following movements of the eye between tertiary positions. *Arch. Ophthal.*, **62**, 804–809.

Westheimer, G. (1954) Mechanism of saccadic eye movements. *Arch. Ophthal.*, **52**, 710–724.

Westheimer, G. (1954) Eye movement responses to a horizontally moving visual stimulus. *Arch. Ophthal.*, **52**, 932–941.

Westheimer, G. & Mitchell, A. M. (1956) Eye movement responses to convergence stimuli. *Arch. Ophthal.*, **55**, 848–856.

Winterson, B. J. & Robinson, D. A. (1975) Fixation by the alert but solitary cat. *Vision Res.*, **15**, 1349–1352.

Young, L. R. & Stark, L. (1963) Variable feedback experiments testing a sampled data model for eye tracking movements. *Int. Elect. Electronics Engineers. Transactions on Human Factors in Electronics.* HFE-4, 38–51.

Zuber, B. L. (1971) Control of vergence eye movements. In *The Control of the Eye Movements.* Ed. Bach-y-Rita, P., Collins, C. C. & Hyde, J. E., pp. 447–471. New York: Academic Press.

Zuber, B. L., Stark, L. & Cook, G. (1965) Microsaccades and the velocity-amplitude relationship of saccadic movements. *Science*, **150**, 1459–1460.

# 17. Nervous control of the eye movements

## GENERAL CONSIDERATIONS

The activities of the muscles of the neck and trunk are ultimately determined by the discharges in the motor neurones of the ventral horns of the spinal cord, which are described as the *final common path*. These neurones are acted upon by sensory neurones, and by inter-neurones localized in all levels of the central nervous system, from the various segments of the cord to the cerebral cortex. Thus the activity of any given muscle is determined by nervous activity in practically the whole of the central nervous system; nevertheless the activity of an organism may be resolved into certain stereotyped patterns of behaviour—the *reflexes*—which may be evoked independently of many regions of the central nervous system; thus the flexor reflex may operate in the spinal animal, i.e. the animal whose spinal cord has been severed from the brain; whilst a large variety of reflex responses may be achieved in the decerebrate animal, the animal whose cerebral hemispheres and basal ganglia have been removed. In general, we may say that these patterns of behaviour may be controlled and modified by activity in brain centres that are located higher in the system than those required for the performance of the reflex activity. This is manifest especially in the voluntary control of muscular activity mediated by the frontal motor cortex and along the pyramidal tracts.

The movements of the eyes form no exception to these general principles; well established reflex patterns of behaviour are present and these may be largely controlled, or at any rate influenced, by voluntary activity initiated, it is presumed, also in the frontal cortex. The final common path is given by motor neurones in the brain, rather than the spinal cord, whilst higher coordinating centres are found in the *superior colliculi* and in the *occipital cortex*. These reflex centres are well established; somewhat more problematical are certain regions of the brain stem, notably the tegmentum of the midbrain, pons and medulla, stimulation of which electrically gives rise to well coordinated patterns of eye movement.

### REFLEXES

In this short, and necessarily somewhat elementary, discussion of the nervous control of the eye movements we shall be concerned with the reflexes and their nervous pathways—the *reflex arcs*. It must be strongly emphasized, however, that most reflexes are only exhibited reproducibly and in the 'pure' form under artificial conditions, for example, in the decerebrate animal; any actual movement in the normal individual represents the final outcome of the antagonistic inter-play of numerous reflexes which, in their turn, may be modified to a greater or lesser extent by the voluntary centres. For example, we shall see that a bright light in the periphery of the visual field evokes a *fixation reflex* whereby the eyes are turned towards the light so as to bring its image on the fovea; this does not mean, how-ever, that the eyes are invariably turned to any light in the peripheral visual field. The reflex, defined as the response of the individual to a stimulus, must be considered, especially in man, as a component in behaviour; under special conditions it may be elicited with the utmost regularity, whilst in the normal intact animal the response to the same stimulus may be greatly different according to the conditions prevailing.

## THE MUSCLE FIBRES AND THEIR INNERVATION

The mammalian extraocular muscles have several structural and physiological features that differentiate them from most other striated muscles, features that they share with some submammalian skeletal muscles and with mammalian intrafusal fibres.

### MOTOR UNIT

The motor unit, i.e. the number of muscle fibres supplied by a single motor axon, is remarkably small being probably of the order of 6, although the difficulty in establishing whether a given fibre is sensory or motor may result in some error in the estimate. This compares with some 100 to 150 for limb muscles (Torre, 1953). Because of its short twitch-time the fusion frequency for an extraocular muscle is very high indeed, of the order of 350 impulses per sec, and this permits a large variation in tension—over a range of tenfold—on passing from a single twitch to a fused tetanus. These two features, namely small size of motor unit and small twitch/tetanus tension ratio, permit a high degree of gradation of the strength of contraction.

# TWITCH AND SLOW FIBRES

Again the muscles of the frog, for example, fall into two classes, characterized by the structure and innervation of their fibres; the *twitch*, or *fast*, fibres are usually large and innervated by large, rapidly conducting, nerve fibres that make typically a single *en plaque* or *sole-plate* ending on each muscle fibre; this type of muscle is concerned with rapid phasic movements and, on electrical stimulation of its nerve, gives a twitch response. The other type of muscle has thinner fibres with multiple *en grappe* nerve terminals distributed over its whole length; this type is associated with the tonic type of sustained contraction; electrical stimulation of its motor nerve leads to graded mechanical responses without the spread of action potentials. Pharmacologically, the fibres are distinguished by their responses to acetylcholine, the slow fibres giving a substantial contraction whilst the twitch fibres give a single twitch. Structurally there are striking differences between the fibres; Krüger pointed out that the slow fibres lacked the M-band whilst their fibrils were irregularly arranged, being clustered together in clumps to give what he called the *Felderstruktur*, which contrasted with *Fibrillenstruktur* of the twitch fibre, in which the myofibrils were regularly arranged in a more abundant sarcoplasm. Subsequent electron-microscopical studies of Peachey and Huxley (1962) and Page (1965) have emphasized these differences; in the fibrillar, or twitch, type of fibre the myofibrils are surrounded by a well defined system of sarcoplasmic reticulum made up of longitudinal and transverse systems giving rise to the *triads*, which are considered to be developments permitting the rapid transmission of the electrical changes occurring on the surface into the depths of the fibre. In the slow fibres the myofibrils lack regular arrangement, the sarcoplasmic reticulum is scarcer and lacks the transverse system, so that the triads are absent.

## MAMMALIAN SKELETAL MUSCLES

In the skeletal muscles of mammals this sharp differentiation into twitch and slow muscles does not occur; nevertheless, it is possible to describe muscles as either fast or slow on the basis of their contraction-times, i.e. the time to reach peak tension after a nerve stimulus; the fast muscles are usually white, capable of rapid action but soon fatiguing, whilst the slow, concerned with sustained postural contractions, are red, having large amounts of myoglobin and usually plentiful mitochondria.* More modern techniques, involving fatiguing the muscle with repetitive stimuli, have permitted the easy identification of the twitch-fibre histologically, and have resulted in a classification based on speed of contraction and fatiguability (Burke *et al.* 1973). According to this, there are three basic types

designated as FF (fast-rapidly fatiguable), FR (fast-fatigue-resistant) and S (slow-fatigue resistant).

# EXTRAOCULAR MUSCLE

Harker (1972) has examined the types of muscle fibre in the sheep's superior rectus and levator palpebrae with histochemical and electron microscopical techniques.

## BASIC STRUCTURE

The basic structure of the superior rectus consists in a central core-layer of mainly large-diameter muscle fibres adjacent to the eyeball, covered dorsally and laterally by an *orbital rim layer* of small muscle fibres; at the origin and insertion ends an additional *peripheral patch layer* surrounds the orbital rim; it contains medium-diameter muscle fibres. (The spindles are distributed peripherally either within the orbital rim or at the junction between this and the peripheral patch layers.)

## FIBRE TYPES

On the basis of fibre diameter and histochemical profiles some six types were differentiated, probably corresponding with the three types just outlined and seen in the gastrocnemius, although a group of the slow, or S fibres, called G, had a distributed grape-like innervation by contrast with the plate-endings of the other types.† According to Harker's classification, there were three types of C-fibre equivalent to the Fast-Fatigue-resistant category, differing in size and their histochemical features; there were A-fibres, equivalent to the Fast-rapidly fatiguable (FF), and finally B fibres, equivalent to the slow fibres (S) of the gastrocnemius. In the superior rectus there were no B-fibres, so that the G fibres with distributed endings, and probably multiply innervated, provided the slow component. The nerve fibres supplying the end plate-innervated twitch-type muscle fibres (A and C) were larger (4·3 $\mu$m) than those supplying the grape-innervated muscle fibres (2·5 $\mu$m). In the levator palpebrae, in which there was

---

*It must be appreciated that many muscles are mixed, containing both twitch and slow fibres; in the mammal, for example, Close (1967) described slow and intermediate types of unit in the soleus, whereas the extensor digitorum longus contained only fast fibres.

† The multiple endings on the slow grape-termination muscle fibres suggest a multiple innervation, although all the grape endings could have been derived from a single fibre; intracellular recordings of these fibres has indicated a multiple innervation (e.g. Hess and Pilar, 1963). Bach-y-Rita and Lennerstrand (1975) have denied the existance of polyneural innervation of cat extraocular muscles, but as Browne points out, their measurements could only apply to the twitch-type fast fibres.

no distinct layering, all fibres were plate-innervated including the slow B-fibres, which were not represented in the rectus.

## Two main types

In general, then, the early observation of two fundamentally differently innervated types of muscle fibres, apparently peculiar to the extraocular muscles in the mammal, but common in submammalian species, has been sustained; and there is little doubt that the two main types of fibre differ functionally as well as morphologically, the plate-innervated belonging to the fast or twitch-type and the grape-innervated to the slow type; apart from this difference, the many gradations of size, and proportions of aerobic to anaerobic enzymes, are found in the extraocular muscles as with the rest of the skeletal musculature.

## CONTRACTILE PROPERTIES

### PHASIC AND TONIC RESPONSES

The Soviet physiologist, Matyushkin (1961), was the first to demonstrate that the mammalian extraocular muscles contained two types of fibre that responded differently to electrical stimulation of the nerve nucleus; Figure 17.1 shows intracellularly recorded responses of muscle fibres of the superior oblique to a single stimulus in the trochlear nucleus; the records in (a) are responses in a phasic fibre, being spikes of 60 to 90 mV amplitude, and duration 1·3 to 1·5 msec, whilst the records in (b) are from a tonic fibre of low amplitude (12 to 28 mV and duration greater than 20 msec). At rest, the phasic fibre was silent whilst the tonic fibre gave small potentials at 20/sec. The longer latent period for the tonic fibre indicated that it was supplied by finer fibres. In a later study (1964), the effects of repetitive stimulation of the nerve supplying the superior oblique indicated the phasic and tonic natures of the two types of fibre; the spikes were associated with the rapid development of tension whilst the small potential changes were only associated with significant tension when repetitive stimulation was maintained.

### TWITCH AND SLOW

In a comparative anatomical and physiological study, Hess and Pilar (1963) showed that the two types of muscle fibre could, indeed, be classed as twitch and slow, as revealed by their responses to repetitive stimulation. Thus, if the nerve to the superior oblique, for example, is stimulated repetitively at 250 to 300 impulses/sec, maximal tension develops, and this soon falls off due to Wedensky block, but the tension does not fall to zero, remaining at about 30 per cent of the maximum value, whilst records of activity in the muscle fibres now show no evidence of spikes but only the slow

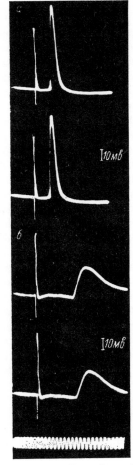

**Fig. 17.1** Intracellular recordings of superior oblique muscle fibres in response to trochlear nerve stimulation. a. Responses in phasic fibre; b. Responses in tonic fibre. (Matyushkin, *Sechenov Physiol. J.*)

monophasic potentials characteristic of the non-propagated type of change occurring in slow muscles (Pilar, 1967).*

### SINGLE MUSCLE UNITS

Lennerstrand (1974) studied single units in the cat's inferior oblique muscle by separating its nerve fibres

---

*Bach-y-Rita and Ito (1966) disputed this difference between slow and fast fibres and considered that both were of the twitch type, but Pilar (1967) has confirmed, with additional experiments, the qualitative difference between the responses in the two types of fibre. It may be, however, that there are three types of fibre: fast and slow twitch-types, and slow graded fibres (Peachey, 1968). In some extraocular muscles, e.g. the superior oblique studied by Hess and Pilar, the *Felderstruktur* fibres are large and plentiful by comparison with the inferior oblique. Hess (1967) has summarized the similarities and differences between submammalian skeletal and mammalian extraocular muscle fibres.

until only one fibre in the remaining portion activated the muscle fibre under examination. Of the three types of fibre, identified by their electrical responses to nerve fibre stimulation, one (MINC) gave only local electrical responses with no propagated action potential and no twitch; this, representing five out of a total of 60 units studied, was clearly the slow grape-innervated unit similar to those in amphibian slow muscle. Of the other types, SI had a single end-plate, and the MIC had two or more end-plates, but they were both of the twitch-type giving action potentials. The MINC fibres gave no mechanical response to a single stimulus, requiring repetitive stimulation to develop tension; the tetanic fusion-frequencies varied from 50 to 350 impulses/sec, being least with the slow MINC and greatest with the fast singly-innervated SI units. An interesting feature of the units studied by Lennerstrand was the small difference in fatiguability between fast and slow units. Lennerstrand suggested that the slow twitch units described by Bach-y-Rita might be the end-plate innervated units described as MIC rather than the MINC with grape innervation.

FUNCTIONAL SIGNIFICANCE

This differentiation of muscle fibre types into twitch, with propagated action potentials, and slow, with non-propagated action potentials, is of obvious significance in the submammalian species, the former being required for rapid activity as in the jumping of a frog, and the latter for sustained postural activity. In the mammal this difference in function is usually achieved by employing the same type of muscle fibre, so far as innervation is concerned, so that all give an all-or-none action potential, but their mechanical features vary, giving a fairly wide spectrum of speed and fatiguability. It has been argued that the requirements for two very different types of movement, as well as posture, namely the fast saccade and the slow pursuit and vergence movements, have led to the retention of the more primitive submammalian muscle fibre with its grape-like terminations and multiple innervation permitting a more precise control over its tension.

TONIC COMPONENT OF CONTRACTION

Subsequent work has not left a completely clear picture as to the functional significance of the morphologically identified two main types of muscle fibre; thus Barmack et al. (1971), studying the effects of two successive stimuli, found no evidence of a significant contribution of slowly contracting fibres to the total muscle tension in the cat's extraocular muscle, arguing that the slowly contracting fibres should show evidence of interaction between two stimuli separated by, say, a twentieth of a second, but in fact there was no interaction until the stimuli were repeated at 40/sec, double the fusion-

frequency for the slow fibres described by Bach-y-Rita (1966). Similarly, Fuchs and Luschei (1971), working on simian muscle, found no measurable tonic component of contraction when stimulating the abducens nerve, supplying the lateral rectus, at low frequencies (30/sec) adequate to cause tetanic fusion of cat's slow fibres. Again, Close and Luff (1974) found only a very small slow component in the response of the rat's inferior rectus giving what they considered to be a negligible contribution to total tension. An important point to appreciate, however, is the large differences in muscular tension required for different types of activity. If these are taken into account, it might very well be that a small development of tension, capable of being sustained for long periods, would escape measurement in many of the muscles studied.

Selective stimulation of slow fibres

Thus Browne (1976) pointed out that the tension in the lateral rectus required to maintain the human eye in the primary position was only 12 to 17 g compared with the 130 g developed in a saccade (Robinson et al. 1969); if the holding tension in the primary position represented activity in slow, slowly-fatiguing fibres, then the demonstration of such tensions in the absence of fast twitch-type muscle activity would confirm the belief in the physiological significance of the obvious morphological difference in muscle fibre-types. Browne profited by the different sizes of nerve fibres supplying the two main types of fibre, employing a technique of electrical stimulation that permitted the blocking of the lower-threshold large nerve fibres supplying the fast fibres, while the high-threshold small fibres transmitted their effects to the muscle. As Figure 17.2 shows, small-fibre stimulation causes a twitch that is slow by comparison with that due to the whole muscle, and developing only a fraction of the tension (note different calibrations for upper and lower records). With selective repetitive stimulation of the small nerve fibres, the tetanic fusion-frequency was 60 Herz, but maximal tension was only developed when the frequency of stimulation was raised to 140 Herz. If it is appreciated that motor neurones to the extraocular muscles are capable of discharging at rates in excess of 700 imp/sec, and when in the primary position the rate is typically 100 imp/sec, it is clear that these slow fibres would be activated nearly maximally. The actual tension developed under optimal conditions and maximal frequency of stimulation was some 8 g, compared with 166 g when the whole muscle was activated at an optimal rate of 280 to 320 imp/sec. Consequently it is very likely that the slow fibres, capable of developing some 8 g of tension compared with the 12 to 17 g required for maintaining the primary position, are capable of contributing a large part of the tension necessary for

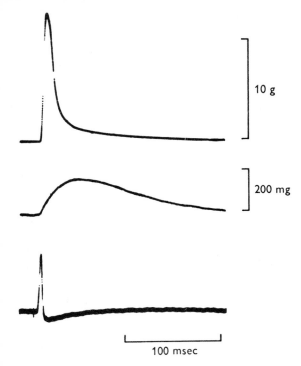

**Fig. 17.2**  Comparison between the time-course of a whole muscle twitch (upper trace) and a slow fibre contraction (middle trace). The bottom trace shows the late monophasic component of the e.m.g. The stimulus was applied 12 msec after the start of the sweep. (Browne, *J. Physiol.*)

sustaining this tension under conditions where fatigue must not be allowed to intervene. Thus these slow fibres showed very little fall in tension when stimulated for 16 sec (9 per cent) compared with that observed in activation of the whole muscle (49 per cent).

## Smoothing function

An additional function of these slow fibres will be to smooth the activities of the twitch-fibres; these have a tetanic fusion-frequency of 350 to 500 imp/sec, so that when the eye is in the primary position and these twitch fibres are being activated along with the slow fibres, their responses will be twitches instead of smooth tetanic contractions. The simultaneous activation of the slow fibres at frequencies to give a smooth tension will tend to reduce the twitch-like character imparted by the fast muscle fibres.

### LENGTH-TENSION CHARACTERISTICS

In general, skeletal muscle can increase its isometric tension, when maximally stimulated, if its length is increased above resting length; this gives the characteristic length-tension curve. The extraocular muscles are no exception, and this means, for example, that

when the eye is in strong adduction, thereby stretching the lateral rectus, the lateral rectus can command much more tension to execute a lateral saccade than if the movement were initiated from the primary position.★

## PHARMACOLOGY

### ACETYLCHOLINE CONTRACTURE

Duke-Elder and Duke-Elder (1930) observed that mammalian extraocular muscles resembled the slow skeletal muscles of lower vertebrates in responding by a contracture to applied acetylcholine.† In the superior rectus of the rabbit, where the slow and fast types of fibre are separated in two layers within the muscle plate, Kern (1965) demonstrated contracture-type responses in the slow fibres. The tension developed during acetylcholine contracture of the superior rectus muscle is about 30 per cent of the maximal tetanic tension, and it is interesting that the tension developed during repetitive stimulation of this muscle settles down to 30 per cent of the maximal tension developed (Pilar, 1967); this corresponds with the situation when the fast fibres have ceased to respond because of Wedensky inhibition, so that the tension finally reached during repetitive stimulation is due to the slow fibres. Suxamethonium is a nicotinic cholinergic drug, and Browne (1976), in his study of the sheep's superior oblique muscle, found that this caused a contracture that developed a tension of 8·0 and 10·6 g in two sheep while the maximal tensions developed by repetitive stimulation of the small nerve were 7·0 and 7·1 g respectively.

### SYMPATHETIC

Paralysis of the sympathetic system produces Horner's syndrome consisting of ptosis, due to paralysis of the adrenergic Muller's muscle, miosis due to paralysis of the dilator pupillae, and enophthalmos. The last effect suggests that the extraocular musculature may be influenced by the sympathetic system. Kern (1968) has found that isoproterenol causes relaxation of tension in

---

★The length-tension diagrams were obtained by holding the muscle of one eye at a fixed length and asking the subject to gaze at points farther and farther in the field of action of the muscle, i.e. more and more nasally for the medial rectus; since the nervous discharge to the muscle increases with increasing size of saccade, this means that graded degrees of 'innervation' of the muscle could be obtained for each artificially maintained length.

†The effects of several cholinomimetics on the cat's extraocular muscles have been examined by Sanghvi and Smith (1969); spike activity certainly occurs, perhaps due to the twitch fibres only. Muscarine has a smaller effect than that of acetylcholine, which is blocked by atropine, so that the receptors are mainly 'nicotinic'.

the monkey's extraocular muscles, an effect that is blocked by DCI, indicating a β-action. Since the amine caused relaxation of tension developed during treatment with acetylcholine, he argued that it was the slow fibres that were sensitive to the sympathomimetic agent. On the same basis, then, the exophthalmos accompanying extreme fear could be due to the β-action of the adrenaline secreted by the adrenal gland under these conditions.

## SENSORY INNERVATION

### MUSCLE SPINDLE

Muscle spindles were described in the sheep's extraocular muscles as early as 1910 by Cilimbaris, but it was not until 1946 that Daniel described spiral nerve endings surrounding fibres of the human extraocular muscle; he surmised that they were sensory in function, responding perhaps to thickening during contraction. Later Cooper and Daniel identified definite muscle spindles in human muscles, a single rectus containing as many as 50, a number comparable with those in a lumbrical muscle of the hand; the spindles were concentrated near the origin of the muscle, and that was why they had not been discovered earlier.

### Structure

In general the structure of the spindle was similar to that classically described in other muscles; 2 to 10 fine muscle fibres (intrafusal) are enclosed in a capsule whilst an adjacent nerve trunk sends fibres into the capsule to end on the intrafusal fibres. The innervation is both motor and sensory; thus the gamma efferents—fine motor fibres—by causing the muscle fibres in the capsule to contract are able to increase the sensitivity of the spindle to stretch; whilst the sensory endings are characteristically of the primary, or annulospiral, and secondary, or flower-spray, types supplied by large Group I and smaller Group II fibres respectively.

### Bag and chain fibres

The sheep's spindle has been described recently by Harker (1972) who has compared the superior rectus with the levator palpebrae. The distributions are illustrated in Figure 17.3, most of them being in the peripheral (orbital rim and peripheral patch) layers. They contained both nuclear bag and nuclear chain fibres and each spindle was innervated by a large Group I primary nerve fibre which divided to supply each intrafusal fibre with annulospiral endings; in addition a spindle could have up to four secondary endings derived from smaller Group II nerves and terminating in irregular spirals mainly on the chain fibres. The nuclear bag fibres have a low alkali-stable ATPase and

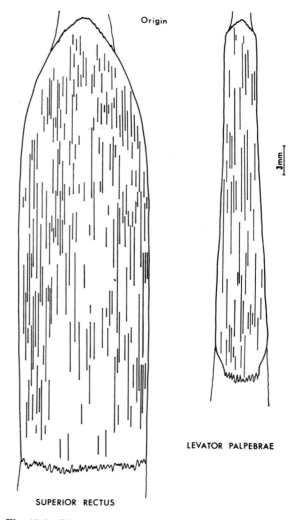

**Fig. 17.3** Distribution of muscle spindle capsules in superior rectus and levator palpebrae muscles of the sheep. The capsules are plotted within the limits of a single representative longitudinal section. (Harker, *Invest. Ophthal.*)

are similar to slow fibres whilst the nuclear chain fibres have high ATPase activity and are similar in this and other respects to twitch fibres.

### Motor innervation

These similarities extend to the mode of motor innervation, the nuclear bag fibres having a grape-ending similar to the G-fibres described above, and the nuclear chain a plate ending similar to the C-fibres (Fig. 17.4). An interesting feature is the collateral motor innervation, so that a fibre that activates an extrafusal fibre also activates a spindle intrafusal fibre—the motor innervation thus corresponds to the β-innervation, which is generally regarded as a more primitive type that fails to provide independent control of the two types of fibre,

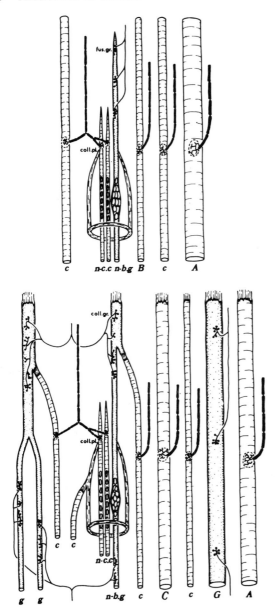

**Fig. 17.4** Top: Schematic representation of the pattern of motor innervation in levator palpebrae. Bottom: Superior rectus. A, large A-fibre; B, small B-fibre; C, intermediate C-fibre; c, small C-fibre; *coll. gr.*, collateral grape endings; *coll. pl.*, collateral plate endings; *fus. gr.*, fusimotor grape endings; G, large G-fibre; g, small G-fibre; *n-b.g.*, nuclear bag G-fibre; *n-c.c*, nuclear-chain C-fibre. (Harker, *Invest. Ophthal.*)

with 181—and the motor innervation was different insofar as the grape innervation of the nuclear bag fibres was not shared with extrafusal fibres (it will be recalled the levator does not contain extrafusal G-fibres).

SPINDLE ANALOGUES

In certain species, notably the cat and monkey, no spindles were found but the richness of the innervation of the muscles certainly suggests the presence of simpler organs, perhaps just spirally wound endings encircling single muscle fibres (Cooper and Fillenz).* Subsequent work, notably that of Wolter, suggests that the sensory apparatus in the human extraocular muscles is even more varied; in all, he has described some six different types of ending, varying in complexity from the spindle to relatively simple terminations, which may be distinguished from motor endings by the circumstance that they end in connective tissue between fibres.

SENSORY PATHWAY

The evidence relating to the pathway for this sensory information has been summarized by Rogers and Whitteridge. There seems little doubt that there are connections between the muscle-nerves and the trigeminal (N V), often within the orbit; recording from these branches, in the goat, Whitteridge (1955) obtained spindle-responses on stretching the eye-muscle. In the sheep, Manni, Bortolami and Desole (1967) found a group of neurones in the Gasserian ganglion with a steady discharge of 10 to 20/sec; on stretching an extraocular muscle, the discharge-rate suddenly increased to up to 300/sec, finally settling down to 150 to 90/sec when the stretch was maintained. On release of the muscle, the discharge ceased abruptly, and later the steady discharge of 10 to 20/sec was reassumed. This is typical of the behaviour of spindle afferent neurones. Subsequent studies in both sheep and pig leave no doubt that in these species the sensory neurones innervating the spindles are located in the Gasserian ganglion, and not in the mesencephalic root of the trigeminal (Manni, Bortolami and Deriu, 1970), nor yet in ganglion cells located along the oculomotor nerve (Manni, Desole and Palmieri, 1970). The fact that the oculomotor nerve in these species contains sensory fibres is due to the passage of the central processes of the ganglion cells of N V into the central portions of the oculomotor nerve (Fig. 17.5) at the oral portion of the cavernous sinus and thus reaching the midbrain through this nerve (N III). As to where these centrally directed processes terminate, Manni *et*

and this doubtless corresponds with the more limited functions of extraocular muscle spindles, which are not involved in any direct servo-type stretch-reflex (p. 414). The levator palpebrae muscle differed from the rectus in having a smaller number of spindles—61 compared

*The responses recorded from N III when the cat's inferior oblique was stretched have been examined in great detail by Bach-y-Rita and Ito (1966), and compared with the responses from true muscle spindles in skeletal muscle. In general, it seems that the receptor-organ of the cat extraocular muscle is simpler, without any gamma-efferent control.

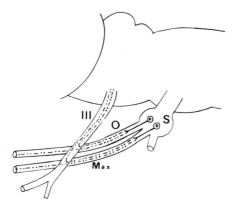

**Fig. 17.5**  Probable arrangement of the trigeminal fibres contained in the oculomotor nerve. III, oculomotor nerve; O, trigeminal ophthalmic branch; Max, trigeminal maxillary branch; S, semilunar ganglion. (Manni *et al.*, *Exp. Neurol.*)

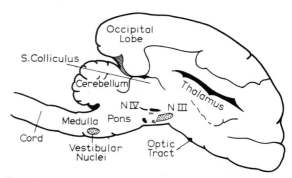

**Fig. 17.6**  Sagittal section of the cat's brain.

*al.* (1976) suggest two possibilities: (i) they could end in the rhombencephalic reticular formation, or (ii) they could run directly to the cervical spinal cord in the trigeminal tract and synapse with motor neurones supplying the muscles of the neck. This latter suggestion arises from the circumstance that stretching eye muscles can affect neck muscle activity as recorded by the EMG (Easton, 1971; Manni *et al.*, 1975).* In the cat, destruction of the eye-muscle nuclei completely denervated the muscles, according to Corbin and Oliver, so that it may well be that in this species the sensory neurones have migrated from N V to the motor nuclei. Certainly Manni, Bortolami and Desole (1968) were unable to find neurones in the Gasserian ganglion responding to stretch of eye muscles, as in the sheep and pig.

## THE FINAL COMMON PATH

### OCULO-MOTOR NUCLEI

The motor neurones activating the eye muscles are derived from the cranial nuclei as follows:

N III. (*Oculomotor nerve*). Superior rectus. Inferior rectus. Medial rectus. Inferior oblique.

N IV. (*Trochlear*). Superior oblique.

N VI. (*Abducens*). Lateral rectus.

These nerves represent essentially the axons of cell bodies collected in the grey matter of the mid-brain and pons—the *eye-muscle nuclei*.

### THE NUCLEUS OF NERVE III

This lies in the tegmentum in the central grey matter in the upper part of the mid-brain below the superior colliculus, beneath and running parallel to the aqueduct of Sylvius (Fig. 17.6). It is some 5 to 6 mm long and

extends between the third and fourth ventricles. It has been conventionally divided into a series of discrete zones each concerned with a pair of muscles; thus according to Brouwer it is divided primarily into a principal (lateral) nucleus providing the motor neurones to the levator palpebrae and four extraocular muscles; the *median nucleus of Perlia*, considered to be a coordinating centre for activating the medial recti during convergence; and the *Edinger-Westphal nucleus*, containing the autonomic motor neurones controlling the ciliary muscle and sphincter pupillae. The principal lateral nucleus was divided into zones from above downwards (rostrocaudally) each zone representing the motor neurones for the levator, superior rectus, medial rectus, inferior oblique and inferior rectus in this order. Subsequent work, notably that of Warwick (1953), has shown that the organization is not so simple. Thus, although the individual eye muscles are indeed operated by localized groups of neurones within the principal nucleus, the simple rostrocaudal arrangement postulated by Brouwer apparently does not pertain. Instead, the arrangement is that shown in Figure 17.7 (*right*); it will be seen that the inferior rectus, inferior oblique and medial rectus possess discrete motor pools arranged along the axis of the somatic nucleus in this *dorsoventral* order. The superior rectus, on the other hand, is innervated by neurones scattered along the medial aspect of the caudal two thirds of the main nucleus. Moreover, Warwick (1955) showed that in monkey and man, the two species in which convergence of the eyes is a prominent feature of the movements, the median nucleus of Perlia is difficult to find, and, when present, its neurones are not internuncial, as one would expect of a coordinating centre, but rather they are motor neurones supplying the superior rectus and inferior

*As summarized by Manni (1974) the central processes of the trigeminal neurones, concerned with proprioception, enter the ipsilateral trigeminal root and run in the descending trigeminal tract to end in the oral part of the descending nucleus and in the main sensory nucleus. Second-order neurones run to the thalamus, tectum and tegmentum.

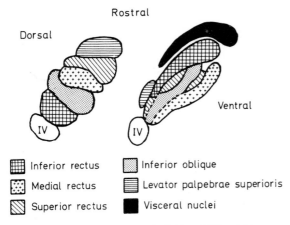

Fig. 17.7 Illustrating Brouwer's (left) and Warwick's (right) views of the organization of the oculomotor nucleus. (After Warwick, *J. comp. Neurol.*)

oblique. Hence we must assume that the act of convergence is organized at a higher level. The Edinger-Westphal nucleus is well established as the autonomic division of the oculomotor nucleus.

### THE NUCLEUS OF NERVE IV

This is a small clump of cells lying caudally to the nucleus of N III, being situated in the central grey matter in the floor of the aqueduct of Sylvius and at the level of the lower part of the inferior colliculus. The nerve fibres, which represent the motor pathway for the superior oblique, undergo an almost complete decussation in the anterior medullary velum, emerging on the dorsal aspect of the brain stem.

### THE NUCLEUS OF NERVE VI

This nucleus lies in the pons, in the floor of the fourth ventricle.

#### MEDIAL LONGITUDINAL FASCICULUS

A conjugate lateral movement requires a co-ordinated response on the part of the medial rectus of one eye, innervated by N III, and the lateral rectus of the other, innervated by N VI. This co-ordination is brought about by interneurones running from the nucleus of N VI or a nearby region—the so-called *parabducens nucleus* of Crosby. These interneurones run to the nucleus of N III in the medial longitudinal fasciculus (also called posterior longitudinal bundle), a tract associated with the brain-stem reticular formation which connects the three motor nuclei bilaterally. Thus the initiation of a conjugate lateral movement may be illustrated by Figure 17.8, the primary stimulus reaching the parabducens and abducens nuclei; an interneurone in the latter passes up in the median longitudinal fasciculus to activate a motor neurone in N III which passes to the medial rectus.*

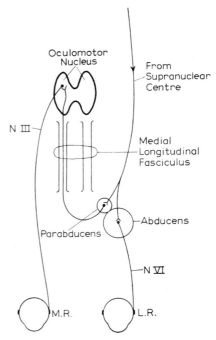

Fig. 17.8 Illustrating the conjugate innervation of the lateral rectus of one side and the medial rectus of the other.

Lesions in the median longitudinal fasciculus may well be responsible for the so-called internuclear ophthalmoplegia; the loss of the power to adduct the eye in conjugate lateral movements with its retention in convergence. Certainly this syndrome may be evoked by bilateral section of the fasciculus (Hyde and Slusher).†

## ELECTROMYOGRAPHIC STUDIES

### ELECTROMYOGRAM

The activity in a muscle may be measured by the action potentials of its muscle fibres recorded from a needle-

---

*Reciprocal innervation demands, of course, that inhibitory impulses pass to the antagonists; thus the parabducens nucleus would have to send inhibitory impulses to N III of the same side besides excitatory impulses to N III of the opposite side. We may note that opinions differ as to whether the fibres related to conjugate deviation cross at once to ascend in the contralateral median longitudinal fasciculus, as indicated in the diagram, or whether they ascend on the same side as the deviation and decussate within the nucleus of N III (see Crosby, 1953, p. 440).

†A possible autonomic control of the extraocular musculature has been discussed by Eakins and Katz (1967); they found that stimulation of the cervical sympathetic, or injection of epinephrine, increased the tension of the superior rectus, an action that was antagonized by α-blockers such as phentolamine. The results were similar to the effects on the nictitating membrane, except that atropine was without effect on the rectus muscle.

electrode inserted into the muscle; the more intense the contraction the greater will be the frequency of the spike discharges in individual units and the larger will be the number of units active. Because of the danger of various artefacts, the results of any study must always be interpreted with caution, but in the hands of experienced workers the EMG has been a valuable tool in assessing the contributions of individual muscles to given eye movements.*

### RECIPROCAL INNERVATION

Björk and Kugelberg showed that, when the eye looked straight ahead, a given extraocular muscle showed a steady tonic discharge, single units firing with frequencies as high as 50/sec.† This is in marked contrast with skeletal muscles which are usually almost or completely silent during relaxation. With needles in the medial and lateral recti, reciprocal innervation was very well demonstrated; thus on abduction, there was an increased discharge in the lateral rectus associated with a simultaneous inhibition of the tonic discharge in the medial rectus. This is illustrated by Figure 17.9; the records marked A are obtained with steady fixation, whilst B shows the discharges in the medial (top) and lateral (bottom) recti during a slow following movement from 40° inward to 40° outward, reading from right to left. The gradual diminution in activity in the medial rectus (top record), accompanied by the increase in activity of the lateral rectus, is well brought out. With this slow type of movement there was never a complete inhibition of the antagonist, by contrast with the rapid 'saccade' (p. 398) where the antagonist becomes quite silent (Fig. 17.10).

### CO-CONTRACTION

The EMG has also permitted a solution of the problem of co-contraction, the simultaneous contraction of agonist and antagonist that is a feature of certain acts involving the skeletal musculature. The phenomenon of reciprocal inhibition shows that in the execution of any movement of the eyes agonist and antagonist do not work against each other, and this applies even to the point at which the movement terminates; thus the move-

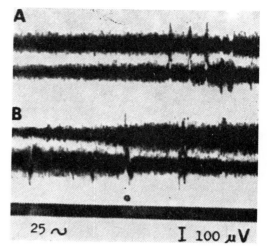

**Fig. 17.9** Electromyograms of human medial and lateral recti during steady fixation (A) and a lateral movement (B). The upper record of each pair is from the lateral rectus, the lower is from the medial rectus. A. The gaze is maintained in the straightforward direction; note intense activity in both muscles. B. Slow following movement from 40° inward to 40° outward showing continuous diminution of activity in medial rectus and corresponding increase in the lateral rectus. The dot below the traces in record B indicates the moment at which the gaze passes the midline. Records read from right to left. (Björk and Kugelberg, *EEG clin. Neurophysiol.*)

ment of abduction, say, is not arrested by activity in the medial rectus, the reciprocal inhibition observed at the beginning and during the course of the movement being maintained to the end, at which time the muscle exhibits a discharge that is characteristic for the particular position of the eye. Thus the movement is not *ballistic*, as a movement arrested by active contraction

---

*The technique and many of the pitfalls in interpretation of the EMG have been described in some detail by Marg, Jampolsky and Tamler (1959).
†This is true of all the extraocular muscles, so that all three antagonistic pairs are constantly pulling against each other at rest; the oblique muscles are less active than the recti. According to Björk and Wahlin (1960) the resting activity is absent in sleep and general anaesthesia.

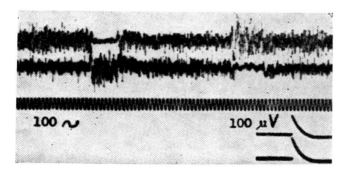

**Fig. 17.10** Electromyograms from lateral rectus (upper tracing) and medial rectus (lower tracing) during quick changes of fixation from 10° outward to straight ahead and back to 10° outward, showing brief acceleration of activity in the agonist accompanied by abrupt cessation of activity in the antagonist. (Björk and Kugelberg, *EEG clin. Neurophysiol.*)

of antagonists would be called (Miller, 1958). Nevertheless it might be argued that the movement would be steadied by co-contraction of the elevator and depressor pairs (Boeder), but Breinin and later workers have shown that there is no increase over the normal tonic resting activity in these muscle-pairs during a lateral movement; similarly, during a pure elevation or depression, the medial and lateral recti fail to show augmented activity.

## THE SENSORY MESSAGES

### EFFECTS OF STRETCH

That sensory information is conveyed from the muscles, during stretch, was shown without doubt by Cooper, Daniel and Whitteridge for the goat where spindles were identified, and also for the cat and monkey in which spindles were not found (Fillenz); action potentials being recorded from the brainstem, or from the nerve supplying the muscle, e.g. the oculomotor nerve (N III), on stretching the inferior oblique (thus proving that this nerve is not purely motor, incidentally). Stretching a muscle activates its spindles and induces a monosynaptic activation of its motor neurons, causing it to contract, thereby resisting the stretch. According to Breinin (1957) the resting activity of a muscle falls to practically nothing on cutting its tendon, suggesting that the normal resting activity is a reflex response to stretch; however, the failure to augment the discharge on pulling the muscle rather discountenances this view of the origin of *tonic* activity, which is therefore probably central in origin.

#### ABSENCE OF STRETCH-REFLEX

Later studies have tended to confirm the absence of a stretch-reflex in the extraocular muscles. Thus Keller and Robinson (1971), in their study of isolated abducens motor neurones in the intact monkey, failed to observe any increase in discharge on rotation of the eye so as to stretch the lateral rectus. Thus the normal tone of the extraocular muscles, as revealed by their tensions when the eye is in the primary position, is due to centrally controlled discharges to the motor neurones.

#### FEED-BACK

The participation of the spindles in a feed-back mechanism is indicated, however, by Breinin's failure to observe the finely graded reciprocal innervation of agonists and antagonists in the muscles of an enucleated eye; thus, under these conditions, the antagonist is abruptly inhibited as soon as the agonist acts instead of showing the gradual decrease illustrated by Figure 17.9. More recently Sears, Teasdall and Stone (1959) have

shown that pulling on the agonist or on the antagonist during a movement causes a decrease in the spike discharge measured electromyographically, whilst mere cutting the insertion of the muscle had no effect on spontaneous activity. On balance, then, the electromyographic evidence suggests that the motor discharge to the muscles is influenced by sensory messages from the muscles themselves, but more work is obviously required.

## MOTOR NEURONE DISCHARGES DURING MOVEMENTS

#### RECORDING TECHNIQUE

Fuchs and Luschei (1970) and Robinson (1970) have described the firing of motor neurones in the abducens and oculomotor nucleus respectively in the waking monkey. To make these measurements, in a preliminary operation a coil of wire was implanted on one globe which, used in conjunction with magnetic fields, permitted the recording of eye movements. At the same time stimulating electrodes were implanted in N III so that when, later, the recording electrode was passed through an implanted guide-tube, the position in the oculomotor nucleus could be ascertained by antidromically conducted impulses evoked by stimulating N III.

#### SACCADE AND FIXATION

Figure 17.11 shows typical discharge activity during fixation and saccadic movements of a unit that was almost certainly one controlling the inferior rectus; with steady fixation there is a steady discharge; with an upward movement (A) this discharge is inhibited, to be resumed at the new fixation point. With a downward movement there is a burst-discharge of very high frequency which increases with the velocity of the saccade; after the saccade the burst did not subside instantaneously but gradually, indicating a continuing contraction of the muscle in spite of the stationary position of the eye and probably necessary to compensate for the slow stress-relaxation of visco-elastic elements in the globe's suspensory tissue. With any given unit we may recognize an ON-direction, such that movement in this direction is associated with a burst-discharge, and an OFF-direction associated with inhibition of the resting fixation discharge; thus with the lateral rectus the ON-direction was an abduction, and so on.

#### DISCHARGE DURING STEADY FIXATION

The steady discharge during fixation depended on the direction of gaze so that, as this was directed from the primary position more and more into the OFF-direction,

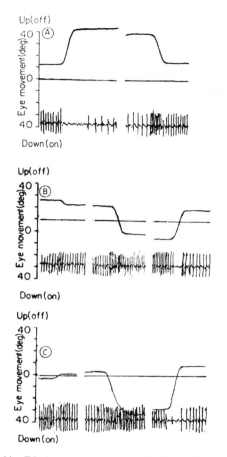

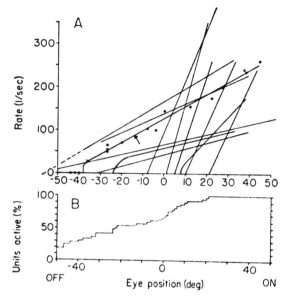

**Fig. 17.12** A. Relation between discharge-rate (impulses/sec) and eye position during steady fixation. Responses in single units are shown to indicate the variety of slopes and thresholds. Experimental points for one curve, marked with arrow, are shown to indicate the typical nature of the fit between experimental points and straight-line approximation. B. Percentage of units active out of the pooled population of 35 units during fixation at any angle of gaze. The ON-direction could be either up or down, left or right, depending on which extraocular muscle was associated with the unit under study. (Robinson, *J. Neurophysiol.*)

**Fig. 17.11** Discharges of single units in the oculomotor nucleus of the intact, alert monkey during fixation and saccadic eye movements. The upper record of each section indicates eye movement, and the lower, unit discharge. The sections were selected from a continuous recording of the activity of a unit associated with downward movements and thought to be a motor neurone of the inferior rectus muscle. (Robinson, *J. Neurophysiol.*)

the steady discharge decreased and finally could be inhibited completely. There was thus a *threshold direction of gaze*. Figure 17.12 shows the behaviour of a variety of units with different ON-directions; also shown is the percentage of units, out of a pooled population of 35, that were active for a given direction of fixation, and it will be seen that, as the eye moved farther in the ON-direction, units were recruited; thus small angles of deviation from the primary position are achieved by contractions of fewer units, and in a given unit, with a lower frequency of discharge. In general the studies of Fuchs and Luschei and of Robinson indicated that, when the eye was in the primary position, a considerable discharge, of the order of 100 imp/sec, was maintained indefinitely. This is very high for a tonic postural type of innervation and demands a muscle fibre with a low degree of fatiguability.[*]

TIME-RELATIONS FOR A SACCADE

These are indicated in Figure 17.13 from Fuchs and Luschei's study of abducens units; the burst-discharge definitely precedes the eye movement and ceases before the end, exhibiting a fast decline ($d_{fast}$) and a slower one, ($d_{slow}$).

HIGH SACCADIC DISCHARGE RATES

It has been noted by several workers that, although the tetanic fusion frequency for fast extraocular muscle fibres is high, of the order of 200 imp/sec, the maximum discharge rates during saccades are very much higher, although maximum tension in the fast muscle units is developed at the fusion frequency. Barmack *et al.* (1971) suggested that the *rate* of development of tension might increase at frequencies beyond the fusion-frequency, and this was true, so that the maximum

[*]The motor neurones to other muscles, such as the gastrocnemius or soleus, usually discharge at rates far below their maximum possible; this 'stabilization' is due, partly at any rate, to the recurrent negative feedback through Renshaw cells. According to Sasaki (1963), the eye-muscle motor neurones do not show this recurrent inhibition; and this may account for the much higher rates of discharge.

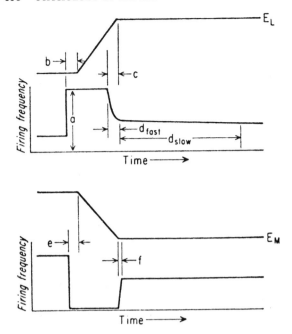

**Fig. 17.13** Schematic representation of changes in firing frequency of abducens units during a lateral saccade ($E_L$) and a medial saccade ($E_M$). Lower case letters designate measurements used to characterize the unit firing patterns. Upper trace represents EOG; lower trace represents firing frequency. (Fuchs and Luschei, *J. Neurophysiol.*)

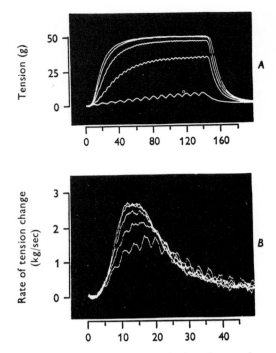

**Fig. 17.14** A, isometric tension in the lateral rectus due to a train of supramaximal shocks applied to the abducens nerve. Stimulation frequencies were varied in 100/sec steps from 100 (lower curve) to 500/sec (upper curve). Maximum tension produced by 400 and 500/sec stimuli was identical. B, rate of change of isometric tension obtained by differentiating records similar to those of A. Stimulation frequencies ranged from 300/sec (lower curve) to 700/sec (upper curve) in 100/sec steps. Maximum rate of change of tension was identical at 600 and 700/sec. (Fuchs and Luschei, *J. Physiol.*)

rate of tension-development was achieved at 600 imp/sec although maximum tension was developed at 400/sec (Fuchs and Luschei, 1971). This is shown by Figure 17.14 where the top curve (A) shows the effects of increasing frequency of stimulation on the development of tension and the lower curve (B) shows the differential of tension with time, i.e. the rate of tension development; and it will be seen that this increased up to 600 imp/sec, so that the maximum rate was about 25 per cent greater at 600/sec than at 400/sec, the fusion frequency, which gave maximum tension.

PURSUIT FOLLOWING MOVEMENT

When the eyes followed a moving target there was an increase in the rate of discharge of a given unit, over that during fixation on a given point, by an amount that was proportional to the velocity of following. Thus the rate of discharge, D, was related to the angle of view, $\theta$, by:

$$D = k\theta + r\, d\theta/dt$$

The ratio $r/k$ is a time-constant, $T$, with a mean of 198 msec. It will be recalled that in terms of force developed, the movement can be characterized by a similar equation:

$$F = K\theta + R\, d\theta/dt$$

where $F$ is the force. $R/K$ is likewise a time-constant and has a value of the same order. Figure 17.15 shows the relation between discharge-rate and eye-velocity; as the velocities of following are so much lower than those in a saccade it is not surprising that the discharge rates are very much less (of the order of 100/sec compared with rates as high as 400/sec in a saccade.

Robinson noted that all the motor neurones studied were involved in any given type of activity—fixation. saccade and pursuit—so that a separate class of motor neurones is apparently not involved in, say, pursuit. The same applies to vergence movements (Keller and Robinson, 1972; Keller, 1973) in spite of the claim that these are mediated by slow muscle fibres (p. 407), all abducens units participating in vergence movements, decreasing their discharge with convergence and increasing it with divergence. Moreover the value of $r$ in the above equation, deduced from vergence movements, was the same.

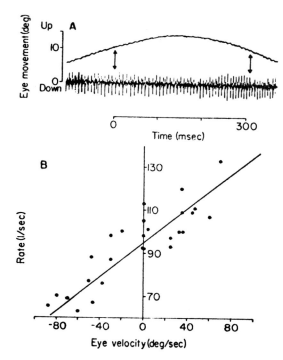

**Fig. 17.15**  Relation between discharge-rate and pursuit eye-velocity.

A. At each arrow the eye position was 10° upwards. The discharge rate for steady fixation at this position was 130/sec but the rate was only 86/sec when the eye was travelling in the OFF-direction at 38°/sec (left arrow) and was 167/sec when travelling in the opposite direction at 49°/sec (right arrow).

B. Discharge-rate of another unit plotted against pursuit velocity at instants when the eye passed through a given position when going in the ON-direction (positive abscissae) and the OFF-direction (negative abscissae). (Robinson, *J. Neurophysiol.*)

ASYMMETRICAL CONVERGENCE

We have seen that on Hering's view, when the eyes converge asymmetrically (Fig. 16.24, p. 397) the apparently motionless eye did, in fact, take part in a convergence which was subsequently balanced by a version to leave its fixation direction the same. Using the electromyograph Tamler *et al.* (1958) showed that the apparently stationary eye in asymmetrical convergence showed activity in both medial and lateral rectus, a *co-contraction* that would balance the version and vergence if the activities were synchronous; this certainly supports Hering's view of conjugate innervation of both medial recti and suggests that, either the tendency for the apparently stationary eye to adduct is balanced by the pull of the lateral rectus, or that the eye does adduct and then abducts in a conjugate version. However, Keller and Robinson found no evidence, from their studies on single units in the waking monkey, of

a co-contraction during asymmetrical convergence and they agreed with Breinin (1957) that the version and vergence commands required for this type of vergence are summed centrally, with the net result appearing as activity in a shared common path, i.e. the opposing commands simply cancel out in each motor nucleus of the horizontal recti of the non-moving eye.

## TENSIONS IN THE EYE MUSCLES

### FIELDS OF ACTION

As we have indicated, a definite tension is developed in the eye muscles even when the eye is in the primary position; this amounts to about 8 g. As the eye moves towards a new fixation point, we may say that it moves into the *field of action* of the muscle that is primarily concerned with the movement, and out of the field of action of its antagonist. For example, the left medial rectus will be active when this eye moves to the right, and the eye would be said to be moving into its field of action and out of the field of the left lateral rectus. Collins et al. (1975) implanted microtransducers into the muscles of human subjects and measured the tensions developed at different fixation points, during following movements, and during saccades. As we should predict from the studies on motor neurone discharges (p. 414), the tension developed as the fixation point is moved into the field of action of the muscle increases; the form of the curve relating tension to fixation position is not linear but parabolic. Tension increases in the antagonist as well, but to a much less extent since this is being inhibited and the tension is the result of passive stretch.

### FOLLOWING MOVEMENTS

During slow following movements the tensions were close to those attained during steady fixation at any position, although there was some hysteresis, in the sense that, when the eye moved nasally, the right medial rectus developed a greater tension for any given fixation point than when it moved in the opposite direction, i.e. out of its field of action (Fig. 17.16).

### SACCADES

During the saccade, the tension rose rapidly and isometrically to a peak before the eye had moved; this was followed by an isotonic phase of lowered tension as the eye moved and fixation at the new point was maintained (Fig. 17.17); the antagonist, although being inhibited by reciprocal innervation, showed a sharp peak in tension due to passive stretch, and finished with a higher tension than before, due to this passive increase in length.

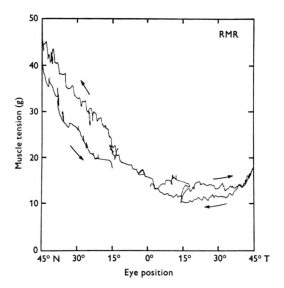

**Fig. 17.16** Tensions recorded in a right medial rectus muscle tendon during unrestrained movements by a patient with 30 prism-dioptres of intermittent exotropia. The eye followed a target moving from the primary position to 45° temporal, then to 45° nasal and back toward 15° nasal at 10°/sec. The lower curve closely approximates the static locus of fixation-tensions. (Collins *et al.*, *J. Physiol.*)

OPERATIONAL ENVELOPE

Collins *et al.* summarized the mechanical features of the muscular contractions during normal movements by the operational envelope shown in Figure 17.18.

The lower heavy line indicates the tension in the muscle during fixation, rising steeply as the eye fixates nasally into the field of action of the medial rectus, and more gradually as the eye moves temporally.

The upper curve is obtained by adding the force-increment necessary to make a saccade to any new position, the upward arrow indicating the increment as the eye fixates 15° medially; after the fixation the tension falls back to the lower curve along the line with the downward pointing arrow. If the eye moved back to the primary (0°) position the medial rectus would be relaxing and only a small passive increase in tension would occur (horizontal arrow).

## CO-ORDINATING CENTRES

The motor nuclei may possibly be regarded as 'centres' integrating, in a limited way, the activities of the individual muscles, so that a given movement is carried out smoothly and efficiently, but it is probably safer to regard them as the final common path for the necessary movements elaborated at higher levels of the nervous system. Certainly the intimate connection of the eye movements with bodily activity, and the dependence of the eye movements on visual impressions, demand centres at a higher level capable of integrating these associated activities on a wider scale. Centres that are considered to be important for these integrative activities are the superior colliculi, certain regions of the brainstem tegmentum, the occipital cortex and the frontal cortes.

## THE SUPERIOR COLLICULI

The functions of these two bodies, often called the *optic tectum*, are not completely known, especially in man, but there is little doubt that they constitute an important

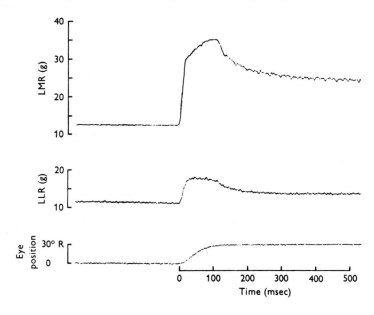

**Fig. 17.17** Tensions recorded in a left medial rectus (upper channel) and a left lateral rectus (middle channel) during an unrestrained saccadic movement from primary position to 30° right (indicated on the lower channel). Note the initial isometric tension rise in the agonist left medial rectus before the eye moves appreciably, then a break in the curve as the eye achieves significant velocity. Note also that the tension in the relaxing antagonist left lateral rectus increases during the early stage of the saccade before it assumes the new steady-state fixation level, which is several grams greater than that for the primary position. (Collins *et al.*, *J. Physiol.*)

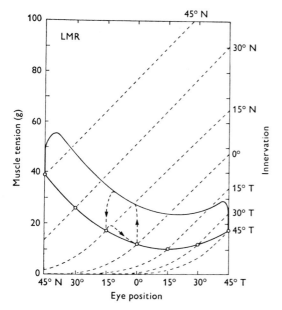

**Fig. 17.18** An operational envelope of the normal ranges of extraocular muscle tension for a typical left medial rectus during fixation, following and saccadic movements. Dashed lines in the background represent a family of length–tension curves for each of the innervations shown. The lower curve represents the static locus or the tension required to maintain the eye in each position of horizontal gaze indicated on the abscissa. The upper curve represents the dynamic locus, or force increment producing a saccadic movement. The area between the static and dynamic loci represents the normal operational envelope. The counterclockwise loop between the zero and 15° nasal positions indicates tension changes in the muscle during a saccadic refixation into the muscle's field of action. The small clockwise loop from left to right indicates the passive tension changes occurring in the muscle as an antagonist during a saccade. (Collins *et al.*, *J. Physiol.*)

centre in the organization of movement, especially those movements of the head and eyes that must be coordinated with movements of the rest of the body.

CONNEXIONS

This is clear from their anatomical connexions since they receive afferent information from the medial and trigeminal lemnisci as well as from the cerebral cortex. Their efferent connexions to the spinal motor neurones constitute the tectospinal tract; connexions with the motor neurones controlling eye movements, however, are not direct (Altman and Carpenter, 1961), so that the pathway is probably through the nuclei of Darkschewitz and Cajal. The anatomical basis for a special role in visually directed movements is provided by the fibres from the optic tract terminating directly on tectal neurones in the superficial* strata of the bodies, together with a projection from Areas 17 and 18 of the cerebral cortex (p. 423). Projections to the thalamus, particularly

the posterior nuclei including the pulvinar, probably permit feedback circuits that allow coordination of all types of sensory information with motor activity through association areas of the cerebral cortex.

PHYLOGENY

In general, the higher the animal phylogenetically, the larger the proportion of optic tract fibres that pass to the lateral geniculate body and thence to the cerebral cortex. In man and the higher apes, therefore, only relatively few fibres terminate in the colliculi.

RESPONSE TO ELECTRICAL STIMULATION

Electrical stimulation of the superior colliculi with electrodes implanted in the waking and freely moving cat, by Hess' technique, causes a movement of the head towards the opposite side; a careful study of the eyes shows that the movement of the head is preceded by a conjugate movement of the eyes (Burgi). Usually there is an upward or downward component in the movement according to the part of the colliculus stimulated; and in the rostro-median region a pure upward movement can be obtained. In the alert monkey, electrical stimulation evoked short-latency saccadic movements (Robinson, 1972; Schiller and Stryker, 1972), and the *amplitude*† and *direction* of the eye movement were largely determined by the point stimulated, rather than intensity, so that there is an upward component if medial regions are stimulated and downward component on lateral stimulation; if the stimulus is applied to a rostral region the movement is small compared with a

*The superficial laminae are exclusively visual in input; the intermediate layers combine visual, somatic and acoustic inputs, whilst the deepest layers are essentially non-visual Stein *et al.*, 1976). Tactile representation in the deeper layers is in register with visual representation in the overlying layers, so that the magnified representation of the visual field overlapped with the tactile representation of the face. In general, the overlap could be described by viewing the visual field as a flexible sheet stretched over the body with the area centralis on the nose and the limbs radiating out at acute angles. Such an overlap would suggest an involvement in coordinating visual input with motor activity in the limbs and face.
† Straschill and Rieger (1973) working on the cat, found, in contradiction to Robinson, that the magnitude of the saccade was not completely determined by the position of stimulation on the colliculus but that it increased with current strength and suggested that Robinson's weakest stimuli had already saturated the system. Continuous stimulation did not produce a series of saccades, but usually just one with subsequent maintenance of fixation. The final position achieved by the eyes was independent of their initial position so that in the cat the stimulus leads to a *goal-directed* response; consequently only in the primary position does the sensory map match the motor map.

caudal stimulus. When a given point was stimulated repeatedly, successive saccades were made from the position reached by the previous one until the limit of movement of the eye was reached.

### FOVEATION

According to Robinson, the electrical stimulation is essentially similar to photic stimulation of a given point on the retina, and the response may be regarded as a 'foveation', but not a goal-directed movement since stimulation of the same collicular point can cause movement to different fixation points whereas in the normal animal allowance would be made for a given movement. Thus the colliculus is responding to a retinal system of coordinates and failing to translate retinal stimuli on to a system of coordinates involving the head.

### TECTAL INFLUENCE ON MOTOR NEURONES

According to Precht *et al.* (1974), the pathway from the colliculus to the motor neurones of the eye muscles is at least disynaptic and many of the responses were trisynaptic, as judged by the latency between stimulating the colliculus and the electrical response in the motor neurone. They consider that the pathway is probably by way of the ipsilateral midbrain reticular formation, with a crossing over to the contralateral motor neurones at some point caudal to the midbrain and certainly not within the collicular commissure.

## RESPONSES TO LIGHT

### EYE MOVEMENTS

In the anaesthetized cat, stimulation of the retina by light does not evoke any eye movements; if, however, the excitability of the superior colliculus is enhanced locally by the application of a crystal of strychnine, then light causes a prompt movement of the eyes directed to a definite spot in space characteristic for the particular region of the colliculus that has been made thus sensitive. By systematically changing the position of the strychninized area, a map of the surface of the colliculus may be plotted indicating regions of the visual field towards which the eyes will be directed when a given point is rendered hypersensitive. Presumably there is a reflex path from the retina through the colliculus that determines that the eyes will move towards the point in the visual field that stimulates the retina; such a concept is the basis of the 'foveation hypothesis' of collicular function (Schiller and Koerner, 1971), a hypothesis that has been questioned by Wurtz and Goldberg (1972).

### DISCHARGES IN COLLICULAR NEURONES

According to the depth of penetration into the colliculus, units may be found that respond to visual stimuli and,

in general, the receptive fields of the units, i.e. the regions on the retina or visual space that evoke responses in the given tectal unit, are located in space where electrical stimulation of the unit would direct the eyes (Robinson, 1972; Cynader and Berman, 1972). Figure 17.19 relates the receptive fields of collicular neurones to position on the colliculus; the periphery of the contralateral half-field is represented on the most posterior part, the upper field on the medial part and the lower on the lateral part, and the fovea, or area centralis, has a disproportionately large representation. In the monkey, Cynader and Berman found that the half-fields were strictly segregated to the contralateral colliculus, whereas in the cat some 10 to 12° ipsilateral to the area centralis could be represented.

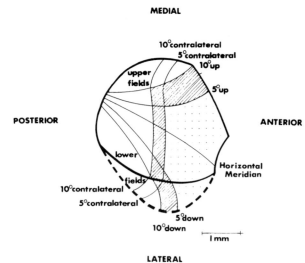

**Fig. 17.19**   Representation of the visual field on the surface of the right colliculus (as viewed from above in the Horsley-Clarke plane). The stippled area represents parts of the contralateral visual field within 5° of the fovea. Striped and stippled areas combined represent parts of the contralateral visual field within 10° of the fovea. (Cynader and Berman, *J. Neurophysiol.*)

## Movement fields

Some units produce discharges in relation to eye movements—*movement-related neurones*—the discharge preceding the movement and having a maximal value for a particular direction and amplitude, so that, in addition to the *receptive field* of the tectal unit we may plot a *movement field*, the region in space to which the eyes must be directed to obtain a discharge in the tectal unit. Figure 17.20 shows the responses as 'burst indices' to movement at different angles from the zero straight-ahead position; four different amplitudes of movement

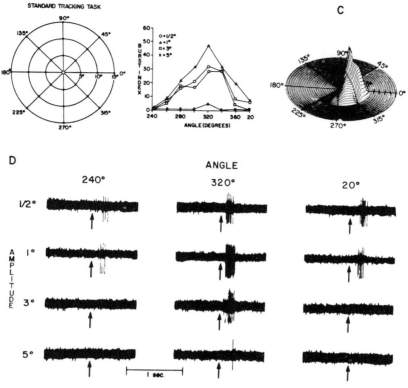

**Fig. 17.20**  A, the visual tracking task. If subject fixated the centre dot for 2 sec, the target was moved to a second position. If the target was acquired within 500 msec and maintained for 2 sec, a reinforcement was given. B, burst index as a function of angle and amplitude of movement. C, three-dimensional representation of burst index as a function of angle and amplitude of eye movement. The maximal burst index value is 48. D, response of a superior colliculus unit to a series of saccades at different angles and amplitudes. The arrow beneath each tracing represents the onset of the target movement. (Sparks *et al., Brain Res.*)

are given. In C a three-dimensional representation of the same results is shown. The maximal discharge of this neurone occurred prior to small right saccades with a downward component (1° in amplitude at an angle of 320°). Movements within the movement-field of greater or smaller amplitude gave smaller responses, and again, deviation from 320° gave a smaller response.★

RECORDING AND STIMULATION AT SAME SITES

Figure 17.21 indicates graphically the correspondence between receptive fields, indicated by contours, of collicular neurones and the directions and magnitudes of saccades evoked by stimulation of the particular collicular unit, so that these units, in the superficial layers, elicited saccades that directed the eyes to the point in space that caused an evoked response.†

NECK AND EYE-MUSCLE AFFERENT INPUTS

If the superior colliculi are to act as integrating centres, controlling movements of both head and eyes, we must expect to find collicular neurones responsive, not only to

---

★Cynader and Berman (1972) point out that the foveal representation of the colliculus, i.e. the region that, when stimulated, directs the eye towards peripheral stimuli that fall on the fovea, receives only a cortical projection (Wilson and Toyne, 1970).

†Robinson and Jarvis (1974) raised the question as to whether the collicular discharge that precedes an eye movement was due to an intended eye movement or, because the head was fixed in these experiments (e.g. Schiller and Stryker, 1972) it was related to an intended head movement. In their study they were able to dissociate head and eye movements and found a better temporal relation of the saccade to the collicular discharge than the head movement.

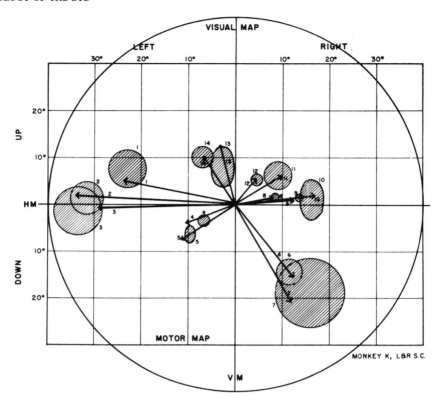

**Fig. 17.21** Effects of recording and stimulation in the superficial layers of the superior colliculus. The visual map with the receptive fields of 14 units is superimposed on the motor map with its arrows representing the electrically elicited saccades at each of the 14 sites. The length of each arrow represents the mean length of 8–14 stimulation-elicited saccades; the direction of each arrow represents the mean direction of saccades. HM = horizontal meridian, VM = vertical meridian. (Schiller and Stryker, *J. Neurophysiol.*)

visual input, but also to proprioceptors from neck and eye-muscles. Thus it is well established that neck-muscle receptors are concerned in tonic neck and eye reflexes in the decerebrate cat (Carpenter, 1972; McCouch, 1951). Abrahams and Rose (1975) found a powerful projection from the neck-muscles of the cat to the superior colliculus, long-latency responses to neck-muscle afferent stimulation being predominantly in the superficial layer, whilst short-latency responses were found in the intermediate and deep layers, predominating in the tegmentum. Projection from the extraocular muscles constituted the single richest projection, long-latency responses being the most common. Table 17.1 indicates the extent of convergence of proprioceptive and visual inputs on to single collicular neurones.

VESTIBULAR COMPETITION

In order that visually and neck-operated responses may act effectively, it might often be necessary to inhibit

vestibular reflexes (p. 435) operating in an opposite sense to that required; there is little doubt from the work of Abrahams (1972) that receptors in the neck can inhibit or excite vestibulospinal output.

**Table 17.1** Convergence of different sensory modalities on single collicular neurones (Abrahams and Rose, 1975)

| No. of stimulus types | | No. of units |
|---|---|---|
| 3 | Extraocular Neck Visual | 61 |
| 2 | Extraocular and visual | 3 |
| | Neck and extraocular | 23 |
| | Visual and neck | 0 |
| 1 | Extraocular | 4 |
| | Neck | 1 |
| | Visual | 1 |
| | Total | 93 |

## RELATION TO CEREBRAL CORTEX

According to the 'foveation hypothesis', we may regard the colliculi as centres that bring about reflex movements of the eyes in response to a visual stimulus. In man and apes the number of visual fibres passing to the colliculi is relatively small; moreover, as we have indicated, the collicular region representing the fovea in the monkey has no direct visual projection, so that visual input is indirect and through the striate and peristriate areas (p. 619). Thus visually evoked reflex activity, bypassing the visual cortex, seems not to occur in man since eye movements cannot be evoked in the absence of the striate area. However, as the number of visual fibres terminating in the colliculi decreases, there is a corresponding increase in the number of fibres running from the occipital cortex to the colliculi, so that on phylogenetic grounds we may assume that the colliculi retain their function as stations for executing eye movements, but they become subordinated to the occipital cortex (Crosby and Henderson, 1948).

### INTEGRATION OF SENSORY MODALITIES

According to Pasik, Pasik and Bender (1966), however, the absence of any permanent oculo-motor deficit in monkeys after ablation of the superior colliculi rules these bodies out as subsidiary motor centres, and they consider that their function is to integrate sensory modalities since afferents from almost every sensory system converge here. The role of these bodies will not be discussed further here, being deferred until the perceptual aspects of collicular and cortical activity are taken up. Let us consider, now, the regions of the cerebral cortex that are concerned with eye movements.

## CORTICAL CENTRES

### OCCIPITAL MOTOR CENTRE

The occipital lobe is differentiated into three areas on histological grounds; area 17, the *striate area*, is where the visual fibres terminate, whilst areas 18 and 19, the *peristriate* and *parastriate areas*, are intimately associated with area 17, neurones passing from area 17 to area 18, and from area 18 to area 19 and back to area 17.

### EFFECTS OF ELECTRICAL STIMULATION

The classical investigators of the cerebral cortex by electrical stimulation observed contralaterally directed

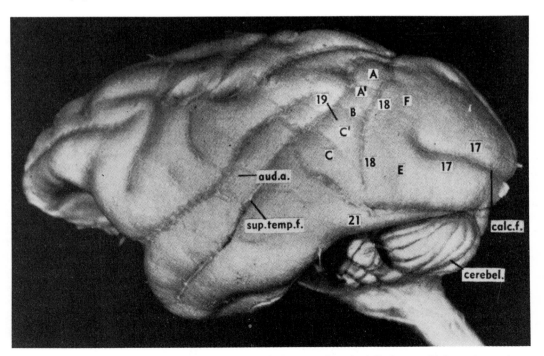

**Fig. 17.22** A photograph of the left side of the brain of *Macaca mulatta* ( × 1·8). On area 19 the various points from which eye movements were elicited are indicated by letters. Stimulation of A produced conjugate upward deviation of the eyes, of A′ conjugate deviation obliquely upward toward the right, of B conjugate horizontal deviation toward the right, of C′ conjugate deviation obliquely downward to the right, and of C, conjugate downward deviation. Stimulation in the region of E (area 17) caused oblique conjugate deviation of the eyes upward and toward the right, and in the region of F, conjugate deviation obliquely downward to the right. (Crosby and Henderson, *J. comp. Neurol.*)

conjugate movements of the eyes in response to stimulation of area 17 in primates, the movements being apparently directed to the parts of the visual field that are projected on to the cortex. More recent work has largely confirmed this; Figure 17.22 illustrates the regions of the occipital cortex of the macaque that gave movements in the experiments of Crosby and Henderson, whilst Figure 17.23 illustrates the results obtained by Wagman, Krieger and Bender. According to these last workers, the occipital cortex may be divided into four quadrants by two planes; a midsagittal vertical plane dividing the cortex into a right and left half, and a roughly horizontal plane through the calcarine fissure. Two of these quadrants are illustrated in Figure 17.23. The movements of the eyes were such as to direct the gaze away from the quadrant stimulated. In man, Foerster obtained eye movements by stimulating area 19.

## Motor area?

Wagman *et al.* raise the point as to whether the movement of the eyes is the result of stimulating a sensory region, evoking a sensation of light to which the animal responds, or whether the stimulus does, indeed, activate a motor centre. This question is probably somewhat academic since the modern tendency is to treat many areas of the brain as both sensory and motor. The finding

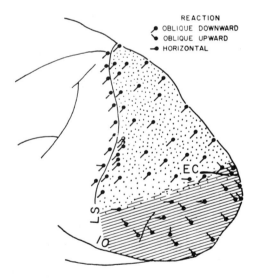

**Fig. 17.23** Contralateral conjugate eye movements elicited upon stimulation of the dorsal, lateral and lateral inferior surfaces of the occipital lobe of *Macaca mulatta*. The solid circles indicate the points that were actually stimulated, whilst the lines originating at these circles indicate the direction of the contralateral eye movement in response to this stimulation. The dots outline the regions from which obliquely downward and contralateral deviations were obtained, whilst the shaded area indicates the region from which obliquely upward and contralateral deviations of the eyes were obtained. (Wagman, Krieger and Bender, *J. comp. Neurol.*)

that certain occipital lesions may impair motor activity without impairing vision (Bender, Postel and Krieger, 1957), and that in human subjects eye movements that are not always associated with visual sensation may be evoked by electrical stimulation, suggests that there are, indeed, motor regions in the occipital cortex (Bickford *et al.*, 1953).

## THE FRONTAL EYE FIELDS

### CLASSICAL STUDIES

In man and other primates there is a region in the frontal cortex, stimulation of which gives rise to movements of the eyes. In man the region is close to the precentral gyrus, stimulation of which gives rise to discrete movements of the parts of the body as indicated in Figure 17.24 (area 8 $\alpha\beta\delta$); in apes the region lies farther forward. Movements of the eyes following stimulation of this area in the chimpanzee were described by Grünbaum and Sherrington in 1903, and similar movements in man by Foerster and Penfield and Boldrey. In the monkey, Graham Brown described movements of two kinds; in the first place, stimulation of the middle frontal convolution caused a conjugate movement of the eyes to the opposite side, the head remaining still; in the second place, stimulation of the superior frontal convolution caused a turning of the head, the eyes remaining fixed in their sockets. In both cases, therefore, the animal 'looked' to the opposite side, but in the former instance the eyes were employed whilst in the latter the muscles of the neck. Interestingly, if the head was prevented from turning, then the eyes moved.

### SACCADES

Since this classical work there have been several studies on the effects of localized stimulation of this frontal area in primates and the cat. As summarized by Robinson and Fuchs (1969), the usual response has been a lateral slow rotation rather than a saccade, although also vergence and nystagmus have been reported. The work of Krieger *et al.* (1958) suggested that the failure to obtain saccades was due to the use of anaesthesia, and in their study of an unanaesthetized monkey they produced saccades. Robinson and Fuchs studied the waking monkey and controlled the stimulus strength whilst recording with precision the eye movements. The great majority of the eye-movements evoked by frontal field stimulation were single contralateral saccades occurring about 25 msec after initiating the stimulation; the amplitude and direction were determined by the position of the stimulating electrode; with continuous stimulation, as with the colliculus, a succession of saccades, spaced about 100 msec apart, took place until the mechanical limit of eye excursion was reached.

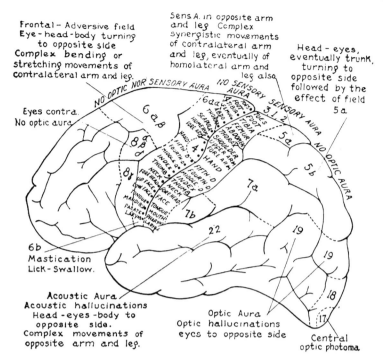

Frontal - Adversive field
Eye - head - body turning
to opposite side
Complex bending or
stretching movements of
contralateral arm and leg.

Sens A. in opposite arm
and leg Complex
synergistic movements
of contralateral arm
and leg, eventually of
homolateral arm and
leg also

Head - eyes,
eventually trunk,
turning to
opposite side
followed by the
effect of field

Eyes contra.
No optic aura.

6b
Mastication
Lick - Swallow.

Acoustic Aura
Acoustic hallucinations
Head - eyes - body to
opposite side.
Complex movements of
opposite arm and leg.

Optic Aura
Optic hallucinations
eyes to opposite side

Central
optic photoma

**Fig. 17.24** Diagram of the human cerebral hemisphere seen from the left side illustrating the areas, electrical stimulation of which is followed by movement or sensation. The numbers of the areas are those of Vogt. (Herrick, *Introduction to Neurology*.)

Directions ranged from horizontal to almost vertical and the amplitudes from 1 to 70° (Fig. 17.25). Centring, smooth pursuit or nystagmus movements were never observed. The saccade had an all-or-none characteristic so that if the stimulus was below a certain strength there was no response; above the critical strength the size and direction of the saccade varied little with further increases.

COMPARISON WITH BRAINSTEM

In general, there was very little influence of eye position so that if a saccade of 10° to the left was produced by a stimulus when the eye was looking straight ahead, the response when the eye was looking 10° to the right would be such as to bring the eyes straight ahead. This is quite different from the goal-directed type of response to brainstem stimulation, described for example by Hyde and Eason (1959), in which stimulation causes the eyes to be directed to the same final point from any initial point.

ANAESTHESIA

Studies on the effects of anaesthesia, which demanded much larger stimulus currents, indicated that responses such as vergence, slow following type movements, and nystagmus, were due to this agent.

TWO-POINT STIMULATION

When two points, producing by themselves saccades of different direction and amplitude, were stimulated

simultaneously the resulting saccade lost its all-or-nothing character, its direction and amplitude depending on the relative stimulus strengths. When bilateral two-point stimulation was employed, the result depended on the time-interval between the two stimuli; if this was less than 30 msec, the two responses interfered with each other, the second tending to cancel, in mid-flight, the first saccade and replacing it by all or a fraction of its own. When the stimuli were simultaneous then a single saccade was evoked which could be graded in size and duration, just as with two-point unilateral stimulation.

**Control of duration of discharge**

The authors emphasize that control over the amplitude of a saccade represents essentially control over the duration of activation of the eye muscles and not over the intensity of stimulation, since the muscular activity is virtually maximum at the beginning of any saccade. The cortical control is determined, not by how many cortical fibres are excited, since once the threshold is exceeded the size of the saccade is unchanged with increasing current-strength and therefore increasing numbers of excited cortical cells. Thus control is governed by *which* cortical cells are excited, and these cells must influence the brainstem neural networks that control the duration of the discharge to the motor neurones. When two cortical points are stimulated, in some way the midbrain circuits are influenced by the two

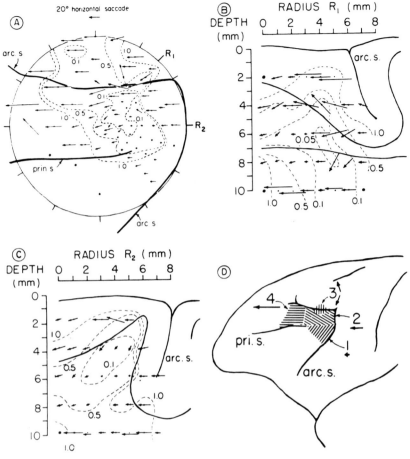

**Fig. 17.25** Maps of amplitude and direction of saccadic eye movements evoked by stimulation of frontal eye field of a rhesus monkey (left cortex shown). Length of each arrow represents amplitude of the evoked saccade (note scale at top); its inclination indicates upward or downward slant of movement to the nearest 15°. Interrupted lines are isopleths of contrast threshold current in milliamperes. Stimulus was a 30- or 60-msec train of 1-msec pulses at 200/sec. A, a cortical section parallel to surface at a depth of 6 mm showing location of chamber (outer circle), arcuate (arc. s.), and principal (prin. s.) sulci as seen at surface of cortex. B, a cross section along the chamber radius $R_1$ showing movements and thresholds in grey and white matter as a function of depth. C, a cross section along radius $R_2$. Filled circles indicate no response below 2 ma. D, a summary of the four regions which exhibited constant saccade characteristics in the four animals for which such maps were made. (Robinson and Fuchs, *J. Neurophysiol.*)

cortical inputs to strike a compromise to produce a pulse width of discharge intermediate between the two evoked separately.

ACTIVITIES OF CORTICAL UNITS

By inserting an electrode into Area 8, Bizzi (1967) recorded from single units that altered their discharge-rate when the eyes moved, the final rate of discharge being a function of the direction of gaze; the interesting feature of these cells was that the change of discharge only occurred *after* the movement had been initiated, so that

Bizzi postulated that this area of the cortex was only concerned with the tonic control of eye position after a movement had been made. He points out that, in the monkey, ablation of Area 8 only leads to paralysis of eye movements for a few days (Pasik and Pasik, 1964). In a later study (1968) these cortical neurones were found to be of two types: *Type I*, which fired during voluntary saccades or the quick phase of nystagmus in a given direction, whilst they were silent during the slow phase; and *Type II*, which discharged during fixation of gaze in a specific direction as well as during smooth pursuit movements. When the head was allowed to move as well as the eyes the same pattern of discharge in Type I and II neurones was observed, so that the slow compensatory following movements that followed the head movement were accompanied by Type II discharge. A further group of neurones were found discharging in association with the head movement, some were specific so far as direction was concerned, others not. Discharge in head neurones occurred *prior* to rotation whilst, as we have seen, the eye neurones discharged after (Bizzi and Schiller, 1970).

VOLUNTARY CENTRE

It is considered that in man, at any rate, the frontal region is the centre for voluntary eye movements in contrast to the occipital regions which are thought to be concerned with purely reflex movements of the eyes in response to a visual stimulus. Thus a stimulus applied to the frontal area is said to be prepotent over a simultaneous one applied to the occipital area; e.g. stimulation of the left occipital lobe causes a conjugate movement to the right; stimulation of the right frontal centre causes a deviation to the left. If both areas are stimulated simultaneously, the eyes move to the left. Surgical removal of the 'ocular responsive cortex' in man leads to the failure of the individual to direct his eyes voluntarily to the side opposite the lesion. In monkeys and apes stimulation of the ocular responsive cortex causes head movements, opening of the palpebral fissures, and dilatation of the pupils, besides eye movements. In general, as with man, the lateral movements are most frequent; if, however, the medial and lateral recti are cut, cortical stimulation leads to upward and downward movements. Simultaneous stimulation of both sides of the cortex can lead to the fixation of the gaze directly forward. As indicated above, however, it may be that Area 8 in monkeys plays a more subsidiary role.

EFFERENT PATHWAYS

The pathways by which cortical impulses determining eye movements reach the eye-muscle motor neurones are not completely worked out. The pathway from the frontal centre is independent of the occipital centre, since ablation of the latter does not abolish voluntary eye movements. Figure 17.26 illustrates Crosby's views. According to this, the pathway from the frontal centre is by way of the corticobulbar tract, i.e. the tract that carries the voluntary impulses to the musculature generally and to which the name *pyramidal* is given in its course through pons, medulla and cord. Thus the fibres run in the internal capsule and cerebral peduncle; at about the midbrain level, fibres to the nucleus of N VI leave the main tract to enter the medial lemniscus, descending in this to the abducens and parabducens nuclei; the remaining eye-muscle nuclei are thought to be indirectly activated by way of the median longitudinal fasciculus. The occipital centres in Areas 17, 18 and 19 are thought to activate the abducens and parabducens nuclei indirectly through the colliculi by way of the internal corticotectal tract and directly through the corticotegmental tract. The reason for imputing this double pathway is the old observation of Graham Brown that damaging the colliculi left eye movements intact but, as Whitteridge (1960) has pointed out, this evidence is not conclusive since the frontal pathway is known to be independent of the colliculi, and the eye

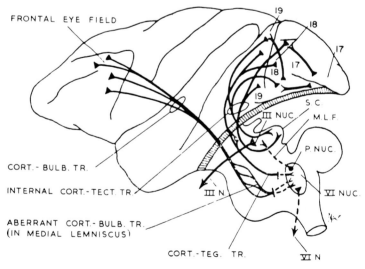

Fig. 17.26 Pathways for the cortical control of the eye movements. (Slightly modified from Crosby, *J. comp. Neurol.*)

movements observed might have been mediated by the frontal centre* (Beckford *et al.* 1953).

## PSYCHO-OPTICAL REFLEXES

The most important group of reflexes involving eye movements in man are those in which the incidence of light on the retina acts as the primary stimulus; it is as a result of these reflexes that the eyes are (*a*) directed to objects of interest and (*b*) maintained in the necessary orientation to ensure that the images formed by the two eyes fall on the foveae. We are accustomed to think of reflexes as being executed without the intervention of awareness on the part of the individual; these visual reactions, however, definitely involve awareness and are therefore classed as *psycho-optical reflexes*; as we shall see, there can be no doubt of their essentially involuntary nature.

## THE FIXATION REFLEX

### SACCADE AND MAINTAINED FIXATION

All psycho-optical reflexes may be regarded as manifestations of the more general *fixation reflex*—the reflex which brings the images of an object on to the foveae of the two eyes and maintains them there. The fixation reflex develops early, since an infant a few days after birth will fixate a bright light; binocular fixation, however, does not develop for some months.

The reflex exhibits itself in two aspects—(*a*) the response to a peripheral stimulus and (*b*) the maintenance of fixation; the mechanism involved in the two processes is nevertheless the same. Thus let us imagine that the eyes are stimulated by a bright object in the peripheral field. Attention is aroused and the eyes move so that the images of the object approach the foveae— suppose that the nervous response grows in intensity as the images approach closer and closer to the foveae and reaches a maximum when they fall actually on them. The eye muscles can then be considered to behave as a tuning system, causing movements of the eyes until a maximum nervous response is obtained, just as the knob of a tuning dial is turned to obtain maximum sound from a wireless set. Fixation is thus achieved when maximum nervous discharge is reached. If the head moves slightly, the images are displaced and the nervous response falls off; immediately the tuning mechanism comes into play and the eyes are readjusted to maintain perfect fixation. Thus the same principle, the movement of the eyes to maintain a maximum nervous response, operates both to initiate and to maintain fixation.

### MOVEMENTS DURING FIXATION

This principle may also explain the neural basis for the frequent small movements of the eyes that take place even during attempts at steady fixation (p. 401). As studies on the stabilized retinal images have shown (p. 322), the perceived image becomes rapidly indistinct in the absence of a shift of this image on the retina, so that the nervous discharge, postulated above, tends to decrease as the eye remains stationary. The eyes presumably respond to this falling off with a movement which shifts the images on the retinae sufficiently to excite a new set of cones and the discharge again becomes maximal. The constant, small movements of the eyes in fixation may thus be regarded as the restless groping of the oculomotor centres for a position that gives maximum nervous discharge.

### OCULAR NYSTAGMUS

This pendular to-and-fro movement of the eyes may be regarded as an exaggeration of this groping; if fixation is difficult, as in amblyopia, the eyes move excessively in a vain attempt to find a clear image. The nystagmus of miners working for long periods in the dark is a complex syndrome; nevertheless an element in the aetiology must be the necessity for using peripheral vision; under these conditions fixation on the fovea reduces the visual acuity rather than enhances it, so that the 'best position' for the retinal image is much more vaguely defined and this might well result in a ceaseless search for it over relatively large areas of retina to give a pendular type of nystagmus.

### SIMPLIFICATION

It must be appreciated, however, that this point of view represents a crude simplification of what must be a more complex process; the factors of attention and meaning cannot be ignored; for example, corrective fusion movements, to be described below, are brought about as a result of a changed orientation of the retinal images, e.g. when one eye rolls, and the intensity of the 'nervous discharge' remains unaltered.

### OTHER MANIFESTATIONS

Other manifestations of the fixation reflex are given by:

1. The following movements in *optokinetic nystagmus*

---

*We may note that the eye movements evoked by occipital cortex and superior colliculus may be affected differently by lesions in the brain stem, once again indicating separate pathways (Hyde, 1962) furthermore, Hyde and Slusher found that discrete lesions in the median longitudinal fasciculus could abolish adductions in response to some forms of stimulation but not to others. This looks as though the concept of a parabducens centre, controlling lateral movements through a final common pathway up the median longitudinal fasciculus, postulated by Crosby, is unsound.

2. The *corrective fusion movements* made in the interests of single vision
3. The movements of the eyes concerned with the *visual righting reflexes*.

OPTOKINETIC NYSTAGMUS

This describes the movements of the eyes when the body is in motion and the subject is gazing at the environment, as when he looks out of a railway carriage (train-nystagmus) or alternatively when the subject is still and his environment, or a prominent part of it, moves around him, as in the clinical test when a rotating striped drum is gazed at steadily. Figure 17.27 is a record of the slow movements in response to a rotating drum at several speeds; it will be noted that the velocity builds up to a plateau. Here the stimulus is the movement of the image of the fixation point on the retina, the eye responding, during the slow phase of the nystagmus, by movements designed to keep the image on the fovea. At a certain moment, fixation of this point ceases and the eyes make a rapid saccade in the opposite direction, to fixate a new stripe on the drum, and the slow movement begins again. The slow movement is obviously a following movement, but the rapid movement is difficult to categorize. It was originally thought that it occurred simply because the eyes were unable to maintain fixation on the fixated stripe, i.e. because the stripe had moved out of the field of binocular fixation. In fact, however, the rapid movement usually takes place long before this limit has been reached; moreover, it is not apparently due to a deliberate attempt to fixate a new stripe appearing in the periphery since optokinetic nystagmus occurs in homonymous hemianopia when, because of a lesion in the visual pathway (p. 542), the eye is blind to half the visual field (Fox and Holmes; Pasik, Pasik and Krieger). Thus the eye makes a rapid movement into its blind side.

NYSTAGMUS CENTRE

We must conclude, therefore, that the unusual stimulus-condition, namely the apparent rotation of the visual environment, activates a *nystagmus centre* that triggers off these involuntary rhythmical movements (Rademaker and Ter Braak).* The involuntary nature of this movement in man is best revealed when he is surrounded by the moving drum so that practically the whole visual field moves; in this case it is impossible to avoid the nystagmus so long as the eyes are open. When only a small part of the visual field moves, as when the subject looks at a small striped drum, then the subject may avoid the nystagmus by 'looking through' the drum, i.e. by refusing to fixate any given stripe (Carmichael, Dix and Hallpike).†

**Cortical and subcortical nystagmus**

In animals, such as the rabbit and dog, an optokinetic nystagmus may be evoked even in the absence of cerebral hemispheres, and for this reason it is called *subcortical*. Visual reactions in man, in the absence of a cortex, are, as we have seen, impossible, presumably because of the relative insignificance of the visual projection to the

---

*The *rhythm* of the nystagmus will be largely determined by the amplitude of the excursions for a given speed of rotation; the fact that this alters under a variety of experimental conditions is further proof that the initiation of the fast movement, which puts an end to the slow movement, is not due to the eyes reaching the maximal extent of the field of fixation. We may note that, according to Ter Braak, the eyes lag slightly behind the movement of the drum during the slow phase; consequently there is a slight drift of the retinal image over the retina.

†When a subject is completely surrounded by the moving field, which thus occupies his whole field of view, within a few seconds of beginning the rotation the perception of movement of the field changes to one of motion of the subject—*circularvection* (Brandt et al. 1973). This persists even if the eyes do not make an optokinetic nystagmus. If only the central part of the visual field rotates, then only optokinetic nystagmus is present, but if only the periphery rotates then circularvection dominates. The phenomenon is interesting since the sensation of motion is obtained in the absence of any vestibular stimulation, but of course, if a subject were rotating at constant velocity for any length of time, vestibular stimulation would subside. Thus the stimulus-conditions are, in fact, indistinguishable from rotation in a chair at constant velocity.

Fig. 17.27 Velocity of slow phases of optokinetic nystagmus (OKN) and optokinetic after-nystagmus (OKAN) induced by full-field rotation at 60°/sec. Note the faster decline of the slow phase velocity during right OKAN. The interrupted lines adjacent to the traces are the exponentials estimating the velocity-declines during OKAN; they were moved by a small amount so as not to be obscured by the experimental records. The time-constants for the declines were 16·2 and 5·3 sec for left and right OKAN respectively. (Cohen et al., *J. Physiol.*)

colliculi. It is unwise, therefore, to transpose studies on lower animals to man, and certainly unwise to speak of a *subcortical nystagmus* as such, although the presence of a nystagmus-centre in the lower parts of the brain is not an unreasonable hypothesis. In man this would be activated by the cortical visual stimuli, but in lower animals the colliculi might suffice.

## Directional preponderance

Under pathological conditions in man, involving lesions in the cerebral hemispheres, optokinetic nystagmus may be impaired, frequently with the appearance of a *directional preponderance*, in the sense that the reflex is normal when the drum moves one way, and reduced when it moves in the other. In the case of optokinetic nystagmus in man the preponderance is seen when the drum rotates away from the affected side, so that the study of this nystagmus may be a valuable diagnostic tool (Fox and Holmes). According to Carmichael *et al.* the regions of the cortex influencing the optokinetic nystagmus are in the supramarginal and angular gyri of the temporal lobe, the right hemisphere, for example, facilitating nystagmus on moving the drum to the left, and inhibiting that on moving the drum to the right.★ A schematic pathway for the nystagmus is illustrated in Figure 17.28; it should be noted, however, that Pasik *et al.* (1959) do not agree with the location of the cortical centres in the supramarginal and angular gyri.

## Single moving point

We may note that when a single object moves across the stationary visual field, there is no optokinetic nystagmus unless the field is otherwise empty; thus according to

Rademaker and Ter Braak, we may think of two opposing tendencies, the shifting of the images of stationary objects on the retina, tending to inhibit the following movement, and the shifting of the image of the single moving point that incites to movement. According to the relative preponderance of the moving or fixed field, one or other of the tendencies will determine the result, i.e. whether the eye remains fixed or follows the moving point. This was well demonstrated by Ter Braak who immobilized one eye of a rabbit and examined the movements of the other, which was blinded. Moving a single point across the visual field of the immobilized eye now caused the other eye to give a genuine nystagmus; because the stimulated eye was immobilized, there was no movement of images of stationary objects over its retina, and so the normal inhibition of the nystagmus was released. A similar experiment on a human subject, with one eye immobilized by paralysis of its extraocular muscles, is described by Mackensen (1958).

---

★ In the macaque Crosby and Henderson abolished optokinetic nystagmus by bilateral destruction of areas 18 and 19; unilateral destruction abolished the nystagmus on rotating the drum away from the injured side, i.e. there was a directional preponderance *towards* the injured side, by contrast with man where it is *away* from the lesion. The frontal eye fields (area 8) exerted an inhibitory action so that on ablation of these the nystagmus was more marked.

We may note the danger of confusion according as the direction of the nystagmus is described in terms of the quick component or the direction of rotation of the drum; thus when Carmichael, Dix and Hallpike (1954) say that nystagmus to the healthy side is suppressed, they are referring to the quick phase of the nystagmus, so that there is suppression of nystagmus when the drum rotates away from the healthy side, i.e. towards the lesion.

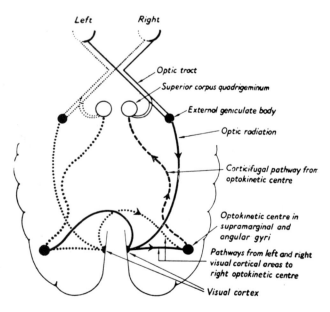

Left   Right

Optic tract
Superior corpus quadrigeminum
External geniculate body
Optic radiation
Corticifugal pathway from optokinetic centre
Optokinetic centre in supramarginal and angular gyri
Pathways from left and right visual cortical areas to right optokinetic centre
Visual cortex

**Fig. 17.28** Suggested schema of optokinetic mechanisms. (Carmichael, Dix and Hallpike, *Brit. med. Bull.*)

## Optokinetic after-nystagmus

When a rotating drum is stopped after the subject viewing it has experienced a nystagmus for some time, there is an after-nystagmus, which is best recorded when the animal is in darkness, after the optokinetic nystagmus has lasted some time, and when fixation is prevented. There are generally two phases, the first with the nystagmus in the same direction as that of the original optokinetic nystagmus; this may give way to a nystagmus in the opposite direction. The significance of the after-nystagmus is probably to be seen in its antagonism to the vestibularly evoked after-nystagmus of rotation, which occurs in the opposite direction (Ter Braak, 1936). This nystagmus has been examined in some detail by Mackensen (1959) and more recently by Cohen *et al.* (1977). The speed of the slow-phase decreases rapidly after the onset of the after-nystagmus, and this deceleration may be promoted by short periods of fixation, so that after 3 sec of fixation the velocity is zero. In general, the development of the after-nystagmus may be likened to a charging process, with a characteristic time-constant, during which a central excitatory state is being built up by the optokinetic nystagmus; if, at any moment, the optokinetic stimulus is stopped, the velocity of the slow movement of the after-nystagmus will have a value governed by the length of exposure to the stimulus. Figure 17.29 illustrates this effect, and the corresponding increase in velocity of the slow phase of the optokinetic nystagmus.

## Torsional nystagmus

Brecher (1934) induced an optokinetic torsional nystagmus by causing a subject to view a rotating sectored disc; the nystagmic jerks were of several degrees in amplitude.

### CORRECTIVE FUSION MOVEMENTS

The orientation of both eyes so as to produce images of the same object on corresponding parts of the two retinae requires continuous adjustments, which are brought about by this reflex. Here the effective stimulus is diplopia.

### VISUAL RIGHTING REFLEXES

In man and the ape the eyes play a very important part in the maintenance of posture. It is quite clear that the movements of the eyes must be closely attuned to the posture of the animal; in lower forms this is achieved by a rigid association of the eye muscles with the vestibular apparatus so that the position of the eyes is completely determined by the orientation of the body in relation to gravity and the parts of the body in relation to each other. Here, then, the labyrinth dominates eye movements. The visual impressions

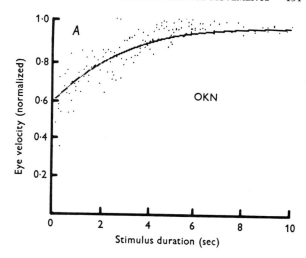

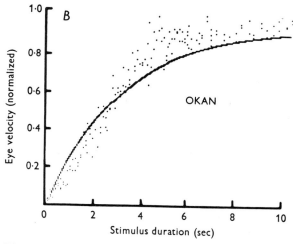

**Fig. 17.29**  Approximation of results from slow rise of OKN (*A*) and charge of OKAN mechanism (*B*) by exponentials with the same time-constant (3 sec). The stimulus was rotation at 60°/sec for both groups of results. Dots are experimental points and continuous lines are exponential fits to them. (Cohen *et al.*, *J. Physiol.*)

alone, however, can clearly provide information as to the orientation of the head, and therefore the body. Thus if a subject is falling forward, his original point of fixation moves upwards and, as a result of the fixation reflex, his eyes move upwards to maintain fixation; the necessary postural adjustments for recovery can be made in the light of impressions provided by the eyes and their adjustments. In this case proprioceptive impulses from the superior recti demand a movement of the head upwards so that the eyes can re-assume the primary position. The proprioceptive impulses from the neck muscles now demand postural changes that will permit the head to re-assume its normal erect position. In man and the ape posture can be maintained quite adequately by means of the visual righting reflexes when the labyrinths have been destroyed.

## VESTIBULAR AND NECK PROPRIOCEPTIVE REFLEXES

The primary stimulus for these reflexes is not visual but arises from the vestibular apparatus or the proprioceptive stretch receptors in the neck muscles. They are classed as *static (or tonic) reflexes* and *kinetic reflexes*; by the first class a definite orientation of the eyes is evoked and maintained for a given position of the head in relation to the trunk and the pull of gravity, whilst by the latter a definite *movement* of the eyes follows a movement or acceleration of the head.

## THE VESTIBULAR APPARATUS

A brief description of the vestibular apparatus would be useful at this point. In man it consists essentially of three semicircular canals in mutually perpendicular planes, and two sacs, the *utricle* and *saccule* (Figs. 17.30 and 17.31). Each canal widens at one end into an *ampulla*

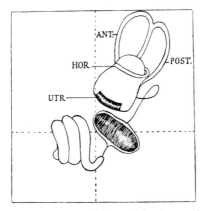

**Fig. 17.30**  The positions of the semicircular canals in the head viewed from the side. (After Quix.)

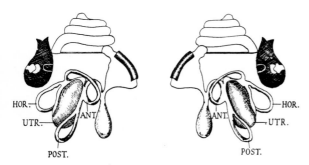

**Fig. 17.31**  Positions of the semicircular canals in the head viewed from above. (After Quix.)

**Fig. 17.32**  The cupula of the crista ampullaris. (After Kolmer.)

containing a sense organ—the *crista ampullaris* or *crista acustica*, which bears as an auxiliary structure a jelly-like *cupula* which encloses the sense hairs protruding from the surface of the crista (Fig. 17.32). The utricle and saccule or otolith organs contain sensory organs called *maculae*, with a similar type of sense cell but with shorter hairs. The sense hairs of the macula are embedded in an otolith membrane which is encrusted with numerous concretions of lime crystals (*otoconia*) (Fig. 17.33). The membranous labyrinth, the cavities of which are filled with endolymph, is surrounded by cartilage or bone, the space between it and the surrounding tissues being filled with perilymph.

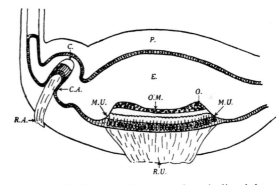

**Fig. 17.33**  Section through the macula utriculi and the ampulla of a vertical semicircular canal in man. *C*, cupula; *C.A.*, crista ampullaris; *E*, endolymph; *M.U.*, macula utriculi; *O*, otoconia; *O.M.*, otolith membrane; *P*, perilymph; *R.A.*, ramus ampullaris; *R.U.*, ramus utricularis of N VIII. (Lowenstein, after Kolmer, *Biol. Rev.*)

### THE UTRICLE AND SACCULE

The maculae of the otolith organs are stimulated by a gravitational pull on the otolith membrane which alters the tension in the hair cells; the macula of the utricle lies normally in a horizontal plane, and under these conditions the tension is at a minimum; if the head is held upside down the stimulus is at a maximum. The utricle is thus a receptor which can respond to alterations in the orientation of the head in relation to the direction of gravity; it can therefore mediate the static class of reflex, i.e. compensating poses of the eyes, head, etc. It is worth noting, however, that a sudden linear acceleration of the head could exert an inertial pull on the hairs of the macula and give rise to a kinetic type of reflex response. The saccule's macula lies in a plane at right-angles to that of the utricle.

## THE SEMICIRCULAR CANALS

The semicircular canals are receptor organs responding to acceleration of the head; opinion on the manner in which the crista ampullaris is stimulated by an acceleration has fluctuated, but the experiments of Steinhausen and Dohlman have now firmly established the original notion of Mach that it is the inertial movements of the endolymph that deform the cupula and stimulate the hair-fibres embedded in it. As Figure 17.34 shows, when the head is accelerated the endolymph within the membranous canal tends to be 'left behind' and so produces an inertial force that deflects the cupula, which is thought to move in much the same way as a rotating door. When the acceleration ceases there is an inertial movement of fluid in the opposite direction.

## Integrating accelerometer

The elements determining the dynamic behaviour of the canals are thus the moment of inertia of the ring of fluid in the canal, the viscous drag of the fluid, and the 'spring-stiffness' of the cupula, which always returns to its neutral position in the absence of external forces. It is apparently the viscous drag that is the dominant force; the velocity of flow is proportional to the angular acceleration of the head, so that the position of the cupula, which is the integral of the fluid velocity, is proportional to the angular velocity of the head. Thus

the sense organ behaves like an integrating accelerometer, with the result that its neural discharge indicates the angular velocity of the head. If the head movements are too short or too long the viscosity no longer predominates over the inertial or spring forces, and the canal ceases to integrate head acceleration properly (Robinson, 1968). Thus, if the head is accelerated to a definite angular speed and then this speed is maintained, as when a subject is rotated in a chair, at the end of the acceleration the deflexion of the cupula indicates the velocity attained, but now the spring-stiffness asserts itself and the cupula gradually returns to its original position, although the velocity remains the same.

## THE HAIR CELLS

The sensitive epithelium of the ampulla is called the crista and consists of a sheet of hair-cells; closely related to these are the terminals of vestibular nerve fibres; the hairs on these cells consist of two types, a single kinocilium and a number of stereocilia arranged in a characteristic pattern. As indicated, the stereocilia increase in size as they approach the kinocilium, and as Lowenstein and Wersäll (1959) pointed out, the arrangement of the cilia constitutes the structural basis for the polarization of the hair cell such that a deflexion of the cilia in the direction of the kinocilium, as indicated by the arrow in Figure 17.35, constitutes an excitatory stimulus whereas a deflexion in the opposite direction constitutes an inhibitory one. In a given ampulla, all the hair-cells are polarized in the same direction so that the response in all the neurones is qualitative uniform.

## Role of kinocilium

Experimentally it has been found that, so far as the horizontal canals are concerned, an ipsilateral acceleration, i.e. an acceleration towards the canal under consideration, causes increased nervous discharge, and a contralateral acceleration a decreased discharge. Examination of Figure 17.34 indicates that an ipsilateral acceleration leads to an inertial flow of fluid towards the ampulla or the utricle—the deflexion of the cupula is

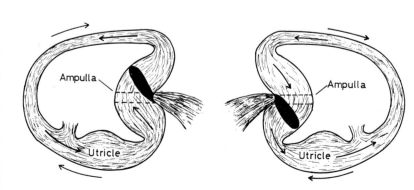

**Fig. 17.34** Illustrating movements of the cupula in the ampullae of the horizontal canals on accelerating the head to the right. In the left canal the inertial flow of fluid is away from the ampulla, whilst in the right canal it is towards the ampulla.

MORPHOLOGICAL  POLARIZATION

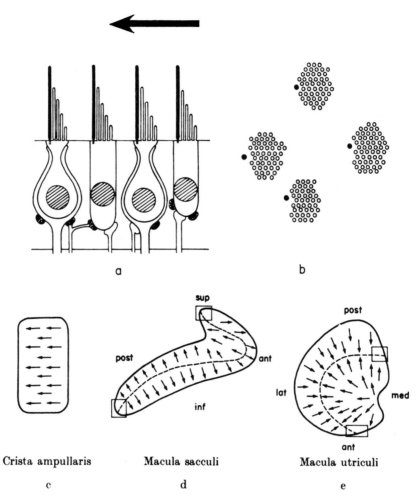

Crista ampullaris          Macula sacculi          Macula utriculi

c                          d                       e

**Fig. 17.35**   Morphological polarization of the sensory cells and the polarization pattern of the vestibular sensory epithelia. The morphological polarization (arrow) of a sensory cell is determined by the position of the kinocilium in relation to the stereocilia. (a) Section perpendicular to the epithelium showing the increasing length of the stereocilium as the kinocilium is approached. (b) Section parallel to the epithelial surface. Sensory cells on the crista ampullaris (c) are polarized in the same direction whereas those on the macula sacculi (d) and macula utriculi (e) are divided by an arbitrary curved line into two areas of opposite morphological polarization, the pars interna and the pars externa. On the macula sacculi the cells are polarized away from the dividing line, on the macula utriculi towards the line. Constant irregularities in the polarization pattern are found in areas corresponding to the peripheral continuation of the striola (rectangles in d and c). (Lindeman, *Acta otolaryngol.*)

thus ampullipetal or utriculopetal—and Lowenstein and Wersäll pointed out that the kinocilium is placed on the hair-cells in such a way that when it is bent towards the utricle it is excited and when bent away from it—utriculofugal—it is inhibited. In the vertical canals the opposite relation pertains, in the sense that movement of fluid towards the ampulla or the utricle causes *inhibition*; and it is found that the orientation of the kinocilium is now opposite, facing away from the utricle, so that it is utriculo*fugal* displacement that causes excitation.

**Polarization in utricle and saccule**

Here the polarization is more complex; in general the otolith membrane shows a characteristic division caused by a denser aggregation of otoconia to form a line, or *striola*; so, as indicated by Figure 17.35, we may speak of a *pars interna* and *pars externa*, and it is found that the polarization of the cilia varies in a characteristic manner in relation to this dividing line; moreover, in the saccule, the polarization is away from the line and in the utricle towards it. This variation in polarization in a given macula means that the organ is sensitive to gravitational pulls in a large variety of directions, in the sense that there will be individual neurones that are maximally excited by a specific direction of pull; in this way the organ can accurately convey information respecting the position of the head in relation to gravity over a wide range of inclinations. If all the hair-cells were polarized in the same direction there would be only one direction of maximal sensitivity, with a corresponding diminution in the range of useful information.

## VESTIBULO-OCULAR REFLEX

Acceleration of a human subject by rotation in a chair causes a slow movement of the eyes in the opposite direction, at about the same angular velocity as that of the rotation, with the result that the eyes are able to maintain fixation on a previously fixated point. The response happens in the dark, but in this case some 'phase-error' may be present, in the sense that the compensatory rotation is not completely adequate to maintain the original direction of gaze. Thus the function of the vestibulo-ocular reflex is to permit a stabilization of the retinal image to the extent that only small corrections of fixation need be made by way of the fixation reflex, utilizing visual input.

### EFFICIENCY

The vestibularly evoked movement serves to keep the image of an external object on the retina stationary, even when the angular velocity is as high as 300°/sec; the control mechanism may therefore be said to be far more effective than the visual tracking movement, which is only effective up to velocities of 30°/sec.

### ARRANGEMENT OF CANALS

As illustrated by Figure 17.31, the canals are arranged in the head in three mutually perpendicular planes. We may distinguish on each side a horizontal or external canal, both lying in the same plane which, in effect, is not horizontal when the head is erect but slopes backwards at 30°, hence they only come into the horizontal plane when the head is tilted forward through this angle. Rotation of the head about a vertical axis has its main effect on this pair of canals, which we may describe as a *synergic pair*. When the head is rotated about its vertical axis, e.g. when the head looks to the right, the

reflex movement of the eyes in the opposite direction that maintains fixation on a point that was originally straight ahead of the eyes will be mediated by the inertial stimuli in the horizontal canals. The anterior vertical canal of one side is in the same plane as the posterior vertical canal of the other, so that we may distinguish two more synergic pairs, namely left anterior-right posterior, and left posterior-right anterior. When the head is moved forwards and backwards, towards and away from the chest, the compensatory rapid movements of the eyes upwards and downwards are mediated by the synergic pair whose planes lie in a sagittal plane, whilst when the head is moved towards and away from the shoulder the other vertical pair, lying in the frontal plane, parallel to the forehead, will mediate the rolling movement that takes place. In fact, as we shall see, the reflex movements involving vertical and torsional rotations are governed by activities in the ampullae of all four vertical canals.

## VESTIBULAR NYSTAGMUS

The mechanism of the semicircular canals is best investigated in the intact animal or man by the study of the nystagmus that occurs on rotating about a given axis. If a human subject is rotated about a vertical axis when he is seated in a rotating chair with his head bending forward to bring his horizontal canals into the plane of rotation, then his eyes go into a *rotatory nystagmus*; it is as though they were making repeated attempts at maintaining fixation of an object in front, but they are not visual responses since they occur when the eyes are closed, the eye-movements being recorded electrically. The nystagmus consists in a slow movement of the eyes in the opposite direction to that of rotation, followed by a rapid saccade in the direction of rotation. The speed of the slow movement is that of the speed of rotation. The direction of the nystagmus is named after the direction of the fast movement; hence the direction of a rotatory nystagmus is the same as that of the rotation. If the rotation is maintained for about 20 to 30 seconds the nystagmus ceases.

### MECHANISM

Why should the eyes go into a nystagmus on rotation? That is, what is there in the stimulus-conditions that results in this repetitive type of motion? We may answer the last question first. It will be recalled that for any abrupt movement of the head there are an acceleration and a deceleration, and therefore there are two sets of messages sent to the central nervous system, determined by two types of inertial fluid movement. The cupula is deflected one way and then the other. By contrast, when the rotation is sustained, the initial acceleration is not followed by a deceleration, and consequently no inertial force is brought into play to return the cupula to its mid-position. Instead the cupula must rely on its own elasticity, and experiments have shown that it behaves as a heavily damped pendulum, requiring some 20 to 30 seconds to return to its mid-position after maximal deflection, i.e. it requires as long as the nystagmus lasts. Thus a sustained deflection of the cupula, with its accompanying pattern of nervous discharges, gives rise to rhythmic nystagmoid movements, whilst a deflection followed by a rapid return gives rise only to a compensatory deflection of the eyes. Experimentally this was demonstrated by Steinhausen who showed that if he applied a transitory pressure to a canal, through a tube inserted into it, the effect was only a simple movement of the eyes; if he maintained a constant pressure, so that the cupula remained deflected, the eyes went into a nystagmus which lasted as long as the pressure was maintained.

## POST-ROTATORY NYSTAGMUS

We may imagine that our human subject has been rotated at a steady speed for some 20 to 30 seconds; the cupulas have returned to their mid-positions and the rotatory nystagmus has ceased. We now abruptly cease the rotation. There is now an inertial flow of endolymph such as to deflect the cupulas in the opposite direction. Moreover, these deflections will be maintained until the natural elasticity brings the cupulas back to their mid-positions. Hence the conditions are such as to induce another nystagmus, lasting another 20 to 30 seconds, but differing only in the direction; thus the slow component will be in the direction of rotation. It is this nystagmus that is easy to demonstrate; the subject is rotated at an even speed for some 20 to 30 seconds and then the rotation is abruptly stopped and the experimenter observes the nystagmoid movements.

## THE VERTICAL CANALS

The effects of stimulating the vertical canals are more complex. Studies of the discharges in the nerve fibres from the ampullae indicate that, whereas rotation in the plane of the horizontal canals only stimulates the horizontal ampullae, rotation in other planes causes a complex group of inhibitory and excitatory responses in the ampullae of *all* the canals. The final compensatory movement of the eyes, resulting from a rotation in a plane other than that of the horizontal canals, represents the integrated response to stimuli arising in all the ampullae. It must be appreciated, moreover, that the macula of the utricle may be stimulated by these rotatory accelerations, and thus make its contribution to the general response of the eyes.*

## REACTION OF LABYRINTH TO ARTIFICIAL STIMULI

A number of artificial stimuli will upset the normal balance between the tonic activities of left and right labyrinths; such an upset, if maintained, usually leads to a nystagmus.

### LABYRINTHECTOMY

Unilateral labyrinthectomy causes a nystagmus for a few days with the slow phase directed towards the operated side; injection of cocaine has a similar effect. It will be recalled that in the normal animal at rest there are tonic discharges from both ampullae of a synergic pair. Accelerations, requiring compensatory movements, cause an imbalance in this system, the discharge from one ampulla increasing and that from the other being inhibited. Clearly, then, extirpation of one labyrinth creates an imbalance that will be interpreted by the central nervous system as an acceleration, and some compensatory movements of the eyes will be initiated. In this way we may explain the effects of extirpation; with irritative lesions in the vestibular apparatus or in its nervous pathway the balance is upset once again, but this time in the opposite sense (see, for example, Cranmer, 1951).† After a time the necessary central compensations take place and the nystagmus subsides. When, after the nystagmus due to a lesion in the labyrinth has subsided, the labyrinth on the opposite side is injured, then a new nystagmus occurs with the slow phase now directed towards the newly injured side—the so-called *Bechterew nystagmus*.

### CALORIC NYSTAGMUS

A current of warm water (40° to 45°) or cold water (22° to 27°) through the auditory meatus causes a nystagmus of vestibular origin; the slow movement is towards the stimulated side, with cold water, and away from this side with warm water. The fact that the nature of the nystagmus depends on the position of the head suggests that the cause is associated with thermal currents in the endolymph.

### ELECTRICAL STIMULATION

Passing a direct current from one ear to the other causes a nystagmus, the slow phase being towards the positive pole. The nature of the nystagmus is independent of the position of the head. In the experimental animal application of repetitive stimuli to an ampulla produces a nystagmus. Thus a single shock, applied to the right vestibular nerve, causes contractions of the left lateral rectus and right medial rectus, leading to a horizontal movement to the left, i.e. away from the stimulated

---

*The effects of electrical stimulation of the ampulla of a single canal have been described by Fluur (1959), but as he points out, this is an exceedingly artificial procedure, since under normal conditions of stimulation, there are reactions in both sides of the head simultaneously. Stimulation of the left lateral canal caused a conjugate movement to the right; of the left anterior vertical canal a conjugate movement upwards, and the left posterior vertical canal a conjugate movement downwards. Combined stimulation of the left anterior and posterior canals caused a conjugate counter-clockwise rolling.
†Because of the involvement of proprioceptive impulses from the neck muscles in eye movements, we might expect irritative lesions in the cervical cord to cause nystagmus; in fact they do. Presumably impulses pass up the cord in the spino-vestibular pathways which connect the proprioceptive centres of the cord with the inferior vestibular nuclei; within the vestibular areas there is a transfer of impulses to pathways connecting with the oculomotor nuclei (Crosby, 1953).

side. If this stimulation is continued the eyes go into a nystagmus with the slow movement away from the stimulated side and a fast movement towards it. Thus the nystagmus may be regarded as a rhythmic interruption in the response to stimulation of one vestibular nerve or ampulla. Experimental reduction in the tonic discharges in a vestibular nerve also induces a nystagmus, this time in the opposite direction.

## STATIC VESTIBULAR REFLEXES

These reflexes, mediated by the utricle and saccule serve to maintain a normally orientated field of vision despite changes in the position of the head. Thus when the body is tilted towards the prone position the eyes move upwards; when the body is inclined sideways, the eyes roll; it should be emphasized, however, that in man the rolling of the eyes by no means compensates entirely for the altered position of the head; psychological adjustments are necessary (p. 506). Movement of the body or the head about a vertical axis will clearly have no influence on the direction of gravitational pull on the otolith membrane, hence any compensatory movement of the eyes that takes place must be mediated by another type of receptor. It is here that the proprioceptive impulses from the neck muscles are involved.

### COMPENSATORY ROLLING

The rolling of the eyes that takes place when the head is tilted towards the shoulder has attracted considerable interest. It is clear that both otolith and semicircular canals may act as receptors, but if the compensatory rolling is to be sustained for any length of time clearly it is only the otolith, or neck proprioception mechanism, that will operate. Merton's (1956) studies have emphasized the much greater effectiveness of the semicircular canals in producing a rolling. When a subject is swung from side to side on a swing hung from a ball-bearing, such that the axis of rotation corresponds to the

visual axis of the eye, then it is found that the eyes roll in the opposite direction to that of the movement of the head; the rolling movement lags behind the head movement, so that the compensation is never complete; in other words, the vertical meridians of the eyes do not remain vertical. The compensation amounts to about a quarter of the actual movement of the head. When a head movement is sustained, however, this partial compensation is not maintained, so that the vertical meridian returns almost completely to the position it would have had if no rotation had occurred. According to Woellner and Graybiel, for example, at a tilt of 15° the rolling is only 1·5°, whilst with a tilt of 70° it is only 4°. As Merton has explained, the partial compensatory movements during rotation are of value in reducing the angular movements of the retina, and thereby permitting a reasonable visual acuity even when the head is swinging through quite large angles.

### THE DOLL-REFLEX

When a subject is moved in the median plane with his head fixed so that movements of the head in relation to the trunk, evoking neck proprioception, are avoided the vestibulo-ocular reflex tends to make the eyes rotate upwards or downwards through utricular stimulation. If the human subject is rotated in this way in the dark and presented with only a series of vertically arranged fixation points we may estimate the degree to which this so-called doll reflex has operated by asking him to state which fixation point is directly ahead. Consistent errors are in fact found (Ebenholtz and Shebilske, 1975) indicating some upward and downward movements; the errors followed a sine-function such as would be anticipated of stimuli governed by the pull of gravity on the otolith membrane assuming this to be at an angle of about 30° backwards when the head is vertical. The results of experiments are illustrated by Figure 17.36;

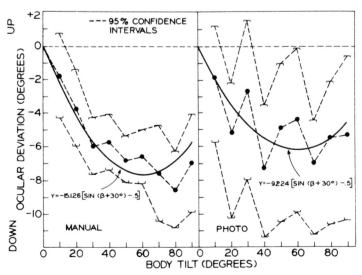

**Fig. 17.36** Ocular deviation due to body-tilt in a median plane. The ordinate represents the difference between the subjectively estimated and the true positions of the fixation point. Dashed lines indicate 95 per cent confidence intervals. 'Manual' and 'photo' indicate different methods of measuring the ocular deviation. (Ebenholtz and Shebilska, *Vision Res.*)

as with the torsional movement the doll-reflex only partially compensates for the tilt, a maximum ocular deviation of about 9° being obtained with a 70° tilt.

### OTOLITH RESPONSES TO ACCELERATION

Although the otolith organs are essentially gravity receptors they are, of course, susceptible to shearing forces developed in other ways than the pull of gravity; thus acceleration of a rabbit forwards should exert shearing forces on the utricular macula and appropriate eye-movements should be recorded, whilst swinging the animal, inducing rhythmic parallel or transverse accelerations, should induce nystagmic types of movement. The responses of a rabbit to simple linear accelerations and to sinusoidal swings have been studied by Baarsma and Collewijn (1975); during transverse swinging vertical eye movements were recorded and during sagittal swinging torsional eye movements. The important feature of these responses was their sluggishness, indicating the unsuitability of the otoliths as acceleration sensors; thus with sinusoidal swings the phase-lag amounted to 180°, and in general the responses took many seconds to develop fully.★

## PROPRIOCEPTIVE NECK REFLEXES

Bending the head forward and backwards towards or away from the chest, besides activating the utricle, will stretch the neck muscles, and it may be shown that the compensatory movements of the eyes that are elicited may be caused by both of these receptive mechanisms. Thus movement of the trunk, keeping the head fixed, gives the same reflex, but by virtue of the stretched neck muscles. Lateral movements in response to moving the head about a vertical axis are mediated exclusively by the neck mechanism. Hence cutting the cord in the cervical region may reduce an animal's ability to maintain orientation (Cohen).

## THE NERVOUS PATHWAYS IN THE LABYRINTHINE REFLEXES

### PRIMARY VESTIBULAR AFFERENTS

The labyrinth is supplied by the vestibular branch of N VIII; as with other sensory nerves, the fibres have their cell bodies outside the central nervous system, in this case the *ganglion of Scarpa* (Fig. 17.37). Some primary sensory fibres travel directly to the cerebellum (p. 454) and the rest pass into the medulla to relay in four *vestibular nuclei*, with functionally different afferent inputs and efferent projections.

### VESTIBULAR NUCLEI

The nuclei are (a) *Deiters'* or the *lateral nucleus*; (b) *Schwalbe's* or the *medial* nucleus; (c) *Bechterew's* or the *superior nucleus* and (d) *Roller's* or the *descending* or *spinal vestibular nucleus*.

### SECONDARY NEURONES

Secondary vestibular nucleus neurones project (a) to the spinal cord as the vestibulo-spinal tracts; thus the lateral vestibulospinal tract is derived entirely from neurones of Deiters' nucleus which thus exerts its descending action on the spinal motor neurones. (b) to the cerebellum and cerebral cortex, (c) to the brainstem eye-muscle nuclei mainly by way of the medial longitudinal fasciculus (MLF); another route is by way of the reticular formation where neurones with monosynaptic connexions with vestibular nuclear neurones may be identified, mostly in *n.r. gigantocellularis* in the medulla and in *n.r. pontis caudalis* in the pons.

### Connexions to motor neurones

According to Tarlov, who made discrete lesions in the vestibular nuclei and followed the resulting degenera-

---

★ The interested reader should consult Baarsma and Collewijn's paper for a valuable discussion of the literature on this subject.

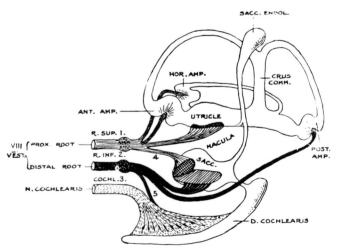

**Fig. 17.37** Innervation of the membranous labyrinth. (De Burlet, *Anat. Rec.*)

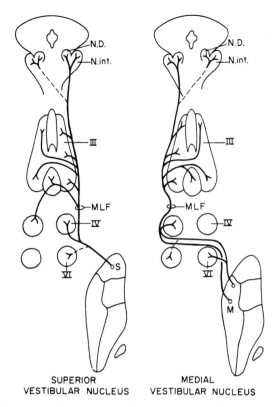

**Fig. 17.38** Courses of vestibulo-oculomotor fibres. Ascending fibres from superior vestibular nucleus (left) pass in ipsilateral medial longitudinal fasciculus (MLF) to extraocular muscle nuclei bilaterally. Note that fibres to contralateral IIIrd and IVth nuclei cross midline within the IIIrd nuclei. In contrast, fibres from medial vestibular nucleus (right) decussate between levels of IVth and VIth nuclei, and ascend entirely in contralateral MLF. Dashed lines indicate sparse projections. (Tarlov, *Progr. Brain Res.*)

tion to the motor nuclei of the eye muscles, only lesions in the superior and rostral portions of the medial vestibular nucleus caused degeneration in the eye motor nuclei, and Figure 17.38 illustrates the courses of the oculo-motor fibres, which pass by way of the medial longitudinal fasciculus. We must note that McMasters

*et al.* considered that the ascending fibres from Deiters' and the medial nucleus were capable of mediating all patterned eye movements, so that Tarlov's study suggests that the connexions from Deiters' (lateral) nucleus involve an extra synapse, possibly by way of the reticular formation, the interstitial nucleus of Cajal and the nucleus of Darkschewitsch. His results are consistent with Highstein and Ito's electrophysiological study, and the observation of Shimazu *et al.* (1971) that most of the Deiters' neurones are activated by the utricle rather than the semicircular canals, and it is therefore likely that utricular responses in the eye muscles are mediated by a more complex pathway than the semicircular canal responses. This is understandable since these last mediate dynamic responses to acceleration whereas the utricle is mainly concerned with static postural responses tending to maintain poses when speed of execution is not so important.

## Additional connexions
The connexions of the vestibular nuclei with the spinal motor neurones, with the eye-muscle motor neurones, with the cerebellum and with the motor nuclei of the reticular formation indicate their possible role as coordinators of eye movements with movements of the head and trunk.

### THE THREE-NEURONE ARC
A variety of anatomical and electrophysiological studies have established the existance of a simple three-neurone arc mediating the vestibulo-ocular reflex, as illustrated in Figure 17.39.

### SPECIFICITY OF INPUT TO EYE MUSCLES
Figure 17.40 from Szentagothai (1950), who stimulated different ampullae and measured the predominant motor action in the individual muscles, shows that the lateral ampulla activates the ipsilateral medial rectus and the contralateral lateral rectus to produce a deviation away from the stimulated side. Stimulation of the superior vertical ampulla activated the ipsilateral

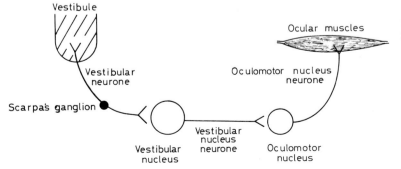

**Fig. 17.39** The three-neurone arc mediating vestibulo-ocular reflexes.

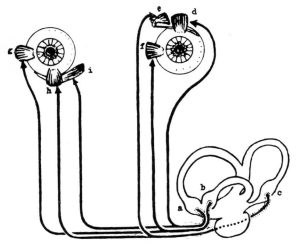

**Fig. 17.40**   Specificity of connexions of semicircular canals. Ampulla of superior canal (a) is in three-neurone connexion with ipsilateral superior rectus (d) and with contralateral inferior oblique (i); ampulla of horizontal canal (b) with ipsilateral medial rectus (f) and contralateral lateral rectus (g); ampulla of posterior canal (c) establishes three-neurone connexions with ipsilateral superior oblique (e) and contralateral inferior rectus (h). (Szentagothai, *J. Neurophysiol.*)

superior rectus and the contralateral inferior oblique; these are a 'linked pair', as we have seen, and thus may be expected to be activated in unison to cause an upward deviation with some rolling. The posterior canal activated the ispilateral superior oblique and the contra-lateral inferior rectus producing downward deviation

and some rolling. An important feature of the responses was the simultaneous inhibition or relaxation of antagonist muscles, e.g. lateral rectus when medial rectus of the same eye contracts, or if linked pairs are concerned, inhibition of the superior rectus of the opposite side when the inferior oblique of the same side is activated.

### Prime and synergic actors

In a more precise study, combining electrical stimulation of individual canal nerves with simultaneous stimulation of pairs, Cohen *et al.* (1964) define the prime actors as well as synergic actors in the response to stimulation of a given nerve; the results are indicated in Table 17.2, where it is seen that a given canal can activate six muscles, three in each eye, and since relaxation of antagonists was also observed this means that even the horizontal canals influence all twelve muscles. As an example we may take the response to stimulating the left anterior vertical canal, having its prime action on the right inferior oblique and left superior rectus; the observed eye movements (Fig. 17.41) were not conjugate, the left eye elevating with a slight counterclockwise rolling and the right giving mainly a counterclockwise rolling and a slight upward movement. When both anterior canal nerves were stimulated conjugate upward eye movements were obtained through activation of the superior recti whilst stimulation of one anterior and one posterior canal of the same side gave a conjugate rolling (Fig. 17.42).

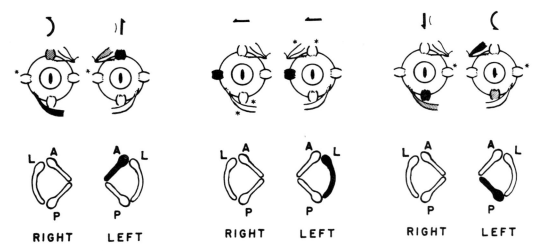

**Fig. 17.41**   Diagram of the eye movements and eye muscle activity elicited by stimulating ampullary nerves from left anterior canal (left), lateral canal (centre), and posterior canal (right). Anterior (A), lateral (L), and posterior (P) semicircular canals are represented in the diagrams below the eyes. The direction of the evoked eye movement is shown by arrows above the eyes. The canal whose nerve is stimulated and the eye muscles primarily activated are shaded in black. Stronger synergistic activity of eye muscles is shown by cross hatching and weaker synergistic activity by asterisks (see text for details). Muscles that are not excited, relax. (Cohen *et al.*, *Ann. Otolaryngol.*)

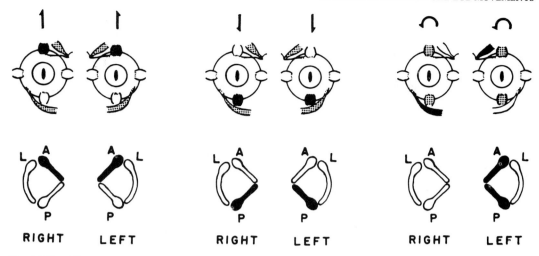

**Fig. 17.42** Diagram of eye movements evoked by simultaneous stimulation of both anterior canal nerves (left), both posterior canal nerves (centre), and the left anterior and posterior canal nerves (right). Scheme as in Figure 17.41 with exception that muscles activated by stimulation of one canal nerve and partially or completely inhibited during stimulation of more than one canal nerve are shown as dotted. (Cohen *et al.*, *Ann. Otolaryngol.*)

## Stimulation of canal pairs

When a pair of canals lying in the same plane were stimulated, e.g. the lateral canals, there was no eye movement; this was true of simultaneous stimulation of the right posterior and left anterior canal; as indicated earlier these lie in the same plane and are normally excited in opposite senses in any movement with a component in this plane; thus activation of one and inhibition of the other is the natural consequence of an acceleration in the plane. Simultaneous activation of both is an artificial situation equivalent to no head movement and thus does not call for any movement of the eyes.

RESPONSES FROM THE OTOLITH ORGANS

Compensatory poses in response to head tilts mediated through the gravity receptors involve elevation and depression, and conjugate rolling; lateral movements are evoked by stimulation of neck receptors. When Suzuki *et al.* (1969) stimulated the left utricular nerve of the cat the response was mainly a conterclockwise rolling, with the strongest increase in tension in the ipsilateral superior oblique and contralateral inferior oblique.

## Utricle

The most thorough study of the electrophysiology of the utricle and saccule is that of Fernandez *et al.*

**Table 17.2** Contractions of extraocular muscles during stimulation of single semi-circular canals. (Cohen *et al.*, 1964)

| Canal nerve stimulated | Primary contractions | Synergistic contractions | |
|---|---|---|---|
| | | *(stronger)* | *(weaker)* |
| left anterior canal | left superior rectus | left superior oblique | left medial rectus |
| | right inferior oblique | right superior rectus | right lateral rectus |
| left posterior canal | left superior oblique | left inferior rectus | left lateral rectus |
| | right inferior rectus | right inferior oblique | right medial rectus |
| left lateral canal | left medial rectus | left superior oblique | |
| | | left superior rectus | |
| | right lateral rectus | right inferior oblique | |
| | | right inferior rectus | |

(1972), based on a study of individual nerve fibres in the superior nerve, which innervates the utricle, and the inferior nerve innervating the saccule. They postulated that acceleration of the animal's head in a given plane and direction should activate those fibres related to hair-cells with maximal response to movement in this direction, so they defined the *functional polarization vector* of the given unit as that direction in space that gave the highest response. So far as the utricle was concerned the polarization vectors lay in the horizontal plane and 70 per cent of the units were most sensitive to an ipsilateral roll of the head sideways, and 30 per cent to a contralateral roll. Of those responding preferentially to pitches, equal numbers responded to nose-up and nose-down accelerations. Thus basically the sensitive cells in the macula utriculae fall into four classes with their discharge affected primarily by ipsilateral or contralateral rolls and forward and backward pitches.

## Saccule

With the inferior nerve, innervating the saccule, the polarization vectors tended to lie in the sagittal plane, and this would conform with the orientation of its macula in this plane.

### PATHWAYS TO THE MOTOR NEURONES

The main pathway from the vestibular nuclei to the motor neurones governing the eye muscles is through the medial longitudinal fasciculus (MLF). Although destruction of the MLF will often abolish vestibular responses in eye muscles, it does not always do so, but only attenuates them. According to Cohen (1971), the semicircular canals may send messages directly through the MLF or alternatively through the reticular formation. An additional route, probably carrying otolith messages, runs via the cerebellum, impulses leaving the inferior vestibular nucleus to reach the fastigial nucleus, whence fastigial neurones send impulses to the motor neurones of the eye muscles.

### INPUT TO NECK MUSCLES

Wilson and Yoshida (1969) found that, whereas a single shock to the labyrinth evoked no monosynaptic responses in motor neurones of the spinal cord activating the fore- and hind-limbs of the cat, responses in neck motor neurones were obtained; and in a more recent study, Wilson and Maeda (1974) obtained EPSP's and IPSP's in all neck motor neurones on stimulating individual ampullary nerves. The responses were quite sterotyped, e.g. the biventer cervicis and complexus motor neurones were excited by stimulation of two anterior canals and inhibited from two posterior canals. Figure 17.43 illustrates the probable connections which involve, once again, both disynaptic excitation and inhibition of the motor neurone through a three-neurone arc.

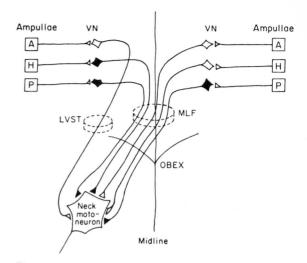

**Fig. 17.43** Schematic diagram showing connexions between ipsilateral and contralateral ampullae and neck motor neurones. A, H, P are anterior, horizontal and posterior ampullae; VN, vestibular nuclei. Inhibitory neurones and their terminals are shown in black; excitatory in white. (Wilson and Maeda, *J. Neurophysiol.*)

## RESPONSES IN PRIMARY AFFERENT NEURONES

### FISH SEMICIRCULAR CANALS

The sensory fibres from the semicircular canals have usually a tonic resting discharge, and this is enhanced when the head is accelerated in one direction and inhibited by acceleration in the opposite direction. A typical response curve for the ray's horizontal semicircular canal is illustrated in Figure 17.44. It will be seen that at rest, or zero acceleration, there is a tonic discharge of 80 imp/sec; this may be increased or decreased according to the direction of acceleration. When different units were examined the same basic feature was revealed, namely reversed type of response with accelerations in opposite directions, so that we may conclude that the hair-cells of the crista ampullaris of any given canal are polarized in the sense that bending in one direction excites its associated axon, and bending in the other inhibits it. The sign of this polarization is the same for all the hair-cells in a given ampulla, and the positions of ampullae in canal pairs are such that excitation of one set of hair cells is accompanied by inhibition of the opposite set. In this way we see the advantage of a high tonic discharge in the labyrinthine neurones; this permits two signals, namely increase and decrease of discharge indicating acceleration in one or the opposite direction; if the neurones, in the absence of acceleration, were silent, then acceleration in one, the

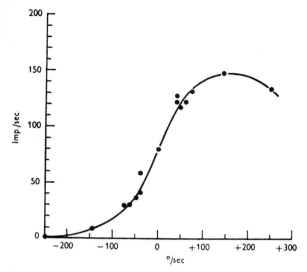

Fig. 17.44 Frequency of discharge of a fibre from the horizontal semicircular canal of the ray plotted as a function of angular acceleration. With zero acceleration there was a resting discharge of 80 imp/sec, which diminished or increased according to the direction of acceleration. (Groen et al., *J. Physiol.*)

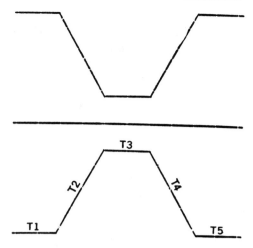

Fig. 17.46 Stimulus pattern used to obtain the responses indicated in Fig. 17.45. Each velocity trapezoid consists of stationary period ($T_1$), constant acceleration period ($T_2$), constant velocity period ($T_3$), constant deceleration period ($T_4$) and final stationary period ($T_5$). The signal is proportional to the angular velocity; the upper and lower traces correspond to counterclockwise and clockwise rotation respectively. (Goldberg and Fernandez, *J. Neurophysiol.*)

inhibitory, direction would result in no signal since there can be no negative discharge in a neurone.

MAMMALIAN UNITS

Goldberg and Fernandez (1971) studied responses of monkey units; Figure 17.45 illustrates the response of a single vestibular nerve fibre to a stimulus-pattern illustrated in Figure 17.46; this consists of a period,

$T_1$, during which the head is at rest, and the record indicates a tonic discharge of 80 imp/sec. A period of abrupt acceleration, $T_2$, indicated by the bar on the record, is accompanied by an abrupt increase in discharge. At the end of the period of acceleration, the velocity is held constant over the period $T_3$, and during this period the discharge steadily wanes, returning to the resting value in about 1 minute. Next the rotation

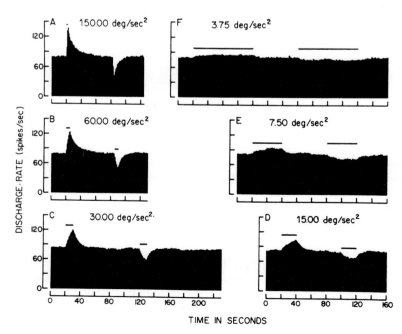

Fig. 17.45 Response of superior canal sensory unit to the velocity trapezoids depicted in Fig. 17.46. Each histogram represents the response to the presentation of a single stimulus. Each histogram column depicts the discharge rate for 1 sec and the horizontal marks denote periods of constant acceleration and deceleration ($T_2$ and $T_4$). Accelerations are as stated, all rotations counterclockwise, and maximum velocity always 300/sec. (Goldberg and Fernandez, *J. Neurophysiol.*)

TIME IN SECONDS

was decelerated over the period $T_4$, indicated by the second bar, and it will be seen that there is a prompt inhibition of the tonic discharge followed by a return to the basal rate when the animal is stationary.

## Torsion-pendulum

The period at constant velocity—zero acceleration—represents the period when the cupula is being restored to its resting position under the influence of its own elasticity, and Goldberg and Fernandez showed that the dynamics of the changes in unit discharge conformed with those predicted on the basis of the movement of a heavily damped torsion-pendulum, as earlier described by Van Egmond *et al.* (1949), when they studied the duration of the sensation of giddiness, or Hallpike and Hood (1953) studying the duration and velocity of the oculogyral illusion, after attaining a steady velocity.

## THE NEURONES OF THE VESTIBULAR NUCLEI

### THE THREE-NEURONE-ARC

Studies on the discharges and membrane potentials of individual neurones in the vestibular nuclei during vestibular stimulation have shown unequivocally that both the excitation of the vestibular neurone, revealed usually as an increase in its basic discharge, and the inhibition, revealed as a reduction in discharge, can be mediated by a monosynaptic connexion between the vestibular nerve and the vestibular nucleus neurone. Since, also, the study of the motor neurones under the same conditions reveals a monosynaptic connexion between these vestibular neurones, the concept of a three-neurone are—labyrinthine sensory neurone in Scarpa's ganglion, vestibular neurone in medulla, and brainstem motor neurone—is well substantiated both for excitation and inhibition.

### RESPONSES OF VESTIBULAR NUCLEUS NEURONES

Duensing and Schaefer (1958) carried out the classical study on responses of these neurones to accelerations of the head, and he defined four types according to whether discharge was accelerated or inhibited by ipsi- or contralateral rotation. Thus Type I gave increased discharge on acceleration to the same side and decrease to the opposite side. Type II gave a reciprocal behaviour, giving decreased discharge on ipsilateral turning and increased on contralateral turning. Type II neurones, which were much rarer, were considered by Shimazu and Precht (1966) to be inhibitory vestibular neurones that could mediate the commissural inhibition of Type I vestibular neurones, as suggested by Figure 17.47, where it is seen that a horizontal ampullary neurone activates Type I neurones, and that one of these activates

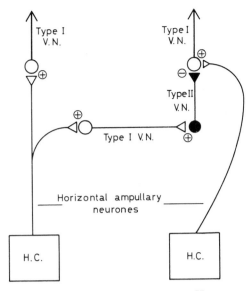

**Fig. 17.47** Illustrating mechanism of Type II neurones. A horizontal ampullary neurone activates Type I neurones, one of which activates a contralateral Type II interneurone which then inhibits a Type I vestibular neurone. (After Shimazu and Precht, *J. Neurophysiol.*)

a contralateral Type II interneurone which then inhibits a Type I vestibular neurone. This inhibition is to be contrasted with the monosynaptic activation of an inhibitory vestibular neurone, $V_i$, that is responsible for the simultaneous activation of agonists and antagonists seen during vestibularly stimulated eye movements, described by Baker and Berthoz (1974) for example. Thus inhibition of a Type II neurone, if this were spontaneously active, would *disinhibit* a Type I neurone and so facilitate the motor neurone with which it is related.

## Specificity of input

Wilson and Felpel (1972), working on the pigeon's labyrinth, found that, of the individual neurones studied in the vestibular nuclei, greater than 55 per cent of the responding cells only fired in response to electrical stimulation of a single ampulla; some were influenced by two ampullae, but when allowance was made for electrical spread from one ampulla to another Wilson and Felpel estimated that at least 81 per cent of the responding vestibular neurones were influenced by only one ampulla.

### INFLUENCE ON MOTOR NEURONES

As an example of the postsynaptic potentials observed on direct stimulation of the vestibular nerve, we may cite the study of Baker *et al.* (1969). Intracellular records of abducens (N VI) motor neurones controlling

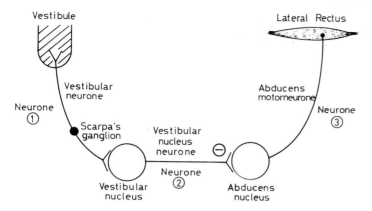

**Fig. 17.48** Reflex arc for inhibition of an abducens motor neurone.

the lateral rectus revealed the development of hyperpolarizing IPSP's in the ipsilateral neurones, which were probably derived from inhibitory vestibular neurones in the ipsilateral medial vestibular nucleus. Thus stimulation of the ipsilateral vestibular nucleus gave slowly developing IPSP's in the abducens motor neurones. The shortest latency from the ipsilateral vestibular nucleus was 0·65 msec, indicating a monosynaptic pathway since conduction-time was some 0·2 to 0·3 msec. The latency from the ipsilateral vestibular nerve to the motor neurone was 1·4 to 2·0 msec, giving a latency-difference of 0·9 msec, on average, which could be the time for activation of the inhibitory neurone in the vestibular nucleus. Thus the IPSP is produced disynaptically from the ipsilateral vestibular nerve Figure 17.48, and the total reflex arc is a 3-neurone one.*

## Trochlear neurones

An essentially similar result was obtained by Baker *et al.* (1973) when studying trochlear (N IV) motor neurones; these receive short-latency excitation (EPSP) from the contralateral labyrinth, and inhibition (IPSP) from the ipsilateral labyrinth. Stimulation of the individual labyrinth nerves from the separate anterior and posterior vertical ampullae revealed a reciprocal activation, the one activating while the other inhibited the trochlear motor neurone. We may note that when Cohen *et al.* (p. 441) stimulated the same pair of canals, there was a strong conjugate torsion caused by powerful activation of the contralateral inferior oblique; at the same time, of course the contralateral superior oblique, innervated by trochlear motor neurones, would be inhibited, so that in a given eye there would be a strong reciprocal activity on a given trochlear motor neurone.

## Obliques and recti

In another study (Baker and Precht, 1972), the combined actions of the posterior canal on the obliques and recti were studied; strong activity was evoked in the ipsilateral superior oblique and the contralateral inferior rectus (linked pair) but a weaker excitation was observed in the ispilateral inferior rectus and the contralateral inferior oblique. The joint action on the eyes was to cause an anti-clockwise rotation and a depression of both eyes.

### MECHANISM OF RELAXATION OF ANTAGONISTS

The development of monosynaptically activated IPSP's in the motor neurones indicates the primary mode of inhibition, or relaxation, of antagonists; another process is described as *disfacilitation*, by commissural fibres running from one vestibular nucleus neurone to another on the opposite side; if the neurone with which this commissural neurone synapses is being tonically excited by a vestibular nerve neurone, then inhibition of this excitation by presynaptic activity would be described as disfacilitation, to distinguish it from the inhibition that results from direct synaptic action producing an IPSP.

### ABDUCENS INTERNEURONES

Horcholle and Tyc-Dumont (1968) found neurones in the cat's abducens nucleus that failed to respond to antidromic stimulation of N VI but were nevertheless involved in the discharges that occurred after vestibular nerve stimulation. These *interneurones* fell into three categories and were described as 2a and 2b, and 3. The Type-3 failed to respond to vestibular nerve shocks but were activated during nystagmus and discharged out of phase with those concerned with motor discharge, i.e. the motor neurones. Figure 17.49 compares the spike discharges of Type 3 interneurone, motor neurone, and Type 2b interneurone with the integrated motor discharges in the abducens nerve during nystagmus. It will be seen that the Type 3 spike discharge is out of phase with the motor discharge. Type 2a interneurone discharges in phase with the motor discharge and the same was true of the other interneurones. It was suggested that the Type 3 interneurones, exclusively concerned in nystagmoid movements, were

---

* Inhibitory nuclear cells, with direct action on motor neurones, are concentrated in the superior vestibular nucleus whilst excitatory cells are contained in the rostral two-thirds of the medial vestibular nucleus (Ito, 1972).

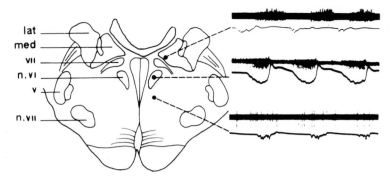

Fig. 17.49    *Left*: diagram showing transverse section of midbrain at the level of the abducens nucleus. *Right*: rhythmic firing pattern (above) and integrated motor discharges (below) for an Interneurone 3 (top records) a motor neurone (middle records) and an Interneurone 2b (bottom records) recorded at the indicated points during nystagmus. (Horscholle and Tyc-Dumont, *Exp. Brain Res.*)

inhibitory interneurones with their activity corresponding to the inhibited phase of the muscle. Thus interneurones 1,[*] 2a and 2b might act as re-excitatory neurones whose activity eventually overcomes the inhibitory activity of interneurones-3 and culminating in a motor neurone burst. These interneurones-3 were situated in a dorsomedial zone of the medial vestibular nucleus, a zone that does not receive direct projection from the labyrinth (Brodal and Walberg, 1962) but receives fibres from the lateral nucleus and projects through the medial longitudinal fasciculus to the eye muscle motor neurones. It could well be that these interneurones-3 terminate in the characteristic manner described by Szentagothai, arborizing around the motor neurones by contrast with the coarser calibre terminations corresponding to the excitatory synapses.

## HYPOTHETICAL NEURAL CIRCUITS

Robinson, largely on the basis of his own studies on the dynamics of eye movements associated with electrical discharges in motor neurones, has elaborated a simple neural circuit that accounts for the main features of the vestibulo-ocular reflex.

### SLOW FOLLOWING MOVEMENT

The basis of this reflex is the slow following movement that permits the retinal image to remain stable during a movement of the head. When the movement of the eyes becomes too extreme, the slow movement is interrupted by a saccade and the following movement is restarted—thus the experimentally studied vestibular nystagmus is simply a manifestation of the vestibulo-ocular reflex. The visually determined saccade may be viewed as a phylogenetically later developed type of movement that simply borrows the quick movement necessary for efficient use of the vestibulo-optic reflex. The sensory basis of the slow following movement is the stimulation of the semicircular canals by the acceleration; experimentally it has been found that the frequency of discharge in the ampullar nerve fibres is proportional to velocity when a constant acceleration is maintained, so that for many head movements it is simply required

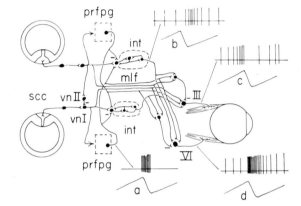

Fig. 17.50    The neural elements necessary and sufficient for the complete dynamic operation of the vestibulo-ocular reflex, both slow and fast components. *scc*, semicircular canals; *vnI* and *II*, vestibular nuclei neurons, types I and II, *mlf*, medial longitudinal fasciculus feeding velocity information directly to the motoneurons; *int*, a brainstem neural network which integrates velocity signals into position signals; *III*, *VI*, oculomotor and abducens nuclei; *prfpg*, pulse generator neural network in the pontine reticular formation, *a*, unit activity of pulse generators; *b*, output of integrator circuits; *c*, *d*, activity seen in agonist and antagonist motoneurons shown in upper traces during slow and fast phases of nystagmus (*lower traces*). (Robinson, *Invest. Ophthal.*)

[*] Type 1 interneurones responding antidromically were considered to be activated by recurrent collaterals of the abducens motor neurones although Sasaki (1963) denies the existence of axon collaterals in eye motor neurones; they were considered to be activated in this way on the grounds that they were not excited in a one-to-one manner by stimuli in the motor nerve, but only after repetitive stimulation. Precht *et al.* (1967) described Type II neurones in the abducens nucleus that increased their discharge with contralateral acceleration and decreased it with ipsilateral; these were antidromically activated by N VI stimulation and were clearly motor neurones. Type I neurones exhibited the reverse behaviour and were not antidromically activated and were possibly interneurones, possibly Horscholle and Tyc-Dumont's Type-3 neurones. They were located in or close to the ventral border of N VI.

that some neural integrator can convert velocity of head movement into a signal representing the position of the head. Thus in Figure 17.50, Robinson postulates an integrator into which the vestibular information is fed. The equation governing motor discharge in relation to an angular movement of the eyes is

$$D = k(\theta - \theta_T) + r\,d\theta/dt$$

and this predicts that the eye will lag behind the motor neurone activity at high frequencies of eye movement, so that a neural phase-lead must be built into the system, since in fact there is no phase-lag in the execution of the total reflex. Thus phase-lead is fed through a monosynaptic pathway by way of the medial longitudinal fasciculus directly to the eye motor neurones, bypassing the integrator. In essence this feed-forward of velocity signals corrects the phase-lag due to the mechanical lag imposed by the orbital contents.

SACCADE

If the smooth movement becomes too large, it is interrupted by a saccade in the opposite direction, a movement achieved by a pulse of high frequency discharge followed by a step-discharge that maintains the new position until the new slow movement is begun. This pulse-step discharge is thought to be generated in a neural circuit in the paramedian pontine reticular formation (p. 448). An important feature is that the pulse should not only be activated, but that it should reset the integrator so that a new slow movement may be begun from a new eye position.

SMOOTH PURSUIT

The smooth visually evoked pursuit system represents a more highly developed apparatus that provides the mechanism with the parameter it has to stabilize, namely the retinal image. In the rabbit, the directionally sensitive ganglion cells could provide the necessary sensory input, and ganglion cells with the necessary range of direction-sensitivity have been described (Oyster et al., 1972). As Collewijn (1972) has shown, in the dark, the rabbit eye wanders about in a random fashion, due to the absence of visual input—the control loop is said to be open. These movements are thought to represent random signals or imbalances generated in the central motor networks (the central noise generator of Figure 17.50), so that the visual input must counteract this noise. The input from the ganglion cells is fed into a central controller network which controls the gain of the reflex; it must command a velocity of eye movement commensurate with the retinal slip velocity, as measured by the direction-sensitive ganglion cells, and this may well be through the same integrator as that employed in the vestibulo-ocular reflex. This system may be described as an image-stabilizer rather than a

tracking mechanism, and as Robinson has emphasized, this is the important feature, a feature obscured by the study of a rabbit's responses to a rotating striped field—a highly unusual stimulus—thus the purpose of the reflex is to make the eyes stay still, not to make them move.

Foveal vision

With the development of the fovea, a new complication appears, since now it is insufficient to keep the eyes still to appreciate a moving point stimulus; tracking must now take place, as in a pursuit, and it has been suggested that this is brought about by a modification of the rabbit's more primitive oculomotor reflex; thus if the foveal directionally sensitive units greatly outnumbered those in the retinal periphery, they would respond to retinal slip the more strongly and would command an eye movement that would increase retinal slip in the periphery. Thus the two retinal slips would be pitted against each other, and the foveal would be the more effective, leading to tracking.

So far as the saccade is concerned, the main feature is the conversion of a spatially discrete visual impulse, e.g. one reaching a topographically organized point in the superior colliculus, into a command that consists of a temporally modulated burst of discharge, the duration and frequency of the burst being the message to the motor neurones that leads to the saccade of almost exactly the correct magnitude for 'foveation'. The correspondence of the motor and sensory maps (Robinson 1972; Cynader and Berman, 1972) forms the anatomical substrate for such a foveation technique, but the neuronal circuitry required to convert this spatially coded information into a burst-discharge is a matter of speculation. As we see, the cerebellum has been invoked, but the relatively small effects of cerebellectomy make it doubtful whether it could fulfil such a vital role.

ACCESSORY OCULO-MOTOR NUCLEI

GAZE CENTRES

The literature contains frequent allusions to *gaze centres*, the one controlling lateral movements and corresponding to the so-called *parabducens nucleus*, the other, near the superior colliculi, supposed to control vertical movements. The parabducens nucleus is best regarded as a relay-station on the motor pathway in conjugate lateral movements, and not as a co-ordinating centre. As Bender and Shanzer (1964) point out, it is too vaguely defined anatomically and physiologically to be treated as a centre controlling lateral eye movements. The movements evoked by stimulation of the colliculi usually have a strong vertical component, and it is doubtless for this reason that the hypothetical gaze centre for vertical movements has been sited near these bodies.

### DIRECTION-SPECIFIC MOVEMENTS

A systematic exploration of the brain stem in anaesthetized *encéphale isolé* cats\* by Hyde has revealed a number of regions, in particular the tegmentum of the midbrain, pons and medulla, from which conjugate movement may be evoked on electrical stimulation. It is unwise to describe these regions as 'centres' controlling the execution of eye movements, however, since they may simply contain the pathways from the cortex to the eye-muscle nuclei. Nevertheless it must be appreciated that, in lower animals at any rate, there are regions of the brain stem in which specific types of movements are organized; these are called by Jung and Hassler *direction-specific movements*, and they cause the animal to turn in a characteristic direction. Thus, stimulation of the interstitial nucleus of Cajal, which is close to the nucleus of N III, causes a rotation of the head about its longitudinal axis as illustrated by Figure 17.51.

**Fig. 17.51**   Direction-specific types of movement resulting from electrical stimulation in the region of the interstitial nucleus of Cajal by means of electrodes implanted in the otherwise intact brain. (Hassler and Hess, *Arch. Psychiat.*)

Associated with the turning movement there is an ocular torsion with contralateral upward deviation of the eyes; and this is consistent with the projection from the nucleus to the motor neurone groups.

## RETICULAR FORMATION

This part of the brainstem, extending from medulla to midbrain, contains coordinating centres for bodily movements; because of the polysynaptic pathways taken through it, activities involving the reticular formation are especially susceptible to anaesthesia.

### PONTINE NEURONES

An obvious involvement of pontine reticular neurones is shown by their responses to accelerations of the head or postural tilts; the study of Duensing and Schaefer (1958) revealed neurones that behaved similarly to vestibular nuclear neurones, showing increased discharge with the fast component of a nystagmus and decreased activity or complete inhibition in the slow phase. Since direct connexions between the primary afferent vestibular nerves with the reticular formation are rare, it is clear that these reticular neurones are being activated by vestibular nucleus neurones. According to Duensing and Schaefer, the rhythmic changes in activity of the reticular neurones were analogous with those taking place in rhythmically discharging respiratory neurones of two types, governing inspiration and expiration, the analogous neurones of the eye muscles being those activating the slow and fast phases of the nystagmus.

### RETICULAR STIMULATION

Cohen and Komatsuzaki (1972) stimulated the pontine reticular formation in waking monkeys, and the interesting finding was that the eye movement was not the quick saccade so commonly the result of stimulating brain centres but a constant-velocity movement. Mere activation of the oculo-motor neurones or the medial longitudinal fasciculus causes a saccade, with the velocity passing through a maximum; therefore it seemed that this part of the brain was carrying out some integration so that the discharge to the motor neurones at any point in the movement was a function of the

\* This preparation is obtained by a section of the brainstem at about the level of C1, leaving intact the nucleus of N V; in this condition the animal has many of the features of the waking state by contrast with the *cerveau isolé* preparation, where the section is higher, at the collicular level.

number of impulses so far discharged.* The movement finished with a period of fixation, the position finally achieved being that which occurs naturally lasting up to several seconds and never less than 100 msec. Thus this region of the posterior pontine reticular formation, close to the abducens nucleus, may represent the supra-nuclear common path for horizontal gaze. When the region in which 'pausing units' were recorded was stimulated *vide infra* there was complete inhibition of further voluntary saccades although the slow phase of vestibular nystagmus was not affected (Keller, 1974).

### RECORDING FROM RETICULAR UNITS

In alert monkeys Cohen and Feldman (1968) recorded potentials in the paramedian zone of the pontine reticular formation that led to rapid eye movements with a delay of some 10 to 20 msec; the eye movements were typical saccades, or fast phases of nystagmus. Interestingly, the same potentials were reflected in potential changes in the lateral geniculate nucleus, and the authors suggested that the same neurones that were commanding eye movements through their descending pathway also sent ascending messages through the mid-brain reticular formation to the lateral geniculate body in a feedback on to the visual input. Of particular interest are the studies of Luschei and Fuchs (1972) and Keller (1974) on alert monkeys with electrodes implanted in the dorsomedial pontine reticular forma-tion close to the abducens (N VI) nucleus. The onset, duration and intensity of the burst firing of these units were tightly correlated with the direction, duration and velocity of saccades, and typically began before the burst of firing in the abducens motor neurones. Thus, unlike activity in frontal eye fields, the unit activity definitely *preceded* the motorneuronal discharge. Altogether four main categories of units were described by Luschei and Fuchs (1972) as follows:

1. *Burst-tonic.* Giving a burst discharge related to a saccade and a tonic discharge related to the steady direction of gaze.
2. *Burst-unit.* Burst related to saccade; usually no tonic activity and if so unrelated to eye position.
3. *Tonic.* Tonic discharge related to eye position. No burst on saccade.
4. *Pausing.* Discharging tonically but pausing when movement occurs in one or several directions. The pause preceded the movement and lasted as long as the saccade.

The lead, defined as the time between first spike of a burst and the beginning of eye movement, varied from as long as 80 msec to only 8 to 9 msec. In some units the burst followed the beginning of the saccade. Keller's (1974) study revealed the same types of unit; however,

he divided the pausing units into two categories, one type clustered just anterior to the rostral pole of N VI, paused with ipsilateral saccades.

In summarizing the results of their study on alert monkeys, Cohen and Henn (1972) concluded that the pontine reticular formation probably represented the most peripheral station capable of generating the orders to the motor neurones requisite for the execution of a saccade and maintenance of the subsequent positions of fixation.

### LESIONS

Lesions in the pontine reticular formation lead to paralysis or paresis of horizontal gaze towards the side ipsilateral to the lesion; stimulation of the labyrinth, designed to cause lateral movements, resulted in a failure to cross the midline (Cohen and Komatsuzaki, 1968).

## INTERSTITIAL NUCLEUS OF CAJAL

The neurones of this brainstem nucleus end on all motor neurones governing the eye muscles except those activating the medial and lateral recti, and electrical stimulation causes conjugate vertical and rolling move-ments. By way of the MLF it connects to the vestibular nuclei and so is able to influence vestibularly evoked eye movements. Markham *et al.* (1966), working on the cat, obtained ipsilateral inhibition of Type I vestibular neurones (p. 444) and activation of Type II neurones. These actions are consistent with a projection from the interstitial nucleus to the medial vestibular nucleus, no other midbrain structure projecting to the vestibular complex. The effect is primarily on the horizontal canals and it is suggested that the interstitial nucleus tends to inhibit lateral canal activity when the vertical canals are active. In the *encéphale isolé* cat, Hyde and Eason (1959) observed a slow movement on electrical stimulation of the nucleus; it was accompanied by rolling. There was no correlation between the initial velocity and the distance to travel.

---

*Keller (1974), during his study of single units of the medial pontine reticular formation in the waking monkey, stimulated the sites of recording and found a single type of movement, whatever the type of unit—burst, tonic, etc. from which he recorded; this was the constant-velocity type described by Cohen and Komatsuzaki, the eyes accelerating smoothly through a transient phase to reach a constant velocity maintained as long as the stimulation, the velocity varying linearly with frequency of stimulation up to 500/sec. He considered that the hypothesis of an integration process during execution of the movement was an oversimplification. Thus the transient phase of acceleration is equivalent to the initial phase of a pursuit movement in which more force is developed than that required to maintain fixation during the subsequent phase, and the two phases are probably pre-programmed.

## NUCLEUS OF DARKSCHEWITSCH

This nucleus has been invoked as a step in the motor pathway from the superior colliculus to the motor neurones, since there is no direct projection; the work of Carpenter et al. (1970), involving discrete lesions, confirmed the connections since these produced lesions in the oculomotor complex bilaterally. Szentagothai and Schab (1956) obtained inhibition of vestibularly evoked eye reflexes by stimulation of the nucleus; this was accompanied by protrusion of the eye-ball suggesting inhibition of tonic contraction of the eye muscles,★ and a similar effect has been described by Scheibel et al. (1961). Carpenter (1972) made lesions in the nucleus that spared the interstitial nucleus of Cajal, and degenerating fibres were found in the interstitial nucleus and the oculomotor complex, so that the pathway may well be through the interstitial nucleus; however, the abducens nucleus showed no degeneration, and the animal showed normal lateral movements with impaired upward movements.

## NYSTAGMUS CENTRES

### THREE-NEURONE ARC

There is by no means any certainty as to the existence of one or more specific centres controlling nystagmus, either optokinetic or vestibular, or both. According to de Kleyn, the vestibular nystagmus may be elicited in a rabbit from which the cerebral hemispheres and cerebellum have been removed, and the brain stem has been sectioned just anterior to the abducens nucleus, so that the nuclei of N III and IV are out of action, whilst a further section through the medulla, just behind the vestibular nuclei, reduces the amount of the brain involved practically to the vestibular nuclei. Under these conditions the lateral rectus exhibits a slow contraction and quick relaxation on stimulating the labyrinth. For vestibular nystagmus to occur, therefore, little more than a three-neurone arc, involving the vestibular nucleus, seems to be necessary.

### RETICULAR FORMATION

Other studies have suggested that the control over the quick phase of the nystagmus is exerted through a more complex pathway than the simple three-neurone arc that may be all that is needed for the slow, or vestibulo-ocular reflex, phase; and the reticular formation has been invoked as a necessary region for eliciting the fast phase and possibly organizing the whole mechanism of vestibular nystagmus. Thus Lorente de Nó found that lesions in the reticular formation abolished the quick phase, but Spiegel was unable to confirm this, obtaining a vestibular nystagmus with large lesions of the reticular

formation so long as the vestibular nuclei and medial longitudinal fasciculus were spared.

## VESTIBULAR INTERACTION

Spiegel suggested that the nystagmus might result from mutual interaction between the contralateral halves of the vestibular nuclei controlling the eye-muscle motor neurones in such a way that the one half dominated the activities of the eye muscles in one phase, only to be dominated by the other half in the other phase. Clearly a study of the electrical events in the motor neurones and vestibular neurones concerned in a given form of nystagmus, e.g. horizontal, would provide information on this point. Maeda et al. (1972) carried out such a study on horizontal nystagmus and Baker and Berthoz (1974) on a rotatory nystagmus involving the oblique muscles. These studies tended to substantiate the vestibular interaction hypothesis.

### MOTOR NEURONE DISCHARGES

Maeda et al. (1972) evoked a horizontal nystagmus in cats by repetitive stimulation of one vestibular nerve, and measured activities in medial and lateral recti neurones of both eyes; if the right vestibular nerve was stimulated, there was a slow deviation to the left associated with increasing discharge in the left lateral rectus and right medial rectus; this discharge was terminated abruptly with the onset of the quick movement to the right associated with a quick burst of activity in the antagonist eye muscles namely the right lateral rectus and left medial rectus (Fig. 17.52).

## EPSP and IPSP

Intracellular records of the motor neurones supplying antagonistic pairs, e.g. left lateral rectus and right lateral rectus, during a deviation indicated that activation of the agonist during the slow movement was achieved by development of a slow depolarizing EPSP, generating spikes, followed by a quick hyperpolarizing potential (IPSP) that suppressed the spikes. Thus electrical stimulation of the contralateral vestibular nerve had this effect on an abducens motor neurone. When the ipsilateral vestibular nerve was stimulated, the same abducens motor neurone was only involved in the quick phase and the rhythmic response was a quick depolarization, generating spikes associated with a quick phase, and this was followed by a slow progression in a hyper-

---

★ Szentagothai and Schab considered that reciprocal inhibition of eye motor neurones was mediated through the nucleus of Darkschewitsch rather than inhibitory vestibular nucleus neurones; there is no doubt, however, that neurones in the vestibular nucleus, activated directly by the labyrinth, mediate this type of inhibition.

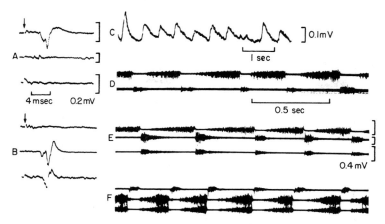

**Fig. 17.52**  *A* and *B*: action potentials of motor nerves innervating left lateral rectus (top), left medial rectus (middle), and right lateral rectus muscle (bottom) in response to a single shock to right (*A*) and left (*B*) vestibular nerve. Arrow indicates the time of stimulus. Calibration and time scale for *A* apply to *B*. *C*: electrooculogram recorded during high-frequency (400/sec) stimulation of right vestibular nerve. Upward deflection indicates ocular deviation to right side. *D*: impulses of motor nerves innervating left lateral rectus (top) and left medial rectus (bottom) during right vestibular nerve stimulation. *E*, *F*: same arrangements of records as in *A* and *B*, but activities induced after repetitive stimulation of right (*E*) and left (*F*) vestibular nerve. Time scale for *D* applies to *E* and *F*, and calibration for *E* applies to *D* and *F*. (Maeda *et al.*, *J. Neurophysiol.*)

polarizing direction, indicating inhibition, while the slow phase was being executed through the antagonist.

## Synchrony

The striking feature was the synchrony between the quick depolarization of one motor neurone and the quick hyperpolarization of the contralateral motor neurone, suggesting the simultaneous activation of excitatory and inhibitory neurones in the vestibular nucleus controlling the rhythmic movement. Maeda *et al.* considered that the delay between cessation of abducens discharge on one side and initiation of a discharge on the other was due to the time required for the quick depolarization to reach firing threshold i.e. to initiate spikes in the motor neurone. Thus an inhibitory hyperpolarization acts quickly to suppress spike activity whereas an excitatory depolarization takes time to build up to firing level. Thus the asynchrony of impulse initiation and suppression in antagonistic pairs, revealed in this and other studies, is not to be attributed to delays elapsing along central pathways in the reticular formation as had been postulated by Lorente de Nó (1938). So far as the mechanism of the rhythmic activities induced in agonists and antagonists is concerned Maeda *et al.* (1971) observed rhythmic discharges in vestibular nucleus neurones in phase with homolateral abducens nerve discharges when the vestibular nerve was stimulated, i.e. when a nystagmus was induced. The activity in the

contralateral abducens nerve had a discharge pattern just the reverse. This behaviour is consistent with the physiological evidence that excitatory and inhibitory fibres projecting to the abducens motor nucleus originate in the contralateral and ipsilateral vestibular nuclei respectively (Baker *et al.* 1969).

### MUTUAL INHIBITION OF NUCLEI

Maeda *et al.* suggested that the vestibular nucleus was itself organizing the rhymic motor discharges, a rhythm that might be achieved by a process of mutual inhibition of the pairs of vestibular nuclei; a commissural inhibitory pathway between vestibular neurones is well established. Such a rhythm, resulting from mutual inhibition, had been postulated long ago by Graham Brown (1911) to describe the generation of rhythmic contraction and relaxation of walking muscles during progression.

### ROLLING NYSTAGMUS

Baker and Berthoz (1974) examined the rotation nystagmus caused by unilateral section of the anterior and posterior canal nerves; the subsequent nystagmus required the intactness of the contralateral labyrinth, and was clearly organized through this structure and was due, as we have argued, to an imbalance between the tonic discharges from both sides. The rolling nystagmus had vertical and horizontal components and the muscles examined were the obliques and their motor nuclei N III (inferior oblique) and N IV (trochlear, superior oblique). The general basis for the experiment is illustrated in Figure 17.53. The fast phase of the nystagmus was directed to the side of the lesion, so that

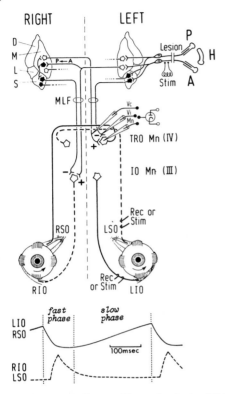

**Fig. 17.53** Schematic diagram illustrating basis of Baker and Berthoz' experiments. Spontaneous vestibular nystagmus is produced by a lesion of the left anterior (A) and posterior (P) canal nerves. The eye movement is rotary with the fast phase (arrows) directed to the side of the lesion. Nerve activities reflecting the fast phase are found in the left superior oblique (LSO) and right inferior oblique (RIO) (dotted lines) and those showing the slow phase are seen in the left inferior oblique (LIO) and right superior oblique (RSO) records (solid lines). Extra- or intracellular records can be obtained from TRO or IO motoneurons and the axons of contralateral, excitatory (Vc) or ipsilateral, inhibitory (Vi) vestibular neurons (solid filled neurons are inhibitory). The left A canal supplies the Vi and Vc neurons projecting to LTRO and RIO motoneurons. A similar set of vestibular neurons are indicated in the right vestibular nucleus but their terminal projection on LIO and RTRO motoneurons is not completed. The left P canal supplies the Vi and Vc neurons projecting to the LIO and RTRO motoneurons (the right vestibular nucleus connection to RIO and LTRO motoneurons is shown). One excitatory vestibular neuron in the left vestibular nucleus is shown to give a collateral to an inhibitory vestibular interneuron in the right medial vestibular nucleus. The lower diagram indicates responses recorded in the indicated oblique muscles during spontaneous nystagmus. Other abbreviations are: M, medial; L, lateral; S, superior; D, descending; H, horizontal; and MLF, medial longitudinal fasciculus. (Baker and Berthoz, *J. Neurophysiol.*)

in this particular case where the lesion was in the left labyrinth, the activities governing the fast phase of contraction were in the left superior oblique (LSO) and the right inferior oblique (RIO), these being the linked pair that would execute the conjugate anticlockwise rotation illustrated in Figure 17.53. The slow phase would be associated with activities in the left inferior oblique (LIO) and right superior oblique (RSO). Thus with this experimental set-up the RIO and LIO are antagonists and fast activity will be found in RIO followed by slow activity in LIO. Figure 17.54 illustrates schematically the neuronal events during a cycle. The spike discharge of the motor neurone is indicated in heavy black, whilst the correlative changes in membrane potential of the motor neurones are indicated by the lines

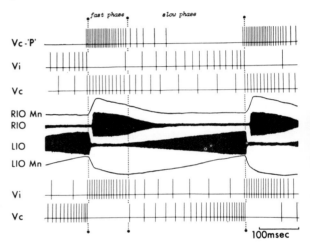

**Fig. 17.54** Summary diagram showing motoneuron, vestibular neuron, and oblique nerve activity during vestibular nystagmus. The left vestibular nerve is presumed lesioned, thus producing a spontaneous nystagmus with the slow phase in the LIO nerve and the fast phase in the RIO nerve. Below the LIO nerve activity is shown the correlative changes in membrane potential (without spike production) in a LIO motoneuron. The activity in the inhibitory (Vi) and excitatory (Vc) vestibular neurons projecting to this LIO motoneuron is shown in the lower part of the diagram. The membrane potential changes in a RIO motoneuron which participates in the fast phase are shown just above the trace depicting the RIO nerve activity. The corresponding activities of the inhibitory (Vi) and excitatory (Vc) vestibular neurons are shown in the uppermost part of the figure. Two classes of excitatory neuron activity are indicated in the diagram: Vc indicates monosynaptic (vestibular neuron) and Vc-'P' polysynaptic activation following stimulation of the left (contralateral) vestibular nerve. The discharge frequencies of the vestibular neurons were chosen as representative from experiments in which the intracellular recorded membrane potential changes and nerve activity were similar to those depicted in the diagram. The two vertical dotted lines on the left are at a 100-msec interval and indicate the approximate duration of the *fast phase* part of the nystagmic cycle. The interval labelled *slow phase* comprises about 300 msec of the nystagmic cycle. (Baker and Berthoz, *J. Neurophysiol.*)

marked *RIO Mn* and *LIO Mn* respectively. The associated discharges in vestibular excitatory and inhibitory neurones, making monosynaptic connexions with the respective motor neurones, are indicated as vertical spikes ($V_c$ and $V_i$).

### Reciprocal activities

The reciprocal activities in the agonist and antagonist motor neurones are clearly evident, as well as the association of motor neurone discharge with the development of motor neurone depolarization that leads to spike activity, whilst the abrupt cessation of discharge in the antagonist LIO is achieved by a hyperpolarizing potential. In this illustration, the LIO is just coming to the end of its slow phase, so that initiation of the fast phase is brought about by simultaneous inhibition of this motor neurone and activation of the RIO. The transition from fast phase to slow is not so sudden, and is associated with a slow repolarization due to combined inhibitory and excitatory input from $V_i$ and $V_c$. The build-up of depolarization in the LIO, which is executing the slow phase, is gradual and reciprocal to the repolarization of the RIO.

### Vestibular interaction

This high degree of reciprocal activation and inhibition of the motor neurones suggests that there is an extensive interaction occurring between the bilateral vestibular nuclei. Baker and Berthoz found nothing in their results inconsistent with Maeda *et al.*'s hypothesis of a vestibular nuclear control centre for nystagmus, and suggested that the undoubted involvement of the reticular formation in organization of eye movements might well, in this instance, be confined to a general facilitation of eye-muscle motor neurone activity, a background upon which the rhythmic vestibular centres play.

## BASIC NYSTAGMIC CIRCUIT

In Figure 17.55 we picture two groups of vestibular nucleus neurones, A and B, controlling the abducens motor nuclei of each side, and, because of their mutual inhibitory action (indicated by the black dots) a rhythmic activation and inhibition of the muscles on each side can be obtained. Thus when Group A is dominant, the ipsilateral lateral rectus is active whilst the contralateral is being inhibited by a direct action on the motor neurones, and less directly through an inhibition of vestibular neurones in B; the inhibition of B also helps the Side I to be dominant since it disinhibits the A group. As activity on the I side diminishes, through say, refractoriness, then Side II tends to gain the ascendency since inhibition of B and D is being reduced,

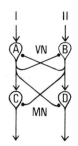

**Fig. 17.55** Simplified schematic representation of the organization of the vestibulo-ocular pathway. I and II, labyrinthine input on each side. A and B, population of neurones in the vestibular nuclei (VN) on each side. C and D, abducens motorneurone (MN) on each side. Excitatory synapse is indicated by arrow terminal and inhibitory synapse by filled circle. In the actual organization A and D, likewise B and C, are on the same side, respectively. A and B contain both excitatory and inhibitory neurones, not individually shown. (Shimazu, *Progr. Brain Res.*)

and any gain in B's activity immediately helps to suppress activity on the I side. Thus the system is delicately poised so that a small change in excitability on one side can be reflected in a positive feedback that magnifies the change and allows the other side to gain the ascendency.

## CENTRAL NYSTAGMUS

Electrical stimulation of the parieto-temporal region of the guinea pig and the temporo-occipital zone of the rabbit (Fig. 17.56) induces a central nystagmus, with the quick phase directed to the opposite side; when the stimulus is over, the nystagmus continues for several seconds as an after-nystagmus in the same direction. The motor pathway uses the same circuitry as that involved in vestibular nystagmus, including the vestibular nuclei, although these are not necessary (Manni and Giretti, 1968).

### SUPERIOR COLLICULI

Fibres from the centre apparently pass through the superior colliculi, so that stimulation of these bodies

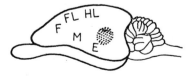

**Fig. 17.56** Lateral view of the cerebral cortex of the rabbit. The nystagmogenic centre is located inside the dotted area. The other motor foci are indicated as follows: F, face; FL, foreleg; HL, hindleg; M, mastication; E, ear. (Manni, Azzena and Desole, *Arch. ital. Biol.*)

evokes a central nystagmus. By using the ponto-mesencephalic preparation illustrated in Figure 17.57, which contains the nuclei of N III, IV and VI, part of the trigeminal nuclear complex, and part of the reticular formation, but lacks the cerebellum and vestibular nuclei, Manni and Giretti (1970) showed that the nystagmus evoked by collicular stimulation was organized by the reticular formation lying between and including the oculomotor and abducens nuclei. When the pontine reticular formation was stimulated in localized regions the most consistent effect was a slow ipsiversive conjugate movement which might lead to a nystagmus, the fast movement being contraversive (Manni *et al.* 1974).

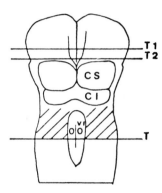

**Fig. 17.57**  A schematic dorsal view of the brain stem of the guinea pig, after removal of the cerebellum, showing the superior (CS) and the inferior (CI) colliculi, the transections (T, T1, T2) delimiting the pontomesencephalic preparation and the lesion on the floor of the fourth ventricle (hatched area) in order to destroy the superior vestibular nuclei and the cranial part of the medial and lateral vestibular nuclei. T is just caudal to the abducens nuclei. T1 is at the level of the posterior thalamus and rostral to the posterior commissure region and to the pretectum. T2 is just precollicular, rostral to the oculomotor nuclei and involves the posterior commissure and the pretectum. (Manni and Giretti, *Exp. Neurol.*)

## THE CEREBELLUM

This region of the brain is off the main ascending and descending pathways from cord to midbrain and cortex, and may be regarded essentially as a regulating device for the smooth execution of the motor activity of the organism. It may be described as a 'feed-back centre', in the sense that it receives information from all regions of the body and from the parts of the brain directly concerned in motor activities; because of its reciprocal connections with these parts, including the regions of the cortex concerned with voluntary movements, it is able to modify their activities continuously in the light of this information. The control of the eye movements, both

reflex and voluntary, therefore involves activity of this part of the brain. The direct connections of the flocculo-nodular lobe with the vestibular nuclei in the medulla suggest a very intimate connection between vestibularly evoked eye reflexes and this, the so-called paleo-cerebellum.

## AFFERENT INPUT FROM EYE MUSCLES

We have seen that the extraocular muscles are well provided with sensory receptors—spindles and comparable bodies—capable of providing information as to the state of tension of their fibres. This information is not utilized directly in a stretch-reflex, and instead is passed to higher integrating centres (Keller and Robinson, 1971). Fuchs and Kornhuber (1969) stretched extraocular muscles of the cat and recorded evoked potentials in the vermian folia of Lobulus V, VI and VII; these had a short latency of some 4 msec, and the location of the evoked responses corresponds with the regions of projection of the visual and auditory cerebral cortex, as well as homologous projections from Area 8 as found in the monkey. Using the more precise techniques of stimulation of N IV, Baker *et al.* (1972) obtained short-latency responses in the cerebellar cortex of the same regions, which were probably carried by mossy fibres; longer-latency responses probably corresponded with impulses carried by the climbing fibres of the olivocerebellar pathway.

## ELECTRICAL STIMULATION

### CAT

Electrical stimulation of the cerebellum has long been known to cause contraction of muscles, organized on a somatotopic basis; Cohen *et al.* (1965) obtained a variety of patterned eye movements in the alert cat, the character depending on the locus stimulated. For example straight up-and-down movements were elicited by stimuli on the midline of the vermis of the anterior and posterior lobes, horizontal movements from the tuber vermis and fastigial nucleus. The responses were very reminiscent of vestibular stimulation.

### MONKEY

In the alert monkey Ron and Robinson (1973) obtained direction-specific saccades by stimulation of vermal and hemispheric regions, as indicated in Figure 17.58, where it will be seen that there were considerable areas that failed to provoke eye movements in contrast to much of the earlier work where it seemed that almost any point on the cerebellum was capable of evoking a movement. As the Figure also indicates, smooth movements

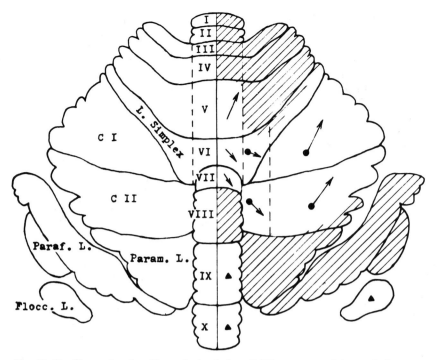

**Fig. 17.58** Illustrating the effects of stimulation of different parts of the cerebellum, so far as eye movements are concerned. Arrows indicate lobes where stimulation elicited saccades, the angle of the saccade above or below the horizontal being indicated by the tilt of the arrow. ● indicates lobes where stimulation evoked smooth movements. A dot and an arrow indicate a structure where stimulation evoked both saccades and smooth movements. The direction of the smooth movements was the same as that of the saccade (the direction of the arrow). ▲ indicates lobes where stimulation evoked nystagmus; shaded areas indicate structures where stimulation did not evoke eye movements. (Ron and Robinson, *J. Neurophysiol.*)

were evoked, the velocity increasing with stimulus current, frequency, and pulse-width up to as high as 150°/sec, according to the site stimulated. With the saccade, the response was independent of the visual field, being the same in complete darkness or with a textured field; by contrast, a textured field exerted a strong inhibitory action on the evoked smooth movement.

**Nystagmus**

Stimulation of the vestibulocerebellum, i.e. the flocculus, uvula and nodulus, evoked a nystagmus; this is shown in Figure 17.59C, where a saccade (A) and a smooth movement interrupted by a saccade, are also shown; the difference between the true nystagmus and the slow movement interrupted by a saccade is obvious, especially as the saccade in B is interrupted by a saccade in the same direction as the slow movement.

**Types of movement**

As Ron and Robinson emphasize, the important feature of an evoked eye movement, when control mechanisms

are considered, is not the direction but the type; thus there are essentially four semi-independent control systems, saccadic, pursuit, vergence and vestibular. Classification on this basis allowed a clear separation of three regions, each involved in a separate task, namely the vestibulocerebellum, the vermal regions of lobules V, VI and VII, giving pure saccades, and the paravermal and hemispheric Crus I and II giving smooth and saccadic movements, both in the same direction. Within these regions all directions of eye movement are included.

## UNIT ACTIVITY DURING MOVEMENTS

Wolfe (1971) implanted electrodes in the cerebellum and studied activity in the waking animal during saccades; an accurate analysis of the time-relations of the field potentials and saccades indicated that these potentials occurred some 25 to 35 msec before the initiation of the eye movement. Once again the activity was confined to a narrow vermal band of folia VI and VII.

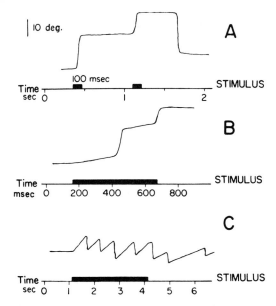

**Fig. 17.59**  A, examples of saccades; B, smooth movements mixed with saccades; C, nystagmus, evoked by cerebellar stimulation. Only the horizontal component of the eye movements is shown. Horizontal bars indicate periods of stimulation. (Ron and Robinson, *J. Neurophysiol.*)

## CEREBELLAR LESIONS

### DYSMETRIA

Lesions of the cerebellar regions associated with oculomotor activity were made by Ritchie (1975), and in confirmation of earlier work he found that the only obvious defect was a dysmetria in saccades, in the sense that they resulted in either overshoot or undershoot. The interesting feature of the defect was its dependence on the original position of the eyes, so that the hypermetria was greatest if movement began from an extreme eccentric position; by contrast, if the movement began from the primary position there was hypometria. Moreover, the lesions caused similar errors when the visual stimulus to the saccade was removed. Thus it is unlikely that the normal cerebellum was processing visual information and thereby modifying the motor activity in the light of this information. It is more likely that, in some way, the cerebellum is utilizing proprioceptive information from the extraocular muscles that permits a more accurate match of the burst-frequency and duration with the required eye movement, but it seems clear that the visual information has already been encoded.

### MAN

Pathologically in humans the oculo-motor errors are similar; as described by Cogan (1954), the ocular

hypermetria is greatest when moving from eccentric points to the primary position. Such dysmetria leads to successive corrective saccades until fixation is achieved, thereby giving rise to oscillatory movements; a similar oscillation was observed by Ritchie.

## CEREBELLAR MECHANISM

### INTEGRATOR

It is beyond the scope of this book to discuss in detail the mechanism of cerebellar control over reflex and voluntary movements, the more so as the subject is by no means clear. Thus, in a nystagmus, the discharges of vestibular neurones, which send their commands directly to the motor neurones, are proportional to the desired eye-velocity, but the discharge in the motor neurones is proportional to the eye-position; clearly some integrating curcuit is necessary for this conversion and Carpenter's (1972) study on cerebellectomy is compatible with removal of an integrating centre of this type. However, such an integrator would be passing commands directly to the motor neurones and therefore cause a sustained deviation, not a nystagmus, whereas stimulation of the vestibular cerebellum causes nystagmus. Thus the integrator would have to be central to the vestibulocerebellum.

### COMPARISON WITH COLLICULUS

An interesting difference between the effects of cerebellar and collicular stimulation, emphasized by Ron and Robinson, is that in the colliculus the direction *and* magnitude of a saccade are coded by position (Fig. 17.21, p. 422) whereas in the cerebellum it is only the direction, the magnitude being governed by intensity of stimulation, i.e. by the number of cerebellar neurones activated.

### PULSE GENERATOR

The authors suggest that the saccade is evoked by activation of a 'pulse generator', which produces a burst of discharge in the motor neurones of pre-programmed frequency and duration; hence continuous stimulation of the pulse-generating region would not produce a saccade of increasing magnitude but a series of pulses leading to successive saccades in the same direction, as found with collicular (p. 419) and cerebellar stimulation. The role of the cerebellum in this pulse generation might be to coordinate proprioceptive information, revealing the position of the eyes and head, together with visual and auditory information, a role compatible with the vermis of lobes V to VII since these receive these sensory inputs. Thus this region would integrate information and send out a signal coded with respect to the direction of movement and the intensity of the burst.

## MODIFICATION OF PROGRAMMED MOVEMENT

The limited effects of cerebellectomy on the execution of eye movements, however, make it very unlikely that the cerebellum occupies such a key role in the execution of movements; these are doubtless programmed in other parts of the central nervous system and it is essentially in the *modification* of these, during their execution, that the role of the cerebellum lies (Eccles, 1973). Thus stimulation of the cerebellum must not be regarded as the activation of cerebellar motor centres controlling specific types of movement but rather the activation of centres in close collateral communication with the cerebellum to produce a figurative 'back-firing' of types of movement pre-programmed in the centres—cortical, collicular and vestibular—that are sending their programmes to the cerebellum for correction.

## MOVEMENTS OF HEAD AND EYES

The great majority of studies of eye movements have been carried out with the head fixed, an unnatural situation and one that obscures the coordination that is necessary to establish fixation when both eyes and head participate in the actual gaze-process.

### SACCADE AND SLOW MOVEMENT

Bizzi *et al.* (1971) noted that, when a light was presented to a monkey in its peripheral field of view, the response was quite stereotyped, consisting of a saccade followed by a slow movement, in the opposite direction to the head movement, that compensated for this and maintained fixation. Although the eyes actually moved first, EMG activity began first in the neck muscles and only some 20 msec later in the eye muscles. It appeared that the movements of eye and head were programmed by a central mechanism that works in a stereotyped fashion activating both eye and head muscles nearly simultaneously, the delay in head movement being due to the greater inertia of the head and slower action of the neck muscles. The slow compensatory movement is not programmed, but is a reflex response to the movement of the head. Its reflex nature was demonstrated by

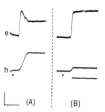

**Fig. 17.60** A, coordinated eye (e) and head (h) response to sudden appearance of a target. B, presentation of the same target is followed by application of the brake, preventing head movement (horizontal bar). Note lack of compensatory eye movement. Dots represent onset of luminous target. Calibration: Horizontal bar equals 500 msec; vertical bar equals 15 degrees. (Bizzi *et al.*, *Science.*)

placing the head in a rotatable frame provided with a brake, so that movement of the head was brought to a halt immediately after or during the saccade; as Figure 17.60 shows, the following movement was inhibited.

### SACCADE DURING HEAD-TURN

Morasso *et al.* (1973) raised the question as to whether, when the head was allowed to move *during* the saccade, so that the goal was reached by a combined saccade and head movement, the saccade would be smaller than if the head had been fixed. Figure 17.61 shows the record of a saccade made with the head fixed (a) and the records of the saccade and head movement when both participated in the gaze-process (b); the Curve G represents the sum of the eye (E) and head (H) displacements, and it is seen that this sum exactly matches the saccade in (a); the head movement reduces the size of the saccade. Experiments indicated that there was no pre-programming, so that if the head movement was suddenly stopped just as the saccade began there was no reduction in the amplitude, suggesting that the reduction was carried out in the light of proprioceptive or vestibular input following the head movement. When the animal was bilaterally vestibulectomized, the saccade-amplitude was the same with head fixed or free, so that there was a remarkable overshoot of gaze; the latency for neck proprioceptive responses was shown to

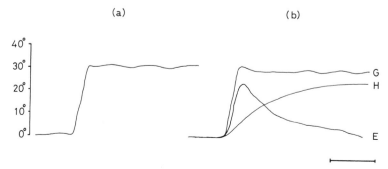

(a)    (b)

**Fig. 17.61** Comparison of saccade with gaze. (a) Eye saccade to a suddenly appearing target with head fixed. (b) Coordinated eye saccade (E) and head movement (H) to the same target with head free. The gaze movement (G) represents the sum of E and H. Note the remarkable similarity of eye saccade in (a) and gaze trajectory in (b) as well as reduced saccade amplitude in (b). Time calibration, 100 msec. (Morasso *et al.*, *Exp. Neurol.*)

be some 50 msec, which was too long to allow neck reflexes to contribute.*

MAN

In man, Gresty (1974) has described head and eye movements during voluntary fixation that lead to comparable stabilization of the gaze; he emphasizes that the head movement is similar to the saccade in being ballistic with an approximately linear relation between peak head velocity and size of target displacement so that the time for executing a 'head saccade' is reasonably constant as with an eye saccade. Both saccades are initiated together and it is only because of the greater inertia of the head that the eyes have nearly fixated a target in the periphery before the head has moved.

RABBIT

Collewijn (1977) employed a technique that permitted the measurement of the absolute position in space of both head and eye at any moment. These experiments revealed a remarkable co-ordination of eye and head movements during normal exploratory behaviour, so that although saccades could be, and often were, initiated before any head movement, the subsequent movement of the head tended to stabilize the direction of gaze or position of the eye in space. In Figure 17.62 records of the positions of head (H) and Eye (E) in space are recorded together with the position of the eye in the head (EH); downward displacements indicate a sideways movement of head or eyes. All the movements of the eyes in space were saccadic and accompanied by a head movement that tended to stabilize the position of the eye in the head. The head movements were mainly smooth, in perfect coordination with the eye movements, and the head was nearly stable in the intervals. Thus, in general, movements of the eye in the head were limited and transient, so that the rabbit, as with other

species, achieves its exploratory and other fixations finally by a changed orientation of the head. The compensatory following movement of the eyes when the head is still moving towards the fixation point, described by Bizzi et al. (1972), was also observed by Collewijn in the rabbit, but the timing may be different so that the compensatory movement of the eye may occur before it has reached the final target, with the result that the eye is still moving in space towards the target, being carried by the head, but it is moving more slowly because, so far as its position in relation to the head is concerned, it is moving backwards. A similar behaviour has been described in man (Bartz, 1966)

## VESTIBULAR AND OPTOKINETIC INTERACTION

We may regard the vestibular and optokinetic reflexes as mechanisms that stabilize the eye-position relative to

*Dichgans et al. (1973) found that this stabilization of the eyes during head movements was just as efficient in the dark as in the light; after labyrinthectomy it was abolished, but there was a remarkable compensation with time, which was retained in spite of de-afferentation of the neck-muscles. In some way, information from the neck-programming centre governing the intended head movement is fed forward to the oculomotor centres. Bizzi et al. (1972) observed, in the monkey, a rather different type of eye-head movement when the event was predicted, as opposed to the response to an unexpected appearance of the target. In this case there was not the ballistic burst of impulses to the eye-muscles but a gradual increase in the agonist's tension and decrease in that of the agonist. The head began to move some 150 to 200 msec before the saccade which was always slower and preceded as well as followed by a compensatory movement in the opposite direction. We may note that the guinea pig's reflex head and eye movements have been studied by Gresty (1975); under swing conditions the guinea pig avoids eye nystagmus, so that stabilization of the visual field relies largely on head movements, which are accompanied by eye movements in the same direction.

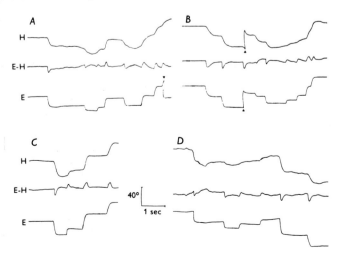

**Fig. 17.62** Examples of spontaneous movements of eye (E) and head (H) in space, and eye in head (E-H). Triangles indicate an automatic resetting of the recording pen. Downward excursion represents clockwise (rightward) rotation for all traces. *A–C*, normal illumination; *D*, darkness. (Collewijn, *J. Physiol.*)

gravity and to the visual surroundings. Maculo-ocular (from utricle and saccule) and neck-proprioceptive reflexes tend to keep the retinal image normally orientated with respect to gravity, whilst the canal-ocular, and the optokinetic reflexes prevent too fast displacements of the image of the external world on the retina. The question of the interaction of the vestibule-based and retina-based reflexes is of obvious interest not only from a functional point of view but also in elucidating the neural mechanisms of control; thus it is of interest to know whether brainstem groups of neurones are involved in both types of reflex or whether the control mechanisms are kept separate until the final common path through the eye-muscle nuclei is reached.

## TORSION-SWING RESPONSES

Baarsma and Collewijn (1974) measured the rabbit's responses to sinusoidal yaw movements on a torsion-swing under conditions where the vestibular response could be examined in isolation (eyes closed), when it co-operated with the optokinetic response (eyes open), and when an artificial situation was created in which the visual environment moved in unison with the swing, so that there was conflict between the vestibular reflex, demanding movement of the eyes away from the direction of movement, and the optokinetic reflex which demanded a stationary position of the eyes, since the visual environment had not moved. The variables measured were the *gain*, which in this case was defined as the amplitude of eye movement/amplitude of swing movement, and the *phase-shift* defined as 0° when the eye movement is shifted 180° with respect to the swing movement and in phase with the visual movement of earth-fixed surroundings, i.e. the phase-shift was zero if there was exact compensation.* The main results are summarized in Figure 17.63; the increase in gain on opening the eyes (B compared with A) is obvious.

### NON-LINEARITY OF VESTIBULAR RESPONSE

An important feature of the purely vestibularly evoked response was its non-linearity with respect to frequency of oscillation, so that at low frequencies, and therefore with low velocities and accelerations, the gain was low, and only at high frequency did it reach a maximum of about 0·8; similarly there was a phase-lead which could be as high as 50° at low velocities but reached zero—perfect compensation—at high frequency. This non-linearity was abolished by opening the eyes with an earth-fixed environment (B of Fig. 17.63) where it is seen that, at low and high frequencies of swing oscillation, there is perfect compensation with respect to both gain and phase.

### VESTIBULECTOMIZED ANIMAL

When the optokinetic system is examined in isolation, i.e. in the vestibulectomized animal, it is found that the gain varies *inversely* with the velocity of movement, in marked contrast to the situation with the normal animal in the dark where the gain increases with velocity; thus the *vestibular system is tuned to operate at high velocities* and the *optokinetic system at low velocities*, so that when they both operate, as in Figure 17.63B, there is a practically constant gain and negligible phase-error. Figure 17.63D illustrates the result with the labyrinthectomized animal; at low swing-frequency, compensation is good but at high frequencies gain is very small.

### ABNORMAL SITUATION

The abnormal stimulus-situation, with swing-fixed visual environment, caused an accentuation of the non-linearities described above for the animal swung in the dark, so that, as Figure 17.63C shows, at low frequency there was virtually no eye movement (gain zero); at the higher frequency gain was improved and phase-lead, which could be as high as 70° at low frequency, was reduced.

## ADAPTIVE CHANGES IN VESTIBULO-OCULAR REFLEX

It is well established that repeated induction of vestibular nystagmus, e.g. by rotation or calorically, leads to a diminution in the amplitude of the response; however, as Gonshor and Melvill Jones[†] have pointed out, such an attentuation, occurring as a result of everyday head movements, would reduce the value of the reflex; and in their experiments, in which human subjects were repeatedly exposed to sinusoidal[‡] rotations at angular velocities that fell within the range expected during normal head movements, there was no attenuation. As a quantitative estimate of the magnitude, or efficacy, of the reflex the gain, $G$, defined as the ratio of peak eye velocity ($\omega_e$) to peak head velocity ($\omega_h$):

$$G = \omega_e \max / \omega_h \max$$

---

*Because the eyes move in the opposite direction to that of the swing, failure of perfect compensation, due to semicircular canal and oculomotor dynamics, is manifest as a phase-*lead*; this amounts in man normally to about +7° (Hixson and Nivon, quoted by Gonshor and Melvill Jones, 1976).

[†] The literature is well summarized by Gonshor and Melvill Jones; Miles and Fuller (1974) caused monkeys to wear telescopic lenses that either reduced or increased the apparent movement of objects in space, and found appropriate changes in gain in the vestibulo-ocular reflex. Thus monkeys that had worn magnifying glasses overcompensated for a head-turn executed in total darkness, so that their eyes lagged progressively behind the target.

[‡] By this is meant a to-and-fro (yaw) oscillation of the body; the peak velocity reached was 40°/sec and the frequency of the oscillation was 1/6 Herz.

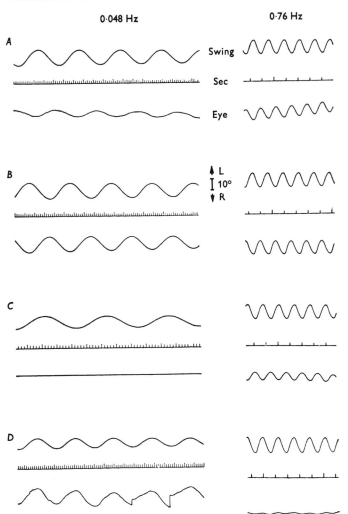

**Fig. 17.63** Examples of compensatory eye movements on the torsion swing, for an amplitude of about 5° and frequencies of 0·048 and 0·76 Hz. *A*: eyes closed. *B*: eyes open, earth-fixed visual surroundings. *C*: eyes open, platform-fixed visual surroundings. *D*: eyes open, earth-fixed surroundings, labyrinthless animal. Calibrations valid for both swing and eye movements, except that polarity of eye movement recording in *D* has been inverted. (Baarsma and Collewijn, *J. Physiol.*)

was employed. In the dark, this normally has a value of 0·7, indicating that the eyes tend to lag behind the head. With eye(s) open, so that the optokinetic reflex can co-operate, the gain becomes about unity, indicating perfect fixation.

REVERSED TRACKING

When subjects were exposed to a reversed optokinetic tracking task, by causing them to view a mirror-image of the environment as the head was rotated, it was found that there was some attenuation of the reflex—decreased gain—when the subjects were subsequently tested in the dark, so that, on average, a reduction of some 23 per cent in gain was achieved by interposing periods of reverse tracking.

When human subjects were exposed to reversed tracking for long periods of time by wearing, during their waking hours, goggles with reversing prisms, the

effects were much more striking. After 7 days the gain fell to a plateau at about 25 per cent of normal, so that when the electro-oculugraphic records of eye movements during rotation were analysed it was difficult to recognize any sinusoidal component in them.

**Reflex reversal**

The more important finding was that there was a phase-change in the eye movements of such a size as to result in a *reflex reversal*, i.e. the vestibular response to a head acceleration was of the opposite character to that normally required, so that the eyes were in fact moving with the direction of rotation (Figure 17.64). This means that the central programming device is able to make use of visual information, translating its false into its true meaning so far as head turning is concerned, and influencing the vestibular reflex mechanism appropriately.

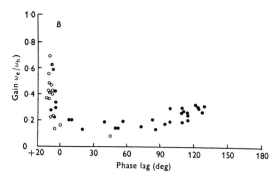

**Fig. 17.64** Changes in vestibulo-ocular reflex gain as a function of phase. Each point gives the mean values of gain and phase for an entire run before (◐) during (●) and after completing (○) the period of maintained reversal of vision. (Gonshor and Melvill Jones, *J. Physiol.*)

### CEREBELLECTOMY

A clue to the mechanism of this adaptive change was provided by Robinson (1976), who showed that it was completely abolished in cats by removal of the vestibular portion of the cerebellum. Thus the cat's gain in the dark was 0·90; after wearing reversing prisms for 8 days the gain decreased by 93 per cent; on removal of the cerebellum the gain rose to 1·17, and now there were no plastic changes with reversing prisms. That visual input can influence activities in cerebellar Purkinje cells is well established, e.g. Maekawa and Simpson (1973).

### SITE OF INTERACTION

According to Ohm, optokinetic and vestibular nystagmus were able to interact in the vestibular nuclei, which were therefore the centre through which both types of nystagmus operated. That the vestibular nuclei are not a necessary stage in the circuitry was suggested by the observation that in cases of streptomycin lesions of the vestibular nuclei, severe enough to abolish caloric nystagmus, optokinetic nystagmus could be elicited normally (Dix, Hallpike and Harrison, 1949). Lachmann, Bergmann and Monnier (1958) discovered a mesencephalic region, stimulation of which would cause a nystagmus; the most sensitive region was near the thalamus and was thus considerably rostral to the vestibular nuclei. By studying the interaction of this 'central nystagmus', evoked by electrical stimulation of the brain, with vestibular nystagmus obtained by mechanical stimulation of the labyrinth, these workers came to the conclusion that the two forms of nystagmus had a common meeting ground in the midbrain, since there was no doubt of their mutual reaction, both subtractive and additive effects being obtained (Bergmann *et al.* 1959). Presumably, then, vestibular

nystagmus is mediated through this mesodiencephalic region or one related to it, although de Kleyn's and other experiments show that this is not necessary. As indicated earlier, the cerebral regions that, when stimulated, give a central nystagmus share a common pathway with vestibular nystagmus, and it seems likely that the mesodiencephalic region of Lachmann *et al.* is closely related to the cortical regions (Manni, Azzena and Atzori, 1965), in which case the locus of interaction is presumably the pontomesencephalic reticular formation.★

### CEREBELLUM

Finally we must note that the abolition of the adaptive change in the vestibulo-ocular reflex, described above, by removal of the vestibular cerebellum suggests that visual and vestibular reflex mechanisms might have a meeting place in the cerebellum, as suggested by Gonsho and Melvill Jones.

## CORTICAL CONTROL OF THE EYE MOVEMENTS

The fixation reflex operates by way of the occipital cortex; that this reflex—although an important component in behaviour—does not dominate the eye movements is clear from the fact that an individual does not respond to every peripheral stimulus by a movement of the eyes; similarly, on turning the head, more often than not the eyes move *with* the head and not against as would happen if purely labyrinthine or proprioceptive impulses determined the eye movements. There is thus an over-riding control over the eye movements capable of inhibiting the more primitive reflex patterns; the centre for this control in man is in the frontal lobe of the cortex.

---

★ When the cortical centre was destroyed the mesodiencephalic centre still gave a nystagmus, so that the effects of stimulation are not merely attributable to the excitation of corticofugal fibres. By the Marchi method, Manni *et al.* (1965) showed that degeneration extended through the internal capsule, cerebral peduncle, the medial and lateral geniculate bodies and superior colliculus, i.e. to structures surrounding the mesodiencephalic centre..Thus this latter centre is a group of relay neurones intercalated in the neural paths connecting the cortical nystagmogenic area with the oculomotor nuclei.

It would seem from the studies of Bergmann and his colleagues (see, for example, Bergmann, Costin and Chaimowitz, 1970) that electrical stimulation of any region of a single visual pathway in the rabbit will cause a nystagmus, and it is interesting that, just as with a unilateral lesion of the labyrinth, so unilateral injury to the lateral geniculate body causes a nystagmus, with the slow phase directed away from the injured side. This nystagmus ceases after a few days, and now injury to the other lateral geniculate body causes a new nystagmus in the opposite direction, the analogue of the 'Bechterew nystagmus'.

## IMPAIRMENT OF VOLUNTARY CONTROL

### DOMINANCE OF FIXATION REFLEX

The importance of the voluntary frontal centre is revealed by the defects in patients suffering from lesions in this area or, more usually, in the projection fibres in the internal capsule or anterior part of the midbrain. As described by Holmes, the patient cannot move his eyes in response to a command although an obvious effort is made to obey, accompanied often by great distress. There is no paralysis of eye movements since a slowly moving object may be followed; if the movement is at all rapid, however, following is unsuccessful. Reading may be carried out slowly provided that the letters are not too far apart. When the eyes are deviated by following a moving point, great difficulty may be experienced in bringing them back to their primary position. When the head moves, the eyes generally move in the opposite direction. Clearly the oculo-motor behaviour of the patient is now determined by reflex patterns—primarily the fixation reflex. The following movements are responses of the eyes to a movement of the image off the fovea; if the point moves too quickly the eyes become the prey to fixation reflexes initiated from other points in the visual field and so cease to follow—the following movements in the normal human subject therefore represent the combined effects of voluntary control and the fixation reflex; the voluntary centre inhibits other reflexes which, if permitted to operate, would prevent the full execution of the primary intention, and it is thus only when this centre is unimpaired that true purposive movements of the eyes can be carried out.

### VESTIBULAR REFLEXES

The powerful nature of the fixation reflex is revealed in lesions of the sort described; the caloric or galvanic nystagmus is a labyrinthine reflex regularly evoked in the normal subject; in these patients, however, such responses can only be elicited when the patient is in the dark—in the light, the fixation reflex, operating without any inhibitory control from the frontal centre, can and does dominate oculo-motor behaviour to the extent of inhibiting vestibular reflexes. Again, when the patient has fixated an object, great difficulty is experienced in removing the gaze from it and many expedients are used to achieve this, such as jerking the head or covering the eyes with the hand.

### READING

The fact that a patient can follow a row of letters in reading indicates that the fixation reflex implies not only a visual stimulus but also interest and attention; after reading the first group of letters interest in them ceases and the occipital centre is ready to respond to the letters in the peripheral field; in the normal subject this response is immediate since the primary reflex response to the first group is readily inhibited by the frontal centre; in the sufferer from a frontal lesion the new response is slow and is only brought about if the new stimulus falls close to the fovea. That the patient reads in the right direction shows the importance of attention and interest even when the voluntary centre is not active.

## IMPAIRMENT OF FIXATION REFLEX

A disturbance in the fixation reflex in which vision is unimpaired is rare; however, cases of lesions of the corticotectal tract—the presumed efferent pathway in the reflex—have been described. In these patients the movements of fixation are normal but the maintenance of fixation is defective, especially if strong deviation of the eyes is necessary. If the eyes are in their position of rest—and fixation is most easily maintained in this position—the voluntary effort is quite adequate to sustain normal fixation; with the eyes deviated, fixation requires a more complex and continuous muscular effort and it is then that the fixation reflex becomes important—the automatic pilot which keeps the plane on its course while the pilot attends to other things. When the head, or the object of fixation is moving, maintenance of fixation becomes likewise a complex act and the patient fails to follow a moving object and complains of being unable to 'focus' objects while he is moving, as in a vehicle. The phenomena of defects in the fixation reflex thus fit in well with the general conception of its function and relation to the voluntary frontal centre.

## ALLOCATION OF CORTICAL CENTRES

Robinson (1968) has suggested that the saccade, pursuit movement and vergence systems are controlled by different cortical regions; he inclines to allocate areas 17 to 19 to the smooth pursuit system, since smooth movements are evoked by stimulation here, but since most investigators have studied the anaesthetized animal, in which all responses are smooth, further work on the unanaesthetized animal is required. Saccades can be evoked by stimulation of area 8, and only saccades, whilst stimulation of areas 19 and 22 produces vergence movements in the lightly anaesthetized monkey (Jampel, 1959).*

---

*The extensive studies of Pasik and Pasik (1964) on cortical ablation and stimulation have led them to be sceptical about cortical motor centres; they conclude: 'We move the eyes with our whole brain. It appears as if oculomotor function is initiated in the entire cortex and uses numerous corticifugal systems that descend and converge toward the tegmentum of the brainstem.'

## OTHER REFLEX RESPONSES

The abrupt turning of the eyes to an unexpected sound—the *acoustic reflex*—or towards a stimulus applied to the body, primitive as it may appear, involves integrations of a high order since spatial localization enters into the response; it would seem that the frontal cortex is involved (i.e. that this part of the brain is not exclusively concerned with voluntary acts). In other movements implying judgment of the nature of the stimulus, such as blinking and aversion of the eyes as a result of a threatening gesture, much wider regions of the cortex are concerned, and the pulvinar of the thalamus— phylogenetically related to the lateral geniculate body— is regarded as a possible 'lower visual centre' intimately associated with these and related responses, since it has close connections with the precuneus, angular gyrus, parieto-occipital and occipito-temporal lobes.

## REFERENCES

Abrahams, V. C. (1972) Neck muscle proprioceptors and a role of the cerebral cortex in postural reflexes in primates. *Rev. canad. Biol.* **31**, Suppl. 115–130.

Abrahams, V. C. & Rose, P. K. (1975) Projections of extraocular, neck muscle, and retinal afferents to superior colliculus in the cat: their connections to cells of origin of tectospinal tract. *J. Neurophysiol.* **38**, 10–18.

Altman, J. & Carpenter, M. B. (1961) Fiber projections of the superior colliculus in the cat. *J. comp. Neurol.* **116**, 157–177.

Apter, J. T. (1945) Projection of the retina on superior colliculus of cats. *J. Neurophysiol.* **8**, 123–134.

Apter, J. T. (1946) Eye movements following strychninization of the superior colliculus of cats. *J. Neurophysiol.* **9**, 73–86.

Baarsma, E. A. & Collewijn, H. (1974) Vestibulo-ocular and optokinetic reaction to rotation and their interaction in the rabbit. *J. Physiol.* **238**, 603–625.

Baarsma, E. A. & Collewijn, H. (1975) Eye movements due to linear accelerations in the rabbit. *J. Physiol.* **245**, 227–247.

Bach-y-Rita, P. & Ito, F. (1966) Properties of stretch receptors in cat extraocular muscles. *J. Physiol.* **186**, 663–688.

Bach-y-Rita, P. & Lennerstrand, G. (1975) Absence of polyneuronal innervation in cat extraocular muscles. *J. Physiol.* **244**, 613–624.

Baker, R. & Berthoz, A. (1974) Organization of vestibular nystagmus in oblique oculomotor system. *J. Neurophysiol.* **37**, 195–217.

Baker, R. G., Mano, N. & Shimazu, H. (1969) Postsynaptic potentials in abducens motoneurons induced by vestibular stimulation. *Brain Res.* **15**, 577–580.

Baker, R. & Precht, W. (1972) Electrophysiological properties of trochlear motoneurons as revealed by IVth nerve stimulation. *Exp. Brain Res.* **14**, 127–157.

Baker, R., Precht, W. & Berthoz, A. (1973) Synaptic connections to trochlear motoneurons determined by individual vestibular nerve branch stimulation in the cat. *Brain Res.* **64**, 402–406.

Baker, R., Precht, W. & Llinas, R. (1972) Mossy and climbing fibre projections of extraocular muscle afferents to the cerebellum. *Brain Res.* **38**, 440–445.

Barmack, N. H., Bell, C. C. & Rence, B. G. (1971) Tension and role of tension development during isometric responses of skeletal muscle. *J. Neurophysiol.* **34**, 1072–1079.

Bartz, A. E. (1966) Eye and head movements in peripheral vision: nature of compensatory eye movements. *Science* **152**, 1644–1645.

Bender, M. B., Postel, D. M. & Krieger, H. P. (1957) Disorders of oculomotor function in lesions of the occipital lobe. *J. Neurol. Neurosurg. Psychiat.* **20**, 139–143.

Bender, M. B. & Shanzer, S. (1964) Oculomotor pathways defined by electric stimulation and lesions in the brainstem of monkey. In *The Oculomotor System*, Ed. M. B. Bender, pp. 81–140. New York: Harper & Row.

Bergmann, F., Costin, A. & Chaimowitz, M. (1970) Influence of lesions in the lateral geniculate body of the rabbit on optic nystagmus. *Exp. Neurol.* **28**, 64–75.

Bergmann, F., Lachmann, J., Monnier, M. & Krupp, P. (1959) Central nystagmus. III. Functional correlations of mesodiencephalic nystagmogenic centre. *Am. J. Physiol.* **197**, 454–460.

Bickford, R. G., Petersen, M. C., Dodge, H. W. & Sem-Jacobsen, C. W. (1953) Observations on depth stimulation of the human brain through implanted electrographic leads. *Proc. Staff Meetings Mayo Clin.* **28**, 181–187.

Bizzi, E. (1967) Discharge of frontal eye field neurons during eye movements in unanesthetized monkeys. *Science* **157**, 1588–1590.

Bizzi, E. (1968) Discharge of frontal eye field neurons during saccadic and following eye movements in unanaesthetized monkeys. *Exp. Brain Res.* **6**, 69–80.

Bizzi, E., Kalil, R. E. & Morasso, P. (1972) Two modes of active eye-head coordination in monkeys. *Brain Res.* **40**, 45–48.

Bizzi, E., Kalil, R. E. & Tagliasco, V. (1971) Eye-head coordination in monkeys: evidence for centrally patterned organization. *Science* **173**, 452–454.

Bizzi, E. & Schiller, P. H. (1970) Single unit activity in the frontal eye fields of unanaesthetized monkeys during eye and head movement. *Exp. Brain Res.* **10**, 151–158.

Björk, A. & Kugelberg, E. (1953) Motor unit activity in the human extraocular muscles. *E.E.G. clin. Neurophysiol.* **5**, 271–278.

Björk, A. & Kugelberg, E. (1953) The electrical activity of the muscles of the eye and eyelids in various positions and during movement. *E.E.G. clin. Neurophysiol.* **5**, 595–602.

Björk, A. & Wahlin, A. (1960) Muscular factor in enophthalmos and exophthalmos. *Acta ophthal. Kbh.* **38**, 701–707.

Boeder, P. (1961) The co-operation of extraocular muscles. *Am. J. Ophthal.* **51**, 469–481.

Brandt, T., Dichgans, J. & Koenig, E. (1973) Differential effects of central versus peripheral vision on egocentric and exocentric motion perception. *Exp. Neurol.* **16**, 476–491.

Brecher, G. A. (1934) Die optokinetische Auslösung von Augenrollung und rotorischem Nystagmus. *Pflüg. Arch. ges. Physiol.* **234**, 13–28.

Breinin, G. M. (1957) Electromyographic evidence for ocular proprioception in man. *Arch. Ophthal.* **57**, 176–180.

Breinin, G. M. (1957) Quantitation of extraocular muscle innervation. *Arch. Ophthal.* **57,** 644–650.

Brodal, A. & Walberg, F. (1962) *The vestibular nuclei and their connections. Anatomy and functional correlations.* Oliver & Boyd, London.

Brown, T. G. (1911) The intrinsic factors in the act of progression in the mammal. *Proc. Roy. Soc. B* **84,** 308–319.

Brown, T. G. (1922) Reflex orientation of the optical axes and the influence upon it of the cerebral cortex. *Arch. néerl. Physiol.* **7,** 571–578.

Browne, J. S. (1976) The contractile properties of slow muscle fibres in sheep extraocular muscle. *J. Physiol.* **254,** 535–550.

Burgi, S. (1957) Das Tectum opticum. Seine Verbindungen bei der Katze und seine Bedeutung beim Menschen. *Deutsch. Z. Nervenheilk.* **176,** 701–729.

Burke, R. E., Levine, D. N., Tsairis, P. & Zajac, F. E. (1973) Physiological types and histochemical profiles in motor units of the cat gastrocnemius. *J. Physiol.* **234,** 723–748.

Carmichael, E. A., Dix, M. R. & Hallpike, C. S. (1954) Lesions of the cerebral hemispheres and their effects upon optokinetic and caloric nystagmus. *Brain* **77,** 345–372.

Carmichael, E. A., Dix, M. R. & Hallpike, C. S. (1956) Pathology, symptomatology and diagnosis of organic affections of the eighth nerve system. *Br. med. Bull.* **12,** 146–152.

Carpenter, M. B. (1971) Central oculomotor pathways. In *The Control of the Eye Movements.* Ed. P. Bach-y-Rita & C. C. Collins. Academic Press: N.Y. pp. 67–103.

Carpenter, M. B., Harbison, J. N. & Peter, P. (1970) Accessory oculomotor nuclei in the monkey: projections and effects of lesions. *J. Comp. Neurol.* **140,** 131–153.

Carpenter, R. H. S. (1972) Cerebellectomy and the transfer function of the vestibulo-ocular reflex in the decerebrate cat. *Proc. Roy. Soc. B* **181,** 353–374.

Chez, R. A., Palmer, R. R., Schultz, S. G. & Curran, P. F. (1967) Effect of inhibitors on alanine transport in isolated rabbit ileum. *J. gen. Physiol.* **50,** 2357–2375.

Close, R. (1967) Properties of motor units in fast and slow skeletal muscles. *J. Physiol.* **193,** 45–55.

Close, R. I. & Luff, A. R. (1974) Dynamic properties of inferior rectus muscle of the rat. *J. Physiol.* **246,** 259–270.

Cogan, D. G. (1954) Ocular dysmetria: flutter-like oscillations of the eyes, and opsoclonus. *Arch. Ophthal.* **51,** 318–335.

Cohen, B. (1971) Vestibulo-ocular relations. In *The Control of Eye Movements.* Ed. P. Bach-y-Rita & C. C. Collins. Academic Press: N.Y. pp. 105–148.

Cohen, B. & Feldman, M. (1968) Relationship of electrical activity in pontiné reticular formation and lateral geniculate body to rapid eye movements. *J. Neurophysiol.* **31,** 806–817.

Cohen, B., Goto, K., Shanzer, S. & Weiss, A. H. (1965) Eye movements induced by electric stimulation of the cerebellum in the alert cat. *Exp. Neurol.* **13,** 145–162.

Cohen, B. & Henn, V. (1972) Unit activity in the pontiné reticular formation associated with eye movements. *Brain Res.* **46,** 403–410.

Cohen, B. & Komatsuzaki, A. (1968) Electroculographic syndrome in monkeys after pontiné reticular formation lesions. *Arch. Neurol.* **18,** 78–92.

Cohen, B. & Komatsuzaki, A. (1972) Eye movements induced by stimulation of the pontiné reticular formation: evidence for integration in oculomotor pathways. *Exp. Neurol.* **36,** 101–117.

Cohen, B., Matsuo, V. & Raphan, T. (1977) Quantitative analysis of the velocity characteristics of optokinetic nystagmus and optokinetic after-nystagmus. *J. Physiol.* **270,** 321–344.

Cohen, B., Suzuki, J.-I. & Bender, M. B. (1964) Eye movements from semicircular canal nerve stimulation in the cat. *Ann. Otol. St. Louis* **73,** 153–169.

Cohen, L. A. (1961) Role of eye and neck proprioceptive mechanisms in body orientation and motor coordination. *J. Neurophysiol.* **24,** 1–11.

Collewijn, H. (1977) Eye and head movements in freely moving rabbits. *J. Physiol.* **266,** 471–498.

Collins, C. C., O'Meara, D. & Scott, A. B. (1975) Muscle tension during unrestrained human eye movements. *J. Physiol.* **245,** 351–369.

Cooper, S. & Daniel, P. D. (1949) Muscle spindles in human extrinsic eye muscles. *Brain* **72,** 1–24.

Cooper, S., Daniel, P. D. & Whitteridge, D. (1955) Muscle spindles and other sensory endings in the extrinsic eye muscles. *Brain* **78,** 564–583.

Cooper, S. & Fillenz, M. (1955) Afferent discharges in response to stretch from the extraocular muscles of the cat and monkey and the innervation of these muscles. *J. Physiol.* **127,** 400–413.

Corbin, K. B. & Oliver, R. K. (1942) The origin of fibres to the grape-like endings in the insertion third of the extraocular muscles. *J. comp. Neurol.* **77,** 171–186.

Cranmer, R. (1951) Nystagmus related to lesions of the central vestibular apparatus and the cerebellum. *Ann. Otol.* **60,** 186–196.

Crosby, E. C. (1953) Relations of brain centres to normal and abnormal eye movements in the horizontal plane. *J. comp. Neurol.* **99,** 437–479.

Crosby, E. C. & Henderson, J. W. (1948) Pathway concerned in automatic eye movements. *J. comp. Neurol.* **88,** 53–91.

Cynader, M. & Berman, W. (1972) Receptive-field organization of monkey superior colliculus. *J. Neurophysiol.* **35,** 187–201.

Daniel, P. D. (1946) Spiral nerve endings in the extrinsic eye muscles of man. *J. Anat. Lond.* **80,** 189–192.

De Burlet, H. M. (1924) Zur Innervation der Macula sacculi bei Säugetieren. *Anat. Anzeiger,* **58,** 26–32.

Dichgans, J., Bizzi, E., Morasso, P. & Tagliasco, V. (1973) Mechanism underlying recovery of eye-head coordination following bilateral labyrinthectomy in monkeys. *Exp. Brain Res.* **18,** 548–562.

Dix, M. R., Hallpike, C. S. & Harrison, M. S. (1949) Some observations upon the otological effects of streptomycin intoxication. *Brain* **72,** 241–245.

Dohlman, G. (1935). Some practical and theoretical points in labyrinthology. *Proc. roy. Soc. Med.* **28,** 1371–1380.

Duensing, F. & Schaefer, K.-P. (1957) Die Neuronenactivität in der Formatio reticularis des Rhombencephalons beim vestibulären Nystagmus. *Arch. f. Psychiat.* **196,** 265–290.

Duensing, F. & Schaefer, K.-P. (1958) Die Activität einzelner Neurone im Bereich der Vestibulariskerne bei Horizontalabsehleunigen unter besonderer Berücksichtigung des vestibularen Nystagmus. *Arch. f. Psychiat.* **198,** 225–252.

Duke-Elder, W. S. & Duke-Elder, P. M. (1930) The contraction of the extrinsic muscles of the eye by choline and nicotine. *Proc. R. Soc. B.* **107,** 332–343.

Eakins, K. E. & Katz, R. L. (1967) The role of the autonomic nervous system in extraocular muscle function. *Invest. Ophthal.* **6,** 253–260.

Easton, T. A. (1971) Inhibition from cat eye muscle stretch. *Brain Res.* **25,** 633–637.

Ebenholtz, S. M. & Shebilske, W. (1975) The doll reflex and ocular counterrolling with head-body tilt in the median plane. *Vision Res.* **15,** 713–717.

Eccles, J. C. (1973) The cerebellum as a computer: patterns in space and time. *J. Physiol.* **229,** 1–32.

Fernandez, C., Goldberg, J. M. & Abend, W. K. (1972) Response to static tilts of peripheral neurons innervating otolith organs of the squirrel monkey. *J. Neurophysiol.* **35**, 978–997.

Fillenz, M. (1955) Responses in the brainstem of the cat to stretch of extrinsic ocular muscles. *J. Physiol.* **128**, 182–199.

Fluur, E. (1959) Influences of semicircular ducts on extraocular muscles. *Acta otolaryngol. Suppl.* **149.**

Foerster, O. (1931) The cerebral cortex in man. *Lancet* (2), 309–312.

Fox, J. C. & Holmes, G. (1926) Optic nystagmus and its value in the localisation of cerebral lesions. *Brain* **49**, 333–371.

Fuchs, A. F. & Kornhuber, H. H. (1969) Extraocular muscle afferents to the cerebellum of the cat. *J. Physiol.* **200**, 713–722.

Fuchs, A. F. & Luschei, E. S. (1970) Firing patterns of abducens neurons of alert monkeys in relationship to horizontal eye movement. *J. Neurophysiol.* **33**, 382–392.

Fuchs, A. F. & Luschei, E. S. (1971) Development of isometric tension in simian extraocular muscle. *J. Physiol.* **219**, 155–166.

Goldberg, J. M. & Fernandez, C. (1971) Physiology of peripheral neurons innervating semicircular canals of the squirrel monkey. *J. Neurophysiol.* **34**, 635–660; 676–684.

Gonshor, A. & Melvill Jones, G. (1976) Extreme vestibulo-ocular adaptation induced by prolonged optical reversal of vision. *J. Physiol.* **256**, 381–416.

Gresty, M. A. (1974) Coordination of head and eye movements to fixate continuous and intermittent targets. *Vision Res.* **14**, 395–403.

Gresty, M. A. (1975) Eye, head and body movements of the guinea pig in response to optokinetic stimulation and sinusoidal oscillation yaw. *Pflüg. Arch. ges. Physiol.* **353**, 201–214.

Groen, J. J., Lowenstein, O. & Vendrik, A. J. H. (1952) The mechanical analysis of the responses from the end organs of the horizontal semicircular canal of the isolated elasmobranch labyrinth. *J. Physiol.* **117**, 329–346.

Grünbaum, A. S. F. & Sherrington, C. S. (1903) Observations on the physiology of the cerebral cortex of the anthropoid ape. *Proc. Roy. Soc. B.* **72**, 152–155.

Hallpike, C. S. & Hood, J. D. (1953) The speed of the slow component of ocular nystagmus induced by angular acceleration of the head: its experimental determination and application to the physical theory of the cupular mechanism. *Proc. Roy. Soc. B.* **141**, 216–230.

Harker, D. W. (1972a) The structure and innervation of sheep superior rectus and levator palpebrae extraocular muscles. I. Extrafusal muscle fibers. *Invest. Ophthal.* **11**, 956–969.

Harker, D. W. (1972b) The structure and innervation of sheep superior rectus and levator palpebrae extraocular muscles. II. Muscle spindles. *Invest. Ophthal.* **11**, 970–979.

Hassler, R. & Hess, W. R. (1954) Experimenteller und anatomische Befunde über die Drehbewegungen und ihre nervösen Apparate. *Arch. Psychiat.* **192**, 488–526.

Hess, A. (1967) The structure of vertebrate slow and twitch muscle fibers. *Invest. Ophthal.* **6**, 217–228.

Hess, A. & Pilar, G. (1963) Slow fibres in the extraocular muscles of the cat. *J. Physiol.* **169**, 780–798.

Highstein, S. M. & Ito, M. (1971) Differential localization within the vestibular nuclear complex of the inhibitory and excitatory cells innervating IIIrd nucleus oculomotor neurons in rabbit. *Brain Res.* **29**, 363–365.

Holmes, G. (1938) The cerebral integration of the ocular movements. *Br. Med. J.* **ii**, 107–112.

Horcholle, G. & Tyč-Dumont, S. (1968) Activités unitaires des neurones vestibulaires et oculomoteurs au cours du nystagmus. *Exp. Brain Res.* **5**, 16–31.

Hyde, J. E. (1962) Effect of hindbrain lesions on conjugate horizontal eye movements in cats. *Exp. Eye Res.* **1**, 206–214.

Hyde, J. E. & Eason, R. G. (1959) Characteristics of ocular movements evoked by stimulation of brainstem of cat. *J. Neurophysiol.* **22**, 666–678.

Hyde, J. E. & Slusher, M. A. (1961) Functional role of median longitudinal fasciculus in evoked conjugate ocular deviations in cats. *Am. J. Physiol.* **200**, 919–922.

Ito, M. (1972) Inhibitory and excitatory relay neurons for the vestibulo-ocular reflexes. *Progr. Brain Res.* **37**, 543–545.

Jung, R. & Hassler, R. (1960) The extrapyramidal motor system. In *Handbook of Physiology.* Ed. Field, Magoun & Hall, vol. II, pp. 863–927. Baltimore: American Physiological Society.

Keller, E. L. (1973) Accomodative vergence in the alert monkey. Motor unit analysis. *Vision Res.* **13**, 1565–1575.

Keller, E. L. (1974) Participation of medial pontine reticular formation in eye movement generation in monkey. *J. Neurophysiol.* **37**, 316–332.

Keller, E. L. & Robinson, D. A. (1971) Absence of a stretch reflex in intra-ocular muscles of the monkey. *J. Neurophysiol.* **34**, 908–919.

Keller, E. L. & Robinson, D. A. (1972) Abducens unit behavior in the monkey during vergence movements. *Vision Res.* **12**, 369–382.

Kern, R. (1965) A comparative pharmacologic-histologic study of slow and twitch fibers in the superior rectus muscle of the rabbit. *Invest. Ophthal.* **4**, 901–910.

Kern, R. (1968) Uber die adrenergischen Receptoren der extraoculären Muskeln der Rhesusaffen. *v. Graefes' Arch. Ophthal.* **174**, 278–286.

de Kleyn, A. (1939) Some remarks on vestibular nystagmus. *Confin. Neurol.* **2**, 257–292.

Krüger, P. (1952) *Tetanus und Tonus der quergestreiften Skelettmuskel der Wirbeltiere und des Menschen.* Leipzig: Acad. Verlag, Geest & Portig.

Lachmann, J., Bergmann, F. & Monnier, M. (1958) Central nystagmus elicited by stimulation of the meso-diencephalon in the rabbit. *Am. J. Physiol.* **193**, 328–334.

Lennerstrand, G. (1974) Electrical activity and isometric tension in motor units of the cat's inferior oblique muscle. *Acta physiol. Scand.* **91**, 458–474.

Lorente, de Nó, R. (1933) Vestibulo-ocular reflex arc. *Arch. Neurol. Psychiat.* **30**, 245–291.

Lorente de Nó, R. (1938) Analysis of the activity of the chains of internuncial neurons. *J. Neurophysiol.* **1**, 207–244.

Lowenstein, O. & Wersäll, J. (1959) Functional interpretation of the electron microscopic structure of the sensory hairs in the cristae of the elasmobranch *Raja clavata* in terms of directional sensitivity. *Nature* **184**, 1807–1808.

Luschei, E. S. & Fuchs, A. F. (1972) Activity of brain stem neurons during eye movements of alert monkeys. *J. Neurophysiol.* **35**, 445–461.

Mackensen, G. (1958) Zur Theorie des optokinetischen Nystagmus. *Klin. Mbl. Augenheilk.* **132**, 769–780.

Mackensen, G. (1959) Untersuchung zur Physiologie des optokinetischen Nachnystagmus *v. Graefes' Arch. Ophthal.* **160**, 497–509.

Maeda, M., Shimazu, H. & Shinoda, Y. (1971) Rhythmic activation of secondary vestibular efferent fibers recorded within the abducens nucleus during nystagmus. *Brain Res.* **34**, 361–365.

Maeda, M., Shimazu, H. & Shinoda, Y. (1972) Nature of

synaptic events in cat abducens motoneurons at slow and quick phase of vestibular nystagmus. *J. Neurophysiol.* **35,** 279–296.

Manni, E., Azzena, G. B. & Atzori, M. L. (1965) Relationships between cerebral and mesodiencephalic nystagmogenic centres in the rabbit. *Arch. ital. Biol.* **103,** 136–145.

Manni, E., Bortolami, R. & Deriu, P. L. (1970) Superior oblique muscle proprioception and the trochlear nerve. *Exp. Neurol.* **26,** 543–550.

Manni, E., Bortolami, R. & Desole, C. (1967) Relationship of Gasserian cells to extraocular muscle proprioception in lamb. *Experientia* **23,** 230–231.

Manni, E., Bortolami R. & Desole, C. (1968) Peripheral pathway of eye muscle proprioception. *Exp. Neurol.* **22,** 1–12.

Manni, E., Bortolami, R., Pettorossi, V. E. & Callegari, E. (1976) Trigeminal afferent fibers in the trunk of the oculomotor nerve of lambs. *Exp. Neurol.* **50,** 465–476.

Manni, E., Desole, C. & Palmieri, G. (1970) On whether eye muscle spindles are innervated by ganglion cells located along the oculomotor nerves. *Exp. Neurol.* **28,** 333–343.

Manni, E. & Giretti, M. L. (1968) Vestibular units influenced by labyrinthine and cerebral nystagmogenic impulses. *Exp. Neurol.* **22,** 145–157.

Manni, E. & Giretti, M. L. (1970) Central eye nystagmus in the pontomesencephalic preparation. *Exp. Neurol.* **26,** 342–353.

Manni, E., Giretti, M. L. & Deriu, P. L. (1974) Eye movements elicited by electric stimulation of the pontiné reticular formation. *Arch. int. Physiol. Biochim.* **82,** 831–842.

Manni, E., Palmieri, G., Marini, R. & Pettorossi, V. E. (1975) Trigeminal influences on extensor muscles of the neck. *Exp. Neurol.* **47,** 330–342.

Marg, E., Jampolsky, A. & Tamler, E. (1959) Elements of human extraocular electromyography. *Arch. Ophthal.* **81,** 258–269.

Markham, C. H., Precht, W. & Shimazu, H. (1966) Effect of stimulation of interstitial nucleus of Cajal on vestibular unit activity in the cat. *J. Neurophysiol.* **29,** 493–507.

Matyushkin, D. P. (1961) Phasic and tonic neuromotor units in the oculomotor apparatus of the rabbit. *Sechenov Physiol. J. USSR* **47,** 960–965.

Matyushkin, D. P. (1964) Motor systems in the oculomotor apparatus of higher animals. *Fed. Proc. Transl. Suppl.* **23,** T 1103–1106.

McCouch, G. P., Deering, I. D. & Ling, T. H. (1951) Location of receptors for tonic neck reflexes. *J. Neurophysiol.* **14,** 191–195.

McMasters, R. E., Weiss, A. H. & Carpenter, M. B. (1966) Vestibular projection to the nuclei of the extraocular muscles. *Amer. J. Anat.* **118,** 163–193.

Merton, P. A. (1956) Compensatory rolling movements of the eye. *J. Physiol.* **132,** 25–27P.

Miles, F. A. & Fuller, J. H. (1974) Adaptive plasticity in the vestibulo-ocular responses of the rhesus monkey. *Brain Res.* **80,** 512–516.

Miller, J. E. (1958) Electromyographic pattern of saccadic eye movements. *Am. J. Ophthal.* **46,** Pt. 2, 183–186.

Morasso, P., Bizzi, E. & Dichgans, E. (1973) Adjustment of saccade characteristics during head movement. *Exp. Brain Res.* **16,** 492–500.

Ohm, J. (1936) Ueber Interferenz mehrerer Arten von Nystagmus. *Proc. Acad. Sci. Amst.* **39,** 549–558.

Oyster, C. W., Takahashi, E. & Collewijn, H. (1972) Direction-selective retinal ganglion cells and control of optokinetic nystagmus in the rabbit. *Vision Res.* **12,** 183–193.

Page, S. G. (1965) A comparison of the fine structures of frog slow and twitch fibres. *J. Cell Biol.* **26,** 477–497.

Pasik, P. & Pasik, T. (1964) Oculomotor functions in monkeys with lesions of the cerebrum and the superior colliculi. In *The Oculomotor System*, Ed. M. B. Bender, pp. 40–80. New York: Harper & Row.

Pasik, T., Pasik, P. & Bender, M. B. (1966) The superior colliculi and eye movements. *Arch. Neurol.* **15,** 420–436.

Pasik, P., Pasik, T. & Krieger, H. P. (1959) Effects of cerebral lesions upon optokinetic nystagmus in monkeys. *J. Neurophysiol.* **22,** 297–304.

Peachey, L. D. (1968) Muscle. *Ann. Rev. Physiol.* **30,** 401–429

Peachey, L. D. & Huxley, A. F. (1962) Structural identification of twitch and slow striated muscle fibers of the frog. *J. Cell Biol.* **13,** 177–180.

Penfield, W. & Boldrey, E. (1937) Somatic motor and sensory representation in the cerebral cortex of man as studied by electrical stimulation. *Brain* **60,** 389–443.

Pilar, G. (1967) Further study of the electrical and mechanical responses of slow fibers in cat extraocular muscles. *J. gen. Physiol.* **50,** 2289–2300.

Precht, W., Gripps, J. & Richter, A. (1967) Effect of horizontal angular acceleration on neurons in the abducens nucleus. *Brain Res.* **5,** 527–531.

Precht, W., Schwindt, P. C. & Magherini, P. C. (1974) Tectal influence on cat ocular motoneurons. *Brain Res.* **82,** 27–40.

Rademaker, G. G. J. & Ter Braak, J. W. G. (1948) On the central mechanism of some optic reactions. *Brain* **71,** 48–76.

Ritchie, L. (1975) Effects of cerebellar lesions on saccadic eye movements. *J. Neurophysiol.* **39,** 1246–1256.

Robinson, D. A. (1968) Eye movement control in primates. *Science* **161,** 1219–1224.

Robinson, D. A. (1970). Oculomotor unit behavior in the monkey. *J. Neurophysiol.* **33,** 393–404.

Robinson, D. A. (1972) Eye movements evoked by collicular stimulation in the alert monkey. *Vision Res.* **12,** 1795–1808.

Robinson, D. A. (1976) Adaptive gain control of vestibulo-ocular reflex by the cerebellum. *J. Neurophysiol.* **39,** 954–969.

Robinson, D. A. & Fuchs, A. F. (1969) Eye movements evoked by stimulation of frontal eye fields. *J. Neurophysiol.* **32,** 637–648.

Robinson, D. A. & Jarvis, C. D. (1974) Superior colliculus neurons studied during head and eye movements of the behaving monkey. *J. Neurophysiol.* **37,** 533–540.

Robinson, D. A., O'Meara, D. M., Scott, A. B. & Collins, C. C. (1969) Mechanical components of human eye movements. *J. appl. Physiol.* **26,** 548–553.

Rogers, K. T. (1957) Ocular proprioceptive neurons in the developing chick. *J. comp. Neurol.* **107,** 427–437.

Ron, S. & Robinson, D. A. (1973) Eye movements evoked by cerebellar stimulation in the alert monkey. *J. Neurophysiol.* **36,** 1004–1022.

Sanghvi, I. S. & Smith, C. M. (1969) Characterization of stimulation of mammalian extraocular muscles by cholinomimetics. *J. Pharmacol.* **167,** 351–364.

Sasaki, K. (1963) Electrophysiological studies on oculomotor neurones of the cat. *Jap. J. Physiol.* **13,** 287–302.

Schiller, P. H. & Koerner, F. (1971) Discharge characteristics of single units in superior colliculus of the alert rhesus monkey. *J. Neurophysiol.* **34,** 920–936.

Schiller, P. H. & Stryker, M. (1972) Single-unit recording and stimulation in superior colliculus of the alert rhesus monkey. *J. Neurophysiol.* **35**, 915–924.

Sears, M. L., Teasdall, R. D. & Stone, H. H. (1959) Stretch effects in human extraocular muscle. *Bull. Johns Hopk. Hosp.* **104**, 174–178.

Shimazu, H. (1972) Vestibulo-oculomotor relations: dynamic responses. *Progr. Brain Res.* **37**, 493–506.

Shimazu, H. & Precht, W. (1966) Inhibition of central vestibular neurons from the contralateral labyrinth and its mediating pathway. *J. Neurophysiol.* **29**, 467–492.

Sparks, D. L., Holland, R. & Guthrie, B. L. (1976) Size and distribution of movement fields in the monkey superior colliculus. *Brain Res.* **113**, 21–34.

Spiegel, E. A. & Price, J. B. (1939) Origin of the quick component of labyrinthine nystagmus. *Arch. Oto-Laryngol.* **30**, 576–588.

Stein, B. E., Magalhâes-Castro, B. & Kruger, L. (1976) Relationship between visual and tactile representations in cat superior colliculus. *J. Neurophysiol.* **39**, 401–419.

Steinhausen, W. (1933) Uber die Beobachtung der Cupula in den Bodengangsampullen des Labyrinths des lebenden Hechts. *Pflüg. Arch.* **232**, 500–512.

Straschill, M. & Rieger, P. (1973) Eye movements evoked by focal stimulation of the cat's superior colliculus. *Brain Res.* **59**, 211–227.

Suzuki, J. I., Todumasu, K. & Goto, K. (1969). Eye movements from singular utricular nerve stimulation in the cat. *Acto Oto-laryngologica*, **68**, 350–362.

Szentágothai, J. (1950) The elementary vestibulo-ocular reflex arc. *J. Neurophysiol.* **13**, 395–407.

Szentágothai, J. & Schab, R. (1956) A midbrain inhibitory mechanism of oculomotor activity. *Acta physiol. Hung.* **9**, 89–98.

Tamler, E., Jampolsky, A. & Marg, E. (1958) An electromyographic study of asymmetric convergence. *Am. J. Ophthal.* **46**, Pt. 2, 174–182.

Tarlov, E. (1972) Anatomy of the two vestibulo-oculomotor projection systems. *Progr. Brain Res.* **37**, 471–491.

Ter Braak, J. W. G. (1936) Untersuchungen ueber optokinetischen Nystagmus. *Arch. néerl. Physiol.* **21**, 309–376.

Torre, M. (1953) Nombre et dimensions des unités motrices dans les muscles extrinsiques de l'oeil et, en général, dans les muscles squelettiques. *Schweiz. Arch. Neurol.* **72**, 362–378.

Van Egmond, A. A. J., Groen, J. J. & Jongkees, L. B. W. (1949) The mechanics of the semicircular canal. *J. Physiol.* **110**, 1–17.

Wagman, I. H., Krieger, H. P. & Bender, M. B. (1958) Eye movements elicited by surface and depth stimulation of the occipital lobe of macaque mulatta. *J. comp. Neurol.* **109**, 169–193.

Warwick, R. (1953) Representation of the extra-ocular muscles in the oculomotor nuclei of the monkey. *J. comp. Neurol.* **98**, 449–503.

Warwick, R. (1955) The so-called nucleus of convergence. *Brain* **78**, 92–114.

Whitteridge, D. (1955) A separate afferent nerve supply from the extraocular muscles of goats. *Quart. J. exp. Physiol.* **40**, 331–336.

Whitteridge, D. (1960) Central control of the eye movements. In *Handbook of Physiology*, Ed. Field, Magoun & Hall, vol. II, pp. 1089–1109. Washington: Am. Physiol. Soc.

Wilson, M. E. & Toyne, M. J. (1970). Retino-tectal and cortico-tectal projections in *Macaca mulatta. Brain Res.* **24**, 395–406.

Wilson, V. J. & Felpel, L. P. (1972) Specificity of semi-circular canal input to neurons in the pigeon. *J. Neurophysiol.* **35**, 253–264.

Wilson, V. J. & Maeda, M. (1974) Connections between semicircular canals and neck motoneurons in the cat. *J. Neurophysiol.* **37**, 346–357.

Wilson, V. J. & Yoshida, M. (1969) Comparison of effects of stimulation of Deiters' nucleus and medial longitudinal fasculus on neck, forelimb, and hindlimb motoneurons. *J. Neurophysiol.* **32**, 743–758.

Woellner, R. C. & Graybiel, A. (1959) Counterrolling of the eyes and its dependence on the magnitude of gravitation or inertial force acting laterally on the body. *J. app. Physiol.* **14**, 632–634.

Wolfe, J. W. (1971) Relationship of cerebellar potentials to saccadic eye movements. *Brain Res.* **30**, 204–206.

Wurtz, R. H. & Goldberg, M. E. (1972) Activity of superior colliculus in behaving monkey. III. Cells discharging before eye movements. *J. Neurophysiol.* **35**, 575–586.

# 18. The pupil

## SPHINCTER AND DILATOR MUSCLES

The aperture of the refracting system of the eye is controlled by the iris which behaves as a diaphragm, contracting or expanding as a result of the opposing actions of two muscles of ectodermal origin, namely the *sphincter pupillae* and the *dilator pupillae*. As classically described, the sphincter is an annular band of smooth muscle, 0·75 to 0·8 mm broad in man, encircling the pupillary border, which on contraction draws out the iris and constricts the pupillary aperture. At the pupillary border the muscle is closely associated with the pigment epithelium, so that, on contraction, the latter tends to be drawn on to the anterior surface of the iris. The sphincter is closely adherent to the adjacent connective tissue so that, after an iridectomy, it does not contract up, and the pupil remains reactive to light. Like the sphincter, the dilator muscle is of ectodermal origin, but in the sphincter the ectodermal cells have been transformed into true muscle fibres whereas the cells of the dilator retain their primitive characteristics and are called *myoepithelial cells*. It will be recalled (p. 12) that the ciliary epithelium consists of two layers of cells, an outer pigmented layer and, inside this, the unpigmented layer; these two layers are continued over the posterior surface of the iris, the pigmented layer becoming the layer of myoepithelial cells and the unpigmented layer acquiring pigment and becoming the posterior epithelium of the iris. The myoepithelial cells are spindle shaped with long processes that run radially to constitute the fibres of the dilator, making a discrete layer anterior to that made up by the bodies of the myoepithelial cells; the layer of fibres is called the *membrane of Bruch*. Close to the edge of the pupil the dilator fibres fuse with the sphincter, and at the other end they continue into the ciliary body where they take origin; between the insertion and origin, 'spurs' make connection with adjacent tissue. When it contracts, the dilator draws the pupillary margin towards its origin and thus dilates the pupil.

### INNERVATION

The sphincter is innervated by parasympathetic fibres from N III by way of the ciliary ganglion and the short ciliary nerves; the dilator is controlled by the cervical sympathetic, the fibres relaying in the superior cervical ganglion; post-ganglionic fibres enter the eye in the short and long ciliary nerves (Fig. 1.6, p. 4).

### FUNCTIONS OF THE PUPIL

A pupil of varying size performs three main functions:

1. It modifies the amount of light entering the eye, thus permitting useful vision over a wide range of luminance levels. The amount of light entering the eye is directly proportional to the area of the pupil; under conditions of night vision the absolute amount of light entering the eye is of great importance and a sixteenfold increase of sensitivity of the eye to light should be obtained by an increase of pupil diameter from 2 to 8 mm. At the other end of the scale, excessive luminance reduces visual acuity; the strong pupillary constriction occurring in these circumstances mitigates this condition and thus extends the range of useful vision. In nocturnal animals this aspect of pupil-size is of greater importance and it is probable that the slit of the fully constricted pupil of the cat is an adaptation permitting a greater restriction of the amount of light penetrating the eye in daylight.

2. As a result of constriction, the pencils of light entering the eye are smaller and the depth of focus of the optical system is increased (p. 487). When the eye is focused for distant objects, the depth of focus is large and the pupil-size is relatively unimportant; when the eye is focused for near objects, on the other hand, the depth of focus becomes small and a constricted pupil contributes materially.

3. By the reduction in the aperture of the optical system aberrations are minimized. In the dark these aberrations are unimportant, since form is only vaguely perceived, so that the advantage accruing from the increased light entering a dilated pupil far outweighs the effects of aberrations; in daylight, however, the reverse is true; and we have already seen how, at varying luminances, the size of the pupil is adjusted to give optimum visual acuity (p. 321).

## THE PUPILLARY REFLEXES

The accurate adjustment of the pupil-size to the optical requirements of the eye implies a reflex response to illumination; this is the basis of the *light reflex* under which term we may include the constriction of the pupil when the luminance is increased, and the dilatation when the luminance is decreased. The other important reflex, from the optical point of view, is the *near reflex*, the constriction of the pupil that takes place during the focusing on a near object. Additional reflexes are the *lid-closure reflex*, a constriction in response to closure of the lid, and the *psycho-sensory reflex*, a dilatation of the pupil in response to a variety of sensory and psychic stimuli.

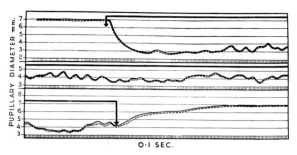

**Fig. 18.1**  Effect of exposure of right eye to a bright light. Ordinate: pupillary diameter in mm. Abscissa: time of exposure. Record in full line applies to right eye; in broken line to left eye.

*First line*: the pupils were large and quiet in darkness. At the arrow the right eye was exposed to a steady illumination.

*Second line*: this shows the pupillary oscillations after the right eye had been exposed for 3 min to the light.

*Third line*: when the light was turned off, at the arrow, the pupils dilated and the oscillations disappeared. (Lowenstein and Loewenfeld, *Amer. J. Ophthal.*)

## EXPERIMENTAL MEASUREMENT—PUPILLOGRAPHY

The experimental study of the size of the pupil—*pupillography*—has been carried out by a variety of techniques; since the pupil responds to visible light the method of measurement must dispense, if possible, with illumination unless this involves flashlight photography where the duration of the flash is so short that the pupil has no time to respond. For continuous studies flash-photography is obviously of no use, since the flashes have their effects, although delayed; by the use of infrared photography, however, a cinematographic record of the pupil in the dark may be obtained and this is the method that has been most commonly employed. More recently Lowenstein and Loewenfeld have developed a method of 'electronic pupillography' based on infrared scanning of the pupil, the reflected light from the scanned pupil being brought on to an infrared sensitive photo-tube. The scanned image is made visible by the usual television principles, if this is required; alternatively the size of the pupil may be recorded less directly on a cathode-ray screen. Thus when the scanning spot crosses from the iris to the pupil there is an abrupt decrease in the reflected light, and this shows up as a sudden fall in the signal on the oscilloscope. The larger the pupil the longer the portion of a given scan will give the low signal.

## THE LIGHT REFLEX

When the light falling on one eye is increased, the pupil of this eye constricts—the *direct light reflex*; at the same time the pupil of the other, unstimulated, eye also contracts—the *indirect* or *consensual light reflex*. A record of the changes in pupil-size in the two eyes during exposure to a bright light is shown in Figure 18.1, where the pupil-diameter is given as the ordinate and is plotted against time of exposure. The record is shown in three parts; at the arrow on the top, the light is switched on to illuminate the right eye, whose pupillary diameter is indicated by the full line. Both pupils contract down to about 3 mm, and the sizes oscillate together whilst, as time progresses, there is an increase in size on which the oscillations are imposed; this is seen by the middle record, the mean pupil-size being about 4 mm. At the arrow in the bottom record the light was turned off, and the pupils return slowly to their original diameters. The increase in diameter that occurs during sustained illumination may be regarded as an effect of light-adaptation. On flashing a light into the eye, then, the response is a constriction of the homolateral and contra-lateral pupils, the extent of the constriction depending on the intensity of the light and the state of adaptation of the eye. If the light-stimulus is prolonged, the pupil dilates slowly and reaches a size which, for a given individual, depends on the intensity of the light. When both eyes are stimulated there is a summation of the individual responses so that both pupils constrict rather more than when one alone is stimulated with the same intensity of light; conversely if, when both are being subjected to a light stimulus, one eye is closed, the contralateral pupil dilates slightly.

### EFFECT OF LUMINANCE

The magnitude of the initial constriction caused by a light-stimulus depends, as we should expect, on the intensity of the light. This is demonstrated by Figure 18.2 where the constriction, in millimetres, is represented by the heights of the black bars, whilst the relative luminances of the stimuli are plotted logarithmically as abscissae. At the arrow marked $a$, the intensity is some millionth of that required to evoke the 3 mm response, and this corresponds to the threshold for colour vision, i.e. the cone threshold. At the arrow marked $b$, the intensity is some $10^{-9}$ to $10^{-10}$ of that required to produce the 3 mm response; this is the threshold for rod vision, and it will be seen that under the conditions of these experiments the threshold for pupillary response is higher than that for the visual response.

### RODS AND CONES

The receptors for the pupillary response to light are obviously the rods or cones or both; in fact, many studies have shown that both receptors may bring about the reflex. At threshold intensities for pupillomotor activity the visual sensation is colourless,

THE PUPIL   469

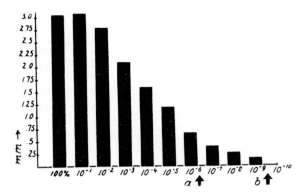

**Fig. 18.2**  Average extent of pupillary reflexes elicited by light-stimuli of decreasing intensity. The heights of the columns indicate the average degree of constriction, whilst the abscissa shows the intensities of the light-stimuli (duration 1 second). The arrow *a* shows the threshold for colour vision and the arrow *b* the absolute threshold. (Lowenstein and Loewenfeld, *Amer. J. Ophthal.*)

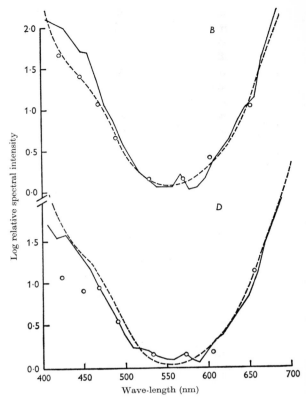

**Fig. 18.3**  Mean spectral sensitivity curve for the pupil response of two subjects. ◯, Differential threshold measurements are plotted for 2 sec flashes of a 2° test patch centrally fixated and seen against a continuous blue background approximately 15° in diameter, which produced a retinal illuminance somewhere between 100 and 200 td. Interrupted line, C.I.E. photopic luminosity curve; solid line, mean results of psychophysical measurements of photopic luminosity (flicker photometry) on the same two subjects with the same apparatus. (Alpern and Campbell, *J. Physiol.*)

proving that the rods are mediating the response; on plotting the sensitivity to the different wavelengths a typical scotopic sensitivity curve is obtained with maximum in the region of 500 nm (Schweitzer, 1956).

SPECTRAL SENSITIVITY

The most thorough investigation of the influence of wavelength on pupillary response is that of Alpern and Campbell (1962), who studied both transient pupillary changes, i.e. the response to rhythmically altered intensities, and steady-state changes. In both cases, the responses to relatively high luminances never gave an uncomplicated photopic luminosity curve comparable with that for vision; instead, the maximum was between 530 and 540 nm, suggesting that, even under highly photopic conditions and foveal stimulation, the rods and cones were both operative, due, presumably, to scatter from the fovea. When the effects of rods were excluded by keeping them adapted to a blue light acting as a background upon which the foveal test-stimulus was imposed, then the action-spectrum gave a curve with a peak at about 560 nm, in reasonable agreement with the C.I.E. standard curve (Fig. 18.3). Analysis of the curve obtained when the maximum was at 530 to 540 nm, on the basis of the different effects of the Stiles-Crawford phenomenon with red and blue lights, suggested that the rod and cone effects were, indeed, simultaneously present, and that their separate contributions to the final pupillary response could be estimated if the measured sensitivity was treated as the weighted mean of the logarithms of the individual rod and cone sensitivities at a given wavelength.* If the response of the pupil is plotted against intensity of the light-stimulus, as in Figure 18.4, there appears an obvious break in the curve when white or blue light is employed, a break corresponding to the transition from rod to cone vision. When red light is employed there is no break, and the threshold is higher because the red light is very ineffective as a rod stimulator (p. 230). In general, the cone response to a flash has a shorter latency (0·2 to 0·3 second) and is more rapid and longer-lasting than the rod response, which has a latency of 0·45 to 0·6 second.

SUMMATION

Schweitzer has studied the effects of increasing the area and duration of the stimulus. It will be recalled that the visual stimulus shows a limited amount of spatial summation, in the sense that if the thresholds for a small patch of light and a larger patch are compared, the threshold is lower for the larger patch; this is due, as we have seen, to the convergence of rods and cones on to ganglion cells (p. 239). The area over which summation of the visual response occurs is relatively small by comparison with that for the pupillographic response; thus Schweitzer found complete summation up to an area subtending 4° in the periphery and 0·5° at the fovea, whilst

---

*Alexandridis and Koeppe (1969) have described the typical Purkinje shift in maximal sensitivity to wavelength on passing from scotopic to photopic stimulation; their sensitivity curves coincided remarkably well with the corresponding sensitivity curves for vision, with maxima at 500 nm and 560 nm.

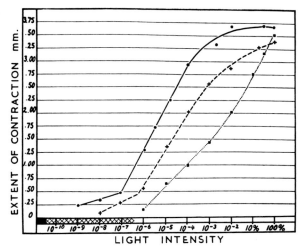

**Fig. 18.4** Pupillary contraction as a function of the intensity of the light-stimulus for three different types of light. Solid line and dots: curve for white light. Broken line and crosses: curve for blue light obtained by placing filter in front of white source; the apparent increase in the threshold is due to the absorption of light by the filter. Dotted line and squares: curve for red light obtained by placing red filter in front of the white source. The black bar above the abscissa indicates values below the absolute threshold; the hatched bar indicates the range of scotopic vision. Note that with red light pupillary responses are only obtained at intensity levels corresponding to photopic vision. (Lowenstein and Loewenfeld, *Amer. J. Ophthal.*)

partial summation extended over much wider areas. This means that to make the most efficient use of light in the pupillomotor response we must stimulate a large area of retina; when this is done we find that the visual and pupillomotor thresholds are equal, but Schweitzer was unable to push the sensitivity of the pupillomotor response beyond this point, i.e. he could not obtain a contraction of the pupil in the absence of visual sensation.* As a result of the operation of the light reflex, the size of the pupil tends to adopt a characteristic value for each level of luminance; according to De Groot and Gebhard the diameter, $d$, in mm, is related to the luminance, $B$, in mL, by the empirical equation:

$$\log d = 0 \cdot 8558 - 0 \cdot 000401 \ (\log B + 8 \cdot 1)^3$$

In the completely dark-adapted eye the mean pupillary diameter is given variously as 6·6 to 7·6 mm.

## THE NEAR REFLEX

By the near reflex is meant the constriction of the pupil accompanying the convergence of the eyes on to a near object; associated with these two actions of convergence and miosis, there is, of course, the act of accommodation, all three being mediated by N III. The response is bilateral and the maximal constriction obtained with this reflex is approximately equal to that obtained with the light reflex; the two reflexes may summate their responses.

## THE LID-CLOSURE REFLEX

This had been classically described as a homolateral pupillary constriction associated with closure of the lid; it was said to be evoked if the lid was held open while the effort to close the eye was made. Lowenstein and Lowenfeld (1969) have pointed out that this latter phenomenon is really an example of the near reflex, since it does not occur if the subject is told to look at a far point all the time. As these authors state, it is best to include all pupillary changes associated with lid-closure in the class of 'lid-closure reflexes', but the variety of conditions under which the lids close means that the pupillary changes will be different and have different mechanisms. Thus, if the lids close to avoid a noxious stimulus, it is probable that the pupils will dilate because of the psycho-sensory reflex; a brief, naturally occurring, blink is associated with dilatation due to the operation of the light-reflex in reverse, i.e. the response to the short period of darkness.

## PSYCHO-SENSORY REFLEX

This is a dilatation of the pupils evoked by stimulation of any sensory nerve or by strong physical stimuli. Pupillary dilatation thus occurs during the induction stage of anaesthesia, in extreme fear and in pain; in sleep, on the other hand, when these stimuli are lacking, the pupils are generally constricted. The mechanism is clearly a cortical one and is apparently mediated both by way of the cervical sympathetic, which activates the dilator pupillae, and by way of N III which inhibits the tonus of the sphincter pupillae. The reaction is not seen in the newborn, but appears in the first few days of life, developing fully at the age of six months.

## NERVOUS PATHWAYS

### REFLEX ARC

The receptors of the light reflex are the rods and cones; the afferent pathway is therefore through the optic

---

* It is commonly stated that the threshold for pupillomotor activity is higher than that for vision, and this is true for small test-fields; thus if this subtended 1° 50′, the ratio of thresholds varied from 8 to 1000; when the Ulbricht integrating sphere was employed, permitting exposure of a very large field, the thresholds became equal because of the greater degree of summation in the pupillomotor response. Temporal summation, whereby two stimuli, following each other rapidly, are more effective than a single one, depends essentially on the photochemical characteristics of the receptors (Bunsen-Roscoe Law) rather than on any neural organization; it is not surprising, therefore, that temporal summation is the same for the pupillomotor and visual responses (Schweitzer, 1956).

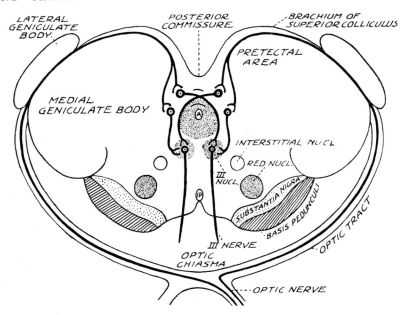

**Fig. 18.5** The path concerned in the pupillary light-reflex, according to Magoun and Ranson (*Arch. Ophthal.*).

nerves and tracts; the pupillary fibres leave the optic tract before the lateral geniculate body is reached, and for many years it was thought that they relayed in the superior colliculus. Magoun and Ranson showed, however, that in cats and monkeys the fibres pass through the medial border of the geniculate body and then run, by way of the brachium of the superior colliculus, into the *pretectal nucleus*, a group of cells occupying a position at the junction of the diencephalon and the tectum of the mid-brain; here they have their first synapse. In the cat the majority of the secondary fibres cross in the posterior commissure and terminate in the Edinger-Westphal nucleus; in primates the crossing is not so extensive. Destruction of the superior colliculus, therefore, leaves the light reflex intact; stimulation of the pretectal area gives a bilateral constriction of the pupils; conversely, destruction of this area gives a dilated and fixed pupil. From the Edinger-Westphal nucleus the pupillo-constrictor fibres pass with the IIIrd nerve fibres to the ciliary ganglion. The reflex arc for the direct and indirect light reflexes is indicated in Figure 18.5; the diagram indicates quite clearly how a light stimulus, falling on one eye, gives rise to pupillary constriction in both eyes.*

## PRETECTAL COMPLEX

The regions of termination of the retinal fibres in the mid-brain have turned out to be complex, so that the simple statement that the pupillo-sensory input relays in the 'pretectal nucleus' is misleading; it is possible furthermore that fibres concerned with accommodation relay in a different nucleus from those mediating pupillary responses (Westheimer

and Blair, 1973). The subject has been discussed in detail by Scalia (1972) in the light of his own studies, and he has constructed a valuable table indicating the equivalent structures in a variety of mammalian species, and the different terminologies employed by the many anatomists who have described one or more of these.

### NUCLEAR MASSES

His own study led to the conclusion that there were four pretectal cell masses or nuclei receiving retinal input, namely: *Nucleus of the Optic Tract (NOT)*; *Anterior Pretectal Nucleus*; *Olivary Pretectal Nucleus (PON)* and *Posterior Pretectal Nucleus*. Some of these nuclei are illustrated in Figure 18.6 which is an outline drawing of a brainstem section through the most compact portion of the posterior commissure. The features common to all species are a heavy retinal projection to the olivary pretectal nucleus and to the nucleus of the optic tract; the posterior pretectal nucleus is present in all species and receives a retinal projection; the retinal projection to the anterior pretectal nucleus is variable amongst species. Of most interest was the re-discovery of the retinal projection to the olivary pretectal nucleus, the high density of retinal terminals indicating an obvious visual function. More recently Pierson and Carpenter (1974) described a similar dense projection to this nucleus in monkeys, and demonstrated, moreover, a projection from the PON to the pregeniculate nucleus. The projections were predominantly contralateral but an ipsilateral

---

* The actual areas of the nucleus of N III concerned with pupillo-constrictor function have been examined by Sillito and Zbrożyna (1970); they include a region containing both the cells of the Edinger-Westphal nucleus and the anteromedian nucleus where they adjoin, and the caudal continuation of this nucleus. The authors emphasize that this pupillary constrictor nucleus is not a paired structure, as represented in Figure 18.5 for example, but is strictly a group of cells with no midline cleavage; in agreement with this, all pupillomotor responses produced by stimulation of the nucleus, or of supranuclear structures, were strictly bilateral.

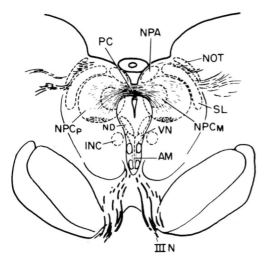

**Fig. 18.6**  Outline drawing of a brain stem section through the most compact portion of the posterior commissure (PC). At this level, the nucleus of the optic tract (NOT), the sublentiform nucleus (SL), the nucleus of the pretectal area (NPA), and the nucleus of the posterior commissure (NPC) are well developed. The anterior median nucleus (AM) is present, but the dorsal visceral nuclei (VN) of the oculomotor complex have not separated into medial and lateral cell columns. Additional abbreviations: INC, interstitial nucleus of Cajal; ND, nucleus of Darkschewitsch; NPCm, nucleus of posterior commissure, pars magnocellularis; NPCp, nucleus of posterior commissure, pars principalis; III N, oculomotor nerve. (Carpenter and Pierson, *J. Comp. Neurol.*)

projection in addition, giving rise to the possibility of binocular visual input, was found in each of the mammals studied by Scalia, but it depended on the nucleus.

PREGENICULATE OR VENTRAL GENICULATE NUCLEUS

The role of this diencephalic nucleus, closely related anatomically and phylogenetically to the thalamus (p. 548), in pupillary or accommodative activity has been discussed by Pierson and Carpenter. There is no doubt that this is a nucleus in which retinal fibres terminate, but these authors were unable to find pupillary misfunction in monkeys with lesions in this nucleus. According to Swanson *et al.* (1974) the ventral nucleus of the lateral geniculate body of cat and rat projects to the pretectal nuclei.

# SINGLE UNIT ACTIVITY

Sillito (1968) and Sillito and Zbrożyna (1970) recorded activities of single neurones in the midbrain of cats which discharged in relation to pupilloconstrictor activity. Thus a given neurone discharged at 1/sec when the pupil diameter was 10 mm, at 2·5/sec at 3 mm, 7·5/sec at 0·5 mm and at 8·5/sec when the pupil was slit-like. During the light-reflex the rate rose to 8·5 to 12·5/sec and could be as high as 25/sec. When the peri-

fornical region of the hypothalamus (a pupillodilator centre) was stimulated, the discharge of the units was completely silenced.

MIDBRAIN LOCATION

When the regions of the midbrain in which these pupilloconstriction-associated neurones were localized were explored they were found to include a region containing both the cells of the Edinger-Westphal nucleus and of the anteromedian nucleus where they adjoin; the caudal part of the Edinger-Westphal nucleus and the rostral part of the anteromedian nucleus were not involved in pupilloconstrictor activity, so that the distribution of neurones cuts across the traditional basis of division that attributed pupilloconstrictor activity entirely to the Edinger-Westphal nucleus. Sillito and Zbrożyna emphasize, moreover, that the pupilloconstrictor centre is not a paired structure, as represented in Figure 18.5, for example, but is strictly a group of cells with no midline cleavage; in agreement with this, all pupillomotor responses produced by stimulation of the nucleus, or of supranuclear structures, were strictly bilateral.*

DILATOR NEURONES

Smith *et al.* (1970) found neurones some 0·5 to 2·0 mm above and below the Edinger-Westphal nucleus which discharged at rates related to the degree of dilatation of the pupil; when the brain was stimulated at the recording sites there was pupillary dilatation, and the same occurred when the cervical sympathetic nerves were cut, so that the response was mediated by inhibition of tonically active pupilloconstrictor neurones in the Edinger-Westphal nucleus with their pathway through the ciliary ganglion.

PRETECTAL UNITS

Cavaggioni *et al.* (1968) recorded responses to light in single units of the nucleus of the optic tract in the midpontine-trigeminal cat and correlated these with pupillary responses. The feature that differentiated the responses from those in the adjacent superior colliculus was their sustained nature, which was usually an increased rate of discharge maintained as long as the light-stimulus; a short phasic component was completed before the pupil began to constrict and might

---

* This view differs a little from that of Jampel and Mindel (1967), who placed an electrode exactly in the midline of the monkey's midbrain and, according to its depth, obtained accommodation only, a mixture of this with pupilloconstriction, or only pupilloconstriction. As soon as the electrode was displaced laterally the responses became ipsilateral, so they concluded that there were paired masses close to the midline responsible for pupillary and accommodative activity, cells controlling accommodation being located dorsally and somewhat rostrally to cells innervating the iris.

have been responsible for activating the Edinger-Westphal neurones to initiate the constrictor response to light, whilst the tonic sustained increase augmented the tonic discharges in the Edinger-Westphal neurones maintaining the pupillary constriction at the value for the new light-intensity. The responses of collicular neurones were easily distinguished from the pretectal units, for example in their sensitivity to moving stimuli. Other pretectal units were suppressed by light and gave a prominent OFF-response, and it was suggested that this OFF-discharge was inhibitory, so that the dilatation of the pupil at onset of darkness was not a mere passive return to a relaxed state but was aided by active inhibition of the tonically active Edinger-Westphal neurones; certainly a 'supranuclear inhibition of the light-reflex', not involving the sympathetic pathway and presumably involving inhibition of the Edinger-Westphal nucleus, has been described (Bonvallet and Zbrożyna, 1963; Lowenstein and Loewenfeld, 1969).*

### SYMPATHETIC UNITS

Passatore and Pettorossi (1976) measured responses in single fibres dissected from the central end of the cat's cervical sympathetic nerve; as Figure 18.7 shows, the

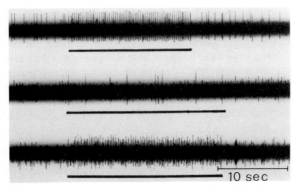

**Fig. 18.7** Different response patterns to ON-OFF light-stimulus in fibres dissected from the central end of the cat's cervical sympathetic nerve. Time signal: 10 sec. (Passatori and Pettorossi, *Exp. Neurol.*)

resting discharge during illumination of the eye was strongly increased with onset and maintenance of darkness, indicating that the response to darkness involves an active process of exciting the dilator pupillae. The effect was stronger in the homolateral eye when the consensual reflex was studied. When painful stimuli were applied to the skin or cornea, discharges were recorded comparable with those obtained on cutting off illumination, and these coincided with pupillary dilatation. Since the same fibres exhibited responses to changes in illumination, we may conclude that the

pathways for the light-reflex and the psycho-sensory reflex converge on the same sympathetic neurones (Passatore, 1976).

## OTHER PUPILLOMOTOR CENTRES

The sphincter pupillae undoubtedly dominates pupil activity so that the pretectal and cortical pupilloconstrictor centres have received most attention. The strong association of dilatation of the pupil with emotional states would indicate the presence of dilator centres in those parts of the brain that have been shown to be strongly associated with emotional states, namely the hypothalamus and the limbic system generally. In fact, electrical stimulation in the hypothalamus, over quite a wide region, causes dilatation of the pupil (Hess). The pathway from the brain in dilator responses involves the sympathetic system, the pupillomotor fibres leaving the cord from the first and second thoracic segments and relaying in the superior cervical ganglion. Stimulation of the cord between the VIth cervical and IVth thoracic segments causes dilatation of the pupil and it is customary to speak of spinal dilator centres, but it seems most likely that the stimulus merely activates the sympathetic motor neurones in the cord.

## THE PERIPHERAL MECHANISMS

### REFLEX DILATATION

Stimulation of the cervical sympathetic trunk in the neck causes a pronounced dilatation of the pupil. If the sympathetic trunk is cut there is an initial dilatation, due to the irritative effect of the cut, and this is followed by a constriction so that the pupil on the cut side is smaller than the other; this is presumably due to the fact that normally the size of the pupil is determined by a balance of forces between sphincter and dilator; putting the dilator out of action allows the sphincter to take control. The same phenomenon is manifest in the reflex response to light; in the sympathectomized eye the response is generally larger. Reflex dilatation, brought about by some sensory stimulus, is mainly due to active

---

* Straschill and Hoffmann (1969) have described units in the pretectum of the cat that are highly reminiscent of collicular units, although those located in the nucleus of the optic tract—the region where Cavaggione *et al.* found their pupillary units—showed no directional sensitivity. They point out that the projection of the colliculus on the pretectum could account for the 'collicular features' of the pretectal units, and of course many of their units might have been recordings from fibres *en passage* rather than from pretectal neurones. The nucleus of the posterior commissure, one of the pretectal nuclei included in their study, and giving direction-sensitive units, sends fibres into the medial longitudinal fasciculus and is therefore obviously concerned with eye movements; lesions here cause paralysis of vertical gaze (Bender and Shanzer, 1964).

dilatation, mediated by the sympathetic, since when the sympathetic trunk is cut there is only a small dilatation, due to inhibition of sphincter tone. Some time after removing the sympathetic supply to the iris it becomes hypersensitive to adrenaline and this gives rise to what has been called 'paradoxical pupillary dilatation'. Thus a sensory stimulus causes, in this sympathectomized eye, a small dilatation due to relaxation of sphincter tone, and this is followed by a much larger dilatation due, presumably, to the liberation into the blood of adrenaline as a consequence of the sensory stimulus.

Stimulation of N III, or the ciliary ganglion, causes constriction of the pupil; damage to the ciliary ganglion or IIIrd nerve reduces the reflex responses to light very considerably, such constriction as occurs being due to inhibition of sympathetic tone.* Weak psycho-sensory stimuli result in reflex pupillary dilatation in both normal and ciliary ganglion-ectomized eyes, the effect being a little larger in the normal eye, as we should expect.

# THE EFFECTS OF DRUGS

The main effects of drugs on the pupil are largely what one would expect of a system innervated by both para-sympathetic and sympathetic fibres.

## PARASYMPATHETIC DRUGS

It will be recalled that parasympathetic activity is mediated by the liberation of acetylcholine, so that this substance and drugs with a similar chemical constitution will behave in the same way as parasympathetic stimulation, causing a constriction of the pupil. Thus acetylcholine, pilocarpine, muscarine and many synthetic compounds are *parasympathomimetic*, by virtue of their chemical similarity to the parasympathetic chemical mediator. Another group mimics the para-sympathetic by virtue of their inhibitory action on cholinesterase, the enzyme that destroys acetylcholine at its site of action and thus prevents the chemical mediator from acting indefinitely. They are said to *potentiate* the action of acetylcholine. Eserine, physo-stigmine and many modern synthetic drugs, such as diisopropylfluorophosphate (DFP), act in this way causing a prolonged pupillary constriction by preserving from destruction the acetycholine, liberated by the tonic and reflex activity of the parasympathetic nerve endings. Another group of substances block the action of acetyl-choline by competing for its sites of activity on the muscle; an example of this *parasympatholytic* action is given by atropine, and various synthetic drugs such as homatropine. These drugs, by blocking the tonic parasympathetic activity, leaving sympathetic activity normal, cause a long-lasting dilatation of the pupil, which is now unresponsive to light.

## Ganglion blockers, etc.

Since the parasympathetic innervation operates through the ciliary ganglion, we may expect drugs that act specifically on autonomic ganglia to influence the pupil; in fact, hexamethonium and pentolinium block trans-mission and cause dilatation of the pupil, due to cutting off of tonic action. Transmission through the ganglion is brought about by the liberation of acetylcholine, hence this transmitter, as well as nicotine and tetramethyl ammonium, causes discharge in the post-ganglionic fibres and thus brings about constriction of the pupil. However, the sympathetic system, causing dilatation of the pupil, also operates through a ganglion, this time the superior cervical, hence the final effect of a blocking agent on the pupil will depend on which division of the autonomic system is having the dominant action at the time of administration of the drug.

## SYMPATHETIC DRUGS

The sympathetic chemical mediator is noradrenaline, whilst adrenaline, liberated by the adrenal medulla in response to stress, has an essentially similar action, in so far as it mimics many of the actions of the sympathetic system. These substances cause dilatation of the pupil.

### Alpha- and beta-actions

Adrenaline has two types of action on tissues, suggesting that there are two types of receptor with which it may react; in accordance with Ahlquist's (1948, 1962) nomenclature, those actions, largely excitatory, that are blocked by such 'classical' adrenaline antagonists as ergotoxin, dibenamine and phentolamine, are the result of reaction with the α-receptors; reaction with the β-receptors, largely inhibitory, is blocked by dichloroiso-proterenol (DCI) and pronethanol. Noradrenaline, the adrenergic transmitter, has a mainly α-action, whilst isoproterenol (isoprenaline) has mainly β-action. In general, then, when the order of activity of the amines is: Adrenaline > Noradrenaline ⩾ Isoproterenol, the influence of α-receptors is involved, whereas the order: Isoproterenol ⩾ Adrenaline > Noradrenaline indicates β-activity. If the tissue has both α- and β-receptors, then any of these amines will activate both, and the response will be determined by the dominant one; by the use of the selective blockers, the individual contributions may be assessed.

### Intraocular injections

Bennett *et al.* (1961) injected sympathetic amines into the posterior chamber of rabbits, and the anterior chamber of dogs, measuring the pupillary effects. If the dilatation caused by adrenaline was put equal to 100, that due to noradrenaline was 37 and to isoproterenol 19;

* In the cat, in contrast to man and monkey, the sympathetic supply to the iris does not pass through the ciliary ganglion; hence in this animal damage to, or removal of the ganglion does not influence sympathetic activity.

the action of isoproterenol was not blocked by DCI, so that we may attribute all three effects to α-activity. In these animals, then, pupillary activity seems to be dominated by α-receptors in the dilator.

## Isolated muscle

The study of isolated muscle-strips allows the assessment of the effects on dilator and sphincter separately. Van Alphen, Robinette and Macri (1964), working on the cat, found that, with the dilator, the order of effectiveness in causing contraction was: Adrenaline > Noradrenaline ≫ Isoproterenol. These effects were blocked by phenoxybenzene so that there is little doubt that the main activity is through α-receptors. Some β-activity was demonstrated by the observation that smaller doses of isoproterenol often produced a relaxation, which was blocked by DCI; again, phenoxybenzene not only blocked the contraction due to isoproterenol but could reverse this into a relaxation. With the sphincter, a relaxation of the muscle, previously caused to contract with acetylcholine, was caused by the amines in the order: Isoproterenol ≫ Adrenaline > Noradrenaline, thus indicating mainly β-activity. The action of adrenaline is indicated by Figure 18.8; at A, contraction was induced with acetylcholine; at E, adrenaline was applied, which caused a relaxation after a brief contraction. When DCI was added (lower trace) the relaxation failed and adrenaline now caused contraction due to its α-action. Thus the cat sphincter is,

indeed, influenced by adrenergic amines, the dominant influence being relaxation through β-receptors. Hence the effects of adrenaline in the cat are synergic, the α-activity on the dilator being assisted by the β-relaxation of the sphincter. Essentially similar conditions were found by Van Alphen, Kern and Robinette (1965) in the rabbit, and by Patil (1969) in the ox; in the monkey, however, α-activity was strongly predominant in both sphincter and dilator, so that under the action of sympathetic amines there should be antagonism.[*]

### ULTRASTRUCTURE OF NERVE TERMINALS

The nature of nerve terminals on smooth muscle may be ascertained with some degree of probability by the types of vesicles within them, as seen in the electron microscope (see, for example, Richardson, 1964). Nishida and Sears (1969) found two types of axon running in the sphincter muscle of the albino guinea pig; 85 per cent belonged to the type containing agranular vesicles characteristic of cholinergic fibres whilst the remaining 15 per cent contained vesicles with a dense core characteristic of adrenergic fibres; these were morphologically identical with the majority of fibres in the dilator muscle. Twenty-four hours after superior cervical ganglionectomy these fibres showed degenerative changes.

### OTHER DRUGS

Certain drugs act directly on smooth muscle, and since the sphincter is the stronger of the two iris muscles, their action on the pupil is largely determined by whether they cause this muscle to relax or contract. Thus *histamine* causes a powerful miosis which overcomes the mydriasis of atropine. *Morphine* and veratrine cause a miosis by direct action on the plain muscle; the pinpoint pupil of morphia addicts is thought to be due, however, to the action of the drug on the central nervous system. The ions *barium*, *strontium* and *potassium* stimulate smooth muscle and cause miosis. We have seen that mechanical stimulation of the peripheral end of the trigeminal nerve causes constriction of the pupil (p. 57) and this is due to the liberation of one or more prostaglandins previously called *irin* by Ambache.

## PUPILLARY UNREST

It is clear from the above that the state of the pupil at any moment is determined by a variety of synergic and antagonistic nervous influences; in general, external

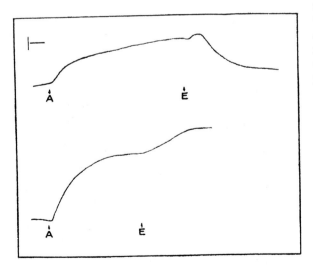

**Fig. 18.8**  Showing effects of epinephrine on the iris sphincter:
*Top*: At A contraction was induced by acetylcholine; at E relaxation is caused by epinephrine.
*Bottom*: After treatment with DCI, the contracture induced by acetylcholine is stronger and the effect of epinephrine is reversed into an additional contraction. (Van Alphen, Robinette and Macri, *Int. J. Neuropharmacol.*)

[*]Electrical stimulation of the excised cat sphincter causes contraction, abolished by atropine; in the presence of the latter, the contraction is replaced by active relaxation, presumably due to electrical stimulation of the adrenergic fibres (Schaeppi and Koella, 1964).

circumstances—light, and proximity of the fixation point—tend to cause constriction whilst the internal factors of sensation, and psychic activity generally, cause dilatation. The continual interplay between these opposing forces results in a constant state of pupillary activity—the pupil is restless—and the size of the pupil under any given conditions is a fluctuating quantity.

The fact that both pupils fluctuate in size in harmony shows that the phenomenon of pupillary unrest is determined centrally, and is not merely an example of the independent rhythmic activity so frequently seen in smooth muscle (Stark, *et al.*, 1958). The pathological condition in which this activity is abnormally great is called *hippus*.

# REFERENCES

Ahlquist, R. P. (1948) A study of the adrenotropic receptors. *Am. J. Physiol.* **153**, 586–600.

Ahlquist, R. P. (1962) The adrenotropic receptor. *Arch. int. Pharmacodyn.* **139**, 38–41.

Alexandridis, E. & Koeppe, E. R. (1969) Die spektrale Empfindlichkeit der für den Pupillenlichtreflex verantwortlichen Photoreceptoren beim Menschen. *v. Graefes' Arch. Ophthal.* **177**, 136–151.

Alpern, M. & Campbell, F. W. (1962) The spectral sensitivity of the consensual light reflex. *J. Physiol.* **164**, 478–507.

Ambachie, N. (1959) Further studies on the preparation purification and nature of irin. *J. Physiol.* **146**, 255–294.

Bender, M. B. & Shanzer, S. (1964) Oculomotor pathways defined by electric stimulation and lesions in the brainstem of the monkey. In *The Oculomotor System*, Ed. Bender, M. B., pp. 81–140. New York: Harper & Row.

Bennett, D. R., Reinke, D. A., Alpert, E., Baum, T. & Vasquez-Leon, H. (1961) The action of intraocularly administered adrenergic drugs on the iris. *J. Pharmacol.* **134**, 190–198.

Cavaggioni, A., Madarasz, I. & Zampollo, A. (1968) Photic reflex and pretectal region. *Arch. ital. Biol.* **106**, 227–240.

De Groot, S. G. & Gebhard, J. W. (1952) Pupil size as determined by adapting luminance. *J. Opt. Soc. Am.* **42**, 492–495.

Hess, W. R. (1957) *The Functional Organization of the Diencephalon.* New York: Grune & Stratton.

Jampel, R. S. (1959) Representation of the near-response on the cerebral cortex of the macaque. *Am. J. Ophthal.* **48**, pt. 2, 573–582.

Jampel, R. S. & Mindel, J. (1967) The nucleus for accommodation in the midbrain of the macaque. *Invest. Ophthal.* **6**, 40–50.

Lowenstein, O. & Loewenfeld, I. E. (1958) Electronic pupillography. *Arch. Ophthal.* **59**, 352–363.

Lowenstein, O. & Loewenfeld, I. E. (1959a) Scotopic and photopic thresholds of the pupillary light reflex in normal man. *Am. J. Ophthal.* **48**, pt. 2, 87–98.

Lowenstein, O. & Loewenfeld, I. E. (1959b) Influence of retinal adaptation upon the pupillary reflex to light in normal man. *Am. J. Ophthal.* **48**, pt. 2, 536–550.

Lowenstein, O. & Loewenfeld, I. E. (1969) The Pupil. In *The Eye*, 2nd ed., Ed. Davson, vol. III, pt. 2. London & New York: Academic Press.

Magoun, H. W. & Ranson, S. W. (1953) The central path of the light reflex. *Arch. Ophthal.* **13**, 791–811.

Nishida, S. & Sears, M. (1969a) Fine structural innervation of the dilator muscle of the iris of the albino guinea pig studied with permanganate fixation. *Exp. Eye Res.* **8**, 292–296.

Nishida, S. & Sears, M. (1969b) Dual innervation of the iris sphincter muscle of the albino guinea pig. *Exp. Eye Res.* **8**, 467–469.

Passatore, M. (1976) Physiological characterization of efferent cervical sympathetic fibers influenced by changes of illumination. *Exp. Neurol.* **53**, 71–81.

Passatore, M. & Pettorossi, V. E. (1976) Efferent fibers in the cervical sympathetic nerve influenced by light. *Exp. Neurol.* **52**, 66–82.

Patil, P. N. (1969) Adrenergic receptors of the bovine iris sphincter. *J. Pharmacol.* **166**, 299–307.

Pierson, R. J. & Carpenter, M. B. (1974) Anatomical analysis of pupillary reflex pathways in the rhesus monkey. *J. comp. Neurol.* **158**, 121–143.

Richardson, K. C. (1964) The fine structure of the albino rabbit iris with special reference to the identification of adrenergic and cholinergic nerves and nerve endings in its intrinsic muscles. *Am. J. Anat.* **114**, 173–184.

Scalia, F. (1972) The termination of retinal axons in the pretectal region of mammals. *J. comp. Neurol.* **145**, 223–245.

Schaeppi, U. & Koella, W. P. (1964) Adrenergic innervation of cat iris sphincter. *Am. J. Physiol.* **207**, 273–278.

Schweitzer, N. M. J. (1956) Threshold measurements on the light reflex of the pupil in the dark adapted eye. *Doc. Ophthal.* **10**, 1–78.

Sillito, A. M. (1968) The location and activity of pupilloconstrictor neurones in the mid-brain of the cat. *J. Physiol.* **194**, 39–40 P.

Sillito, A. M. & Zbrozyna, A. W. (1970) The localization of pupilloconstrictor function within the mid-brain of the cat. *J. Physiol.* **211**, 461–477.

Smith, J. D., Masek, G. A., Ichinose, L. Y., Watanabe, T. & Stark, L. (1970) Single neuron activity in the pupillary system. *Brain Res.* **24**, 219–234.

Stark, L., Campbell, F. W. & Atwood, J. (1958) Pupil unrest: an example of noise in a biological servomechanism. *Nature, Lond.* **182**, 857–858.

Straschill, M. & Hoffmann, K. P. (1969) Response characteristics of movement-detecting neurons in pretectal region of the cat. *Exp. Neurol.* **25**, 165–176.

Swanson, L. W., Cowan, W. M. & Jones, E. G. (1974) An autoradiographic study of the efferent connections of the ventral lateral geniculate nucleus in the albino rat and the cat. *J. comp. Neurol.* **156**, 143–163.

Van Alphen, G. W. H. M., Kern, R. & Robinette, L. (1965) Adrenergic receptors of the intraocular muscles. *Arch. Ophthal. N.Y.* **74**, 253–259.

Van Alphen, G. W. H. M., Robinette, S. L. & Macri, F. J. (1964) The adrenergic receptors of the intraocular muscles of the cat. *Int. J. Neuropharmacol.* **2**, 259–272.

Westheimer, G. & Blair, S. M. (1973) The parasympathetic pathways to internal eye muscles. *Invest. Ophthal.* **12**, 193–197.

# 19. Accommodation

The emmetropic eye is defined as one whose optical system is such as to cause images of distant objects to be focused on the retina, which is therefore at the second principal focus of the dioptric system. Images of near objects will necessarily be formed behind the retina (Fig. 6.7, p. 183) so that in order to bring these on to the retina an increase in dioptric power is necessary, and this increase is described as accommodation.

## AMPLITUDE OF ACCOMMODATION

As an object is moved closer and closer to the eye, at a certain point—*the near point of accommodation*—the eye can no longer increase its dioptric power and the object becomes indistinct. The far point of the eye has been defined as the position of an object such that its image is formed on the retina of the completely relaxed eye. Thus in the emmetrope this is infinity. We may define the *amplitude of accommodation*, therefore, as the difference in refracting power of the eye in the two states of complete relaxation and maximal accommodation. Thus if Q is the refracting power of the eye in maximum accommodation, and S its power when completely relaxed, the amplitude, A, is given by:

$$A = Q - S$$

Since the amplitude is a *difference* in refracting power of the eye in two conditions, it is not necessary to know its actual power in either; the amplitude may be calculated from the near and far points quite simply, as follows.

In Figure 19.1 the near and far points of a myopic eye are shown. It is required to find how much the power of the eye is increased when an object, formerly at the far point, is seen distinctly at the near point. To do this we can imagine that the eye, instead of accommodating, has remained in the relaxed condition and that the increased dioptric power to permit distinct vision of the object at the near point has been provided by a converging lens. The power of this lens would represent the increase in dioptric power of the eye in maximum accommodation. The completely relaxed eye sees distinctly an object at the far point; the lens must be chosen, therefore, of such a strength that it will form an image of an object, situated at the near point, at the far point. In the lens formula:

$$1/v - 1/u = 1/f$$

$v = r$, where $r$ is the distance of the far point in metres,

$u = p$, where $p$ is the distance of the near point in metres.

Thus $$1/r - 1/p = 1/f.$$

If we let $1/r$ in metres be R; $1/p$ be P; the power of the lens, A, is given by:

$$R - P = A$$

R may be called the *dioptric value of the far point distance*; P the *dioptric value of the near point distance*, and A is the amplitude of accommodation, in dioptres.

This relationship between R, P, and A has been derived from the lens formula, consequently it will only hold if the usual sign convention is applied. Thus in the case of the myope R and P are both negative.

EMMETROPE AND HYPERMETROPE

In the emmetrope the Far Point is infinity, so that R is zero and the formula becomes:

$$A = -P,$$

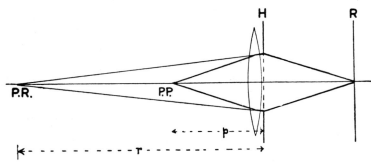

**Fig. 19.1** Illustrating the amplitude of accommodation in a myopic subject. At maximum accommodation the extra dioptric power exerted is equivalent to that given by placing a thin lens of focal length p in front of the eye.

so that if the Near Point is, say, 10 cm or 0·1 m, the amplitude is $1/0·1 = 10$ dioptres. In the hypermetrope the Far Point is a fictitious point behind the eye, so that R is positive and the same Near Point-distance of 10 cm would correspond to a greater amplitude than 10 dioptres. This becomes clear when we realize that, in order to see distant objects distinctly, the hypermetrope has to make use of a part of his total amplitude.

## THEORY OF ACCOMMODATION

According to the modern theory of accommodation, which is largely based on Helmholtz's ideas, the capsule has sufficient elasticity to mould the lens into a more strongly curved system than is necessary for distant vision. The elasticity of the capsule is held in check by the normal tension in the zonule, so that accommodation consists in a relaxation of the tension in the zonule which permits the capsule to mould the lens into a more strongly curved system. The relaxation of the zonule follows a *contraction* of the ciliary muscle.

## THE CILIARY MUSCLE

### THE FIBRE TYPES

The ciliary muscle is classically described as containing three types of unstriped muscle fibre:

1. *Meridional fibres* arising from the epichoroid and attaching to the scleral spur.
2. *Radial fibres*, situated more internally and anteriorly, arranged in the manner of a fan.
3. *Circular fibres*, continuous with, and inseparable from the radial fibres, running round the free edge of the ciliary body just behind the root of the iris.

However, Fincham has shown that this formulation is rather too rigid; the fibres take origin from the epichoroid, and in the outer part of the muscle (nearest the sclera) they run a purely meridional course, to be inserted in the scleral spur. More internally, the fibres make up a branching meshwork and it is this part that has been incorrectly called the radial portion—the fibres are not arranged in a regular radial manner but, as indicated, in a reticular meshwork. As a result of the branching of the bundles of fibres, a number acquire a more or less circular direction, a tendency that becomes more pronounced towards the inner edge of the muscle. Thus the innermost edge consists essentially of circular fibres—the so-called *sphincter muscle of Müller*—but they are not to be regarded as making up a separate muscle. The circular elements give off short tufts of fibres which are inserted into a ring of elastic tissue (A of Fig. 19.2) at the base of the iris. Another layer of elastic tissue (B) is situated between the muscle and the pigment epithelium of the ciliary body. The elastic elements presumably resist the tension of the zonule.

### EFFECTS OF CONTRACTION

The general anatomical characteristics of the ciliary muscle indicate that its contraction must relax the

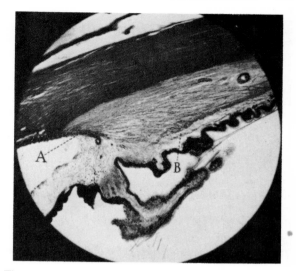

**Fig. 19.2** The ciliary muscle as described by Fincham. A and B indicate elastic tissue. (*Brit. J. Ophthal.*)

zonule; the meridional fibres pull on the epichoroidal tissue and drag the ciliary body forward. The circular fibres, probably the most important in accommodation, must, on shortening, decrease the radius of the circle that they constitute; this can only be achieved by a radial extension of the ciliary body which must reduce the tension in the zonule (Fig. 19.3). These anatomical considerations are confirmed by the observation that during accommodation the external surface of the ciliary body moves forward by 0·5 mm. Moreover, careful observation of the lens during accommodation shows that it becomes tremulous as though its normal taut suspension had been relaxed; again, the equator of the capsule of the unaccommodated lens shows fine wrinkles, proving that it is held under tension; during accommodation these wrinkles are smoothed out.

## THE ZONULAR SYSTEM

### STRUCTURE

The actual mechanism whereby contraction of the ciliary muscle affects the tension in the zonular fibres is not com-

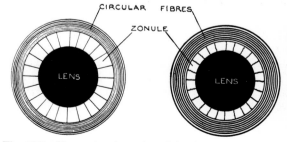

**Fig. 19.3** Illustrating the action of the circular fibres in relaxing tension in the zonule.

pletely clear, since the muscle has no direct connections with the fibres. The structure of the human zonule has been examined in some detail by Rohen and Rentsch (1969) with a view to elucidating this point. According to this study, the fibres fall into two classes, the separation into two groups occurring in the middle third of the ciliary body; behind this, the system is a uniform mat of fibres. As indicated in Figure 19.4 the main system of 'holding fibres', H, runs forward towards the pars orbicularis where the individual fibres split at the *zonular fork*, ZF, and finally both branches insert into the lens capsule. Branching off from the holding fibres is the second system of *tension fibres*, T, inserting into the ciliary epithelium in the orbicularis region of the ciliary body. It seems that it is because of the close relation of these tension fibres to the ciliary epithelium on the ciliary processes, and in the valleys between, that effects of changes in volume of the ciliary body, due to contraction of the muscle, are transmitted to the lens.

### CHANGES WITH ACCOMMODATION

Thus Figure 19.4 illustrates the postulated changes; below, the muscle is relaxed in the unaccommodated state, and the holding fibres are stretched, maintaining tension on the capsule. Above, the muscle has contracted and the changed shape of the ciliary body has caused the tension fibres to become stretched, and this has permitted the main holding fibres to relax. It is essentially the tension fibres that cause the important change in tension in the capsule, and they do this by virtue of their attachment to the ciliary epithelium, the changed shape of the ciliary body causing a forward and axial pull.

## CHANGES IN THE LENS

### PURKINJE IMAGES

The actual changes in the lens during accommodation were studied by Helmholtz by means of the Purkinje images; during accommodation image III, formed by the anterior surface of the lens, became smaller, indicating an increase in the curvature of this surface (Fig. 19.5). Studies on individuals with congenital aniridia have shown that the lens does become smaller (1·25 mm) and thicker (1·3 mm) as Helmholtz's theory would demand.

### ANTERIOR SURFACE

Fincham's work suggested that the central zone of the anterior surface became more convex in relation to the

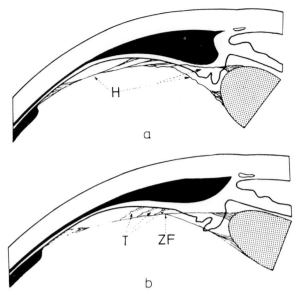

**Fig. 19.4** Illustrating Rohen and Rentsch' concept of the structure of the zonular fibres. (a) The accommodated state. (b) The relaxed state. H are the *holding fibres* which split at the *zonular fork*, ZF. Branching from the holding fibres are the *tension fibres*, T, inserting into the ciliary epithelium. (After Rohen and Rentsch, *v. Graefes' Arch. Ophthal.*)

peripheral parts, the surface becoming what Fincham called 'conoid', but Fisher's (1969) measurements of the lens in both accommodated and unaccommodated states have shown that the anterior surface may be described accurately as that of an ellipsoid; Fisher also calculated that the energy required for this postulated conoid transformation could not be supplied by the capsule.

### REARRANGEMENT OF LENS SUBSTANCE

Whatever the exact changes may be, they must involve a rearrangement of the lens substance since this is quite incompressible; the general increase in curvature must mean a decrease in the surface of the lens and this must mean that the shape approaches more closely to that of a sphere (in the sphere we have the condition of maximum

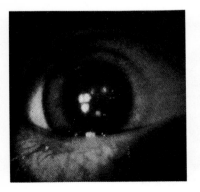 

**Fig. 19.5** The change in the size of the Purkinje image, formed by the anterior surface of the lens, following accommodation. The object, whose image was formed by the reflecting surfaces of the eye, consisted of three bright lamps arranged in a triangle. Left: unaccommodated eye. Right: accommodated eye. (Fincham, *Brit. J. Ophthal.*)

volume for minimum area; if we decrease the surface of a body and maintain its volume the same, we must make it more nearly spherical). The effects of this rearrangement may be observed with the slit-lamp microscope when there are mobile opacities in the lens; the movement consists generally of an axial flow from the periphery, as we should expect.

### ZONULAR RIDGES

The attachment of the zonular fibres at the equator gives rise to an anterior and a posterior equatorial ridge, the posterior one being visualized by gonioscopy and the anterior in subjects with iris colobomata (Brown, 1974). The changes in these ridges during stimulation and paralysis of accommodation support strongly the Helmholtz-Fincham hypothesis of accommodation, the ridges becoming more pronounced in the relaxed phase when the tension is expected to be greatest.

### EXCISED LENS

It follows from the theory of accommodation that the excised lens must be in a state of maximum accommodation. Hartridge and Yamada estimated the dioptric power of the cat's eye *in situ* by measuring the magnification of a portion of the fundus observed through a modified ophthalmoscope. On removal of the lens its curvatures were measured and its dioptric power calculated. They showed that the combined power of the cornea and excised lens amounted to 12 dioptres more than the power of the eye *in vivo*, i.e. the lens, with the curvatures it possessed in the excised state, would provide the eye with an extra 12 dioptres of refractive power. However, the experimentally determined amplitude of accommodation in the cat is only about 2 D (Ripps *et al.*, 1961); moreover Fisher (1971), from his measurements of Young's modulus of the cat lens, together with the capsular energy, concluded that any change in shape of the lens would be out of the question, and the same was true of the rabbit.* Again Vakkur *et al.* (1963), in their valuable study of the dioptrics of the cat's eye, concluded that there was little or no accommodative power, any change that did occur being more likely to be due to a shift in position of the lens than a change in shape, and they showed that Hartridge and Yamada's conclusion was based on a miscalculation of the power of the eye *in situ*. In fact O'Neill and Brodkey (1969) have measured a linear movement of the anterior pole of the cat's lens associated with increased power when stimulating the ciliary ganglion; the average maximum accommodation exerted was 1·72 D, corresponding to a movement of 0·57 mm.

## THE CHANGE IN ACCOMMODATION WITH AGE

### CILIARY MUSCLE

The amplitude of accommodation decreases progressively with age (Fig. 19.6) falling from some 14 dioptres in a child of ten years to about zero at the age of 52 (Hamasaki *et al.*, 1956). It has been argued that the failure is due to atrophy of the ciliary muscle, but the study of Swegmark (1969), in which he measured ciliary muscular activity by means of its changes in electrical impedance, measured through four electrodes on the perilimbal region of the eye. Through one pair an

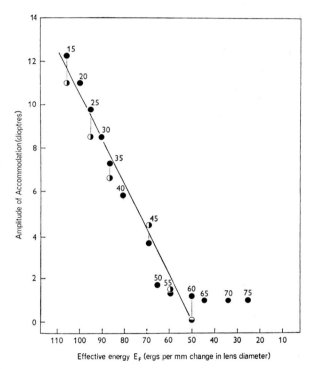

**Fig. 19.6** Showing presbyopic changes in amplitude of accommodation and effective energy of the lens capsule. ● Duane's data; ◑ Bruckner's data mean per decade; ◓ Hamasaki's data. Age in years (15 to 75). (Fisher, *J. Physiol.*)

---

*With this technique, which avoids the damaging procedure of compression, Fisher (1971) showed that the lens is a truly elastic body and therefore repeated deformations, as in accommodation, will not tend to 'set' the lens in a given shape. The technique permitted the calculation of Young's modulus of elasticity in the polar and equatorial regions. A necessary condition for the attainment of ellipsoidal shapes at different degrees of flattening (accommodation) is that the strain at the pole should be twice that at the equator, and the finding that the ratio of polar to equatorial elasticities was, in man, 0·62 indicates that the human lens is adapted to permit this; by contrast, the cat and rabbit had ratios of about 3.

alternating current was passed and through the other impedance was measured. The ciliary contractile power, measured in this way, remained the same up to the age of 60 years; Swegmark emphasized, moreover, that a disuse atrophy of the ciliary muscle was extremely unlikely to occur, since the ciliary muscle would be activated during convergence of the eyes—part of the near response (p. 489)—whether or not accommodation took place.

## LENS CAPSULE

Helmholtz's view, that there is a failure of the lens capsule to modify the shape of its contents receives far more support. Thus Fincham observed that the curvature of the excised senile lens was considerably less than that of a juvenile one. This failure could be due to a hardening of the lens material, *sclerosis*, to a decrease in modulus of elasticity, or to a decrease in thickness, of the capsule, or to all three of these factors.

### CAPSULAR SURFACE TENSION

In an attempt to assess the importance of the various factors, Fisher (1969b) measured the capsular surface tension, i.e. the energy that unit area the capsule can exert, at rest and during accommodation; this was computed from the measured changes in outline of the anterior capsule and the force required to stretch it. In Figure 19.6 the effective energy of the capsule is plotted against the amplitude for lenses taken from different aged eyes; there is a remarkable correspondence between the two since the capsule actually increases in thickness with age up to 60 years; this suggests that the modulus of elasticity, which decreases from some $6 \times 10^7$ dynes/cm$^2$ in childhood to $1 \cdot 5 \times 10^7$ dynes/cm$^2$ in extreme old age, plays an important role. However, this parallelism does not rule out a significant contribution of increased stiffness of the lens material; it does show, however, that the earlier tendency to attribute presbyopia entirely to lens sclerosis is unjustified.*

When the effects of altered elasticity and lens shape† were taken into account, the measurements indicated that, in the absence of changed power of the ciliary muscle, about 50 per cent of the loss of accommodative power would be due to lens hardening.

## FISHER'S RECENT STUDIES

### MUSCLE FORCE

In a more elaborate study in which lenses of freshly enucleated cadaver eyes were studied with their zonular and ciliary attachments intact, Fisher (1977) was able to deduce, not only the changes in the ability of the lens to alter its shape passively in response to a change in diameter of the ciliary ring, but also to compute the probable changes in muscle tension that took place at different degrees of accommodative effort; and he concluded that the basic concept of Helmholtz was, indeed, correct, namely that presbyopic changes were due to altered physical characteristics of the lens, whilst the actual power of the ciliary muscle to develop tension actually increased with age up to about 45 years old (Fig. 19.7).

### CILIARY RING

Fisher placed the lens-ciliary body preparation in an apparatus that enabled him to increase or decrease the

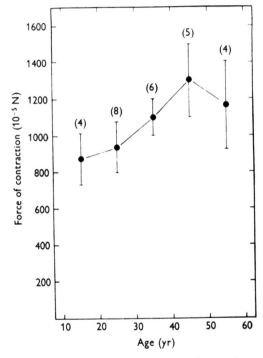

**Fig. 19.7**   Maximum force of contraction of the entire ciliary muscle plotted against age. Bracketed numbers indicate numbers of subjects. Vertical lines standard deviation per decade. (Fisher, *J. Physiol.*)

---

* The difficulty in accepting lens sclerosis as the sole cause of presbyopia is that the condition begins essentially in childhood, when it is unlikely that 'senile' changes in the lens fibres would have occurred.
† The effective energy of the capsule, in relation to its power of altering refractive power, depends on the shape of the lens as well as on the modulus of elasticity; the flatter human lens allows the capsule to exert a greater moulding pressure for a given modulus of elasticity than the more spherical lenses of the cat and rabbit; to some extent, then, the differences in accommodative power of these species are attributable to this since their elasticities are not greatly different: 5, 0·9 and $2 \cdot 5 \times 10^7$ dynes/cm$^2$ respectively.

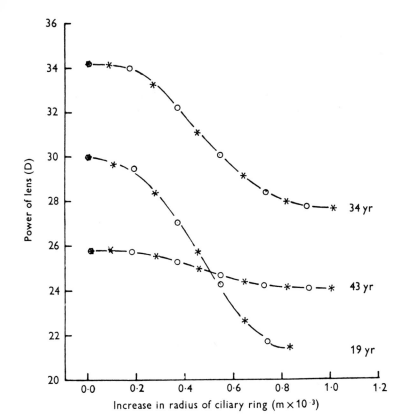

**Fig. 19.8**  Typical dioptric power/ciliary ring strain curves of human lenses. *Note*: 0·0 radius of ciliary ring corresponds to the unaccommodated state of the lens. *, increase in radius; ○, decrease in radius. (Fisher, *J. Physiol.*)

radius of the ciliary ring by gentle traction; Figure 19.8 shows the measured changes in dioptric power of the lens with a given increase in radius of the ciliary ring, and it is immediately evident that in the old lens the dioptric power shows little increase with decreasing radius of the ring, so that contractions of the ciliary muscle that reduce the size of the ring would contribute little to the dioptric power of the eye.

ESTIMATED MUSCLE FORCE

Fisher measured the force required to be exerted on the equator of the capsule to produce a given change of shape, and thus of dioptric power; this was done by spinning the lens around its antero-posterior axis, which has the effect of flattening the lens in a manner comparable to that brought about by tension in the zonule (Fisher, 1971), and by an appropriate calculation he was able to relate a given dioptric change with a change in force, and in this way he computed the force of contraction exerted by the entire ciliary body. The equation used was:

$$F_{CB} = \tfrac{2}{3}\pi a_s^2 \{a_u t_u \Omega_u^2 - a_a t_a \Omega_a^2\}p,$$

where  $\Omega_a \rightarrow \Omega_u$ is such $st_s = st_a$.

$F_{CB}$ = radial force exerted by the entire ciliary body (N).
  $a_a$ = equatorial radius of accommodated lens (m)
  $a_u$ = equatorial radius of unaccommodated lens (m)
  $a_s$ = equatorial radius of spinning lens (m)
  $t_u$ = polar thickness of unaccommodated lens (m)
  $t_a$ = polar thickness of accommodated lens (m)
  $st_a$ = change in thickness of lens under zonular tension (m)
  $st_s$ = change in thickness of lens when spinning (m)
  $\rho$ = density of lens substance (kg m$^{-3}$)
  $\Omega_a$ = equivalent speed of rotation of lens when eye is accommodated (rad sec)
  $\Omega_u$ = equivalent speeds of rotation of lens when eye is unaccommodated (rad sec$^{-1}$).

Thus the important experimental parameters were the equivalent speeds of rotation of the lens to produce the shape when the eye is accommodated and unaccommodated. In this way the points in Figure 19.9 were calculated. Thus to predict the amplitude of accommodation, D, from the force that the ciliary muscle can exert and the forces required to produce a change in shape Fisher gave the equation:

$$D = K_{df} \sqrt{F_{CB}}$$

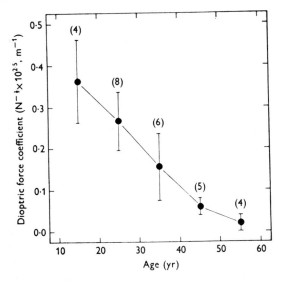

**Fig. 19.9** Dioptric force coefficient plotted against age. Bracketed numbers indicate numbers of subjects. Vertical lines standard deviation per decade. (Fisher, *J. Physiol.*)

$K_{df}$ is a dioptric force-coefficient and $F_{CB}$ is the force of contraction of the ciliary muscle; as Figure 19.9 shows, this coefficient reaches a value of zero between ages 50 and 60.★

DEPTH OF FOCUS

Fisher points out that the computed amplitudes are some 1·25 D less at each age than those obtained *in vivo*, when amplitude of accommodation was measured by the 'blurring technique'. However, if amplitude is measured objectively by retinoscopy a smaller value is obtained (Hamasaki *et al.*, 1956; Miles, 1953), and, as Figure 19.6 shows, accommodative power, measured in this way, becomes zero at age 50 about. Thus the 'accommodation' apparently exerted by subjects over 52 years is in fact fictitious and a reflexion of the subject's depth of focus, i.e. the degree of blurring of the optical image he is prepared to accept.

VALUE OF MAGNIFICATION

Hamasaki *et al.* pose the problem as to why subjects, who have reached the age of 52, with no accommodative power still require stronger presbyopic corrections as they get older. The answer is to be sought in the larger image provided by the increased spectacle power; thus the subject is using the spectacles as a younger subject would use a magnifying glass; in both situations distinct vision is possible with the eye closer to the object being viewed, and this means an effective increase in magnification.

ACCOMMODATION FOR NEAR VISION

The amount of accommodative power for near vision, as in reading, is about 3·5 D when measured by the blur technique, and accepting this value, we may calculate the percentage of the total ciliary muscle force available required for near vision at different ages; in the young person this is small, namely about 5 per cent, but with age this percentage increases rapidly to 100, indicating that, with onset of presbyopia, it is likely that the subject is using maximal ciliary muscle power, a condition that might produce 'eye-strain' rendering the use of spectacles desirable before near vision is impossible. However, as Fisher emphasizes, the extra muscle force that becomes available due to hypertrophy with age, indicated by Figure 19.7, is only sufficient to provide 0·6 D of extra accommodative power.

## THE NERVOUS MECHANISM

PARASYMPATHETIC

Contraction of the ciliary muscle is brought about by the activity of the parasympathetic fibres of N III, with their origin in the Edinger-Westphal nucleus; the fibres relay in the ciliary ganglion. Stimulation of N III causes an increase in the refractive power of the eye (the eye becomes myopic), an effect that is inhibited by blocking the ciliary ganglion or by the application of atropine to the eye. Eserine, which potentiates the action of acetylcholine, and pilocarpine, which mimics its action, cause a spasm of accommodation, that by the former being due to failure to hydrolyse the transmitter liberated as a result of tonic action of the nerve.

### Alternate pathway

Westheimer and Blair (1973) found, in the monkey, that painting the ciliary ganglion with nicotine reduced, but did not abolish, the accommodative response to stimulation of N III intra-orbitally; furthermore, when they stimulated N III at high frequency (440 Herz) they obtained full accommodation although at this frequency transmission through the ganglion would be impaired or blocked. They concluded, therefore, that fibres to the ciliary muscle passed through the ciliary ganglion without relaying.

---

★An earlier study (Fisher 1973) led to the same conclusions regarding the cause of presbyopia; here a 'lens coefficient' was computed from the elasticities of capsule and lens substance and the dimensions of the lens in accommodated and unaccommodated states; this gives a measure of the ability of the capsule to overcome the forces preventing the lens from assuming a more spherical shape. The line relating this coefficient to age coincided well with that relating accommodation to age. The three factors leading to presbyopia are thus the decreased elastic modulus of the capsule, the increased modulus of the lens substance, and a flattening of the lens.

## SYMPATHETIC

Opinions have varied as to whether the sympathetic plays any role in accommodation (see, e.g. Cogan, 1937); the action of the ciliary muscle is antagonized by the elasticity of the capsule, and fine adjustments are therefore possible by means of the interplay of these opposing forces; an antagonistic muscle, innervated by the sympathetic, whose contraction tightens the zonule and therefore flattens the lens, may be thought to be unnecessary in man on this account. Some studies of Olmsted and Morgan and their colleagues have suggested, however, that the sympathetic nerves might cause an active flattening of the lens, insults likely to cause generalized sympathetic stimulation, such as a tap on the nose of a rabbit, being said to cause an increase of the animal's normal hypermetropia by about 1 D; stimulation of the cervical sympathetic in the anaesthetized cat and dog caused large changes in refractive condition, especially if the parasympathetic division to the eye was cut.

In primates Törnqvist (1966, 1967) has shown that stimulation of the cervical sympathetic causes a small decrease in the dioptric power of the eye; this becomes greater if the stimulation is carried out against a background of parasympathetic activity, either by pretreatment of the eye with pilocarpine or by stimulation of the oculomotor nerve; it was never greater than 1·5 to 2·0 D, however. He pointed out that the responses were so slow, taking some 5 to 10 seconds to begin and 10 to 40 seconds to reach a maximum, that they are unlikely to have physiological significance. These effects could, conceivably, be due to the presence of genuine radial fibres, innervated by the sympathetic, whose action would be, in contraction, to tighten the zonule, although it has been argued that the effects are due to constriction of the blood vessels in the eye with a consequent decrease in volume of the ciliary body leading to an increase in the pull on the zonule. Thus Fleming and Olmsted (1955) observed that the transient myopia that occurs on cutting the cervical sympathetic runs a parallel course to the transient dilatation of blood vessels. However, Törnqvist (1966) found that the effects of sympathetic stimulation were not blocked by a typical sympathetic α-blocker, phentolamine, and this would be expected if the action depended on the vascular effects of sympathetic stimulation; again, the effect on accommodation was inhibited by a β-blocker, propanolol, which had no effect on the vasoconstriction due to sympathetic stimulation.*

## Excised muscle

Studies on excised strips of ciliary muscle leave little doubt that sympathetic amines are active, but the effect varies with the species; in the cat the order of increasing effectiveness in causing relaxation is Isoproterenol ≫

Noradrenaline > Adrenaline, an order indicating a β-action of the sympathetic, especially since the relaxation due to isoproterenol is blocked by DCI. That there are also α-receptors is indicated by the fact that the relaxation caused by adrenaline and noradrenaline is converted to contraction when the β-activity is blocked by DCI (Bennett et al., 1961). In the monkey, sympathetic activity is exclusively β whilst in the rabbit the action is mainly α with a little β.

## THE STIMULUS FOR ACCOMMODATION

It might be thought that the effective stimulus for accommodation would be the blurring of the retinal image, but if this were the only clue, the centre responsible would be unable to determine whether the object had come nearer or moved farther away, since in both cases the image would be blurred. We should therefore expect the eye to 'grope' for the correct change in power, whereas it was found by Fincham (1951) that when the eye observes an object, and the vergence† of the light is altered, the accommodative effort is correct; that is, if the vergence is changed so that the object is apparently closer, the eye increases its power, and *vice versa*. There is no oscillation, as we should expect in an instrument unable to distinguish the nature of the vergence-change.

### ABERRATIONS

What then is the stimulus? Fincham's experiments indicated that in many subjects the chromatic aberration of the retinal image provided the necessary clue, the aberration rings being theoretically different according to the nature of the out-of-focus image. Thus, if the object were too close, the image of a point would be a light disc surrounded by a red fringe, whilst if it were too far away the disc would be surrounded by a blue fringe. By using monochromatic light many subjects were unable to make correct adjustments. A certain number of subjects were able to do so, however, and it

---

*Biggs, Alpern and Bennett (1959) found a decrease in accommodative power in humans after subconjunctival injections of adrenaline.
†By *vergence*, in this context, is meant the degree of divergence or convergence of the rays of light in a pencil as they strike the principal plane of a refracting system; the closer an object, the greater the vergence of its rays, hence the reciprocal of its distance from the principal plane may be used as a measure of vergence; if this is in metres the unit of vergence is the dioptre; thus the vergence of the light from an object is the *dioptric value of the object-distance*. We may note that the term vergence is also applied to the degree of convergence or divergence of the fixation axes; this often leads to ambiguities and confusion. Thus by 'vergence-induced accommodation' a writer could mean accommodation induced by convergence or by changed light-vergence.

would seem that spherical aberration provides a clue (Campbell and Westheimer, 1959) since reducing the spherical aberration of the eye reduced the subject's accuracy in accommodating in monochromatic light.*

## FLUCTUATIONS

When a subject views a near object steadily his accommodation fluctuates through a range of about 0·25 D with a dominant frequency of the order of 2 cycles/sec (Campbell, Robson and Westheimer, 1959); this is presumably analogous with pupillary unrest, and with the high-frequency tremor in the eye movements although here the frequency is in the range of 75 cycles/sec. Since the unrest is considerably reduced when the subject views an empty field it is possible that these oscillations are determined by variations in the clarity of the image—retinal feedback.

# DEPTH OF FOCUS OF THE EYE

In a lens system there is theoretically only one position of the object that will give an image at a definite distance from the principal plane; hence with the eye, where the position of the retina is fixed, there is only one position, for a given state of the dioptric power, where an object will produce a distinct image. For the emmetropic eye, this is at infinity. When the object is brought nearer to the eye, the image of a point on it becomes a 'blur-circle' because the true image is behind the retina. If the blur-circle is not large, however, it is not recognized as differing from a point and the object may thus be brought within a finite distance of the eye before the blur-circle becomes so large that the image is appreciated as 'indistinct'. The extent to which the object may be moved before the points on it appear out of focus is called the *depth of focus*. As Figure 19.10 shows, the depth of focus obviously depends on the pupil-size. It will also depend on the distance of the fixation point from the eye; if this is large (e.g. 100˙m) the point may be moved many metres closer to the eye before it goes out of focus; if it is small (e.g. 25 cm) a movement of a centimetre or two is enough to blur the image. For the human eye the figures obtained by Campbell were as follows:

| Pupil diameter | Depth at infinity | Depth at 1 metre |
|---|---|---|
| 1 mm | From infinity to 1·25 m | 5·0 m to 56 cm |
| 2 mm | 2·33 m | 1·8 m to 70 cm |
| 3 mm | 2·94 m | 1·5 m to 75 cm |
| 4 mm | 3·57 m | 1·4 m to 78 cm |

## VISUAL ACUITY

Ogle and Schwartz presented the subject with a chart and changed the vergence of light from it by means of a

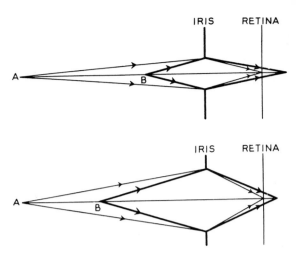

**Fig. 19.10**   Illustrating the influence of the size of the pupil on the depth of focus. With the small pupil (above) B produces a blur-circle of fixed size when it is closer to the eye than with the large pupil (below). The depth of focus (AB) is smaller with the larger pupil.

lens, keeping the image-size constant. At a certain point the visual acuity was decreased and the change of vergence was recorded. This amounted to 0·6 dioptres and Ogle and Schwartz defined the depth of focus in terms of this amount of vergence. The depth of field, on the other hand, could be defined as the equivalent distance through which the chart would have been moved to produce this change of vergence. The *hyperfocal distance* was defined as the distance of the fixation point at which objects beyond this all appear in focus.

### TOLERATED CHANGE OF VERGENCE

Adamson and Fincham measured the change of vergence tolerated by the eye; i.e. if an object was made to appear closer or farther away, by means of lenses, a point was reached when the eye adjusted its accommodation, and this was indicated by a change of its refracting power as measured with an optometer. The average change in vergence tolerated by the eye was found to be ±0·25 D. An object 1 m away (vergence −1 D) may be moved so that the vergence changes to

---

*Fincham (1951) observed that if a subject, who failed to accommodate correctly in monochromatic light, was allowed to scan the object in view, correct accommodation became possible, and he pointed out that when the eye turns through a small angle from the point of fixation the rays of light at the opposite sides of the blur-circles on the retina acquire different degrees of obliquity to the surface, this difference being opposite in direction when the image is out of focus due to being too close and too far away. The different degrees of obliquity will be translated into different sensations of brightness by the operation of the Stiles-Crawford effect.

−0·75 D or to −1·25 D without blurring of the image; i.e. it can be moved to 1·33 m or 0·80 m, giving a depth of focus of 53 cm. If the object is 25 cm away (vergence −4 D) the limits of tolerance are −3·75 D and −4·25 D, representing a depth of focus of 3·0 cm (26·6 −23·6 cm). The difference between the values found by Ogle and Schwartz and Fincham depends on their different criteria.⋆

## THE RESTING POSITION OF ACCOMMODATION

On the classical view, e.g. that of Helmholtz, the emmetropic subject's relaxed state of accommodation, when there is no stimulus for distinct vision, is zero, i.e. adjusted for infinity. An alternative suggestion, that the eye tends to adopt an intermediate condition, being focussed at about 1 metre, has been put forward. Experimentally, Hennessy et al. (1976) have approached the problem by allowing subjects to view objects, at different distances, through artificial pupils of smaller and smaller size. The increased depth of focus provided by the small pupils reduces the necessity for accommodation on nearer objects, so that by extrapolation it could be estimated what state of accommodation the eye would take up when there was no necessity to accommodate. Figure 19.11 illustrates some results; the theoretical line indicates the amounts of accommodation required for the different distances, and the filled circles are the experimental values for an emmetropic subject with normal pupils; the slopes are similar indicating, as we should expect, the adoption of appropriate increases in dioptric power with proximity of the target, and extrapolating to about 6 metres for zero accommodation. As the pupil is reduced in size, there is the expected flattening of the line, indicating smaller accommodation for a given fixation-distance; and at the lowest pupillary size the state of accommodation is nearly constant at the value required for 1 metre viewing-distance, suggesting that this, indeed, is the true relaxed condition.†

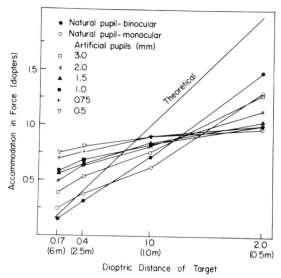

**Fig. 19.11** Accommodation as a function of dioptric distance for various sized artificial pupils. The theoretical line indicates values expected if accommodation corresponded to target distance. (Hennessy et al., Vision Res.)

⋆Campbell and Westheimer (1958) prevented their subjects from accommodating with atropine and they were asked to recognize the rhythmic blurring of the image of an object brought nearer to and farther from the eye; when this rhythmic excursion was too small, no change was experienced, and on increasing it a point was reached when the rhythmical changes were noted. The average 'sensitivity to difference of focus' was 0·2 D. Oshima (1958) found a much higher sensitivity, 0·02–0·06 D.

† The relation of this finding to 'instrument myopia' is discussed by Hennessy et al. This is the unnecessary accommodation exerted by the viewer through, say, a microscope, where the position of the image is normally adjusted to infinity. The exit-pupil of the microscope is about 2 mm or smaller, so that the depth of focus is large and the eye adjusts its focus to that in the resting state (Hennessy, 1975). Charman and Tucker (1977) have examined the effects of the spatial frequency of a grating on the accommodation stimulus; in the extreme case of zero frequency we have the 'empty field' situation and the myopia was some 3 D.

## REFERENCES

Adamson, J. & Fincham, E. F. (1939) The effect of lenses and convergence upon the state of accommodation of the eye. Trans. Ophthal. Soc. U.K. 59, 163–179.

Biggs, R. D., Alpern, M. & Bennett, D. R. (1959) The effect of sympathomimetic drugs upon the amplitude of accommodation. Am. J. Ophthal. 48, pt. 2, 169–172.

Brown, N. (1974) The shape of the lens equator. Exp. Eye Res. 19, 571–576.

Campbell, F. W., Robson, J. G. & Westheimer, G. (1959) Fluctuations in accommodation under steady viewing conditions. J. Physiol. 145, 579–594.

Campbell, F. W. & Westheimer, G. (1958) Sensitivity of the eye to difference in focus. J. Physiol. 143, 18 P.

Campbell, F. W. & Westheimer, G. (1959) Factors influencing accommodation responses of the human eye. J. opt. Soc. Am. 49, 568–571.

Charman, W. N. & Tucker, J. (1977) Dependence of accommodation response on the spatial frequency spectrum of the observed object. Vision Res. 17, 129–139.

Cogan, D. G. (1937) Accommodation and the autonomic nervous system. Arch. Ophthal. 18, 739–766.

Fincham, E. F. (1937) The mechanism of accommodation. Br. J. Ophthal., Mon. Suppl. No. 8.

Fincham, E. F. (1951) The accommodation reflex and its stimulus. Br. J. Ophthal. 35, 381–393.

Fisher, R. F. (1969a) Elastic constants of the human lens capsule. J. Physiol. 201, 1–19.

Fisher, R. F. (1969b) The significance of the shape of the

lens and capsular energy changes in accommodation. *J. Physiol.* **201**, 21–47.

Fisher, R. F. (1971) The elastic constants of the human lens. *J. Physiol.* **212**, 147–180.

Fisher, R. F. (1973) Presbyopia and its changes with age in the human crystalline lens. *J. Physiol.* **228**, 765–779.

Fisher, R. F. (1977) The force of contraction of the human ciliary muscle during accommodation. *J. Physiol.* **270**, 51–74.

Fleming, D. G. & Olmsted, J. M. D. (1955) Influence of cervical ganglionectomy on the lens of the eye. *Am. J. Physiol.* **181**, 664–668.

Hamasaki, D., Ong, J. & Marg, E. (1956) The amplitude of accommodation in presbyopia. *Amer. J. Optom.* **33**, 3–14.

Hennessy, R. T. (1975) Instrument myopia. *J. opt. Soc. Am.* **65**, 1114–1120.

Hennessy, R. T., Iida, T., Shina, K. & Leibowitz, H. W. (1976) The effect of pupil size on accommodation. *Vision Res.* **16**, 587–589.

Morgan, M. W., Olmsted, J. M. D. & Watrus, W. G. (1940) Sympathetic action in accommodation for far vision. *Am. J. Physiol.* **128**, 588–591.

Ogle, K. N. & Schwartz, J. T. (1959) Depth of focus of the human eye. *J. opt. Soc. Am.* **49**, 273–280.

Olmsted, J. M. D. & Morgan, M. W. (1939) Refraction of the rabbit's eyes in the unexcited and excited state. *Am. J. Physiol.* **127**, 602–604.

Olmsted, J. M. D. & Morgan, M. W. (1941) The influence of the cervical sympathetic nerve on the lens of the eye. *Am. J. Physiol.* **133**, 720–723.

O'Neill, W. D. & Brodkey, J. S. (1969) Linear regression of lens movement with refractive state. *Arch. Ophthal.* **82**, 795–799.

Oshima, S. (1958) Studies on the depth of focus of the eye. *Jap. J. Ophthal.* **2**, 63–72.

Ripps, H., Breinin, G. M. & Baum, J. L. (1961) Accommodation in the cat. *Trans. Am. Ophthal. Soc.* **59**, 176–193.

Rohen, J. W. & Rentsch, F. J. (1969) Der konstruktive Bau des Zonulaapparatus beim Menschen und dessen funktionelle Bedeutung. *v. Graefes' Arch. Ophthal.* **178**, 1–19.

Swegmark, G. (1969) Studies with impedance cyclography on human ocular accommodation at different ages. *Acta ophthal.* **47**, 1186–1206.

Törnqvist, G. (1966) Effect of cervical sympathetic stimulation on accommodation in monkeys. *Acta physiol. Scand.* **67**, 363–372.

Törnqvist, G. (1967) The relative importance of the parasympathetic and sympathetic nervous systems for accommodation in monkeys. *Invest. Ophthal.* **6**, 612–617.

Vakkur, G. J., Bishop, P. O. & Kozak, W. (1963) Visual optics in the cat, including posterior nodal distance and retinal landmarks. *Vision Res.* **3**, 289–314.

Westheimer, G. & Blair, S. M. (1973) The parasympathetic pathways to internal eye muscles. *Invest. Ophthal.* **12**, 193–197.

# 20. The near response

## CONVERGENCE, ACCOMMODATION AND MIOSIS

When a near object is focused, three events occur simultaneously—convergence, accommodation and pupillary constriction (miosis). The motor pathway for all three activities is through the nucleus of N III, accommodation and miosis being mediated by the autonomic division of this motor nucleus—the Edinger-Westphal nucleus; and there is no doubt that these pathways, along the oculomotor nerve, are separate since the three reactions may be abolished selectively by appropriate treatment of the nerve or ciliary ganglion. As we have seen, the central coordination of this triad of responses is probably at a higher level of the brain, in the cerebral cortex. The question naturally arises as to the extent to which these responses to the near stimulus are independent of each other. Is it possible to converge without accommodating? Can one accommodate without converging? If the answers to these questions are yes, is pupillary constriction tied preferentially to the convergence or the accommodation? These questions, so important for the theory of squint, have been examined and discussed intensively, whilst in recent years Fincham in England and Alpern in America have submitted them to exact quantitative study.

### ACCOMMODATION CONVERGENCE

Suppose first we examine a normal young subject and measure his accommodation, by means of an objective optometric test, whilst he converges on an object that is brought closer and closer to his eyes. Clearly, if he is to see single and distinctly, the convergence and accommodation must 'match', 1 dioptre of accommodation occurring with 1 metre-angle of convergence. The results of such a measurement are shown by the line marked 'normal' in Figure 20.1; it will be seen that accommodation does not match exactly; and since diplopia does not occur, we may attribute this to an 'accommodation lag', the eye tolerating a certain degree of blurring of the image that is important for depth of focus (p. 486).* The line marked 'normal' of Figure 20.1 thus gives us a standard of reference, telling us how convergence and accommodation go together when the stimuli for these responses match, and presumably reinforce each other.

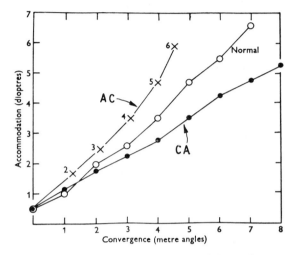

**Fig. 20.1** The relation between accommodation and convergence. The curve marked AC represents accommodative convergence, i.e. the convergence that takes place when the eyes accommodate under conditions where altered convergence does not lead to diplopia. The curve marked CA represents the accommodation that takes place when the eyes converge, the arrangement being such that altered accommodation does not lead to blurring of the image. (Fincham and Walton, *J. Physiol.*)

### ACCOMMODATIVE CONVERGENCE

Next we may allow the two eyes to view separate objects independently by a haploscopic device; under these conditions there is no compulsion for fusion and the eyes may adopt any degree of convergence without experiencing diplopia. The apparent nearness of the test-chart, seen by the right eye, is now varied by means of lenses, so that in order for it to be seen distinctly the right eye must accommodate. The line in Figure 20.1 marked AC shows that the eyes converge as a result of the accommodation of the right eye; this *accommodative convergence* matches the accommodation over a certain range; but not exactly, so that with 6 dioptres of

---

*It has been argued, e.g. by Ogle, that a certain degree of 'fixation disparity' is tolerated, in which case, of course, the failure of accommodation to match convergence could be due to errors in binocular fixation, the resultant physiological diplopia (p. 519) being in some way suppressed or tolerated. Not all workers are willing to accept the existence of a significant fixation disparity, however.

accommodation the convergence is only a little over 4 metre-angles.★

## CONVERGENCE-INDUCED ACCOMMODATION

Next, we may measure the changes of accommodation that take place when the eyes are *made* to converge whilst the object remains at the same apparent distance—the *vergence* of the light (p. 485) is said to be fixed.† The *convergence-induced accommodation* is indicated in the line of Figure 20.1 marked CA. Once again the linkage between the two processes is manifest, and in this case, too, the accommodation lags behind the convergence. This is because of the age of the subject who was 32 years old, and the lag is a sign of presbyopia, subjects under 24 years old showing an almost perfect match between convergence and accommodation. *In a young subject, therefore, the stimulus to convergence automatically induces an accommodation of the right order of magnitude for distinct vision.*

## CONFLICT BETWEEN REFLEXES

In the experimental arrangements so far employed there has been no conflict between the two stimuli, an alteration in the degree of convergence, for example, being compatible with single vision over the whole range. If we cause a subject to fixate a test-chart some 0·33 metres away, his accommodation for distinct vision is 3 D whilst his convergence is 3 metre-angles. By means of prisms, or other devices, we may make it necessary for him to converge further, or diverge, in order to maintain single vision. We may ask: What happens to his accommodation? If the linkage were still maintained, an increase in convergence would be accompanied by an increased accommodation, but since the vergence of the light has not been altered the chart would become blurred by the unnecessary increase in power. There would be thus a conflict between the accommodation reflex, which demands that the accommodation remain unchanged, and the convergence-induced accommodation that operates through the link. Figure 20.2 shows that the link has some flexibility. The line through the circles shows the convergence-induced accommodation that occurred under the earlier conditions where there was no compulsion to maintain any degree of accommodation. The curve through the crosses shows the convergence-induced accommodation during the present experiment; the arrow indicates the initial condition, the subject converging on an object 0·33 m away, showing 3 metre-angles of convergence and 3 D of accommodation. As convergence increases, accommodation does indeed follow it, but to a much smaller extent, an increase of convergence from 3 to 7 metre-angles being associated with an increase of only about 1 D, by contrast with an increase of 3 D that would have occurred under the

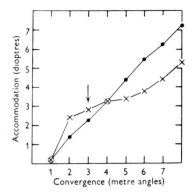

**Fig. 20.2**   Illustrating the convergence-induced accommodation that occurs when alterations of accommodation will lead to blurring of the image. The line through the circles shows the normal relationship between accommodation and convergence, whilst the line through the crosses shows the convergence-induced accommodation taking place when the subject views an object 0·33 m away whilst the degree of convergence is varied by a haploscopic device. (Fincham and Walton, *J. Physiol.*)

'free conditions'. Thus some blurring of the image has occurred, but this has acted as a 'brake' on the convergence-induced accommodation. This experiment thus reveals, once again, the presence of a link between convergence and accommodation, but also demonstrates that the link is not rigid and may be relaxed when its effects militate against distinct vision.

## Relative amplitude of accommodation

The converse experiment is carried out by fixing the convergence and placing lenses in front of the eyes to alter the degree of accommodation. Any accommodative convergence, under these conditions, will cause diplopia, so that the compulsion for fusion operates as an inhibitor of this linked convergence. In fact we do indeed find that convergence is maintained to give single vision over quite a wide range of accommodation, but the conflict between the two requirements leads to a restriction of the amount of accommodation that can be exercised, in other words the *relative amplitude of accommodation* is less than the amplitude when the eyes are free to converge and diverge in accordance with the nearness or farness of the object.

---

★ The ratio of the accommodative convergence, indicated by AC, over the accommodation, indicated by A, is called the AC/A ratio, and is considered a valuable diagnostic measurement. It has been examined in great detail by Alpern and his collaborators.
† Convergence was stimulated by bringing the object closer, but the effective vergence of the light was held constant by the use of very narrow pencils of light so that the rays entering the eye were effectively parallel for all distances of the object from the eye.

## CONVERGENCE-ACCOMMODATION LINK

In general, then, we may conclude that accommodation and convergence are linked together by a common nervous mechanism that activates the two together. Because the stimulus of diplopia is more effective than that of blurred vision, we may consider that the response to nearness is primarily one of convergence; in a young person the associated activation of the accommodation mechanism is accurate enough to make the image fairly distinct, whilst the fine adjustment is provided by the accommodation reflex *per se*, which adjusts the focal power of the eye more exactly, using as clues the alterations of chromatic aberration etc. that take place when the image is not accurately focused. In older persons the convergence-induced accommodation is less adequate and the reflex response must be larger.

## PUPILLARY RESPONSE

Finally we may ask how the pupillary reflex is involved.

When the eyes are made to converge, with accommodation kept fixed, there is usually some pupillary constriction, but very small by comparison with that which would occur were accommodation permitted to occur to match the convergence (Knoll, 1949). Thus the inhibition of the convergence-induced accommodation has been associated with inhibition of the near pupillary reflex. By contrast, the relationship with accommodation is very strong (Marg and Morgan, 1950), in fact, according to Alpern, Ellen and Goldsmith (1958), the pupil continues to constrict even when the amplitude of accommodation has been exceeded; thus the increased effort to accommodate is apparently sufficient to cause pupillary constriction. In a similar way, with presbyopic subjects, the stimulus to accommodation is sufficient to cause pupillary constriction, so that presenting a near object, even if accommodation is inadequate, causes a constriction of a magnitude that would be expected had the accommodation taken place (Alpern, Mason and Jardinico, 1961).

## REFERENCES

Alpern, M. (1958) Vergence and accommodation. I and II. *Arch. Ophthal.*, **60**, 355–357, 358–359.

Alpern, M., Ellen, P. & Goldsmith, R. I. (1958) The electrical response of the human eye in far-to-near accommodation. *Arch. Ophthal.*, **60**, 595–602.

Alpern, M. (1969) Movements of the eyes. In *The Eye*. Ed. Davson, vol. III, 2nd ed. pt. 1. London and New York: Academic Press.

Alpern, M., Mason, G. L. & Jardinico, R. E. (1961) Vergence and accommodation. V. Pupil size changes associated with changes in accommodative convergence. *Am. J. Ophthal.*, **52**, pt. 2, 762–766.

Breinin, G. M. (1957) Relationship between accommodation and convergence. *Trans. Am. Acad. Ophthal.*, **61**, 375–382.

Fincham, E. F. & Walton, J. (1957) The reciprocal actions of accommodation and convergence. *J. Physiol.*, **137**, 488–508.

Knoll, H. A. (1949) Pupillary changes associated with accommodation and convergence. *Am. J. Optom.*, **26**, 346–357.

Marg, E. & Morgan, M. W. (1950) The effect of accommodation, fusional convergence and the proximity factor on pupillary diameter. *Am. J. Optom.*, **27**, 217–225.

# 21. The protective mechanism

The eyes are protected from mechanical insults by two mechanisms—blinking and the secretion of tears; by the latter process irritating particles and fumes are washed away from the sensitive cornea, and the surface of the globe is maintained in a normally moist condition. Blinking, besides its obvious protective function, also operates to prevent dazzle by a blinding light, to maintain the exposed surface of the globe moist by spreading the lacrimal secretions and, during sleep, by preventing evaporation. Finally the act of blinking assists in the drainage of tears. The pathological condition of *keratitis sicca* may thus result from either defective lid closure or defective lacrimal secretion.

## BLINKING

ORBICULARIS OCULI

Closure of the eyes is brought about by contraction of the orbicularis oculi muscle, associated with a reciprocal inhibition of the levator palpebri of the upper lid. The orbicularis oculi is an oval sheet of concentric striped muscle fibres extending from the regions of the forehead and face surrounding the orbit into the lids (Fig. 21.1). Although a single muscle, with its origin mainly in the medial canthal region of the orbit, it is customary to divide it into a number of separate muscles, or parts, the main divisions being the *pars orbitalis* and *pars palpebralis*. The pars palpebralis is divided itself into the

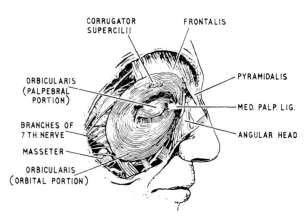

**Fig. 21.1**   Illustrating the orbicularis oculi muscle. (After Wolff, *Anatomy of the Eye and Orbit.*)

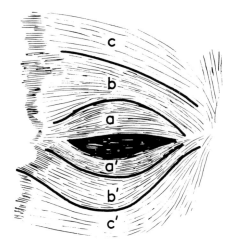

**Fig. 21.2**   Illustrating the subdivisions of the orbicularis muscle. *c* and *c'* represent the pars orbitalis; *b* and *b'* are the pretarsal, and *a* and *a'* the preseptal muscles. (After Jones, *Trans. Amer. Acad. Ophthal.*)

upper and lower pretarsal and upper and lower preseptal muscles (Fig. 21.2). All the palpebral muscles insert into the lateral palpebral raphe and thus, by contracting, act as tensor muscles to the lids, moving the nasal end of each lid towards the nose.

**Orbital portion**

The orbital portion of the muscle is not normally concerned with blinking, which may be carried out entirely by the palpebral portion; however, when the eyes are closed tightly, for example in the blepharospasm consequent on a painful stimulus, then the skin of the forehead, temple and cheek is drawn towards the medial side of the orbit; the radiating furrows caused by this action of the orbital portion eventually lead to the 'crow's feet' of elderly persons. The two portions can be activated independently; thus the orbital portion may contract, causing a furrowing of the brows that reduces the amount of light entering from above, e.g. in strong sunshine, whilst the palpebral portions remains relaxed and allows the eyes to remain open.

LEVATOR PALPEBRAE

Opening of the eye is not just the result of passive relaxation of the orbicularis muscle; the upper lid contains the *levator palpebrae superioris*, which takes origin with the extraocular muscles at the apex of the orbit as a narrow tendon and runs forward into the upper lid as a very broad tendon, the *levator aponeurosis*, which inserts into the anterior surface of the tarsus and the skin covering the upper lid. Contraction of the muscle will obviously cause elevation of the upper eye-lid, and it is interesting that the nervous connections of this muscle are closely related to those of the superior rectus, required to elevate the eye, so that when the eye looks upwards the upper eyelid moves up in unison.

### MÜLLER'S MUSCLE

The orbicularis and levator are striped muscles under voluntary control; the lids contain, in addition, unstriped muscle fibres deep to the septum orbitale; they take origin among the fibres of the levator in the upper lid and the prolongation of the inferior rectus in the lower, whilst they are inserted into the attached margins of the tarsal plates. They are described as the *superior* and *inferior palpebral muscles* respectively. The muscles are activated by the sympathetic division of the autonomic system and tend to widen the palpebral fissure by elevation of the upper, and depression of the lower, lid.

### ACCESSORY MUSCLES OF LID-CLOSURE AND -OPENING

In addition to the muscles already described, we may note that other, facial, muscles often cooperate in the acts of lid-closure or opening; thus the *corrugator supercilii* muscles pull the eyebrows towards the root of the nose, making a projecting roof over the medial angle of the eye and producing characteristic furrows in the forehead; they are used primarily to protect the eye from the glare of the sun. The *pyramidalis*, or *procerus*, muscles occupy the bridge of the nose; they arise from the lower portion of the nasal bones and, being inserted into the skin of the lower part of the forehead on either side of the midline, they pull the skin into transverse furrows. In lid-opening, the *frontalis muscle*, arising midway between the coronal suture of the scalp and the orbital margin, is inserted into the skin of the eyebrows; contraction therefore causes the eyebrows to rise and opposes the action of the orbital portion of the orbicularis; the muscle is especially used in gazing upwards. It is also brought into action when vision is rendered difficult, either by distance or the absence of sufficient light.

### REFLEXES

Reflex blinking may be caused by practically any peripheral stimulus but the two functionally significant reflexes are (*a*) that resulting from stimulation of the endings of N V in the cornea, lid or conjunctiva—the *sensory blink reflex*, or *corneal reflex*, and (*b*) that caused by bright light—the *optical blink reflex*. The corneal reflex is rapid (0·1 second reflex time) and is the last to disappear in deepening anaesthesia; it is mediated by N V, impulses being relayed from the nucleus of this nerve to N VII. The reflex is said to be under the control of a medullary centre. The optical reflex is slower and, in man, the nervous pathway includes the visual cortex; the reflex is absent in children of less than nine months.

The blink- or wink-responses to a tap on the face, for example on the supra-orbital region, have been described as a group of *facial reflexes* to which various names, after their discoverers, have been attached (McCarthy, v. Bechterew, etc.). The name *orbicularis oculi reflex* has been suggested to cover them all since their reflex paths are essentially the same, and they apparently consist of two phases—a *proprioceptive* response due to stretch of the facial musculature, followed by a *nociceptive* response which may be regarded as protective in function (Rushworth, 1962).

### NORMAL RHYTHM

In the waking hours the eyes blink fairly regularly at intervals of two to ten seconds, the actual rate being a characteristic of the individual. The function of this is to spread the lacrimal secretions over the cornea, and it might be thought that each blink would be reflexly determined by a corneal stimulus—drying and irritation. As a result of extensive studies by Ponder and Kennedy (1928) on human subjects, it would appear that this view is wrong; the normal blinking rate is apparently determined by the activity of a 'blinking centre' in the globus pallidus of the caudate nucleus. This is not to deny, however, that the blink-rate is modified by external stimuli, particularly through N V and N II, and also by the emotional state of the individual; Ponder's work shows this most clearly—for example, the blink-rate of an individual in the witness-box went up markedly during cross-examination; that of another decreased on ingestion of a large dose of alcohol, and so on; the fact remains, however, that when any given route for sensory impulses is blocked, e.g. by cocainization of the cornea, the blink-rate is, on the average, unchanged, i.e. characteristic of the individual.

### MOVEMENTS OF EYES

In conclusion, attention should be drawn to the strong association between blinking and the action of the extraocular muscles. A movement of the eyes is generally accompanied by a blink and it is thought that this aids the eyes in changing their fixation point. It will be remembered that in lesions of the frontal motor centre the fixation reflex asserts itself so strongly that a change in the fixation point becomes very difficult but may be brought about by cutting off the visual stimulus by closing the eyes or jerking the head. It is possible that blinking in the normal individual is an aid, although not a necessary one, in inhibiting the fixation reflex preparatory to the adoption of a new point of fixation.

## LACRIMATION

The surface of the globe is kept moist by the tears, secreted by the lacrimal apparatus, together with the mucous and oily secretions of the other secretory organs and cells of the conjunctiva and lids.

## LACRIMAL GLANDS

The lacrimal gland, proper (Fig. 21.3), lies in the upper and outer corner of the orbit, just within the orbital

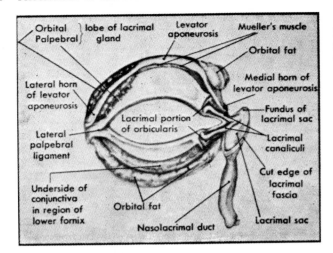

**Fig. 21.3**  Tarsal plates and lacrimal system. (Kronfeld *et al.*, *The Human Eye in Anatomical Transparencies*.)

margin. It is divided by the lateral horn of the aponeurosis of the levator muscle into two lobes, *orbital* and *palpebral*. About twelve ducts gather the secretion of the entire gland and open through the palpebral conjunctiva, some 4 to 5 mm above the upper border of the tarsus. It has a typical serous gland structure, and is supplied mainly by the lacrimal branch of the ophthalmic artery, the lacrimal vein draining the blood into the superior ophthalmic vein. The glands are supplied by the *lacrimal nerve*, which is the smallest terminal twig of the ophthalmic division of the trigeminal (Fig. 21.6, p. 498).

### HISTOLOGY

The microscopical appearance of the gland has been described by Scott and Pease (1959), Kühnel (1968) and Ruskell, (1975). The basic structure, as with the salivary glands with which it has been compared, is the acinus, opening into the tubule. The cells constituting the acinus are of two main types, as characterized by the types of granule within their cytoplasm, and on this basis they have been described as mucous (with clear granules, Type A or K) and another serous (dense granules, Type B or G).* According to the histochemical study of Allen *et al.* (1972), the type A, or mucous cell, granules stain similarly to the granules in the Paneth cell, which contain lysozyme, whilst the type B or serous cell granules contain either neutral or acid glycoprotein. Thus the mucous and serous cell types are responsible for secretion of the glycoprotein and lysozyme of tears respectively. The myoepithelial cells are the third type, and these presumably serve to compress the acini, expelling the secretion into the ducts.

### ACCESSORY LACRIMAL GLANDS

In addition to the lacrimal gland, there are numerous microscopical *accessory lacrimal glands* in the conjunctiva—the glands of Krause and Wolfring—structurally organized in an essentially similar fashion but on a very much smaller scale.

### SEBACEOUS AND MUCOUS SECRETIONS

An oily fluid is secreted by the glands of Zeis and the Meibomian glands, in the eyelids, whilst goblet cells in the conjunctiva secrete mucus.

## PRECORNEAL FILM

The fluid layer on the cornea is called the precorneal film and is said by Wolff to consist of an innermost layer of mucus, a middle layer of lacrimal secretions and an outer oil film which serves to reduce evaporation of the underlying watery layer (Mishima and Maurice, 1961) as well as preventing overflow at the lid-margin. Ehlers (1965) absorbed the film on a sponge and estimated, from its weight, that it would be some 8·5 to 4·5 $\mu$m thick according to the length of time between blinks; its chemistry suggested that the lipid was predominantly from the Meibomian glands, containing as it did mainly cholesterol esters.

### WETTING

The factors governing the wetting of the cornea by the tears, i.e. the spreading of a coherent film over the corneal surface, have been discussed by Holly (1973); they are characterized by a *spreading coefficient, S*, determined by the surface tensions indicated in Figure 21.4. Thus, when *S* is greater than zero, spontaneous spreading will occur. Alternatively, the concept of *wettability* is employed, a surface being wettable by a

---

*Ruskell's study of the human lacrimal gland revealed three types of cell, dark, medium, and light, according to the granulation; he was satisfied that the different appearances were not due to successive stages in maturation of the cell, their nuclei, for example, being of different shape. Completely clear cells were probably lymphocytes.

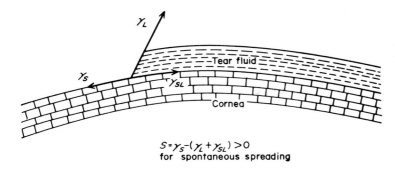

$$S = \gamma_S - (\gamma_L + \gamma_{SL}) > 0$$
for spontaneous spreading

**Fig. 21.4** Interfacial tensions determining the spread of tear fluid over the cornea. (Holly, *Exp. Eye Res.*)

liquid if the critical surface tension of the solid is greater than the surface tension of the liquid, the *critical surface tension* being an empirical parameter. In general, then, a low surface tension of the tears will favour spreading, but too low a tension will cause foaming instead of even spreading, so that conventional surfactants, such as soaps, would be useless, whilst the mucin, which is a universal concomitant of mucous membranes, fulfils this role. Thus Teflon, which cannot be wetted by commercially available surfactant can be wetted when exposed to dissolved mucins. The mucins achieve this by maintaining a low value of interfacial tension between cornea and tears ($\gamma_{SL}$ of Fig. 21.4), without solubilizing the lipids, and the main source is the glycoproteins secreted by the conjunctival goblet cells.

RE-FORMING THE FILM

When the eyes close, the superficial layer of lipid is actually swept off the mucinous surface and compressed to the small area between the lid edges. When the eyes open, a new lipid film must be spread over the surface, and it is important that this remain stable for some time. According to Holly, the breakdown of the new film provides the stimulus for the next blink. Thus, when blinking is prevented, the tear-film ruptures in 15 to 40 sec to produce dry spots. When patients with various 'dry-eye' syndromes were studied, no very significant change in the character of the tears, sufficient to affect the stability of the tear-film, was observed, so that it is more likely that defective mucus secretion by the goblet cells is the prime factor.

RATE OF EVAPORATION

The importance of the precorneal film in reducing the evaporation from the surface of the eye was demonstrated by Mishima and Maurice (1961a) who used, as a measure of evaporation, the thinning of the cornea that took place as a result of loss of water; in the normal eye no such thinning took place since the fluid lost from the surface was made good by the aqueous humour; by replacement of the aqueous humour with oil, however, there was a steady thinning; the rate was increased greatly by washing the surface of the cornea with saline, i.e. by removing the precorneal film but this could be brought back to the normal rate by allowing the rabbit to blink, a procedure that presumably replaced the lipid precorneal film, since preliminary destruction of the Meibomian glands rendered lid-closure ineffective. Removal of the epithelium had no obvious effect so that Mishima and Maurice concluded that it offered no obstruction to evaporation of water. Iwata *et al.* (1969) measured the evaporative rate more directly and confirmed the striking increase in rate when the precorneal film was washed away; as Figure 21.5 shows, this rapid loss does not last for long, and the rate settles down to a much slower value similar to that from the intact eye. The initial rapid loss occurs because of the presence of a layer of tears enclosed by the pre-corneal film; when this has evaporated, the loss is now apparently determined by the rate at which new fluid can pass from aqueous humour to the surface of the eye; thus preliminary blotting off of this layer gives a slow evaporative loss equal in rate to that reached with the lipid film removed. The limiting factor under these conditions seems to be the epithelium, since removal of this allowed the high rate of evaporation to be maintained indefinitely (Fig. 21.5).

NORMAL OUTPUT

The normal daily output of tears was estimated by Schirmer in 1903 in subjects with their lacrimal ducts and sacs removed; the time for a drop to form was measured; the amount of evaporation that had occurred was estimated by a separate experiment, and knowing the weight of the drop the formation during this period was computed. Schirmer estimated a secretion of 0·5 to 0·75 g in 16 hours. He concluded that there was no secretion during sleep.* Schirmer also devised tests for normal lacrimation; these consisted of hooking a strip of filter paper over the lower lid margin and measuring the length of wetting that occurred in a fixed time. In

---

*More recently Ehlers *et al.* (1972) have estimated a basal rate of 1 to 2 $\mu$l/min or 2 to 3 ml/24 hr.

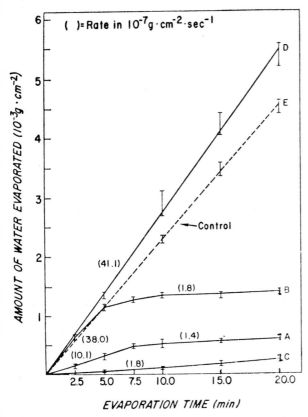

**Fig. 21.5** The weight of water evaporated, and evaporation rates, from the precorneal and corneal surfaces as measured over a 20 minute period. Line A shows the results from the intact tear film. Two discrete phases with differing rates are noted. Line B shows the results from the tear film after the superficial lipid layer was removed. Again, two discrete phases with differing rates are seen. The rate of the second phase in Line B parallels that of the second phase in Line A. Line C shows the results measured from the epithelial surface. Only one phase with a rate similar to the second phases of A and B is noted. Line D shows the evaporation from the stromal surface which occurs at a rate similar to that from a pure water surface (Line E). (Iwata, Lemp, Holly and Dohlman, *Invest. Ophthal.*)

some tests the presence of the filter paper was the stimulus for secretion, whilst in adaptations of this test the nasal mucosa was stroked with a brush, or the subject was asked to look into the sun. Applying this test de Roetth observed a considerable decrease in the magnitude of this lacrimation-response with age, the secretion at 80 years being only about a fifth that at 20 years.

## CHEMISTRY OF THE LACRIMAL SECRETION

### OSMOLALITY

The fluid collecting in the lacus lacrimalis is a mixture of mucus and secretions of the lacrimal glands—the tears proper—and is thus slightly opalescent. According to Krogh, Lund and Pedersen-Bjergaard, the tears are isosmolal with blood plasma, but the precise relationship between the two fluids will obviously depend on the degree of evaporation that the fluid undergoes before collection, and, presumably, on the rate of secretion of the fluid. Thus Giardini and Roberts found the concentration of chloride to fall from 135 mM to 110 mM as the rate of secretion was increased by exposing the human eye to larger and larger concentrations of teargas. Essentially similar changes were found by Balik for chloride, sodium, urea and phosphate, whilst Thaysen and Thorn found the concentrations of urea, sodium, potassium and chloride to be independent of rate of secretion.

Whilst there is little doubt that evaporation will tend to increase the concentrations of solutes in the tear-film, and hence the osmolality of the fluid, the extent to which this happens will be governed by the slow evaporative rate, discussed above, and the replacement from the aqueous humour. v. Bahr (1956) observed that the cornea did not change in thickness when a 1 per cent solution of NaCl was placed on it, whereas higher concentrations caused it to come thinner; hence the tear-film probably has an equivalent osmolality; Mishima and Maurice (1961b) found a value of 0·9 per cent NaCl if the eye was closed and 1·0 per cent if open.[*]

### SECRETION RATIO

Balik defined a 'secretion ratio', F, being the ratio of the concentrations of a given substance in tears over plasma; with urea, sodium and phosphate this was greater than unity with normal rates of flow, suggesting possible active accumulation in the tears, although evaporation as the main cause of these ratios cannot be ruled out. The concentration of glucose in the tears is remarkably low, being only a few per cent of that in plasma (Giardini and Roberts). The pH is about 7·2 becoming more alkaline with continued lacrimation (Swan, Trussell and Allen).

### PROTEINS

The concentration of proteins in the tears is low, being of the order of 0·1 to 0·6 per cent by comparison with 6 to 7 per cent for plasma; they have been resolved into

---

[*] The authors point out that the thinning of the cornea when the eye is open, and its thickening when the eye is kept closed, indicated that under normal conditions the aqueous humour loses water, and they found a difference of 1·5 per cent in the estimated osmolalities of the fluid in the open and closed eyes.

a number of components by the technique of paper-electrophoresis.*

## Lysozyme

Thus Brunish described an albumin, representing some 39 per cent of the total, 3 globulins representing 45 per cent, whilst the rest was made up of *lysozyme*, an antibacterial substance first described by Fleming in 1922 which is responsible for the low bacterial count of normal human tears. Lysozyme is a strongly basic protein, of molecular weight 14 000 to 25 000; with an iso-electric point of 10·5 to 11·0 it has a net positive charge at physiological pH and thus migrates to the cathode by contrast with the other proteins which have a net negative charge. It is an enzyme, and owes its bactericidal activity to its power of dissolving the mucopolysaccharidic coats of certain bacteria. It is not peculiar to human tears, being widely distributed in the plant and animal kingdoms, egg-white and fig-tree latex being rich sources, for example.

## Other enzymes

Van Haeringen and Glasius (1974; 1976 a,b) have described a variety of enzymes in human tears, notably the hydrolases associated with breakdown of tissue, such as acid and alkaline phosphatase. It seems likely that when enzymes concerned in carbohydrate metabolism, such as lactate dehydrogenase, are found in the tears collected from the conjunctival sac they are derived from the corneal epithelium through trauma, so that when fluid is collected by a glass cannula these enzymes are absent.

## Globulins

Although it is usual to designate the globulins as α-, β- and γ-, it is unlikely that the protein fractions of tears are identical with their plasma analogues, so that it is better to indicate them by non-committal names; thus U. Krause isolated six fractions which he denoted by arabic numerals, 1 to 5 in descending order of speed of migration in the electric field. His fraction 1 would correspond to 'tear-albumin' of Erickson (1956); fraction 2 to a globulin with a speed of migration close to that of $\alpha_2$-globulin, and so on.

It would seem that the tear proteins differ from their plasma analogues by virtue of the larger amounts of mucopolysaccharide in combination with them; at any rate, according to McEwan, Kimura and Feeney, they stain strongly with the PAS reagent (p. 95). Iwata and Kabasawa (1971) have separated three glycosaminoglycan fractions, of molecular weights 400 000, 50 000 and 14 000 daltons; these contain fucose, mannose, galactose, N-acetylglucosamine and sialic acid.

### LIPIDS

These have been examined by Andrews (1970), Ehlers (1965), and Van Haeringen and Glasius (1975); remarkable is the large concentration of cholesterol (3 to 6·6 mM); as Van Haeringen and Glasius emphasize, it is unlikely that such large concentrations would be derived from the lacrimal gland, especially as the transport requires a β-lipoprotein, the low protein concentration in lacrimal secretion being incompatible with this. It is likely that the cholesterol is derived from the Meibomian secretion, as suggested by Ehlers (1965), who found no evidence of lipids in the lacrimal glands, goblet cells, or glands of Wolfring, whereas the lipids of the Meibomian secretion corresponded with those in the precorneal film.

## DRAINAGE OF TEARS

When the secretion of tears exceeds the loss due to evaporation the excess fluid, if it is not to overflow—*epiphora*—must be drained away. The passage for this drainage consists of the *canaliculi*, the *lacrimal sac* and the *naso-lacrimal duct*. The lacrimal *puncta* are the openings of the canaliculi in the lid-borders of the inner canthus; the canaliculi course at first vertically and then more or less horizontally to the lacrimal sac, which is continued inferiorly as the nasolacrimal duct emptying into the inferior meatus of the nose (Fig. 21.3). The sac and the duct may be regarded as essentially a continuous structure, the term 'sac' really applying to *that part of the duct lying within the orbit*. Any theory of the mechanism of conduction of the tears must take into account the common experience that the act of blinking favours drainage and prevents overflow; certainly the dimensions of the canaliculi and puncta are so small (about 0·5 mm diameter) as to make it unlikely that an adequate drainage could be maintained by the unassisted action of gravity (siphoning would of course be necessary for the upper canaliculus).

### THE LACRIMAL PUMP

The mechanism has been analysed most recently by Jones, and is described by him as the *lacrimal pump*. Essentially the pumping action consists in an alternate

*The literature on the protein fractionation of tears is large and has been summarized by U. Krause (1959) and McEwan and Goodner (1969). The first description is that of Miglior and Pirodda (1954); other outstanding contributions are from the laboratories of McEwen, Erickson and Brunish, to name only a few. We may note that François (1959) has shown that, by contrast with aqueous humour and lens, the use of agar in place of paper for electrophoresis is contra-indicated; at any rate the fractions obtained by this technique are so different as to suggest interaction of the proteins with the agar.

negative and positive pressure in the lacrimal sac caused by the contraction of the orbicularis muscle. Because of the anatomical relationships of this muscle, in particular that part called Horner's muscle, to the palpebral ligament and lacrimal sac, contraction during blinking causes a dilatation of the sac. At the same time, the muscular contraction also causes the canaliculi to become shorter and broader. Finally contraction of the orbicularis also tends to invert the lower lid, thus ensuring that the punctum dips in the lacus lacrimalis.* Negative pressure in the nose during inhalation, and gravity, are also factors in emptying the sac.

## THE NERVOUS MECHANISM

Tears are secreted reflexly in response to a variety of stimuli, e.g. irritative stimuli to the cornea, conjunctiva, or nasal muscosa; thermal stimuli, including peppery foods, applied to the mouth and tongue; bright lights, and so on. In addition it occurs in association with vomiting, coughing and yawning. The secretion associated with emotional upset is described as *psychical weeping*. Section of the sensory root of the trigeminal prevents all reflex weeping, leaving psychical weeping unaffected; similarly the application of cocaine to the surface of the eye, which paralyses the sensory nerve endings, inhibits reflex weeping, even when the eye is exposed to potent tear-gases. The afferent path in the reflex is thus by way of N V. The motor pathway to the main lacrimal gland is more complex.

### PARASYMPATHETIC

The parasympathetic supply is from the *facial nerve* (N VII); preganglionic fibres arise from the *lacrimal nucleus* in the pons; they run in the *nervus intermedius*, through the *geniculate ganglion* (without relaying), to the *sphenopalatine ganglion* as the *greater superficial petrosal nerve*. Post-ganglionic fibres enter the *zygomatic nerve* and thence the *lacrimal* (Fig. 21.6). The sympathetic pathway is by way of the *deep petrosal nerve* which arises from the cervical sympathetic plexus; it joins the greater

superficial petrosal nerve to form the *nerve of the ptery-goid canal*, or *Vidian nerve*, which ends in the spheno-palatine ganglion; thence the fibres run with the para-sympathetic as described above. Other fibres reach the gland by way of the lacrimal artery.

### NERVE TERMINALS

The nerve terminals in the gland have been examined by Ruskell (1968, 1975) in the monkey and Tsukahara and Tanishima (1974) in man, with a view to their identification as either adrenergic or cholinergic. Although Tsukahara and Tanishima found cate-cholamine fluorescence, they were unable to identify the characteristically cored vesicles of adrenergic terminals, and they concluded that, in man, there was little adrenergic innervation. Cholinergic terminals were identified, and these were related to the myoepithelium. In the monkey, Ruskell found both sympathetic and parasympathetic terminals, but the sympathetic in-nervation seemed to be confined to the interstices of the gland, whilst the parasympathetic fibres ended in relation to acini and ducts, most being adjacent to myoepithelial cells, as in man. If the myoepithelial cells are responsible for forcing the secretion along the ducts, then the fibres ending in relation to these would be responsible for this mechanical aspect of secretion of tears, whilst the actual elaboration in the acini would be controlled by the fibres making relation to acinus cells. †

### NEWBORN

Innervation of the lacrimal gland is not always complete at the time of birth; thus Sjögren found a complete lack of reflex secretion of tears in 13 per cent of normal full-term babies and in 37 per cent of premature babies

---

*According to Brienen and Snell (1969), however, the punctum always opens into a strip of tear fluid, so that a 'dipping into the tear lake' during eye-closure is out of the question. These authors attribute drainage entirely to the pressure developed in the conjunctival sac.

†In the sheep's gland Yamauchii and Burnstock (1967) found that the myoepithelial cells covering the acini were very prominent, and bare axons could be seen in grooves on their cytoplasm. These authors suggest that all axons might well be innervating myoepithelial, rather than acinar, cells.

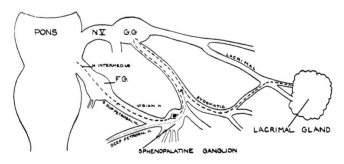

**Fig. 21.6**  Innervation of the lacrimal gland. G.G., Gasserian ganglion; F.G., facial ganglion. (After Mutch, *Brit. J. Ophthal.*)

during the first three days of extra-uterine life. Since this condition was not associated with any dryness of the cornea—keratitis sicca—it might be suspected that normally the secretions of the accessory lacrimal glands, which are not involved in reflex secretion, are adequate to keep the cornea moist. This suggestion is supported by the relative absence of corneal symptoms after complete removal of the lacrimal gland. Thus the reflex secretion may be regarded as an emergency response.

## SYMPATHETIC

The function of the sympathetic in the secretion of tears is problematical; section of the superficial petrosal nerve—a procedure often concomitant with operative removal of the Gasserian ganglion—or blocking of the sphenopalatine ganglion, causes a failure of reflex and psychical lacrimation, suggesting that the para-sympathetic only is concerned with reflex secretion. Furthermore, there is no apparent interference with weeping as a result of bilateral section of the cervical sympathetic. According to Whitwell (1961), therefore, it is the sympathetic innervation that is responsible for a tonic secretion adequate to keep the eye moist, but the possibility must be considered that the main gland is normally at rest, whilst the accessory glands operate continuously and independently of any innervation. In this event the sympathetic would play no role in the formation of tears.

## DRUGS

As one would expect of a parasympathetic-dominated process, intravenous pilocarpine produces a large and prolonged increase in flow-rate of tears; according to a study of Goldstein, de Palau and Botelho (1967) there may be an eleven-fold increase, in the rabbit, lasting for 33 minutes. Intravenous noradrenaline increased flow in this animal some threefold for two minutes; it inhibited the increased flow due to pilocarpine, however, perhaps because of its vasoconstrictive action.

## CROCODILE TEARS

Some human subjects lacrimate strongly when eating, and the syndrome has been given the name of 'crocodile tears'. The mechanism has been discussed by Golding-Wood (1963), and he concludes that the prime cause is a degeneration of the greater superficial petrosal nerve which provokes sprouting of fibres from the lesser superficial that normally subserves salivary secretion; as a result, the salivary secretory fibres stimulate the lacrimal gland. He found that intracranial section of the glosso-pharyngeal nerve relieved the condition.

# REFERENCES

Allen, M., Wright, P. & Reid, L. (1972) The human lacrimal gland. *Arch. Ophthal.* **88**, 493–497.

Andrews, J. S. (1970) Human tear film lipids. I. Composition of the principal non-polar component. *Exp. Eye Res.* **10**, 223–227.

v. Bahr, G. (1956) Corneal thickness. Its measurement and changes. *Amer. J. Ophthal.* **42**, 251–263.

Balik, J. (1959) Uber die Ausscheidung von Natrium in die Tränen. *Ophthalmologica, Basel* **137**, 95–102.

Balik, J. (1959) Uber die Ausscheidung von Harnstoff in die Tränenflüssigkeit bei Keratoconjunctivitis sicca. *Klin. Mbl. Augenheilk.* **135**, 533–537.

Balik, J. (1959) Uber die Ausscheidung von Natrium in die Tränen bei trockener Keratoconjunctivitis. *v. Graefes' Arch. Ophthal.* **160**, 633–657.

Balik, J. (1960) Secretion of inorganic phosphorus in tears. *Am. J. Ophthal.* **49**, 941–945.

Brienen, J. A. & Snell, C. A. R. D. (1969) The mechanism of the lacrimal flow. *Ophthalmologica, Basel* **159**, 223–232.

Brunish, R. (1957) The protein components of human tears. *Arch. Ophthal.* **57**, 554–556.

Ehlers, N. (1965) The precorneal film. *Acta Ophthal., Kbh.* Suppl. 81.

Ehlers, N., Kessing, S. V. & Norn, M. S. (1972) Quantitative amounts of conjunctival mucous secretion and tears. *Acta. Ophthal.* **50**, 210–213.

Erickson, O. F. (1956) Albumins in lacrimal protein patterns. *Stanf. med. Bull.* **14**, 124–125.

Fleming, A. (1922) On a remarkable bacteriolytic element found in tissues and secretions. *Proc. Roy. Soc. B.* **93**, 306–317.

François, P. (1959) Micro-electrophorese sur gélose des larmes humaines. *Bull. Soc. belge d'Ophthal.* **122**, 343–351.

Giardini, A. & Roberts, J. R. E. (1950) Concentration of glucose and total chloride in tears. *Br. J. Ophthal.* **34**, 737–743.

Golding-Wood, P. H. (1963) Crocodile tears. *Br. med. J.* **(i),** 1518–1521.

Goldstein, A. M., de Palau, A. & Botelho, S. Y. (1967) Inhibition and facilitation of pilocarpine-induced lacrimal flow by noradrenaline. *Invest. Ophthal.* **6**, 498–511.

Holly, F. J. (1973) Formation and rupture of the tear film. *Exp. Eye Res.* **15**, 515–525.

Iwata, S. & Kabasawa, I. (1971) Fractionation and chemical properties of tear mucoids. *Exp. Eye Res.* **12**, 360–367.

Iwata, S., Lemp, M. A., Holly, F. J. & Dohlman, C. H. (1969) Evaporation rate of water from the precorneal tear film and cornea in the rabbit. *Invest. Ophthal.* **8**, 613–619.

Jones, L. T. (1958) Practical fundamental anatomy and physiology (of lacrimal apparatus). *Trans. Am. Acad. Ophthal.* **62**, 669–678.

Krause, U. (1959) A paper electrophoretic study of human tear proteins. *Acta ophthal. Kbh.* **53**, Suppl., pp. 67.

Krogh, A., Lund, C. G. & Pedersen-Bjergaard, (1945) The osmotic concentration of human lacrymal fluid. *Acta physiol. Scand.* **10**, 88–90.

Kühnel, W. (1968) Menschliche Tränendrüse. *Z. Zellforsch.* **89**, 550–572.

McEwan, W. K. & Goodner, E. K. (1969) Secretion of tears and blinking. In *The Eye*, Ed. Davson, 2nd ed. vol. III, pt. 3. London & New York: Academic Press.

McEwen, W. K., Kimura, S. J. & Feeney, M. L. (1958) Filter-paper electrophoresis of tears. III. *Am. J. Ophthal.* **45**, 67–70.

Miglior, M. & Pirodda, A. (1954) Indagini elettroforetiche sulla composizione proteica delle lacrime umane normali. *Giorn. ital. Oftal.* **7**, 429–439.

Mishima, S. & Maurice, D. M. (1961a) The oily layer of the tear film and evaporation from the corneal surface. *Exp. Eye Res.* **1**, 39–45.

Mishima, S. & Maurice, D. M. (1961b) The effect of normal evaporation on the eye. *Exp. Eye Res.* **1**, 46–52.

Mutch, J. R. (1944) The lacrimation reflex. *Br. J. Ophthal.* **28**, 317–336.

Ponder, E. & Kennedy, W. P. (1928) On the act of blinking. *Quart. J. exp. Physiol.* **18**, 89–110.

Rushworth, G. (1962) Observations on blink reflexes. *J. Neurol. Neurosurg. Psychiat.* **25**, 93–108.

Ruskell, G. L. (1968) The fine structure of nerve terminations in the lacrimal glands of monkeys. *J. Anat.* **103**, 65–76.

Ruskell, G. L. (1969) Changes in nerve terminals and acini of the lacrimal gland and changes in secretion induced by autonomic denervation. *Z. Zellforsch.* **94**, 261–281.

Ruskell, G. L. (1975) Nerve terminals and epithelial cell variety in the human lacrimal gland. *Cell Tiss. Res.* **158**, 121–136.

Schirmer, O. (1903) Studien zur Physiologie und Pathologie der Tränenabsonderung und Tränenabfuhr. *v. Graefes' Arch. Ophthal.* **56**, 197–291.

Scott, B. L. & Pease, D. C. (1959) Electron microscopy of the salivary and lacrimal glands of the rat. *Amer. J. Anat.* **104**, 115–161.

Sjögren, H. (1955) The lacrimal secretion in newborn, premature and fully developed children. *Acta ophthal. Kbh.* **33**, 557–560.

Swan, K. C., Trussell, R. E. & Allen, J. H. (1939) pH of secretion in normal conjunctival sac determined by glass electrode. *Proc. Soc. exp. Biol. Med., N.Y.* **42**, 296–298.

Thaysen, J. H. & Thorn, N. A. (1954) Excretion of urea, potassium and chloride in human tears. *Am. J. Physiol.* **178**, 160–164.

Tsukahara, S. & Tanishima, T. (1974) Adrenergic and cholinergic innervation of the human lacrimal gland. *Jap. J. Ophthal.* **18**, 70–77.

Van Haeringen, N. J. & Glasius, E. (1974) Lactate dehydrogenase in tear fluid. *Exp. Eye Res.* **18**, 345–349.

Van Haeringen, N. J. & Glasius, E. (1975) Cholesterol in human tear fluid. *Exp. Eye Res.* **20**, 271–274.

Van Haeringen, N. J. & Glasius, E. (1976a) The origin of some enzymes in tear fluid, determined by comparative investigation with two collection methods. *Exp. Eye Res.* **22**, 267–272.

Van Haeringen, N. J. & Glasius, E. (1976b) Characteristics of acid hydrolases in human tear fluid. *Ophthalmic Res.* **8**, 367–373.

Whitwell, J. (1961) Role of the sympathetic in lacrimal secretion. *Br. J. Ophthal.* **45**, 439–445.

Yamauchii, A. & Burnstock, G. (1967) Nerve-myoepithelium and nerve-glandular epithelium contacts in the lacrimal gland of the sheep. *J. Cell Biol.* **34**, 917–919.

# Visual perception

# 22. Introduction

## SENSATION AND PERCEPTION

In Section II we have investigated the 'mechanism of vision' in the limited sense of attempting to understand the elementary processes in the retina that result in discharges in the optic nerve, and of attempting to correlate these with the sensations. To this end, very simple visual phenomena were chosen—the sensation of luminosity, the discrimination of flicker, colour and form. In respect to this last category, we confined ourselves to the very element of form discrimination—the resolution of points and lines—and in visual acuity studies we indicated that the recognition of letters was an unsuitable basis for the study of resolution since it involved activities at a higher level than those we were immediately concerned with. In this section we must consider some of the manifestations of this higher form of cerebral activity.

Essentially we shall be concerned with the modes in which visual sensations are interpreted; the eyes may gaze at an assembly of objects and we know that a fairly accurate image of the assembly is formed on the retinae of the two eyes, and this is 'transferred' by a 'point-to-point' projection to the occipital cortex. This is the basis for the visual sensation evoked by the objects but there is a large variety of evidence, derived from everyday experience, which tells us that the final awareness of the assembly of objects involves considerably more nervous activity than a mere point-to-point projection of the retina on the cortex would require. As a result of this activity we appreciate a *perceptual pattern*; the primary visual sensation, resulting from stimulating the visual cortex, is integrated with sensations from other sources presented simultaneously and, more important still, with the memory of past experience. For example, a subject looks at a box; a plane image is produced on each retina which may be represented as a number of lines, but the perceptual pattern evoked in the subject's mind is something much more complex. As a result of past experience the lines are interpreted as edges orientated in different directions; the light and shade and many other factors to be discussed later are all interpreted in accordance with experience; its hardness is remembered and its uses, and the original

visual sensation is *interpreted as a box*, not a series of lines. It is impossible, in practice, to separate the sensation of seeing from the perceptual process; the sensation is essentially an abstraction—what we think would be the result of a visual stimulus if this higher nervous integrative activity had not happened. In the case of a flash of light the influence of higher nervous activity is very small and the sensation is virtually identical with the final perception. With a written word, on the other hand, the final perceptual pattern presented to consciousness is something far different from the bare visual sensation.

We may begin by illustrations of the perceptual process, as revealed by psycho-physical studies, and later we shall consider some electrophysiological studies that illustrate the possible neural basis for some of these higher integrative functions.

## HIGHER INTEGRATIVE ACTIVITY

### INADEQUATE REPRESENTATION

*Schroeder's staircase* (Fig. 22.1) shows how the same visual stimulus may evoke different meanings in the same subject; to most people the first impression created by Figure 22.1 is that of a staircase; on gazing at it

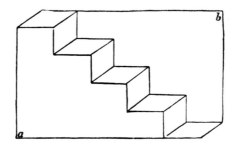

**Fig. 22.1**   Schroeder's staircase.

intently, however, the observer may see it as a piece of overhanging masonry. (If the reader has any difficulty in seeing Figure 22.1 as a piece of overhanging masonry, he has only to turn the page upside down by slowly rotating the book, keeping the eyes fixed on the drawing.) *Rubin's vase* (Fig. 22.2) may be seen either as

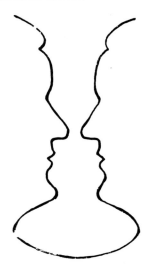

**Fig. 22.2** Rubin's vase.

a vase or two human profiles. In general, diagrams of this sort are called *inadequate representations* of the object they are intended to copy, or *ambiguous figures*. The lines, as such, can be interpreted with more or less ease in two different ways and the fact that they are perceived in one or other definite manner emphasizes the arbitrary element of interpretation in the visual process. Helmholtz quotes in this connection the case of a windmill, observed against the night sky; according as the observer fancied that the sails were in front of, or behind, the body of the mill he saw them revolving in different directions. Brown (1962) has described an experimental set-up that simulates the situation described by Helmholtz, the subject observing the shadows cast by a set of six vertical pins arranged in a circle and rotating at constant speed about a vertical axis through the centre of the circle. The stimulus on the screen was two-dimensional, with one set of pins moving left and the other right, but the subject perceived it three-dimensionally, with the apparent direction of rotation of the pins varying. This is because the basic stimulus contains strong depth cues, but without decisive ones concerning the depth relation between the two sets of pins moving in right and left directions.

Careful experimentation with ambiguous figures has shown that, over a period of time, the fluctuations in meaning exhibit a periodicity characteristic of the observer. This is not to deny, however, that with an effort of will, by moving the eyes, blinking, etc., the meaning of the pattern may be made to change. Nevertheless, if the subject merely regards the figure for several minutes and the times during which it is seen in its two meanings are recorded, it is found that the fluctuations are definitely not random in nature. It is interesting that the periodicity may be affected by drugs.

Brown's experimental arrangement described above is particularly suitable for investigation of the rate of change of apparent movement; he found that it increased with viewing time. If this increase had been achieved by monocular observation, then the other eye showed the same increase, so that we may say that there was *interocular transfer* of this adaptational effect, indicating that the process occurs at a higher central level than the retina, namely at some point in the visual pathway common to both eyes.

### READING

The reading process is an instance of the arbitrary attachment of meaning to symbols; with practice the meaning becomes associated not with the complete picture of a word but with its general shape and thus the practised reader can be imposed on easily with mis-spelt words; the context leads him to expect a certain word and he actually sees it even though it is not on the page. Here the perceptual process is imposing a meaning on the visual stimulus that does not correctly belong to it, and many examples of this arbitrary activity of the higher centres may be cited; for example most people read the word 'vicegerent' as 'viceregent' because the latter is the word they expect even though it does not occur in the dictionary. The filling up of the blind spot and the failure of the hemianopic subject to realize his defect are further instances.

## COLOUR PERCEPTION

### COLOUR CONSTANCY

By this term we mean the tendency of the subject to perceive the colour of an object as being the same in spite of large variations in the actual chromaticity of the light emanating from it. The colour becomes one of an object's attributes—its *memory colour*—and there is a strong tendency to perceive this even when the physical conditions have been chosen to make this difficult. For example, a red letterbox may be illuminated with blue light so that far more blue light is reflected off it than red, yet the subject continues to call it red, and his *perception* is that of a red object. If the object, so illuminated, is looked at through a tube, so that its colour may be divorced from its shape, it is said to appear blue. Again, we may place a green filter at arm's length and note that the part of the field of vision seen through the filter acquires a distinct green colour—*colour conversion*—but when the filter is brought close to one eye and the other is closed, the natural colours of the objects viewed are identified readily—*colour constancy*. Thus, as Helmholtz pointed out: '. . . with all coloured surfaces without distinction, wherever they are in the sphere of the coloured illumination, we get accustomed to subtracting the illuminating colour from them in order to find the colour of the object.' The basis on which the 'discounting of the illuminant colour' is achieved must be an unconscious recognition of the chromaticity of the light reflected off the more highly reflecting objects in the scene, e.g. a piece of white paper.

If we call the colour perceived in an object as the *object colour*, we may refer to these phenomena as examples of *object-colour constancy*. When a coloured light is viewed independently of any object, as through a hole in a screen, then we may call this an *aperture colour*, but it must be appreciated that the perceived aperture colour is by no means invariant for a given chromaticity of its emitted light, and this is largely because of adaptational and contrast effects. Thus, as Helson (1938) showed, it is easily possible to devise a scene in which an object reflecting light of any chromaticity whatsoever is perceived as grey, the main basis for this being the tendency for the subject to see the colour of the illuminant, the colour of the after-image complementary to the illuminant, or achromaticity, depending on the relative reflectances of the surface being viewed and the adaptation reflectance.

These principles, which lead to modifications of the perceived colours of objects, have been formulated quantitatively by Judd (1940), the process of discounting the spectral quality of the illuminant being achieved by defining, on the Maxwellian triangle, a point to correspond with the perception of grey in the object-mode of viewing, this point being close to the chromaticity of the illuminant. The prediction of the hue reported by an observer was given by the vector extending from this achromatic point to the point defining the chromaticity of the light actually emanating from the object.

## TWO PRIMARY PROJECTIONS

Interest in colour constancy, and its related phenomena, was revived with the description by Land (1959) of a series of experiments in which he produced, with the aid of only two illuminants, pictures that seemed to give a faithful reproduction of the colours of various scenes—pictures that, on the basis of trichromatic theory, should have required three independent sources of illumination. In general, Land produced two transparent positives of the scene by photographing it first through a red-transmitting filter (585 to 700 nm) and then through a filter transmitting the middle third of the spectrum (490 to 600 nm). The image of one of these positives was usually projected on to a white screen with red light (590 to 700 nm), whilst that of the other usually by an incandescent lamp, the two images being adjusted carefully to be in register. Space will not permit of a summary of the findings, but his Experiment 3 may be

quoted as an example: By combining the long-wave picture shown in red light with the middle-wave picture shown in incandescent light, the subject perceived the full colour in a portrait of a blonde girl, i.e. blonde hair, pale blue eyes, red coat, blue-green collar, and strikingly natural flesh tones. For adequate colour representation through photography, a three-primary system is ordinarily employed, and the question to be asked is how this apparently true representation of colours is achieved with only two positives. It must be appreciated, however, that the use of white light in the incandescent illuminant provides the viewer with all colours of the spectrum, and it is essentially the fact that the subject ignores the exact chromatic content of the light reflected off the various parts of the screen that permits him to perceive a fairly true representation of the original. According to Judd's analysis, the perceived colours in this, and in other examples described by Land, are, in fact, predictable on the basis of his earlier quantitative formulation, so that there is no reason to abandon the trichromatic theory of colour perception on the basis of these experiments of Land.

## TONE CONSTANCY

Essentially similar phenomena are described for the blacks, greys and whites. A grey wall may be compared with a grey piece of paper on a table reflecting the same amount of light; the wall appears white whilst the paper appears grey, presumably because we associate lightness with the walls of a room. These phenomena are classified under the heading of *tone constancy*, but it must be appreciated that, just as with chromatic vision, the final perception depends on inductive effects taking place at the retinal level. For example, it is argued that the black of this type, seen on a fine day, reflects more light than does the white of the paper in dim illumination, and yet the black is seen as black in both circumstances. The psychologist 'explains' this as tone constancy; the physiologist, however, would merely state that the black sensation is evoked by a certain area of the retina being stimulated less intensely than another, adjoining, area; he would be the last to suggest that there is an absolute black sensation depending on some critical level of retinal stimulation. We know, moreover, that the sensitivity of the retina is enormously affected by light-adaptation (p. 366), so that it is impossible to expect that the same sensation will be evoked by the same stimulus under widely differing states of adaptation. We must therefore be careful of attributing to the activities of higher centres phenomena that properly belong to the retina.

Having clarified our position in relation to the difference between a perception and a sensation, we can now proceed to analyse some of the simple mechanisms concerned in the perceptual processes.

# MONOCULAR PERCEPTION

## THE PROJECTION OF THE RETINA

Objects are perceived in definite positions in space—positions definite in relation to each other and to the percipient. Our first problem is to analyse the physiological basis for this spatial perception or, as it is expressed, the *projection of the retina into space*.

### RELATIVE POSITIONS OF OBJECTS

The perception of the positions of objects in relation to each other is essentially a geometrical problem. Let us confine the attention, for the present, to the perception of these relationships by one eye—*monocular perception*. A group of objects, as in Figure 22.3, produces images

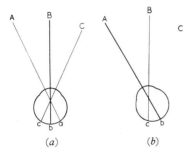

**Fig. 22.3** (*a*) A group of objects, A, B, C, produces images on the retina, *a*, *b*, *c*; the retinal images are projected outwards in space towards the points evoking them (*b*) The eye has moved to the left so that the image of A falls on *b*. *b* was previously projected to B but it is now projected to A.

on the retina in a certain fixed geometrical relationship; for the perception of the fact that A is to the left of B, that B is to the left of C, etc., it is necessary that the incidence of images at *a*, *b*, *c* on the retina be interpreted in a similar, but, of course, inverted geometrical relationship. The neural requirements for this interpretation are (*a*) that the retina should be built up of elements which behave as units throughout their conducting system to the visual cortex and (*b*) that the retinal elements should have 'local signs'. The local sign could represent an innate disposition or could result from experience—the association of the directions of objects in space, as determined by touch, etc., with the retinal pattern of stimulation. However they arise, the local signs of the retinal elements provide the complete information as to the relative positions of A, B and C.

The retinal stimuli at *a*, *b* and *c* are appreciated as objects outside the eye—the retina is said to be *projected* into space, and the field of vision is thus the projection of the retina through the nodal point. It will be seen that the geometrical relationship between objects and retinal stimuli is reversed; in the retina *a* is to the right of *b*,

and so on; it is customary to speak of the 'psychological erection of the retinal image'; this is thoroughly misleading since it suggests an awareness, on the part of the higher centres, of the actual points on the retina stimulated. This of course is untrue; *a* and *c* are projected to the left and right of B, not because the higher centres have learnt the laws of geometrical optics but because such a projection corresponds with experience.

### POSITION IN RELATION TO OBSERVER

The recognition of the directions of objects in relation to the observer is more complex. If the eye (Fig. 22.3*b*) is turned to the left, the image of A falls on the retinal point *b*, so that if *b* were always projected into the same direction in space, A would appear to be in B's place. In practice we know that A is perceived as fixed in space in spite of the movements of the eye; hence the direction of projection of the retinal point is constantly modified to take into account movements of the eye—we may call this '*psychological compensation*'. It will be seen that correct projection is achieved by projecting the stimulated retinal point through the nodal point of the eye. Movements of the eye due to movements of the head must be similarly compensated. As a result, any point in space remains fixed in spite of movements of the eye and head. Given this system of 'compensated projection,' the recognition of direction in relation to the individual is now feasible. B may be said to be due north, or more vaguely 'over there'; when the head is turned, since B is perceived to be in the same place, it is still due north or 'over there'.

### FALSE PROJECTION

Under some conditions the human subject will make an error in projecting his retinal image so that the object giving rise to this image appears to be in a different place from its true one; the image is said to be *falsely projected*. If the eye is moved passively, for example by pulling on the conjunctiva with forceps, the subject has the impression that objects in the outside world are moving in the opposite direction to the movement of the eye. The exact nature of this false projection is illustrated in Figure 22.4. The eye is fixated on an object F, whose image falls on the fovea, *f*.* The eye is moved forcibly through 30°, so that the image of F falls on a new point, *q* (Fig. 22.4*b*). The correct projection of *q* would be through the nodal point out to F; however, projection takes place as though the eye had not moved. The image falls 30° to the right of the fovea; with the eye in its original position, such a retinal stimulus would correspond to an object 30° to the left, the point Q in

---

* In this chapter the term 'fovea' will be used very loosely to indicate the position of the image of a centrally fixated point; it will be the intersection of the visual axis with the retina.

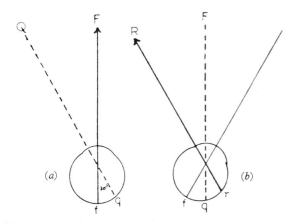

**Fig. 22.4** Illustrating false projection resulting from the forcible movement of the eye. (*a*) The point F is fixated. (*b*) The eye has been moved forcibly 30° to the right. The image of F falls on *q*, and the projection of *q* is made on the assumption that the eye has not moved, i.e. towards **R**.

Figure 22.4*a*, so that the stimulus on *q*, in the new position of the eye, is projected 30° to the left, along the line *r*R. Consequently, as the eye is moved to the right, the retinal image of F is projected progressively to the left.

## Information from motor discharge

Quite clearly, under these abnormal conditions, the necessary information as to the position of the eye is not being sent to what we may call a '*space representation centre*' (Ludvigh), so that interpretation of the retinal image is made on the basis of inadequate information. The tension in the spindles and tendon-organs will undoubtedly be altered as a result of this passive movement, and the fact that no use is made of this proprioceptive information indicates that normally the space representation centre employs a different type of information on which to base its estimates of the position of the eye. This information could be provided by the motor discharge from the voluntary and reflex centres controlling the movements of the eyes, as suggested by Helmholtz; thus when the subject moves his eyes to the left to fixate point A in Figure 22.3, the centre that initiates the motor discharges to the medial and lateral recti, besides sending its impulses to the motor neurones of the nuclei of N III and VI, may also send a message to the space representation centre indicating the intensities of these discharges, and thus the approximate extents of the eye movements. The space representation centre thus interprets the shifting images of A, B and C over the retinae not as a movement of these points but as a movement of the eyes to fixate A. The proprioceptive messages from the muscle spindles and tendon-organs operate on a completely unconscious level, simply serving to moderate the strengths of the contractions of

individual fibres in accordance with variations in local conditions—the so-called *parametric feedback* (Ludvigh). In the case of the eye that has been moved passively, the space representation centre has obviously not been informed of the eye movement, and the interpretation of the shifting retinal images is now what would be expected had the eyes remained stationary, i.e. the interpretation is that of a moving of the external objects from right to left, with its consequent false projection. Before examining evidence directly concerning the existence of a collateral supply of information derived from the motor commands—*corollary discharge* of Sperry (1950) or *Efferenzkopie* of v. Holst and Mittelstaedt (1950)—we may consider some phenomena closely related to the simple case of false projection described above.

### PERCEPTUAL ERRORS DURING PURSUIT MOVEMENTS

A number of studies on perceived position of stationary objects while the eye is tracking a moving point suggest that there is very little compensation for the eye movements; for example Stoper (1967) flashed successively two lines while the observer tracked a moving target on a homogeneous ground; if the intervals were short—up to about 300 msec—the subject failed to take account of the eye movement intervening when judging the relative positions of the two lines, which were therefore determined by the retinal pattern with no allowance for the intervening motion; with longer intervals of up to 1·7 sec compensation became manifest but it was on average only 64 per cent. More recently Festinger *et al.* (1976) compared the perceived with the actual paths of spots of light in harmonic motion during the tracking of another spot and concluded that there was little compensation for the tracking movement. Thus, perceptually, the subject is ignorant of the actual speed of his eye movements during slow tracking, but of course practically the control over the eye muscles indicates an accurate adjustment in accordance with speed. Festinger suggested that initially, on the basis of perceived motion, tracking commands are initiated and the interpretation of retinal images is based on this primary command, which assumes that a relatively low speed is being utilized almost independently of the actual speed of the target; the actual adjustment of speed to permit accurate tracking is achieved by visual input that is apparently ignored by the perceptual process, thereby leading to considerable perceptual errors.★

★Festinger *et al.* (1976) quote a number of illusions based on failure to correct for eye movements; for example Johannsen reported that if an observer follows a target moving horizontally in simple harmonic motion, a vertically moving spot, 90° out of phase with the horizontally moving spot, is perceived to move in a circle, and this is the path it would sweep out on the retina.

CONTROL OF EYE POSITION IN THE DARK

Early studies of the behaviour of the eyes in the dark, when no fixation points are available, indicated that within a few seconds of extinguishing a luminous target the eyes rapidly wandered from the target; however, when Skavenski and Steinman (1970) examined behaviour during periods of up to two minutes they found that the error in 'fixating' the extinguished target, after increasing rapidly after the target was extinguished increased more slowly and eventually the position of the eye stabilized with an error of about 2°, suggesting that there was some 'extraretinal' memory and ability to direct the eyes in the light of this memory. In a later study Skavenski (1971) made a correlative study of the apparently random movements under the same conditions and showed that there was a corrective element that led to stabilization of the fixation error. When the position of an eye was changed artificially, without the subject being aware of the manipulation (a pull was exerted on a contact lens attached to the eye), the subject could report the movement correctly more often than not, with a probability of less than 0·001 that the answers could have occurred by chance. Moreover, the subject could also report the position of a target previously seen without the load on the eye that was causing an artificial deviation. Thus it seemed that the subject was correcting for the passive deflexion of the eye in the light of extraretinal *inflow* of information, since the outflow correlative discharge would not provide sufficient information as to the passive deflexion. These experiments indicate that some position-sense remains when the eyes are in the dark, but it is clearly of a very crude nature, and any accurate interpretation of visual images requires more precise information than this extraretinal inflow.

## Constant drift

Working over longer periods in the dark, Becker and Klein (1973) described a constant drift, tending to bring the eyes back to the primary position when attempting to maintain any direction of gaze; small corrective saccades were made, of which the subject was unaware, and also larger consciously programmed saccades when the subject felt that the eyes had drifted away from the target. When a subject tried repeatedly to make a given saccade, the size depended on the drift that had taken place during the preliminary fixation in the dark, the larger the drift the smaller the movement, so that the saccade took into account that part of the movement that had already been made by the drift.

## VISION DURING A SACCADE

It was shown by Dodge in 1900 that vision was apparently suppressed during a saccade; a convincing demonstration of

this was that of Ditchburn (1955) in which a subject fixated a bright spot on a cathode-ray oscillograph, the position of which moved in advance of the eye movement (positive feedback); as a result the spot made rapid flicks whenever the eyes moved, and the subject was unable to see these movements although an observer beside him could. Volkmann (1962) presented 20 μsec flashes to a subject who was told to move his eyes at given times; she found that, during a fixation movement, there was a 0·5 logunit rise in the threshold to the flashes, the change beginning some 40 msec before the beginning of the movement and lasting 80 msec after the end. In addition, the finest gratings that could be resolved by the moving eye were about 1·4 times coarser than those resolved by the fixating eye. Beeler (1967), in confirming the 0·5 logunit rise in threshold, pointed out that this decrease in visual sensitivity would not have been sufficient to prevent Ditchburn's subject from seeing the illuminated spot, so that there must be some other factor operating to prevent detection of movement. He presented a subject with a light-spot and made this move sideways 15 minutes of arc, the movement of the spot being arranged to occur at random intervals after an involuntary saccade; the movement of the spot was too small in extent to cause a fixation movement by the subject. Beeler determined the frequency of seeing the spot when it moved at different intervals after the saccade; and this increased from zero, during the saccade, to 93 per cent if the movement was separated from the flick by more than 100 msec. Thus, during a saccade there is apparently a suppression of target information; Beeler suggested that the cortex investigates the oculomotor system when a movement of the retinal images occurs, for example after a flick; during this time there is no awareness of further motions. If the efferent system has operated, i.e. if the retinal image-movement is due to motor activity and not to actual object movement, then all image-motion is ignored.*

## 'RETINAL SMEAR'

The obvious interpretation of this retinal 'suppression' is that it is responsible for the absence of 'retinal smear' during eye movements. Such a suppression could be achieved by corollary discharge, the command signals to

---

*Michael and Stark (1967) measured the visually evoked response (VER, p. 601) with scalp electrodes and found that, when saccadic suppression of light-flashes occurred, there was suppression or reduction of some waves on this record. Again, Zuber, Stark and Lorber (1966) found that the reduction in pupillary area in response to a flash could be as little as 10 per cent of the control value if the flash occurred during a saccade. Mitrani, Mateeff and Yakimoff (1970) have demonstrated very neatly that 'retinal smear' is a factor of some significance by showing that the depression of visual acuity was large for vertically orientated gratings, when the eye made a horizontal saccade, but small for horizontally orientated gratings. On the other hand, Collewijn (1969) found depression of the VER in rabbit cortex and colliculus evoked by a flash in one, immobile, eye when the saccade occurred only in the other, non-seeing, eye. The objective measurement of the suppression is well illustrated by the work of Riggs et al. (1974) which showed that the phosphene evoked by electrical stimulation of the eye in human subjects required a higher current-intensity during the execution of a saccade; by a suitable calibration of current versus subjective sensation they computed that the threshold rose by some 0·5 log units, or threefold, as found by others using photic stimulation.

the motor neurones being accompanied by discharge to the visual centres enabling them to interpret the shift in retinal images in terms of eye motion. To have value, of course, such a suppression must distinguish between movement of the image due to movement of the eyes and movement due to movement of the visual environment, since the latter could be of significance and should therefore not be suppressed. MacKay (1970) showed that the increase in threshold to a flash of light could occur by displacement of the visual images on the retina however they occurred, and he suggested that the elevation of threshold was purely determined by the movement of the image, so that the information could not have been provided by corollary discharge. MacKay's results have been abundantly confirmed, e.g. by Brooks and Fuchs (1975) who have discussed the significance of saccadic suppression in some detail in the light of their studies defining the stimulus parameters necessary for eliciting the phenomenon.*

## BEHAVIOUR OF VISUAL UNITS

It is of interest to examine the discharges of neurones in the visual pathway during eye-movements, and to distinguish, if possible, the effects of eye movements from those of movements of the objective world.

### GANGLION CELLS

Noda and Adey (1974) examined the responses of ganglion cells in the waking cat to movements of the eyes over a grating, or to moving targets. An obvious difference between sustained and transient ganglion cells was manifest, the transient units giving a burst discharge when scanning the target whilst the sustained gave a sustained discharge which was related to the luminance of the target. In all units, moving the target produced a corresponding type of discharge, and it was once again clear that it was the transient units that were specifically responsive to the dynamic aspect of the stimulus-situation.

### LATERAL GENICULATE BODY

Adey and Noda (1973) measured field potentials in the geniculate and striate regions during eye movements in waking cats; in the geniculate there was a suppression of postsynaptic events lasting some 150 msec associated with eye movements, but this was not observed in complete darkness and so could not have been due to a corollary discharge from the motor apparatus. In the striate region, however, there was an enhancement lasting for some 200 msec, which occurred in complete darkness but not when the visual field moved, so this cortical event might be the manifestation of a motor signal.

## Sustained and transient units

When the relay cells in the lateral geniculate body were studied, it was found that their excitabilities, as measured by the firing probability of cells being stimulated through the optic tract, were profoundly depressed as a result of an eye movement, but only if this occurred in the light, as found by Büttner and Fuchs (1973); this depression also occurred if there was a movement of a patterned eye-field rather than a movement of the eyes. Therefore, the lateral geniculate body is not the first place where visual and motor impulses meet. It will be recalled that ganglion cells fall into two main classes, transient or Y-cells and sustained or X-cells, and the same distinction is preserved in the geniculate nucleus where Noda referred to them as T and S cells respectively.

When tested with a moving grating, the Sustained (S) cells responded in a simple fashion with a discharge as each line moved across the receptive field, whereas the Transient (T) cells showed the same (primary) response to a slow motion of the grating but when the speed was increased a non-specific discharge occurred unrelated to the striped pattern. The general failure of the T cells to respond to patterned information, as opposed to transient changes, indicates that it is these cells that are relaying information regarding the dynamic aspects of the visual field, responding in bursts to a quick shift in retinal image during a saccade or object-movement. By contrast, the sustained (S) cells fire during movements and maintain a sustained discharge at the end of the movement, related to the stimulus-pattern; clearly their function is analysis of the spatial aspects of the visual field as discussed earlier. The actual discharges of the two types of geniculate cell during a saccade in the light and in darkness are shown in Figure 22.5, where it is seen that the T cell in the light gives a burst discharge whilst the resting discharge of the sustained S cell is completely inhibited; corresponding with the suppression of activity there was a depression of excitability, but interestingly the same reduction in firing probability in response to electrical stimulation of the optic tract was observed in the T cell following the saccade in the light but not in darkness. As before, the same responses were obtained by moving a grating in front of the stationary eyes. The impairment of transmission through both S and T cells may, as Noda has argued, be the important feature of the response to eye- or image-movement, and be closely related to the elevation of visual threshold obtained under the same conditions. An important discrepancy, however, is that the electrophysiological events have not been shown to precede the onset of the eye movement (Volkmann et al.,

---

*Of significance is the failure of thresholds to rise during saccades if these are made in the dark.

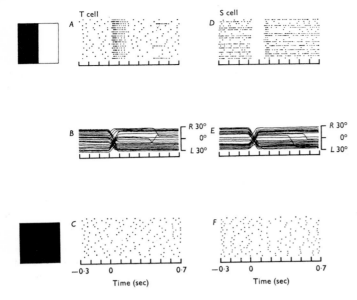

**Fig. 22.5**   Responses of a T cell (*A*) and an S cell (*D*) to saccadic eye movements in a visual field where only the right half of the tangent screen was diffusely illuminated. Discharges of the cells in response to saccades are shown for a period from − 300 to 700 msec in reference to the onset of saccades. The discharges are represented by horizontal rows of dots and the responses are aligned to the start of saccades (time 0). The time course of the corresponding saccades is shown below the responses by superimposed twenty consecutive sweeps of horizontal e.o.g. (*B* and *E*). In the e.o.g. calibrations *R* 30° represents 30° to the right and *L* 30° means 30° to the left seen from the cat. When tested in complete darkness, the changes seen in the light disappeared (*C* and *F*). *C* is for the T cell and *F* is for the S cell. (Noda, *J. Physiol.*)

1968) although the rise in threshold may be as much as 40 msec before the movement and extend up to 80 msec after.

Thus the lateral geniculate body fails to provide evidence for any interaction between oculo-motor output and visual input, and it seems likely that this will be found more definitively in the superior colliculi (p. 622).\*

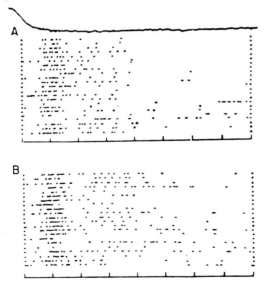

**Fig. 22.6**   Response of a cortical neurone to, A, movement of the eye through 20° and, B, movement of the visual field through the same angle. The solid line in A is the electro-oculogram, and the rows of dots are spike responses resulting from a single movement, successive rows corresponding to successive movements. Note similarity in responses in spite of the different ways of stimulating the receptive field of the neurone. (Wurtz, *J. Neurophysiol.*)

CORTICAL NEURONES

So far as the activity of cortical neurones is concerned, there is no doubt that this, in the presence of an appropriate stimulus falling on their receptive fields, may be markedly altered when the eyes make a rapid movement; this may be an increase in unit activity or a suppression; in others, however, there is no alteration at all (Kawamura and Marchiafava, 1968; Wurtz, 1969a). However, Wurtz (1969b) was able to match any response to a saccade by a movement of a patterned stimulus across the retinal field at 900°/sec (Fig. 22.6) so that once again we have no evidence here of corollary discharge.

**Oculogyral illusion**

Another example of false projection is given by the oculogyral illusion described by Graybiel and Hupp in 1946. A subject is seated in a rotatable chair in the dark, and he fixates a small source of light which is attached to the chair and rotates with it. When the chair is rotated

---

\* The reader's attention may profitably be directed to pages 817 and 818 of Adey and Noda's (1973) paper for a discussion of the conflicting evidence associating lateral geniculate activity with eye movements; it would seem that many of the results obtained in anaesthetized or encéphale isolé animals are not repeatable in the alert cat; moreover, the changes observed during vestibular nystagmus, e.g. by Jeannerod and Putkonen (1971), may be peculiar to this form of eye movement. Again, the well established bursts in the lateral geniculate nucleus associated with rapid eye movements (REM) of sleep (Bizzi, 1966), which have been suggested to be corollary discharges, are certainly not observed during spontaneous eye movements in alert cats. It should be mentioned, however, that Feldman and Cohen found negative potentials in the alert monkey's lateral geniculate nucleus associated with rapid eye movements in the dark; slow movements failed to produce these.

the light appears to move in the same direction. In this case the rotation has stimulated the semicircular canals, and the space representation centre has been informed that the body has rotated. The subject is unaware that the fixation light has also moved, so that the stationary retinal image is reinterpreted; clearly, if the head moves whilst the image of a point remains stationary, the point must have moved to the same angular extent as the head. Hence the retinal image is projected as though the object had moved in relation to the observer. Thus, if the chair is rotated to the right, the light will be falsely projected to the right of the observer, although it has remained fixed in relation to him.

## Autokinetic illusion

The explorer Alexander von Humboldt in 1799 noticed that after watching a faint star for a short time it would start to move around in various directions. An obvious explanation for this apparent movement may be the occurrence of involuntary eye movements, the space representation centre being inadequately informed of them whilst the relatively uniform background against which the light is viewed fails to provide adequate retinal information of the eye movements. There is still some disagreement as to whether such movements could always account for the phenomenon, so that other factors have been invoked as well. The subject has been reviewed by Grosvenor (1959), whilst the contribution of eccentric vision to the phenomenon has been described by Crone and Verduyn Lunel (1969).

### APPARENT MOVEMENT OF AN AFTER-IMAGE

The apparent movement of an after-image, when the eye moves, is an excellent illustration of psychological compensation. A retinal stimulus, being normally projected through the nodal point, is projected into different points in space as the eye moves; an after-image can be considered to be the manifestation of a continued retinal impulse, and its projection changes as the eye moves. The after-image thus appears to move in the same direction as that of the movement of the eye. Whether or not the gradual drift of an after-image across the field of view is entirely due to eye movements, it is difficult to say. One certainly has the impression that the eye is 'chasing' the after-image.

### FIGURAL AFTER-EFFECTS

This is the term given by psychologists to a variety of illusions having as their basis the adaptation to prolonged observation of a given figure; thus a subject fixating a curved line for some time sees it as less and less curved, whilst a straight line presented immediately after this prolonged inspection of the curved line appears curved in a direction opposite to that of the original curvature (see, for example, Gibson, 1933). Particularly

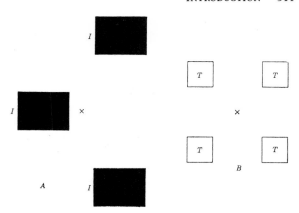

**Fig. 22.7** Demonstrating displacement effects. Regard the fixation cross of A at reading distance for about 40 seconds, and then look at the fixation cross of B. After A has been observed, the right-hand squares of B appear nearer the central horizontal axis of this figure than do the left-hand squares. Inspection of the left-hand I-figure in A causes the left-hand squares of B to 'move apart vertically'; and inspection of the right-hand I-figures of A causes the right-hand squares of B to 'move together'. Subjects say that test figures are displaced from the positions of previous inspection figures. (Graham, *Vision and Visual Perception.*)

instructive is Figure 22.7 from Köhler and Wallach (1944); the subject inspects the fixation cross of the left-hand ensemble for about 40 seconds and then inspects the fixation cross of the right-hand ensemble. As a result of the first period of fixation the right-hand pair of squares look closer to each other in the vertical direction than do the left-hand pair. Thus there is a tendency for the perceived object to be displaced as a result of previous viewing, and the direction of displacement is away from the previously inspected object.

## VISUAL ESTIMATES

### THE DIRECTIONS OF LINES

We have so far considered the problem of estimating the positions of points, in relation to each other and to the percipient. The estimate of the directions of lines involves no really new principles, since if two points, A and B, are exactly localized, the direction of the line AB can be appreciated. The perspective distortion, however, and the torsion of the eye resulting from a change in the direction of fixation, can modify the estimate of the direction of a line, referred to a fixed line in space—e.g. the horizontal or vertical—unless some psychological compensatory process is introduced.

### Retinal meridians

A horizontal line—fixated in the primary position of the eye—is recognized to be horizontal because its image falls on a set of receptor elements on the retina lying on the horizontal meridian. Similarly a vertical line is recognized because its image falls on the vertical meridian. When the eye moves to the right, into a secondary position, a horizontal line still falls on the meridian because there has been no torsion. On looking up and to the right, however, the image of a horizontal line is inclined at a certain fixed angle, depending on the position of the eye, to the horizontal meridian of the retina. Because of the torsion a psychological compensation must be made so that an image, inclined across the horizontal meridian, is perceived as the image of a truly horizontal line. The full significance of Donders' Law (p. 392) now becomes evident. Movements of the eye involve torsion—this cannot be avoided—and torsion involves a disturbance of orientation of lines unless psychologically compensated. The higher centres, in order to compensate, must know the degree of torsion, and this information becomes physiologically simple to acquire if the torsion is characteristic of the position of the eye. Thus the motor innervation bringing about a given position of the eyes is 'reported' to the space representation centre, and this centre, in interpreting the retinal image, works on the assumption that this movement has brought about a characteristic degree of torsion. Donders' Law states that the torsion is a definite quantity for a given position of the eyes; and this regularity is achieved by the eyes rotating about an axis in Listing's plane (Listing's Law).

**Vestibular influences**

Movements of the head are associated with compensatory reflex movements of the eyes, mainly of vestibular origin, designed to keep the visual fields normally orientated. In man, however, the rolling necessary to compensate for an inclination of the head towards the shoulder is quite inadequate, amounting to only one-tenth to one-fifth of the inclination of the head; since horizontal lines appear horizontal even when the head is strongly inclined, psychological compensation must be involved here; presumably the information required is provided by the labyrinth. That this labyrinthine information is not entirely adequate is shown, however, by *Aubert's phenomenon*. A subject regards a bright vertical slit in a completely dark room; on bending the head over to the side, the slit appears to incline in the opposite direction. When the lights are switched on the slit becomes vertical. If, in the dark, the head is inclined suddenly the slit remains vertical for a while but finally tilts over; it would seem that the abrupt movement stimulates the semicircular canals (p. 433) which cause adequate compensation. If the higher centres have to depend on the tonic receptors only—utricle and proprioceptive neck impulses—the compensatory mechanism becomes inadequate

(Merton). It would appear, therefore, that the tonic receptors must be reinforced by the visual sensation; the observer knows that the bars of the window, for example, are not usually tilted over and so a compensation is initiated.

**Accuracy of estimates**

The power of appreciating whether a line is horizontal or vertical is very highly developed in man. Volkmann describes experiments in which he repeatedly set a line to what he considered to be the true horizontal; his mean error was only 0·2° with his left eye, the left-hand end of the line was too low; with his right eye the right-hand end was too low. We may say that Volkmann's '*apparently horizontal meridians*' were slightly inclined to the true horizontal. With vertical lines his error was greater—more than 1°—but it was a constant error indicating an inclination of his '*apparently vertical meridians*'. More recent work, notably that of Gibson, bears out the findings of Volkmann; vertical lines are, on the average, estimated to within 0·28° and horizontal lines to within 0·52°; when an attempt is made to adjust a line to a fixed angle with the vertical, e.g. 60°, the error is much greater (1·74°) a finding which suggests that the vertical and the horizontal act as a frame of reference to which all other directions are referred. A modification of this type of experiment consists in presenting the subject with a short line, AB (Fig. 22.8)

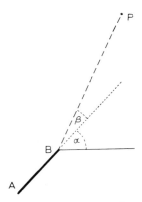

**Fig. 22.8** The subject views the line AB and must adjust the position of P such that it appears to coincide with the prolongation of the line AB. (After Bouma and Andriessen, *Vision Res.*)

at a given angle of slant, α, from the horizontal, and asking him to adjust a dot, some distance away, to a position such that an extension of the line would pass through it. The error is indicated by the angle, β, and the angle (α + β) is called the perceived direction. Bouma and Andriessen (1968) found a tendency for the perceived orientation to be closer to the nearest horizontal or vertical than the geometrical slant, $\alpha$. It was found that $\beta$ decreased with the length of the short line, AB, up to a subtense of about 1°, so that a foveal line-segment was employed in the estimate; two dots were as effective as a line-segment.

## COMPARISON OF LENGTHS

The influence of the movements of the eyes in the estimation of length was emphasized by Helmholtz. An accurate comparison of the lengths of the lines AB and CD (Fig. 22.9) can be made, whereas if an attempt is

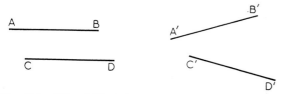

**Fig. 22.9**   AB and CD, being parallel, can be compared in length with accuracy whereas A′B′ and C′D′ cannot.

made to compare the lines A′B′ and C′D′, quite large errors occur. According to Helmholtz, the eye fixates first the point A and the line AB falls along a definite row of receptors, thereby indicating its length. The eye is now moved to fixate C, and if the image of CD falls along the same set of receptors the length of CD is said to be the same as that of AB. Such a movement of the eye is not feasible with the lines A′B′ and C′D′. Similarly the parallelism, or otherwise, of pairs of lines can be perceived accurately because on moving the eye over the lines the distance between them must remain the same.

Fairly accurate estimates of relative size may be made, nevertheless, without movements of the eyes. If two equal lines are observed simultaneously, the one with direct fixation and the other with peripheral vision, their images fall, of course, on different parts of the retina; if the images were equally long it could be stated that a certain length of stimulated retina was interpreted as a certain length of line in space. It is probable that this is roughly the basis on which rapid estimates of length depend; a white line stimulating, say, twenty-five retinal elements being considered to have the same length as another line stimulating the same number of elements in another part of the retina. It is easy to show, however, that the matter is not quite so simple; the retina is curved so that lines of equal length in different parts of the visual field do not produce images of equal length on the retina; the distortion in the image resulting from the use of oblique pencils of light also complicates the matter, so that it is unlikely that such a simple relationship between size of image and projected size is achieved. The problem of the estima-

tion of absolute size is closely linked with that of the estimation of distance and will be discussed under that heading (p. 515).

## ESTIMATES OF ANGLES

Two angles may be compared with accuracy if eye movements are permitted and if the sides are parallel. The errors in comparison increase greatly if the sides are not parallel. Essentially we are dealing with the same problem—the superimposition of the retinal pattern on another. In the estimation of right-angles some characteristic errors are encountered. Thus the cross made by the lines *ab* and *cd* (Fig. 22.10) appeared to Helmholtz

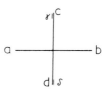

**Fig. 22.10**   The cross *abcd* appeared correct to Helmholtz, whereas the true cross is given by the line-segments $\gamma$ and $\delta$.

to be truly rectangular, whereas it is correctly given by the line-segments $\gamma$ and $\delta$. This error is an expression of the fact that the apparently horizontal and vertical meridians of the retina (p. 387) are not exactly at right-angles. When the cross was turned at an angle of 45° the error in assessing the right-angles became much greater; the true right-angles to the right and left appeared to be 92° whilst those lying above and below appeared to be 88°.

## OPTICAL ILLUSIONS

There are innumerable instances in the psychological literature of well-defined and consistent errors in visual estimates under special conditions; there is probably no single hard-and-fast rule by which they can be explained, but an important element may be the tendency for distinctly perceptible differences to appear larger than those more vaguely perceived. Thus the line AB of Figure 22.11 has been correctly bisected at C, but because AC is divided into easily perceptible parts, whilst the line CB is not, the former appears longer. The triangles ABC and A′B′C′ are both equilateral but, because of the division of the lines A′B′ and A′C′, the triangle A′B′C′ appears taller and therefore the angles at the base appear greater than 60°. Similarly, ABCD and A′B′C′D′ are squares yet the height of the latter figure appears to be greater than its breadth.

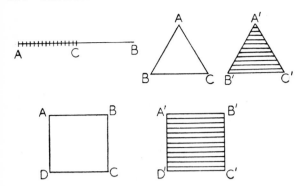

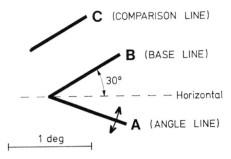

Fig. 22.11   Some optical illusions. AC appears greater than CB. The angles B′ and C′ of the equilateral triangle A′B′C′ appear to be greater than A′. The sides A′D′ and B′C′ of the square A′B′C′D′ appear to be greater than the sides A′B′ and C′D′.

Fig. 22.13   The 'base line' and the 'angle line' meet at the centre of a screen to form an angle so that they appeared to the subject like the hands of a clock. There was a similar 'comparison line', C, with its centre on the normal through the middle of B. During the experiments, B was fixed in the upper right quadrant of the screen at an angle of 30° to the horizontal. Line A was rotated to form various angles with B. (Blakemore *et al.*, *Nature*.)

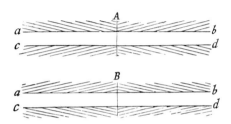

Fig. 22.12   Hering's illusion.

In Figure 22.12 (after Hering) the lines *ab* and *cd* are truly parallel and yet they appear to converge on each other on moving away from A (above) and diverge on moving away from B (below). It may be shown that the illusion results from the tendency of an angle to appear greater when lines are drawn through it parallel to one of the sides (thus the angles B′ and C′ in the triangle A′B′C′ of Figure 22.11 appear too large). Blakemore *et al.* (1970) have described a simple experiment in which the influence of one line on the estimated orientation of another is measured; thus the subject views the *Base Line* of Figure 22.13 and moves the *Comparison Line* so as to appear parallel. The *Angle Line* is moved by the experimenter, and it is found that the subjects make systematic errors in estimating the orientation of the *Bar Line* according to the angle made by the *Angle Line*. The tendency is for the acute angle to appear larger. Blakemore *et al.* have argued that such a result is predictable if the estimate of direction depends on the excitation of orientation-specific cortical neurones whilst other orientation-specific neurones, sensitive to a different orientation, exert an inhibitory action. Such an inhibition would modify the orientation-selectivity of

the neurones responsible for determining the orientation of the Bar Line, angling it away from its correct orientation.

SUMMARY

To summarize, we may say that the pattern of objects in the outside world is perceived as a result of the interpretation of the geometrical pattern of the image on the retina. It is not sufficient, however, to know merely the pattern of objects in space, it is useful also to relate *direction* on the retina with certain fixed directions in space. Primarily these are the vertical and the horizontal, the most frequently recurring directions of contours in the outside world. The retina has therefore become specialized by the development of apparently vertical and horizontal meridians; these act as a pair of rectangular co-ordinates through the fovea to which the position of any point or line is referred. In order that the interpretation of the retinal pattern be independent of the movements of the fixation axis, and of the inclination of the apparently horizontal meridian (torsion), a compensated system of projection into space is necessary. As a result of this compensation, the *absolute* position of an object in space is always correctly interpreted and the directions of lines, in relation to the horizontal and vertical, are also accurately estimated.

There is a great deal of evidence, that the tendency to project the retina is 'innate' although few would deny that accurate projection is only acquired with experience. Presumably the development in evolution of an innate tendency to project the retina in accordance with certain geometrical rules had survival value; the 'rules' turn out to be similar to those that would be followed if the individual learnt to project the retina in accordance with his own experience.

## THE MONOCULAR PERCEPTION OF DEPTH

From purely geometrical considerations, the estimation of the relative positions of objects in three dimensions by a single eye is not feasible; the points A, B and C in space (Fig. 22.14) produces images, *c*, *b* and *a* on the

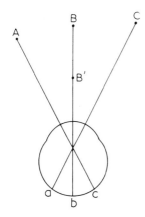

**Fig. 22.14** With uniocular projection, the relative positions of B and B′ cannot be distinguished.

retina which are projected outwards along the lines *c*A, *b*B, and *a*C, but there is nothing in the geometry of this projection system to enable it to differentiate between the points B and B′ for example. The projection of the retinal image of a single eye is thus two-dimensional and objects in space are presented to consciousness as a plane pattern. Nevertheless, the individual may *perceive* depth with a single eye by making use of a variety of clues, whose significance has been ascertained by accumulated experience. These clues are:

### 'AERIAL PERSPECTIVE'

As a result of the scattering of light by the atmosphere the colours of distant objects gave some clue to their distance; the scattered light is predominantly blue and violet and the eye views a distant object through a 'wall' of blue light. The thicker this wall, the bluer will the object appear to be; distant mountains thus appear blue, a fact exemplified in the well-known lines:

'T'is distance lends enchantment to the view
And robes the mountain in its azure hue.'[*]

It must be appreciated, however, that a self-luminous object, such as a lamp, is visible because its own emitted radiations reach the eye; since blue light is scattered

preferentially to red, a distant white light appears yellow or red when seen through a hazy atmosphere; the red light penetrates the atmosphere whilst the blue is scattered on the way. It might be objected that a mountain reflects light and that we see it because of this reflected light; this reflected light suffers scattering and so the mountain should appear yellow or red. Distant objects, however, are seen largely through contrast with the brighter sky; it is because they reflect less light than that emitted by the background that they are discriminated; the reflected light is therefore not important in perception and it is the wall of scattered light that determines the colour sensation. (When the brightness of 'wall' plus the brightness of the mountain due to reflected light becomes equal to the brightness of the background, the mountain becomes invisible. For this reason, the visibility of distant points in the landscape constitutes a good measure of the atmospheric haze.)

The scattering of light also blurs the outlines of objects and thus gives another clue to their distance from the observer; the apparent closeness of mountains in a very clear atmosphere is an illusion based on the normal association of clearly defined objects with proximity to the observer.

### RELATIVE SIZES OF OBJECTS

The apparent size of an object may be represented by the angle it subtends at the nodal point of the eye; thus AB and A′B′ (Figure 22.15) appear to have the same

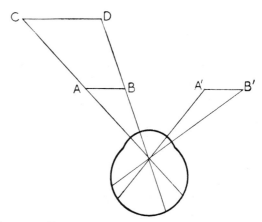

**Fig. 22.15** Illustrating the value of relative size in estimating depth.

size because they subtend the same angle and they are the same distance away; CD, moreover, appears as big as AB because it subtends the same angle although its real size is twice as great as that of AB. If the observer knows that CD is twice as large as AB, e.g. if AB is a

[*]Thomas Campbell: *Pleasure of Hope*; pt. i, l. 7–8.

bicycle and CD is a car, he can say that CD is twice as far away from him as AB, i.e. he can estimate their *relative* distances in the third dimension. A fair estimate of the *absolute* distance away can be made if the object viewed is familiar, e.g. a man or a cottage. When a man walks away from an observer the image on the latter's retina becomes smaller and smaller; this is interpreted as a more and more distant, not as a smaller and smaller, man. In gunnery, the visual estimation of range is made almost entirely on this basis.

The association of size with distance can give rise to common illusions; in a thick fog a figure 'looming up' looks very large because the figure is assumed to be farther away than it actually is. Similarly objects seen through binoculars appear to be very close whereas their images, seen through the instrument, are at infinity.

*Absolute size*
The determination of *absolute size* is thus closely interwoven with that of distance; any rigid formulation of this process on the basis of the number of retinal elements stimulated by the image of an object is therefore doomed to failure. A given retinal distance represents an absolute size of an object in space only for one distance of the object from the observer. A change in this distance must necessitate a revaluation of retinal distance. When an object is recognized to be far away, a small retinal distance is interpreted as a certain unit of 'absolute size' in space; if the same object is brought much closer, its image becomes larger so that a larger retinal distance now becomes equivalent to the same unit of 'absolute size' in space. The close linkage between this revaluation process and the acts of convergence and accommodation is shown by the condition of '*paresis micropsia.*' As a result of a defective convergence mechanism, for example, a strong effort of convergence is made in order to fixate a near object; this effort, being out of all proportion to the resulting convergence, is associated with much too great a change in retinal dimension values; the object converged upon appears smaller because it is treated as being much closer and the retinal dimension values have been scaled down too far.

*The moon illusion*
The assignment of a size to objects very far away, such as the sun and moon, must be entirely arbitrary, and so it is not surprising that the assigned size varies, although the reasons for the variations are not apparent. Thus the sun and moon, when seen on the horizon appear very much larger than when viewed at the zenith. It is interesting that the magnitude of the discrepancy between the apparent sizes of a disc viewed horizontally and vertically is much greater in children than in adults; thus in 4 to 5-year-olds a 10-inch disc viewed horizontally was estimated to be of the same size as a 20-inch disc seen vertically; in adults it was considered equivalent to a 16-inch disc (Leibowitz and Hartman, 1959).

FORM
The contours of objects give some idea as to their relative positions in the third dimension. Thus in Figure 22.16 the cottage is in front of the tree which is in front of the windmill which itself is perceived to be in front of the hill, entirely because of the overlapping of the contours.

PERSPECTIVE
The projected retinal image of an object in space may be represented as a series of lines on a plane, e.g. a box; these lines, however, are not a unique representation of the box since the same lines could be used to convey the

**Fig. 22.16**   The value of overlapping contours in the estimation of depth.

impression of a perfectly flat object with the lines drawn on it, or of a rectangular, but not cubical, box viewed at a different angle. In order that a three-dimensional object be correctly represented to the subject on a two-dimensional surface, he must know what the object is, i.e. it must be familiar to him. Thus a bicycle is a familiar object; viewed flat-on it has the appearance of Figure 22.17; if it is turned away from the observer it appears as in Figure 22.18; the wheels, for example, are elliptical

**Fig. 22.17**    Full side view.

**Fig. 22.18**    Three-quarters view. The front wheel appears smaller than the back; both appear elliptical.

and the front one is smaller than the rear one. Because the observer knows that the wheels are actually circular and of the same size he can state that the bicycle is directed into the third dimension, i.e. he perceives depth in a two-dimensional pattern of lines. The perception of depth in a two-dimensional pattern thus depends greatly on experience—the knowledge of the true shape of things when viewed in a certain way. The limitations of two-dimensional representation become evident when we look at the photograph of an unfamiliar object, such as a complex crystal; the picture is almost meaningless because we have no idea of the actual shapes of its component parts.

## LIGHT AND SHADE

A sphere always appears in a two-dimensional pattern as a circle, however viewed; perspective can therefore give no clue to its solidity. The light and shade on its surface, however, immediately give it a three-dimensional appearance (Fig. 22.19); in general, it is

**Fig. 20.19**    Illustrating the value of light and shade in giving the appearance of depth to a sphere.

important that the origin of the source of light be known and then the relations of the shadows of objects to each other indicate their mutual arrangement in space. The protective colouring of a large number of animals is designed to offset the appearance of solidity resulting from uneven illumination. Thus the back of an animal receives more light than the belly and if the back reflected the light just as efficiently as the belly it would appear much brighter and convey the impression of solidity (Fig. 22.20). However, the white belly of the donkey, for example, increases the amount of light reflected by it so as to make it approximate in brightness to its back, which is grey. As a result the animal appears, at a distance, to be fairly uniformly bright over its whole surface and therefore appears flat (Fig. 22.20), a characteristic that helps it to merge with its background. Under certain conditions of illumination, light and shade can give valuable clues to the length of a room, for example, in the third dimension; if the source of illumination is close to the observer, the luminance of the walls decreases progressively with their distance away. As a result of the process that results in 'tone constancy,' the walls generally appear to be evenly illuminated, the actual change in luminance being interpreted as an *increase in distance* from the observer. In this case the phenomenon of 'tone constancy' is essentially similar to that of 'size constancy'—the tendency to interpret a diminishing retinal image as an object of constant size moving away from the observer.

## PARALLACTIC DISPLACEMENT

On movement of the head from side to side, the objects nearer to the observer appear to move, in relation to more distant points, in the opposite direction. Similarly,

**Fig. 22.20**    Protective shading of an animal (see text).

when the observer moves forward, the nearer objects appear to move past him whilst the more distant objects appear to move with him. Helmholtz remarks that if one sits in a wood with one eye closed and the head motionless, the picture presented is quite flat; on moving the head, however, the environment springs to life due to the parallactic displacements, which indicate the essentially three-dimensional nature of the scene.

ACCOMMODATION

Within a limited range, the accommodative effort required to see an object clearly might be expected to give a clue as to its absolute distance from the observer, but although opinions differ, it seems safe to conclude that the kinaesthetic sensations from the ciliary muscle do not contribute materially to the power to assess distance, and the same would appear to be true, in binocular vision, for the process of convergence (Zajac, 1960). The

extreme difficulty in separating the factor of accommodation from all the other factors concerned reveals that depth perception is a complex process whose components are so interwoven as to make it a somewhat academic exercise to attempt to separate them and measure their individual contributions.

This description of the monocular factors in depth perception shows that there are many ways in which the one-eyed person can perceive depth in his environment; nevertheless his perceptions are grossly inadequate when compared with those of a normal subject, and we may now inquire into the general characteristics of binocular vision.

## BINOCULAR PERCEPTION

The two eyes behave, in so far as their motor taxis is concerned, as a single unit; binocular vision is likewise integrated, so that in many respects the perception can be described in terms of a single eye.

'THE CYLOPEAN EYE'

If its two images on the retinae were projected independently into space, an object would appear to be differently orientated in relation to the individual according as its direction was referred to the primary position of one or the other eye. For example, in Figure 22.21, the point A is fixated by both eyes, and it is

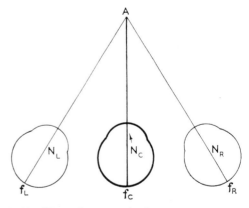

**Fig. 22.21**    The cyclopean projection.

correctly projected into space at A; to the left eye, however, it is to the right of the fixation axis in its primary position, whilst to the right eye it is to the left. Since it is symmetrically disposed in regard to the two eyes, the object is considered to be straight ahead of the observer—neither to the right nor to the left—so that binocular projection can be represented by the *'Cyclopean eye'*, as in Figure 22.21, an imaginary eye in

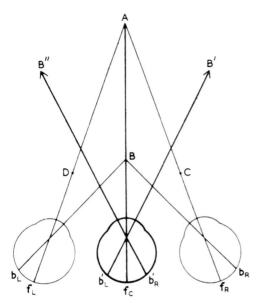

**Fig. 22.22** Heteronymous diplopia. B is closer to the eyes than the fixation point. By the left eye the image is projected to B′; by the right eye to B″.

the forehead to which all directions are, in fact, referred. The point A is considered to be straight ahead of the observer because it lies on the fixation axis of the Cyclopean eye in its 'primary' position.

BINOCULAR DOUBLE VISION

The Cyclopean projection is represented by transposing the images on the two retinae to the single, median, eye as in Figure 22.21. When the images on the two retinae are not symmetrically disposed about the fixation point, however, such a transposition leads to a double projection, and it is found, in fact, that when the same object can be projected in two directions by the Cyclopean eye double vision does actually occur. In Figure 22.22, the two eyes fixate the point A, as before. A point B, closer to the eye, on the median line, produces images at $b_L$ and $b_R$; the problem is to determine how the observer projects these images. In Figure 22.22, the Cyclopean eye, occupying a mean position between the two eyes, is drawn in. The image of A falls on the fovea of each eye and hence its projection by the Cyclopean eye is given by projecting $f_c$ through the nodal point, i.e. it coincides with the true direction of A. The image of B on the left retina, $b_L$, occurs to the left of $f_L$ and must be put in a similar position on the Cyclopean retina; $b_L$ is thus projected to B′. On the right retina, B forms an image, $b_R$, to the right of $f_R$, which, when transposed to the Cyclopean eye, gives a projection to B″. Since B is projected to B′ and B″ at the same time, it is seen double; the diplopia is called *heteronymous*

*diplopia* because on closing one eye the image appears to move away from the fixing eye, e.g. on closing the left eye the image of B appears to be at B″, i.e. away from the right, fixing, eye. The *homonymous diplopia*, caused by an object farther away from the fixation point, is indicated in Figure 22.23. The normal presence of double images in the binocular field of vision is not usually noticed; they are easily demonstrated, however, by placing a pencil close to the eyes in the median line; when a distant object is fixated, the pencil is seen double.

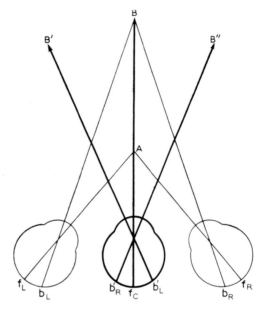

**Fig. 22.23** Homonymous diplopia. B is farther away from the eyes than the fixation point, A. By the left eye the image of B is projected to B′; by the right eye to B″.

BINOCULAR ALIGNMENT

On a binocular projection, the point B is aligned with the point A, so that if an observer places a pencil in front of him so as to coincide with a distant object when both eyes are open, he really puts the pencil in such a position that its double images lie at equal distances to the left and right of the distant object—B′ and B″ of Figure 22.22. Such a Cyclopean alignment is, in general, not accurate; the careful alignments necessary for sighting a rifle, for example, are made with one eye; if the alignment were made with the right eye a point C, on the fixation axis, would be correctly aligned with A; for the left eye, the point D (Fig. 22.22). (The reader should try rapidly pointing with a pencil at a distant object; keeping the pencil fixed, he should inspect the double images of the pencil while he fixates the distant object; if the latter lies between the double images, the pointing has been carried out on a Cyclopean projection; if, on the other hand, one of the double images coincides with the distant object, the pointing has been carried out on the basis of a monocular projection.)

DIPLOPIA DUE TO SQUINT OR PRISMS

In Figure 22.22 the point A was seen single, on a Cyclopean projection, because the lines joining it to the nodal point of each eye met the retinae in the foveae. B was seen double because the lines from it through the nodal points met the retinae on opposite sides of the fovea. A similar state of affairs occurs in squint. Figure 22.24 represents a divergent squint. A is fixated by the right eye, but the left visual axis diverges from A; the image of A on the left retina thus falls at $a_L$. If the images on the two retinae are transposed to the Cyclopean eye, the image of A is projected towards A and towards A', to the right of A. A heteronymous diplopia is therefore caused by this divergent squint.

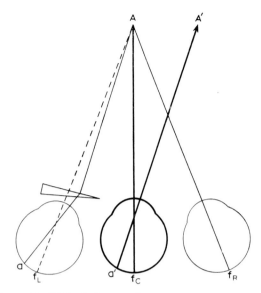

Fig. 22.25   Diplopia due to prism.

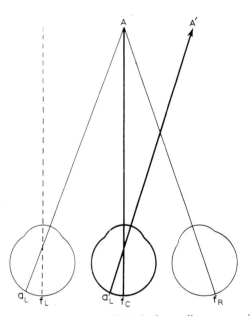

**Fig. 22.24**   Heteronymous diplopia due to divergent squint. The image of A is correctly projected by the right, fixating, eye to A; by the left eye it is projected to A'.

## Prism

A prism may cause diplopia, provided that the eyes do not compensate for the change of apparent direction of the object caused by the prism. In Figure 22.25, a base-out prism has been placed before the left eye. The image of A falls on $f_R$ the fovea of the right eye, but in the left eye it falls on $a$, to the left of the fovea $f_L$. On the Cyclopean projection, the images of A are projected in two directions, A and A'. In Figure 22.26, the eye has compensated for the prism by adducting; both images

now fall on the foveae, and the Cyclopean projection indicates single vision. It is interesting that a single projection is obtained in this instance; the left eye has adducted, so that on the basis of our earlier arguments (p. 507) an image falling on the fovea should be projected through the nodal point—to the right of A. If the left eye, only, were being used, such a projection would occur—a *false projection*, in the sense that the image of A is not projected towards A. The fact that with binocular vision the two images of A are correctly projected to a single point indicates that the Cyclopean projection is determined by the projection of the eye that has not been interfered with.

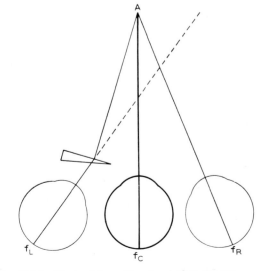

**Fig. 22.26**   Single vision with prism before one eye.

When prisms are placed in front of both eyes, if the necessary oculo-motor adjustments are possible, single vision is retained, but this time the retinal images of objects are erroneously projected, either closer to, or farther away from, their true position.

## THE 'FALSE MACULA'

The diplopia due to squint is an instance of false projection; the squinting eye fails to project a retinal image through the nodal point, e.g. the stimulated point $a_L$ of Figure 22.24 is not projected through the nodal point to A but its projection may be described as though the point stimulated were to the left of $a_L$. The fovea of the squinting eye is likewise projected falsely, an image falling on $f_L$ is projected towards A, the fixation point of the other eye. In many cases the false projection may be corrected by the development of a 'false macula', that portion of the retina on which the image of the fixation point, A, of the other eye falls, acquiring some of the characteristics of the fovea. The change is a cortical one—it is impossible for the anatomical characteristics of the retina to alter; the development of a false macula represents a change in the significance of the stimulus falling on it; the stimulus becomes the centre of attention, and that falling on the true macula is suppressed; moreover, the false macula is correctly projected into space. We have seen that projection is essentially a matter of localization in space of a retinal stimulus; to do this correctly the positions of the eye and head must be known and psychological compensations for movements of the eye and head are an integral part of the projection mechanism. The development of the false macula requires a re-adjustment of this psychological compensation to take into account the abnormal relationship of the visual axes of the two eyes.*

## INNATE PROJECTION

The existence of false projection following squint reveals that the projection of the eye through the nodal point is an innate characteristic, developed doubtless during the early years of life, but nevertheless an inherited quality. The possibility of a development of a false macula reveals, on the other hand, a certain plasticity in the function of the visual cortex. We have already remarked on the rigidity of the relationship between stimulus and sensation in this area of the cortex, a rigidity which differentiates it from other parts of the cerebral mantle, and it is interesting that the relationship may be modified by experience and is therefore not absolute as with processes occurring at lower levels in the brain.

A condition of *monocular diplopia* may arise from the development of a false macula. Thus the stimulus falling on $a_L$ (Fig. 22.24) is originally projected to A'; after the development of a false macula it is correctly projected to A. Sometimes both projections take place simultaneously, so that, as a result entirely of cortical activity (there is clearly only one image on the retina), the point A is seen simultaneously at A and A'.

The development of a 'pseudo-fovea' may result from hemianopia; in this condition one half of the foveal image is seen, a condition so annoying that the affected person tends to use an adjacent portion of the retina for fixation. With time this point acquires the characteristics of the true fovea—it is correctly projected into space and the images of objects falling on it become the centre of attention and interest.

## CONVERGENCE

With two eyes we have the geometrical basis for an estimate of the absolute distance of a point from an observer, since the converging power necessary for binocular fixation is uniquely determined by its distance away in the third dimension of space. The relative distances of different objects are likewise determined by the relative degrees of convergence necessary for binocular fixation. That variations in convergence may cause three-dimensional illusions was shown long ago by Meyer in the so-called *wallpaper phenomenon*. As described by Helmholtz, the illusion is brought about by looking at a wallpaper with a vertical line pattern; by making his eyes cross, so that the two eyes were fixating adjacent lines, Helmholtz found that the wallpaper appeared to float out of the wall, whilst if he made his eyes diverge the paper receded into the wall in the region of the fixation points. The crossing or divergence of the eyes caused no apparent diplopia because the fields of view consisted of only vertical lines. As Helmholtz and others have interpreted the phenomenon, the changed convergence *per se* was responsible for the apparent three-dimensionality of the wallpaper; in other words, the act of convergence contributed to the sensation of depth. However, more modern workers are inclined to deny to the convergence *per se* any significant contribution to depth perception (Stevenson Smith, 1946; 1949; Zajac, 1960); it is because alterations in convergence, by altering the fixation point, change the relative sizes and positions of other objects around the fixation point, that the retinal images are modified and thus the appearance of the fused binocular retinal images. Nevertheless, there is every reason to believe that the act of convergence provides signals to the brain that, in the absence of other clues, are adequate to indicate nearness or farther-awayness with some accuracy; thus Hofsten (1976) exposed two small spots of light to the eyes separately, varying the retinal disparities, and the degree of convergence necessary to fuse the identical images on the retinae provided the cue to apparent depth. The subjects indicated their estimates

---

*There is unfortunately a serious danger of confusion arising out of the different senses in which 'false projection' is regarded. To the neurologist the macula of the squinting eye is falsely projected because the individual grasps wide of the object whose image falls on it (he 'past-points'). False projection thus means the projection of a retinal image to some other point in space than that occupied by the object evoking it, and this is the sense maintained in this book. To some ophthalmologists, on the other hand, false projection is synonymous with what is called 'abnormal retinal correspondence'. With the development of a 'false macula', an image, not on the macula of the squinting eye, is projected to the same point in space as the image on the macula of the normal eye; the retinal correspondence (p. 522) has been readjusted to give single binocular vision of an object whose images do not fall on the innate corresponding points of the two retinae. Thus the neurologist would say that the 'false macula' is correctly projected into space, whilst some ophthalmologists would say that it is falsely projected.

on a metre-scale seen at the same time. Considerable accuracy was achieved.

## STEREOSCOPIC DEPTH PERCEPTION

We have so far considered the geometrical consequences of the use of the two eyes in estimating distance—the system has been treated as a range-finder—and the problem of the fusion of the two images in the separate eyes has not been considered. Not only does the fusion of the images provide a unitary perception of the binocular field of view, but it presents the objects in this field in a three-dimensional pattern that is quite impossible of achievement with monocular vision. In the following pages we shall inquire into those characteristics of the two retinal images that (a) permit or favour fusion and (b) contribute to the three-dimensional percept.

## CORRESPONDING POINTS AND THE HOROPTER

### CORRESPONDING POINTS

We have seen, in our discussion of binocular projection, that a single object in space is projected in two different directions when its images, formed by the two eyes, do not both fall on the foveae (Fig. 22.22); this results in diplopia, and single vision is only achieved when the retinal stimuli are projected to the same point in space. The foveae may therefore be described as *corresponding points*; they are points on the two retinae normally projected to the same point in space.

Every point on the retina of each eye is projected into space to give the field of vision; where the projections overlap we have the binocular field of vision; the problem is to determine which points in the binocular field of vision are seen single or, more correctly, which pairs of points in the retinae of the two eyes are projected to the same point in space. The matter is subject to experimental investigation and the following conclusions were drawn by Helmholtz:

1. The apparently horizontal meridians of the two retinae correspond.
2. The apparently vertical meridians correspond.
3. Points on the apparently horizontal meridians, equally distant from the foveae, are pairs of corresponding points.
4. Points equally distant from the apparently horizontal meridians, which are on the apparently vertical corresponding lines, are a pair of corresponding points.

In so many words, points equidistant from the apparently horizontal and vertical meridians of the retinae were said to be corresponding points.

### THE VIETH-MÜLLER HOROPTER CIRCLE

Figure 22.27 represents a horizontal section through the two eyes, fixating the point F. The points $a_L$ and $a_R$ are equally distant from the vertical meridians of the two eyes and are therefore projected to the same point in space through the respective nodal points, namely, to A. The points $b_L$ and $b_R$, also symmetrically disposed about the vertical meridians, are projected to B. From the geometry of this binocular projection it is evident that A, F and B lie on a circle passing through the fixation point, F, and the nodal points of the eyes. All points in space, lying on this circle, so long as they produce images on the two retinae, produce them on corresponding points provided that the symmetry, postulated by Helmholtz, exists. The circle is called the *Müller* or *Vieth-Müller horopter circle*: it is the locus of points, in a horizontal plane, that produce images falling on corresponding points of the two retinae when the fixation axes are horizontal, and are directed on the point F. Moreover, if the eyes may be considered to rotate about the nodal points instead of their centres of rotation, the fixation point may move to any point on the circle and all other points will still produce images on corresponding points. With this approximation the Müller horopter circle becomes the locus of points, in a horizontal plane, that produce images on corresponding points for a given degree of convergence of the

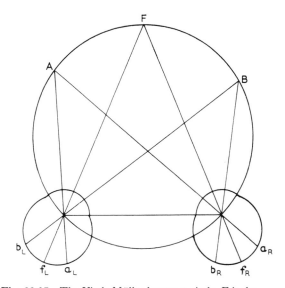

**Fig. 22.27** The Vieth-Müller horopter circle. F is the fixation point. If corresponding points are symmetrically distributed about the foveae the points in space, in the fixation plane, whose images fall on corresponding points lie on the circle.

eyes when the retinal horizons are horizontal. The Müller circle only indicates the points in space *in a single plane* that produce images on corresponding points. If the circle is rotated about a line joining the nodal points, the surface so produced gives the locus of all points in space whose images fall on corresponding points, for a fixed degree of convergence of the eyes. This is only true if the two retinae possess strict spherical symmetry and if the corresponding points are symmetrically disposed around the foveae. As the apparently vertical meridians of the retinae are not parallel to each other (p. 513) there is no such perfect symmetry and this Müller horopter surface, as we may call it, is only an approximation.

### THE HOROPTER

The *horopter* as originally defined is the *locus of points in space whose images fall on corresponding points of the two retinae*, i.e. no proviso is made as to the fixation point or the degree of convergence. The Müller horopter circle must therefore be regarded as a special case of the horopter when the fixation point has been defined. The number of points in space that will produce images on corresponding points, *however the eyes move*, is strictly limited; thus in Figure 22.28, when the eyes move to a new fixation point not lying on the original Müller circle, e.g. the point F′, the images of A do not lie symmetrically with regard to the foveae and are therefore not on corresponding points. Hence, although A lies on the Müller horopter circle when the eyes fixate F, it does not lie on the horopter. The horopter was calculated

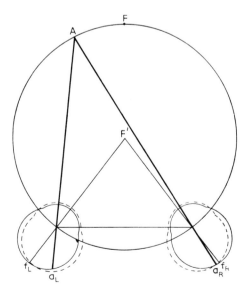

**Fig. 22.28** When F was the fixation point, A lay on the horopter-circle; when F′ becomes the fixation point A no longer produces images on corresponding points, so that A is not on the more generally defined horopter.

by Helmholtz from experimental data on the positions of corresponding points and the laws of the eye movements; it is a curve of the third degree (a curve that cuts a plane in three points). Under certain conditions the horopter acquires a simpler form; e.g. if the eyes are looking straight ahead in the primary position, the horopter becomes a horizontal plane passing through the feet, all points on this plane being seen single. When the point of fixation is either in the median plane of the head or in a horizontal plane with the fixation axes horizontal, the horopter becomes a Müller circle and a vertical line, intersecting in a point.

### Longitudinal horopter

For practical purposes, the so-called *longitudinal horopter* is the most useful. For a given fixation point it is the locus of points in a horizontal plane whose images fall on corresponding points of the two retinae. Thus, if Müller's assumptions were correct, it would be his horopter circle. In fact, however, the postulated symmetry about the vertical meridians does not hold good, so that the longitudinal horopter is not a circle.

### HERING-HILLEBRAND DEVIATION

Thus if we wish to find the point corresponding to $a_L$ (Fig. 22.27) it is, in general, not at $a_R$ such that $f_L a_L = f_R a_R$, but closer to, or farther away from $f_R$. This so-called *Hering-Hillebrand deviation* causes a characteristic change in the longitudinal horopter, so that not only does it deviate from a circle but its shape changes with the fixation distance. With strong convergence it was found to be concave to the face; at about 2 metres it became approximately a straight line in the frontal plane, whilst beyond this distance it was a curve, convex to the face.

### Modern analysis

Interest in the horopter, up to 1932, was largely academic; in this year, however, Ogle and his associates drew attention to the implications of the horopter from the point of view of the relative sizes of the retinal images. On the basis of certain assumptions, Ogle showed that the shape of the longitudinal horopter should be an index to the degree of discrepancy between the sizes of the retinal images. His interest in the problem of aniseikonia led to a very thorough investigation of the horopter, both from a practical and theoretical aspect; it is impossible in a brief space to do justice to the results or the conclusions of this study. The mathematical analysis confirmed the general conclusions of Hillebrand on the variation of the shape of the horopter with distance. Thus at a certain distance it coincided with the frontal plane whilst at other distances it was curved. The significance of altered image-size, i.e. of *aniseikonia*, will be discussed later (p. 532).

**Fig. 22.29**    Stereographic photographs of a book.

## THE STEREOSCOPE AND DEPTH PERCEPTION

The essential feature of stereoscopic perception is the existence on the two retinae of *different images of the same object*; the object is seen in two aspects because of the different viewpoints of the eyes. Thus Figure 22.29 represents two photographs of a book as it would be seen by the left and right eyes; if these photographs are viewed separately by the two eyes and their images fused, the book is perceived as a three-dimensional object.

### THE STEREOSCOPE

The fusion of the two images can be carried out, with a little practice, by holding them directly in front of the eyes, a piece of cardboard in the median plane being used to keep the fields separate. The conditions are, however, artificial in that the eyes normally converge when near objects are viewed, and a convenient instrument, which permits this convergence when the objects viewed are close to the eyes and directly in front of them, is the *stereoscope*, the optical properties of which are shown in Figure 22.30. Two pictures, representing different aspects of the same object, AB and CD, are placed about 10 cm from the eyes. They are viewed separately    through    base-out    prisms,    so    that symmetrically disposed points on the stereograms are projected to the same point, e.g. *a* and *b* are projected to *c*.

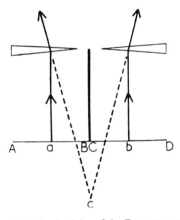

**Fig. 22.30**    Optical principles of the Brewster stereoscope.

### DISCREPANT IMAGES

If the two identical circles of Figure 22.31*a* are placed in the stereoscope, their images are fused but no sensation of depth is achieved precisely because the pictures are identical and are such as would, in effect, be presented to the eyes were a flat circle viewed. If the dots are placed in the circles, as in Figure 22.31*b*, fusion of the images creates the impression that the dot is floating in space in front of the circle, whereas if the dots are placed as in Figure 22.31*c*, the dots appear to be behind the plane of the circle. The stereograms imitate the appearances to the two eyes of a three-dimensional pattern; thus let us imagine that there is a dot on the

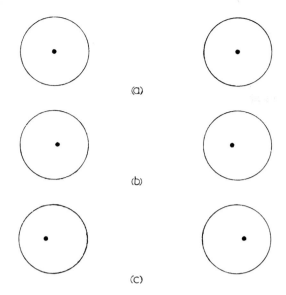

**Fig. 22.31** (*a*) The spots are in the centres of the two circles; the fused image is two-dimensional. (*b*) The spots are de-centred nasally. The spot appears, on fusion of the images, to be nearer the observer than the plane of the circle. (*c*) The spots are de-centred temporally. The spot appears, on fusion of the images, to be farther from the observer than the plane of the circle.

surface of a sphere placed symmetrically with regard to the two eyes; viewed with the left eye, the dot appears to the right of the centre of a circle—the outline of the sphere—whilst viewed by the right eye it is to the left. The combined image, as seen in the stereoscope, gives the impression of a point on the surface of the sphere. It will be noted that a sphere, viewed from any angle, appears as a circle; the two aspects of a sphere are therefore identical circles and binocular perception of a uniformly bright sphere can give no impression of solidity. It is essentially because a sphere is normally unevenly illuminated, and has light and dark spots on it, that the aspects presented to the eyes are different; the fusion of these aspects places the lights and shades in their true relationships in three dimensions and so creates the appearance of solidity. The importance of light and shade in stereoscopic vision may be seen by examining photographs of unfamiliar objects which reflect light to varying extents on different parts; stereoscopic vision puts the various high lights in their correct positions in the third dimension and so contributes greatly to the intelligibility of the presentation.

PARALLAX

The difference in the two aspects of the same object (or group of objects) is measured as the *instantaneous parallax*. In Figure 22.32, B is closer to the observer than


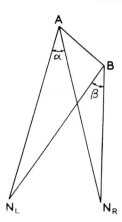

**Fig. 22.32** Binocular and instantaneous parallax.

A; the fact is perceived stereoscopically because the line AB subtends different angles at the two eyes, and the instantaneous parallax is measured by the difference between the angles $\alpha$ and $\beta$. The *binocular parallax* of any point in space is given by the angle subtended at it by the line joining the nodal points of the two eyes; hence the binocular parallax of A is $\alpha$; that of B is $\beta$; the instantaneous parallax is thus the difference of binocular parallax of the two points considered.

ACCURACY AND LIMITS OF STEREOSCOPIC PERCEPTION

The accuracy of stereoscopic perception is remarkable. Andersen and Weymouth and later, Ten Doesschate, placed three wires in front of an observer; the distance the middle wire had to be moved to appear out of the plane of the other two was measured; the parallax was of the order of four seconds of arc, corresponding to a disparity of the retinal images far smaller than the diameter of a single cone. Other values in the literature range from 1·6 to 24 seconds.

With two editions of the same book, it is not possible, by mere inspection, to detect that a given line of print was not printed from the same type as the same line in the other book. If the two lines in question are placed in the stereoscope, it is found that some letters appear to float in space, a stereoscopic impression created by the minute differences in size, shape, and relative position of the letters in the two lines. The stereoscope may thus be used to detect whether a bank-note has been forged, whether two coins have been stamped by the same die, and so on.

It may be shown, on geometrical grounds, that beyond a certain distance from the eyes, the difference of parallax of objects becomes effectively zero; beyond this point, then, objects in the environment appear to be in a single plane; the distance has been computed to be about

450 metres but of course it depends on the figure taken for the smallest perceptible instantaneous parallax.

## EFFECT OF DISTANCE

Several studies of the effects of distance on relative depth discrimination have been described. Under these conditions the observer is presented, in the distance, with two similar objects, separated laterally but initially at the same distance from him; one object is moved nearer or farther away until the observer states that he can detect the fact; and the distance between the two objects represents the *linear* depth limen, which may be converted to an angular disparity or parallax. According to Teichner, Kobrick and Wehrkamp, there is an actual improvement in the power to discriminate as the distance increases, although on theoretical grounds we should not expect any change, when the limen is expressed in terms of an angle of parallax. Ogle (1958) has pointed out, however, that the monocular clue given by difference in perceived size, when the two objects are not equidistant from the observer, becomes relatively more important in depth discrimination as distance increases; and it is for this reason that there is an improvement of depth discrimination, the eye being more sensitive to changes of size than to changes of parallax at large distances. This view was confirmed by Jameson and Hurvich.

## DURATION OF THE STIMULUS

Dove in 1841 observed that stereopsis could be experienced by viewing pictures in a stereoscope when they were illuminated by an electric spark, proving that the experience did not depend on movements of the eyes, such as those involved in convergence. A more systematic study by Ogle and Weil has shown that the accuracy* of stereopsis increases by a factor of four on increasing the exposure-time from 4 to 1000 msec. This suggested to Ogle and Weil that small nystagmoid movements, by contributing to visual acuity, improved stereoscopic acuity, since there is no doubt that these two parameters are related. That this is not the correct explanation, however, is shown by Keesey's study on visual acuity with different exposure-times; this did, indeed, improve as the time increased from 20 to 200 msec, but the same improvement was found when the image on the retina was 'stabilized' in such a way that it always remained on the same portion of the retina. Thus the small involuntary scanning movements do not contribute to visual acuity, and the improvement of both visual acuity and stereoscopic acuity with viewing-time implies only that the neural processes concerned with the discrimination require a certain minimum time for their best performance.

## DISPARATE POINTS AND STEREOSCOPIC PERCEPTION

Stereoscopic perception results from the fusion of dissimilar images; it follows from this that the fused percept is made up by the integration of stimuli falling on disparate points on the two retinae, i.e. of points which, on a Cyclopean projection, would be seen double. Thus in Figure 22.33, A′B′C′D′ and A″B″C″D″

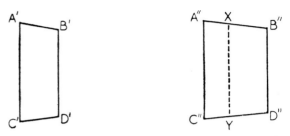

**Fig. 22.33**   Illustrating the importance of the incidence of images on non-corresponding points in stereoscopic perception.

are the two aspects of a slab, ABCD, inclined to the median plane. If the lines B′D′ and B″D″ correspond, it is quite clear that the lines A′C′ and A″C″ *cannot*, the line XY corresponding approximately with A′C′. Hence, on a purely Cyclopean projection, if the edge of the slab, BD, is seen single, the near edge, AC, is seen double. The fused percept is, nevertheless, that corresponding to a single entity—a slab 'floating in space'. The existence of images of the same point in space on disparate points of the two retinae is, therefore, just as much a characteristic of stereoscopic perception as the existence of images on corresponding points; in fact, if all points on an object produced images on corresponding points, there would be no stereoscopic perception.

## CORRESPONDING POINTS DEFINED

The facts of stereoscopic perception tell us, therefore, that a pair of corresponding points on the two retinae is not completely defined by the statement that their images are seen single, since disparate points may also be seen single. Let us consider the points A and B of Figure 22.34. They are on the horopter and their images fall on corresponding points, $a_L$ and $a_R$; $b_L$ and $b_R$. Both points are seen single in binocular vision. Let us now consider the points A and B′; if A is the fixation point,

---

*The accuracy of stereopsis is determined by measuring the threshold, i.e. the smallest instantaneous parallax that is just detected with stereoscopic vision. As with other thresholds, of course, there is a range of uncertainty over which we must speak of a *probability of detection*, and the threshold must be assessed on a statistical basis (Ogle, 1950; 1962).

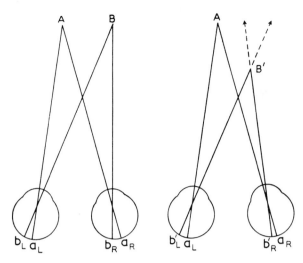

**Fig. 22.34** Illustrating the definition of corresponding points.

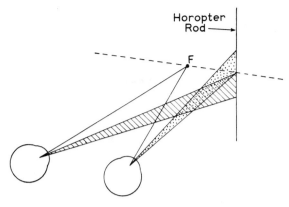

**Fig. 22.35** Nonius method for determining corresponding points. Both eyes are able to fixate F, the fixation point, whilst the right eye sees only the upper half of the horopter rod and the left eye only the bottom half. When the rod is on the horopter its upper and lower halves appear in line. (After Ogle.)

B′ is not on the horopter and, on a Cyclopean projection, would be seen double. If, however, A and B′ are points on an object, seen single as a stereoscopic percept, B′ must be seen single although it is not on the horopter; we must find what it is that differentiates its images $b'_L$ and $b'_R$ from the pair $b_L$ and $b_R$ which enables us to say that they are not corresponding points. It will be seen that when A and B are on the horopter, closing either eye fails to alter their apparent relative positions; on the other hand, when A is on the horopter but B′ is not, B′ is projected in different directions in relation to A according as one or other eye is closed. Thus with the left eye closed, B′ appears closer to A than when the right eye is closed. The actual binocular, or stereoscopic, projection of B′ in relation to A is, therefore, different from either of the separate projections. It is essentially this difference that is made use of in the most reliable experimental method for assessing the positions of corresponding points (e.g. the so-called *nonius* method). If the fixation point is used for reference, another point will produce images on corresponding points if its binocular projection is identical with the separate projections of the two eyes.

**Nonius method**
Experimentally the points on the longitudinal horopter may be determined by an apparatus illustrated schematically by Figure 22.35. The observer views a fixation point and a vertical horopter-rod through an arrangement of screens that permits both eyes to fixate the fixation point, F, whilst the right eye is able to see only the upper half of the vertical rod, and the left eye only the bottom half. The position of this rod is adjusted

so that when the eyes are fixating F, upper and lower halves of the rod appear above and below each other. Under these conditions the retinal direction values of the points stimulated by the rod are the same.

PANUM'S AREA
If the stereograms of Figure 22.33 are modified so that the lines A′C′ and B′D′ progressively get closer, whilst the equivalent lines of the right stereogram get progressively farther apart, a point is reached where fusion becomes impossible and double vision takes the place of a stereoscopic percept. There is thus a limit to the disparity of retinal points, beyond which they cannot be fused. Panum attempted to define the limiting degree of disparity beyond which fusion becomes impossible. He stated in general that 'contours resembling each other, depicted on approximately corresponding points, must be fused together'. The circumference that contained those points on the other retina that could be fused with a given point on the first, he called the *corresponding circle of sensation*. This formulation is probably rather too rigid; the ability to fuse contours is determined largely by the possibility of interpreting them as the right- and left-eyed views of a three-dimensional object, and thus the meaning of the final fused percept contributes to some extent to the ability to fuse disparate images. Nevertheless there is reason to suppose that there is a certain contour of finite size such that any point within it will invariably fuse with a single point on the other retina. It is to be noted that the two aspects of any object in space, as seen by the left and right eyes, differ principally in respect to their width, their height being very nearly the same; it is not surprising there-

fore that it is possible to fuse pairs of vertical parallel lines with quite large discrepancies in their separation, whereas only very small discrepancies in the separation of horizontal lines are tolerated. Panum's 'circle' should thus be described as an ellipse with its long axis horizontal. According to Ogle (1950) the length of the horizontal axis of this ellipse varies from one subject to another and, in a given subject, with the position on the retina at which the measurement is made. In one subject it increased from 5 to 30 minutes of arc on passing from central to peripheral vision, and in another from 10 to 40 minutes. Mitchell (1966), using central vision and flash-presentation of the stimuli so as to preclude fusional eye movements, found that horizontal disparities of 13 to 23 minutes of arc were necessary if two points were to be seen as double; the required vertical disparities were 8·9 to 15·9 minutes of arc.

## DOUBLE IMAGES IN STEREOSCOPIC PERCEPTION

If a pencil is placed vertically at arm's length, an inch or two to the side and in front of a thin bar, a unitary stereoscopic percept of the two is obtained, with no double images. The images would be said to fall within the Panum circles. On bringing the pencil closer to the eyes, with the bar as a fixation point, a stage is reached at which the double images of the pencil can be distinguished; nevertheless the pencil and the bar may still be perceived as a unitary pattern because one of the double images is ignored. When the pencil is brought very close to the eyes, no accurate perception of it in relation to the bar may be obtained because the double images have now become too obvious. In this experiment the pencil was not on the horopter, which passes through the fixation point; the farther removed from the horopter the pencil becomes, the less accurate and well defined is the stereoscopic percept; fortunately, however, for the unitary perception of the outside world, the more distant an object from the horopter, the more vaguely is it seen, owing to the limitations in the depth of focus of the eye. Thus objects in the environment may produce retinal images of such gross disparity that, if they were distinctly perceived, a unitary perception would be impossible. The vagueness of the double images, however, permits the ready suppression of one, so that all objects in the binocular field of view appear normally single and it requires a careful examination to reveal the actual existence of double images.

In general, the closer the position of an object to the horopter, the more accurate is depth perception; for example if one pin is placed on the horopter and another very close to it, but off the horopter, the eyes can estimate with extraordinary accuracy the displacement in the third dimension. The farther the pins are off the horopter, the grosser the estimate, a fact following from Weber's psychophysical law.

## RANGE AND SCOPE OF BINOCULAR DEPTH DISCRIMINATION

We may ask whether stereoscopic depth perception is achieved when the disparity of retinal images is greater than Panum's area so that an object closer to the fixation point is presented as a pair of double images. Helmholtz was in no doubt that objects could be quite accurately located as being nearer to, or farther from, a given point by virtue of their double images; and it was presumably the crossed disparity in the one case (cf. Fig. 22.22) and the uncrossed disparity in the other (cf. Fig. 22.23) that permitted this. Ogle (1962) investigated the effects of greater and greater disparities in the retinal images on the perception of depth; he confirmed the efficacy of double images in permitting estimates of depth, and divided stereoscopic depth into several parts; for small retinal disparities where fusion takes place, the sensation of depth is compelling and the subject is able to make accurate quantitative estimates of relative depth in accordance with the degree of retinal disparity; beyond this region, where double images may be shown to occur, i.e. where fusion is not real, the experience of depth remains, but quantitative estimates of relative depth are poor or non-existent. Finally, when the disparity is still larger, the sense of depth disappears and the two images are indefinitely localized.

More recently Blakemore (1970) has made a thorough investigation of the limits of retinal disparity within which estimates of depth can be reliably made, and these turned out to be very large; thus with objects immediately in front of the observer, a convergent disparity as high as 7° could be interpreted as nearness to the observer, and a divergent disparity of 9° could be interpreted as farther awayness. In the periphery the same observer perceived depth with disparities as high as 13 to 14°. Blakemore confirmed that the accuracy of stereopsis, as indicated by the threshold disparity required to permit discrimination of depth, was greatest at the horopter; this is illustrated by Figure 22.36, where lines indicate contours of isothreshold disparity, the actual values in minutes of arc being indicated by the figures. The filled circles represent the subject's longitudinal horopter, and it is interesting how the contours follow the shape of this horopter and how the range is compressed in the middle of the visual field.

### Binocularly stabilized vision

Fender and Julesz (1967) have emphasized the dynamic character of Panum's fusional area; thus when fusion of disparate images was brought about using stabilized images, these images could be pulled apart very much farther than the limits of Panum's fusional area, without the appearance of double images. The fact that the images were stabilized means that no disjunctive move-

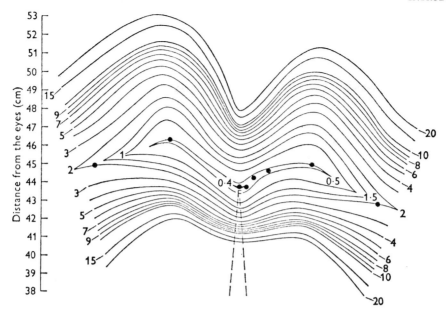

**Fig. 22.36** Illustrating retinal disparities that may be fused, as a function of position of the images in relation to the fixation point, F. The disparities are represented on ordinates as distances of the object point from the eyes, causing the disparities. Each smooth line is a contour of iso-threshold-disparity, and the magnitude of that threshold, in minutes, is shown at the end of the contour line. (Blakemore, *J. Physiol.*)

ments of the eyes occurred, as would have happened in normal vision. During this pulling apart of the retinal images there was no change in the stereoscopic perception, so that if the images were of a vertical line, its localization in space was unaltered. Thus Fender and Julesz speak of first, a labelling process, operative in Panum's fusional region, which establishes corresponding areas in the left and right images having various disparities, and second a cortical registration process that preserves the labelling of the retinal images of a given point on an object in spite of considerable movements of these images in relation to each other, i.e. in spite of considerable variations in their disparity.

CYCLOPEAN AND STEREOSCOPIC PROJECTION

The study of stereoscopic perception has revealed a defect in the Cyclopean system of representing the projection of the two eyes in binocular vision. According to this, the point B of Figure 22.22 (p. 519) is seen double however close it is to A, whereas we know that, if the separation is not too large, B is seen single. On a *stereoscopic* projection system, therefore, the points $b_L$ and $b_R$ are correctly projected through their respective nodal points to B. The Cyclopean projection of the images of B is unreal, in the sense that it projects the images of a single object to two different points in space simultaneously; the stereoscopic visual process imposes a truer meaning on the two retinal stimuli, not by

suppressing one or the other of the two projected images, but by establishing a single projection that corresponds to the true position of B in relation to A. This process extends to retinal images that fall outside Panum's circle, and, even when double images become apparent, it is found that they tend to be projected into space in a direction much closer to the true position of the point giving rise to them than would be expected on a Cyclopean projection.

The Cyclopean projection is not, nevertheless, to be treated as a useless and academic mode of representing the projections of the two eyes; it tells us how the eyes would combine their separate projections if this combination were a simple additive process; thus in Figure 22.23 (p. 519) the right eye alone would project its image of B to the right of A, whilst the left eye would project it to the left. The Cyclopean projection is nothing more than an addition of these two projections. By comparing the Cyclopean with the actual stereoscopic projection we can form an idea of the fundamental revaluation of retinal points that must take place to permit single vision as a result of the stimulation of disparate points.

**Experimental demonstration**
An interesting demonstration of this shift in visual directions has been given by Burian and is illustrated in Figure 22.37; the subject views a stereogram consisting

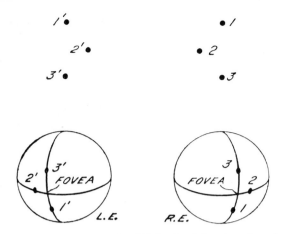

**Fig. 22.37** Illustrating the shift in visual direction of a point by virtue of viewing it stereoscopically. The subject views the stereogram consisting of the sets of dots; on fusion the dots appear in a vertical line. (Burian, *Documenta Ophthalmologica*.)

of a set of three vertically arranged dots. The images of the upper and lower dots fall on corresponding points, whilst those of the middle dots, 2 and 2', do not, being displaced nasally to give a crossed disparity. When viewed, the appearance will be that of the dots 1 and 3 lying in a frontal plane with dot 2 closer to the observer (cf. Fig. 22.31*b*). What is remarkable, is that the three dots will all appear to be in the same vertical line; consequently the visual directions of the retinal points, on which the points marked 2 fall, have been altered because of the stereoscopic effect, since we know that if, in the stereograms, the points 2 were placed immediately under points 1 and 3 they would still appear in the same vertical line, but this time, of course, also in the same fronto-parallel plane.

### STEREOPSIS WITHOUT FAMILIARITY CUES

It has been argued that retinal disparity is not the only, or even the determining, factor in stereopsis; thus with most stereograms familiarity with the figure represented is thought to be an important element, whilst in addition there are often monocular depth cues in the individual stereograms. However, Julesz (1964) has shown that it is possible for the monocularly viewed stereograms to be such that the object, seen in stereopsis, is completely invisible in the individual stereograms. He devised stereograms composed of random dots such that neither had any significance on its own, yet, on viewing these in the stereoscope, the observer saw a central square which could be in front of, or behind, the rest of the field; in Figure 22.38 it is in front of the background.* The manner in which Figure 22.38 was constructed is too elaborate to permit a detailed description here; suffice it to say that the stereograms are divided into $9 \times 10$ picture-elements composed of dots; some of these elements occupy corresponding points whilst others have disparity. It is the organized disparities of certain of these picture-elements that allow the stereoscopic appearance on fusion. This ingenious technique of Julesz allows, now, an unequivocal verification of the interpretation of Dove's experiments with flash-illumination of the stereograms, since it has been argued that fusion under Dove's conditions was only possible because the subjects knew what they were expected to see; with Julesz' stereograms there is 'nothing to see' except a random array of dots. The experiments also dispose of Hering's claim that it is the awareness of double images that determines stereopsis; even if the random elements in Figure 22.38 were giving double images there could be no awareness of this. Finally, the experiment shows that discrete contours in the left- and right-eye images are not a necessary feature of stereopsis.

---

*When the pictures are rotated through 90°, so that the disparities become vertical, stereopsis is lost but the perception of the central square remains, although unrecognizable in either monocular presentation. Thus we must distinguish between the perception of form in depth and form without depth (Kaufman, 1964a).

**Fig. 22.38** Random-dot stereograms constructed so that the individual pictures reveal no form, yet when viewed in the stereoscope the fused percept is one of a square in front of the background. (Julesz, *Science*.)

## Role of eye movements

Although flash-presentation still permits stereoscopic perception, it is important to note that when Julesz-type patterns are presented to the eyes, i.e. pictures without monocular contours, then perception is greatly reduced and even abolished by flash-presentation. Presumably, when eye movements are allowed, the obscure disparity cues of the Julesz patterns have a better chance of being picked up. As Richards (1977) emphasizes, the Julesz patterns, even under the most favourable conditions, do not permit significant estimates of magnitude of depth implied by the actual disparities of the dots.

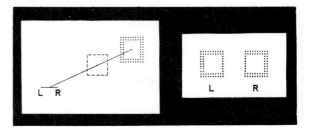

**Fig. 22.39**    Illustrating depth perception with anomalous contours. The stereogram shown at the right represents the actual situation illustrated at the left, where the view of the squares of dots is interrupted by the white square. (After Lawson and Gulick, *Vision Res.*)

## STEREOPSIS AND PERCEPTION

It has been customary to treat stereopsis as being the result of a higher perceptual process that interprets the separate fields of view of the two eyes. Thus it has been argued that the two fields 'are presented separately to consciousness and that the final percept results from a psychic act'. As Ogle has emphasized, however, the experience resulting from fused disparate images is so immediate and compelling that it is difficult to separate it from a visual sensation, as such, and it might well be more correct to describe the experience simply as the response to the falling of disparate images on the two retinae. Interpretation, in the sense of conscious analysis of the two retinal images, is of course not an element in the experience. The same may be said of stereoscopic lustre (p. 535); for explanatory purposes we may say that the experience results because it is the only way of interpreting the two images of different luminosity, yet once again the experience is so compelling and immediate that we must treat it as the 'response to retinal images of different luminosity'.

### ANOMALOUS CONTOUR

The perception of depth created by observing a distant object binocularly, when nearer objects are interposed between the distant object and the observer, provides an example of what the psychologist has called *anomalous contour*; and its contribution to depth perception, apart from the fusion of disparate images, may be investigated by constructing appropriate stereograms that simulate the actual objects viewed. Thus Figure 22.39, *right* (Lawson and Gulick, 1967), when viewed in the stereoscope, gives rise to the impression of a white square in front of the dots, which lie in a single plane behind. The two pictures are identical except that the right vertical row of dots in the left picture was omitted and the left vertical line of the right picture. The real situation corresponding to the left- and right-eye views could consist of a square frame of dots with a white

opaque screen interposed between it and the observer, as illustrated to the left, and this is the 'interpretation' put on the left- and right-eye views. In this case there is no fusion of disparate images, of dots since these all appear to lie in a single plane.

## REVERSAL OF RELIEF

It will be noted by comparing the stereograms of Figure 22.31*b* and *c*, that the dot in the circle appears to be in front or behind according as it is displaced to the right or left of the centre as viewed by the left eye, and according as the dot is displaced to the left or right of the centre as viewed by the right eye. Thus the pair of stereograms, *c*, is really obtained from the pair, *b*, by interchanging the pictures presented to the left and right eyes. In any pair of stereograms a *reversal of relief* may be obtained by interchanging the pictures presented to the right and left eyes. Such a reversal implies a knowledge by the higher perceptual centres of the contributions of each retina to the final fused image, and yet it is interesting that we are normally not aware as to which eye sees any object not seen by both eyes in the binocular field of view. Thus Rogers describes the following experiment. A tube, about 2 inches in diameter, is held in front of the right eye and directed towards the far left corner of the room. A piece of white paper is placed in front of the left eye so as to screen this eye from the part of the room seen by the right eye. The impression is obtained of seeing the objects in the room with the *left eye* through a hole in the paper.

## SIZE EFFECTS

### HORIZONTAL MAGNIFICATION

The importance of the sizes of the retinal images in determining space perception can best be described with the aid of an example, in which the size of one retinal image is increased with the aid of a lens that modifies the

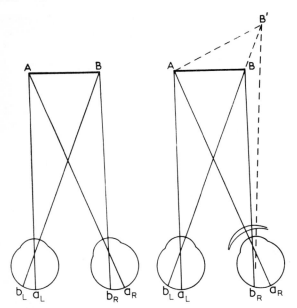

**Fig. 22.40** The effect of an aniseikonia glass on depth perception. *Left*: AB is in the frontal plane and may be said to lie on the longitudinal horopter. *Right*: AB is now projected to AB′.

image-size but leaves the dioptric power of the eye almost unchanged (an aniseikonia glass). In Figure 22.40 (*left*) we have a line on the longitudinal horopter; it appears to be in the frontal plane and the images on the two retinae are of the same size. In Figure 22.40 (*right*) a lens has been put in front of the right eye; the lens magnifies in the horizontal meridian only, i.e. it is a cylinder with axis vertical. The retinal image is now formed at $a_R$ and $b'_R$ and is projected through the nodal point. The intersections of the projections of both eyes give the apparent positions of A and B in stereoscopic single vision, namely, at A and B′. It is clear that a considerable distortion of space values has resulted, AB appearing to lie along AB′. AB therefore no longer lies on the 'horopter' of the individual with the aniseikonia glass in front of one eye, and the horopter has thus changed its shape. It has been found that the eye is sensitive to as little as a 0·25 per cent change in the size of one retinal image. This spatial distortion can only be evoked by a magnification in the horizontal meridian; if an ordinary spherical magnifying lens is employed, magnifying in both horizontal and vertical meridians, then the distortion disappears.

VERTICAL MAGNIFICATION

If the magnification is confined to the vertical meridian, then we have what Ogle calls the *induced size effect*. The space dimensions are changed, but in the sense that would have been produced by horizontal magnification

of the image in the other eye. It is for this reason that simple over-all magnification of the image in one eye, by a spherical lens, has no effect with regard to spatial distortion. Thus the horizontal magnification in the right eye would give the distortion illustrated by Figure 22.40, with B appearing at B′; the vertical magnification, by its induced effect, would cause A to move back into the plane of B′, and so the distortion disappears.

**Induced size effect**

The induced effect, which is essentially the effect of producing *vertical disparities* in the retinal images, is a most striking phenomenon. The interpretation put forward by Ogle is difficult to present in a short space and this will not be attempted here; suffice it to say that, as with so many stereoscopic and other illusions, the experience may be regarded as a matter of putting the best interpretation on an anomalous set of stimuli. The experience in ordinary viewing conditions where vertical disparities become prominent is in asymmetrical convergence, one eye fixating a point immediately in front of it, whilst the other converges (see, for example, Fig. 16.24, p. 397). Under these conditions certain readjustments in spatial orientation are made; if, now, the vertical disparities are introduced artificially, then the adjustments in spatial orientation that would have been appropriate to asymmetrical convergence are carried out, leading to distortion of space values.

## THE PULFRICH PHENOMENON

If a swinging pendulum with a luminous bob is observed simultaneously through a green glass by one eye and a dark red glass by the other, instead of moving in a plane the pendulum appears to move through an elliptical path. Thus if the left eye wears the red glass, the pendulum on its left-to-right swing appears to move away from the observer and on its reverse swing to move towards him. If the colours are interchanged, the direction of apparent rotation is reversed. Different coloured glasses are unnecessary to evoke the effect, and it is sufficient if the light entering one eye is reduced by placing a neutral filter before it.

INTERPRETATION

The cause of the illusion is the difference in reaction-times of the two eyes in response to the different colours or different luminances. In Figure 22.41 let us assume for simplicity that the right eye responds instantaneously to the image of the bob at any point in its path, whilst the left eye responds after a delay. In effect, therefore, when the two eyes are fixating the bob, indicated in its true position in black at A, the left eye is seeing it as though it were at A′, i.e. as though its image on the left

The importance of the *meaning* of the final percept is evident when one attempts to fuse two pictures in the stereoscope; it may be found impossible for a time, but if fusion once takes place to give a three-dimensional representation of some object, then the significance of the two aspects has become clear and the stereograms may be fused rapidly at will.

If two pictures that cannot possibly be related as two aspects of the same three-dimensional object are presented to the two eyes, single vision *may*, under some conditions, be obtained but the phenomenon of *retinal rivalry* enters. Thus the two letters **F** and **L** of the stereogram in Figure 22.42, when presented to the two

**Fig. 22.42** The letters F and L in the stereoscope give simultaneous perception to create the impression of the letter E.

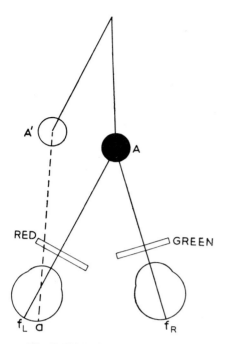

**Fig. 22.41** The Pulfrich phenomenon.

retina were at *a* instead of at $f_L$. The disparate images are fused to give a stereoscopic effect, so that the bob is actually projected to a point outside the plane of the pendulum. The disparity is uncrossed, so that the bob on its left-to right swing appears farther from the observer. The degree of disparity will obviously depend on the speed of the bob, and this passes from zero at the extreme of its swing through a maximum at the bottom of its swing; hence the bob will appear in the plane of the pendulum at the extremes of its swing, and closest to and farthest from the observer at the bottom of its right-to-left and left-to-right swings respectively. In other words, its motion will be elliptical. Careful experimental studies of the Pulfrich phenomenon, for example those of Lit, have confirmed this explanation.*

## FURTHER ASPECTS OF BINOCULAR VISION

### RETINAL RIVALRY AND OCULAR DOMINANCE

Stereoscopic vision is essentially the response to disparate images on the retinae of the two eyes; if the images are too disparate, as under the artificial conditions of the stereoscope, great difficulty is experienced in obtaining a single fused percept, the double images being so far apart as to force themselves on the awareness and so destroy the unity of the percept.

eyes separately, can be fused to give the letter **E**; the letters **F** and **L** cannot, however, by any stretch of the imagination be regarded as left and right aspects of a real object in space, so that the final percept is not three-dimensional and, moreover, it is not a unitary percept in the sense used in this discussion; great difficulty is experienced in retaining the appearance of the letter **E**, the two separate images, **F** and **L**, tending to float apart. We are here dealing with a mode of binocular vision that may be more appropriately called *simultaneous perception*; the two images are seen simultaneously, and it is by *superimposition*, rather than fusion, that the illusion of the letter **E** is created.

With the stereograms of Figure 22.43, if both patterns were fused or superimposed, the result would be a criss-cross pattern; in actual fact, however, no such perfect superimposition is achieved and the phenomenon of retinal rivalry becomes strongly evident. In different parts of the field one or other set of lines predominates

*Wist *et al.* (1977) have shown that, if the swinging object is a grating, then the apparent depth during a swing varies with the spatial frequency of the grating, the higher the spatial frequency of the moving grating the smaller the angular speed required to produce a given displacement of its apparent plane of motion. The authors consider that this finding is difficult to reconcile with the interpretation discussed here, but the situation with a vertically striped object must be complex, since now stereopsis is possible by convergence or divergence of the eyes to bring a given stripe on corresponding points, as in the wallpaper phenomenon (p. 521).

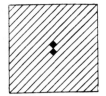 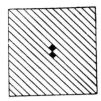

**Fig. 22.43** The stereograms do not give a regular criss-cross pattern in the stereoscope.

whilst the opposing set is suppressed; sometimes for a brief interval one set of lines will appear by itself over the entire field; in general the impression is one of fluctuation and Helmholtz stated that the appearance at any moment was largely determined by the attention paid to any aspect, e.g. if the observer started to count the lines of one set the others, in the opposite direction, tended to disappear. Helmholtz deduced from his observation that 'the content of each separate field comes to consciousness without being fused with that of the other field by means of organic mechanisms, and therefore the fusion of the two fields in one common image, when it does occur, is a psychic act'. More recent studies of retinal rivalry, however, indicate that there is a periodicity in the recurrence of one or other field in awareness which appears to be independent of volition; the distinction between fusion by a 'psychic act' and an 'organic mechanism' is, moreover, not one that would be countenanced by the modern physiologist; as we shall see, the fusion of retinal images, when these occur within Panum's area, is brought about by the projection of the evoked nervous discharges from the retinae on to the same cortical neurones. When retinal disparity is so large as to evoke rivalry this projection on to single cortical cells presumably does not occur.

OCULAR DOMINANCE

Retinal rivalry may be viewed as the competition of the retinal fields for attention; such a notion leads to the concept of *ocular dominance*—the condition when one retinal image habitually compels attention at the expense of the other. Whilst there seems little doubt that a person may use one eye in preference to the other in acts requiring monocular vision, e.g. in aiming a rifle, it seems doubtful whether, in the normal individual, ocular dominance is really an important factor in the final awareness of the two retinal images. Where the retinal images overlap, stereoscopic perception is possible and the two fields, in this region, are combined into a single three-dimensional percept. In the extreme temporal fields, entirely different objects are seen by the two eyes and as to which dominates the awareness at any moment depends largely on the interest it arouses; as a result, the complete field of view is filled in and one is not aware of what objects are seen only by one eye.

Where the fields overlap, and different objects are seen by the two eyes, e.g. on looking through a window the bars may obscure some objects as seen by one eye but not as seen by the other, the final percept is determined by the need to make something intelligible out of the combined fields. Thus the left eye may see a chimney-pot on a house whilst the other eye sees the bar of a window in its place; the final perceptual pattern involves the simultaneous awareness of both the bar and the chimney-pot because the retinal images only have meaning if both are present in consciousness. So long as the individual retinal images can be regarded as the visual tokens of an actual arrangement of objects, it is possible to obtain a single percept, and there seems no reason to suppose that the final percept will be greatly influenced by the dominance of one or other eye. When a single percept is impossible, retinal rivalry enters; this is essentially an alternation of awareness of the two fields—the subject apparently makes attempts to find something intelligible in the combined presentation by suppressing first one field and then the other—and certainly it would be incorrect to speak of ocular dominance as an absolute and invariable imposition of a single field on awareness, since this does not occur.*

When the field of one eye is consistently dominant over that of the other, this generally occurs as a result of a general law that the predominant field along an edge is that field in which the edge lies. Thus if a vertical bar is presented to the left eye and a horizontal bar to the right eye the combined impression may be represented by the cross in Figure 22.44, although it must be appreciated that the result is by no means so clear-cut as a simple diagrammatic representation would indicate. At the extremes of the fused cross one eye sees only white whilst the other sees the contour, or edge; this eye is 'dominant' and the black bar is seen in preference to the white space. Where the bars overlap there is a mixed state of affairs. To the right eye the field above and below the overlap is white and, since this eye is dominant along the horizontal edge, we may expect white above and below the bar; to the left eye the field to the left and right of the overlap is white and since, in this region, the left eye is also 'dominant' we expect white to the left and right.

OPTOKINETIC NYSTAGMUS AND DOMINANCE

An interesting situation is provided by allowing the eyes to view a rotating striped pattern through Dove prisms,

---

* An interesting observation of Creed's should be recorded here; in studying the retinal rivalry caused by observation in the stereoscope of two postage stamps, differing in colour as well as slightly in design, he noted that the dominance of one colour was not necessarily associated with the dominance of the design belonging to it—thus at a given moment it was quite possible for the design of one stamp to prevail with the colour of the other.

**Fig. 22.44**   Schematic representation of the simultaneous perception of a vertical and horizontal bar in the stereoscope. (Helmholtz, *Physiological Optics*.)

such that to one eye the pattern is moving, say, to the left and to the other to the right. With the aid of the electro-oculogram it is easy to measure objectively the form of the ensuing nystagmus. This changes repeatedly and the changes are accurately correlated with the subjective dominance of one or other movement-pattern (Fox *et al.* 1975).

DOMINANCE AND HANDEDNESS

Experimentally the subject of ocular dominance has received more attention than it probably deserves, and attempts have been made to relate it to 'handedness'. Dolman's peephole test consists of a $13 \times 20$ cm card with a round central hole in it of 3 cm diameter. A target 6 metres away is looked at through the hole held at arm's length. Under these conditions only one eye can actually see the target, and the one that is found to be doing so is adjudged the dominant eye. Walls has summarized the various tests and placed them in five categories and has stated that in effect they reduce to only two, namely those based on motor control of the eyes and those based on the assessment of subjective visual direction. According to Walls, for instance, the motor taxis of the eyes may be likened to the steering apparatus of a motor car; one wheel is directly attached to the steering rod whilst the other is guided by the track rod. It is difficult to translate this analogy into physiological terms, however; we know that the muscles to both eyes are innervated at the same time in any movement, and it would be quite incorrect to think of the one eye dragging the other along with it, as this view of ocular dominance would require.

## STEREOSCOPIC LUSTRE

If the stereograms of Figure 22.45 are fused, the impression of lustre is created. By lustre is meant the

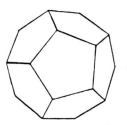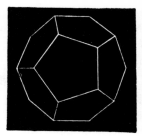

**Fig. 22.45**   Stereoscopic lustre.

appearance of a surface when some of its reflected light seems to come from within it; it is as though one sees some of the light from the object *through* the rest of the reflected light. The stereograms of Figure 22.45 represent the two appearances of a crystal when viewed by the two eyes separately, but the two differ also in being bright and dark at corresponding places. Stereoscopic lustre is therefore the result of stimulating corresponding points of the retinae to different extents, and the phenomenon suggests an explanation for the lustre in real objects. If the two stereoscopic pictures are to be interpreted as tokens of the same object, the surface of the latter must be such that a small change in viewpoint (from left eye to right eye) makes a profound change in the amount of light reflected into the eye. An object with a diffusely reflecting surface could not produce such a change and it therefore appears 'flat'; a strongly reflecting surface, on the other hand, would provide the reflected light with a directional quality such that the amount of light entering the eye varied considerably with the viewpoint; such a surface is, in general, lustrous. The chromatic lustre of a bird's plumage, of a thin film of oil, and so on, is due to the variation in the quantity and *quality* of the light entering the eye with change of viewpoint. It must be pointed out, however, that lustre in objects can be appreciated by *one eye only*; Helmholtz has shown, nevertheless, that the *illusion* of lustre may likewise be produced with a single eye; in these circumstances the illusion is created by rapid changes in the brightness of the points in the retinal image, associated with either movements of the eye or of the object; thus on Helmholtz's view essentially the same impression can be gained in one eye by successive stimuli as is gained in two eyes by simultaneous stimuli. Bartley has confirmed this viewpoint; he has shown that successive monocular presentation of two stereograms of different luminance creates the sensation of lustre. The psychological issues involved in such an equivalence of time and space need not concern us here; the important feature of stereoscopic lustre resides in its demonstration that, wherever this is possible, the higher perceptual centres will read such a meaning into two retinal images as to make them the tokens of a single object in space; the simultaneous presentation to the two eyes of a light and a dark view of an object is *interpreted* as the single presentation of a lustrous object in three dimensions, and this happens even though both the light and dark views reflect light diffusely. If the two views cannot be regarded as the tokens of a single object, retinal rivalry takes the place of this perceptual process; in the case of stereoscopic lustre this can be shown by presenting differently coloured views of the same object; if the difference in colour is not very pronounced the impression of lustre is gained on fusion; when the difference is made great lustre gives way to retinal rivalry.

## THE ESSENTIAL BASIS OF STEREOPSIS

The studies of Julesz have prompted several investigations into the nature of stereopsis; for example, the question has been asked whether fusion of disparate images is necessary; whether contours are 'irrelevant', and so on. Thus Kaufman (1964, 1965), in a series of papers has described stereoscopic effects using, instead of random dots, rectangles composed of randomly selected typewriter letters; in one stereogram a block of these letters is shifted to the left or right so as to be out of phase with the identical block in the other stereogram, and as a result this central block appears in depth in relation to the background of surrounding letters. Kaufman concluded that retinal rivalry was present using this type of pattern as well as Julesz' dot patterns, since the binocular percept did not give rise to the density of dots or letters that would have been expected by a simple physical superposition of the two monocular stereograms; instead, the density corresponded with that of a single stereogram, as though the percept of depth had been evoked by alternate suppression of one or other stereogram. Kaufman (1965) generalizes correctly when he says that stereopsis will occur when correlated stimuli are out of spatial phase with respect to some reference system. The reference system can be another set of correlated objects or point-arrays, or it may be the edge of the overall half-fields. Any mechanism that can detect the correlation between the binocular stimuli, and also detect a difference in their phase, can yield a representation of depth.

### VISUAL AND OBJECTIVE UNREALITY

When attempting to steer a way through the rather complex arguments in this field, it is best to keep clearly in mind that the binocular percept, resulting from the presentation of separate and different pictures to the two eyes, will be governed by the relation of these two pictures to reality. How can they be interpreted as left- and right-eyed views of the external world? As indicated earlier, rivalry is the response to *visual* unreality; no object when viewed by left and right eyes simultaneously can appear as a vertical bar and a horizontal bar. *Objective* unreality is given by the formation of double images, e.g. of a pencil held close to the face when viewing a distant point; the double images are real visually in the sense that real objects in space have provoked them, but they are objectively unreal because our sense of touch provides contradictory evidence. Stereopsis is achieved under these conditions, and it is assisted by the suppression of objectively unreal parts of the retinal images. It is probably this ability to suppress parts of the retinal images that permits the development of stereopsis when

differently coloured monocular stereograms are presented to the two eyes; according to Treisman (1962), stereopsis is achieved although the colour of one of the fields may be suppressed; thus the brain, under these conditions, was accepting the contours of the stereograms but suppressing the colour of one. Similarly, with many of the random dot and letter stereograms devised by Julesz and Kaufman, the situations presented are often so unreal visually that stereopsis is only achieved by virtue of suppression of parts of the visual fields.

## BINOCULAR ASPECTS OF SOME VISUAL FUNCTIONS

Stereoscopic vision reveals the existence of processes integrating the responses in the two eyes to give a unitary percept quite different from that predictable on the basis of a summation of the separate impressions. Retinal rivalry, on the other hand, provides a condition in which the separate impressions appear, at certain moments, to be unaffected by each other, whilst between these two states we may have a more or less perfect overlap of the visual impressions with, however, dominance of one or other field in certain parts. It is worth inquiring to what extent the simpler aspects of vision discussed in Section II are affected by the use of two eyes. There is, unfortunately, disagreement on many points so that the whole subject merits further investigation.

### LIGHT THRESHOLD

If there is the possibility of summation between the two eyes, we may expect that the absolute light threshold, i.e. the smallest light stimulus necessary to evoke the sensation of light in the completely dark-adapted eye, will be smaller when the two eyes are employed to view the stimulating patch. There are conflicting claims regarding this possibility; Pirenne has shown that the threshold is definitely lower in binocular vision, but this *does not* indicate a summation between the two eyes. The use of two eyes increases the chance of one eye receiving the requisite number of quanta from a flash of light to stimulate vision. The decrease in threshold, to be expected on this basis of quantum fluctuations, can be calculated; it corresponds to that actually found.

### BRIGHTNESS SENSATION

At higher levels of luminance the position is more complex; when the fields presented to the two eyes are equal, there is no obvious summation—this is evident from everyday experience, since the closing of one eye does not make a previously binocularly viewed piece of white paper appear significantly darker. When there is a gross discrepancy in the luminance of the two fields, retinal rivalry appears, although it may most frequently manifest itself as ocular dominance, the bright field determining the binocular sensation. (Under special conditions, it will be remembered, the sensation of lustre may be created.) Under intermediate conditions there is a definite fusion of the separate sensations, so that the binocular sensation is compounded of the separate monocular sensations or, to speak more precisely, the binocular sensation is different from that evoked by either eye separately, and is intermediate between these two.

## Weighting coefficients

Recent experimental studies have been devoted to quantifying the relation between the monocular and binocular stimuli. Thus Levelt (1965) caused subjects to view binocularly two equally illuminated fields; this was the *comparison stimulus*. The sensation of brightness evoked in this way was compared with the *test stimulus*, where the subject one again viewed two monocular fields binocularly, but this time the luminance of one was fixed by the experimenter and that of the other was adjusted by the subject until the binocular sensation of brightness equalled that evoked by the comparison stimulus. An example of the results is given in Figure 22.46, where the luminance of the left field is plotted as ordinate against that of the right field, the plotted points representing the left and right luminances which, on fusion, gave a binocular sensation of brightness equal to that obtained when both eyes viewed fields of the

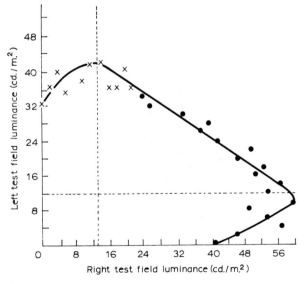

**Fig. 22.46** The subject views test fields of different luminances through left and right eyes separately; one field is fixed and the other is varied by the subject so that the combined impression is one of equality with a standard binocularly viewed luminance of 30 cd/m². Thus a field of 40 cd/m² viewed in conjunction with a right field of about 16 cd/m² gave the same subjective sensation as that obtained by viewing fields of 30 cd/m² through both eyes. (After Levelt, *Brit. J. Psychol.*)

same luminance of 30 cd/m². Over part of the range there is a linear relation between the two field luminances, indicating a weighted averaging of the two; e.g. if the left field was 40 cd/m², the right was 16; if the left was 36 the right was 24; if the left was 12 the right was 60, and so on. When the discrepancy between the luminances presented to the two eyes was large, the relation became non-linear and provides the basis for some well known paradoxes. Thus in *Fechner's paradox*: a monocularly viewed white surface appears brighter than when it is viewed binocularly in such a way that one eye views it directly and the other through a dark glass. Thus if we look at Figure 22.46 we see that if the subject was presented through his left eye with a field of about 40 cd/m² it would provoke a considerably brighter sensation than if this field was seen binocularly with a field of about 12 cd/m², since the combined sensation for binocular vision is equivalent to that evoked by 30 cd/m². In general, Levelt showed that over the linear range, the brightness sensation could be predicted from the weighted mean of the left and right stimuli:

$$W_R E_L + W_R E_R = C$$

where the weighting coefficients, $W_R$ and $W_L$, add up to unity and are constant for a given observer. If a contour was present in one of the binocular fields and not in the corresponding region of the other, then the weighting coefficient for the eye with the contour increased at the expense of that for the other eye. This is demonstrated by viewing Figure 22.47 stereoscopically; when fused, A appears much brighter than C, although at the centres of the discs the luminances are the same; on the other hand no difference between B and C is observed. Usually, contours in non-corresponding parts of the field give rise to rivalry, and this means, essentially that the contours of one eye dominate the weighting coefficient of this eye when dominant; at another moment, the contour of the other field will become dominant.*

### FLICKER

The classical experiments of Sherrington would indicate a more or less complete independence of the two eyes in respect to the discrimination of flicker. Thus under conditions where flicker is perceptible at 20 cycles/sec, the presentation of separate flickering patches to the two eyes in such a way that the dark period of one eye corresponded to the bright period of the other, did not give fusion although a flicker rate of 40 cycles/sec in one eye would have done. Smith has shown that when one

---

*Engel (1967) has examined in some detail the manner in which the sensations are combined; his formulation carries that of Levelt farther, invoking a cross-correlation process between the two eyes that allows of an improved prediction of the effects of combining monocular fields.

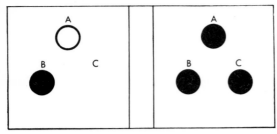

**Fig. 22.47** When the left and right figures are fused in the stereoscope, A appears brighter than C although no difference between B and C is noticed. (Levelt, *Brit. J. Psychol.*)

eye views a steady light source and the other an intermittent one, the sensation is one of binocular flicker, i.e. it is impossible to discriminate between the sensation so evoked and that which would be produced by two flickering sources presented separately to the two eyes. In the region of the critical fusion frequency, however, retinal rivalry occurs, so that the flicker sensation alternates with the impression of a stationary source. The critical fusion frequency, under these conditions, is not entirely determined by the luminance of the flickering patch but on the ratio of this luminance to that of the stationary source. The experimental conditions for the study of binocular flicker are clearly essentially the same as those under which retinal rivalry and ocular dominance are studied; where a unitary percept is possible, as in the special case of the two flickering patches being synchronous, we may expect the binocular flicker sensation to be identical with that evoked in either eye separately. When conditions are different in the two eyes, but not too discrepant, retinal rivalry becomes obvious. When the discrepancy is very strong, rivalry gives place to the dominance of one eye. A study of the tendency for dominance, when different pictures are presented in the stereoscope, shows that the picture with the most contours tends to become dominant; in the case of flicker, it is the flickering field that imposes itself on awareness at the expense of the stationary.

BINOCULAR COLOUR MIXING

When a pair of stereograms consists merely of two coloured pieces of paper, the condition most frequently encountered is that of retinal rivalry, one or other of the colours dominating awareness for a time to give place to the other. A true binocular mixing can be obtained, however, conforming approximately to the laws of monocular colour mixing, if the conditions are suitable. According to some studies by Johansen, conditions favouring mixture are low absolute luminance and a small separation of the two colours in the spectrum. When rivalry occurs, the longer wavelength tends to

dominate, and where there is a difference in luminosity between the colours, dominance occurs with the brighter of the two. The fact that so careful an experimenter as Helmholtz denied the existence of binocular colour mixing reveals that there are large individual differences in the capacity to mix colours binocularly. When red and green stereograms of a three-dimensional object are presented to the eye, it is seen in depth but one or other colour is usually suppressed (Treisman, 1962).

COLOUR STEREOSCOPY

The stereoscopic appearance obtained by regarding two differently coloured, but otherwise identical, plane pictures with the two eyes separately, is probably due to chromatic differences of magnification. If the left eye, for example, views a plane picture through a red glass and the right eye views the same picture through a blue glass, an illusion of solidity results.

ADAPTATION

It would seem from the work of Mandelbaum that the adaptive state of one eye is unaffected by that of the other; thus the absolute threshold of one eye was not altered by adapting the other to a variety of luminance levels. This is an important finding, since it justifies the use of one eye as a control in adaptation studies, e.g. those of Wright (p. 192).

## THE PERCEPTION OF MOTION

The perception of motion has two elements in it; there is the direct appreciation of movement in consequence of the gliding of the image over the retina when the eye is still (or as a result of the movement of the eye when the image of the object fixated remains stationary on the retina), and there is the more 'intellectual' recognition of the fact of movement deduced from the observation that an object is projected to a certain point in space at one moment and at another point after a certain interval of time. Thus we recognize that the large hand of a watch is in motion because at one minute it points at 'ten' and at the next minute at 'eleven'; there is, however, no real perception of motion, the threshold rate of movement being too low. In experimental studies the measurements are concerned with the direct perception of movement. The sensitivity of the retina to movement is highest at the fovea, as one might expect, since it is here that discrimination, generally, reaches its acme; nevertheless the periphery shows a high sensitivity in comparison with the low order of visual acuity attainable in this region; in the peripheral retina, so great is the difference between visual acuity and motion acuity, that the existence of objects may often only enter awareness when they are moved; as soon as they become stationary they are no longer perceived.*

## THRESHOLDS

The threshold sensitivity to movement can be measured in a number of ways as follows:

1. *The minimum excursion of a moving body necessary for its being in motion to be appreciated.* This minimum depends on the speed of movement, but under favourable conditions it is of the order of 10 seconds of arc. This is an extremely small displacement, so small that if two stationary points were separated by this amount they could not be discriminated as separate. The image of the moving point thus need not move over a whole cone before its movement is appreciated. As a result of diffraction and chromatic aberration, the image of a point covers several cones; these are stimulated to different degrees, since the brightness of the image falls off as the periphery of the blur-circle is approached; a very slight movement of the whole image modifies the distribution of light on the several cones and thus modifies the pattern of discharge in their nerve fibres. A movement much smaller than the diameter of a cone could modify the distribution of light on the cones and so provide the peripheral basis for this high degree of acuity. According to McColgin, the threshold increases with increasing distance from the fovea.

2. *The minimum angular speed of a point.* In the absence of stationary reference points, this of the order of one to two minutes of arc per second.

3. *The minimum duration for a given excursion.* If a point is moved between two stationary points, there is a speed above which the motion cannot be appreciated, i.e. the observer sees the point at all points in its path simultaneously; the moving luminous point becomes a 'streak'. With an excursion of 10°, the minimum duration is of the order of 50 msec, corresponding to a speed of movement of 200°/sec.

## APPARENT MOVEMENT

A common experimental method of inducing the appearance of movement is to present the subject with a given figure or spot for a short period of time and then to present the same figure at a different position in space; if the times and distances fall within certain ranges, the subject reports the perception of movement. This tachistoscopic type of presentation is, of course, the basis of motion pictures. According to Wertheimer's (1912) classical study, when the intervals are less than about 30 msec the two stimuli are reported as occurring simultaneously so that no motion is perceived; with 60 msec, movement is optimal whilst with intervals of 200 msec or more the two stimuli are perceived in succession.

## ILLUSIONS

A variety of illusions of movement have been described; for example, the apparent movement of the moon behind fast-flying clouds; the apparent backward motion of the

---

* This is a somewhat loose way of stating the fact; one becomes aware of an object when it moves in the peripheral field, but this may be an expression of visual acuity rather than motion acuity; owing to adaptation, the awareness of a stationary object in the peripheral field soon ceases, and it requires the stimulation of a new set of receptors to come again into consciousness. The sudden awareness of this fresh stimulation is *interpreted* as a movement of the object, yet experiments seem to show that if the movement of the object is continued there need be no real perception of motion. According to Warden, the direct comparison of visual with movement acuity is beset with difficulties.

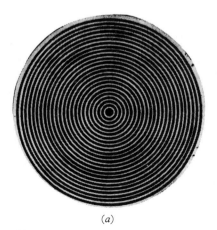

(a)                   (b)

**Fig. 22.48** Illustrating illusions of movement. For details see text. (After Bowditch, *J. Physiol.*)

water near the shores of a large river if the current is rapid at the middle; the apparent ducking forward of the masts of a ship in passing under a bridge from which it is observed, and so on. If the series of concentric lines in Figure 22.48a is subjected to a slight but rapid circular motion, similar to that used in rinsing a circular dish half-filled with water, an interesting 'propeller-like' motion will appear. Again, if Figure 22.48b is rotated about an axis through its centre, then, if the direction of rotation is that in which the spiral line approaches the centre, the entire surface appears to expand during rotation and to contract after it has ceased, and *vice versa*

if the rotation is in the opposite direction (Bowditch and Hall, 1880–2). Alternatively, the prolonged viewing of an object in motion gives rise to a movement-after-image in the opposite direction—the so-called *waterfall phenomenon*. Barlow and Brindley (1963) have shown that there is interocular transfer of the adaptation effect, since if only one eye is exposed to the moving figure, and this eye is subsequently blinded temporarily by exerting pressure on the globe, the after-effect appears in the other eye. The possible retinal basis for the phenomenon will be mentioned later (p. 565 footnote).

## REFERENCES

Adey, W. R. & Noda, H. (1973) Influence of eye movements on geniculate-striate excitability in the cat. *J. Physiol.* **235,** 805–821.

Andersen, E. E. & Weymouth, F. W. (1923) Visual perception and the retinal mosaic. *Am. J. Physiol.* **64,** 561–594.

Barlow, H. B. & Brindley, G. S. (1963) Inter-ocular transfer of movement after-effects during pressure blinding of the stimulated eye. *Nature* **200,** 1347.

Becker, W. & Klein, H.-M. (1973) Accuracy of saccadic eye movements and maintenance of eccentric eye positions in the dark. *Vision Res.* **13,** 1021–1034.

Beeler, G. W. (1967) Visual threshold changes resulting from spontaneous saccadic eye movements. *Vision Res.* **7,** 769–775.

Bizzi, E. (1966) Changes in the orthodromic and antidromic response of optic tract during eye movements of sleep. *J. Neurophysiol.* **29,** 861–870.

Blakemore, C. (1970) The range and scope of binocular depth discrimination in man. *J. Physiol.* **211,** 599–562.

Blakemore, C., Carpenter, R. H. S. & Georgeson, M. A. (1970). Lateral inhibition between orientation detectors in the human visual system. *Nature* **228,** 37–39.

Bowditch, H. P. & Hall, G. S. (1880–82) Optical illusions of motion. *J. Physiol.* **3,** 297–307.

Burian, H. M. (1951) Stereopsis. *Doc. Ophthal.* **5–6,** 169–183.

Buttner, U. & Fuchs, A. F. (1973) Influence of saccadic eye movements on unit activity in simian lateral geniculate and pregeniculate nuclei. *J. Neurophysiol.* **36,** 127–141.

Collewijn, H. (1969) Optokinetic eye movements in the rabbit: input-output relations. *Vision Res.* **9,** 117–132.

Creed, R. S. (1935) Observations on binocular fusion and rivalry. *J. Physiol.* **84,** 381–391.

Crone, R. A. & Verduyn Lunel, H. F. E. (1969) Autokinesis and the perception of movement: the physiology of eccentric fixation. *Vision Res.* **9,** 89–101.

Ditchburn, R. W. (1955) Eye-movements in relation to retinal action. *Optica Acta* **1,** 171.

Engel, G. R. (1967) The visual process underlying binocular brightness summation. *Vision Res.* **7,** 753–767.

Feldman, M. & Cohen, B. (1968) Electrical activity in the lateral geniculate body of the alert monkey associated with eye movements. *J. Neurophysiol.* **31,** 455–466.

Festinger, L., Sedgwick, H. A. & Holtzman, J. D. (1976) Visual preception during smooth pursuit eye movements. *Vision Res.* **16,** 1377–1386.

Fox, R., Todd, S. & Bettinger, L. A. (1975) Optokinetic

nystagmus as an objective indicator of binocular rivalry. *Vision Res.* **15,** 849–853.

Graham, C. H. (1965) Ed. *Vision and Visual Perception.* New York: Wiley.

Graybiel, A. & Hupp, D. (1946) The oculogyral illusion: a form of apparent motion which may be observed following stimulation of the semicircular canals. *J. aviat. Med.* **17,** 3–27.

Grosvenor, T. (1959) Eye movements and the autokinetic illusion. *Am. J. Optom.* **36,** 78–87.

Helmholtz, H. von. (1925) *Physiological Optics* (translated J. P. C. Southall). Optical Soc. Am.

Jameson, D. & Hurvich, L. M. (1959) Note on factors influencing the relation between stereoscopic acuity and observation distance. *J. opt. Soc. Am.* **49,** 639.

Jeannerod, M. & Putkonen, P. T. S. (1971) Lateral geniculate unit activity and eye movements: saccade-locked changes in dark and light. *Exp. Brain Res.* **13,** 533–546.

Julesz, B. (1964) Binocular depth perception without familiarity cues. *Science* **145,** 356–362.

Kaufman, L. (1964a) Suppression and fusion in viewing complex stereograms. *Am. J. Psychol.* **77,** 193–205.

Kaufman, L. (1964b) On the nature of binocular disparity. *Am. J. Psychol.* **77,** 393–402.

Kaufman, L. (1965) Some new stereoscopic phenomena and their implications for the theory of stereopsis. *Am. J. Psychol.* **78,** 1–20.

Kaufman, L. & Pitblado, C. (1965) Further observations on the nature of effective binocular disparities. *Am. J. Psychol.* **78,** 379–391.

Keesey, U. T. (1960) Effects of involuntary eye movements on visual acuity. *J. opt. Soc. Am.* **50,** 769–774.

Lawson, R. B. & Gulick, W. L. (1967) Stereopsis and anomalous contour. *Vision Res.* **7,** 271–297.

Levelt, W. J. M. (1965) Binocular brightness averaging and contour information. *Brit. J. Psychol.* **56,** 1–13.

Lit, A. (1960) The magnitude of the Pulfrich stereo-phenomenon as a function of target velocity. *J. exp. Psychol.* **59,** 165–175.

Ludvigh, E. (1952) Control of ocular movements and visual interpretation of environment. *Arch. Ophthal.* **48,** 442–448.

Mandelbaum, J. (1941). Dark adaptation. Some physiologic and clinical considerations. *Arch. Ophthal.* **26,** 203–239.

McColgin, F. H. (1960) Movement thresholds in peripheral vision. *J. opt. Soc. Am.* **50,** 774–779.

Merton, P. A. (1956) Compensatory rolling movements of the eye. *J. Physiol.* **132,** 25–27P.

Michael, J. A. & Stark, L. (1967) Electrophysiological

correlates of saccadic suppression. *Exp. Neurol.* **17**, 233–246.

Mitchell, D. E. (1966) Retinal disparity and diplopia. *Vision Res.* **6**, 441–451.

Mitrani, L., Mateeff, St. & Yakimoff, N. (1970) Smearing of the retinal image during voluntary saccadic eye movements. *Vision Res.* **10**, 405–410.

Noda, H. (1975). Depression in the excitability of relay cells of lateral geniculate nucleus following saccadic eye movements in the cat. *J. Physiol.* **249**, 87–102.

Noda, H. & Adey, W. R. (1974) Retinal ganglion cells of the cat transfer information on saccadic eye movement and quick target motion. *Brain Res.* **70**, 340–345.

Ogle, K. N. (1950) *Researches in Binocular Vision.* Philadelphia: Saunders.

Ogle, K. N. (1953) Precision and validity of stereoscopic depth perception from double images. *J. opt. Soc. Am.* **43**, 906–913.

Ogle, K. N. (1958a) Note on stereoscopic acuity and observation distance. *J. opt. Soc. Am.* **48**, 794–798.

Ogle, K. N. (1958b) Present state of our knowledge of stereoscopic vision. *Arch. Ophthal.* **60**, 755–774.

Ogle, K. N. (1962) The optical space sense. In *The Eye*, Ed. Davson, vol. IV, chapters 11–18. London & New York: Academic Press.

Ogle, K. N. & Weil, M. P. (1958) Stereoscopic vision and the duration of the stimulus. *Arch. Ophthal.* **59**, 4–17.

Panum, P. L. (1858) *Physiologische Untersuchungen uber das Sehen mit zwei Augen.* Kiel.

Pirenne, M. H. (1943) Binocular and uniocular thresholds of vision. *Nature* **152**, 698–699.

Richards, W. (1977) Stereopsis with and without monocular contours. *Vision Res.* **17**, 967–969.

Riggs, L. A., Merton, P. A. & Morton, H. B. (1974) Suppression of visual phosphenes during saccadic eye movements. *Vision Res.* **14**, 997–1011.

Skavenski, A. A. (1971) Extraretinal correction and memory for target position. *Vision Res.* **11**, 743–746.

Skavenski, A. A. & Steinman, R. M. (1970) Control of eye position in the dark. *Vision Res.* **10**, 193–203.

Smith, S. (1946) The essential stimuli in stereoscopic depth perception. *J. exp. Psychol.* **36**, 518–521.

Smith, S. (1949) A further reduction of sensory factors in stereoscopic depth perception. *J. exp. Psychol.* **39**, 393–394.

Teichner, W. H., Kobrick, J. L. & Wehrkamp, R. F. (1955) The effects of terrain and observation distance on relative depth discrimination. *Am. J. Psychol.* **68**, 193–208.

Ten Doesschate, G. (1955) Results on an investigation of depth perception at a distance of 50 metres. *Ophthalmologica, Basel* **129**, 56–57.

Treisman, A. (1962) Binocular rivalry and stereoscopic depth perception. *Quart. J. exp. Psychol.* **14**, 23–37.

Volkmann, A. W. (1863) *Physiologische Untersuchungen in Gebiet der Optik.* Leipzig.

Volkmann, F. C. (1962) Vision during voluntary saccadic movements. *J. opt. Soc. Amer.* **52**, 571–578.

Walls, G. L. (1951) A theory of ocular dominance. *Arch. Ophthal.* **45**, 387–412.

Widén, L. & Ajmone-Marsan, C. (1961) Action of afferent and corticofugal impulses on single elements of the dorsal lateral geniculate nucleus. In *The Visual System.* Ed. Jung & Kornhuber, pp. 125–133. Berlin: Springer.

Wist, E. R., Brandt, T., Diener, H.-C. & Dichgans, J. (1977) Spatial frequency effect on the Pulfrich phenomenon. *Vision Res.* **17**, 391–397.

Wurtz, R. H. (1969a) Response of striate cortex neurons to stimuli during rapid eye movements in the monkey. *J. Neurophysiol.* **32**, 975–986.

Wurtz, R. H. (1969b) Comparison of effects of eye movements and stimulus movements on striate cortex neurons of the monkey. *J. Neurophysiol.* **32**, 987–994.

Zajac, J. L. (1960) Convergence, accommodation and visual angle as factors in perception of size and distance. *Am. J. Psychol.* **73**, 142–146.

Zuber, B. L., Stark, L. & Lorber, M. (1966) Saccadic suppression of the pupillary light reflex. *Exp. Neurol.* **14**, 351–370.

# 23. Neurophysiology of perception

## THE PRIMARY VISUAL PATHWAY TO THE CEREBRAL CORTEX

### OPTIC NERVE AND TRACT

The ganglion cells of the retina (p. 167) are the second-order neurones in the transmission of retinal impulses to the cerebral cortex; the axons of these cells constitute the *optic nerves* (N II) and *optic tracts*, passing either to the *lateral geniculate bodies* or the *superior colliculi* (the pupillary fibres, it will be remembered, pass to the pretectal region of the midbrain).

#### DECUSSATION

The general plan of the visual pathway is indicated in Figure 23.1; the fibres from the nasal halves of the retinae decussate in the *optic chiasma*, so that each optic tract contains fibres from the temporal half of one retina and the nasal half of the other. A visual impulse, arising from the right half of the field, is therefore conveyed exclusively along the left optic tract and thus to the left cerebral hemisphere. Section of the optic tract thus gives rise to what is called *homonymous hemianopia*, or half-blindness—one half of the visual field of each eye, the temporal half of the opposite side and the nasal half of the other, being obliterated. The degree of decussation varies from species to species and is obviously a function of binocular vision; thus in the guinea pig some 99 per cent of the optic nerve fibres cross to the opposite side, leaving only 1 per cent uncrossed; in the rat some 10 per cent remain uncrossed; in the opossum 20 per cent, whilst in the cat and primates, with frontally directed eyes, about equal numbers of fibres are crossed and uncrossed.

### Phylogeny

Phylogenetically considered, then, the uncrossed fibres are the 'younger' ones and may be expected to subserve primarily stereoscopic perception of depth (Blakemore and Pettigrew, 1970). Thus in an animal with frontally developed eyes, as Figure 23.1 shows, the temporal half of the retina of one eye is receiving an image of the same region of space as does the nasal half of the other. Hence with both laterally and frontally directed eyes, each cerebral hemisphere receives information from the contralateral half of the visual world.

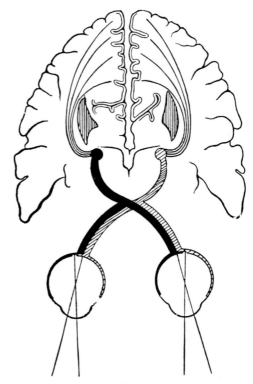

**Fig. 23.1** The visual pathway. Fibres from the nasal retina, including the nasal half of the macula, decussate in the chiasma to join the uncrossed fibres of the temporal half of the retina. (Wolff, after Traquair.)

#### ARRANGEMENT OF FIBRES

The fibres are grouped in the nerve in three bundles according to the retinal fields from which they originate as follows:

1. Uncrossed temporal fibres.
2. Crossed nasal fibres.
3. Macular fibres.

Half of the macular fibres undergo decussation in the chiasma, so that half of the macula is represented in each tract. The relative positions of the bundles in the optic nerve and tract are indicated in Figure 23.2.

The fibres vary in size, being divided roughly into coarse and fine; since the speed of conduction in a nerve

 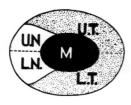  

**A. In the optic nerve (distal).**    **B. In the optic nerve (proximal).**    **C. In the optic tract.**    **D. In the lateral geniculate body.**

The crescents below U.P. and L.P. are the uniocular fibres.

**Fig. 23.2** Distribution of the fibres in optic nerve, optic tract, and lateral geniculate body. U.T., upper temporal; U.N., upper nasal; L.N., lower nasal; L.T., lower temporal; U.P., upper peripheral; L.P., lower peripheral. (Wolff, *Anatomy of the Eye and Orbit.*)

fibre increases with its diameter, this variation in size must be associated with different conduction rates in the visual pathway. It is interesting that in the cat the fine, slowly conducting, fibres pass to the midbrain visual centres whilst the coarse fibres go to the visual cortex. So great is the difference in calibre of the fibres, that a visual impulse can be transmitted to the lateral geniculate body, relayed here to the cortex, and the cortical response may pass to the midbrain before the impulse reaches this region by way of the fine fibres directly. It is theoretically possible, therefore, for midbrain reflex activities to be anticipated and modified by cortical influences initiated by the same stimulus. In fact Doty states that the evoked potential recorded from the midbrain, in response to a light stimulus, appears some 12 msec after that recorded from the cerebral cortex of the cat.*

## NON-GENICULATE TARGETS FOR RETINOFUGAL INPUT

In mammals and marsupials there are additional targets for the retinal fibres in addition to those relaying in the lateral geniculate body and thence passing to the cerebral cortex. These may be enumerated as follows:

1. Superior colliculus.
2. Accessory optic system.
3. Hypothalamus.
4. Pretectal complex.†

## COLLICULAR FIBRES

In all species, including man, a portion of the optic tract fibres avoids the lateral geniculate body and goes to the superior colliculus in what has been called the *mesencephalic root of the optic tract* to distinguish it from the *diencephalic root* running to the lateral geniculate body. According to the position of the animal in the phylogenetic scale, the proportion of optic fibres travelling to the midbrain station changes; in birds there is very little cortical representation of retinal impulses,

the vast majority of the optic fibres running to the optic tectum, or superior colliculus; in rodents the collicular projection is less, but still very significant, whilst in apes and man it is smaller. The fact that there is a projection of retinal fibres on the superior colliculus in man might suggest that visual responses to light would occur in the absence of an occipital cortex; in fact, however, only pupillary responses, which, as we have seen, are mediated by fibres relaying in the pretectal nucleus, are obtained in the absence of a functioning cortex. We shall discuss some aspects of collicular physiology later, in so far as it bears on the interpretation of the retinal image. Its possible role as a motor centre in control of the eye movements has been discussed earlier.

## ACCESSORY OPTIC SYSTEM

As described by Hayhow (1960), the accessory optic system consists of a group of crossed finely medullated

---

* Studies with the electron microscope have shown that the number of small unmyelinated fibres in the optic nerve is vastly greater than had been estimated on the basis of light-microscopical measurements, so many of these fibres being beyond the limits of resolution of the light-microscope. For example, Maturana has counted some 470 000 unmyelinated fibres in the frog's optic nerve and 12 000 thicker myelinated fibres, whereas earlier counts had shown a total of only about 14 000. Hughes and Wässle (1976) found 193 000 fibres in the cat's optic nerve, with diameters ranging from 0·5 to 13·5 $\mu$m; there was a high-density core of fibres, corresponding to the area centralis. The number of ganglion cells in the cat's retina has been estimated at 190 000 by Wässle *et al.* (1974) and 217 000 by Hughes *et al.* (1975). In the rabbit's optic nerve, Vancy and Hughes (1976) found 349 000 fibres, 98 per cent being myelinated. Diameters ranged from 0·25 to 7 $\mu$m with a peak at 0·75 $\mu$m. The number of ganglion cells in the retina was 455 000 to 547 000, and the authors suggest that some of these might be displaced amacrine cells.

† The lateral geniculate body, viewed as a relay-station for visual input to the cerebral cortex, should be described as the *dorsal nucleus of the lateral geniculate body* (DLGB); the ventral nucleus of lower mammals (VLGB) and its analogue, the pregeniculate nucleus of primates (PGN), receive retinal input that is not relayed directly to the cerebral cortex (p. 548).

axons of retinal origin which is divided into *inferior* and *superior accessory optic tracts*. The inferior fasciculus is a group of fibres deviating from the course of the main optic tract and terminating in the dorsal portion of a prominent *terminal nucleus* in the midbrain tegmentum (Fig. 23.3); Hayhow *et al.* (1960) call this the *medial terminal nucleus* of the accessory optic system.

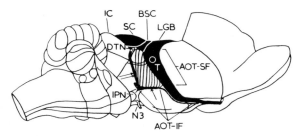

**Fig. 23.3** A right inferolateral view of a dissection of the rat brain, showing the course taken by the degenerating contralateral optic axons (shown in black) resulting from section of the left optic nerve 7 days previously. The course of the inferior fasciculus has been indicated by fine stipple. The course of the superior fasciculus has been indicated semi-diagrammatically by unbroken lines. The single arrow indicates the termination in the anterior extremity of the medial terminal nucleus of the slender fibre bundle formed by the most anterior fibres of the superior fasciculus. The double arrows indicate the course of the more posterior fibres of the superior fasciculus (the classical tractus peduncularis transversus) as observed with the aid of a stereo-microscope. The location of the dorsal terminal nucleus has been depicted by white dots. (Hayhow *et al.*, *J. comp. Neurol.*)

The superior fasciculus is larger and more diffuse and terminates in the basal portion of the medial terminal nucleus, whilst more posterior fibres of the same fasciculus terminate additionally in two superficially located nuclei, which were called the *dorsal* and *lateral terminal nuclei of the optic tract*. This superior fasciculus includes the classically described *tractus peduncularis transversus* and the *posterior accessory optic tract* of Bochenegg, this latter also being described as *the basal optic root*. The application of the technique of labelling axoplasm, and measuring its flow towards nerve-endings by autoradiography, has permitted a much more precise definition of the visual pathways; this, coupled with electron-microscopical examination of regions of high radioactivity after section of nerves, has revealed terminal degeneration and has often permitted the decision as to whether the fibres end in a particular nucleus or not. The application of these techniques has generally confirmed the existence of these accessory optic tracts and their termination in the terminal nuclei.

SUPRACHIASMATIC NUCLEI

In addition a hypothalamic termination in the suprachiasmatic nucleus has been described, reached by what Moore and Klein describe as the retino-hypothalamic tract. For example, Moore and Lenn (1972) injected [3]H-leucine into the rat's eye and subsequently found label in this nucleus bilaterally; the labelling was confined to this hypothalamic nucleus so that the supraoptic nucleus, for example, had no activity. Electron microscopy after removal of the eye showed degenerating terminals on the dendrites of the neurones in the nucleus. The bilaterality of the projection is interesting and occurs even in the albino rat where all the remaining visual input is completely crossed; these authors showed that the hypothalamic projection was common to rat, guinea pig, rabbit, cat and monkey.

CIRCADIAN RHYTHMS

The significance of this hypothalamic projection becomes clear when we consider that, in many mammals, certain bodily functions, controlled by the hypothalamus, such as oestrus, exhibit a circadian rhythm dependent on the light-dark cycle. For example, rearing a rat in continuous light, produces a permanent state of oestrus, and the classical studies of Critchlow (1963) and Butler and Donovan (1971) have shown that these effects may be abolished by a lesion in the suprachiasmatic nucleus. The pineal gland, or epiphysis, is involved in circadian rhythms, but its anatomical location within the skull precludes its being influenced directly by light. Instead, the influence of light must be mediated through the central nervous system, culminating in a sympathetic supply from the superior cervical ganglion. Thus, interruption of this pathway—for example by removing the superior cervical ganglion—completely blocks the responses of the pineal to light or darkness (Wurtman *et al.*, 1964). The gland secretes a hormone, *melanotonin*; and the cyclical activity of the gland depends on the cyclical synthesis or activation of the two key enzymes involved in its synthesis, namely hydroxy-indole-O-methyltransferase (HIOMT) and serotonin-N-acetyltransferase. Thus the pineal exerts its action on oestrus and other activities through the secretion of melanotonin, which inhibits the secretion of pituitary gonadotrophic hormones, the effect of light being to inhibit the secretion of pineal hormones and thus to release the pituitary from inhibitory control.

RETINO-HYPOTHALMO-PINEAL PATHWAY

The pathway to the hypothalamus apparently leaves the primary optic tracts separately from the accessory optic tracts, as indicated in Figure 23.4. Moore and Klein (1974) made lesions that sectioned the primary

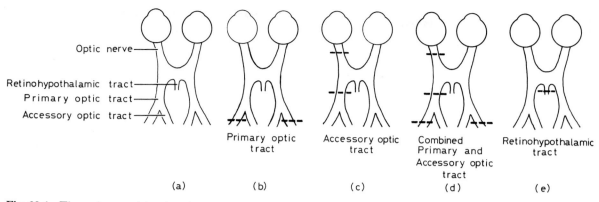

**Fig. 23.4**   The optic tracts (a) and sections made to elucidate the different pathways. (After Moore and Klein, *Brain Res.*)

optic tracts (*b*), or the the accessory optic tracts (*c*),★ or sectioned both the primary and accessory optic tracts (*d*) leaving the retinohypothalamic tract intact, and finally one that destroyed the retinohypothalamic

tract at the same time as the suprachiasmatic nuclei (*e*). Lesions that involved the suprachiasmatic nuclei and retinohypothalamic pathway completely abolished the rhythmic changes in N-acetyltransferase activity associated with the light-dark cycle. On the contrary, bilateral lesions involving the primary or accessory optic tracts or both were without significant effects. The probable pathway from the suprachiasmatic nucleus to the superior cervical ganglion is indicated in Figure 23.5.

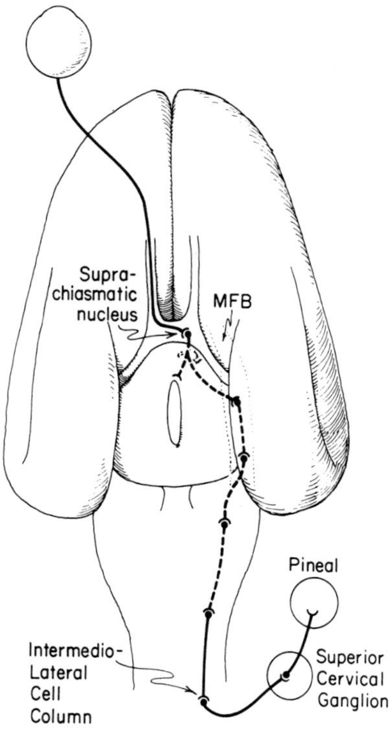

TONIC AND RHYTHMIC LIGHT EFFECTS

If it is the retinohypothalamic system that provides the light-information to the circadian rhythm centre, we may ask what function the accessory optic system, terminating in the nuclei of the optic tract, fulfils? Stephan and Zucker (1972) emphasized the distinction between tonic effects of light on bodily function and the rhythmic or phasic effects manifest in circadian rhythms. According to this distinction, the circadian rhythm would be influenced by the light-input through the suprachiasmatic nuclei and pineal gland, whereas the non-rhythmic effects of continuous exposure to light or dark might be influenced by the accessory system.†

PRETECTAL FIBRES

These leave the primary optic tract before it enters the lateral geniculate body and relay in a group of midbrain

**Fig. 23.5**   Diagram of proposed central neural pathway regulating pineal N-acetyltransferase activity. Photic input from the eyes traverses the retinohypothalamic tracts to the suprachiasmatic nuclei. From these nuclei information passes through the lateral hypothalamus and brain stem, in unidentified pathways, to the spinal cord and the interomediolateral cell column whose neurons provide preganglionic fibres to the superior cervical ganglia. The dashed lines indicate areas where the exact fibres transmitting information have not been identified with certainty. (Moore and Klein, *Brain Res.*)

★ The lesion in (*c*) designed to cut off input through the accessory optic tracts relies on the complete decussation of these accessory fibres in the chiasma; the same is true for the combined lesions in (*d*).

† The circadian adrenal corticosterone rhythm of the rat was not abolished by section of the primary and accessory optic tracts; only when the terminal nucleus of the retinohypothalamic pathway was destroyed—namely the suprachiasmatic nucleus—was the rhythm abolished and the level of the hormone remained the same at night as in the morning (Moore and Eichler, 1972).

nuclei. The fibres carry the visual input to the pupillomotor and accommodative muscles of the eye and have been discussed in these contexts.

## LATERAL POSTERIOR NUCLEUS, OR PULVINAR

A direct pathway from retina to this 'lower visual centre' has been the subject of debate and it may well be that the situation varies with the species; thus there is definitely no direct projection in the squirrel monkey but one has been reported for the phalanger (Rockel et al., 1972), tree shrew (Laemle, 1968) and the rhesus monkey and baboon (Campos-Ortega et al., 1970). In the two marsupial opossums studied by Royce et al. (1976), employing the radioautographic axon-flow technique, there were small but distinct retinal projections. As we shall see, there is strong electrophysiological evidence for visual input, but this may well be secondary to input to the superior colliculus, the tecto-pulvinar pathway being the major subcortical input to the pulvinar (Glendenning et al., 1975).*

## LATERAL GENICULATE BODY

### LAMINATION

In primates including man the cells of the dorsal nucleus of the lateral geniculate body are arranged in six laminae, and the fibres of the optic tract make synaptic connections with these cells. It was shown by Glees and LeGros Clark that the crossed optic fibres end in layers one, four and six, whilst the uncrossed, temporal fibres synapse with cells in layers two, three and five (Fig. 23.6). There thus seems to be a rigid separation of the fibres arriving from the two eyes, so that the fusion of the two retinal images required by binocular vision cannot apparently take place at this level. In this respect, then, we may regard the lateral geniculate body as a sorting centre that rearranges the crossed and uncrossed fibres that had become quite mixed in the optic tract. By noting the distribution of transneuronal atrophies† following small retinal lesions, Penman and LeGros Clark showed that there is a virtual 'point-to-point' representation of the retina in the lateral geniculate body, thereby ensuring that the one-to-one relationship between cones of the fovea, bipolar and ganglion cells is carried through as far as this stage in the visual pathway.

### OPTIC TRACT TERMINATIONS

Each nerve fibre of the optic tract, after entering its lamina, breaks up into a number of terminals, each ending in a 'bouton' applied to a lateral geniculate

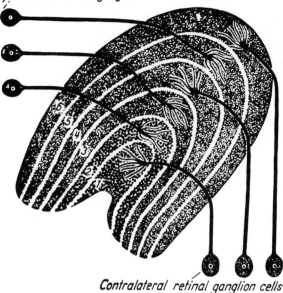

*Ipsilateral retinal ganglion cells*

*Contralateral retinal ganglion cells*

**Fig. 23.6** Illustrating the segregation of the optic tract fibres into the separate laminae of the lateral geniculate body in primates. Fibres from the contralateral retina, i.e. the crossed fibres, relay in layers 1, 4 and 6. (Glees, In *The Visual System.*)

body cell; each cell has only one bouton applied to it, so that there is no overlapping of impulses from different retinal receptors. (In the cat, on the other hand, there are a large number of boutons to each cell—in this animal visual acuity is subservient to sensitivity to light.) The terminal branches of each optic tract fibre end on up to thirty lateral geniculate cells. Since each geniculate cell receives a bouton from only one optic tract fibre, there should be some 10 to 30 times as many geniculate cells as optic tract fibres, yet the numbers appear to be about equal.

### SOMATOTOPY

As we should expect from the orderly projections of fibres from the retina through the optic tracts and their ending points in the laminae of the lateral geniculate bodies, there is a strict

---

*It is worth noting that, according to Gilbert and Kelly (1975), there is a connexion between the midbrain reticular nuclei and the lateral geniculate body, and this corresponds with the physiological studies of Singer (1974) on the effects of midbrain nuclei on the activities of geniculate neurones.

†When a neurone is injured, the degeneration is not necessarily confined to this cell; thus the destruction of primary visual neurones in the retina causes the cell bodies of geniculate cells to degenerate, a phenomenon described as *transneuronal degeneration*; apparently the geniculate neurone depends for its viability on receiving impulses from the ganglion cells of the retina.

somatotopy in the relation of retinal points to given points in the dorsal lateral geniculate nucleus (Bishop *et al.*, 1962). This regularity extends to the types of retinal ganglion cells. Thus *classified morphologically and their mode of projection.* Kelly and Gilbert (1975) have exploited the technique of injecting horseradish peroxidase into the presumed regions of termination of the ganglion cell axons—lateral geniculate body and superior colliculus—and have identified the cells of origin in the retina after the tracer has been carried retrogradely back. Kelly and Gilbert identified three classes of retinal ganglion cell according to their destinations, and these classes could also be related to the morphology of the cells. Thus there was a class of small cells labelled almost exclusively after injections into the superior colliculus; a second class of intermediate size that was labelled much more extensively from the lateral geniculate nucleus than after collicular injection, and lastly there were large cells labelled after injections in either structure.

The central terminations of these large cells suggested that their axons branched to supply both colliculus and geniculate body, a branching that has been established electrophysiologically. The authors compared the morphology of the retinal cells, identified in this manner, with the classification of Boycott and Wässle (p. 290) and concluded that the large cells were equivalent to their alpha-cells, the medium to their betas, and the small cells to their gammas. These ganglion cell types fall into separate classes on the basis of their responses to light (p. 288).

LAMINATION IN THE CAT

The classical study of Thuma (1928) differentiated three layers, described as A, $A_1$ and B; A and $A_1$ had strong morphological similarities whilst B was different from the other two; the transneuronal degeneration following removal of one eye suggested and A and B received retinal fibres from the contralateral eye, whilst layer $A_1$ received fibres from the ipsilateral eye; thus A and $A_1$ would be concerned with the binocular field and would provide a basis for keeping the inputs from the two eyes separate. Subsequent work, notably that of Guillery (1970), has indicated a more complex lamination, as indicated in Figure 23.7 where the layers A and $A_1$ are shown corresponding with Thuma's layers, whilst layer B is replaced by three layers, C, $C_1$ and $C_2$ differing morphologically mainly by the size of their cells, although transitions between layers are not sharp. Degeneration experiments indicated that layers C and $C_1$ were analogous with A and $A_1$, in the sense that they received retinal inputs from contralateral and ipsilateral eyes respectively; and it seemed that $C_2$ received no retinal input.

**Subdivision**

By employing the axonal transport of radioactive tracer as a marker for fibre terminations, a technique that gives much better definition of terminals so that fibres of passage are easily distinguishable from those of termination (Hickey and Guillery 1974), it was possible to differentiate four laminae in addition to A and $A_1$ namely C, $C_1$, $C_2$, and $C_3$, this last having no retinal projection.* Laminae A and C and $C_2$ received afferents from the contralateral eye and laminae $A_1$

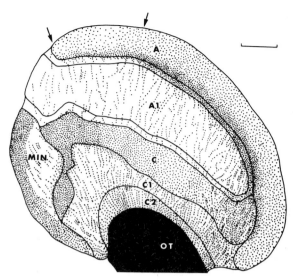

**Fig. 23.7**  Frontal section through the rostral third of the *right* dorsal lateral geniculate nucleus. The rat survived 12 days after removal of the left eye. The unlabelled arrows indicate the part of lamina A in which the degeneration is least dense. The continuous lines show the borders of the cell laminae. The dots, which show the distribution of the degenerating fibres, have been drawn to indicate the density of the degeneration and the orientation of the degenerating fibres as accurately as possible. A, A1, C, C1, and C2 indicate the cell laminae; OT, optic tract; MIN, medial interlaminar nucleus. Scale: 500 $\mu$m. (Guillery, *J. comp. Neurol.*)

and $C_1$ received afferents from the ipsilateral eye. There was no evidence of binocular overlap within the nucleus. As indicated by Figure 23.1, the parts of the retina that are concerned with monocular vision are the extreme nasal areas corresponding to the temporal fields, and electrophysiological evidence indicates that this essentially monocular input is confined to a separate part of the lateral geniculate body, which might be lamina C of Guillery, and might constitute a part of lamina B of Thuma and others.

**Medial interlaminar nucleus**

The lateral geniculate body contains, in addition to its laminated regions, the medial interlaminar nucleus (Fig. 23.8). It would seem that the laminated regions project only to areas 17 and 18 in the cat, whilst the medial interlaminar nucleus projects to area 19 (see, for example, Rosenquist *et al.*, 1974).

SIGNIFICANCE OF LAMINATION

The basic function of lamination is obviously concerned in maintaining the inputs from the two eyes separate; this means that integration of the inputs from the two eyes to give a single field is a function of cerebral cortical cells, and it is only here that neurones driven by both eyes are found. As to why there are six layers in the primate there has been some speculation. LeGros Clark

---

* According to Gray (quoted by LeVay and Gilbert, 1976) $C_3$ receives an input from the superior colliculus.

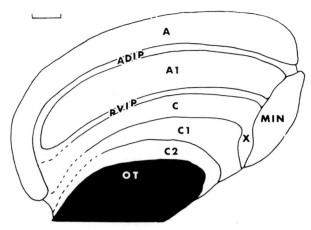

**Fig. 23.8** Sketch indicating lamination of cat's dorsal lateral geniculate nucleus as seen in a frontal section through the rostral third. MIN, medial interlaminar nucleus; X shows the part of the lamina, C, that is continuous with the lateral aspect of the medial interlaminar nucleus. ADIP and PVIP indicate the anterodorsal and posteroventral leaflets of the interlaminar plexus. Scale: 500 μm. (Guillery, *J. comp. Neurol.*)

pointed out that the smallest retinal lesion always caused degeneration in three layers of the lateral geniculate body, suggesting that the unit of conduction to the cortex is a three-neurone one, and he thought that this might be related to the trichromatic nature of vision (p. 190) whereby colour responses are determined by mixtures of three primary sensations. Certainly the electrophysiological studies of De Valois (1960) and Hubel and Wiesel (1961) lend support to the notion of a functional separation of neurones according to layer. Thus De Valois found that the responses of single cells of the monkey's lateral geniculate body depended on the layer in which they were situated; thus the dorsal pair of layers gave responses when the light was flashed on to the retina (ON-responses, p. 248), whilst the cells of the intermediate layer gave responses either at ON or at OFF, acording to the wavelength of the light employed; the ventral pair of layers gave responses only when the light was switched off (OFF-responses), so that light projected on to the retina caused only a cessation of spontaneous activity followed by a spike discharge when the light was switched off.

## VENTRAL NUCLEUS

The function of this much smaller and morphologically distinct portion of the lateral geniculate body is by no means clear (Swanson *et al.*, 1974). It receives input from the optic tract through collaterals relaying in the dorsal nucleus, from the superior colliculus (Gilbert and Kelly, 1975) and from the visual and juxta-visual areas of the cerebral cortex, and finally from the dorsal nucleus.

### EFFERENT CONNEXIONS

Its efferent connexions were studied by Swanson *et al.*, using the transport of radioactive leucine as a marker;

these proved to be the pretectal nuclei, the lateral terminal nuclei of the accessory optic system, the ventral portion of the suprachiasmatic nuclei of the hypothalamus, and the superior colliculus; all these regions receive retinal projections and, except for the colliculus, the ventral geniculate input coincides with the retinal. (The ventral geniculate input to the colliculus was confined to the deeper layers.) There was no evidence for a cortical projection.

### PREGENICULATE NUCLEUS

The ventral nucleus of the lateral geniculate body of mammals, such as the cat, is clearly separated from the dorsal nucleus by a neurone-free lamina. The equivalent structure in the primate is the pregeniculate nucleus. In man, Preobrazhenskaya (1964) followed the changes in the lateral geniculate body during embryonic and fetal development and described, in the early stages, the separation into ventral and dorsal nuclei; at later stages, after 5 months, the ventral nucleus is no longer seen and its place is taken by a group of small faintly staining cells, the *pregeniculate nucleus* (PrGN), lying over the rostral part of the lateral geniculate body, which has now become a layered structure. In the infant the connexion between the PrGN and the zona incerta is clear, and the author considers that the PrGN, developing as it does in close association with the LGN, is, indeed, part of the visual system and also part of the thalamic reticular formation.

### Possible function

The similarity in projections of the retina and the ventral geniculate or pregeniculate nucleus has suggested to Swanson *et al.* that the ventral nucleus might play a centrifugal role, comparable with that played by the isthmo-optic nucleus on the amacrine cells of the pigeon retina (p. 175); in the case of the mammal, however, the centrifugal effect would be directed to higher brain centres than the retina, such as the superior colliculi, pulvinar, and so on.

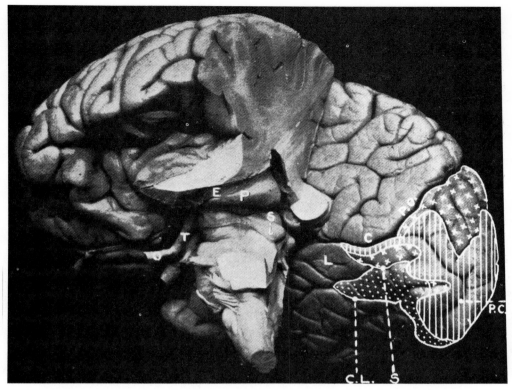

**Fig. 23.9**  The cortical visual areas. Striate area marked with vertical lines; the parastriate area with crosses and the peristriate area with dots. C, calcarine fissure; C.L., collateral sulcus; S, sagittal sulcus of the gyrus lingualis (L); O, optic nerve; T, optic tract; E, lateral geniculate body; P, pulvinar; S, superior colliculus; *i*, inferior colliculus. (Whitnall, *Anatomy of the Human Orbit*.)

## THE VISUAL CORTEX

STRIATE AREA

From the lateral geniculate body, the third-order neurones pass as the *optic radiation* to the surface of the occipital pole, relaying with the cortical neurones in Brodmann's area 17, or the *striate area*, so called because the axons of the optic radiation form a well-defined *white line of Gennari* on their way to their synapses with the subjacent layer IV, together with axons of association fibres passing to adjacent cortex. The region over which the striate area extends is illustrated in Figures 23.9 and 23.10. The grey matter of the cerebral cortex is divided into some six layers numbered I to VI from above downwards, and it is layer IV that contains the main sensory input in the receiving areas. In general, the more superficial layers are the source of commissural and association connexions, while the deeper layers are the sources of corticofugal connexions, for example, the cells of layer V send axons to the superior colliculus (Palmer and Rosenquist, 1974) and those of layer VI to the lateral geniculate body (Gilbert and Kelly 1975).

## RETINAL PROJECTION ON THE CORTEX

Small lesions in the striate area of man lead to blindness in well-defined regions of the visual fields, and a systematic study of the effects of gun-shot wounds in this region of the cortex has led to the definite picture of the projection of the retina on the visual cortex reproduced in Figure 23.11 from Gordon Holmes. In general the projection may be represented by picturing one half of the retina spread over the surface of one striate area, the macular region being placed posteriorly, the periphery anteriorly, the upper margin along its upper edge and the lower on its inferior border. The macula, as one might expect, is projected over an area out of all proportion to its size, just as with the motor representation of area 4, where the cortical centres concerned with the hand are much larger than those connected with movements of the proximal segments of the arm. More recent studies of head injuries by Spalding has largely confirmed the picture presented by Gordon Holmes.

'POINT-TO-POINT PROJECTION'

Anatomical and physiological investigation has con-

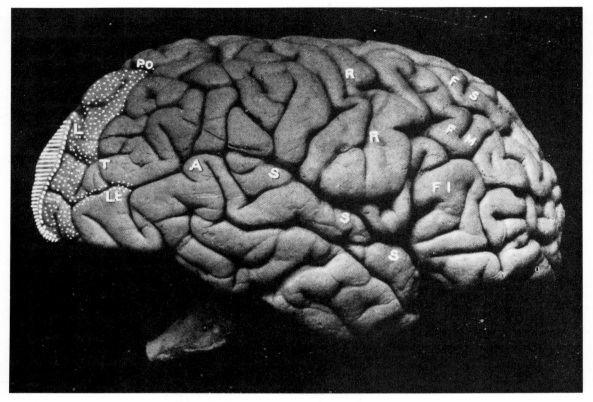

**Fig. 23.10**   Lateral view of the right hemisphere. Striate area marked with cross lines; the parastriate and peristriate areas by large and small dots respectively. PO, parieto-occipital; L, lunate; T, transverse occipital; Lt, lateral sulcus. A is placed on the angular gyrus, which surrounds the upturned end of the superior temporal sulcus. S, S, S is the lateral cerebral fissure (fissure of Sylvius). R, R is the central sulcus (fissure of Rolando). FS, FM, FI are placed on the posterior parts of the superior, middle and inferior frontal convolutions respectively; between them and the sulcus are the pre-central sulcus and convolution. (Whitnall, *Anatomy of the Human Orbit.*)

firmed this general plan of the retinal projection and has brought out the existence of a virtual 'point-to-point' relationship between retinal stimulus and cortical representation. For example, in the monkey, a definite projection map can be plotted by stimulating the retina with a point source of light and determining the exact locus of the electrical potential evoked in the cortex (Talbot and Marshall). From a strictly anatomical point of view this point-to-point projection, in the sense of a single cone being connected to a single bipolar cell, which is connected to a single ganglion cell, and so on, is not possible; in the retina the one-to-one relationship between cone and optic nerve fibre only holds good in the central fovea, and even here it is only a functional relationship since impulses from a single cone have subsidiary paths which permit a spread of their responses outwards to several ganglion cells. Similarly, as we have just seen, in the lateral geniculate body, although there are no internunciary neurones, each fibre of the tract ends on a number of geniculate cells, and a further

dispersion is possible in the cortex by intracortical association cells. The point-to-point relationship is therefore a functional one, in that there is a preferential path from one cone, say, to one cortical cell. This point-to-point relationship between retinal light-stimulus and cortical evoked response may be found, moreover, only under highly artificial conditions, involving deep anaesthesia. In the intact animal, with implanted electrodes in different regions of the cortex, it is found that the evoked response to discrete stimuli is by no means so highly localized. In a similar way, the more elaborate study of visual sensitivity in human subjects with lesions in their visual pathway has shown that portions of the visual field that are apparently 'spared', in the sense that they respond to light, are not necessarily completely normal (Battersby *et al.*, 1960).

CORTICAL MAGNIFICATION FACTOR

If the retina is projected on to the visual cortex in the way these studies suggest, we may speak of a *cortical magnification*

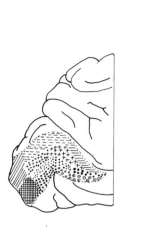
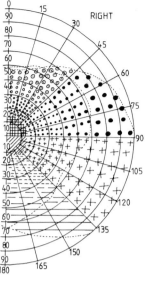

**Fig. 23.11** Projection of the retina on the visual cortex. On the left the striate area of the left hemisphere of the brain is shown, the calcarine fissure being opened up to reveal that portion of it that lines its walls. On the right is the right half of one visual field. The correspondence of the markings indicates the representation of different segments of the visual field on the cortex. (Gordon Holmes, *Proc. Roy. Soc.*)

*factor*; for example, if 1 mm of retina were represented by 3 mm of cortex, the magnification would be 3, or, since 1 mm of retina corresponds to a visual field of about 3.3°, the magnification would be about 1 mm/degree. Daniel and Whitteridge (1959) have extended Talbot and Marshall's studies on visual cortex of the monkey and baboon and have shown that the magnification decreases evenly from a value of 5.6 mm/degree at the fovea to about 0.1 mm/degree at a visual angle of 60°. Having measured the magnifications along different meridians of the visual field, Daniel and Whitteridge made a map of the striate cortex by applying these magnifications; it was obviously not the simple surface of a sphere, because of the distortion imposed by the variation in magnification. The area of the calculated surface came out at 1320 mm². If this was folded twice, once along the horizontal meridian for its anterior half, and vertically at the junction of the posterior and middle thirds, the resulting folded surface corresponded with the actual surface of the calcarine cortex in monkeys and baboons.

### GENICULATE MAPS

In a similar way Choudhury and Whitteridge (1965) have plotted a map of retinal projection on the rabbit's lateral geniculate body; in this animal there is a retinal region—the streak—in which the density of ganglion cells is very high, and it probably corresponds with the fovea of primates. This streak—in which the density of ganglion cells is very high, tion to other retinal areas. In the rabbit, a small number of optic nerve fibres project ipsilaterally, and it was found that the ipsilateral and contralateral projection areas overlapped, indicating the possibility of binocular interaction at this level in this non-laminated relay station.

## PROJECTION OF THE LATERAL GENICULATE BODIES ON CORTEX

### CORTICAL LESIONS

Lesions in the visual cortex are reflected in regions of degeneration in the lateral geniculate body of the same side; the smallest cortical lesion, leading to degeneration, always involves cells in all six layers of the geniculate body; the conducting unit of the optic radiation is therefore apparently six fibred, and impulses from corresponding regions of the six cell layers are brought into relation in the cortex; fusion of the temporal field of one eye with the nasal field of the other is therefore possible at this level.

### GENICULATE LESIONS

Wilson and Cragg (1967) examined the projection of the lateral geniculate body on the visual cortex by the location of degeneration in the cortex following lesions placed in the geniculate bodies. In the monkey they confirmed that the projection was confined to area 17, but in the cat degeneration extended to area 18 (Visual II); and this agrees with the finding of short-latency responses to light recorded from this area by Berkley, Wolf and Glickstein, responses that survived cutting of the corpus callosum and removal of area 17. A further projection was found on the suprasylvian gyrus. Lesions medial to the lateral geniculate body (pulvinar) caused some degeneration in area 19 (Visual III), and this may correlate with Suzuki and Kato's (1969) description of cells in the posterior thalamus of the cat responding to photic stimulation of the retina.

### PROJECTION TO CORTICAL LAMINAE

The specific sensory afferents ramify principally within layer IV of the sensory cortex, and the visual cortex is no exception; the extent to which other laminae are involved in direct projection is a matter on which there is not complete agreement, and there may well be species differences. The modern technique of injecting labelled leucine into a region of the brain, and observing its axonal transport to the projection areas of the neurones taking up the leucine, has enabled more accurate studies of projection. Thus LeVay and Gilbert (1976) injected into the main separate layers A, A₁ and C of the cat's geniculate body and examined the distribution of radioactivity in the visual cortex. With lamina A injections they found uptake in areas 17 and 18 confined to two bands, one occupying layer IV and the bottom of layer III and the other layer VI; injection of layer A₁ gave a similar pattern. No attempt was made to discriminate between the C layers; projection extended not only to areas 17 and 18 but also area 19 and the suprasylvian gyrus. In area 17 the main projection split into two bands, one at the IV/V border and one in the upper part of

IVab extending well into layer III; in addition, layer I received a clear projection but there was no projection to layer VI. Thus, according to this study in the cat, the visual projection is distributed more widely than had been suspected, with only the upper parts of layers II + III and the lower part of layer V completely free.

## Branched projection
Stone and Dreher (1973) found that antidromic activation of a given Y-cell in the lateral geniculate body of the cat could be achieved by stimulation of both area 17 and 18, indicating a branching of the Y-cell's axon. The slower-conducting X-cells apparently only projected to area 17.

## Feedback
Modern neurophysiology has brought out the reciprocal nature of the connections between any two stations in the central nervous system. Traditionally the lateral geniculate body has been regarded as a relay-station from retina to cerebral cortex, and, as such, doing little more than pass the visual messages on. However, there is some evidence that the message is modified, and this modification is effected not only by activity on the part of geniculate cells, but also by *corticofugal fibres*, i.e. by fibres originating in the occipital cortex which, by either excitatory or inhibitory activity, modify the responses of the geniculate cells to the retinal impulses (Guillery, 1967). By this feedback arrangement the cortex is able to modify and control the messages it receives. Thus Hull (1968) found that the responses of some geniculate neurones to photic stimulation of the retina were facilitated by reversible ablation of the cortex through cooling; others were inhibited.*

### The 'sparing of the macula'
The fact that lesions in the optic radiation, or in the visual cortex, frequently do not involve the macula (the 'sparing of the macula') has led many to suppose that fibres from the lateral geniculate body do not pass exclusively to the cortex of the same side, but that some macular fibres cross in the corpus callosum, or alternatively by way of mesencephalic connections, and therefore give rise to a bilateral cortical representation of the macula. In primates and man, however, complete destruction of one occipital lobe is followed by complete cellular atrophy of the homolateral geniculate body, hence the cells of this body project exclusively on to the occipital cortex of the same side. Moreover, such damage leads to loss of half the retinal fields including half of each macula (Spalding; LeGros Clark). The apparent sparing of the macula is probably essentially a result of its large cortical representation by comparison with that of the peripheral retina; moreover, the part of the cortex involved in this projection lies in the margins of distribution of the middle and posterior cerebral arteries so that, when one of these is occluded, a portion of it, at least, may receive sufficient blood from the other to remain functional. The possibility that macular sparing may be due to a crossing of macular fibres in the chiasma is ruled out by Gordon Holmes on the grounds that lesions in the optic tract cause a complete hemianopia extending into the fovea.

We shall see that the integration of the responses of the two eyes in stereoscopic binocular vision demands that some cortical cells of a given hemisphere receive impulses from the 'wrong side' of its appropriate retina; this could be achieved by the failure of all the fibres from the nasal half to cross, or by the crossing of some temporal fibres; and this may well happen in the cat, but in man such integration of the naso-temporal region of the retina is almost certainly brought about by neurones transmitting influences from one side of the cortex to the other in the corpus callosum (Hubel and Wiesel, 1967; Berluchi, Gazzaniga and Rizzolatti, 1967).

## PROJECTIONS FROM THE CORTICAL LAYERS

The pathway from cerebral cortex to lower visual centres was investigated by Gilbert and Kelly (1975), who injected horseradish peroxidase into different parts of the brain concerned with vision; this tracer is taken up by nerve endings and passes *retrogradely* to the cell-body and dendrites, where it is identified in the electron microscope. When injected into layers A and $A_1$ of the lateral geniculate body, the tracer was found in layer VI of the visual cortex, the labelled cells being pyramids. Areas 18 and 19 also contained the tracer, indicating cortico-geniculate pathways from these prestriate areas. Using the opposite technique of axonal transport of $^3$H-proline from cell body to terminals, Updyke (1975) described projections from areas 17, 18 and 19 to the various laminae of the cat's lateral geniculate body, the most dense projection being from area 17.

## THE PRESTRIATE CORTEX

Area 17, the striate cortex, is surrounded by cortex concerned with vision, and this was delineated by Brodmann on a morphological basis; this *prestriate cortex* was designated areas 18 and 19 or the *parastriate* and *peristriate areas* respectively, but there are no obvious cytoarchitectonic features to differentiate the two areas from each other. In order to determine how

---

* Inhibition of geniculate cells, following light stimuli, may also be achieved by recurrent activity on the part of the geniculate cells themselves. According to Vastola (1960), a geniculate neurone, activated by a ganglion cell axon, sends its message to the cerebral cortex but also, by a collateral branch, activates short-axon neurones in the lateral geniculate body which inhibit the activities of neighbouring geniculate neurones. The cells that are most strongly inhibited are those that receive the weakest excitatory stimuli from the optic tract. Such an inhibition of the weakly stimulated cells by the more strongly excited ones is of obvious value in sharpening up the retinal image. The lateral geniculate body thus continues a process that has already begun in the retina itself, and the projection from the cortex on the geniculate may represent a further continuation of the process.

these areas were interrelated, v. Bonin, Garol and McCulloch (1942) applied strychnine to localized portions of any one area of the monkey's cortex and studied the 'strychnine spikes' evoked in other regions.* Applied to area 17, there was remarkably little spread within this area; spikes appeared in area 18, but not in 19 and never in the contralateral hemisphere, so that the fusion of the two halves of the visual field in any one eye does not take place by direct interhemispheric connections between areas 17. When applied locally to area 18, the parastriate area, there was widespread activity within this area and also in areas 19 and 17. Moreover, activity spread to symmetrical loci in area 18 of the opposite lobe, and also to the temporal lobe (middle and inferior convolutions). Strychninization of area 19 caused only local activity and apparently caused a spreading depression so that spontaneous activity in large areas of the cortex was finally suppressed. Thus area 18, the parastriate area, stands out as the part of the cortex most closely concerned with integrating visual activity in the two hemispheres, and also with activity in the temporal lobe.

## PROJECTION IN THE CAT

More recent studies of projections to the cortex, and of intercortical projections, notably those between the striate and prestriate areas, have established four visual areas in the cat and have shown that, unlike the situation in primates, area 18 in the cat receives a direct geniculate projection. The projections may be summarized:

1. Dorsal lateral geniculate to areas 17 and 18 (Vision I and II).
2. Probably the medial interlaminar nucleus of the dorsal lateral geniculate to area 19 or Vision III.
3. Lateral geniculate nucleus plus pulvinar to lateral suprasylvian gyrus, or Vision IV.†

By employing the more precise methods of retrograde movement of horseradish peroxidase, Gilbert and Kelly, in the study already alluded to, were able to establish the interconnexions of the cat's visual cortical areas with more precision. These interconnexions seem to be more widespread; thus area 17 was said to project to areas 18, 19 and the Clare-Bishop area; area 18 projected to area 19 and the Clare-Bishop area, and area 19 projected to the Clare-Bishop area. Each of the projections to the Clare-Bishop area was reciprocal since this area projected back to areas 17, 18 and 19.

## TOPOGRAPHICAL PROJECTION FROM STRIATE AREA

The projection from the retina to the striate area is highly organized, so that a given point on the retina is represented by a definite region on the cortex (Fig. 23.11, p. 551). In a similar way, the study of Bilge et al.

(1967) on the cat indicated an orderly projection of re-representation of the retina on areas 18 and 19, as indicated by Figure 23.12. Of great interest is the circumstance that the vertical meridian, indicated by the heavy line marked 270°–90°, separates Visual I (area 17) from Visual II (area 18); it is essentially in the vertical meridian that integration between the two eyes is so important for stereopsis; thus this 'juxtastriate region' integrates the retinal points near the vertical meridians of both eyes by virtue of the callosal connexions that are absent in area 17 but present in area 18.

## PRESTRIATE PROJECTION IN THE MONKEY

The projections in the monkey have been examined in detail by Choudhury et al. (1965), who confirmed the importance of area 18 in the bilateral response to a visual stimulus; section of the corpus callosum, or cooling of the contralateral projection point on the cortex, extinguished the ipsilateral cortical response in animals with one optic tract cut. More detailed analyses of the primate are those of Meyers and Cragg (1969) and Zeki (e.g. 1969, 1971) based on lesions in the striate area and following degeneration by the Nauta method into the prestriate areas. As we have indicated, there is no clear definition cytoarchitectonically between areas 18 and 19 so that there is some ambiguity as to the limit between the two. As defined by Zeki (e.g. 1974) they are roughly illustrated by Figure 23.13. Cragg found projection from the striate area to the posterior wall of the lunate sulcus, but concluded that this did not extend to area 19, although this area received input from the pulvinar and from the corpus callosum, and he emphasized that area 19, as separately defined in the monkey and the cat, could not be considered homologous. Zeki (1969) also placed lesions in the striate area and showed that there are three areas of re-representation of the visual field: at the opercular lip,

---

* Strychnine applied locally to the cortex makes the underlying neurones highly sensitive to any afferent impulse; as a result, the neurone discharges, apparently spontaneously, and the occurrence of 'strychnine-spikes' at a distance from the region at which the strychnine was applied indicates the likely point of termination of these artificially activated neurones. Thus the appearance of spikes in area 18 and not in area 19, after strychninization of area 17, suggests that neurones of area 17 make direct connections with area 18 but not with area 19. It was considered earlier that the action of strychnine in the cortex was to abolish inhibition, by analogy with its action in the cord, but studies on cortical neurones rather oppose this view (Krnjević, Randic and Straughan, 1966).

† This cortical area in the cat has been given various names, such as Clare-Bishop area, visual area of suprasylvian sulcus (Kawamura, 1973). This author has described the intercortical projections relating to the visual areas of the cat in some detail and emphasizes the possible role of the middle suprasylvian gyrus in integrating somatic, auditory and visual sensory modalities.

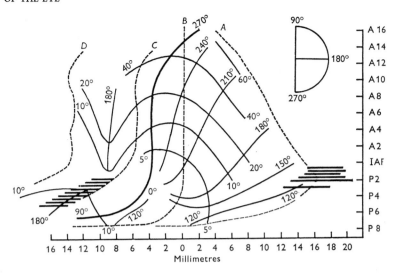

**Fig. 23.12**   An extended cortical surface to show the visual areas of the cat. A, The bottom of the splenial sulcus; B, the medial edge of the left hemisphere continuing as a section 1 mm in front of the posterior pole; C, the bottom of the post-lateral sulcus; D, the bottom of the lateral sulcus. The heavy line 270°–90° is the vertical meridian of the visual field and separates Visual I medially from Visual II laterally. Between C and D the 180° meridian separates Visual II from Visual III. Meridians are shown 30° apart. Incomplete semicircles of latitude are shown 5°, 10°, 20°, 40° and 60°. Heavy horizontal lines indicate areas in which folding of cortex makes reliable representation of the cortical surface impossible. I.A.P., interaural plane. Ordinates, Clarke-Horsley anterior and posterior planes; Abscissae, extended cortical surface in millimetres. (Bilge *et al.*, *J. Physiol.*)

in the lunate sulcus, and finally the posterior bank of the superior temporal sulcus. Thus any lesion made above the line of representation of the horizontal meridian in the striate cortex leads to a degeneration in dorsolateral prestriate cortex, and any lesion made below the horizontal meridian in the striate cortex leads to degeneration in ventrolateral prestriate cortex.

**Vertical meridian**

The scheme of projection is illustrated schematically by Figure 23.14 and anatomically by Figure 23.15. It will be seen that, as in the cat, the vertical meridian in prestriate cortex is represented by the junction between areas 17 and 18; similarly Zeki concluded that the vertical meridian was represented by another line delimiting area 19 from adjacent temporal cortex; as we should expect, these regions are interhemispherically connected.

**Foveal representation**

The foveal area appears to claim a large field of prestriate cortex extending into prelunate gyrus and the later third of the anterior bank of the lunate sulcus, and, in addition, extending ventrally into the inferior occipital gyrus. The large areas of re-representation in prestriate

areas suggests that there is some overlap between projections, so that a lesion in area 17 need not necessarily result in two distinct lesions in areas 18 and 19; and this was often found by Zeki (1969) whilst Cowey (1964) applying punctate stimuli to the retina, found prestriate receptive fields larger than those of the striate area and that adjacent points in the striate area projected to overlapping areas in prestriate cortex.

**Superior temporal sulcus**

In addition to the re-representations of the retina on areas 18 and 19, Zeki observed degeneration in the posterior bank of the superior temporal sulcus but the mode of projection was not clear; frequently lesions in the striate area producing lesions in this area were not accompanied by lesions in areas 18 and 19; there appeared to be no obvious topographical organization so that cortical neurones here might analyse features of the peripheral stimulus independently of the position in space, e.g. on the basis of its colour (p. 588).

TERMINOLOGY

Zeki compares his subdivisions of the prestriate area with those proposed by Myers (1965); Myers defined four areas, depending on whether they received fibres from striate cortex and from opposite prestriate cortex. His *juxtastriate 18*, the

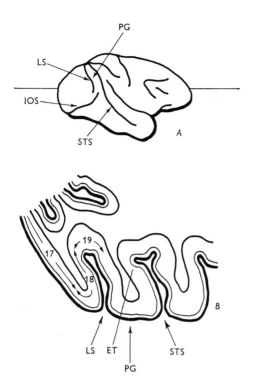

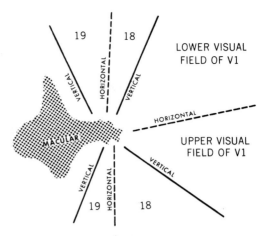

**Fig. 23.14** A schematic representation of prestriate cortex and its relation to striate cortex. The prestriate cortex may be completely divided into upper and lower sections, corresponding to the upper and lower retinal quadrants, by an imaginary prolongation of the line of representation of the horizontal meridian in the primary visual cortex (V1). The horizontal meridian re-representation in prestriate cortex is almost at right angles to its representation in striate cortex. This horizontal re-representation may be considered to form the boundary between the areas 18 and 19, as defined in this study. Note that the areas 18 and 19 of this study are within the lunate sulcus. Points further away dorsally (or ventrally) along the vertical meridian in striate cortex project to more dorsal (or ventral) points in prestriate cortex. Points further peripheral along the horizontal line project to more dorsal or ventral levels, depending on whether they are in the upper or lower visual field representation, the projection from the 2 fields of representation being separate. Not shown is the third vertical and horizontal re-representation in the superior temporal sulcus. (Zeki, *J. Physiol.*)

**Fig. 23.13** Tracing of a horizontal section through the brain of the rhesus monkey, at the level indicated in *A*. Only the posterior and lateral part of the section is shown in *B*, to indicate the position of the cortex of the posterior bank of the superior temporal sulcus relative to areas 17, 18 and 19 as these have been determined anatomically for regions of central visual field representation in the rhesus monkey. ET = electrode track; LS = lunate sulcus; IOS = inferior occipital sulcus; STS = superior temporal sulcus; PG = prelunate gyrus. (Zeki, *J. Physiol.*)

striate-prestriate border, receives striate and contralateral prestriate projections, and projects to the opposite juxtastriate region. In his scheme, the lunate sulcus receives fibres from striate cortex only, with no connexions with the opposite hemisphere, and forms his *striate receptive 19*. This is followed on the surface of the prelunate gyrus by *area 18 proper*, with connexions to the opposite area 18 proper and situated in the posterior part of the prelunate gyrus, and *area 19 proper*, with no connexions with the opposite hemisphere and situated in the anterior part of the prelunate sulcus. On Zeki's scheme, the lunate sulcus and inferior temporal sulcus are divided into at least two areas in which there is an orderly representation from vertical to horizontal and then back to vertical; the areas in which the midline is represented are connected with the opposite hemisphere. Thus the distinction between the two prestriate areas 18 and 19 is made on the basis of the re-representation of the visual fields and not on the basis of whether they are in connexion with the opposite hemisphere or with striate cortex only. Zeki also comments on terminology, suggesting that the terms 'secondary visual cortex' and 'peri-' and 'parastriate cortex' be abandoned. The terms OA and OB should be used when referring to the subdivisions proposed by Von Bonin and Bailey (1947) but

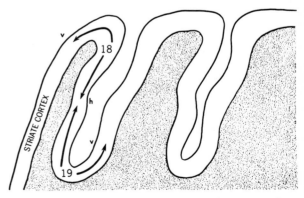

**Fig. 23.15** Diagrammatic representations of areas 18 and 19, as defined by Zeki, as they would appear in a horizontal cross-section passing through the lunate (or inferior occipital) sulcus. *v*, area of representation of the vertical meridian; *h*, area of representation of the horizontal meridian. (Zeki, *Brain Res.*)

should not be used as synonyms for areas 18 and 19 of Brodmann (1905). The term prestriate cortex may be used as a general term to indicate the cortical field receiving fibres from striate cortex. Areas 18 and 19 defined by Brodmann have been equated with two physiologically distinct areas V2 and V3 in the cat (Hubel and Wiesel, 1965) and it seems likely that the areas 18 and 19 defined by Zeki will have corresponding physiologically defined identities, but he emphasizes that these differ from those defined by Brodmann.

AREAS V4 AND V4A

The posterior bank of the superior temporal sulcus (Fig. 23.13, p. 555) receives a convergent projection from area 17, in the sense that lesions in widely distributed points in area 17 produce a single field of degeneration in this region. When examining the projections from areas 18 and 19 to neighbouring parts of the cortex, Zeki (1971b) found projections to the posterior bank of the superior temporal sulcus as though there were two routes from striate cortex to here, a direct and more complex. As a result of each projection, the topographical organization becomes blurred. In addition to this projection Zeki found two other areas indicated in Figure 23.16, which he named V4 and V4A rather than areas 20 and 21; in A the projection from the upper part of V2 and V3 (areas 18 and 19) is shown, and in B the projections from the lower part; owing to the changed folding of the cortex V4 and V4A appear as invaginations into V2 just beyond the lip of the operculum, due to the obliteration of the lunate sulcus and the opening up of the inferior occipital sulcus. In general, the organized topographic projection of the striate area to V2 and V3 breaks down, so that for example, the areas corresponding to horizontal and vertical meridians in V2 and V3 overlap in V4 and V4A.

RE-REPRESENTATIONS

Zeki's work emphasizes the complex series of steps in analysis of visual input that may take place through these 're-representations' of the visual field; the simple classical picture of projection from area 17 to area 18, from area 18 to area 19 and from area 19 to the so-called 'inferior temporal area' will have to be abandoned. In his study he found no projections from V2 and V3 (areas 18 and 19) to the 'inferior temporal areas', so that their involvement in visually directed activity must be through other connexions.

EFFECTS OF EXTIRPATION ON LEARNING

In the monkey, bilateral extirpation of areas 18 and 19 caused the irrecoverable loss of a learned habit. Thus Ades and Raab (1949) presented monkeys with two doors leading to chambers in one of which was placed a pellet of food. By placing a letter F on one door, which led to the compartment containing food, and an inverted leter F on the other, which did not contain food, the monkeys could be trained to distinguish between the upright and inverted F. In other words they were taught to discriminate visual form. After bilateral extirpation of areas 18 and 19 the habit was lost, but could be re-learned in about the same time as it was originally acquired. If one area only was removed, there was no loss; moreover, if the second area was removed after a delay of two weeks, there was also no loss. This indicates that the learned exercise had become bilaterally organized between operations, the initial excision having initiated some compensatory activity in other parts of the cortex that not only compensated for the loss of the cortex removed on the one side but 'in anticipation' for the removal on the other. A similar bilateral compensation for extirpation of the motor area has already been described.

**Temporal lobe**

If the temporal lobes were also removed, all possibility of learning, or re-learning, was lost although removal of the temporal lobes alone was without influence either on the learned habit nor yet on the power of the animal to acquire it. Thus the temporal lobes, in some way, are able to act for the visual 'association areas' in the learning of visual discriminations involving form. It is worth noting that none of the animals with temporal lobectomy showed the signs of psychic blindness described by Klüver and Bucy (1939); in other words, they were quite capable of selecting edible from inedible articles from a miscellaneous collection. Since Area 17, the striate area, is not directly connected with the temporal lobe, the learning in the absence of Areas 18 and 19 is presumably brought about by the mediation of subcortical links, e.g. by way of corticofugal fibres from the striate area to the pulvinar of the thalamus, and thence from the pulvinar to the temporal cortex; at any rate Jasper, Ajmone-Marsan and Stoll (1952) have shown that cortical stimulation leads to evoked activity in the pulvinar and also the lateral geniculate body.

When visual input to one side of the brain was cut off, by section of the left optic tract, then learning could be impaired by removing the right temporal lobe; further impairment could be achieved by section of the corpus callosum (Ettlinger, 1958). It would appear, then, that co-operation between the visual cortex of one side and the temporal lobe of the other, in the performance of learned tasks, may take place through the corpus callosum, but is not adequate to replace the co-operation between visual and temporal areas of the same side.

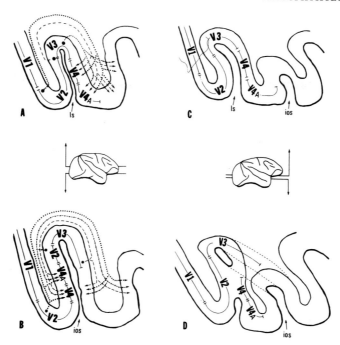

**Fig. 23.16**  Projections from areas 18 (V2) and 19 (V3) revealed by Zeki's microelectrode study. The projection pattern from upper prestriate V2 and V3 is shown in A and that from lower prestriate V2 and V3 is shown in B. All parts of areas V2 and V3 reported here appear to project to Visual 4 (V4) and Visual 4a (V4a). V4 and V4a of lower prestriate cortex appear as invaginations into lower V2, just beyond the lip of the operculum. Sections C and D show the probable steps by which such a transformation in the position of V4 and V4a occurs, based upon evidence presented in this and previous studies. The obliteration of the lunate sulcus and the opening up of the inferior occipital sulcus bring about this change. The dotted lines in section D show the position along which the inferior occipital sulcus will open up. With such an opening and with the obliteration of the lunate sulcus, V4 and V4a appear in the posterior bank of the inferior occipital sulcus, as may be seen in section B. Beyond the projection to V4 and V4a, areas V2 and V3 send a weak projection to the anterior part of the prelunate gyrus (from upper V2 and V3—section A) and the anterior bank of the inferior occipital sulcus (from lower V2 and V3—section B). The change in the position of this field from the anterior part of the prelunate gyrus to the anterior part of the inferior occipital sulcus occurs because of the changes in the sulcal and gyral patterns from dorsal to ventral (see text for further details). The regions of areas V2 and V3 studied also project to the cortex of the posterior bank of the superior temporal sulcus. Not shown are the following connections: local connections between V2 and V3 and within V2 and V3 and the connections between upper and lower parts of V3. (Zeki, *Brain Res.*)

## CORPUS CALLOSUM

This is an interhemispherical tract that provides connections between a given point on one hemisphere with a symmetrical point on the other. By stimulating discrete points on the cortex, Curtis (1940) recorded electrical responses on symmetrical points of the cortex of the other hemisphere. In primates, the only area that did not give a response was Area 17, showing the probable absence of interhemispherical connections between these two areas, and the anatomical studies of Myers (1962) and Ebner and Myers (1965) leave no doubt that it is at the junction of Areas 18 and 17 that the great majority of the posterior callosal fibres project.* In the cat Choudhury, Whitteridge and Wilson (1965) found that the receptive fields of cortical units in the medial edge of Area 18, just adjacent to Area 17, were within a few degrees of the vertical meridian; these responses were recorded when the optic tract of the same side was cut, so that a callosal route from the opposite hemisphere was probably involved, and this was confirmed by cooling the corresponding point on the opposite hemisphere or by complete section of the callosum.

## CALLOSAL NEURONES

Again, Berlucchi, Gazzaniga and Rizzolatti (1967) inserted electrodes into the posterior callosum of cats and found axons that gave discharges when the eyes were stimulated with patterns corresponding to those described by Hubel and Wiesel, so that they obtained simple, complex and hypercomplex units (p. 566). In general, it was only when the stimulus was presented close to the vertical meridian that callosal units were excited; three out of seventeen units were driven by both eyes by stimuli located in the same part of the visual field; nine were driven by one eye, and with the remaining five a clear response was obtained with one eye and a weak one with the other. Thus, as Berlucchi *et al.* say, the representation of the visual world is on a continuum, the neurones associated with the vertical meridian and their callosal connections being the hyphen necessary to bring together the two half-fields. Thus the cortical cells connected in this way are at the junction of Areas 17 and 18 and it is here that the vertical meridian projects (Choudhury, *et al.*, 1965; Bilge *et al.*, 1967), whilst it is here, too, that the great majority of the callosal fibres project (Ebner and Myers, 1965).

### BINOCULAR FIELDS

When the chiasma of the cat was split, so that the visual input from a given eye was entirely homolateral, Berlucchi and Rizzolatti (1968) found 9/70 units driven by both eyes in the visual cortex at the boundary of Areas 17 and 18; both receptive fields were very similar and covered the vertical meridian. Thus the representation of the visual fields of the two eyes may be indicated by Figure 23.17; the central 20° is bilaterally represented through callosal connections; the remaining parts of the binocular field, making up a total of 120°, are represented by cortical neurones driven by both eyes, but a given visual stimulus in this field will only drive cortical cells in one hemisphere. Fields of 80° on each side are uniocular, whilst a final 80° is blind.

## CALLOSAL TRANSFER OF VISUAL HABIT

If a cat is placed in a box from which there are two exits, it may be trained to discriminate between two signs, for

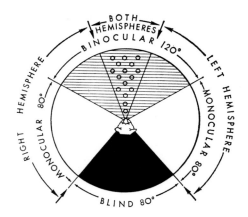

**Fig. 23.17** The visual field of the cat and its relationship with the eyes and the cerebral hemispheres. The visual callosal mechanisms allow both hemispheres to receive projections from an area of the visual field extending 20° on either side of the vertical meridian. The extension of the bihemispheric region of the visual field is inferred from earlier results of Berlucchi *et al.* (1967) showing that some visual receptive fields of callosal units are as large as 20°. (Berlucchi and Rizzolatti, *Science*.)

example between a vertical and a horizontal bar, if these are placed over the respective exits and the animal is rewarded with food when choosing the one exit and not when choosing the other. If one eye is covered while the animal is trained in this pattern-discrimination, it is found that when the animal is made to use the other eye for the discrimination it succeeds as well as if both eyes had been open during the training. In other words, we may say that there has been transfer of the learned habit from one eye to the other; if the left eye is trained the right eye can also execute the task, and *vice versa*. If the optic chiasma is sectioned by a sagittal incision, so that all visual impulses from one eye are only carried to one hemisphere, it is found that the animal is still able to make the transfer, in the sense that the task learnt with the left eye can still be executed by the right eye. Finally, if both the chiasma and corpus callosum are separated by sagittal incisions, the animal must now learn this habit separately with each eye.

### SPLIT-BRAIN CAT

Thus, one cat, operated on in this way, required

---

* Myers (1962) examined the degeneration of fibres in the opposite cortex when the occipital lobe of one side was removed; in addition to area 18, areas 7 and 23 of the parietal lobe and Area 20 of the temporal lobe, in the region where they border Area 19, showed degeneration; thus, the striate cortex, receiving no afferents from the corpus callosum, is completely surrounded by cortex that does. We may note that Area 19 was free of degeneration as well as Area 17; in the cat, Ebner and Myers (1965) found essentially the same picture, earlier claims of callosal connections between Areas 19 being due to the inclusion of part of Area 18 in Area 19.

respectively 80, 170 and 850 trials to learn three different types of visual discrimination with the left eye, whilst on covering this and testing it with the other eye open it required 70, 160 and 830 trials in order to relearn the discriminations. Furthermore, the animals could be taught to make contradictory discriminations with the two eyes; for example, the left eye could be trained to respond positively to the vertical bar whilst the right eye could be trained to respond negatively to this, and positively to the horizontal bar. Hence its choice of exit depended on which eye it was using (Sperry, Stamm and Miner).

However, if the split-brain cats were trained using a shock-avoidance technique instead of a food-reward one, pattern discrimination was successfully transferred, and it was suggested that this was made possible by the meeting of visual and pain stimuli in the superior colliculi (Sechzer, 1963). In these experiments the cats were discriminating patterns; when the simpler discrimination of light-intensity was required, it was found that interocular transfer occurred in the presence of split chiasma and corpus callosum (Meikle and Sechzer, 1960), and only when additional commissures were cut, namely the anterior and posterior, the massa intermedia, habenular, hippocampal and superior collicular commissures, was the interocular transfer prevented (Meikle, 1964). Finally we may train a cat to respond positively to two sets of lines with their directions at right-angles, i.e. to ||| ≡, and negatively to two sets with their directions parallel, i.e. to ||| |||; if the stimuli are presented in such a way that one eye receives the vertical, and the other the horizontal, lines of the positive stimulus, then if the chiasma and the dorsal two-thirds of the corpus callosum are split, the power to discriminate is grossly impaired; only if the rostral third of the callosum is also cut does the performance degenerate to chance level (Voneida and Robinson, 1970).

In monkeys, too, with their optic chiasma sectioned, there is interhemispheral transfer of learned visual discrimination, which is abolished by section of the commissures (Downer 1959). This author studied the 'handedness' of his experimental monkeys under these conditions and showed that, according to which eye was open, they employed the contralateral arm, i.e. the arm controlled by that side of the brain to which the visual information was restricted by the section of the chiasma. If they were prevented from doing this, by binding the arm, then the animal groped with its free arm as though it were blind. Unlike cats, split-brain monkeys could not make an interocular transfer of a discrimination based on brightness (Hamilton and Gazzaniga, 1964).★

## Mirror-image stimuli

A very revealing study on transfer of learned habit in the monkey is that of Noble (1966) who used as positive and negative stimuli objects that were not bilaterally symmetrical; thus the usual stimuli would be + and ○ or ||| and ≡, the choice of one being rewarded and the other not. When a split-chiasma animal was trained to react positively to an object and negatively to its mirror-image, e.g. to ⌐ and ⌐, with one eye occluded, it was found that when the animal was tested for transfer to the opposite eye it consistently chose the negative stimulus. Thus, if it had been trained to react positively to ⌐ with its right eye, then it reacted positively to ⌐ with its left eye. Noble considered that this was to be expected on the basis of a mirror-image reversal of the projection of the retinal image on the opposite cortex through Area 18 and the corpus callosum (Fig. 23.18).

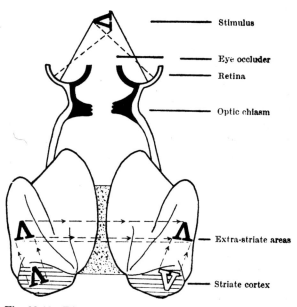

**Fig. 23.18**  Diagram to explain mirror-image reversal. With right eye occluded, the input to left striate cortex is propagated to extrastriate areas (Λ) and is laterally reversed in crossing the midline via the forebrain commissures (stippled). When the left eye is occluded the input to the right striate area (⩔) does not match the stimulus (Λ) transposed from the left hemisphere; it does, however, match its mirror-image. (Noble, *Nature*.)

## Mechanism of transfer

Berlucchi and Rizzolatti asked whether the learned process was coded and transferred to the opposite hemisphere, or whether the information necessary for the learning process was received in both hemispheres, through the corpus callosum if necessary, and the

---

★ The normal rabbit behaves similarly to the split-brain split-chiasma cat so far as inability to transfer pattern discrimination is concerned; there is no doubt, however, that the two visual areas are linked by callosal connections (Van Hof, 1970).

information was subsequently processed into memory patterns separately in the two halves of the brain. Certainly their studies on information from the vertical meridian of the visual field indicates that primary information passes virtually simultaneously to the two hemispheres.

## PHYLOGENETIC SIGNIFICANCE OF VISUAL AREAS

The phylogenetic significance of the striate and associated cortical areas has been well summarized by Holmes:

'In the evolution of the brain, vision is the first sense organ to obtain a representation in the cortex of the forebrain. Even in birds, fibres carrying retinal impulses reach what is regarded as the neopallium; it is only later in evolution that other sense organs attain cortical connections and it is still later that so-called 'motor centres' appear. With vision, too, localization in the cortex reached its acme, for every point of the retinae is represented rigidly in it; then the brain, if we may speak teleologically, decided that such a rigid machine was unsuitable to its further development and adopted a more plastic organization for its later-evolved functions. It is consequently not surprising that cortical visual reactions are more highly organized, and possibly organized on a simpler plan, than other cerebral activities. Even in birds, the original visual cortex is connected with the roof of the mid-brain by efferent fibres—the *tractus occipito-mesencephalicus* of Kappers —which are probably the homologue of the occipito-mesencephalic fibres of mammals, and by means of these the cortex may be able to influence reactions, including ocular movements and postures, in response to retinal stimuli. As the cortex became the main, and finally the exclusive organ for visual perceptions, the primary visual reactions—as reflex direction of the eyes to light, accommodation, fusion when binocular vision was acquired, and fixation—were transferred to it till in man they can be evoked through it only. The next step in development was the appearance of associational areas around the visual cortex which enable it to co-operate with other sense impressions and to elaborate further the faculties of spatial perception, discrimination, and recognition.'

## LOWER VISUAL CENTRES

In man removal of the striate areas causes complete blindness; thus awareness of light, and the ability to respond to it by motor activity,* are cortical functions.

In lower animals, however, this is not true; for example, in reptiles and birds vision is barely affected by removal of the cortex, so that a pigeon will fly and avoid obstacles as well as a normal one. In rodents, such as the rabbit, removal of the occipital lobes causes some impairment of vision, but the animal can avoid obstacles when running, recognize food by sight, and so on.

### MONKEYS

In monkeys it was thought for many decades that the animal was totally and permanently blind after bilateral occipital lobectomy, and, as Klüver pointed out, this conclusion was easily reached from casual observation of the animal provided it was in unfamiliar surroundings; under these conditions the monkey bumps into walls or objects in the room and makes no attempt to avoid obstacles. It does not seem to be able to find food or other objects except by sense of touch, and it sits for hours in its cage with the door wide open. By contrast, when the monkey is in familiar surroundings its movements are so quick and efficiently performed that on casual observation it passes for normal, and it requires careful experimental study to assess the degree of impairment of visual function. Such a monkey is said to exhibit an optokinetic nystagmus some months after the operation (Pasik and Pasik, 1964), and it may be trained to respond to light and to changes in the total luminous flux entering the eye; up till recently, it was considered that this was the total basis of its visual discrimination. However, Humphrey and Weiskrantz (1967) have shown that, under the proper conditions, a monkey with bilateral striate removal can not only detect objects but locate these accurately in space, i.e. is able to discriminate on the basis of position as well as luminous flux. Thus they found that, in order that monkeys could be made to notice an object, e.g. a cube, it had to be moved, either from side to side or by rotation; once this condition for notice had been fulfilled, the monkey could be trained easily to grasp the object accurately when it appeared at a given point in the visual field, the accuracy being greatest when it was in the central field and poor at 40° of eccentricity. Eventually it became difficult to hold an object still enough for the monkey not to perceive the natural tremor of the experimenter's hand, and finally, after a longer period, stationary objects could be noticed and the monkey trained to localize them accurately.

In the dark, a stationary luminous object was noticed, but in ordinary illumination a neon light, for example, was not, unless it was flashed repetitively; at the

---

* Except, of course, the pupillary response to light, mediated by optic tract fibres that pass to the pretectal nucleus.

frequency required for human fusion the monkey failed to notice the light.*

Subsequent work on the monkey by Pasik and his colleagues has emphasized the ability of the monkey, with striate area removed, to discriminate form as well as differences in light-intensity (Pasik *et al.*, 1969; Schilder *et al.*, 1971); and it was suggested that the more severe deficits, described by Klüver (1942) were due to removal of more cortex than that in Area 17.

### SUBCORTICAL MECHANISMS

As to the mechanism of these discriminations, it might be argued that, because the geniculate projection—at any rate in primates—is exclusively to Area 17, the pathway would be subcortical, for example, by way of the superior colliculus or pretectum. As indicated earlier, however, both these subcortical centres have a projection to the cerebral cortex through the thalamus, so that a purely subcortical mechanism need not be invoked. In the past there has been a tendency to regard the superior colliculus as a purely motor centre, but Sprague and Meikle (1965), on the basis of their studies of lesions in the colliculus and in the afferent and efferent pathways, argue that the body must be regarded as an integrating centre rather than as a simple relay station for cortically directed eye movements.

Before reviewing in some detail the modern studies on the superior colliculus, which involve recording of unit behaviour in its different layers, we must consider the basic electrophysiology of the higher stages in the primary visual pathway, namely the lateral geniculate body and striate and prestriate areas.

## ELECTROPHYSIOLOGY OF THE HIGHER VISUAL PATHWAY

### GANGLION CELLS

#### CENTRE-SURROUND ORGANIZATION

The organization of the receptive fields of the ganglion cells of the retina has been described earlier; to recapitulate, the basic organization is circular, with a central spot and a surrounding annulus responding in opposite fashions; thus we may have a field in which an ON-response occurs when the light stimulus is projected onto this region of retina whilst the surrounding annulus gives an OFF-response, i.e. switching the light on does not excite, in fact it causes inhibition of the background spontaneous discharge if this is present. On switching the light off, there is a prominent OFF-discharge. This unit would be called an ON-centre unit. Figure 23.19, A and B, illustrates both ON- and OFF-centre fields. Quite clearly, then, by the time the responses of the cones or rods have been transmitted through the ganglion cells there has been a considerable elaboration, of which a prominent feature is inhibition. We have seen how this type of unit, acting in conjunction with others, can signal movement and even the specific direction of the movement; in addition, too, the signals of many of the units have become colour-coded in a

---

* Pasik, Pasik and Schilder (1969) have shown that monkeys, with striate cortex removed, can discriminate different sized figures illuminated in such a way that the total light-flux entering the eye is the same; they thus apparently were able to discriminate on the basis of size.

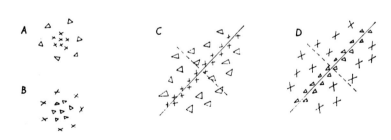

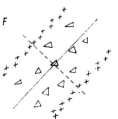

**Fig. 23.19** Types of receptive field. Crosses indicate excitatory or ON-responses; triangles indicate inhibitory or OFF-responses. A: ON-centre response typically seen in retina and lateral geniculate body. B: OFF-centre response. C-G: Various arrangements of 'simple' cortical fields. (Hubel and Wiesel, *J. Physiol.*)

different manner from the coding found in the receptors themselves, so that a given 'dominator-type' for example, can indicate by its frequency of discharge the prevailing luminosity independently of the colour of the light, whilst others, modulators, will respond only to a relatively narrow band of wavelengths.

## GENICULATE NEURONES

### COMPARISON WITH OPTIC TRACT FIBRE

A study of lateral geniculate neurones, e.g. by Hubel and Weisel (1961), has revealed essentially the same basic organization of receptive fields, characterized by ON-centre—OFF-surround, or *vice versa*. This does not represent a mere copy of the ganglion cell input, however, since a given geniculate cell will have been influenced by many hundreds of ganglion cells excited within the receptive field; moreover, there is considerable accentuation of the inhibitory actions of the centre and surround of any given field. This is illustrated by Figure 23.20 where the microelectrode was recording from an optic tract fibre simultaneously with a geniculate cell. A shows the responses to both units when a small

light-spot falls on the centre of the receptive field; the optic tract fibre, being an ON-centre one, gives a burst of spikes which are distinguishable from those of the geniculate fibre by their being monophasic and small. The response of the geniculate fibre appears at OFF as large diphasic spikes. Increasing the size of spot, so that the stimulus now falls on the surrounding annulus of opposite response characteristics, has almost completely inhibited the geniculate response, but had only slight effect on the tract fibre (B). In C the spot has been made much larger, and the geniculate response is virtually abolished and the tract response is reduced. In D only the surround has been stimulated, and now the large diphasic spikes appear first at ON, and the small tract spikes at OFF. Thus the geniculate cells differ quantitatively in the degree of 'surround antagonism', to such an extent indeed does this occur that diffuse illumination, because it stimulates both centre and surround of the receptive field, has little effect on geniculate cells.

### VALUE OF INHIBITION

Thus, by application of inhibition, Nature has reduced the tremendous bombardment of the visual cortex

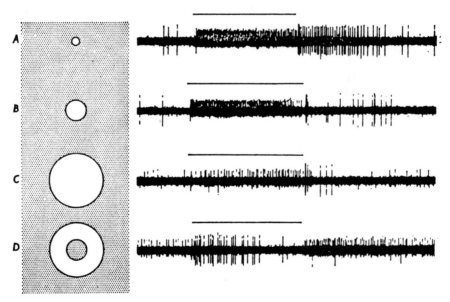

**Fig. 23.20**  Simultaneous recording from optic tract fibre and from geniculate neurone; the optic tract fibre gives small monophasic spikes.
A. The small light-stimulus falls on the centre of the receptive field; the optic tract fibre is excited at ON (it is ON-centre) whilst the geniculate one is excited at OFF (it is OFF-centre).
B. The larger stimulus, covering both central region and surrounding region of opposite characteristics, causes almost complete inhibition of the geniculate response.
C. The stimulus is larger and has now reduced the optic tract response as well.
D. Only the surround is stimulated, and the geniculate neurone has fired at ON and the tract neurone at OFF. Horizontal lines above records indicate duration of light stimulus. (Hubel and Wiesel, *J. Physiol.*)

that would occur continuously as long as the eyes were open in the light; instead, many neurones remain silent and are ready to respond to what is important in the environment, namely the contrast involved in contours and the effects of a moving stimulus, be it light or dark, which, passing across the concentric receptive fields, has a definite effect.

### SUSTAINED AND TRANSIENT TYPES

As with the ganglion cells, the geniculate units can be separated into two main types, sustained or X-, and transient or Y-cells. Thus Figure 11.7, page 288, may be compared with Figure 23.21, both records being obtained with a spot stimulus at the centre of the receptive field. Simultaneous recording from ganglion and geniculate cells showed that there was a functional segregation, so that a given sustained geniculate cell received input from one or more sustained ganglion cells. The similarity extended to conduction-velocities, so that fast-conducting axons from ganglion cells synapsed with fast-conducting geniculate cells (Cleland et al., 1971); this is understandable, since Y-axons conduct more rapidly than the X-axons.*

### CLELAND CLASSIFICATION

When Cleland et al. (1976) applied their more elaborate classification to ganglion cells, in an examination of input to lateral geniculate neurones, they were able to extend this classification to the geniculate neurones, identifying not only the sustained and transient (X- and Y-) categories but also more rarely encountered sluggish sustained and sluggish transient neurones. As the scatter-diagram of Figure 23.22 indicates, there is a fair correlation between the cortico-geniculate antidromic latency and the retino-geniculate latency, suggesting that relay to the cortex is carried out by geniculate neurones having the same features as those of the ganglion cells that transmit to them. There were some exceptions, but these could have been due to mixed inputs from retina to geniculate neurone. Finally, we may note that Wilson et al. (1976), retaining their W-X-Y-classification, derived the input relations illustrated by Figure 23.23.

---

* Stone and Fukuda (1974) have described the central projections of different types of retinal ganglion cells, including the W-cells first described by Stone and Hoffmann (1972).

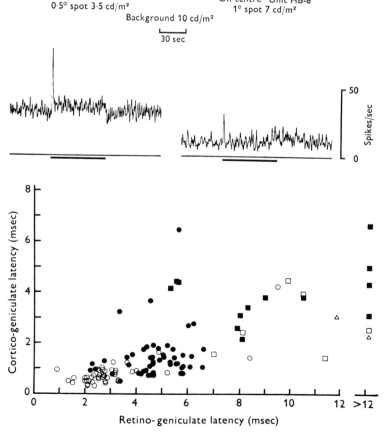

**Fig. 23.21** Sustained and transient responses to standing contrast at the geniculate level. In each case the centred spots of added light were about 10 × threshold for the respective cells, and fell within the limits of the central zones. On the left there was a definite though small sustained response for the duration of the stimulus (black line below). With the transient cell on the right, there was no combination of area or luminance of the spot, including that illustrated, which could produce a sustained response. (Cleland et al., J. Physiol.)

**Fig. 23.22** Scatter diagram of cortico-geniculate latency (antidromic) against retino-geniculate latency for 115 LGN cells. Symbols indicate class of cell: ○ 'brisk-transient'; ● 'brisk-sustained'; □ 'sluggish-transient'; ■ 'sluggish-sustained'; △ 'local-edge-detector'. (Cleland et al., J. Physiol.)

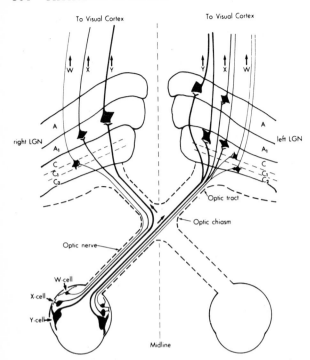

**Fig. 23.23** Schematic diagram of the relay of W-, X-, and Y-cells through the LGN. Many of the features represented are derived from earlier studies. Two new features are shown. First, retinal W-cells relay to the visual cortex via cells in laminae C, $C_1$, and $C_2$. Second, each type of retinal cell relays through one lamina of the ipsilateral LGN ($A_1$ or $C_1$) and through two laminae of the contralateral LGN (X- and Y-cells via laminae A and C, W-cells through C and $C_2$). (Wilson et al., *J. Neurophysiol.*)

LAMINATION

The types of neurone were not regularly distributed in the laminae of the lateral geniculate body, so that whereas brisk-sustained units were found in A and $A_1$, they were poorly represented in C and $C_1$, and so on.

SPATIAL ORGANIZATION OF THE GENICULATE FIELD

A simple model for the production of the geniculate cell's centre-surround field is that suggested by Singer and Creutzfeld (1970) by which an ON-centre geniculate cell receives input from the centre of a retinal ON-centre unit, and its OFF-surround receives input from several OFF-centre ganglion cells. On this basis it could be predicted that the receptive field of an ON-centre geniculate cell would have an annular zone—or outer surround—beyond the inhibitory surround formed from the antagonistic surrounds of retinal ganglion cell inputs whose centres constitute the inhibitory surround (Fig. 23.24). Hammond (1972) projected annuli of larger and larger radii centred on a

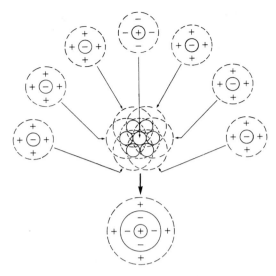

**Fig. 23.24** Exploded diagram of a model for convergence of retinal fibre inputs on to a single LGN cell. The resultant geniculate receptive field (lower), consisting of centre, surround and outer surround, takes account of the receptive field centres *and* surrounds of the constituent retinal inputs (top), spatially juxtaposed on the retina as in the composite diagram (centre). The model is equally applicable to OFF-centre geniculate cells. (Hammond, *J. Physiol.*)

geniculate cell's receptive field; as Figure 23.25 shows, the response decreased and eventually became inhibitory, and finally there was a region of weak disinhibition. With optic nerve fibres, this disinhibitory surround was not observed.* As Hammond pointed out, the model of the geniculate cell field illustrated by Figure 23.24 implies a basic difference between the centres of ganglion and geniculate fields since, if Rodieck and Stone are correct (p. 285), the surround field of the ganglion cell extends into the centre of its field, the response being the algebraic sum of the opposing excitatory and inhibitory responses; by contrast, the geniculate surround need not produce a significant influence on the centre and, if it does, it is in the same sense as that exerted by the centre.

**Analysis of contrast**

As Maffei and Fiorentini (1972) have emphasized, the absence of a simple relay of centre and surround from ganglion to geniculate cell means that the analysis of contrast, which the centre-surround organization favours, is not carried out at the ganglion cell level, but at the geniculate. These authors showed that the

---

* Stevens and Gerstein (1976) have questioned this model and, on the basis of a spatio-temporal analysis of geniculate cell discharges, and the relations between them, have proposed a model based on inhibitory interneuronal action in the lateral geniculate nucleus.

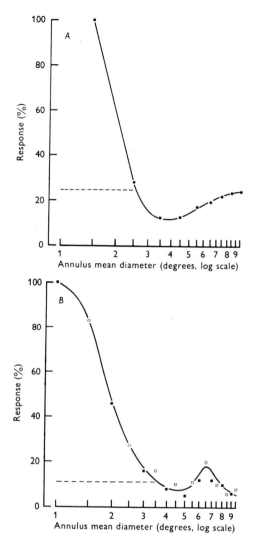

**Fig. 23.25**   Response variation as a function of mean diameter of the stimulating annulus. Comparison between optic tract fibre (A) and LGN cell (B) with approximately the same field centre size and eccentricity. Stimulus wavelength was 525 nm presented against an achromatic adapting field. Mean spike counts are presented as percentage of maximum. (■ increasing diameters; □ decreasing diameters.) Equivalent spontaneous firing levels are indicated by dashed lines. (Hammond, *J. Physiol.*)

to movement of a spot of light in the 'preferred' direction and showing inhibition of spontaneous discharge in the 'null' direction (p. 285). In other species, this phenomenon cannot be detected in the retina, and it is in the lateral geniculate body, or superior colliculus, that such directional-sensitivity is elaborated. In the rat lateral geniculate body, for example, Montero and Brugge (1969) have described neurones with essentially the features described by Barlow *et al.* (1964) for retinal neurones; they showed that the basis of this discrimination was both inhibition of adjacent receptors in the null direction and facilitation in the preferred direction. About 80 per cent of the cells were sensitive to up-and-down movement, and most were concentrated in the area of central vision.*

## VISUAL CORTICAL NEURONES

### LINEAR FIELDS

When recordings are taken from neurones in the striate cortex, an entirely different organization of receptive fields appears, the dominant feature being the linear organization of the ON- and OFF-regions, as opposed to the concentric circular organization of the lower neurones. The types of field that are described as 'simple' are illustrated in Figure 23.19 C to G, A and B being the geniculate fields described earlier. Here the ON or excitatory responses are indicated by crosses and the OFF by triangles. The receptive field represented by C consists of a central slit of retina orientated in a definite direction; light falling on this slit, or a spot of light moving along it, produces ON-responses; light falling on the narrow areas to each side gives OFF-responses, whilst light falling on both central slit and the parallel surrounds leads to complete cancellation, so that diffuse illumination of the retina produces no response in these cortical neurones. Summation occurs along the slits, in the sense that a slit of half the length gives a smaller response than one occupying the whole length of the field.

---

* Barlow and Brindley (1963) studied the effects of repeatedly stimulating a direction-sensitive ganglion cell by moving a spot of light over the retina in the preferred direction; before stimulation, there was a maintained discharge of 7/min; the discharge rose to 60/min and finally settled down to 25/min. When the stimulation ceased, the discharge fell to zero and gradually rose, over a period of a minute, to the steady level of 7/min. They consider that this fall to zero and slow return of the background discharge give rise to the illusion of movement, the normal balance of steady discharges between ganglion cells of opposite directions of preferred movement being upset.

surround-sensitivity of the geniculate cell, unlike that of the ganglion cell (Barlow *et al.*, 1957) did not disappear during dark-adaptation, and thus the surround of the geniculate cell was unlikely to be the relayed surround of the ganglion cell.

### DIRECTIONAL SENSITIVITY

In many species, e.g. the frog and rabbit, retinal ganglion cells show directional specificity, responding

## ORIENTATION SELECTIVITY

Illumination of both flanks simultaneously produced a larger response than separate illumination of either. The necessary orientation of the slit stimulus to obtain maximal response was quite critical, so that a variation of more than 5 to 10° had a measurable effect. We may assume that these cortical neurones receive their inputs from a number of geniculate cells, each with a circular field, and that the effect of this combination is to give a linear array of these fields that would behave effectively as a unit with a linear receptive field as illustrated in Figure 23.26. By appropriately arranging geniculate cells the different receptive fields of other types of 'simple' cortical neurones, illustrated in Figure 23.19, E to G, may be constructed.

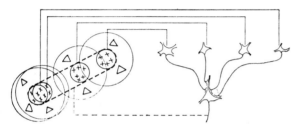

**Fig. 23.26**  Scheme to explain the organization of the linear cortical receptive field on the basis of summation of circular fields at the geniculate level. (Hubel and Wiesel, *J. Physiol.*)

To return to Figure 23.19, D illustrates the mirror-image of field C, the outer flanks being excitatory and the inner slit inhibitory. In both C and D the flanks of the field were more diffuse than the central regions. F illustrates a cell with a more diffuse centre and narrow flanks, and the optimal response was obtained by simultaneously illuminating the two flanks with parallel slits. In G, only two regions could be distinguished; the most effective stimulus, here, was two areas of different luminance placed so that the line separating them fell exactly over the boundary of the fields; this was called by Hubel and Wiesel an 'edge stimulus'. With all these 'simple' units a moving stimulus was always very effective, but it is important that the correct orientation of the slit be maintained, the movement being executed at right-angles to the direction of the slit. Usually the response was the same with forward or backward movement, but in others the responses were very unequal, due to an asymmetry in the flanking regions; thus with unit G the difference was very marked.

## SIMPLE UNITS

The fields of these units were described as *simple* because, like retinal and geniculate fields, they were subdivided into distinct excitatory and inhibitory regions; there was summation within the separate excitatory and inhibitory parts; there was antagonism between excitatory and inhibitory regions; and finally because it was possible to predict responses to stationary or moving stimuli of varying shapes from a map of the excitatory areas.

## COMPLEX UNITS

Another class of units showed similarities with the simple ones, but they were called *complex* because their responses to variously shaped stationary or moving forms could not be predicted from a map made with circular spot-stimuli. In general, the fields were larger than those of simple cells, responding poorly to flashes as opposed to movement. The activating region showed no subdivision into spatially separate zones and, when a cell responded to flashes, these elicited responses at both ON and OFF throughout the field. Similarly, a response was obtained to both a bright and dark edge of a moving stimulus. The cells were more often driven binocularly, had greater spontaneous activity, and responded to higher velocities of movement. Hubel and Wiesel suggested that these complex units received input from several simple units, thereby exhibiting a hierarchy of cortical cells (Fig. 23.27).

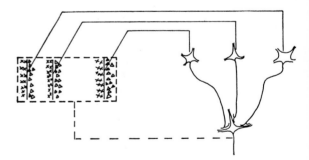

**Fig. 23.27**  Possible scheme for explaining the organization of complex receptive fields. A number of cells with simple fields, of which three are shown schematically, are imagined to project to a single cortical cell of higher order. Each projecting neurone has a receptive field arranged as shown to the left: an excitatory region to the left and an inhibitory region to the right of a vertical straight-line boundary. The boundaries of the fields are staggered within an area outlined by the interrupted lines. Any vertical-edge stimulus falling across this rectangle, regardless of its position, will excite some simple-field cells, leading to excitation of the higher-order cell. (Hubel and Wiesel, *J. Physiol.*)

## HYPERCOMPLEX UNITS

When Hubel and Wiesel explored Areas 18 and 19 (II and III) they found neurones entirely of the complex type, as might be expected were the visual input to these areas exclusively from Area 17.* Furthermore, new degrees of complexity in the receptive fields emerged, so that a further class of 'hypercomplex' cells had to be made, and within this group 'higher order' and 'lower order' cells were described. Visual Area II contained almost entirely complex cells similar to those found in Area I, the remainder being hypercomplex, whilst in Area III the majority were hypercomplex. An example of a hypercomplex cell was one that, like the complex cells, could be excited by a line of specific orientation, but this line had to be limited in length in one or both directions to obtain maximal effect. The adequate stimulus was thus a critically orientated line falling within a given region of retina (the activating region), provided that a similarly orientated line did not fall over an adjacent, antagonistic, region. This hypercomplex cell behaved as though it were activated by two complex cells, one excitatory to the cell with a receptive field occupying the activating portion and one, inhibitory to the cell, having its field in the antagonistic portion.

UNORIENTATED CELLS

Striate cortex also contains units that show no specific orientation-specificity, and these have been divided into several subgroups (see, for example, Dow, 1974).

## ORIENTATIONAL SELECTIVITY

The characteristic feature of the simple cortical unit is its requirement for a line-stimulus at a definite angle to one of the meridians of the retina. This orientational

selectivity was quantified by Campbell *et al.* (1969) by varying the angle of a striped pattern which was moved over the retina. As Figure 23.28 shows, the impulse-frequency decreases linearly with the obliquity of the grating from the preferred orientation. The angular selectivity of a unit was defined by drawing the line AB through the value of half-maximum amplitude of response. Bisecting the angle determined by the length of AB gives the angular selectivity, which in this unit was $\pm 19°$. Angular selectivities ranged from $10°$ to $50°$, but it is interesting that there was no preference for the horizontal or vertical as might be suggested by the finding that grating-acuity is highest when the lines are either vertical or horizontal (p. 318).

## DIRECTIONAL SENSITIVITY

Having established the preferred *orientation* of a simple cell's receptive field, we may discover frequently another preference, namely the direction in which this line is moved across the retina. Thus with many simple cells the response to movement is the same whether it is from left to right or from right to left, but in many others Bishop *et al.* (1973) found a very pronounced directional sensitivity. For example, they described a cell that was responsive to a bar flashed on to its linear receptive field at the appropriate angle. A moving edge crossing the field at the same angle was effective when moved in one direction but in the

---

*It must be emphasized, however, that Area 18 in the cat receives direct input from the lateral geniculate body, although this is not true of the primate.

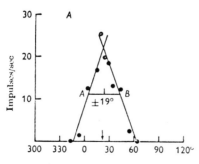
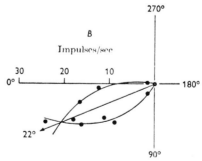

**Fig. 23.28**  Illustrating sensitivity of a cortical cell to direction of movement of a grating.

A. Linear plot of response amplitude (of cortical cell) versus the direction of movement (angle) of the stimulus grating. Their point of intersection is taken as the best estimate of the maximum response amplitude and the preferred angle (arrowed). The horizontal line is drawn at half the maximum response amplitude. Spatial frequency: 0·24 c/deg. Drift frequency: 3·8 c/s. Contrast: 0·7.

B. Polar plot of results in A. The arrowed line indicates the preferred angle. $0° =$ a movement from right to left in the visual field; $90° =$ a downward movement. (Campbell, Cleland, Cooper and Enroth-Cugell, *J. Physiol.*)

opposite direction there was no response at all—the field had a preferred direction.* Bishop *et al.* showed that it was not necessary for the edge to cross the boundary between the excitatory and inhibitory areas of the receptive field to produce this directional selectivity, very small movements of 1 minute of arc eliciting a response in the preferred direction and none in the opposite direction.

### DIRECTIONAL SELECTIVITY. GENERAL FEATURES

Some 72 per cent of simple cells described by Bishop and his colleagues were directionally selective, 51 per cent responding to both light and dark edges in the one—*preferred*—direction and 21 per cent to only one edge. For any given cell the selectivity did not depend on the length of the bar so that a spot was just as direction-selective. A further point to appreciate is that, although directional selectivity is usually studied by moving a bar at right-angles to its preferred orientation, the selectivity is still manifest if the bar is angled to this preferred orientation, so that the response to movement fails before directional selectivity fails.

## DISCHARGE CENTRES

Hubel and Wiesel analysed the receptive fields of their simple cells largely in terms of responses to stationary flashes; Bishop *et al.* (1971) emphasized the importance of movement of the stimulus-pattern, revealing as it does the directional sensitivity of many units; a further aspect revealed by moving stimuli was the response to leading and trailing edges, and the contrast of the

edges. Thus a bright vertical slit, moving from left to right across a unit with a vertically directed receptive field presents two stimuli as it passes over the field; first the leading edge of bright contrast and next the trailing edge of dark contrast; on moving the slit back from right to left the stimuli are repeated with movement in the opposite direction (i.e. preferred or null). An example of such a movement is shown diagrammatically in Figure 23.29. Clearly, if the width of the slit is increased, the spatial separation of the successive responses will increase and if the 'discharge-centres' responsible for the responses coincide, i.e. if the same retinal region discharges to bright and dark contrast edges, then the spatial separation will increase by the width of the slit. If the discharge-centres are not the same—and in many units this was the case—then the discharge peaks in the histograms are separated by a larger distance (Fig. 23.29).

### CORTICAL INHIBITION

A systematic analysis of the effects of slit-widths and other parameters of these moving stimuli led Bishop *et al.* (1971, 1973) to a fundamentally different concept of organization of simple cell receptive fields, through input from geniculate receptive fields, from that suggested by Hubel and Wiesel, the important feature of their model being the presence of inhibitory interactions at a cortical level. A tonic inhibition of cortical neurones is to be expected, in view of the absence of any resting

---

* A cell is said to be directionally sensitive if movement in one direction produces a smaller or larger response than movement in the opposite direction; the extreme case is when movement in one direction produces no response.

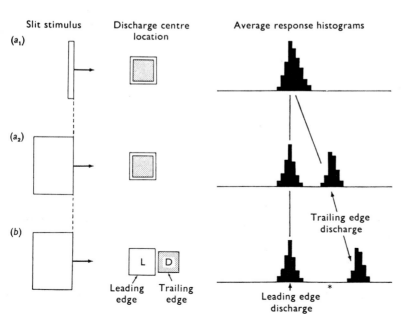

**Fig. 23.29** Diagrammatic representation of average response histograms showing the expected locations of discharge peaks in response to the leading and trailing edges of a moving slit of light. With the discharge centres coincident, there is a single discharge peak when the slit is narrow ($a_1$) and when it is widened there are two peaks separated by a distance equal to the width of the slit ($a_2$). With spatially separated discharge centres ($b$), the discharge peaks are separated by the width of the slit (i.e. from vertical arrow to asterisk) plus the distance between the discharge centres. Starting position of the leading edge held constant throughout. (Bishop *et al.*, *J. Physiol.*)

discharge in these in spite of a tonic input from geniculate cells. The authors emphasize that the characteristics of the receptive field are not usually predictable from the effects of flashing stationary slits; moreover, the regions of ON and OFF, defined in this way, are not necessarily to be equated with regions of excitation and inhibition, so that on moving a slit across the field, discharges can be obtained in both ON and OFF regions. In general, there are relatively wide inhibitory bands of about 2° to either side of the discharge-centre of a unit, even when the slit is moving in the preferred direction and with the optimum orientation. Another feature was the increase in discharge as the length of the slit was increased up to a maximum, beyond this there was no further increase in discharge.

LATERAL GENICULATE INPUT

Dreher and Sanderson (1973) have examined the receptive field characteristics of geniculate units in the light of Bishop *et al*.'s studies on simple cells, and they have shown that, like these, the field is, in effect, organized into discharge centres which are often spatially separate, responding to moving edges in a similar way to that described by Bishop *et al*. For example, in 50 per cent of the simple cells described by Bishop *et al*., with elongated stimuli moving along an axis perpendicular to the optimal orientation, the spatial arrangement of light and dark edge discharge-centres was similar to that found in geniculate ON-centre cells and OFF-centre cells, the dark edge discharge being closer to the starting point of stimulus movement than the light edge discharge-centre (Fig. 23.30). Dreher and Sanderson concluded that the discharge-centre organization of the simple cells was a direct result of the geniculate cell input with little cortical modification; furthermore, simple cells would receive a direct excitatory input from either ON-centre or OFF-centre geniculate cells, but not from both. Orientation- and direction-specificity, however, require more than a simple relay of geniculate field characteristics.

## BASIS OF CORTICAL UNIT'S DIRECTIONAL SENSITIVITY

### RADIAL INHIBITION

The basis of the selectivity of the ganglion cell is, as we have seen, an inhibitory process that prevents the excitatory process from getting through the ganglion cell when movement is in the preferred direction (p. 287). Goodwin *et al*. (1975) have emphasized that the geniculate cell has a pronounced directional sensitivity in that an OFF-centre unit, for example, shows a pronounced burst of discharges when a moving light bar

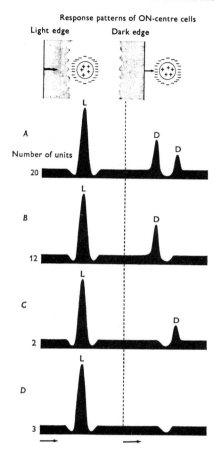

**Fig. 23.30** Schematic diagrams of the four types of response patterns of ON-centre LGN units to moving single light (L) and dark (D) edges. The left halves of the diagrams illustrate the responses to the light edges while the right halves illustrate the responses to the dark edges. Only one direction of movement is illustrated since the responses were virtually identical for all directions of movement. For details see text. (Dreher and Sanderson, *J. Physiol*.)

moves across the receptive field away from the centre (Fig. 23.31), i.e. there is a *radial* directional selectivity, and the question is how this could be converted into the lateral selectivity occurring in a row of geniculate fields corresponding to the postulated simple cell's field. In other words, how can the simple cortical cell, with its receptive field made up of a linear array of geniculate receptive fields as postulated in Figure 23.26, convert this radial directional selectivity in its component geniculate fields into a lateral directional sensitivity.

### INHIBITORY AREA

Goodwin *et al*. showed that the directional sensitivity was not a function of the centre-surround organization since a very small movement (one or two minutes of

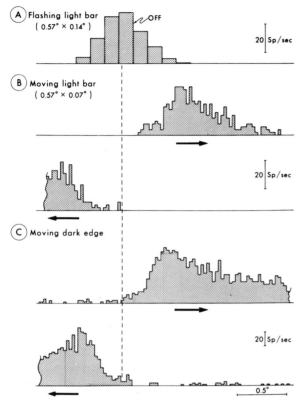

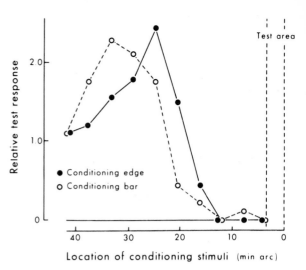

**Fig. 23.32** Spread of inhibition from a conditioning bar at different distances from the test area. Note that at about 10 minutes of arc separation inhibition is very strong; at greater distances it is less and reverses to become a facilitation. (Goodwin *et al.*, *J. Neurophysiol.*)

**Fig. 23.31** Responses of a sustained (X) OFF-centre neurone recorded in the lateral geniculate nucleus showing radial direction selectivity to stimuli moving away from the receptive field centre. *A*: receptive field OFF-centre plotted with a short light bar flashed on and off at successive locations 0·14° apart. OFF-responses averaged over initial 200 ms. Flash frequency 0·2 Hz. The vertical dashed line indicates the midpoint of the OFF-centre. *B, C*: responses to a light bar (*B*) and to a dark edge (*C*; 0·57° long), both moving forward and backward over the receptive field at 1·14°/s. (Goodwin *et al.*, *J. Neurophysiol.*)

arc) of a slit within the field centre was directionally sensitive. They showed that a thin slit, or an edge, exerted a strong inhibitory action on the adjacent part of the receptive field as tested by a second edge or bar stimulus; this is illustrated by Figure 23.32, where it is seen that if the conditioning stimulus—a standing light edge—is within about 10 minutes of arc from the test-area, the response is suppressed; with increasing distance, inhibition decreases and eventually is converted to a facilitation. The important feature is that the inhibitory effect can only be provoked from one side of the test area—the side that would be illuminated by a single edge as it moved in the non-preferred direction. Thus light on the opposite side of the test-area fails to produce inhibition, as indicated by the firing that comes from the test area when the edge

moves in the preferred direction. The inhibition takes time to develop, but if an edge is maintained in the field the inhibition, once developed, remains, so that the first movement of the edge is in fact direction-selective; hence so long as a cell responds to a moving edge it will retain its directional selectivity however much the speed of movement is reduced.

SUSTAINED INHIBITION

In comparing their results with those on the rabbit ganglion cell, Goodwin *et al.* emphasize the sustained nature of the inhibition established by an edge in the receptive field of the simple cortical cell, by contrast with its transience in the ganglion cell; and they suggest that the basis is the sustained, or X-type, of ganglion and geniculate cells making up the receptive fields of the direction-selective simple cell (Hoffmann *et al.*, 1972).

MODEL

Figure 23.33 illustrates the model of directional selectivity proposed by Goodwin *et al.* The column of circles represents a linear array of geniculate receptive fields; each field has a radial directional selectivity in that an edge, moving away from the centres, excites. The postulated inhibitory mechanism effectively blocks off half of these fields, so that moving edges will only cause a response on moving to the right, i.e. away from the centres. On moving to the left in the uninhibited regions the edge would be moving towards the centres of the fields, and when moving away from them

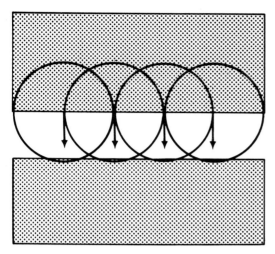

**Fig. 23.33** Simple cell receptive field: model for directional selectivity. The simple cell receptive field is formed by a row of geniculate fields. Cortical directional selectivity is based on the radial directional selectivity of retinal and geniculate receptive fields. Dotted areas indicate inhibitory regions. For details see text. (Goodwin *et al.*, *J. Neurophysiol.*)

it would be in the inhibited zone. The postulated blanking inhibition is applied by intracortical circuits (see, for example, Benevento *et al.*, 1972).

SIMPLE AND COMPLEX CELL MECHANISMS

Goodwin and Henry (1975) concluded that, although directional specificity of the simple cell was due to a blanketing inhibition of the cortical neurone, as indicated above, the specificity of complex cells was based on a different mechanism, the complex cell receiving inputs from other neurones that were already directionally sensitive. This was deduced from the observation that, whereas movement in the non-preferred direction abolished both spontaneous and driven activity in the simple cortical neurone, the same movement only abolished driven activity, leaving spontaneous activity unaffected. In an attempt to assess the role of inhibition in determining directional specificity in cortical neurones, Sillito (1977) applied the anti-GABA drug, bicuculline, locally to simple and complex cells of the cat's visual cortex, and assessed the changes in directional sensitivity. With simple cells directional sensitivity was rapidly reduced or abolished by iontophoretic application of bicuculline, as indicated by Figure 23.34. With complex cells the results were different and in order to characterize them it was necessary to classify the units in three categories— Types 1 to 3. Type 1 behaved similarly to simple cells, and so presumably received, like these, a primary non-directional excitatory input that was modified by intracortical inhibition. Type 2 were unaffected by bicu-

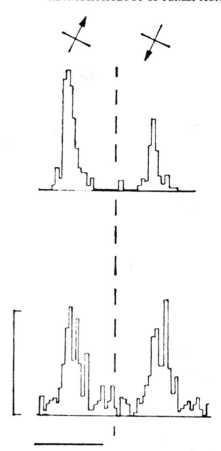

**Fig. 23.34** Effects of local application of an anti-GABA drug, bicuculline, on the directional sensitivity of a Type-1 complex cell. Upper record, before application. (Sillito, *J. Physiol.*)

culline, and it was concluded that they received a directionally specific input, and so did not have to rely on intracortical inhibition for their directional sensitivity. Type 3 apparently corresponded with the corticotectal neurones described by other workers (p. 621) and their direction-specificity was unaffected by bicuculline. Hypercomplex cells were unaffected, so they, too, apparently receive a directionally specific input.*

BINOCULAR FIELDS

The visual fields of animals with frontally placed eyes overlap, and the phenomena of binocular vision indicate

* The characteristic 'end-stopping' or length preference of hypercomplex cells of layers II and III was not greatly affected by bicuculline, suggesting that this feature of the hypercomplex cell's receptive field is built in by its input from length-sensitive units, perhaps cells in layer V (Sillito and Versiani, 1977).

that at some stage, or stages, in the visual neuronal pathway, from retina to striate and adjacent cortical areas, there is interaction between the impulses arriving from the separate eyes. When lateral geniculate neurones were examined by Hubel and Wiesel (1961), they found units that could be driven by one or other eye, but they never found one that could be activated by both eyes. With striate neurones the position was different, some 85 per cent responding to stimulation of either eye (Hubel and Wiesel, 1962); furthermore, as we might expect, the receptive fields in the two eyes that stimulated a single cortical neurone were similar in shape and orientation, and they occupied symmetrical positions in the two retinae, i.e. what are called corresponding areas. Summation and antagonism occurred between the receptive fields of the two eyes. For example, if the ON-area of the left eye and the OFF-area of the right eye were stimulated together, the effects tended to cancel each other. In some cases a cell could only respond if both eyes were stimulated together by an appropriately orientated slit.

DOMINANCE

We may speak of the *dominance* of one or other eye, according to its power of exciting a given cortical neurone by comparison with the other eye; thus a neurone excited only by the left eye, i.e. showing no measurable binocular interaction, would be said to be completely left-dominant; between this condition and right-dominance there were many stages. In general, there was a preponderance of contralateral dominance, presumably corresponding with the preponderance of crossed over uncrossed fibres in the cat's optic tracts. An illustration is in Figure 23.35, the response being heavily dominated by the contralateral eye. Each cell, then, receives a somewhat greater excitatory input from the contralateral eye than from the ipsilateral eye. The significance of this will be discussed in connection with stereoscopic vision.

## CORTICAL ARCHITECTURE

Mountcastle (1957) discovered that the somatosensory cortex was divided by a system of vertical columns extending from surface to white matter, with cross-sectional widths of some 0·5 mm, the responses to neurones in these columns being similar with respect to modality of stimulation, the one responding to superficial and the other to deep sensation.

ORIENTATION AND DOMINANCE COLUMNS

In the visual area, too, Hubel and Wiesel (1963) found that their units responding to a given orientation of a slit were grouped together in columns; when units were classified in accordance with eye-dominance, a similar

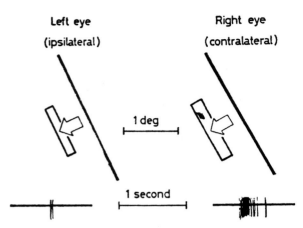

**Fig. 23.35** The binocular neurone whose responses are illustrated here belongs to Hubel and Wiesel's eye-dominance group 2: it is heavily dominated by the input from the contralateral eye. The receptive fields for the left and right eyes are shown as rectangles on the left and right, respectively. Below each field is an oscilloscope record of the action potentials typically evoked from the neurone by a single, slow sweep of a thin bright slit across the receptive field in that eye. Clearly the unit is far more strongly activated by a stimulus to the right (contralateral) eye than by one to the left. (Blakemore and Pettigrew, *Nature*.)

vertical distribution of units was found, overlapping with those based on line preference. In the monkey, too, Hubel and Wiesel (1968) showed that the units were organized on a vertical basis in so far as orientation of receptive field and eye preference, or ocular dominance, was concerned. When a slightly oblique penetration was made it was interesting to observe the step-by-step changes in orientation of the receptive fields as the electrode passed through and along adjacent orientation columns (Fig. 23.36).

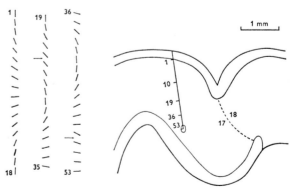

**Fig. 23.36** Reconstruction of a penetration through striate cortex about 1 mm from the border of areas 17 and 18, near the occipital pole of the spider monkey. To the left of the figure the lines indicate orientations of columns traversed; each line represents one or several units recorded against a rich unresolved background activity. Arrows indicate reversal of direction of shifts in orientation. (Hubel and Wiesel, *J. Physiol.*)

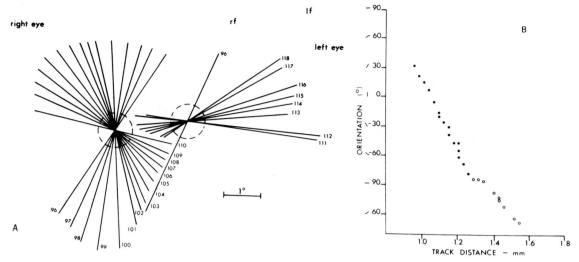

**Fig. 23.37**  (*A*) Observations during an oblique microelectrode penetration through monkey's striate cortex. The drawing was made on a sheet of paper affixed to the tangent screen 146 cm from the monkey. The eyes converged slightly so that the visual axes crossed, with the left and right foveas ophthalmoscopically projected as shown (lf and rf). The receptive fields of 22 cells or clusters of cells were distributed through the regions marked with dashed circles, 3° below and 3° to the left of the foveas. The first cell in the sequence, No. 96 in the experiment, had an orientation 32° clockwise to the vertical and was influenced from both eyes but more strongly from the right (ipsilateral). Cells 97–110 were likewise strongly dominated by the right eye; at No. 111, there was an abrupt switch to the left (contralateral) eye, which dominated for the rest of the sequence.

(*B*)  Graph of orientation vs track distance for the sequence of *A*. In plotting orientations, 0° is vertical, angles clockwise up to 90° are positive, and those counterclockwise to 89°, negative. (●) Ipsilateral (right) eye; (○) contralateral (left) eye. (Hubel *et al.*, *Cold Spring Harbor Symposia*.)

OBLIQUE PENETRATIONS

When the electrode penetrated approximately parallel with the surface of the cortex then a very regular shift in preferred orientation was observed, every 25 to 50 μm shift giving rise to a change in orientation of about 10° clockwise or counter-clockwise with occasional shifts in direction of rotation (Fig. 23.37). During this penetration, illustrated by Figure 23.37, there was a change in ocular dominance so that cells 96 to 110 were driven preferably by the right eye, but from 111 onwards they were driven by the left eye. This shift in ocular dominance occurred with regularity and at distances of about 0·5 mm. As mentioned above, in a vertical penetration all cells have the same dominance, so that we may envisage the cortex as being made up of 'hypercolumns' representing both ocular dominance and orientation-specificity as in Figure 23.38.

**Layer IV**

When horizontal penetrations were made at different depths, it was found that in layer IVc the same alternating pattern occurred, but now the cells were strictly monocular, with the non-dominant eye silent; in this layer the geniculate cells make synapse with cortical cells, and thus, according to the geniculate layers from which the fibres are derived, the cortical cells will be driven by one or the other eye.

HISTOLOGICAL DEMONSTRATION

The columnar arrangement of the cortical cells in dominance columns was demonstrated histologically; this is illustrated by Figure 23.39 from Hubel and Wiesel (1969), who made use of the fact that the lateral geniculate cells were carrying messages from either the contralateral eye (layers 6, 4, 1) or the ipsilateral eye (layers 5, 3, 2). Thus a lesion in layer 6 of the left lateral geniculate body should represent a lesion in the right eye, and the distribution of degenerating axons and terminals in the left hemisphere should indicate how messages from the right eye were being organized. It was found, from the degenerating fibres, that they passed mainly to layer IV of the cortex, where they relayed with cortical neurones; some passed vertically up into layer III also. Figure 23.39 shows the pattern of degeneration, which occurred in patches, as viewed from above, so that shaded areas represent those parts of layer IV receiving input from the contralateral eye, and unshaded portions from the ipsilateral eye. Later, Wiesel *et al.* (1974) injected radioactive proline plus

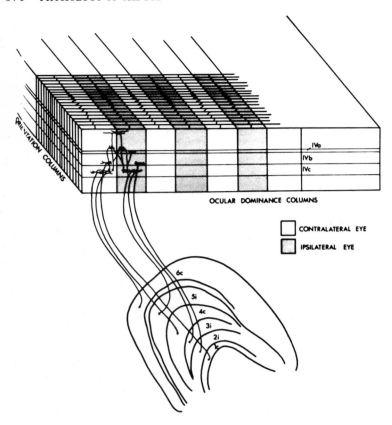

**Fig. 23.38** Diagram showing a possible relationship between ocular-dominance columns and orientation columns assuming the orientation columns are also long narrow parallel slabs, and that their arrangement is very orderly. Note that the width of these orientation slabs is much less than that of the ocular-dominance columns. A complex cell in an upper layer is shown receiving input from two neighbouring ocular-dominance columns, but from the same orientation column. (Hubel and Wiesel, *J. comp. Neurol.*)

ORIENTATION COLUMNS

OCULAR DOMINANCE COLUMNS

☐ CONTRALATERAL EYE
▦ IPSILATERAL EYE

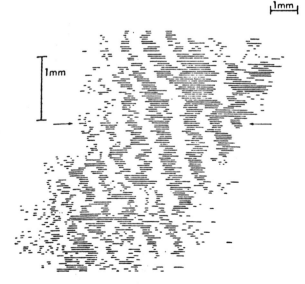

**Fig. 23.39** Reconstructed map of regions of fourth layer degeneration, following localized injury to the lateral geniculate body, as they would appear from above. Each horizontal line represents areas of terminal degeneration from one parasagittal section. Shaded areas thus represent those parts of Layer IV receiving input from the contralateral eye; unshaded areas, ipsilateral eye. (Hubel and Wiesel, *Nature.*)

fucose into one or other eye; these isotopes are taken up by ganglion cells and pass into the lateral geniculate body and thence into the cortex, and permit the visualization of the terminals in layer IV of geniculate fibres from the injected eye. Radioautography showed alternating patches of light and dark in layer IV, the light patches indicating the deposition of silver grains (Fig. 23.40). These patches are, in effect, sections through parallel bands.

DOMINANCE HISTOGRAM

The cells in layers above and below layer IV are not monocular, but receive input from both eyes, with a preference for one or other; this binocularity results from receiving input from many cells in layer IV, so that if, as the anatomical connexions indicate, a cortical neurone receives input from an area of about 1 $\mu$m or more in horizontal extent, it will receive input from probably two dominance columns; according to the cell's proximity to one or other dominance column, it will have a predominant input from one or the other eye. Experimentally the degrees of dominance may be classified in a rough quantitative manner in accordance with the relative magnitudes of ipsi- and contralateral responses; in this way Hubel and Wiesel (1962) divided responses into seven groups ranging from complete

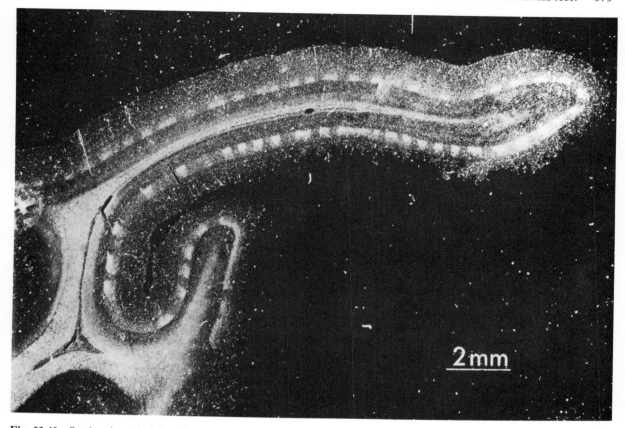

**Fig. 23.40** Section through right striate cortex of a normal 3-year-old monkey whose right eye had been injected 2 weeks before perfusion with a mixture of tritiated proline and fucose. Plane of section is perpendicular to the sagittal plane, tipped back about 45° from the coronal. The upper bank of the fingerlike gyrus forms part of the exposed surface of the occipital lobe; the lower part is the superior-posterior bank of the calcarine fissure. (The opposite bank has fallen away from this section.) Separating these folds of cortex is the relatively densely labelled optic radiation, which shows up as white in this dark-field photograph. Layer IVc is seen prominently as a series of alternating dark and light patches running about midway between the surface and the depths; about 56 pairs of columns can be counted. (Hubel *et al.*, *Cold Spring Harbor Symposia.*)

dominance by the contralateral eye (Group 1) through cells that are equally driven from each eye (Group 4) to neurones that are only excited through the ipsilateral eye (Group 7). Figure 23.41 is a typical dominance histogram for a normal animal, representing the numbers of cells of given groups encountered during electrode penetrations to increasing depths of the cortex.

SIGNIFICANCE OF COLUMNAR ARRANGEMENT

This has been discussed by Hubel and Wiesel (1974); probably the main significance is the economy of connexions from geniculate to cortical cells. Thus a cortical receptive field is brought about by inputs from several geniculate units whose ON- and OFF-field centres are arranged in a line; by dropping a few geniculate inputs and adding a few, the orientation of the line-field may be changed slightly, but the inputs to the old and new lines will be approximately the same. Secondly, as suggested by Blakemore and Tobin (1972),

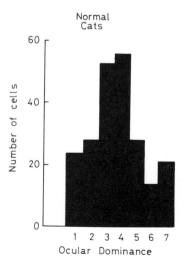

**Fig. 23.41** Typical dominance histogram for a normal cat. (Blakemore from Hubel and Wiesel, *Brit. med. Bull.*)

orientation-selectivity may be sharpened by a cell's receiving inhibitory input from cells with slightly different orientation-selectivity; if this is true, this could be best achieved by having units with similar orientation-selectivity close together. Finally, during development, it may be that the order reflects a mechanism guaranteeing that all orientations be represented once or only a few times in each part of the visual field, with no omissions and a minimum of redundancy; and it is interesting that these orientation columns are laid down before birth and are not affected by visual deprivation (Hubel and Wiesel, 1974b).

## QUANTITATIVE ANALYSIS OF MONKEY'S CORTICAL UNITS

### S-CELLS

In an elaborate study of cortical neurones in the monkey, Schiller et al. (1976, a–e) have distinguished some seven types of simple, or S-cells. The differences are illustrated in Figure 23.42, and may be described in terms of differing sub-fields, in the sense that the receptive field of any S-cell can be described in terms of multiple activating regions. $S_1$ is the simplest type of orientated cell, responding to only one sign of contrast change, in this case a dark edge moving in one

direction only. $S_2$ is the double-field cell, in which one sub-field is excited by a light-edge and the other by a dark-edge; both respond to movement in the same direction and some 2/3 were undirectional. $S_3$ cells show interaction between contrast and direction of movement; they have two subfields and each responds only to one direction of movement, but in opposite directions to each other; and so on. An example of the analysis is shown in Figure 23.43 for an $S_7$ cell which is

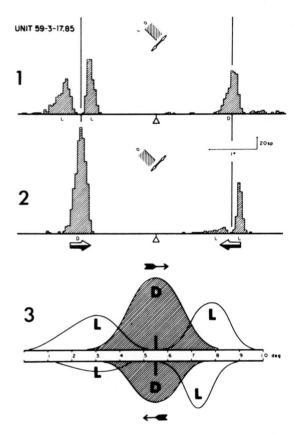

**Fig. 23.43**  Bidirectional multiple subfield S-7 type cortical cell. 1 and 2, responses to moving single edges; 3, schematic drawing of receptive field. (Schiller et al., J. Neurophysiol.)

directional with multiple subfields. When a single edge is passed from left to right, three well defined subfields are manifest; on moving in the opposite direction the first subfield is barely observable.

Essentially similar subfield organization had been described by Bishop et al. (1971) in the cat. Schiller et al. suggested that the more complex fields of S-cells, exhibited by $S_2$ to $S_7$, indicate a hierarchical ordering, so that $S_2$ receives input from two $S_1$ units of opposite contrast-response, but the same orientation and direction.

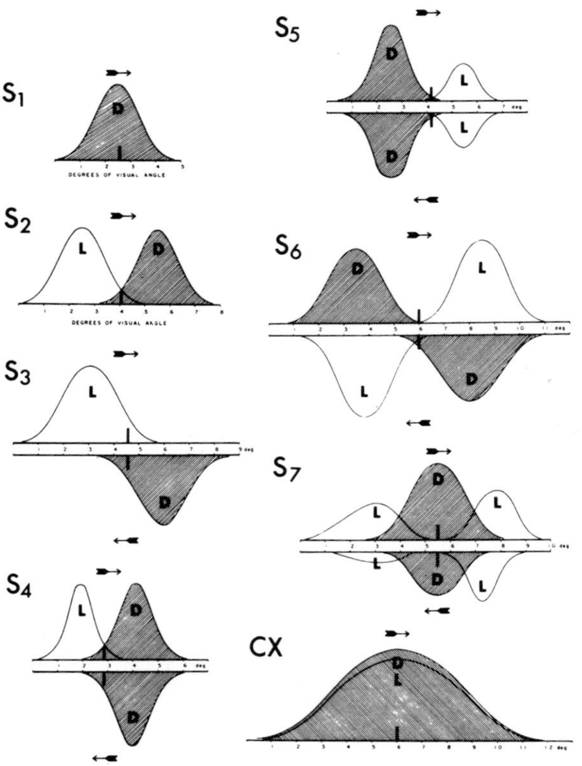

### CX-CELLS

Schiller et al. examined a second group of cells which they indicated by CX; these were equivalent to the complex cells of Hubel and Wiesel and defined as those cells with a unified activating region within which a response can be elicited by both light- and dark-increment; the receptive fields were

**Fig. 23.42**  Schematic drawings for seven S-type striate cells and one CX-type. (Schiller et al., J. Neurophysiol.)

larger, but there was some overlap with S-cells; they could be divided tentatively into two subgroups according to whether they were uni- or bidirectional.

### T-TYPE CELLS

These showed no orientation-selectivity, and a higher percentage were colour-coded than amongst S- and CX-cells; they doubtless corresponded with Dow's class I cells, with little transformation of the geniculate input.

### END-STOPPING

An important feature of Schiller's quantitative analysis was the failure of the hypercomplex cells to emerge as a distinct class; the basic feature of this type of cell is the 'end-stopping' manifest as a reduction in discharge when the length of the slit exceeds a certain value. Thus many of the S-cells and CX-cells exhibited end-stopping, so that in so far as hypercomplex cells in Area 17 of the monkey are concerned, they are not completely defined by this feature of the stimulus.

This feature of the cortical cell was further examined by Rose (1977) who studied the effects of increasing the length of a bar-stimulus from a very short to a long bar, i.e. from a small spot to a bar subtending 16°. Figure 23.44 shows some examples; at first there is a region of facilitation which only later may lead to inhibition as length increases beyond a certain point. The bottom records are for hypercomplex cells and it is seen that they only represent extreme examples of a 'flanking inhibition' that is by no means peculiar to this class of cell. In general, when different cells are compared, the effects of the flanking area are graded along a continuum of profound inhibition to marked facilitation, passing through a null-point where no flanks are apparent.

### ORIENTATION AND DIRECTION TUNING

In general, S-type cells exhibited sharper direction-tuning than CX-cells; moreover, the sharpness was the same at all cortical depths for S-cells, whereas CX-cells showed broader tuning at greater depths. So far as dominance was concerned

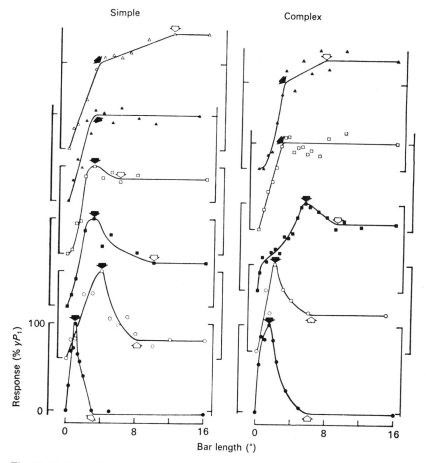

**Fig. 23.44** Length-tuning in visual cortical cells. Evoked spike counts have been normalized to 100 per cent at the point marked by black arrow where facilitation ceases; white arrows, where appropriate, indicate end of inhibitory phase. Simple cells on the left, complex on the right. The bottom pair of curves (filled circles) are from cells classed qualitatively as hypercomplex; the pair above (open circles) are from two dubious hypercomplex cells which could not be classified by ear with certainty as hypercomplex; the remaining seven curves are from simple and complex cells. (Rose, *J. Physiol.*)

more CX-type cells (88 per cent) were binocularly activated than S-cells (49 per cent). Schiller *et al.* found that, whereas orientation-selectivity exhibited the columnar organization in the cortex described by Hubel and Wiesel, the direction-sensitivity did not; this and other observations led to the inference that separate cortical mechanisms governed the two features. A possible basis for these two types of cortical interaction is illustrated by Figure 22.45, based on the supposition

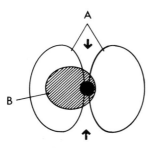

**Fig. 23.45**    Schematic basis for cortical interaction leading to directional and orientational sensitivity. A, apical dendritic field; B, basal dendritic field. For details see text. (Schiller *et al.*, *J. Neurophysiol.*)

that two separate inhibitory systems operate on the same cortical neurone through different dendritic systems. The black central disc is the cell-body, and is the site of excitatory input to the S1-type cell. The cross-hatched oval (B) represents the dendritic field involved in direction-selectivity, and the bilobed oval fields (A) are responsible for orientation-selectivity. Stimuli moving from left to right fail to activate the cell, due to inhibition through field-B, whilst edges or bars with orientations significantly different from the projection line between the two arrows produce inhibition from A. Hence the cell responds best to a bar or edge that moves from right to left and is orientated parallel with the arrows.

### TRANSFORMATION OF INPUT

As summarized by Schiller *et al.* (1976), the transformation of the input from lateral geniculate nucleus to simple cortical cells involves five features, the first producing *specificity for orientation of contours*; the second, *direction-selectivity*; the third involves *combination of responses of opposite sign*; thus whereas most LGN cells in the centre of their receptive field are excited by light increment or light decrement, but not by both, the cortical cell responds at the centre of its active region to both types of contrast; fourthly, the *inputs from both eyes are combined*, and finally, the cortical cells have a *greater spatial frequency-selectivity* than that found in the geniculate cells (Maffei and Fiorentini, 1973).

## VELOCITY TUNING

The assessment of the velocity of a moving object is an important and very necessary feature of the visual

system; Movshon (1975) has studied cat's cortical neurones from this aspect and has shown that, just as for spatial frequency of gratings (p. 583) and orientation of lines, the cortical neurones are, in fact, tuned to different bands of velocity, covering a range of velocities of some two-hundredfold from some 0·25 to 64° per sec. Figure 23.46 shows how different types of cortical

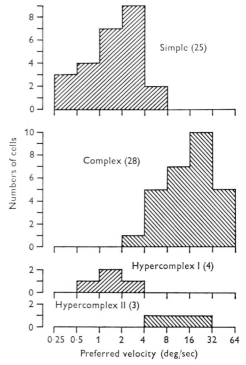

**Fig. 23.46**    Histograms showing the distributions of preferred velocities for sixty cortical cells of the four major types. (Movshon, *J. Physiol.*)

cells—simple, complex, and hypercomplex—are apparently tuned to different ranges, the complex and some hypercomplex cells being adapted to especially high velocities. Movshon points out that there are two basic ways of developing a system for signalling movement-velocity; units may simply increase rate of discharge monotonically over a wide range of movement, and such seems to be the method employed by the visual neurones of lower vertebrates, and this is also a feature of velocity-sensitive Y-neurones of the cat's lateral geniculate body (Dreher and Sanderson, 1973); the second method is that of multiple-channel analysis, whereby velocity-information is abstracted by determining the most active neural group in a population of such channels, each selective to a narrow range of velocity; and this is the analysis carried out by the cortical units, simple and Type I* hypercomplex cells forming a slow group responding optimally to 0·25 to 7·5°/sec, while the fast group is formed by complex and Type II hypercomplex units with preferred velocities of 4 to 50°/sec.

## STRUCTURAL CORRELATES

Kelly and Van Essen (1974) established the type of receptive field, i.e. simple, complex or hypercomplex, of cortical neurones in the different layers and, after establishing this, they injected Procion yellow through the microelectrode to permit morphological identification. The majority of simple units were stellate cells and the majority of complex and hypercomplex were pyramidal cells; and this accounts for the earlier observations that these last two categories tend to be concentrated in the superficial and deep layers of cortex, whilst the simple cells are more concentrated in the middle layers, where the input from the lateral geniculate body takes place. The identification of the complex cell with the pyramidal cell is reasonable, since these pyramidal cells are those with axons leading out of the cortex; we should expect the neurones in which visual information had been processed to a high degree to be those that would influence other parts of the brain or the periphery. Thus the corticotectal neurones are found predominantly, if not exclusively, in layer V and, according to Kelly and Van Essen, these are pyramids. Although simple cells are concentrated in layer IV and although stellate cells are also concentrated here, Kelly and Van Essen emphasize that they have not provided unequivocal proof that the geniculate input is, indeed, selectively to stellate rather than pyramidal cells; such a situation is demanded by the hierarchical concept of neuronal organization, a concept that has been questioned on several occasions (p. 566).

## LAMINATION OF CORTEX

Gilbert (1977) has emphasized the importance of assigning cortical cells of a given receptive field type to a given lamina of the cortex. His study in the cat showed that simple cells were not only concentrated in layer IV but in the adjacent part of layer III and also in layer VI, these being the regions of termination of the dorsal lateral geniculate nucleus in the cat. Complex cells were rare in layer IV but were frequent in all other layers. Layer VI has a strong projection to the lateral geniculate nucleus (Gilbert and Kelly, 1975) and it is of some interest that, when units in layer VI were antidromically activated from the lateral geniculate body, their characteristics were similar to those of all the units found in this layer, i.e. there seemed to be no specialization of receptive fields in these cortical cells feeding back to the geniculate.

The concentration of complex cells in layers II and III is quite consistent with an input from simple cells in layer IV; certainly the cells in these upper layers do not send dendrites to layer IV† and so probably receive no geniculate projection; as suggested by Hubel and Wiesel (1962), a considerable convergence of simple on complex cells would be required. As others had found, there was a strong similarity between the receptive field characteristics of the complex cells in layer V and the visually activated collicular cells (p. 617).

In general, there is not only a correlation between receptive field type and the cortical cell's morphology, but also between the field-type and the layer, and this in turn reflects the connexions the cortical cell makes with other parts of the nervous system, each layer receiving a unique set of afferents and projecting to a particular area (Gilbert, 1977).

## DIRECT CORTICAL FEEDBACK

By making lesions in various parts of the cat's visual cortex—Areas 17, 18, 19, 20, 21 and lateral suprasylvian area of Clare-Bishop (CB) Kawamura et al. (1974) identified a very considerable corticofugal input to lower visual centres. Thus Areas 17, 18 and 19 project to the dorsolateral geniculate nucles, Area 17 to the laminar part whilst Area 18 projects to both the laminar and interlaminar nuclei, and Area 19 to NIM alone. This corticofugal feedback is not peculiar to the geniculate system; thus the pretectum, besides being a nucleus in the pupillary reflex pathway, also acts as a relay station for visual input by way of the lateral pulvinar nucleus and thence to Areas 19, 20, 21 and the lower band of the splenial sulcus, so that the superior colliculus is able to relay its visual input to Areas 19, 20 and CB through the medial pulvinar. All these lower visual centres receive corticofugal inputs; the corticotectal input has been discussed elsewhere, and is of striking importance in controlling responsiveness of this centre to visual stimuli (p. 621); Areas 17, 18 and 19 all project doubly to the colliculus; in the superficial layers (chiefly lamina II) the projection is retinotopically organized, whereas in the deeper layers this is more diffuse. All other cortical regions project only to the deep layers in a diffuse manner. In the pulvinar, cortical projection is from all visual areas.

The cortical input to the pretectal nuclei provides a means of influencing pupillary reflexes; according to Kawamura et al., all the visual cortical areas project to the various pretectal nuclei.

## CONNECTIONS OF STRIATE NEURONES

There have been several studies seeking to correlate the types of receptive field of a cortical neurone with its afferent or efferent connexions, especially to determine whether simple and complex neurones can be related to their sustained or transient geniculate input.

---

* Dreher (1972) divided hypercomplex cells into two subclasses; the units are hypercomplex, because they are selectively responsive to stimuli limited in length at one or both ends, but apart from this they resemble simple or complex cells, and are classified as Type I (simple type) and Type II (complex type).

† Except for the large pyramids at the bottom of Layer III, and these may well receive a direct geniculate input.

AFFERENT AND EFFERENT CONNECTIONS

Singer *et al.* (1975) sought to correlate the type of receptive field of a striate unit with its afferent and efferent connexions, which they assessed by electrical stimulation at various points; they took account, too, of the X- or Y-classification of the geniculate or retinal input. So far as this aspect is concerned, they were unable to distinguish simple and complex units on the basis of their geniculate input, X and Y units connecting just as frequently with either main class.\* However, they confirmed that cortico-tectal units were mainly activated by Y-type geniculate input, and it is interesting that retinal Y-units are those that send branching axons to the tectum and geniculate. Cortical feedback to the geniculate took place through cortical cells that received both X- and Y-type input. In general, Singer *et al.* found that the afferent input to a given type of cortical unit—simple, complex, etc.—was not by any means peculiar to the type, so that cortical units with direct geniculate input and therefore supposed to be executing the more elementary analysis or processing of information, could belong to the complex categories, and, moreover, could have efferent output to subcortical structures and therefore might be considered to have completed their processing. On the hierarchical model, the complex and hypercomplex units would not be expected to receive specific geniculate input directly, but only after this had been processed by simple cells. Thus Singer *et al.* conclude that it is unlikely that serial 'filter operations' take place in Area 17, but rather that two separate operations are carried out in parallel.

## Area 18

A similar study of Area 18 in the cat, which unlike the primate region, receives direct geniculate input, lent further support to the notion of parallel processing. There were numerous similarities between the two areas with regard to their afferent and efferent connexions and their intrinsic organization, so that the receptive fields and numerical distribution were similar too. One fundamental difference was obvious, however, namely the response to much higher stimulus-velocities together with larger receptive fields in Area 18, which might well be due to the fact that the subcortical input to Area 18 is of the transient, Y-type, whereas the dominant input to Area 17 is of the sustained or X-type. Thus Area 17 would be analysing information largely on the basis of its spatial frequency which, as we have seen, requires units of the sustained type, whereas Area 18 would be analysing the information on a temporal basis, requiring units of the transient type. Of interest in this connexion is the projection to Area 18 from the superior colliculus by way of the thalamic lateralis posterior nucleus; and there is little doubt that the colliculus is concerned in localizing moving targets in space.

SERIAL AND PARALLEL PROCESSING

Ikeda and Wright pointed out that a scheme of geniculo-cortical connexions based on a sustained and complex pattern, such that simple cells receive sustained input and complex cells receive transient input is inconsistent with the hierarchical serial processing scheme of Hubel and Wiesel; the alternate scheme whereby we may have sustained and transient simple cells and sustained and transient complex cells is consistent with this, but also it is consistent with a parallel processing scheme. According to this latter, two separate processes, are carried out in parallel; the first extracts certain luminance characteristics from the stimulus, giving rise to simple cells, which may pass their information to other regions of the cortex or back to geniculate neurones in a feedback. A second activity, carried out in parallel, would be to correlate or integrate activities in the simple cells with activity derived from subcortical centres and also from cortico-cortical association systems. Such integration would be accomplished by complex cells.†

SUSTAINED AND TRANSIENT UNITS

Ikeda and Wright (1975a) examined the receptive fields of striate neurones and were able to separate both simple and complex units on the basis of their sustained or transient responses; examples of responses are presented in Figure 23.47 which shows the response histograms

(a) when an optimally orientated bar at the receptive field centre is reversed in contrast at 0·1 Hz
(b) when an optimal sine-wave grating drifted across the field at 1 Hz
(c) Plots of the receptive field with a stationary bar reversing in contrast at 1 Hz, where + is response to bright bar and − response to black bar
(d) the receptive field profile obtained from the receptive field plots.

Clearly the sustained responses belong to the two upper cells; when the response to a sine-wave grating is examined, it is seen that responses of cells 1 and 3 are modulated in phase with the stimulus, whereas 2 and 4 give continuous discharge throughout the cycle. This unmodulated response is characteristic of complex

---

\* It should be noted, however, that Wilson and Sherman (1976) have found a good correlation between the proportions of complex to simple cells in the cat's striate cortex and the proportions of Y cells in the lateral geniculate nucleus to total X + Y, when these were examined as a function of retinal eccentricity. Again, Stone and Dreher (1973) found that in Area 18 of the cat, which has a direct geniculate projection, cortical units were only activated by Y-type or transient geniculate units and not by X-type. Thus the lack of simple cells in the cat's Area 18 could be due to lack of X-type input.
† The article by Zeki (1976) may be read for a discussion on the merits of the parallel-processing theory.

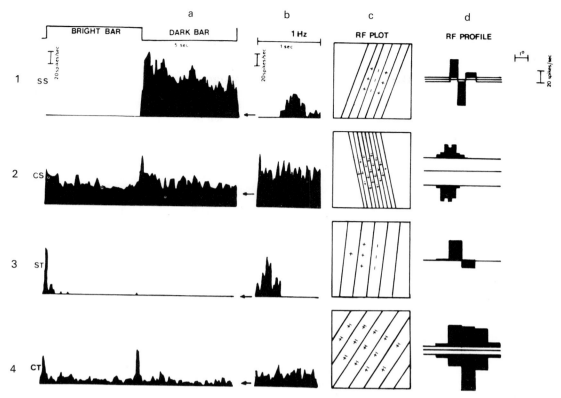

**Fig. 23.47** Classification of cortical cells into four types. First column (a) Post stimulus histograms obtained from 'simple-sustained' (SS), 'complex-sustained' (CS), 'simple-transient' (ST) and 'complex-transient' (CT) cortical cells. Each histogram is collected over 16 presentations of an optimal bar at the receptive field centre, reversing in contrast at 0·1 Hz. Contrast of the bar above and below background = 0·2. Arrows— spontaneous firing rate. Second column (b) Post stimulus histograms obtained from the same four cells over 16 stimulation cycles using an optimal sine-wave grating drifting across the receptive field at 1 Hz. Grating contrast, 0·4. Third column (c) Receptive field plots of the same cells obtained with a stationary bar, optimally orientated and with a width equal to $\frac{1}{2}$ cycle of the optimal spatial frequency grating, reversing in contrast (0·4) at 1 Hz. Responses to bright bar +, responses to dark bar −. Fourth column (d) Receptive field profiles obtained from these receptive field plots. Ordinate: cell response (spikes/sec); responses to the bright bar are above the abscissa and responses to the dark bar below the abscissa. The spontaneous firing rate is indicated by a horizontal line cutting the histogram above and below the abscissa. The abscissa is distance across the receptive field at right angles to the preferred orientation (calibration 1°). (Ikeda and Wright, *Exp. Brain Res.*)

cells while the modulated response is characteristic of the simple cell (Maffei and Fiorentini, 1973). Again the cells with separate ON and OFF responses in their receptive fields (1 and 3) are characterized as simple cells, whereas those with ON-OFF responses at the receptive field centre are complex.

### Spatial and temporal frequency tuning

When the spatial frequency-sensitivity of the neurones was examined it was found that it was the cells with sustained responses that exhibited the sharp tuning required for high spatial resolution (Fig. 23.48), as with ganglion cells; by contrast, the transient units were sensitive only to low spatial frequencies. Estimates of receptive field centre size for retinal ganglion cells are 0·15° to 2·9° for sustained cells and 0·5° to 7° for transient cells; if the receptive field widths of cortical units were estimated on the basis of the peak spatial frequency (half-cycle width) a similar difference between sustained (0·13° to 1·7°) and transient (0·3° to 5°) was obtained. Finally, the temporal frequency-tuning, based on a flickering stimulus, indicated parellelism between cortical and ganglion cells of the transient and sustained type. Thus the features of the ganglion cell that fit it for analysis of spatial frequency on the one hand, or of temporal frequency on the other (Sustained and Transient respectively) are transmitted to cortical units irrespective of their simple or complex nature.

## SPATIAL FREQUENCY TUNING CURVES

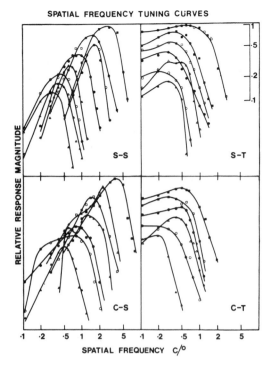

**Fig. 23.48** Spatial frequency tuning curves of 'simple-sustained' (top left), 'complex-sustained' (bottom left), 'simple-transient' (top right), and 'complex-transient' (bottom right) cortical neurones. The maximum response for each cell is scaled to 1 and each curve displaced downwards for clarity. Both axes are on log scales. The tuning curves represent cell responses as a function of spatial frequency of gratings of optimum orientation, contrast 0·4, drifted at 1 Hz, i.e., 1 grating cycle traversed a given point in 1 sec. (Ikeda and Wright, *Exp. Brain Res.*)

### Retinotopic distribution

Ikeda and Wright (1975b) found that sustained cortical units were more frequent in the part of Area 17 on which the fovea was represented, as we might expect of units concerned with the precise spatial-frequency resolution required for high visual acuity. Complex and simple cells were evenly distributed over the visual cortex. When cortical neurones were examined for the sharpness of their orientation tuning (i.e. the minimum change in orientation required to elicit a change in discharge) then there was no difference between sustained and transient units, although there was a significant difference between simple and complex cells, the simple cells being the more sharply tuned, as Rose and Blakemore (1974) had found. Thus orientation-selectivity is a peculiar cortical attribute depending on intracortical inhibition, and input from either type of ganglion cell is adequate.

### FAMILIES OF CELLS IN VISUAL CORTEX

Rose (1977) when summarizing his studies on 'length-tuning' or end-stopping of cortical cells emphasized the gradual mergence of specific features rather than the sharp segregation of types into simple, complex and hypercomplex. Thus we may speak of two groups, or families, of simple and complex cells which may derive some of their basic properties from the geniculate inputs and which pass on these properties to other cells in the same family. Within a family there will be a continuum, from cells dominated by geniculate input to cells dominated by input from other cortical units.

## ELECTROPHYSIOLOGICAL ASPECTS OF VISUAL ACUITY

Although the resolving power of the retina depends in the last analysis on the size and density of packing of the receptors in the retina, we must remember that it is the neural organization of these receptors, to give receptive fields of higher order neurones such as the ganglion cell, that determines whether or not the maximal theoretical resolving power is attainable.

### CONTRAST-SENSITIVITY FUNCTION

It is of some interest, therefore, to determine how a ganglion cell, and neurones higher in the visual pathway, respond to the presence of a grating form of stimulus in its receptive field. The subject has been examined in some detail by Enroth-Cugell and Campbell (1966) in the cat. When a stationary series of black and white lines is projected on to the retina, the responses of ganglion cells fall into two classes, X- and Y-cells. The X-cells behaved as though the responses to individual units of area in the field were linearly additive, the units in the centre adding to give a combined Centre-response, and those of the periphery to give a combined Surround-response, the final response being determined by subtraction of the combined Centre and Surround responses. With the Y-cells, no such linearity was present.

#### X-CELLS

To confine attention to the X-cells, it is interesting that two positions of the grating could be found that gave no response, the positions depending on the phase of the grating in relation to the centre of the field (Fig. 23.49); the existence of these null-points means that, when the illuminations over both halves of the total receptive field are equal, their responses just balance. When the grating was caused to drift over the retina in

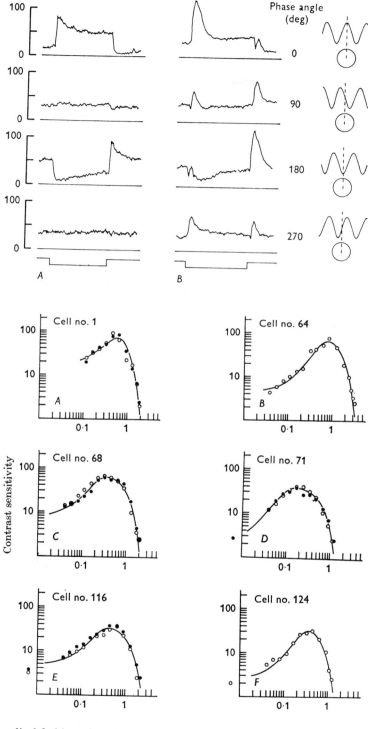

**Fig. 23.49** Illustrating the responses of two types of ganglion cell to the presence of a grating on their receptive fields. A is the response of an OFF-centre X-cell and B that of an OFF-centre Y-cell. The phases of the grating in relation to the centre of the field are indicated at the right. In order to evoke a response, the contrast of the grating was turned on and off, as indicated in the lowest traces where a downward deflection indicates turning off. Note that with the X-cell there are two positions of the grating that give a null response. Ordinates at the left give pulses per sec of the ganglion cell discharge. (Enroth-Cugell and Robson, *J. Physiol.*)

**Fig. 23.50** Contrast-sensitivity functions for five ON-centre X-cells (A-E) and one OFF-centre X-cell (F). Gratings of different frequencies drifted over the receptive field at a frequency of 1 cycle/sec, and the contrast was altered until the observer could detect a modulation of the discharge frequency at 1 cycle/sec (i.e. synchronously with the temporal luminance modulation). (Enroth-Cugell and Robson, *J. Physiol.*)

a cyclical fashion, then the ganglion cell responded in a cyclical fashion too, in phase with the drift, and it was found that this was the best way of stimulating a ganglion cell in order to examine the effects of contrast and grating-width, or, its reciprocal, the *grating-frequency* in cycles/degree. In this way the *contrast-sensitivity* of a given unit may be determined as the change in contrast of the drifting grating that can produce a measurable effect on the discharge. When contrast-sensitivity was plotted against grating-width, characteristic curves—

called contrast-sensitivity functions—were obtained with an exponential falling off at high grating-frequencies. This falling off at high frequencies corresponds to the limit of resolving power of the ganglion cell as the grating becomes fine; and a theoretical treatment, based on summation[*] of effects in the linear fashion indicated above, permitted the derivation of curves that could be made to match these 'contrast sensitivity functions' (Fig. 23.50).

### Central field

The contrast-sensitivity functions for different ganglion cells varied, indicating selective sensitivity to different ranges of grating-frequency, and it followed from theory that these were determined by the size of the central/region of the receptive field which, in the cat, was estimated to range from 0·5 to 4·4 degrees.

## SUSTAINED AND TRANSIENT GANGLION CELLS

Earlier we have seen that the basic classification by Enroth-Cugell and Robson (1966) of ganglion cells into X and Y, characterized primarily by their sustained response (X-cells) to a spot stimulus on their centre and their transient response (Y-cells) to the same sustained stimulus, has been established as a fundamental differentiation of the majority of ganglion cells with concentrically organized fields, be they ON- or OFF-centre. Additional criteria have been established for the differentiation, namely conduction-velocity of their axons, optimal size of target, existence of a periphery effect, and so on. These criteria stem from the basic differences, namely transience or sustained response and size of centre, the transient units having a larger centre—1° or greater—than that of the sustained units—½° to 1°—(Cleland *et al.*, 1971); and the same applies to the diameters of the surrounds (Cleland *et al.*, 1973).

### MOVEMENT VS PATTERN

This broad difference suggests that the transient (Y) cells are synaptically organized to transmit messages regarding movement, whilst those giving sustained responses, having small receptive fields, are better adapted to respond to patterns, as manifest in local difference in retinal illumination. Thus, according to Cleland *et al.* (1971), the transient type of ganglion cell represents the initial stage in the development of a specific sensitivity to motion, by virtue of its responsiveness to movement of large objects in the periphery of its receptive field—*periphery effect*—and to objects of any size suddenly crossing or moving within the

receptive field. With small objects moving slowly, both types of neurone can provide information that *localizes* the stimulus—in the sense that the response-magnitude increases as movement is towards the centre of the field—and *identifies* the stimulus, in the sense that the response depends on lightness or darkness of its image.

Kulikowski and Tolhurst (1973) have carried this distinction further with pyschophysical studies on humans in which grating stimuli were modulated spatially—i.e. the spatial frequency was altered, as in the determination of the contrast-sensitivity function of Figure 23.50, and the stimuli were also modulated temporally, either being flashed on and off at varying frequencies, or being 'alternated', in the sense that bars that were black at one moment became white, and *vice versa*; in this latter case the change in contrast of the adjacent bars was twice that when the grid was just switched on and off. Figure 23.51 shows a typical experiment where contrast-sensitivity has been plotted against spatial frequency as before. The interesting point is that at high spatial frequencies—i.e. with fine stripes—there is no difference in sensitivity to the two modes of stimulation, although, on the basis of temporal sequence of events, the retina is being exposed to twice the contrast during alternation. At lower spatial frequencies (wider gratings) a difference in sensitivity becomes manifest, and below 2·5 c/degree the predicted doubling of sensitivity is obtained. The results might suggest that different ganglion cells were being utilized for conveying the message; at the high spatial frequencies (fine lines) the retina would not be taking notice of the temporal sequence of stimuli and would be responding as though the receptors were being stimulated continuously with the contrast appertaining to a static presentation. Thus, it might be the sustained type of ganglion cell, with small receptive field, that was responsive. With the wider gratings the retina would be taking note of the alternation (which is equivalent to a movement of the grating backwards and forwards 3·5 times per sec) and this permits a greater effective contrast-sensitivity, since the retina is working on a basis of doubled contrast.

### FLICKER SENSATION

The difference is manifest in a difference of sensation at threshold; thus, with alternating spatial frequencies of greater than about 4/sec, and in the region where alternation was not being recognized, the gratings appeared stationary, whereas at the lower spatial frequencies—broader stripes—there was a definite sensation of motion, or flicker, i.e. the subject was recognizing

---

[*] The assumed basis of this summation is through a 'weighting function' by which points more distant from the centre contribute less than points closer to it; the actual function chosen was a Gaussian one (Rodieck, 1965).

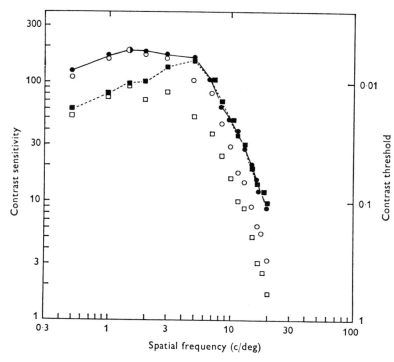

**Fig. 23.51**  Contrast sensitivity to sinusoidal gratings alternated in phase and switched on and off at 3·5 Hz. The filled symbols show the sensitivity for detecting a stimulus irrespective of its appearance: alternating (●) and on/off (■). The open symbols show the sensitivity for detecting that the stimuli were flickering: alternating (○) and on/off (□). The alternating grating was seen better than the corresponding on/off grating only when the alternating grating appeared to be flickering. Mean luminance 7 cd/m² . (Kulikowski and Tolhurst, *J. Physiol.*)

the temporal, as well as the spatial, features of the stimulus. In the experiment illustrated by Figure 23.51, the subject was required to say whether or not he considered that the stimulus appeared to be flickering at his threshold; if not, he was required to increase the contrast until the flicker was just apparent—the *flicker-threshold*. It is clear from Figure 23.51 that the flicker-threshold and stimulus-detection threshold are the same for wide stripes, but differ with fine gratings so that the contrast must be increased in order to detect flicker as well as contrast. It is evident from Figure 23.51, as well, that when the contrast has been raised so as to appreciate flicker it must be more for ON-OFF presentation than for alternation. Thus, under these conditions, namely recognition of flicker, the retina is taking notice of alternation and utilizing the greater contrast provided by this method of stimulation. As indicated, the ratio of contrast-sensitivities for the two types of presentation is theoretically 2, but varies from this to unity according to the spatial frequency, or fineness, of the pattern.

FLICKER VS PATTERN

When a subject is asked to use, as threshold, the contrast necessary to appreciate pattern and, again, the contrast necessary to detect the stimulus either through its flickering or its pattern, two types of graph are obtained (Fig. 23.52) where the ordinate is the sensitivity-ratio for the two types of presentation, i.e. for ON-OFF

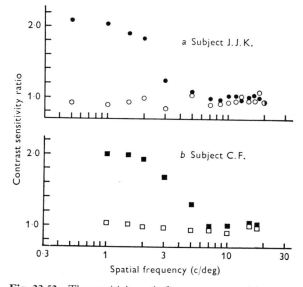

**Fig. 23.52**  The sensitivity ratio for pattern recognition as a function of spatial frequency. The filled symbols represent the sensitivity ratio for detecting the stimulus irrespective of the appearance. The open symbols show the ratio for the pattern recognition task; it is close to 1 over the whole range of spatial frequency. (Kulikowski and Tolhurst, *J. Physiol.*)

flashing and alternation. With pattern recognition, the ratio is unity at all spatial frequencies, but when the temporal features of the stimulus are recognized, the ratio rises to two at low spatial frequencies (broad stripes).

These studies emphasize, then, the operation of two separate mechanisms for recognition of the spatial and temporal characteristics of the stimulus and Kulikowski and Tolhurst argue that they are mediated through sustained and transient ganglion cell types.

### LINE STIMULUS

The above results applied to sinusoidal gratings; Keesey (1972) observed the same phenomenon with a flickering line-stimulus subtending 4 minutes or arc at the eye on a background of a 1° diameter field; at the visual contrast-threshold, only a diffuse flickering sensation was perceived and only at a higher contrast could the pattern perception of a well localized line be perceived. King-Smith and Kulikowski (1975) have examined the respective features of the pattern and flicker detectors; for example the receptive field of the most sensitive flicker detector is about twice as broad as that for pattern detection. They noticed that the vertical test line, modulated at frequencies of 3 Hz or more, produced a strong sensation of lateral motion at threshold comparable with that described for modulated gratings. Thus, it is likely that the flicker-sensitive cells are really motion sensitive; as we shall see, there are cells in the cerebral cortex that are specifically sensitive to moving targets (Zeki, 1974).

### ADAPTATION

Blakemore and Campbell's study on adaptation by gratings of different spatial frequency indicated the existence of separate 'channels' for analysis of this type of image. Tolhurst (1973) has extended his studies on flickering gratings to the effect of adaptation and shown that, at low spatial frequencies, where Blakemore and Campbell had been unable to detect specific channels on the basis of stationary targets, these could be detected when the target was moved; then it was found that the contrast-sensitivity was unaffected by adaptation to stationary targets.

### VALUE OF GRATING ADAPTATION

Adaptation to luminance has the value of extending the working range of the receptors over a very wide range of luminance so that the response of retinal ganglion cells to a fractional change in luminance, $\Delta I / I$, is roughly the same whether the eye is exposed to a luminance of 1 or 100. This is achieved by a decrease in contrast-sensitivity as retinal illumination increases, so that the maintained discharge in the ON-centre ganglion cells remains fairly constant. Lateral inhibition is another feature of the adaptive process, increasing sensitivity to high spatial frequencies compared with low, and finally there is the greater sensitivity to flicker or temporal resolution at high luminances. All these effects of adaptation have obvious value in enabling the eye to convey information to the brain. Barlow et al. (1976) pose the question as to the value, if any, of the adaptive effects of gratings; since these adaptive effects on one eye transfer to the other it is concluded that the mechanism is cortical rather than retinal. Their studies indicated, however, that there was no improvement in discriminating changes of spatial frequency or orientation of gratings.

## LATERAL GENICULATE AND CORTICAL UNITS

Campbell et al. (1969) compared geniculate and cortical units in the cat and obtained characteristic contrast-sensitivity functions with sharp cut-off points such that a small change in spatial frequency of the grating caused an abrupt decrease in sensitivity; the 'attenuation points' varied from about 0·2 to 1·5 cycles/degrees for cortical cells and 0·3 to 4 cycles/degrees for geniculate cells (Fig. 23.53). By human standards the grating

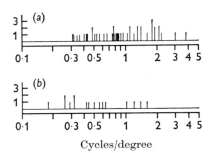

**Fig. 23.53** High frequency 'attenuation points' for (a) lateral geniculate fibres recorded in the cortex, and (b) cortical cells. (Campbell et al., J. Physiol.)

discrimination displayed by these cells, is low, indicating a low visual acuity in the cat. Later Maffei and Fiorentini (1973) measured the amplitude of response of simple cortical units as a function of the spatial frequency of gratings moved in the preferred direction; Figure 23.54 shows that the units responded over separate and narrow ranges of spatial frequency; and comparison with curves for ganglion cells and lateral geniculate units indicated that there was a progressive narrowing of the range of sensitivity, indicating the development of spatial frequency analysers that may well constitute the basis for the psychophysical results of Campbell and his colleagues (p. 317). The maximal resolution, of which these cat cortical analysers were capable, was 0·2, which is comparable with the resolving power of the cat determined through the visually evoked response (Campbell et al., 1973).* Complex cells did

---

* Campbell et al. (1973) emphasize that the falling-off in contrast-sensitivity with wide gratings found in the human (p. 317) is shifted towards wider gratings in the cat, so that, although its resolution of fine gratings is far inferior to man's, the cat is able to discriminate gross patterns better. In an extremely interesting comparison, these authors point out that the cat's vision is not comparable with the blurred vision a man would have if his acuity were reduced artificially with a blurring lens of +3 D, so that the comparison might be better with a man viewing through binoculars the wrong way round, i.e. distinct vision with lower resolution.

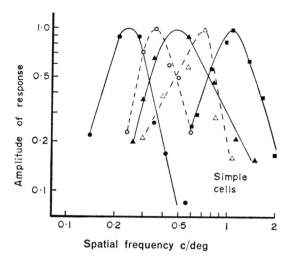

**Fig. 23.54**  Response of simple cortical cells to a sinusoidal grating, as a function of spatial frequency. Filled circles: unit with an ON-centre region of 1·2° width, flanked by two OFF-regions. Total width of the receptive field 5°. Open circles: unit with a receptive field of the same type as the previous one, but with very weak OFF-flanks. ON-centre region of 1° width. Filled triangles: unit with a bipartite receptive field consisting of an ON-region of 1·3° flanked by one OFF-region of 0·7° width. Open triangles: unit with a bipartite receptive field of total width 0·9°. Filled squares: unit with a bipartite receptive field of total width 0·6°. (Maffei and Fiorentini, *Vision Res.*)

not show such narrow ranges of frequency-selection, so that resolution of spatial detail was never greater than 1°.

### DOUBLE-FIELD SIMPLE CELL

A possible candidate for spatial frequency selectivity, according to Bishop *et al.* (1973) is the double-field simple cell in which both sub-fields respond to the same direction of movement, with one region responsive to a dark edge and the other to a light edge (S₂ type of Schiller *et al.*). The optimal response would be obtained when the moving bar had a width equal to the centre-to-centre separation of the two subfields, when the two subfields would be excited simultaneously by the opposite contrasts. An alternate, or additional, mechanism, in which inhibitory discharges arising when edges of non-optimal bar-widths impinged on inhibitory regions of the receptive field would probably be far more effective in governing selectivity, since it is possible to envisage no response at all to bars of widths exceeding a certain critical value.

### SPATIAL-FREQUENCY TUNING IN COMPLEX CELLS

The tuning of the simple cell's receptive field, so that it responds maximally to a certain spatial-frequency with a sharp cut-off at higher frequencies (Fig. 23.51, p. 585), seems

to be a feature of the width of the receptive field, units with narrow fields being able to signal higher spatial frequencies (thinner bars) than those with wider fields; in fact, as we have seen, the visual acuity of a cat can be assessed in terms of the width of its simple cell receptive fields. Pollen and Ronner (1975) examined the receptive fields of complex cells of the cat by moving a fine slit at the correct orientation across the field; they found a periodic series of peaks in the response-histograms, most having five principal peaks (Fig. 23.55), but one cell located near the border of Areas 17 to 18 had as many as thirteen. The response-pattern to the moving slit remained the same while velocity was varied, although the magnitudes of the responses showed some dependence. Thus these cells are capable of analysing a spatial pattern of stripes moving across the visual field, independently of its movement, and they could well be responsible for the spatial frequency-analysis examined psychophysically by Campbell and his colleagues. Moreover, if the complex cell received input from simple cells, the receptive field characteristics could be the result of this input.

The spatial frequencies of the individual cells varied from 0·4 to 5·0 cycles per degree, so that a half-period as small as 0·1 degree could be achieved; in general, the periodicity of the receptive field, in cycles per degree, was inversely related to the width of the receptive field, a wide field representing a coarser pattern. As we should expect from earlier studies, simple cells showed no such periodicity in response. Pollen and Ronner projected a slit of varying width symmetrically on the receptive field of a simple cell; the response increased as slit-width increased to about 1°, where it peaked and fell to about zero at 2°; each width of slit could be represented as a spatial-frequency by taking the reciprocal of twice the width, and so a spatial-frequency-response curve comparable with those described by Maffei and Fiorentini (1973) was obtained (Fig. 23.56).*

### CORTICAL COLUMNS

It will be recalled that, during a vertical penetration, at right-angles to the surface of the cortex, the preferred orientation of cortical units remains the same within a few degrees, whereas with tangential penetration there are successive changes in preferred orientation. So far as spatial frequency-tuning of cortical cells is concerned,

---

* Maffei and Fiorentini reported broader-band tuning of complex cells using their grating stimuli by contrast with the much more selective tuning revealed by Pollen and Ronner's study; these latter authors suggest that Maffei and Fiorentini were, in fact, measuring an average receptive field *shape*, rather than the periodic component revealed here. In their more recent and exhaustive study Schiller *et al.* (1976c) found very little difference between the two types of cortical cell; however, a very clear difference was manifest in the temporal modulation of the cell's discharge as a grating was moved over the field; the S-cells showed a sharp burst to each cycle that traversed the field whereas the CX cell responded in a rather continuous fashion, and this feature provides a reliable means of classifying S and CX cells. An interesting feature is the much greater selectivity for a sine-wave grating than for the classical black-and-white or square-wave grating; this is surprising since the sine-wave grating is similar to an out-of-focus image without attenuation of contrast, but it may be that these cortical neurones are concerned with the analysis of contours lacking great contrast.

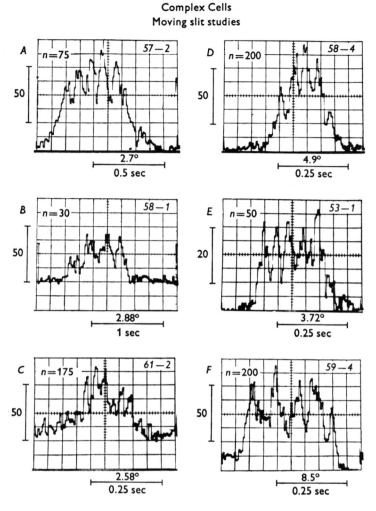

Complex Cells
Moving slit studies

**Fig. 23.55** Average response histograms representing receptive field shape for complex cells as a narrow slit is moved across the receptive field. *n* equals the number of stimulus presentations. Periodic changes in excitability across the receptive field may be noted for cells in area 17 (*A* and *B*), area 18 (*D–F*) and for one cell near the 17–18 border (*C*). In *A*, *D–F*, the zero level is indicated by the lowermost horizontal line. In *B* and *C* the zero level is indicated by the lowest filled bin. In *C*, the spontaneous level had fallen to zero during the period in which the histogram was taken. (Pollen and Ronner, *J. Physiol.*)

there is an opposite relationship, so that during a depth-penetration Maffei and Fiorentini (1977) found a changing spatial-frequency optimum with different depths (Fig. 23.57A), whereas with a tangential penetration the spatial frequency optima were the same within ±20 per cent over a length of 2 mm (Fig. 23.57B). Thus cells in a column have the same preferred direction; they also have the same area of receptive field so that they 'look' at the same area of visual field. If we assume that the cortex analyses the spatial features of the visual field, we may expect an orientation-column to contain all the tools required for processing visual information contained in the area of visual space the column is 'looking' at. Each column should therefore have cells tuned to a variety of spatial frequencies spread over a sufficiently wide range of frequency spectra, and in fact this was found, a range of about ten-fold, or 1 log-unit, in spatial frequencies being measured in a depth-penetration. Thus the neurones in a column contain all

the spatial information necessary to describe any object in a given orientation.

## COLOUR-CODING OF CORTICAL UNITS

EARLY STUDIES

Motokawa, Taira and Okuda (1962) recorded from single neurones in the striate cortex of the monkey; many of these gave simple ON-responses to illumination with maximal responses to one part of the spectrum; the regions of maximum response were 600 to 640, 520 to 540 and 460 nm, i.e. the red, green and blue; others gave opponent type responses, similar to those described by De Valois for the lateral geniculate body, the wavelengths for maximum ON- and OFF-responses being in different regions, e.g. in the red and blue. Some units gave ON-OFF responses to light over the whole range of wavelengths with maximum effect at 500 nm; these

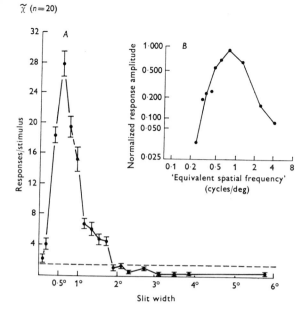

$\tilde{\chi}$ (n=20)

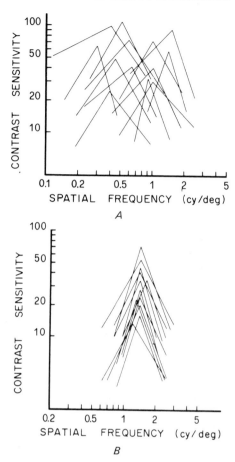

**Fig. 23.56** *A*, average response per stimulus for the first 500 msec of the ON-response is plotted against slit-width for a number of slits presented symmetrically around the receptive field centre in a random sequence for an ON-centre simple cell. The spontaneous level is indicated by the dashed line. Stimulus duration is 500 msec.

*B*, the response amplitudes are normalized for a peak response of 1 and are plotted on a log scale against the reciprocal of twice the width to give an 'equivalent spatial frequency' for one cycle of a square-wave grating. (Pollen and Ronner, *J. Physiol.*)

**Fig. 23.57** *A*, contrast sensitivity curves for a sample of 13 out of 27 neurones recorded in a penetration perpendicular to the cortical surface in area 17. All the cells had approximately the same preferred stimulus orientation (vertical).

*B*, contrast sensitivity curves of 12 neurones recorded in a tangential penetration perpendicular to the medial plane in area 17 (layer IV). (Maffei and Fiorentini, *Vision Res.*)

were presumably related to rods, but their sensitivity to light was low suggesting an inhibitory activity in the pathway. With other units there was no clear-cut peak over the whole spectrum, whilst with still others there were peaks at the two ends, suggesting connections with two types of cone; as the intensity of the stimulus was reduced the peaks diminished and a new peak at 500 nm emerged, indicating that rods and cones were converging on the same cortical neurone; in the light of the studies on geniculate cells this was to be expected.

IMPORTANCE OF RECEPTIVE FIELD

These pioneering experiments on cortical cells were carried out before the work of Hubel and Wiesel had drawn attention to the special spatial characteristics of the receptive fields of cortical neurones. Thus, in order to excite most of these neurones with white light it is necessary to choose a stimulus of a particular orientation, a particular length, a particular direction of movement, and so on. Hence a thorough examination of the chromatic characteristics of cortical cells would require that first the special spatial and temporal features of the receptive field be determined with white light, and

then, having categorized the cell in terms of these—simple, complex, direction-sensitive, and so on—to ascertain the effects of different wavelengths on the responses.

SIMPLE VS COMPLEX CELLS

Hubel and Wiesel found that some 6 out of 25 simple cells showed chromatic sensitivity; they responded to long narrow fields with greatest sensitivity to long wavelengths; these fields were flanked by a more inhibitory area with blue-green sensitivity. Thus the cortical cell behaved as though it were connected with a set of Type I red-ON-centre green-OFF-surround geniculate cells organized in a linear fashion. Twelve out of 177 complex cells showed chromatic preference; these could be stimulated by a blue slit orientated in a specific direction; they did not respond to white; other units were similar but favoured longer wavelengths. Since this study, several workers, notably Dow and Gouras, and Poggio *et al.*, have examined the monkey's cortical neurones from this aspect, the monkey possessing good colour vision characteristic of a human trichromat (De Valois and Morgan, 1974) by contrast with the cat on which so many studies of cortical neurones have been carried out.

### PARALLEL PROCESSING

Dow and Gouras (1973) emphasized that a simple cortical cell, responding to a slit or edge orientated in a definite direction, but also showing chromatic specificity, is not really specific for either colour or shape. Thus a Red-ON-Centre-Green-OFF-surround cell is not specific for long wavelengths since a small white light is, if anything, more effective than a small red light. Further, it is not specific for small spots since a large red spot is powerfully excitatory. The cell shows partial specificity in that it is not excited by large white or blue spots, so that the message is ambiguous: 'either a long-wavelength light or a small spot'. In general, so far as the colour opponence mechanisms are not co-extensive—being organized on a centre-surround or line-flank basis—the cells are not completely colour-specific, the effects of large coloured stimuli being mimicked by small white spots; and, insofar as the spatial opponent mechanisms have different chromatic sensitivities, the cells lack complete spatial selectivity. A similar ambiguity, of course, applies to the ganglion cells of the retina, and it may be, as Dow and Gouras suggest, that the information from these cells is processed in parallel, so that the spatial features are emphasized to produce cortical cells specific for orientation of contours, and the chromatic features are processed to produce cortical cells with a high degree of chromatic specificity. On this basis we might expect to find cortical cells without orientation-specificity but chromatically sensitive, and certainly the rarity of chromatic sensitivity amongst simple cells, described by Hubel and Wiesel, seems to confirm this.

### MONKEY'S FOVEAL UNITS

Dow and Gouras, in their examination of the monkey's striate area, paid special attention to the region of foveal representation, and they found some 34 per cent that showed specific chromatic sensitivity whilst 42 per cent were highly orientation-selective but without chromatic selectivity. These unselective units had, usually, broad action-spectra, suggesting that they were equally driven by all three types of cone; Figure 23.58 illustrates one with a fairly sharp peak of sensitivity at about 550 nm, and, when this was selectively adapted to a blue-green or red light, the action-spectrum altered, indicating the

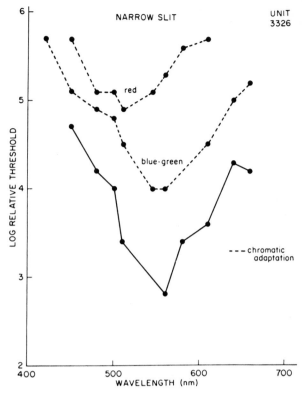

**Fig. 23.58** Wavelength-response curve for a cortical unit without chromatic selectivity showing broad action-spectrum. The effects of adaptation red and blue-green lights indicate the operation of at least two cone mechanisms. (Dow and Gouras, *J. Neurophysiol.*)

operation of red and green cone mechanisms. With a yellow background adaptation, no responses to monochromatic stimuli were obtained, suggesting the absence of a blue-mechanism.

### ORIENTATION-SELECTIVITY

When responses to slits of a single colour were examined, there was no difference in the orientation-selectivity, the line-flank arrangement being unaltered and sensitivity to changed orientation being the same with all colours. This means that the spatial inhibition, responsible for the orientation-specificity, involved interaction within each cone mechanism. Thus, if the two mechanisms contributed to give the inhibition necessary to inhibit the response to a horizontal line, as opposed to a vertical one, then adapting with a light so as to put one of the mechanisms out of action would seriously impair this inhibitory process and thus restrict orientation-selectivity.

### COLOUR-OPPONENT CELLS

24 per cent of the units examined were colour-opponent,

in the sense that they could be excited by one colour, e.g. red, and inhibited by another, e.g. green. So far as could be determined, the receptive fields for the opponent responses were co-extensive; thus Figure 23.59 shows the responses to a red stimulus presented on a blue-green background, and thus activating primarily the red cone mechanism, and to a green stimulus on a red background. The red response is at ON, and may be described as excitatory, and the green response is at OFF, and is essentially a release from inhibition, so that the green stimulus is inhibitory. The responses occur at the same point of stimulation of the retina, indicating overlap of the excitatory red- and inhibitory green-sensitive regions. As indicated, the responses are the same whether movement is vertical or horizontal, so that these units have no orientation-selectivity.

## SPATIAL COLOUR CELLS

Another group, representing 10 per cent, combined chromatic and spatial specificities and were described as *spatial colour cells*; these were typically discovered by moving white slits at an appropriate orientation; when stationary chromatic slits were substituted there was a centre-channel opponent response, with ON-responses from one colour-channel and OFF-responses from the other—usually red-ON and green-OFF. When a very narrow slit was employed, it appeared that a pure red or green response could be obtained, suggesting that the opponent response derived mainly from the flanking region, but the flanking region is not necessarily sensitive to a different colour-channel, e.g. a cell responding to a thin red line could exhibit flanking red inhibition. Dow and Gouras were unable to identify double-opponent cells with centre and flanks each exhibiting opponent responses such as those seen lower in the usual pathway.

The existence in the cortex of opponent colour cells, having a centre-surround antagonism *involving the same colour-channel* is apparently a new feature, emerging in the cortex and not present in the lateral geniculate body where the centre-surround cells receive excitation exclusively from one cone mechanism and inhibition exclusively from another so that, effectively, the geniculate cells have spectral, but lack spatial, opponency at least at one end of the spectrum; this makes them poorly sensitive to spatial contrast and completely insensitive to colour-contrast (Gouras, 1974). This centre-surround antagonism, utilizing the same cone-mechanism, might be brought about if the same type of opponent colour geniculate cells mediated both the centre and surround of the cortical cell; such a cell

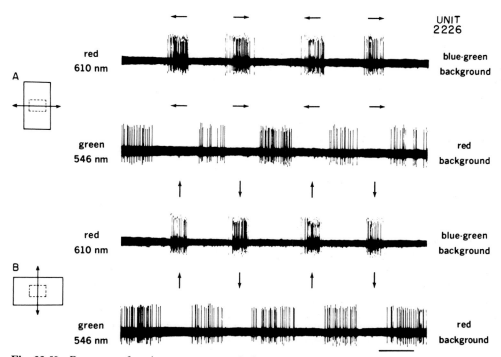

**Fig. 23.59** Responses of a colour-opponent cortical neurone to orthogonally orientated moving monochromatic slits in the presence of chromatic adaptation. Time-mark, 1 sec. (Dow and Gouras, *J. Neurophysiol.*)

would be more sensitive to spatial contrast than the opponent colour geniculate cell. It could be, therefore, that this particular feature of the chromatic sensitivity of the cortical cell contributed to the detection of spatial contrast. They could be sensitive to simultaneous contrast because appropriate wavelengths in the surround might disinhibit the centre.

### LUMINOSITY AND TUNED UNITS

The group of chromatically unselective units described by Dow and Gouras have been characterized by Poggio *et al.* as *luminosity units* which they consider to be analogous with the dominators of Granit. They constituted nearly half of the cortical units studied by these workers and they presumably received inputs from all three cone mechanisms to give an action-spectrum similar to the photopic luminosity function of the monkey, which is similar to man's with a maximum in the yellow-green. These units are useless for hue-discrimination since, by appropriate variation of the intensity of a single wavelength, the same response may be obtained. Thus if the cell is stimulated by different wavelengths, $\lambda_2$ and $\lambda_1$ with intensities related by:

$$I(\lambda_2) = \frac{S(\lambda_1)}{S(\lambda_2)} \cdot I(\lambda_1)$$

the observation that the responses are the same means that the unit is a luminosity unit; if this relationship fails, then the unit shows some chromatic specificity indicating a preponderance of input from one or two of the three cone mechanisms. Poggio *et al.* described such units as *tuned*; Figure 23.60 illustrates action-spectra for a luminosity unit (11-h), a blue-green tuned unit (11-d) and a red tuned unit (11-e) whilst Figure 23.61 gives the actual responses of two tuned units to different coloured stimuli adjusted in intensity to give equal sensation of brightness to the photopically stimulated eye.

### OPPONENT NEURONES

Another class described by Poggio *et al.* were opponent neurones, stimuli of different spectral composition evoking qualitatively different responses (ON and OFF) from the same or adjacent regions of their receptive fields. Essentially these belong to the opponent types of Dow and Gouras and their features need not be recapitulated here in detail; the receptive fields were either uniform or else were bipartite, containing a central slit and flanking surrounds, or centre-surround concentricity. In those with uniform fields, short-wavelength stimuli evoked one type of response and long-wavelength stimuli the opposite; lights about the middle of the spectrum were either ineffective or evoked ON-OFF-responses of high threshold. Units

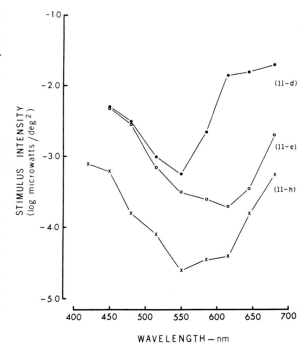

**Fig. 23.60** Action spectra for a luminosity-unit (11-h), a blue-green tuned unit (11-d) and a red tuned unit (11-e) in the monkey's visual cortex. (Poggio *et al.*, *Brain Res.*)

with bipartite receptive fields usually had different spectral sensitivities for the two parts of the field, e.g. centre response to blue-green and flank response to red. Some of these bipartite units also exhibited an opponency within the individual parts of the field so that the centre might be red-ON, blue-OFF, and a reciprocal arrangement could be inferred for the surround (red-OFF-blue-ON). Such units have been called *double opponent*. Poggio *et al.* consider that their tuned units may well be the same as some of the opponent units described by Dow and Gouras, the tuned neurones reflecting the interaction of two systems, one of which does not manifest its action directly, either because of its intrinsic weakness or because it is inhibited by an opponent system. Hence the two categories of tuned and opponent neurones could cover the whole continuum of spectrally selective properties from evident and balanced opponence, through opponence with one system dominant, to tuned.

### VERVET MONKEY

Bertulis *et al.* (1977), working on cortical cells of the vervet monkey, *Cerpithoecus Aethiops*, have been unable to confirm some of the features described by Gouras. Thus, studying cells with foveal representation, they found that, of the cells responding to a narrow range of spectral hues, three quarters were orientation-specific and half of these were of the complex 'stopped-end' variety. Cells that responded over only a narrow

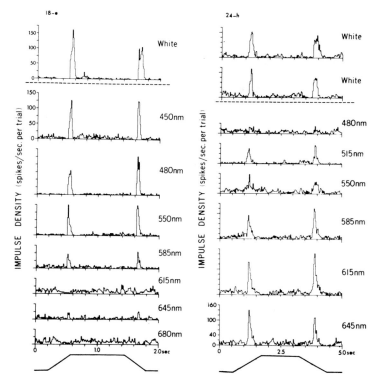

**Fig. 23.61**  Responses of two spectrally tuned cortical neurones to luminous slits of different wavelengths adjusted to equal luminance. *Left*: Blue-green tuned neurone with uniform receptive field 1° from visual axis. *Right*: Red tuned neurone with complex receptive field 1·8° from visual axis. (Poggio *et al.*, *Brain Res.*)

range of wavelengths had maxima at 590, 560 and 495 nm; thus the cells with sensitivity in the longest wavelengths were 'yellow-sensitive' and the others green- and blue-sensitive. Some broad-band units could not be equated completely with 'luminosity units', since, when the stimuli were adjusted to equal-energy on the basis of the C.I.E. photopic sensitivity curve, they showed greater responses at middle and long wavelengths. Some units, however, showed little or no change in response over the whole spectrum when this was adjusted to equal energy.

## LAMINATION AND RESPONSE FEATURES

### LAYER IV

The lateral geniculate input to Area 17 is to Layer IV, the terminals being not only in the stria of Gennari (Layer IVb) but also immediately above (Layer IVa) and below (Layer IVc).\* If, as seems likely, processing takes place on a vertical or columnar basis, cells exhibiting more specialized receptive fields—complex and hypercomplex—may be expected to predominate in the more superficial and deeper layers; and this seems to be true (Kelly and Van Essen, 1974; p. 579). It is of some interest to see how processing of colour information is associated with processing of spatial information. Gouras (1974) has established some correlations with the colour opponence of cortical units and their spatial characteristics, when recording from different layers.

### COLOUR OPPONENCE IN IVb

A striking finding in the monkey is that the majority of cells in Layer IVb have opponent colour properties, indicating that this form of chromatic organization is continued from the geniculate input, although, by contrast with geniculate cells, some striate units receive centre-surround opponent inputs from the same cone mechanism, a feature that might militate against chromatic analysis of the environment, the centre and surround responses tending to cancel each other. Another interesting feature of Layer IVb is that the majority of cells did not show orientation-selectivity, i.e. they would not fall into Hubel and Wiesel's simple category.

### SPATIAL VS COLOUR OPPONENCE

On moving away from Layer IVb, the proportion of orientated cells increased and that of colour-opponent cells decreased, so that it would seem that, during spatial processing, some of the chromatic specificity is lost, provided that colour opponence, as such, may be regarded as the result of a processing that enables the cortical cell to distinguish colours more effectively than, say, the geniculate or ganglion cell could do.

---

\* Visually driven cortical neurones are also found in Layers II, III, V and VI, but are presumably driven secondarily by the cortical neurones receiving the primary projection.

VALUE OF COLOUR OPPONENCE

Thus, if we consider the receptive field illustrated by C in Figure 14.37, p. 353, it is opponent in that, when red light is flashed on to its field, it gives an ON-response; when blue light falls on its field there is no excitation, and instead, any tonic discharge is inhibited (OFF-response). Such a cell discriminates between red and blue with a high degree of accuracy because the responses are qualitatively opposite, and it is likely that it functions better than a single red or a single blue receptor, which would give signals that differed only quantitatively in response to red and blue lights, signals, moreover, that could be made equal by appropriate variation of intensities. If, as is found, the colour-opponent unit retains its opponence in the face of large variations in intensities of the two coloured stimuli, then clearly the colour opponence represents a processing that enables the single cell to discriminate more accurately than the two cones working on their own.

SPATIAL OPPONENCE

As to whether spatial opponence may be regarded as a further stage in chromatic processing is another matter; the advantage of spatial opponence for spatial frequency analysis is obvious, so that an estimate of the animal's visual acuity can, in fact, be made on the basis of the width of a simple cell's receptive field (Maffei and Fiorentini, 1973). Since high spatial resolution runs parallel with highly developed colour resolution, we must expect spatial opponence to be manifest chromatically, in fact if this were not so, the development of the centre-surround organization might obliterate the chromatic features of the signals, so that if cones ultimately converged on ganglion cells irrespective of their colour specificity, red, green and blue converging on the centre and red, green and blue on to the surround, the ganglion cells would exhibit spatial opponence but any colour opponence that resulted from imbalance of inputs to centre and surround would fail to signal the chromatic features of the stimulus with any clarity.

## CORTICAL AREA V4

We have indicated earlier that Zeki found that this region, lying in the anterior bank of the lunate sulcus, dorsally, and emerging ventrally on the posterior bank of the inferior occipital sulcus, contained neurones all of which were colour-coded, responding vigorously to one wavelength and grudgingly or not at all to another and to white light; they were organized in columns responding to the same band of wavelengths. In some there were no opponent colour surrounds but with others there were, such as green Centre ON-red

Surround-OFF. Some were described as *successive contrast units*, responding to a change from one colour to another. The main projection to this area is from Areas 18 and 19, which receive from Area 17; only a few Area 17 units are colour-specific and so far no colour-specific units have been described in Areas 18 and 19, so that the input to V4, leading to an apparently exclusive colour-specificity, is of some interest.

## NEUROPHYSIOLOGICAL BASIS OF STEREOPSIS

In looking for electrophysiological correlates of binocular vision involving stereopsis, we must expect to find, as we have indeed, cortical neurones that respond to stimuli in both eyes. Furthermore, as Barlow, Blakemore and Pettigrew (1967) have put it, the neurone must select those parts of the two images that belong to each other, in the sense that they are images of the same point; secondly, for stereopsis, they must assess the small displacements from exact symmetry that give the binocular parallax.

MINIMUM RESPONSE FIELDS

Barlow et al. recorded from neurones in Area 17 of the cat and found, as Hubel and Wiesel had earlier, that binocular stimulation was more effective in stimulating many units than uniocular; if the positioning of the targets was wrong, however, one eye could veto the response of the other as in Figure 23.62; here a bar moved over the receptive field of the left eye and right eye separately; acting together there was powerful facilitation. When the separate fields were mapped out for each eye, it was found that these 'minimum response fields' did not occupy exactly symmetrical positions in the two eyes, i.e. they were not corresponding areas; instead, for maximal binocular facilitation there had to be a discrepancy; for the unit illustrated by the left part of Figure 23.62 this was 5·7°; in this case the cat's eyes were divergent by 6·4°, so that the actual disparity corresponded to a convergence of 0·7° for the left unit and 3·1° for the right unit. In general, different neurones required different disparities to obtain maximal facilitation between the two eyes.*

---

* Nikara, Bishop and Pettigrew (1968) have made comparable studies in the cat, studying the vertical and horizontal disparities between receptive fields of binocularly represented units of Area 17. The disparities were in the range of ±1·2° in both horizontal and vertical directions. The significance of vertical disparity is that it permits single vision of points when the directions of regard of the two eyes are in error due to random movements of the fixation axes in relation to each other.

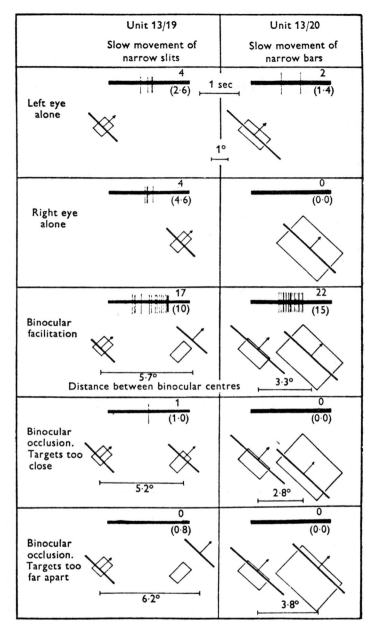

|  | Unit 13/19<br>Slow movement of<br>narrow slits | Unit 13/20<br>Slow movement of<br>narrow bars |
|---|---|---|
| Left eye<br>alone | 4<br>(2·6) | 2<br>(1·4) |
| Right eye<br>alone | 4<br>(4·6) | 0<br>(0·0) |
| Binocular<br>facilitation | 17<br>(10) | 22<br>(15) |
| Binocular<br>occlusion.<br>Targets too<br>close | 1<br>(1·0) | 0<br>(0·0) |
| Binocular<br>occlusion.<br>Targets too<br>far apart | 0<br>(0·8) | 0<br>(0·0) |

**Fig. 23.62** Illustrating responses of two binocularly driven units. Maximal facilitation in binocular stimulation is obtained only when there has been a certain disparity between the stimulated regions in the two retinae. (Barlow, Blakemore and Pettigrew, *J. Physiol.*)

## DISPARITY

To estimate the disparities of the fields the optimum orientation of a slit was first found, and this was the same for both eyes. The slit was moved over the retina and the limits beyond which no response was obtained were drawn on a screen in front of the eye. This gave the primary borders of the field; the lateral borders were determined by moving the continuously oscillating slit, maintaining the axis of orientation, until the end of the slit had moved out of the field and no constant response could be elicited. This is called the *minimum response field*, and is indicated in the rectangles of Figure 23.62. The fields for the two eyes were not necessarily of the same size, the dominant eye having the larger. It seems likely that the centres of these fields represent the centres of retinal areas connecting to a single cortical neurone. For obtaining the binocular disparities the two fields must be examined together; usually the centres of their separately determined fields give a good clue to this, but to find the exact measure the two slits are oscillated in synchrony over their respective fields to obtain maximal facilitation. At a critical separation this is obtained and the positions of the optimally separated slits are marked with a line on the respective monocularly determined fields, as in Figure 23.62. The point on this line where the normal to it would pass through the centre of the monocular minimum response field is called the *binocular centre*. It is the angle between these centres that is used as measure of retinal disparity.

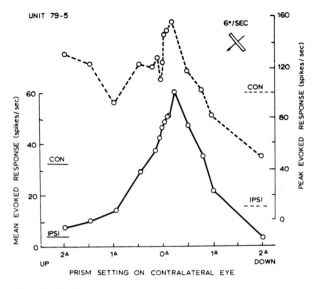

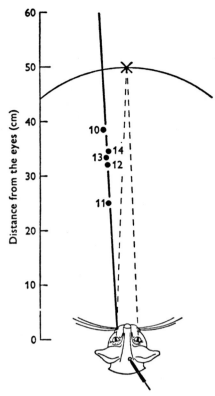

**Fig. 23.63** Responses of a cat's cortical neurone to a bar-stimulus applied to both eyes when the degree of retinal disparity was varied by placing prisms in front of the contralateral eye. The bars marked IPSI and CON indicate the responses of ipsi- and contralateral eyes when stimulated alone. Full lines indicate mean evoked response and dashed lines the peak evoked response. (Pettigrew, Nikara and Bishop, *Exp. Brain Res.*)

**Fig. 23.64** Illustrating the positions in space of points that would give rise to the observed retinal disparities of cortical neurones in a constant direction column. (Blakemore, *J. Physiol.*)

### DETECTION OF DISPARITY

Pettigrew, Nikara and Bishop (1968) recorded the summated action potentials from binocularly driven cortical units of the cat when the disparity between left and right-eyed fields was changed by means of a prism. Figure 23.63 shows the evoked response of a unit as a function of prism setting, 0° corresponding to perfect superimposition. The responses of the ipsi- and contra-lateral eyes are indicated by horizontal bars, and it will be seen that over a certain range of disparities there is facilitation through binocular stimulation. Over the remainder of the range of disparities there is mutual inhibition. The sensitivity of the unit to disparity is high, of the order of 3 minutes of arc $(0.1\Delta)$.

### DEPTH AND DIRECTION COLUMNS

It will be recalled that Hubel and Wiesel found that the neurones of the striate area were organized in columns according to eye-dominance and according to the direction-preference of the slits required to excite; Blakemore (1970) examined the effects of recording from striate neurones at successive depths in a column, and he found that, in certain columns, the neurones all had optimal retinal disparities of about the same direction and magnitude; all these neurones were responding, then, to a strip of space at a definite distance from the

fixation point. Adjacent columns had disparities differing by about 0.6°. Blakemore called these *constant depth* columns to distinguish them from *constant direction* columns whose neurones showed enormous variations in disparity from unit to unit at successive depths. When the left- and right-eye fields of these units were compared, it was found that for one eye, e.g. the left, the fields of successive units were all superimposed, so that the variation in disparity was due to the variations in the positions of the fields of the right eye. In Figure 23.64 a spatial plot of the positions of points in space that would give rise to these disparities is shown, and it becomes clear that these units have the common parameter of constant direction in relation to the left eye. Thus the column of neurones is 'peering along a tube of visual space that is lined up along one oculocentric visual direction'. Examination of several direction-columns gave a similar pattern, namely that the superimposable response fields were found in the hemisphere contra-lateral to the retina in which they were evoked, so that those responsible for the disparities were derived from the ipsilateral retina. It has been pointed out earlier that the ipsilateral projection of the retina is the phylogenetically newer projection since in lower vertebrates the decussation tends to be complete. It is interesting therefore that it is the ipsilaterally projecting neurones that show the binocular disparities that are essential for stereoscopic vision.* Thus in non-stereoscopic vision it would be the direction-columns that were of absolute importance; with the requirements of stereopsis information both with respect to direction and disparity of retinal images is required, and this is provided by the constant disparity (depth) columns.

## Cortical neurones required

Blakemore estimated an average of 0.6° difference of horizontal disparity between neighbouring depth-columns, and about 4° of oculocentric visual direction between direction-columns. If the mosaic were maintained across the whole binocular field, and if the total range of disparities were 6°, this would require 500 columns for one orientation of the target. If there were ten to fifteen optimal orientations of the target, this would make a grand total of 8000 to 12 000 columns to encode every orientation and every spatial position, a number that is not impossible anatomically (Blakemore, 1970).

### STEREOPSIS IN THE MONKEY

By the use of random-dot patterns, as illustrated in Figure 22.38, p. 530, such that the individual pictures presented to the eyes in a stereoscope are indistinguishable, it is possible to study the stereoscopic vision of the monkey, training it to respond only when the two pictures give a required stereoscopic depth perception on fusion. In this way Bough (1970) demonstrated that monkeys could discriminate between pictures that, when fused, would give no stereopsis, and those that would. Hubel and Wiesel (1970) have examined the cortical cells of Areas 17 and 18 of macaque monkeys and have shown that in Area 18 binocularly responding cells are found reacting similarly to complex and hyper-complex cells in Area 17; these they called 'ordinary cells of Area 18'; they had receptive fields in anatomically corresponding areas; they responded to separate stimulation of the eyes and showed moderate summation when stimulated together. The other half were called 'binocular depth cells', the most common of which gave no response to stimulation of the separate eyes but a brisk response to simultaneous stimulation. Some responded well to stimuli in corresponding points of the retinae, and others exhibited some disparity in their receptive fields. By contrast with the cat, no convincing evidence was found of binocular depth cells in Area 17. Thus Area 18 has two functions in the monkey, namely linking the two half-fields (as also in the cat), and the elaboration of stereoscopic depth perception.

## BILATERAL REPRESENTATION OF RETINA

Blakemore (1969) has emphasized that if the decussation of the optic nerve fibres splits the visual field exactly into two halves across the vertical meridian, so that all stimuli

from the temporal side of the field remain ipsilateral, and all on the nasal side cross over, then, when an object produces disparate images on the retinae, as in Figure 22.34, $a_L$ and $a_R$, the images can fall on the nasal halves of each retina and so their messages are carried to opposite hemispheres of the brain. In this event, the single cortical neurones that receive messages from both of these points on the retinae will have to be connected to a pair of cortical neurones in opposite hemispheres, perhaps through the corpus callosum. Alternatively, of course, we may postulate that over the small regions adjacent to the vertical meridians, where retinal disparities are utilized in binocular stereoscopic vision, the decussation may be incomplete, so that this part of the fovea will be represented, in effect, bilaterally. This might account for the 'sparing of the macula' discussed earlier, but in the primate the evidence against such a bilateral representation through incomplete decussation is preponderant, and callosal connections mediate the necessary integration between the two hemispheres.

### AREAS 17 AND 18

Hubel and Wiesel (1967) recorded from separate units simultaneously in the two hemispheres of the cat, and found some that had receptive fields that overlapped in the middle; these were invariably in the boundary between Areas 17 and 18 where the representations of the central vertical meridian in the primary and secondary visual areas are side by side (Fig. 23.12, p. 554). Blakemore has examined the response fields of binocularly driven units at the junctions of Areas 17 and 18, and has confirmed that some of these overlap, in fact their binocular centres (p. 595) could be on the 'wrong' side of the estimated vertical meridian by a degree or two, as illustrated by Figure 23.65, where the centres of the response-fields of right eyes have been plotted against azimuth from the vertical meridian. The filled circles indicate responses in the right, or ipsilateral, hemisphere which, in the absence of bilateral representation, should be confined to the left of the Figure, and the open circles indicate respones in the left, contralateral, hemisphere and should be confined to the right of the Figure. The 'wrong fields' are indicated by the line joining the points, and this indicates the degree of overlap. This central strip of retina, with a width of about 1.5° and equivalent to horizontal

---

*Blakemore and Pettigrew (1970) have shown that the receptive fields of the ipsilaterally projecting neurones are much more scattered than those of the contralaterally, when penetrations to successive depths in a column are made. Thus the ipsilaterally projecting neurones can give less precise information regarding actual direction in space, but of course the stereoscopic information they can convey is what is important.

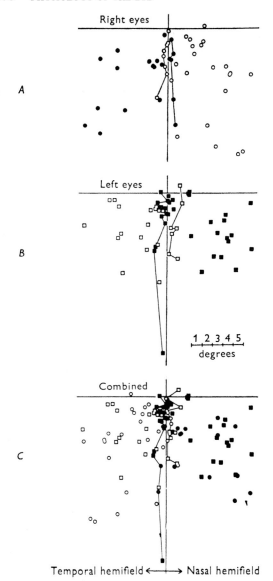

Fig. 23.65    A. Response field centres for four right eyes are
pooled by superimposing the true vertical through the middle
of the overlap and the horizontal through the estimated area
centralis. The open circles are the centres for units recorded
in the left (contralateral) hemisphere and the filled circles for
units in the right (ipsilateral) hemisphere.

B. The field centres for five left eyes are pooled in the
same way. The open squares are centres for right
(contralateral) hemisphere units and the filled squares for
neurones in the left (ipsilateral) hemisphere.

C. The data from A and B are combined by lateral
inversion of the right eyes' field-centres and the
superposition of the verticals and horizontals. Now temporal
visual hemifield lies to the left of the vertical meridian and
nasal hemifield to the right. The symbols are those used in A
and B. All field-centres lying on the 'wrong' side of the
vertical meridian (i.e. in the ipsilateral hemifield), are joined
to give an indication of the width of the strip of bilateral
projection. (Blakemore, *J. Physiol.*)

disparities of about twice this, doubtless is the region of
bilateral representation that is required for stereopsis in
the vertical meridian.

CORPUS CALLOSUM

The mechanism of this overlap could be through the
corpus callosum, in which event monocularly driven
callosal units should be found, as they apparently are
(Berlucchi, Gazzaniga and Rizzolatti 1967), whilst
cutting the corpus callosum abolishes stereopsis
(Mitchell and Blakemore, 1970), of objects in the midline but not, of course, in the periphery since in this
event both images are projected to the same hemisphere.
The situation is illustrated by Figure 23.66, which

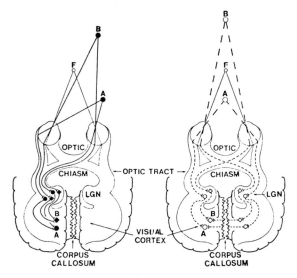

Fig. 23.66    The possible neural system for binocular depth
perception in the split-brain human. On the left is the
arrangement of neurones which enabled the subject to
recognize the depth of the peripheral objects A and B,
relative to the fixation point, F. For both objects the image in
the right eye falls on nasal retina and that in the left upon
temporal, and all the information projects to the left optic
tract. The messages from the two eyes remain segregated at
the lateral geniculate nucleus (LGN). Solid circles represent
neurones within whose receptive field the image of A falls.
Solid diamonds are cells to which object B projects. The
two binocular cells, A and B, shown as large symbols in the
cortex, encode the disparities of the objects. In this case the
sectioned corpus callosum is of no hindrance. The binocular
information is processed entirely in the left hemisphere and
the judgements can therefore be vocalized.

In the right diagram objects A and B lie directly in the
midline and, therefore, their images fall on temporal or nasal
retina, respectively, in *both* eyes. The interrupted lines and
open symbols in the visual pathway show the neurones
normally responsible for the recognition of the disparity of
these objects. The binocular cells A and B in the cortex
receive one of their inputs from a fibre from the opposite
visual cortex, crossing in the corpus callosum. Section of the
commissure has severed this link and binocular integration is
impossible. (Mitchell and Blakemore, *Vision Res.*)

describes the neural arrangements in a human subject with sectioned corpus callosum.* To the left, the subject, fixating F, will appreciate the stereopsis created by the positions of A and B because the images fall on the temporal and medial retinae, cortical neurones being binocularly driven by disparate points on the retinae. To the right, the disparities of A and B are such that both the images fall on nasal or temporal halves of the retinae, so that a cortical neurone must receive messages from the contralateral hemisphere to integrate the disparate messages. Mitchell and Blakemore found that, whereas the corpus callosum-sectioned subject could discriminate the peripherally placed points A and B in depth, he could not do so when they were in the midline (Fig. 23.66 *right*).

## PRESTRIATE CORTICAL NEURONES

Zeki (see, for example, Zeki 1976) has emphasized the role of different prestriate regions in a parallel processing of the information reaching the striate Area 17, a role suggested by Hubel and Wiesel's studies on receptive fields, which indicated that those with complex and hypercomplex fields, and presumably the result of convergence of simple cells, were much more numerous in Area 18. Moreover, the ability to discriminate depth through disparity-sensitivity belongs to cortical neurones in Area 18 (Hubel and Wiesel, 1970).

### ELECTRODE PENETRATIONS

By penetrating through different areas of striate and prestriate cortex Zeki (see, for example, 1976) was able to establish some striking differences in the characteristics of cortical neurones, according as the electrode passed through the several visual areas defined in his earlier anatomical studies. Figure 23.67 illustrates the changes on passing into V2 (Area 18) across the lunate sulcus to V4 and finally to V5 the posterior bank of the superior temporal sulcus.

---

*This operation is carried out to prevent spread of an epileptic focus from one hemisphere to the other.

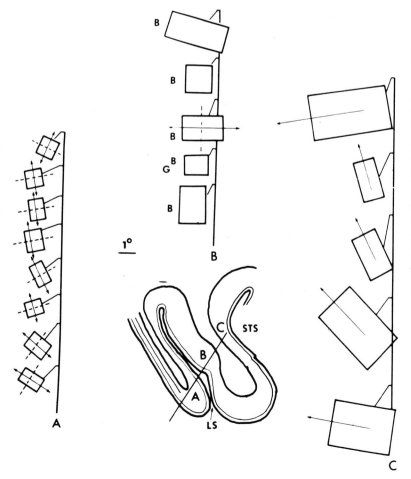

**Fig. 23.67** Illustrating changes in character of cortical units during penetration in the manner illustrated. The first cells recorded from (A), (to left of figure) were in V2; they were all binocularly driven, orientation-selective cells without any colour preference. As the electrode crossed the lunate sulcus and hit V4 (B), the cells encountered were colour-coded. They were binocularly driven. In the cortex of the posterior bank of the superior temporal sulcus, the character of the cells changed once again (C, to right of figure). The cells met here were binocularly driven, directionally sensitive for which the form, orientation or colour of the stimulus were irrelevant. Arrows indicate the preferred direction of motion of the stimulus. B, blue; G, green; LS, lunate sulcus; STS, superior temporal sulcus. (Zeki, *Cold Spring Harbor Symposia*.)

## AREA V2

Cells in V2 were, as Hubel and Wiesel (1970) had found, binocularly driven with well defined orientational preference. Others were disparity-selective (p. 594) and some colour-coded cells were also found. It seemed that among the functions emphasized in V2 is analysis of the visual fields in terms of binocular disparity, i.e. in depth perception.

## AREA V4

When the electrode passed into V4 all units examined were colour-coded (Zeki, 1973) responding vigorously to one colour and grudgingly if at all to other wavelengths and to white light; some required a field of a definite shape and position but others were independent of these parameters. In a given penetration successive cells responded to the same wavelength, suggesting an organization in columns, so that with an oblique penetration successive cells responded to different colours. Since colour-coded units in Areas 17 and 18 are rare compared with Area V4 it may be inferred that V4 is a region for analysis of the stimulus in terms of wavelength. Area 4A contained a wide variety of cells, including colour-coded ones, complex, hypercomplex, and so on. Because of this great variety it is not yet clear what its prime function is.

## AREA V5

When the electrode passed into the posterior bank of the superior temporal sulcus, V5, the majority of cells were driven from the two eyes with similar receptive fields for either eye, and most cells were equally driven by either eye—i.e. there was little indication of eye-dominance. Of interest were some cells that were similar to the pan-direction units of the superior colliculus their responses being essentially to movement in any direction, rather than to contours. Combined with this appearance of movement-sensitivity, *per se*, there was an increased proportion of direction-sensitive cells compared with Areas 17 and 18, so that a specialization for the analysis of movement is clearly a feature of this region. As with Area 17, oblique penetrations indicated a columnar organization with an orderly shift in preferred direction of motion when direction-sensitive cells were encountered.

### Opposed movement cells

In a further study, Zeki (1974b) identified cells that probably signalled movement of an external object closer to or farther away from the subject; thus, as illustrated by Figure 23.68, movement of the large bar closer to the observer produces two types of signal; as far as an individual eye is concerned the image will expand and can be mimicked by bringing two edges farther apart; as far as both eyes are concerned, the movements of the

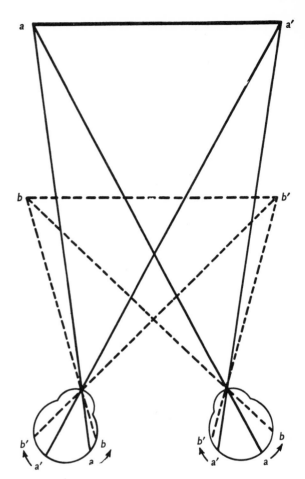

**Fig. 23.68**   Diagram to show the displacement of the retinal images when a large bar, *aa'*, is moved to *bb'*. The images will move in opposite directions in each eye. (Zeki, *J. Physiol.*)

images are in opposite directions. Cells responding maximally to these types of movement were in fact found, being either monocularly or binocularly driven. Figure 23.69 illustrates the response of a binocularly driven 'opposed movement' hypercomplex cell on stimulation of the ipsilateral eye. The cell responded maximally when two edges moved towards each other along a definite orientation; reducing the length of the edges reduced the response, indicating hypercomplex behaviour. The disparities involved in this type of movement are much larger than those involved in the responses of the disparity-sensitive cells discussed earlier, these latter responding to a fixed disparity whereas the present ones respond to changing disparity. All cells in this region were not selective of wavelength.

### PARALLEL PROCESSING

In general, Zeki's studies indicate a parallel processing of visual input in different prestriate regions; since Area

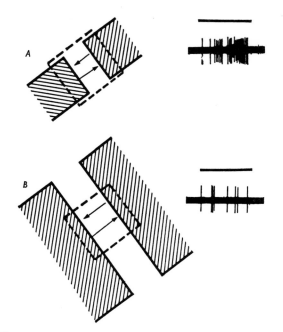

**Fig. 23.69** The response of a binocularly driven, opposed movement hypercomplex cell in the cortex of the posterior bank of the superior temporal sulcus to stimulation of the right (ipsilateral) eye. The cell responded when two edges of the appropriate orientation were moved towards each other within the receptive field. Increasing the length of the edges diminished the response (*B*). The cell was 5° × 4° and was located in the lower contralateral quadrant and crossed the midline. (Zeki, *J. Physiol.*)

17 can make direct connexions with these areas, as well as indirectly, there is the possibility of convergence of input; thus V5 receives a divergent output from Area 17; the input from Area 18 (V2) converges on this, so that it may be that the properties of most of the cells in V5 are derived from an input from Area 17, whilst the opposed movement feature, signalling changing disparity, might be supplied by Area 18 (V2) which would be an economical procedure. Thus, instead of sending multiple inputs to V2 for the analysis of fixed disparities, and another set of inputs to V5 for the analysis of changing disparities, it is only necessary to build up fixed disparity cells in V2 by a direct input from Area 17 and then build up changing disparity cells in the cortex of V5 by a direct input from V2.

### INFERO-TEMPORAL CORTEX

Bilateral extirpation of this region impairs visual learning (p. 556), perhaps by virtue of its input from the pulvinar; neurones driven by visual stimuli have been examined by Gross *et al.* (1972); the receptive fields are large, being driven by both eyes, the ipsilateral only, or the contralateral eye, with sensitivities to contrast, wavelength, size, shape and direction of movement. The latencies of response were consistent with a striate-prestriate-inferotemporal pathway (Gross *et al.*, 1967).

## THE VISUALLY EVOKED RESPONSE

The arrival of afferent impulses at the sensory cortex gives rise to a series of electrical changes recorded from the surface of the exposed cortex—the electrocorticogram (ECoG). These are of sufficient magnitude to be distinguished from the background activity, especially when this has been depressed by anaesthesia. In man, attempts have been made to record similar changes through scalp electrodes, but their attenuation through this method of recording demands the use of special 'averaging techniques' that allow a separation of the evoked response from the background, i.e. the signal-to-noise ratio must be increased. The summation technique is based on the principle that the average amplitude of a signal, time-locked to a reference point in time, increases in direct proportion to the number of samples, whilst recurrent activity randomly related in time to the reference increases as the square root of the number. The principle, then, is to stimulate with repeated flashes of light and to add the responses by a computer of average transients (CAT).

### AMPLITUDE AND LATENCY

Figure 23.70 illustrates schematically the cycle of changes of potential that may be obtained by this technique; as will be seen, as the intensity of the stimulus is decreased, the latency of the response increases and the amplitudes of the different waves change in a complex manner, reflecting the algebraic summation of positive and negative components; thus the increase in amplitude of the first positive wave, $P_1$, as intensity of stimulus decreases, is due to the failure of the negative wave, $N_1$. Consequently latency is more closely related to intensity than amplitude. The fact that the Stiles-Crawford effect is prominent in the VER indicates that cones have a dominant influence, and when a luminosity curve was constructed, on the basis of the reciprocal of the latency versus wavelength of stimulating light, this had the typical photopic maximum (De Voe, Ripps and Vaughan, 1968).[*]

---

[*] Ennever *et al.* (1967) have shown that the potential recorded from the human scalp can be seriously contaminated by non-cortical activity; thus the act of blinking produces a record very similar to the VER, but this is due to modulation of the electric field over the scalp caused by the corneo-retinal potential, movements of scalp, neck and face muscles altering the electrical resistance of the scalp and so altering the recorded potential. We must note, on the other hand, that Cobb and Dawson's (1960) studies of the latencies of the ERG and VER have shown quite conclusively that the VER is not merely the electrotonically transmitted record of the ERG.

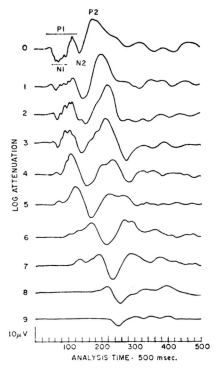

**Fig. 23.70**   VER as a function of stimulus. 4° stimulus, 10 μsec duration, 2·5 + 10⁷ mL peak luminance (0 log attenuation). Electrode placement: mid-occipital to linked ears. (Vaughan in *Clinical Electroretinography*.)

SCOTOPIC INFLUENCE

However, when modern averaging techniques were employed the existance of visually evoked responses to scotopic stimuli could be demonstrated; thus Klingaman (1976) obtained a characteristic dark-adaptation curve, with the typical break after 10 minutes in the dark, when he plotted the reciprocal of the amplitude of the late positive wave in response to a weak blue light against time in the dark. The time-course of the VER-adaptation corresponded well with that of the psychophysically determined threshold. When different wavelengths were employed, Fujimura *et al.* (1975) obtained a characteristic scotopic sensitivity curve with maximum at about 500 nm.

APPLICATION

The VER has been employed in the study of a number of phenomena of human vision; for example, in retinal rivalry it seems that the amplitude of the flash-evoked VER recorded from a given side of the head is depressed if the contralateral visual field is being suppressed during rivalry (Lansing, 1964; Lawwill and Biersdorf,* 1968). Again, the presence of orientation- and size-selective neurones in the human cortex has been surmised on the basis of studies of the VER (Campbell and Maffei, 1970).

## DEVELOPMENTAL ASPECTS OF VISUAL FIELD CHARACTERISTICS

### VISUAL DEPRIVATION

The consequences of failure to use an eye in childhood are well known as clinical manifestations of squint or of the operated congenital cataract (Von Senden, 1960). In the human infant, the deprivation must apparently extend for several years, and the loss of vision exhibits some degree of reversibility if the disuse has not lasted too long. In the cat, Hubel and Wiesel carried out pioneer investigations into the requirements of previous visual experience for the features of the receptive field of cortical neurones; and the subject has been subsequently investigated in great depth by Blakemore and his colleagues.

DOMINANCE AND ORIENTATION SPECIFICITY

The two features that have excited most interest are the degree of dominance by a given eye, revealed by the dominance histogram (Fig. 23.41, p. 575), and the orientation-specificity, of simple cells, the question posed being, essentially, whether these features of the cortical cells are innately determined at birth and, if so, to what extent they may be modified by visual experience as soon as the eyes are capable of responding to light and a patterned environment. In this respect it must be appreciated that about seven days must elapse in the kitten before the eyes open and even then the optical quality of the retinal image is poor until the pupillary vascular membrane disintegrates (Freeman and Marg, 1975), whereas the eye of the newborn monkey is open and optically clear. A newborn kitten, kept in the dark or with its eyes sutured for several weeks after birth, is completely blind; but this does not mean that its cortical cells are unresponsive, so that the orientation-specificity and ocular dominance of given cortical cells may be determined, and compared with those of normally reared animals or with those in the newborn.

## DOMINANCE HISTOGRAMS

BILATERAL DEPRIVATION

In general, it is established that, at the time the kitten first opens its eyes, most of its cortical neurones have bilateral input giving an adult histogram (Hubel and

---

*Donchin and Cohen (1970) question this interpretation of the depression of the VER in rivalry; they point out that if the subject's attention is drawn to the flash rather than the rivalry patterns a prominent VER is obtained, whereas the same flash gives no VER if the subject concentrates on the rivalry pattern. This supports the contention of Ennever *et al.* that the VER is a record of eye-movement rather than of cortical responses.

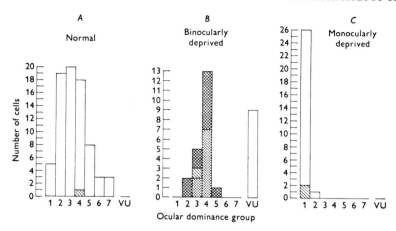

**Fig. 23.71**   Histograms, showing the number of neurones, their ocular dominance groups and the types of receptive fields classified on the basis of responses in the eye dominating the cell. Cells in groups 1 and 7 are monocularly driven, being excitable only through the contralateral (left) eye or ipsilateral (right) eye respectively. Cells of group 4 are equally driven by the two eyes. Group 3 and 5 cells are slightly more strongly driven by the contralateral and ipsilateral eyes respectively; group 2 and 6 neurones are very strongly dominated by the contralateral and ipsilateral eyes respectively. Receptive field types: *orientation selective* = open blocks; *orientational bias* = cross-hatched blocks; *pure direction selective* = diagonally striped blocks; *non-oriented* = stippled blocks; *visually unresponsive* = open blocks under a separate column labelled V.U. *A*, seventy-six neurones from three normal kittens aged 3 to 5 months. *B*, thirty neurones from a kitten binocularly deprived until recording at 8 weeks. *C*, twenty-seven neurones from a kitten monocularly deprived by suturing the right (ipsilateral) eyelids until recording at 5 weeks. (Blakemore and Van Sluyters, *J. Physiol.*)

Wiesel, 1963); and this binocularity survives rearing in the dark or *bilateral* lid-suturing (Fig. 23.71B).

UNILATERAL DEPRIVATION

When only one eye is sutured, however, there is a striking change in the dominance histogram, as seen in Figure 23.71 which shows the histogram for normal cats (A) and for kittens that had been reared with the ipsilateral eye occluded (C). It will be seen that a negligible number of cells are driven by the ipsilateral eye (Group 1) and practically all by the contralateral eye. Clearly, this robbing the animal of visual experience has resulted in effective destruction of cortical synapses, and it would seem that there is a state of competition between the two eyes for synaptic connexions with cortical cells during development.

PERIOD OF SUSCEPTIBILITY

INDUCTION INDEX

Occlusion of a single eye in the adult cat has no influence on ocular dominance, as measured in electrode penetrations, so that it is of interest to determine the period during which the synaptic connexions remain plastic, and the extent to which any changes, due to modifying the kitten's visual environment, are reversible. Figure 23.72 summarizes the results of Hubel and Wiesel, where the ordinate is the *induction index*, defined as the ratio of the number of cells dominated by the experienced eye over the total number of cells; thus the value of about 0·3, represented by the dashed horizontal line, corresponds to the normal adult value; the horizontal bars indicate the periods during which the contralateral eye was held closed; and it will be seen that, to achieve an index of unity, the closure must occur at an early state in development, whilst the total sensitive period is of the order of fifteen weeks. The period of susceptibility begins rather suddenly at about the beginning of the fourth week; it is very high during that week, and it begins to fall between the sixth and eighth week and has disappeared at the end of the third month.

RAPIDITY OF EFFECT

Hubel and Wiesel (1970), examining kittens at the most sensitive period to monocular deprivation (3 to 5 weeks old), found definite abnormalities in the histogram after only three days whilst after 6 days the picture was

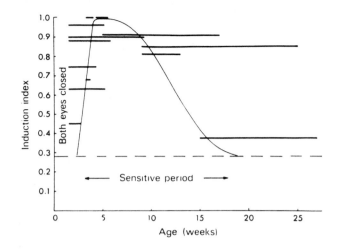

**Fig. 23.72** Time-course of the sensitive period for changes in ocular dominance in kitten cortical cells, induced by monocular visual deprivation. (Blakemore from Hubel and Wiesel, *Brit. med. Bull.*)

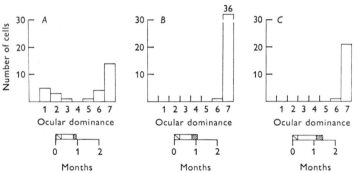

**Fig. 23.73** Ocular-dominance histograms for cells in the left visual cortex in three kittens deprived by right-eye closure during days (*A*) 23 to 26; (*B*) 23 to 29; (*C*) 30 to 39. Kittens *B* and *C* were litter-mates. (Hubel and Wiesel, *J. Physiol.*)

similar to that of an animal monocularly deprived for a long time (Figure 23.73). Again, Olson and Freeman (1975) found definite changes in the ocular dominance-pattern after 1 day of monocular deprivation of a 4-week-old kitten; they pointed out that it was un-necessary to keep the kittens in the dark before imposing the monocular deprivation. When the sensitivity of the cortex to exposure to stripes of a given orientation was examined, Blakemore and Mitchell (1973) found that exposure of as little as 1 hour to vertical stripes produced cortical neurones with receptive fields pre-dominantly sensitive to the vertical by comparison with control animals similarly reared in the dark but with no visual experience. When the criterion was a change in the dominance histogram, then 6 to 20 hours of monocular vision on the 29th day of lid suturing was sufficient to shift dominance towards the experienced eye (Peck and Blakemore, 1975).

## STRABISMUS

When squints were induced from birth by section of a rectus muscle, there was no obvious visual defect, but when the distribution of ocular dominance amongst

cortical cells was examined, this was found to be grossly abnormal (Fig. 23.74) Groups 1 and 7 predominating at the expense of bilaterally driven neurones; under these conditions each eye is receiving experience, and this experience is permitting it to pre-empt cortical neurones

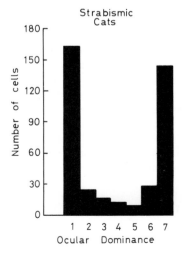

**Fig. 23.74** Ocular dominance histogram for kittens reared with artificial strabismus. (Blakemore from Hubel and Wiesel, *Brit. med. Bull.*)

at the expense of the other. Thus, for a neurone to be binocularly driven, both eyes must apparently view the same object or patterned environment at the same time, and on corresponding parts of the two retinae. Hence any method of robbing the cortical neurones of congruent signals from the two eyes may be expected to reduce binocularity.

## STRIPED ENVIRONMENTS

Blakemore established squints in the young kittens surgically, and reared these in the dark except at intervals when they were exposed to a striped environment. Under these conditions there was no significant loss of binocularity since, although the eyes were squinting, the retinal images overlapped sufficiently to permit simultaneous activation of cortical cells by the two eyes. If, however, a squinting kitten was exposed to horizontal stripes through one eye and vertical stripes through the other, binocularity was impaired (Hirsch and Spinelli, 1970, 1971; Blakemore, 1976).

## ALTERNATING MONOCULAR OCCLUSION

Hubel and Wiesel found that exposing only one eye at a time to a patterned environment, by alternately occluding one eye, had a similar effect to that produced by strabismus. The essential feature of the treatment is, once again, robbing the cortical cells of simultaneous congruent signals from both eyes. The importance of simultaneity was demonstrated by exposing dark-reared kittens for periods to a striped environment with alternating occlusion; thus they received the same retinal image through each eye but never simultaneously; and binocularity was reduced to the same extent as through simple alternating monocular occlusion. Thus the necessary condition for the maintenance of binocularity is that cortical neurones should be stimulated by patterned retinal images with contours of approximately the same orientation, falling simultaneously on the receptive fields of the two eyes (Blakemore, 1976).

## IMPORTANCE OF PATTERN

Blakemore (1976) placed a transparent neutral filter in front of one eye of a kitten, thereby reducing the light-flux entering the eye but not affecting the distinctness of the images of external objects; in front of the other eye he placed a translucent opal glass that not only reduced the light-flux entering the eye but also impaired image-formation. The ocular dominance was shifted to the eye with the patterned experience. Thus mere reduction of retinal illumination of one eye, with maintained patterned vision, has only a very small effect, if any, on ocular dominance.

# DEVELOPMENT OF PATTERN SENSITIVITY

## VISUAL ACUITY

The visual acuity of the kitten, measured in terms of resolvable spatial frequency, increases rapidly from about the start of the fourth week, at about the same time as the sensitive period for exhibiting effects of deprivation comes to an end (Freeman and Marg, 1975). At the 23rd day it was about 0·3 cycles/degree of 20/200; at Day 33 it was 1 cycle/degree of 20/600, and between Days 33 to 100 it reached 3 cycles/degree or 20/200.

## REQUIREMENTS FOR PATTERNED VISION

The development of patterned vision requires the specialization of cortical cells so that they respond to, or are triggered by, highly specific visual stimuli, such as the orientation of a line, the direction and velocity of movement, to the polarity of an edge, to the angular width of a bar, and so on; and there is no doubt that these qualities, as well as the ability to respond to visual stimuli in both eyes, can be considerably modified by visual experience or its absence.

## DEVELOPMENT OF SPECIFIC SENSITIVITIES

Pettigrew (1974), in his study of disparity-tuning (p. 596), made observations on orientation- and direction-specificity of cortical cells in the developing kitten, and compared the relative proportions of sensitive cells in the adult with those in the visually deprived animal. It is clear from Figure 23.75 that the picture in the deprived kitten is not much different from that of the normal kitten in its second week, when good pattern vision first becomes feasible, whereas in the fifth week the visually experienced kitten has a nearly normal pattern of sensitive neurones, by contrast with the binocularly deprived animal. In general, Pettigrew concluded that, whereas binocularity and motion-sensitivity are innately specified, visual experience is essential for development of orientation- and disparity-specificities. Hirsch and Spinelli (1970) reared kittens in an environment such that one eye viewed vertical stripes and the other horizontal stripes, and found that the preferred directions of simple cortical cells matched the direction to which the eye had been exposed. Again, Blakemore and Cooper (1970) exposed kittens to a variety of directional stripes and, after five months, examined their behaviour; they were obviously blind to lines at right-angles to those of their environment if tested by a visual placing reaction on to a lined surface; Figure 23.76 shows the preferred directions of the cortical neurones examined some 7–8 months after the visual testing.

## RECEPTIVE FIELD TYPES

In a more elaborate study of the effects of visual experience on cortical neurones Blakemore and Van

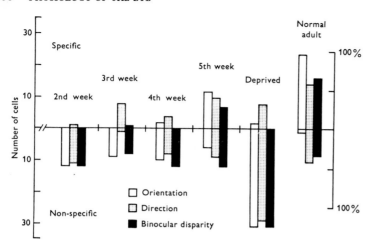

**Fig. 23.75** The numbers of cortical cells responding specifically (above horizontal) to orientation, direction and binocular disparity are shown during development of the normal kitten; included in the diagram are the numbers in binocularly deprived kittens aged 2 to 6 weeks. The proportion of specific cells increased with age and visual experience, particularly with respect to orientation and disparity. No cells were found in binocularly deprived cortex that were specific for disparity and very few showed orientation selectivity. Directional sensitivity was seen at all ages irrespective of visual experience. (Pettigrew, *J. Physiol.*)

Sluyters (1975)* have made a study of receptive fields of over 700 cortical neurones in normal kittens with ages ranging from 9 days to 22½ weeks, and the effects of depriving the kitten of visual experience on the receptive fields. The receptive fields were defined as follows: *Non-orientated*, responding to moving stimuli but with no clear preference for direction of movement or any orientation of edge. *Pure direction-sensitive*, the receptive field having a clear axis, movement in one (preferred) direction exciting and in the reverse (null) direction, not; they are comparable to Barlow's direction-selective ganglion cells in the rabbit; *Orienta-*

*tion-bias*; here the cell responds to movement in all directions but shows a preference for one axis; this cell

---

* The requirement of experience for the development of orientation-selectivity in cortical neurones has been questioned in more than one study; Hubel and Wiesel, in their original studies on binocular deprivation, found orientation-selective neurones, but Blakemore and Mitchell (1973) and Imbert and Buisseret (1975) were unable to do so. Sherk and Stryker (1976) sutured kittens from first lid-opening and examined them at 22 to 29 days old. Ninety out of 98 cortical cells responded to a moving bar stimulus at an appropriate orientation, and a similar columnar organization of the orientation-selective units to that in the adult was found.

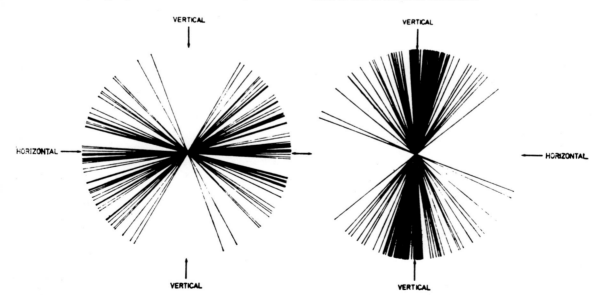

**Fig. 23.76** These polar histograms show the distributions of optimal orientations for fifty-two neurones from a horizontally experienced cat on the left, and seventy-two from a vertically experienced cat on the right. The slight torsion of the eyes, caused by the relaxant drug, was assessed by photographing the pupils before and after anaesthesia and paralysis. A correction has been applied for torsion, so the polar plots are properly orientated for the cats' visual fields. Each line shows the optimal orientation for a single neurone. For each binocular cell the line is drawn at the mean of the estimates of optimal orientation in the two eyes. No units have been disregarded except for one with a concentric receptive field and hence no orientational selectivity. (Blakemore and Cooper, *Nature*.)

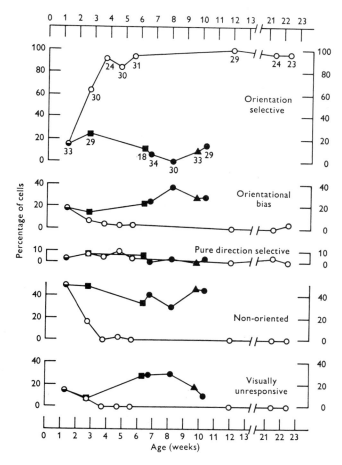

**Fig. 23.77**  Comparison of the effects of binocular deprivation with those of normal visual experience. The five pairs of curves show the percentages of each type of receptive field for 397 cells recorded from seven normal kittens (open circles) and six binocularly deprived kittens (filled symbols) of various ages. In each case the two curves originate from points representing the data for the 9-day-old kitten recorded at the time of natural eye opening. For the binocularly deprived kittens: *Lid suture* = filled circles; *dark reared* = filled squares; *nictitating membrane suture* = filled triangle. The numbers beneath the data points for the top pair of curves indicate the total number of cells recorded in each animal. (Blakemore and Van Sluyters, *J. Physiol.*)

is rare in the adult cat, but has been identified in the rabbit (Van Sluyters and Stewart, 1974); *Orientation-selective* with a clear preferred orientation for a linear stimulus, with no discharge for movement along the orthogonal axis; these are the cells described by Huber and Wiesel (p. 566) and are subdivided into simple, complex and hypercomplex categories.

ORIENTATION-SELECTIVITY

Blakemore and Van Sluyters found that kittens older than about 4 weeks, brought up in a normal environment, were virtually adult in their cortical physiology, and even the youngest studied (19 days) was surprisingly mature, so that despite the cloudy optics, the majority of neurones were distinctly orientation-selective, narrowly tuned for orientation, and binocularly driven. This orientation-selectivity could be demonstrated by flashing a bright bar with an appropriate orientation, and so was differentiated from the *direction-sensitive* neurone that required motion to excite. The dependence on visual experience is illustrated by Figure 23.77 where the initial point represents a nine-day-old kitten that had just opened its eyes and was, indeed, visually in-

experienced. The curves indicate changes due to visual experience, but the existence of orientation-selectivity in the nine-day-old kitten is remarkable and is illustrated in more detail in Figure 23.78, which reconstructs the penetration of an electrode. The ocular dominance pattern for this kitten was normal.*

**Requirements of visual experience**

Blakemore and Van Sluyters varied the type of visual experience in a variety of ways and measured the proportions of cortical neurones exhibiting given receptive field-types. The results emphasized the importance of lines, and, as had been found earlier, if the lines were in a given orientation the orientation-selective units

---

* An interesting exception amongst the visually deprived kittens was one that had been kept in a diffusely illuminated cylinder which showed a strong preference for vertical or near-vertical lines, in contrast with another kitten with its nictitating membranes sutured and thus similarly exposed to diffuse uncontoured light; however, it turned out that there were, in fact, some low-contrast contours visible to the kitten in the cylinder, e.g. a distinct vertical seam where the sheet of white card lining the walls of the cylinder was joined together.

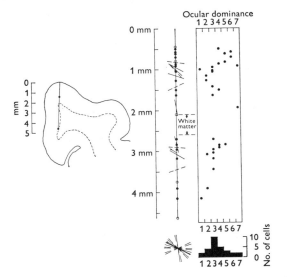

**Fig. 23.78** Illustrating direction-sensitivity in the 9-day-old kitten. This penetration, as is shown in the coronal section, briefly passed through white matter. The points at which the electrode entered and left the white matter are indicated by horizontal interrupted lines beside the schematic reconstruction. *Orientation selective* = continuous line at preferred orientation; *orientational bias* = interrupted line at the optimal orientation; *pure direction selective* = arrow in the preferred direction; *non-oriented* = filled circle; *visually unresponsive* = open circle. (Blakemore and Van Sluyters, *J. Physiol.*)

were sensitive to this direction (Fig. 23.79). As their summary-diagram indicates (Fig. 23.80) the main change in the character of the receptive field after birth was in orientation-selectivity, a characteristic that is poorly developed at birth and requires a patterned environment for its development.

### ANATOMICAL FEATURES

Cragg (1972) found that only 1·5 per cent of the adult complement of synaptic terminals is present at day 8 of the kitten, the youngest animal studied, and dendritic spines did not appear at all until day 7 to 10, so that the responsiveness and stimulus-specificity of the young kitten are surprising; such synapses as are present in the very young kitten occur in the deeper cortical layers where, in fact, the orientation-responsive neurones are found; and it is only later, in weeks 2 to 4, that large increases in synaptic density appear in the more superficial layers, coinciding with the almost adult character of the kitten cortex at 4 weeks old revealed by Figure 23.77.

## DEVELOPMENT OF DISPARITY SPECIFICITY

### DISPARITY TUNING

We have seen (p. 596) that cortical neurones show an optimal binocular response when the points on the two retinae that are excited have a certain disparity; within this range there is facilitation of responses from either eye, but when this range is exceeded, greater disparities lead to mutual inhibition, as revealed by Figure 23.63 (p. 596). In very young kittens, the optic axes diverge strongly (as much as 30° in a 2-week-old animal), but during the critical period there is a rapid decrease in this retinal disparity running parallel with other features already discussed. In these very young kittens it is possible to evoke binocular responses in cortical neurones, provided the optic axes are brought into alignment by means of prisms, but there is none

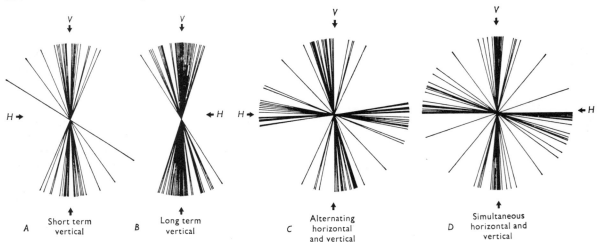

**Fig. 23.79** Polar diagrams summarizing the preferred orientations in the dominant eye for all orientation selective and orientational bias cells recorded from animals reared in striped environments. *A*, 'short term vertical' kitten; *B*, one of the 'long term vertical' kittens; *C*, one of the 'alternating horizontal and vertical' kittens and *D*, 'simultaneous horizontal and vertical' kitten. (Blakemore and Van Sluyters, *J. Physiol.*)

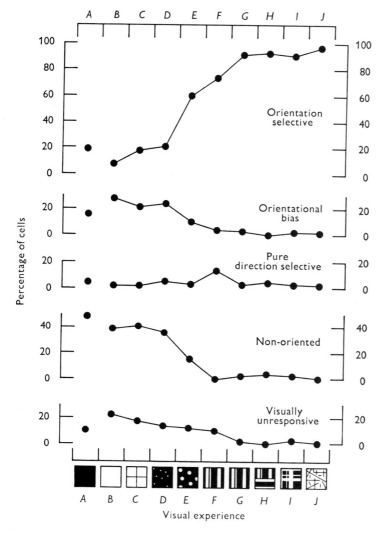

Percentage of cells

Visual experience

**Fig. 23.80** Summarizing the effects of different kinds of visual experience on the selectivity of cortical cells. A represents two very young visually inexperienced kittens; B, five binocularly deprived kittens; C, one kitten reared in a diffusely illuminated tube and given minimal experience of horizontal and vertical contour; D, two kittens reared in 'small spots'; E, two kittens reared in 'large spots'; F, one kitten given 'short term' experience of vertical contour; G, two kittens given 'long term' experience of vertical contour; H, two kittens exposed alternately to horizontal and vertical contour; I, one kitten exposed simultaneously to horizontal and vertical contour; J, four normally reared kittens all of whom were older than five weeks at the time of recording. (Blakemore and Van Sluyters, *J. Physiol.*)

of the disparity-selectiveness, or tuning, that is characteristic of the adult where 63 per cent of cortical neurones exhibit this sharp tuning (Pettigrew *et al.*, 1968). Thus in kittens of 16 days, for example, a neurone could be found sensitive to binocular stimulation so that nearly three times more spikes were elicited on stimulating both eyes than by stimulating the contralateral eye; however, this facilitation of one eye by the other extended over a wide disparity (over 6 degrees in the ipsilateral retina), and the characteristic inhibition at high disparities, and peak response at a minimum disparity, were not present.

## Maturation

The changes leading to maturity are illustrated in Figure 23.81; here the adult response in this instance exhibits *no* facilitation, so that the binocular response at the tuned disparity is equal to that of the contralateral eye alone; increases of disparity lead to reduced

responses indicating mutual inhibition. Thus development to adult is manifest as a reduction in *facilitation*, and an increase in *inhibition*, to produce the highly tuned binocular disparity-selector.

## Effects of deprivation

That the development of this selectivity depended on visual experience rather than on an inherent tendency to maturation, was shown by binocular deprivation experiments; an example is given in Figure 23.82 which shows the disparity-response curve for a 6-week-old kitten reared normally (lower curve) and for a kitten with lids sutured during this period; the binocularly deprived neurone shows facilitation over a wide range of disparity similar to an immature neurone, although the optimum is fairly well defined. In contrast, the normal neurone exhibits facilitation over a narrow range and, at non-optimal disparities, the binocular response is depressed below monocular levels.

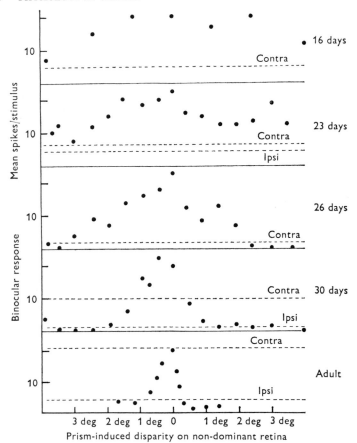

**Fig. 23.81** Showing changes in response to disparity with increasing age. Each point represents the mean number of spikes ($n = 8$ or 10 repetitions) elicited by the movement of the preferred stimulus over both receptive fields of a binocularly activated neurone. The mean response elicited monocularly is given by the dotted levels (ipsilateral or contralateral eye); absence of the monocular level for the ipsilateral eye in two cases, signifies that no response could be elicited from that eye. Stimuli were 6 degrees × 0·2 degree lines of preferred orientation and direction moving at speeds from 2 to 4 degrees/sec. Immature neurones were characterized by binocular facilitation which was relatively insensitive to changes in the retinal disparity of stimuli presented to both eyes. The response to a target passing over both fields was much greater than any response which could be elicited by stimulating each field alone, even when the two fields were not optimally aligned. With increasing age and visual experience narrower limits are placed on the zone of facilitation, the binocular response actually being depressed below monocular levels unless the receptive fields are exactly aligned in the plane of stimulation. (Pettigrew, *J. Physiol.*)

## Development of inhibition

Thus inexperienced binocular neurones tolerate a wide range of retinal disparity, and this range is narrowed by age and visual experience; hence disparity-detection, which is the basis of stereoscopic vision, results from a refinement of the interaction of the two eyes on a given cortical neurone, a refinement that would be very difficult to develop in an innate fashion in view of the variations in the directions of the optic axes of growing kittens. The development is one of inhibition, and another example of inhibitory interactions in visually experienced animals has been described by Sherman and Sanderson (1972) whilst Lund and Lund (1972) have shown, in the visually inexperienced rat, the failure of development in the superior colliculus of F-type terminals, presumed intrinsic inhibitory connections.

### STEREOPSIS

The absence or severe reduction in the number of binocularly driven cells found in the monocularly deprived kitten suggests that stereopsis would be poor or non-existent in this experimental animal. By rearing kittens in the dark and allowing them only uniocular vision in alternating eyes for short periods, Packwood and Gordon (1975) destroyed all power of stereopsis,

as measured by performance tests, although visual acuity was normal (resolution of about 6 minutes of arc). The Siamese cat is abnormal in having almost complete decussation of its optic pathway, with the result that very few of its striate neurones are binocularly driven (Hubel and Wiesel, 1971; Kaas and Guillery, 1973); like the kitten submitted to alternating monocular deprivation, the Siamese cat showed no stereopsis.*

---

* Siamese cats are often cross-eyed and this may follow from an unusual projection from the retina on to the lateral geniculate body, since Guillery (1970) has shown that many fibres that should, in the normal animal, have crossed in the chiasm, failed to do so, with the result that parts of the ipsilateral visual field were represented on the lateral geniculate body and thence on to the cortex. Hubel and Wiesel (1971) showed that this projection from the contralateral eye relayed in layer $A_1$, which in the normal cat receives input from the ipsilateral eye. This aberrant projection resulted in the appearance on the cortex of regions represented by the ipsilateral field inserted between regions representing the contralateral field. Interestingly, no cortical cells were driven by both eyes, as found with kittens brought up with artificially created squints. Later Kaas and Guillery (1973) described two different types of abnormality in American bred Siamese cats; in one type the relays from layers $A_1$ and $C_1$ were suppressed, whereas in the other there was a rearrangement with a reversal of input from these segments.

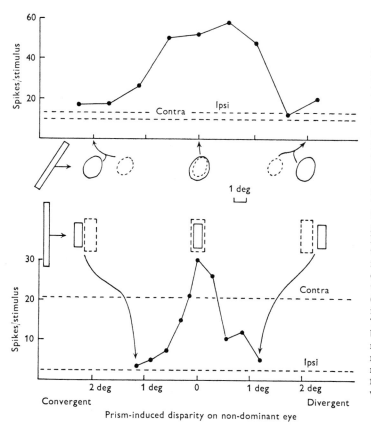

**Fig. 23.82** Most selective binocular response curves found in visual cortex of binocularly deprived (upper curve) and of normal (lower) kittens at 6 weeks. Both neurones were well matched in terms of receptive field size and both were located less than three degrees from the area centralis. Despite comparable accuracy in their detection of the retinal position of a stimulus presented to one eye alone, the two cells differ markedly in their ability to detect the relative retinal positions of the two images of the same stimulus presented to both eyes. The binocularly deprived neurone gave a binocular response which is much greater than the monocular response levels (indicated by dashed lines) and it was independent of disparity over a wide range; binocular facilitation was still evident when the stimulus passed over the two receptive fields successively (contralateral field—continuous outline, ipsilateral field—interrupted outline). The normal neurone gave a facilitatory binocular response over a very narrow range of disparity; the binocular response was suppressed below the monocular level when the two receptive fields were offset by a fraction of their width. (Pettigrew, *J. Physiol.*)

## THE VISUAL DEFECTS FOLLOWING DEPRIVATION

### BILATERAL DEPRIVATION

The kitten deprived of visual experience bilaterally for some months after birth is said to be behaviourally blind (Ganz *et al.*, 1972), but according to Sherman (1974) there is improvement after eye-opening, and the initial behaviour may be the consequence of complete absence of visuomotor experience during deprivation. Sherman used a perimetric technique, whereby he introduced new objects into a definite segment of the visual field and observed the visuomotor response. Both his binocularly deprived cats seemed totally blind at first, then visual placing, and the following of moving objects and reaching with a forepaw, returned after 17 days, in one, and 5 days in the other. The perimetric tests indicated that the binocular field of vision was fairly normal, but the responses to appearance of a new object indicated that the cats were ignoring the contra-lateral field of view when tested monocularly, respond-ing only to objects in the ipsilateral hemifield.★

### MONOCULAR DEPRIVATION

Dews and Wiesel (1970) compared the behaviour of kittens monocularly deprived, when the deprived eye

was open and the undeprived eye was closed, and *vice versa*. They concluded that, for kittens sutured after birth and opened after 4 months, the animal had no visuomotor control when using only the deprived eye. However, Sherman (1973, 1974), using his perimetric technique, showed that monocularly deprived cats responded to objects in the monocular segment of the ipsilateral visual field and ignored the binocular seg-ments, as illustrated by Figure 23.83. We may presume, then, that the monocularly deprived kitten makes use of such monocularly driven cortical neurones as are avail-able. When Dews and Wiesel deferred lid-suture until days 8 to 15 after birth, and lid-opening was allowed at about 110 to 145 days, there was some evidence of response to light, but not to patterned stimuli. As date of lid-suture was advanced, the response to perform-ance-tests improved and, in general, the effects of timing of lid-suture on the degree and permanence of visual defects paralleled the changes in the cortical neurones with respect to ocular dominance.

---

★ Blakemore and Van Sluyters (1974), in discussing the amblyopia in a squinting eye, have pointed out that children with extreme astigmatism that defocusses one meridian, by contrast with the orthogonal one, are often left with a meri-dional amblyopia, acuity for contours being poor when these are in the out-of-focus meridian.

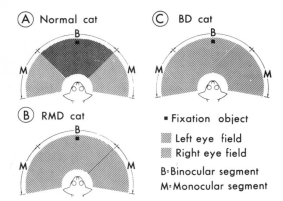

Fig. 23.83 Monocular and binocular segments of cat's visual field when both eyes are aligned on the fixation object. A, *Normal cats.* The binocular field extends between about 100° on either side. The monocular fields each extend from about ipsilateral 100° to contralateral 45° which delineates the *binocular segment* of visual field as bounded bilaterally by about 45° on either side and the *monocular segment* on each side as extending from about 45° to about 100°. B, *Monocularly deprived (MD) cats.* The binocular field has a normal extent as in A. Also, the non-deprived eye monocular field is normal but the deprived eye monocular field includes only the monocular segment. C, *Binocularly deprived (BD) cats.* The binocular field has a normal extent as in A, but each monocular field includes only the ipsilateral hemifield. (Sherman, *Brain Res.*)

## DEFECTS IN GENICULATE UNITS OF SQUINTING EYE

Hubel and Wiesel found no obvious visual defects in their kittens raised with squints due to section of a rectus muscle; however, there is no doubt that central visual acuity is markedly impaired in the squinting eye, manifest in the human as *amblyopia ex anopsia.* Ikeda and Wright (1976) studied the receptive field characteristics of lateral geniculate neurones, representing central vision, in the cat reared with a convergent squint, the spatial resolution of these cells being poor compared with geniculate cells representing the non-squinting eye. Ikeda *et al.* (1976) correlated the defective responsiveness of the geniculate cells with morphological changes, similar to those observed in cells representing the monocularly deprived eye, but they also observed that these morphological defects extended beyond the regions of representation of the area centralis of the squinting eye, suggesting that the visual defects were not confined to central vision.

### NASAL VISUAL FIELD

In a more elaborate study, Ikeda *et al.* (1977) measured the spatial resolution of cells in different parts of the lateral geniculate body of kittens reared with a convergent strabismus; the regions were chosen so as to represent central and more and more peripheral parts

of the retina, and the units responding were compared with corresponding areas for the non-squinting eye. In general it was found that, in each lateral geniculate body of the normal kitten, cells with normal receptive fields were equally common in layers A and A1, whereas in kittens with over 23·5° of squint no briskly functional cells could be recorded from layer A1 of the left lateral geniculate body where the squinting eye's *nasal visual field* projects. In addition to this restriction in the visual field, it was found that cells representing the area centralis of the squinting eye, although responsive, had much lower spatial resolution; the degree of loss did not depend on the degree of squint, whereas the restriction of the nasal field did do so. In general, there was a graded deterioration in the function of cells fed by the squinting eye, ranging from minor alterations in the receptive field properties in the temporal visual field projection zones, and the loss of spatial resolution at the central retinal projection zone leading to complete loss of function at the extreme nasal field projection zone. Morphologically a gradation in defect could also be established; thus Figure 23.84 shows the mean perikaryal size of cells in different zones, in *A* as we pass from that fed by the central visual field (1) temporally, and in *B* in layer A1 passing from the zone fed by the central visual field (9) nasally. Obviously shrinkage occurs in all regions represented by the squinting eye, and the effect is greatest in the extreme nasal field. An examination of the visual fields of the normal and

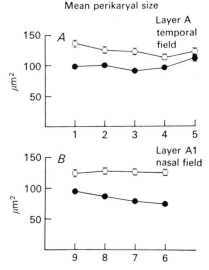

Fig. 23.84 Effect of strabismus on mean perikaryal size of lateral geniculate cells. A, Cells sampled at the five different visual field projection zones in layer A representing the area centralis (zone 1) to progressively peripheral temporal fields (zones 2–5). B, Cells sampled at the four zones in layer A1, representing the area centralis (zone 9) to progressively peripheral nasal fields (zones 8, 7, 6). Normal eye, open circles; squinting eye, closed circles. (Ikeda *et al.*, *J. Physiol.*)

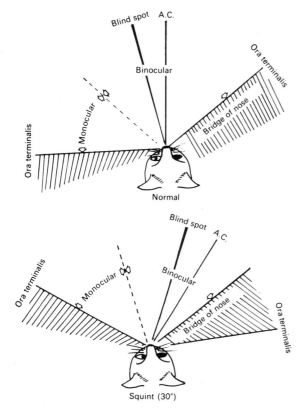

**Fig. 23.85**  The limit of the visual field of a cat seen by a normal left eye (upper diagram) and that seen by a left eye with a convergent squint of 30° (lower diagram). Note that the convergent squint of 30° results in approximately 30° of the temporal retina, which should receive inputs from the peripheral nasal field, being hidden behind the bridge of the nose, thus preventing stimulation. (Ikeda *et al.*, *J. Physiol.*)

squinting left eyes, as in Figure 23.85 reveals that a portion of the nasal field of the squinting eye is restricted by the bridge of the nose thus preventing stimulation. Thus the defects are essentially those due to disuse, and they will be graded because the shadow of the bridge of the nose does not create a sudden occlusion, the effects depending on the degree of eye movement and the angle of incident light. The defects in the neurones fed by the area centralis, and giving rise to low spatial resolution in central vision, are not due to disuse but to inappropriate stimulation of non-corresponding areas of the two retinae during development.

## Behavioural defects

Ikeda and Jacobson (1977) extended these studies to the responses of kittens to objects presented at different positions in the visual field and found restrictions in the visual field that correlated with the neurophysiological and neuro-anatomical findings, the defect being graded and located mainly in the nasal visual field.

## SPECIES DIFFERENCES

### MONKEY

Hubel and Wiesel (1970) found, in monkeys, that two weeks of monocular closure during the first five weeks produced at least temporary blindness in the eye that had been closed, and an absence of responses of cortical cells to stimulation of that eye. Again, Von Noorden *et al.* (1970) deprived monkeys of form-vision, but not of light, and found a critical period of up to three months of deprivation from birth when no recovery of form vision was obtained, as tested with a Landolt-C. If deprivation occurred after 3 months of form experience there was no effect. Later Baker *et al.* (1974) showed that the sensitive period begins at birth and may continue up to 8 weeks. They were able to correlate the low visual acuity following deprivation or strabismus with the paucity of cortical cells driven by the deprived eye.

### RABBIT

In the rabbit Van Hof (1969) found no abnormality in the ability of young animals to discriminate patterns when they were deprived of vision for six weeks after birth. So far as binocularly driven cortical neurones were concerned, monocular deprivation had only a small effect on dominance, skewing it towards the ipsilateral eye when the contralateral eye was sutured.[*] Deprivation did affect some other characteristics of the receptive fields such as the clustering of preferred directions around horizontal and vertical meridians and increasing the number of binocularly driven cells with non-identical receptive fields (Van Sluyters and Stewart, 1974).[†]

---

[*] The receptive fields of rabbit cortical neurones have been described by Chow, Masland and Stewart (1971); they found a large variety of types including simple, complex and hypercomplex cells; simpler types, found also in the retina and lateral geniculate body, were those with concentric fields, those with no surround, directionally sensitive, movement-selective responding better to moving than stationary stimuli, with no preferred direction. The simple and complex cells were the only exclusively cortical type. There is an area of visual cortex representing the limited region of binocular overlap, and the dominance distribution of the neurones has been described by Van Sluyters and Stewart (1974a), showing a skew towards the contralateral eye with no ipsilaterally driven neurones. Only rarely did the corresponding receptive fields project to the same point in visual space, so that it is unlikely that these binocular neurones provided reliable information regarding position in space.

[†] In general, it would appear that the ontogenesis of the rabbit's visual responses to flashes of light is governed by passive maturation of the retina; Chow and Spear (1974) have reviewed the information, and in their own experiments have shown that cortical units, which can be driven by optic nerve stimulation from birth, are not driven by light till day 8. Cortical cells with symmetrical types of receptive field, e.g. concentric, uniform with no surround, and those responsive to any direction of motion, are present at eye-opening (days 10 to 11) but more complex asymmetrical fields, e.g. units responsive to motion in a particular direction, appear some days later. By day 18 all types of receptive field are found, and at day 25 the proportions of cells with the various fields have adult values. This contrasts with the superior colliculi where the same is found at days 10 to 11; at this time the retinal receptors have developed and ganglion cells show evoked responses (Noell, 1958).

## EFFECTS OF CORTICAL LESIONS ON RECOVERY

Such recovery as takes place in deprived eyes could be due to utilization of the few cortical neurones that remain responsive to light, or it could be due to the employment of subcortical pathways. Spear and Ganz (1975) found that, after lesions that destroyed all the visual cortex in the cat (Areas 17, 18 and 19), there was no recovery from monocular deprivation. Removal of the monocular segment of Area 17 was without effect on recovery. These authors suggested that the superior colliculus might have been adequate to permit orientating and approach responses in the cat, whilst visual discrimination tests, which were more sensitive to cortical lesions, might have required visual cortex.

## REVERSAL OF SOME EFFECTS OF DEPRIVATION

### REVERSE SUTURING

Blakemore and Van Sluyters (1974) sutured one eye of newborn kittens for five weeks and established the binocular histogram which, as we have seen, indicates complete monocularity. With other kittens the sutured eyes were opened after periods ranging from 5 to 14 weeks, and the previously open eyes were sutured (*reverse suturing*) and, after allowing the kitten to use its previously deprived eye for 9 weeks, its cortical neurones were examined. When reverse suturing was carried out at 5 weeks, there was a complete reversal of the dominance pattern, the previously sutured eye losing its dominance and transferring it to the newly opened eye, whilst few or no cells were binocularly driven. As the time for reverse suturing was delayed, the reversal became less and less effective so that reversal at 14 weeks left the dominance distribution similar to that in the animal deprived for five weeks without reverse suturing; at intermediate periods, e.g. reverse suturing at 6 weeks, a fair number of cells were binocularly driven.

### PREFERRED ORIENTATION

In these cells, which had never had binocular visual experience, there was a remarkable difference in preferred orientation for line-stimuli according to the eye being stimulated, preferred orientations differing by as much as 70° by contrast with normal kittens where there is a strong correlation between preferred orientations in the two eyes when binocularly driven neurones are studied. This change is shown by Figure 23.86 from a later study by Movshon (1976), which compares

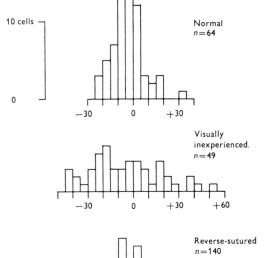
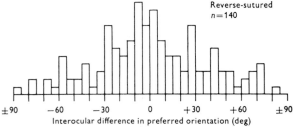

**Fig. 23.86** Showing the effects of binocular deprivation and subsequent reverse suturing on the differences in preferred orientation between the two receptive fields of given binocular neurones. In the normal eye the preferred orientations are close to each other in the majority of cells (top histogram). In the visually inexperienced kittens much larger differences occur, widening the spectrum; reverse suturing widens the spectrum still further so that direction sensitive units could have preferred orientations differing by as much as 90° in the two eyes. Differences of positive sign occur when the receptive field of the right eye is tilted clockwise with respect to that in the left eye; negative differences are anticlockwise. (Movshon, *J. Physiol.*)

histograms of preferred orientation-differences in the two eyes of binocularly driven cortical neurones; visual inexperience alone widens the spectrum, and reverse suturing not only fails to re-establish the normal situation but widens the spectrum still further. This emphasizes the importance of binocular visual experience for maintaining this similarity in orientation-preference, so essential for fusion of retinal images.

### SENSITIVE PERIOD

The sensitive period during which reversal of the effects of deprivation can be effected coincides fairly closely with the sensitive period for producing some effect of deprivation, so that the sensitive period is one of plasticity during which synaptic connexions may be broken and re-established.

RECOVERY OF VISUAL FUNCTION

Behaviourally, the kitten after 5 weeks of monocular deprivation was blind in the sutured eye, but after reverse suturing at 5 to 8 weeks there was considerable recovery of visual function in the newly opened eye; when reverse suturing was delayed to 8 to 14 weeks there was some vision through both eyes with progressively less in the newly opened eye.*

DOMINANCE COLUMNS

Monocular deprivation, as we have seen, makes a striking change in ocular dominance so that on a given side of the cortex virtually all neurones are driven by the eye contralateral to lid-suture; the same columnar arrangement of neurones is, however, maintained, so that ocular dominance-columns become larger when projected to the surface of the cortex; Movshon (1976) showed that, during reversal of dominance, there was a strong correlation between the reversal index† and the column size (Table 23.1). In the normal cat there is

**Table 23.1** Reconstructed column sizes in reverse-sutured kittens. The extent of each defined ocular dominance column is given in millimetres projected to the cortical surface in twelve kittens. The animals are arranged in order of increasing reversal index, their rearing conditions are indicated in the left-hand column as: age (weeks) at reverse-suturing + days of reversed suture. The left eye was the contralateral, initially experienced eye in all cases (Movshon 1976).

| Animal | Reversal index | Column sizes | |
|---|---|---|---|
| | | Left-eye | Right eye |
| 7 + 6 | 0·25 | 1·6 | 0·3 |
| 5 + 3 | 0·26 | 0·4 | 0·5, 0·5 |
| 7 + 9 | 0·32 | 1·2 | 0·6 |
| 6 + 6 | 0·36 | 0·5 | 0·5 |
| 5 + 6 | 0·38 | 1·4, 0·9 | 0·5 |
| 7 + 24 | 0·40 | 0·9 | 0·6 |
| 7 + 63 | 0·48 | 0·6, 0·4 | 0·5 |
| 6 + 9 | 0·49 | 0·4 | 0·4 |
| 4 + 3 | 0·63 | 0·1, 0·4 | 0·5 |
| 6 + 63 | 0·65 | 0·7, 0·3 | 0·7 |
| 6 + 12 | 0·75 | 0·2, 0·3 | 0·7 |
| 5 + 9 | 0·77 | 0·3, 0·6 | 1·4 |

no obvious change in the character of the orientation-preference columns when dominance is grossly changed by strabismus, so that the overlapping systems of dominance- and orientation-columns seem to be independent of each other; this means that orientation is coded across the cortex without regard to the eye from which the visual information comes. With reverse suturing one might expect the same, so that as the opened eye took over dominance columns it might also take over their orientation sequences, but this was not true, the cortical representation of orientation being

disrupted at the borders of dominance columns. Thus it seems not to be possible to account for the observed patterns of orientation columns and interocular orientation-differences, by assuming that visual experience, following reverse-suturing, simply reactivates synapses made ineffective by the initial period of deprivation. According to Movshon, the new visual experience seems rather to stimulate the development of a new visual cortex, similar to the old one but differing in points of detail.

**Recapture of control**

The question arises as to whether re-innervation, leading to a recapture of control over a cortical neurone, represents regrowth of a synaptic structure that had been lost, or whether it represents something more subtle, e.g. a change in the cortical neurone that enables it to respond postsynaptically in a different fashion, or to a re-functioning of synapses that had gone silent during deprivation. So far as the effects of deprivation are concerned, there is little doubt that, in monkeys, the input from the deprived eye, as measured by autoreadiographic techniques, is strongly reduced (Hubel et al., 1976), so that re-establishment of dominance must be the consequence of regrowth of axons, although the computed rate of growth of 100 μm a day is high. Such a regrowth would probably lead to the new pattern of cortical organization found, rather than precisely reproduce the old pattern.

**Movement of horseradish peroxidase**

As Movshon points out, it could be that both processes operate but at different times during the sensitive period. Thus if deprivation in the kitten begins at birth, horseradish peroxidase, injected into the visual cortex, is transported mainly to layers of the lateral geniculate nucleus driven by the experienced eye, i.e. the cortical terminals of the geniculate neurones activated by the deprived eye have become ineffective. If deprivation is begun at 4 weeks, i.e. still within the sensitive period, the horseradish peroxidase is retrogradely transported to all layers of the geniculate. Corresponding with this, when suture-reversal is carried out after deprivation begun during the fifth week, the re-innervation of the cortex leads to a similar connexion pattern to that

---

* In a more elaborate study of the recovery of visual function after cross-suturing, Movshon (1976b) established that, unless the previously deprived eye had recovered control over an appreciable proportion of cortical neurones, it was unable to perform reliably the more demanding visually controlled tasks such as placing or following. He points out that Spears and Ganz (1975) have shown that recovery depends on the presence of the visual cortex, so that ablation prevents this.
† By this is meant the proportion of visually responsive neurones dominated by the initially deprived eye.

found normally, the orientation patterns overlapping nearly as exactly as they do in the normal (Blakemore et al., 1976).

### CORTICAL STRIPE-SIZES

Hubel et al. (1976) injected tritiated fucose and proline into the experienced eye of a monocularly deprived monkey and made autoradiographs of the cortex; the average total width of the left- plus right-eye stripes, indicating terminations in the cortex, was unchanged, but the relative sizes of the two indicated expansion by the experienced eye at the expense of the deprived eye. Since the cell-bodies had not shrunk, the altered width of the stripes indicated changes in synaptic development. That the change is due to competition is indicated by the fact that, in regions of the cortex representing the extreme temporal crescent area, and thus receiving input from only one eye, there was no change in the width of the stripes representing the deprived eye.

### LATERAL GENICULATE BODY

So far as the lateral geniculate body is concerned, bilateral closure has no effect on its morphology, but after uniocular closure there is a large shrinkage of the laminae belonging to the deprived eye, due to diminution in cell size (Wiesel and Hubel, 1963; Garey et al., 1973). Thus the geniculate neurones do not disappear but shrink in consequence of their reduced synaptic connexions in the cerebral cortex. Reverse suturing during the critical period reverses the difference in cell size, so that the layers receiving from the previously closed eye now have the larger effective cell cross-sectional area (Dürsteler et al., 1976). When the binocular and uniocular segments of the cat's lateral geniculate body are compared, the importance of competition for control becomes even more evident. Thus in the ipsilateral portion of lamina A, thinning is restricted to the medial portion, which has a binocular input; no thinning occurs in the lateral part of lamina A which is uniocular in representation, receiving its input from the extreme nasal crescent of the retina (Guillery and Stelzner, 1970).

### POST-CRITICAL PERIOD REVERSAL

Kratz et al. (1976) obtained a considerable reversal in dominance by the previously deprived eye, in spite of waiting till after the critical period, by enucleating the normal eye at the end of the monocular deprivation period. Thus at the end of some 4 months of monocular deprivation the number of cortical cells driven by the deprived eye was of the order of 5 per cent, as found by others, but when kittens deprived in this way were enucleated, the percentage increased to some 35 per cent immediately after enucleation of the normal eye, and there was no change if testing was carried out 3 months later or longer (Fig. 23.87). The immediate effects of enucleation suggest strongly that inhibitory actions were at play from the normal eye, acting along the geniculo-striate pathway, and preventing the cortical neurones from responding to the deprived eye. A similar inhibition has been described for control over collicular neurones (p. 619).

## SIGNIFICANCE OF MATURATION AND DEPRIVATION

### RABBIT VS CAT

In the rabbit the more complex type of receptive field, typified by the cat's simple and complex cells, appears later—between 10 and 25 days after birth—by contrast with the kitten where some are present at birth. The maturation process is passive in so far as it is not modified by visual deprivation (Chow and Spear, 1974), and the same is essentially true for ocular dominance (Van Sluyters and Stewart, 1974). Thus the receptive field characteristics of the rabbit's cortical (and retinal ganglion) cells are laid down before birth and reach their maturity as a result of processes that are independent of visual experience.

### REQUIREMENTS FOR STEREOPSIS

This is in marked contrast with the cat, and the differ-

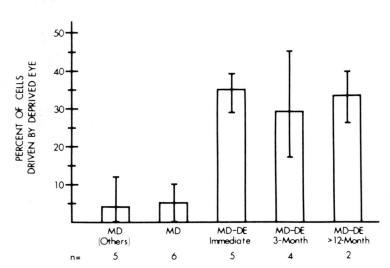

**Fig. 23.87** Effects of enucleation of the normal eye on the dominance of the monocularly deprived (MD) eye. The bar-graphs indicate percentage of striate cortex cells responding to the deprived eye in each rearing condition. DE refers to enucleation. 'Others' refers to earlier studies of Wiesel and Hubel. (Kratz et al., J. Neurophysiol.)

ence is probably to be sought in the virtual absence of stereoscopic vision in the rabbit, so that a precise correspondence between events happening on the two retinae is unnecessary. Such a correspondence would be extremely difficult to predict, in view of the mobility of the eyes, so that a learning process, imposed on a genetically determined synaptic field, is probably the only manner in which precise binocular—and stereoscopic—vision can be achieved. Thus the plasticity revealed in the kitten enables genetically specified simple cells to impose similar orientation-selectivity on the cells around them, creating the columnar system that is virtually absent in the rabbit; it also ensures that most cortical cells remain binocularly driven and acquire similar preferred orientations in the two eyes★ (Blakemore and Van Sluyters, 1974), a necessary condition for fusion of the two retinal images; and finally it can improve the orientation-tuning, converting the 'non-orientated' units to orientation-selective ones.

## ELECTROPHYSIOLOGY OF THE LOWER VISUAL CENTRES

## THE SUPERIOR COLLICULI

The probable role of this paired nucleus in the control of visually directed movements has already been indicated (p. 418) and here we may consider in some detail the types of receptive fields of the collicular neurones, and the interaction of these neurones with cortical and other non-retinal inputs.

### STRUCTURE

Each colliculus is composed of several layers, from outside there is the fine *stratum zonale*, the *superior gray*, *optic gray*, *intermediate gray* and *intermediate white*, *deep gray* and *deep white* layers. Fibres from the contralateral optic tract terminate in the superficial, optic and intermediate gray layers.

### RETINAL INPUT

Hoffmann and Stone (1973) using their W, X, Y-classification of retinal ganglion cells, described three main pathways from retina to colliculus:

1. A direct retinotectal pathway carried by axons of retinal W-cells, transmitting to some 73 per cent of collicular cells.
2. A direct retinotectal pathway carried by axons of Y-cells, transmitting to some 9 per cent of collicular cells.
3. An indirect pathway involving a cortical loop consisting of retinal Y-cell, geniculate Y-cell, and a cortical complex cell sending its axon through the corticotectal pathway.

At least a part of the retinotectal pathway results from branching of the retinogeniculate fibres since electrical stimulation of the superior colliculus produces antidromic spikes in the optic tract which interfere with spikes antidromically activated from the lateral geniculate body (Hayashi *et al.*, 1967).

### TECTAL VISUAL FIELDS

In primates the tectal visual fields are organized similarly to those of the lateral geniculate body and visual cortex, in the sense that the partial decussation of retinal fibres leads to the representation of a hemifield in each colliculus. A similar projection was described in the cat by Apter (1945), but subsequent studies on this species and the squirrel and tree-shrew have shown that a portion of the ipsilateral hemifield, in addition to the whole contralateral hemifield, is represented in each colliculus, indicating an ipsilateral input from a part of the nasal retina (Lane *et al.*, 1973; Casagrande and Diamond, 1974). Thus McIlwain and Buser (1968) found many collicular neurones in the cat with receptive fields extending across the midline into the ipsilateral visual field.†

### RECEPTIVE FIELDS

Sterling and Wickelgren (1969) examined the receptive fields of cat collicular neurones and compared them with those of cortical, geniculate and retinal units. The similarity was strongest with complex or hypercomplex cortical units, in so far as the fields were elliptical, rather than circular, and were surrounded usually by an inhibitory area so that an optimal size was manifest when a light or dark bar just did not overlap with the inhibitory area. The best orientation was not nearly so critical as with cortical units, so that a change of at least 30° was necessary to evoke a reduction in discharge. Again, cortical units responded well to fine jittery movements of a bar whereas the collicular units responded best to large sweeping movements over most of the receptive field. In cortical units the preferred orientations are distributed equally through the

---

★ The preferred orientations of a cortical cell's receptive field tend to be similar in the two eyes when a binocularly driven unit is studied. However, after visual deprivation, such that the animal has never had binocular vision, the preferred orientations can differ by as much as 70°.
† In the Siamese cat there is an abnormally large ipsilateral projection of visual input to the superior colliculus, as with the lateral geniculate body (p. 610). Thus, in normal cats this consists of a vertical strip in the midline about 15 to 30° wide; in the Siamese cat this extends to some 40°. Removal of the visual cortex, which, in normal cats, skews the ocular dominance-histogram strongly towards the contralateral eye, has no effect in the Siamese cat (Berman and Cynader, 1972).

360°, whereas collicular orientations were largely confined to the horizontal.

## MOTION SENSITIVITY

When the neurone was direction-sensitive, giving a strong response in one preferred direction and a small or no response in the opposite (null) direction, the best response was usually obtained when the moving stimulus passed away from the centre of the field into the periphery. The velocity of movement was important, so that the collicular neurones would be capable of providing information regarding both rate and direction of movement of objects in the external world. The response to movement from centre to periphery of the field is what would be required of a sensor brought into play to follow a moving object. Thus, when an object moves from the periphery to the centre of the visual field no movement of the eye is necessary, but when it crosses the mid-point and moves into the periphery, following movements become more important, and this is when the units are activated best; and it is interesting that the size of movement required to elicit a following eye movement is fairly large and comparable with that required to excite a collicular neurone. Thus Sterling and Wickelgren's study, and later ones by Cynader and Berman, (1972, 1973) emphasize the sensitivity to motion, direction of motion and the speed of this motion,* so that the responses to stationary flashes are relatively weak by comparison with retinal, geniculate or cortical cells. As Hoffmann and Dreher showed, the best stimulus for a neurone was a bar of optimal width passing across the receptive field.

## LOCATION AND DOMINANCE

Most of the receptive fields were in the contralateral half of the visual field; and the retinotopic projection on the colliculus was similar to that derived earlier by Apter, most of the units being confined to the superficial gray and optic layers. The ocular dominance histogram (Fig. 23.88), was similar to the cortical one but more skewed to the contralateral eye, 97 per cent of the cells being driven by both eyes. As with the cortex, the type of field was similar for both eyes.

## CORTICAL RELATIONS

Wickelgren and Sterling (1969) were impressed with the similarity between collicular neurones and cortical hypercomplex neurones, and considered the possibility that the collicular field was being largely imposed on it by its cortical projection, a projection that is from the ipsilateral cortex. Thus the colliculus could be responding to the contralateral eye directly, by the retino-

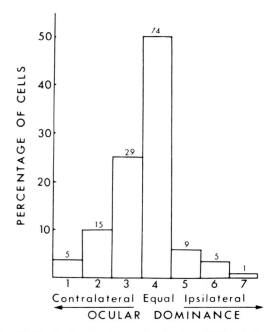

**Fig. 23.88** Ocular dominance distribution of 138 units in superior colliculus. Number over each bar gives absolute number of cells in that category. Group 1 cells are driven only by the contralateral eye; Group 4 are driven about equally by both eyes; Group 7 respond only to ipsilateral eye. (Sterling and Wickelgren, *J. Neurophysiol.*)

tectal pathway, and to the ipsilateral eye indirectly via the geniculo-striato-collicular pathway (Fig. 23.89). Removal of the ipsilateral visual cortex produced a great change in the character of the collicular units; they now responded well to stationary targets, had lost their directional sensitivity, and were driven only by the contralateral eye. Cooling of the visual cortex was sufficient to produce this change, and this was reversed by rewarming.

## DEEPER LAYERS

The intermediate and deep gray layers of the colliculus receive input from visual, auditory and somatosensory systems, and single units responding to several modalities of sensation have been described; it is reasonable to assume that these units are concerned with the integration of visual and other sensory information. Gordon (1973) has described the receptive fields of collicular units in these deeper layers;

---

* The units of Drener and Hoffmann fell into three groups of speed preferences, slow, medium and fast. So far as direction-sensitivity was concerned, they confirmed the importance of moving away from the centre of the visual field; cells responding to stimuli in the upper visual quadrants responded best to upward motion, those in the lower to downward motion. Another feature was the presence in some units of two or more centres of activity, so that an edge moving through the receptive field produced two or more bursts of activity.

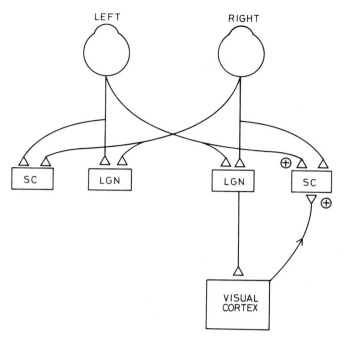

**Fig. 23.89**    The right superior colliculus responds directly to the contralateral eye and indirectly to the ipsilateral eye by way of the lateral geniculate nucleus (LGN) and visual cortex.

they are much larger, with some extending over the entire contralateral field; many responded to a wide range of stimulus-sizes and shapes and were directionally sensitive, responding only to a movement with a horizontal component towards the periphery of the contralateral visual field. Many units responded to both auditory and visual stimuli and a few to both somatic and visual stimuli. Somatic receptive fields showed an interesting correlation with the visual; thus the rostral colliculus contained units responsive to the area centralis of the retina and the face, whilst the caudal colliculus had units responsive to the peripheral retina and the forelimbs, shoulders and trunk. Such a correlation would be expected of a system controlling eye and head movements in response to a moving visual stimulus.

### EFFECTS OF MONOCULAR OCCLUSION

Since monocular occlusion in the kitten makes the contralateral cortex unresponsive to visual stimuli, we may expect the superior colliculi to lose any influence from the contralateral cerebral cortex and therefore to be influenced only by the normal, experienced eye, directly by the optic tract and indirectly through the cerebral cortex of the side opposite to the normal eye.

In other words, the effects of monocular occlusion should be similar to those of removal of cerebral cortex on the side opposite to the occlusion. In fact this was found, most of the collicular neurones being responsive only to the experienced eye.

### CORTICAL INHIBITION

The effect of cortical ablation on collicular neurones suggests an inhibitory action on the ipsilateral colliculus that normally prevents its neurones from responding to the ipsilateral eye. The directional sensitivity of the collicular neurones is not like that in Barlow's rabbit ganglion cells, which is due to an asymmetrical inhibition (p. 285), since very small movements in the excitatory field are direction-sensitive, i.e. movements that do not involve crossing the excitatory-inhibitory interface, so that directional sensitivity is due to the sequence of activation of the collicular units, a sensitivity that could be imposed by the cortex through inhibition. Thus, according to Dreher and Hoffmann (1973), complex cortical cells with preferred directions in the null would inhibit superior collicular cells, and those with preferred directions in the same directions as those of the collicular cells would excite. Rosenquist and Palmer (1971) confirmed the effects of cortical ablations and showed that they had to include Area 17 so that lesions in Areas 18 and 19, sparing Area 17, were without effect. Acute lesions were just as effective as long-standing ones, indicating an immediate release phenomen (Berman and Sterling, 1974.[*]

### RECEPTIVE FIELDS IN DECORTICATE CAT

Berman and Cynader (1975) found, in general, that besides the change in ocular dominance and direction-sensitivity that the collicular cells undergo with removal

---

[*]Hoffmann and Straschill (1971) were unable to confirm the effects of lesions in Areas 17 to 19 on direction-specificity or ocular dominance. Stimulation of the cortex could inhibit a collicular cell's background discharge and also the response to a moving light stimulus. Other collicular units could be excited. We may note, here, that in the rabbit Stewart *et al.* (1973) found no effects of large visual cortical ablations on the receptive fields or trigger features of collicular units.

of visual cortex, the receptive fields of some units become similar to those of ganglion cells; thus one group had concentrically organized ON-OFF fields; another group, sensitive to a flickering stimulus, were similar to the phasic units described by Cleland *et al.* (1971) and Fukada and Saito (1971), whilst the third group, constituting some 90 per cent, had fields similar to W-cells which, according to Hoffmann (1972), constitute the dominant retino-collicular projection. These retinal units have mixed ON-OFF receptive fields, relatively low velocity-tuning and slow conduction-time to the colliculus, and it was suggested that they played a dominant role in control of the collicular receptive fields in the decorticate cat, and that the cortex exerts an inhibitory action, so that sensitivity to flickering stimuli is normally suppressed.

BINOCULAR DEPRIVATION

In binocularly deprived animals, the W-pathway to the superior colliculi remained normal but there was a striking reduction in the number of Y-cells in both the monocular and binocular segments of the lateral geniculate body, and, correlated with this, there was a loss in the indirect Y-pathway via the cortical loop (p. 617) to the superior colliculi. The loss of cortical control over the colliculi was not due to a defect in the visual cortex since stimulation of this activated collicular units normally. Thus the effects of deprivation seem to be due to failure in the retino-geniculate pathway to the cortex (Hoffmann and Sherman, 1975).

# CORTICAL ABLATION AND MONOCULAR DEPRIVATION

The loss of directional sensitivity is, in effect, the loss of the power to inhibit the response to a stimulus moving in the null direction; consequently, in the absence of cortical activity—responsible for inhibition of collicular neurones—we may expect the collicular response to a moving stimulus to be the same when it moves in both the null and preferred directions. Berman and Sterling (1976) investigated this point in monocularly deprived kittens. As we have indicated earlier, the effects of monocular deprivation are to cause collicular units to be activated mainly by the experienced eye. So far as the superior colliculus contralateral to the deprived eye is concerned, the situation is that shown in Figure 23.90, the collicular neurones being activated by the experienced eye mainly through the visual cortex, whilst those activated by the deprived eye are activated directly. On this basis, it should be possible to find neurones responding equally well in the null and preferred directions through the deprived eye, and showing directional

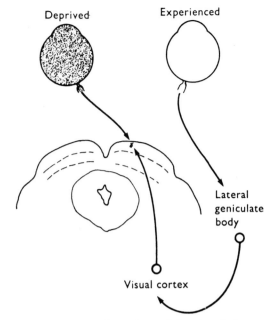

**Fig. 23.90** Diagram of experimental situation. Left eye, deprived from birth by lid suture, provides direct retinal input to right superior colliculus. Right eye (experienced) provides ipsilateral input to colliculus via geniculo-cortical pathway. In the intact animal most neurones in both colliculus and cortex are activated almost exclusively from the experienced eye. (Berman and Sterling, *J. Physiol.*)

sensitivity through the experienced eye. Berman and Sterling did find a few, presumably activated by the two pathways indicated by Figure 23.90.

OCULAR DOMINANCE

We have seen earlier that removal of the visual cortex produces a striking change in the ocular dominance pattern of collicular neurones; whereas in the normal animal most collicular neurones are driven by both eyes, after removal of the visual cortex the collicular neurones were only driven by the contralateral eye.

A similar effect may be shown in the kitten reared with monocular occlusion. In this animal the collicular neurones are driven almost entirely by the experienced eye. Berman and Sterling (1976) carried out a preliminary monocular deprivation in kittens, and the shift in ocular dominance in favour of the experienced eye is shown in Figure 23.91 (*left*). After appropriate testing of the collicular neurones, a lesion was made in Areas 17 and 18, and recording was begun within 15 to 60 min after the lesion. Figure 23.91 (*right*) shows the change in the ocular dominance pattern, the majority of collicular neurones now being driven by the deprived eye; this happens so rapidly that it is impossible to account for it on the basis of sprouting of nerve terminals. It is difficult to escape the conclusion that

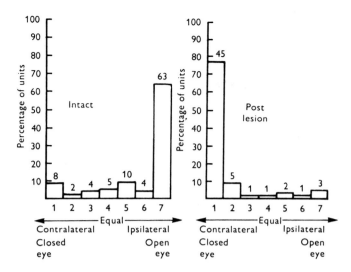

**Fig. 23.91**  Ocular dominance distribution of units in the superior colliculus contralateral to the deprived eye in five animals before and immediately following removal of visual cortex. Groups 1–7 on the abscissa represents a contralateral to ipsilateral trend in ocular dominance with units in group 1 driven only by the contralateral eye, units in group 4 driven equally by either eye, and units in group 7 driven only by the ipsilateral eye. The numbers above each bar represent the number of units in that group. (Sherman and Sterling, *J. Physiol.*)

the visual cortex had been suppressing the input from the deprived eye.

**Exceptions**

Berman and Sterling found, in three cats, an exceptional behaviour in that many collicular cells were quite effectively driven by the deprived eye and often dominated over the experienced eye; none of the collicular neurones that were driven by the deprived eye were directionally sensitive, so that it would seem that, here, the cortex was not competing effectively with the retinal input; this was not due to failure of cortical responsiveness, since in Area 17 simple and complex cells were found, all driven exclusively by the experienced eye. The extent to which the retinal input dominated over the cortical input varied, so that it appeared that there had been competition for capture of collicular cells.

## DEVELOPMENT OF RECEPTIVE FIELDS

Kittens appear blind directly after their eyes open, and their visually guided behaviour only develops over a period of several weeks. Norton (1974) found no direction-selective collicular cells in the very young kitten, most being driven by the contralateral eye *exclusively*, whereas in the adult most collicular neurones are activated by either eye. Between days 15 and 25 there were rapid changes in the receptive field characteristics, leading to the adult picture; and visual following movements could be elicited at about days 16 to 18. More complex visual placing reactions could only be achieved after 26 days, when the receptive fields were completely adult in their pattern. Norton commented that the collicular receptive fields of the young kitten are similar to those in the adult after cortical ablation, so that the development presumably represents acquisition of collicular control by the visual cortex.*

## CORTICOTECTAL NEURONES

Palmer and Rosenquist (1974) found in the cat that all cortical neurones that responded antidromically to stimulation of the superior colliculus belonged to the complex and hypercomplex categories of Hubel and Wiesel; the cells were found in layer 5 and exhibited very pronounced binocularity and direction-sensitivity but were more broadly tuned to stimulus orientation, and they postulated that the changes in collicular neurones following visual cortex ablation could be accounted for by a loss of properties imposed by these cortical cells. In the monkey, however, Schiller *et al.* (1974) found that the pattern of response to cooling or ablation of the visual cortex was more subtle, only slight changes in response and dominance in the superficial collicular cells being observed, whereas in the deeper layers the units lost their powers to respond to visual stimuli.† Finlay *et al.* (1976) found the corticotectal units in the monkey to be confined to layers 5 and 6 of Area 17, and they were usually classified as CX-cells but with broader orientation-tuning, larger receptive fields and

---

* Goldberg and Wurtz (1972) examined the receptive fields of collicular neurones of the monkey, using chronically implanted electrodes in the waking animal. Most of the units were pandirectional and only 10 per cent were direction-sensitive, by contrast with about 70 per cent in the cat (Cynader and Berman, 1972), and usually a moving stimulus was necessary, the response increasing with speed. Interestingly, no cell gave excitatory discharges with eye movements in the dark, and over half were inhibited. Perhaps in this way the colliculus can inform higher centres that a movement has occurred.
† The region of the striate cortex representing foveal projection sends fibres to the rostral third of the superior colliculus, but Wilson and Toyne (1970) found no retinal fibres ending here, the predominant projection, as determined anatomically, being to the caudal two-thirds of the contralateral colliculus. Evoked responses to light stimuli falling on the fovea might therefore be expected to be brought about through the cortical projection; however, Schiller *et al.* (1974) found that cooling or removal of striate cortex did not prevent foveal stimuli from producing responses in the rostral colliculus.

greater binocularity than CX-cells have in general; a third were direction-sensitive. In general the corticotectal units of monkey and cat were remarkably similar in spite of the fact that the effects of cortical ablation were so different in the two species, the cat's collicular neurones losing binocularity, direction-sensitivity and response to movement. The strong effects on the deep collicular units suggested to Finlay *et al.* that the cortical input exerted a grating action on the flow of information from the superficial to the deep units of the colliculus, rather than imposing receptive field characteristics.

## TYPES OF RESPONSE AND LOCATION OF NEURONES

Mohler and Wurtz (1976) have summarized the variation in response-type as an electrode penetrates deeper into the monkey's colliculus. Thus the neurones in the most superficial layers respond, as we have seen, to visual input but they do not discharge in association with eye movements; the next, deeper, type is also visually responsive but it does discharge *before* movements of the eyes to a restricted part of the visual field—*the movement field.**** Deeper still, cells respond before eye movements but are usually visually non-responsive. Thus the visually responsive cells are located in the superficial layers, and the movement-related cells in the deeper layers. As we might expect, the receptive fields of the superficially located visually responsive neurones are similar to the movement fields of adjacent movement-related neurones in the intermediate layers. These relations clearly suggest an organization in depth, in the sense that visual input evokes responses in superficial cells which then activate more deeply located movement-related neurones, which then activate oculomotor and neck neurones. Mohler and Wurtz worked on trained monkeys with implanted electrodes in the superior colliculi, and eye movements were recorded with the EOG. Confirming their earlier study, they found movement-related cells primarily in the intermediate grey and white layers; these discharged before an eye movement, and a *movement field* could be defined, representing a limited area in the visual field, movements to which were preceded by this vigorous discharge. The most dorsal units were found just below the junction between the optic layer and the intermediate gray layer. As the electrode penetrated deeper amongst these movement-related cells, the interval between the initiation of a saccade and the discharge increased from some 50 msec to 100 to 150 msec. The deeper the cells, moreover, the longer the duration of the discharge. Another feature was that the size of the movement-field increased with depth. It must be emphasized that movement-related cells could also be responsive to visual stimuli, and that the number with both characteristics was highest in the superficial layers.

### VISUALLY TRIGGERED MOVEMENT-RELATED CELLS

At the zone of transition of the two main types of cell, Mohler and Wurtz found units that discharged before eye movements and which were visually triggered, i.e., they discharged when the monkey made a saccade towards an object presented to the visual field, but they failed to make this discharge when the movement was spontaneous. These cells were thus visually triggered, but they showed two bursts of discharge, namely an initial burst in response to the appearance of the target—the *visual response*—and the *movement-related discharge*. These visually triggered cells had the smallest movement-fields and had the shortest lead-times; the movement-fields were of about the same size as the visual receptive fields. Experimentally

it was possible to dissociate the burst of discharge from the occurrence of an actual eye movement. Thus the monkey would be looking at a fixation point, and a new visual stimulus would be presented at a different point, as in the usual tests. If the fixation point was extinguished, and the visual target was turned on for only 100 to 200 msec and then the fixation point was turned on again, the monkey sometimes made saccades to the target and sometimes not. Even when no saccade occurred, there was a strong discharge which could not have been a simple visual response since this cell showed only a weak visual response. Thus these eye-movement cells could discharge at the time appropriate for association with a saccade, even though no saccade occurred; so that the discharge could easily be looked upon as a command for eye movements. It appeared, too, that the discharge was prominent if the monkey anticipated the requirement for a saccade rather than if this occurred randomly. Thus the characteristics of the movement-cells seem more closely related to the monkey's *readiness* to make a movement than to its actual execution.

### Co-ordinators

Mohler and Wurtz suggested that these visually triggered movement-cells were, in fact, co-ordinating visual and movement information, providing some higher centre with this combined information rather than being concerned in the execution of the movement as such. Their anatomical location at the junction between the regions where the purely visually responsive and the 'ordinary' movement-related cells occurred would favour this function; and the similarity in magnitude of receptive fields would be consonant with such a function. It is possible, then, that the visually triggered movement-cell represents the culmination of one line of collicular processing, and it may be an important efferent neurone from the superior colliculus.

This processing would be from above—visual—and below—movement-related—and would represent a convergence of information, the source of the visual input being the retinal and cortical inputs to the superficial layers; the source of the movement-related input is not clear, however, but the fact that their activity begins before that in more superficial layers rules out the superficial collicular cells as the origin. A final possibility is that of parallel inputs and outputs without vertical interaction in the colliculus.

### IMPORTANCE OF ATTENTION

In studying the visually responsive cells of the superficial layer, Wurtz and Goldberg (1972) had concluded that the eye movement itself was not correlated with the collicular neurone's discharge, but rather the animal's attention to the peripheral spot to which the movement was directed, since the discharge preceding an eye movement was enhanced if the movement was directed towards a light-spot rather than being made without this stimulus. This *enhancement* was studied in greater detail by Wurtz and Mohler (1976a). It was shown to be highly specific for the direction and extent of the saccade, so that a maximal enhancement was obtained when the eyes fixated the centre of the receptive field of the collicular neurone, and fell off with distance away from here. Wurtz and Mohler concluded that it resulted from an input from the deeper, movement-related, cells; thus if the monkey was trained to make a hand, rather than an eye, response to the visual stimulus the enhancement was not present.

---

* The movement-field is defined as the outline in space where movement from the fixation point to another point in the visual field is preceded by an increase in rate of discharge.

**Agent for readiness**

The main problem, of course, is the relation between the processing within the colliculus, as revealed by these visually and movement-related discharges, and the actual discharges to the motor neurones that determine the movement. The visual input apparently induces a readiness to respond in the superficial collicular neurones, mediated possibly through a parallel input to the deeper cells; and as a result, the eyes are finally directed towards a specific region of the visual field. Thus the colliculus is acting as an agent for readiness-to-respond rather than as the initiator, and as such it facilitates eye movements rather than initiates them, so that lesions in the colliculus may increase latency of movements and only slightly decrease the accuracy.

## EFFECTS OF LESIONS

The effects of lesions in the superior colliculi are not easily defined, and they doubtless depend critically on the extent of the lesion since the projections to the superficial and deep layers are fundamentally different. The effects have been summarized by Sprague (1972). In the cat, for example, total extirpation produces an inattention to visual stimuli, but the ability to follow objects and localize them visually is not seriously impaired. In the monkey, too, no serious defects in visual guidance were evident, and the emphasis has been on inattention rather than any precise defect. Wurtz and Goldberg (1972) have pointed to the grossness of the movement-fields of collicular neurones, so that these neurones are unlikely to be responsible for the highly accurate movements of fixation. These fields were of the order of 5°, whereas errors in a saccade are or of the order of 0·5°. Casagrande and Diamond (1974) made elaborate studies of visually directed behaviour in the spider monkey after lesions that invaded different extents of the superior colliculi. Many of the visual defects could not have been accounted for by failure of attention, and they concluded that the superior colliculus should be regarded as a double organ, its superficial layers, receiving the direct retinal and visual cortical inputs, might be responsible for some degree of pattern vision, whilst the deeper layers, with their more extensive and complex inputs, would be responsible for visual attentiveness, lesions here causing a remarkable degree of inattention to visual stimuli. Thus the projections from the deep layers descend to motor areas of the brainstem and ascend to thalamic areas not usually associated with vision, such as the posterior nuclei, the intralaminar complex, and regions of the subthalamus. The main problem, however, is how the deeper layers receive the visual input necessary to permit the visually guided behaviour that was clearly present in the absence of the superficial layers; Casagrande and Diamond quote some work indicating a visual input to the deeper layers.*

In general, then, it seems that the superior colliculus acts at a very early stage in eye-movement processing, whilst other inputs must be available to the oculomotor system that permit adequate control in the presence of collicular lesions.

## COMPARISON WITH STRIATE CORTEX

Wurtz and Mohler (1976b) emphasize the great power of the striate cortex to analyse the visual stimulus in terms of contour, motion and colour, and they contrast this with the poor behaviour of the colliculi in these respects but their excellent ability to assess the significance of the visual stimulus in relation to the individual. Thus the cortical neurones do not distinguish between movement of the object in space from the similar movement of the retinal image that takes place when the eye moves (Wurtz, 1969), although the collicular neurones do (Robinson and Wurtz, 1975). The enhancement of the collicular neurone is selective to the movement, in the sense that it is greatest when the movement is directed towards the centre of the receptive field of the collicular neurone and falls off with distance from here; this selectivity clearly allows the collicular cell to signal movement to a specific point in space, whereas such enhancement as was seen in striate cortical neurones was unspecific and might have been related to general alerting. Hence, as Wurtz and Mohler state, 'at these early stages in processing, the cortical and the collicular areas stand in a nearly reciprocal relation: the cortex is an excellent analyser of stimulus characteristics but a poor evaluator of stimulus significance, while the colliculus is an excellent evaluator but a poor analyser'.

## THE PULVINAR

Phylogenetically the pulvinar, or lateral posterior nucleus of the thalamus, is remarkable for its great expansion in primates, its close connexion with the lateral geniculate bodies, and its relations to parts of the cerebral cortex that may be described as visual, acoustic and somato-sensory association areas. It has therefore been suggested that it might act as a lower visual centre.

VISUALLY EVOKED RESPONSES

Experimentally, light-evoked responses have been recorded from the pulvinar, most in the inferior nucleus

---

*The reader is referred to Casagrande and Diamond's discussion of the phylogenetic aspects of collicular and striate lesions. Thus in the cat ablation of Area 17 has little or no effect on learned pattern discrimination, whereas ablation of the superior colliculus affects pattern discrimination.

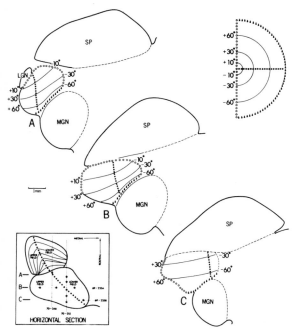

**Fig. 23.92** The representation of the visual field in the inferior pulvinar of the owl monkey. In the box in the lower left corner is a drawing of a horizontal section through the inferior pulvinar and the lateral geniculate nucleus. The horizontal meridian (indicated by the heavy dashed line) divides the inferior pulvinar into a rostromedial portion in which the lower visual quadrant is represented and a caudolateral portion in which the upper visual quadrant is represented. The horizontal meridian continues rostrolaterally and divides the lateral geniculate nucleus in a similar fashion. The light dashed lines correspond to the intersection with the horizontal plane of the parasagittal sections from experiments 70–346 and 70–315 and the coronal sections from experiment 69–235. The stars indicate the positions of the microelectrode penetrations in these experiments, as seen in the horizontal plane. The levels of coronal sections A, B, and C are indicated to the left of the horizontal section. In the upper right corner is a perimeter chart which serves as a key for the projection of the visual field in the inferior pulvinar as shown in coronal sections A, B, and C. The vertical meridian is indicated by small circles; the horizontal meridian by the heavy dashed line, and the extreme temporal periphery of the visual field by the small black triangles. (Mathers and Rapisardi, *Brain Res.*)

but many also in the lateral nucleus (Mathers and Rapisardi, 1973); most of the neurones were monocularly driven from the contralateral visual hemifield, with large receptive fields (100 square degrees compared with about 10 in the lateral geniculate nucleus). Interestingly, there were no polymodal units, although the connexions of the pulvinar with the areas adjacent to the auditory and somato-sensory areas are reciprocal. The projection of the visual field on to the inferior pulvinar is visuo-topically organized; as Figure 23.92 shows, the horizontal meridian divides this into a rostromedial portion, in which the lower visual quadrant is represented, and a caudo-lateral portion, representing the upper visual quadrant. The horizontal meridian continues rostrally to divide the lateral geniculate nucleus similarly.

### VISUAL PATHWAY

As to the visual pathway, it seems unlikely that this is retinal, since there is no retinal projection in the squirrel monkey, and that in the owl monkey could not account for the well defined visuotopic organization. Alternative or additional inputs could be by way of the well established collicular pathway and a cortical input which, in the macque, includes projections from both striate and prestriate regions. An input from the lateral geniculate nucleus is unlikely.[*] Mathers and Rapisardi have discussed the matter, when comparing the receptive fields of pulvinar with collicular and prestriate neurones, expecting that similarities would indicate the main source of visual input. Most surprising in this context is the large amount of ocular dominance, by contrast with the collicular and prestriate neurones, which are mostly binocularly driven.[†]

---

[*] Glendenning *et al.* (1975) found that, in the bush baby, *Galago senegalensis*, lesions in the superior colliculus were accompanied by degeneration in the pulvinar. They concluded that the pathway from the inferior pulvinar was to the temporal cortex and not to Area 18. The superior pulvinar, with no collicular projection, does project to Area 18.

[†] Lesions in the inferior pulvinar impair visual pattern discrimination in monkeys, such as the differentiation of the letters N from Z (Chalupa *et al.*, 1976).

## REFERENCES

Ades, H. W. & Raab, D. H. (1949) Effect of preoccipital and temporal decortication on learned visual discrimination in monkeys. *J. Neurophysiol.* **12**, 101–108.

Allman, J. M., Kaas, J. H., Lane, R. H. & Miezin, F. M. (1972) A representation of the visual field in the inferior nucleus of the pulvinar in the owl monkey. *Brain Res.* **40**, 291–302.

Apter, J. T. (1945) Projection of the retina on superior colliculus of cats. *J. Neurophysiol.* **8**, 123–134.

Baker, F. H., Grigg, P. & von Noorden, G. K. (1974) Effects of visual deprivation and strabismus on the response of neurons in the visual cortex of the monkey, including studies on the striate and peristriate cortex in the normal animal. *Brain Res.* **66**, 185–208.

Barlow, H. B., Blakemore, C. & Pettigrew, J. D. (1967) The neural mechanism of binocular depth perception. *J. Physiol.* **193**, 327–342.

Barlow, H. B. & Brindley, G. S. (1963) Evidence for a physiological explanation of the waterfall phenomenon and figural after-effects. *Nature* **200**, 1345–1347.

Barlow, H. B., Fitzhugh, R. & Kuffler, S. W. (1957) Change of organization in the receptive fields of the cat retina during dark adaptation. *J. Physiol.* **137**, 338–354.

Barlow, H. B., Hill, R. M. & Levick, W. R. (1964) Retinal ganglion cells responding selectively to direction and speed of image motion in the rabbit. *J. Physiol.* **173**, 377–407.

Battersby, W. S., Wagman, I. H., Karp, E. & Bender, M. E. (1960) Neural limitations of visual excitability: alterations produced by cerebral lesions. *Arch. Neurol.* **3**, 24–42.

Benevento, L. A., Creutzfeldt, O. D. & Kuhnt, U. (1972) Significance of intracortical inhibition in the visual cortex. *Nature, New Biol.* **238**, 124–126.

Berkley, M., Wolf, E. & Glickstein, M. (1967) Photic evoked potentials in the cat: evidence for a direct geniculate input to visual II. *Exp. Neurol.* **19**, 188–198.

Berlucchi, G., Gazzaniga, S. & Rizzolatti, G. (1967) Microelectrode analysis of transfer of visual information by the corpus callosum. *Arch. ital. Biol.* **105**, 583–596.

Berlucchi, G. & Rizzolatti, G. (1968) Binocularly driven neurons in visual cortex of split-chiasm cats. *Science* **159**, 308–310.

Berman, N. & Cynader, M. (1972) Comparison of receptive field organization of the superior colliculus in Siamese and normal cats. *J. Physiol.* **224**, 363–389.

Berman, N. & Cynader, M. (1975) Receptive fields of cat superior colliculus after visual cortex lesions. *J. Physiol.* **245**, 261–270.

Berman, N. J. & Sterling, P. (1974) Immediate reversal of eye dominance in the superior colliculus of the monocularly deprived cat following cortical removal. *Anat. Rec.* **178**, 310.

Berman, N. & Sterling, P. (1976) Cortical suppression of the retino-collicular pathway in the monocularly deprived cat. *J. Physiol.* **255**, 263–273.

Bertulis, A., Gold, G. C. & Lennox-Buchtal, M. A. (1977) Spectral and orientation specificity of single cells in foveal striate cortex of the vervet monkey, *Cercopithecus aethiops*. *J. Physiol.* **268**, 1–20.

Bilge, M., Bingle, A., Seneviratne, K. N. & Whitteridge, D. (1967) A map of the visual cortex in the cat. *J. Physiol.* **191**, 116–117P.

Bishop, P. O., Coombs, J. S. & Henry, G. H. (1971) Responses to visual contours: Spatio-temporal aspects of excitation in the receptive fields of simple striate neurones. *J. Physiol.* **219**, 625–657.

Bishop, P. O., Coombs, J. S. & Henry, G. H. (1973) Receptive fields of simple cells in the cat striate cortex. *J. Physiol.* **231**, 31–60.

Bishop, P. O., Goodwin, A. W. & Henry, G. H. (1974) Direction-sensitive sub-regions in striate simple cell receptive fields. *J. Physiol.* **238**, 25–27P.

Bishop, P. O., Kozak, W. & Vakkur, G. J. (1962). Some quantitative aspects of the cat's eye: axis and plane of reference, visual field co-ordinates and optics. *J. Physiol.* **163**, 466–502.

Blakemore, C. (1969) Binocular depth discrimination and the nasotemporal division. *J. Physiol.* **205**, 471–497.

Blakemore, C. (1970a) The representation of three-dimensional visual space in the rat's striate cortex. *J. Physiol.* **209**, 155–178.

Blakemore, C. (1970b) The range and scope of binocular depth discrimination in man. *J. Physiol.* **211**, 599–622.

Blakemore, C. (1974) Development of functional connexions in the mammalian visual system. *Brit. med. Bull.* **30**, 152–157.

Blakemore, C. (1976) The conditions required for the maintenance of binocularity in the kitten's visual cortex. *J. Physiol.* **261**, 423–444.

Blakemore, C. & Campbell, F. W. (1969) On the existence of neurones in the human visual system selectively sensitive to the orientation and size of retinal images. *J. Physiol.* **203**, 237–260.

Blakemore, C. & Cooper, G. F. (1970) Development of the brain depends on the visual environment. *Nature* **228**, 477–478.

Blakemore, C. & Mitchell, D. E. (1973) Environmental modification of the visual cortex and the neural basis of learning and memory. *Nature* **241**, 467–468.

Blakemore, C. & Pettigrew, J. D. (1970) Eye dominance in the visual cortex. *Nature* **225**, 426–429.

Blakemore, C. & Van Sluyters, R. C. (1974) Experimental analysis of amblyopia and strabismus. *Brit. J. Ophthal.* **58**, 176–182.

Blakemore, C. & Van Sluyters, R. C. (1974) Reversal of the physiological effects of monocular deprivation in kittens: further evidence for a sensitive period. *J. Physiol.* **237**, 195–216.

Blakemore, C. & Van Sluyters, R. C. (1975) Innate and environmental factors in the development of the kitten's visual cortex. *J. Physiol.* **248**, 663–716.

Blakemore, C., Van Sluyters, R. C. & Movshon, J. A. (1976) Synaptic competition in the kitten's visual cortex. *Cold Spr. Harb. Symp. quant. Biol.* **40**, 601–609.

v. Bonin, G. N., Garol, H. W. & McCulloch, W. S. (1942) The functional organization of the occipital lobe. *Biol. Symp.* **7**, 165–192.

v. Bonin, G. & Bailey, P. (1947) *The Necocortex of Macaca Mulatta.* Univ. of Illinois Press, Urbana.

Brodmann, K. (1905) Beiträge zur histologischen Lokalisation der Grosshirnrinde. *J. Psychol. Neurol. (Leipzig)*, **4**, 176.

Campbell, F. W., Cooper, G. F. & Enroth-Cugell, C. (1969) The spatial selectivity of the visual cells of the cat. *J. Physiol.* **203**, 223–235.

Campbell, F. W., Maffei, L. & Piccolino, M. (1973) The contrast sensitivity of the cat. *J. Physiol.* **229**, 719–731.

Campos-Ortega, J. A., Hayhow, W. R. & Cluver, P. F. de V. (1970) A note of the problem of retinal projections to the inferior pulvinar nucleus of primates. *Brain Res.* **22**, 126–130.

Casagrande, V. A. & Diamond, I. T. (1974). Ablation study of the superior colliculus in the tree shrew *(Tupaia glis)*. *J. comp. Neurol.* **156**, 207–238.

Chalupa, L. M., Coyle, R. S. & Lindsley, D. B. (1976) Effects of pulvinar lesions on visual pattern discrimination in monkeys. *J. Neurophysiol.* **39**, 334–369.

Choudhury, B. P. & Whitteridge, D. (1965) Visual field projections on the dorsal nucleus of the lateral geniculate body. *Quart. J. exp. Physiol.* **50**, 104–111.

Choudhury, B. P., Whitteridge, D. & Wilson, M. E. (1965) The function of the callosal connections of the visual cortex. *Quart. J. exp. Physiol.* **50**, 214–219.

Chow, K. L., Masland, R. H. & Stewart, D. L. (1971) Receptive field characteristics of striate cortical neurons in the rabbit. *Brain Res.* **33**, 337–352.

Chow, K. L. & Spear, P. D. (1974) Morphological and functional effects of visual deprivation on the rabbit visual system. *Exp. Neurol.* **42**, 429–447.

Clark, W. E. LeGros. (1959) The anatomy of cortical vision. *Trans. Ophthal. Soc. U.K.* **79**, 455–461.

Cleland, B. G., Dubin, M. W. & Levick, W. R. (1971) Sustained and transient neurones in the cat's retina and lateral geniculate nucleus. *J. Physiol.* **217**, 473–497.

Cleland, B. G., Levick, W. R., Morstyn, R. & Wagner, H. G. (1976) Lateral geniculate relay of slowly conducting retinal afferents to cat visual cortex. *J. Physiol.* **255**, 299–320.

Cleland, B. G., Levick, W. R. & Sanderson, K. J. (1973) Properties of sustained and transient ganglion cells in the cat retina. *J. Physiol.* **228**, 649–680.

Cobb, W. A. & Dawson, G. D. (1960) The latency and form in man of the occipital potentials evoked by bright flashes. *J. Physiol.* **152**, 108–121.

Cowey, A. (1964) Projection of the retina on to striate and prestriate cortex in the squirrel monkey, *Samiri sciureus. J. Neurophysiol.* **27**, 366–396.

Cragg, B. G. (1969) The topography of the afferent projections in the circumstriate visual cortex of the monkey studied by the Nauta method. *Vision Res.* **9**, 733–747.

Critchlow, V. (1963) The role of light in the neuroendocrine system. *Adv. Neuroendocrinol.* pp. 377–402.

Curtis, H. J. (1940) Intercortical connections of corpus callosum as indicated by evoked potentials. *J. Neurophysiol.* **3**, 407–413.

Cynader, M. & Berman, N. (1972) Receptive field organization of monkey superior colliculus. *J. Neurophysiol.* **35**, 187–201.

Daniel, P. M. & Whitteridge, D. (1959) The representation of the visual field on the calcarine cortex of baboons and monkeys. *J. Physiol.* **148**, 33–34P.

De Valois, R. L. (1960) Colour vision mechanisms in the monkey. *J. gen. Physiol.* **43**, No. 6 (Suppl.), 115–128.

De Valois, R. L. & Morgan, H. C. (1974) Psychophysical studies of monkey vision. I. Macaque luminosity and color vision tests. *Vision Res.* **14**, 53–67.

De Voe, G. G., Ripps, H. & Vaughan, H. G. (1968) Cortical responses to stimulation of the human fovea. *Vision Res.* **8**, 135–147.

Dews, P. B. & Wiesel, T. N. (1970) Consequences of monocular deprivation on visual behaviour in kittens. *J. Physiol.* **206**, 437–455.

Donchin, E. & Cohen, L. (1970) Evoked potentials to stimuli presented to the suppressed eye in a binocular rivalry experiment. *Vision Res.* **10**, 103–105.

Dow, B. M. (1974) Functional classes of cells and their laminar distribution in monkey visual cortex. *J. Neurophysiol.* **37**, 927–946.

Dow, B. M. & Gouras, P. (1973) Color and spatial specificity of single units in rhesus monkey striate cortex. *J. Neurophysiol.* **36**, 79–100.

Downer, J. L. de C. (1959) Changes in visually guided behaviour following midsagittal division of optic chiasma and corpus callosum in monkey. *Brain* **82**, 251–259.

Dreher, B. (1972) Hypercomplex cells in the cat's striate cortex. *Invest. Ophthal.* **11**, 355–356.

Dreher, B. & Hoffmann, K.-P. (1973) Properties of excitatory and inhibitory regions in the receptive fields of single units in the cat's superior colliculus. *Exp. Brain Res.* **16**, 333–353.

Dreher, B. & Sanderson, K. J. (1973) Receptive field analysis: responses to moving visual contours by single lateral geniculate neurones in the cat. *J. Physiol.* **234**, 95–118.

Dürsteler, M. R., Garey, L. J. & Movshon, J. A. (1976) Reversal of the morphological effects of monocular deprivation in the kitten's lateral geniculate nucleus. *J. Physiol.* **261**, 189–210.

Ebner, F. F. & Myers, R. E. (1965) Distribution of corpus callosum and anterior commissure in cat and raccoon. *J. comp. Neurol.* **124**, 353–365.

Ennever, J., Gartside, I. B., Lippold, O. C. J., Novotny, G. E. K. & Shagass, C. (1967) Contamination of the human cortical evoked response with potentials of intra-orbital origin. *J. Physiol.* **191**, 5–6P.

Enroth-Cugell, C. & Robson, J. G. (1966) The contrast sensitivity of retinal ganglion cells in the cat. *J. Physiol.* **187**, 517–552.

Ettlinger, G. (1958) Visual discrimination following successive temporal excisions in monkeys. *J. Physiol.* **140**, 38–39P.

Feldman, M. & Cohen, B. (1968) Electrical activity in the lateral geniculate body of the alert monkey associated with eye movements. *J. Neurophysiol.* **31**, 455–466.

Fender, D. & Julesz, B. (1967) Extension of Panum's fusional area in binocularly stabilized vision. *J. opt. Soc. Am.* **57**, 819–830.

Fukada, Y. & Saito, H. (1971) The relationship between response characteristics to flicker stimulation and receptive field organization in the cat's optic nerve fibres. *Vision Res.* **11**, 227–240.

Ganz, L. & Hirsch, H. V. B. (1972) The nature of visual deficits in visually deprived cats. *Brain Res.* **44**, 547–568.

Garey, L. J., Fisher, R. A. & Powell, T. P. S. (1973) Effects of experimental deafferentation on cells in the lateral geniculate nucleus of the cat. *Brain Res.* **52**, 363–369.

Gilbert, C. D. (1977) Laminar differences in receptive field properties of cells in cat primary visual cortex. *J. Physiol.* **268**, 391–421.

Gilbert, C. D. & Kelly, J. P. (1975) The projections of cells in different layers of the cat's visual cortex. *J. comp. Neurol.* **163**, 81–106.

Glees, P. & Clark, W. E. LeGros. (1941) The termination of optic fibres in the lateral geniculate body of the monkey. *J. Anat.* **75**, 295–308.

Glendenning, K. K., Hall, J. A., Diamond, I. T. & Hall, W. C. (1975) The pulvinar nucleus of *Galago Senegalensis. J. comp. Neurol.* **161**, 419–458.

Goldberg, M. E. & Wurtz, R. H. (1972) Activity of superior colliculus in behaving monkey. I. Visual receptive fields of single neurons. *J. Neurophysiol.* **35**, 542–559.

Goodwin, A. W. & Henry, G. H. (1975) Direction sensitivity of complex cells in a comparison with simple cells. *J. Neurophysiol.* **38**, 1524–1540.

Goodwin, A. W., Henry, G. H. & Bishop, P. O. (1975) Direction selectivity of simple striate cells: properties and mechanism. *J. Neurophysiol.* **38**, 1500–1523.

Gordon, B. (1973) Receptive fields in deep layers of cat superior colliculus. *J. Neurophysiol.* **36**, 157–178.

Gouras, P. (1974) Opponent-colour cells in different layers of foveal striate cortex. *J. Physiol.* **238**, 583–602.

Gross, C. G., Rocha-Miranda, C. E. & Bender, D. B. (1972) Visual properties of neurons in inferotemporal cortex of macaque. *J. Neurophysiol.* **35**, 96–111.

Gross, C. G., Schiller, P. H., Wells, C. & Gerstein, G. L. (1967) Single-unit activity in temporal association cortex of the monkey. *J. Neurophysiol.* **30**, 833–843.

Guillery, R. W. (1970) An abnormal retinogeniculate projection in Siamese cats. *Brain Res.* **14**, 739–741.

Guillery, R. W. (1970) The laminar distribution of retinal fibres in the dorsal lateral geniculate nucleus of the cat: a new interpretation. *J. comp. Neurol.* **138**, 339–357.

Guillery, R. W. & Stelzner, D. J. (1970) The differential effects of unilateral lid closure upon monocular and binocular segments of the dorsal lateral geniculate nucleus in the cat. *J. comp. Neurol.* **139**, 413–422.

Hamilton, C. R. & Gazzaniga, M. S. (1964) Lateralization of learning of colour and brightness discriminations following brain bisection. *Nature* **201**, 220.

Hammond, P. (1972) Spatial organization of receptive fields of LGN neurones. *J. Physiol.* **222**, 53–54P.

Hayashi, Y., Sumitomo, I. & Iwama, K. (1967) Activation of lateral geniculate neurons by electrical stimulation of superior colliculus in cats. *Jap. J. Physiol.* **17**, 638–651.

Hayhow, W. R., Webb, C. & Jervie, A. (1960) The accessory optic fiber system in the rat. *J. comp. Neurol.* **115**, 187–215.

Helmholtz, H. von. (1925) *Physiological Optics* (translated J. P. C. Southall). Optical Soc. Am.

Hickey, T. L. & Guillery, R. W. (1973) An autoradiographic study of the efferent connections of the ventral lateral geniculate nucleus in the cat and fox. *J. comp. Neurol.* **156**, 239–253.

Hirsch, H. V. B. & Spinelli, D. N. (1970) Visual experience modifying distribution of horizontally and vertically oriented receptive fields in cats. *Science* **168**, 869–871.

Hirsch, H. V. B. & Spinelli, D. N. (1971) Modification of the distribution of receptive field orientation in cats by selective visual exposure during development. *Exp. Brain Res.* **12**, 509–527.

Hoffmann, K.-P. (1972) The retinal input to the superior colliculus in the cat. *Invest. Ophthal.* **11**, 467–470.

Hoffmann, K.-P. & Dreher, B. (1973) The spatial organisation of the excitatory region of receptive fields in the cat's superior colliculus. *Exp. Brain Res.* **16**, 354–370.

Hoffmann, K.-P. & Sherman, S. M. (1974) Effects of early monocular deprivation on visual input to cat superior colliculus. *J. Neurophysiol.* **37**, 1276–1286.

Hoffmann, K.-P., Stone, J. & Sherman, S. M. (1972) Relay of receptive-field properties in dorsal lateral geniculate nucleus of the cat. *J. Neurophysiol.* **35**, 518–531.

Hoffmann, K.-P. & Straschli, M. (1971) Influences of cortico-tectal and intertectal connections on visual responses in the cat's superior colliculus. *Exp. Brain Res.* **12**, 120–131.

Holmes, G. (1945) Organization of the visual cortex in man. *Proc. Roy. Soc. B* **132**, 348–361.

Hubel, D. H. & Wiesel, T. N. (1961) Integrative action in the cat's lateral geniculate body. *J. Physiol.* **155**, 385–398.

Hubel, D. H. & Wiesel, T. N. (1962) Receptive fields, binocular interaction and functional architecture in the cat's visual cortex. *J. Physiol.* **160**, 106–154.

Hubel, D. H. & Wiesel, T. N. (1963) Receptive fields of cells in striate cortex of very young, visually inexperienced kittens. *J. Neurophysiol.* **26**, 994–1002.

Hubel, D. H. & Wiesel, T. N. (1965a) Receptive fields and functional architecture in two non-striate areas (18 and 19) of the cat. *J. Neurophysiol.* **28**, 229–289.

Hubel, D. H. & Wiesel, T. N. (1965b) Binocular interaction in striate cortex of kittens reared with artificial squint. *J. Neurophysiol.* **28**, 1041–1059.

Hubel, D. H. & Wiesel, T. N. (1967) Cortical and callosal connections concerned with the vertical meridian of visual fields in the cat. *J. Neurophysiol.* **30**, 1561–1573.

Hubel, D. H. & Wiesel, T. N. (1968) Receptive fields and functional architecture of monkey striate cortex. *J. Physiol.* **195**, 215–243.

Hubel, D. H. & Wiesel, T. N. (1969) Anatomical demonstration of columns in the monkey striate cortex. *Nature* **221**, 747–750.

Hubel, D. H. & Wiesel, T. N. (1970) Stereoscopic vision in macaque monkey. *Nature* **225**, 41–42.

Hubel, D. H. & Wiesel, T. N. (1970) The period of susceptibility to the physiological effects of unilateral eye closure in kittens. *J. Physiol.* **206**, 419–436.

Hubel, D. H. & Wiesel, T. N. (1971) Aberrant visual projections in the Siamese cat. *J. Physiol.* **218**, 33–62.

Hubel, D. H. & Wiesel, T. N. (1972) Laminar and columnar distribution of geniculo-cortical fibers in the macaque monkey. *J. comp. Neurol.* **146**, 421–450.

Hubel, D. H. & Wiesel, T. N. (1974) Sequence regularity and geometry of orientation columns in the monkey striate cortex. *J. comp. Neurol.* **158**, 267–293.

Hubel, D. H., Wiesel, T. N. & Le Vay, S. (1976) Functional architecture of Area 17 in normal and monocularly deprived macaque monkeys. *Cold Spr. Harb. Symp. quant. Biol.* **40**, 581–589.

Hughes, A. (1975) A quantitative analysis of the cat retinal ganglion cell topography. *J. comp. Neurol.* **163**, 107–128.

Hughes, A. & Wässle, H. (1976) The cat optic nerve: fibre total count and diameter spectrum. *J. comp. Neurol.* **169**, 171–184.

Hull, E. M. (1968) Corticofugal influence in the macaque lateral geniculate nucleus. *Vision Res.* **8**, 1285–1298.

Humphrey, N. K. & Weiskrantz, L. (1967) Vision in monkeys after removal of the striate cortex. *Nature* **215**, 595–597.

Ikeda, H. & Jacobson, S. G. (1977) Nasal field loss in cats reared with convergent squint: behavioural studies. *J. Physiol.* **270**, 367–381.

Ikeda, H., Plant, G. T. & Tremain, K. E. (1977) Nasal field loss in kittens reared with convergent squint: neuro-physiological and morphological studies of the lateral geniculate nucleus. *J. Physiol.* **270**, 345–366.

Ikeda, H. & Wright, M. J. (1975a) Spatial and temporal properties of 'sustained' and 'transient' neurones in Area 17 of the cat's visual cortex. *Exp. Brain Res.* **22**, 363–383.

Ikeda, H. & Wright, M. J. (1975b) Retinotopic distribution, visual latency and orientation tuning of 'sustained' and 'transient' cortical neurones in Area 17 of the cat. *Exp. Brain Res.* **22**, 385–398.

Ikeda, H. & Wright, M. J. (1976) Properties of LGN cells in kittens reared with convergent squint: a neurophysiological demonstration of amblyopia. *Exp. Brain Res.* **25**, 63–77.

Imbert, M. & Buisseret, P. (1975) Receptive field characteristics and plastic properties of visual cortical cells in kittens reared with or without visual experience. *Exp. Brain Res.* **22**, 25–36.

Jasper, H., Ajmone-Marsan, C. & Stoll, J. (1952) Cortico-fugal projections to the brain stem. *Arch. Neurol. Psychiat.* **67**, 155–171.

Kaas, J. H. & Guillery, R. W. (1973) The transfer of abnormal visual field representations from the dorsal lateral geniculate nucleus to the visual cortex in Siamese cats. *Brain Res.* **59**, 61–95.

Kawamura, K. (1973) Corticocortical fiber connections of the cat cerebrum. II. The parietal region. *Brain Res.* **51**, 23–40.

Kawamura, S., Sprague, J. M. & Niimi, K. (1974) Cortico-fugal projections from the visual cortex to the thalamus, pretectum and superior colliculus in the cat. *J. comp. Neurol.* **158**, 339–362.

Keesey, U. (1972) Flicker and pattern detection: a comparison of thresholds. *J. opt. Soc. Amer.* **62**, 446–448.

Kelly, J. P. & Gilbert, C. D. (1975) The projections of different morphological types of ganglion cells in the cat retina. *J. comp. Neurol.* **163**, 65–80.

Kelly, J. P. & Van Essen, D. C. (1974) Cell structure and function in the visual cortex of the cat. *J. Physiol.* **238**, 515–547.

King-Smith, P. E. & Kulikowski, J. J. (1975) The detection of gratings by independent activation of line detectors. *J. Physiol.* **247**, 237–271.

Klingaman, R. L. (1976) The human visual evoked cortical potential and dark adaptation. *Vision Res.* **16**, 1471–1477.

Klüver, H. (1942) Functional significance of the geniculostriate system. *Biol. Symp.* **7**, 253–299.

Klüver, H. & Bucy, P. C. (1939) Preliminary analysis of functions of the temporal lobes in monkeys. *Arch. Neurol. Psychiat.* **42**, 979–1000.

Kratz, K. E., Spear, P. D. & Smith, D. C. (1976) Post critical period reversal of effects of monocular deprivation on striate cortex cells in the cat. *J. Neurophysiol.* **39**, 501–511.

Krnjević, K., Randić, M. & Straughan, D. W. (1966) Pharmacology of cortical inhibition. *J. Physiol.* **184**, 78–105.

Kulikowski, J. J. & Tolhurst, D. J. (1973) Psychophysical evidence for sustained and transient detectors in human vision. *J. Physiol.* **232**, 149–162.

Laemle, J. K. (1968) Retinal projections of *Tupaia glis.* *Brain, Behav. and Evol.* **1**, 473–499.

Lane, R. H., Allman, J. M., Kaas, J. H. & Miezin, F. M. (1973) The visuotopic organization of the superior colliculus of the owl monkey (*Aotus trivirgatus*) and the bush baby (*Galago Senegalensis*). *Brain Res.* **60**, 335–349.

Lansing, R. W. (1964) Electroencephalographic correlates of binocular rivalry in man. *Science* **146**, 1325–1327.

Lawwill, T. & Biersdorf, W. R. (1968). Binocular rivalry and visual evoked responses. *Invest. Ophthal.* **7**, 378–385.

Le Vay, S. & Gilbert, C. D. (1976) Laminar patterns of geniculocortical projection in the cat. *Brain Res.* **113**, 1–19.

Lund, R. D. & Lund, J. S. (1972) Development of synaptic patterns in the superior colliculus of the rat. *Brain Res.* **42**, 1–20.

Maffei, L. & Fiorentini, A. (1972) Retinogeniculate convergence and analysis of contrast. *J. Neurophysiol.* **35**, 65–72.

Maffei, L. & Fiorentini, A. (1973) The visual cortex as a spatial frequency analyser. *Vision Res.* **13**, 1255–1267.

Maffei, L. & Fiorentini, A. (1977) Spatial frequency rows in the striate visual cortex. *Vision Res.* **17**, 257–264.

Mathers, L. H. & Rapisardi, S. C. (1973) Visual and somatosensory receptive fields of neurons in the squirrel monkey pulvinar. *Brain Res.* **64**, 65–83.

McIlwain, J. T. & Buser, P. (1968) Receptive fields of single cells in the cat's superior colliculus. *Exp. Brain Res.* **5**, 314–325.

Meikle, T. H. (1964) Failure of interocular transfer of brightness discrimination. *Nature* **202**, 1243–1244.

Meikle, T. H. & Sechzer, J. A. (1960) Interocular transfer of brightness discrimination in 'split-brain' cats. *Science* **132**, 734–735.

Mitchell, D. E. & Blakemore, C. (1970) Binocular depth perception and the corpus callosum. *Vision Res.* **10**, 49–54.

Mohler, C. W. & Wurtz, R. W. (1976) Organization of monkey superior colliculus: intermediate layer cells discharging before eye movements. *J. Neurophysiol.* **39**, 722–744.

Montero, V. M. & Brugge, J. F. (1969) Direction of movement as the significant stimulus parameter for some lateral geniculate cells in the rat. *Vision Res.* **9**, 71–88.

Moore, R. Y. & Eichler, V. B. (1972) Loss of a circadian adrenal corticosterone rhythm following suprachiasmatic lesion in the rat. *Brain Res.* **42**, 201–206.

Moore, R. Y. & Klein, D. C. (1974) Visual pathways and the central neural control of a circadian rhythm in pineal serotonin N-acetyltransferase activity. *Brain Res.* **71**, 17–33.

Moore, R. Y. & Lenn, N. J. (1972) A retinohypothalamic projection in the rat. *J. comp. Neurol.* **146**, 1–9.

Motokawa, K., Taira, N. & Okuda, J. (1962) Spectral responses of single units in the primate visual cortex. *Tohoku J. exp. Med.* **78**, 320–337.

Mountcastle, V. B. (1957) Modality and topographic properties of single neurones of cat's somatic sensory cortex. *J. Neurophysiol.* **20**, 408–434.

Movshon, J. A. (1975) The velocity tuning of single units in cat striate cortex. *J. Physiol.* **249**, 445–468.

Movshon, J. A. (1976a) Reversal of the physiological effects of monocular deprivation in the kitten's visual cortex. *J. Physiol.* **261**, 125–174.

Movshon, J. A. (1976b) Reversal of the behavioural effects of monocular deprivation in the kitten. *J. Physiol.* **261**, 175–187.

Myers, R. E. (1962) Commissural connections between occipital lobes of the monkey. *J. comp. Neurol.* **118**, 1–10.

Myers, R. E. (1965) Organization of visual pathways. In *Function of the Corpus Callosum.* Ed. Ettlinger, E. G. Churchill, London, pp. 133–142.

Nikara, T., Bishop, P. O. & Pettigrew, J. D. (1968) Analysis of retinal correspondence by studying receptive fields of binocular single units in cat striate cortex. *Exp. Brain Res.* **6**, 353–372.

Noble, J. (1966) Mirror-images and the forebrain commissures of the monkey. *Nature* **211**, 1263–1266.

Norton, T. T. (1974) Receptive-field properties of superior colliculus cells and development of visual behavior in kittens. *J. Neurophysiol.* **37**, 674–690.

Ogle, K. N. (1950) *Researches in Binocular Vision.* Philadelphia: Saunders.

Olson, C. R. & Freeman, R. D. (1975) Progressive changes in kitten striate cortex during monocular vision. *J. Neurophysiol.* **38**, 26–32.

Packwood, J. & Gordon, B. (1975) Stereopsis in normal domestic cat, and cat raised with alternating monocular occlusion. *J. Neurophysiol.* **38**, 1485–1499.

Palmer, L. A. & Rosenquist, A. C. (1971) Visual receptive field properties of cells of the superior colliculus after cortical lesions in the cat. *Exp. Neurol.* **33**, 629–652.

Palmer, L. A. & Rosenquist, A. C. (1974) Visual receptive fields of single striate cortical units projecting to the superior colliculus in the cat. *Brain Res.* **67**, 27–42.

Pasik, P. & Pasik, T. (1964) Oculomotor functions in monkeys with lesions in the cerebrum and the superior colliculi. In *The Oculomotor System*, pp. 40–80. Ed. M. B. Bender. New York: Harper & Row.

Pasik, P., Pasik, T. & Schilder, P. (1969) Extrageniculostriate vision in the monkey: discrimination of luminous flux-equated figures. *Exp. Neurol.* **24**, 421–437.

Peck, C. K. & Blakemore, C. (1975) Modification of single neurons in the kitten's visual cortex after brief periods of monocular visual experience. *Exp. Brain Res.* **22**, 57–68.

Penman, G. G. (1934) The representation of the areas of the retina in the lateral geniculate body. *Trans. Ophthal. Soc. U.K.* **54**, 232–270.

Pettigrew, J. D. (1974) The effect of visual experience on the development of stimulus specificity by kitten cortical neurones. *J. Physiol.* **237**, 49–74.

Pettigrew, J. D., Nikara, T. N. & Bishop, P. O. (1968a) Responses to moving slits by single units in cat's striate cortex. *Exp. Brain Res.* **6**, 373–390.

Pettigrew, J. D., Nikara, T. & Bishop, P. O. (1968b) Binocular interaction on single units in cat striate cortex: simultaneous stimulation by single moving slit with receptive fields in correspondence. *Exp. Brain Res.* **6**, 391–410.

Poggio, G. F., Baker, F. H., Mansfield, R. J. W., Sillito, A. & Grigg, P. (1975) Spatial and chromatic properties of neurons subserving foveal and parafoveal vision in rhesus monkey. *Brain Res.* **100**, 25–59.

Pollen, D. A. & Ronner, S. F. (1975) Periodic excitability

changes across the receptive fields of complex cells in the striate and parastriate cortex of the cat. *J. Physiol.* **245**, 667–697.

Preobrazhenskaya, N. S. (1964) Development and functional significance of pregeniculate nucleus in man. *Fed. Proc.* **23** (transl. supp.) 715–718.

Rodieck, R. W. (1965) Quantitative analysis of cat retinal ganglion cell response to visual stimuli. *Vision Res.* **5**, 583–601.

Rose, D. (1977) Responses of single units in cat visual cortex to moving bars of light as a function of bar length. *J. Physiol.* **271**, 1–23.

Rose, D. & Blakemore, C. (1974) An analysis of orientation selectivity in the cat's visual cortex. *Exp. Brain Res.* **20**, 1–17.

Rosenquist, A. C. & Palmer, L. A. (1971) Visual receptive field properties of cells of the superior colliculus after cortical lesions in the cat. *Exp. Neurol.* **33**, 629–652.

Royce, G. J., Ward, J. P. & Harting, J. K. (1976) Retinofugal pathways in two marsupials. *J. comp. Neurol.* **170**, 391–413.

Schilder, P., Pasik, T. & Pasik, P. (1971) Extrageniculostriate vision in the monkey. II. Demonstration of brightness discrimination. *Brain Res.* **32**, 383–398.

Schiller, P. H., Finlay, B. L. & Volman, S. F. (1976a–e) Quantitative studies of single-cell properties in monkey striate cortex. I. Spatiotemporal organization of receptive fields. II. Orientation specificity and ocular dominance. III. Spatial frequency. IV. Corticotectal cells. V. Multivariate statistical analyses and models. *J. Neurophysiol.* **39**, 1288–1319; 1320–1333; 1334–1351; 1352–1361; 1362–1374.

Schiller, P. H., Stryker, M., Cynader, M. & Berman, N. (1974) Response characteristics of single cells in the monkey superior colliculus following ablation or cooling of visual cortex. *J. Neurophysiol.* **37**, 181–194.

Sechzer, J. A. (1963) Successful interocular transfer of pattern discrimination in 'split-brain' cats with shock-avoidance motivation. *J. comp. physiol. Psychol.* **58**, 76–83.

Sherk, H. & Stryker, M. P. (1976) Quantitative study of cortical orientation selectivity in visually inexperienced kitten. *J. Neurophysiol.* **39**, 63–70.

Sherman, S. M. (1973) Visual field defects in monocularly and binocularly deprived cats. *Brain Res.* **49**, 25–45.

Sherman, S. M. (1974) Permanence of visual perimetry deficits in monocularly deprived cats. *Brain Res.* **73**, 491–501.

Sherman, S. M. & Sanderson, K. J. (1972) Binocular interaction on cells of the dorsal lateral geniculate nucleus of visually deprived cats. *Brain Res.* **37**, 126–131.

Sillito, A. M. (1977) Inhibitory processes underlying the directional specificity of simple, complex and hypercomplex cells in the cat's visual cortex. *J. Physiol.* **271**, 699–720.

Sillito, A. M. & Versiani, V. (1977) The contribution of excitatory and inhibitory inputs to the length preference of hypercomplex cells in layers II and III of the cat's striate cortex. *J. Physiol.* **273**, 775–790.

Singer, W. (1974) The effect of mesencephalic reticular stimulation on intracellular potentials of cat lateral geniculate neurons. *Brain Res.* **61**, 35–54.

Singer, W. & Creutzfeldt, O. D. (1970) Reciprocal lateral inhibition of ON- and OFF-center neurones in the lateral geniculate body. *Exp. Brain Res.* **10**, 311–330.

Singer, W., Tretter, F. & Cynader, M. (1975) Organization of cat striate cortex: a correlation of receptive field properties with afferent and efferent connections. *J. Neurophysiol.* **38**, 1080–1098.

Spalding, J. M. K. (1952a) Wounds of the visual pathway. I. The visual radiation. *J. Neurol.* **15**, 99–107.

Spalding, J. M. K. (1952b) Wounds of the visual pathway. II. The striate cortex. *J. Neurol.* **15**, 169–183.

Spear, P. D. & Ganz, L. (1975) Effects of visual cortex lesions following recovery from monocular deprivation in cat. *Exp. Brain Res.* **23**, 181–201.

Sperry, R. W., Stamm, J. S. & Miner, N. (1956) Relearning tests for interocular transfer following division of optic chiasma and corpus callosum in cats. *J. comp. physiol. Psychol.* **49**, 529–533.

Sprague, J. M. (1966) Interaction of cortex and superior colliculus in mediation of visually guided behaviour in the cat. *Science* **153**, 1544–1547.

Sprague, J. M. (1972) The superior colliculus and pretectum in visual behavior. *Invest. Ophthal.* **11**, 473–482.

Sprague, J. M. & Meikle, T. H. (1965) The role of the superior colliculus in visually guided behaviour. *Exp. Neurol.* **11**, 115–146.

Stephan, F. K. & Zucker, I. (1972) Circadian rhythms in drinking behavior and locomotor activity of rats are eliminated by hypothalamic lesions. *Proc. Nat. Acad. Sci. Wash.* **69**, 1583–1586.

Sterling, P. & Wickelgren, B. G. (1969) Visual receptive fields in the superior colliculus of the cat. *J. Neurophysiol.* **32**, 1–15.

Stevens, J. K. & Gerstein, G. L. (1976a) Spatiotemporal organization of cat lateral geniculate receptive fields. *J. Neurophysiol.* **39**, 213–238.

Stevens, J. K. & Gerstein, G. L. (1976b) Interactions between cat lateral geniculate neurons. *J. Neurophysiol.* **39**, 239–256.

Stewart, D. L., Birt, D. & Towns, L. C. (1973) Visual receptive-field characteristics of superior colliculus neurons after cortical lesions in rabbit. *Vision Res.* **13**, 1965–1977.

Stewart, D. L. & Van Sluyter, R. C. (1974) Binocular neurons of the rabbit's visual cortex: receptive field characteristics. *Exp. Brain Res.* **19**, 166–195.

Stone, J. & Dreher, B. (1973) Projection of X- and Y-cells of the cat's lateral geniculate nucleus to areas 17 and 18 of visual cortex. *J. Neurophysiol.* **36**, 551–567.

Stone, J. & Fukuda, Y. (1974) Properties of cat retinal ganglion cells: a comparison of W-cells with X- and Y-cells. *J. Neurophysiol.* **37**, 722–749.

Stone, J. & Fukuda, Y. (1974) The naso-temporal division of the cat's retina re-examined in terms of Y-, X- and W-cells. *J. comp. Neurol.* **155**, 377–394.

Suzuki, H. & Kato, H. (1969) Neurons with visual properties in the posterior group of the thalamic nuclei. *Exp. Neurol.* **23**, 353–365.

Swanson, L. W., Cowan, W. M. & Jones, E. G. (1974) An autoradiographic study of the efferent connections of the ventral lateral geniculate nucleus in the albino rat and the cat. *J. comp. Neurol.* **156**, 143–164.

Talbot, S. A. & Marshall, W. H. (1941) Physiological studies on neural mechanisms of visual localization and discrimination. *Am. J. Ophthal.* **24**, 1255–1264.

Thuma, B. D. (1928) The cytoarchitecture of the corpus geniculatum laterale. *J. comp. Neurol.* **46**, 173–200.

Tolhurst, D. J. (1973) Separate channels for the analysis of the shape and movement of a moving visual stimulus. *J. Physiol.* **231**, 385–402.

Updyke, B. V. (1975) The patterns of projection of cortical areas 17, 18, and 19 onto the laminae of the dorsal lateral geniculate nucleus in the cat. *J. comp. Neurol.* **163**, 377–396.

Vancy, D. I. & Hughes, A. (1976) The rabbit optic nerve: fibre diameter spectrum, fibre count, and comparison with a retinal ganglion cell count. *J. comp. Neurol.* **170**, 241–251.

Van Hof, M. W. (1969) Discrimination of striated patterns of different orientation in rabbits deprived of light after birth. *Exp. Neurol.* **23,** 561–565.

Van Hof, M. W. (1970) Interocular transfer in the rabbit. *Exp. Neurol.* **26,** 103–108.

Van Sluyters, R. C. & Stewart, D. L. (1974a) Binocular neurons of the rabbit's visual cortex: receptive field characteristics. *Exp. Brain Res.* **19,** 166–195.

Van Sluyters, R. C. & Stewart, D. L. (1974b) Binocular neurons of the rabbit's visual cortex: effects of monocular sensory deprivation. *Exp. Brain Res.* **19,** 196–204.

Vastola, E. F. (1960) Monocular inhibition in the lateral geniculate body. *E.E.G. clin. Neurophysiol.* **12,** 399–403.

Vaughan, H. G. (1966) The perceptual and physiologic significance of visual evoked responses recorded from the scalp in man. In *Clinical Electroretinography*, pp. 203–223. Ed. H. M. Burian & J. H. Jacobson. New York: Pergamon.

Volkmann, A. W. (1863) *Physiologische Untersuchungen in Gebiet der Optik*. Leipzig.

Volkmann, F. C., Schick, A. M. L. & Riggs, L. A. (1968) Time course of visual inhibition during voluntary saccades. *J. opt. Soc. Amer.* **58,** 562–569.

Voneida, T. J. & Robinson, J. S. (1970) Effects of brain bisection on capacity for cross comparison of patterned visual input. *Exp. Neurol.* **26,** 60–71.

von Noorden, G. K., Dowling, J. E. & Ferguson, J. E. (1970) Experimental amblyopia in monkeys. I. *Arch. Ophthal.* **84,** 206–214.

Wässle, H., Levick, W. R. & Cleland, B. G. (1975) The distribution of the alpha type of ganglion cell in the cat retina. *J. comp. Neurol.* **159,** 419–438.

Wickelgren, B. G. & Sterling, P. (1969) Influence of visual cortex on receptive fields in the superior colliculus of the cat. *J. Neurophysiol.* **32,** 16–23.

Wiesel, T. N. & Hubel, D. H. (1963a) Effects of visual deprivation on morphology and physiology of cells in the cat's lateral geniculate body. *J. Neurophysiol.* **26,** 978–993.

Wiesel, T. N. & Hubel, D. H. (1963b) Single cell responses in striate cortex of kittens deprived of vision in one eye. *J. Neurophysiol.* **26,** 1003–1017.

Wiesel, T. N. & Hubel, D. H. (1965) Comparison of the effects of unilateral and bilateral eye closure on cortical unit responses in kittens. *J. Neurophysiol.* **28,** 1029–1040.

Wiesel, T. N. & Hubel, D. H. (1966) Spatial and chromatic interactions in the lateral geniculate body of the rhesus monkey. *J. Neurophysiol.* **29,** 1115–1156.

Wiesel, T. N. & Hubel, D. H. (1974) Ordered arrangement of orientation columns in monkeys lacking visual experience. *J. comp. Neurol.* **158,** 307–318.

Wiesel, T. N., Hubel, D. H. & Lam, D. (1974) Autoradiographic demonstration of ocular dominance columns in the monkey striate cortex by means of transsynaptic transport. *Brain Res.* **79,** 273–279.

Wilson, J. R. & Sherman, S. M. (1976) Receptive-field characteristics of neurons in cat striate cortex: changes with visual field eccentricity. *J. Neurophysiol.* **39,** 512–523.

Wilson, M. E. & Cragg, B. G. (1957) Projections from the lateral geniculate nucleus in the cat and monkey. *J. Anat.* **101,** 677–692.

Wilson, M. E. & Toyne, M. J. (1970) Retino-tectal and cortico-tectal projections in *Macaca mulatta*. *Brain Res.* **24,** 395–406.

Wilson, P. D., Rowe, M. H. & Stone, J. (1976) Properties of relay cells in cat's lateral geniculate nucleus: a comparison of W-cells with X- and Y-cells. *J. Neurophysiol.* **39,** 1193–1209.

Wurtman, H. J., Axelrod, J., Chu, E. W. & Fischer, J. E. (1964) Mediation of some effects of illumination on the rat estrous cycle by the sympathetic nervous system. *Endocrinology* **75,** 266–272.

Wurtz, R. W. & Goldberg, M. E. (1972) Activity of superior colliculus in behaving monkey. III and IV. *J. Neurophysiol.* **35,** 575–586; 587–596.

Wurtz, R. W. & Goldberg, M. E. (1972) The primate superior colliculus and the shift of visual attention. *Invest. Ophthal.* **11,** 441–450.

Wurtz, R. W. & Mohler, C. W. (1976a) Organization of monkey superior colliculus: enhanced visual response of superficial layer cells. *J. Neurophysiol.* **39,** 745–765.

Wurtz, R. W. & Mohler, C. W. (1976b) Enhancement of visual response in monkey striate cortex and frontal eye fields. *J. Neurophysiol.* **39,** 766–772.

Yates, J. T. (1974) Chromatic information processing in the foveal projection (area striata) of unanesthetized primate. *Vision Res.* **14,** 163–173.

Zeki, S. M. (1969) Representation of central visual fields in prestriate cortex of monkey. *Brain Res.* **14,** 271–291.

Zeki, S. M. (1971a) Convergent input from the striate cortex (area 17) to the cortex of the superior temporal sulcus in the rhesus monkey. *Brain Res.* **28,** 338–340.

Zeki, S. M. (1971b) Cortical projections from two prestriate areas in the monkey. *Brain Res.* **34,** 19–35.

Zeki, S. M. (1973) Colour coding in rhesus monkey prestriate cortex. *Brain Res.* **53,** 422–427.

Zeki, S. M. (1974a) Functional organization of a visual area in the posterior bank of the superior temporal sulcus of the rhesus monkey. *J. Physiol.* **236,** 549–573.

Zeki, S. M. (1974b) Cells responding to changing image size and disparity in the cortex of the rhesus monkey. *J. Physiol.* **242,** 827–841.

Zeki, S. M. (1976) The functional organization of projections from striate to prestriate visual cortex in the rhesus monkey. *Cold Spr. Harb. Symp. quant. Biol.* **40,** 591–600.

# Index